FEATURES AND BEN...
Mathematics: Applications and Co...

Integration	... of mathematical topics helps students to see mathematics as a whole. Algebra lessons prepare students for first-year algebra.	11
	Other integrated topics: • geometry	38
	• measurement	504
	• statistics	546
	• probability	253
	• proportional reasoning	215
Applications	... show students how mathematics relates to the real world around them.	
	• Lessons open with an application or connection relevant to teenagers	62
	• *"When am I ever going to use this?"* answers the age-old question.	318
	• *School to Career* demonstrates careers in which mathematics is used.	395
	• *Math in the Media* helps students interpret mathematics in print.	289
Connections	... to other subject areas help students appreciate the role of mathematics in the other courses they are taking, such as science, social studies, and literature.	
	• *Chapter Projects*	329
	• Connection examples and exercises	19, 284
	• *Interdisciplinary Investigations*	274-275
Problem Solving	... activities and applications are integrated into every chapter.	
	• problem-solving strategies through *Thinking Labs*	90-91
	• 4-step problem-solving plan used throughout each chapter	4
	• Study Hints	297
Labs and Investigations	... provide students with an opportunity to explore, create mathematical models, and work cooperatively with a partner or group.	
	• Optional Hands-On Labs and Technology Labs provide a preview of the next lesson or can extend the concepts in the preceding lesson.	205, 118
	• Mini-labs introduce or reinforce concepts within a lesson.	301
	• Activities from **Interactive Mathematics: Activities and Investigations** provide additional options for extension activities in real-world settings.	2d
Ample Practice and Review	... reinforces new skills and concepts.	
	• Extra Practice for each lesson in the back of the text.	600-646
	• Mixed Review exercises in each exercise set	95
	• *Let the Games Begin* helps students maintain skills in previous lessons	285
Test Preparation and Assessment	... provides practice for local, state, and national tests.	
	• Test Practice questions in each exercise set	341
	• Standardized Test Practice at the end of each chapter includes multiple-choice and free-response questions.	326-327
	• Alternative Assessment options in the Study Guide and Assessment	99
	• A free-response test for each chapter in the back of the text.	647-659
Technology	... strand prepares students to function in a technological society through a variety of instruction and activities, including Technology Labs.	
	• scientific calculators	9
	• graphing calculators	16
	• spreadsheets	162
	• Internet Connections	316
Teacher Support	... in the **Teacher's Wraparound Edition** makes it easy for you to organize, present, and enhance the content. The extensive set of resource materials denoted in the Teacher's Edition helps you increase each student's chance for success.	

Teacher's Wraparound Edition

Glencoe
Mathematics
Applications and Connections

Course 3

Glencoe McGraw-Hill

New York, New York   Columbus, Ohio   Woodland Hills, California   Peoria, Illinois

Mathematics: Applications and Connections
Course 3

Student Edition
Teacher's Wraparound Edition
Spanish Student Edition

Applications

Classroom Games
Diversity Masters
Family Letters and Activities
Investigations and Projects Masters
School to Career Masters
Science and Mathematics Lab Manual

Assessment/Evaluation

Assessment and Evaluation Masters

 MindJogger Videoquizzes

 Test and Review Software

Meeting Individual Needs

Enrichment Masters
Investigations for the Special Education Student
Practice Masters
Spanish Study Guide and Assessment
Study Guide and Practice Workbook
Study Guide Masters
Transition Booklet

Manipulatives/Modeling

Hands-On Lab Masters
Glencoe Mathematics Classroom Manipulative Kit
Overhead Manipulative Resources
Glencoe Mathematics Student Manipulative Kit

Technology/Multimedia

 CD-ROM Program

 Interactive Mathematics Tools Software

Technology Masters

Teaching Aids

Answer Key Masters
Block Scheduling Booklet

 5-Minute Check Transparencies
Teaching Transparencies

Lesson Planning Guide
Solutions Manual

Electronic Teacher's Classroom Resources

 All blackline masters on a single CD-ROM

Glencoe/McGraw-Hill
A Division of The **McGraw·Hill** Companies

Copyright © 1999 by the McGraw-Hill Companies, Inc. All rights reserved. Printed in the United States of America. Except as permitted under the United States Copyright Act, no part of this publication may be reproduced or distributed in any form or by any means, or stored in a database or retrieval system, without prior written permission of the publisher.

Send all inquiries to:
Glencoe/McGraw-Hill
936 Eastwind Drive
Westerville, OH 43081-3374

ISBN: 0-02-833052-8 (Student Edition)
       0-02-833055-2 (Teacher's Wraparound Edition)

1 2 3 4 5 6 7 8 9 10 67/043 07 06 05 04 03 02 01 00 99 98

LETTER FROM THE AUTHORS

Dear Students, Teachers, and Parents,

Middle School math students are very special to us! That's why we wrote **Mathematics: Applications and Connections,** the first middle school math program designed specifically for you. The exciting, relevant content and up-to-date design will hold your interest and answer the question "When am I ever going to use this?"

As you page through your text, you'll notice the variety of ways math is presented for you. You'll see real-world applications as well as connections to other subjects like science, history, language arts, and music. You'll have opportunities to use technology tools such as the Internet, CD-ROM, graphing calculators, and computer applications like spreadsheets.

You'll appreciate the easy-to-follow lesson format. Each new concept is introduced with an interesting application or connection followed by clear explanations and examples. As you complete the exercises and solve interesting problems, you'll learn a great deal of useful math. You'll also have the opportunity to complete relevant Chapter Projects, Hands-On Labs, and Interdisciplinary Investigations. Test Practice, Test-Taking Tips, and Reading Math Study Hints will help you improve your test-taking skills.

Each day, as you use **Mathematics: Applications and Connections,** you'll see the practical value of math. You'll quickly grow to appreciate how often math is used in ways that relate directly to your life. If you don't already realize the importance of math in your life, you soon will!

Sincerely, The Authors

Kay McClain
Patricia S. Wilson
Patricia Frey
Linda Dritsas
Barbara Smith
Jack M. Ott
Ron Pelfrey
David Molina
Beatrice Moore-Harris

Meet Our Authors

William Collins
*Director of The Sisyphus
  Mathematics Learning Center
W.C. Overfelt High School
San Jose, CA*

William Collins provides mathematics support services to the students of Overfelt and the East Side community. He also has many years of experience as a mathematics instructor and mathematics department chairperson. Mr. Collins received his B.A. in Mathematics and his B.A. in Philosophy from Herbert H. Lehman College, CUNY, and his M.S. in Mathematics Education from California State University, Hayward. He is active in several professional mathematics organizations.

Linda Dritsas
*Mathematics Coordinator
Fresno Unified School District
Fresno, CA*

Linda Dritsas is highly involved with the Fresno Urban Systemic Initiative. She received her B.A. and M.A. from California State University, Fresno, where she also taught. Ms. Dritsas has published numerous articles, workbooks, and other supplementary materials. She received the Edward Begle Award from the California Mathematics Council. Ms. Dritsas has is an active member of NCTM and the Association for Supervision and Curriculum Development.

Patricia Frey
*Mathematics Department
  Chairperson
Buffalo Academy for Visual and
  Performing Arts
Buffalo, NY*

Patricia Frey received her B.A. from D'Youville College in Buffalo, and her M.Ed. from the State University of New York at Buffalo. Her publications include articles in mathematics education journals as well as mathematics, computer, and calculator curriculum journals. Ms. Frey is a Woodrow Wilson fellow and a New York State Math Mentor. She is an active member of NCTM and other professional mathematics organizations.

Arthur C. Howard
*Program Director for Secondary
  Mathematics
Aldine Independent School District
Houston, TX*

Arthur Howard is also an adjunct mathematics teacher at North Harris College and a member of the Executive Board of the Rice University School Mathematics Project. He received his B.S. in Mathematics and M.Ed. in Mathematics Education from the University of Houston. Mr. Howard is active in NCTM and has made presentations at local, state, regional, and national mathematics conferences.

Kay McClain
*Education Specialist
George Peabody College
Vanderbilt University
Nashville, TN*

Kay McClain is currently working on a Ph.D. at Vanderbilt University. While a teacher at Mountain Brook Middle School in Birmingham, Alabama. she received the Presidential Award for Excellence in the Teaching of Mathematics and was a Woodrow Wilson Fellow. Ms. McClain received her B.A. for Auburn University and her Educational specialist degree from the University of Montevallo. She is an active member of NCTM.

David Molina
*Adjunct Professor of Mathematics
  Education
The University of Texas at Austin
Austin, TX*

Dr. David Molina is the Associate Director of the Charles A. Dana Center. He earned his B.S. in Mathematics from the University of Notre Dame and his M.A. and Ph.D. in Mathematics Education from The University of Texas at Austin. Dr. Molina is active in every area of mathematics education. He has published numerous articles about professional development and educational technology.

Beatrice Moore-Harris
*Staff Development Specialist
Bureau of Education and Research
Houston, TX*

Beatrice Moore-Harris is a former middle school mathematics teacher and Mathematics Supervisor for the Houston and Fort Worth Independent School Districts. She received her B.A. from Prairie View A&M University. Ms. Moore-Harris is also a consultant and one of four official spokespersons for NCTM. She is also an active member of several professional mathematics organizations.

Jack M. Ott
*Distinguished Professor of
    Mathematics Education
University of South Carolina
Columbia, SC*

Jack Ott has taught grades 5-12 and college. He recently received the South Carolina Council of Teachers of Mathematics Award for Outstanding Contributions in Mathematics Education. Dr. Ott received his A.B. from Indiana Wesleyan University, his M.A. from Ball State University, and Ph.D. from The Ohio State University. He has written articles for The Mathematics Teacher and The Arithmetic Teacher and has been a speaker at national and state mathematics conferences.

Ronald Pelfrey
*Mathematics Consultant
Fayette County Schools
Lexington, KY*

Ronald S. Pelfry is a former middle school and high school mathematics teacher and district-level mathematics supervisor. He received his B.S., M.A., and Ed.D. from the University of Kentucky. Dr. Pelfrey has been active with NCTM and its local and state affiliates in Kentucky. He has been an author of several articles on curriculum and assessment, a speaker, and both a regional and state conference chair.

Jack Price
*Co-Director of Science and
    Mathematics
California State Polytechnic
    University
Pomona, CA*

Past-president of the National Council of Teacher of Mathematics, Dr. Price teaches mathematics and mathematics methods classes to prospective teachers. He received his B.A. from Eastern Michigan University and his M.Ed. and Ed. D. from Wayne State University in Detroit. Dr. Price also serves on the Expert Panel on Mathematics and Science Education of the U.S. Department of Education and the standing mathematics committee for NAEP.

Barbara Smith
*Mathematics Consultant
Unionville-Chadds Ford School
    District
Kennett Square, PA*

Barbara Smith is a former mathematics teacher with 13 years experience at the middle school level and 3 years at the high school level. She received her B.S. from Grove City College and her M.Ed. from the University of Pittsburgh. Ms. Smith is an active member of NCTM and has held offices in several state and local mathematics organizations

Patricia S. Wilson
*Associate Professor of Mathematics
    Education
University of Georgia
Athens, GA*

Patricia Wilson, a former middle school mathematics teacher, is currently working with preservice and inservice teachers. She received the Excellence in Teaching Award from the College of Education at the University of Georgia. Dr. Wilson received her B.S. from Ohio University and her M.A. and Ph.D. from The Ohio State University. She is an active member of NCTM and other professional mathematics organizations.

Academic Consultants and Teacher Reviewers

Each of the Academic Consultants read all 39 chapters in Courses 1, 2, and 3, while each Teacher Reviewer read two chapters. The Consultants and Reviewers gave suggestions for improving the Student Editions and the Teacher's Wraparound Editions.

ACADEMIC CONSULTANTS

Richie Berman, Ph.D.
Mathematics Lecturer and Supervisor
University of California, Santa Barbara
Santa Barbara, California

Robbie Bonneville
Mathematics Coordinator
La Joya Unified School District
Alamo, Texas

Cindy J. Boyd
Mathematics Teacher
Abilene High School
Abilene, Texas

Gail Burrill
Mathematics Teacher
Whitnall High School
Hales Corners, Wisconsin

Georgia Cobbs
Assistant Professor
The University of Montana
Missoula, Montana

Gilbert Cuevas
Professor of Mathematics Education
University of Miami
Coral Gables, Florida

David Foster
Mathematics Director
Robert Noyce Foundation
Palo Alto, California

Eva Gates
Independent Mathematics
   Consultant
Pearland, Texas

Berchie Gordon-Holliday
Mathematics/Science Coordinator
Northwest Local School District
Cincinnati, Ohio

Deborah Grabosky
Mathematics Teacher
Hillview Middle School
Whittier, California

Deborah Ann Haver
Principal
Great Bridge Middle School
Virginia Beach, Virginia

Carol E. Malloy
Assistant Professor, Math Education
The University of North Carolina,
   Chapel Hill
Chapel Hill, North Carolina

Daniel Marks, Ed.D.
Associate Professor of Mathematics
Auburn University at Montgomery
Montgomery, Alabama

Melissa McClure
Mathematics Consultant
Teaching for Tomorrow
Fort Worth, Texas

TEACHER REVIEWERS

Course 1

Carleen Alford
Math Department Head
Onslow W. Minnis, Sr. Middle School
Richmond, Virginia

Margaret L. Bangerter
Mathematics Coordinator K-6
St. Joseph School District
St. Joseph, Missouri

Diana F. Brock
Sixth and Seventh Grade Math Teacher
Memorial Parkway Junior High
Katy, Texas

Mary Burkholder
Mathematics Department Chair
Chambersburg Area Senior High
Chambersburg, Pennsylvania

Eileen M. Egan
Sixth Grade Teacher
Howard M. Phifer Middle School
Pennsauken, New Jersey

Melisa R. Grove
Sixth Grade Math Teacher
King Philip Middle School
West Hartford, Connecticut

David J. Hall
Teacher
Ben Franklin Middle School
Baltimore, Maryland

Ms. Karen T. Jamieson, B.A., M.Ed.
Teacher
Thurman White Middle School
Henderson, Nevada

David Lancaster
Teacher/Mathematics Coordinator
North Cumberland Middle School
Cumberland, Rhode Island

Jane A. Mahan
Sixth Grade Math Teacher
Helfrich Park Middle School
Evansville, Indiana

Margaret E. Martin
Mathematics Teacher
Powell Middle School
Powell, Tennessee

Diane Duggento Sawyer
Mathematics Department Chair
Exeter Area Junior High
Exeter, New Hampshire

Susan Uhrig
Teacher
Monroe Middle School
Columbus, Ohio

Cindy Webb
Title 1 Math Demonstration Teacher
Federal Programs LISD
Lubbock, Texas

Katherine A. Yule
Teacher
Los Alisos Intermediate School
Mission Viejo, California

Course 2

Sybil Y. Brown
Math Teacher Support Team-USI
Columbus Public Schools
Columbus, Ohio

Ruth Ann Bruny
Mathematics Teacher
Preston Junior High School
Fort Collins, Colorado

BonnieLee Gin
Junior High Teacher
St. Mary of the Woods
Chicago, Illinois

Larry J. Gonzales
Math Department Chair
Desert Ridge Middle School
Albuquerque, New Mexico

Susan Hertz
Mathematics Teacher
Revere Middle School
Houston, Texas

Rosalin McMullan
Mathematics Teacher
Honea Path Middle School
Honea Path, South Carolina

Mrs. Susan W. Palmer
Teacher
Fort Mill Middle School
Fort Mill, South Carolina

Donna J. Parish
Teacher
Zia Middle School
Las Cruces, New Mexico

Ronald J. Pischke
Mathematics Coordinator
St. Mary of the Woods
Chicago, Illinois

Sister Edward William Quinn I.H.M.
Chairperson Elementary Mathematics
Curriculum
Archdiocese of Philadelphia
Philadelphia, Pennsylvania

Marlyn G. Slater
Title 1 Math Specialist
Paradise Valley USD
Paradise Valley, Arizona

Sister Margaret Smith O.S.F.
Seventh and Eighth Grade Math Teacher
St. Mary's Elementary School
Lancaster, New York

Pamela Ann Summers
Coordinator, Secondary Math/Science
Lubbock ISD
Lubbock, Texas

Dora Swart
Teacher/Math Department Chair
W. F. West High School
Chehalis, Washington

Rosemary O'Brien Wisniewski
Middle School Math Chairperson
Arthur Slade Regional School
Glen Burnie, Maryland

Laura J. Young, Ed.D.
Eighth Grade Mathematics Teacher
Edwards Middle School
Conyers, Georgia

Susan Luckie Youngblood
Teacher/Math Department Chair
Weaver Middle School
Macon, Georgia

Course 3

Beth Murphy Anderson
Mathematics Department Chair
Brownell Talbot School
Omaha, Nebraska

David S. Bradley
Mathematics Teacher
Thomas Jefferson Junior High School
Salt Lake City, Utah

Sandy Brownell
Math Teacher/Team Leader
Los Alamos Middle School
Los Alamos, New Mexico

Eduardo Cancino
Mathematics Specialist
Education Service Center, Region One
Edinburg, Texas

Sharon Cichocki
Secondary Math Coordinator
Hamburg High School
Hamburg, New York

Nancy W. Crowther
Teacher, retired
Sandy Springs Middle School
Atlanta, Georgia

Charlene Mitchell DeRidder, Ph.D.
Mathematics Supervisor K-12
Knox County Schools
Knoxville, Tennessee

Ruth S. Garrard
Mathematics Teacher
Davidson Fine Arts School
Augusta, Georgia

Lolita Gerardo
Secondary Math Teacher
Pharr San Juan Alamo High School
San Juan, Texas

Donna Jorgensen
Teacher of Mathematics/Science
Toms River Intermediate East
Toms River, New Jersey

Statha Kline-Cherry, Ed.D.
Director of Elementary Education
University of Houston – Downtown
Houston, Texas

Charlotte Laverne Sykes Marvel
Mathematics Instructor
Bryant Junior High School
Bryant, Arkansas

Albert H. Mauthe, Ed.D.
Supervisor of Mathematics
Norristown Area School District
Norristown, Pennsylvania

Barbara Gluskin McCune
Teacher
East Middle School
Farmington, Michigan

Laurie D. Newton
Teacher
Crossler Middle School
Salem, Oregon

Indercio Abel Reyes
Mathematics Teacher
PSJA Memorial High School
Alamo, Texas

Fernando Rosa
Mathematics Department Chair
Edinburg High School
Edinburg, Texas

Mary Ambriz Soto
Mathematics Coordinator
PSJA I.S.D.
Pharr, Texas

Judy L. Thompson
Eighth Grade Mathematics Teacher
Adams Middle School
North Platte, Nebraska

Karen A. Universal
Eighth Grade Mathematics Teacher
Cassadaga Valley Central School
Sinclairville, New York

Tommie L. Walsh
Teacher
S. Wilson Junior High School
Lubbock, Texas

Marcia K. Ziegler
Mathematics Teacher
Pharr-San Juan- Alamo North High School
Pharr, Texas

Student Advisory Board

The Student Advisory Board gave the editorial staff and design team feedback on the design, content, and covers of the Student Editions. We thank these students from Crestview Middle School in Columbus, Ohio, and McCord Middle School in Worthington, Ohio, for their hard work and creative suggestions in making *Mathematics: Applications and Connections* more student friendly.

Table of Contents

CHAPTER 1: Problem Solving and Algebra

Chapter Project — Theme: Technology
Home Page Bound 3

1-1	A Plan for Problem Solving..............................	4
1-2	Powers and Exponents	8
1-3	Variables, Expressions, and Equations	11
1-3B	**Technology Lab:** Graphing Calculators Evaluating Expressions........................	16
1-4	Solving Subtraction and Addition Equations ..	17
1-5	Solving Division and Multiplication Equations	21
	Mid-Chapter Self Test	25
1-6	Writing Expressions and Equations.................	27
1-7A	**Thinking Lab:** Problem Solving Work Backward	30
1-7	Solving Two-Step Equations............................	32
1-7B	**Hands-On Lab:** Cooperative Learning Function Machines.............................	37
1-8	**Integration:** Geometry Perimeter and Area	38
1-9	Solving Inequalities	43
	Study Guide and Assessment	48
	Standardized Test Practice.............................	52

Applications, Connections, and Integration Index, pages xxii–1.

Let the Games Begin
- You're the Greatest, **47**

- Technology, **26**

interNET CONNECTION
- Chapter Project, **3**
- Data Update, **15**
- School to Career, **26**
- Let the Games Begin, **47**

Test Practice
10, 15, 20, 25, 29, 31, 36, 42, 47, 52–53

MATH IN THE MEDIA
- Fill It Up!, **20**

CHAPTER 2
Algebra: Using Integers

	CHAPTER Project	Theme: Endangered Species	
		Where the Wild Things Are	55
2-1		Integers and Absolute Value	56
2-2		Comparing and Ordering Integers	59
2-3		Adding Integers	62
2-4		More About Adding Integers	66
2-5		Subtracting Integers	69
		Mid-Chapter Self Test	72
2-6	**Integration:** Statistics	Matrices	73
2-6B	**Technology Lab:** Graphing Calculators	Matrices	77
2-7		Multiplying Integers	78
2-8		Dividing Integers	81
2-9A	**Hands-On Lab:** Cooperative Learning	Solving Equations	84
2-9		Solving Equations	86
2-9B	**Thinking Lab:** Problem Solving	Eliminate Possibilities	90
2-10	**Integration:** Geometry	The Coordinate System	92
		Study Guide and Assessment	96
		Standardized Test Practice	100

Let the Games Begin
- Absolutely!, **61**
- Matrix Madness, **76**

interNET CONNECTION
- Chapter Project, **55**
- Let the Games Begin, **61, 76**
- Data Update, **62**

Test Practice
58, 61, 65, 68, 72, 76, 80, 83, 89, 91, 95, 100–101

MATH IN THE MEDIA
- Broadway Bound?, **89**

CHAPTER 3
Using Proportion and Percent

CHAPTER Project   **Theme: Travel**
Pack Your Bags **103**

3-1	Ratios and Rates	**104**
3-2	Ratios and Percents	**107**
3-3	Solving Proportions	**111**
3-4	Fractions, Decimals, and Percents	**114**
	Mid-Chapter Self Test	**117**
3-4B	**Hands-On Lab:** Cooperative Learning The Golden Ratio	**118**
3-5	Finding Percents	**120**
3-6A	**Thinking Lab:** Problem Solving Reasonable Answers	**124**
3-6	Percent and Estimation	**126**
	Study Guide and Assessment	**130**
	Standardized Test Practice	**134**

Applications, Connections, and
Integration Index, pages xxii–1.

What's For Dinner?, **136**

- Per-fraction, **123**

interNET CONNECTION

- Chapter Project, **103**
- Data Update, **122**
- Let the Games Begin, **123**
- Interdisciplinary Investigation, **137**

Test Practice

106, 110, 113, 117, 122, 125, 129, 134–135

CHAPTER 4
Statistics: Analyzing Data

CHAPTER Project	**Theme: Museums**	
	Oldies But Goodies!	**139**
4-1A	**Thinking Lab:** Problem Solving	
	Make a Table	**140**
4-1	Bar Graphs and Histograms	**142**
4-1B	**Technology Lab:** Graphing Calculators	
	Histograms	**147**
4-2	Circle Graphs	**148**
4-3	Line Plots	**153**
4-3B	**Hands-On Lab:** Cooperative Learning	
	Maps and Statistics	**156**
4-4	Measures of Central Tendency	**158**
4-4B	**Technology Lab:** Spreadsheets	
	Finding a Mean	**162**
4-5	Measures of Variation	**163**
	Mid-Chapter Self Test	**166**
4-5B	**Technology Lab:** Graphing Calculators	
	Box-and-Whisker Plots	**167**
4-6	**Integration:** Algebra	
	Scatter Plots	**168**
4-7	Choosing an Appropriate Display	**171**
4-8	Misleading Graphs and Statistics	**174**
	Study Guide and Assessment	**178**
	Standardized Test Practice	**182**

Let the Games Begin
- You're the Winner...Bar None!, **146**

- History, **152**

MATH IN THE MEDIA
- Computer Comparisons, **151**

interNET CONNECTION
- Chapter Project, **139**
- Let the Games Begin, **146**
- School to Career, **152**
- Data Update, **170**

Test Practice
141, 146, 151, 155, 161, 166, 170, 173, 177, 182–183

CHAPTER 5
Geometry: Investigating Patterns

	CHAPTER Project	**Theme: Advertising**	
		Be True to Your School.................	**185**
5-1A	Hands-On Lab: Cooperative Learning Measuring and Constructing Line Segments and Angles		**186**
5-1	Parallel Lines...................................		**188**
5-1B	Hands-On Lab: Cooperative Learning Constructing Parallel Lines		**193**
5-2A	Thinking Lab: Problem Solving Use a Venn Diagram.........................		**194**
5-2	Classifying Triangles		**196**
5-3A	Hands-On Lab: Cooperative Learning Polygons as Networks.......................		**200**
5-3	Classifying Quadrilaterals		**201**
	Mid-Chapter Self Test		**204**
5-4A	Hands-On Lab: Cooperative Learning Reflections		**205**
5-4	Symmetry		**206**
5-5	Congruent Triangles		**210**
5-5B	Hands-On Lab: Cooperative Learning Constructing Congruent Triangles ...		**213**
5-6A	Hands-On Lab: Cooperative Learning Dilations		**214**
5-6	Similar Triangles............................		**215**
5-7	Transformations and M.C. Escher................		**220**
	Study Guide and Assessment......................		**224**
	Standardized Test Practice		**228**

Let the Games Begin

- Shape Relations, **218**

School to Career
- Advertising, **219**

interNET CONNECTION
- Chapter Project, **185**
- Data Update, **212**
- Let the Games Begin, **218**
- School to Career, **219**

Test Practice
192, 195, 199, 204, 209, 212, 218, 223, 228–229

Applications, Connections, and Integration Index, pages xxii–1.

CHAPTER 6: Exploring Number Patterns

CHAPTER Project — **Theme: Sports**
Take Me Out to the Ball Game **231**

6-1	Divisibility Patterns	**232**
6-2	Prime Factorization	**235**
6-2B	**Hands-On Lab:** Cooperative Learning Basketball Factors	**239**
6-3A	**Thinking Lab:** Problem Solving Make a List	**240**
6-3	Greatest Common Factor	**242**
6-4	Rational Numbers	**245**
	Mid-Chapter Self Test	**248**
6-5	Rational Numbers and Decimals	**249**
6-6	**Integration:** Probability Simple Events	**253**
6-7	Least Common Multiple	**257**
6-8A	**Hands-On Lab:** Cooperative Learning Density Property	**260**
6-8	Comparing and Ordering Rational Numbers	**261**
6-9	Scientific Notation	**265**
	Study Guide and Assessment	**268**
	Standardized Test Practice	**272**

What in the World is "WYSIWYG?", **274**

- Collecting Factors, **238**

interNET CONNECTION

- Chapter Project, **231**
- Let the Games Begin, **238**
- Data Update, **247**
- Interdisciplinary Investigation, **275**

Test Practice

234, 238, 241, 244, 248, 252, 256, 259, 264, 267, 272–273

MATH IN THE MEDIA

- Fox Trot, **267**

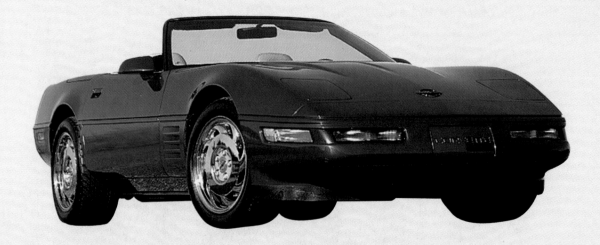

xiv Table of Contents

CHAPTER 7: Algebra: Using Rational Numbers

Chapter Project — **Theme: Nature**
Patterns in Nature 277

7-1	Adding and Subtracting Like Fractions	278
7-2	Adding and Subtracting Unlike Fractions	281
7-3	Multiplying Fractions	286
7-4	Properties of Rational Numbers	290
7-5A	**Thinking Lab:** Problem Solving Look for a Pattern.............................	294
7-5	**Integration:** Patterns and Functions Sequences	296
	Mid-Chapter Self Test	299
7-5B	**Hands-On Lab:** Cooperative Learning The Fibonacci Sequence	300
7-6	**Integration:** Geometry Area of Triangles and Trapezoids......	301
7-6B	**Hands-On Lab:** Cooperative Learning Area and Pick's Theorem	305
7-7A	**Hands-On Lab:** Cooperative Learning Graphing Pi...	307
7-7	**Integration:** Geometry Circles and Circumference	309
7-8	Dividing Fractions	312
7-9	Solving Equations...	315
7-10	Solving Inequalities......................................	318
	Study Guide and Assessment	322
	Standardized Test Practice	326

Applications, Connections, and Integration Index, pages xxii–1.

Let the Games Begin
- Fraction Track, **285**

School to Career

- Health, **293**

Math in the Media
- Close to Home, **289**

interNET CONNECTION
- Chapter Project, **277**
- Let the Games Begin, **285**
- School to Career, **293**
- Data Update, **316**

Test Practice
280, 284, 289, 292, 295, 299, 304, 311, 314, 317, 321, 326–327

CHAPTER 8: Applying Proportional Reasoning

	CHAPTER Project Theme: Television	
	Stay Tuned!	**329**
8-1	Using Proportions	**330**
8-1B	**Technology Lab:** Spreadsheets	
	Proportions	**334**
8-2	The Percent Proportion	**335**
8-3	**Integration:** Algebra	
	The Percent Equation	**339**
8-3B	**Thinking Lab:** Problem Solving	
	Solve a Simpler Problem	**342**
8-4	Large and Small Percents	**344**
8-5	Percent of Change	**348**
8-5B	**Technology Lab:** Spreadsheets	
	Discounts	**352**
8-6	Simple Interest	**353**
	Mid-Chapter Self Test	**356**
8-7	**Integration:** Geometry	
	Similar Polygons	**357**
8-8	Indirect Measurement	**361**
8-8B	**Hands-On Lab:** Cooperative Learning	
	Trigonometry	**365**
8-9	Scale Drawings and Models	**366**
8-10	**Integration:** Geometry	
	Dilations	**370**
	Study Guide and Assessment	**374**
	Standardized Test Practice	**378**

Let the Games Begin
- What's Missing?, **333**
- Map Sense, **369**

interNET CONNECTION
- Chapter Project, **329**
- Let the Games Begin, **333, 369**
- Data Update, **356**

Test Practice
333, 338, 341, 343, 347, 351, 356, 360, 364, 369, 373, 378–379

MATH IN THE MEDIA
- B.C., **364**

xvi Table of Contents

CHAPTER 9
Algebra: Exploring Real Numbers

CHAPTER Project   **Theme: Traffic Safety**
Proceed with Caution **381**

9-1	Square Roots..	**382**
9-2A	Hands-On Lab: Cooperative Learning Estimating Square Roots	**385**
9-2	Estimating Square Roots	**386**
9-3	The Real Number System	**390**
	Mid-Chapter Self Test	**394**
9-4A	Hands-On Lab: Cooperative Learning The Pythagorean Theorem.................	**396**
9-4	The Pythagorean Theorem.............................	**398**
9-5A	Thinking Lab: Problem Solving Draw a Diagram	**402**
9-5	Using the Pythagorean Theorem	**404**
9-5B	Hands-On Lab: Cooperative Learning Graphing Irrational Numbers	**408**
9-6	Integration: Geometry Distance on the Coordinate Plane	**410**
9-7	Integration: Geometry Special Right Triangles......................	**414**
	Study Guide and Assessment	**418**
	Standardized Test Practice	**422**

Applications, Connections, and Integration Index, pages xxii–1.

Math at the Mall, **424**

- Estimate and Eliminate, **389**

- Law Enforcement, **395**

interNET CONNECTION
- Chapter Project, **381**
- Let the Games Begin, **389**
- Data Update, **394**
- School to Career, **395**
- Interdisciplinary Investigation, **425**

Test Practice

384, 389, 394, 401, 403, 407, 413, 417, 422–423

Table of Contents   **xvii**

CHAPTER 10 Algebra: Graphing Functions

CHAPTER Project	Theme: Computer Games	
	Games People Play	**427**
10-1	Functions	**428**
10-1B	**Technology Lab:** Graphing Calculators Function Tables..................................	**432**
10-2	Using Tables to Graph Functions	**433**
10-3	Equations With Two Variables	**437**
10-4A	**Hands-On Lab:** Cooperative Learning Graphing Relationships	**441**
10-4	Graphing Linear Functions............................	**442**
10-4B	**Technology Lab:** Graphing Calculators Linear Functions................................	**445**
10-5	Graphing Systems of Equations	**446**
	Mid-Chapter Self Test	**449**
10-6A	**Thinking Lab:** Problem Solving Use a Graph...	**450**
10-6	Graphing Quadratic Functions	**452**
10-7	**Integration:** Geometry Translations...	**456**
10-8	**Integration:** Geometry Reflections..	**460**
10-9	**Integration:** Geometry Rotations..	**464**
	Study Guide and Assessment	**468**
	Standardized Test Practice	**472**

Let the Games Begin
- Guess My Rule, **431**
- Parabola Hit or Miss, **455**

School to Career
- Computer Programming, **436**

MATH IN THE MEDIA
- Herman, **440**

interNET CONNECTION
- Chapter Project, **427**
- Data Update, **430**
- Let the Games Begin, **431, 455**
- School to Career, **436**

Test Practice
431, 435, 440, 444, 449, 451, 455, 459, 463, 467, 472–473

CHAPTER 11
Geometry: Using Area and Volume

CHAPTER Project   **Theme: Kites**
Up, Up, and Away.......................... **475**

11-1	Area of Circles...............................	**476**
11-2A	**Thinking Lab:** Problem Solving Make a Model.................................	**480**
11-2	Three-Dimensional Figures	**482**
11-3	Volume of Prisms and Cylinders....................	**486**
11-4	Volume of Pyramids and Cones	**490**
	Mid-Chapter Self Test	**493**
11-5A	**Hands-On Lab:** Cooperative Learning Nets...	**494**
11-5	Surface Area of Prisms................................	**495**
11-6	Surface Area of Cylinders	**499**
11-6B	**Hands-On Lab:** Cooperative Learning Surface Area and Volume	**503**
11-7	**Integration:** Measurement Precision and Significant Digits........	**504**
	Study Guide and Assessment	**508**
	Standardized Test Practice	**512**

Applications, Connections, and Integration Index, pages xxii–1.

Let the Games Begin
- Architest, **498**

interNET CONNECTION
- Chapter Project, **475**
- Let the Games Begin, **498**
- Data Update, **506**

Test Practice
479, 481, 485, 489, 493, 498, 502, 507, 512–513

MATH IN THE MEDIA
- Hi and Lois, **507**

CHAPTER 12: Investigating Discrete Math and Probability

CHAPTER Project Theme: Athletes
Consider the Probabilities **515**

12-1A	**Hands-On Lab:** Cooperative Learning Fair and Unfair Games	**516**
12-1	Counting Outcomes	**518**
12-2	Permutations	**521**
12-3	Combinations	**524**
12-4	Pascal's Triangle	**528**
	Mid-Chapter Self Test	**531**
12-4B	**Hands-On Lab:** Cooperative Learning Patterns in Pascal's Triangle	**532**
12-5	Probability of Compound Events	**534**
12-6A	**Thinking Lab:** Problem Solving Act It Out	**538**
12-6	Experimental Probability	**540**
12-7A	**Hands-On Lab:** Cooperative Learning Punnett Squares	**544**
12-7	**Integration:** Statistics Using Sampling to Predict	**546**
	Study Guide and Assessment	**550**
	Standardized Test Practice	**554**

Extra! Extra! Newspapers May Take Over the Planet!, **556**

- Win the Lottery, **543**

- Health, **549**

interNET CONNECTION

- Chapter Project, **515**
- Data Update, **537**
- Let the Games Begin, **543**
- School to Career, **549**
- Interdisciplinary Investigation, **557**

Test Practice

520, 523, 527, 531, 537, 539, 543, 548, 554–555

CHAPTER 13 Algebra: Exploring Polynomials

CHAPTER Project Theme: Basketball
Hit or Miss 559

- **13-1A** Hands-On Lab: Cooperative Learning
  Algebra Tiles 560
- **13-1** Modeling Polynomials 561
- **13-2** Simplifying Polynomials 565
- **13-3** Adding Polynomials 570
- **13-4** Subtracting Polynomials 573
  Mid-Chapter Self Test 576
- **13-5A** Hands-On Lab: Cooperative Learning
  Modeling Products 577
- **13-5** Multiplying Monomials and Polynomials 578
- **13-6** Multiplying Binomials 583
- **13-7A** Thinking Lab: Problem Solving
  Guess and Check 586
- **13-7** Factoring Polynomials 588
  Study Guide and Assessment 592
  Standardized Test Practice 596

Let the Games Begin

- Factor Challenge, **591**

SCHOOL to CAREER

- Sports Management, **582**

- Chapter Project, **559**
- Data Update, **561**
- School to Career, **582**
- Let the Games Begin, **591**

Test Practice

564, 569, 572, 576, 581, 585, 587, 591, 596–597

Student Handbook 599
  Extra Practice 600–646
  Chapter Tests 647–659
  Getting Acquainted with the Graphing
    Calculator 660–661
  Getting Acquainted with Spreadsheets 662–663
  Selected Answers 664–699
  Photo Credits 700–701
  Glossary 702–710
  Spanish Glossary 711–721
  Index 722–731
  Symbols, Formulas, and Measurement
    Conversions .... **inside back cover**

Applications, Connections, and
Integration Index, pages xxii–1.

Applications, Connections, and Integration Index

APPLICATIONS

Agriculture, 75, 83, 181, 384, 488, 543

Air-Conditioning, 317

Animals
Pets, 295, 484, 522
Wildlife Management, 181

Architecture, 199, 217, 227, 363, 364, 462, 483, 493, 581

Business, 80, 345, 349, 448, 491, 526, 527, 539, 553, 587
Advertising, 195, 207, 350, 463, 548
Apparel Sales, 73, 76
Economics, 448
Inventory, 333
Manufacturing, 259, 280, 311, 314
Marketing, 42, 174, 176, 195, 467, 546
Profit, 471
Rental Agency, 36
Retail Sales, 28, 587
Sales, 181
Shipping, 10

Calendar, 234

Camp, 241

Camping, 496

Careers, 177
Farmer, 325
Engineering, 362
Social Work, 24
See also Index on pages 722-731

Carpentry, 244, 572

Celebrations, 41

Census, 534

Charities, 151

Clothing, 506
Fashion, 176, 451, 539

Communication, 289, 471, 520

Community Service, 501

Computers,
See Index on pages 722-731

Construction, 38, 223, 227, 299, 415–416, 421, 500

Crafts, 244

Demographics, 168–169

Design, 204, 212, 359, 383, 384, 401, 421

Economics, 541

Education, 347
Home Schooling, 76
Test Scores, 338

Employment, 83, 172
After-school Jobs, 91

Energy, 80, 82, 488

xxii Applications, Connections, and Integration Index

Entertainment, 9, 165, 310, 377, 403, 467, 506
  Amusement Parks, 263, 363
  Concerts, 547
  Drive-in Theaters, 15, 34
  Fireworks, 454
  Game Shows, 64
  Movies, 64, 68, 347, 367, 368, 371
  Party Planning, 241
  State Fairs, 505
  Television, 7, 151, 175
  Theater, 247
  Videos, 70, 161, 542

Environment, 106, 489
  Natural Resources, 146

Flags, 466

Flooring, 325

Food, 15, 287, 288, 313, 444, 479, 511, 553, 567, 575, 591
  Baking, 338
  Cooking, 113, 296, 313, 403
  Food Design, 223

Forestry, 311, 364, 394

Framing, 587

Fund-raising, 36, 332, 356, 530

Furniture, 191

Games, 47
  Backgammon, 541
  Clue, 520
  Contests, 543
  Cribbage, 527
  Dominoes, 537
  Monopoly, 31, 177, 537
  Puzzles, 35
  Scrabble, 255
  Sorry!, 536
  Tic Tac Toe, 523
  Totolospi, 520
  Trouble, 271
  UNO, 255

Gardening, 421, 580, 595

Health, 51, 88, 122, 137, 169, 330, 333, 343, 451, 489
  Medicine, 20, 23, 51
  Nutrition, 29, 155, 165, 321

Hobbies, 15, 113, 133, 222, 241, 360, 369

Home Decorating, 279

Home Improvement, 47

Horticulture, 477, 497

Interior Design, 243, 485, 590, 595

Internet, 15, 26, 47, 55, 61, 62, 76, 113, 122, 123, 137, 139, 146, 152, 170, 185, 212, 218, 219, 231, 238, 247, 275, 277, 285, 293, 316, 329, 333, 369, 381, 389, 394, 395, 425, 427, 430, 431, 436, 455, 475, 498, 506, 515, 537, 543, 549, 557, 559, 561, 582, 591

Inventions, 5

Jewelry, 338

Journalism, 120

Landscaping, 304, 388

Applications, Connections, and Integration Index

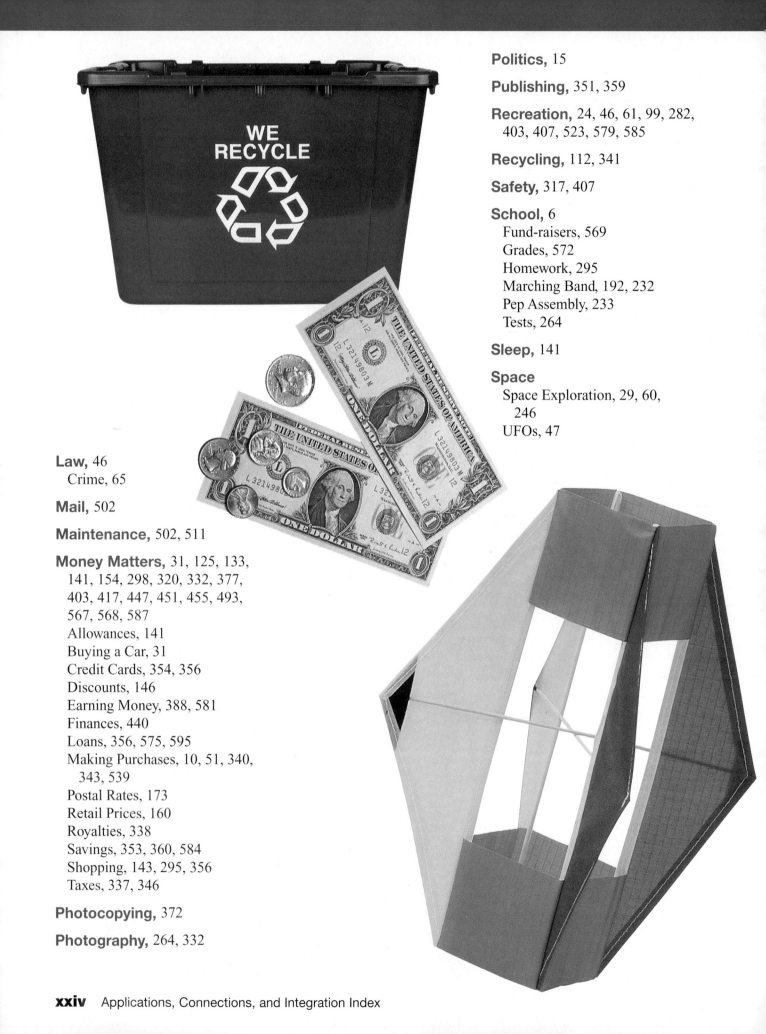

Law, 46
  Crime, 65

Mail, 502

Maintenance, 502, 511

Money Matters, 31, 125, 133, 141, 154, 298, 320, 332, 377, 403, 417, 447, 451, 455, 493, 567, 568, 587
  Allowances, 141
  Buying a Car, 31
  Credit Cards, 354, 356
  Discounts, 146
  Earning Money, 388, 581
  Finances, 440
  Loans, 356, 575, 595
  Making Purchases, 10, 51, 340, 343, 539
  Postal Rates, 173
  Retail Prices, 160
  Royalties, 338
  Savings, 353, 360, 584
  Shopping, 143, 295, 356
  Taxes, 337, 346

Photocopying, 372

Photography, 264, 332

Politics, 15

Publishing, 351, 359

Recreation, 24, 46, 61, 99, 282, 403, 407, 523, 579, 585

Recycling, 112, 341

Safety, 317, 407

School, 6
  Fund-raisers, 569
  Grades, 572
  Homework, 295
  Marching Band, 192, 232
  Pep Assembly, 233
  Tests, 264

Sleep, 141

Space
  Space Exploration, 29, 60, 246
  UFOs, 47

Spreadsheets, 71, 252, 295
World Wide Web, 25
See also Computers and Index on pages 722–731

Time, 325

Transportation, 7, 14, 117, 125, 129, 321, 444
  Aviation, 88, 170, 449, 454
  Bicycles, 199, 289
  Bus Routes, 259
  Driver Safety, 434
  Highway Maintenance, 493
  Highway Safety, 161
  Traffic, 227

Sports, 79, 91, 123, 302, 435, 443, 452, 575
  Baseball, 110, 159–160, 241, 262, 271, 316, 400, 523, 564
  Basketball, 128, 336, 485
  Bowling, 431
  Daytona 500, 58
  Football, 67, 155, 209, 248, 347, 405
  Golf, 80, 99, 471
  Gymnastics, 407
  Horse Racing, 95
  Injuries, 125, 141
  In-line Skating, 164
  Marathon, 23
  Olympics, 173
  Skiing, 520
  Soccer, 519
  Softball, 553
  Tee-ball, 41
  Tennis, 299, 539
  Track, 493
  Water Skiing, 497
  WNBA, 19

Technology, 142, 241
  Cable TV, 110, 351
  Calculators, 7, 91, 507
  E-mail, 13
  Graphing Calculators, 413, 542
  Internet, 141
  Memory, 332

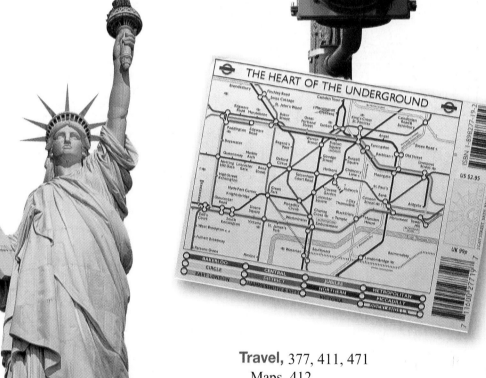

Travel, 377, 411, 471
  Maps, 412
  Signs, 416
  Tourism, 31, 361, 366

Weather, 547
  Arizona, 247

Urban Planning, 192

Writing, 133

CONNECTIONS

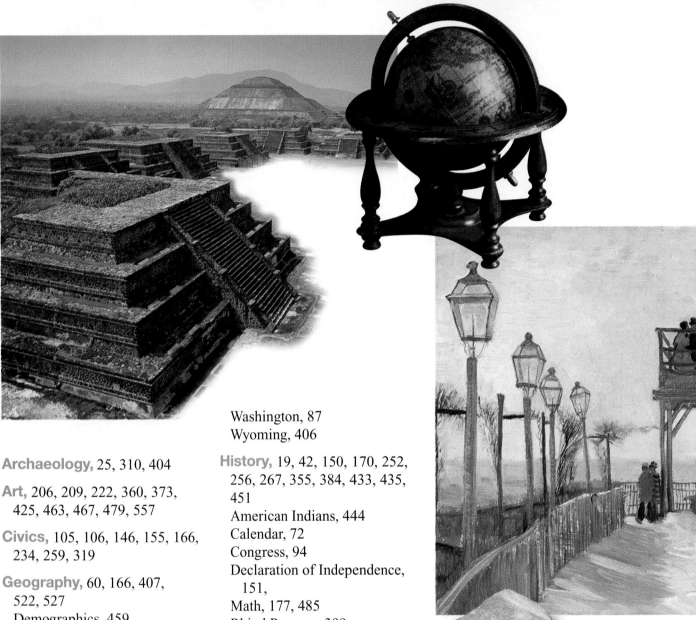

Archaeology, 25, 310, 404

Art, 206, 209, 222, 360, 373, 425, 463, 467, 479, 557

Civics, 105, 106, 146, 155, 166, 234, 259, 319

Geography, 60, 166, 407, 522, 527
Demographics, 459
El Salvador, 449
Great Lakes, 72
Hawaii, 150
Massachusetts, 68
Mexico City, 369
Midway, 95
Minnesota, 343
Netherlands, 368
Niagara Falls, 29
North America, 250
Population, 122, 351
United Kingdom, 208
Virginia, 303
Washington, 87
Wyoming, 406

History, 19, 42, 150, 170, 252, 256, 267, 355, 384, 433, 435, 451
American Indians, 444
Calendar, 72
Congress, 94
Declaration of Independence, 151,
Math, 177, 485
Rhind Papyrus, 399

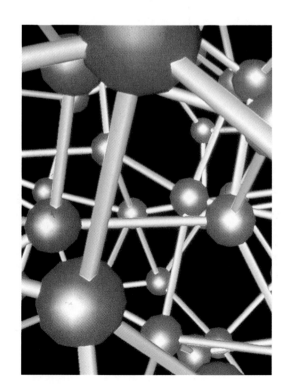

Music, 67, 150, 284, 341, 440, 451, 511, 531

Science, 137, 557
 Astronomy, 9, 364
 Chemistry,
 Earth Science, 31, 57, 58, 61, 65, 116, 125, 127, 129, 144, 172, 217, 311, 332, 369, 430, 438
 Life Science, 10, 20, 28, 75, 80, 81, 95, 105, 113, 116, 129, 145, 166, 170, 171–172, 173, 209, 252, 259, 265, 266, 284, 292, 331, 332, 333, 341, 343, 368, 429, 430, 435, 463, 525, 529, 531, 548, 563
 Physical Science, 7, 267, 291, 295, 316, 387, 388, 393, 440, 454, 498, 505, 539, 562, 574

Social Studies, 425

Home Economics, 155

Language, 72

Language Arts, 137, 148–149, 425, 557

Literature, 44

Meteorology, 51, 392

Applications, Connections, and Integration Index **xxvii**

INTEGRATION

Geometry and Spatial Sense, 38–42, 184–227, 256, 288, 373, 384, 401, 403, 431, 439, 458, 474–511, 525, 564, 581
  Angles, 20, 199
  Area, 68, 95, 110, 117, 233, 292, 321, 337, 523, 569, 585, 589, 590
  Circumference, 314, 347

Algebra, 58, 54–99, 168–170, 250, 289, 297, 304, 314, 383, 393, 440, 457, 459, 461, 498, 531, 543, 572, 576, 585
  Evaluating Expressions, 25, 29, 61, 64, 67, 71, 80, 93, 236, 237, 280, 281, 283, 287, 291, 569, 581
  Simplifying Expressions, 283, 590, 591
  Solving Equations, 25, 29, 36, 42, 47, 58, 61, 65, 72, 83, 89, 95, 106, 113, 189, 191, 197, 198, 199, 203, 204, 209, 211, 216, 259, 321, 341, 384, 387, 401, 444, 489
  Solving Inequalities, 58, 117
  Solving Proportions, 112, 117, 161, 192, 252
  Writing Equations, 36, 42, 202–203, 256
  Writing Expressions, 68, 106
  Writing Inequalities, 238, 523
  See also Index on pages 722-731

Data Analysis, Statistics, and Probability
  Data Analysis, 195
  Probability, 253–256, 280, 299, 321, 478, 479, 507, 531, 548, 576, 581
  Statistics, 73–76, 109, 122, 129, 138–181, 234, 248, 263, 267, 284, 299, 343, 351, 384, 394, 527, 537, 539, 564

xxviii   Applications, Connections, and Integration Index

Congruent Figures, 218, 238
The Coordinate System, 92–95
Dilations, 370–373
Distance on the Coordinate
  Plane, 410–413
Graphing, 94, 106, 192, 373
Heron, 386
Perimeter, 95, 110, 284, 343,
  571, 576
Probability, 254, 271
Rectangles, 295
Reflections, 460–464
Scale Drawings, 373
Similar Figures, 223
Surface Area, 502, 507
Symmetry, 212, 244
Triangles, 252, 280
Volume, 10, 317, 498, 572
See also Index on pages
  722-731

Measurement,
  Customary, 251
  Metric, 520
  See also Index on pages
    722-731

Number and Operations
  Sense, 311, 317
  Theory, 237, 403, 587

Patterns and Functions,
  296–299
  Functions, 428–431, 569
  Number Patterns, 237, 304,
    405, 451
  Patterns, 42, 99, 141

Problem Solving
  Act It Out, 538–539
  Draw a Diagram, 402–403
  Eliminate Possibilities, 90–91
  Guess and Check, 586–587
  Look for a Pattern, 294–295
  Make a List, 240–241
  Make a Model, 480–481
  Make a Table, 140–141
  Reasonable Answers, 124–125
  Solve a Simpler Problem,
    342–343
  Use a Graph, 450–451
  Use a Venn Diagram, 194–195
  Working Backward, 30–31

Proportionality
  Area, 479
  Percent, 115–116, 335–338
  Perimeter, 360
  Probability, 478
  Surface Area, 503
  Volume, 486, 503

Vist the Glencoe Mathematics internet site for
Mathematics: Applications and Connections at
www.glencoe.com/sec/math/mac/mathnet

You'll find:

Group Activities

Games

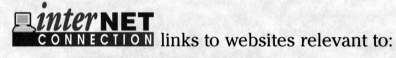 links to websites relevant to:
- Chapter Projects
- Interdisciplinary Investigations
- exercises

and much more!

About the Cover
The optimum viewing angle for the hologram on the cover of this textbook is a 45° angle. For best results, view the hologram at this angle under a direct light source, such as sunlight or incandescent lighting.

TEACHER'S WRAPAROUND EDITION

Table of Contents

Mathematics: Applications and Connections, Course 3

Pages T2-T31 provide a brief overview of the latest trends and research in mathematics education as well as to illustrate ways in which Glencoe has responded to these.

Pathways to Success T2
Bridging the Gap T4
Mathematics: Applications and Connections Exemplifies the NCTM Standards .T6
Motivating Middle School Students T8
Developing Problem Solving T10
Using Technology T12
Assessment T14
Using Projects T16
Using Games in the Math Classroom T19
Multiple Learning Styles T20
Involving Parents and Family T21
Classroom Management T22
Classroom Vignettes T23
Mathematics: Applications and Connections Research Activities T24
Scope and Sequence T25
Planning Your Course of Study T30

Chapter Overviews

Chapter 1 ....2a-2f
Chapter 2 ....54a-54f
Chapter 3 ....102a-102f
Chapter 4 ....138a-138f
Chapter 5 ....184a-184f
Chapter 6 ....230a-230f
Chapter 7 ....276a-276f
Chapter 8 ....328a-379f
Chapter 9 ....380a-380f
Chapter 10 ....426a-426f
Chapter 11 ....474a-474f
Chapter 12 ....514a-514f
Chapter 13 ....558a-558f

Student Handbook

Extra Practice 600-646
  (Basic Skills, Lesson by Lesson, Mixed Problem Solving)
Chapter Tests 647-659
Getting Acquainted with the Graphing Calculator 660-661
Getting Acquainted with Spreadsheets 662-663
Selected Answers 664-699
Photo Credits 700-701
Glossary 702-710
Spanish Glossary 711-721

Teacher's Wraparound Edition Index 722-733
Additional Answers AA1-AA40
Symbols, Formulas, and Measurement Conversions ......inside back cover

PATHWAYS TO SUCCESS

For the last several years, it's been difficult to find a newspaper or magazine that does not have an article about how poorly U.S. students score on international math exams compared to students from other countries such as Japan or Germany. For example, in the **Third International Mathematics and Science Study (TIMSS),** U.S. eighth grade students scored *below* the average for industrialized countries.

This result should not be surprising given the well-known study by James Flanders. He found that only 30-40% of the content in grades 6-8 of the most widely used K-8 mathematics series was new while algebra textbooks contained about 88% new content. (*Arithmetic Teacher*, "How Much of the Content in Mathematics Textbooks is New?", September 1987). This lack of new content in middle school combined with almost all new content in algebra 1 often leads to student frustration and failure in algebra 1. Flanders' research and the TIMSS data were both obtained before the first edition of ***Mathematics: Applications and Connections*** was published.

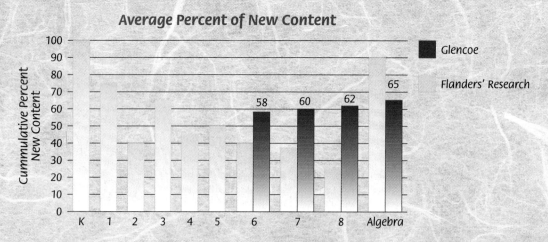

Note that the TIMMS data was collected when the programs described in the Flanders study were **very** widely used.

Glencoe changed this pattern with the publication of ***Mathematics: Applications and Connections.***

Glencoe understands that the middle school experience is critical in preparing students for success in algebra 1 and geometry. *Mathematics: Applications and Connections* is designed to smooth the path to algebra and geometry by creating a program that has about the same amount of new material in Grade 6, Grade 7, and Grade 8 as well as in Glencoe's *Algebra 1*.

Here are some of the key findings from TIMSS and Glencoe's response.

TIMMS Finding

Achievement U.S. eighth-grade students test at about the international average in algebra, fractions, statistics, and probability. U.S. students do not do as well in geometry, measurement, and proportionality.

Curriculum The content taught in U.S. eighth-grade classrooms is at a seventh-grade level in comparison to other countries.

Curriculum Topic coverage in U.S. eighth-grade mathematics classes is not as focused as in Germany and Japan. (The U.S. curriculum is "a mile wide and an inch deep.)

Glencoe's Response

All mathematics topics are integrated throughout Course 1-3 of *Mathematics: Applications and Connections.*

Mathematics: Applications and Connections meets or exceeds international standards in grades 6-8. It emphasizes geometry, measurement, and proportionality, as well as algebra, fractions, and statistics.

Mathematics: Applications and Connections follows a structured scope and sequence that introduces, reinforces, and extends topics needed for success in algebra 1 and geometry.

In addition to Glencoe's progressive response to the Flanders' research and the TIMSS, here are some other key features and benefits that middle school mathematics educators asked us to include in *Mathematics: Applications and Connections.*

Feature

Integrated Content There is an emphasis on integrating algebra, geometry, measurement, proportional reasoning, statistics, probability, technology, and problem solving.

Applications and Connections Relevant, real-life applications and interdisciplinary connections are a part of every lesson. Every application and connection is written in an engaging style and often accompanied by colorful, high-interest photos, graphs, tables, and charts.

Test Preparation Every lesson has at least one multiple-choice *Test Practice* question. Every chapter has two pages of *Standardized Test Practice,* which includes multiple-choice test items, open-ended test items, and a *Test-Taking Tip.*

Benefit

Students will be prepared for success in algebra 1 and geometry, the gateway courses for success in college and careers.

Students are motivated to learn mathematics when it is interesting and related to their lives. Every page is designed to help you answer the question, "When am I ever going to use this?".

Test scores will increase because students will be prepared for success on state, national, and international standardized tests as well as classroom tests and end-of-course examinations.

As you examine *Mathematics: Applications and Connections,* you will see that this program will help you prepare your students for success in school and in their lives. As students use the program, they will repeatedly receive this message: "Math is for everyone...You can do it...You'll use it every day."

Mathematics: Applications and Connections— something good just got better!

BRIDGING THE GAP

From Elementary Mathematics to High School Mathematics

Students' experiences in middle school mathematics are crucial in preparing them for success in algebra and geometry.

Glencoe's *Mathematics: Applications and Connections* prepares all students for success in algebra and geometry. How? By introducing a variety of new concepts at the right time in the right way and by integrating them appropriately into all three courses. For example, integers are introduced in Chapter 11 of Course 1, in Chapter 5 of Course 2, and in Chapter 2 of Course 3. Because algebra and geometry are introduced in Chapter 1 in all three courses – and because both are reinforced throughout – students are much better prepared to take these courses in high school.

Manipulatives help students bridge the gap from the concrete to the abstract.

How does *Mathematics: Applications and Connections* help students bridge from the concrete, number-oriented mathematics in elementary school to abstract, symbol-centered algebra and geometry in high school? Many middle school students' learning styles lend themselves to concrete operations. They may not be able to grasp abstract concepts easily, but they can understand abstract concepts on a concrete level.

Hands-On Labs and Mini-Labs help students discover concepts on their own.

Glencoe's *Mathematics: Applications and Connections* encourages students to **do** mathematics. **Hands-On Labs** give students hands-on experiences, with a partner or group, in discovering mathematical concepts for themselves. The **Hands-On Lab Masters** provide students with a way to record what they observe and discover in the Hands-On Labs. Students may also participate in shorter **Hands-On Mini-Labs** in which they investigate mathematical concepts within a lesson. **Hands-On Math** exercises provide even more opportunities for concrete learning.

The **Overhead Manipulative Resources** include a wealth of transparent manipulatives for use with overhead projectors such as a compass, spinner, counters, geoboard, and coordinate planes. A complete Teacher's Guide, correlated to the Hands-On Labs and Hands-On Mini-Labs, provides suggestions for demonstrations by the teacher or students.

The Glencoe **Mathematics Manipulative Kit** offers the tools students need to work through the Hands-On Labs and Hands-On Mini-Labs. A Teacher's Guide provides a correlation to all three courses.

> "Hands-on experiential learning in which students take an active role and assume responsibility for their own learning is an integral part of the instructional practices of this program."
>
> Beatrice Moore-Harris, Author

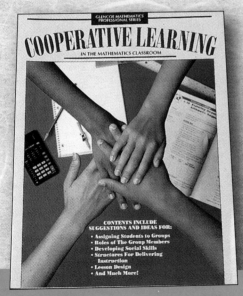

Cooperative Learning in the Mathematics Classroom, part of the Glencoe Mathematics Professional Series, provides suggestions for implementing cooperative learning techniques in your classroom.

Cooperative learning helps students with academic achievement and interpersonal skills.

In cooperative learning, small groups of students work as teams to share ideas, solve problems, and justify conclusions. At the same time, they develop vital skills in cooperating with others toward a common goal. The increased communication that takes place in cooperative learning experiences clarifies and strengthens individual students' understanding.

Glencoe's *Mathematics: Applications and Connections* offers an abundance of ways for students to learn cooperatively. **Chapter Projects** provide open-ended activities that often include collecting and organizing data. Students work together to explore, analyze, and report their results. **Interdisciplinary Investigations** relate mathematics to other content areas students are studying. These long-term projects are ideal for cooperative learning groups. The **Investigations and Projects Masters** provide students with ways to organize their explorations in the Chapter Projects and Interdisciplinary Investigations.

Hands-On Labs, Technology Labs, Thinking Labs, and **Mini-Labs** throughout all three courses of Mathematics: Applications and Connections also provide excellent opportunities for cooperative learning experiences. **Let the Games Begin** features interesting and fun games to help students reinforce and extend concepts.

Mathematics: Applications and Connections Exemplifies the NCTM Standards

Mathematics: Applications and Connections thoroughly integrates all thirteen curriculum standards for grades 5-8 as outlined in the NCTM Standards. For more information on the Standards, refer to pages 64-119 of the *Curriculum and Evaluation Standards for School Mathematics* © 1989 by the National Council of Teachers of Mathematics.

STANDARD 1: Mathematics as Problem Solving

Course 1 (Chapters 1-13)
4-37, 46-85, 94-121, 132-155, 157-169, 178-219, 228-257, 268-283, 285-301, 310-343, 352-385, 396-425, 434-467, 476-503, 514-539

Course 2 (Chapters 1-13)
4-35, 44-79, 88-121, 132-175, 184-217, 226-257, 268-307, 316-351, 360-397, 408-441, 450-481, 490-517, 528-553

Course 3 (Chapters 1-13)
4-47, 56-95, 104-129, 140-177, 186-223, 232-267, 278-321, 330-373, 382-417, 428-467, 476-507, 516-548, 560-591

STANDARD 2: Mathematics as Communication

Course 1 (Chapters 1-13)
4-37, 46-85, 94-121, 132-169, 178-219, 228-257, 28-301, 310-343, 352-385, 396-425, 434-467, 476-503, 514-539

Course 2 (Chapters 1-13)
4-35, 44-79, 88-121, 132-175, 184-217, 226-257, 268-307, 316-351, 360-397, 408-441, 450-481, 490-517, 528-553

Course 3 (Chapters 1-13)
4-47, 56-95, 104-129, 140-177, 186-223, 232-267, 278-321, 330-373, 382-417, 428-467, 476-507, 516-548, 560-591

STANDARD 3: Mathematics as Reasoning

Course 1 (Chapters 1-13)
4-37, 46-85, 94-121, 132-169, 178-219, 228-257, 268-283, 285-301, 310-343, 352-385, 396-425, 434-467, 476-503, 514-539

Course 2 (Chapters 1-13)
4-35, 44-79, 88-121, 132-175, 184-217, 226-257, 268-307, 316-351, 360-397, 408-441, 450-481, 490-517, 528-553

Course 3 (Chapters 1-13)
4-15, 17-47, 56-95, 104-129, 140-177, 186-223, 232-267, 278-321, 330-373, 382-417, 428-467, 476-507, 516-548, 560-591

STANDARD 4: Mathematical Connections

Course 1 (Chapters 1-13)
4-37, 46-85, 94-121, 133-139, 141-169, 178-180, 182-190, 193-219, 228-241, 243-257, 268-270, 273-283, 285-301, 312-343, 352-361, 364-366, 370-373, 375-382, 396-414, 416-425, 434-439, 441-461, 464-467, 476-487, 492-493, 496-503, 514-539

Course 2 (Chapters 1-13)
4-35, 44-79, 88-121, 132-175, 184-217, 226-257, 268-307, 316-351, 360-397, 408-441, 450-481, 490-517, 528-553

Course 3 (Chapters 1-13)
4-15, 17-47, 56-95, 104-129, 140-177, 186-192, 194-212, 214-223, 232-267, 278-321, 330-373, 382-417, 428-467, 476-507, 516-548, 560-591

STANDARD 5: Number and Number Relationships

Course 1 (Chapters 1-13)
8-11, 20-25, 28-37, 46-49, 54-85, 94-121, 132-148, 152-169, 178-180, 182-184, 188-219, 228-241, 243-257, 268-301, 310-321, 324-343, 352-366, 370-373, 375-382, 396-414, 416-424, 434-461, 464-467, 476-503, 515-534, 536-539

Course 2 (Chapters 1-11, 13)
4-20, 24-35, 44-79, 88-121, 132-175, 184-217, 226-257, 268-307, 316-351, 366-367, 370-373, 382-385, 408-411, 450-459, 464-468, 474-481, 528-533, 538-553

Course 3 (Chapters 1-13)
4-10, 17-25, 30-36, 56-65, 69-95, 104-129, 140-155, 158-177, 232-252, 257-267, 278-280, 286-300, 307-314, 318-321, 330-333, 335-350, 352-369, 382-389, 396-417, 433-444, 456-467, 476-479, 504-507, 532-533, 586-587

STANDARD 6: Number Systems and Number Theory

Course 1 (Chapters 1-13)
28-31, 34-37, 50-53, 60-63, 94-121, 132-148, 150-163, 167-169, 178-180, 182-219, 228-241, 243-257, 268-301, 312-321, 324-343, 352-366, 370-373, 375-382, 396-414, 416-424, 434-461, 464-467, 476-503, 515-529, 531-534, 536-539

Course 2 (Chapters 1-11)
4-7, 44-53, 56-73, 77-79, 132-164, 169-175, 184-209, 212-214, 226-231, 234-238, 249-252, 268-287, 292-307, 316-329, 332-335, 339-341, 376-379, 410-422, 432-435, 450-455, 474-480

Course 3 (Chapters 1, 2, 6, 7, 9)
11-15, 17-25, 30-36, 44-47, 66-72, 78-83, 86-89, 232-238, 240-260, 281-285, 307-308, 382-384, 390-394

STANDARD 7: Computation and Estimation

Course 1 (Chapters 1-13)
4-37, 46-67, 71-77, 112-121, 132-143, 145-148, 152-163, 167-169, 178-180, 182-190, 193-196, 217-219, 228-241, 243-257, 268-301, 312-343, 352-366, 370-373, 375-382, 396-414, 416-425, 434-461, 464-467, 476-503, 514-534, 536-539

Course 2 (Chapters 1-13)
4-16, 21-23, 28-35, 50-63, 66-79, 88-121, 132-175, 197-200, 202-205, 207-217, 228-237, 239-241, 246-248, 253-257, 268-279, 284-307, 317-351, 362-365, 369-373, 376-379, 382-385, 408-417, 419-426, 428-441, 450-458, 460-467, 469-481, 503-506, 510-517, 528-533, 538-545, 547-549, 551-553

Course 3 (Chapters 1-13)
4-16, 21-31, 37-47, 56-95, 104-129, 140-146, 148-167, 188-192, 194-199, 201-204, 206-212, 214-218, 232-238, 240-264, 278-299, 301-306, 309-321, 330-373, 382-417, 428-440, 42-444, 446-467, 476-481, 486-493, 495-507, 518-543, 546-548, 561-591

STANDARD 8: Patterns and Functions

Course 1 (Chapters 1, 2, 4-5, 7-13)
8-11, 28-31, 46-49, 54-57, 64-85, 133-139, 145-155, 178-184, 217-219, 280-283, 285-288, 296-301, 322-327, 334-336, 356-357, 384-385, 402-409, 416-417, 425, 440, 456-467, 476-486, 514-534, 536-539

Course 2 (Chapters 1, 3-7, 9, 11)
24-27, 94-97, 132-149, 207-211, 215-217, 249-257, 288, 360-361, 388-397, 459-467, 478-480

Course 3 (Chapters 1-10, 12, 13)
11-15, 37, 73-76, 107-110, 118-119, 140-147, 153-157, 162-177, 194-195, 200, 205-209, 220-223, 232-234, 239-244, 257-260, 278-280, 294-300, 305-308, 318-321, 334, 342-343, 352, 365, 370-373, 385-389, 396-407, 414-417, 428-467, 528-533, 586-587

STANDARD 9: Algebra

Course 1 (Chapters 1, 2, 4-13)
16-31, 34-37, 46-57, 64-85, 133-139, 141-143, 145-148, 150-155, 161-163, 167-169, 178-180, 182-184, 197-201, 214-216, 243-245, 250-253, 273-283, 285-294, 296-301, 317-321, 324-327, 330-333, 337-343, 370-373, 398-409, 416-420, 425, 434-461, 464-467, 476-503, 515-518

Course 2 (Chapters 1, 2, 4-9, 11-13)
8-23, 28-33, 56-63, 66-69, 138-141, 197-200, 202-205, 207-209, 212-214, 226-231, 234-245, 253-257, 284-287, 297-304, 317-320, 325-328, 332-335, 346-348, 362-365, 369-373, 376-379, 382-385, 450-453, 456-458, 469-481, 514-517, 538-541

Course 3 (Chapters 1-11, 13)
11-29, 32-47, 62-72, 78-89, 92-95, 107-117, 142-147, 153-155, 158-162, 168-177, 188-192, 196-199, 201-204, 210-212, 215-218, 242-252, 257-259, 278-298, 301-321, 330-341, 344-373, 390-394, 398-401, 414-417, 428-440, 4422-449, 452-467, 476-479, 490-493, 560-585, 588-591

STANDARD 10: Statistics

Course 1 (Chapters 1-13)
4-7, 12-15, 28-31, 34-37, 46-85, 10-5-108, 112-115, 133-139, 141-148, 150-151, 157-159, 182-184, 188-190, 193-196, 206-219, 242-245, 250-253, 268-270, 289-291, 298-301, 316, 330-343, 358-361, 410-411, 434-439, 445-448, 456-461, 484-487, 500-503, 514-525, 530, 535-539

Course 2 (Chapters 2-6, 8, 11)
44-63, 66-73, 88-121, 172-175, 212-214, 253, 336-338, 342-345, 459-467, 469-472

Course 3 (Chapters 2-6, 8, 12)
56-65, 73-76, 104-113, 120-123, 126-129, 140-177, 194-195, 260-264, , 330-333, 441, 445, 450-451, 546-548

STANDARD 11: Probability

Course 1 (Chapters 1-4, 8, 11, 13)
34-37, 54-57, 64-67, 105-108, 132, 316, 434-436, 514-518, 522-539

Course 2 (Chapters 4, 8, 10, 11, 13)
165-168, 317-320, 436-441, 464-467, 528-533, 542-545, 547-549, 551-553

Course 3 (Chapters 4-6, 11, 12)
142-146, 200, 253-256, 476-479, 516-545

STANDARD 12: Geometry

Course 1 (Chapters 1, 3-8, 10-13)
34-37, 102-111, 141-143, 145-155, 161-163, 178-181, 188-190, 198-201, 228-234, 238-241, 246-249 , 268-272, 277-283, 296-301, 310-315, 317-320, 322-327, 334-336, 340-343, 398-425, 434-436, 445-448, 456-467, 484-487, 500-503, 522-534, 536-539

Course 2 (Chapters 1, 2, 5-10, 12)
17-20, 28-33, 56-59, 66-69, 191-195, 215-217, 254-257, 292-300, 317-320, 362-365, 369-385, 388-397, 410-441, 490-517

Course 3 (Chapters 1-11, 13)
8-10, 17-25, 38-42, 81-83, 92-95, 107-110, 118-123, 126-129, 148-151, 186-193, 196-223, 239, 245-248, 253-256, 296-299, 301-314, 342-343, 357-373, 382-384, 386-389, 396-401, 404-417, 433-435, 442-444, 446-449, 452-467, 476-503, 506

STANDARD 13: Measurement

Course 1 (Chapters 1, 3-13)
4-7, 12-15, 28-31, 34-37, 95-104, 109-112, 118-121, 133-136, 145-155, 161-166, 188-190, 197-213, 228-234, 238-241, 246-249, 254-257, 268-283, 292-301, 312-315, 317-327, 334-336, 340-343, 352-355, 358-366, 368-373, 379-382, 434-436, 449-452, 459-461, 464-467, 484-487, 522-529, 531-534

Course 2 (Chapters 1-4, 7-9, 11-12)
4-7, 74-76, 118, 148-149, 272-279, 284-287, 289-300, 305-307, 325-328, 360-385, 388-397, 469-472, 490-517

Course 3 (Chapters 1-9, 11)
38-47, 56-58, 62-72, 92-95, 118-119, 126-129, 148-151, 186-193, 196-199, 201-205, 213-215, 249-252, 265-267, 278-280, 307-311, 331-373, 396-401, 404-417, 482-485, 499-507

Motivating Middle School Students

"When are we ever going to use it?"

"Why do I have to learn this?"

Although these may be just complaints from our middle school students, they are legitimate questions that deserve answers. We at Glencoe realize that effective mathematics programs that really motivate middle school students must have more than strong content. They must demonstrate the usefulness of mathematics in a way that relates to student interests.

Mathematics: Applications and Connections makes math a part of students' daily school lives.

- **Applications** begin lessons with a real-life situation that points out how mathematics is used in students' everyday lives as well as in the world about them.
- **When am I ever going to use this?** gives an example of when the math concepts of the lesson will be useful in students' lives.
- **Connection** examples and exercises show how mathematics is used in other courses they may be taking such as Language Arts, Life Science, Geography, Music, Health, Physical Education, and Art.
- **Integration** examples and exercises illustrate how the math they are learning relates to mathematics courses they may take later such as Algebra, Geometry, and Probability and Statistics.

> "The curriculum must go beyond the basics—to be relevant, it must be of interest to students and emphasize the usefulness of mathematics."
>
> Ron Pelfrey, Author

Mathematics: Applications and Connections invites students to look beyond their textbook to see mathematics in the world.

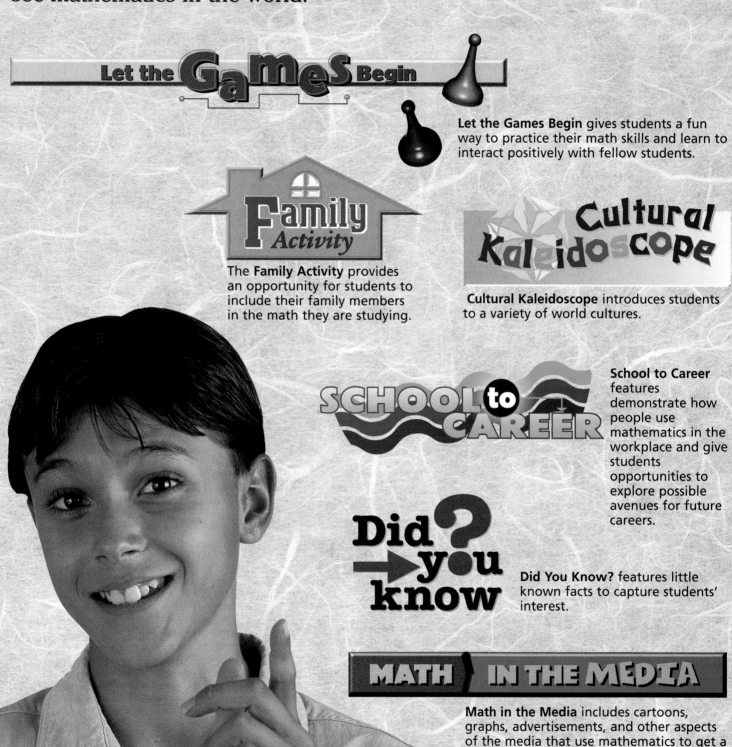

Let the Games Begin gives students a fun way to practice their math skills and learn to interact positively with fellow students.

The **Family Activity** provides an opportunity for students to include their family members in the math they are studying.

Cultural Kaleidoscope introduces students to a variety of world cultures.

School to Career features demonstrate how people use mathematics in the workplace and give students opportunities to explore possible avenues for future careers.

Did You Know? features little known facts to capture students' interest.

Math in the Media includes cartoons, graphs, advertisements, and other aspects of the media that use mathematics to get a point across.

Developing Problem Solving

According to the NCTM Standards, "Problem solving is the process by which students experience the power and usefulness of mathematics in the world around them. It is also a method of inquiry and application ... to provide a consistent context for learning and applying mathematics. Problem situations can establish a 'need to know' and foster the motivation for the development of concepts."

Problem solving is an integral part of every lesson in every course of Glencoe's *Mathematics: Applications and Connections*. How is this accomplished?

The **first chapter** of each course focuses on problem solving.
- Course 1:
  Problem Solving, Numbers, and Algebra
- Course 2:
  Problem Solving, Algebra, and Geometry
- Course 3: **Problem Solving and Algebra**

Thinking Labs in every chapter focus on problem-solving strategies, such as solving a simpler problem and working backward, and present opportunities to solve nonroutine problems.

Frequent **Problem-Solving Study Hints** suggest using various problem-solving strategies to investigate and understand mathematical content and apply strategies to new problem situations.

Reading Math Study Hints provide tips on how to read and interpret the language or symbolism of mathematics.

Study Hint
Reading Math The symbols < and > always point to the lesser of the two numbers.

Critical Thinking exercises in every lesson give students practice in developing and applying higher-order thinking skills.

Glencoe's *Mathematics: Applications and Connections* links practical problem solving to students' real-life interests. Mathematics becomes a vital force in the lives of middle school students as their eyes are opened to the relationship between mathematics and sports, shopping, and other teen interests. They "take ownership" of their skills by writing their own problems and presenting class projects connected to real life.

> "Problem solving is an integral component of this program. It requires students to think critically, examine new concepts, and then extend or generalize what they already know."
>
> Linda Dritsas, Author

Most lessons begin with either a real-world **Application,** an interdisciplinary **Connection,** or content **Integration** that provides students with a reason to learn mathematics. Application, connection, and integration examples also occur throughout the texts to give students the opportunity to study completely worked-out problems.

Applications and Problem Solving exercises in every lesson directly link mathematics to real-world topics like entertainment, and to art, history, science, and other subjects.

The **Chapter Projects** and **Interdisciplinary Investigations** enable students to become more deeply engaged in a problem situation.

USING TECHNOLOGY

Is there a day that goes by when you don't encounter a computer-operated machine? The future world of our students will be even more involved in high technology. Clearly, technology is changing the workplace and the home at an increasingly rapid pace. Without technical mathematical skills, today's students will have little or no chance of finding good jobs.

Mathematics: Applications and Connections provides many opportunities for you to introduce your students to the world of technology as they learn mathematics.

Internet Connections throughout the Student Edition refer students to the Glencoe *Mathematics: Applications and Connections* site **www.glencoe.com/sec/math/mac/mathnet** that provides links to other websites on the Internet. These sites provide more information on the topics that students are studying.

Technology Labs and **Mini-Labs** give students hands-on opportunities to use a computer or a graphing calculator to solve problems. Students learn how to use a spreadsheet to organize data and make predictions. Graphing calculator programs provide an introduction to logic and a way to perform calculations in a timely manner.

Technology Study Hints provide helpful suggestions on how students can use scientific calculators, graphing calculators, or software in studying the topics of the lesson.

In the **Chapter Projects** and **Interdisciplinary Investigations,** students are encouraged to use the Internet and application software like word processing, publishing software, and spreadsheets.

Additional materials in the Teacher's Classroom Resources provide other ways for you to incorporate technology in your lesson planning.

 The **CD-ROM Program** for *Mathematics: Applications and Connections* contains a variety of activities correlated to each chapter. These include a Chapter Introduction and a Resource Lesson for each lesson in the Student Edition. The activities also include Interactive Lessons and Extended Activities for some of the lessons in the chapter. Every chapter includes an Assessment Game that is a cumulative review of skills they have learned.

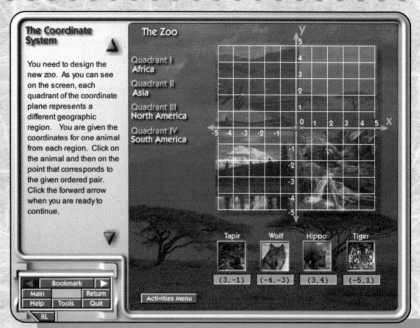

CD-ROM Program; Course 2, Chapter 7

MindJogger Videoquizzes feature a game show format that provides an alternative assessment and an entertaining way for your students to review chapter concepts and problem-solving skills.

Interactive Mathematics Tools Software helps students gain mathematical power through interactive activities that combine video, sound, animation, graphics, and text.

All of the blackline masters for *Mathematics: Applications and Connections* are available on the **Electronic Teacher's Classroom Resources** CD-ROM. There's no need to carry a large array of booklets from room to room—just print out the masters you wish to use from a single CD-ROM that works on both Macintosh and Windows formats.

Ass(e)ssment

The assessment tools built into *Mathematics: Applications and Connections* are designed to assess traditional basic skills as well as those skills that will be required for success in the 21st century. In addition to basic skills, *Mathematics: Applications and Connections* also helps you assess students' ability to organize information, apply previously-learned information, and make conjectures based on gathered data. The curriculum alignment in *Mathematics: Applications and Connections* also enables students to perform well on standardized testing at both state and national levels.

The following features and components help you accurately assess each student's achievement.

In the Student Editions...

- Every lesson has a **Mixed Review** that includes a **Test Practice** item.
- Every chapter has **Math Journals**, a **Chapter Project**, a **Mid-Chapter Self Test**, **Standardized Test Practice**, and a **Chapter Test**.
- Every chapter has a **Study Guide and Assessment** that includes Vocabulary, Understanding and Using Vocabulary, Examples and Review Exercises for each Objective, Applications and Problem Solving, a **Performance Task**, and a **Portfolio** suggestion.
- Since practically every mathematics test is also a reading test, several **Reading Math** study hints are included in every chapter.

In the Teacher's Wraparound Editions...

- Every lesson is correlated to the major national standardized tests: CAT, CTBS, ITBS, MAT, SAT, and Terra Nova.
- Every lesson includes a **5-Minute Check** that covers the previous lesson or chapter.
- Every lesson has a **Closing Activity** that involves writing, speaking, or modeling.

In the supplementary materials...

Assessment and Evaluation Masters include:

1. For each chapter: three Multiple Choice tests (Basic, Average, Honors), three Free-Response Tests (Basic, Average, Honors), Performance Assessment (includes a Scoring Guide), Standardized Test Practice, a Cumulative Review, four quizzes, and a Mid-Chapter Test.
2. Also Included: a Placement Test, two Semester Tests and a Final Test

MindJogger Videoquizzes (VHS) review each chapter by using a game show format. As students compete on teams, they hear and see each review problem as it is presented and then completely solved.

Test and Review Software (Windows & Macintosh) combines a test generator and test bank. You can easily create your own free-response, multiple choice, and open-ended tests for honors, average, and basic students.

5-Minute Check Transparencies provide a quick review of the previous lesson or chapter. There is one full-color transparency for every lesson. The 5-Minute Check is also printed in the Teacher's Wraparound Edition to make your lesson plans easier to prepare.

The CD-ROM Program includes guided practice and an Assessment Game that is similar to Trivial Pursuit.

Alternative Assessment in the Mathematics Classroom, part of the Glencoe Mathematics Professional Series, gives an overview of the latest trends in assessment.

Using Projects

Doing mathematics is so much more effective than memorizing mathematics!

Mathematics: Applications and Connections gives you and your students several opportunities to engage in projects of varying lengths that put mathematics into motion.

The **Chapter Project** is introduced at the beginning of the chapter and sets in motion several activities that culminate with **Completing the Chapter Project** in the Chapter Study Guide and Assessment. Students apply the mathematics they learn in the chapter to various real-life situations. They are often asked to represent their results as a working model or in statistical graph form. Throughout the chapter, **Working on the Chapter Project** exercises remind students of the next steps in their project. These activities not only implement the mathematics student are learning, but give students the opportunity to work together as a team.

The **Interdisciplinary Investigations** found at the ends of Chapters 3, 6, 9, and 12 show how the mathematics students have learned can be applied to other courses they may be taking. Each investigation uses the skills from the previous three chapters. In addition to the problem they are assigned to investigate, they are given writing assignments that explore other areas such as Language Arts, Foreign Language, Science, Health, Physical Education, and Social Studies.

Interactive Mathematics: Activities and Investigations

offers an innovative approach to teaching and learning middle school mathematics. Each of the 18 units that make up this comprehensive, activity-based program may be used to enhance chapters in *Mathematics: Applications and Connections.*

The interleaf pages of each chapter in the Teacher's Wraparound Edition highlight two activities from *Interactive Mathematics: Activities and Investigations* that are appropriate for that lesson. However, there are many other activities that may be useful.

The chart below summarizes the mathematical focus of each unit of *Interactive Mathematics: Activities and Investigations.*

	Unit	Title	Mathematical Focus
Course 1	1	From the Beginning	Building Math Power
	2	A Million to One	Number Sense
	3	Just the Right Size	Scale Drawings and Proportional Reasoning
	4	Through the Looking Glass	Spatial Visualization
	5	Get a Clue	Logical Reasoning
	6	The Road Not Taken	Graph Theory and Networks
Course 2	7	Take It From the Top	Building Math Power
	8	Data Sense	Statistics and Data Analysis
	9	Don't Fence Me In	Area and Perimeter
	10	Against the Odds	Probability
	11	Cycles	Algebra Patterns
	12	Treasure Island	Geometry and Measurement
Course 3	13	Start Your Engines	Building Math Power
	14	Run for Cover	Surface Area and Volume
	15	On the Move	Graphing and Functions
	16	Growing Pains	Linear and Exponential Growth
	17	Infinite Windows	Fractals and Chaos Theory
	18	Quality Control	Applied Data Analysis

Glencoe has correlated *Mathematics: Applications and Connections (MAC)* with the *Interactive Mathematics: Activities and Investigations (IMAI)* units to provide a unique opportunity for you to tailor a program to your middle school students. Integrating the two programs allows you to emphasize areas and mathematical ideas that interest and fit the unique needs of the students in your class.

MAC Course 1 Chapter	IMAI Units 1-6	Other IMAI Units	MAC Course 2 Chapter	IMAI Units 7-12	Other IMAI Units	MAC Course 3 Chapter	IMAI Units 13-18	Other IMAI Units
1	1, 5		1	7		1	15	8
2	6	8	2	7	2	2	15	
3	2		3	8		3	16	3
4	2	7, 9	4	11	18	4	18	8
5	2, 3		5	11	16	5	14	4
6	3	11	6	11	15	6	16, 17	
7	3	10	7	9, 10		7	16, 17	
8	3	10	8	10	3	8	16, 17	3, 12
9	4	12	9	12		9	17, 18	10
10	4	14	10	9		10	15, 16	
11		11	11	10		11	14	
12		11	12	9	4, 14	12	17	6, 10
13	6	10	13	10	6	13	16	5

Interactive Mathematics: Activities and Investigations is available as individual units or as a three book series that groups Units 1-6, 7-12, and 13-18. Each unit also has a Classroom Instructional Resources binder that includes teaching instructions, blackline masters, transparencies, and guidance on using projects in the classroom.

For more information on *Interactive Mathematics: Activities and Investigations,* contact your nearest Glencoe Regional Office or call 1-800-334-7344.

Using Games in the Math Classroom

During the past few years, the use of games in the mathematics classroom has increased dramatically. These activities tend to be very motivational, and they often help improve students' attitudes toward mathematics in general. Because games are enjoyable, they allow students to learn and review mathematical skills and concepts while they unknowingly increase their problem-solving, logic, and computational skills.

Why Use Games? While mathematical games are not the only method you can use to provide individualized instruction or enrichment, they may be a very effective strategy. According to NCTM, most mathematical games serve one or more of the following functions.
- To develop concepts.
- To provide drill and reinforcement.
- To develop perceptual abilities.
- To provide opportunities for logical thinking and/or problem solving.

Therefore, mathematical games can be used in a number of different ways. You may want to use a game to introduce a new topic, thus encouraging students to employ discovery learning. You may want to use a game within the framework of your lesson instruction, for example, in the last few moments of a class period. Finally, you may want to use a game to review a concept previously taught.

The games presented in **Let the Games Begin** are designed to be useful and enjoyable to all students. Feel free to develop new and fun variations that will make each game your own.

More information on games is available from Glencoe on the Internet at **www.glencoe.com/sec/math/mac/mathnet**.

In addition to the games presented in the Student Edition, more games are available in the **Classrooms Games** booklet.

Multiple Learning Styles

People learn in many different ways. There are several different learning styles that help us approach and solve problems. Everyone possesses varying degrees of each of these learning styles, but the ways in which they combine and blend are as varied as the personalities of the individuals. Glencoe's *Mathematics: Applications and Connections* provides you with ways to accommodate students with these diverse learning styles.

Learning Style	Characteristics of Students	Activities in Student Edition
verbal/linguistic	read regularly, write clearly, and easily understand the written word	**Communicating Mathematics** exercises ask students to tell, write, and explain mathematical concepts. Students also express what they have learned in their **Math Journals**.
logical	use numbers, logic, and critical thinking skills	Clearly-written **Examples** present important concepts, and **Critical Thinking** exercises extend those concepts. **Thinking Labs** encourage students to practice their logical thinking skills by using various strategies.
visual/spatial	think in terms of pictures and images	**Integration** and **Hands-On Lab** exercises ask students to draw or show mathematical concepts through modeling, coordinate grids, charts, and graphs.
auditory/musical	have "good ears" and can produce rhythms and melodies	Multimedia software, such as the **CD-ROM Program**, **MindJogger Videoquizzes**, and **Interactive Mathematics Tools Software**, can be easily incorporated into lessons.
kinesthetic	learn from touch, movement, and manipulating objects	**Hands-On Labs** and **Mini-Labs** provide for physical involvement in learning.
interpersonal	understand and work well with other people	**Chapter Projects, Interdisciplinary Investigations,** as well as **Hands-On Labs, Technology Labs,** and **Mini-Labs,** allow students to collaborate with others.
intrapersonal	have a realistic understanding of their strengths and weaknesses	**Math Journals, Portfolios,** and **Family Activities** help students personalize mathematics.
naturalist	can distinguish among, classify, and use features of the environment	**Application** and **Connection** examples and exercises show students how mathematics relates to the world around them.

As mathematics teacher, you may want to assign activities to students that accommodate their strongest learning styles, but frequently ask them to use their weakest learning styles. Additional activities are provided in the bottom margins of the lesson notes in the Teacher's Wraparound Edition.

The resources available in *Mathematics: Applications and Connections* guarantee that your classroom will be a multisensory environment, providing multiple paths for student learning.

Involving Parents and the Community

When children enter 6th grade, for a variety of reasons, it may become difficult for parents or guardians to remain on top of what's going on at school. The curricula grow more specialized, the typical middle school student becomes more independent of his or her parents, and multiple teachers replace the primary teacher of the elementary school years.

When parents do have the opportunity to meet with teachers, they often ask what they can do to motivate their children. Here are some ideas to share with parents. Developed by Reginald Clark of the Academy for Educational Development in Washington, D.C., they are designed to foster positive attitudes and boost learning.

- Share the fact that there is an inverse correlation between excessive TV viewing and high achievement in school.
- Stress the importance of seeing that their children complete homework assignments.
- Urge them to provide time, space, and materials needed for homework and reading.
- Remind them how important it is to listen to their children read and/or to read to their children.

Each chapter in Glencoe's *Mathematics: Applications and Connections* contains a suggested **Family Activity** for your students to complete with a family member. The **Family Letters and Activities** provide a letter and activity for each chapter to encourage students' families to become active participants in their students' learning.

Fostering Community Involvement

There are many ways to involve the community in your mathematics classroom and to share your learning with them.

- Send photos and press releases to your local newspaper to inform the community about the exciting things your students are learning.
- Have a "math career day" and invite local people to describe to students how they use math in their jobs.
- Set up a "shadow day" in which students spend a half-day "shadowing" people in the workplace who use math in their careers.
- Find out about – or implement your own – math fairs and competitions that give your students a chance to shine in the "outside world."

Involving Parents and the Community in the Mathematics Classroom, one of the booklets in Glencoe's Mathematics Professional Series, presents additional suggestions on how parents and the community can be active participants in supporting mathematics instruction.

Classroom Management

Meet Susan Uhrig. Mrs. Uhrig has taught in the Columbus Public Schools in Columbus, Ohio, for more than 25 years. She currently uses *Mathematics: Applications and Connections* in her classroom at Monroe Middle School.

Just one look at her classroom tells you that Mrs. Uhrig enjoys teaching middle school students. Their work is displayed throughout her bright classroom. Positive sayings on the bulletin boards encourage students to do their best. Manipulatives are kept out so they are easy to use—evidence that Mrs. Uhrig encourages her students to do mathematics.

Mrs. Uhrig says that she enjoys the diverse personalities of her students. You never know what challenges lie ahead each day! She organizes her classroom in ways that really get her students involved and keep their parents informed. Here are some of her ideas.

- Write the **assignment** and **objective** for the day on the chalkboard for students to copy when they come into class.
- Keep a **master notebook** where assignments are written so students can refer to the notebook if they miss a class.
- Have students keep assignment **logs** throughout the year. This helps students get and stay organized.
- Keep **manipulatives** on a table so they are easy to find and use. Mrs. Uhrig assigns several students to help distribute and collect manipulatives and calculators.
- Use "homework coupons" as a **reward** for special behavior.
- Have students write in their **journals** every day. Mrs. Uhrig has her students do the Check for Understanding exercises and 5-Minute Checks in their journals and allows them to refer to their journals during tests.
- Communicate with **parents and guardians** early and often. At the beginning of the year, Mrs. Uhrig sends each parent a letter that indicates her classroom policies. Throughout the year, she communicates with parents though student logs, interim reports, and grade cards.

There are many teachers like Mrs. Uhrig who, through years of experience, have acquired classroom management techniques that help their classrooms run smoothly. They also possess a wealth of knowledge about how to teach mathematics. Glencoe allows you to network with these teachers through the **Classroom Vignettes** in the Teacher's Wraparound Edition. Teachers from across the country share their ides with you so that you may add to your own list of classroom management techniques.

Classroom Vignettes

Glencoe's unique **Classroom Vignettes** are classroom-tested teaching suggestions made by experienced mathematics educators. Each vignette was written by a teacher who uses Glencoe's ***Mathematics: Applications and Connections*** or by a Glencoe reviewer, consultant, or author. Glencoe thanks these outstanding mathematics educators for their unique contributions.

Robert Allen (p. 171)
Explorer Middle School
Phoenix, AZ

Beth Murphy Anderson (p. 382)
Brownell Talbot School
Omaha, NE

Richie Berman (p. 39)
University of California,
Santa Barbara, CA

Cindy J. Boyd (pp. 465, 566)
Abilene High School
Abilene, TX

Anne B. Breda (p. 115)
Hambrick Middle School
Houston, TX

Mary Jo Deschene (p. 142)
St. Francis Junior High School
St. Francis, MN

Mark Fisher (p. 62)
Clearview Regional Middle School
Mullica Hill, NJ

Denis J. Fogaroli (p. 398)
Eisenhower Middle School
Succasunna, NJ

Suetta Gladfelter (p. 206)
Caroline Middle School
Milford, VA

Cindy Hansard (p. 357)
North Forsyth School
Cumming, GA

Mrs. Sandy Hardy (p. 94)
Taylor Street Middle School
Griffin, GA

Berchie Gordon-Holliday (p. 540)
Northwest Local School District
Cincinnati, OH

Dennis Iffland (p. 278)
East Toledo Junior High School
Toledo, OH

Lorraine Jefferson (p. 363)
Ridgeview Middle School
Columbus, OH

Mary Judy (p. 17)
Westmoor Middle School
Columbus, OH

Janine Lopez (p. 237)
Broomfield Heights Middle School
Broomfield, CO

Jerry W. Murkerson (p. 281)
West Bainbridge Middle School
Bainbridge, GA

Clare L. Polzin (p. 196)
Red Rock Central Junior High
  School
Lamberton, MN

Susan Stier (p. 232)
Belle Plaine Junior High School
Belle Plaine, MN

Steve Werges (p. 482)
LaSalle Springs Middle School
Glencoe, MD

Research Activities

Glencoe's **Mathematics: Applications and Connections**, as well as the entire Glencoe Mathematics Series, is the product of ongoing, classroom-oriented research that involves students, teachers, curriculum supervisors, administrators, parents, and college-level mathematics educators.

The programs that make up the Glencoe Mathematics Series are currently being used by millions of students and tens of thousands of teachers. The key reason that Glencoe Mathematics programs are so successful in the classroom is the fact that each Glencoe author team is a mix of practicing classroom teachers, curriculum supervisors, and college-level mathematics educators. Glencoe's balanced author teams help ensure that Glencoe Mathematics programs are both practical and progressive.

Prior to publication of a Glencoe program, typical research activities include:
- a review of educational research and recommendations made by groups such as NCTM
- mail surveys of mathematics educators
- discussion groups involving mathematics teachers, department heads, and supervisors
- focus groups involving mathematics educators
- face-to-face interviews with mathematics educators
- telephone surveys of mathematics educators
- in-depth analyses of manuscript by a wide range of reviewers and consultants
- field tests in which students and teachers use pre-publication manuscript in the classroom

Feedback from teachers, curriculum supervisors, and even students who currently use Glencoe Mathematics programs is also incorporated as Glencoe plans and publishes new and revised programs. For example, Classroom Vignettes, which are printed in the *Teacher's Wraparound Editions*, are one result of this feedback.

All of this research and information is used by Glencoe's authors and editors to publish the best instructional resources possible.

Scope and Sequence

Mathematics: Applications and Connections is a comprehensive, well-balanced, three-course program that prepares middle school students for success in algebra and geometry. Through a carefully planned scope and sequence of mathematical topics, students encounter, practice, and extend their knowledge of mathematics to promote confidence and mastery.

The chart below shows the chapter titles for the three courses in *Mathematics: Applications and Connections.* The following pages present a detailed chart describing the depth to which each topic is covered.

Chapter	Course 1	Course 2	Course 3
1	Problem Solving, Numbers, and Algebra	Problem Solving, Algebra, and Geometry	Problem Solving and Algebra
2	**Statistics:** Graphing Data	Applying Decimals	**Algebra:** Using Integers
3	Adding and Subtracting Decimals	**Statistics:** Analyzing Data	Using Proportion and Percent
4	Multiplying and Dividing Decimals	Using Number Patterns, Fractions, and Percents	**Statistics:** Analyzing Data
5	Using Number Patterns, Fractions, and Ratios	**Algebra:** Using Integers	**Geometry:** Investigating Patterns
6	Adding and Subtracting Fractions	**Algebra:** Exploring Equations and Functions	Exploring Number Patterns
7	Multiplying and Dividing Fractions	Applying Fractions	**Algebra:** Using Rational Numbers
8	Exploring Ratio, Proportion, and Percent	Using Proportional Reasoning	Applying Proportional Reasoning
9	**Geometry:** Investigating Patterns	**Geometry:** Investigating Patterns	**Algebra:** Exploring Real Numbers
10	**Geometry:** Understanding Area and Volume	**Geometry:** Exploring Area	**Algebra:** Graphing Functions
11	**Algebra:** Investigating Integers	Applying Percents	**Geometry:** Using Area and Volume
12	**Algebra:** Exploring Equations	**Geometry:** Finding Volume and Surface Area	Investigating Discrete Math and Probability
13	Using Probability	Exploring Discrete Math and Probability	**Algebra:** Exploring Polynomials

Mathematics: Applications and Connections
Scope and Sequence

Legend: I = Introduce, D = Develop, R = Reinforce

Course	1	2	3
PROBLEM SOLVING			
Develop a plan	D	D	R
Strategies			
Look for a pattern	D	D	R
Solve a simpler problem		D	R
Act it out		D	R
Guess and check	D	D	R
Draw a diagram	D	D	R
Make a table	D	D	R
Work backward	D	D	R
Choose the method of computation	D	D	R
Make a list	D	D	R
Eliminate the possibilities	D	D	R
Determine reasonable answers	D	D	R
Make a model	D	D	R
Use a graph	D	D	R
Use an equation	D	D	R
Use logical reasoning	D	D	R
Use the Pythagorean Theorem		D	R
Use a Venn diagram	D	D	R
Use a frequency table	D	D	R
Use a spreadsheet	I	D	R
Use proportional reasoning			R
NUMBER AND OPERATIONS			
Number Relationships			
Decimals			
Decimal concepts	D	R	R
Reading and writing	D	R	R
Decimal place value	D	R	R
Comparing and ordering	D	R	R
Rounding	D	R	R
Relating decimals and fractions	D	R	R
Relating decimals, ratios, and percents	D	R	R
Terminating and repeating decimals	D	R	R
Scientific notation		D	R
Powers of ten		D	R
Fractions			
Fraction concepts	D	R	R
Writing mixed numbers as fractions	D	R	R
Mixed numbers and improper fractions	D	R	R

Course	1	2	3
Equivalent fractions	D	R	R
Comparing and ordering fractions	D	R	R
Simplifying fractions	D	R	R
Least common denominator (LCD)	D	R	R
Rounding and estimating fractions	D	R	R
Relating fractions and decimals	D	R	R
Relating fractions and percents		D	R
Proportional Reasoning			
Ratio			
Concept of ratio	D	D	R
Reading and writing ratios	D	D	R
Simplifying ratios	D	D	R
Relating ratios and fractions	D	D	R
Relating ratios and rates	D	D	R
Ratio and probability	D	D	R
The Golden Ratio			R
Proportion			
Concept of proportion	D	D	R
Solving proportions	D	D	R
Property of proportion (cross product)	D	D	R
Scale drawings	D	D	R
Similar figures	D	D	R
Dilations		D	R
Indirect measurement		D	R
Percent			
Concept of percent	D	D	R
Writing fractions and decimals as percent	D	D	R
Percents greater than 100% or less than 1%	D	D	R
Find percent of a number	D	D	R
Percent one number is of another		D	R
Finding number when percent is known		D	R
Percent proportion		D	R
Relating percent and ratio		D	R
Percent equation		D	R
Capture/recapture		D	
Non-proportional relationships			R
Computations and Estimation			
Order of operations	D	D	R
Decimals			
Adding and subtracting	D	R	R
Multiplying by a whole number	D	R	R

Course	1	2	3
Multiplying two decimals	●	●	●
Dividing by a whole number	●	●	●
Dividing by decimals	●	●	●
Dividing with zeros in the quotient	●		
Fractions			
Adding and subtracting	●	●	●
Subtracting with renaming	●	●	●
Multiplying and dividing	●	●	●
Add and subtract mixed numbers	●	●	●
Multiply and divide mixed numbers	●	●	●
Percents			
Discount	●	●	●
Sales tax		●	
Simple interest		●	●
Percent of change		●	●
Integers			
Adding and subtracting		●	●
Multiplying and dividing		●	●
Estimation			
Whole numbers			
Rounding	●	●	●
Sums and differences	●	●	●
Products and quotients	●	●	●
Decimals			
Rounding	●	●	●
Sums and differences	●	●	●
Products and quotients	●	●	●
Fractions			
Sums and differences	●	●	●
Products and quotients	●	●	●
Percents	●	●	●
Use equivalent fractions, decimals, and percents			●
Strategies for estimating			
Rounding	●	●	●
Compatible numbers	●		
Capture-recapture		●	
Clustering	●	●	
Square roots		●	●
Area or volume	●	●	●
Mental math			
Divisibility patterns	●	●	●
Compatible numbers	●		
Solving equations mentally		●	
Finding percents			●
Powers of ten		●	
Using formulas	●	●	
Number Systems and Number Theory			
Reading and writing whole numbers	●	●	●

Course	1	2	3
Place value of whole numbers	●	●	●
Place value of decimals	●	●	●
Comparing and ordering			
Whole numbers	●	●	●
Decimals	●	●	●
Fractions	●	●	●
Integers	●	●	●
Rationals			●
Positive exponents	●	●	●
Negative exponents			●
Divisibility patterns	●	●	●
Prime and composite numbers	●	●	●
Relative primes			●
Prime factorization	●	●	●
Greatest common factor (GCF)	●	●	●
Least common multiple (LCM)	●	●	●
Scientific notation		●	●
Square roots		●	●
Factorials		●	●
Properties			
Properties of numbers		●	●
Distributive property	●	●	●
Property of proportions (cross products)	●	●	●
Properties of equality		●	●
Density property			●

PATTERNS AND FUNCTIONS

	1	2	3
Numeric patterns			
Sequences	●	●	●
Fibonacci sequence			●
Pascal's triangle		●	●
Divisibility patterns	●	●	●
Geometric patterns			
Recognizing geometry patterns	●	●	●
Tessellations	●	●	●
Fractals		●	
Represent relationships			
Tables	●	●	●
Graphs		●	●
Function rules		●	●
Analyze functional relationships			●
Use patterns and functions to solve problems	●	●	●

ALGEBRA

	1	2	3
Integers			
Reading and writing integers	●	●	●
Graphing integers on a number line	●	●	●
Comparing and ordering integers	●	●	●
Adding and subtracting integers	●	●	●

Course	1	2	3
Multiplying and dividing integers	●	●	●
Absolute value		●	●
Rational numbers			
Identify and simplify rational numbers			●
Properties of rational numbers			●
Density property			●
Rational numbers and decimals			●
Scientific notation			●
Comparing and ordering			●
Solving equations with rational number solutions			●
Real numbers			
Identify and classify real numbers			●
Square roots		●	●
Irrational numbers			●
Density property			●
Functions			
Function machine	●		●
Function tables	●	●	●
Linear functions			●
Analyze tables and graphs	●	●	●
Equations and expressions			
Concepts of variable, expression, equation	●	●	●
Order of operations		●	●
Evaluate algebraic expressions	●	●	●
Write algebraic expressions and equations		●	●
Solve addition and subtraction equations	●	●	●
Solve multiplication and division equations	●	●	●
Solve two-step equations	●	●	●
Solve equations with two variables		●	●
Solve inequalities		●	●
Solve equations with concrete methods	●	●	
Solve equations with informal methods	●	●	●
Solve equations with formal methods		●	●
Graphing			
Integers on a number line	●	●	●
Irrational numbers on a number line			●
Inequalities on a number line		●	●
Points on a coordinate plane	●	●	●
Transformations on a coordinate plane	●	●	●
Functions	●	●	●
Linear functions			●
Quadratic functions			●
Equations		●	●
Systems of equations		●	●
Using graphing calculators		●	●
Polynomials			
Model with algebra tiles			●
Represent and simplify polynomials			●
Add, subtract, and multiply polynomials			●

Course	1	2	3
Factor polynomials			●
Multiply binomials			●
Apply algebra to real-world and math problems	●	●	●
Use spreadsheets and formulas	●	●	●

STATISTICS

Course	1	2	3
Taking a survey	●	●	●
Analyzing survey data	●	●	●
Organizing data			
Using a table to organize data	●	●	●
Frequency tables	●	●	●
Using tables to solve problems	●	●	●
Using matrices to organize data			●
Constructing and interpreting graphs			
Bar graphs	●	●	●
Circle graphs	●	●	●
Line graphs	●	●	●
Stem-and-leaf plots	●	●	●
Box-and-whisker plots		●	●
Line plots	●		●
Histograms			●
Scatter plots	●		●
Maps that show statistics			●
Choosing an appropriate display			●
Interpreting data			
Clusters		●	●
Mean, median, and mode	●	●	●
Range and quartiles		●	●
Misleading graphs and statistics		●	●
Making predictions from statistics	●	●	●
Making predictions from graphs	●	●	●
Making predictions from a sample	●	●	●

PROBABILITY

Course	1	2	3
Outcomes	●	●	●
Simple event	●	●	●
Independent events	●	●	●
Dependent events		●	●
Complementary events	●		
Experimental probability	●	●	●
Theoretical probability		●	●
Tree diagrams		●	●
Counting principle		●	●
Permutations and combinations	●	●	●
Probability and ratio		●	●
Fair and unfair games		●	●
Simulations or experiments		●	●
Area models	●	●	●
Capture-recapture		●	

Course	1	2	3
Punnett squares			●

GEOMETRY

Course	1	2	3
Constructions			
Congruent segments	●		●
Perpendicular lines		●	●
Parallel lines		●	●
Segment bisectors	●		
Congruent angles	●		●
Angle bisectors	●		
Polygons, inscribed	●	●	●
Congruent triangles			●
Angles			
Classify and measure angles	●	●	●
Sum of angle measures		●	●
Parallel lines and transversal		●	●
Polygons			
Identify polygons	●	●	●
Classify triangles and quadrilaterals	●	●	●
Identify congruent figures	●	●	●
Using polygons as networks			●
Triangles			
Determine congruent triangles			●
Right triangle relationships (trigonometry)			●
Pythagorean Theorem		●	●
Special right triangles			●
Similarity			
Corresponding parts of similar figures	●	●	●
Identify similar figures	●	●	●
Scale drawings	●	●	●
Dilations	●	●	
Circles			
Circumference (radius, diameter)	●	●	●
Area	●	●	●
Perimeter	●	●	●
Area			
Rectangles	●	●	●
Parallelograms (base, height)	●	●	●
Trapezoids		●	●
Triangles	●	●	●
Circles	●	●	●
Square roots and area of squares		●	●
Pick's Theorem			●
Area and probability	●	●	
Transformations			
Translations, reflections, and rotations	●	●	●
Dilations		●	●
On the coordinate plane	●	●	●
Tessellations	●	●	●

Course	1	2	3
Symmetry	●	●	●
Solids			
Identify, draw three-dimensional figures	●	●	●
Nets	●	●	●
Surface area		●	●
Volume	●	●	●
Coordinate Geometry			
Graphing ordered pairs	●	●	●
Distance in the coordinate plane			●
Transformations on the coordinate plane	●	●	●
Patterns			
Recognizing geometric patterns	●	●	●
Symmetry	●	●	●
Tessellations	●	●	●
Fractals		●	
Trigonometry			●
Inductive and deductive thinking			●

MEASUREMENT

Course	1	2	3
Metric System			
Units of length, capacity, and mass	●	●	●
Changing units within the metric system	●	●	●
Customary system			
Units of length, capacity, and weight	●	●	●
Change units within the customary system	●	●	●
Time	●		
Non-standard units	●		
Perimeter and circumference	●	●	●
Area			
Irregular figures	●	●	●
Rectangles	●	●	●
Parallelograms	●	●	●
Triangles	●	●	●
Circles	●	●	●
Trapezoids		●	●
Surface area			
Rectangular prisms		●	●
Triangular prisms			●
Cylinders		●	●
Volume			
Rectangular prisms	●	●	●
Cylinders		●	●
Pyramids and circular cones		●	●
Relating area and perimeter	●		
Relating surface area and volume			●
Precision and significant digits			●
Indirect measurement		●	●

Legend: ● Introduce   ● Develop   ● Reinforce

Planning Your

The charts on these two pages contain course planning guides for three types of classes. A Pacing Chart included in the interleaf of each chapter shows in greater detail the number of days suggested for all three types of pacing and which lessons are appropriate. This same pacing is repeated in the margin notes for each lesson for your convenience.

40- TO 50-MINUTE CLASS PERIODS

The chart below gives suggested pacing guides for two options, Standard and Honors, for four 9-week grading periods. The Standard option covers Chapters 1-12 while the Honors option covers Chapter 1-13.

The total number of days suggested for each option is 165 days. This allows for teacher flexibility in planning due to school cancellation or shortened class periods.

Grading Period	Standard		Honors	
	Chapter	Days	Chapter	Days
1	1	13	1	13
	2	15	2	14
	3	11	3	10
	4-1 to 4-2	4	4-1 to 4-3	5
2	4-3 to end	9	4-3B to end	8
	5	15	5	15
	6	15	6	13
			7-1 to 7-5A	6
3	7	15	7-5 to end	9
	8	15	8	15
	9	13	9	11
			10-1 to 10-4	6
4	10	14	10-5 to end	8
	11	13	11	11
	12	13	12	11
			13	10

Course of Study

BLOCK SCHEDULE, 90-MINUTE CLASS PERIODS

The chart below gives a suggested pacing guide for teaching Course 3 using block scheduling in one semester (classes meet every day) or during the entire year (classes meet every other day). A total of 85 class periods is suggested to cover Chapters 1-12.

Chapter	Class Periods
1	7
2	8
3	6
4	7
5	8
6	8
7	8
8	7
9	7
10	7
11	6
12	6

For more detailed descriptions for lesson planning and pacing, please refer to the **Lesson Planning Guide** and the **Block Scheduling Booklet** for **Course 3** of *Mathematics: Applications and Connections*.

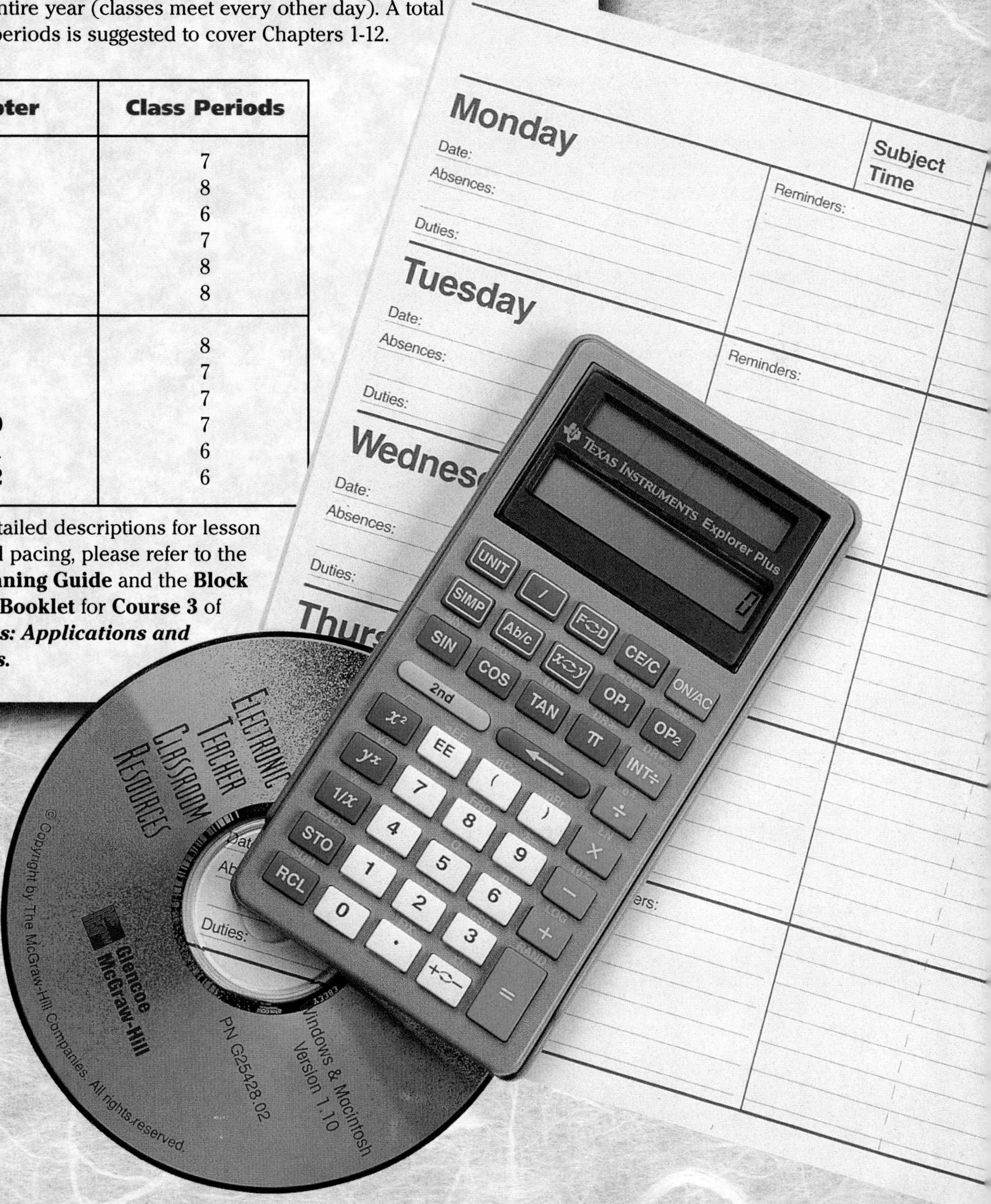

CHAPTER 1: Problem Solving and Algebra

Previewing the Chapter

Overview

This chapter provides a solid foundation for problem solving throughout the text. It begins with a detailed look at the four-step problem-solving plan. Students learn the meaning of a variable and how to use variables to write algebraic expressions and equations. Methods for solving various types of equations and inequalities are developed. Students solve problems by working backward and using function machines. Finally, students apply the methods shown in the chapter to problems involving the perimeters and areas of rectangles, squares, and parallelograms.

Lesson (pages)	Lesson Objectives	NCTM Standards	Standardized Tests	State/Local Objectives
1-1 (4–7)	Solve problems by using the four-step plan.	1–5, 7	MAT	
1-2 (8–10)	Use powers and exponents in expressions.	1–5, 7, 12	CAT, SAT	
1-3 (11–15)	Evaluate expressions and find the solutions of equations.	1–4, 6–9	CTBS, MAT, SAT, TN	
1-3B (16)	Evaluate expressions by using a graphing calculator.	1, 2, 7, 9		
1-4 (17–20)	Solve equations by using the subtraction and addition properties of equality.	1–6, 9, 12	CTBS, ITBS, MAT, SAT, TN	
1-5 (21–25)	Solve equations by using the division and multiplication properties of equality.	1–7, 9, 12	CTBS, ITBS, MAT, SAT, TN	
1-6 (27–29)	Write algebraic expressions and equations from verbal phrases and sentences.	1–4, 7, 9	SAT	
1-7A (30–31)	Solve problems by working backward.	1–7	MAT	
1-7 (32–36)	Solve two-step equations.	1–6, 9	ITBS, MAT, SAT	
1-7B (37)	Use function machines to find output from a given input and vice versa.	1–4, 7–9	CTBS, SAT, TN	
1-8 (38–42)	Find the perimeters and areas of rectangles, squares, and parallelograms.	1–4, 7, 9, 12, 13	CAT, CTBS, ITBS, MAT, SAT, TN	
1-9 (43–47)	Write and solve inequalities.	1–4, 6, 7, 9, 13	CTBS, SAT, TN	

CAT = California Achievement Tests, CTBS = Comprehensive Tests of Basic Skills, ITBS = Iowa Tests of Basic Skills, MAT = Metropolitan Achievement Tests, SAT = Stanford Achievement Tests, TN = Terra Nova

Organizing the Chapter

CD-ROM
All of the blackline masters in the Teacher's Classroom Resources are available on the **Electronic Teacher's Classroom Resources** CD-ROM.

LESSON PLANNING GUIDE

Lesson	Extra Practice (Student Edition)	Blackline Masters (Page Numbers)										
		Study Guide	Practice	Enrichment	Assessment & Evaluation	Classroom Games	Diversity	Hands-On Lab	School to Career	Science and Math Lab Manual	Technology	Transparencies A and B
1-1	p. 605	1	1	1								1-1
1-2	p. 605	2	2	2							53	1-2
1-3	p. 605	3	3	3	15							1-3
1-3B												
1-4	p. 606	4	4	4		1–2						1-4
1-5	p. 606	5	5	5	14, 15		27					1-5
1-6	p. 606	6	6	6						49–52		1-6
1-7A	p. 607											
1-7	p. 607	7	7	7	16							1-7
1-7B								40				
1-8	p. 607	8	8	8				68				1-8
1-9	p. 608	9	9	9	16				27		54	1-9
Study Guide/ Assessment					1–13, 17–19							

Other Chapter Resources

Student Edition
Chapter Project, pp. 3, 7, 25, 41, 51
Math in the Media, p. 20
School to Career, p. 26
Let the Games Begin, p. 47

Technology
 CD-ROM Program
 Interactive Mathematics Tools Software

Applications
Family Letters and Activities, pp. 53–54
Investigations and Projects Masters, pp. 17–20
Meeting Individual Needs
Transition Booklet, pp. 7–12
Investigations for the Special Education Student, pp. 1–2, 7

Teacher's Classroom Resources
Teaching Aids
Answer Key Masters
Block Scheduling Booklet
Lesson Planning Guide
Solutions Manual

Professional Publications
Glencoe Mathematics Professional Series

Planning the Chapter

 MindJogger Videoquizzes provide a unique format for reviewing concepts presented in the chapter.

Assessment Resources

Student Edition
Mixed Review, pp. 10, 15, 20, 25, 29, 36, 42, 47
Mid-Chapter Self Test, p. 25
Math Journal, pp. 6, 9, 40
Study Guide and Assessment, pp. 48–51
Performance Task, p. 51
Portfolio Suggestion, p. 51
Standardized Test Practice, pp. 52–53
Chapter Test, p. 647

Assessment and Evaluation Masters
Multiple-Choice Tests (Forms 1A, 1B, 1C), pp. 1–6
Free-Response Tests (Forms 2A, 2B, 2C), pp. 7–12
Performance Assessment, p. 13
Mid-Chapter Test, p. 14
Quizzes A–D, pp. 15–16
Standardized Test Practice, pp. 17–18
Cumulative Review, p. 19

Teacher's Wraparound Edition
5-Minute Check, pp. 4, 8, 11, 17, 21, 27, 32, 38, 43
Building Portfolios, p. 2
Math Journal, pp. 16, 37
Closing Activity, pp. 7, 10, 15, 20, 25, 29, 31, 36, 42, 47

Technology
Test and Review Software
MindJogger Videoquizzes
CD-ROM Program

Materials and Manipulatives

Lesson 1-3B
graphing calculator

Lesson 1-4
counters*†
cups*†
equation mat*†

Lesson 1-5
counters*†
cups*†
equation mat*†

Lesson 1-7
counters*†
cups*†
equation mat*†

Lesson 1-8
grid paper†
scissors*

Lesson 1-9
index cards
spinner*†

*Glencoe Manipulative Kit †Glencoe Overhead Manipulative Resources

Pacing Chart

See pages T25–T27 for the Course Planning Calendar.

COURSE	DAY 1	DAY 2	DAY 3	DAY 4	DAY 5	DAY 6	DAY 7
Standard	Chapter Project	Lesson 1-1	Lesson 1-2	Lesson 1-3	Lesson 1-4	Lesson 1-5	Lesson 1-6
Honors	Chapter Project & Lesson 1-1	Lesson 1-2	Lessons 1-3 & 1-3B	Lesson 1-4	Lesson 1-5	Lesson 1-6	Lesson 1-7A
Block	Chapter Project & Lesson 1-1	Lessons 1-2 & 1-3	Lessons 1-4 & 1-5	Lessons 1-6 & 1-7A	Lesson 1-7	Lessons 1-8 & 1-9	Study Guide and Assessment, Chapter Test

The *Transition Booklet* (Skills 2–4) can be used to practice adding, subtracting, multiplying, and dividing with decimals.

Interactive Mathematics:
Activities and Investigations

is an activity-based program that may be used as an enhancement for chapters in *Mathematics: Applications and Connections*.

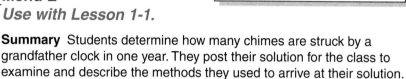

Unit 11, Activity One, Menu E
Use with Lesson 1-1.

Summary Students determine how many chimes are struck by a grandfather clock in one year. They post their solution for the class to examine and describe the methods they used to arrive at their solution.

Math Connection Students solve a nonroutine problem involving patterns and elapsed time. They are given information on how many times a grandfather clock strikes at each 15-minute interval. Using this information, they determine how many chimes are struck by the clock in one year.

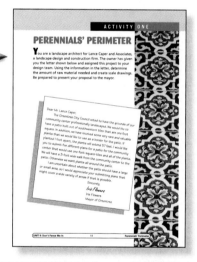

Unit 9, Activity One
Use with Lesson 1-8.

Summary Students work in groups as landscape design teams. Each team receives a request from a potential client to submit five different plans for landscaping a community center. Each team develops a detailed plan and submits drawings to accompany its proposal.

Math Connection Students develop scale drawings of different patios that have the same area and perimeter. This activity involves estimation, measurement, geometry, spatial visualization, accuracy, and the use of number.

DAY 8	DAY 9	DAY 10	DAY 11	DAY 12	DAY 13	DAY 14	DAY 15
Lesson 1-7A	Lesson 1-7	Lesson 1-8	Lesson 1-9	Study Guide and Assessment	Chapter Test		
Lessons 1-7 & 1-7B		Lesson 1-8	Lesson 1-9	Study Guide and Assessment	Chapter Test		

Chapter 1 **2d**

Enhancing the Chapter

APPLICATIONS

Classroom Games, pp. 1–2
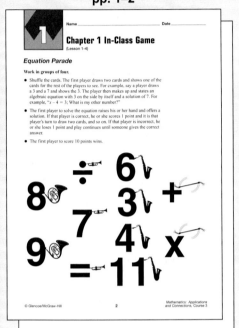

Diversity Masters, p. 27
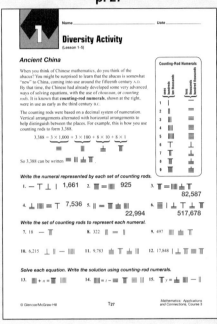

School to Career Masters, p. 27
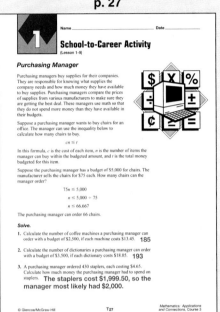

Family Letters and Activities, pp. 53–54
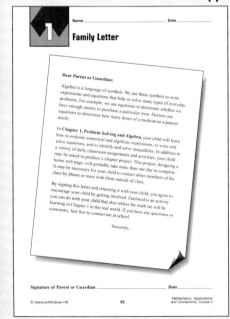
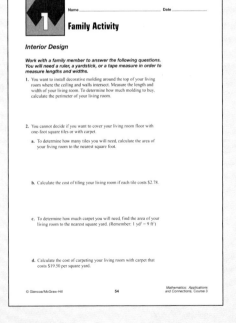

Science and Math Lab Manual, pp. 49–52
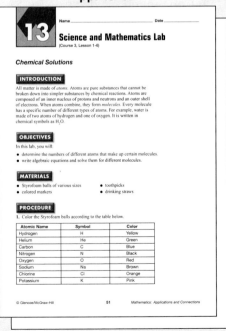

MANIPULATIVES/MODELING

Hands-On Lab Masters,
p. 68

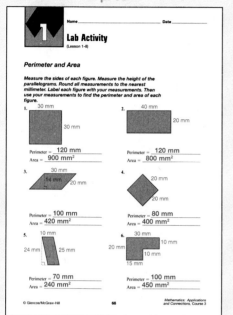

ASSESSMENT/EVALUATION

Assessment and Evaluation Masters,
pp. 14–16

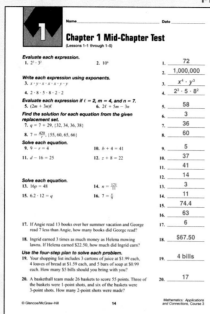

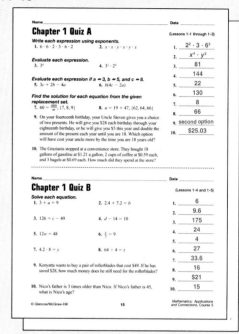

TECHNOLOGY/MULTIMEDIA

Technology Masters,
pp. 53–54

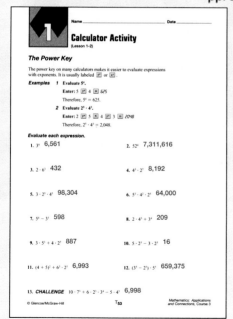

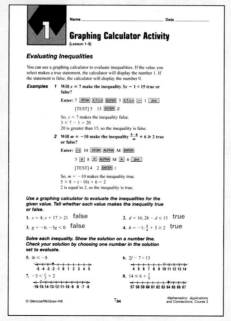

MEETING INDIVIDUAL NEEDS

Investigations for the Special Education Student, pp. 1–2, 7

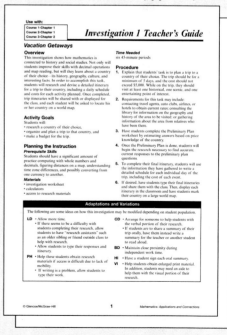

Chapter 1 **2f**

CHAPTER 1 NOTES

Theme: Technology

The World Wide Web is a system of documents available on the Internet that are written in HTML—HyperText Markup Language. HTML codes translate into text, pictures, voice, music, and even video. Because the Web's popularity is increasing, work has begun on the development of Internet II, which will transmit information 45,000 times as fast as 1997 modems.

Question of the Day Like HTML, algebra is a type of code. How would you use variables to express the number of web pages designed by a class if each student made 3? $p = 3 \times s$

Assess Prerequisite Skills

Ask students to read through the list of objectives presented in "What you'll learn in Chapter 1." You may wish to ask them what each of the objectives means or if they have experienced or used any of these math concepts before.

 Building Portfolios

A portfolio is each student's collection of work that expresses how the student has grown as a mathematics student. It is not necessarily a collection of the best work done by a student, but rather the portfolio should demonstrate how the student has improved. You may want to have students attach an explanation to each paper in their portfolios to state why they chose to add it to the collection. Students should update their portfolios with the completion of each chapter.

 Math and the Family

In the *Family Letters and Activities* booklet (pp. 53–54), you will find a letter to the parents explaining what students will study in Chapter 1. An activity appropriate for the whole family is also available.

CHAPTER 1 — Problem Solving and Algebra

What you'll learn in Chapter 1

- to solve problems by using the four-step problem solving plan,
- to evaluate numerical and algebraic expressions,
- to write and solve equations,
- to solve problems by working backward,
- to find the areas and perimeters of rectangles, squares, and parallelograms, and
- to identify and solve inequalities.

 CD-ROM Program

Activities for Chapter 1
- Chapter 1 Introduction
- Interactive Lessons 1-4, 1-5, 1-7, 1-8
- Assessment Game
- Resource Lessons 1-1 through 1-9

CHAPTER Project

HOME PAGE BOUND

The Internet has changed the way we communicate. One way we can communicate to others about ourselves is through a home page on the World Wide Web. Software is available that makes designing a web page a fairly simple task. In this project, you will design your own home page for the World Wide Web and organize your results into a report.

Getting Started

- Find out about software that can be used to design a home page that is available for your computer. Decide on the software you will use.
- Learn how to use the software. Be sure to follow any instructions carefully. Experiment to learn how to best use the capabilities of your software. Work towards designing a page that is attractive and one to which people will be attracted.
- Write a report about what you learned about designing a home page and how you designed your page.

Technology Tips

- Use **graphing software, paint programs,** or **photo editing software** to create graphics and to design your home page.
- Use a **word processor** to write your report.
- Surf the **Internet** to study other home pages.

interNET CONNECTION For more information on creating a home page, visit:
www.glencoe.com/sec/math/mac/mathnet

Working on the Project

You can use what you'll learn in Chapter 1 to help you design your own home page.

Page	Exercise
7	7
25	34
41	21
51	Alternative Assessment

interNET CONNECTION

Glencoe has made every effort to ensure that the website links for *Mathematics: Applications and Connections* at www.glencoe.com/sec/math/mac/mathnet are current and contain appropriate content. However, these website links are not under Glencoe's control.

Instructional Resources ▶▶▶

A recording sheet to help students organize their data for the Chapter Project is shown at the right and is available in the *Investigations and Projects Masters*, p. 20.

CHAPTER Project NOTES

Objectives
Students should
- use the four-step problem-solving plan to design and organize their home pages.
- gain an awareness of the technology available for designing home pages.
- extend their ability to express themselves verbally and graphically.

Project Pointer You may suggest that students begin a *Project Folder* to keep their work as they complete each stage of the Chapter Project. The completed project may also be added to their portfolios.

Software for surfing the Internet and for designing web pages may be available from your school or your local library. Many titles are also available on the World Wide Web for purchase, free trial, or even free downloading. Image and sound software, as well as public domain images and sounds, are available for free downloading. Advise students on the limits concerning copyrighted materials.

Investigations and Projects Masters, p. 20

Chapter 1 Project

Home Page Bound

Page 7, Working on the Chapter Project, Exercise 7
Explore

Plan

Solve

Examine

Page 25, Working on the Chapter Project, Exercise 34
Interests:

Page 41, Working on the Chapter Project, Exercise 21
Home Page Layout:

1-1 Lesson Notes

Instructional Resources
- *Study Guide Masters*, p. 1
- *Practice Masters*, p. 1
- *Enrichment Masters*, p. 1
- Transparencies 1-1, A and B
- CD-ROM Program
  - Resource Lesson 1-1

Recommended Pacing	
Standard	Day 2 of 13
Honors	Day 1 of 13
Block	Day 1 of 7

1 FOCUS

5-Minute Check

1. Find 70 × 20. **1,400**
2. Find 1,426 + 93.4 + 375.8. **1,895.2**
3. How much greater is 5,000 than the sum of 856 and 2,063? **2,081**
4. Find 5,008 ÷ 16. **313**
5. A $75 jacket was marked down $25.05 for clearance. Find the sale price. **$49.95**

 The 5-Minute Check is also available on **Transparency 1-1A** for this lesson.

Motivating the Lesson

Problem Solving Four people meeting for the first time each want to shake hands with everyone else. How many handshakes will occur? **6** How can you figure this out? **Sample answer: diagram or act it out**

1-1 A Plan for Problem Solving

What you'll learn
You will learn to solve problems by using the four-step plan.

When am I ever going to use this?
You will graph data in science class and look for a pattern in the graph to help analyze the data.

The London Underground in England is the oldest subway system in the world. The first lines in the system opened in 1863. Now the network contains 272 stations and 267 miles of track!

Part of the map displayed in the Underground stations is shown. Each interchange is shown as a circle or connected circles, but the map is not drawn to scale. How can you get from Bond Street to Tower Hill with the fewest train changes?

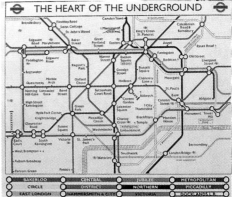

Where would you begin to solve this problem? In mathematics, we have a four-step plan to solve problems like this.

1. *Explore* Determine what information is given in the problem and what you need to find. Do you have all the information you need to solve the problem? Is there too much information?

2. *Plan* After you understand the problem, select a strategy for solving it. There may be several strategies that you can use. It is usually helpful to make an estimate of what you think the answer should be.

3. *Solve* Solve the problem by carrying out your plan. If your plan doesn't work, try another, and maybe even another.

4. *Examine* Finally, examine your answer carefully. See if it fits the facts given in the problem. Compare it to your estimate. You may also want to check your answer by solving the problem again in a different way. If the answer is not reasonable, make a new plan and start again.

4 Chapter 1 Problem Solving and Algebra

 Cross-Curriculum Cue

Inform the other teachers on your team that your students are using a four-step plan for problem solving. Share the plan with them. Ask them to use it occasionally to solve problems in their disciplines.

Suggestions for curriculum integration are:
Physics: velocity
Health: public health case studies
Geography: map exploration

Example Refer to the beginning of the lesson. Find the route from Bond Street to Tower Hill with the fewest train changes.

Explore *What do you know?*
There are several lines that run through the city in the area of Bond Street and Tower Hill.

What are you trying to find?
A route from Bond Street to Tower Hill with the fewest train changes.

Plan You could find every route between Bond Street and Tower Hill. However, since you could change trains in many interchanges to get on and off of different lines, this method would take a great deal of time. Another alternative is to find a route and then look for one with fewer train changes until no more changes can be eliminated.

Solve There are two lines that run through the Bond Street station – Jubilee and Central. Look for the route with the fewest train changes using Jubilee. Then look for the route with the fewest train changes using Central and compare.

Jubilee

Using the Jubilee line, one possible route is Bond Street – Green Park – Victoria – Tower Hill. This involves two train changes.

Another option on the Jubilee line is Bond Street – Charring Cross – Embankment – Tower Hill, which also has two train changes.

There are no routes that use fewer than two train changes using the Jubilee line.

Central

On the Central line, a possible route is Bond Street – Bank – Tower Hill. This route uses only one train change.

There are no routes on the Central line that use fewer than one train change.

So, the route from Bond Street to Tower Hill with the fewest train changes is Bond Street – Bank – Tower Hill.

Examine Is your answer reasonable? Only a route using one line has fewer train changes than the one found. Since there is no line that stops at both Bond Street and Tower Hill, the best possible route must have at least one train change.

Throughout this textbook, you will be solving many kinds of problems. Some can be solved easily by adding, subtracting, multiplying, or dividing. Others can be solved by using a strategy like finding a pattern, solving a simpler problem, making a model, drawing a graph, and so on. No matter which strategy you use, you can always use the four-step plan to solve a problem.

Lesson 1-1 A Plan for Problem Solving **5**

---

2 TEACH

Transparency 1-1B contains a teaching aid for this lesson.

Modeling Mathematics Have students estimate the number of pennies needed to cover their desktops without overlap. Have them describe their estimation methods. **Sample answers: Place pennies across the top and down one side of the desk, then count and multiply; divide the area of the desktop by the area of one penny.**

In-Class Example

For Example 1
Using the map on page 4 of the Student Edition, design a route from Notting Hill Gate to Liverpool Street that allows you to travel on at least 3 of the 13 lines. The route will not be direct. **Sample answer: Notting Hill Gate to Oxford Circus to Baker Street to Liverpool Street**

Teaching Tip Students can learn strategies for solving problems by working together. Have students work in pairs to share strategies and ways of graphing data.

In-Class Example

For Example 2
Joe is swimming to improve his physical fitness. After 2 weeks, he begins to lose weight. Use the table to determine when he will weigh 220 pounds if he weighed 228 pounds before he began exercising. **week 21**

Week	Weight Loss (lb)
1	0
2	0
3	0.5
4	0.5
5	0.5
6	0.5
7	0
8	0.5
9	0.5
10	0.5
11	0.5
12	0

3 PRACTICE/APPLY

Check for Understanding

If students need additional practice or instruction after completing Exercises 1–5, one of these options may be helpful.
- Extra Practice, see p. 605
- Reteaching Activity
- *Transition Booklet*, pp. 7–8
- *Study Guide Masters*, p. 1
- *Practice Masters*, p. 1

***Study Guide Masters*, p. 1**

Example 2 APPLICATION

Medicine The table shows how the amount of a medicine in the bloodstream is related to the time since it was taken. If the doctor wants the concentration to be at least 0.30 mg/L all of the time, how often should the medicine be taken?

Time (h)	Concentration (mg/L)
0	0
1	0.60
2	0.75
3	0.69
4	0.60
5	0.52
6	0.45

Explore You know the concentrations for 0 to 6 hours after the medication is taken. You need to predict when the concentration reaches 0.30 mg/L.

Plan One good way to find a pattern in a set of data is to show the data on a graph. In this case, make a graph that shows time on the horizontal axis and the concentration on the vertical axis. Then look for a pattern in the graph.

Solve

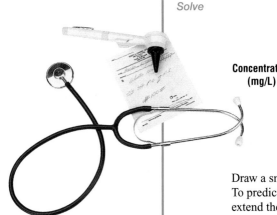

Draw a smooth curve through the points on the graph. To predict when the concentration will reach 0.30 mg/L, extend the curve. The extended line is shown in red. This corresponds to about 9 on the horizontal axis. Therefore, the medicine should be taken about every 9 hours.

Examine Look for another pattern in the data. The time 9 hours seems reasonable.

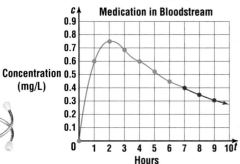

CHECK FOR UNDERSTANDING

Communicating Mathematics

Read and study the lesson to answer each question. 2. See margin.

1. **Tell** what each step in the four-step plan means.   **See Answer Appendix.**
2. **Explain** what to do when your plan to solve the problem doesn't work.
3. **Write** two or three sentences in your journal that describe what you expect to learn in this course.   **See students' work.**

6   Chapter 1 Problem Solving and Algebra

■ Reteaching the Lesson ■

Activity Have students use historic weather data to decide whether weather in a region is changing. Students should decide what data to use and where to find them. This data can be found on the Internet or in almanacs.

Additional Answer
2. Try another strategy.

Guided Practice Use the four-step plan to solve each problem.

4. *Physical Science* A chemist pours sodium chloride, or table salt, into a beaker. If the beaker plus the sodium chloride have a mass of 84.8 grams, and the beaker itself has a mass of 63.3 grams, what is the mass of the sodium chloride that was poured into the beaker? **21.5 grams**

5. *Transportation* Refer to the beginning of the lesson. Find a route from Oxford Circus to Old Street using the fewest train changes. **Sample answer: Oxford Circus – Euston – Old Street**

EXERCISES

Applications and Problem Solving

Use the four-step plan to solve each problem. **6a. See Answer Appendix.**

6. *Look for a Pattern* The table shows the cost of mailing a first-class letter.
   a. Make a graph of the data.
   b. Predict the cost of mailing a letter that weighs 6 ounces. **$1.57**

Weight (ounces)	Cost (1997)
1	$0.32
2	$0.57
3	$0.82
4	$1.07
5	$1.32

7. *Working on the* **CHAPTER Project**
   Designing a home page is like solving a problem. Use the plan for problem solving to help you devise a plan for your home page. Record your plan and follow it as you work on the Chapter Project. **See Answer Appendix.**

8a. Thurs.; Mon.; Wed.; Tues.; Sun.; Fri.; Sat.

8b. about 600 million

8. *Television* The graph shows the number of prime-time television viewers in millions for each night of the week.
   a. List the nights of the week from most viewers to least viewers.
   b. Estimate the total number of viewers for the entire week.
   c. About how many more viewers watch television on Thursday than on Friday? **about 16 million**

Prime-time Viewers (millions)

Monday	91.9
Tuesday	89.8
Wednesday	90.6
Thursday	93.9
Friday	78.0
Saturday	77.1
Sunday	87.7

Source: Nielsen Media Research, 1995

9. *Use a Model* Suppose you had 100 sugar cubes. What is the largest cube you could build with the sugar cubes? **4 by 4 by 4**

10. 7,000 × 5 = 35,000

10. *Technology* Armando used his calculator to find 7,129 + 6,859 + 7,523 + 6,792 + 6,928. How can he tell if the answer shown is reasonable?

    7129 [+] 6859 [+] 7523 [+] 6792 [+] 6928 [=] 35231

11. *Critical Thinking* Replace each ■ in the figure at the right so that the total of the numbers along each side is the same.
    Sample answer: clockwise from top – 17, 10, 45

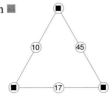

Lesson 1-1 A Plan for Problem Solving **7**

Extending the Lesson

Enrichment Masters, p. 1

Activity Have students use the four-step plan for problem solving to determine a direct route from the school to a park, mall, or other local attraction. Refer students to maps of the neighborhood.

Assignment Guide

Core: 6, 8–11
Enriched: 6, 8–11

CHAPTER Project

Exercise 7 asks students to advance to the next stage of work on the Chapter Project. You may have students work in pairs to plan a home page.

4 ASSESS

Closing Activity

Writing Have pairs of students make up their own word problems. You may want to give them a list of topics from which to choose. Pairs should then exchange problems, solve the problems they receive by using the four-step plan, and explain their solutions to the students who wrote the problems.

Practice Masters, p. 1

1-2 Lesson Notes

Instructional Resources
- *Study Guide Masters,* p. 2
- *Practice Masters,* p. 2
- *Enrichment Masters,* p. 2
- *Transparencies 1-2, A and B*
- *Technology Masters,* p. 53

CD-ROM Program
- Resource Lesson 1-2

Recommended Pacing	
Standard	Day 3 of 13
Honors	Day 2 of 13
Block	Day 2 of 7

1 FOCUS

5-Minute Check
(Lesson 1-1)

Use the four-step plan to solve this problem.
A car rents for $28 per day plus $0.09 per mile. Mrs. Kent rented a car for three days and drove 560 miles. Find the rental charge. **$134.40**

The 5-Minute Check is also available on **Transparency 1-2A** for this lesson.

Motivating the Lesson
Problem Solving Every person has 2 biological parents. Each parent has 2 parents, so every person has $2 \cdot 2$ or 4 grandparents. Each grandparent has 2 parents, so every person has $2(2 \cdot 2)$ or 8 great grandparents. How many great-great-great-grandparents does every person have?
$2 \cdot 2 \cdot 2 \cdot 2 \cdot 2 = 32$ or 2^5

2 TEACH

Transparency 1-2B contains a teaching aid for this lesson.

Reading Mathematics This lesson includes mathematic terms also used in other areas. Encourage students to compare the common and mathematical definitions of *factor, power, base,* and *evaluate.*

1-2 Powers and Exponents

What you'll learn
You'll learn to use powers and exponents in expressions.

When am I ever going to use this?
You can use exponents to determine how much money will be in your savings account after time.

Word Wise
exponent
factor
power
base
associative property
commutative property
evaluate

In the movie *I.Q.*, Walter Matthau portrays the famous physicist Albert Einstein. He tries to set up his niece, mathematician Catherine Boyd, played by Meg Ryan, with Ed Walters, an auto mechanic played by Tim Robbins. When Ed asks Catherine "How many stars do you think there are in the sky?" she replies "10 to the twelfth plus 1."

Catherine's answer uses an **exponent**. An exponent is a way of writing a repeated multiplication in a simple way.

When two or more numbers are multiplied, these numbers are called **factors**. When the same factor is repeated, you may use an exponent.

$$32 = 2 \cdot 2 \cdot 2 \cdot 2 \cdot 2 \rightarrow 2^5 \quad \textit{5 is the exponent.}$$

An expression like 2^5 is called a **power**. The 2 in this expression is called the **base**. The base names the factor being repeated.

Example Write the expression $3 \cdot 3 \cdot 3 \cdot 3$ using exponents.

There are four factors of 3.
$3 \cdot 3 \cdot 3 \cdot 3 = 3^4$

Study Hint
Reading Math To read a power, say the base and then the exponent. For example, 3^4 is read as 3 to the fourth power.

You can use these properties to help write expressions involving exponents.

Property	Arithmetic	Algebra
Associative	$3 + (4 + 10) = (3 + 4) + 10$ $5 \cdot (2 \cdot 12) = (5 \cdot 2) \cdot 12$	$a + (b + c) = (a + b) + c$ $a \cdot (b \cdot c) = (a \cdot b) \cdot c$
Commutative	$1 + 2 = 2 + 1$ $3 \cdot 2 = 2 \cdot 3$	$a + b = b + a$ $a \cdot b = b \cdot a$

Examples Write each expression using exponents.

2 $2 \cdot 2 \cdot 5 \cdot 5 \cdot 7 \cdot 7 \cdot 7$

Use the associative property to group the factors that are the same.

$2 \cdot 2 \cdot 5 \cdot 5 \cdot 7 \cdot 7 \cdot 7$
$= (2 \cdot 2) \cdot (5 \cdot 5) \cdot (7 \cdot 7 \cdot 7)$
$= 2^2 \cdot 5^2 \cdot 7^3$

3 $3 \cdot 4 \cdot 3 \cdot 3 \cdot 4 \cdot 6 \cdot 6$

Use the commutative property to rearrange the factors. Then use the associative property to group them.

$3 \cdot 4 \cdot 3 \cdot 3 \cdot 4 \cdot 6 \cdot 6$
$= 3 \cdot 3 \cdot 3 \cdot 4 \cdot 4 \cdot 6 \cdot 6$
$= (3 \cdot 3 \cdot 3) \cdot (4 \cdot 4) \cdot (6 \cdot 6)$
$= 3^3 \cdot 4^2 \cdot 6^2$

8 Chapter 1 Problem Solving and Algebra

When you find the value of a power, you are **evaluating** the power. The y^x key on your calculator allows you to evaluate one or more powers.

Examples

4 Evaluate 7^3.

Method 1 Use pencil and paper.
$7^3 = 7 \cdot 7 \cdot 7$
$= 49 \cdot 7$
$= 343$

Method 2 Use a calculator.
7 y^x 3 = *343*

5 Evaluate $2^4 \cdot 3^2$.

Method 1 Use pencil and paper.
$2^4 \cdot 3^2 = 2 \cdot 2 \cdot 2 \cdot 2 \cdot 3 \cdot 3$
$= 16 \cdot 9$
$= 144$

Method 2 Use a calculator.
2 y^x 4 × 3 y^x 2 = *144*

APPLICATION

6 **Entertainment** Refer to the beginning of the lesson. Evaluate the expression for the number of stars in the sky.

Write an expression for *ten to the twelfth plus 1*. Then evaluate.
$10^{12} + 1 = 10 \cdot 10 \cdot 10 \cdot 10 \cdot 10 \cdot 10 \cdot 10 \cdot 10 \cdot 10 \cdot 10 \cdot 10 \cdot 10 + 1$
$= 1,000,000,000,000 + 1$
$= 1,000,000,000,001$

This number is read as one trillion one.

Cultural Kaleidoscope
Many constellations were named in Greece around 3000 B.C. But other civilizations named star patterns based on stories and people that were important to them.

CHECK FOR UNDERSTANDING

Communicating Mathematics
Read and study the lesson to answer each question.

1. **Tell** what the 3 represents in 2^3. **Use 2 as a factor 3 times**
2. **Express** *four to the 5th power* using exponents. **4^5**
3. **Write** a few sentences about why scientists working with very large numbers often use exponents. **See students' work.**

Guided Practice
Write each expression using exponents. 5. $16^2 \cdot 20^2$ 6. $2^3 \cdot 3^2 \cdot 7$
4. $7 \cdot 7 \cdot 7$ **7^3**
5. $16 \cdot 16 \cdot 20 \cdot 20$
6. $2 \cdot 3 \cdot 3 \cdot 2 \cdot 2 \cdot 7$

Evaluate each expression. 10. **1,512,000**
7. 4^3 **64**
8. 9^2 **81**
9. $6^2 \cdot 2^3$ **288**
10. $7 \cdot 6^3 \cdot 10^3$

11. **Astronomy** The average distance from Jupiter to the Sun is 4.84×10^8 miles.
   a. Write an expression for 10^8. **$10 \cdot 10 \cdot 10 \cdot 10 \cdot 10 \cdot 10 \cdot 10 \cdot 10$**
   b. Evaluate 4.84×10^8. **484,000,000**

Lesson 1-2 Powers and Exponents **9**

In-Class Examples

For Example 1
Write the expression $4 \cdot 4 \cdot 4 \cdot 4$ using exponents. **4^4**

For Example 2
Write the expression $2 \cdot 2 \cdot 2 \cdot 9 \cdot 9 \cdot 11 \cdot 11 \cdot 11$ using exponents. **$2^3 \cdot 9^2 \cdot 11^3$**

For Example 3
Write the expression $4 \cdot 8 \cdot 9 \cdot 4 \cdot 4 \cdot 9 \cdot 9$ using exponents. **$4^3 \cdot 8 \cdot 9^3$**

For Example 4
Evaluate 6^3. **216**

For Example 5
Evaluate $5^3 \cdot 2^3$. **1,000**

For Example 6
How many people would you include if you researched your family back 10 generations before your own? Write an expression and then evaluate. **2^{10}; 1,024**

Teaching Tip Remind students that on their calculators, x^2 gives the square immediately, while y^x can evaluate powers greater than 2 but takes more steps.

Reteaching the Lesson

Activity Pose the following problem.
How many times must you multiply each first number by itself to obtain the second number?
a. 3; 81 **4**
b. 2; 64 **6**
c. 5; 125 **3**
d. 4; 1,024 **5**
e. 13; 169 **2**
f. 7; 2,401 **4**

Error Analysis
Watch for students who multiply base times exponent to evaluate an exponential expression.
$2^5 = 2 \cdot 5 = 10$
Prevent by stressing that an exponent tells how many times to multiply the base by itself. $2^5 = 2 \cdot 2 \cdot 2 \cdot 2 \cdot 2 = 32$

3 PRACTICE/APPLY

Check for Understanding
If students need additional practice or instruction after completing Exercises 1–11, one of these options may be helpful.
- Extra Practice, see p. 605
- Reteaching Activity, see p. 9
- *Transition Booklet,* pp. 9–10
- *Study Guide Masters,* p. 2
- *Practice Masters,* p. 2
- Interactive Mathematics Tools Software

Assignment Guide
Core: 13–37 odd, 38–40
Enriched: 12–34 even, 35–40

4 ASSESS

Closing Activity
Speaking Write several products for students to see. Have students state each as a power. **Example: 25 equals 5 to the second power.**

Additional Answer
38. When a value increases exponentially, its exponent increases by 1 each time. For example, 2^1, 2^2, 2^3, and so on.

Practice Masters, p. 2

EXERCISES

Practice Write each expression using exponents.   17. $12 \cdot 14^2 \cdot 5^3$

12. $6 \cdot 6 \cdot 6 \cdot 6$ 6^4 13. $10 \cdot 10$ 10^2 14. $8 \cdot 8 \cdot 8 \cdot 8 \cdot 8 \cdot 8$ 8^6

15. $4 \cdot 4 \cdot 4 \cdot 8 \cdot 8$ $4^3 \cdot 8^2$ 16. $2 \cdot 2 \cdot 2 \cdot 5 \cdot 5$ $2^3 \cdot 5^2$ 17. $12 \cdot 14 \cdot 14 \cdot 5 \cdot 5 \cdot 5$

18. $5 \cdot 5 \cdot 4 \cdot 2 \cdot 5 \cdot 4$ 19. $18 \cdot 5 \cdot 5 \cdot 5 \cdot 18$ 20. $a \cdot b \cdot b \cdot a \cdot b$
    $5^3 \cdot 4^2 \cdot 2$ $18^2 \cdot 5^3$ $a^3 b^3$

Evaluate each expression. 26. 43,200 27. 1,000,000 32. 17,496

21. 6^4 1,296   22. 8^3 512   23. 1^{10} 1   24. 10^1 10

25. $4^2 \cdot 5^3$ 2,000   26. $5^2 \cdot 8^2 \cdot 3^3$ 27. $1,000^2$ 28. 16^4 65,536

29. $7^2 - 3^3$ 22   30. $2 \cdot 2^2 \cdot 2^3$ 64   31. $3 \cdot 2^2 \cdot 4^3$ 768   32. $9 \cdot 6^2 \cdot 2 \cdot 3^3$

33. Write the product $10 \cdot 10 \cdot 10 \cdot 24 \cdot 24 \cdot 50$ using exponents. $10^3 \cdot 24^2 \cdot 50$

34. Evaluate $2 \cdot 2^3 \cdot 3^2$. **144**

Applications and Problem Solving

36. 193,000,000

35. *Geometry* To find the volume of a cube, multiply its length, its width, and its depth. **b. 12.167 cubic meters**
    a. Write an expression for the volume of a cube whose edges are each 12 centimeters long. 12^3
    b. If each edge of a cube measures 2.3 meters, what is the volume of the cube?

36. *Shipping* The busiest port in the United States is the Port of South Louisiana. In 1993, it handled 1.93×10^8 tons of cargo. Express this number without exponents.

37. *Life Science* Every person has 2 biological parents. Each of their parents has 2 parents, so every person has $2 \cdot 2$ or 4 grandparents, and so on. How many great-great grandparents does every person have? **16**

38. *Critical Thinking* What does it mean to say that a value increases exponentially? **See margin.**

Mixed Review

39. *Money Matters* At the school bookstore, a pen costs $0.28, and a small writing tablet costs $0.23. What combination of pens and tablets could you buy for exactly $0.74? *(Lesson 1-1)* **1 pen and 2 tablets**

40. **Test Practice** Shelby selected 6 items at the craft store that ranged in price from $3.97 to $5.31. Which is a reasonable total for her purchases? *(Lesson 1-1)* **C**

   A less than $20
   B between $20 and $25
   C between $25 and $30
   D between $30 and $35
   E more than $35

10 Chapter 1 Problem Solving and Algebra

Enrichment Masters, p. 2

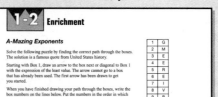

Extending the Lesson

Activity Have students express all numbers between 1 and 64 as powers, if possible. Some numbers may be powers of more than one base. **Sample answers: $8 = 2^3$; $16 = 2^4$ or 4^2; $64 = 8^2$ or 4^3 or 2^6**

1-3 Variables, Expressions, and Equations

What you'll learn

You'll learn to evaluate expressions and to find the solutions of equations.

When am I ever going to use this?

You can use equations to enter information into a spreadsheet.

Word Wise

numerical expression
substitute
order of operations
variable
algebraic expression
equation
open sentence
solution
replacement set

Study Hint

Reading Math
Grouping symbols include:
- parentheses ( ),
- brackets [ ], and
- fraction bars, as in $\frac{5+3}{2}$, which means $(5 + 3) \div 2$.

In 1996, the Chicago Bulls became the first team in the history of the National Basketball Association to win 72 games. Almost everyone knows that Michael Jordan led the Bulls in their record-breaking season, but did you know that Toni Kukoc was the NBA's sixth man award winner? The sixth man award is given to the league's best substitute player. A substitute enters a game to replace one of the starters.

In mathematics, the concept of substitution is just as important as it is in sports. Consider the **numerical expression** $8 + 4$. It has a value of 12. However, the expression $x + 4$ does not have a value until a value for x is given. Suppose you let $x = 13$. To find the value of the expression, you must **substitute** 13 for x, or put 13 in place of x, in the expression. It becomes the numerical expression $13 + 4$, which has a value of 17. Therefore, if $x = 13$, then $x + 4 = 17$.

When you evaluate a numerical expression, you need to follow an order of operations. That is, you need to know which operations to do first when there is more than one operation in the expression. The following rules are used when evaluating numerical expressions.

Order of Operations	1. Do all operations within grouping symbols first; start with the innermost grouping symbols. 2. Do all powers before other operations. 3. Multiply and divide in order from left to right. 4. Add and subtract in order from left to right.

Example

1 Evaluate $(9 + 6) \div 5 \times 4 + (2^3 - 3)$.

Follow the order of operations. Do operations inside the parentheses first.

$(9 + 6) \div 5 \times 4 + (2^3 - 3)$
$= 15 \div 5 \times 4 + (8 - 3)$ $\quad 9 + 6 = 15$ and $2^3 = 8$
$= 15 \div 5 \times 4 + 5$ $\quad 8 - 3 = 5$
$= 3 \times 4 + 5$ $\quad 15 \div 5 = 3$
$= 12 + 5$ $\quad 3 \times 4 = 12$
$= 17$

Lesson 1-3 Variables, Expressions, and Equations **11**

1-3 Lesson Notes

Instructional Resources
- Study Guide Masters, p. 3
- Practice Masters, p. 3
- Enrichment Masters, p. 3
- Transparencies 1-3, A and B
- Assessment and Evaluation Masters, p. 15

 CD-ROM Program
- Resource Lesson 1-3

Recommended Pacing	
Standard	Day 4 of 13
Honors	Day 2 of 13
Block	Day 2 of 7

1 FOCUS

 5-Minute Check
(Lesson 1-2)

1. Write $2 \cdot 4 \cdot 2 \cdot 5 \cdot 5 \cdot 6$ using exponents. $2^2 \cdot 4 \cdot 5^2 \cdot 6$

Evaluate each expression.
2. 3^3 27
3. 10^3 1,000
4. $3.45 \cdot 10^3$ 3,450
5. $3 \cdot 4^3 \cdot 5^4$ 120,000

 The 5-Minute Check is also available on **Transparency 1-3A** for this lesson.

Motivating the Lesson

Problem Solving Ask students what $25 + 15 \times 3$ equals. 70 Then ask students what answer they would get if they first added 25 and 15 and then multiplied by 3. 120

Multiple Learning Styles

 Kinesthetic Prepare this set of seven cards, 12, 6, 3, +, −, ×, ÷, for each group. Tell students to place the cards in the order 12, 6, and 3. Then direct them to use two operation cards to form expressions that equal 6, 10, and 75. $12 \div 6 \times 3$; $12 - 6 \div 3$; $12 \times 6 + 3$

Then add an exponent card, ?², and ask students to create expressions for 24 and 18. $6^2 \div 3 + 12$; $3^2 \times 12 \div 6$

Lesson 1-3 **11**

2 TEACH

 Transparency 1-3B contains a teaching aid for this lesson.

Thinking Algebraically Describe to students the difference between arithmetic and algebraic equations. Arithmetic equations deal only with real numbers and represent a specific occurrence, such as $2 + 3 = 5$. Algebraic equations use numbers and letters and represent a variety of occurrences that fit a certain pattern, such as $x + 3 = y$.

Teaching Tip Before Example 1, emphasize that the rules for the order of operations are indeed necessary. Use $10 - 6 - 2$ as an example. If $10 - 6$ is evaluated first, the answer is 2. In contrast, subtracting 2 from 6 first yields a wrong answer of 6.

In-Class Examples

For Example 1
Evaluate $(60 - 20) \div 4 + (3^2 \times 5)$. **55**

For Example 2
Evaluate $m + p \cdot 3$ if $m = 5$ and $p = 2$. **11**

For Example 3
Evaluate $4x - 2y$ if $x = 8$ and $y = 6$. **20**

For Example 4
Evaluate $\frac{(h + k^2)}{3h}$ if $h = 2$ and $k = 4$. **3**

Study Hint

Technology If your calculator does not follow the order of operations, you can evaluate expressions with it by entering the numbers as you would evaluate the expression with pencil and paper. Press $=$ after each operation, and continue without clearing.

Algebra is a language of symbols. In algebra, letters, called **variables**, are used to represent unknown quantities. In the expression $x + 6$, x is a variable. Expressions that contain variables are called **algebraic expressions**. In order to evaluate numerical and algebraic expressions, you must know how to read algebraic expressions.

$3(2)$ means 3×2

$5x$ means $5 \times x$

xy means $x \times y$

$4 \cdot 7a$ means $4 \times 7 \times a$

$8ab^2$ means $8 \times a \times b \times b$

$a[b(cd)]$ means $a \times [b \times (c \times d)]$

$\frac{x}{10z}$ means $x \div (10 \times z)$

$a\left(\frac{b}{3}\right)$ means $a \times (b \div 3)$

Examples

2 Evaluate $a - b + 7$ if $a = 15$ and $b = 9$.

Begin by using substitution. Replace each variable in the expression with its value. Then use the order of operations.

$a - b + 7 = 15 - 9 + 7$ *Replace a with 15 and b with 9.*
$ = 6 + 7$ *Follow the order of operations.*
$ = 13$ *Check your answer mentally.*

3 Evaluate $3a + 4b$ if $a = 5$ and $b = 17$.

$3a + 4b = 3(5) + 4(17)$ *Replace a with 5 and b with 17.*
$ = 15 + 68$ *Do multiplication before addition.*
$ = 83$ *Add.*

4 Evaluate $\frac{m^3}{2n}$ if $m = 6$ and $n = 9$.

The fraction bar, which means division, is also a grouping symbol. Evaluate the expressions in the numerator and denominator separately before dividing.

$\frac{m^3}{2n} = \frac{6^3}{2 \cdot 9}$ *Replace m with 6 and n with 9.*
$\phantom{\frac{m^3}{2n}} = \frac{216}{18}$ *Evaluate the numerator and the denominator separately.*
$\phantom{\frac{m^3}{2n}} = 216 \div 18$ *Then divide.*
$\phantom{\frac{m^3}{2n}} = 12$

12 Chapter 1 Problem Solving and Algebra

 Investigations for the Special Education Student

This blackline master booklet helps you plan for the needs of your special education students by providing long-term projects along with teacher notes. Investigation 1, *Vacation Gateways*, and Investigation 3, *Dining Out*, may be used with this chapter.

A mathematical sentence that contains an "=" is called an **equation**. Some examples of equations are $5 + 9 = 14$, $12(3) = 36$, and $8 - (4 \div 2) = 6$. An equation that contains a variable is an **open sentence**. When a number is substituted for the variable in an open sentence, the sentence may be true or false.

$$\begin{aligned}
\textbf{Equation:} \quad & 34 + 16 = c \\
\text{Replace } c \text{ with 40.} \quad \rightarrow \quad & 34 + 16 = 40 \\
& \text{This equation is false.} \\
\text{Replace } c \text{ with 50.} \quad \rightarrow \quad & 34 + 16 = 50 \\
& \text{This equation is true.}
\end{aligned}$$

The values of the variable that make the equation true are called the **solutions** of the equation. In the equation $34 + 16 = c$, the solution is 50. The process of finding a solution is called *solving the equation*.

In-Class Examples

For Example 5
Find the solution of $71 - n = 29$ if the value of n can be selected from the set {41, 42, 43}. **42**

For Example 6
Abbey earns $6.75 an hour at her weekend job. One Saturday she earned $47.25. How many hours did she work? Use the equation $\$6.75 \times a = \$47.25$ and the replacement set {5, 6, 7, 8}. **7 hours**

Example 5 Find the solution of $82 + a = 96$ if the value of a can be selected from the set {10, 12, 14}.

Try each value from the solution set to see which is a solution.

Try 10.	Try 12.	Try 14.
$82 + a = 96$	$82 + a = 96$	$82 + a = 96$
$82 + 10 \stackrel{?}{=} 96$	$82 + 12 \stackrel{?}{=} 96$	$82 + 14 \stackrel{?}{=} 96$
$92 = 96$ *false*	$94 = 96$ *false*	$96 = 96$ *true*

14 is the solution of $82 + a = 96$.

When you are given a set of numbers from which to choose the value of the variable, the set is called the **replacement set** for the equation.

Example 6 APPLICATION

Technology E-mail is the most widely used and rapidly growing online activity among Internet subscribers. Suppose an e-mail message contains 72,000 bytes, and 24,000 of this is used to include a picture. How many bytes of the message were used to write the text? Use the equation $24,000 + t = 72,000$ and the replacement set {48,000, 52,000, 56,000}.

$24,000 + t = 72,000$

$24,000 + 48,000 \stackrel{?}{=} 72,000$ *Replace t with 48,000.*

$72,000 = 72,000$ *true*

The solution is 48,000. The text uses 48,000 bytes.

Lesson 1-3 Variables, Expressions, and Equations **13**

3 PRACTICE/APPLY

Check for Understanding
If students need additional practice or instruction after completing Exercises 1–13, one of these options may be helpful.
- Extra Practice, see p. 605
- Reteaching Activity
- *Study Guide Masters*, p. 3
- *Practice Masters*, p. 3

Assignment Guide
Core: 15–43 odd, 44–48
Enriched: 14–40 even, 42–48

Additional Answers
1. Numerical expressions contain only numbers and algebraic expressions contain numbers and variables.
2. A replacement set is the given set of numbers from which to choose the correct value of the variable. The solution is part of the replacement set.

Study Guide Masters, p. 3

CHECK FOR UNDERSTANDING

Communicating Mathematics

Read and study the lesson to answer each question. 1–2. See margin.
1. *Tell* the difference between a numerical expression and an algebraic expression.
2. *Write* a definition for a replacement set. Explain how the replacement set is related to the solution.

Guided Practice

Evaluate each expression.
3. $17 + 2 \cdot 8$ **33**
4. $3(7) - 4 \div 2$ **19**
5. $\frac{36}{3^2 - 3}$ **6**

Evaluate each expression if $a = 5$, $b = 6$, $c = 3$, and $d = 4$.
6. $2a + 3b - 4c$ **16**
7. $bdc \div 12$ **6**
8. $(3a + 2c)d$ **84**

Find the solution for each equation from the given replacement set.

10. 1.80

9. $x = 16 + 28$, $\{42, 44, 46\}$ **44**
10. $2.30 - y = 0.50$, $\{1.40, 0.90, 1.80\}$
11. $4m = 60$, $\{20, 15, 10\}$ **15**
12. $15 = \frac{480}{p}$, $\{32, 30, 28\}$ **32**

13. *Transportation* There were 7,000 miles of concrete roads in the United States in 1917. Just ten years later in 1927, a network of 50,000 miles of roads had been built. Find the number of miles of concrete roads built between 1917 and 1927. Use the equation $7,000 + b = 50,000$ and the replacement set $\{41,000, 43,000, 45,000\}$. **43,000**

EXERCISES

Practice

Evaluate each expression.
14. $(3 + 6)^2$ **81**
15. $5 \cdot (4 + 2^2)$ **40**
16. $(8 - 3)^2 + \frac{8 \cdot 3}{12}$ **27**
17. $27 \div 9 + 5$ **8**
18. $(5^2 + 3) \div 7$ **4**
19. $[(6 + 5)2] \div 11$ **2**
20. $4^2 - 2 \cdot 5 + (8 - 2)$ **12**
21. $\frac{28}{4^2 - 2}$ **2**
22. $3 \cdot (18 - 7) + \frac{25 - 9}{8}$ **35**

Evaluate each expression if $w = 4$, $x = 7$, $y = 6$, and $z = 3$.
23. $3x + 4y - 2w$ **37**
24. $xyz \div 21$ **6**
25. zw^2 **48**
26. $(zw)^2$ **144**
27. $(3y + 2z) \cdot w$ **96**
28. $(2z \cdot w) + 3y$ **42**
29. $\frac{3yz}{w}$ **13.5**
30. $\frac{x^2}{2z + 1}$ **7**
31. $\frac{(wz)^2}{y + 10}$ **9**

Find the solution for each equation from the given replacement set.
32. $r + 35 = 80$, $\{35, 40, 45, 50\}$ **45**
33. $216 = 127 + n$, $\{81, 89, 91, 99\}$ **89**
34. $w - 18 = 62$, $\{80, 85, 90, 95\}$ **80**
35. $4x = 124$, $\{29, 31, 33, 35\}$ **31**
36. $108 = 6h$, $\{20, 18, 16, 14\}$ **18**
37. $\frac{576}{w} = 72$, $\{4, 6, 8, 12\}$ **8**
38. $2x - 6 = 18$, $\{8, 10, 12, 14\}$ **12**
39. $9 = 25 - 4a$, $\{0, 1, 3, 4\}$ **4**

14 Chapter 1 Problem Solving and Algebra

Reteaching the Lesson

Activity Students can use a calculator to test values from the replacement set in an equation. For example, to test values for x in $177 + x = 315$, students should first store 177 in memory. They can then enter a value from the replacement set, add the stored value by recalling it from memory, and see if the sum is 315.

Error Analysis
Watch for students who confuse the order of operations in expressions like $(2^3 + 8)$ and $(2 + 8)^3$.
Prevent by insisting that students write out each step.
$(2^3 + 8) = (8 + 8) = 16$
$(2 + 8)^3 = (10)^3 = 1,000$

40. $4.67

40. Solve $d = \$7.00 - \$2.33$ if the replacement set is $\{\$5.67, \$4.67, \$4.33\}$.

41. Find the solution for $6m = 42$. Use the replacement set $\{6, 7, 8, 9\}$. **7**

Applications and Problem Solving

For the latest information on drive-in theaters, visit:
www.glencoe.com/sec/math/mac/mathnet

42a. the approximate number of theaters now

42. Entertainment Today there are 2,170 fewer drive-in theaters in the United States than there were in the 1980s. Find the number of drive-in theaters now if there were approximately 3,000 drive-in theaters in the 1980s. Use the equation $3,000 - 2,170 = x$ to represent the situation.

a. What does the variable x represent?

b. Use the replacement set $\{830, 850, 880, 930\}$ with the equation to find the approximate number of drive-in theaters that remain. **830**

Carol Moseley Braun

43. Politics In 1992, Carol Moseley Braun became the first African-American woman elected to the U.S. Senate. Shirley Chisholm had become the first African-American woman to enter the U.S. House of Representatives in 1968. The equation $1968 + y = 1992$ can be used to represent the situation.

a. What does the variable y represent? **See margin.**

b. Use the replacement set $\{19, 24, 29, 34\}$ with the equation to find the number of years between Ms. Chisholm's and Ms. Braun's elections. **24**

Shirley Chisholm

44. Critical Thinking Write two different open sentences whose solution is 4. **Sample answers:** $4x = 16$; $x + 8 = 12$

Mixed Review

45. Find the value of the expression 2^6. *(Lesson 1-2)* **64**

46. **Test Practice** How can $5 \cdot 5 \cdot 7 \cdot 7 \cdot 7 \cdot q \cdot q$ be written in exponential notation? *(Lesson 1-2)* **B**

A $5 \cdot 12^2 \cdot q^2$
B $5^2 \cdot 7^3 \cdot q^2$
C $35^2 \cdot q^2$
D $70q^2$

47. 6 pounds

47. Hobbies Jerome put 4 pounds of sunflower seeds in his bird feeder on Sunday. On Friday, the bird feeder was empty, so Jerome put 4 more pounds of seed in it. The following Sunday, the seeds were half gone. How many pounds of sunflower seeds were consumed by the birds that week? *(Lesson 1-1)*

48. Food On a typical day, 2 million gallons of ice cream are produced in the United States. About how many gallons of ice cream are produced each year? *(Lesson 1-1)* **about 730 million gallons**

Lesson 1-3 Variables, Expressions, and Equations **15**

Extending the Lesson

Enrichment Masters, p. 3

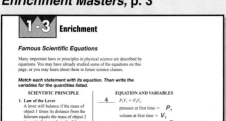

Activity The sum of the solutions in each row, column, and diagonal equals the length (in miles) of one of the longest stretches of straight highway in North America. **45 mi**

$x + 6 = 19$	$2p = 46$	$n - 5 = 4$
$4y = 44$	$5 = k \div 3$	$20 - m = 1$
$h + 4 = 25$	$a = 28 \div 4$	$25 = d + 8$

1-3B LAB Notes

GET READY

Objective Students evaluate expressions by using a graphing calculator.

Technology Resources
- graphing calculator (TI-83, TI-82, TI-81 or TI-80)

Optional Resources
Hands-On Lab Masters
- Calculator Layout (TI-83), p. 33

MANAGEMENT TIPS

Recommended Time
50 minutes

Getting Started The Technology Labs in this text use TI-83 keystrokes. Other TI calculators function in a similar manner. For other calculators, please refer to the user's manual for specific keys. Allow students plenty of time to explore the various keys of their TI-83 calculators. It may be beneficial to have students work in pairs to assist each other in the exploration and discovery process.

Using a Scientific Calculator If a graphing calculator is not available, show students how to use a simple scientific calculator to evaluate expressions. Values for variables need to be entered by hand. Have students test whether the calculator follows the order of operations by entering the expression $3 + 2 \times 2$. A calculator that does not use the order of operations will give $(3 + 2) \times 2$ or 10 as a solution. A calculator that uses the order of operations will give $3 + (2 \times 2)$ or 7 as a solution.

ASSESS

After students answer Exercises 1–5, ask them to determine how they would change the way they enter the expressions when a graphing calculator is not available.

Technology Lab — GRAPHING CALCULATORS

1-3B Evaluating Expressions

A Follow-Up of Lesson 1-3

graphing calculator

Graphing calculators follow the order of operations. So there is no need to perform each operation separately. To evaluate an expression, enter it just as it is written. If an expression contains parentheses, enter them in the calculator just as they are written, and the expression will be evaluated correctly.

On TI calculators, the multiplication and division signs do not appear on the screen as they do on the keys. Instead, the calculator displays the symbols used in computer language. That is, * means multiplication and / means division.

TRY THIS

Work with a partner.

Evaluate $3(x - 6) \div 2 + (x^2 - 15)$ for $x = 8$ and for $x = 12$.

You can replace each x with 8 as you enter the expression. However, if you will evaluate an expression for more than one value of x, it is helpful to enter the variables so that you can reevaluate quickly.

Step 1 Use the following keystrokes to evaluate for $x = 8$.

8 [STO▶] [X,T,θ,n] [ALPHA] [:] *Stores 8 as the value of x.*

3 [(] [X,T,θ,n] [−] 6 [)] [÷] 2 [+] [(] [X,T,θ,n] [x²] [−] 15 [)] [ENTER] 52

Step 2 You do not need to reenter the whole expression to reevaluate for $x = 12$. Just change the value for x.

[2nd] [ENTRY] *Redisplays the last entry.*

Use the arrow keys to scroll to the beginning of the display. Press [DEL] to delete the 8. Then press [2nd] [INS] 12 to insert the new value for x. Reevaluate the expression by pressing [ENTER]. When $x = 12$, the value of the expression is 138.

ON YOUR OWN

Use a graphing calculator to evaluate each expression for $x = 3$, $x = 6$, and $x = 15$.

1. $x^2 - 9$ 0; 27; 216
2. $2x^2 + 10$ 28; 82; 460
3. $\dfrac{16x^2}{3}$ 48; 192; 1,200
4. $x(20 - x)$ 51; 84; 75

5. How would you evaluate xy^2 for $x = 4$ and $y = 7$ on a graphing calculator? (*Hint:* Enter Y by pressing [ALPHA] [Y].) See margin.

16 Chapter 1 Problem Solving and Algebra

Ask students to write a paragraph about how evaluating expressions with a graphing calculator varies from evaluating expressions with pencil and paper.

Additional Answer

5. 4 [STO▶] [X,T,θ,n] [ALPHA] [:] 7 [STO▶] [ALPHA] [Y] [ALPHA] [:] [X,T,θ,n] [ALPHA] [Y] [x²] [ENTER] 196

1-4 Solving Subtraction and Addition Equations

What you'll learn
You'll learn to solve equations by using the subtraction and addition properties of equality.

When am I ever going to use this?
You can use addition and subtraction equations to find increases and decreases in prices.

Word Wise
addition property
subtraction property
inverse operation

In World War II, the Navajo language was used to devise a secret code. It is one of the few unbroken codes in history. There were 29 original "code talkers," Navajo Marines who used the code. By the end of the war, 420 Navajos served as code talkers. How many recruits were added to the original Navajo code talkers?

Suppose we let r represent the number of recruits added to the code talkers. Then we could use the equation $29 + r = 420$ to find the number of recruits. *This problem will be solved in Example 3.*

In Lesson 1-3, you learned that the value of a variable that makes an equation true is different for different equations. You used the values in a given replacement set to find the solutions.

Often when you need to find the solution to an equation, a replacement set is not given or is too large to try all of the numbers. You can then use the properties of algebra to find the solution of the equation.

MINI-LAB

Work with a partner.

Solve $x + 3 = 7$ using cups and counters.

🥤 cups and counters

▯=▯ equation mat

Try This
- Let a cup represent x. Put a cup and 3 counters on one side of the mat and 7 counters on the other side. These two quantities are equal.

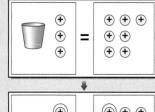

- The goal is to get the cup all by itself on one side of the mat. Take 3 counters away from each side. What you have left is the value of the cup, which is the value of x.

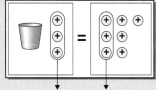

Lesson 1-4 Solving Subtraction and Addition Equations **17**

1-4 Lesson Notes

Instructional Resources
- *Study Guide Masters*, p. 4
- *Practice Masters*, p. 4
- *Enrichment Masters*, p. 4
- Transparencies 1-4, A and B
- *Classroom Games*, pp. 1–2
- CD-ROM Program
  - Resource Lesson 1-4
  - Interactive Lesson 1-4

Recommended Pacing	
Standard	Day 5 of 13
Honors	Day 4 of 13
Block	Day 3 of 7

1 FOCUS

5-Minute Check
(Lesson 1-3)
Evaluate each expression.
1. $2^5 - 4 \cdot 6 + (8 + 10)$ **26**
2. $(13 - 3^2)^2 + \frac{10 \cdot 7}{5}$ **30**

Evaluate each expression if $a = 2$, $b = 3$, and $c = 5$.
3. $2a + 3b + 4c$ **33**
4. $(4a + 4b)c$ **100**
5. Find the solution for $x + 27 = 41$ if the replacement set is $\{12, 14, 16\}$. **14**

The 5-Minute Check is also available on **Transparency 1-4A** for this lesson.

Motivating the Lesson

Hands-On Activity Place equal weights on each side of a two-pan balance. Add weight to one pan and ask students what can be done to bring the scale back into balance. **Add the same weight to the other pan.**

Begin again with equal weights on each side. Remove weight from one side, and ask what can be done to bring the scale back into balance. **Remove an equal weight from the other pan.**

Classroom Vignette

"You may want to use bingo chips as counters instead of using paper models. I also use golf tees instead of cups. Both manipulatives are inexpensive, more durable, easy to store, and take less set-up time."

Mary L. Judy, Teacher
Westmoor Middle School
Columbus, OH

Mary L. Judy

Lesson 1-4 **17**

2 TEACH

 Transparency 1-4B contains a teaching aid for this lesson.

Using the Mini-Lab After modeling the equation with the cup and counters, ask students why the goal is to get the cup by itself on one side of the mat. *The cup represents x, and we want to find the value of x.* Ask students why the same number of counters must be removed from each side of the mat. *to keep both sides of the equation in balance*

In-Class Examples

For Example 1
Solve $46 + n = 99$. Check your solution. **53**

For Example 2
Solve $k - 1.29 = 2.36$. **3.65**

For Example 3
When a park was established, it consisted of 53 acres. Over time, people donated acreage until the park had a total area of 676 acres. Use the equation $53 + a = 676$ to find the number of acres added to the original park. **623**

Teaching Tip In the Examples, remind students of the meanings of *equal* and *equation*. *Equal* means "of the same quantity." An equation is a statement that includes an equals sign. In an equation, the expression on one side of the equals sign is the same quantity as the expression on the other side.

---

2. 5; The goal is to get the cup by itself on one side of the mat.

Talk About It
1. What is the value of x? **4**
2. If there were 5 counters on the side with the cup and 8 on the other side of the mat, how many would you have taken from each side to solve for x? Explain how you know.
3. In the equation $x + 3 = 7$, 3 is added to x. To solve it, you subtract 3. Suppose you were solving the equation $y - 4 = 8$. What operation would you use with 4 to find the value of y? **addition**

The Mini-Lab suggests that the following two properties can help us solve equations.

Addition Property of Equality	**Words:** If you add the same number to each side of an equation, then the two sides remain equal.	
	Symbols: Arithmetic	Algebra
	$3 = 3$	$a = b$
	$3 + 5 = 3 + 5$	$a + c = b + c$
	$8 = 8$	
Subtraction Property of Equality	**Words:** If you subtract the same number from each side of an equation, then the two sides remain equal.	
	Symbols: Arithmetic	Algebra
	$3 = 3$	$a = b$
	$3 - 2 = 3 - 2$	$a - c = b - c$
	$1 = 1$	

To solve an equation in which a number is added to or subtracted from the variable, you can use the opposite, or **inverse**, operation. Remember, it is always wise to check your solution.

Examples

1 Solve $y - 27 = 18$. Check your solution.

27 is subtracted from y. To solve, add 27 to each side.

Solve the equation.
$$y - 27 = 18$$
$$y - 27 + 27 = 18 + 27$$
$$y = 45$$

Check the solution. Replace y with 45.
$$y - 27 = 18$$
$$45 - 27 \stackrel{?}{=} 18$$
$$18 = 18 \checkmark$$

The solution is 45.

2 Solve $9.84 = 5.75 + m$.

5.75 is added to m. To solve, subtract 5.75 from each side.

$$9.84 = 5.75 + m$$
$$9.84 - 5.75 = 5.75 + m - 5.75$$
$$4.09 = m$$

Check: $9.84 = 5.75 + m$
$9.84 \stackrel{?}{=} 5.75 + 4.09$
$9.84 = 9.84 \checkmark$

The solution is 4.09.

18 Chapter 1 Problem Solving and Algebra

Example 3 CONNECTION

History Refer to the beginning of the lesson. Use the equation $29 + r = 420$ to find the number of recruits added to the code talkers after their formation.

$$29 + r = 420$$
$$29 + r - 29 = 420 - 29$$
$$r = 391$$

There were 391 recruits added.

Check: $29 + r = 420$
$29 + 391 \stackrel{?}{=} 420$
$420 = 420$ ✓

Did you know? According to William McCabe, one of the code designers, there were no code books for the Navajo code. The code talkers used the 411-word code from memory.

CHECK FOR UNDERSTANDING

Communicating Mathematics

Read and study the lesson to answer each question. 1–3. See margin.

1. *Tell* how you determine whether to add or subtract to solve an equation.
2. *Explain* how the addition and subtraction properties of equality are used to solve equations.
3. *You Decide* Taina says that you need to use an inverse operation to solve $y = 8 + 12$. Tess disagrees. Who is correct and why?

HANDS-ON MATH

4. *Model* and solve the equation $x + 5 = 8$ by using cups and counters. See Answer Appendix.

Guided Practice

Solve each equation. Check your solution.

5. $2 + x = 9$ **7**
6. $30 + n = 55$ **25**
7. $p - 1 = 8$ **9**
8. $y = 3.4 + 5.9$ **9.3**
9. $m - 24 = 19$ **43**
10. $8.3 - 4.6 = z$ **3.7**

11. **Sports** The Women's National Basketball Association (WNBA) began play in 1997. Lisa Leslie and Rebecca Lobo were among the league stars who earned as much as $250,000 that year. That salary was $2,500 more than the minimum salary in the men's National Basketball Association (NBA). Use the equation $s + 2{,}500 = \$250{,}000$ to find the minimum salary in the NBA for 1997. **$247,500**

EXERCISES

Practice

Solve each equation. Check your solution.

12. $y = 29 - 7$ **22**
13. $q + 3 = 7$ **4**
14. $21 = 17 + s$ **4**
15. $m + 40 = 75$ **35**
16. $p = 17 - 8$ **9**
17. $z - 1 = 0$ **1**
18. $16 + t = 23$ **7**
19. $s - 9.6 = 17.8$ **27.4**
20. $q - 62 = 153$ **215**
21. $115 - 32 = r$ **83**
22. $0.7 + x = 2.62$ **1.92**
23. $36 = s - 16$ **52**
24. $c + 2.4 = 11.3$ **8.9**
25. $25 + t = 52$ **27**
26. $6.01 - 5.28 = b$ **0.73**
27. $w = 8.26 + 6.2$ **14.46**
28. $143 = h - 98$ **241**
29. $183 = y + 97$ **86**

Lesson 1-4 Solving Subtraction and Addition Equations **19**

---

3 PRACTICE/APPLY

Check for Understanding

If students need additional practice or instruction after completing Exercises 1–11, one of these options may be helpful.
- Extra Practice, see p. 606
- Reteaching Activity
- *Study Guide Masters*, p. 4
- *Practice Masters*, p. 4
- Interactive Mathematics Tools Software

Assignment Guide

Core: 13–31 odd, 33–37
Enriched: 12–28 even, 30–37

Additional Answers

1. Add when a number is subtracted from the variable and subtract when a number is added to the variable.
2. These properties help to isolate the variable while maintaining the equality of the original equation.
3. Tess is correct. You need only to use an inverse operation to solve if there is an operation performed on the side with the variable.

Study Guide Masters, p. 4

Reteaching the Lesson

Activity Working in groups of four, each student should do one of the following: write an equation, do the inverse operation, find the solution, or check the solution. Repeat the activity three times so that each student does each step once.

Error Analysis
Watch for students who choose the number to add or subtract by looking at the side of the equation without a variable.
Prevent by stressing that they should use the number added to or subtracted from the variable.

4 ASSESS

Closing Activity
Modeling Have students use manipulatives as in the Mini-Lab to solve a variety of equations such as $x + 4 = 9$, $x + 2 = 8$, and $y + 3 = 10$. **5, 6, 7**

Additional Answer for Math in the Media
2. Sample answer: The information in the article and the information in the table do not agree as they are written. The averages were probably found, then the average increase was found by subtracting. Then the prices were rounded to the nearest cent to be shown in the table.

Practice Masters, p. 4

Applications and Problem Solving

30. *Medicine* The first kidney transplant was performed in 1954. In 1968, the first transplant of a pancreas was performed. Use the equation $1954 + y = 1968$ to find the number of years between these two medical breakthroughs. **14 years**

31. *Life Science* An African elephant calf weighs approximately 250 pounds at birth. Male adult elephants can weigh up to 12,000 pounds and adult females up to 8,000 pounds.
    a. Use the equation $250 + w = 12,000$ to find the amount of weight gained by a male elephant between birth and adulthood. **11,750 pounds**
    b. Find the difference in weight between an adult female elephant and an adult male using the equation $12,000 - d = 8,000$. **4,000 pounds**

32. *Geometry* Suppose the measure of angle $X = x$ and the measure of angle $Y = y$. If angle X and angle Y are complementary, then $x + y = 90°$. Angle X has a measure of $30°$. Use the equation to find the measure of angle Y. **60°**

33. *Critical Thinking* Solve $2x + 3 = x + 7$. **4**

Mixed Review

34. Evaluate the expression $6 \cdot (18 - 14) + \frac{22 - 4}{9}$. *(Lesson 1-3)* **26**

35. Solve $456 = 76y$ if the replacement set is $\{4, 6, 7, 9\}$. *(Lesson 1-3)* **6**

36. **Test Practice** The Perfect Painting Company uses the formula $A = \ell w$ where A is the area of the wall, ℓ is the wall length, and w is the wall width. What is the area of a wall 13 feet long and 9 feet wide? *(Lesson 1-3)* **C**
    A 22 sq ft **B** 52 sq ft **C** 117 sq ft **D** 127 sq ft

37. Write $5 \cdot 5 \cdot 8 \cdot 8 \cdot 8$ using exponents. *(Lesson 1-2)* $5^2 \cdot 8^3$

MATH IN THE MEDIA

Fill It Up!

The portion of the article below appeared in *USA TODAY* on August 13, 1997.

After a long, steady decline, gasoline prices are suddenly soaring at the peak of the summer driving season.

Nationally, the average price of a gallon of unleaded self-serve regular spurted 5.4 cents in one week to $1.24 a gallon in the latest check by Lundberg Survey, an industry price-taking service.

"To see things pop up this quickly is very unusual," says Mike Morrissey, spokesman for the American Automobile Association. "It caught people off guard."

Pumped-up Prices
Average retail prices for one-gallon of regular unleaded self-serve gasoline:

City	Aug 9, '96	July 25	Aug 8
Atlanta	$1.09	$1.02	$1.07
Chicago	$1.32	$1.31	$1.36
Long Island, N.Y.	$1.31	$1.28	$1.32
Los Angeles	$1.33	$1.24	$1.32
U. S. Average	$1.24	$1.19	$1.24

Source: Lundberg Survey

1. The equation $p + 0.054 = 1.24$ could be used to find the previous week's average price per gallon. Solve for the price. **$1.186**

2. Does the information in the table match the information in the article? Explain. **See margin.**

20 Chapter 1 Problem Solving and Algebra

■ Extending the Lesson ■

Enrichment Masters, p. 4

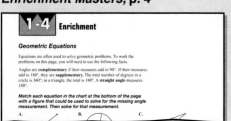

MATH IN THE MEDIA

Ask students to rewrite the information in the story to match the table. Make certain that the students use all of the data in the table. Students should note the gas increase over two weeks in 1997 and compare the price on August 9, 1996 with August 8, 1997.

1-5 Solving Division and Multiplication Equations

What you'll learn
You'll learn to solve equations by using the division and multiplication properties of equality.

When am I ever going to use this?
You can use division and multiplication equations to find how long you can expect a car trip to take.

Word Wise
division property of equality
multiplication property of equality

On April 21, 1997, Lameck Aguta of Kenya won the 26.2-mile Boston Marathon with a time of 2 hours 10 minutes 33 seconds. Many of the 11,000 athletes who ran the race also had hopes of winning, but some athletes had a goal of posting their personal best time. On average, how fast would you have had to run each mile if your goal was to complete the race in 3 hours?

The formula $d = rt$ can be used to solve this problem. In the formula, d represents the distance traveled, r represents the rate, and t represents the time. Substitute 26.2 for d and 3 for t. The formula becomes $26.2 = 3r$. *You will solve this equation in Example 3.*

In Lesson 1-4, you learned that equations could be solved by using the inverse operation. This is also true for multiplication and division equations. The Mini-Lab below shows how this works.

HANDS-ON MINI-LAB

Work with a partner. 🥤 cups and counters ▭ equation mat

Solve $4y = 12$ using cups and counters.

Try This
- Each cup represents y or $1y$. So $4y$ means 4 cups. Put 4 cups on one side of the mat. On the other side of the mat, put 12 counters.
- Arrange the counters into 4 equal groups to correspond to the cups.

Talk About It
1. What number of counters matches with each cup? **3**
2. If $4y = 12$, what is the value of y? **3**
3. What operation does the lab suggest you could use to find the value of y? **Division; separating into equation groups suggests division.**

1-5 Lesson Notes

Instructional Resources
- *Study Guide Masters*, p. 5
- *Practice Masters*, p. 5
- *Enrichment Masters*, p. 5
- Transparencies 1-5, A and B
- *Assessment and Evaluation Masters*, pp. 14, 15
- CD-ROM Program
  - Resource Lesson 1-5
  - Interactive Lesson 1-5

Recommended Pacing	
Standard	Day 6 of 13
Honors	Day 5 of 13
Block	Day 3 of 7

1 FOCUS

5-Minute Check (Lesson 1-4)
Solve each equation. Check your solution.
1. $98 + x = 131$ **33**
2. $3.6 + 4.9 = k$ **8.5**
3. $3.6 = h - 12.7$ **16.3**
4. $m - 254 = 76$ **330**
5. $y = 37 - 15.3$ **21.7**

The 5-Minute Check is also available on **Transparency 1-5A** for this lesson.

Motivating the Lesson
Communication Ask students to describe the difference between the equations $4 + x = 20$ and $4x = 20$. The first equation states that the sum of 4 and a number is 20. The second equation states that the product of 4 and a number is 20.

2 TEACH

Transparency 1-5B contains a teaching aid for this lesson.

Using the Mini-Lab Remind students that their goal is to find the value of y (1 cup). They can do this by dividing the counters into 4 equal groups. Ask students how they would solve $3y = 12$. Use 3 cups and divide the counters into 3 equal groups.

Multiple Learning Styles

Intrapersonal Have students think of ways they encounter multiplication and division in their lives. Examples: One sweater costs $21. How much do 3 sweaters cost? $s = \$21 \cdot 3$, or $63 If you have to have a 288-page book read in 12 hours, how many pages do you have to read each hour? $x = \frac{288}{12}$, or 24 pages

Teaching Tip Before introducing Example 1, stress that when solving an equation, the goal is to isolate the variable on one side of the equation by using inverse operations to eliminate numbers that appear with the variable.

In-Class Examples

For Example 1
Solve $207 = 9n$. Check your solution. **23**

For Example 2
Solve $\frac{x}{7} = 3$. Check your solution. **21**

In the Mini-Lab, you solved the multiplication equation $4y = 12$ by separating, or dividing, the counters into 4 equal groups. You would get the same result if you simply divided each side of the equation by 4. When you divide each side of an equation by the same number, you are using the **division property of equality** to solve the equation. There is also a **multiplication property of equality**.

Division Property of Equality	Words:	If you divide each side of an equation by the same number, then the two sides remain equal.	
	Symbols:	Arithmetic $8 = 8$ $8 \div 2 = 8 \div 2$ $4 = 4$	Algebra $a = b$ $\frac{a}{c} = \frac{b}{c}, c \neq 0$
Multiplication Property of Equality	Words:	If you multiply each side of an equation by the same number, then the two sides remain equal.	
	Symbols:	Arithmetic $8 = 8$ $8 \cdot 2 = 8 \cdot 2$ $16 = 16$	Algebra $a = b$ $ac = bc$

To solve an equation in which the variable is multiplied or divided by a number, use the opposite, or inverse operation. Multiplication and division are inverse operations.

Examples

1 Solve $183 = 3m$. Check your solution.

m is multiplied by 3. To solve, divide each side by 3.

Solve the equation.

$183 = 3m$

$\frac{183}{3} = \frac{3m}{3}$ *Divide each side by 3 to undo multiplication by 3.*

$61 = m$

The solution is 61.

Check the solution.

Check: $183 = 3m$

$183 \stackrel{?}{=} 3 \cdot 61$

$183 = 183$ ✓

2 Solve $\frac{y}{5} = 3$. Check your solution.

y is divided by 5. To solve, multiply each side by 5.

Solve the equation.

$\frac{y}{5} = 3$

$\frac{y}{5} \cdot 5 = 3 \cdot 5$ *Multiply each side by 5 to undo division by 5.*

$y = 15$

The solution is 15.

Check the solution.

Check: $\frac{y}{5} = 3$

$\frac{15}{5} \stackrel{?}{=} 3$

$3 = 3$ ✓

Chapter 1 Problem Solving and Algebra

Example 3 — APPLICATION

Sports Refer to the beginning of the lesson. Use the equation $26.2 = r \cdot 3$ to find the average rate needed to run a marathon in 3 hours.

$26.2 = r \cdot 3$

$\dfrac{26.2}{3} = \dfrac{r \cdot 3}{3}$ *Divide each side by 3 to undo the multiplication.*

$\dfrac{26.2}{3} = r$ 26.2 3 8.7333333333

In order to run a marathon in 3 hours, you would have to run about 8.7 miles per hour.

In-Class Example
For Example 3
The sum of the measures of the interior angles of a pentagon is 540°. The five angles all have the same measure. Solve the equation $5x = 540$ to find the measure of each angle. **108°**

3 PRACTICE/APPLY

Check for Understanding
If students need additional practice or instruction after completing Exercises 1–10, one of these options may be helpful.
- Extra Practice, see p. 606
- Reteaching Activity
- *Transition Booklet*, pp. 11–12
- *Study Guide Masters*, p. 5
- *Practice Masters*, p. 5
- Interactive Mathematics Tools Software

Additional Answers
1. Multiply; to undo division, you multiply.
2. Sample answer: 138,336 ÷ 7833 = 17.66066641

CHECK FOR UNDERSTANDING

Communicating Mathematics

Read and study the lesson to answer each question.   1–2. See margin.

1. *Tell* whether you would first multiply or divide to solve the equation $\dfrac{x}{4} = 20$. Explain how you know.

2. *Show* how to use a calculator to solve $138{,}336 = 7{,}833r$ to determine Lameck Aguta's average rate in feet per second. Round to the nearest tenth.

HANDS-ON MATH

3. *Write* the equation shown by the model at the right. Then solve. $3x = 12$; $x = 4$

Guided Practice

Solve each equation. Check your solution.

4. $16x = 64$ **4**
5. $40 = 8p$ **5**
6. $n = \dfrac{280}{70}$ **4**
7. $\dfrac{n}{6} = 13$ **78**
8. $3.7 \cdot 4 = g$ **14.8**
9. $23.8 \div 7 = s$ **3.4**

10. **Medicine** In 1997, the American Medical Association (AMA) elected its first female president, Texas physician Dr. Nancy Dickey. About 32,120 or 0.11 of the doctors in the AMA are women. Use the equation $0.11w = 32{,}120$ to find the number of doctors in the AMA. Round to the nearest whole number.   **292,000 members of the AMA**

Lesson 1-5 Solving Division and Multiplication Equations

Reteaching the Lesson

Activity Give 12 counters to each student. Direct students to solve $2x = 12$, $3x = 12$, and $6x = 12$ by dividing their counters into equal groups. **6 per group, $x = 6$; 4 per group, $x = 4$; 2 per group, $x = 2$**

Error Analysis
Watch for students who use the wrong operation when applying the inverse. **Prevent by** reminding them that if they divide on one side of the equation, they must divide on the other. The same applies to multiplying. Students should always check their solutions.

Assignment Guide

Core: 11–33 odd, 35–38
Enriched: 12–30 even, 31–33, 35–38
All: Self Test 1–10

Chapter Project

Exercise 34 asks students to advance to the next stage of work on the Chapter Project. You may have students work in groups of 4 to 6 to share their ideas of what interests to mention on their home pages.

EXERCISES

Practice

Solve each equation. Check your solution. 28. 28

11. $9 \cdot 25 = a$ **225**
12. $13m = 182$ **14**
13. $78 = 13b$ **6**
14. $6 = \frac{k}{50}$ **300**
15. $56 \times 8 = n$ **448**
16. $\frac{z}{0.7} = 4.2$ **2.94**
17. $\frac{q}{9} = 6$ **54**
18. $245 \div 5 = f$ **49**
19. $3.18 \div 6 = q$ **0.53**
20. $6 = \frac{p}{35}$ **210**
21. $8 = \frac{x}{25}$ **200**
22. $322 = 14h$ **23**
23. $43.4 = 6.2t$ **7**
24. $\frac{n}{0.19} = 8$ **1.52**
25. $16.8 = 0.4r$ **42**
26. $350 = 14d$ **25**
27. $\frac{t}{3} = 61$ **183**
28. $\$7.56 = \$0.27b$

29. Solve $\frac{d}{0.11} = 5$.   **0.55**
30. Find the value of x for $\$5.44 = \$0.34x$.   **16**

Applications and Problem Solving

31. *Recreation* The Boundary Waters of Minnesota and Canada is a popular area for canoeing and camping. The outfitters in nearby Ely will arrange for your permits and rent all of the equipment for a trip.
    a. Cliff Wold's offers a three-day package for two people for $322. Use the equation $2c = 322$ to find the per-person cost.   **$161**
    b. Wilderness Outfitters rents canoes on a daily basis. If the daily charge for a rental is $32, use the equation $\frac{t}{32} = 7$ to find the total charge for a seven-day rental.   **$224**

32. *Social Work* Dr. Pedro José Greer Jr. has founded four free clinics to provide health care for homeless people. One of those, the Camillus Health Concern, in Miami, Florida, gets about 30,000 visits per year. Use the equation $365v = 30,000$ to find the average number of visits per day. Round to the nearest tenth.   **82.2 visits per day**

33. *Critical Thinking* In order for a bill that has been vetoed by the president to become law, it must be passed by two-thirds of both the House of Representatives and the Senate. That means that 290 members of the House of Representatives would have to pass a bill to override a veto. Use the equation $\frac{2}{3}x = 290$ to find the number of members of the House of Representatives.   **435 representatives**

24 Chapter 1 Problem Solving and Algebra

Study Guide Masters, p. 5

34. **Working on the CHAPTER Project** Your home page should include information about your interests. For example, if you are interested in music, you might want to include some information about your favorite artists. Make a list of your interests and choose a few to incorporate into that plan you have made for your home page. **See students' work.**

Mixed Review

35. **Archaeology** *Nature* magazine reports that stone tools found in Ethiopia are estimated to be 2.5 million years old. That is about 700,000 years older than similar tools found in Tanzania. Use the equation $2,500,000 = a + 700,000$ to find the age of the tools found in Tanzania. *(Lesson 1-4)* **1.8 million years**

36. **Algebra** Solve $t - 9 = 23$. Check your solution. *(Lesson 1-4)* **32**

37. Write $4 \cdot 4 \cdot 8 \cdot 8 \cdot 4$ as an expression using exponents. *(Lesson 1-2)* **$4^3 \cdot 8^2$**

38. **Test Practice** One centimeter is about 0.3937 inch. If Dale is 68 inches tall, then his height in centimeters must be — *(Lesson 1-1)* **E**
    A less than 140 centimeters.
    B between 140 and 150 centimeters.
    C between 150 and 160 centimeters.
    D between 160 and 170 centimeters.
    E more than 170 centimeters.

CHAPTER 1 Mid-Chapter Self Test

1. **Transportation** A Boeing 747-300 jet carries 342 passengers with 36 in first-class and the rest in coach. Suppose a first-class ticket to fly from Los Angeles to Chicago costs $750 and a coach ticket costs $450. Use the four-step problem solving plan to find the greatest possible ticket sales on one flight. *(Lesson 1-1)* **$164,700**

2. Write the product $5 \cdot 5 \cdot 5$ using exponents. *(Lesson 1-2)* **5^3**

3. Evaluate the expression $3^3 \cdot 4^4$. *(Lesson 1-2)* **6,912**

4. **Algebra** Find the value of $a^2 + b^2 - c^2$ if $a = 10$, $b = 5$, and $c = 4$. *(Lesson 1-3)* **109**

5. **Algebra** Solve $y + 45 = 60$ if the replacement set is {10, 15, 20, 25}. *(Lesson 1-3)* **15**

Solve each equation. Check your solution. *(Lessons 1-4 and 1-5)*

6. $m + 30 = 100$ **70**
7. $s - 5.8 = 14.3$ **20.1**
8. $16x = 48$ **3**
9. $18 = \frac{y}{7}$ **126**

10. **Technology** In 1996, shares of Yahoo!, the company that created one of the most popular world wide web search engines, were offered for sale. At $33 each, the shares owned by company co-founder Jerry Yang were valued at $132 million. Use the equation $33p = 132,000,000$ to find how many shares Mr. Yang owned. *(Lesson 1-5)* **4 million**

Lesson 1-5 Solving Division and Multiplication Equations 25

Motivating Students

As students learned in the Chapter Project on page 3, there are many opportunities to explore the Internet. They may be interested in a career in web-page design or in a related field. To start the discussion, you may ask students questions about the history of the Internet.
- When did the first part of the Internet begin and what was its purpose? **In 1969, the ARPANET was started by the U.S. Department of Defense to link military and academic researchers with remote computers.**
- When did the first businesses provide commercial access to the Internet? **1995**

Making the Math Connection

Mathematical Markup Language, or MathML, developed in 1997, allows mathematical symbols to be presented on the Web, enabling students, teachers, and researchers to do more math on the web than was previously possible.

Working on *Your Turn*

Students may want to work in groups to research the future of careers in Internet or software development. Make sure they understand that the World Wide Web is just one part of the computer industry and that jobs are wide-ranging and available in many places around the world.

*An additional School to Career activity is available on page 27 of the **School to Career Masters**.*

TECHNOLOGY

Tim Berners-Lee
WORLD WIDE WEB INVENTOR

Tim Berners-Lee had a problem he needed to solve. In 1989, he was a physicist in Switzerland and wanted to exchange information with colleagues all over the world. So, he created the World Wide Web to link him with this information. His invention has opened up a wealth of information for people all over the world. He now heads the W3 Consortium at MIT's Laboratory for Computer Science in Boston, Massachusetts.

Most companies prefer that Internet developers have college degrees. Courses in computer science, graphics design, and mathematics are helpful in preparing for a career as an Internet developer.

For more information:
American Electronics Association
5201 Great American Pkwy.
Suite 520
Santa Clara, CA 95054

inter CONNECTION
www.glencoe.com/sec/math/mac/mathnet

Someday, I'd like to develop a computer product to help people solve problems.

Your Turn

Use the World Wide Web to gather information about the future of careers in Internet or software development. Will more developers be needed in the next decade? What trends are expected in the industry over the next 10 to 20 years?

More About Tim Berners-Lee
- Mr. Berners-Lee has received numerous awards and honorary degrees. In 1997, he won the Duddell Medal of the Institute of Physics and the MCI Computerworld/Smithsonian Award for Leadership in Innovation.
- Mr. Berners-Lee earned a bachelor of arts degree with honors in physics from The Queen's College, Oxford University, England, which he attended from 1973 to 1976. While there, he built his first computer with a soldering iron, an old television, and spare computer parts.

1-6 Writing Expressions and Equations

What you'll learn
You'll learn to write algebraic expressions and equations from verbal phrases and sentences.

When am I ever going to use this?
You can use an equation to change Fahrenheit temperatures to Celsius.

Digital television will become common as technology improves. There will be better sound and video quality and Internet access through your television. Have you ever wondered how digital television works? Signals are sent through the air in a computer-based language of 0s and 1s. Then a digital converter box in the television translates the signals back into the images you see.

You can act as a translator in mathematics, interpreting words and ideas and translating them into mathematical expressions and equations. There are many words and phrases that suggest arithmetic operations. Any variable can be used to represent a number.

Verbal Phrase	Algebraic Expression
five less than a number	$x - 5$
a number increased by 12	$y + 12$
twice a number decreased by 3	$2c - 3$
the quotient of a number and 4	$\frac{n}{4}$

The chart shows common phrases that usually indicate the four operations.

Addition or Subtraction		Multiplication or Division	
plus	minus	times	divided
sum	difference	product	quotient
more than	less than	multiplied	separated
increased by	subtract	each	
total	decreased by	of	
in all			

Examples

Write each sentence or phrase as an algebraic expression.

1 The number of e-mail users on a server increased by 300.

Let u represent the number of e-mail users before the increase. The word *increased* means there are more e-mail users.

The algebraic expression is $u + 300$.

2 three times the number of points required for an A

Let $a =$ the number of points required for an A.

Three times means multiply by 3. → $3 \cdot a$ or $3a$

The algebraic expression is $3a$.

Lesson 1-6 Writing Expressions and Equations **27**

Multiple Learning Styles

Auditory After completing Examples 1–3, test students' understanding. Call out "Four times a number is 40." Have students translate the phrase into an algebraic equation. Make up progressively more complicated examples.

1-6 Lesson Notes

Instructional Resources
- *Study Guide Masters*, p. 6
- *Practice Masters*, p. 6
- *Enrichment Masters*, p. 6
- *Science and Math Lab Manual*, pp. 49–52
- Transparencies 1-6, A and B

CD-ROM Program
- Resource Lesson 1-6

Recommended Pacing	
Standard	Day 7 of 13
Honors	Day 6 of 13
Block	Day 4 of 7

1 FOCUS

5-Minute Check
(Lesson 1-5)
Solve each equation. Check your solution.
1. $\frac{x}{3} = 6$ **18**
2. $91 = 13k$ **7**
3. $2.4(1.8) = w$ **4.32**
4. $\$8.46h = \$54.99$ **6.5**
5. When the number q is divided by 18, the quotient is 8. Find the number. **144**

The 5-Minute Check is also available on **Transparency 1-6A** for this lesson.

Motivating the Lesson
Communication Have students read the two opening paragraphs of the lesson. Ask for volunteers to state a number in English and then to say it in another language, identifying the language used. Guide students to understand that number phrases can be translated into words and vice versa.

2 TEACH

Transparency 1-6B contains a teaching aid for this lesson.

Reading Mathematics Write an algebraic expression or equation on the chalkboard. Have students translate it into a phrase or sentence.

Lesson 1-6 **27**

In-Class Examples

For Example 1
Tony added 15 CDs to his collection. Let c represent the number of CDs in Tony's original collection. Write an algebraic expression for the new number of CDs. $c + 15$

For Example 2
Write the phrase as an algebraic expression: 7 times the cost of one book.
$7 \times b$ or $7b$

For Example 3
Meg scored 11 points. That was 4 points less than the number of points Sumi scored. Let s represent Sumi's score. Write an equation to represent this problem. $11 = s - 4$

3 PRACTICE/APPLY

Check for Understanding
If students need additional practice or instruction after completing Exercises 1–10, one of these options may be helpful.
- Extra Practice, see p. 606
- Reteaching Activity
- *Study Guide Masters*, p. 6
- *Practice Masters*, p. 6

Study Guide Masters, p. 6

You can also translate a verbal sentence into an equation. Then the equation can be used to solve a problem.

Verbal Sentence	Algebraic Equation
24 is 6 more than a number.	$24 = n + 6$
Five times a number is 60.	$5n = 60$

Example 3 — APPLICATION

Retail Sales In 1997, NikeTown opened in New York City. Shoppers are surrounded by sports monuments and video screens showing their favorite sports heroes. There are 6,600 more customers each day than there are full- or part-time employees. If there are 7,000 customers each day, write an equation to find the average number of employees.

Let n = the number of employees.

6,600 more than the number of employees → $n + 6{,}600$
The number of customers, 7,000, equals this. → $7{,}000 = n + 6{,}600$

CHECK FOR UNDERSTANDING

Communicating Mathematics

Read and study the lesson to answer each equation. 1–3. See margin.

1. **Write** two different verbal phrases that could be represented by the algebraic expression $y + 3$.
2. **State** one or more words that tell you that a verbal sentence can be written as an equation.
3. **You Decide** Kathy says that the phrases *7 less than y* and *7 less y* are translated into the same algebraic expression. Chapa disagrees. Who is correct and why?

Guided Practice

Write each phrase or sentence as an algebraic expression or equation.

4. 15 more than x $x + 15$
5. the product of 7 and n $7 \cdot n$ or $7n$
6. the quotient of m and 8 $m \div 8$ or $\frac{m}{8}$
7. 9 less than w is 15. $w - 9 = 15$
8. The difference between 29 and t is 8. $29 - t = 8$
9. The desks in the classroom can be separated into 7 groups of 4. $d \div 4 = 7$
10. **Life Science** Did you know that a mosquito has 15 more teeth than an adult human? If t represents the number of teeth a human has, write an expression for the number of teeth in a mosquito. $t + 15$

28 Chapter 1 Problem Solving and Algebra

■ **Reteaching the Lesson** ■

Activity Copy the chart on the chalkboard. Ask students to copy this chart and supply verbal phrases for each symbol. Sample answers are given.

Operation	Phrase
+	more than; in all
−	difference; less than
×	each; of
÷	quotient; separated

Additional Answers
1. Sample answer: three more than a number; a number increased by 3
2. Sample answer: is, equal, total
3. Chapa is correct. *7 less than y* is translated as $y - 7$ and *7 less y* is translated $7 - y$.

EXERCISES

Practice Write each phrase or sentence as an algebraic expression or equation.

15. $16r = 80$
19. $j + 10$
22. $a + 5$

11. the sum of k and 9 $k + 9$
12. 28 less n is 25. $28 - n = 25$
13. 5 more than a number $n + 5$
14. 7 less than y is 25. $y - 7 = 25$
15. The product of 16 and r is 80.
16. thirteen fewer than the total $t - 13$
17. the sum of 6 and b $6 + b$
18. Four times w is 72. $4w = 72$
19. 10 more than Juanita scored
20. half of Tyrone's sales $\frac{t}{2}$
21. 48 less v is 15. $48 - v = 15$
22. 5 points more than the class average
23. two times as many touchdowns as the Cowboys $2c$
24. Five more than m is 7. $m + 5 = 7$
25. The difference between a number and 8 is 25. $n - 8 = 25$
26. The product of a number and 7 is 56. $7n = 56$
27. The number of desks divided into 3 classrooms is 25. $n \div 3 = 25$
28. $100 more than the amount made yesterday $y + 100$

Family Activity

Have family members tell you phrases to write as expressions and equations. For example, your brother may say "If you are 3 years older than I am and you are 14, how old am I?"

Applications and Problem Solving

31b. $2.3a = 103.5$

29. **Geography** Niagara Falls is one of the most visited waterfalls in North America, but it is not the tallest. Yosemite Falls is 2,249 feet taller!
    a. If n is the height of Niagara Falls, write an expression for the height of Yosemite Falls. $n + 2,249$
    b. Yosemite Falls is 2,425 feet high. Write an equation to find the height of Niagara Falls. $n + 2,249 = 2,425$

30. **Space Exploration** In 1983, Sally Ride became the first woman in space, traveling aboard the *Challenger*. Three years later, *Challenger* tragically exploded in the first seconds of a flight. Write an equation to find the year in which the *Challenger* accident occurred. $3 = n - 1983$

31. **Nutrition** Use the information in the table to answer each question.
    a. Write an expression for the amount of vitamin C in a serving of orange if there are s milligrams in 5.5 ounces of strawberries. $s - 9$
    b. There is 2.3 times the recommended daily allowance of vitamin C in 5.5 ounces of kiwifruit. Write an equation for the amount of vitamin C recommended for each day.

Fruit	Vitamin C (mg in 5.5 oz)
Orange	52
Strawberries	63
Kiwifruit	103.5

Source: Food and Drug Administration

32. **Critical Thinking** Write an equation to represent *5 times the sum of twice x and 4 is 40*. $5(2x + 4) = 40$

Mixed Review

33. **Algebra** Solve $7b = 105$. *(Lesson 1-5)*   15
34. **Test Practice** Sydney's packages weighed 12.5 pounds, 9 pounds, and 11.25 pounds. What was the total weight of the packages? *(Lesson 1-4)*   E
    A 35.25 lb   B 22.75 lb   C 16.25 lb   D 12.50 lb   E Not Here
35. **Algebra** Evaluate $6a^2 + b$ if $a = 4$ and $b = 1$. *(Lesson 1-3)*   97
36. Find the value of the expression 2^6. *(Lesson 1-2)*   64

Lesson 1-6 Writing Expressions and Equations   **29**

Extending the Lesson

Enrichment Masters, p. 6

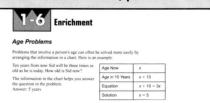

Activity Ask students to search newspapers and magazines for stories containing numerical information. Students can then select the relevant portions of the stories, choose appropriate variables, and translate newspaper phrases into algebraic expressions or equations.

Assignment Guide

Core: 11–31 odd, 32–36
Enriched: 12–28 even, 29–36

Family Activity

From each student, collect phrases written as expressions and equations for the Family Activity. Create a book that members of the class can use for extra practice.

4 ASSESS

Closing Activity

Writing Ask students to write the steps they would follow to translate this sentence into an equation.
Four more than k is the product of 3 and 7.
1. Translate "four more than k" as $k + 4$.
2. Translate "is" as =.
3. Translate "the product of 3 and 7" as $3 \cdot 7$.
4. Rewrite the sentence as $k + 4 = 3 \cdot 7$.

Practice Masters, p. 6

1-6 **Practice**

Writing Expressions and Equations

Write each phrase or sentence as an algebraic expression or equation.
1. 8 more than x $x + 8$
2. 12 less than b $b - 12$
3. the product of 6 and y $6y$
4. the quotient of a and 4 $\frac{a}{4}$
5. the sum of 9 and c $9 + c$
6. the difference of q and 12 $q - 12$
7. 7 times d $7d$
8. 20 less n $20 - n$
9. 6 less than x is 18. $x - 6 = 18$
10. 4 more than y is 17. $y + 4 = 17$
11. The product of a and 7 is 21. $7a = 21$
12. The sum of 1 and w is 12. $1 + w = 12$
13. the sum of 8 and 6 times y $8 + 6y$
14. eight dollars less than Joni earned $j - 8$
15. Pak's salary minus a $223 deduction $p - 223$
16. twice as many flowers as Susan picked $2s$
17. 6 less than the product of 8 and c is 58. $8c - 6 = 58$
18. The cost of the tea plus 10 cents tax is $2.09. $c + 0.10 = 2.09$
19. 8 more than the number of meals served on Tuesday $T + 8$
20. 18 less than the number of gameboard squares is 126. $s - 18 = 126$
21. 63 is 1 more than twice the number of miles Timothy drove. $63 = 2m + 1$
22. The sum of 9 and the quotient of x and 7 is 11. $9 + \frac{x}{7} = 11$
23. 12 less than twice the number of cows is 36. $2c - 12 = 36$
24. 17 inches less than 3 times Maria's height is 169 inches. $3m - 17 = 169$
25. 5 more than the number of paper clips divided into 4 groups $\frac{n}{4} + 5$
26. 18 more than 3 times Tony's wages $3t + 18$
27. 8 less than the number of apples divided into 5 groups is 32. $\frac{a}{5} - 8 = 32$

Lesson 1-6   **29**

1-7A THINKING LAB Notes

Objective Students solve problems by working backward.

Recommended Pacing	
Standard	Day 8 of 13
Honors	Day 7 of 13
Block	Day 4 of 7

1 FOCUS

Getting Started Ask students how they would solve this problem: *The price of 6 notebooks is $14.50 including $0.70 tax. Find the cost of 1 notebook.* Subtract $0.70 from $14.50 and divide by 6; $2.30.

2 TEACH

Teaching Tip Advise students to be careful to undo operations in the correct order. Have them write out the sequence of operations in the problem and then work backward, undoing operations one by one.

In-Class Example
When Cindy multiplies her age by 5, subtracts 12, and divides by three, the result is 16. How old is Cindy? 12

Additional Answers
1. They started with the money they had left and added on the amounts they spent in reverse order.
2. Two-thirds of the money was $40. $\frac{2}{3}n = 40$. So $n = \$60$. They do have the correct amount of change.
3. It is easier to work backward than forward when you know a result and need to know the start.

THINKING LAB PROBLEM SOLVING

1-7A Work Backward
A Preview of Lesson 1-7

Alonso and Ron have just finished buying supplies for their camping trip. They think they may have gotten the wrong change back at one of the stores. Let's listen in!

Alonso: I think we should have more money left. We started with $60 and I only have about $2 left.

Ron: Wow, it does seem like we should have more than that left!

Alonso: Well, we just left the grocery. The receipt says we spent $15.89. With the $2 or so we have left, that makes about $18.

Before that we were at the sporting goods store. We got all that equipment. The total there was $21.91. Adding that on, that's about $40.

And the first thing we did was use a third of the money as a deposit on the campsite.

So do we have the right change or not?

THINK ABOUT IT

Work with a partner. 1–4. See margin.

1. **Tell** how Alonso and Ron were working backward to solve the problem.
2. **Work backward** to finish the problem and determine whether Alonso and Ron have the correct change.
3. **Explain** when is it easier to work backward to solve a problem than it is to work forward.
4. **Show** the order that you would need to use to solve this problem by **working backward**.

   Start with a number. Multiply it by 4, divide by 10, and add 14. Then double. If the result is 124, what was the starting number?

30 Chapter 1 Problem Solving and Algebra

■ Reteaching the Lesson ■

Activity Read this example to students: *If I add 3 to my number, then divide by 6, the answer is 2. Guess my number.* 9 One student states a problem involving two operations, like the example. The student who correctly guesses the number scores one point. Each member of the group takes a turn making up a problem.

Additional Answer
4. Divide by 2. → 124 ÷ 2 = 62
   Subtract 14. → 62 − 14 = 48
   Multiply by 10. → 48 × 10 = 480
   Divide by 4. → 480 ÷ 4 = 120
   The starting number was 120.

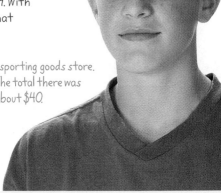

ON YOUR OWN

5. The last step of the 4-step plan for problem solving is to *examine* the solution. **Explain** how you can examine the solution when you solve a problem by working backward. **See margin.**

6. *Write a Problem* that could be solved by working backward. **See students' work.**

7. *Look Ahead* If following the order of operations is working forward, how is the solution to Example 1 on page 33 working backward? **See margin.**

MIXED PROBLEM SOLVING

Strategies
Look for a pattern.
Solve a simpler problem.
Act it out.
Guess and check.
Draw a diagram.
Make a chart.
Work backward.

Solve. Use any strategy.

8. *Tourism* Fantasyland in Iraan, Texas, features giant statues of comic strip characters. If you multiply Dinny the Dinosaur's height by 4 and add 1 foot, you will find his length. If the statue of Dinny is 65 feet long, how tall is it? **16 feet**

9. *Money Matters* Mrs. Navarro's class bought food for the Adopt-A-Family program. They spent $127.68. The 19 students divided the cost equally. How much did each student contribute? **$6.72**

10. *Earth Science* According to *Health* magazine, the average American produces 110,000 pounds of garbage by the time he or she is 75 years old. How many pounds of garbage does the average American produce each year? **about 1,467 pounds**

11. *Money Matters* Shelley Green just bought her first new car. She will pay $400 a month for five years. The actual amount of the loan is $15,900. How much extra will Ms. Green pay in order to spread the payment out over five years? **$8,100**

12. **Test Practice** In one month, Bryant's father spent $51.90, $22.78, $33.11, and $45.08 on groceries. A good estimate of the total amount spent on groceries during the month is — **D**
  A $100. B $120. C $140.
  D $150. E $200.

13. *Games* Parker Brothers began selling Monopoly in 1935. The rules haven't changed since then. The chart shows how the game would have changed if it had kept up with inflation.

	1935	1995
Pool of Money	$15,140	$184,794
Dollars per Player	$1,500	$18,308
Park Place Rent	$35	$428
Passing GO	$200	$2,441

Source: Parker Brothers

  a. How much greater would the rent be on Park Place in 1995 than in 1935? **$393**
  b. It costs $400 to buy Boardwalk. How much would it be if it was adjusted for inflation? **$4,882**

14. *Money Matters* A store tripled the price it paid for a pair of jeans. After a month, the jeans were marked down $5. Two weeks later, the price was divided in half. Finally, the price was reduced by $3 and the jeans sold for $14.99.
  a. How much did the store pay for the pair of jeans? **$13.66**
  b. Did the store make or lose money on the jeans? **The store made $1.33.**

Lesson 1-7A THINKING LAB **31**

Extending the Lesson

Activity Have students write problems involving three or more steps that can be solved by working backward. Be sure to have them to check the answers. Then have students try to stump the class with their problems.

Sample problem: *Khadija accidentally spills $2\frac{1}{3}$ cups of sugar, which she discards. She then makes 9 dozen cookies using a recipe that calls for 2 cups of sugar per 3-dozen batch. If there is no sugar left, how much sugar was there in the beginning?* $8\frac{1}{3}$ **cups**

---

3 PRACTICE/APPLY

Check for Understanding
Use the results from Exercise 4 to determine whether students comprehend the process for working a problem backward.

Extra Practice If students need additional practice in problem solving, extra practice is available on the following pages.
- Work Backward, see p. 607
- Mixed Problem Solving, see pp. 645–646

Assignment Guide

All: 5–14

4 ASSESS

Closing Activity
Modeling Use blocks of four different sizes to represent the four arithmetic operations. Have students stack the blocks, in any order. Then have them dismantle the stack from the top down, describing how each operation can be undone.

Additional Answers
5. Work the problem forward and make sure the ending number is correct.
7. Undoing the subtraction first then undoing the multiplication is backward from the order of operations.

Lesson 1-7A **31**

1-7 Lesson Notes

Instructional Resources
- *Study Guide Masters*, p. 7
- *Practice Masters*, p. 7
- *Enrichment Masters*, p. 7
- Transparencies 1-7, A and B
- *Assessment and Evaluation Masters*, p. 16

 CD-ROM Program
- Resource Lesson 1-7
- Interactive Lesson 1-7

Recommended Pacing	
Standard	Day 9 of 13
Honors	Days 8 & 9 of 13
Block	Day 5 of 7

1 FOCUS

5-Minute Check
(Lesson 1-6)

Write each phrase or sentence as an algebraic expression or equation.
1. three times n $3n$
2. five less than m $m - 5$
3. the quotient of 5 and h $\frac{5}{h}$
4. The difference between m and 9 is 5. $m - 9 = 5$
5. The product of y and 7 is 12. $7y = 12$

 The 5-Minute Check is also available on **Transparency 1-7A** for this lesson.

Motivating the Lesson

Problem Solving Ask students to identify the error in the following solution.

$$2x + 6 = 14$$
$$\frac{2x}{2} + 6 = \frac{14}{2}$$
$$x + 6 = 7$$
$$x = 1$$

In the first step, the entire equation must be divided by 2; $x = 4$.

1-7 Solving Two-Step Equations

What you'll learn
You'll learn to solve two-step equations.

When am I ever going to use this?
Doctors use two-step equations to determine how many doses of medication a patient needs.

Word Wise
distributive property

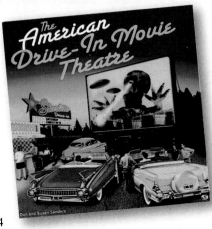

Have you ever been to a drive-in theater? Richard Hollingshead, Jr. opened the first drive-in theater in Camden, New Jersey, in 1933. He first tested his idea for a drive-in theater by setting up a projector and screen in his driveway. He even turned on the sprinkler to see if he could still watch the movie in the rain!

Admission to Mr. Hollingshead's theater was $1 per carload or $0.25 per person. If there were 214 carload admissions one night and the total sales was $285.50, how many individual tickets were sold? *This problem will be solved in Example 3.*

In order to find the number of individual tickets sold, you will need to solve a two-step equation. You can investigate solving two-step equations using cups and counters.

HANDS-ON MINI-LAB

Work with a partner.

cups and counters

equation mat

Solve $4 + 2x = 10$ using cups and counters.

Try This
- Place 4 counters and 2 cups on one side of the mat to represent $4 + 2x$. On the other side of the mat, place 10 counters.
- Remove 4 counters from each side of the mat.
- Separate the remaining counters into two equal groups to correspond to the two cups.

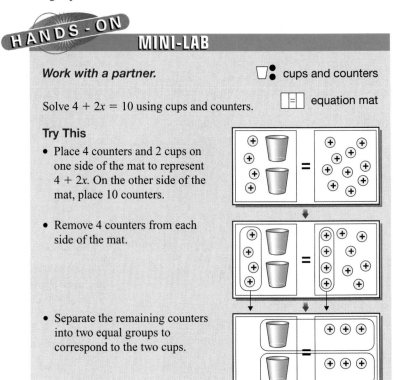

32 Chapter 1 Problem Solving and Algebra

Talk About It
1. What number of counters matches with each cup? 3
2. What operation is represented by removing the four counters from each side? **subtracting 4**
3. What operation is represented by separating the counters into two groups? **division by 2**
4. What is the solution of $4 + 2x = 10$? 3

In the Mini-Lab, you solved the equation $4 + 2x = 10$. Notice that in the equation, the x is first multiplied by 2, and then 4 is added. When you solved the equation, you first "undid" the addition by subtracting 4. Then you "undid" the multiplication by dividing by 2. Notice that this is the opposite order than the order of operations. Many equations can be solved algebraically in a similar manner.

Examples

Study Hint
Problem Solving You can use the work backward strategy to solve two-step equations.

1 Solve $6m - 4 = 38$. Check your solution.

$$6m - 4 = 38$$
$$6m - 4 + 4 = 38 + 4 \quad \text{Add 4 to each side to undo the subtraction of 4.}$$
$$6m = 42$$
$$\frac{6m}{6} = \frac{42}{6} \quad \text{Divide each side by 6 to undo the multiplication by 6.}$$
$$m = 7$$

The solution is 7.

Check: $6m - 4 = 38$
$6(7) - 4 \stackrel{?}{=} 38$ *Replace m with 7.*
$42 - 4 \stackrel{?}{=} 38$ *Multiply before subtracting.*
$38 = 38$ ✓   *It checks.*

2 Solve $16.45 = \frac{c}{3.6} - 5.2$. Check your solution.

$$16.45 = \frac{c}{3.6} - 5.2$$
$$16.45 + 5.2 = \frac{c}{3.6} - 5.2 + 5.2 \quad \text{Add 5.2 to each side to undo the subtraction of 5.2.}$$
$$21.65 = \frac{c}{3.6}$$
$$21.65 \cdot 3.6 = \frac{c}{3.6} \cdot 3.6 \quad \text{Multiply each side by 3.6 to undo the division by 3.6.}$$
$$77.94 = c$$

The solution is 77.94.

Check: $\frac{c}{3.6} - 5.2 = 16.45$
$\frac{77.94}{3.6} - 5.2 \stackrel{?}{=} 16.45$
$21.65 - 5.2 \stackrel{?}{=} 16.45$
$16.45 = 16.45$ ✓   *The solution of 77.94 checks.*

Lesson 1-7 Solving Two-Step Equations   **33**

---

2 TEACH

 Transparency 1-7B contains a teaching aid for this lesson.

Using the Mini-Lab Ask students why cups and counters should not be separated into two groups in Step 1. **This separates the equation into two equations, neither of which is solved. The cups must be by themselves on one side of the mat before separating the counters into groups.**

Teaching Tip Before Example 1, write several two-step equations on the chalkboard. Ask students to state which operation they should undo first when solving each equation.

In-Class Examples

For Example 1
Solve $5y + 9 = 24$. Check your solution. **3**

For Example 2
Solve $\frac{n}{3.1} - 11.52 = 8.48$. Check your solution. **62**

Lesson 1-7   **33**

In-Class Examples

For Example 3
Your doctor has written you a prescription and instructs you to take the 29 pills until they are gone. You're supposed to take 4 the first two days and 3 per day thereafter. How long will you take the pills after the first two days? **7 days**

For Example 4
Solve $15 + 2(x - 7) = 35$. **17**

Example 3
APPLICATION

Entertainment Refer to the beginning of the lesson. Find the number of individual tickets sold.

Explore   You know that admission was $1 per carload or $0.25 per person. There were 214 carload admissions and the total sales was $285.50. You need to find how many individual tickets were sold.

Plan   Write and solve an equation.

Solve   Let a represent the number of individual tickets.

$$\underbrace{\text{sales from}}_{\text{carload admissions}} \text{ and } \underbrace{\text{sales from}}_{\text{individual admissions}} \text{ is total sales}$$

$$\$1 \cdot 214 \quad + \quad \$0.25 \cdot a \quad = \quad \$285.50$$

$$\$214 + \$0.25a = \$285.50$$

$\$214 + \$0.25a - \$214 = \$285.50 - \$214$ *Subtract 214 from each side.*

$$\$0.25a = \$71.50$$

$$\frac{\$0.25a}{\$0.25} = \frac{\$71.50}{\$0.25} \quad \text{\textit{Divide each side by 0.25.}}$$

$$a = 286$$

There were 286 individual tickets sold.

Examine   The sales from 214 carload admissions and 286 individual tickets is $214 \cdot \$1 + 286 \cdot \$0.25$ which is $\$214 + 71.50$ or $\$285.50$. The solution checks.

You can use the distributive property to help solve two-step equations.

Distributive Property	Words:	The sum of two addends multiplied by a number is the sum of the products of each addend and the number.
	Symbols:	Arithmetic $6 \cdot (5 + 7) = (6 \cdot 5) + (6 \cdot 7)$ Algebra $a \cdot (b + c) = (a \cdot b) + (a \cdot c)$

Example 4 Solve $3(x - 5) = 27$. Check your solution.

$$3(x - 5) = 27$$
$$3 \cdot x - 3 \cdot 5 = 27 \quad \text{Apply the distributive property.}$$
$$3x - 15 = 27$$
$$3x - 15 + 15 = 27 + 15 \quad \text{Add 15 to each side.}$$
$$3x = 42$$
$$\frac{3x}{3} = \frac{42}{3} \quad \text{Divide each side by 3.}$$
$$x = 14 \quad \text{The solution is 14.}$$

Check: $3(x - 5) = 27$
$3(14 - 5) \stackrel{?}{=} 27$
$3(9) \stackrel{?}{=} 27$
$27 = 27$ ✓ The solution checks.

CHECK FOR UNDERSTANDING

Communicating Mathematics

Read and study the lesson to answer each question. 1–3. See margin.

1. **Tell** what step you would do first to solve $3x + 5 = 17$.
2. **Explain** how the order of operations is used to solve two-step equations.
3. **Show** how to use the model to solve the equation $4x + 3 = 15$. How can you check your solution?

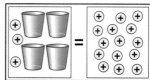

HANDS-ON MATH

Guided Practice

Solve each equation. Check your solution.

4. $8c - 3 = 13$ **2**
5. $9 + 3d = 45$ **12**
6. $7 + \frac{m}{3} = 9$ **6**
7. $\frac{m}{8} - 6 = 2$ **64**
8. $\frac{y}{7} + 0.8 = 2.3$ **10.5**
9. $5(x + 1) = 25$ **4**

10. Write *four more than twice a number is 38* as an equation and solve.

11. **Puzzles** The first crossword puzzle was published in 1913. Sixty-six years later, the world's largest crossword puzzle was created with 20 more than 384 times the number of clues that the first puzzle had! If the world's largest puzzle has 12,308 clues, how many clues were in the first puzzle? **32 clues**

10. $4 + 2x = 38$; 17

EXERCISES

Practice

Solve each equation. Check your solution. 23. 0.3

12. $4c - 3 = 25$ **7**
13. $3r + 7 = 16$ **3**
14. $\frac{m}{16} - 2 = 10$ **192**
15. $9 + 11s = 53$ **4**
16. $\frac{r}{4} - 1 = 12$ **52**
17. $\frac{g}{8} + 2 = 21$ **152**
18. $4(x + 2) = 20$ **3**
19. $\frac{r}{8} + 3 = 27$ **192**
20. $6f - 15 = 3$ **3**
21. $5g + 1.8 = 4.3$ **0.5**
22. $2.7 + 0.8n = 4$ **1.625**
23. $0.68 = 0.17 + 1.7k$
24. $2.3 + 6t = 10.1$ **1.3**
25. $2b + 7.5 = 8.0$ **0.25**
26. $11y - 5.3 = 5.7$ **1**
27. $4(w - 3) = 36$ **12**
28. $18 = 2(t + 1)$ **8**
29. $1.3 = \frac{n}{9} - 0.8$ **18.9**

Lesson 1-7 Solving Two-Step Equations **35**

3 PRACTICE/APPLY

Check for Understanding

If students need additional practice or instruction after completing Exercises 1–11, one of these options may be helpful.
- Extra Practice, see p. 607
- Reteaching Activity
- *Study Guide Masters*, p. 7
- *Practice Masters*, p. 7
- Interactive Mathematics Tools Software

Assignment Guide

Core: 13–37 odd, 38–43
Enriched: 12–34 even, 35–43

Additional Answers

1. Subtract 5 from each side.
2. You can use the order of operations in reverse to solve two-step equations.
3. Remove 3 counters from each side. Then divide the remaining counters into four groups. The solution is $x = 3$. Check by substituting 3 counters for each cup and verifying that the result is a true equation.

Study Guide Masters, p. 7

Reteaching the Lesson

Activity Have students complete the following statements by inserting the words *addition, subtraction, multiplication*, and *division*.

To solve a two-step equation, first undo the <u>addition</u> or <u>subtraction</u>. Then undo the <u>multiplication</u> or the <u>division</u>.

Error Analysis
Watch for students who have difficulty turning words into algebraic equations.
Prevent by helping students write the equation in words first as shown in Example 3.

Lesson 1-7 **35**

4 ASSESS

Closing Activity

Modeling Use 4 cards to model the equation $3p + 4 = 13$.

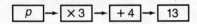

Students should write the appropriate inverse operations on blank cards and solve the equation. **3**

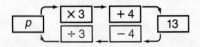

Chapter 1, Quiz C (Lessons 1-6 and 1-7) is available in the *Assessment and Evaluation Masters*, p. 16.

Practice Masters, p. 7

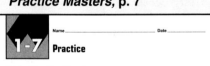

30. Find the solution of $3x - 4 = 7$. $3\frac{2}{3}$
31. What is the value of g for $7 = 4g - 9$? **4**

Write each sentence as an equation. Then solve.

32. Eight more than three times a number is 59. $8 + 3x = 59$; **17**
33. The quotient of a number and four decreased by one is 7. $\frac{n}{4} - 1 = 7$; **32**
34. Ruben scored 12 points less than half Jean's total in Laser Tag. Ruben scored 103 points. What was Jean's score? $\frac{J}{2} - 12 = 103$; **230**

Applications and Problem Solving

35. *Business* A rental agency charges $52 per day for a car. The first 100 miles are free, but any miles after that cost $0.32 each. Ms. Misel rented a car for three days, and the rental fee was $237.60. How many miles did Ms. Misel drive? **355 miles**

36. *Fund-raising* The Live Aid Concert on July 13, 1985, was the longest television concert ever, with more than 60 bands performing in Philadelphia and London. It raised $65 million for food for African countries experiencing a drought. Ticket sales for the Philadelphia concert totaled $3,375,000, with 15,000 of these tickets sold at $50 each.
    a. If general admission tickets were $35 each, write an equation to find how many general admission tickets were sold. $3,375,000 = 50(15,000) + 35t$
    b. Find the number of general admission tickets sold. **75,000 tickets**

37. See students' work.

37. *Write a Problem* that can be solved using the equation $2x + 3 = 15$.

38. *Critical Thinking* In diving, seven judges score each dive, and the sum of the three middle scores is multiplied by the degree of difficulty for the dive. After eleven dives, the scores for the dives are added to find the diver's final score.
    a. Beatriz's score is 366.5 going into the last dive. The degree of difficulty is 2.7, and the leader's score is 439.4. Write an equation to find the total of the judges' scores for her to tie for first.
    b. What must Beatriz score to tie for first? **27** 38a. $2.7s + 366.5 = 439.4$

Mixed Review

39. *Algebra* Write an equation to represent *three more than the number of cars is fifteen*. (Lesson 1-6) $c + 3 = 15$

40. **Test Practice** A day-care center's thermostat used to be set on 70° for 24 hours each day. To save money, the owner decided to turn down the furnace for 9 hours during the night between the hours of 8 P.M. and 5 A.M. If she saves $1.70 per day on her heating bill, how much will she save during a 30-day month? (Lesson 1-5) **C**

    A $15.30 **B** $33.50 **C** $51.00 **D** $72.80 **E** Not Here

41. *Algebra* Solve $b - 19 = 73$. (Lesson 1-4) **92**
42. Evaluate $\frac{45 - 9}{3^2 + 3}$. (Lesson 1-3) **3**
43. Evaluate $2^2 \cdot 7^2$. (Lesson 1-2) **196**

36 Chapter 1 Problem Solving and Algebra

Enrichment Masters, p. 7

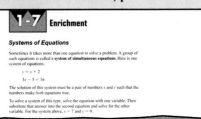

Extending the Lesson

Activity Have pairs of students use cups and counters to solve equations where variables appear on both sides of the equation. Point out that when modeling the equations, there will be cups on both sides of the mat and that a solution is represented when the cups remain on just one side.

COOPERATIVE LEARNING

1-7B Function Machines

A Follow-Up of Lesson 1-7

Cacao beans

Do you like Hershey's Chocolate Kisses? The machines at the factory in Hershey, Pennsylvania, make about 33 million kisses each day! It takes seven steps to turn the input of cacao beans in the machines to the output of chocolate.

In mathematics, we have *functions* that work like machines. In a function, you *input* a number. Since the *output* depends on the input, it is *a function of* the input number. If the input number is represented by x, then the function of the input number can be represented by $f(x)$, which is read "f of x."

TRY THIS

Work in pairs.

Put a number in the top of the "function machine" at the right. At each stage, an operation is performed. Then the result moves to the next stage.

- What number is being used as the input? **4**
- What is the first operation performed? What is the result after the first operation? **multiply by 7; 28**
- What is the second operation? What is the result after the second operation? **subtract 2; 26**
- What is the final output of the function? **26**

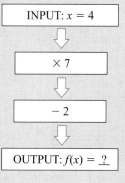

INPUT: $x = 4$

$\times 7$

$- 2$

OUTPUT: $f(x) = \underline{?}$

ON YOUR OWN

1. Write a sentence or two to describe the operations performed on the input by the function machine. **Multiply the input by 7. Then subtract 2 from the result.**
2. Each stage of the function machine represents a mathematical operation. If the input is represented by x, write an expression for the value of the output. **$7x - 2$**
3. Suppose you were given the output. How could you find the number that was the input? **Work backward to solve the two-step equation.**

Find the missing input or output for each function machine.

4.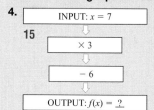

INPUT: $x = 7$
15
$\times 3$
$- 6$
OUTPUT: $f(x) = \underline{?}$

5. INPUT: $x = ?$
12
$\div 2$
$+ 1$
OUTPUT: $f(x) = 7$

6.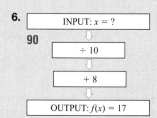

INPUT: $x = ?$
90
$\div 10$
$+ 8$
OUTPUT: $f(x) = 17$

Lesson 1-7B HANDS-ON LAB **37**

Math Journal

Have students write a paragraph about how using a function machine makes it easier to understand the order of operations and to solve problems by working backward.

1-8 Lesson Notes

Instructional Resources
- *Study Guide Masters*, p. 8
- *Practice Masters*, p. 8
- *Enrichment Masters*, p. 8
- *Transparencies 1-8, A and B*
- *Hands-On Lab Masters*, p. 68

CD-ROM Program
- Resource Lesson 1-8
- Interactive Lesson 1-8

Recommended Pacing	
Standard	Day 10 of 13
Honors	Day 10 of 13
Block	Day 6 of 7

1 FOCUS

5-Minute Check
(Lesson 1-7)

Solve each equation. Check your solution.
1. $6 + 4x = 38$ **8**
2. $\frac{n}{9} - 3 = 2$ **45**
3. $30 = 13k - 9$ **3**
4. $4(m + 7) = -8$ **−9**
5. Write eight more than half a number is 15 as an equation and solve.
   $\frac{x}{2} + 8 = 15; x = 14$

The 5-Minute Check is also available on **Transparency 1-8A** for this lesson.

Motivating the Lesson

Communication Have students read the first two paragraphs of the lesson. Show students these two arrangements of four squares.

Figure 1   Figure 2

- Ask students to compare the perimeters of the two figures. The perimeter of Figure 1 (8 units) is 2 units less than the perimeter of Figure 2 (10 units).
- Ask students to draw an arrangement of five squares whose perimeter is 12 units.

38 Chapter 1

1-8 Integration: Geometry
Perimeter and Area

What you'll learn
You'll learn to find the perimeters and areas of rectangles, squares, and parallelograms.

When am I ever going to use this?
You can estimate the amount of paint needed for your room by finding the areas of the walls.

Word Wise
rectangle
perimeter
square
parallelogram
area
base
altitude
height

Stop the presses! The big news is that the Newseum in Arlington, Virginia, is open. The Newseum is the first museum dedicated to news. Interactive displays allow visitors to the Newseum to imagine themselves as reporters, editors, and news anchors. One display includes a wall that is a **rectangle** 126 feet long and 10 feet high with video screens that carry news broadcasts. How much material was needed to build the frame around the video screens? *This problem will be solved in Example 1.*

The **perimeter** of a geometric figure is the sum of the measures of all of its sides. The formula for the perimeter of any rectangle is $P = 2\ell + 2w$, where ℓ represents the length and w represents the width.

$P = 2\ell + 2w$

Example **APPLICATION**

① Construction Refer to the beginning of the lesson. Find the amount of material needed to build the frame around the video screens.

The length of the wall is 126 feet, and the width is 10 feet.

$P = 2\ell + 2w$
$P = 2(126) + 2(10)$ *Replace ℓ with 126 and w with 10.*
$P = 252 + 20$
$P = 272$

The frame around the video screens required 272 feet of material.

A rectangle whose sides are all equal is called a **square**. In a square, the values for ℓ and w are the same. So, the formula for the perimeter of a square is often written as $P = 4s$, where s is the length of a side. The perimeter of the square at the right is 24 inches.

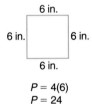

$P = 4(6)$
$P = 24$

38 Chapter 1 Problem Solving and Algebra

Squares and rectangles are special types of **parallelograms**. In a parallelogram, each pair of opposite sides are parallel and have the same length. To find the perimeter of a parallelogram, add the lengths of the sides. So, the formula for the perimeter of a parallelogram can be written as $P = 2a + 2b$, where a and b are the lengths of adjacent sides. The parallelogram at the right has a perimeter of 26 meters.

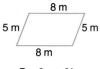

$P = 2a + 2b$
$P = 2(8) + 2(5)$
$P = 16 + 10$
$P = 26$

Perimeter Formulas for Rectangles, Squares, and Parallelograms

rectangle	square	parallelogram
$P = 2\ell + 2w$	$P = 4s$	$P = 2a + 2b$

In addition to the perimeter, we often solve problems by using the **area** of a geometric figure. The area is the measure of the surface enclosed by the figure. The area of any rectangle can be found by multiplying the width and the length.

Examples

Find the area of each rectangle.

2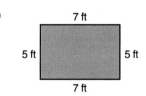

$A = \ell w$
$A = 7 \cdot 5$
$A = 35$

The area is 35 square feet.

3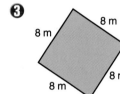

$A = s \cdot s$ or s^2
$A = 8 \cdot 8$
$A = 64$

The area is 64 square meters.

The formula for the area of a parallelogram is *not* the product of the sides. However, it is related to the formula for the area of a rectangle.

Lesson 1-8 Integration: Geometry   Perimeter and Area   **39**

---

2 TEACH

 Transparency 1-8B contains a teaching aid for this lesson.

In-Class Examples

For Example 1
The exterior of the John Hancock Tower in Boston, Massachusetts, contains 10,334 rectangles of glass that are each 4 feet by 11 feet. Calculate the perimeter of each pane. **30 ft**

For Example 2
Find the area of the rectangle.

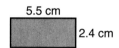

13.2 cm²

For Example 3
Find the area of the square.

144 m²

Teaching Tip Before Example 3, point out that the orientation of a figure does not affect its classification.

Classroom Vignette

"For students who have difficulty remembering $P = 2\ell + 2w$, I find it helpful to show them the perimeter in expanded form, $\ell + w + \ell + w$. Then they can see there are two ℓs and two ws and recall the formula more easily."

Richie Berman, Ph.D., Consultant
University of California, Santa Barbara
Santa Barbara, CA

Richie Berman

Lesson 1-8   **39**

Using the Mini-Lab Be sure students distinguish between the altitude and the side of a parallelogram. An altitude is drawn at a right angle to each of the bases. A side does not form right angles with the base unless the parallelogram is a rectangle.

In-Class Example

For Example 4
Find the area of the parallelogram.

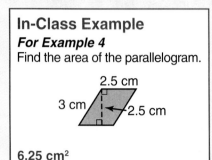

6.25 cm²

---

3 PRACTICE/APPLY

Check for Understanding
If students need additional practice or instruction after completing Exercises 1–8, one of these options may be helpful.
- Extra Practice, see p. 607
- Reteaching Activity
- *Study Guide Masters,* p. 8
- *Practice Masters,* p. 8

Additional Answers
1.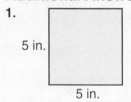

   $P = 20$ in.

2. A square is a rectangle with all sides equal; a rectangle is a special parallelogram.

3. Sample answer: Area is the amount of space a figure takes up; the perimeter is the distance around the figure.

 **MINI-LAB**

Work with a partner.   grid paper   scissors

Try This
- Copy the parallelogram onto a piece of grid paper. Cut it out.
- Make a cut along the dashed line. Move the parts so that they form a rectangle.

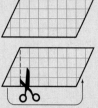

Talk About It

1. 32 square units

1. Find the area of the rectangle you formed.
2. Consider the parts labeled on the figure at the right. Compare the length and width of the rectangle with the **base** and **altitude** of the parallelogram. base = length, altitude = width

   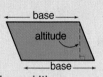

3. The length of the altitude is the **height** of a parallelogram. If b represents the length of the base and h represents the height, write a formula for the area of a parallelogram. $A = bh$

Example ④ Find the area of the parallelogram.

The height is 2.2 meters, and the base is 4.6 meters.

$A = bh$
$A = 4.6(2.2)$
$A = 10.12$ The area is 10.12 square meters.

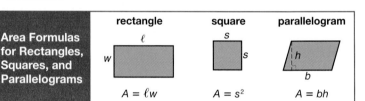

Area Formulas for Rectangles, Squares, and Parallelograms

rectangle: $A = \ell w$
square: $A = s^2$
parallelogram: $A = bh$

CHECK FOR UNDERSTANDING

Communicating Mathematics

Read and study the lesson to answer each question. 1–3. See margin.

1. **Draw** and label a square that has an area of 25 square inches. What is its perimeter?
2. **Tell** how parallelograms, rectangles, and squares are related.
3. **Write** in your own words how the area of a figure differs from its perimeter.

40 Chapter 1 Problem Solving and Algebra

■ Reteaching the Lesson ■

Activity Have students find several rectangular objects in the classroom such as textbook covers and posters. Then have them use rulers to measure the dimensions of each object in order to compute its perimeter and area.

Guided Practice

Find the perimeter and area of each figure.

4. 7.5 m × 7.5 m square
$P = 30$ m; $A = 56.25$ m²

5. parallelogram, 5 yd × 8 yd
$P = 26$ yd; $A = 40$ yd²

6. parallelogram, 11 ft top, $5\frac{1}{2}$ ft, 6 ft
$P = 34$ ft; $A = 60\frac{1}{2}$ ft²

7. A rectangle has a length that is twice its width. Find the area and perimeter of the rectangle if the width is 8 centimeters.

8. *Celebrations* In celebration of Flag Day in 1997, Kraft Foods made a giant flag cake at the base of the Statue of Liberty. The cake was $62\frac{1}{2}$ feet by 88 feet. Find the perimeter and area of the cake.
$P = 301$ ft; $A = 5,500$ ft²

7. $P = 48$ cm; $A = 128$ cm²

EXERCISES

Practice Find the perimeter and area of each figure.

9. $P = 16$ units; $A = 15$ units²

10. $P = 22.2$ m; $A = 17.5$ m²

11. $P = 22.6$ m; $A = 29.2$ m²

12. $P = 32$ yd; $A = 64$ yd²

13. $P = 20$ m; $A = 23.5$ m²

14. $P = 16$ m; $A = 12.4$ m²

15. $P = 16$ units; $A = 12$ units²

16. $P = 40$ in.; $A = 60$ in²

17. $P = 60$ in.; $A = 185$ in²

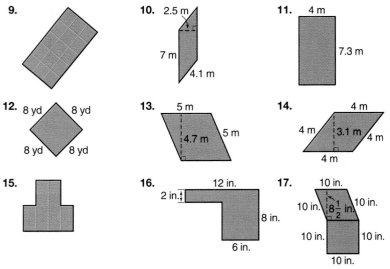

18. Write an expression for the perimeter of a rectangle whose length is three times its width. $P = 8w$

19. Use an equation to find the base of a parallelogram whose height is 8 yards and whose area is the same as a square with sides each 12 yards long.

19. $144 = 8b$; $b = 18$ yards

20. The area of a parallelogram is 91 square meters. Find the height of the parallelogram if its base is 13 meters long. **7 m**

Applications and Problem Solving

21. See students' work.

21. *Working on the CHAPTER Project* Part of designing a home page is planning the layout. The different parts of the page, such as text, art, and photographs, should be placed so that the page is attractive and easy to read. Use what you have learned about area to help you lay out your home page.

Lesson 1-8 Integration: Geometry Perimeter and Area **41**

Assignment Guide
Core: 9–19 odd, 23, 25–28
Enriched: 10–20 even, 22–28

CHAPTER Project

Exercise 21 asks students to advance to the next stage of work on the Chapter Project. You may wish to have the students exchange layouts in order to critique them. Encourage them to first state the layout's best feature. Then have them suggest improvements.

Study Guide Masters, p. 8

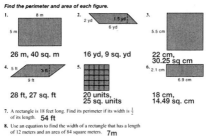

4 ASSESS

Closing Activity

Speaking Sketch a rectangle and a parallelogram, including the dimensions, on the chalkboard. Have students describe how to find the perimeter and the area of each figure.

Additional Answer

25. a — $\frac{1}{4}$ unit^2; b — $\frac{1}{4}$ unit^2;

c — $\frac{1}{16}$ unit^2; d — $\frac{1}{8}$ unit^2;

e — $\frac{1}{16}$ unit^2; f — $\frac{1}{8}$ unit^2;

g — $\frac{1}{8}$ unit^2

Practice Masters, p. 8

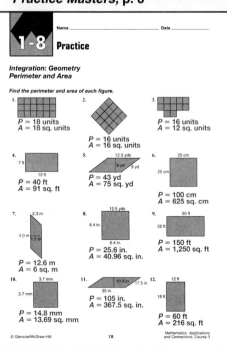

22. **History** Did you know that the poem that became our national anthem, "The Star Spangled Banner," was inspired by the flag flying over Fort McHenry in Maryland in 1814? The flag now hangs in the Smithsonian Institution.

   a. Mary Pickersgill was paid $405.60 to make the Fort McHenry flag in 1813. When commissioned, it was 42 feet wide and 30 feet high. What was the area and perimeter of the flag? **P = 144 ft; A = 1,260 ft^2**

   b. In 1997, a project to clean and restore the flag was started. Experts found that the area of the flag had been reduced by 270 square feet because of battle, souvenir hunters, and deterioration. What is the remaining area to be restored? **990 ft^2**

23. **Marketing** If you had bought a box of Quaker cereal in 1955, you would own a tiny piece of land in Canada's Yukon Territory. Quaker gave deeds to 21 million one-inch by one-inch plots of land in boxes of cereal!

   a. What is the area of one plot of the land given away? **1 in^2**

   b. One man claimed he had collected the deeds to 75 square feet of land. How many boxes of cereal did he have to buy? (*Hint:* He bought more than 900 boxes.) **10,800 boxes**

24. **Patterns** Copy and complete the table of perimeters and areas of rectangles.

Length	Width	Perimeter	Area
1	2	6	2
2	4	12	8
4	8	24	32

   a. Notice that the dimensions of each rectangle are doubled. How do the perimeters of the rectangles compare? **The perimeter is doubled.**

   b. How are the areas of the rectangles changed when the length and width are doubled? **The area is multiplied by 4.**

25. **Critical Thinking** The figure is an ancient Chinese tangram puzzle. The square is cut into seven pieces: five triangles, one square, and one parallelogram. Suppose the area of the entire tangram is 1 square unit. Find the area of each piece. **See margin.**

Mixed Review

26. **Algebra** Write an equation to represent *three times a number less five is 16.* Then solve the equation. (*Lesson 1-7*) **3n − 5 = 16; 7**

27. **Test Practice** Lenora has saved $2.26, Karen has saved $3.47, and Kevin has saved $2.93 toward the purchase of a new video game that costs $25.95. What is the total the friends have saved? (*Lesson 1-4*) **A**

   A $8.66   **B** $8.56   **C** $7.66   **D** $7.56   **E** Not Here

28. **Algebra** Solve $x − 32 = 20$ if the replacement set is {47, 52, 57, 63}. (*Lesson 1-3*) **52**

42 Chapter 1 Problem Solving and Algebra

Enrichment Masters, p. 8

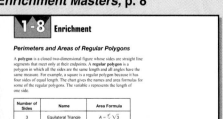

Extending the Lesson

Activity Organize a class contest. Prepare a variety of rectangles, squares, and parallelograms on notecards with their areas and perimeters given. For each turn, call out the name of the figure with its area and perimeter. Students then have to draw the figure correctly. You may have them compete individually or in teams of 2, 3, or 4.

1-9 Solving Inequalities

What you'll learn
You'll learn to write and solve inequalities.

When am I ever going to use this?
You can find the minimum score you need on a test to get a certain grade in the course.

Word Wise
inequality

Have you wondered what it feels like to jump from an airplane? At Paramount's *Great America* in Santa Clara, California, you can get the feel of skydiving in the world's tallest free-fall ride "The DROP ZONE Stunt Tower." On this ride, up to four people at a time plummet from a height of 224 feet reaching speeds in excess of 60 miles per hour! While there is no minimum age requirement to go on this ride, you must be greater than 54 inches tall.

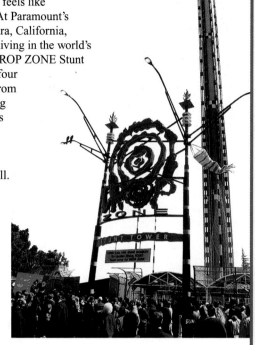

If h represents your height, then $h > 54$ represents the height you must be to go on the ride. The sentence $h > 54$ is called an **inequality**. Inequalities are sentences that compare quantities. They contain symbols like $>$ and $<$.

		Words	Symbols
Arithmetic		6 is greater than 4.	$6 > 4$
		2 is less than 15.	$2 < 15$
Algebra		$2x + 9$ is greater than 11.	$2x + 9 > 11$
		$\frac{m}{3} - 5$ is less than 8.	$\frac{m}{3} - 5 < 8$

Study Hint
Reading Math Recall that $>$ is read as *is greater than* and $<$ is read as *is less than*. The symbol $\geq$ is read as *is greater than or equal to* and $\leq$ is read as *is less than or equal to*.

Some inequalities use the symbols $\geq$ or $\leq$. They are combinations of the equals sign and the inequality symbols.

		Words	Symbols
Arithmetic		8 is greater than or equal to 7.	$8 \geq 7$
		16 is less than or equal to 16.	$16 \leq 16$
Algebra		11 is greater than or equal to $5x$.	$11 \geq 5x$
		$\frac{t}{3} + 4$ is less than or equal to 12.	$\frac{t}{3} + 4 \leq 12$

Equations with one variable often have one solution. Unlike an equation, an inequality may have many solutions. The solution can be written as a set of numbers.

Lesson 1-9 Solving Inequalities **43**

1-9 Lesson Notes

Instructional Resources
- *Study Guide Masters*, p. 9
- *Practice Masters*, p. 9
- *Enrichment Masters*, p. 9
- Transparencies 1-9, A and B
- *Assessment and Evaluation Masters*, p. 16
- *School to Career Masters*, p. 27
- *Technology Masters*, p. 54

CD-ROM Program
- Resource Lesson 1-9

Recommended Pacing	
Standard	Day 11 of 13
Honors	Day 11 of 13
Block	Day 6 of 7

1 FOCUS

5-Minute Check
(Lesson 1-8)
Find the perimeter and area of each figure.

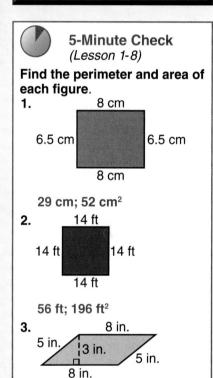

1. [square: 8 cm × 6.5 cm, with 6.5 cm sides]
   29 cm; 52 cm²

2. [square: 14 ft on all sides]
   56 ft; 196 ft²

3. [parallelogram: 8 in. base, 5 in. sides, 3 in. height]
   26 in.; 24 in²

The 5-Minute Check is also available on **Transparency 1-9A** for this lesson.

Lesson 1-9 43

Motivating the Lesson

Hands-On Activity Have students draw a number line from 0 to 10. Then ask them to:
- mark with a long, black, vertical line where $x = 5$;
- shade with red the area where x is less than 5; and
- shade with blue the area where x is greater than 5.

2 TEACH

 Transparency 1-9B contains a teaching aid for this lesson.

Using the Mini-Lab To show that the solution set includes whole numbers greater than 10, have students test 11 and 20. You may also ask them to test numbers such as 5.3 and $4\frac{1}{2}$ so they see that not only whole numbers may be solutions.

Teaching Tip In both parts of Example 1, make certain that students understand that you are indicating possible positions on the number line for c and s.

In-Class Example

For Example 1
At least 6 extra beds will be needed to house visitors for a wedding.

a. Write an inequality to express this. Then graph the inequality. $n \geq 6$

```
  +--+--+--+--+--+--●──●──●──●──●→
  0  1  2  3  4  5  6  7  8  9
```

b. You can house up to two of these visitors. Write an inequality to express this. Then graph the inequality. $a \leq 2$

```
←●──●──●──+--+--+--+--+--+--+→
  0  1  2  3  4  5  6  7  8  9
```

 **MINI-LAB**

Work with a partner.

Try This
- Copy the number line.

- Use the whole numbers 0 through 10 as the replacement set. Color in each circle that represents a solution for the inequality $3a \geq 12$.

Talk About It
1. What does the number line suggest about the solution set?
2. If all real numbers are included in the replacement set, are there solutions that are not marked on the number line? If so, name one.
3. Solve the equation $3a = 12$. How is the solution of this equation related to the solution of the inequality?

1. All of the numbers greater than or equal to 4 are solutions.
2. yes; 4.5
3. It is the least number that is a solution.

There are many situations in real life that can be described using an inequality. The table below shows some common phrases and corresponding inequalities.

<	>	≤	≥
• less than • fewer than • up to	• greater than • more than • exceeds • in excess of	• less than or equal to • no more than • at most	• greater than or equal to • no less than • at least

Example 1

CONNECTION

Literature *Alice In Wonderland* is one of the most popular children's books ever. But its author, Lewis Carroll, said he would be surprised if it sold more than 2,000 copies. Before Mr. Carroll died in 1898, *Alice In Wonderland* had sold more than 180,000 copies.

a. Write an inequality to show the number of copies Mr. Carroll expected to sell. Then graph the inequality.

$$\underbrace{\text{copies expected to sell}}_{c} \quad \underbrace{\text{no more than}}_{\leq} \quad \underbrace{2{,}000}_{2{,}000}$$

The copies Mr. Carroll expected to sell can be represented by $c \leq 2{,}000$.

```
←●━━━━━●──+──+──+→
 0   1,000  2,000  3,000  4,000
```

When ≤ or ≥ is used, the circle on the graph is solid to show that the number is included in the solution.

44 Chapter 1 Problem Solving and Algebra

b. Write an inequality to show how many books were actually sold before Mr. Carroll died. Then graph the inequality.

$$\underbrace{\text{copies sold}}_{s} \quad \underbrace{\text{were more than}}_{>} \quad \underbrace{180{,}000}_{180{,}000}$$

The copies sold before Mr. Carroll died can be represented by $s > 180{,}000$.

When $<$ or $>$ is used, the circle on the graph is open to show that the number is not included in the solution.

As we learned in solving equations, the guess-and-check method is not always the quickest way to solve an equation. Likewise, it is usually not the best way to solve an inequality. You can use your knowledge of solving equations to solve an inequality.

Example 2 Solve $5x - 15 > 17$. Graph the solution on a number line.

Solve the related equation, $5x - 15 = 17$. Remember to add to undo the subtraction. Then divide to undo the multiplication.

$$5x - 15 = 17$$
$$5x - 15 + 15 = 17 + 15 \quad \textit{Add 15 to each side.}$$
$$5x = 32$$
$$\frac{5x}{5} = \frac{32}{5} \quad \textit{Divide each side by 5.}$$
$$x = 6.4$$

The solution will either be numbers greater than 6.4 or numbers less than 6.4. Let's test a number to see which is correct.

numbers greater than 6.4	*numbers less than 6.4*
Try 9.	Try 5.
$5(9) - 15 > 17$	$5(5) - 15 > 17$
$45 - 15 \stackrel{?}{>} 17$	$25 - 15 \stackrel{?}{>} 17$
$30 > 17$ *true*	$10 > 17$ *false*

The numbers greater than 6.4 make up the solution set. So, $x > 6.4$.

To graph the solution, draw an empty circle at 6.4. Then draw a large arrow to indicate the numbers that are solutions.

Try other numbers in the set to check your solution.

Lesson 1-9 Solving Inequalities **45**

In-Class Example

For Example 2
Solve $2x - 9 > 1$. Graph the solution on a number line. $x > 5$

3 PRACTICE/APPLY

Check for Understanding

If students need additional practice or instruction after completing Exercises 1–11, one of these options may be helpful.
- Extra Practice, see p. 608
- Reteaching Activity
- *Study Guide Masters*, p. 9
- *Practice Masters*, p. 9

Assignment Guide

Core: 13–33 odd, 35–39
Enriched: 12–32 even, 33–39

Additional Answers

1. "In excess of" indicates that the speeds are more than 60 mph.
2. One number is greater than, less than, or equal to another.
3.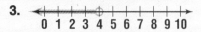
   0 1 2 3 4 5 6 7 8 9 10
4.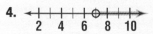
   2 4 6 8 10
5.
   4 6 8 10 12
6.
   12 14 16 18 20
7.
   4 6 8 10 12

Study Guide Masters, p. 9

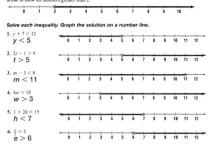

CHECK FOR UNDERSTANDING

Communicating Mathematics

Read and study the lesson to answer each question. 1–3. See margin.

1. *Tell* what is meant by the phrase "in excess of" in describing the speed of the DROP ZONE Stunt Tower in the beginning of the lesson.
2. *Describe* the three different ways two numbers can compare.
3. *Draw* a number line for a solution set that is all numbers less than 4.

HANDS-ON MATH

Guided Practice
4–9. See margin for number lines.

Solve each inequality. Graph the solution on a number line.

4. $a - 4 > 3$ $a > 7$
5. $21 > 12 + d$ $d < 9$
6. $x - 3 \leq 14$ $x \leq 17$
7. $3x < 27$ $x < 9$
8. $\frac{x}{8} \geq 5$ $x \geq 40$
9. $7 < 2c - 3$ $c > 5$

10. Write an inequality for *five times a number is greater than sixty*. Then solve the inequality. $5x > 60$; $x > 12$

11. **Recreation** *Women's Sports and Fitness* magazine recommends that when you go hiking, three times the weight of your backpack and its contents should be no more than your body weight.
    a. If you weigh 90 pounds and your empty backpack weighs 12 pounds, write an inequality to find how much the contents of your backpack should weigh. $3(x + 12) \leq 90$
    b. Solve the inequality. $x \leq 18$

EXERCISES

Practice

Solve each inequality. Graph the solution on a number line.

12–29. See Answer Appendix for number lines.

12. $x - 5 < 18$ $x < 23$
13. $y + 9 \geq 34$ $y \geq 25$
14. $8 < k - 7$ $15 < k$
15. $3d \leq 12$ $d \leq 4$
16. $28 \leq 4x$ $x \geq 7$
17. $\frac{m}{5} > 2$ $m > 10$
18. $30 \geq 6a$ $5 \geq a$
19. $18 > 2h$ $9 > h$
20. $5x \geq 25$ $x \geq 5$
21. $5r \leq 35$ $r \leq 7$
22. $8 < \frac{b}{3}$ $24 < b$
23. $\frac{k}{2} \leq 8$ $k \leq 16$
24. $7 < 2x - 3$ $5 < x$
25. $\frac{y}{3} + 8 \leq 11$ $y \leq 9$
26. $7 < 5x - 8$ $x > 3$
27. $2 > \frac{r}{5} - 8$ $r < 50$
28. $9q + 4 \leq 22$ $q \leq 2$
29. $\frac{m}{6} + 1 \leq 4$ $m \leq 18$

Write an inequality for each sentence. Then solve the inequality.

30. Three times a number is less than sixty. $3x < 60$; $x < 20$
31. Four more than a number is greater than fifteen. $4 + x > 15$; $x > 11$
32. Six less than four times a number is greater than thirty. $4x - 6 > 30$; $x > 9$

Applications and Problem Solving

33. **Law** In many states, a person has to be greater than 16 years old to obtain a driver's license. Write an inequality that tells the ages at which a person can obtain a driver's license in those states. $a > 16$

46 Chapter 1 Problem Solving and Algebra

■ Reteaching the Lesson ■

Activity Have students work in groups of three. The first student writes a sentence containing the phrase *less than* or *more than*, the second student translates the sentence into an inequality, and the third student graphs the inequality.

Additional Answers

8.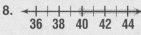
   36 38 40 42 44

9.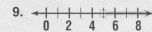
   0 2 4 6 8

34. **Space** More than 50,000 UFO sightings have been reported in the last 20 years. Between 1947 and 1969, the U.S. Air Force collected and studied reports of UFO sightings. In more than 10,800 cases, more than 90% of those studied, they were able to conclude that these were meteors, stars, satellites, aircraft, clouds, or jokes! a. $s > 50{,}000$
    a. Write an inequality for the UFO sightings in the last 20 years.
    b. Write and solve an inequality for the number of cases the Air Force studied. $0.90c < 10{,}800;\ c < 12{,}000$

35. **Critical Thinking** Find the possible values of x for the inequality $10 < 4x + 2 \leq 30$. $2 < x \leq 7$

Mixed Review

36. **Home Improvement** Mr. Delgado decided to extend his living room to add a sunroom to his home. If the length of the room is increased by 10 feet and the width of the room is 16 feet, find the area of the new sunroom. *(Lesson 1-8)*

    36. 160 ft²

37. **Algebra** Solve $\frac{n}{2} + 31 = 45$. Check your solution. *(Lesson 1-7)* $n = 28$

38. **Test Practice** Last week Carlos ran 3.75 kilometers on Monday, 5 kilometers on Tuesday, and 4.5 kilometers on Wednesday. How many kilometers did Carlos run on those three days? *(Lesson 1-4)* **B**

    A 5.75 km   **B** 13.25 km   **C** 17.25 km   **D** 20.75 km   **E** Not Here

39. Evaluate $3[14 - (8 - 5)^2] + 20$. *(Lesson 1-3)* **35**

You're the Greatest

Get Ready This game is for two players.
- 31 index cards
- spinner

Get Set Copy the numbers 0 to 30 onto the cards, one number per card. Shuffle the cards and place them facedown in a pile. Label the sections of a spinner with the following inequalities.

$3x - 5 > 11$ $17 + c < 25$ $4p > 16$

$\frac{d}{2} + 15 < 20$ $2f + 60 > 75$ $42 - m > 21$

Go
- Decide which player will go first.
- In turn, each player selects a card and spins the spinner. If the number on the card is in the inequality's solution, the player gets 1 point.
- The first player to score 10 points is the winner.

 Visit www.glencoe.com/sec/math/mac/mathnet for more games.

4 ASSESS

Closing Activity

Speaking Sketch several number lines on the chalkboard, each showing the solution set of an inequality. Have students state the corresponding inequalities.

Chapter 1, Quiz D (Lessons 1-8 and 1-9) is available in the *Assessment and Evaluation Masters*, p. 16.

■ **Extending the Lesson** ■

Enrichment Masters, p. 9

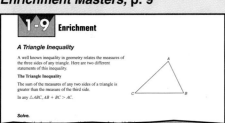

Let the Games Begin

Depending on students' levels of ability, decide whether players can use calculators, pencils and paper, number lines, or mental math to solve the inequalities.
Additional resources for this game can be found on page 46 of the **Classroom Games**.

Study Guide and Assessment

Vocabulary

This section provides a listing of the new terms, properties, and phrases that were introduced in this chapter. Have students define each term and provide an example or two, if appropriate.

Understanding and Using the Vocabulary

These exercises check students' understanding of the terms by using a variety of verbal formats including matching, completion, and true/false.

Glossaries A complete glossary of terms appears on pages 702–710. The glossary also appears in Spanish on pages 711–721.

CHAPTER 1 Study Guide and Assessment

Vocabulary

After completing this chapter, you should be able to define each term, concept, or phrase and give an example or two of each.

Numbers and Operations
base (p. 8)
equation (p. 13)
evaluate (p. 9)
exponent (p. 8)
factor (p. 8)
numerical expression (p. 11)
order of operations (p. 11)
power (p. 8)

Geometry
altitude (p. 40)
area (p. 39)
base (p. 40)
height (p. 40)
parallelogram (p. 39)
perimeter (p. 38)
rectangle (p. 38)
square (p. 38)

Algebra
addition property of equality (p. 18)
algebraic expression (p. 12)
associative property (p. 8)
commutative property (p. 8)
distributive property (p. 34)
division property of equality (p. 22)
inequality (p. 43)
inverse operation (p. 18)
multiplication property of equality (p. 22)
open sentence (p. 13)
replacement set (p. 13)
solution (p. 13)
substitute (p. 11)
subtraction property of equality (p. 18)
variable (p. 12)

Problem Solving
four-step plan (p. 4)
work backward (p. 30)

Understanding and Using the Vocabulary

State whether each sentence is *true* or *false*. If false, replace the underlined word or number to make a true sentence.

1. In the expression 6^3, the <u>base</u> is 3. **false; exponent**
2. $9 + 12$ is an <u>algebraic</u> expression. **false; numerical**
3. The sentence $x > 4$ is called an <u>equation</u>. **false; inequality**
4. In the expression $t - 8$, t is a <u>variable</u>. **true**
5. The <u>first</u> step in solving $4y + 3 = 19$ is to divide each side of the equation by 4. **false; second**
6. 4 is the <u>solution</u> of $x + 6 = 10$. **true**
7. The <u>area</u> of a rectangle is found by multiplying its width and length. **true**

In Your Own Words

8. *Write* a sentence that explains the difference between an algebraic expression and an equation. **Sample answer: An equation contains an algebraic expression and an equals sign.**

48 Chapter 1 Problem Solving and Algebra

 MindJogger Videoquizzes

MindJogger Videoquizzes provide an alternative review of concepts presented in this chapter. Students work in teams to answer questions, gaining points for correct answers. The questions are presented in three rounds.
Round 1 Concepts–5 questions
Round 2 Skills–4 questions
Round 3 Problem Solving–4 questions

Study Guide and Assessment Chapter 1

Objectives & Examples

Upon completing this chapter, you should be able to:

● solve problems using the four-step plan *(Lesson 1-1)*

Explore	What do you know? What are you trying to find?
Plan	How will I go about solving this?
Solve	Carry out your plan. Does it work? Do you need another plan?
Examine	Does your answer seem reasonable? If not, check your solution.

● use powers and exponents in expressions *(Lesson 1-2)*

Find $4^2 \cdot 3^3$.

$4^2 \cdot 3^3 = 4 \cdot 4 \cdot 3 \cdot 3 \cdot 3$
$ = 16 \cdot 27 \text{ or } 432$

● evaluate expressions *(Lesson 1-3)*

Evaluate $5x + y$ if $x = 3$ and $y = 2$.

$5x + y = 5(3) + 2$
$ = 15 + 2 \text{ or } 17$

● find the solutions of equations *(Lesson 1-3)*

Find the solution for $9 + m = 15$. Use the replacement set $\{6, 7, 8, 9\}$.

Let $m = 6$. $\quad\quad 9 + 6 = 15$
6 is the solution. $\quad\quad 15 = 15$ ✓

● solve equations by using the subtraction and addition properties of equality *(Lesson 1-4)*

Solve $d - 14 = 20$.

$d - 14 + 14 = 20 + 14 \quad$ Add to undo
$d = 34 \quad\quad\quad\quad\quad\;$ subtraction.

Review Exercises

Use these exercises to review and prepare for the chapter test.

Use the four-step plan to solve each problem.

9. **Time** The distance between Alicia's house and Paquita's house is 1,600 feet. If it takes Alicia 3 seconds to walk 10 feet, how long will it take her to walk to Paquita's house? **8 minutes**

10. **Construction** The concrete slab for a floor is 40 feet long, 30 feet wide, and 2 feet deep. If the volume of the concrete is measured in cubic feet and is found by multiplying the length, width, and height, find the volume of the concrete used. **2,400 ft³**

Write each expression using exponents.

11. $5 \cdot 5 \cdot 6 \cdot 6 \cdot 6$ $\mathbf{5^2 \cdot 6^3}$
12. $2 \cdot 4 \cdot 2 \cdot 2 \cdot 4 \cdot 7$ $\mathbf{2^3 \cdot 4^2 \cdot 7}$

Evaluate each expression.

13. 5^4 **625**
14. $4^2 \cdot 2^3$ **128**

Evaluate each expression if $a = 1$, $b = 2$, $c = 5$, and $d = 8$.

15. $cd^2 + (3b - 4)$ **322**
16. $abd + 3a$ **19**
17. $(4c) \div (ab)$ **10**
18. $7a + 2c$ **17**

Find the solution for each equation from the given replacement set.

19. $3z = 480, \{160, 140, 110\}$ **160**
20. $x + 10 = 58, \{68, 58, 48\}$ **48**
21. $87 - b = 37, \{124, 50, 40\}$ **50**

Solve each equation. Check your solution.

22. $n + 40 = 90$ **50**
23. $s - 13 = 62$ **75**
24. $143 = a + 35$ **108**
25. $7.6 = 4.7 + t$ **2.9**
26. $r - 12 = 27$ **39**
27. $1.6 = p - 2.5$ **4.1**

Chapter 1 Study Guide and Assessment **49**

Objectives & Examples

This section reviews the skills and concepts of the chapter and shows completely worked examples.

Review Exercises

These exercises provide practice for the corresponding objectives.

Assessment and Evaluation Masters, pp. 3–4

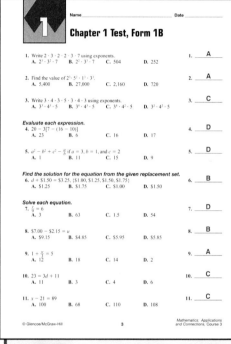

Assessment and Evaluation

Six forms of Chapter 1 Test are available in the *Assessment and Evaluation Masters* as shown in the chart.

Chapter 1 Test, Form 1B, is shown at the right. Chapter 1 Test, Form 2B, is shown on the next page.

1A	Multiple Choice	Honors
1B	Multiple Choice	Average
1C	Multiple Choice	Basic
2A	Free Response	Honors
2B	Free Response	Average
2C	Free Response	Basic

Study Guide and Assessment

Additional Answers
44. $r < 5$

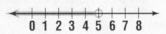

45. $a > 12$

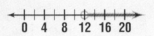

Assessment and Evaluation Masters, pp. 9–10

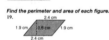

Chapter 1 Study Guide and Assessment

Objectives & Examples

• solve equations by using the division and multiplication properties of equality *(Lesson 1-5)*

Solve $5f = 45$.

$\frac{5f}{5} = \frac{45}{5}$ *Divide to undo multiplication.*

$f = 9$

• write algebraic expressions and equations from verbal phrases and sentences *(Lesson 1-6)*

Write an algebraic expression to represent *the sum of a number and 7*.

The algebraic expression is $n + 7$.

• solve two-step equations *(Lesson 1-7)*

Solve $3d + 3 = 30$.

$3d + 3 - 3 = 30 - 3$

$3d = 27$

$\frac{3d}{3} = \frac{27}{3}$ *Divide each side by 3.*

$d = 9$

• find the perimeters and areas of rectangles, squares, and parallelograms *(Lesson 1-8)*

Perimeter and Area Formulas		
Figure	Perimeter	Area
rectangle	$P = 2\ell + 2w$	$A = \ell w$
square	$P = 4s$	$A = s^2$
parallelogram	$P = 2a + 2b$	$A = bh$

• write and solve inequalities *(Lesson 1-9)*

Solve $4x - 5 > 15$.

Solve $4x - 5 = 15$. → $x = 5$

Test a number greater than 5.

Try 6. $4(6) - 5 \stackrel{?}{>} 10$

$19 \stackrel{?}{>} 15$ true

The solution to $4x - 5 > 15$ is all numbers greater than 5.

Review Exercises

Solve each equation. Check your solution.

28. $2.7m = 8.1$ **3**
29. $\frac{s}{7} = 42$ **294**
30. $60 = 15t$ **4**
31. $3 = \frac{d}{24}$ **72**
32. $\frac{r}{0.2} = 6.6$ **1.32**
33. $6.72 = 0.56k$ **12**

Write each phrase or sentence as an algebraic expression or equation.

34. the sum of 9 and z **$9 + z$**
35. the product of 6 and x **$6x$**
36. Ten less a number is 25.   **$10 - y = 25$**
37. Four times a number is 48.   **$4x = 48$**

Solve each equation. Check your solution.

38. $5f - 12 = 18$ **6**
39. $\frac{h}{6} + 30 = 41$ **66**
40. $16 = \frac{t}{2} - 8$ **48**
41. $3(m + 1) = 27$ **8**

Find the perimeter and area of each figure.

42.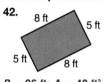

$P = 26$ ft; $A = 40$ ft^2

43.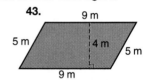

$P = 28$ m; $A = 36$ m^2

Solve each inequality. Graph the solution on a number line.   44–45. See margin.

44. $r - 4 < 1$
45. $9 < \frac{a}{4} + 6$

Write an inequality for each sentence. Then solve the inequality.   46. $3n < 90$; $n < 30$

46. Three times a number is less than ninety.
47. Five more than a number is greater than sixteen.   **$n + 5 > 16$; $n > 11$**

50 Chapter 1 Problem Solving and Algebra

Test and Review Software

You may use this software, a combination of an item generator and item bank, to create your own tests or worksheets. Types of items include free response, multiple choice, short answer, and open ended.

CD-ROM Program

The CD-ROM Program contains an Assessment Game whose questions review the concepts in this chapter.

Study Guide and Assessment   Chapter 1

Applications & Problem Solving

48. Medicine The doctor gave Mitsu 30 capsules for strep throat. She was to take two capsules with every meal for the first two days and then one capsule with every meal after that until the capsules are gone. If Mitsu starts the medication with lunch on Monday and eats three meals a day, when will she take the last capsule? *(Lesson 1-1)*

49. Money Matters The Village Produce Market charges $1.79 a pound for red grapes. Mr. Franklin paid $7.16 for a bag of red grapes. How many pounds of grapes did he buy? *(Lesson 1-5)* **4 lbs**

50. Meteorology A degree day is the difference between 64°F and the day's average temperature. Write an equation to find the degree day, if you are given the average temperature. Then find the degree day for a day when the average temperature is 83°F. *(Lesson 1-6)* $D = a - 64$; 19°

48. the following Tuesday at breakfast

51. Work Backward A cup, two saucers, and three bowls cost $44. One bowl costs as much as one saucer and one cup. If a saucer costs $4, how much is each bowl and cup? *(Lesson 1-7A)* **cup: $6; bowl: $10**

52. Health Neal lives on the block shown in the diagram below. If Neal jogs around the block four times, how many feet has he jogged? *(Lesson 1-8)* **6,400 ft**

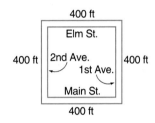

Alternative Assessment

● **Performance Task**
Your family plans to rent a car on your next vacation. Company A charges a flat rate of $39 a day plus $0.10 per mile. Company B charges a daily fee of $45 plus $0.05 for each mile. Suppose you plan to rent the car for 4 days. How many miles would you need to drive for Company A to be less expensive?
less than 480 miles
Create your own car rental charges other than the ones given. Choose a flat rate to start with and then a per day or per mile charge.
See students' work.

A practice test for Chapter 1 is provided on page 647.

● **Completing the CHAPTER Project**
Use the following checklist to make sure your home page design is complete.
☑ The information included on the page is accurate and easy to read.
☑ The graphics on the page are attractive and clear.
☑ The layout of the page is well-organized.
Add any finishing touches that will make your home page more attractive.

 Select a problem you solved in this chapter and place it in your portfolio. Attach a note explaining how the problem illustrates an important concept covered in this chapter.

Performance Assessment

Additional performance assessment tasks for this chapter are included in the *Assessment and Evaluation Masters* on page 13. A scoring guide is also provided on page 25.

Applications & Problem Solving
This section provides additional practice in solving real-world problems that involve the skills of this chapter.

Alternative Assessment
The **Performance Task** provides students with a performance assessment opportunity to evaluate their work and understanding.

CHAPTER Project
Students should complete the final stages of their project and prepare a class demonstration of their results. A scoring guide for the project is available in the *Investigations and Projects Masters*, p. 19.

 Students should add to their portfolios at this time.

Assessment and Evaluation Masters, p. 13

Standardized Test Practice

The Standardized Test Practice may be used to help students prepare for standardized tests. The test items are written in the same style as those in state proficiency tests and standardized tests like CAT, CTBS, ITBS, MAT, SAT, and Terra Nova. The test items cover skills and concepts covered up to this point in the text.

The pages can be used as an overnight assessment. After students have completed the pages, discuss how each problem can be solved, or provide copies of the solutions from the *Solutions Manual*.

Assessment and Evaluation Masters, p. 19

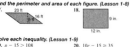

CHAPTER 1
Standardized Test Practice
Assessing Knowledge & Skills

Section One: Multiple Choice

There are twelve multiple-choice questions in this section. Choose the best answer. If a correct answer is *not here,* choose the letter for Not Here.

1. The scale weighs a shoe in pounds. What is the weight of the shoe to the nearest pound? **C**

   A 3 pounds
   B $3\frac{1}{2}$ pounds
   C 4 pounds
   D $4\frac{1}{2}$ pounds

2. $(2 \times 8) + (3 \times 8)$ is equivalent to — **J**
   F 6×8.
   G 6×64.
   H $8(2 + 5)$.
   J $8(2 + 3)$.

3. Evaluate $(3 + 4)^2 + (6 - 2)^2$. **D**
   A 11
   B 121
   C 49
   D 65

4. How is the product $4 \times 4 \times 4$ expressed using exponents? **G**
   F 4×3
   G 4^3
   H 3^4
   J 4^4

52 Chapter 1 Standardized Test Practice

◀◀◀ **Instructional Resources**
Another cumulative review is shown at the left and is available in the *Assessment and Evaluation Masters*, p. 19.

Please note that Questions 5–12 have five answer choices.

5. Mrs. Numkena and 4 of her students planned to leave a 15% tip when they had lunch. Two of the students had hamburgers, two of the students had salads, and Mrs. Numkena bought a salad and soup. What other information is needed to determine how much to leave for a tip? **D**
   A the cost of a salad
   B the cost of hamburgers
   C where they had lunch
   D the cost of the meals
   E the time the waiter was at the table

6. During a 3-day time period, Lindsay spent $3.00, $3.25, and $2.98 for her lunches. A good estimate for the total amount Lindsay spent on lunches during the 3 days is — **H**
   F $7   G $8
   H $9   J $10
   K $11

7. The graph shows how often families eat together at breakfast.

   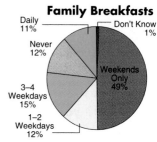

   Family Breakfasts
   Daily 11%
   Don't Know 1%
   Never 12%
   Weekends Only 49%
   3–4 Weekdays 15%
   1–2 Weekdays 12%

   What percent of the families did *not* say that they eat together 1–4 weekdays? **B**
   A 27%   B 73%
   C 60%   D 13%
   E 64%

Standardized Test Practice Chapter 1

8. Chet had $20 to spend at the fair. His admission was $3.50, and he gave the cashier a $5 bill. How much change should he receive? **F**
   - F $1.50
   - G $2.50
   - H $3.50
   - J $16.50
   - K Not Here

9. Robbie has saved $11.42, Beth has saved $12.00, and Monsa has saved $6.78. They want to purchase a collectible baseball card that costs $40. What total have they saved? **B**
   - A $23.42
   - B $30.20
   - C $33.68
   - D $40.20
   - E Not Here

10. If 33% was deducted from a paycheck for income taxes and 5% was deducted for a charitable contribution, what is the total percent deducted from the paycheck? **H**
    - F 28%
    - G 33%
    - H 38%
    - J 43%
    - K Not Here

11. When three bunches of bananas are weighed at the check-out, they weighed 2.5 pounds, 3.04 pounds, and 1 pound. What was the total weight of the bananas? **E**
    - A 3.5 lb
    - B 5.54 lb
    - C 4.04 lb
    - D 6.04 lb
    - E Not Here

12. Wesley's movie ticket cost $5.75, and his snacks cost $4.35. If he had $23, how much does Wesley have now? **G**
    - F $8.55
    - G $12.90
    - H $14.50
    - J $17.25
    - K Not Here

Test-Taking Tip

Most standardized tests have a time limit, so you must budget your time carefully. Some questions will be much easier than others. If you cannot answer a question within a few minutes, go on to the next one. If there is still time left when you get to the end of the test, go back and work the questions that you skipped.

Section Two: Free Response

This section contains seven questions for which you will provide short answers. Write your answers on your paper.

13. Bonnie is reading a 186-page book. She needs to read twice as many pages as she has already read. How many pages has she read? **62 pages**

14. A certain number is divided by 4 and then 5 is subtracted from the result. The final answer is 25. What is the number? **120**

15. What is the solution to the inequality $12 > t + 8$? **$t < 4$**

16. Write an equation to represent *four more than twice a number is 12*. **$4 + 2n = 12$**

17. If $c = 10$ and $p = 2$, what is the value of cp? **20**

18. If you add 30 centimeters to Rosa's height and take a third of the sum, you have half of her younger brother Hugo's height. Hugo is 140 centimeters tall. How tall is Rosa? **180 cm**

19. Find the area and perimeter of a rectangle that is 2 inches wide and 5 inches long. **$A = 10$ in^2; $P = 14$ in.**

CHAPTER 2
Algebra: Using Integers

Previewing the Chapter

Overview
In this chapter, students extend their knowledge of numbers less than zero. The first lesson defines integers and shows where negative integers fall on the number line. Methods for comparing and ordering positive and negative integers are given in the next lesson. A thorough treatment of addition, subtraction, multiplication, and division of integers follows. Students learn to solve equations involving integers using counters and equation mats. They also learn to locate ordered pairs on the coordinate plane.

Lesson (pages)	Lesson Objectives	NCTM Standards	Standardized Tests	State/Local Objectives
2-1 (56–58)	Graph integers on a number line and find absolute value.	1–7, 10, 13		
2-2 (59–61)	Compare and order integers.	1–7, 10	CAT, CTBS, MAT, SAT, TN	
2-3 (62–65)	Add integers.	1–7, 9, 10, 13	CAT, CTBS, SAT, TN	
2-4 (66–68)	Add three or more integers.	1–4, 6, 7, 9, 13	CAT, CTBS, SAT, TN	
2-5 (69–72)	Subtract integers.	1–7, 9, 13	CAT, CTBS, SAT, TN	
2-6 (73–76)	Use matrices to organize data.	1–10		
2-6B (77)	Perform matrix operations with a graphing calculator.	7, 9, 10		
2-7 (78–80)	Multiply integers.	1–7, 9	CAT, CTBS, SAT, TN	
2-8 (81–83)	Divide integers.	1–7, 9, 12	CAT, CTBS, SAT, TN	
2-9A (84–85)	Solve equations by using models.	1–5, 7, 9		
2-9 (86–89)	Solve equations with integer solutions.	1–7, 9	CTBS, MAT, SAT, TN	
2-9B (90–91)	Solve problems by eliminating possibilities.	1–5, 7		
2-10 (92–95)	Graph points on the coordinate plane.	1–5, 7, 9, 12, 13	CTBS, SAT, TN	

CAT = California Achievement Tests, CTBS = Comprehensive Tests of Basic Skills, ITBS = Iowa Tests of Basic Skills, MAT = Metropolitan Achievement Tests, SAT = Stanford Achievement Tests, TN = Terra Nova

Organizing the Chapter

CD-ROM
All of the blackline masters in the Teacher's Classroom Resources are available on the **Electronic Teacher's Classroom Resources** CD-ROM.

LESSON PLANNING GUIDE

BLACKLINE MASTERS (PAGE NUMBERS)

Lesson	Extra Practice (Student Edition)	Study Guide	Practice	Enrichment	Assessment & Evaluation	Classroom Games	Diversity	Hands-On Lab	School to Career	Science and Math Lab Manual	Technology	Transparencies A and B
2-1	p. 608	10	10	10			28					2-1
2-2	p. 608	11	11	11								2-2
2-3	p. 609	12	12	12	43							2-3
2-4	p. 609	13	13	13								2-4
2-5	p. 609	14	14	14	42, 43							2-5
2-6	p. 610	15	15	15								2-6
2-6B												
2-7	p. 610	16	16	16					28	89–92		2-7
2-8	p. 610	17	17	17	44	3–6						2-8
2-9A								41				
2-9	p. 611	18	18	18							55, 56	2-9
2-9B	p. 611											
2-10	p. 611	19	19	19	44			69				2-10
Study Guide/ Assessment					29–41, 45–47							

OTHER CHAPTER RESOURCES

Student Edition
Chapter Project, pp. 55, 65, 83, 95, 99
Math in the Media, p. 89
Let the Games Begin, pp. 61, 76

Technology
 CD-ROM Program
 Interactive Mathematics Tools Software

Teacher's Classroom Resources

Applications
Family Letters and Activities, pp. 55–56
Investigations and Projects Masters, pp. 21–24
Meeting Individual Needs
Investigations for the Special Education Student, pp. 9–12

Teaching Aids
Answer Key Masters
Block Scheduling Booklet
Lesson Planning Guide
Solutions Manual

Professional Publications
Glencoe Mathematics Professional Series

Planning the Chapter

MindJogger Videoquizzes provide a unique format for reviewing concepts presented in the chapter.

Assessment Resources

Student Edition
Mixed Review, pp. 58, 61, 65, 68, 72, 76, 80, 83, 89, 95
Mid-Chapter Self Test, p. 72
Math Journal, pp. 57, 88, 93
Study Guide and Assessment, pp. 96–99
Performance Task, p. 99
Portfolio Suggestion, p. 99
Standardized Test Practice, pp. 100–101
Chapter Test, p. 648

Assessment and Evaluation Masters
Multiple-Choice Tests (Forms 1A, 1B, 1C), pp. 29–34
Free-Response Tests (Forms 2A, 2B, 2C), pp. 35–40
Performance Assessment, p. 41
Mid-Chapter Test, p. 42
Quizzes A–D, pp. 43–44
Standardized Test Practice, pp. 45–46
Cumulative Review, p. 47

Teacher's Wraparound Edition
5-Minute Check, pp. 56, 59, 62, 66, 69, 73, 78, 81, 86, 92
Building Portfolios, p. 54
Math Journal, pp. 77, 85
Closing Activity, pp. 58, 61, 65, 68, 72, 76, 80, 83, 89, 91, 95

Technology
Test and Review Software
MindJogger Videoquizzes
CD-ROM Program

Materials and Manipulatives

Lesson 2-2	**Lesson 2-4**	**Lesson 2-6B**	**Lesson 2-9A**
index cards	scientific calculator	graphing calculator	counters*†
scissors*			cups*†
	Lesson 2-5	**Lesson 2-7**	equation mat*†
Lesson 2-3	counters*†	counters*†	
counters*†	integer mat*†	integer mat*†	
integer mat*†		calculator	

*Glencoe Manipulative Kit   †Glencoe Overhead Manipulative Resources

Pacing Chart

See pages T25–T27 for the Course Planning Calendar.

COURSE	DAY 1	DAY 2	DAY 3	DAY 4	DAY 5	DAY 6	DAY 7
Standard	Chapter Project	Lesson 2-1	Lesson 2-2	Lesson 2-3	Lesson 2-4	Lesson 2-5	Lesson 2-6
Honors	Chapter Project & Lesson 2-1	Lesson 2-2	Lesson 2-3	Lesson 2-4	Lesson 2-5	Lessons 2-6 & 2-6B	
Block	Chapter Project & Lessons 2-1 & 2-2	Lessons 2-3 & 2-4	Lesson 2-5	Lesson 2-6	Lessons 2-7 & 2-8	Lessons 2-9A & 2-9	Lessons 2-9B & 2-10

Interactive Mathematics:
Activities and Investigations

is an activity-based program that may be used as an enhancement for chapters in *Mathematics: Applications and Connections*.

Unit 11, Activity Three, *Back to the Future*
Use with Lesson 2-8.

Summary Students work in groups to calculate times and dates in the past and future. Each student then writes a note to another student summarizing their methods of arriving at their solutions.

Math Connection Students use clocks in sizes 12 and 7 to solve time measurement problems involving times of the day, months of the year, and days of the week. This activity involves calculating with time measurements, operations with integers, and summarizing methods of solution.

Unit 6, Activity Eight
Use with Lesson 2-9B.

Summary Students are given three alternative ways to commute between a person's residence and a work site. They examine cost, time, and environmental issues to determine which route is the best. They then report their approaches, considerations, and conclusions.

Math Connection Students solve a network problem by determining the best method of transportation for commuting between two locations. They use a variety of mathematical tools, including making lists, graphs, maps, or tables.

DAY 8	DAY 9	DAY 10	DAY 11	DAY 12	DAY 13	DAY 14	DAY 15
Lesson 2-7	Lesson 2-8	Lessons 2-9A & 2-9		Lesson 2-9B	Lesson 2-10	Study Guide and Assessment	Chapter Test
Lesson 2-7	Lesson 2-8	Lesson 2-9	Lesson 2-9B	Lesson 2-10	Study Guide and Assessment	Chapter Test	
Study Guide and Assessment, Chapter Test							

Chapter 2 **54d**

Enhancing the Chapter

APPLICATIONS

Classroom Games, pp. 3–6
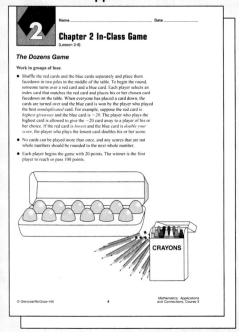

Diversity Masters, p. 28
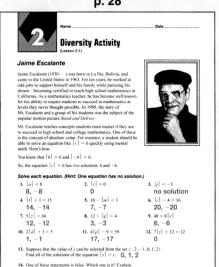

School to Career Masters, p. 28
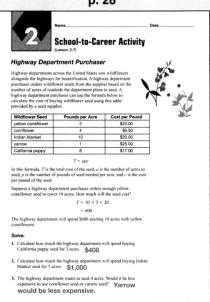

Family Letters and Activities, pp. 55–56

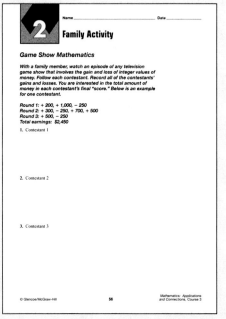

Science and Math Lab Manual, pp. 89–92
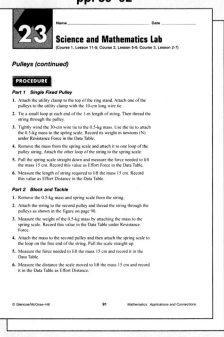

MANIPULATIVES/MODELING

Hands-On Lab Masters, p. 69

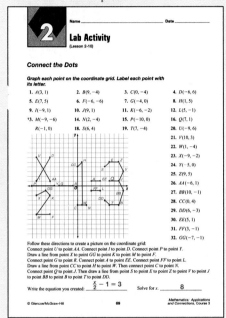

ASSESSMENT/EVALUATION

Assessment and Evaluation Masters, pp. 42–44

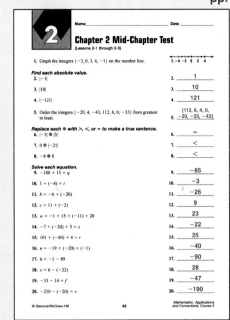

TECHNOLOGY/MULTIMEDIA

Technology Masters, pp. 55–56

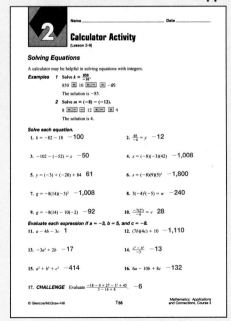

MEETING INDIVIDUAL NEEDS

Investigations for the Special Education Student, pp. 9–12

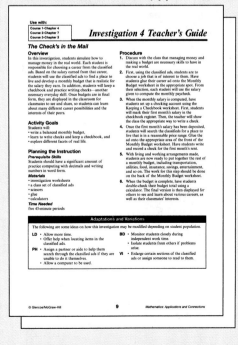

Chapter 2 **54f**

CHAPTER 2 NOTES

Theme: Endangered Species

Sea otters live in coastal waters from California to Alaska and around the Pacific Rim to northern Japan. Until the 1700s, sea otters abounded—150,000 to 300,000. Then fur traders hunted them to near extinction. By 1911, the California sea otter population was about 50.

The Endangered Species Act and other protection have contributed to an increase in sea otters. In 1995, there were more than 2,300 otters along the California coast.

Question of the Day If the California population of sea otters doubles every ten years, approximately how many sea otters will there be in 2005? **4,600**

Assess Prerequisite Skills

Ask students to read through the list of objectives presented in "What you'll learn in Chapter 2." You may wish to ask them what each of the objectives means or if they have experienced or used any of these math concepts before.

 Building Portfolios

Encourage students to revise their portfolios as they study this chapter. Students can save results from a variety of activities—number line exercises, graphing coordinate pairs, and drawings of solving equations with counters. These activities can help students retain important math concepts.

 Math and the Family

In the *Family Letters and Activities* booklet (pp. 55–56), you will find a letter to the parents explaining what students will study in Chapter 2. An activity appropriate for the whole family is also available.

CHAPTER 2 — Algebra: Using Integers

What you'll learn in Chapter 2

- to compare, order, and compute with integers,
- to organize and use data in matrices,
- to solve problems by eliminating possibilities,
- to solve problems involving integers, and
- to graph points on the coordinate plane.

CD-ROM Program

Activities for Chapter 2
- Chapter 2 Introduction
- Interactive Lessons 2-3, 2-5, 2-8
- Extended Activity 2-9
- Assessment Game
- Resource Lessons 2-1 through 2-10

CHAPTER Project

WHERE THE WILD THINGS ARE

In this project, you will work in a group to investigate a threatened or endangered species in the United States or elsewhere in the world and report on changes in its population. You will present your findings as a script for a 1- or 2-minute television public service announcement and illustrate your points using algebra, graphs, and statistics.

Getting Started

- Research an endangered species of your choice.
- Make a list of information that you will include in the script for your television public service announcement. The tables show population information on four endangered or threatened species.

Bald Eagle	
Year	Number in Lower 48 States
1782	25,000
1963	900
1994	9,000

Source: U.S. Fish & Wildlife Service

California Condor	
Year	Number in World
1979	35
1987	27
1997	119

Source: California Dept. of Fish & Game

Grizzly Bear	
Year	Number in Lower 48 States
1800	50,000
1975	1,000
1995	665

Source: U.S. Fish & Wildlife Service

Whooping Crane	
Year	Number in North America
1870	1,400
1939	18
1997	180

Source: Texas Parks & Wildlife Dept.

Technology Tips

- Use an **electronic encyclopedia** to research your selected species.
- Use a **word processor** to write the script for your television public service announcement.
- Use a **spreadsheet** to organize any data that you collect.

 interNET CONNECTION For up-to-date information on endangered species, visit: www.glencoe.com/sec/math/mac/mathnet

Working on the Project

You can use what you'll learn in Chapter 2 to help you make a television public service announcement script about your endangered species.

Page	Exercise
65	38
83	44
95	48
99	Alternative Assessment

Chapter 2 **55**

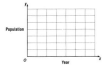 **interNET CONNECTION**

Glencoe has made every effort to ensure that the website links for *Mathematics: Applications and Connections* at www.glencoe.com/sec/math/mac/mathnet are current and contain appropriate content. However, these website links are not under Glencoe's control.

Instructional Resources ▶▶▶
A recording sheet to help students organize their data for the Chapter Project is shown at the right and is available in the *Investigations and Projects Masters*, p. 24.

CHAPTER Project NOTES

Objectives Students should
- use the library, CD-ROMs, and websites to research animal populations.
- use spreadsheets to organize information.
- understand how animal populations fluctuate and how these fluctuations can indicate trends.

Project Pointer You may suggest that students begin a *Project Folder* to keep their work as they complete each stage of the Chapter Project. The completed project may also be added to their portfolios.

Using the Tables Students should note that the data in these tables skip over time periods. For example, the bald eagle data jump from 1782 to 1963—an interval of 181 years. The next information comes from only 11 years later. By the end of this chapter, students should understand that information in these tables would be difficult to graph.

Investigations and Projects Masters, p. 24

Name _____ Date _____

Chapter 2 Project

Where the Wild Things Are
Page 55, Getting Started

Page 65, Working on the Chapter Project, Exercise 38
a. The number of bald eagles in 1990 was:

b.

Page 83, Working on the Chapter Project, Exercise 44

Page 95, Working on the Chapter Project, Exercise 48

Chapter 2 Project **55**

2-1 Lesson Notes

Instructional Resources
- *Study Guide Masters*, p. 10
- *Practice Masters*, p. 10
- *Enrichment Masters*, p. 10
- Transparencies 2-1, A and B
- *Diversity Masters*, p. 28
- CD-ROM Program
  - Resource Lesson 2-1

Recommended Pacing	
Standard	Day 2 of 15
Honors	Day 1 of 14
Block	Day 1 of 8

1 FOCUS

5-Minute Check
(Chapter 1)

1. Evaluate $3a^2 - (b - c)$ if $a = 2$, $b = 5$, and $c = 3$. **10**

Solve each equation or inequality.

2. $3.7p = 22.2$ **6**
3. $\frac{p}{6} + 13 = 21$ **48**
4. $2.4x - 4.8 < 8.4$ **$x < 5.5$**
5. Find the perimeter and area of the figure. **56 cm; 180 cm²**

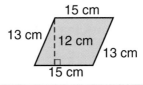

The 5-Minute Check is also available on **Transparency 2-1A** for this lesson.

2 TEACH

Transparency 2-1B contains a teaching aid for this lesson.

Using Calculators Have students find the 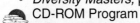 key on their calculators. Demonstrate that pressing the key changes the sign of the displayed number. Compare the 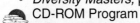 key with the ⊟ key, which performs subtraction.

2-1 Integers and Absolute Value

What you'll learn
You'll learn to graph integers on a number line and find absolute value.

When am I ever going to use this?
Integers are used to describe scores in sports and to measure temperature.

Word Wise
integer
graph
coordinate
absolute value

Study Hint
Reading Math The periods (...) in the set of integers are called an *ellipsis*. They indicate that the set continues without end, following the same pattern.

What is your favorite way to beat the heat? People in the Daloi Danakil Depression in Ethiopia, Africa, really have to learn to tolerate hot days. They have the hottest average annual temperature in the world: 95°F! On the other end of the spectrum, Plateau Station, Antarctica, has the coldest average annual temperature, −71°F.

Negative numbers like −71, as well as positive whole numbers like 95, are part of the set of **integers**. The set of integers can be written as $\{..., -4, -3, -2, -1, 0, +1, +2, +3, +4, ...\}$.

Positive integers can be written with or without the + sign. So, +95 and 95 are the same.

You can graph integers on a number line. 0 is considered the starting point on the number line, with positive numbers to the right and negative numbers to the left. Zero is neither negative nor positive.

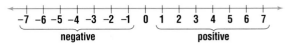

The arrows show that the numbers continue without end.

To **graph** an integer, locate the number and draw a dot at that point on the line. Sometimes letters are used to name points on a number line. The integer that corresponds to a letter is called the **coordinate** of the point.

Examples

① Name the coordinates of each point graphed on the number line.

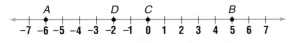

The coordinate of *A* is −6. The coordinate of *B* is 5.
The coordinate of *C* is 0. The coordinate of *D* is −2.

② Graph the following points on a number line.
E has the coordinate −3. *F* has the coordinate 4.
G has the coordinate −6. *H* has the coordinate 0.

Find each number. Draw a dot there. Write the letter above the dot.

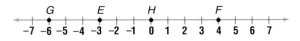

56 Chapter 2 Algebra: Using Integers

Motivating the Lesson
Hands-On Activity Draw a thermometer large enough to label −100°F and 100°F. Ask students to copy it. Then have students read the opening paragraph of the lesson and label 95°F and −71°F on their thermometers.

Example 3

Earth Science The table at the right shows the average January temperatures in Celsius for selected cities. Graph each temperature on a number line.

You can use the first letter of each city name to label the points.

City	Temperature (°C)
Boston	2
Chicago	-2
Detroit	-1
Milwaukee	-3
New York	3
Pittsburgh	1

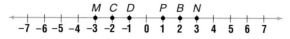

The distance from a number to 0 on the number line is called the **absolute value** of the number. We write *the absolute value of* -4 as $|-4|$. Let's find $|-4|$. Graph -4.

The graph of -4 is 4 units from 0. So, $|-4| = 4$.

Study Hint

Reading Math

$|-4|$ is read as *the absolute value of negative four.*

Example 4

4 Find $|-5|$ and $|5|$.

First graph each number on a number line.

Then count how many units each number is from 0.

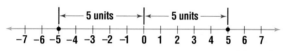

5 is 5 units from 0, and -5 is 5 units from 0. So, $|5|$ and $|-5|$ are both 5.

CHECK FOR UNDERSTANDING

Communicating Mathematics

Read and study the lesson to answer each question. 1–2. See Answer Appendix.

1. *Write* how to graph an integer on a number line.
2. *Draw* a number line from -8 to 8. Graph two points that are the same distance from 0. Write a number sentence about these two points.
3. *Write a Problem* about a real-life situation involving a negative integer. See students' work.

Guided Practice

4. Name the coordinate of each point graphed on the number line.

4. $C, -7; D, 0; B, 1$

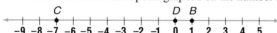

Graph each set of points on a number line. 5–7. See Answer Appendix.

5. $\{-3, 1, 4\}$ 6. $\{5, 0, -2\}$ 7. $\{6, 4, -4\}$

Lesson 2-1 Integers and Absolute Value **57**

In-Class Examples

For Example 1
Name the coordinates of each point graphed on the number line.
$K: -5; M: -3; E: 0; R: 2$

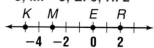

For Example 2
Graph the set of points $\{1, -2, -6\}$ on a number line.

For Example 3
Graph each yardage on a number line.

Players	Yardage
Rob (R)	-10
Tyrone (T)	7
Lyle (L)	13
Stan (S)	16
P.J. (P)	-6
Darnay (D)	-12

For Example 4
Find $|-2|$ and $|2|$. **Both are 2.**

Study Guide Masters, p. 10

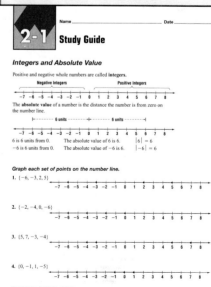

Reteaching the Lesson

Activity Working in pairs, students take turns rolling a number cube, tossing a coin for positive and negative, and graphing the number on a number line—positive for heads, negative for tails.

Error Analysis
Watch for students who confuse the absolute value of an integer with the integer itself.
Prevent by describing absolute value as a distance in units. The number of units is always more than zero.

Lesson 2-1   **57**

PRACTICE/APPLY

Check for Understanding
If students need additional practice or instruction after completing Exercises 1–13, one of these options may be helpful.
- Extra Practice, see p. 608
- Reteaching Activity, see p. 57
- *Study Guide Masters*, p. 10
- *Practice Masters*, p. 10

Assignment Guide
Core: 15–41 odd, 42–46
Enriched: 14–38 even, 39–46

4 ASSESS

Closing Activity
Speaking Ask this question: *What are the coordinates of two points whose absolute values are 4? Describe their locations on the number line.* −4 and 4; 4 units left and 4 units right of 0

Practice Masters, p. 10

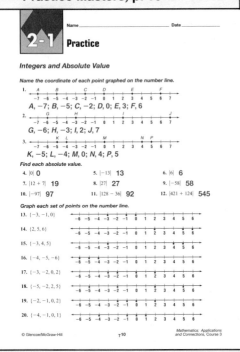

Find each absolute value.

8. $|-7|$ 7   9. $|23|$ 23   10. $|-44|$ 44   11. $|0|$ 0

12. −10

12. Write an integer to describe a low temperature that is 10 degrees below zero.

13. **Earth Science** The Caspian Sea between Europe and Asia is the only major natural lake in the world below sea level. It is 92 feet below sea level. Write an integer to describe the level of the Caspian Sea. −92

EXERCISES

Practice

14. K, −10; L, −9; N, −6; M, −3; Q, −1; P, 2

14. Name the coordinate of each point graphed on the number line.

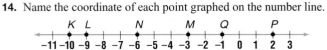

Graph each set of points on a number line. 15–23. See Answer Appendix.

15. {4, 5, 7}   16. {−1, −3, 5}   17. {−3, 5, 6, 0}
18. {−6, −8, 0, 1}   19. {9, −5, −7, 12}   20. {3, −6, −8, 1}
21. {9, 5, 8, −2}   22. {2, 5, −6, −1}   23. {3, −2, 2, −3, 0}

Find each absolute value.

24. $|-4|$ 4   25. $|5|$ 5   26. $|-16|$ 16   27. $|12|$ 12
28. $|-15|$ 15   29. $|45|$ 45   30. $|-65|$ 65   31. $|88|$ 88
32. $|4+8|$ 12   33. $|13-9|$ 4   34. $|5|+|-9|$ 14   35. $|-17|-|12|$ 5

Write an integer to describe each situation.

36. Amparo finished the race 3 seconds ahead of the 2nd place finisher. 3
37. Myron ended his round of golf 3 under par. (Par is a score of 0.) −3
38. Denver, the mile high city, is 5,280 feet above sea level. +5,280

Applications and Problem Solving

Jeff Gordon

39. **Sports** Jeff Gordon won the 1997 Daytona 500. But Mark Martin led the pack for 12 laps more than Gordon. Express 12 laps as an integer. 12

40. **Algebra** The sum of two numbers is 0. What must be true about their absolute values? Explain. **See Answer Appendix.**

41. False. For example, $|-4| > |2|$ but −4 < 2; and $|-8| > |1|$, but −8 < 1.

41. **Critical Thinking** a and b are integers. Is it *true* or *false* that if $|a| > |b|$, then $a > b$? Use examples to prove your answer.

Mixed Review

42. **Algebra** Solve $4s < 36$. Graph the solution on a number line. (*Lesson 1-9*) $s < 9$; See Answer Appendix for graph.

43. **Test Practice** Shanté wants to paint a wall that is 8 feet by 12 feet. A door that measures 7 feet by 3 feet is in the middle of the wall and will not be painted. How much surface area will be painted? (*Lesson 1-8*) **C**
A 20 sq ft   B 21 sq ft   C 75 sq ft   D 96 sq ft   E Not Here

44. **Algebra** Solve $4y = 196$. Check your solution. (*Lesson 1-5*) 49
45. **Algebra** Solve $30 = k - 141$. Check your solution. (*Lesson 1-4*) 171
46. Write $5 \cdot 5 \cdot 8 \cdot 8 \cdot 8$ using exponents. (*Lesson 1-2*) $5^2 \cdot 8^3$

58 Chapter 2 Algebra: Using Integers

Enrichment Masters, p. 10

Extending the Lesson

Activity Have students think of situations that use positives, negatives, or neither positive nor negative integers.
Sample answer: money—profit, loss, break even; altitude—above sea level, below sea level, sea level

2-2 Comparing and Ordering Integers

What you'll learn
You'll learn to compare and order integers.

When am I ever going to use this?
You can compare and order integers to determine the winner in a miniature golf game.

Did you know The original snowboards were called "snurfers" because the sport was compared to surfing on snow.

Have you ever gone snowboarding? Snowboarding is a rather new sport, which began in the 1960s. It is one of several sports that is increasing in popularity. The table shows the percent of increase or decrease of people participating in different sports from 1992 to 1993. Which sport has had the greatest percent of increase? Which had the greatest percent of decrease?

SPORTING TRENDS

Sport	Percent Increase or Decrease
In-line Skating	27
Racquetball	−18
Snowboarding	50
Step Aerobics	15
Windsurfing	−22

Source: National Sporting Goods Association

Let's sketch a number line for the percents and graph each number.

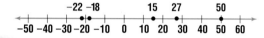

Use the number line to list the numbers from greatest to least. They would be in order from *right to left* on the number line.

$$50 \quad 27 \quad 15 \quad -18 \quad -22$$

You can use one of three symbols to compare any two numbers.

$=$ is equal to $>$ is greater than $<$ is less than

Use the number line to compare integers. Replace each ● with $>$, $<$, or $=$ to make a true sentence.

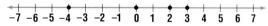

① $-4 \bullet 3$
-4 is to the left of 3 on the number line.
So, $-4 < 3$.

② $|-2| \bullet 0$
$|-2| = 2$
2 is to the right of 0 on the number line.
So, $|-2| > 0$.

Lesson 2-2 Comparing and Ordering Integers **59**

2-2 Lesson Notes

Instructional Resources
- *Study Guide Masters*, p. 11
- *Practice Masters*, p. 11
- *Enrichment Masters*, p. 11
- Transparencies 2-2, A and B
- CD-ROM Program
  - Resource Lesson 2-2

Recommended Pacing	
Standard	Day 3 of 15
Honors	Day 2 of 14
Block	Day 2 of 8

1 FOCUS

5-Minute Check
(Lesson 2-1)

1. Name the coordinate of each point graphed on the number line. **A, −8; B, −6; C, 3; D, −2; E, 1; F, −5**

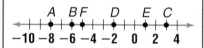

2. Graph the set of numbers {4, 1, 0, −2, −3} on a number line.

Find each absolute value.
3. $|-112|$ **112**
4. $|72 - 59|$ **13**
5. The Fishinger Middle School girls' basketball team scored 62 points in its final game of the season. Write an integer to express the situation. **62**

The 5-Minute Check is also available on **Transparency 2-2A** for this lesson.

2 TEACH

Transparency 2-2B contains a teaching aid for this lesson.

Thinking Algebraically Ask students to think algebraically about how they would use a number line to represent the degrees above or below the average temperature for each day of the week.

Lesson 2-2 **59**

Motivating the Lesson
Problem Solving Ask this question: *A snail on a number line crawled to the left from −5, beginning at noon. Was the coordinate of the snail's 1 P.M. location greater or less than −5?* **less than**

In-Class Examples

Replace each ● with >, <, or = to make a true sentence.

For Example 1
−5 ● 2   <

For Example 2
−1 ● 0   <

For Example 3
A class bought a mutual fund. The table shows how it fared compared to the Dow Jones industrial average.

Week	% Difference from Dow
1	−5
2	−6
3	+3
4	0
5	+4
6	+1

Order the differences by graphing them on a number line. from least to greatest: −6, −5, 0, +1, +3, +4

Study Guide Masters, p. 11

Example 3 CONNECTION

Geography Many western cities are increasing in population, while some eastern cities are decreasing. The table lists six of the largest counties that had changes in population from 1990 to 1995. Order the changes by graphing the integers on a number line.

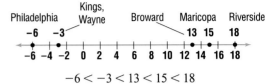

County	Percent of Population Change
Broward, FL	13
Kings, NY	−3
Maricopa, AZ	15
Philadelphia, PA	−6
Riverside, CA	18
Wayne, MI	−3

Source: Bureau of the Census

$-6 < -3 < 13 < 15 < 18$

CHECK FOR UNDERSTANDING

Communicating Mathematics

Read and study the lesson to answer each question.

1. *Write* a comparison of −3 and 4 using an inequality sign. Write a second comparison using a different inequality sign.   −3 < 4; 4 > −3

2. *Tell* which of the following are true and which are false.
   - 2a. true
   - 2b. true
   - 2c. false
   - 2d. false

   a. −5 < −3   b. −5 < |−3| c. |−5| < −3   d. |−5| < |−3|

3. *You Decide* Mariano says that the inequality symbol always points to the lesser number. Judy says it points to the greater number. Who is correct? Explain.   Mariano is correct.

Guided Practice

Replace each ● with >, <, or = to make a true sentence.

4. −3 ● 2   <   5. 2 ● −3   >   6. |−23| ● 23   =   7. 34 ● 16   >

8. −37, −23, −12, 0, 45, 55

8. Order the integers {45, −23, 55, 0, −12, −37} from least to greatest.

9. *Space Exploration* In 1997, NASA established websites to allow the public to follow the Pathfinder Mission to Mars. The internal temperature of the Rover fluctuated between 40C° and −22C°. Write two inequalities to show the relationship between these temperatures.   40 > −22; −22 < 40

EXERCISES

Practice

Replace each ● with >, <, or = to make a true sentence.   21. =

10. 4 ● −4   >   11. 5 ● |−5| =   12. 0 ● |3| <   13. −6 ● −12   >
14. −35 ● −16   <   15. 0 ● |−12| <   16. |−8| ● 0   >   17. |3| ● |−3| =
18. −19 ● −22   >   19. 90 ● 21   >   20. |−34| ● |−9| >   21. |−821| ● |821|

22. −34, −22, −12, 12, 34, 56

22. Order the integers {−12, 12, −34, 56, −22, 34} from least to greatest.

23. 564, 254, −100, −356, −450

23. Order the integers {−450, 564, −356, −100, 254} from greatest to least.

24. Order the integers {−467, −237, 1,276, −3,456, −943} from greatest to least.   1,276, −237, −467, −943, −3,456

60 Chapter 2 Algebra: Using Integers

■ Reteaching the Lesson ■

Activity Give each of five students a card with an integer or absolute value on it. Then have the students line up from least to greatest. Repeat with five new students and cards and line up from greatest to least.

Applications and Problem Solving

25. *Recreation* Hippos dance and broomsticks march at Disney's 36-hole *Fantasia Gardens Miniature Golf Course*. On the first hole, Nate scored a 1, and his sister Jamila scored −1. Write two inequalities to compare their scores.

26. *Earth Science* The table shows the record low temperatures in the U.S. Graph the temperatures on a number line. Order them from least to greatest.

City	Temp. (°F)	Date
Island Park Dam, ID	−60	1/18/43
McIntosh, SD	−58	2/17/36
Moran, WY	−63	2/9/33
Prospect Creek Camp, AK	−80	1/23/71
Rogers Pass, MT	−70	1/20/54
Seneca, OR	−54	1/18/43

Source: National Climatic Data Center

25. $1 > -1; -1 < 1$
26. See Answer Appendix for graph. $\{-80, -70, -63, -60, -58, -54\}$

Mixed Review

27. *Critical Thinking* Consider the inequality $|x| < 3$.
   a. Is 3 a solution to this inequality? Explain.   No; $|3| = 3$ and $3 \not< 3$.
   b. Graph all integer solutions for the equation on a number line.   See Answer Appendix.

28. *Algebra* Evaluate $2 + |1 - 0|$. *(Lesson 2-1)*   3

29. *Algebra* Solve $102 = 17p$. Check your solution. *(Lesson 1-5)*   6

30. **Test Practice**   Mrs. Acosta buys 2 flag pins for each of the 168 band members. Pins cost $0.09 each. Estimate the cost of the pins. *(Lesson 1-1)*   **C**
   A $8   **B** $20   **C** $30   **D** $50   **E** $70

Absolutely!

Get Ready This game is for two players.   index cards   scissors

Get Set Cut the index cards in half. Then copy the integers below, one integer onto each card. Shuffle the cards and divide them into two equal facedown piles, one for each player. Each player writes "absolute value" on two other cards and places these cards aside.

$-17 \quad -3 \quad 0 \quad 19 \quad 16 \quad 5 \quad 25 \quad -10 \quad 3 \quad -2 \quad -8 \quad -7$
$7 \quad 6 \quad 9 \quad 22 \quad 11 \quad 12 \quad 1 \quad 14 \quad -20 \quad -13 \quad -16 \quad -18$

Go • Each player places the top card from his or her pile faceup. The partner with the greater card takes both cards and puts them facedown in a separate pile. When there are no more cards in your original pile, shuffle the cards in the second pile and use them.

• Twice during the game, each player can use an "absolute value" card after the two other cards have been played. When an absolute value card is played, players compare the absolute values of the integers on the cards. When there is a tie, continue play; the first player with a greater integer takes all of the upturned cards.

• The winner is the player who takes all of the cards.

 Visit www.glencoe.com/sec/math/mac/mathnet for more games.

Lesson 2-2 Comparing and Ordering Integers   **61**

3 PRACTICE/APPLY

Check for Understanding

If students need additional practice or instruction after completing Exercises 1–9, one of these options may be helpful.
• Extra Practice, see p. 608
• Reteaching Activity
• *Study Guide Masters*, p. 11
• *Practice Masters*, p. 11

Assignment Guide

Core: 11–27 odd, 28–30
Enriched: 10–24 even, 25–30

4 ASSESS

Closing Activity

Writing Have students write a paragraph explaining how to compare two numbers graphed on a number line. Make certain that students use the terms *absolute value*, *greater than*, and *less than*.

Practice Masters, p. 11

Extending the Lesson

Enrichment Masters, p. 11

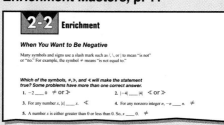

Inform students that this game closely resembles the card game War. It differs from War in that it recognizes positive and negative integers.

Lesson 2-2   **61**

2-3 Lesson Notes

Instructional Resources
- *Study Guide Masters*, p. 12
- *Practice Masters*, p. 12
- *Enrichment Masters*, p. 12
- Transparencies 2-3, A and B
- *Assessment and Evaluation Masters*, p. 43
- CD-ROM Program
  - Resource Lesson 2-3
  - Interactive Lesson 2-3

Recommended Pacing	
Standard	Day 4 of 15
Honors	Day 3 of 14
Block	Day 2 of 8

1 FOCUS

5-Minute Check
(Lesson 2-2)

Replace each ● with >, <, or = to make a true sentence.
1. 13 ● −11 >
2. −4 ● −9 >
3. |−4| ● |−6| <
4. |8| ● |−8| =
5. Order the integers {18, −36, −12, 0, 12} from least to greatest.
   −36, −12, 0, 12, 18

The 5-Minute Check is also available on **Transparency 2-3A** for this lesson.

Motivating the Lesson

Hands-On Activity Borrow a pan balance from the science department. Label one pan "+" and the other "−". Place one penny on the "−" pan and label the pointer position −1. Add one penny at a time and label the pointer positions for −2 through −5. Remove all the pennies. Repeat the process on the "+" pan for positions 1 through 5. Find these sums using the labeled balance and pennies: 2 + 3 = 5, (−2) + (−3) = −5, −1 + 4 = 3, and 2 + (−5) = −3.

62 Chapter 2

2-3 Adding Integers

What you'll learn
You'll learn to add integers.

When am I ever going to use this?
You can add integers to find your profit at a bake sale.

Word Wise
zero pair

interNET CONNECTION
For the latest *American Girl You Said It Poll*, visit: www.glencoe.com/sec/math/mac/mathnet

"Are you a spender or a saver?" *American Girl* magazine asked its website visitors this question in its weekly *You Said It* poll. They found that chances are you have a full piggy bank at home. Of those who answered the poll, 53 percent called themselves savers. About 27 percent like to save half and spend half when they get some money. And for 20 percent, money gets spent right away.

You can use integers to keep track of your savings and spending.
- Receiving money can be modeled as a positive number. For example, earning $10 from mowing the lawn can be written as +10.
- When you spend money or give it to someone else, that can be modeled as a negative number. So if you spent $3 on a magazine, that is written as −3.

Suppose Amy borrowed some money from her friend Yoko. She borrowed $5 for lunch and later borrowed $7 for a movie. How much does Amy owe Yoko? Let a = the total amount Amy owes.

$$\underbrace{-5}_{\text{lunch \$}} \; \underbrace{+}_{\text{plus}} \; \underbrace{(-7)}_{\text{movie \$}} \; \underbrace{=}_{\text{is}} \; \underbrace{a}_{\text{total owed}}$$

Addition can be modeled on a number line. First graph −5. Since −7 is negative, you will move 7 units to the left of −5.

$-5 + (-7) = a$

$-12 = a$ Amy owes Yoko $12.

You can also use counters to add integers.

Example

1 Solve $-9 + (-6) = s$.

Put 9 negative counters on a mat. Add 6 more negative counters.

There are 15 negative counters. Therefore, $s = -15$. The solution is −15.

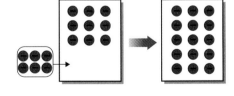

62 Chapter 2 Algebra: Using Integers

Classroom Vignette

"When adding and subtracting integers or decimals, we include an activity that deals with making a budget or running a company. Students practice their mathematical skills and experience the idea of what accounting is all about."

Mark Fisher, Teacher
Clearview Regional
Middle School
Mulica Hill, NJ

You already know how to add two positive integers. For example, to solve $r = 23 + 34$, you simply add 23 and 34. The solution is 57. In Example 1, you found that the sum $-9 + (-6)$ is -15. Notice that the value of $-(9 + 6)$ is the same. In each case, you add the absolute values of the addends. The sum has the same sign as the integers.

Adding Integers with Same Sign	To add integers with the same sign, add their absolute values. Give the result the same sign as the integers.

If you add a negative integer and a positive integer, what do you suppose happens? Let's use counters to find a rule.

MINI-LAB

Work with a partner. ◐ two colors of counters

Find the sum $5 + (-4)$ using counters. ▭ integer mat

Try This

- Let one color counter represent positive and the other color represent negative. Here we used yellow counters for positive and red for negative.
- Place 5 positive and 4 negative counters on the mat.
- When a positive counter is paired with a negative counter, the result is called a **zero pair**. Since a zero pair has a value of zero, you can add or remove zero pairs without changing the value of the set. Remove all of the zero pairs from the mat.
- The counters you have left represent the solution.

Talk About It

1. Is the sum of $5 + (-4)$ positive or negative? **positive**
2. Which number, 5 or -4, has the greater absolute value? **5**
3. Use counters to find the sum $-5 + 4 = t$. Compare the sign of the sum to the sign of the number with the greater absolute value. -1; **The signs are the same.**
4. Make a conjecture about the sign of the sum $-9 + 8 = m$.

4. The sum is negative.

Lesson 2-3 Adding Integers **63**

---

2 TEACH

 Transparency 2-3B contains a teaching aid for this lesson.

In-Class Example
For Example 1
Solve $-6 + (-5) = s$. -11

Using the Mini-Lab Integers with the same absolute value but opposite signs are called *opposites*. Point out that the sum of a number and its opposite is zero. Ask this question: *Why does a zero pair result when you pair a positive counter and a negative counter?* The counters represent opposites, -1 and 1, so their sum is zero.

Teaching Tip After doing the Mini-Lab with counters, use a large number line taped to the classroom floor to act out the Mini-Lab. Have a student move 5 places from 0 to $+5$ and then 4 units to the left. The student will be at the sum, 1.

Cross-Curriculum Cue

Inform the other teachers on your team that your classes are studying integers. Suggestions for curriculum integration are:
Health: weight gain and loss
Geography: elevation of mountains and oceanic trenches
Earth Science: temperature
Language Arts: synonyms and antonyms

Lesson 2-3 **63**

In-Class Examples

For Example 2
Solve $k = (-13) + 10$. -3

For Example 3
Solve $31 + (-41) = p$. -10

For Example 4
If you have $43 in a savings account but owe $53 for long distance calls made last month, what is your balance now?
$-$10$

3 PRACTICE/APPLY

Check for Understanding

If students need additional practice or instruction after completing Exercises 1–12, one of these options may be helpful.
- Extra Practice, see p. 609
- Reteaching Activity
- *Study Guide Masters*, p. 12
- *Practice Masters*, p. 12
- Interactive Mathematics Tools Software

Study Guide Masters, p. 12

The results of the Mini-Lab suggest the following rule for adding two integers with different signs.

Adding Integers with Different Signs	To add integers with different signs, subtract their absolute values. Give the result the same sign as the integer with the greater absolute value.

Examples

2 Solve $f = (-17) + 20$.
$|20| > |-17|$, so the sum is positive.
The difference of 20 and 17 is 3, so $f = 3$.

3 Solve $52 + (-60) = m$.
$|-60| > |52|$, so the sum is negative.
The difference of 60 and 52 is 8, so $m = -8$.

APPLICATION

4 Entertainment Each question in *Jeopardy!* has a dollar value. When a contestant answers a question correctly, they score that dollar amount. If a contestant answers incorrectly, he or she loses that dollar amount. If Rogelio has a score of $200 and misses a $400 question, what is his new score?

The problem can be represented by the equation $s = 200 + (-400)$.
$|200| < |-400|$, so the sum is negative.

The difference of 400 and 200 is 200.

Therefore, $s = -200$. Rogelio's score is $-$200$.

CHECK FOR UNDERSTANDING

Communicating Mathematics

Read and study the lesson to answer each question.   1–2. See margin.

1. **Demonstrate** how to find the sum of two negative numbers on a number line.

2. **Explain** how the absolute values of numbers are used when adding two integers.

HANDS-ON MATH

3. **Find** the sums $(-3) + 2$ and $2 + (-3)$ using counters. Does the order of the addends make a difference? Explain. -1; no, addition is commutative.

Guided Practice

Solve each equation.

4. $t = 4 + 6$ **10**
5. $-4 + (-3) = q$ -7
6. $6 + (-3) = s$ **3**
7. $g = 3 + (-8)$ -5
8. $w = (-32) + 44$ **12**
9. $-4 + 4 = y$ **0**

10. Find the sum $45 + (-67)$. -22

11. **Algebra** Evaluate $w + (-7)$ if $w = -3$. -10

12. **Entertainment** Did you know that *The Father of the Bride* with Steve Martin was a remake of a movie made 41 years earlier? The remake was made in 1991.
   a. Let $y = $ the year the original movie was made. Write an addition equation for y involving integers. $y = 1991 + (-41)$
   b. Solve the equation to find when the original movie was made. **1950**

64 Chapter 2 Algebra: Using Integers

■ Reteaching the Lesson ■

Activity Working in pairs, students take turns rolling a number cube, tossing a coin, and graphing the number (positive for heads, negative for tails) on a number line labeled from -6 to 6. If a number is already graphed, 1 point is scored. When all the numbers are graphed, the player with the low score wins.

Additional Answers

1. Start at the first addend. Move to the left to add a negative number.

2. To add numbers with the same sign, add the absolute values of the numbers. Then use the sign of the numbers. To add numbers with different signs, subtract the absolute values of the numbers. Then use the sign of the number with the greater absolute value.

EXERCISES

Practice

Solve each equation. 16. −59 22. −59 27. −130

13. $-45 + (-5) = p$ **−50**
14. $5 + 17 = r$ **22**
15. $y = -13 + 13$ **0**
16. $s = (-31) + (-28)$
17. $v = 71 + (-60)$ **11**
18. $k = 5 + (-12)$ **−7**
19. $a = 34 + (-60)$ **−26**
20. $d = -34 + 75$ **41**
21. $(-9) + 29 = c$ **20**
22. $-18 + (-41) = u$
23. $h = 45 + 63$ **108**
24. $m = 35 + (-32)$ **3**
25. $56 + (-34) = m$ **22**
26. $-19 + (-37) = z$ **−56**
27. $b = -98 + (-32)$
28. $t = 56 + (-2)$ **54**
29. $-60 + 30 = r$ **−30**
30. $-23 + (-456) = n$ **−479**
31. Find the sum $76 + (-45)$. **31**
32. What is the value of $-319 + (-100)$? **−419**

Evaluate each expression if $a = -3$, $b = -5$, and $c = 5$.

37b. **−9,406 million**
33. $c + b$ **0**
34. $a + |b|$ **2**
35. $|a + b|$ **8**

Applications and Problem Solving

36. **Earth Science** The highest waterfall in the world is Angel on the Carrao River in Venezuela. The longest single drop is 2,648 feet! The water falls another 268 feet in the rest of the drops.
    a. Let $f =$ the change in the water level. Write an addition equation for f involving integers. $f = -2,648 + (-268)$
    b. Solve the equation to find length that the water falls. **−2,916**

37. **Crime** In 1992, thefts in the United States totaled $14,608 million. Police were able to recover goods worth $5,202 million.
    a. Let $v =$ the value of the unrecovered goods. Write an addition equation for v involving integers. $v = -14,608 + 5,202$
    b. Solve the equation to find the value of the unrecovered goods.

38. **Working on the CHAPTER Project** Refer to the data on bald eagles on page 55. There were 3,000 fewer bald eagles in 1990 than in 1994.
    a. Use $p + 3,000 = 9,000$ to find the number of bald eagles in 1990. **6,000**
    b. Write an equation to compare the populations of your species in two different years. **See students' work.**

39. **Critical Thinking** Are values of $|-3| + |5|$, $|-3 + 5|$, and $-3 + 5$ the same? Explain. no; $|-3| + |5| = 8$, $|-3 + 5| = 2$; $-3 + 5 = 2$

Mixed Review

40. Order the integers $\{226, -3, 18, -157, 2, -28\}$ from least to greatest. *(Lesson 2-2)* $\{-157, -28, -3, 2, 18, 226\}$

41. **Test Practice** Jen had $15 and wanted to spend no more than that to buy 4 notebooks for school. Which inequality could she use to find the price, p, of each notebook? *(Lesson 1-9)* **D**
    A $p + 4 \leq 15$
    B $4p \geq 15$
    C $15 > 4p$
    D $4p \leq 15$
    E $15 - p \leq 4$

42. **Algebra** Solve $6x = 42$. Check your solution. *(Lesson 1-5)* **7**

43. Evaluate $[3(18 - 2)] - 4^2$. *(Lesson 1-3)* **32**

44. Write $4 \cdot 4 \cdot 6 \cdot 6 \cdot 6$ using exponents. *(Lesson 1-2)* $4^2 \cdot 6^3$

Lesson 2-3 Adding Integers **65**

Extending the Lesson

Enrichment Masters, p. 12

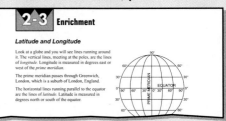

Activity Have students collect examples of integers and addition in newspapers and magazines. Articles on business, sports, or weather often use negative and positive integers. Then have students work in groups to write word problems involving integer addition.

2-4 Lesson Notes

Instructional Resources
- *Study Guide Masters*, p. 13
- *Practice Masters*, p. 13
- *Enrichment Masters*, p. 13
- Transparencies 2-4, A and B
- CD-ROM Program
  - Resource Lesson 2-4

Recommended Pacing	
Standard	Day 5 of 15
Honors	Day 4 of 14
Block	Day 2 of 8

1 FOCUS

5-Minute Check
(Lesson 2-3)

Solve each equation.
1. $-53 + (-71) = k$ **−124**
2. $13 + (-21) = n$ **−8**
3. $x = 270 + 130$ **400**
4. Evaluate $w + z$ if $w = 14$ and $z = -6$.   **8**
5. A shoe store bought shoes for $49.70 wholesale. It increased the price $35 to make a profit. What is the selling price?   **$84.70**

The 5-Minute Check is also available on **Transparency 2-4A** for this lesson.

Motivating the Lesson

Communication Ask students to suggest an easy way to add $9 + 437 + 91$ mentally. **Because $9 + 91 = 100$, then $9 + 437 + 91 = 100 + 437 = 537$.** Ask, *Which properties were used?* **commutative and associative**

2 TEACH

Transparency 2-4B contains a teaching aid for this lesson.

Mastering Basic Skills While students are using the associative and commutative properties, they are practicing adding integers.

2-4 More About Adding Integers

What you'll learn
You'll learn to add three or more integers.

When am I ever going to use this?
You can add integers to find the total yardage gained by a football team in a series of plays.

You have probably been in or seen a tug-of-war. But did you know that the history of the sport goes back to the ancient Egyptians? Teams of three men would hold each other by the waist and try to pull the other team across a mark. No rope was used then. From there the sport spread to ancient Greece, China, Scandinavia, and western Europe. Now the Tug-of-War International Federation organizes championship tournaments for men's and women's teams.

The progress of a tug-of-war team can be modeled by adding integers. Pulling the opponent toward the mark is a positive number, and being pulled is a negative number.

Suppose a United States team is competing with a Canadian team and the U.S. moves at the end of each of the first five minutes of competition are $+1, -1, +1, +2, -1$. If each integer represents a number of feet, what is the position of the mark at the end of the first five minutes?

LOOK BACK
You can refer to Lesson 1-2 to review the commutative and associative properties.

Let p = the position of the mark.
$p = 1 + (-1) + 1 + 2 + (-1)$
$p = [1 + (-1)] + (1 + 2) + (-1)$ *Use the associative property to group the addends.*
$p = 0 + 3 + (-1)$
$p = (0 + 3) + (-1)$
$p = 3 + (-1)$
$p = 2$ The United States team has pulled the Canadian team 2 feet.

Sometimes using both the associative and commutative properties allows you to find the result in fewer steps. One way is to group all of the positive numbers and all of the negative numbers.

$p = 1 + (-1) + 1 + 2 + (-1)$ → $p = (1 + 1 + 2) + (-1 + (-1))$
$p = 4 + (-2)$
$p = 2$

Solve $d = 7 + 20 + (-5)$.

Use the associative property to group the first two addends.
$d = (7 + 20) + (-5)$
$d = 27 + (-5)$
$d = 22$

Check: *Use the associative property to group the last two addends.*
$d = 7 + (20 + (-5))$
$d = 7 + 15$
$d = 22$ ✓

66 Chapter 2 Algebra: Using Integers

Multiple Learning Styles

Visual/Spatial Separate students into groups to investigate how a checking account works. Assign each group a $1,000 income per month, and give them a list of expenses to deduct. Each month's expenses should be different. Let the group construct a large number line from −$500 to +$1,000 and locate each month's balance along this line.

Examples

2 Solve $w = -4 + 3 + (-3) + 4$.

Look for groupings that make the addition simpler. Since adding 0 is easy and $-4 + 4 = 0$, group these.

$w = -4 + 3 + (-3) + 4$ **Check:** $w = -4 + 3 + (-3) + 4$
$w = (-4 + 4) + (-3 + 3)$ $w = (-4 + (-3)) + (3 + 4)$
$w = 0 + 0$ $w = -7 + 7$
$w = 0$ $w = 0$ ✓

APPLICATION

3 **Music** They say that rock and roll is here to stay! But the sales of rock music are not the same from one year to the next. In 1991, 35% of the music sold was rock. The table shows the change in sales in the successive years. What percent of the music sold in 1995 was rock?

Years	Change
1991-1992	-3
1992-1993	-1
1993-1994	+5
1994-1995	-1

Source: Recording Industry Assn. of America

$t = 35 + (-3) + (-1) + 5 + (-1)$
$t = (35 + 5) + (-3 + (-1) + (-1))$
$t = 40 + (-4 + (-1))$
$t = 40 + (-5)$
$t = 35$

Check:  ✓

Study Hint

Technology To enter -54 in a scientific calculator, press 54 and then press the [+/−] key. A negative sign will appear.

In-Class Examples

For Example 1
Solve $h = 19 + (-22) + 5$. **2**

For Example 2
Solve $f = -8 + 2 + (-2) + 8$. **0**

For Example 3
The table compares the number of skis sold at Hilltown Sports to the number sold the year before. If 200 pairs were sold in 1993–1994, how many pairs of skis were sold in 1997–1998? **270**

Year	Change
1994–95	-110
1995–96	+85
1996–97	-25
1997–98	+120

3 PRACTICE/APPLY

Check for Understanding

If students need additional practice or instruction after completing Exercises 1–8, one of these options may be helpful.
- Extra Practice, see p. 609
- Reteaching Activity
- *Study Guide Masters*, p. 13
- *Practice Masters*, p. 13

Study Guide Masters, p. 13

---

1. Sample Answer: Add $-13 + 13 = 0$; $-45 + (-55) = -100$; and $15 + 25 = 40$. Then $0 + (-100) + 40 = -60$.
2. Sample Answer: $(-3 + 4) + (-5) + 6 = [1 + (-5)] + 6$; $(-3 + 4) + (-5 + 6)$; $[-3 + (-5)] + (4 + 6)$

CHECK FOR UNDERSTANDING

Communicating Mathematics

Read and study the lesson to answer each question.

1. **Write** a sentence to tell how you might use the commutative and associative properties to solve $x = -13 + (-45) + 15 + (-55) + 25 + 13$ mentally.

2. **Show** three different ways you can solve $t = -3 + 4 + (-5) + 6$.

Guided Practice

Solve each equation. Check by solving another way. 6. -105

3. $h = 8 + (-4) + 9$ **13**
4. $n = -3 + 8 + 9$ **14**
5. $16 + (-33) + (-14) + 33 = q$ **2**
6. $-235 + 613 + (-844) + 361 = y$

7. **Algebra** Evaluate the expression $d + (-8) + 4 + 3$ if $d = 5$. **4**

8. **Sports** During a fourth-quarter possession in Super Bowl XXXI, the Green Bay Packers gained or lost the following yards in 10 plays. What was the net gain on the series? **39 yards**

Play	1	2	3	4	5	6	7	8	9	10
Yards Gained/Lost	0	7	18	7	1	-5	7	3	1	0

■ **Reteaching the Lesson** ■

Activity Have students create equations containing three addends, some with negative integers. Then have students find each sum in two different ways, using the associative and commutative properties.

Assignment Guide

Core: 9–27 odd, 28–32
Enriched: 10–24 even, 26–32

4 ASSESS

Closing Activity

Modeling Have students use counters to demonstrate that the commutative and associative properties do not alter the sum. Ask them to create equations containing three addends. Then have them find each sum in two different ways.

Additional Answer

28. No; $6 + (-9) + 14 = 11$. Beng evaluated $-6 + 9 + 14$, and $-6 + 9 + 14 = 17$.

Practice Masters, p. 13

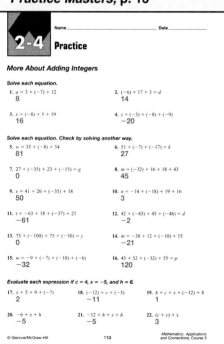

EXERCISES

Practice Solve each equation. Check by solving another way.

9. $d = -3 + 5 + (-9)$ **−7**
10. $x = 3 + 21 + (-6)$ **18**
11. $-7 + 12 + 9 = v$ **14**
12. $3 + (-2) + (-10) + 6 = b$ **−3**
13. $m = 4 + (-8) + 12 + (-11)$ **−3**
14. $g = 23 + 19 + (-8) + 12$ **46**
15. $9 + 50 + 3 + (-50) = k$ **12**
16. $14 + 7 + (-23) + 10 = p$ **8**
17. $8 + (-4) + 10 + (-2) = s$ **12**
18. $-6 + 12 + (-11) + 1 = r$ **−4**
19. $t = 92 + 73 + (-51) + 100$ **214**
20. $y = 17 + (-21) + 10 + (-17)$ **−11**

Evaluate each expression if $x = -4$, $y = -5$, and $z = 4$.

21. $2 + (-6) + x + 10$ **2**
22. $-7 + y + z$ **−8**
23. $z + 4 + (-9)$ **−1**
24. $-144 + x + z + 2$ **−142**

25. *Write a Problem* that can be more easily solved by using at least two mental math strategies. **See students' work.**

Applications and Problem Solving

26. *Geography* In 1970, the population of Worcester, Massachusetts, was 177 thousand. By 1980, the population had decreased by 12 thousand. From 1980 to 1990, the population increased 8 thousand. What was the population of Worcester in 1990? **173 thousand**

27. *Entertainment* The table shows the top ten money-losing movies. **a. $389,100,000 loss**

  a. How much money was lost by these movies?
  b. Columbia Pictures produced both *The Adventures of Baron Münchhausen* and *Ishtar*. Suppose they are able to make $30,000 on video sales of *Ishtar* and $120,000 on television showings for *The Adventures of Baron Münchhausen*. What would their total gain or loss be on these movies? **−$95,250,000**

Film	Revenue
The Adventures of Baron Münchhausen	−$48,100,000
Ishtar	−$47,300,000
Hudson Hawk	−$47,000,000
Inchon	−$44,100,000
The Cotton Club	−$38,100,000
Santa Claus – The Movie	−$37,000,000
Heaven's Gate	−$34,200,000
Billy Bathgate	−$33,000,000
Pirates	−$30,300,000
Rambo III	−$30,000,000

Source: *The Top Ten of Everything*

28. *Critical Thinking* Beng was trying to evaluate $6 + (-9) + 14$ using a calculator. He accidentally pressed the 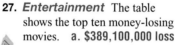 key before 9 instead of after. Is the result still correct? Explain. **See margin.**

Mixed Review

29. **Test Practice** A stock on the New York Stock Exchange opened at $52 on Monday morning. During the week, the stock lost $2, gained $1, gained $3, lost $1, and lost $4. What was the stock worth at the close of business on Friday? *(Lesson 2-3)* **B**

  A $41   **B** $49   **C** $57   **D** $63   **E** Not Here

30. Find $|4^2|$. *(Lesson 2-1)* **16**

31. *Geometry* Find the area of a square with sides of 8 meters. *(Lesson 1-8)* **31. 64 m²**

32. *Algebra* Write an expression for *17 more than p*. *(Lesson 1-6)* **17 + p**

68 Chapter 2 Algebra: Using Integers

Extending the Lesson

Enrichment Masters, p. 13

Activity Ask students to make a diagram of a football field with yardage lines marked. Students should treat the marked football field as a variation of a number line. Then give students yardage gained and lost during several drives by a team and have them map it. Use different colored pencils for each drive.

2-5 Subtracting Integers

What you'll learn
You'll learn to subtract integers.

When am I ever going to use this?
You can subtract integers to find differences in altitudes.

Word Wise
opposite
additive inverse

How do you spend your leisure time? The graph shows how the way people spend leisure time has changed from 1990 to 1994. What is the difference between the change in percent watching videos and the change of percent watching TV? *This problem will be solved in Example 3.*

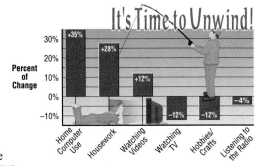
Source: Market Directions

In order to solve this problem, you will need to subtract integers. You can use counters to show subtraction.

Cultural Kaleidoscope

Made in England in 1929, the Baird "Model B Televisor" was the world's first television to be mass produced. Its picture was the size of a postage stamp.

MINI-LAB — HANDS-ON

Work with a partner. two colors of counters
 integer mat

Try This

Solve $x = -3 - (-1)$.
- Put 3 negative counters on the mat.
- Remove 1 negative counter.
- What is the value of x?

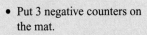

Solve $z = 4 - (-2)$.
- Put 4 positive counters on the mat.
- You need to remove 2 negative counters, but there are none on the mat. Add 2 zero pairs to the mat. Now remove the 2 negative counters.
- What is the value of z?

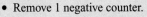

Talk About It
1. Use counters to solve $a = -3 + 1$ and compare the result to $x = -3 - (-1)$. See students' work. The results are both -2.
2. How does the solution of $c = 4 + 2$ compare to $z = 4 - (-2)$?

2. The results are both 6.

2-5 Lesson Notes

Instructional Resources
- *Study Guide Masters*, p. 14
- *Practice Masters*, p. 14
- *Enrichment Masters*, p. 14
- Transparencies 2-5, A and B
- *Assessment and Evaluation Masters*, pp. 42, 43
- CD-ROM Program
  - Resource Lesson 2-5
  - Interactive Lesson 2-5

Recommended Pacing	
Standard	Day 6 of 15
Honors	Day 5 of 14
Block	Day 3 of 8

1 FOCUS

5-Minute Check (Lesson 2-4)

Solve each equation. Check by solving another way.
1. $m = (-7) + 6 + 15$ **14**
2. $k = -9 + (-5) + 9 + 13$ **8**
3. $80 + (-50) + (-70) = w$ **-40**
4. Evaluate $15 + a + (-8) + b$ if $a = 11$ and $b = -15$.   **3**
5. The number of births in Texas in 1993 was about 331,000. In 1994, births decreased about 9,000. In 1995, births increased approximately 7,000 over 1994. About how many babies were born in Texas in 1995?   **329,000**

 The 5-Minute Check is also available on **Transparency 2-5A** for this lesson.

Motivating the Lesson
Problem Solving Have students solve this problem. *The lowest point on Earth is the Mariana Trench at 35,840 feet below sea level. The highest point is Mount Everest at 29,028 feet above sea level. What is the difference between these two points?*
$d = 29,028 - (-35,840)$; 64,868 ft

Multiple Learning Styles

 Interpersonal Ask students to discuss why it is difficult to think about subtracting negative numbers. Then suggest that they come up with a real life example that matches the equation $z = 4 - (-2)$ from the Mini-Lab. Sample answer: I had $4 after I gave you $2. Then you returned it to me. I have $z = \$4 - (-\$2) = \$6$.

Lesson 2-5 Subtracting Integers

2 TEACH

 Transparency 2-5B contains a teaching aid for this lesson.

Using the Mini-Lab At the conclusion of the lab, ask:
- In $x = -3 - (-1)$, why were no zero pairs formed? **Zero pairs need both positive and negative counters. There are no positive counters on the mat.**
- Why were two zero pairs added to solve the equation $z = 4 - (-2)$? **There were no negative counters to remove. Zero pairs were added so the equation mat contained negative counters to remove.**
- Why did the addition of a zero pair not increase the value of z? **A zero pair adds zero to any number.**

In-Class Examples

For Example 1
Solve $p = -4 - 3$. **-7**

For Example 2
Solve $34 - (-25) = s$. **59**

For Example 3
A coal mine elevator starts from 132 feet below ground level. It descends 256 feet, goes up 195 feet, and then descends 57 feet. Where does it end up?
-250 feet

Teaching Tip For In-Class Example 3, use a vertical number line to show the changes in elevation of the coal mine elevator.

Each integer has an opposite. The **opposite** is the number that is the same distance from zero but in the opposite direction. For example, the opposite of 3 is -3. The opposite of an integer is called its **additive inverse**. Additive inverses can be used to subtract integers.

Additive Inverse	Words:	The sum of an integer and its additive inverse is 0.	
	Symbols:	**Arithmetic**	**Algebra**
		$4 + (-4) = 0$	$a + (-a) = 0$

In the Mini-Lab, you compared the result of subtracting an integer with the result of adding its inverse.

Subtracting   *Adding the Additive Inverse*

$-3 - (-1) = -2$ ⟹ $-3 + 1 = -2$

$4 - (-2) = 6$ ⟹ $4 + 2 = 6$

Notice that adding the additive inverse of an integer produces the same result as subtracting the integer.

Subtracting Integers	To subtract an integer, add its additive inverse.

Examples

1 Solve $r = -5 - 2$.

$r = -5 - 2$

To subtract 2, add -2.

$r = -5 + (-2)$

$r = -7$

2 Solve $25 - (-45) = t$.

$25 - (-45) = t$

$25 + 45 = t$ *To subtract -45, add 45.*

$70 = t$

APPLICATION

3 Entertainment Refer to the beginning of the lesson. What is the difference between the change in percent watching videos and the change of percent watching TV?

The change in percent of time watching videos is 12%, and the change in percent of time watching TV is -12%. The difference can be expressed by an equation. Let d represent the difference.

$d = 12 - (-12)$

$d = 12 + 12$ *To subtract -12, add 12.*

$d = 24$

The difference is 24%.

70 Chapter 2 Algebra: Using Integers

CHECK FOR UNDERSTANDING

Communicating Mathematics

Read and study the lesson to answer each question. 1–2. See margin.

1. *Write* how you find the additive inverse of an integer. Use the term *opposite* in your explanation. Give an example of an integer and its additive inverse.
2. *Tell* how you know whether a subtraction problem will have a positive or a negative answer.

HANDS-ON MATH

3. *Draw* a model showing $3 - (-1) = a$. Then solve for a. See Answer Appendix.
4. *Write* the additive inverse of -3. 3

Guided Practice

12. $534 - 50 = b$, $b = 484$; $484 + 30 = b$, $b = 514$; $514 - 20 = b$, $b = 494$

Solve each equation.

5. $s = -4 - (-3)$ -1
6. $r = 3 - (-2)$ 5
7. $t = 4 - (-5)$ 9
8. $-23 - (34) = y$ -57
9. $b = 3 - 3$ 0
10. $0 - (-4) = d$ 4

11. *Algebra* Find the value of $68 - a$ if $a = -7$. 75

12. *Spreadsheets* Millions of people use *Quicken* to organize their finances. The spreadsheet is part of a *Quicken* page. The program finds the balance after each transaction is entered. Write an addition or subtraction equation and find the balance for each transaction.

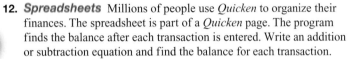

Date	Num	Payee/Category/Memo	Payment	Ck	Deposit	Balance
1/27/98		Opening Balance			534 00	534 00
		(Checking)				
1/30/98	234	Food King	50 00			
		Groceries				
2/01/98		Deposit			30 00	
		Rebate				
2/03/98	ATM	Cash	20 00			
		Entertainment				

EXERCISES

Practice

Write the additive inverse of each integer.

13. 7 -7
14. -37 37
15. 0 0

Solve each equation.

16. $h = 4 - (-7)$ 11
17. $-23 - (-2) = g$ -21
18. $14 - 14 = d$ 0
19. $x = 53 - 78$ -25
20. $y = 17 - (-26)$ 43
21. $w = -43 - 88$ -131
22. $-78 - (-98) = z$ 20
23. $k = 44 - (-11)$ 55
24. $u = 4 - (-89)$ 93

26. 1,313
29. -412
30. -9
32. -1

25. $n = 56 - (-22)$ 78
26. $435 - (-878) = b$
27. $w = 63 - 92$ -29
28. $r = -9 - (-4)$ -5
29. $-345 - 67 = t$
30. $s = -34 - (-25)$
31. $v = 823 - (-19)$ 842
32. $-13 - (-12) = x$
33. $d = 0 - (-6)$ 6

34. Find the value of s for $s = 56 - (-78)$. 134
35. What value of f makes $-18 - 0 = f$ true? -18

Evaluate each expression if $a = 9$, $b = -6$, and $c = -2$.

36. $78 - b$ 84
37. $c - a$ -11
38. $12 - a - b$ 9

Lesson 2-5 Subtracting Integers **71**

3 PRACTICE/APPLY

Check for Understanding

If students need additional practice or instruction after completing Exercises 1–12, one of these options may be helpful.
- Extra Practice, see p. 609
- Reteaching Activity
- *Study Guide Masters*, p. 14
- *Practice Masters*, p. 14
- Interactive Mathematics Tools Software

Assignment Guide

Core: 13–41 odd, 42–47
Enriched: 14–38 even, 40–47
All: Self Test, 1–5

Additional Answers

1. Sample answer: The additive inverse of an integer is called its opposite. The opposite of an integer is the same distance from 0, but has the opposite sign. For example, 4 an -4 are additive inverses.
2. Sample answer: Rewrite the subtraction equation as an addition equation using the additive inverse. Then use the same rules as for addition.

Study Guide Masters, p. 14

Reteaching the Lesson

Activity Have students draw a chart like the one below. Give them subtraction expressions to compute. Example:

Expression	$-5 - (-2)$
Opposite of 2nd Integer	$+2$
Expression as an Addition	$-5 + (+2)$
Sum	-3

Error Analysis
Watch for students who subtract -2 from -5 and come up with the solution of -7.
Prevent by using a number line. Show how to add -2 by moving along the line from -5 to -7 and to subtract -2 by moving from -5 to -3.

Lesson 2-5 **71**

4 ASSESS

Closing Activity
Speaking Ask students these questions.
- What is the additive inverse of an integer? **the integer with the same absolute value but the opposite sign**
- How do you subtract one integer from another? **Add the additive inverse of the second integer to the first integer.**

Chapter 2, Quiz B (Lessons 2-4 and 2-5) is available in the *Assessment and Evaluation Masters,* p. 43.

Mid-Chapter Test (Lessons 2-1 through 2-5) is available in the *Assessment and Evaluation Masters,* p. 42.

Mid-Chapter Self Test
The Mid-Chapter Self Test reviews the concepts in Lessons 2-1 through 2-5. Lesson references are given so students can review concepts not yet mastered.

Additional Answer
42. No, for 0 the integer and its additive inverse both have no sign. 0 is its own additive inverse.

Practice Masters, p. 14

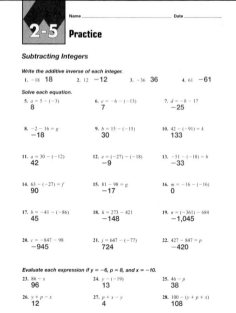

---

39. **Write a Problem** involving subtraction of integers for which the answer is -3. **See students' work.**

Applications and Problem Solving

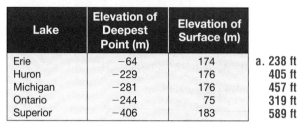

40. **Geography** The table shows the elevations above sea level of the surface and the deepest point of each of the Great Lakes.

Lake	Elevation of Deepest Point (m)	Elevation of Surface (m)	
Erie	-64	174	a. 238 ft
Huron	-229	176	405 ft
Michigan	-281	176	457 ft
Ontario	-244	75	319 ft
Superior	-406	183	589 ft

Source: National Ocean Service

a. How far below the surface is the deepest part of each lake?
b. How does the deepest part of Lake Ontario compare with the deepest part of Lake Superior? **Lake Superior is 162 feet deeper than Lake Ontario.**
c. Find the difference between the deepest part of Lake Erie and the deepest part of Lake Superior. **Lake Superior is 342 feet deeper than Lake Erie.**

41. **History** The calendar that astronomers use began on Jan 1, 4713 B.C. On that day, the Julian calendar, the lunar calendar, and the Roman tax system calendar all coincided. This won't happen again until A.D. 3267!
  a. How many years ago was the astronomer's calendar started? (*Hint:* There was no year 0.) **Answers will vary. For 1998, it was 6,710 years ago.**
  b. Find the number of years between times that the three calendars coincide.

41b. **7,979 years**

42. **Critical Thinking** Do an integer and its additive inverse always have different signs? Is there an integer that is its own inverse? **See margin.**

Mixed Review

44. **>**

43. Solve $44 + 8 + (-20) + 15 = s$. Check your solution. *(Lesson 2-4)* **47**
44. Replace ● with >, <, or = to make $10 ● -10$ a true sentence. *(Lesson 2-2)*
45. **Test Practice** If $12 + 5d = 72$, what is the value of d? *(Lesson 1-7)* **B**
  A 10 B 12 C 14 D 16
46. **Language** There are 999 million people who speak Mandarin. This is 512 million more than speak English. Write an equation to find the number of people who speak English. *(Lesson 1-6)* **Sample answer: $999 - x = 512$**
47. **Algebra** Solve $c + 9 = 27$ if the replacement set is $\{12, 17, 18, 22\}$. *(Lesson 1-3)* **18**

CHAPTER 2 Mid-Chapter Self Test

1. Graph $\{8, 4, -3, 2, 0, -8, -1\}$ on a number line. *(Lesson 2-1)* **See Answer Appendix.**
2. Write $\{-7, 8, 0, -3, -2, 5, 6\}$ in order from least to greatest. *(Lesson 2-2)*
   $\{-7, -3, -2, 0, 5, 6, 8\}$

Solve each equation. *(Lessons 2-3, 2-4, and 2-5)*

3. $x = -5 + 3$ **-2** 4. $514 - 600 = s$ **-86** 5. $y = 2 + (-4) + (-6) + 8$ **0**

72 Chapter 2 Algebra: Using Integers

Extending the Lesson

Enrichment Masters, p. 14

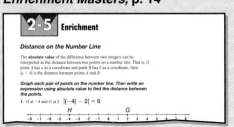

Activity Have students use the *nomograph* below to subtract integers. Example: For $2 - 3$, draw a line from 3 on number line A through 2 on B, and read the answer, -1, where the line crosses C.

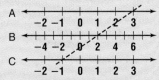

2-6 Integration: Statistics
Matrices

What you'll learn
You'll learn to use matrices to organize data.

When am I ever going to use this?
You can use a matrix to record sales of different items in a bookstore.

Word Wise
matrix
row
column
element

Keeping up with the fashion fads is hard work. And for the corporations that make the fashions, being behind the times can mean going out of business. The table lists the sales of the top three clothing makers in the United States for 1995 and 1996.

Corporation	1995 Sales (millions)	1996 Sales (millions)
Nike, Inc.	$4,761	$6,471
VF Corporation	$5,061	$5,137
Reebok International, Ltd.	$3,518	$3,483

Source: Fortune

Another way to organize information is by using a **matrix**. A matrix is a rectangular arrangement of numbers in **rows** and **columns**. Each number in a matrix is called an **element** of the matrix.

The matrix at the right has 2 rows and 3 columns. Elements are named by telling the row and the column in which they appear.

$$\begin{bmatrix} 3 & 8 & 0 \\ 5 & -2 & -3 \end{bmatrix}$$

column 3
row 2
element (2, 1)

Did you know
VF Corporation is the United States' largest jeans maker with the Lee, Rustler, and Wrangler brands.

Example 1 APPLICATION

Business Refer to the beginning of the lesson. Write a matrix for the apparel sales data.

The matrix will have a row for each corporation and a column for each year. Only the numbers are part of the matrix. The labels are written outside of the matrix.

Nike, Inc.
VF Corporation
Reebok International, Ltd.
$$\begin{bmatrix} 4{,}761 & 6{,}471 \\ 5{,}061 & 5{,}137 \\ 3{,}518 & 3{,}483 \end{bmatrix}$$
1995 Sales   1996 Sales

Lesson 2-6 Integration: Statistics Matrices **73**

Multiple Learning Styles

Kinesthetic Arrange students in the class into two matrices of the same dimensions. Give each student a card with a positive or negative integer written on it. Then have students from each matrix, proceeding element by element, "add" themselves and walk over in pairs to form a new matrix, standing in the spaces corresponding with their original ones. When this is finished, write the sum matrix on the chalkboard. Any additional students can be assigned the tasks of passing out numbers, directing traffic, and writing the matrix on the chalkboard.

2-6 Lesson Notes

Instructional Resources
- *Study Guide Masters*, p. 15
- *Practice Masters*, p. 15
- *Enrichment Masters*, p. 15
- Transparencies 2-6, A and B
- CD-ROM Program
  - Resource Lesson 2-6

Recommended Pacing	
Standard	Day 7 of 15
Honors	Days 6 & 7 of 14
Block	Day 4 of 8

1 FOCUS

5-Minute Check
(Lesson 2-5)

1. Write the additive inverse of 8. -8

Solve each equation.

2. $m = 25 - (-14)$ 39
3. $-6 - (-4) = x$ -2

4. Evaluate $12 - a - b - c$ if $a = -9$, $b = -6$, and $c = 2$. 25

5. Patrice starts with $5 at the beginning of the week, pays $2 for a magazine and $1 for gum, earns $4.50, and spends $3 on food. How much money does she have left? $3.50

The 5-Minute Check is also available on **Transparency 2-6A** for this lesson.

Motivating the Lesson

Problem Solving Read the first two paragraphs of the lesson with the class. Then ask students how they could arrange all of the class scores on four quizzes. **in a matrix** Ask students how they could compare these scores with the scores from the next four quizzes they take.
Write the scores from the next four quizzes in a matrix and subtract one matrix from the other.

Lesson 2-6 **73**

2 TEACH

 Transparency 2-6B contains a teaching aid for this lesson.

Reading Mathematics Help students understand why matrices might be added together by giving word problems that could be represented by the matrices in Example 2. Example: The matrix might represent the number of full and daytime-only memberships sold by a health club in the first, second, and third quarters of the year. Adding the matrices together gives total sales for each quarter.

In-Class Examples

For Example 1
Write a matrix for this information. In 1890, the population of Connecticut was 746,258; Florida's was 391,422; and Oregon's was 317,704. In 1990, Connecticut's was 3,287,116; Florida's was 12,938,071; and Oregon's was 2,842,337.

$$\begin{bmatrix} 746{,}258 & 3{,}287{,}116 \\ 391{,}422 & 12{,}938{,}071 \\ 317{,}704 & 2{,}842{,}337 \end{bmatrix}$$

For Example 2
Find $\begin{bmatrix} 4 & -3 & -1 \\ -4 & 4 & -2 \end{bmatrix} + \begin{bmatrix} -1 & 3 & 7 \\ -1 & 4 & 5 \end{bmatrix}$.

$\begin{bmatrix} 3 & 0 & 6 \\ 5 & 8 & 3 \end{bmatrix}$

For Example 3
Find $\begin{bmatrix} 7 & 4 \\ 1 & -5 \\ 3 & 2 \end{bmatrix} - \begin{bmatrix} -3 & 8 \\ -4 & 6 \\ 0 & -3 \end{bmatrix}$.

$\begin{bmatrix} 10 & -4 \\ 5 & -11 \\ 3 & 5 \end{bmatrix}$

For Example 4
Find $\begin{bmatrix} 2 & 3 \\ 11 & -4 \end{bmatrix} + \begin{bmatrix} 4 & -2 \\ 3 & -8 \\ 7 & -7 \end{bmatrix}$.

impossible

You can add or subtract matrices that have the same number of rows and columns. You add and subtract matrices by adding or subtracting the corresponding elements.

Examples

2 Find $\begin{bmatrix} 5 & -2 & 0 \\ -3 & 4 & 1 \end{bmatrix} + \begin{bmatrix} 3 & -1 & 4 \\ 3 & 2 & -2 \end{bmatrix}$.

$\begin{bmatrix} 5 & -2 & 0 \\ -3 & 4 & 1 \end{bmatrix} + \begin{bmatrix} 3 & -1 & 4 \\ 3 & 2 & -2 \end{bmatrix} = \begin{bmatrix} 5+3 & -2+(-1) & 0+4 \\ -3+3 & 4+2 & 1+(-2) \end{bmatrix}$

$= \begin{bmatrix} 8 & -3 & 4 \\ 0 & 6 & -1 \end{bmatrix}$

Study Hint
Reading Math The plural of matrix is matrices. It is pronounced MAY tra cees.

3 Find $\begin{bmatrix} 8 & 3 \\ 0 & -1 \\ 2 & 2 \end{bmatrix} - \begin{bmatrix} -3 & 4 \\ 1 & 2 \\ -2 & 1 \end{bmatrix}$.

$\begin{bmatrix} 8 & 3 \\ 0 & -1 \\ 2 & 2 \end{bmatrix} - \begin{bmatrix} -3 & 4 \\ 1 & 2 \\ -2 & 1 \end{bmatrix} = \begin{bmatrix} 8-(-3) & 3-4 \\ 0-1 & -1-2 \\ 2-(-2) & 2-1 \end{bmatrix}$

$= \begin{bmatrix} 11 & -1 \\ -1 & -3 \\ 4 & 1 \end{bmatrix}$

4 Find $\begin{bmatrix} 11 & -3 \\ 1 & 1 \end{bmatrix} + \begin{bmatrix} 4 & 12 \\ 2 & -5 \\ -5 & 8 \end{bmatrix}$.

The first matrix has 2 rows and 2 columns, and the second matrix has 3 rows and 2 columns, so these matrices cannot be added or subtracted.

2. You can add or subtract two matrices that have the same number of rows and columns.

CHECK FOR UNDERSTANDING

Communicating Mathematics
Read and study the lesson to answer each question. 1. 3 rows, 2 columns
1. **Tell** how many rows and columns there are in the matrix at the right. $\begin{bmatrix} 4 & -1 \\ 1 & 3 \\ 2 & -4 \end{bmatrix}$
2. **Explain** when you can add or subtract two matrices.

Guided Practice
Find each sum or difference. If there is no sum or difference, write impossible. 6. impossible

4. $\begin{bmatrix} 3 & 1 & -2 \\ -4 & 0 & 10 \end{bmatrix}$

3. $\begin{bmatrix} 1 & 4 \\ 3 & -2 \end{bmatrix} + \begin{bmatrix} -2 & 3 \\ 2 & 0 \end{bmatrix} \begin{bmatrix} -1 & 7 \\ 5 & -2 \end{bmatrix}$

4. $\begin{bmatrix} 0 & 6 & 3 \\ -12 & -7 & 9 \end{bmatrix} - \begin{bmatrix} -3 & 5 & 5 \\ -8 & -7 & -1 \end{bmatrix}$

5. $\begin{bmatrix} 0 & 4 & 1 \\ -2 & -3 & -2 \\ 3 & 0 & 3 \end{bmatrix}$

5. $\begin{bmatrix} -1 & 2 & 0 \\ 2 & 1 & 0 \\ 4 & 1 & 3 \end{bmatrix} - \begin{bmatrix} -1 & -2 & -1 \\ 4 & 4 & 2 \\ 1 & 1 & 0 \end{bmatrix}$

6. $\begin{bmatrix} 3 & 7 \\ -2 & -4 \\ 6 & -6 \end{bmatrix} + \begin{bmatrix} 4 & 16 & 8 \\ -9 & 14 & 5 \end{bmatrix}$

74 Chapter 2 Algebra: Using Integers

7. Life Science Manatees have been living off of the coast of Florida for centuries. A decline in population has lead to preservation and protection programs. The table shows the number of manatees that died from January to April of 1996. Write a matrix of the data.

Cause of Death	J	F	M	A
Watercraft	8	4	7	6
Weather	13	2	0	0
Natural Causes	7	5	47	24
Undetermined	13	12	47	36

Source: Florida Department of Environmental Protection

$\begin{bmatrix} 8 & 4 & 7 & 6 \\ 13 & 2 & 0 & 0 \\ 7 & 5 & 47 & 24 \\ 13 & 12 & 47 & 36 \end{bmatrix}$

EXERCISES

Practice Find each sum or difference. If there is no sum or difference, write *impossible*. 9. impossible   12. impossible

8. $\begin{bmatrix} 3 & -2 \\ 4 & -1 \end{bmatrix} + \begin{bmatrix} -3 & 8 \\ 0 & 6 \end{bmatrix}$ $\begin{bmatrix} 0 & 6 \\ 4 & 5 \end{bmatrix}$ 9. $\begin{bmatrix} 7 & 20 & 6 \\ -1 & 15 & 1 \end{bmatrix} - \begin{bmatrix} 4 & 3 \\ -5 & -3 \\ -1 & 0 \end{bmatrix}$

10. $\begin{bmatrix} -1 & 0 \\ 2 & 6 \\ 5 & -2 \end{bmatrix}$

10. $\begin{bmatrix} 12 & -7 \\ -5 & 9 \\ 7 & 11 \end{bmatrix} - \begin{bmatrix} 13 & -7 \\ -7 & 3 \\ 2 & 13 \end{bmatrix}$

11. $\begin{bmatrix} 6 & -9 \\ -11 & 17 \\ 8 & 10 \end{bmatrix} + \begin{bmatrix} 5 & -3 \\ 12 & 9 \\ -5 & 18 \end{bmatrix}$

11. $\begin{bmatrix} 11 & -12 \\ 1 & 26 \\ 3 & 28 \end{bmatrix}$

12. $\begin{bmatrix} 1 & 8 \\ -2 & 16 \end{bmatrix} - \begin{bmatrix} 8 & -5 & 0 \\ -6 & 1 & 1 \\ 1 & -3 & 0 \end{bmatrix}$

13. $\begin{bmatrix} 14 & 23 & 9 \\ -5 & -7 & 4 \end{bmatrix} - \begin{bmatrix} 9 & 18 & 6 \\ -6 & -4 & -6 \end{bmatrix}$

13. $\begin{bmatrix} 5 & 5 & 3 \\ 1 & -3 & 10 \end{bmatrix}$

14. $\begin{bmatrix} 4 & 3 & -10 \\ 0 & -1 & 4 \\ 0 & 0 & -6 \end{bmatrix} - \begin{bmatrix} -4 & 5 & 2 \\ -4 & 9 & 6 \\ 1 & 0 & 0 \end{bmatrix}$

15. $\begin{bmatrix} 7 & -3 & -1 \\ 5 & -4 & 14 \\ 7 & 22 & -6 \end{bmatrix} - \begin{bmatrix} -9 & -7 & 6 \\ 0 & -3 & 9 \\ 6 & 13 & 6 \end{bmatrix}$

14. $\begin{bmatrix} 8 & -2 & -12 \\ 4 & -10 & -2 \\ -1 & 0 & -6 \end{bmatrix}$

16. $\begin{bmatrix} 1 \\ 2 \\ 9 \end{bmatrix} - \begin{bmatrix} -5 & -2 & -7 \end{bmatrix}$

17. $\begin{bmatrix} 1 & 23 & -11 \\ 0 & -8 & 6 \\ 0 & 2 & -9 \end{bmatrix} + \begin{bmatrix} -1 & -15 & 7 \\ 0 & -1 & 4 \\ 1 & 19 & 6 \end{bmatrix}$

15. $\begin{bmatrix} 16 & 4 & -7 \\ 5 & -1 & 5 \\ 1 & 9 & -12 \end{bmatrix}$

16. impossible

18. $\begin{bmatrix} 18 & -3 & 10 \\ 14 & 28 & 7 \\ -9 & -6 & -7 \end{bmatrix} + \begin{bmatrix} 5 & 17 & 1 \\ -6 & -15 & 0 \\ -8 & 14 & 4 \end{bmatrix}$ $\begin{bmatrix} 23 & 14 & 11 \\ 8 & 13 & 7 \\ -17 & 8 & -3 \end{bmatrix}$

17. $\begin{bmatrix} 0 & 8 & -4 \\ 0 & -9 & 10 \\ 1 & 21 & -3 \end{bmatrix}$

19. $\begin{bmatrix} 3 & 12 & 22 \\ 4 & 8 & 21 \\ 9 & -7 & -6 \\ -5 & 18 & -9 \end{bmatrix} + \begin{bmatrix} -3 & -2 & -2 \\ 5 & -9 & 27 \\ -6 & -4 & -2 \\ -1 & 16 & 18 \end{bmatrix}$ $\begin{bmatrix} 0 & 10 & 20 \\ 9 & -1 & 48 \\ 3 & -11 & -8 \\ -6 & 34 & 9 \end{bmatrix}$

Applications and Problem Solving

20. Agriculture The table shows the number of thousands of livestock on farms in the United States in 1994 to 1996. Write a matrix for the data. **See margin.**

	Livestock on U.S. Farms (thousands)		
Year	Beef Cattle/ Milk Cows	Sheep	Hogs
1994	110,516	9,742	57,904
1995	112,252	8,886	59,900
1996	113,231	8,457	58,700

Lesson 2-6 Integration: Statistics Matrices **75**

■ Reteaching the Lesson ■

Activity Have students use the associative and commutative properties to show $\begin{bmatrix} a & b \\ c & d \end{bmatrix} + \begin{bmatrix} -a & -b \\ -c & -d \end{bmatrix} = \begin{bmatrix} 0 & 0 \\ 0 & 0 \end{bmatrix}$.

$\begin{bmatrix} a + (-a) + b + (-b) \\ c + (-c) + d + (-d) \end{bmatrix} = \begin{bmatrix} 0 & 0 \\ 0 & 0 \end{bmatrix}$

---

3 PRACTICE/APPLY

Check for Understanding
If students need additional practice or instruction after completing Exercises 1–7, one of these options may be helpful.
- Extra Practice, see p. 610
- Reteaching Activity
- *Study Guide Masters*, p. 15
- *Practice Masters*, p. 15

Assignment Guide
Core: 9–21 odd, 22–25
Enriched: 8–18 even, 20–25

Additional Answer
20. $\begin{bmatrix} 110{,}516 & 9{,}742 & 57{,}904 \\ 112{,}252 & 8{,}886 & 59{,}900 \\ 113{,}231 & 8{,}457 & 58{,}700 \end{bmatrix}$

***Study Guide Masters*, p. 15**

Lesson 2-6   **75**

21. **Business** The table shows the expenses for the top three apparel corporations for 1995 and 1996.

   a. Write a matrix for the expenses data.

   b. Find the profit matrix by subtracting the matrix in part a from the sales matrix in Example 1.

21a. $\begin{bmatrix} 4{,}362 & 5{,}917 \\ 4{,}904 & 4{,}838 \\ 3{,}353 & 3{,}344 \end{bmatrix}$

21b. $\begin{bmatrix} 399 & 554 \\ 157 & 299 \\ 165 & 139 \end{bmatrix}$

Corporation	1995 Expenses (millions)	1996 Expenses (millions)
Nike, Inc.	$4,362	$5,917
VF Corporation	$4,904	$4,838
Reebok International, Ltd.	$3,353	$3,344

Source: Fortune

22. **Critical Thinking** In which row and column of the matrix will the number 100 occur? **row 6, column 9**

$\begin{bmatrix} 1 & 3 & 6 & 10 & 15 & \cdots \\ 2 & 5 & 9 & 14 & 20 & \cdots \\ 4 & 8 & 13 & 19 & 26 & \cdots \\ 7 & 12 & 18 & 25 & 33 & \cdots \\ 11 & 17 & 24 & 32 & 41 & \cdots \\ 16 & 23 & 31 & 40 & 50 & \cdots \\ \vdots & \vdots & \vdots & \vdots & \vdots & \end{bmatrix}$

Mixed Review

23. **Test Practice** Solve $b = 65 - (-87)$. *(Lesson 2-5)* **A**

   A 152   **B** 22   **C** -22 **D** -152 **E** Not Here

24. Solve $h = -8 + 4 - (-3)$. *(Lesson 2-4)* **−1**

25. **Education** Do you or someone you know attend school at home? According to the U.S. Department of Education, more than fifty-six times as many students are being taught at home now as there were in the 1970s. If 12,500 students were home-schooled in the 1970s, how many are home-schooled now? *(Lesson 1-9)* **more than 700,000**

Let the Games Begin
Matrix Madness

Get Ready This game is for two players.

Get Set Draw an empty matrix with four rows and four columns like the one at the right. Write the integers from 1 to 16 below the matrix as shown.

Go
- Players take turns crossing a number off of the list and writing it in the matrix. A number cannot be placed so that an adjacent element is a consecutive number. For example, a 4 cannot be placed so that a 3 or a 5 is next to it vertically, horizontally, or diagonally.

$\begin{bmatrix} - & - & - & - \\ - & - & - & - \\ - & - & - & - \\ - & - & - & - \end{bmatrix}$

1 2 3 4 5 6 7 8
9 10 11 12 13 14 15 16

- The winner is the last player who is able to write a number.

interNET CONNECTION Visit www.glencoe.com/sec/math/mac/mathnet for more games.

76 Chapter 2 Algebra: Using Integers

GRAPHING CALCULATORS

2-6B Matrices

A Follow-Up of Lesson 2-6

graphing calculator

You can use most graphing calculators to perform operations with matrices. On a TI-83, the [MATRX] key accesses the matrix operations.

TRY THIS

Work with a partner.

Enter matrix $A = \begin{bmatrix} -5 & 1 \\ 2 & -1 \\ 0 & 4 \end{bmatrix}$ and $B = \begin{bmatrix} 8 & -1 \\ -4 & 3 \\ 2 & -2 \end{bmatrix}$. Find $A + B$.

Step 1 Begin by entering matrix A into the calculator's memory.

Enter: [MATRX] [◄] [ENTER] — *Choose the edit option and matrix A.*
3 [ENTER] 2 [ENTER] — *Enter number of rows and columns.*
[(−)] 5 [ENTER] 1 [ENTER] — *Enter each matrix element.*
2 [ENTER] [(−)] 1 [ENTER]
0 [ENTER] 4 [ENTER]

Step 2 Next enter matrix B.

Enter: [MATRX] [◄] 2 — *Choose the edit option and matrix B.*
3 [ENTER] 2 [ENTER] — *Enter number of rows and columns.*
8 [ENTER] [(−)] 1 [ENTER] — *Enter each matrix element.*
[(−)] 4 [ENTER] 3 [ENTER] 2 [ENTER] [(−)] 2 [ENTER]

Step 3 Find the sum of matrices A and B.

Enter: [2nd] [QUIT] — *Exit EDIT screen.*
[MATRX] 1 — *Choose matrix A.*
[+] [MATRX] — *Add matrix B.*
2 [ENTER]

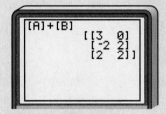

ON YOUR OWN

Enter the matrices into a graphing calculator. Then find each sum or difference.

$D = \begin{bmatrix} 4 & 0 & 2 \\ 2 & 1 & 1 \\ 3 & -1 & 4 \end{bmatrix}$ $E = \begin{bmatrix} 1 & 5 & 3 \\ -3 & 10 & 6 \\ 6 & -7 & -2 \end{bmatrix}$ $F = \begin{bmatrix} 5 & -3 \\ -3 & 1 \\ 11 & 0 \end{bmatrix}$ 1. $\begin{bmatrix} 5 & 5 & 5 \\ -1 & 11 & 7 \\ 9 & -8 & 2 \end{bmatrix}$ 3. $\begin{bmatrix} 3 & -5 & -1 \\ 5 & -9 & -5 \\ -3 & 6 & 6 \end{bmatrix}$

1. $E + D$
2. $D + F$ **impossible**
3. $D - E$
4. $D + E$ $\begin{bmatrix} 5 & 5 & 5 \\ -1 & 11 & 7 \\ 9 & -8 & 2 \end{bmatrix}$

5. Observe the solutions for Exercises 1 and 4. Do you think that addition of matrices is commutative? Explain. **Yes; $D + E = E + D$.**

Lesson 2-6B TECHNOLOGY LAB

2-7 Lesson Notes

Instructional Resources
- *Study Guide Masters*, p. 16
- *Practice Masters*, p. 16
- *Enrichment Masters*, p. 16
- Transparencies 2-7, A and B
- *School to Career Masters*, p. 28
- *Science and Math Lab Manual*, pp. 89–92
- CD-ROM Program
  - Resource Lesson 2-7
  - Interactive Lesson 2-6

Recommended Pacing	
Standard	Day 8 of 15
Honors	Day 8 of 14
Block	Day 5 of 8

1 FOCUS

5-Minute Check
(Lesson 2-6)

Find each sum or difference. If there is no sum or difference, write *impossible*.

1. $\begin{bmatrix} 2 & -6 \\ 5 & -1 \end{bmatrix} + \begin{bmatrix} -2 & 8 \\ 0 & 4 \end{bmatrix} \begin{bmatrix} 0 & 2 \\ 5 & 3 \end{bmatrix}$

2. $\begin{bmatrix} 1 & -3 \\ -7 & 2 \\ 0 & 9 \end{bmatrix} - \begin{bmatrix} -6 & 3 \\ -5 & -2 \\ -4 & 8 \end{bmatrix}$

$\begin{bmatrix} 7 & -6 \\ -2 & 4 \\ 4 & 1 \end{bmatrix}$

3. Make a matrix for these data.

U.S. Crop Production (1992–1995) (in bushels)			
Year	Corn	Oats	Barley
1992	9,476,698	294,229	455,090
1993	6,336,470	206,770	389,041
1994	10,102,735	229,008	374,862
1995	7,378,876	161,847	359,102

$\begin{bmatrix} 9,476,698 & 294,229 & 455,090 \\ 6,336,470 & 206,770 & 389,041 \\ 10,102,735 & 229,008 & 374,862 \\ 7,378,876 & 161,847 & 359,102 \end{bmatrix}$

The 5-Minute Check is also available on **Transparency 2-7A** for this lesson.

2-7 Multiplying Integers

What you'll learn
You'll learn to multiply integers.

When am I ever going to use this?
You can multiply integers to find temperature changes over time.

The first occupied hot air balloon carried a duck, a rooster, and a sheep on an eight-minute flight on September 19, 1783. Ballooning has come a long way since then. Now, six teams are attempting to be the first to fly a balloon around the world.

One of the most serious challenges the balloonists must overcome is weather. The temperature drops about 2°F for each rise of 530 feet. The teams expect to travel about 40,000 feet high on an around-the-world trip. About how many degrees difference will there be between the ground temperature and the temperature at 40,000 feet? *This problem will be solved in Example 5.*

This problem can be solved by multiplying integers. Let's investigate how to multiply integers using counters.

HANDS-ON MINI-LAB

Work with a partner. • two colors of counters

Try This ☐ integer mat

Solve $y = 3 \cdot (-2)$.
- Begin with an empty mat.
- The 3 means put 3 sets of counters on the mat. The -2 means that each set contains 2 negative counters.

Solve $z = -2 \cdot (-5)$.
- Begin with an empty mat.
- The -2 means *remove* 2 sets of counters from the mat. The -5 means that each set contains 5 negative counters.
- Since the mat contains no counters, add zero pairs to the mat. Add just enough so that you can remove 2 sets of 5 negative counters.

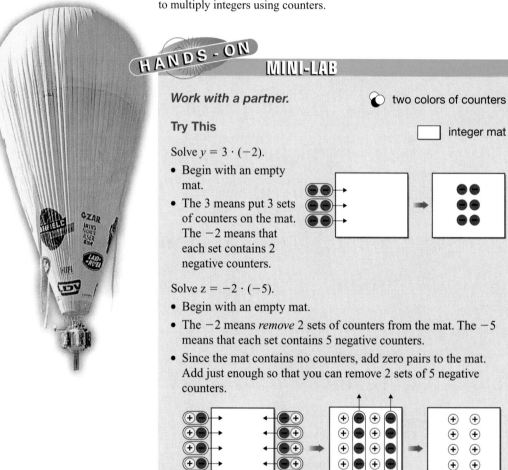

78 Chapter 2 Algebra: Using Integers

Motivating the Lesson

Communication Ask the following questions.
- How can you write 4(3) as a sum of 3s? $3 + 3 + 3 + 3$
- How can you write 4(−3) as a sum of −3s? $-3 + -3 + -3 + -3$
- Add to find the product 4(−3). -12

2. Both factors have the same sign.
3. Factors have different signs.

Talk About It
1. What are the values of y and z? **−6; 10**
2. Describe the factors when the product was positive.
3. Describe the factors when the product was negative.

You can generalize the results of the Mini-Lab as follows.

| Multiplying Integers | The product of two integers with the same sign is positive. The product of two integers with different signs is negative. |

Examples

1 Solve $r = (-3)(-2)$.
The two integers have the same sign. The product will be positive.
$r = (-3)(-2)$
$r = 6$

2 Solve $(-4)(7) = t$.
The two integers have different signs. The product will be negative.
$(-4)(7) = t$
$-28 = t$

3 Solve $s = (-3)(-2)(-5)$.
Group the factors by using the associative property.
$s = [(-3)(-2)](-5)$
$s = 6(-5)$
$s = -30$

4 Solve $a = (-4)^2$.
The exponent says there are two factors of -4.
$a = (-4)(-4)$
$a = 16$

APPLICATION

5 **Sports** Refer to the beginning of the lesson. About how many degrees difference will there be between the ground temperature and the temperature at 40,000 feet?

Explore You know the height and the temperature change per 530 feet. You need to find the change for 40,000 feet.

Plan Write and solve an equation to find the number of 2° temperature changes. Then solve for the temperature change at 40,000 feet.

Solve Let n represent the number of 2° temperature changes.
$n = 40,000 \div 530$
$40000 \; \boxed{\div} \; 530 \; \boxed{=} \; 75.47169811$

The temperature t will drop 2°F about 75 times.
$t \approx 75(-2)$
$t \approx -150$
The difference in temperature is 150°F.

Examine Check the solution by finding the number of 2° decreases it would take to change the temperature by 150°. Then multiply by 530 feet to verify the distance.

Lesson 2-7 Multiplying Integers **79**

■ **Reteaching the Lesson** ■

Activity Give groups of three students bags with three number cubes of one color (positive) and three of another color (negative). One player draws and rolls two cubes, a second states the sign of the product, and a third states the product and records it as a score. The group with the highest total after five rounds wins.

Teaching Tip After working Example 5, point out to students that the check in the "Examine" step does not work out to exactly 40,000 feet because numbers were rounded in the "Solve" step.

2 TEACH

 Transparency 2-7B contains a teaching aid for this lesson.

Using the Mini-Lab The first number in a product tells how many sets of counters to place on the mat or remove. The second number specifies the number and type of counters in each set. For more practice, have students reverse the order of the factors in each example and use counters to model the multiplication.

In-Class Examples

For Example 1
Solve $s = (-2)(-5)$. **10**

For Example 2
Solve $(-5)(7) = p$. **−35**

For Example 3
Solve $r = (-3)(-2)(-4)$. **−24**

For Example 4
Solve $c = (-3)^2$. **9**

For Example 5
Every day Tarek and his friends scuba dive observing sea life at different depths. If they dive 3 feet farther each day for 2 weeks, how deep are they on the last day? **−42 feet**

Study Guide Masters, p. 16

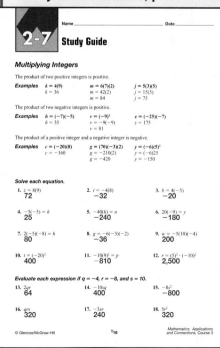

Lesson 2-7 **79**

3 PRACTICE/APPLY

Check for Understanding

If students need additional practice or instruction after completing Exercises 1–11, one of these options may be helpful.
- Extra Practice, see p. 610
- Reteaching Activity, see p. 79
- Study Guide Masters, p. 16
- Practice Masters, p. 16
- Interactive Mathematics Tools Software

Assignment Guide

Core: 13–35 odd, 36–38
Enriched: 12–32 even, 33–38

4 ASSESS

Closing Activity

Speaking Have students complete each statement using the word *positive* or *negative*.
- The product of two positive integers is ? . **positive**
- The product of two negative integers is ? . **positive**
- The product of a positive integer and a negative integer is ? . **negative**

Practice Masters, p. 16

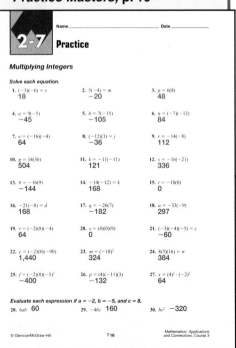

CHECK FOR UNDERSTANDING

Communicating Mathematics

Read and study the lesson to answer each question.   1–3. See Answer Appendix.
1. *Show* two different ways to find the product $(-35)(45)(-2)$.
2. *Write a Problem* that requires the multiplication of 2 and -3. Write about another situation that requires the multiplication of -2 and 3. Are the problems the same? Are the products the same? Explain.

HANDS-ON MATH

Guided Practice

3. *Draw* a model using counters to illustrate $c = (-2)(-4)$. Then solve for c.

Solve each equation. 7. -54

4. $(4)(5) = k$ **20**
5. $(-3)(-2) = y$ **6**
6. $n = 14(-7)$ -98
7. $p = (-2)(-3)(-9)$
8. $q = (4)(-67)(0)$ **0**
9. $(-4)^2 = x$ **16**

10. Evaluate $4xy$ if $x = -5$ and $y = 2$. -40

11. *Life Science* Did you know that most people lose 100 to 200 hairs per day? If you were to lose 120 hairs each day in a week, what is the change in the number of hairs you have? -840

EXERCISES

Practice

Solve each equation. 12. $-8,575$ 26. -156

12. $k = (-35)(245)$
13. $t = 8(6)$ **48**
14. $9(-9) = y$ -81
15. $-8(-3) = m$ **24**
16. $-9(12) = n$ -108
17. $-6(-13) = r$ **78**
18. $55(-11) = s$ -605
19. $-5(-14) = t$ **70**
20. $v = (-8)^2$ **64**
21. $a = 7(-6)(-12)$ **504**
22. $z = 50(-4)(-1)$ **200**
23. $c = 5(23)(7)$ **805**
24. $(-2)(12)(3) = d$ -72
25. $(-21)^2 = f$ **441**
26. $g = (6)(-2)(13)$
27. $h = (9)(-8)(6)^2$ $-2,592$
28. $k = 15(-3)^2$ **135**
29. $m = (5)^2 \cdot (-4)^2$ **400**

Evaluate each expression if $a = -5$, $b = -9$, and $c = 10$.

30. $6cb$ -540
31. $-4ac$ **200**
32. $5abc$ **2,250**

Applications and Problem Solving

33. *Energy* Degree-days help heating companies determine the needs of their customers. The number of degree–days for a day is the temperature minus 65°. If the mean temperature in St. Louis has been 55 degrees for 6 days, what is the total number of degree-days? -60 or 60 heating degree–days

34. *Business* Euro Disney is the number one tourist attraction in France. But in 1993, its revenue was about $-\$930,000,000$! If this continued, what would the revenue at Euro Disney have been after five years? $-\$4,650,000,000$

35. *Critical Thinking* What is the sign of y if $y = x^2$? Explain. See Answer Appendix.

Mixed Review

36. Write two matrices that have no sum. *(Lesson 2-6)* See Answer Appendix.

37. -13

37. *Sports* Jeff Sluman won the 1997 Tucson Classic for golf. His scores for the four rounds were 3, -4, -7, and -5. What was his total score? *(Lesson 2-4)*

38. **Test Practice** Find the area of a parallelogram whose height is 6.5 feet and whose base is 9 feet. *(Lesson 1-8)* **A**

   A 58.5 ft²   **B** 31 ft²   **C** 42.5 ft²   **D** 56.5 ft²

80   Chapter 2 Algebra: Using Integers

Extending the Lesson

Enrichment Masters, p. 16

Activity Find the pair of integers whose sum and product are given.

	Sum	Product	
1.	-7	10	$-2, -5$
2.	11	24	3, 8
3.	-2	-24	4, -6
4.	8	-9	9, -1
5.	-5	-14	$-7, 2$

2-8 Dividing Integers

What you'll learn
You'll learn to divide integers.

When am I ever going to use this?
You can divide integers to find the average amount of decrease in prices over time.

Are the tigers among your favorite animals to visit at the zoo? The natural habitat of tigers is the steamy hot jungles or the icy forests of Asia. Experts estimate that there may have been 100,000 tigers living 100 years ago. Now there are only about 6,000. What was the average change in tiger population each of the last 100 years? *This problem will be solved in Example 3.*

You can solve this problem by dividing integers. Since division is related to multiplication, it uses the same rules of signs.

| **Dividing Integers** | The quotient of two integers with the same sign is positive. The quotient of two integers with different signs is negative. |

Examples

1 Solve $r = -56 \div 8$.
   $r = -56 \div 8$ *The signs are different.*
   $r = -7$ *The quotient is negative.*

2 Solve $(-42) \div (-7) = t$.
   $(-42) \div (-7) = t$ *The signs are the same.*
   $6 = t$ *The quotient is positive.*

CONNECTION

3 **Life Science** Refer to the beginning of the lesson. Find the average change in tiger population each of the last 100 years.

The change in tiger population can be expressed as $6{,}000 - 100{,}000$ or $-94{,}000$. The change occurred over 100 years. The average change can be found by dividing $-94{,}000$ by 100. Let c represent the average change.

   $c = -94{,}000 \div 100$ *The signs are different.*
   $c = -940$ *The quotient is negative.*

The average change in population was -940 tigers per year. This means that each year there were about 940 fewer tigers in the world than the year before.

Remember that fractions are also a way of showing division. Another way to show $c = a \div b$ is $c = \frac{a}{b}$.

Lesson 2-8 Dividing Integers **81**

In-Class Examples

For Example 1
Solve $x = -48 \div 6$. **−8**

For Example 2
Solve $(-36) \div (-6) = w$. **6**

For Example 3
After the stock market crash of 1929, the percent of employed people fell from 97% in 1929 to 76% in 1932. What was the average change of employment per year? **−7%**

For Example 4
Solve $a = \dfrac{108}{-6}$. **−18**

3 PRACTICE/APPLY

Check for Understanding
If students need additional practice or instruction after completing Exercises 1–13, one of these options may be helpful.
- Extra Practice, see p. 610
- Reteaching Activity
- *Study Guide Masters*, p. 17
- *Practice Masters*, p. 17

Assignment Guide

Core: 15–43 odd, 45–50
Enriched: 14–40 even, 41–43, 45–50

Study Guide Masters, p. 17

Example 4
Solve $r = \dfrac{96}{-12}$.

$r = \dfrac{96}{-12}$ *The signs are different.*
$r = -8$ *The quotient is negative.*

CHECK FOR UNDERSTANDING

Communicating Mathematics

Read and study the lesson to answer each question.

1. *Tell* whether the dividend is positive or negative if the divisor is positive and the quotient is negative. **negative**

2. Multiply the quotient and the divisor. If the result is the dividend, then the answer checks.

2. *Explain* how to check a solution of an equation solved by dividing integers.

3. *You Decide* Lesharo says that the next division sentence in the pattern is $-15 \div 3 = -5$. Rachel says the next division sentence is $-15 \div 5 = -3$. Who is correct and why? **Rachel is correct. The divisor is 5 in all of the problems.**

$5 \div 5 = 1$
$0 \div 5 = 0$
$-5 \div 5 = -1$
$-10 \div 5 = -2$

Guided Practice Solve each equation. 6. **−4**

4. $-12 \div (-4) = m$ **3**
5. $320 \div (-8) = n$ **−40**
6. $p = -240 \div 60$
7. $b = \dfrac{-56}{8}$ **−7**
8. $s = 90 \div 10$ **9**
9. $\dfrac{365}{-5} = v$ **−73**

Evaluate each expression if $x = 3$, $y = -10$, and $z = 9$.

10. $\dfrac{800}{y}$ **−80**
11. $z \div (-3)$ **−3**
12. $5yz \div x$ **−150**

13. *Energy* Many homes built before 1960 were heated by coal. In 1950, about 115 million tons of coal were used for fuel each year. By the year 2000, it is estimated that coal use will be only about 5 million tons each year. What was the average change in coal use per year? **about 2 million tons**

EXERCISES

Practice Solve each equation. 22. **31** 25. **−6** 31. **−98**

14. $q = \dfrac{400}{-10}$ **−40**
15. $t = -26 \div (-13)$ **2**
16. $295 \div 5 = h$ **59**
17. $-63 \div (-9) = w$ **7**
18. $c = \dfrac{-88}{44}$ **−2**
19. $z = 49 \div (-7)$ **−7**
20. $112 \div (-4) = f$ **−28**
21. $-56 \div (-2) = r$ **28**
22. $x = -930 \div (-30)$
23. $g = \dfrac{245}{-5}$ **−49**
24. $62 \div 2 = u$ **31**
25. $90 \div (-15) = h$
26. $p = -76 \div (-4)$ **19**
27. $s = 216 \div (-18)$ **−12**
28. $\dfrac{224}{-32} = b$ **−7**
29. $-143 \div 11 = d$ **−13**
30. $f = -195 \div 65$ **−3**
31. $k = 588 \div (-6)$

82 Chapter 2 Algebra: Using Integers

■ Reteaching the Lesson ■

Activity Help students see the relationship between division and multiplication by having them solve several division problems following the format shown.
$24 \div (-6) = \blacksquare \rightarrow \blacksquare \cdot (-6) = 24$
The boxes must contain the same integer.

Evaluate each expression if $a = -8$, $b = 4$, and $c = -16$.

32. $\frac{72}{a}$ -9
33. $\frac{-32}{b}$ -8
34. $-48 \div c$ 3
35. $52 \div b$ 13
36. $c \div (-2)$ 8
37. $-424 \div a$ 53
38. $ac \div (-32)$ -4
39. $bc \div a$ 8
40. $(abc)^2 \div 64$ $4,096$

Applications and Problem Solving

41. *Employment* There were 7,404,000 unemployed American workers in 1995. In 1980, there were 7,637,000 Americans unemployed. a. about $-15,533$
    a. What was the average change in unemployment for each of these 15 years?
    b. If this decline continues at the same rate, how many Americans will be unemployed in the year 2055? about 6,472,000

42. *Agriculture* Texas is the number one agricultural state in the United States. The graph shows the number of farms and the average size of farms in Texas in 1987 and 1992.
    a. Find the average annual change in number of farms of 50 to 179 acres.
    b. There were 35,610,951 acres of land farmed in Texas in 1987. In 1992, 36,381,847 acres were farmed. Explain how this is related to the graph. See margin.

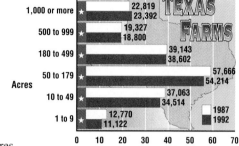

42a. -690.4

43–44. See students' work.

45a. Division is not commutative. For example, $9 \div 3 \neq 3 \div 9$.

43. *Write a Problem* about the photograph at the left that requires division of integers in the solution.

44. *Working on the* CHAPTER Project
    Research your species. Then use the procedure in Example 3 on page 81 to find an average change in population.

45. *Critical Thinking* Use examples to support your answers.
    a. The commutative property of multiplication is $a \cdot b = b \cdot a$. Is division commutative?
    b. The associative property of multiplication is $a(b \cdot c) = (a \cdot b) \cdot c$. Is division associative?
    Division is not associative. For example, $(16 \div 4) \div 2 \neq 16 \div (4 \div 2)$.

Mixed Review

46. Solve $t = -4(12)$. *(Lesson 2-7)* -48
47. Solve $h = -28 + 15 + 6 + (-30)$. Check your solution. *(Lesson 2-4)* -37
48. Graph $\{-3, -1, 0, 2\}$ on a number line. *(Lesson 2-1)* See Answer Appendix.
49. **Test Practice** ReadyRent charges $24 to rent a power saw for 4 hours. The rent is $4.50 for each hour after the first 4. Which sentence could be used to find c, the cost for keeping the saw 10 hours? *(Lesson 1-6)* **B**
    A $c = \$24 + 10(\$4.50)$ B $c = \$24 + 6(\$4.50)$ C $c = 10(\$4.50)$
    D $\$24 + c = 6(\$4.50)$ E $\$4.50 + 6(\$24) = c$
50. *Algebra* Evaluate $6a^2 + b$ if $a = 4$ and $b = 1$. *(Lesson 1-3)* 97

Lesson 2-8 Dividing Integers **83**

Enrichment Masters, p. 17

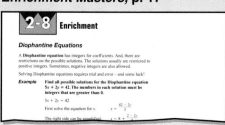

Activity Have small groups solve this problem: "Yesterday was very cold," said the meteorologist. "If you multiply the low temperature in degrees Fahrenheit by -4, add -4, divide by -4, and subtract -4, the result is -4." What was the temperature? $-9°F$

CHAPTER Project

Exercise 44 asks students to advance to the next stage of work on the Chapter Project. You may wish to have students work together to calculate the average change in each animal population studied by the students.

4 ASSESS

Closing Activity

Writing Refer to the Closing Activity on page 80 and change *product* to *quotient*. Have students compare their responses for each activity.

Chapter 2, Quiz C (Lessons 2-6 through 2-8) is available in the *Assessment and Evaluation Masters*, p. 44.

Additional Answer

42b. Sample answer: Some smaller farms were bought and combined to form larger ones. Also, new acreage was acquired to make some smaller farms larger.

Practice Masters, p. 17

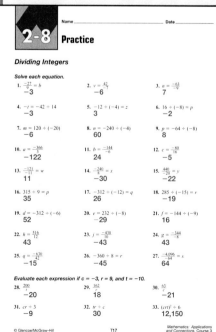

Lesson 2-8 **83**

2-9A LAB Notes

GET READY

Objective Students solve equations by using models.

Optional Resources
Hands-On Lab Masters
- integer counters, p. 6
- pattern for a cup, p. 7
- equation mat, p. 9
- worksheet, p. 41

Overhead Manipulative Resources
- counters
- cup
- equation mat

Manipulative Kit
- counters
- equation mat
- cups

MANAGEMENT TIPS

Recommended Time
30 minutes

Getting Started Before you start, have students refer back to the Mini-Labs on pages 69 and 78. Review the concept of zero pairs of counters and how a zero pair can be added or subtracted from the mat. Also review the definition of an additive inverse.

In **Activity 1,** students will be able to visualize why a negative number on one side of an equals sign becomes a positive number on the other.

HANDS-ON LAB

COOPERATIVE LEARNING

2-9A Solving Equations

A Preview of Lesson 2-9

 two colors of counters

 cups

equation mat

You have used counters to model operations with integers. You can also use counters to solve equations that involve integers.

TRY THIS

Work with a partner.

1. Solve $x + (-2) = 7$.
   - Start with an empty mat.
   - Let a cup represent the unknown x value. Put the cup and 2 negative counters on one side of the mat. Place 7 positive counters on the other side of the mat.
   - The goal is to get the cup by itself on one side of the mat. Then the counters on the other side will be the value of the cup, or x.

   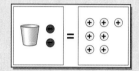

   - Add 2 positive counters to each side of the mat to eliminate the 2 negative counters on the side with the cup.

   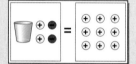

   - Group the counters to form zero pairs. Then remove all of the zero pairs.

   - Now the cup is by itself on one side of the mat. The counters are on the other side.

ON YOUR OWN

1. What is the solution of $x + (-2) = 7$? **9**
2. How do you know what type of counter to add to each side of the mat in order to be able to get the cup by itself? **Add the opposite type of counter than the type that is already there.**

84 Chapter 2 Algebra: Using Integers

TRY THIS

Work with a partner.

② Solve $2x - (-3) = 7$.

Before using the counters, rewrite the expression using the additive inverse.
$2x - (-3) = 7$ becomes $2x + 3 = 7$.

- Start with an empty mat.
- Place 2 cups on one side of the mat to represent $2x$. Add 3 positive counters on that side of the mat. Place 7 positive counters on the other side of the mat.

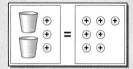

- Add 3 negative counters to each side of the mat.

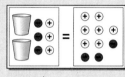

- Remove all of the zero pairs that can be formed.

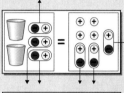

- Arrange the remaining counters on the right side of the mat into 2 equal groups so that they correspond to the 2 cups.

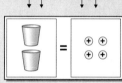

ON YOUR OWN

3. How many counters correspond to each cup? **2**
4. What is the solution of $2x - (-3) = 7$? **2**

Write an equation for each model.

5.  $x - 5 = 8$

6. 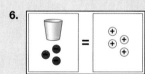 $x - 3 = 4$

Solve each equation by using models.

7. $x + 2 = 8$ **6**
8. $x - 4 = 6$ **10**
9. $x + (-3) = 2$ **5**
10. $x - (-2) = 4$ **2**
11. $x + 6 = -2$ **-8**
12. $x + (-4) = 6$ **10**
13. $4x = 8$ **2**
14. $3x = -15$ **-5**
15. $2x + (-1) = 11$ **6**

Lesson 2-9A HANDS-ON **LAB** 85

2-9 Lesson Notes

Instructional Resources
- *Study Guide Masters*, p. 18
- *Practice Masters*, p. 18
- *Enrichment Masters*, p. 18
- Transparencies 2-9, A and B
- *Technology Masters*, pp. 55, 56
- CD-ROM Program
  - Resource Lesson 2-9
  - Extended Activity 2-8

Recommended Pacing	
Standard	Days 10 & 11 of 15
Honors	Day 10 of 14
Block	Day 6 of 8

1 FOCUS

5-Minute Check
(Lesson 2-8)

1. Solve $p = -64 \div 8$. **-8**
2. Solve $(-27) \div (-3) = t$. **9**
3. Solve $116 \div (-4) = d$. **-29**
4. Evaluate $\frac{96}{a}$ and $\frac{160}{a}$ if $a = -8$. **$-12, -20$**
5. The Wester Savings and Loan invested $3 million. After 7 weeks, the investment value dropped to $2,989,080. About how much did Wester lose on average each week? **$1,560**

The 5-Minute Check is also available on **Transparency 2-9A** for this lesson.

Motivating the Lesson

Hands-On Activity Have students use counters to solve $x + 4 = 8$. Students should write the equation on a blank sheet of paper, lay out a cup and counters appropriately beneath the equation, and then solve the equation by removing the correct number of counters to isolate the x on one side of the equals sign.

2-9 Solving Equations

What you'll learn
You'll learn to solve equations with integer solutions.

When am I ever going to use this?
You can solve an equation to determine how to place the seeds in a garden for best growth.

In the 1800s, forests in the United States were cleared to provide firewood and lumber for housing and to make way for farms. Now that the country is settled, trees are being replanted. Other countries in the growth stage of development are clearing forests. In Brazil, an area just 802 square miles less than the state of Washington was cleared in five years!

You can use an equation to find the average amount of land cleared each year in Brazil. Equations involving integers are solved in the same way that you solved equations involving whole numbers. *This problem will be solved in Example 5.*

Examples

1 Solve $r + 5 = -10$.

$$r + 5 = -10$$
$$r + 5 - 5 = -10 - 5 \quad \text{Subtract 5 from each side.}$$
$$r = -15$$

Check: $\quad r + 5 = -10$
$\quad\quad -15 + 5 \stackrel{?}{=} -10 \quad \text{Replace } r \text{ with } -15.$
$\quad\quad -10 = -10 \checkmark$

2 Solve $p - (-3) = -6$.

$$p - (-3) = -6$$
$$p + 3 = -6 \quad \text{Rewrite using the additive inverse.}$$
$$p + 3 - 3 = -6 - 3 \quad \text{Subtract 3 from each side.}$$
$$p = -9$$

Check: $\quad p - (-3) = -6$
$\quad\quad -9 - (-3) \stackrel{?}{=} -6 \quad \text{Replace } p \text{ with } -9.$
$\quad\quad -6 = -6 \checkmark$

3 Solve $2t = -98$.

$$2t = -98$$
$$\frac{2t}{2} = \frac{-98}{2} \quad \text{Divide to undo multiplication.}$$
$$t = -49$$

Check: $\quad 2t = -98$
$\quad\quad 2(-49) \stackrel{?}{=} -98 \quad \text{Replace } t \text{ with } -49.$
$\quad\quad -98 = -98 \checkmark$

86 Chapter 2 Algebra: Using Integers

Some equations involving integers require two steps to solve. You can work backward to solve for the variable.

Examples

4 Solve $-5x + 8 = -7$.

$$-5x + 8 = -7$$
$$-5x + 8 - 8 = -7 - 8 \quad \textit{Subtract 8 from each side.}$$
$$-5x = -15$$
$$\frac{-5x}{-5} = \frac{-15}{-5} \quad \textit{Divide each side by } -5.$$
$$x = 3$$

Check: $-5x + 8 = -7$
$$-5(3) + 8 \stackrel{?}{=} -7 \quad \textit{Replace x with 3.}$$
$$-15 + 8 \stackrel{?}{=} -7$$
$$-7 = -7 \checkmark$$

CONNECTION

5 **Geography** Refer to the beginning of the lesson. If the area of the state of Washington is 71,302 square miles, what was the average amount of land cleared in Brazil each year?

Explore You know the area of Washington and that the area cleared in Brazil was 802 square miles less than the area of Washington. You need to find the average area cleared each year.

Plan Let a = the amount of land cleared each year. Therefore, the amount cleared in five years was $5a$. Write an equation to solve the problem.

land cleared in five years plus 802 sq mi is area of Washington
$$5a \qquad + \quad 802 \quad = \quad 71{,}302$$

Solve $5a + 802 = 71{,}302$
$$5a + 802 - 802 = 71{,}302 - 802 \quad \textit{Subtract 802 from each side.}$$
$$5a = 70{,}500$$
$$\frac{5a}{5} = \frac{70{,}500}{5} \quad \textit{Divide each side by 5.}$$
$$a = 14{,}100$$

There was an average of 14,100 square miles of forest cleared each year.

Examine Check the solution against the words of the problem. If 14,100 square miles were cleared each year, then $5 \cdot 14{,}100$ or 70,500 square miles were cleared. This is $71{,}302 - 70{,}500$ or 802 square miles less than the size of Washington.

Lesson 2-9 Solving Equations **87**

2 TEACH

 Transparency 2-9B contains a teaching aid for this lesson.

Reading Mathematics The word *equation* is defined algebraically as "a mathematical sentence that contains an equals sign." The word *equation* comes from the Latin word meaning "even."

In-Class Examples

For Example 1
Solve $s + 6 = -11$. -17

For Example 2
Solve $x - (-9) = -9$. -18

For Example 3
Solve $4m = -104$. -26

For Example 4
Solve $-3x + 11 = -10$. 7

For Example 5
Write and solve an equation for this problem: Isaac borrowed $63 from his parents to buy a video game. Then he borrowed money twice to go to the movies. He now owes his parents $87. If he borrowed an equal amount each time he went to the movies, how much did he borrow each time?
$-63 + (-2m) = -87$; $m = 12$

Teaching Tip In Example 4, emphasize that any addition and subtraction must be done before division.

Investigations for the Special Education Student

This blackline master booklet helps you plan for the needs of your special education students by providing long-term projects along with teacher notes. Investigation 4, *The Check's in the Mail*, may be used with this chapter.

3 PRACTICE/APPLY

Check for Understanding

If students need additional practice or instruction after completing Exercises 1–10, one of these options may be helpful.
- Extra Practice, see p. 611
- Reteaching Activity
- *Study Guide Masters*, p. 18
- *Practice Masters*, p. 18

Assignment Guide

Core: 11–33 odd, 34–38
Enriched: 12–30 even, 32–38

Additional Answers

1. For $3r + 5 = -10$: Subtract 5 from each side ($3r + 5 - 5 = -10 - 5$); simplify ($3r = -15$); divide each side by 3 $\left(\frac{3r}{3} = \frac{-15}{3}\right)$; simplify ($r = -5$). For $3r - (-5) = -10$, rewrite using the additive inverse ($3r + 5 = -10$). Then follow the same steps as above.

3. Sample answer: Your mistake could be in the solution of the equation or could be in your check. Check both.

Study Guide Masters, p. 18

CHECK FOR UNDERSTANDING

Communicating Mathematics

Read and study the lesson to answer each question. 1. See margin.

1. **Compare** the steps used to solve $3r + 5 = -10$ and $3r - (-5) = -10$.
2. **Write** a two-step equation that has a solution of 0. **Sample answer:** $3t + 5 = 5$
3. **Write** a sentence to explain why it is a good idea to check your work. What should you do if your check does not agree? **See margin.**

Guided Practice

Solve each equation. Check your solution. 7. −275 9. −20

4. $3b = -36$ **−12**
5. $v + 35 = 32$ **−3**
6. $\frac{x}{-14} = 32$ **−448**
7. $x - (-35) = -240$
8. $8c - 12 = 36$ **6**
9. $75 = -3y + 15$

10. The sum of two integers is −24. One of the integers is −13. Write an equation to find the other integer. Then find the integer. $x + (-13) = -24$; **−11**

EXERCISES

Practice

Solve each equation. Check your solution. 15. −318 26. −6

11. $x - 13 = -22$ **−9**
12. $2y = -90$ **−45**
13. $-30 = 42 + k$ **−72**
14. $\frac{y}{15} = 22$ **330**
15. $w - (-350) = 32$
16. $-4,968 = -69n$ **72**
17. $y - 13 = 45$ **58**
18. $-200 = \frac{r}{3}$ **−600**
19. $\frac{x}{-7} = -5$ **35**
20. $n + 34 = 16$ **−18**
21. $4c = -60$ **−15**
22. $v - 16 = -64$ **−48**
23. $300 = 120 + 5h$ **36**
24. $-15 + 4m = 45$ **15**
25. $6b - (-3) = 105$ **17**
26. $4x - (-7) = -17$
27. $\frac{x}{6} - 5 = -13$ **−48**
28. $12 + \frac{n}{4} = 0$ **−48**

Write an equation for each problem and solve.

29. The difference of two integers is 8. The greater integer is −2. What is the other integer? $-2 - x = 8$; **−10**
30. Two integers have a product of 35. One of the integers is −7. Find the other integer. $-7x = 35$; **−5**
31. Two times a number plus 7 is −21. What is the number? $2x + 7 = -21$; **−14**

Applications and Problem Solving

32. **Health** Would you like to brush your teeth with hogs hair? Toothbrushes were invented in China in the late 1400s and were made of hogs hair! The first nylon toothbrush was made in 1938, 23 years before the first electric toothbrush. Write an equation for the year that the electric toothbrush was invented. Then find the year. $x - 23 = 1938$; **1961**

33. **Aviation** Wilbur Wright made the first flight in Kitty Hawk, North Carolina, on December 17, 1903. His brother Orville made the second flight that same day. Orville flew 4 yards more than 6 times as far as Wilbur. If Orville's flight was 244 yards, how far did Wilbur fly? **40 yd**

34. **Critical Thinking** Solve $6x + 7 = 8x - 13$. **10**

Reteaching the Lesson

Activity Have pairs of students solve $2x + 3 = 11$ as a means of reviewing the steps involved in solving an equation. They should write out each step to solve the equation.

$2x + 3 = 11$
$2x + 3 - 3 = 11 - 3$ Subtract 3 from each side.
$2x = 8$ Simplify.
$\left(\frac{2x}{2}\right) = \left(\frac{8}{2}\right)$ Divide each side by 2.
$x = 4$ Simplify.

Mixed Review

35. Solve $-325 \div 25 = q$. *(Lesson 2-8)* **−13**

36. Order the integers $\{-219, -52, 18, 3, -24, 120, -186\}$ from greatest to least. *(Lesson 2-2)* **{120, 18, 3, −24, −52, −186, −219}**

37. **Algebra** Solve $5 = \frac{k}{2} + 3$. Check your solution. *(Lesson 1-7)* **4**

38. **Test Practice** Rey has $12.64 in his wallet. He wants to buy a soccer ball that is $17.95. How much more will Rey need to save? *(Lesson 1-1)* **D**

    A $30.59 **B** $20.59 **C** $15.31 **D** $5.31 **E** Not Here

MATH IN THE MEDIA

Broadway Bound?

The financial comparison shown below appeared in the *New York Times* on November 21, 1994.

Basic Economics: Calculating Against Theatrical Disaster

Neil Simon plans to open his newest play, "London Suite," Off Broadway. His producer Emanuel Azenberg, provided this financial comparison.

MONEY NEEDED TO OPEN		BROADWAY 1,000 seats $55 top ticket price	OFF BROADWAY 500 seats $40 top ticket price
What the producer spends	Sets, costumes, lights	$357,000	$87,000
	Loading in (building set, etc.)	175,000	8,000
	Rehearsal salaries	102,000	63,000
	Director and designer fees	126,000	61,000
	Advertising	300,000	121,000
	Administration	235,000	100,000
		$1,295,000	**$440,000**
THE WEEKLY BUDGET			
Revenues	Weekly receipts at 75 percent capacity	$250,000	$109,000
Minus expenses	Theater rent, house crew	45,000	12,000
	Salaries	54,000	20,000
	Advertising	30,000	15,000
	Lights and sound rental	6,000	3,500
	Administration	32,000	11,000
	Royalties	23,500	14,000
	Extra rent, salaries based on ticket sales	16,000	6,500
		−206,500	−82,000
Equals weekly profit		**= 43,500**	**= 27,000**

1. Broadway: 29.7 weeks; Off Broadway: 16.3 weeks
1. Use an equation to find the number of performances that it would take to make up the money needed to open on Broadway and Off Broadway.

2. Do you agree with Neil Simon's decision to produce the play Off Broadway? Explain why or why not. **See students' work.**

Lesson 2-9 Solving Equations **89**

Extending the Lesson

Enrichment Masters, p. 18

MATH IN THE MEDIA

Have students find out how much it costs to stage a play in their school or at the community theater.

4 ASSESS

Closing Activity

Modeling Have students solve the equation $3x + (-2) = 7$ using counters and then solve it using algebra. In a few sentences, have them compare and contrast the two methods.

Practice Masters, p. 18

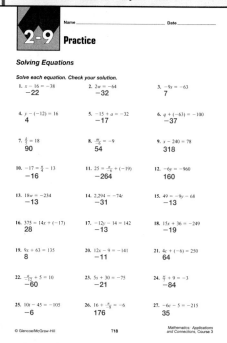

Lesson 2-9 **89**

2-9B THINKING LAB Notes

Objective Students solve problems by eliminating possibilities.

Recommended Pacing	
Standard	Day 12 of 15
Honors	Day 11 of 14
Block	Day 7 of 8

1 FOCUS

Getting Started With a partner, ask students to act out the opening dialogue. Students should follow through with creating a table and crossing out busy slots. Partners should work through Exercises 1–4. Then discuss how the table could have been different if Crystal worked Saturday morning but not Saturday afternoons. **Because the girls would have more free time on the weekend, Saturday and Sunday could both be divided into morning, afternoon, and evening columns. This way, if Crystal works Saturday mornings, she would still be open to play in the afternoon or evening.**

2 TEACH

Teaching Tip When assigning Exercise 3, elicit suggestions for real-life examples to stimulate student thinking. **Possible ideas: Eliminate jobs that conflict with school times, recipes with unavailable ingredients, or gifts that exceed the money limit.**

In-Class Example
What are the most coins, excluding pennies, that you can use to make 45¢? **9 nickels** the fewest coins? **1 quarter, 2 dimes** the most coins for 95¢? **19 nickels** the fewest? **1 half dollar, 1 quarter, 2 dimes**

PROBLEM SOLVING
2-9B Eliminate Possibilities
A Follow-Up of Lesson 2-9

Crystal and Jackie are choosing a time to meet their friends Opa and Ellen for doubles tennis. Let's listen in!

Jackie: I have to work Monday nights and Saturday mornings, so that's out.

Crystal: Ok. And I volunteer at the library on Wednesdays and Thursdays after school.

Jackie: I know that Ellen baby-sits for her neighbors after school on Monday, Wednesday, and Friday. And Sundays are out for her too.

Crystal: This isn't leaving many options is it? I think Opa said that she works on Saturdays, too.

Jackie: Let's organize all of this. I'll use a table with the days on the top and each of our names on the side. Then we can mark out the times we can't play and see what's left.

	Mon.	Tues.	Wed.	Thurs.	Fri.	Sat.	Sun.
Crystal	X					X	
Jackie			X	X			
Ellen	X		X		X		X
Opa						X	

Crystal: Looks like Tuesday is the day! Let's go practice so we're ready!

THINK ABOUT IT

Work with a partner.

1. *Tell* how Crystal and Jackie eliminated possibilities to solve the problem.
2. *Eliminate possibilities* to solve the multiple-choice problem. **C**
   $16,340 \div 19 =$
   A 80 B 86
   C 860 D 8,600
3. *Describe* a real-life situation in which you could eliminate possibilities to solve a problem.
4. *Tell* what day the girls could play tennis if Ellen's baby-sitting job changed to Tuesdays, Thursdays, and Saturdays. **Fridays**

1. Crystal and Jackie eliminated the days that they could not play to see which day or days they could.
3. See students' work.

90 Chapter 2 Algebra: Using Integers

■ Reteaching the Lesson ■

Activity Write 45.6, 85.9, 63.1, 105.7, and 74.7 on the chalkboard. Ask students whether 49, 75, or 97 is the average of these numbers. Students should not calculate but instead eliminate possibilities. Ask students for their reasoning in selecting the average. **Sample answer: Eliminate 49 because it is only slightly greater than the least number in the list.**

ON YOUR OWN

5. The second step of the 4-step plan for problem solving is to *plan* the solution. *Tell* what is involved in planning to solve a problem by eliminating possibilities. **See margin.**

6. *Write a Problem* that could be solved by eliminating possibilities.

7. *Reflect Back* Explain how you could use eliminating possibilities to solve Exercise 38 on page 89. **6–7. See students' work.**

MIXED PROBLEM SOLVING

Strategies
Look for a pattern.
Solve a simpler problem.
Act it out.
Guess and check.
Draw a diagram.
Make a chart.
Work backward.

Solve. Use any strategy.

8. **Technology** The price of calculators has been decreasing. A 4-function calculator sold for $12.50 in 1980. A similar calculator sold for $8.90 in 1990. If the price decrease continues at the same rate, what would be the price in 2010? **$1.70**

9. Trey and Matias are going to take a bus to Baltimore for the day. The buses run every hour on the hour from 7:00 A.M. to 11:00 P.M. It will take one hour to get to Baltimore. They don't want to leave home before 7:30 A.M. and they want to return before 10:00 P.M. While they are in Baltimore, they want to spend $2\frac{1}{2}$ hours at the Aquarium and at least 2 hours at the Science Center. If they have time, they would like to take the 30-minute submarine tour.

   a. If they take an hour to eat lunch at the Inner Harbor, what times might they take the bus to and from Baltimore? **See margin.**

   b. Will they have time for the tour of the submarine? **yes**

10. **Employment** Three after-school jobs are posted on the job board. The first job pays $5.15 per hour for 15 hours of work each week. The second job pays $10.95 per day for 2 hours of work, 5 days each week. The third job pays $82.50 for 15 hours of work each week. If you want to apply for the best-paying job, which job should you choose? **the $82.50 per 15-hour week job**

11. **Test Practice** A human heart beats an average of 72 times in one minute. Estimate the number of times a human heart beats in one year. **A**

    A 37,800,000   B 378,000
    C 37,800   D 3,780

12. **Test Practice** The table below shows the results of a survey of popular cookie flavors. How many flavors were chosen by fewer than 10 students? **A**

    A 3
    B 4
    C 5
    D 6

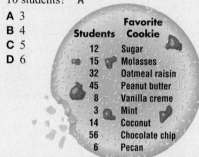

Students	Favorite Cookie
12	Sugar
15	Molasses
32	Oatmeal raisin
45	Peanut butter
8	Vanilla creme
3	Mint
14	Coconut
56	Chocolate chip
6	Pecan

13. **Sports** What non-motor sport do you think is the fastest? It's snow skiing. The record was set at 142.165 mph in 1992. The second fastest record is for speed skating. It was set at 30.68 mph in 1988. How much faster is the skiing record? **111.485 mph**

14. **Test Practice** The Meadowlands Arena seats 20,039 people. If each ticket for an event sells for $12.75, how much should the receipts for a sell-out be? **B**

    A $2,554,972   B $255,497.25
    C $25,549.72   D $255,497.20

Lesson 2-9B THINKING LAB **91**

Extending the Lesson

Activity Have groups solve this problem by eliminating possibilities. They may need outside resources. *Ada, Bob, Cal, and Dot live in Austin, Boise, Chicago, and Phoenix, though not in that order. Have students use clues to decide who lives in each city: Ada lives the farthest north; Bob's city is not in Texas or Idaho; Cal's city is not a state capital; Dot's state borders Mexico.* **Ada, Boise; Bob, Phoenix; Cal, Chicago; Dot, Austin.** Have students make up similar problems for other groups to solve.

---

3 PRACTICE/APPLY

Check for Understanding
Use Exercise 2 to show students how thinking through problems in a logical way can help eliminate possible answers in a multiple-choice test.

Extra Practice If students need additional practice in problem solving, extra practice is available on the following pages.
- Eliminate Possibilities, see p. 611
- Mixed Problem Solving, see pp. 645–646

Assignment Guide
All: 5–14

4 ASSESS

Closing Activity
Writing Tell students that they have been given three possible answers to a math problem, one of which is correct. Have them write a few sentences describing how they could decide which answer is correct without actually solving the problem.

Additional Answers

5. To plan the solution, you need to identify the possibilities for the solution. Then when you solve, you can eliminate the ones that are not the solution.

9a. They could take the following buses to Baltimore and home: 8:00 A.M. and 3:00 P.M.; 9:00 A.M. and 4:00 P.M.; 10:00 A.M. and 5:00 P.M.; 11:00 A.M. and 6:00 P.M.; 12:00 P.M. and 7:00 P.M.; 1:00 P.M. and 8:00 P.M.

Thinking Lab 2-9B **91**

2-10 Lesson Notes

Instructional Resources
- *Study Guide Masters*, p. 19
- *Practice Masters*, p. 19
- *Enrichment Masters*, p. 19
- Transparencies 2-10, A and B
- *Assessment and Evaluation Masters*, p. 44
- *Hands-On Lab Masters*, p. 69
- CD-ROM Program
  - Resource Lesson 2-10
  - Extended Activity 2-10

Recommended Pacing	
Standard	Day 13 of 15
Honors	Day 12 of 14
Block	Day 7 of 8

1 FOCUS

5-Minute Check
(Lesson 2-9)

Solve each equation. Check your solution.
1. $x - 11 = -37$ -26
2. $\frac{y}{21} = -7$ -147
3. $x - 7 = -119$ -112
4. $2b - (-8) = 102$ 47
5. The difference between two integers is 15. The greater integer is -3. Find the other integer using an equation. $-3 - x = 15, x = -18$

 The 5-Minute Check is also available on **Transparency 2-10A** for this lesson.

2 TEACH

 Transparency 2-10B contains a teaching aid for this lesson.

Thinking Algebraically Compare graphing a point on a number line and graphing a point on a coordinate system. A point graphed on a number line has one coordinate and the point lies on the number line. **A point graphed on a coordinate system has two coordinates, and the point may or may not lie on one or both of the axes.**

92 Chapter 2

2-10

Integration: Geometry
The Coordinate System

What you'll learn
You'll learn to graph points on the coordinate plane.

When am I ever going to use this?
You can graph points to look for trends in data collected in science class.

Word Wise
coordinate system
origin
x-axis
y-axis
quadrant
ordered pair
x-coordinate
y-coordinate

Most of us have used a globe to locate places in the world. The first globe was made in about 300 B.C., by the Greek scientist Dicaearchus. Since then, globes have changed quite a bit. New continents have been discovered, and the longitude and latitude system has been perfected.

When reading a globe, you can use the longitude and latitude lines to pinpoint a location on Earth. In mathematics, you can locate a point precisely by using a **coordinate system** similar to the longitude and latitude system used on a globe. The coordinate system is formed by two number lines that intersect at their zero points. This intersection point is called the **origin**. The horizontal number line is called the **x-axis**, and the vertical number line is the **y-axis**. The two axes separate the coordinate plane into four sections called **quadrants**.

Any point on the coordinate plane can be graphed by using an **ordered pair** of numbers. The first number in an ordered pair is called the **x-coordinate**. The second number is the **y-coordinate**. The coordinates are your directions to find the point.

Example 1 Graph the point whose coordinates are (4, −3).

Start at the origin. The *x*-coordinate is 4. This tells you to go 4 units to the right of the origin.

The *y*-coordinate is −3. It tells you to go down 3 units.

Draw a dot. The dot is the graph of the point whose coordinates are (4, −3).

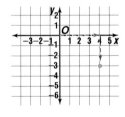

Letters are often used to name points. The symbol B(−3, 2) means point B has an *x*-coordinate of −3 and a *y*-coordinate of 2.

92 Chapter 2 Algebra: Using Integers

Motivating the Lesson
Hands-On Activity Pass out state or city maps to groups of students. Give pairs of coordinates and have students use their maps to determine what is located at each spot. Then name several places on the map, and have students find the coordinates that correspond to each place.

Example 2 Name the ordered pair for point *D*.

Move left on the *x*-axis to find the *x*-coordinate of point *D*. The *x*-coordinate is -1.

Move up along the *y*-axis to find the *y*-coordinate. The *y*-coordinate is 5.

The ordered pair for point *D* is $(-1, 5)$.

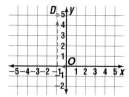

You can use ordered pairs to show how data are related.

Example 3 INTEGRATION

Algebra Evaluate the expression $3x - 2$ for $x = -2, 0, 1,$ and 3. Graph the ordered pairs formed by *x* and the corresponding value of the expression for *x*.

Use a table to evaluate the expression.

x	3x − 2	(x, y)
−2	3(−2) − 2 = −6 − 2 or −8	(−2, −8)
0	3(0) − 2 = 0 − 2 or −2	(0, −2)
1	3(1) − 2 = 3 − 2 or 1	(1, 1)
3	3(3) − 2 = 9 − 2 or 7	(3, 7)

Now graph each ordered pair.

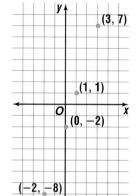

Study Hint

Reading Math In this book, when no numbers are given on the *x*- and *y*-axis, you can assume that each grid square is one unit.

CHECK FOR UNDERSTANDING

Communicating Mathematics

Read and study the lesson to answer each question. 1–2. See Answer Appendix.

1. **Tell** how you know whether to go left or right for the *x*-coordinate and whether to go up or down for the *y*-coordinate when graphing a point.
2. **Describe** how you would graph the points whose coordinates are (0, 3) and (3, 0).
3. **Write** how you can remember whether the first number of the set of ordered pairs is the *x*-coordinate or the *y*-coordinate. **Sample answer: x comes before y in the alphabet.**

Guided Practice

Name the ordered pair for the coordinates of each point graphed on the coordinate plane.

4. *A* (4, 2)
5. *B* (0, 3)
6. *C* (−1, −2)
7. *D* (−3, 2)

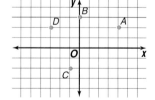

Lesson 2-10 Integration: Geometry The Coordinate System 93

In-Class Examples

For Example 1
Graph the point whose coordinates are (3, −1) and label it *A*.

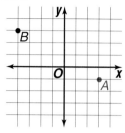

For Example 2
Name the ordered pair for point *B* on the coordinate plane above. (−4, 3)

For Example 3
Evaluate the expression $2x - 1$ for $x = -1, 0, 1,$ and 2. Graph the ordered pairs formed by *x* and the corresponding value of the expression for *y*.

x	2x − 1	y	(x, y)
−1	2(−1) − 1	−3	(−1, −3)
0	2(0) − 1	−1	(0, −1)
1	2(1) − 1	1	(1, 1)
2	2(2) − 1	3	(2, 3)

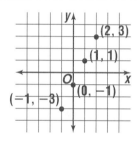

3 PRACTICE/APPLY

Check for Understanding
If students need additional practice or instruction after completing Exercises 1–14, one of these options may be helpful.
- Extra Practice, see p. 611
- Reteaching Activity
- *Study Guide Masters*, p. 19
- *Practice Masters*, p. 19
- Interactive Mathematics Tools Software

Reteaching the Lesson

Activity Graph points on a coordinate grid and label them with the letters of the alphabet. Write a series of ordered pairs corresponding to the graphed points to form a coded message. Have students decode the message by finding the letter that corresponds to each ordered pair.

Error Analysis
Watch for students who read or graph the coordinates of points in reverse order.
Prevent by stressing the phrase "*x* before *y*," that is, relate the coordinates of a point first to the *x*-axis and then to the *y*-axis.

Assignment Guide
Core: 15–47 odd, 49–55
Enriched: 16–44 even, 45–47, 49–55

As a classroom project, have all students find the coordinates of their homes on a map of the area served by the school district. Students should flag each location with their names (or initials) and the coordinates.

Additional Answer
45.
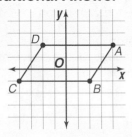

Study Guide Masters, p. 19

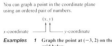

Graph each point on the same coordinate plane. 8–13. See Answer Appendix.
8. $E(-4, 3)$ 9. $F(-3, -9)$ 10. $G(9, -7)$
11. $H(5, 0)$ 12. $I(0, 0)$ 13. $J(7, 7)$

14. **History** Since gaining the right to vote in 1920, women have made slow gains in Congress. The table shows the number of females in the U.S. Senate for several years since 1975. Use the years as the x-coordinates and the numbers of female Senators as the y-coordinates to graph the data. **See Answer Appendix.**

Year	Female Senators
1975	0
1977	2
1979	1
1981	2
1983	2
1985	2
1987	2
1989	2
1991	3
1993	7
1995	9
1997	9

Source: National Women's Political Caucus

EXERCISES

Practice

Name the ordered pair for the coordinates of each point graphed on the coordinate plane.

15. Z $(-5, 5)$ 16. X $(3, 3)$
17. W $(2, 1)$ 18. Y $(-3, 2)$
19. T $(0, 5)$ 20. V $(-3, -2)$
21. U $(-6, -2)$ 22. S $(1, -4)$
23. Q $(0, -7)$ 24. R $(6, 0)$
25. P $(2, -1)$ 26. M $(-5, -7)$

Find a street map that includes your neighborhood. Also find maps that include the homes of two of your relatives or friends who do not live with you. List the coordinates of your home and the other two homes.

Graph each point on the same coordinate plane.   27–44. See Answer Appendix.
27. $A(4, 7)$ 28. $C(1, 0)$ 29. $B(0, 7)$
30. $E(-1, -2)$ 31. $D(-4, -7)$ 32. $F(-10, 3)$
33. $G(9, 9)$ 34. $J(7, -8)$ 35. $K(-6, 0)$
36. $H(0, -3)$ 37. $I(4, 0)$ 38. $M(2, 7)$
39. $N(8, -1)$ 40. $L(-1, -1)$ 41. $P(3, 3)$
42. $Q(-2, 0)$ 43. $S(-3, -3)$ 44. $R(5, -4)$

45. See margin for graph. It is not a rectangle because the angles are not right.

Applications and Problem Solving

45. **Geometry** Graph the points $A(4, 2)$, $B(2, -1)$, $C(-4, -1)$, and $D(-2, 2)$ on the same coordinate plane. Draw the line segments from A to B, from B to C, from C to D, and from D to A. Is the shape formed a rectangle? Explain.

94 Chapter 2 Algebra: Using Integers

Classroom Vignette

"I allow my students to play Battleship. Each player graphs 3 battleships, 3 destroyers, and 3 PT boats. Each pair of players takes turns calling out coordinates to sink the ships."

Mrs. Sandy Hardy, Teacher
Taylor Street Middle School
Griffin, GA

Sandy Hardy

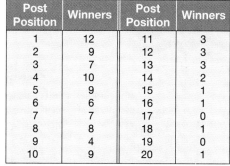

46. **Sports** And they're off! Post position, where a horse begins a race, may give a horse an edge on the victory. The table shows how many horses at the various post positions have won the Kentucky Derby since 1900. Use the post positions as the *x*-coordinates and the number of winners as the *y*-coordinates to graph the data. **See Answer Appendix.**

Post Position	Winners	Post Position	Winners
1	12	11	3
2	9	12	3
3	7	13	3
4	10	14	2
5	9	15	1
6	6	16	1
7	7	17	0
8	8	18	1
9	4	19	0
10	9	20	1

47. **Life Science** At 100 feet long and 150 tons, the blue whale is the largest living mammal today. At full speed, a blue whale can travel 36 feet per second. If *x* represents the number of seconds, the expression 36*x* gives the total distance a whale can travel. **b. See Answer Appendix.**
   a. Evaluate the expression to find the distances traveled in 0, 1, 3, 4, and 6 seconds. **0, 0; 1, 36; 3, 108; 4, 144; 6, 216**
   b. Graph the ordered pairs formed by the times and distances.

48. **Working on the CHAPTER Project** Use the populations that you found for your species. Graph each population on the same coordinate plane, using the year as the *x*-coordinate and the population as the *y*-coordinate. Write a description of the population changes and add the graph and the description to your television public service announcement script. **See students' work.**

49. **Critical Thinking** The points at (−3, 3), (−3, −3), and (−6, 3) are three of the vertices, or corners, of a square. What are the coordinates of the fourth vertex? **(−6, −3)**

Mixed Review

50. **Algebra** Solve 6*d* + 18 = 36. *(Lesson 2-9)* **3**

51. Find $\begin{bmatrix} -5 & -1 & 0 \\ 2 & 3 & -1 \end{bmatrix} + \begin{bmatrix} -1 & -1 & 3 \\ 2 & 0 & 6 \end{bmatrix}$. *(Lesson 2-6)* $\begin{bmatrix} -6 & -2 & 3 \\ 4 & 3 & 5 \end{bmatrix}$

52. **Test Practice** Solve *g* = −56 − 77. *(Lesson 2-5)* **B**
   A −21   B −133   C 21   D 133   E Not Here

53. Replace ● with >, <, or = to make −114 ● −97 a true sentence. *(Lesson 2-2)* **<**

54. **Geometry** Find the perimeter and area of the figure. *(Lesson 1-8)* **P = 20 cm, A = 25 cm²**

   5 cm square

55. **Geography** Do you live in Midway? Midway is the most common place name in the United States. There are 15 more occurrences of the name Midway than of the second most common name, Fairview. If there are 192 Fairviews in the U.S., how many Midway's are there? *(Lesson 1-6)* **207**

Lesson 2-10 Integration: Geometry The Coordinate System

Extending the Lesson

Enrichment Masters, p. 19

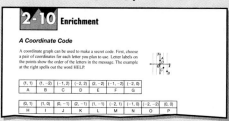

Activity Have students use a globe, atlas, or world map to find the coordinates of these cities: Buenos Aires 35°S, 58°W Karachi, 25°N, 67°E Durban, 30°S, 30°E and Winnipeg. 50°N, 97°W Compare this to a coordinate system in mathematics.

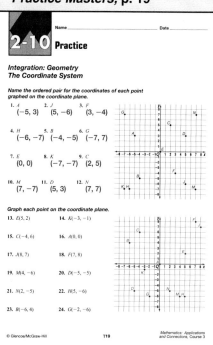

Study Guide and Assessment

Vocabulary

This section provides a listing of the new terms, properties, and phrases that were introduced in this chapter. Have students define each term and provide an example or two of it, if appropriate.

Understanding and Using the Vocabulary

These exercises check students' understanding of the terms by using a variety of verbal formats including matching, completion, and true/false.

Glossaries A complete glossary of terms appears on pages 702–710. The glossary also appears in Spanish on pages 711–721.

CHAPTER 2 Study Guide and Assessment

Vocabulary

After completing this chapter, you should be able to define each term, concept, or phrase and give an example or two of each.

Numbers and Operations
absolute value (p. 57)
additive inverse (p. 70)
integer (p. 56)
opposite (p. 70)
zero pair (p. 63)

Algebra
column (p. 73)
element (p. 73)
graph (p. 56)
matrix (p. 73)
row (p. 73)

Geometry
coordinate (p. 56)
coordinate system (p. 92)
ordered pair (p. 92)
origin (p. 92)
quadrant (p. 92)
x-axis (p. 92)
x-coordinate (p. 92)
y-axis (p. 92)
y-coordinate (p. 92)

Problem Solving
eliminate possibilities (p. 90)

Understanding and Using the Vocabulary

Choose the letter of the term that best matches each statement or phrase.

1. the distance a number is from zero on a number line   e
2. what you add to subtract an integer   f
3. the sign of the product of two integers with the same sign   a
4. the sign of the quotient of two integers with different signs   b
5. one of the four parts that the two axes separate a coordinate plane into   g
6. the first number in an ordered pair   d
7. one of the numbers in a matrix   i
8. the point where the x- and y-axes intersect   h
9. the number that corresponds to a point graphed on a number line   c

a. positive
b. negative
c. coordinate
d. x-coordinate
e. absolute value
f. additive inverse
g. quadrant
h. origin
i. element

In Your Own Words

10. *Write* a sentence that explains why the additive inverse of zero is zero.   The only number you can add to zero to get zero is zero.

96   Chapter 2 Algebra: Using Integers

MindJogger Videoquizzes

MindJogger Videoquizzes provide an alternative review of concepts presented in this chapter. Students work in teams to answer questions, gaining points for correct answers. The questions are presented in three rounds.
Round 1 Concepts–5 questions
Round 2 Skills–4 questions
Round 3 Problem Solving–4 questions

Study Guide and Assessment   Chapter 2

Objectives & Examples

Upon completing this chapter, you should be able to:

● graph integers on a number line and find absolute value *(Lesson 2-1)*

Graph −3 and 2.
Find each number on a number line.
Draw a dot at that point.

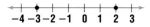

● compare and order integers *(Lesson 2-2)*

Compare −5 and −10.
−5 is to the right of −10 on a number line, so −5 > −10.

● add integers *(Lesson 2-3)*

Solve −14 + (−20) = x.
$|-14| + |-20| = 14 + 20$ or 34
So, x = −34.

● add three or more integers *(Lesson 2-4)*

Solve y = 22 + (−17) + 3.
y = [22 + (−17)] + 3   *Associative property*
y = 5 + 3
y = 8

● subtract integers *(Lesson 2-5)*

Solve z = −30 − 13.
z = −30 − 13
z = −30 + (−13)   *To subtract 13, add −13.*
z = −43

Review Exercises

Use these exercises to review and prepare for the chapter test.

Graph each set of points on a number line.

11. {3, −2, 6, 0}   11–12. See Answer Appendix.
12. {−5, 4, −1, 2}

Find each absolute value.

13. |11| **11**
14. |−88| **88**

Replace each ● with >, <, or = to make a true sentence.

15. −49 ● −340   **>**
16. −13 ● 2   **<**
17. |−483| ● 483   **=**
18. −6 ● |−6| **<**

Solve each equation.

19. 125 + (−75) = a   **50**
20. −66 + (−119) = b   **−185**
21. c = −46 + 38   **−8**
22. y = −54 + 21   **−33**

Solve each equation. Check by solving another way.

23. 20 + 16 + (−5) + 6 = k   **37**
24. −14 + 37 + (−20) + 2 = x   **5**
25. v = 52 + (−78) + 8   **−18**
26. 28 + (−50) + 12 + (−15) = p   **−25**

Solve each equation.

27. 45 − (−63) = b   **108**
28. y = −58 − 34   **−92**
29. m = −76 − (−56)   **−20**
30. −16 − 24 = n   **−40**

Objectives & Examples

This section reviews the skills and concepts of the chapter and shows completely worked examples.

Review Exercises

These exercises provide practice for the corresponding objectives.

Assessment and Evaluation Masters, pp. 31–32

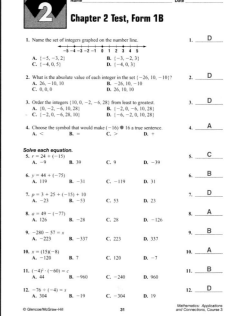

Assessment and Evaluation

Six forms of Chapter 2 Test are available in the *Assessment and Evaluation Masters* as shown in the chart.

Chapter 2 Test, Form 1B, is shown at the right. Chapter 2 Test, Form 2B, is shown on the next page.

1A	Multiple Choice	Honors
1B	Multiple Choice	Average
1C	Multiple Choice	Basic
2A	Free Response	Honors
2B	Free Response	Average
2C	Free Response	Basic

Study Guide and Assessment

Chapter 2 Study Guide and Assessment

Assessment and Evaluation Masters, pp. 37–38

Objectives & Examples

● use matrices to organize data *(Lesson 2-6)*

Find the sum.

$\begin{bmatrix} 4 & -2 \\ 3 & 0 \end{bmatrix} + \begin{bmatrix} -3 & 2 \\ 5 & -4 \end{bmatrix}$

$= \begin{bmatrix} 4 + (-3) & -2 + 2 \\ 3 + 5 & 0 + (-4) \end{bmatrix}$

$= \begin{bmatrix} 1 & 0 \\ 8 & -4 \end{bmatrix}$

● multiply integers *(Lesson 2-7)*

Solve $a = (-7)(3)(5)$.

$a = [(-7)(3)](5)$ *Associative property*

$a = (-21)(5)$

$a = -105$

● divide integers *(Lesson 2-8)*

Solve $b = -48 \div (-6)$.

$b = -48 \div (-6)$

$b = 8$

● solve equations with integer solutions *(Lesson 2-9)*

Solve $c - (-30) = 255$.

$c - (-30) = 255$

$c + 30 = 255$

$c + 30 - 30 = 255 - 30$

$c = 225$

● graph points on the coordinate plane *(Lesson 2-10)*

Graph the point whose coordinates are $(3, -4)$.

Review Exercises

Find each sum or difference. If there is no sum or difference, write *impossible*.

31. $\begin{bmatrix} 2 & -1 \\ -3 & 9 \end{bmatrix} + \begin{bmatrix} 10 & -3 \\ 8 & -3 \end{bmatrix}$ $\begin{bmatrix} 12 & -4 \\ 5 & 6 \end{bmatrix}$

32. $\begin{bmatrix} 3 & -2 \\ 4 & -1 \end{bmatrix} + \begin{bmatrix} -3 & 8 \\ 0 & 6 \\ -4 & 2 \end{bmatrix}$ impossible

33. $\begin{bmatrix} -4 & -2 \\ 5 & 6 \\ 7 & -5 \end{bmatrix} - \begin{bmatrix} 2 & 2 \\ 4 & -4 \\ 9 & 0 \end{bmatrix}$ $\begin{bmatrix} -6 & -4 \\ 1 & 10 \\ -2 & -5 \end{bmatrix}$

Solve each equation.

34. $u = -7(11)$ -77

35. $w = -4(-25)$ 100

36. $(-20)(-2)(3) = q$ 120

37. $(-15)(-4)(-1) = g$ -60

Solve each equation.

38. $-88 \div 8 = c$ -11 39. $\frac{-210}{7} = h$ -30

40. $z = 170 \div 5$ 34 41. $b = \frac{180}{-15}$ -12

Solve each equation. Check your solution.

42. $-3s = -51$ 17

43. $3p - (-7) = 52$ 15

44. $24 + m = -93$ -117

45. $\frac{r}{4} - 12 = -14$ -8

Graph each point on the same coordinate plane. 46–49. See Answer Appendix.

46. $A(-1, -6)$

47. $B(-3, -2)$

48. $C(3, 6)$

49. $D(5, -4)$

98 Chapter 2 Algebra: Using Integers

Test and Review Software

You may use this software, a combination of an item generator and item bank, to create your own tests or worksheets. Types of items include free response, multiple choice, short answer, and open ended.

CD-ROM Program

The CD-ROM Program contains an Assessment Game whose questions review the concepts in this chapter.

Study Guide and Assessment Chapter 2

Applications & Problem Solving

50. Recreation The Chess Club at Mason Middle School is holding a tournament in which each player earns +1 point for a win, −1 point for a loss, and 0 for a draw. The scoreboard below shows the final standings of five players. *(Lessons 2-1 and 2-4)*

Player	Wins	Losses
Neva	6	0
Kaitlin	4	2
Antoine	2	4
Trenna	3	3
Sarah	0	6

a. **See margin.**
a. Determine each player's total points.
b. Graph the total scores on a number line. Let the first letter of each player's name be the letter for each coordinate. **See Answer Appendix.**

51. Sports Alberto scored −2 on five miniature golf holes. What was his score for these 5 holes? *(Lesson 2-7)* **−10**

52. Patterns Third grader Koleka Smith of Chehalis, Washington, discovered a new way of multiplying by 5. To multiply an even number by 5, divide by 2 and then add a 0 to the end. To multiply an odd number by 5, subtract 1, divide by 2, and then add a 5 to the end. *(Lesson 2-8)*
a. Will Koleka's method work with negative integers?
b. If not, give another pattern that will work. **a–b. See Answer Appendix.**

53. Eliminate Possibilities Yuria and Evan have the same favorite color. It is either blue, red, or green. Yuria does not like green. Evan does not like red. What is each person's favorite color? *(Lesson 2-9B)*
blue

Alternative Assessment

● *Performance Task*
You just received a statement from your bank showing that your account is overdrawn by $25.60, which is shown as −25.60. Your records show a balance of $15.90, or +15.90, in your account. You balance your checkbook and decide that your balance is correct. After discussing the error with the bank, you found that they made just one error. What error could the bank have made? **See margin.**

You discover that you withdrew $10.00, recorded as −10.00, from your account that you forgot to record. How does this change the mistake made by the bank? Is your account overdrawn now? Explain the amount of the mistake now.
                                  See margin.

A practice test for Chapter 2 is provided on page 648.

● Completing the **Chapter Project**
Use the following checklist to make sure that your television public service announcement script is complete.

☑ Summarize your research on the species that you chose. Why is it endangered or threatened?

☑ Include population figures for various years and discuss how the population has changed.

☑ Add descriptions of video clips to accompany the audio.

 Select one of the problems you solved in this chapter and place it in your portfolio. Attach a note explaining how your problem illustrates one or more of the important concepts covered in this chapter.

Additional Answers for the Performance Task
- Sample answer: The bank subtracted a check for $41.50 twice (recorded as −41.50), or neglected to record a deposit of $41.50 (recorded as +41.50).
- Your balance is now $5.90, so your account is not overdrawn. The bank's mistake is now $31.50.

Performance Assessment
Additional performance assessment tasks for this chapter are included in the *Assessment and Evaluation Masters* on page 41. A scoring guide is also provided on page 53.

Standardized Test Practice

The Standardized Test Practice may be used to help students prepare for standardized tests. The test items are written in the same style as those in state proficiency tests and standardized tests like CAT, CTBS, ITBS, MAT, SAT, and Terra Nova. The test items cover skills and concepts covered up to this point in the text.

The pages can be used as an overnight assessment. After students have completed the pages, discuss how each problem can be solved, or provide copies of the solutions from the *Solutions Manual*.

Assessment and Evaluation Masters, p. 47

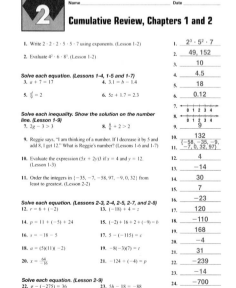

CHAPTERS 1–2 Standardized Test Practice
Assessing Knowledge & Skills

Section One: Multiple Choice

There are twelve multiple-choice questions in this section. Choose the best answer. If a correct answer is *not here*, choose the letter for Not Here.

1. How is $2 \cdot 2 \cdot 7 \cdot a \cdot a \cdot a$ written in exponential notation? **C**
   A $2 \cdot 5 \cdot a$
   B $2^7 \cdot 7^2 \cdot a$
   C $2^2 \cdot 7 \cdot a^3$
   D $14^2 \cdot a^3$

2. What is the perimeter of the rectangle? **G**

   F 11.1 cm
   G 22.2 cm
   H 33.3 cm
   J 30.5 cm

3. If $b = 5$ and $c = 8$, what is the value of $b(10 - c)$? **A**
   A 10
   B 42
   C 50
   D 52

4. There are approximately 6,479,000 bicycles in North Carolina. What is this number rounded to the nearest million? **F**
   F 6,000,000
   G 6,500,000
   H 6,480,000
   J 7,000,000

Please note that Questions 5–12 have five answer choices.

5. The emergency squad responded to 120 calls for help in the first three months of the year. They responded to 42 calls the fourth month. Which would be a reasonable answer for the number of calls that they responded to? **B**
   A less than 35 calls per month
   B between 35 and 45 calls per month
   C between 45 and 55 calls per month
   D between 55 and 65 calls per month
   E more than 65 calls per month

6. Manuela wanted to spend no more than $20 at the grocery. If she bought 9 bottles of sports drink, which inequality could be used to find the price, p, of each bottle so that the total would be at or below $20 excluding tax? **J**
   F $p + 9 < 20$
   G $p + 9 \leq 20$
   H $9p < 20$
   J $9p \leq 20$
   K $9 - p < 20$

7. On a math test, Lee scored 10 points less than twice the lowest score. If his score was 96, what was the lowest score? **B**
   A 86
   B 53
   C 48
   D 43
   E 33

8. It costs approximately $60 per month to heat a 2,000 square-foot home. At that rate, how much should it cost per month to heat a 3,000 square-foot home? **H**
   F $40
   G $70
   H $90
   J $110
   K $120

100 Chapters 1–2 Standardized Test Practice

◄◄◄ **Instructional Resources**
Another cumulative review is shown at the left and is available in the *Assessment and Evaluation Masters*, p. 47.

Standardized Test Practice Chapters 1–2

9. Victor bought a stereo that cost $532.16. He paid for the stereo in 12 equal payments. The best estimate for the amount of each payment is — **D**
   - A less than $10.
   - B between $10 and $20.
   - C between $25 and $35.
   - D between $40 and $50.
   - E between $55 and $65.

10. Three friends will equally share the cost of a $28.95 board game. Which is a *not* a reasonable estimate of each person's share? **H**
    - F $9.90
    - G $9.65
    - H $8.80
    - J $9.50
    - K All are reasonable.

11. Missy bought 3 sweaters for $108. If each sweater cost the same amount, how much did each sweater cost? **C**
    - A $30.24
    - B $32.40
    - C $36.00
    - D $54.00
    - E Not Here

12. The Lewis Center 4-H Club sold 312 pocket calendars last year. This year they sold 447 pocket calendars. How many more calendars did they sell this year than last year? **F**
    - F 135
    - G 125
    - H 155
    - J 189
    - K Not Here

Test-Taking Tip

Most basic formulas that you need to answer the questions on a standardized test are usually given to you in the test booklet. However, it is a good idea to review common formulas before the test. Make sure to familiarize yourself with how to use formulas. A quick review may allow you to answer more questions correctly.

Section Two: Free Response

This section contains seven questions for which you will provide short answers. Write your answers on your paper.

13. Solve $-64 \div 8 = q$. **−8**

14. Find the product of 20 and -9. **−180**

15. If $\frac{c}{-3} = 6$, what is the value of c? **−18**

16. Simplify $(3a + 1) + (2a + 5)$. **$5a + 6$**

17. Find the product $4(3x - 2)$. **$12x - 8$**

18. The product of a number and itself is 196. What is the number? **14 or −14**

19. What is the area of the parallelogram? $41\frac{7}{16}$ ft²

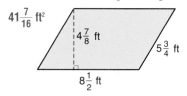

Test-Taking Tip

When taking standardized tests, it is very important to use formulas correctly. This tip advises students to practice solving problems involving common formulas before a test.

Assessment and Evaluation Masters, pp. 45–46

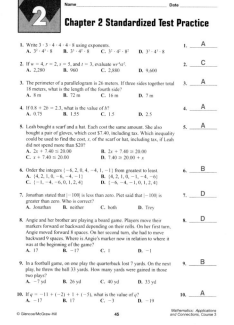

Instructional Resources ▶▶▶

Additional standardized test practice is shown at the right and is available in the *Assessment and Evaluation Masters*, pp. 45–46.

CHAPTER 3
Using Proportion and Percent

Previewing the Chapter

Overview
This chapter introduces ratios, rates, percents, and proportions. Students learn the distinction between ratio and rate. Several facets of percent are explored, including relationships with fractions and decimals, and estimation. Students develop strategies to estimate and compute mentally using percents and to determine reasonable answers to real world problems.

Lesson (pages)	Lesson Objectives	NCTM Standards	Standardized Tests	State/Local Objectives
3-1 (104–106)	Express ratios as fractions in simplest form and determine unit rates.	1–5, 7, 10	CTBS, ITBS, TN	
3-2 (107–110)	Express ratios as percents and vice versa.	1–5, 7, 10, 12	CTBS, TN	
3-3 (111–113)	Solve proportions.	1–5, 7–10	CAT, CTBS, MAT, SAT, TN	
3-4 (114–117)	Express percents as fractions and decimals and vice versa.	1–5, 7, 9	CTBS, TN	
3-4B (118–119)	Find the value of the golden ratio.	1–5, 7, 8, 12, 13		
3-5 (120–123)	Compute mentally with percents.	1–5, 7, 10, 12		
3-6A (124–125)	Determine whether answers to problems are reasonable.	1–5, 7	CTBS, MAT, TN	
3-6 (126–129)	Estimate by using equivalent fractions, decimals, and percents.	1–5, 7, 10, 12, 13		

CAT = California Achievement Tests, CTBS = Comprehensive Tests of Basic Skills, ITBS = Iowa Tests of Basic Skills, MAT = Metropolitan Achievement Tests, SAT = Stanford Achievement Tests, TN = Terra Nova

102a Chapter 3

Organizing the Chapter

CD-ROM: All of the blackline masters in the Teacher's Classroom Resources are available on the **Electronic Teacher's Classroom Resources** CD-ROM.

LESSON PLANNING GUIDE

Lesson	Extra Practice (Student Edition)	Blackline Masters (page numbers)										
		Study Guide	Practice	Enrichment	Assessment & Evaluation	Classroom Games	Diversity	Hands-On Lab	School to Career	Science and Math Lab Manual	Technology	Transparencies A and B
3-1	p. 612	20	20	20							58	3-1
3-2	p. 612	21	21	21	71							3-2
3-3	p. 612	22	22	22	70, 71		29		29		57	3-3
3-4	p. 613	23	23	23			7–9	70				3-4
3-4B								42				
3-5	p. 613	24	24	24	72					53–56		3-5
3-6A	p. 613											
3-6	p. 614	25	25	25	72					85–88		3-6
Study Guide/ Assessment					57–69, 73–75							

Other Chapter Resources

Student Edition
Chapter Project, pp. 103, 106, 113, 129, 133
Let the Games Begin, p. 123

Technology
CD-ROM Program
 Interactive Mathematics Tools Software

Teacher's Classroom Resources
Applications
Family Letters and Activities, pp. 57–58
Investigations and Projects Masters, pp. 25–28
Meeting Individual Needs
Transition Booklet, pp. 9–16, 29–30
Investigations for the Special Education Student, pp. 21–28

Teaching Aids
Answer Key Masters
Block Scheduling Booklet
Lesson Planning Guide
Solutions Manual

Professional Publications
Glencoe Mathematics Professional Series

Planning the Chapter

 MindJogger Videoquizzes provide a unique format for reviewing concepts presented in the chapter.

Assessment Resources

Student Edition
Mixed Review, pp. 106, 110, 113, 117, 122, 129
Mid-Chapter Self Test, p.117
Math Journal, pp. 105, 109, 112
Study Guide and Assessment, pp. 130–133
Performance Task, p. 133
Portfolio Suggestion, p. 133
Standardized Test Practice, pp. 134–135
Chapter Test, p. 649

Assessment and Evaluation Masters
Multiple-Choice Tests (Forms 1A, 1B, 1C), pp. 57–62
Free-Response Tests (Forms 2A, 2B, 2C), pp. 63–68
Performance Assessment, p. 69
Mid-Chapter Test, p. 70
Quizzes A–D, pp. 71–72
Standardized Test Practice, pp. 73–74
Cumulative Review, p. 75

Teacher's Wraparound Edition
5-Minute Check, pp. 104, 107, 111, 114, 120, 126
Building Portfolios, p. 102
Math Journal, p. 119
Closing Activity, pp. 106, 110, 113, 117, 123, 125, 129

Technology
Test and Review Software
MindJogger Videoquizzes
CD-ROM Program

Materials and Manipulatives

Lesson 3-1
coordinate grid*

Lesson 3-2
grid paper

Lesson 3-3
calculator

Lesson 3-4
calculator

Lesson 3-4B
calculator
grid paper
scissors*
tape measure*

Lesson 3-5
calculator
index cards
scissors*
markers

Lesson 3-6
calculator
compass*†
markers

*Glencoe Manipulative Kit   †Glencoe Overhead Manipulative Resources

Pacing Chart

See pages T25–T27 for the Course Planning Calendar.

COURSE	DAY 1	DAY 2	DAY 3	DAY 4	DAY 5	DAY 6	DAY 7
Standard	Chapter Project	Lesson 3-1	Lesson 3-2	Lesson 3-3	Lessons 3-4 & 3-4B		Lesson 3-5
Honors	Chapter Project & Lesson 3-1	Lesson 3-2	Lesson 3-3	Lessons 3-4 & 3-4B		Lesson 3-5	Lesson 3-6a
Block	Chapter Project & Lesson 3-1	Lessons 3-2 & 3-3	Lesson 3-4	Lesson 3-5	Lessons 3-6A & 3-6	Study Guide and Assessment, Chapter Test	

The *Transition Booklet* (Skills 3–6, 13) can be used to practice working with fractions and decimals and measuring length in the customary system.

Interactive Mathematics:
Activities and Investigations

is an activity-based program that may be used as an enhancement for chapters in *Mathematics: Applications and Connections*.

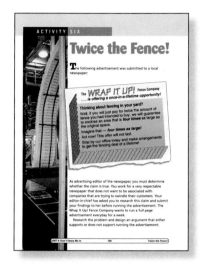

Unit 9, Activity Six
Use with Lesson 3-1.

Summary Students explore the meaning of ratio by determining whether a fence company's ad is legitimate. They examine the relationship between changes in the perimeter of a rectangle and the resulting changes in its area. Then each student prepares a report.

Math Connection Students are introduced to the concept of ratio. Students investigate how the ratio of the perimeters of two rectangles relates to the ratio of their respective areas.

Unit 7, Activity Three
Use with Lesson 3-6A.

Summary Students work in groups to solve problems in which each member of the group receives a clue to the solution of the problem. Students draw diagrams or make models of their solutions and write about their problem-solving processes.

Math Connection Students work cooperatively in groups and follow directions in order to solve problems. Students should determine if their models represent a reasonable solution to the problem.

DAY 8	DAY 9	DAY 10	DAY 11	DAY 12	DAY 13	DAY 14	DAY 15
Lesson 3-6A	Lesson 3-6	Study Guide and Assessment	Chapter Test				
Lesson 3-6	Study Guide and Assessment	Chapter Test					

Chapter 3 **102d**

Enhancing the Chapter

APPLICATIONS

Classroom Games, pp. 7–9

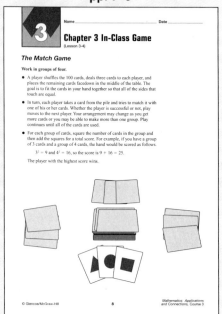

Diversity Masters, p. 29

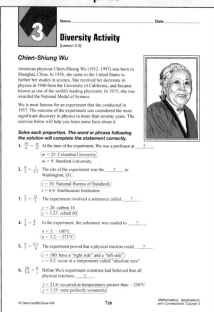

School to Career Masters, p. 29

Family Letters and Activities, pp. 57–58

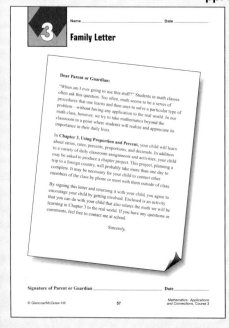

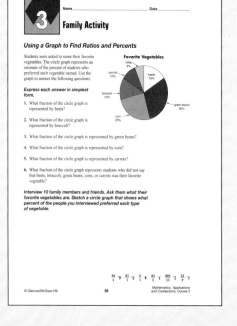

Science and Math Lab Manual, pp. 53–56 and 85–88

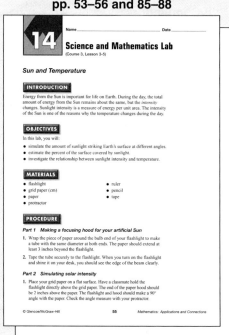

102e Chapter 3

MANIPULATIVES/MODELING

Hands-On Lab Masters, p. 70

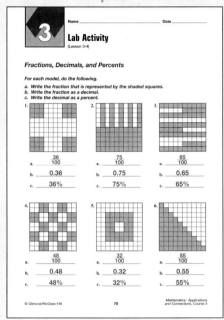

ASSESSMENT/EVALUATION

Assessment and Evaluation Masters, pp. 70–72

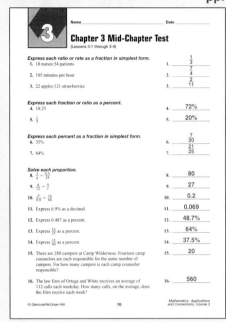

TECHNOLOGY/MULTIMEDIA

Technology Masters, pp. 57–58

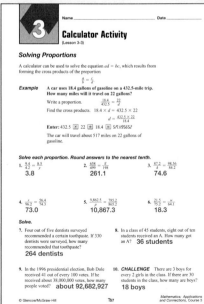

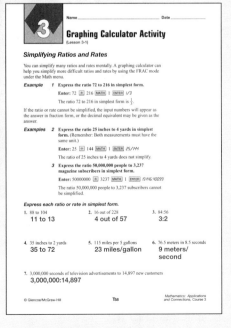

MEETING INDIVIDUAL NEEDS

Investigations for the Special Education Student, pp. 21–28

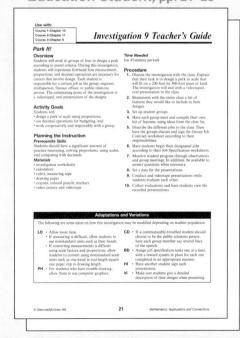

Chapter 3 **102f**

CHAPTER 3 NOTES

Theme: Travel

The European Monetary Union consists of several nations that are working to develop one form of currency, called the *euro*, that will be used throughout Europe. By the year 2002, the euro is expected to replace the money currently used in the Netherlands, France, Germany, and many other countries. Someday you will need to convert dollars only into euro in order to spend money anywhere in Europe.

Question of the Day In Ireland, the U.S. dollar is worth about 1.5 Irish pounds. How many U.S. dollars would you need to pay for a five-night stay at a hotel costing 150 Irish pounds per night? $500 (U.S.)

Assess Prerequisite Skills

Ask students to read through the list of objectives presented in "What you'll learn in Chapter 3." You may wish to ask them what each of the objectives means or if they have experienced or used any of these math concepts before.

 Building Portfolios

Encourage students to revise their portfolios as they study this chapter. They may wish to include examples illustrating a concept in this chapter that was new to them.

 Math and the Family

In the *Family Letters and Activities* booklet (pp. 57–58), you will find a letter to the parents explaining what students will study in Chapter 3. An activity appropriate for the whole family is also available.

CHAPTER 3
Using Proportion and Percent

What you'll learn in Chapter 3

- to express ratios as fractions and determine unit rates,
- to solve problems using proportions,
- to express percents as fractions and decimals,
- to estimate and compute mentally with percents, and
- to determine reasonable answers in real-world problems.

 CD-ROM Program

Activities for Chapter 3
- Chapter 3 Introduction
- Interactive Lessons 3-1, 3-3, 3-6
- Assessment Game
- Resource Lessons 3-1 through 3-6

CHAPTER Project

PACK YOUR BAGS

In this project, you will use ratios, proportions, and percents to help you plan a week-long trip to a foreign country of your choice. You will write a report describing your plan and the cost of your trip.

Getting Started

- Choose a foreign country you would like to visit. Then make a list of the cities and attractions you would like to visit.
- Estimate how much money your trip will cost. It will be helpful to organize your expenses into these categories: transportation, housing, food, and souvenirs.
- When you travel to a foreign country, you must exchange U.S. dollars for that country's currency. Find the exchange rate for your chosen country. Some exchange rates are shown in the table.

Country	Currency Name	Dollar Equivalent as of 7/7/97
Brazil	Real	1.0777
Canada	Dollar	1.3767
China	Yuan	8.3214
Egypt	Pound	3.3938
Ireland	Punt	0.6573
Japan	Yen	112.77
Mexico	Peso	7.9210
Pakistan	Rupee	40.01
South Africa	Rand	4.5365
Switzerland	Franc	1.4610

Technology Tips

- Use an **electronic encyclopedia** to do your research.
- Use a **spreadsheet** to organize your expenses.
- Use a **word processor** to write your report.

interNET CONNECTION For up-to-date information on foreign currency rates, visit: www.glencoe.com/sec/math/mac/mathnet

Working on the Project

You can use what you'll learn in Chapter 3 to help you make your travel plan.

Page	Exercise
106	28
113	29
129	35
133	Alternative Assessment

Chapter 3 **103**

CHAPTER Project NOTES

Objectives Students should
- learn to convert exchange rates to fractions, decimals, and percents.
- estimate the amount of money needed for travel.
- understand how to sketch circle graphs.

Project Pointer You may suggest that students begin a *Project Folder* to keep their work as they complete each stage of the Chapter Project. The completed project may also be added to their portfolios.

Students should include circle graphs and exchange charts in their reports. Both charts and graphs should include how much money students plan to spend on hotels, food, souvenirs, and travel and what percent of their budget each of these represents.

Using the Table Exchange rates for the U.S. dollar can also be shown in terms of the other currency. For example, other sources may show that one dollar equals 0.9279 Brazilian reales. Caution students not to confuse the two forms of comparison when researching their countries.

Investigations and Projects Masters, p. 28

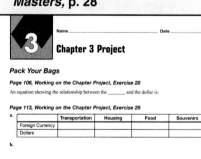

---

interNET CONNECTION

Glencoe has made every effort to ensure that the website links for *Mathematics: Applications and Connections* at www.glencoe.com/sec/math/mac/mathnet are current and contain appropriate content. However, these website links are not under Glencoe's control.

Instructional Resources ▶▶▶
A recording sheet to help students organize their data for the Chapter Project is shown at the right and is available in the *Investigations and Projects Masters,* p. 28.

Chapter 3 Project **103**

3-1 Lesson Notes

Instructional Resources
- *Study Guide Masters*, p. 20
- *Practice Masters*, p. 20
- *Enrichment Masters*, p. 20
- *Transparencies 3-1*, A and B
- *Technology Masters*, p. 58
- CD-ROM Program
  - Resource Lesson 3-1
  - Interactive Lesson 3-1

Recommended Pacing	
Standard	Day 2 of 11
Honors	Day 1 of 10
Block	Day 1 of 6

1 FOCUS

5-Minute Check
(Chapter 2)

Solve each equation.
1. $k = 9 + (-9)$ **0**
2. $15 + (-12) - 20 + 12 = n$ **−5**
3. $x = (-2)(-14)(5)$ **140**
4. $\frac{m}{3} - 6 = -2$ **12**
5. Graph $A(1, -2)$ on a coordinate plane.

 The 5-Minute Check is also available on **Transparency 3-1A** for this lesson.

2 TEACH

 Transparency 3-1B contains a teaching aid for this lesson.

Thinking Algebraically Stress that multiplying or dividing both the numerator and denominator of a fraction by any number forms an equivalent fraction. Show that an expression such as $\frac{a \div c}{b \div c}$ is equivalent to $\frac{a}{b} \div \frac{c}{c}$ and $\frac{c}{c} = 1$. Division by 1 does not affect the value of the number. Likewise, $\frac{a \cdot c}{b \cdot c} = \frac{a}{b} \cdot \frac{c}{c}$ or $\frac{a}{b} \cdot 1$.

104   Chapter 3

3-1 Ratios and Rates

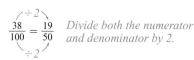 **you'll learn**
You'll learn to express ratios as fractions in simplest form and determine unit rates.

When am I ever going to use this?
Ratios and rates are frequently used in advertising and commercials.

Word Wise
ratio
rate
unit rate

The characters that appear in many television commercials are entertaining as well as effective. One example is the Pillsbury Dough Boy. In a 1996 Louis Harris Poll, about 38 out of 100 people who were surveyed rated the Pillsbury Dough Boy commercials very effective. The expression *38 out of 100* is an example of a **ratio**.

Ratio	A ratio is a comparison of two numbers by division.

The ratio that compares 38 to 100 can be expressed as follows.

Say: 38 to 100, 38 out of 100

Write: 38:100, $\frac{38}{100}$ *A fraction bar is used to indicate division.*

Since a ratio can be written as a fraction, it can be simplified.

$$\frac{38}{100} \overset{\div 2}{\underset{\div 2}{=}} \frac{19}{50}$$ *Divide both the numerator and denominator by 2.*

The equation $\frac{38}{100} = \frac{19}{50}$ indicates that the two ratios are equivalent. This means that 38 out of 100 is equivalent to 19 out of 50.

Examples

Express each ratio in simplest form.

1 number of shaded squares : total number of squares

15 out of 25 squares are shaded.

$\frac{15}{25} = \frac{3}{5}$ *Divide the numerator and denominator by 5.*

The ratio in simplest form is $\frac{3}{5}$ or 3 out of 5.

Study Hint
Reading Math In Example 1, the ratio 3 out of 5 means that for every 5 squares in the grid, 3 are shaded.

2 24 inches to 1 yard

Both measurements must have the same unit.

$\frac{24 \text{ inches}}{1 \text{ yard}} = \frac{24 \text{ inches}}{36 \text{ inches}}$

24 inches

1 yard = 36 inches

$\frac{24}{36} = \frac{2}{3}$ *Divide the numerator and denominator by 12.*

The ratio in simplest form is $\frac{2}{3}$ or 2:3.

104   Chapter 3 Using Proportion and Percent

Motivating the Lesson
Problem Solving On the chalkboard, write the number of right-handed students in the class and the number of left-handed students. Ask students to suggest ways to compare numbers.

Multiple Learning Styles

 Intrapersonal Have students keep track of how they spend their time for one week. Categories could include sleeping, eating, attending school, exercising, and so on. Have the class create a list of word ratios involving these categories. Students can then determine their personal ratios from their own data.

A **rate** is a special kind of ratio. It is a comparison of two measurements with different units, such as miles to gallons or cents to pounds. For example, suppose you spend $9.75 on 2.5 pounds of candy. The rate $\frac{\$9.75}{2.5 \text{ pounds}}$ compares the money spent to the number of pounds of candy.

Rate	A rate is a ratio of two measurements with different units.

When a rate is simplified so it has a denominator of 1, it is called a **unit rate**. This type of rate is frequently used when referring to statistics.

Example 3

Civics Membership in the U.S. House of Representatives is based on population in the preceding census. In 1990, the population of the United States was about 248,000,000. There are 435 members in the House. On average, how many people are represented by each member of the House?

Write the rate that compares the population to the number of members of the House. Then divide both the numerator and the denominator by 435.

$$\frac{248{,}000{,}000 \text{ people}}{435 \text{ members}} \approx \frac{570{,}000 \text{ people}}{1 \text{ member}}$$

The symbol ≈ means about equal to.

(÷435 in numerator and denominator)

Each member of the House of Representatives represents about 570,000 people.

Study Hint

Problem Solving
You can also find the unit rate by dividing the numerator by the denominator.

In-Class Examples

For Example 1
Express the ratio *number of states that begin with letter "A":total number of states* in simplest form. $\frac{2}{25}$ or 2:25

For Example 2
Express the ratio *9 eggs out of 3 dozen eggs* in simplest form. $\frac{1}{4}$ or 1:4

For Example 3
A 1.8-pound wedge of cheddar cheese sells for $6.83. For how much does the cheese sell per pound? **about $3.54 per pound**

---

3 PRACTICE/APPLY

Check for Understanding
If students need additional practice or instruction after completing Exercises 1–9, one of these options may be helpful.
- Extra Practice, see p. 612
- Reteaching Activity
- *Transition Booklet,* pp. 11–12, 15–16, 29–30
- *Study Guide Masters,* p. 20
- *Practice Masters,* p. 20
- Interactive Mathematics Tools Software

CHECK FOR UNDERSTANDING

Communicating Mathematics
Read and study the lesson to answer each question.

1. *Explain* how ratios and rates are alike and how they are different. **See Answer Appendix.**
2. *Write* a ratio about your class. **See students' work.**
3. *Draw* a figure in which the ratio of shaded squares to total number of squares is 3:8. **See students' work.**

Guided Practice
Express each ratio or rate in simplest form.

4. 2 cups:18 cups **1:9**
5. $20 in 25 days **4 to 5**
6. shaded squares in the figure at the right to total number of squares **1 to 2**

Express each rate as a unit rate.

7. 24 pounds in 8 weeks **3 lb/week**
8. 3 inches of rain in 2 hours **1.5 in./h**

Lesson 3-1 Ratios and Rates 105

Reteaching the Lesson

Activity Have pairs of students use a yardstick to measure the length, width, and height of the room to the nearest foot. Then have them calculate the following ratios:

$\frac{\text{width}}{\text{length}}, \frac{\text{height}}{\text{length}}, \frac{\text{height}}{\text{width}}$

Error Analysis
Watch for students who write ratios in the incorrect order.
Prevent by having them include descriptive words as well as numbers when writing a ratio or rate.

Study Guide Masters, p. 20

Assignment Guide

Core: 11–27 odd, 29–33
Enriched: 10–24 even, 26, 27, 29–33

Exercise 28 asks students to advance to the next stage of work on the Chapter Project. You may wish to have students work in groups to develop tables that compare the different exchange rates.

4 ASSESS

Closing Activity
Modeling Have students count several types of classroom objects—books, writing instruments, and so on—and compare the quantities of each using rates and ratios.

Additional Answer
30.

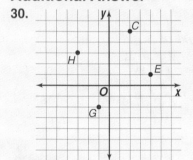

Practice Masters, p. 20

9. *Environment* In 1996, six California condors were released into the Vermillion Cliffs in Arizona. The rarest birds in North America, they have a wingspan of about 9 feet. The common Barn Owl has a 3-foot wingspan. Use a ratio in simplest form to compare the wingspan of a California condor to the wingspan of a Barn Owl. **3:1**

EXERCISES

Practice Express each ratio or rate in simplest form.

10. shaded squares in the figure at the right to total number of squares **3 to 4**
11. 18 brown-eyed students:12 blue-eyed students **3:2**
12. 6 absences in 180 school days **1 to 30**
13. 1 foot:1 yard **1:3**
14. 99 wins:99 losses **1:1**
15. 15 out of 45 **1:3**
16. 45 minutes per hour **3 to 4**
17. 36 to 24 **3 to 2**
18. 150 miles per 6 gallons **25 to 1**

Express each rate as a unit rate.

21. 15 students/teacher
22. 9.5 meters/second

19. $1.75 for 5 minutes **$0.35/minute**
20. $25 for 10 disks **$2.50/disk**
21. 300 students to 20 teachers
22. 100 meters in 10.5 seconds
23. $8.80 for 11 pounds **$0.80/pound**
24. $0.96 per dozen **$0.08/item**
25. Express the ratio *25 out of 40* as a fraction in simplest form. $\frac{5}{8}$

Applications and Problem Solving

26. 37,000 people
27. 3.8 dandelion plants, 17.7 grass plants
28. See students' work.

26. *Civics* In the 1780s, the House of Representatives had 68 members. At that time, the population of the United States was about 2,500,000. On average, about how many people were represented by each member of the House?
27. *Life Science* A science class is studying a section of weeds that has an area of 12 square meters. They count 46 dandelion plants and 212 grass plants. On average, how many of each kind of plant are in one square meter?
28. *Working on the* **CHAPTER Project** Refer to the foreign exchange rate you found on page 103. Write an equation that shows the relationship between the unit of currency used in your foreign country and 1 U.S. dollar.
29. *Critical Thinking* Find the value of *x* so that the ratios $\frac{2}{x}$ and $\frac{x}{18}$ are equal. **6 or −6**

Mixed Review

30. *Geometry* Graph the points $C(2, 5)$, $H(-3, 3)$, $E(4, 1)$, and $G(-1, -2)$ on the same coordinate plane. *(Lesson 2-10)* **See margin.**
31. *Algebra* Solve $-2y + 15 = 55$. *(Lesson 2-9)* **−20**
32. *Algebra* Write an algebraic expression to represent *eight million less than four times the population of Africa.* *(Lesson 1-6)* **4x − 8,000,000**
33. **Test Practice** Tonio uses his calculator to divide 685,300 by 86.3. Which is the best estimate of the result? *(Lesson 1-1)* **C**

A 100 B 800 C 8,000 D 10,000 E 80,000

106 Chapter 3 Using Proportion and Percent

Extending the Lesson

Enrichment Masters, p. 20

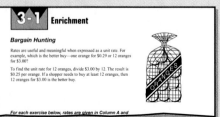

Activity Ask students to write a ratio describing the relationship between a team's win-loss record and the total number of games played. Have students use an almanac to find the year-end records of a sports league of their choosing. Ask them to make a chart of the win-loss ratios of all the teams in the league.

3-2 Ratios and Percents

What you'll learn
You'll learn to express ratios as percents and vice versa.

When am I ever going to use this?
Knowing how to express ratios as percents will help you estimate the sale price of your next pair of jeans.

Word Wise
percent

Where is your ideal vacation spot? In 1996, the George Gallup International Institute reported that about 75 out of 100 U.S. teenagers would like to visit the Grand Canyon. The ratio 75 out of 100 can be expressed as a **percent**.

Percent	A percent is a ratio that compares a number to 100.	
	Ratio: 75 out of 100	Model:
	Words: seventy-five percent	
	Symbol: 75%	

Example

1 Express each ratio as a percent.
a. Nine out of 100 U.S. Senators are women.
   9 out of 100 = 9%
b. He answered 87.5 questions out of 100 correctly.
   87.5 out of 100 = 87.5%

Ratios like 1:2, $\frac{3}{10}$, and 4 out of 5 can also be expressed as percents.

HANDS-ON MINI-LAB

Work with a partner. grid paper

You can use grid paper to express the ratio 1:2 as a percent.
- Since 1:2 = $\frac{1}{2}$, shade $\frac{1}{2}$ of the squares in the decimal model.
- $\frac{1}{2} = \frac{50}{100}$
   = 50% So, $\frac{1}{2}$ = 50%.

Try This 1–3. See Answer Appendix.
Use decimal models to express each ratio as a percent.
1. $\frac{3}{10}$ 30%   2. 4:5 80%   3. 1 out of 3 $33\frac{1}{3}$%

Talk About It
4. Use what you have learned in this lab to express the ratio 7 out of 10 as a percent without using models. **70%**

Did you know In the Gallup survey, 58% of teenagers mentioned the White House and 41% mentioned Graceland as desirable vacation spots.

Lesson 3-2 Ratios and Percents **107**

3-2 Lesson Notes

Instructional Resources
- *Study Guide Masters*, p. 21
- *Practice Masters*, p. 21
- *Enrichment Masters*, p. 21
- Transparencies 3-2, A and B
- *Assessment and Evaluation Masters*, p. 71
- CD-ROM Program
  - Resource Lesson 3-2

Recommended Pacing	
Standard	Day 3 of 11
Honors	Day 2 of 10
Block	Day 2 of 6

1 FOCUS

5-Minute Check
(Lesson 3-1)
Express each ratio or rate in simplest form.
1. 8 out of 12 $\frac{2}{3}$
2. 12 girls:20 students $\frac{3}{5}$
3. 9 inches per foot $\frac{3}{4}$

Express each rate as a unit rate.
4. $17.40 in 3 hours **$5.80 per hour**
5. $0.56 for 16 radishes **$0.035 per radish**

 The 5-Minute Check is also available on **Transparency 3-2A** for this lesson.

Motivating the Lesson
Hands-On Activity Display a collection of ten objects—for example, 4 pencils, 1 ruler, 2 erasers, and 3 books. Ask students to calculate what percent of the total each number of objects represents. Have them combine objects into three groups such as erasers and pencils, books and rulers, and so on. Then ask them to calculate the percent of the total that each group now represents.

Multiple Learning Styles

 Visual/Spatial Make six circle graphs that are divided into five equal sections each. Color some of the sections of each graph, and display them where students can easily see them. Separate the class into small groups, and have them work together to determine what percent of each graph is shaded.

Lesson 3-2 **107**

2 TEACH

 Transparency 3-2B contains a teaching aid for this lesson.

Using the Mini-Lab Remind students that the decimal models provide a visual representation of fractions. Stress that there are no right squares to shade and that it is the number of squares, not the particular squares shaded, that is important.

In-Class Examples

For Example 1
Express each ratio as a percent.
a. Six out of 40 people walk to work. **15%**
b. One out of every 4 people in the United States is younger than 18. **25%**

For Example 2
Express each ratio or fraction as a percent.
a. 6 to 8 **75%**
b. $\frac{4}{5}$ **80%**

For Example 3
Express each percent as a fraction.
a. 20% $\frac{2}{5}$
b. 28% $\frac{7}{25}$

Teaching Tip In Example 2a, point out that there are many different equivalent fractions that equal 20%.

Teaching Tip Before discussing Example 4, explain to students that statistical information is often presented as percents. Students can work together to create their own circle graphs based on data they collect such as the number of classmates who walk, ride the bus, or ride a bike to school.

One way to express a fraction or a ratio as a percent is by finding an equivalent fraction with a denominator of 100.

Example 2 Express each ratio or fraction as a percent.

a. In 1995, 2 out of 25 households had a fax machine.

$$\frac{2}{25} \overset{\times 4}{\underset{\times 4}{=}} \frac{8}{100}$$

So, $\frac{2}{25} = 8\%$.

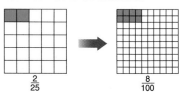

Study Hint
Mental Math Here's another way to express a fraction as a percent. First, divide 100 by the denominator of the fraction. Then, multiply the result by the numerator.

b. About $\frac{1}{5}$ of all adults have taken dance lessons.

$$\frac{1}{5} \overset{\times 20}{\underset{\times 20}{=}} \frac{20}{100}$$

So, $\frac{1}{5} = 20\%$.

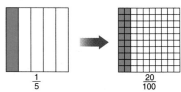

You can express a percent as a fraction by writing it as a fraction with a denominator of 100. Then write the fraction in simplest form.

Example 3 Express each percent as a fraction.

a. 70%

$$70\% = \frac{70}{100} = \frac{7}{10}$$

So, $70\% = \frac{7}{10}$.

b. 45%

$$45\% = \frac{45}{100} = \frac{9}{20}$$

So, $45\% = \frac{9}{20}$.

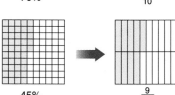

108 Chapter 3 Using Proportion and Percent

✱ Cross-Curriculum Cue

Inform other teachers on your team that your students are studying proportion and percent. Suggestions for curriculum integration are:
Economics: inflation rates
Language: literacy rates
Life Science: extinction rates

The parts of a circle graph are often labeled with percents. The sum of the percents in a circle graph is 100%.

Example 4 INTEGRATION

Statistics The circle graph shows an estimate of the percent of people who traveled to their vacations by car, plane, bus, or other methods. What fraction of the circle graph is represented by each section?

Car: $65\% = \frac{65}{100}$ or $\frac{13}{20}$

Plane: $30\% = \frac{30}{100}$ or $\frac{3}{10}$

Bus: $3\% = \frac{3}{100}$

Other: $2\% = \frac{2}{100}$ or $\frac{1}{50}$

How We Get to Our Summer Vacation

Car 65%, Other 2%, Bus 3%, Plane 30%

Source: Daniel J. Edelman Inc., 1995

CHECK FOR UNDERSTANDING

Communicating Mathematics

Read and study the lesson to answer each question.

1. **Write** the percent and the fraction in simplest form for the model shown at the right. $25\%, \frac{1}{4}$

2. **Explain** how to express the ratio 3 out of 50 as a percent. **See margin.**

3. **Use** grid paper to show 17%. **See margin.**

HANDS-ON MATH

Guided Practice

Express each ratio or fraction as a percent.

4. $\frac{3}{5}$ **60%**   5. 1:10 **10%**

6. Nearly $\frac{90}{100}$ of students ages 9 to 13 play video games. **90%**

Express each percent as a fraction in simplest form.

7. 75% $\frac{3}{4}$ 8. 8% $\frac{2}{25}$ 9. 100% **1**

10. **History** In 1860, Abraham Lincoln was elected president with less than 40% of the popular vote. Express 40% as a fraction in simplest form. $\frac{2}{5}$

EXERCISES

Practice

Express each ratio or fraction as a percent.

11. 2 out of 5 **40%**   12. $\frac{3}{4}$ **75%**   13. $\frac{9}{10}$ **90%**

14. 19:100 **19%**   15. 21 out of 25 **84%**   16. 12.5:100 **12.5%**

Lesson 3-2 Ratios and Percents **109**

In-Class Example

For Example 4
Refer to Example 4 in the Student Edition. Ask students to determine what fraction of people did not use a car to travel to their vacations. $\frac{7}{20}$

3 PRACTICE/APPLY

Check for Understanding

If students need additional practice or instruction after completing Exercises 1–10, one of these options may be helpful.
- Extra Practice, see p. 612
- Reteaching Activity
- *Transition Booklet,* pp. 13–14
- *Study Guide Masters,* p. 21
- *Practice Masters,* p. 21

Assignment Guide

Core: 11–33 odd, 35–39
Enriched: 12–32 even, 33–39

Additional Answers

2. Express $\frac{3}{50}$ as $\frac{6}{100}$, which is 6%.

3.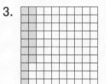

Study Guide Masters, p. 21

■ Reteaching the Lesson ■

Activity Have students find examples of percents used in articles and advertisements in newspapers and magazines. Write each example on the chalkboard. Ask students to explain each percent's meaning in their own words and express the percents as fractions in simplest form.

4 ASSESS

Closing Activity

Modeling Have students use multiple representations (circle graphs, decimal models, mathematical sentences) to show how two ratios are equivalent to the same percent.

Chapter 3, Quiz A (Lessons 3-1 and 3-2) is available in the *Assessment and Evaluation Masters*, p. 71.

Practice Masters, p. 21

Enrichment Masters, p. 21

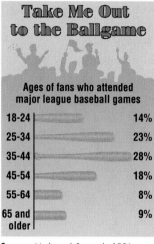

Express each ratio as a percent.

17. Ninety-nine out of 100 U.S. households have at least one radio. **99%**
18. About 1 out of 4 mammals in the world are bats. **25%**
19. In 1995, 23 out of 50 girls ages 12 to 17 played basketball. **46%**

Express each percent as a fraction in simplest form.

20. 40% $\frac{2}{5}$ 21. 5% $\frac{1}{20}$ 22. 10% $\frac{1}{10}$ 23. 25% $\frac{1}{4}$
24. 95% $\frac{19}{20}$ 25. 1% $\frac{1}{100}$ 26. 60% $\frac{3}{5}$ 27. 80% $\frac{4}{5}$

28. Nearly 55% of U.S. households own at least one UNO card game. $\frac{11}{20}$
29. Every day, 77% of Americans age 12 or older listen to the radio. $\frac{77}{100}$
30. There is a 20% chance of rain tomorrow. $\frac{1}{5}$
31. Which is greater, $\frac{3}{5}$ or 70%? **70%**
32. Which is less, $\frac{1}{4}$ or 30%? $\frac{1}{4}$

Applications and Problem Solving

Ages of fans who attended major league baseball games

Age	%
18–24	14%
25–34	23%
35–44	28%
45–54	18%
55–64	8%
65 and older	9%

Source: Mediamark Research, 1996

33. **Baseball** The graph shows the ages of adults who went to baseball games in 1996. Out of every 100 people who went to baseball games, how many would you expect to be between the ages of 25 and 34? **23 people**

34. **Technology** According to *Cablevision*, eleven out of 20 cable TV subscribers decide what program to watch by surfing the channels. Two out of 5 decide by looking in a weekly publication like *TV Guide*. Which way of finding a program represents the greater percent? **surfing**

35. **Critical Thinking** Suppose you choose a blue marble from a bag of different-colored marbles 7 of every 10 times. What is the probability that you will *not* choose a blue marble? **30%**

Mixed Review

36. **Test Practice** A car traveled 828.8 miles on 28 gallons of gas. How many miles per gallon did the car average? *(Lesson 3-1)* **C**
    A 17.96 mpg   B 28.6 mpg   C 29.6 mpg
    D 62.42 mpg   E Not Here

37. Solve $f = 320 \div (-40)$. *(Lesson 2-8)* **−8**

38. Order the set {52, −3, 128, 4, −22, 15, 0, −78} from greatest to least. *(Lesson 2-2)* **{128, 52, 15, 4, 0, −3, −22, −78}**

39. **Geometry** Find the perimeter and area of the parallelogram. *(Lesson 1-8)* **P = 36 in., A = 60 in²**

(parallelogram: 12 in. base, 6 in. height)

110 Chapter 3 Using Proportion and Percent

Extending the Lesson

Activity Have students write at least five questions for a survey on a topic that interests them. Then have students conduct a survey of the class. Tell the students to express their results as percents of the total number of students in the class.

3-3 Solving Proportions

What you'll learn
You'll learn to solve proportions.

When am I ever going to use this?
Knowing how to solve proportions can help you halve a recipe when you are cooking.

Word Wise
proportion
cross products

When a marching band goes by, you can easily recognize the brass section because of its brilliant color. Trumpets, trombones, and baritone horns are made of an alloy called brass, which is a mixture of copper and zinc. The copper and zinc are combined in a ratio of 80 parts copper to 20 parts zinc.

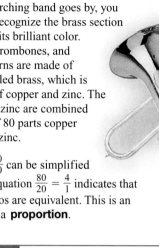

The ratio $\frac{80}{20}$ can be simplified to $\frac{4}{1}$. The equation $\frac{80}{20} = \frac{4}{1}$ indicates that the two ratios are equivalent. This is an example of a **proportion**.

Proportion	Words:	A proportion is an equation that shows that two ratios are equivalent.
	Symbols:	**Arithmetic** $\frac{80}{20} = \frac{4}{1}$ **Algebra** $\frac{a}{b} = \frac{c}{d}, b \neq 0, d \neq 0$

In a proportion, the two **cross products** are equal. The cross products in the proportion below are 2×12 and 3×8.

$$\frac{2}{3} \times \frac{8}{12} \qquad 2 \times 12 = 24, 3 \times 8 = 24$$

Property of Proportions	Words:	The cross products of a proportion are equal.
	Symbols:	If $\frac{a}{b} = \frac{c}{d}$, then $ad = bc$.

You can use cross products to determine whether a pair of ratios forms a proportion. If the cross products of two ratios are equal, then the ratios form a proportion. If the cross products are *not* equal, the ratios do *not* form a proportion.

Study Hint
Mental Math If both ratios simplify to the same fraction, they form a proportion.
$\frac{8}{12} = \frac{2}{3}, \frac{12}{18} = \frac{2}{3}$

Example 1 Determine whether the ratios $\frac{9}{12}$ and $\frac{18}{24}$ form a proportion.

Find the cross products. $\frac{9}{12} \times \frac{18}{24}$

9 ⨯ 24 = 216   12 ⨯ 18 = 216

Since the cross products are equal, the ratios form a proportion.

Lesson 3-3 Solving Proportions **111**

3-3 Lesson Notes

Instructional Resources
- *Study Guide Masters*, p. 22
- *Practice Masters*, p. 22
- *Enrichment Masters*, p. 22
- Transparencies 3-3, A and B
- *Assessment and Evaluation Masters*, pp. 70, 71
- *Diversity Masters*, p. 29
- *School to Career Masters*, p. 29
- *Technology Masters*, p. 58
- CD-ROM Program
  - Resource Lesson 3-3
  - Interactive Lesson 3-3

Recommended Pacing	
Standard	Day 4 of 11
Honors	Day 3 of 10
Block	Day 2 of 6

1 FOCUS

5-Minute Check
(Lesson 3-2)

Express each ratio or fraction as a percent.
1. Thirteen out of 100 cars are green. **13%**
2. $\frac{3}{5}$ **60%**

Express each percent as a fraction in simplest form.
3. 35% $\frac{7}{20}$
4. 6% $\frac{3}{50}$
5. Which is greater, $\frac{2}{3}$ or 65%? $\frac{2}{3}$

The 5-Minute Check is also available on **Transparency 3-3A** for this lesson.

2 TEACH

Transparency 3-3B contains a teaching aid for this lesson.

Reading Mathematics The introduction to this lesson includes mathematical concepts presented in verbal form. Make sure that students can pronounce the word and understand the meaning of *proportion* and how it is used to express equality between two ratios.

Motivating the Lesson

Hands-On Activity Give students 24 counters and ask them to separate the counters into two groups that form the ratios 2:4 and 6:12. Ask students to determine the relationships between these two ratios. **Both ratios simplify to the same fraction of $\frac{1}{2}$.**

Lesson 3-3 **111**

In-Class Examples

For Example 1
Determine whether the ratios 3:18 and 6:36 form a proportion. **Yes. Both cross products equal 108.**

For Example 2
Solve $\frac{5}{8} = \frac{7.5}{z}$. **12**

For Example 3
If you travel 540 miles in 3 days during your summer vacation, how many miles could you travel in 10 days? **1,800 miles**

Teaching Tip Point out that in a proportion, the unit rates are equal.

proportion: $\frac{3}{50} = \frac{1{,}220}{20{,}000}$

unit rates: $\frac{0.06}{1} = \frac{0.06}{1}$

3 PRACTICE/APPLY

Check for Understanding
If students need additional practice or instruction after completing Exercises 1–10, one of these options may be helpful.
- Extra Practice, see p. 612
- Reteaching Activity
- *Transition Booklet,* pp. 9–10
- *Study Guide Masters,* p. 22
- *Practice Masters,* p. 22

Study Guide Masters, p. 22

In a proportion like $\frac{3}{7} = \frac{2.1}{d}$, one of the terms is not known. You can use cross products to find the value of d. This is known as *solving the proportion*.

Examples

INTEGRATION

LOOK BACK
You can refer to Lesson 1-5 to review how to solve equations.

② **Algebra** Solve $\frac{3}{7} = \frac{2.1}{d}$.

$\frac{3}{7} = \frac{2.1}{d}$

$3 \times d = 7 \times 2.1$ *Find the cross products.*

$3d = 14.7$ *Simplify.*

$\frac{3d}{3} = \frac{14.7}{3}$ *Solve the equation by dividing each side by 3.*

$d = 4.9$ *The solution is 4.9.*

You can check the solution by substituting the value of d into the original proportion and checking the cross products.

APPLICATION

③ **Recycling** When 2,000 pounds of paper are recycled or reused, 17 trees are saved. How many trees would be saved if 5,000 pounds of paper are recycled?

Write a proportion. Let t represent the number of trees.

paper → $\frac{2{,}000}{17} = \frac{5{,}000}{t}$ ← paper
trees →   ← trees

Find the cross products. Then solve for t.

Method 1 Use paper and pencil.

$2{,}000 \times t = 17 \times 5{,}000$

$2{,}000t = 85{,}000$

$\frac{2{,}000t}{2{,}000} = \frac{85{,}000}{2{,}000}$

$t = 42.5$

Method 2 Use a calculator.

17 $\times$ 5000 $\div$ 2000 $=$ *42.5*

If 5,000 pounds of paper are recycled, 42.5 trees are saved.

CHECK FOR UNDERSTANDING

Communicating Mathematics

Read and study the lesson to answer each question.   1–3. See Answer Appendix.

1. **Tell** two methods for determining whether two ratios form a proportion.
2. **List** two ratios that do not form a proportion. Explain why they do not.
3. **Write** a paragraph explaining how you use algebra when solving a proportion.

112   Chapter 3 Using Proportion and Percent

Reteaching the Lesson

Activity Have students find the prices of one, two, three, and four tapes at $8 per tape. Then have them write the ratio of the number of tapes to the cost. $\frac{1}{8}, \frac{2}{16}, \frac{3}{24}, \frac{4}{32}$ Show that any pair of these ratios forms a proportion.

Error Analysis
Watch for students who multiply numerators and multiply denominators to find cross products.
Prevent by stressing the term *cross product* and depicting it as a pair of crossing lines, indicating how the products are to be found.

Guided Practice

Determine whether each pair of ratios forms a proportion.

4. $\frac{75}{100}, \frac{3}{4}$ yes 5. $\frac{16}{12}, \frac{12}{9}$ yes 6. $\frac{3}{11}, \frac{55}{200}$ no

Solve each proportion. 9. 0.2

7. $\frac{2}{34} = \frac{5}{x}$ 85 8. $\frac{3}{5} = \frac{c}{10}$ 6 9. $\frac{a}{0.9} = \frac{0.6}{2.7}$

10. **Life Science** When a robin flies, it beats its wings an average of 23 times in 10 seconds. How many times will its wings beat in two minutes? **276 times**

EXERCISES

Practice

Determine whether each pair of ratios forms a proportion.

11. $\frac{7}{6}, \frac{6}{7}$ no   12. $\frac{8}{12}, \frac{10}{15}$ yes   13. $\frac{10}{12}, \frac{5}{6}$ yes   14. $\frac{6}{9}, \frac{8}{12}$ yes

15. $\frac{10}{16}, \frac{25}{45}$ no   16. $\frac{18}{14}, \frac{54}{42}$ yes   17. $\frac{0.2}{0.6}, \frac{0.5}{1.5}$ yes   18. $\frac{0.6}{2.4}, \frac{0.4}{0.16}$ no

Solve each proportion.

19. $\frac{4}{100} = \frac{12}{n}$ 300 20. $\frac{9}{1} = \frac{n}{14}$ 126 21. $\frac{5}{4} = \frac{y}{12}$ 15 22. $\frac{120}{b} = \frac{24}{60}$ 300

23. $\frac{2}{3} = \frac{7}{y}$ 10.5 24. $\frac{10}{6} = \frac{y}{26}$ $43\frac{1}{3}$ 25. $\frac{0.35}{3} = \frac{c}{18}$ 2.1 26. $\frac{n}{2} = \frac{0.7}{0.4}$ 3.5

Applications and Problem Solving

27. **Cooking** Ruth Wakefield invented the recipe for Original Toll House Chocolate Chip Cookies in 1933. It calls for 2.25 cups of flour and 0.75 cup of granulated sugar. Suppose you only have 1.5 cups of flour. How much granulated sugar should you use when you bake the cookies? **0.5 cup**

28. **Hobbies** Model railroads are scaled-down replicas of real trains. One of the most popular modeling sizes is HO, where 1 inch on the model represents 87 inches on a real train. An HO model of a modern diesel locomotive is 8 inches long. How long is the real locomotive? **696 inches or 58 feet**

29. **Working on the CHAPTER Project** Refer to the foreign exchange rate you found on page 103.
    a. Estimate the amount of U.S. currency you will need to exchange for transportation, housing, food, and souvenirs. **See students' work.**
    b. Write and solve a proportion to find the amount of foreign currency you will receive for each category. **See students' work.**

30. **Critical Thinking** In the proportion $\frac{10}{x} = \frac{y}{5}$, how does the value of y change as the value of x gets larger? **The value of y decreases.**

Mixed Review

31. Express 35% as a fraction in simplest form. *(Lesson 3-2)* $\frac{7}{20}$

32. **Algebra** Solve $-460 = -16a + 52$. *(Lesson 2-9)* **32**

33. Find $-7(-31)$. *(Lesson 2-7)* **217**

34. **Test Practice** What is the value of the expression $3^2 + 4 \cdot 2 - 6 \div 2$? *(Lesson 1-3)* **A**

    A 14   B 10   C 5.5   D -26

Extending the Lesson

Enrichment Masters, p. 22

Activity If the numerator of one fraction and the denominator of the other fraction in a proportion are equal, the equal terms are called *mean proportionals*: $\frac{1}{6} = \frac{6}{36}$. Thus, 6 is the mean proportional between 1 and 36. Have students find the mean proportionals between 2 and 8 and between 8 and 18. **4, 12**

Assignment Guide

Core: 11–27 odd, 30–34
Enriched: 12–26 even, 27, 28, 30–34

CHAPTER Project

Exercise 29 asks students to advance to the next stage of work on the Chapter Project. You may want to separate students into groups of four or five to continue work on their foreign exchange charts and to solve the percent problems.

4 ASSESS

Closing Activity

Speaking Write simple proportions on the chalkboard with one term missing, for example, $\frac{1}{2} = \frac{x}{6}$. Have students find the missing term using mental math. $x = 3$

Chapter 3, Quiz B (Lesson 3-3) is available in the *Assessment and Evaluation Masters*, p. 71.

Mid-Chapter Test (Lessons 3-1 through 3-3) is available in the *Assessment and Evaluation Masters*, p. 70.

Practice Masters, p. 22

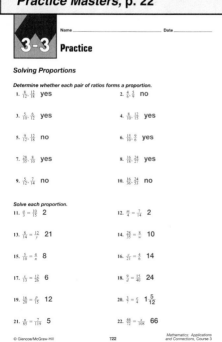

3-4 Lesson Notes

Instructional Resources
- *Study Guide Masters*, p. 23
- *Practice Masters*, p. 23
- *Enrichment Masters*, p. 23
- *Transparencies 3-4, A and B*
- *Classroom Games*, pp. 7–9
- *Hands-On Lab Masters*, p. 70
- CD-ROM Program
  - Resource Lesson 3-4

Recommended Pacing	
Standard	Days 5 & 6 of 11
Honors	Days 4 & 5 of 10
Block	Day 3 of 6

1 FOCUS

5-Minute Check
(Lesson 3-3)

Determine whether each pair of ratios forms a proportion.
1. $\frac{4}{3}, \frac{6}{8}$ **no**
2. $\frac{4}{10}, \frac{10}{25}$ **yes**

Solve each proportion.
3. $\frac{10}{n} = \frac{6}{9}$ **15**
4. $\frac{15}{35} = \frac{k}{14}$ **6**
5. If 12 people at a baseball game eat 17 hot dogs, how many hot dogs will 24 people eat? **34**

The 5-Minute Check is also available on **Transparency 3-4A** for this lesson.

Motivating the Lesson

Communication As students read the lesson opener, ask them to think about how often they see important information expressed as percents. Explain that being able to understand and compile data in percents will help when analyzing information and ideas.

3-4 Fractions, Decimals, and Percents

What you'll learn
You'll learn to express percents as fractions and decimals and vice versa.

When am I ever going to use this?
Fractions, decimals, and percents are used interchangeably in situations like the stock market and department store sales.

What would you do without your blow dryer? How about the telephone? The table shows that 42% of adults say they can't live without a telephone.

In order to compute with percents, you must first express them as fractions or decimals. In Lesson 3-2, you expressed percents as fractions by writing the percent as a fraction with a denominator of 100. The word percent also means *hundredths* or *per hundred*.

Famous Inventions
% of adults who say they can't live without them

automobile	63%
light bulb	54%
telephone	42%
television	22%
aspirin	19%
microwave	13%
blow dryer	8%

Source: *American Demographics*, 1996

You can express percents as decimals by dividing by 100 and expressing the answer as a decimal.

TECHNOLOGY MINI-LAB

Work with a partner.   calculator

You can use the percent key on your calculator to express 42% as a decimal.
- Enter 42 [2nd] [%].
- The display shows 0.42. So, 42% = 0.42.

Try This

Use a calculator to express each percent as a decimal.
1. 16% **0.16**   2. 55% **0.55**   3. 93% **0.93**   4. 100% **1**
5. 40% **0.4**   6. 2% **0.02**   7. 12.5% **0.125**   8. 66.7% **0.667**

Talk About It

9. Explain what happened to the number in the display when you pressed the percent key.
10. Use what you have learned in this lab to express 5% as a decimal without using a calculator. **0.05**

9. The decimal point was moved two places to the left.

Example 1 Express 60% as a fraction and as a decimal.

THINK: 60 hundredths

$60\% = \frac{60}{100}$ or $\frac{3}{5}$

$60\% = 0.60$ or 0.6

114   Chapter 3 Using Proportion and Percent

Multiple Learning Styles

Logical Have students work in pairs to research statistical data presented in publications like *USA Today* or *Newsweek*. Have each pair use percents converted to decimals to explain the data to the class.

To express a decimal as a percent, first express the decimal as a fraction with a denominator of 100. Then express the fraction as a percent.

Example 2 Express each decimal as a percent.

a. 0.78

$$0.78 = \frac{78}{100}$$
$$= 78\%$$

b. 0.3

$$0.3 = \frac{3}{10} \text{ or } \frac{30}{100}$$
$$= 30\%$$

You can use these shortcuts when working with decimals and percents.

Study Hint

Mental Math To multiply by 100, move the decimal point two places to the right. To divide by 100, move the decimal point two places to the left.

Decimals and Percents
- To write a decimal as a percent, multiply by 100 and add the % symbol.
  $$0.24 = 0.24 = 24\%$$
- To write a percent as a decimal, divide by 100 and remove the % symbol.
  $$24\% = 24\% = 0.24$$

You have learned to express a fraction as a percent by finding an equivalent fraction with a denominator of 100. This method works well if the denominator of the fraction is a number like 2, 4, 5, 10, or 25. That is, the denominator is a factor of 100. If the denominator is *not* a factor of 100, you can solve a proportion.

Example 3 Express each fraction as a percent.

a. $\frac{1}{8}$

$$\frac{1}{8} = \frac{n}{100} \quad \text{Write a proportion.}$$
$$1 \times 100 = 8 \times n \quad \text{Find the cross products.}$$
$$100 = 8n$$
$$\frac{100}{8} = \frac{8n}{8} \quad \text{Solve the equation by dividing each side by 8.}$$
$$12.5 = n$$

So, $\frac{1}{8} = 12.5\%$.

b. $\frac{2}{3}$

$$\frac{2}{3} = \frac{n}{100} \quad \text{Write a proportion.}$$
$$2 \times 100 = 3 \times n \quad \text{Find the cross products.}$$
$$2 \;\boxed{\times}\; 100 \;\boxed{\div}\; 3 \;\boxed{=}\; 66.66666667$$

So, $\frac{2}{3} \approx 66.67\%$.

Lesson 3-4 Fractions, Decimals, and Percents **115**

---

2 TEACH

Transparency 3-4B contains a teaching aid for this lesson.

Using the Mini-Lab Ask students to use their calculators to express the following percents as decimals: 82%, 0.5%, 3%. 0.82, 0.005, 0.03

In-Class Examples

For Example 1
Express each percent as a fraction and as a decimal.
a. 30% $\frac{3}{10}$, 0.30
b. 85% $\frac{17}{20}$, 0.85

For Example 2
Express each decimal as a percent.
a. 0.63 63%
b. 0.42 42%

For Example 3
Express each fraction as a percent.
a. $\frac{1}{16}$ 6.25%
b. $\frac{5}{6}$ 83.33%

Classroom Vignette

"I use a container of colored cubes to help students understand the relationship among fractions, decimals, and percents. Students must sort the cubes by color, find the total, then find the percent each color is of the total. As an added feature, students may graph their results."

Anne B. Breda, Department Chair
Hambrick Middle School
Houston, TX

In-Class Example

For Example 4
If there are 128 people living in an apartment building and 46 of them own pets, what percent of the people are pet owners? **35.9%**

Teaching Tip In Example 4, remind students that $27 \div 250 \times 100$ produces the same answer as $27 \times 100 \div 250$.

3 PRACTICE/APPLY

Check for Understanding

If students need additional practice or instruction after completing Exercises 1–13, one of these options may be helpful.
- Extra Practice, see p. 613
- Reteaching Activity
- *Study Guide Masters*, p. 23
- *Practice Masters*, p. 23

Have students share with the class some of the advertisements that use ratios. Have them categorize the ads, for example, according to the type of product being sold.

Study Guide Masters, p. 23

Example 4 — **Life Science** There are 250 known species of sharks, and of that number, only 27 species have been involved in attacks on humans. What percent of known species of sharks have attacked humans?

Express the fraction $\frac{27}{250}$ as a percent by solving a proportion. Let n represent the percent of sharks.

$$\frac{27}{250} = \frac{n}{100}$$

27 × 100 ÷ 250 = *10.8*

So, 10.8% of the known species of sharks have attacked humans.

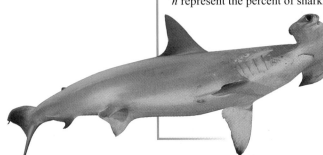

CHECK FOR UNDERSTANDING

Communicating Mathematics

Read and study the lesson to answer each question.

1. *Write* a fraction, decimal, and percent to represent the model at the right. **$\frac{7}{20}$, 0.35, 35%**

2. *Explain* two different methods you can use to express fractions as percents. Use $\frac{13}{20}$ and $\frac{13}{19}$ in your explanation. **See Answer Appendix.**

HANDS-ON MATH

3. *Use models* to show which is greater, 15% or $\frac{1}{8}$. **See Answer Appendix.**

Guided Practice

Express each percent as a decimal.

4. 58% **0.58**   5. 9% **0.09**   6. 23.8% **0.238**

Express each decimal as a percent.

7. 0.29 **29%**   8. 0.7 **70%**   9. 0.042 **4.2%**

Express each fraction as a percent.

10. $\frac{3}{4}$ **75%**   11. $\frac{5}{8}$ **62.5%**   12. $\frac{1}{6}$ **16.67%**

13. *Earth Science* About $\frac{3}{20}$ of the total land surface of Earth is covered by deserts. What percent is this? **15%**

EXERCISES

Practice

Express each percent as a decimal.

14. 26% **0.26**   15. 74% **0.74**   16. 6% **0.06**   17. 1% **0.01**
18. 80% **0.80 or 0.8**   19. 15.5% **0.155**   20. 3.2% **0.032**   21. 3.8% **0.038**

Express each decimal as a percent.

22. 0.19 **19%**   23. 0.02 **2%**   24. 0.9 **90%**   25. 0.026 **2.6%**
26. 0.375 **37.5%**   27. 0.667 **66.7%**   28. 0.083 **8.3%**   29. 0.0625 **6.25%**

Reteaching the Lesson

Activity Make charts like this one for students to complete.

Fraction	Decimal	Percent
? $\frac{1}{4}$	0.25	25%
$\frac{7}{10}$	? 0.7	70%
$\frac{1}{20}$	0.7	? 5%

Error Analysis
Watch for students who express fractions as percents incorrectly and vice versa.
Prevent by having students prepare a fraction-percent equivalencies chart that includes the most commonly used numbers. This will guide them in estimating their answers.

Express each fraction as a percent.

30. $\frac{23}{50}$ 46%   31. $\frac{7}{8}$ 87.5%   32. $\frac{19}{20}$ 95%   33. $\frac{5}{8}$ 62.5%
34. $\frac{1}{3}$ 33.3%   35. $\frac{5}{16}$ 31.25%   36. $\frac{27}{30}$ 90%   37. $\frac{15}{40}$ 37.5%

Replace each ● with <, >, or = to make a true sentence.

38. 45% ● 4.5 <   39. 9.3 ● 93% >   40. 0.05 ● 50% <
41. 57.8% ● 0.578 =   42. 0.3 ● 30% =   43. 1% ● 0.09 <

44. Express the ratio 1:40 as a percent. **2.5%**

45. Which is greater: $\frac{3}{8}$ or 40%? **40%**

Applications and Problem Solving

46. *Geometry* What percent of the area of the square at the right is shaded? **31.25%**

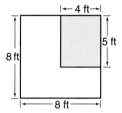

47. *Transportation* According to the Association of American Railroads, about 2 out of every 5 trains in use in 1995 were built in the 1970s. Express 2 out of 5 as a percent. **40%**

48. *Critical Thinking* Which is greater: 2% of 98 or 98% of 2? **They are equal.**

Mixed Review

49. *Algebra* Solve $\frac{3}{7} = \frac{n}{28}$. *(Lesson 3-3)* **12**

50. What percent of the squares at the right are shaded? *(Lesson 3-2)* **30%**

51. Express the ratio *20 out of 50 students* in simplest form. *(Lesson 3-1)* **2 out of 5**

52. *Algebra* Solve $4m > 24$. Graph the solution on a number line. *(Lesson 1-9)* **$m > 6$; See margin for graph.**

53. Evaluate 3^4. *(Lesson 1-2)* **81**

54. **Test Practice** Kyung had $17. His dinner cost $5.62, and he gave the cashier a $10 bill. How much change should he receive from the cashier? *(Lesson 1-1)* **B**

   A $4.32   B $4.38   C $11.38   D $15.62   E Not Here

CHAPTER 3 Mid-Chapter Self Test

1. Express the rate $420 for 15 tickets as a unit rate. *(Lesson 3-1)* **$28/ticket**
2. Express the ratio 3 out of 20 as a percent. *(Lesson 3-2)* **15%**
3. *Algebra* Solve $\frac{x}{36} = \frac{15}{24}$. *(Lesson 3-3)* **22.5**
4. *Physical Science* Light travels approximately 1,860,000 miles in 10 seconds. How long will it take light to travel the 93,000,000 miles from the Sun to Earth? *(Lesson 3-3)* **500 s or about 8.3 min**
5. Express $\frac{3}{50}$ as a percent using a proportion. *(Lesson 3-4)* **6%**

Lesson 3-4 Fractions, Decimals, and Percents **117**

Extending the Lesson

Enrichment Masters, p. 23

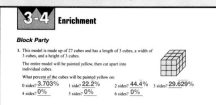

Activity Have students brainstorm to create a list of book or movie titles, slogans, or well-known phrases that contain fractions, decimals, or percents. Have them rewrite each phrase using the two other forms of the value.

Assignment Guide

Core: 15–47 odd, 48–54
Enriched: 14–44 even, 46–54
All: Self Test 1–5

4 ASSESS

Closing Activity
Writing Have students write a fraction, a decimal, and a percent and explain how to express each in the other two forms.

Mid-Chapter Self Test
The Mid-Chapter Self Test reviews the concepts in Lessons 3-1 through 3-4. Lesson references are given so students can review concepts not yet mastered.

Additional Answer
52.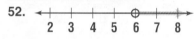

Practice Masters, p. 23

Lesson 3-4 **117**

3-4B Hands-On Lab Notes

GET READY

Objective Students find the value of the golden ratio.

Optional Resources
Hands-On Lab Masters
- grid paper, p. 10
- worksheet, p. 42

Overhead Manipulative Resources
- centimeter grid

Manipulative Kit
- scissors
- tape measure

MANAGEMENT TIPS

Recommended Time
45 minutes

Getting Started Use a 50-square centimeter grid transparency on the overhead projector to illustrate examples of rectangles. Ask students to record the length and width of each rectangle and to determine the ratio of length to width and whether it can be called a golden rectangle.

Teaching Tip Some students may perceive length and width according to the position of the rectangle. Remind them that in this situation, length refers to the longest side.

The **Activity** demonstrates how to use a model to determine a sequence involving the golden ratio. As an extension, ask students to use their tables to determine the dimensions of the next larger golden rectangle after the 34 units by 21 units rectangle they began with. **55 units by 34 units**

Hands-On Lab

COOPERATIVE LEARNING

3-4B The Golden Ratio

A Follow-Up of Lesson 3-4

 grid paper

 scissors

 calculator

 tape measure

What do a cereal box, a skyscraper, and a famous painting have in common? One thing is their rectangular shape.

Mathematicians have studied rectangles for centuries. Many ancient Greek mathematicians and philosophers looked for a special rectangle that was the most perfect, visually appealing rectangle. Some rectangles are shown below. Take an informal survey of your classmates to find which one they like best.

 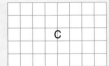

Most people choose rectangle C. It's not too skinny and not too square. This rectangle is called a *golden rectangle*. Notice that the ratio of the length to width is $\frac{8}{5}$ or 1.6.

TRY THIS

Work in groups of three.

Step 1 Cut a rectangle out of grid paper that is 34 squares long and 21 squares wide. The ratio of length to width is $\frac{34}{21}$ or about 1.62.

Step 2 Cut the rectangle into two parts, in which one part is the largest possible square and the other part is a rectangle. Measure the rectangle and record its length and width. Write the ratio of length to width and then express it as a decimal. Record your data in a table like the one shown below.

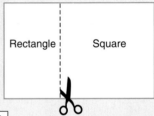

length	34	21	?	?	?	?	?
width	21	13	?	?	?	?	?
ratio	$\frac{34}{21}$	?	?	?	?	?	?
decimal	1.62	?	?	?	?	?	?

Step 3 Use the same method to cut the smaller rectangle into two parts until the remaining rectangle is 1 square by 2 squares. Measure each rectangle to the nearest tenth of a centimeter, and find the ratio of length to width.

Additional Answer for Step 2

length	34	21	13	8	5	3	2
width	21	13	8	5	3	2	1
ratio	$\frac{34}{21}$	$\frac{21}{13}$	$\frac{13}{8}$	$\frac{8}{5}$	$\frac{5}{3}$	$\frac{3}{2}$	$\frac{2}{1}$
decimal	1.62	1.62	1.63	1.6	1.67	1.5	2

ON YOUR OWN

1. Look for a pattern in the ratios you recorded. Describe the pattern.   **Except for 2, all are near 1.6.**

2. If the rectangles you cut out are described as golden rectangles, what do you think the value of the *golden ratio* is?   **about 1.6**

3. Write a definition of *golden rectangle*. Use the word *ratio* in your definition.   **See margin.**

4. The ancient Greeks considered the value of the golden ratio to be 1.618.
   a. Measure the height of your face from the top of your head to your chin. Then measure the width of your face at your cheekbone. Find the ratio for each person in your group.   **See students' work.**
   b. Find the average ratio for your group. Is it close to the golden ratio?   **See margin.**

5. There are many examples of the golden ratio in music, architecture, art, and nature. Some are shown below.

Composition with Grey, Red, Yellow, and Blue by Piet Mondrian

The Parthenon (Greece, 450 B.C.)

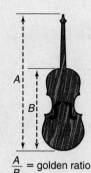
$\dfrac{A}{B}$ = golden ratio

   Do research to find another example of the golden ratio.   **Sample answer: United Nations building in New York City**

6. The spiral of the chambered nautilus follows the pattern of the golden ratio.   **a–b. See margin.**
   a. Trace the figure at the right. It is a smaller version of the golden rectangle you cut apart at the beginning of this lab.
   b. An arc connects the opposite corners of the large square. Starting at the endpoint of the first arc, sketch an arc that connects the opposite corners of the next largest square. Continue the process.
   c. What do the combined arcs resemble?   **spiral**

7. The *Fibonacci sequence* is a famous mathematical sequence. It begins 1, 1, 2, 3, 5, 8,… .
   a. Find the next six numbers in the sequence.   **13, 21, 34, 55, 89, 144**
   b. What is the relationship between the numbers in the Fibonacci sequence and the golden ratio?   **See margin.**

Lesson 3-4B **HANDS-ON LAB** 119

3-5 Lesson Notes

Instructional Resources
- *Study Guide Masters*, p. 24
- *Practice Masters*, p. 24
- *Enrichment Masters*, p. 24
- Transparencies 3-5, A and B
- *Assessment and Evaluation Masters*, p. 72
- *Science and Math Lab Manual*, pp. 53–56
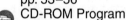
  CD-ROM Program
  - Resource Lesson 3-5

Recommended Pacing	
Standard	Day 7 of 11
Honors	Day 6 of 10
Block	Day 4 of 6

1 FOCUS

 5-Minute Check
(Lesson 3-4)

1. Which is greater, $\frac{1}{25}$ or 5%?
   5%
2. Express 0.2 as a percent.
   20%
3. Express $\frac{13}{20}$ as a percent.
   65%
4. Express 8% as a decimal.
   0.08
5. If there are 20 houses on a block and 4 of them are tan, what percent are *not* tan?
   80%

The 5-Minute Check is also available on **Transparency 3-5A** for this lesson.

2 TEACH

Transparency 3-5B contains a teaching aid for this lesson.

Reading Mathematics Students may not remember that *of* means to multiply. Remind them that 10% of a number means $\frac{10}{100} \times$ the number and that, because 10% is less than 1, their answer will be less than the original amount.

3-5 Finding Percents

What you'll learn
You'll learn to compute mentally with percents.

When am I ever going to use this?
You'll use mental math strategies to help estimate the amount of a tip to leave at a restaurant.

The idea of percent may have originated when the Roman emperor Augustus (63 B.C.–A.D. 14) levied a tax on all goods sold at auction. The tax rate was $\frac{1}{100}$ or 1%.

Look for a pattern in the following problems to find 1% of a number mentally.

Number	Computation	Result
48	1% of 48 → 1 [2nd] [%] [×] 48 [=]	0.48
6	1% of 6 → 1 [2nd] [%] [×] 6 [=]	0.06
125	1% of 125 → 1 [2nd] [%] [×] 125 [=]	1.25

These and other examples suggest that you can find 1% of a number mentally by moving the decimal point two places to the left.

Examples

1 Find 1% of 235.

1% of 235 is 2.35.

2 Find 1% of 12.8.

1% of 12.8 is 0.128.

You can also find 10% of a number mentally.

10% of 82 → 10 [2nd] [%] [×] 82 [=]   8.2

10% of 235 → 10 [2nd] [%] [×] 235 [=]   23.5

These and other examples suggest that you can find 10% of a number mentally by moving the decimal point one place to the left.

Example APPLICATION 3

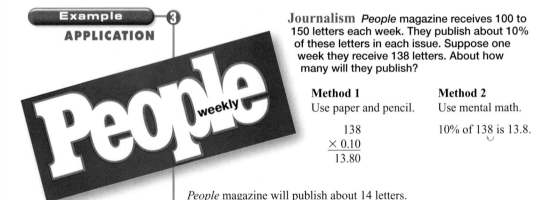

Journalism *People* magazine receives 100 to 150 letters each week. They publish about 10% of these letters in each issue. Suppose one week they receive 138 letters. About how many will they publish?

Method 1
Use paper and pencil.

$$\begin{array}{r} 138 \\ \times\, 0.10 \\ \hline 13.80 \end{array}$$

Method 2
Use mental math.

10% of 138 is 13.8.

People magazine will publish about 14 letters.

120 Chapter 3 Using Proportion and Percent

Motivating the Lesson
Problem Solving Ask students how much a 15% tip would be on a meal costing $45.00. **$6.75** A 10% tip? **$4.50** A 20% tip? **$9.00**

When you compute with common percents like 25%, 12.5%, or $33\frac{1}{3}$%, it may be easier to use the fraction form of a percent. This number line shows some fraction-percent equivalents.

Examples

4 Find 25% of 32.

25% of 32 = $\frac{1}{4}$ of 32 or 8

5 Find 50% of 120.

50% of 120 = $\frac{1}{2}$ of 120 or 60

6 Find 80% of $45.

80% of 45 = $\frac{4}{5}$ of 45

THINK: $\frac{1}{5}$ of 45 is 9. So, $\frac{4}{5}$ of 45 is 4 × 9 or 36.

80% of $45 is $36.

Cultural Kaleidoscope

The percent sign may have come from a symbol in an anonymous Italian manuscript written in 1425. Instead of *per 100*, the author used P. By about 1650, had become %.

Some percents are used much more frequently than others. For this reason, it's a good idea to memorize the equivalent fractions, decimals, and percents listed in the table below.

Equivalent Fractions, Decimals, and Percents			
$\frac{1}{2}$, 0.5, 50%	$\frac{2}{3}$, 0.66$\frac{2}{3}$, 66$\frac{2}{3}$%	$\frac{4}{5}$, 0.8, 80%	$\frac{7}{8}$, 0.875, 87.5%
$\frac{1}{4}$, 0.25, 25%	$\frac{1}{5}$, 0.2, 20%	$\frac{1}{8}$, 0.125, 12.5%	$\frac{3}{10}$, 0.3, 30%
$\frac{3}{4}$, 0.75, 75%	$\frac{2}{5}$, 0.4, 40%	$\frac{3}{8}$, 0.375, 37.5%	$\frac{7}{10}$, 0.7, 70%
$\frac{1}{3}$, 0.33$\frac{1}{3}$, 33$\frac{1}{3}$%	$\frac{3}{5}$, 0.6, 60%	$\frac{5}{8}$, 0.625, 62.5%	$\frac{9}{10}$, 0.9, 90%

CHECK FOR UNDERSTANDING

Communicating Mathematics

Read and study the lesson to answer each question.

1. *Explain* how to find 1% of a number mentally. **See margin.**
2. *Name* the fraction you would use to find 62.5% of 16 mentally. $\frac{5}{8}$
3. *You Decide* Lawana thinks 10% of 95 is 0.95. Elisa thinks 10% of 95 is 9.5. Who is correct? Explain your reasoning. **Elisa; you move the decimal point one place to the left.**

Guided Practice

Compute mentally.

4. 10% of 25 **2.5**
5. 1% of $1,355 **$13.55**
6. 75% of $120 **$90**
7. 37.5% of 16 **6**

Lesson 3-5 Finding Percents **121**

Assignment Guide
Core: 11–29 odd, 30–34
Enriched: 10–26 even, 28–34

4 ASSESS

Closing Activity
Writing Have students explain in writing how to mentally compute 8% of $35,675. **$2,854**

Chapter 3, Quiz C (Lessons 3-4 and 3-5) is available in the *Assessment and Evaluation Masters*, p. 72

Additional Answer
30. Sample answer: Find 10% of the number and then find half of that. Add them together.

Practice Masters, p. 24

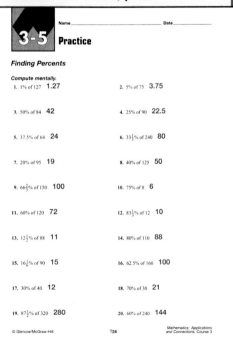

8. Which is greater, 25% of 16 or 5? **5**

9. **Health** Many health authorities recommend that a healthy diet contains no more than 30% of its Calories from fat. If you consume 1,500 Calories each day, what is the maximum number of Calories you should consume from fat? **450 Calories**

EXERCISES

Practice Compute mentally.

10. 10% of $105 **$10.50**
11. 1% of 17.2 **0.172**
12. 20% of 5 **1**
13. 87.5% of 160 **140**
14. 50% of $36 **$18**
15. 12.5% of 48 **6**
16. $33\frac{1}{3}$% of 24 **8**
17. 90% of 1,000 **900**
18. 30% of 50 **15**
19. 40% of 75 **30**
20. 1% of 15,389 **153.89**
21. 10% of $38.41 **$3.84**

Replace each ● with <, >, or = to make a true sentence.

22. 10 ● 10% of 90 **>**
23. 75% of 20 ● 16 **<**
24. $66\frac{2}{3}$% of 18 ● 60% of 15 **>**
25. 1% of 120 ● 10% of 12 **=**

26. Find 1% of $36,700 mentally. **$367**
27. Find 10% of $12.95 mentally. **$1.30**

Applications and Problem Solving

28. **Statistics** Refer to the graph. Suppose 1,600 women were surveyed.
    a. How many women said they bought more than seven pairs of shoes? **400 women**
    b. How many women responded that they didn't know how many pairs of shoes they had purchased? **16 women**

29. **Population** In 1996, there were about 15,167,000 people in the United States who were ages 14 to 17. That section of the population is expected to increase by about 10% by the year 2010. Estimate how many people in the United States will be ages 14 to 17 in 2010. **about 16,683,700 people**

Source: Naturalizer Shoes Trends Survey/Gallup, 1994

30. **Critical Thinking** Devise a method to find 15% of a number mentally. **See margin.**

Mixed Review

31. **Sports** In 1996, 39.5% of high school varsity athletes were girls. In 1973, only 17.8% of varsity athletes were girls. Express both percents as decimals. *(Lesson 3-4)* **0.395, 0.178**

32. **Test Practice** Solve the equation $2t - (-65) = 47$. *(Lesson 2-9)* **B**
    A −18 B −9 C 56 D 112

33. Find $-15 + (-2)$. *(Lesson 2-3)* **−17**

34. **Geometry** Find the perimeter and area of a rectangle that is 2 centimeters wide and 5 centimeters long. *(Lesson 1-8)* **P = 14 cm; A = 10 cm²**

Internet Connection For the latest population statistics, visit: www.glencoe.com/sec/math/mac/mathnet

122 Chapter 3 Using Proportion and Percent

Extending the Lesson

Enrichment Masters, p. 24

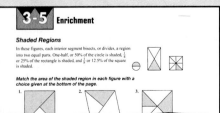

Activity Scott, Amie, and Chip eat lunch at a restaurant. Their bill totals $25.00. They determine that Scott and Chip each owe $\frac{2}{5}$ of the total while Amie owes $\frac{1}{5}$. What amount and percent of the bill should each person pay? **Scott: $10, 40%; Chip: $10, 40%; Amie: $5, 20%**

Let the Games Begin

Per-fraction

Get Ready This game is for two, three, or four players.

 32 index cards   scissors   markers

Get Set Cut two index cards in half for a total of four pieces. On one piece, write a fraction from the table on page 121. On another piece, write any fraction that is equivalent to the first fraction. On the third piece, write the equivalent decimal. On the last piece, write the equivalent percent. Here's a sample.

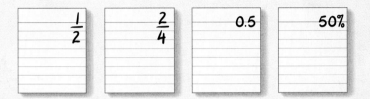

Repeat these steps until you have used all 16 fractions from the table. You will have 64 cards in all.

Go
- Shuffle the cards and deal seven cards to each player. Place the remaining cards facedown in the middle of the table. Take the top card and place it face up next to the deck, forming the discard pile.

Scoring Set

- All players check their cards for scoring sets. A scoring set consists of three equivalent numbers. Scoring sets should be placed face up on the table.

- The player to the left of the dealer draws the top card from the deck or the top card in the discard pile. If the player has a scoring set, he or she should place it faceup on the table. The player may also build onto another player's scoring set by placing a card faceup on the table and announcing the set on which the player is building. The player's turn ends when he or she discards a card.

- Play continues until one player has no cards remaining in his or her hand. That player is the winner.

 Visit www.glencoe.com/sec/math/mac/mathnet for more games.

Let the Games Begin

It may be helpful for students to play a practice round to be sure they understand the rules of the game. Students might be permitted to use the Equivalent Fractions, Decimals, and Percents chart on page 121 as a reference while they play the game.

3-6A LAB Notes

Objective Students should be able to determine whether answers to problems are reasonable.

Recommended Pacing	
Standard	Day 8 of 11
Honors	Day 7 of 10
Block	Day 5 of 6

1 FOCUS

Getting Started Have students act out the situation presented at the beginning of the lesson. Then have them work in pairs to answer Exercises 1 and 2. Have the class discuss the results. Then have them solve Exercise 3.

2 TEACH

Teaching Tip After students complete the opening problem, they may find it helpful to work in groups to formulate problems based on similar situations.

In-Class Example

Martin wants to buy a bike that costs $155.00 plus 8% sales tax. He looks at his savings passbook and finds that he has $200.00. Does it seem reasonable that he has enough money to buy both a bike and a helmet that costs $24.95 if there is 8% sales tax? **Yes; total cost is $194.35.**

THINKING LAB — PROBLEM SOLVING

3-6A Reasonable Answers

A Preview of Lesson 3-6

Tanya wants to buy in-line skates at a sporting goods store. She and her friend Krista are trying to decide whether she has enough money for the skates and the sales tax.

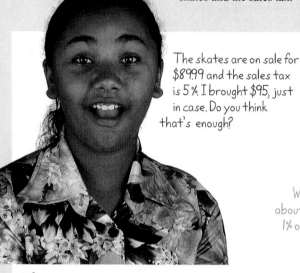

Tanya: The skates are on sale for $89.99 and the sales tax is 5%. I brought $95, just in case. Do you think that's enough?

Krista: Well, the skates cost about $90. I know that 1% of $90 is $0.90. Will that help?

Tanya: Sure! $0.90 times 5 is $4.50. Since $90 and $4.50 is $94.50, I have just enough money!

THINK ABOUT IT

Work with a partner.

1. *Explain* how Krista knows that 1% of $90 is $0.90. **Sample answer: 0.01 × 90 = 0.9**

2. *Think* of another way to help Tanya decide whether she has enough money. **Sample answer: 10% of $90 is $9. One-half of $9 is $4.50.**

3. *Apply* the strategy of determining **reasonable answers** to solve this problem.

   Lucia knows that it is customary to leave a tip for the server at a restaurant. The tip is usually 15% of the value of the meal. How much should Lucia leave for a meal that costs $24.90? **about $3.75**

124 Chapter 3 Using Proportion and Percent

■ Reteaching the Lesson ■

Activity Ask students to work together to write and solve problems that are similar to those in the lesson. It might be helpful for one group to present its problems to another group to solve.

3 PRACTICE/APPLY

Check for Understanding
Use the results from Exercise 6 to determine whether students understand the concept of the lesson and can conclude whether or not an answer is reasonable.

Extra Practice If students need additional practice in problem solving, extra practice is available on the following pages.
- Determine Reasonable Answers, see p. 613
- Mixed Problem Solving, see pp. 645–646

Assignment Guide
All: 4–14

4 ASSESS

Closing Activity
Writing Ask students to imagine that they have been offered a paper route that will pay them $5.75 per hour for 2 hours of work each day. At the end of one week, they expect to receive a paycheck of $100. Ask them to determine whether this is a reasonable amount to expect and to explain their answers.

Additional Answer
4. Sample answer: Mental math skills can help estimate a solution.

ON YOUR OWN

4. See margin.
4. *Explain* how you can use mental math skills to help plan the solution to a problem.
5. Do research to find the state sales tax rate for your state. *Write a Problem* using the tax rate. **See students' work.**
6. *Look Ahead* Explain whether the statement *11% of 98 is 0.99* is reasonable or not. If not, rewrite the statement so it is reasonable. **No; 11% of 98 is about 10.**

MIXED PROBLEM SOLVING

STRATEGIES
Look for a pattern.
Solve a simpler problem.
Act it out.
Guess and check.
Draw a diagram.
Make a chart.
Work backward.

Solve. Use any strategy.

7. *Money Matters* You spend $5.55 plus $0.44 tax at the store for makeup and pay with a $10 bill. Would it be more reasonable to expect about $3.00 or $4.00 in change? **$4**

8. *Ecology* In a survey of 1,413 shoppers, 6% said they would be willing to pay more for environmentally safe products. Is 8.4, 84, or 841 a reasonable estimate for the number of shoppers willing to pay more? **84**

9. *Transportation* Yesterday you noted that the mileage on the family car read 60,094.8 miles. Today it reads 60,099.1 miles. Was the car driven about 4 or 40 miles? **4 miles**

10. *Agriculture* An orange grower harvested 1,260 pounds of oranges from one grove, 874 pounds from another, and 602 pounds from a third. What is a reasonable number of crates to have on hand if each crate holds 14 pounds of oranges? **about 200 crates**

11. *Money Matters* Della has only nickels in her pocket. Ayita has only dimes in hers. Kareem has only quarters in his. Marta approached them for a donation for their school fundraiser. What is the least each person could donate so that each one gives the same amount? **50 cents**

12. *Earth Science* Geothermal energy is heat from inside Earth. Underground temperatures generally increase 9° C for every 300 feet of depth. How deep would you have to dig so that the underground temperature is 90° C greater than the ground temperature? **3,000 feet**

13. *Sports* The graph shows the number of injuries treated in emergency rooms for the top seven sports. About how many injuries were treated in all? **2.7 million people**

Sports Injuries
Top sports for number of injuries treated in emergency rooms

Sport	Injuries
Basketball	693,933
Cycling	599,874
Football	390,180
Snow skiing	330,289
Skating (all types)	322,311
Baseball	219,023
Soccer	157,251

Source: Consumer Product Safety Commission, American Academy of Orthopaedic Surgeons, 1995

14. **Test Practice** Diego wants to buy a pair of shoes that cost $59.95. The sales tax is 6%. Which amount is reasonable for the total cost? **B**
   A $60.31
   B $63.55
   C $65.95
   D $89.95
   E $95.95

Lesson 3-6A THINKING LAB 125

Extending the Lesson

Activity Tell students that they have been given three possible answers to a problem, one of which is correct. Have them work together in small groups to write a few sentences describing how to determine the most reasonable answer without actually solving the problem.

Sample problem: Mei spent $7.56 plus $0.38 tax for food. She paid with a $10 bill. What is a reasonable amount of change to expect?
a. about $1
b. about $2
c. about $3 **b. about $2.00**

3-6 Lesson Notes

Instructional Resources
- *Study Guide Masters,* p. 25
- *Practice Masters,* p. 25
- *Enrichment Masters,* p. 25
- Transparencies 3-6, A and B
- *Assessment and Evaluation Masters,* p. 72
- *Science and Math Lab Manual,* pp. 85–88
- CD-ROM Program
  - Resource Lesson 3-6
  - Interactive Lesson 3-6

Recommended Pacing	
Standard	Day 9 of 11
Honors	Day 8 of 10
Block	Day 5 of 6

1 FOCUS

5-Minute Check
(Lesson 3-5)

1. Find 1% of 79. **0.79**
2. Find 10% of 13.4. **1.34**
3. Find 60% of 140. **84**
4. Which is greater, 9 or 30% of 24? **9**
5. Name the fraction you would use to find 37.5% of 84. $\frac{3}{8}$

The 5-Minute Check is also available on **Transparency 3-6A** for this lesson.

Motivating the Lesson
Communication After reading the lesson opener, ask students to estimate the percent of area damaged when only 10 squares are shaded. **20%** Discuss other ways that estimation helps solve problems in real-life situations.

3-6 Percent and Estimation

What you'll learn
You'll learn to estimate by using equivalent fractions, decimals, and percents.

When am I ever going to use this?
Knowing how to estimate with percents can help you determine prices in a 20%-off sale.

Word Wise
compatible numbers

Fire fighters use geometry and aerial photography to estimate how much of a forest has been damaged by fire. A grid is superimposed on a photograph of the forest.

For example, suppose the tan part of the figure at the right represents the area damaged by a forest fire. About 25 small squares out of 49 squares are shaded tan.

$\frac{25}{49} \approx \frac{25}{50}$ or $\frac{1}{2}$

$\frac{1}{2} = 50\%$

So, about 50% of the area has been damaged by the fire.

You can estimate a percent of a number by using **compatible numbers**. Compatible numbers are two numbers that are easy to divide mentally.

Examples

1 Estimate 19% of 60.

19% is about 20% or $\frac{1}{5}$.   *5 and 60 are compatible numbers.*

$\frac{1}{5}$ of 60 is 12.

So, 19% of 60 is about 12.

2 Estimate 25% of 78.

25% = $\frac{1}{4}$, and 78 is about 80.   *4 and 80 are compatible numbers.*

$\frac{1}{4}$ of 80 is 20.

So, 25% of 78 is about 20.

3 Estimate 65% of 34.

65% is about $66\frac{2}{3}\%$ or $\frac{2}{3}$, and 34 is about 33.   *3 and 33 are compatible numbers.*

$\frac{2}{3}$ of 33 is 22.

So, 65% of 34 is about 22.

Circle graphs are one way to display statistical data. You can use estimation to sketch a circle graph.

126 Chapter 3 Using Proportion and Percent

Investigations for the Special Education Student

This blackline master booklet helps you plan for the needs of your special education students by providing long-term projects along with teacher notes. Investigation 9, *Park It!,* may be used with this chapter.

126 Chapter 3

Example 4 **Earth Science** Seventy percent of Earth's surface is covered by water. The remaining 30% is land. Sketch a circle graph of this data.

Study Hint
Technology Hint You can also use software like *Microsoft Word* to construct a circle graph.

Explore — Circle graphs show how parts are related to a whole. The circle represents the whole quantity. In this case, it represents the surface of Earth.

Plan — Use estimation to find the size of each part. 70% is a little less than 75% or $\frac{3}{4}$. So, the part representing water should be a little less than $\frac{3}{4}$ of the circle. The remaining part will represent land.

Solve — Draw a circle and divide it into fourths using dotted lines to visualize $\frac{3}{4}$ of the circle. Using solid lines, sketch the graph to represent 70% and 30%. Give the graph a title and label the parts.

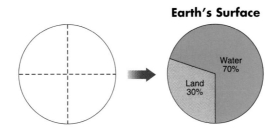

Examine — Since 70% is a little more than twice 30%, the part representing 70% should be more than twice as large as the part representing 30%. The graph is reasonable.

HANDS-ON MINI-LAB

Work with a partner.   calculator   compass

Try This   markers

1. Choose a topic of interest to you and conduct a survey of 20 of your classmates. Here are some suggestions.   **1–3. See students' work.**
   a. number of siblings   b. state where born
   c. kinds of pets   d. favorite sports team
2. Show your data as percents.
3. Sketch a circle graph of the data.

Talk About It

4. Name an advantage of displaying data in a graph instead of a table.

4. Sample answer: In a graph you can visually determine how the parts relate to each other.

---

2 TEACH

Transparency 3-6B contains a teaching aid for this lesson.

Teaching Tip In Example 1, point out that because 19% is less than 20%, 19% of 60 is less than 12.

In-Class Examples

For Example 1
Estimate 9% of 40. about 4

For Example 2
Estimate 20% of 64. about 13

For Example 3
Estimate 32% of 89. about 30

For Example 4
Sketch a circle graph of the following data: In a crowd of people, 10% had red hair, 20% had blond hair, 30% had black hair, and 40% had brown hair.

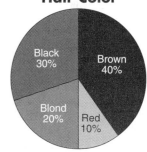

Using the Mini-Lab You may wish to discuss with students what kind of question might be asked in a survey and how they might choose to tally the results. Encourage them not to use a question that could have multiple responses from one interviewee. Have each pair display their circle graph and explain the method they used to convert their data.

Lesson 3-6 Percent and Estimation **127**

3 PRACTICE/APPLY

Check for Understanding

If students need additional practice or instruction after completing Exercises 1–10, one of these options may be helpful.
- Extra Practice, see p. 614
- Reteaching Activity
- *Study Guide Masters*, p. 25
- *Practice Masters*, p. 25

Assignment Guide

Core: 11–33 odd, 36–40
Enriched: 12–30 even, 32–34, 36–40

CHAPTER Project

Exercise 35 asks students to advance to the next stage of work on the Chapter Project. You may have students work in groups to complete their circle graphs which can be included in their portfolios or displayed in the classroom.

Additional Answer

2. Sample answer:

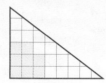

Study Guide Masters, p. 25

128 Chapter 3

CHECK FOR UNDERSTANDING

Communicating Mathematics

Read and study the lesson to answer each question.

1. *Tell* how you can estimate 23% of $98.95 using fractions and compatible numbers. $\frac{1}{4}$ of $100 is $25.
2. *Draw* a triangle on graph paper. Shade about $\frac{1}{3}$ of the area. See margin.
3. *You Decide* Darnell thinks 46% of 80 is less than 40. Elaina thinks it is greater than 40. Who is correct? Explain your reasoning. Darnell; 46% is less than 50%, so the answer will be less than half of 80.

Guided Practice Estimate the percent of the area shaded. 4–10. Sample answers given.

4. $\frac{6}{25} \approx \frac{5}{25}$ or 20%
5.   about 20%

Estimate.

6. 29% of 50 $\frac{3}{10}$ of 50 is 15
7. 88% of 64 $\frac{7}{8}$ of 64 is 56

Estimate the percent.

8. 7 out of 16 $\frac{7}{16} \approx \frac{8}{16}$ or 50%
9. 22 out of 60 $\frac{22}{60} \approx \frac{20}{60}$ or 33.3%

10. *Sports* Estimate Shaun's foul shooting percent in basketball if he made 13 foul shots in 22 attempts. $\frac{13}{22} \approx \frac{15}{25}$ or 60%

EXERCISES

Practice Estimate the percent of the area shaded. 11–16. Sample answers given.

11. $\frac{15}{49} \approx \frac{15}{50}$ or 30%
12. $\frac{7}{25}$ or 28%
13. $\frac{64}{100}$ or 64%

14. about 51%
15. about 12%
16. about 67%

Estimate. 17–22. Sample answers given.

17. $\frac{1}{3}$ of 90 or 30
18. $\frac{3}{4}$ of 64 or 48
19. $\frac{1}{5}$ of 70 or 14
20. $\frac{9}{10}$ of 40 or 36
21. $\frac{2}{3}$ of 9 or 6
22. $\frac{1}{4}$ of 120 or 30

17. 32% of 89
18. 73% of 65
19. 21% of 72
20. 92% of 41
21. 68% of 9.4
22. 26.5% of 125

128 Chapter 3 Using Proportion and Percent

Reteaching the Lesson

Activity Have students shade irregular regions on sections of graph paper and tell how to estimate the percent of each section that is shaded.

Error Analysis
Watch for students who have difficulty in determining fractional estimates for percents; for example, similar numbers such as 3% and 30%.
Prevent by referring students to the equivalence chart in Lesson 3-5 when in doubt.

Estimate the percent. 23–28. See margin.

23. 8 out of 13
24. 4 out of 25
25. 12 out of 60
26. 7 out of 57
27. 9.2 out of 11
28. 16 out of 45.8

29. What compatible numbers could you use to estimate 32% of 154? $\frac{1}{3}$, 150
30. Estimate what percent 8 out of 15 represents. $\frac{8}{15} \approx \frac{8}{16}$ or 50%
31. *Write a Problem* in which the solution is about 50%. See students' work.

Applications and Problem Solving

32. *Earth Science* There are about 90 elements that occur naturally in Earth's crust. The four most common elements are listed in the table. Sketch a circle graph of this data. **See Answer Appendix.**

Elements in Earth's Crust	
Oxygen	47%
Silicon	28%
Aluminum	8%
Iron	5%
Other	12%

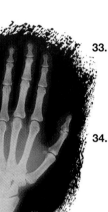

33. *Life Science* The adult skeleton has 206 bones. Sixty of them are in the arms and hands. Estimate the percent of bones that are in the arms and hands. $\frac{60}{206} \approx \frac{60}{200}$ or 30%

34. *Statistics* Refer to the data at right. a–c. **See Answer Appendix.**
    a. Sketch this data in a circle graph and a bar graph.
    b. What advantage does the circle graph have over the bar graph?
    c. What advantage does the bar graph have over the circle graph?

Source: The NPD Group for Book Industry Study Group, 1995

35. *Working on the* **CHAPTER Project** Refer to the budget you planned on page 103. a–b. See students' work.
    a. Of your total expenses, estimate the percent that you'll spend on transportation, housing, food, and souvenirs.
    b. Sketch a circle graph of your expenses.

36. *Critical Thinking* How would you determine whether 8% of 400 is greater than 4% of 180?

36. Using compatible numbers, 10% of 400 is 40 and 5% of 200 is 10, so 8% of 400 is greater than 4% of 180.

Mixed Review

37. Compute 10% of 1,358 mentally. *(Lesson 3-5)* **135.8**

38. **Test Practice** Wesley can run 3.5 miles in 40 minutes. How long would it take him to run 8 miles at this same rate? *(Lesson 3-3)* **B**
    A 182 minutes   B 91 minutes   C 17 minutes   D 0.7 minutes

39. Evaluate $(17 - 8) \div 3 + 5^2$. *(Lesson 1-3)* **28**

40. *Transportation* A service station along I-64 in Lexington, Virginia, charges $1.39 a gallon for diesel fuel and $1.10 a quart for oil. How much will 38 gallons of diesel fuel and 2 quarts of oil cost? *(Lesson 1-1)* **$55.02**

Lesson 3-6 Percent and Estimation **129**

Extending the Lesson

Enrichment Masters, p. 25

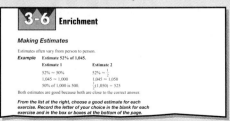

Activity Have students read about mapmaking or speak to a cartographer to learn the methods used to estimate the areas of irregular regions on the surface of Earth.

---

4 ASSESS

Closing Activity

Speaking Have students explain how to use compatible numbers to estimate solutions to percent problems.

Chapter 3, Quiz D (Lesson 3-6) is available in the *Assessment and Evaluation Masters*, p. 72.

Additional Answers

23–28. Sample answers are given.

23. $\frac{8}{13} \approx \frac{8}{12}$ or 66.7%
24. $\frac{4}{25} \approx \frac{5}{25}$ or 20%
25. $\frac{12}{60} = \frac{1}{5}$ or 20%
26. $\frac{7}{57} \approx \frac{7}{58}$ or 12.5%
27. $\frac{9.2}{11} \approx \frac{9}{10}$ or 90%
28. $\frac{16}{45.8} \approx \frac{16}{48}$ or 33.3%

Practice Masters, p. 25

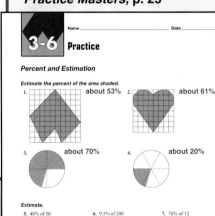

CHAPTER 3 Study Guide and Assessment

Vocabulary

After completing this chapter, you should be able to define each term, concept, or phrase and give an example or two of each.

Number and Operations
compatible numbers (p. 126)
cross products (p. 111)
percent (p. 107)
proportion (p. 111)
rate (p. 105)
ratio (p. 104)
unit rate (p. 105)

Problem Solving
reasonable answers (p. 124)

Understanding and Using the Vocabulary

Choose the correct term or number to complete each sentence.

1. A (<u>ratio</u>, proportion) is a comparison of two numbers by division.
2. The expression $25/5 hours is an example of a (unit rate, <u>rate</u>).
3. An equation that indicates that two ratios are equivalent is a (<u>proportion</u>, percent).
4. A (<u>unit rate</u>, ratio) always has a denominator of 1.
5. The model at the right represents (<u>60%</u>, 6%).
6. The decimal 0.03 written as a percent is (30%, <u>3%</u>).
7. The fraction $\frac{1}{2}$ written as a percent is ($\frac{1}{2}$%, <u>50%</u>).
8. Numbers that are easy to divide mentally are called (<u>compatible</u>, rates).
9. You can estimate 26% of 16 by using the fraction ($\frac{1}{4}$, $\frac{2}{5}$).

10. Sample answer: A ratio is a comparison of any two numbers by division, while a rate is a special kind of ratio that compares two measurements with different units.

In Your Own Words

10. *Explain* the differences between a ratio and a rate.

 MindJogger Videoquizzes

MindJogger Videoquizzes provide an alternative review of concepts presented in this chapter. Students work in teams to answer questions, gaining points for correct answers. The questions are presented in three rounds.
Round 1 Concepts–5 questions
Round 2 Skills–4 questions
Round 3 Problem Solving–4 questions

Study Guide and Assessment

Vocabulary

This section provides a listing of the new terms, properties, and phrases that were introduced in this chapter. Have students define each term and provide an example or two of it, if appropriate.

Understanding and Using the Vocabulary

These exercises check students' understanding of the terms by using a variety of verbal formats including matching, completion, and true/false.

Glossaries A complete glossary of terms appears on pages 702–710. The glossary also appears in Spanish on pages 711–721.

Study Guide and Assessment   Chapter 3

Objectives & Examples

Upon completing this chapter, you should be able to:

● express ratios as fractions in simplest form and determine unit rates *(Lesson 3-1)*

Express the ratio 9 out of 15 in simplest form.

$$\frac{9}{15} \stackrel{\div 3}{=} \frac{3}{5}$$

16. 50 students per bus
17. 22 miles per gallon

Review Exercises

Use these exercises to review and prepare for the chapter test.

Express each ratio or rate in simplest form.

11. 12 peaches:18 pears   **2:3**
12. 40 centimeters per meter   **2 to 5**
13. 25 to 9   **25 to 9**
14. 5 girls:40 people   **1:8**

Express each rate as a unit rate.

15. $5 for 2 minutes   **$2.50 per minute**
16. 150 students to 3 buses
17. 110 miles on 5 gallons

● express ratios as percents and vice versa *(Lesson 3-2)*

Express $\frac{4}{5}$ as a percent.

$$\frac{4}{5} \stackrel{\times 20}{=} \frac{80}{100}$$

So, $\frac{4}{5} = 80\%$.

Express each ratio or fraction as a percent.

18. 3 out of 5   **60%**
19. 16.5:100   **16.5%**
20. $\frac{1}{4}$ **25%**
21. $\frac{14}{25}$ **56%**

Express each percent as a fraction in simplest form.

22. 20% $\frac{1}{5}$
23. 7% $\frac{7}{100}$
24. 85% $\frac{17}{20}$
25. 90% $\frac{9}{10}$

● solve proportions *(Lesson 3-3)*

Solve $\frac{7}{4} = \frac{n}{2}$.

$7 \times 2 = 4 \times n$ *Cross products*

$14 = 4n$

$\frac{14}{4} = \frac{4n}{4}$ *Divide each side by 4.*

$3.5 = n$

Solve each proportion.

26. $\frac{5}{8} = \frac{n}{72}$ **45**
27. $\frac{3}{r} = \frac{6}{8}$ **4**
28. $\frac{9}{x} = \frac{4}{18}$ **40.5**
29. $\frac{30}{0.5} = \frac{y}{0.25}$ **15**

Objectives & Examples

This section reviews the skills and concepts of the chapter and shows completely worked examples.

Review Exercises

These exercises provide practice for the corresponding objectives.

Assessment and Evaluation Masters, pp. 59–60

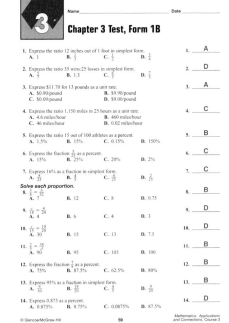

Assessment and Evaluation

Six forms of Chapter 3 Test are available in the *Assessment and Evaluation Masters* as shown in the chart.

Chapter 3 Test, Form 1B, is shown at the right. Chapter 3 Test, Form 2B, is shown on the next page.

1A	Multiple Choice	Honors
1B	Multiple Choice	Average
1C	Multiple Choice	Basic
2A	Free Response	Honors
2B	Free Response	Average
2C	Free Response	Basic

Chapter 3 Study Guide and Assessment   **131**

Study Guide and Assessment   **131**

Study Guide and Assessment

Chapter 3 Study Guide and Assessment

Objectives & Examples

● express percents as fractions and decimals and vice versa *(Lesson 3-4)*

Express 0.18 as a percent.
$0.18 = \frac{18}{100}$ or 18%

Express $\frac{1}{3}$ as a percent.
$\frac{1}{3} = \frac{n}{100}$
$3n = 100$
$n \approx 33.3\%$

● compute mentally with percents *(Lesson 3-5)*

Find 10% of 65.
10% of 65 is 6.5.

Find 25% of 8.
25% of 8 = $\frac{1}{4}$ of 8 or 2

● estimate by using equivalent fractions, decimals, and percents *(Lesson 3-6)*

Estimate 40% of 83.
$40\% = \frac{2}{5}$
$\frac{2}{5}$ of 80 is 32.   *Round 83 to 80.*
So, 40% of 83 ≈ 32.

Review Exercises

Express each percent as a decimal.
30. 13%   **0.13**
31. 90%   **0.90 or 0.9**
32. 24%   **0.24**
33. 4.3% **0.043**

Express each decimal as a percent.
34. 0.35   **35%**
35. 0.7   **70%**
36. 0.04   **4%**
37. 0.655 **65.5%**

Express each fraction as a percent.
38. $\frac{3}{50}$ **6%**
39. $\frac{7}{20}$ **35%**
40. $\frac{9}{25}$ **36%**
41. $\frac{1}{6}$ **16.67%**

Compute mentally.
42. 10% of 18.3   **1.83**
43. 20% of 60   **12**
44. 90% of 100   **90**
45. 50% of $42   **$21**
46. $66\frac{2}{3}\%$ of 24   **16**

Estimate.   47–51. Sample answers given.
47. 8% of 104 $\frac{8}{100}$ **of 100 or 8**
48. 62% of 50 $\frac{3}{5}$ **of 50 or 30**
49. 99% of 35 $\frac{1}{1}$ **of 35 or 35**

Estimate the percent.
50. 11 out of 24 $\frac{11}{24} \approx \frac{12}{24}$ **or 50%**
51. 20 out of 52 $\frac{20}{52} \approx \frac{20}{50}$ **or 40%**

132 Chapter 3 Using Proportion and Percent

Test and Review Software

You may use this software, a combination of an item generator and item bank, to create your own tests or worksheets. Types of items include free response, multiple choice, short answer, and open ended.

CD-ROM Program

The CD-ROM Program contains an Assessment Game whose questions review the concepts in this chapter.

Study Guide and Assessment

Assessment and Evaluation Masters, pp. 65–66

Chapter 3 Test, Form 2B

Express each ratio or rate as a fraction in simplest form.
1. 8 inches per foot — $\frac{2}{3}$
2. 100 menus for 155 diners — $\frac{20}{31}$
3. 20 ounces per 140 loaves of bread — $\frac{1}{7}$
4. 15 girls:36 students — 5:12
5. Express $480 for 12 tickets as a unit rate. — $40/ticket
6. Express the fraction $\frac{12}{25}$ as a percent. — 48%
7. Express the fraction $\frac{27}{40}$ as a percent. — 67.5%
8. Express the ratio 17.5 out of 100 patients as a percent. — 17.5%
9. Express 16% as a fraction in simplest form. — $\frac{4}{25}$
10. Express 98% as a fraction in simplest form. — $\frac{49}{50}$

Solve each proportion.
11. $\frac{3}{36} = \frac{5}{y}$ — 22.5
12. $\frac{12}{f} = \frac{3}{29}$ — 116
13. $\frac{p}{4} = \frac{0.6}{1.6}$ — 1.5
14. $\frac{5}{33} = \frac{c}{99}$ — 15
15. $\frac{2}{3} = \frac{d}{36}$ — 24
16. Express 82% as a decimal and as a fraction. — 0.82, $\frac{41}{50}$
17. Express 0.112 as a percent. — 11.2%
18. Express $\frac{15}{120}$ as a percent. — 12.5%
19. Express $\frac{17}{25}$ as a percent. — 68%
20. Which is greater: $\frac{3}{7}$ or 45%? — 45%

© Glencoe/McGraw-Hill 65 *Mathematics: Applications and Connections, Course 3*

Chapter 3 Test, Form 2B (continued)

Compute mentally.
21. 10% of 592 — 59.2
22. 40% of 65 — 26
23. 25% of 96 — 24
24. 12.5% of 16 — 2
25. $33\frac{1}{3}\%$ of 30 — 10
26. Of 25, 40, and 50, what is the best estimate for 19.5% of 203? — 40

Estimate. *Accept all reasonable answers.*
27. 65% of 147 — $\frac{2}{3}$ of 150 = 100
28. 28% of 34 — $\frac{1}{4}$ of 36 = 9
29. 82% of 126 — $\frac{4}{5}$ of 125 = 100
30. 11% of $129.50 — 10% of $130 = $13
31. 91% of 161 — $\frac{9}{10}$ of 160 ≈ 4
32. On Bat Day, 45,000 fans attended the baseball game at the stadium. The first 1,250 fans received bats. On average, for each bat given away, how many people attended? — 36
33. At 3:00 P.M., a tree 20 meters tall casts a 30-meter shadow. A utility pole next to the tree casts a 12-meter shadow. How tall is the pole? — 8 m

© Glencoe/McGraw-Hill 66 *Mathematics: Applications and Connections, Course 3*

132 Chapter 3

Study Guide and Assessment Chapter 3

Applications & Problem Solving

52. *Reasonable Answers* At a restaurant, entrees cost from $9.95 to $18.95, appetizers cost from $4.95 to $6.95, a salad costs $3.25, and soup costs $2.50. A full meal includes soup, salad, appetizer, and an entree. Without tax and tip, would a couple expect to pay more or less than $40 for a full meal at this restaurant? Explain. *(Lesson 3-6A)*
more than $40

53. *Hobbies* Javier has 130 baseball cards in his collection. Forty-eight of the cards are of Cleveland Indian players. Find the ratio of the Indians cards to all cards. *(Lesson 3-1)*
24 to 65

54. *Money Matters* Penelope spends 42% of her monthly income on housing. If her monthly income is $1,900, about how much does she spend on housing? *(Lesson 3-6)*
$\frac{2}{5}$ **of $2,000 or $800**

55. *Writing* Americans spend about $1.7 billion each year on writing instruments. Refer to the graph. Express each percent as a decimal. *(Lesson 3-4)* **0.43, 0.16, 0.14, 0.11, 0.06, 0.04, 0.03**

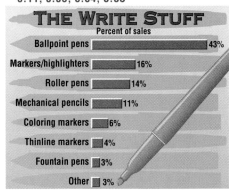

THE WRITE STUFF
Percent of sales
- Ballpoint pens 43%
- Markers/highlighters 16%
- Roller pens 14%
- Mechanical pencils 11%
- Coloring markers 6%
- Thinline markers 4%
- Fountain pens 3%
- Other 3%

Source: Writing Instrument Manufacturers Association, 1996

Alternative Assessment

Performance Task
Sam estimated that Sarah made 75% of the 2-point shots she attempted in a basketball game. If Sarah attempted 18 shots, how many shots could she have made for Sam to arrive at this estimate? Explain your reasoning. **See margin.**

In the next game, Sam estimated that Sarah made 50% of the 2-point shots she attempted. Did Sarah score more or fewer points in the second game? Explain your reasoning. **See margin.**

Completing the CHAPTER Project
Use the following checklist to make sure your report is complete.
- ☑ The foreign exchange rate is included.
- ☑ The computation of the number of U.S. dollars you will need to take on your trip is accurate.
- ☑ Your circle graph is accurate.

Add any finishing touches that you would like to make your report more attractive.

 Select one of the assignments from this chapter and place it in your portfolio. Attach a note to it explaining why you selected it.

A practice test for Chapter 3 is provided on page 649.

Chapter 3 Study Guide and Assessment **133**

Applications & Problem Solving
This section provides additional practice in solving real-world problems that involve the skills of this chapter.

Alternative Assessment
The *Performance Task* provides students with a performance assessment opportunity to evaluate their work and understanding.

CHAPTER Project
Students should complete the final stages of their project and prepare a class demonstration of their results. A scoring guide for the project is available in the *Investigations and Projects Masters*, p. 27.

 Students should add to their portfolios at this time.

Assessment and Evaluation Masters, p. 69

Chapter 3 Performance Assessment

Instructions: Demonstrate your knowledge by giving a clear, concise solution to each problem. Be sure to include all relevant drawings and justify your answers. You may show your solutions in more than one way or investigate beyond the requirements of the problems.

1. Ratios are used to make concrete. A recommended concrete mix is 1 part cement, 2 parts sand, 4 parts gravel, and water to moisten.
   a. Explain what is meant by a *ratio*.
   b. Write three ratios that describe the concrete mix. Write each ratio in a different form. Then write each ratio as a percent.
   c. Explain what is meant by a *proportion*.
   d. Suppose 400 pounds of sand were loaded into a mixer. Tell how much cement and gravel should be added. Show your work.
   e. Circle graphs show how parts are related to the whole. Draw a circle graph that shows the parts (not including the water) that go into the concrete mixture. Use estimation to find the size and percent of each part on the circle graph.

2. An understanding of percent and its relationship to fractions and decimals is essential in shopping for the best price.
   a. Explain the meaning of *percent*.
   b. A car dealer gives employees 25% off the list price of any car in stock or special ordered. A close-out sale of cars in stock advertises them as 70% of the list price. Which is the greater rate of discount? Explain your reasoning.
   c. Give circumstances under which an employee in part b may wish to take the lesser rate of discount.
   d. If the employee discount in part b were changed to $\frac{1}{3}$ off, would the employee or advertised discount rate be greater? Explain your reasoning.
   e. A dealership employee has a choice of a $2,000 rebate or a 25% discount when buying a car. Explain which discount rate is greater or what additional information is needed.
   f. Look at parts b, d, and e. In which case is it easiest to compare discount rates? Why?

© Glencoe/McGraw-Hill   69   Mathematics: Applications and Connections, Course 3

Additional Answers for the Performance Task

Sample answer: about 13 or 14 shots; $\frac{12}{18} \approx 67\%$ and $\frac{15}{18} \approx 83\%$

Sample answer: You cannot tell in which game she scored the most points because you do not know how many baskets Sarah attempted in the second game.

Performance Assessment
Additional performance assessment tasks for this chapter are included in the *Assessment and Evaluation Masters* on page 69. A scoring guide is also provided on page 81.

Study Guide and Assessment **133**

Standardized Test Practice

The Standardized Test Practice may be used to help students prepare for standardized tests. The test items are written in the same style as those in state proficiency tests and standardized tests like CAT, CTBS, ITBS, MAT, SAT, and Terra Nova. The test items cover skills and concepts covered up to this point in the text.

The pages can be used as an overnight assessment. After students have completed the pages, discuss how each problem can be solved, or provide copies of the solutions from the *Solutions Manual*.

Assessment and Evaluation Masters, p. 75

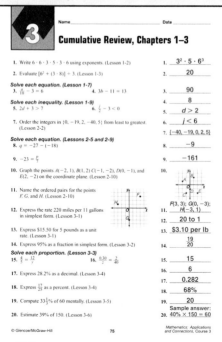

CHAPTERS 1-3 Standardized Test Practice
Assessing Knowledge & Skills

Section One: Multiple Choice

There are ten multiple choice questions in this section. Choose the best answer. If a correct answer is *not here,* choose the letter for Not Here.

1. What is the perimeter of the rectangle? **C**

A 24.5 cm
B 34.5 cm
C 49 cm
D 145 cm

2. Which integers are graphed on the number line? **H**

F $\{-5, -2\}$
G $\{-5, -4, -3, ...\}$
H $\{-5, -2, 0, 1\}$
J $\{-5, 0, 5\}$

3. In a random sample of 150 students, 60 ride the bus to school, 54 ride in car pools, and 36 walk. If there will be 800 students next year, about how many will need bus transportation? **A**
A 320
B 160
C 80
D 60

134 Chapters 1-3 Standardized Test Practice

Please note that Questions 4–10 have five answer choices.

4. Including tax, the cost of renting a steam cleaner is $24.99 for the first day and $5.99 for each day after that. Which number sentence could be used to find C, the cost in dollars for keeping the cleaner for 4 days? **J**
F $C = \$24.99 + \$5.99$
G $C = 4(\$24.99) + \$5.99$
H $C = 4(\$5.99) + \$24.99$
J $C = 3(\$5.99) + \$24.99$
K $C = 4(\$5.99) + 4(\$24.99)$

5. Markers ordered for the art room cost $0.04 per marker for the first 100. After that the cost drops to $0.02 per marker. Which is a reasonable total cost for 250 markers? **B**
A The cost is less than $5.
B The cost is between $5 and $10.
C The cost is between $10 and $15.
D The cost is between $15 and $20.
E The cost is greater than $20.

6. Mallory works 4 hours a day for 4 days each week. She earns $5.25 per hour. Which problem *cannot* be solved using the information given? **H**
F How many days must Mallory work to earn $600?
G How much more would Mallory earn each week if she earns $5.75 per hour?
H How many CDs can Mallory buy with the money she earns in 2 weeks?
J How much does Mallory earn each week?
K How much money would Mallory earn in a week in which she was ill and didn't work one day?

◀◀◀ **Instructional Resources**

Another cumulative review is shown at the left and is available in the *Assessment and Evaluation Masters*, p. 75.

Standardized Test Practice Chapters 1–3

7. The lengths of the sides of a five-sided field are 524 feet, 498 feet, 519 feet, 502 feet, and 486 feet. Estimate the amount of fencing needed to enclose the field. **C**
   A 2,000 feet
   B 2,300 feet
   C 2,500 feet
   D 2,800 feet
   E 3,000 feet

8. The recycling center pays 5¢ for every two aluminum cans turned in. If a group of students brought in 1,200 cans, how much money would they get back? **H**
   F $3
   G $6
   H $30
   J $300
   K Not Here

9. Mr. Mitchell drove his car 1,048 miles in 8 months. What is the average distance he drove per month? **A**
   A 131 miles
   B 1,040 miles
   C 1,056 miles
   D 8,384 miles
   E Not Here

10. The Perfect Painting Company determines how much paint they will need for a wall by finding the area of the wall. What is the area of the wall if its length is 13 feet and its width is 9 feet? **H**
    F 22 ft^2
    G 52 ft^2
    H 117 ft^2
    J 127 ft^2
    K Not Here

Test-Taking Tip

You can prepare for taking a standardized test by working through practice tests like this one. The more you work with questions similar in format to the actual test, the better you become at test taking. Do not wait until the night before the test to review. Allow yourself plenty of time to review the basic skills and formulas needed.

Section Two: Free Response

This section contains three questions for which you will provide short answers. Write your answers on your paper.

11. Max is 4 years younger than three times Carl's age. Carl is 14. Write an equation you can use to find Max's age. **Sample answer: $m + 4 = 3(14)$**

12. $(3 \times 8) + (17 \times 8) =$ **160**

13. The scale measures the mass of an item in grams. To the nearest gram, what is the mass of the calcium carbonate? **9 grams**

Chapters 1–3 Standardized Test Practice 135

Instructional Resources ▶▶▶

Additional standardized test practice is shown at the right and is available in the *Assessment and Evaluation Masters*, pp. 73–74.

Test-Taking Tip

Preparation is the key to performing well on standardized tests. Remind students it is always wise to get a good night's sleep the night before the test.

Assessment and Evaluation Masters, pp. 73–74

Standardized Test Practice 135

Interdisciplinary Investigation

GET READY

This optional investigation is designed to be completed by groups of 2 or 4 students over several days or several weeks.

Mathematical Overview
This investigation utilizes the concepts from Chapters 1–3.
- solving problems
- adding integers
- comparing and ordering integers
- calculating percents

Time Management	
Gathering Data	30 minutes
Calculations	30 minutes
Creating Graphs	30 minutes
Summarizing Data	15 minutes
Presentation	10 minutes

Instructional Resources
- *Investigations and Projects Masters*, pp. 1–4

Investigations and Projects Masters, p. 4

Interdisciplinary Investigation

WHAT'S FOR DINNER?

Do you love to go out to eat? Isn't it great to order whatever you want and get it within minutes? Today, more than 500 fast-food chains are operating in the United States.

Some people have claimed that fast food is unhealthy because it contains too much fat and sodium. What do you think? Can such easy meals be good for you?

What You'll Do

In this investigation, you will research data about fast food and plan a healthy menu from your favorite fast-food restaurants.

Materials   nutritional guides from various fast-food restaurants

 calculator

Procedure

1. Work individually. Suppose you were going to eat at fast-food restaurants for an entire day. List items you would order for breakfast, lunch, and dinner.

2. The number of daily Calories needed by a teenager is about 2,400. Nutrition experts say that a teenager should have about 46 grams of protein per day. In addition, the number of milligrams of sodium should be less than 2,400. For your menu, find the total number of Calories, grams of protein, and milligrams of sodium. How did your menu compare to the recommendations of nutrition experts?

3. Work with a partner. Another recommendation is that the total Calories from fat should not exceed 30% of the total daily Calories. One gram of fat contains 9 Calories. Find the percent of daily Calories from fat in your menus from Step 1.

4. Plan a one-day menu that meets the requirements in Steps 2 and 3.

Technology Tips
- Use a **spreadsheet** to calculate totals and percents.
- Surf the **Internet** to find nutrition information.

◀◀◀ Instructional Resources
A recording sheet to help students organize their data for this investigation is shown at the left and is available in the *Investigations and Projects Masters*, p. 4.

Cooperative Learning
This investigation offers an excellent opportunity for using cooperative learning groups. For more information on cooperative learning strategies and group management, see *Cooperative Learning in the Mathematics Classroom*.

Making the Connection

Use the data collected about fast food to help in these investigations.

Language Arts

Do you think that eating fast food is healthy or unhealthy? Write a newspaper article supporting your opinion. Use facts from the investigation.

Health

Choose several other nutritional aspects of food such as calcium, iron, or vitamins. Did your menu meet the daily requirements of these nutrients?

Science

Research the effects of fast-food containers on the environment. For example, explain why McDonalds stopped using chlorofluorocarbons in their cartons.

Go Further

- Estimate the cost for your family to eat out every day for one year.
- Survey students in your school or families in your neighborhood. How many times per week do they eat at fast-food restaurants? Which restaurants are their favorites? Prepare a display to present your findings to the class.

interNET CONNECTION For current information on fast-food restaurants, visit: www.glencoe.com/sec/math/mac/mathnet

PORTFOLIO You may want to place your work on this investigation in your portfolio.

Interdisciplinary Investigation What's for Dinner?

CHAPTER 4
Statistics: Analyzing Data

Previewing the Chapter

Overview

This chapter emphasizes and explores the different ways to display and interpret data. Students learn how to construct and interpret frequency tables, bar graphs, histograms, circle graphs, statistical maps, and scatter plots. They discover how to find the mean, median, and mode as well as quartiles, range, and interquartile range of a set of data. Lessons on measures of central tendency, measures of variation, and misleading statistics focus on analyzing data.

Lesson (pages)	Lesson Objectives	NCTM Standards	Standardized Tests	State/Local Objectives
4-1A (140–141)	Solve problems by organizing data into a table.	1–5, 7, 8, 10	CTBS, MAT, TN	
4-1 (142–146)	Construct and interpret bar graphs and histograms.	1–5, 7–11	CAT, CTBS, MAT, SAT, TN	
4-1B (147)	Use a graphing calculator to make a histogram for a set of data.	1–5, 8–10		
4-2 (148–151)	Construct and interpret circle graphs.	1–5, 10, 12, 13	ITBS	
4-3 (153–155)	Construct and interpret line plots.	1–5, 8–10		
4-3B (156–157)	Use maps to display statistical data.	1–4, 8, 10		
4-4 (158–161)	Find the mean, median, and mode of a set of data.	1–5, 7, 9, 10	CTBS, ITBS, MAT, TN	
4-4B (162)	Use a spreadsheet to find the mean of a set of data.	1–5, 7–10		
4-5 (163–166)	Find the range and quartiles of a set of data.	1–5, 7, 10		
4-5B (167)	Use a graphing calculator to construct and interpret box-and-whisker plots.	1–5, 7, 8, 10		
4-6 (168–170)	Construct and interpret scatter plots.	1–5, 8–10	SAT	
4-7 (171–173)	Choose an appropriate display for a set of data.	1–5, 8–10	CTBS, TN	
4-8 (174–177)	Recognize when graphs and statistics are misleading.	1–5, 8–10		

CAT = California Achievement Tests, CTBS = Comprehensive Tests of Basic Skills, ITBS = Iowa Tests of Basic Skills, MAT = Metropolitan Achievement Tests, SAT = Stanford Achievement Tests, TN = Terra Nova

Organizing the Chapter

CD-ROM
All of the blackline masters in the Teacher's Classroom Resources are available on the **Electronic Teacher's Classroom Resources** CD-ROM.

LESSON PLANNING GUIDE

Lesson	Extra Practice (Student Edition)	Blackline Masters (page numbers)										Transparencies A and B
		Study Guide	Practice	Enrichment	Assessment & Evaluation	Classroom Games	Diversity	Hands-On Lab	School to Career	Science and Math Lab Manual	Technology	
4-1A	p. 614											
4-1	p. 614	26	26	26			30					4-1
4-1B												
4-2	p. 615	27	27	27	99			71				4-2
4-3	p. 615	28	28	28								4-3
4-3B								43–44				
4-4	p. 615	29	29	29	98, 99				30	73–76	59	4-4
4-4B												
4-5	p. 616	30	30	30								4-5
4-5B												
4-6	p. 616	31	31	31	100					57–60	60	4-6
4-7	p. 616	32	32	32		11–12						4-7
4-8	p. 617	33	33	33	100							4-8
Study Guide/ Assessment					85–97, 101–103							

OTHER CHAPTER RESOURCES

Student Edition
Chapter Project, pp. 139, 146, 161, 173, 181
Math in the Media, p. 151
School to Career, p. 152
Let the Games Begin, p. 146

Technology
CD-ROM Program
Interactive Mathematics Tools Software

Teacher's Classroom Resources

Applications
Family Letters and Activities, pp. 59–60
Investigations and Projects Masters, pp. 29–32

Meeting Individual Needs
Transition Booklet, pp. 5–8
Investigations for the Special Education Student, pp. 3–5

Teaching Aids
Answer Key Masters
Block Scheduling Booklet
Lesson Planning Guide
Solutions Manual

Professional Publications
Glencoe Mathematics Professional Series

Planning the Chapter

 MindJogger Videoquizzes provide a unique format for reviewing concepts presented in the chapter.

ASSESSMENT RESOURCES

Student Edition
Mixed Review, pp. 146, 151, 155, 161, 166, 170, 173, 177
Mid-Chapter Self Test, p. 166
➡ Math Journal, pp. 144, 172
Study Guide and Assessment, pp. 178–181
Performance Task, p. 181
📓 Portfolio Suggestion, p. 181
Standardized Test Practice, pp. 182–183
Chapter Test, p. 650

Assessment and Evaluation Masters
Multiple-Choice Tests (Forms 1A, 1B, 1C), pp. 85–90
Free-Response Tests (Forms 2A, 2B, 2C), pp. 91–96
Performance Assessment, p. 97
Mid-Chapter Test, p. 98
Quizzes A–D, pp. 99–100
Standardized Test Practice, pp. 101–102
Cumulative Review, p. 103

Teacher's Wraparound Edition
5-Minute Check, pp. 142, 148, 153, 158, 163, 168, 171, 174
📓 Building Portfolios, p. 138
➡ Math Journal, pp. 147, 157, 162, 167
Closing Activity, pp. 141, 146, 151, 155, 161, 166, 170, 173, 177

Technology
Test and Review Software
MindJogger Videoquizzes
CD-ROM Program

MATERIALS AND MANIPULATIVES

Lesson 4-1
number cubes

Lesson 4-1B
graphing calculator

Lesson 4-3
tape measure*
masking tape
markers

Lesson 4-3B
colored pencils
outline map of the U.S.

Lesson 4-4B
computer spreadsheet software

Lesson 4-5B
graphing calculator

Lesson 4-6
tape measure*
ruler*
grid paper
centimeter grid†

*Glencoe Manipulative Kit   †Glencoe Overhead Manipulative Resources

PACING CHART

See pages T25–T27 for the Course Planning Calendar.

COURSE	DAY 1	DAY 2	DAY 3	DAY 4	DAY 5	DAY 6	DAY 7
Standard	Chapter Project	Lesson 4-1A	Lesson 4-1	Lesson 4-2	Lessons 4-3 & 4-3B		Lesson 4-4
Honors	Chapter Project & Lesson 4-1A	Lessons 4-1 & 4-1B		Lesson 4-2	Lessons 4-3 & 4-3B	Lessons 4-4 & 4-4B	Lessons 4-5 & 4-5B
Block	Chapter Project & Lesson 4-1A	Lessons 4-1 & 4-2	Lessons 4-3 & 4-4	Lesson 4-5	Lesson 4-6	Lessons 4-7 & 4-8	Study Guide and Assessment, Chapter Test

The *Transition Booklet* (Skills 1, 2) can be used to practice basic operations with decimals.

Interactive Mathematics: Activities and Investigations

is an activity-based program that may be used as an enhancement for chapters in *Mathematics: Applications and Connections.*

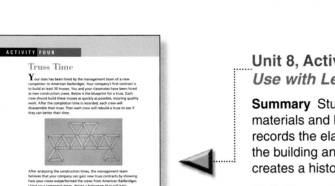

Unit 8, Activity Four
Use with Lesson 4-1.

Summary Students work in pairs to build a bridge truss with the materials and blueprint given to them. Each pair builds their truss and records the elapsed construction time. After disassembling the truss, the building and recording process is repeated. Each pair of students creates a histogram on poster board using data from all student pairs.

Math Connection Students construct a histogram from data created in a simulation. They should determine if the intervals they have chosen are appropriate to represent the data.

Unit 8, Activity Two
Use with Lesson 4-4.

Summary Students work in groups to complete the activity. Each member of the group chooses one of the five different stations that require the students to determine which measure of central tendency best represents the data shown. Students write a report defending their selection and present it to their group. The group then decides which situation they will present to the class.

Math Connection Students analyze the data using the measures of central tendency. A calculator may prove helpful.

DAY 8	DAY 9	DAY 10	DAY 11	DAY 12	DAY 13	DAY 14	DAY 15
Lesson 4-5	Lesson 4-6	Lesson 4-7	Lesson 4-8	Study Guide and Assessment	Chapter Test		
Continue from Day 7	Lesson 4-6	Lesson 4-7	Lesson 4-8	Study Guide and Assessment	Chapter Test		

Chapter 4 **138d**

Enhancing the Chapter

APPLICATIONS

Classroom Games, pp. 11–12
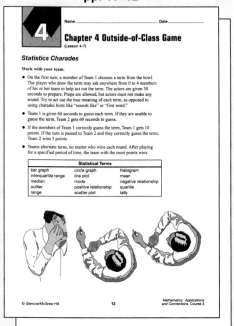

Diversity Masters, p. 30
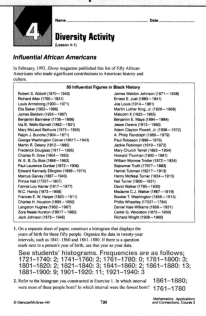

School to Career Masters, p. 30
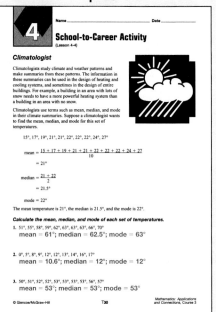

Family Letters and Activities, pp. 59–60
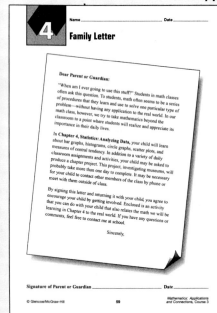

Science and Math Lab Manual, pp. 57–60 and 73–76
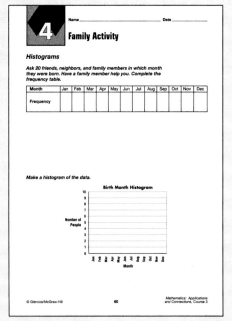

MANIPULATIVES/MODELING

Hands-On Lab Masters,
p. 71

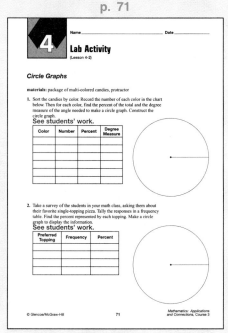

ASSESSMENT/EVALUATION

Assessment and Evaluation Masters,
pp. 98–100

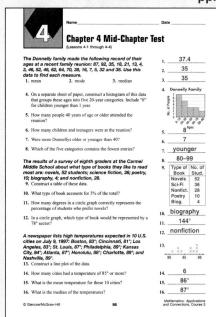

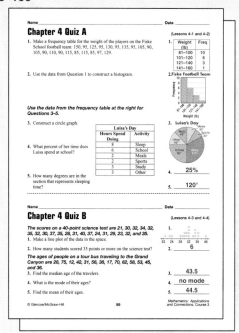

TECHNOLOGY/MULTIMEDIA

Technology Masters,
pp. 59–60

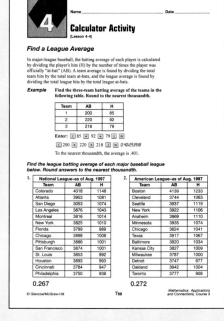

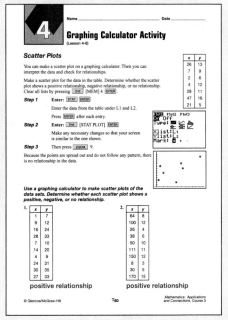

MEETING INDIVIDUAL NEEDS

Investigations for the Special Education Student, pp. 3–5

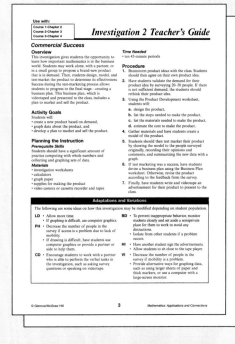

Chapter 4 **138f**

CHAPTER 4 NOTES

Theme: Museums
On September 2, 1995, the Rock and Roll Hall of Fame and Museum opened in Cleveland, Ohio. Students may not know that the term *rock and roll* was popularized in the 1950s by Cleveland deejay Alan Freed.

The Rock and Roll Hall of Fame and Museum contains a visual and audio history of more than 40 years of rock and roll performers and their music.

At the museum, visitors can listen to 500 songs that shaped rock and roll, view the original stage costumes of performers such as Elton John and Madonna, and see Janis Joplin's 1965 Porsche.

Question of the Day In 1996, 800,000 people visited the Rock and Roll Hall of Fame and Museum. What is the mean number of visitors per month? 66,667

Assess Prerequisite Skills
Ask students to read through the list of objectives presented in "What you'll learn in Chapter 4." You may wish to ask them what each of the objectives means or if they have experienced or used any of these math concepts before.

 Building Portfolios
Encourage students to revise their portfolios as they study this chapter. They may wish to include examples of each type of statistical display they learn to make in this chapter.

 Math and the Family
In the *Family Letters and Activities* booklet (pp. 59–60), you will find a letter to the parents explaining what students will study in Chapter 4. An activity appropriate for the whole family is also available.

CHAPTER 4

Statistics: Analyzing Data

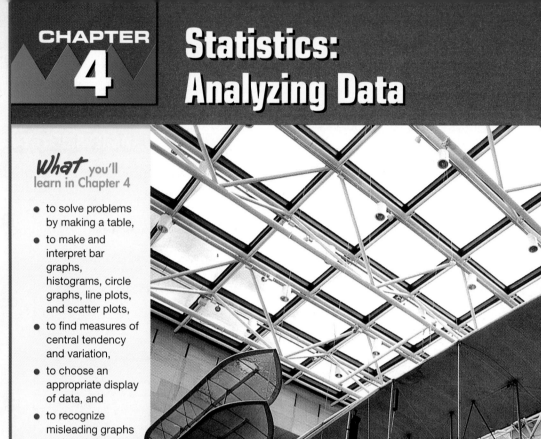

What you'll learn in Chapter 4

- to solve problems by making a table,
- to make and interpret bar graphs, histograms, circle graphs, line plots, and scatter plots,
- to find measures of central tendency and variation,
- to choose an appropriate display of data, and
- to recognize misleading graphs and statistics.

 CD-ROM Program

Activities for Chapter 4
- Chapter 4 Introduction
- Interactive Lessons 4-1A, 4-3, 4-4B
- Extended Activity 4-8
- Assessment Game
- Resource Lessons 4-1 through 4-8

CHAPTER Project

OLDIES BUT GOODIES!

In this project, you will work in a group to investigate how to run a museum and present your findings in a brochure or poster.

Getting Started

- Choose a museum that houses things of interest to your group. For example, the National Air and Space Museum in Washington, D.C., records the history of flight with items like the Wright Brothers' airplane, and the National Museum of American Art in Washington, D.C., collects art from all regions, cultures, and traditions in the United States.
- Make a list of information that you would like to know if you were considering giving a donation to your museum. Research the museum and its collections and other museums that have similar collections.

Top 5 Art Museums in the United States by Number of Annual Visitors	
Museum	**Annual Visitors**
National Gallery of Art, Washington, D.C.	7,500,000
Metropolitan Museum of Art, New York, NY	3,700,000
Art Institute of Chicago, IL	1,800,000
Museum of Modern Art, New York, NY	1,600,000
Hirshhorn Museum and Sculpture Garden, Washington, D.C.	1,300,000

Source: *The Top Ten of Everything*

Technology Tips

- Use a **spreadsheet** to keep track of the data you gather about your museum and others like it.
- Use a **word processor** to write a brochure or poster about your museum.

interNET CONNECTION For up-to-date information on the Smithsonian museums, visit: www.glencoe.com/sec/math/mac/mathnet

Working on the Project

You can use what you'll learn in Chapter 4 to help you make a brochure or poster about your favorite museum.

Page	Exercise
146	11
161	22
173	21
181	Alternative Assessment

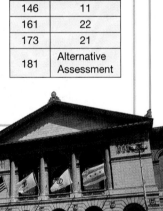

CHAPTER Project NOTES

Objectives Students should
- be able to see the difference in the numbers of visitors that visit each art museum.
- gain an understanding of how to track data and how to graph the results of their research.

Project Pointer You may suggest that students begin a *Project Folder* to keep their work as they complete each stage of the Chapter Project. The completed project may also be added to their portfolios.

If students would like to present their findings in a brochure, suggest they use spreadsheet, word-processing, or booklet-making software to create their own brochures.

Using the Table Students may not understand the word *annual*. If they know that *annual* means "covering the period of one year," they can find the average number of visitors per month and even per day. Students may not realize that the figures in the table are rounded to make the table easier to read and work with.

Investigations and Projects Masters, p. 32

interNET CONNECTION

Glencoe has made every effort to ensure that the website links for *Mathematics: Applications and Connections* at www.glencoe.com/sec/math/mac/mathnet are current and contain appropriate content. However, these website links are not under Glencoe's control.

Instructional Resources ▶▶▶

A recording sheet to help students organize their data for the Chapter Project is shown at the right and is available in the *Investigations and Projects Masters*, p. 32.

4-1A LAB Notes

Objective Students will learn to solve problems by organizing data into a table.

Instructional Resources
CD-ROM Program
- Resource Lesson 4-1A
- Interactive Lesson 4-1A

Recommended Pacing	
Standard	Day 2 of 13
Honors	Day 1 of 13
Block	Day 1 of 7

1 FOCUS

Getting Started After reading the introduction and studying the table, ask students which play the club should consider performing if there are not enough copies of *The Diary of Anne Frank*. Discuss their choice.

2 TEACH

Teaching Tip In addition to working Exercises 1–4, it may help students to develop their own frequency tables based on a familiar situation.

In-Class Example

John and Georgia are planning a fund-raiser to benefit the local humane society. They conducted a survey to find out what type of event their classmates would attend. The results were: 24 votes for a car wash, 14 votes for a bake sale, and 16 votes for a garage sale. Create a frequency table based on the information that includes tally marks for each event. Interpret the results of the survey.

PROBLEM SOLVING
4-1A Make a Table

A Preview of Lesson 4-1

Orlando and Michelle are helping their drama teacher Mrs. Moesley choose the next play the club will perform. Mrs. Moesley made a list of plays to consider. Now Orlando and Michelle are discussing how to poll the club members. Let's listen in!

Orlando: Well, our club meeting is tonight and we have lots of things to talk about. So how can we vote fast and make sure that we keep track of the votes?

Good idea. Let's see. I'll write the play names in a *frequency table* like this. Then they can make tally mark votes. I'll add up the votes after the voting sheet goes around and we'll put on the play that gets the most votes.

Michelle: I wrote down descriptions of the plays, which we can read out loud. Then we can pass around a sheet for people to vote for the play they like best.

Play	Tally	Frequency
The Diary of Anne Frank	|||| |||| ||	12
I Know Why the Caged Bird Sings	|||| ||	7
The Music Man	|||	3
You're a Good Man, Charlie Brown	|||| ||||	9

THINK ABOUT IT

Work with a partner.

1. **Tell** why you think Orlando used tally marks to record the votes. **See margin.**
2. **Brainstorm** types of information you have seen recorded in a table.
3. **Find** survey information that is recorded in a table, in a newspaper, or on the Internet. Which category received the most votes? Which received the fewest votes? **2–3. See students' work.**
4. **Compare and contrast** the table Orlando made with the table at the right. **See margin.**

Sample Tree Diameters from Cumberland National Forest		
Diameter (in.)	Tally	Frequency
2–4	||||		6
4–6	|||| |||| |||| |||| |||| ||||	30
6–8	|||| |||| |||| |||| |||| |||| |||| |||	38
8–10	|||| |||| |||| |||| |||| |||| |||	33
10–12	|||| ||||	9
12–14	||||	4

140 Chapter 4 Statistics: Analyzing Data

■ Reteaching the Lesson ■

Activity Tell students you are going to survey the class to find the popularity of these sports: baseball, basketball, football, tennis, and soccer. Discuss ways to organize the data as the survey is taken. Then conduct the survey, and ask students to interpret the results.

Additional Answers

1. Sample answer: Tally marks can be made quickly, and each person can add his or her mark without having to erase and add the total.

4. Sample answers: Orlando's table has categories, and this one has ranges of numbers; Orlando's table has fewer categories.

ON YOUR OWN

5. The first step of the 4-step plan for problem solving is to *explore* the problem. **Describe** two things you would need to explore in a set of data before you could **make a table** of the data. Explain your answer. **See margin.**

6. *Write a Problem* that can be answered using a table in this lesson. **See students' work.**

7. Refer to the frequency table on page 142. How many states have 501–1,000 malls? **15**

MIXED PROBLEM SOLVING

Strategies
Look for a pattern.
Solve a simpler problem.
Act it out.
Guess and check.
Draw a diagram.
Make a chart.
Work backward.

Solve. Use any strategy.

8. **Technology** The average Internet user spends $6\frac{1}{2}$ hours on-line each week. What percent of the week does the average user spend on-line? **3.9%**

9. *Money Matters* The list shows weekly allowances for a group of 13- and 14-year-olds.

$2.50	$3.00	$3.75	$4.25	$4.25
$4.50	$4.75	$4.75	$5.00	$5.00
$5.00	$5.00	$5.50	$5.50	$5.75
$5.80	$6.00	$6.00	$6.00	$6.50
$6.75	$7.00	$8.50	$10.00	$10.00
$12.00	$15.00			

   a. Make a frequency table of the allowances using $1.00 intervals. **See margin.**
   b. What is the most common interval of allowance amounts? **$4.01–$5.00**

10. **Sports** About 17% of American high-school athletes are injured each year. If there are 133 student athletes at Richmond High School, how many would you expect to be injured this year? **23**

11. *Patterns* I am thinking of a number. If I divide it by 2, then multiply the result by itself, I have 5,184. What was my original number? **144**

12. *Money Matters* Kwan has $12 to spend at the movies. After she pays the $4.75 admission, she estimates that she can buy a tub of popcorn that costs $4.25 and a medium drink that is $2.25. Is this reasonable? **yes; 5 + 4 + 2 < 12**

13. *Sleep* Use the frequency table. Which age group has the greatest percent of people who talk in their sleep? **18–24 years old**

GOOD NIGHT!

Age	Percent who talk in their sleep
18-24	29
25-34	23
35-49	15
50+	9

Source: The Better Sleep Council

14. **Test Practice** If $|a| = 4$, choose the possible values of *a*. **C**
   A −4
   B 0 and 4
   C −4 and 4
   D −16 and 16

15. **Test Practice** A number divided by 0.4 equals 20. What is the number? **B**
   A 4   B 8
   C 10   D 12

Lesson 4-1A THINKING LAB 141

Extending the Lesson

Activity On the chalkboard, write these intervals: January–April, May–August, and September–December. Have each student place a tally mark next to the interval containing the month in which they were born. Then have students create a frequency table based on the information from the birthday survey.

3 PRACTICE/APPLY

Check for Understanding
Use the results from Exercise 6 to determine whether students understand how to use a table to solve problems or make information easier to comprehend.

Extra Practice If students need additional practice in problem solving, extra practice is available on the following pages.
• Make a Table, see p. 614
• Mixed Practice Solving, pp. 645–646

Assignment Guide
All: 5–15

4 ASSESS

Closing Activity

Modeling Prepare a stack of 40 index cards numbered 1 through 40. Have students draw 10 cards. Have them place the cards in piles according to intervals of 5 (that is, 1–5, 6–10, 11–15, and so on). Determine the frequency of each interval.

Additional Answers

5. You would need the lowest and highest numbers in the data set. Knowing those numbers allows you to choose the range of the set.

9a.

Allowance	Number of 13- and 14-year olds
$2.01–3.00	2
$3.01–4.00	1
$4.01–5.00	9
$5.01–6.00	7
$6.01–7.00	3
$7.01–8.00	0
$8.01–9.00	1
$9.01–10.00	2
$10.01–11.00	0
$11.01–12.00	1
$12.01–13.00	0
$13.01–14.00	0
$14.01–15.00	1

Thinking Lab 4-1A 141

4-1 Lesson Notes

Instructional Resources
- *Study Guide Masters*, p. 26
- *Practice Masters*, p. 26
- *Enrichment Masters*, p. 26
- Transparencies 4-1, A and B
- *Diversity Masters*, p. 30
- CD-ROM Program
  - Resource Lesson 4-1

Recommended Pacing	
Standard	Day 3 of 13
Honors	Days 2 & 3 of 13
Block	Day 2 of 7

1 FOCUS

5-Minute Check
(Chapter 3)

1. Express 25:45 as a fraction in simplest form. $\frac{5}{9}$
2. Express 4 feet:144 inches as a percent. $33\frac{1}{3}\%$
3. Solve $\frac{4}{7} = \frac{6.2}{d}$. 10.9
4. Express 29% as a decimal. 0.29
5. Find 6% of $830. $49.80

The 5-Minute Check is also available on **Transparency 4-1A** for this lesson.

Motivating the Lesson
Hands-On Activity Have students use stacks of counters to build a model for the number of states in each category in the table at the beginning of the lesson.

4-1 Bar Graphs and Histograms

What you'll learn
You'll learn to construct and interpret bar graphs and histograms.

When am I ever going to use this?
Bar graphs and histograms are often used to display information in newspapers and magazines.

Word Wise
frequency table
statistics
bar graph
histogram

Do you like to shop till you drop? If so, you're not alone. Thousands of malls across America are filled with shoppers. The **frequency table** shows the number of states with different numbers of malls.

Statistics involves collecting, organizing, and analyzing data. Statisticians sometimes use **bar graphs** and special bar graphs, called **histograms** to display data like that in the table. A bar graph compares different categories of data by showing each as a bar whose length is related to the frequency.

Number of malls	Number of States*
1–500	24
501–1,000	15
1,001–1,500	6
1,501–2,000	3
2,001–2,500	0
2,501–3,000	1
3,001–3,500	1
3,501–4,000	0
4,001–4,500	0
4,501–5,000	0
5,001–5,500	1

* Includes Washington D.C.
Source: National Research Bureau

Example APPLICATION 1

Technology The table shows the percent of people in the United States who own different consumer electronic products. Make a bar graph of the data.

First draw a horizontal axis and a vertical axis. Label the axes as shown. Be sure to include the title, Electronics at Home.

Each category has a bar to represent it. The vertical scale is the percent of the population.

Product	Percent Who Own
Television	98
Color, stereo television	52
CD player	48
Rack or component audio system	34
Camcorder	23
Car CD player	15
Laserdisc player	2

Source: Electronic Industries Association

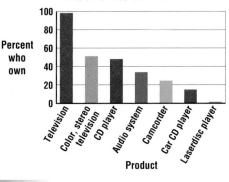

142 Chapter 4 Statistics: Analyzing Data

Classroom Vignette

"I like to have my students create a survey to measure the characteristics of the average 8th grader. We mail them out to other schools and then use spreadsheet software to analyze the results and make graphs."

Mary Jo Deschene, Math Instructor
St. Francis Jr. High School
St. Francis, MN

A histogram uses bars to display numerical data that have been organized into equal intervals. The intervals cover the entire range of the data with no overlapping intervals.

Example **APPLICATION**

2 Shopping Refer to the beginning of the lesson.

a. Construct a histogram of the data.

First draw a horizontal axis and a vertical axis. Label the axes as shown. Be sure to include the title, America's Malls.

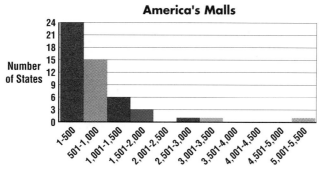

- Equal intervals of 500 are shown on the horizontal axis.
- Because the intervals are equal, all of the bars have the same width.
- As in a bar graph, the frequency of data in each interval is shown by the bar height.
- The vertical scale on bar graphs and histograms is often a factor of the greatest frequency. In this case, the greatest frequency is 24, and the vertical scale is 3.
- Intervals with a frequency of 0, like 2,001–2,500, have no bar.

b. How does the number of states with 1–500 malls compare to the number of states with 1,001–1,500 malls?

Compare the heights of the bars. The bar for 1–500 malls is about four times as tall as the bar for 1,001–1,500 malls. There are about four times as many states with 1–500 malls as there are states with 1,001–1,500 malls.

Sometimes when the data in a graph do not start at zero, a jagged line is shown on one of the axes of the histogram.

2 TEACH

 Transparency 4-1B contains a teaching aid for this lesson.

Reading Mathematics Make sure that students know the difference between horizontal and vertical. Emphasize that a histogram is a type of bar graph.

In-Class Examples

For Example 1
Ask students to explain how an additional entry, *88% own VCRs*, would change the graph. **A bar that reaches 88 on the vertical axis would be added after *Television*.**

For Example 2
a. Construct a histogram of the data.

Number of Public Libraries	Number of States
1–100	22
101–200	11
201–300	7
301–400	5
401–500	2
501–600	1
601–700	1
701–800	1

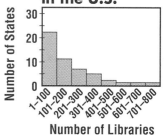

b. How does the number of States with 1–100 libraries compare to the number of states with 201–300 libraries? **There are more than three times as many states with 1–100 libraries as there are with 201–300 libraries.**

Lesson 4-1 Bar Graphs and Histograms **143**

Multiple Learning Styles

Auditory/Musical Separate students into groups of four or five and refer them to the histogram of record high temperatures in Example 3. Ask them to imagine that each bar on the histogram represents an individual musical note or drum beat with the tallest bar being the loudest note or heaviest beat. Have students compose a short song that uses the rhythm of the histogram.

In-Class Example

For Example 3
Refer to the graph in Example 3 of the Student Edition. Ask students to explain how the histogram would change if 8 states had a high temperature of 95–99° and only 9 states had a high temperature of 110–114°. A bar labeled 95–99 and reaching to 8 on the vertical axis would be inserted before the first bar; the 110–114° bar would reach only to 9 on the vertical axis.

Teaching Tip Point out that in Example 3, the temperature is used as the horizontal axis and the number of states as the vertical axis.

Teaching Tip In Lesson 4-1B Students learn to use a graphing calculator to make a histogram for a set of data.

3 PRACTICE/APPLY

Check for Understanding

If students need additional practice or instruction after completing Exercises 1–4, one of these options may be helpful.
- Extra Practice, see p. 614
- Reteaching Activity
- *Study Guide Masters*, p. 26
- *Practice Masters*, p. 26
- Interactive Mathematics Tools Software

Additional Answers

1. Sample answer: Because it is more visual, a histogram is more useful than a table when you are trying to show a general trend. Because the individual numbers are shown, a table is more useful when you need to know exact numbers.

2. Koko is correct. The intervals in a histogram cover all possibilities, but they do not overlap.

4a. the number of states that have different levels of spending per student each year

---

Example 3 CONNECTION

Earth Science The table below shows the record high temperatures for each state. Make a histogram of this data.

CHECK FOR UNDERSTANDING

Communicating Mathematics

Read and study the lesson to answer each question.

1. *Describe* when a histogram is more useful than a table. When is a table more useful than a histogram? **See margin.**

2. *You Decide* Kayla says that a histogram has no spaces between the bars because the intervals overlap. Koko says there are no spaces because the intervals cover every possible data value. Who is correct? Explain your reasoning. **See margin.**

3. *Write* a few sentences about when to use a bar graph and when to use a histogram. **See students' work.**

Guided Practice

4. Use the histogram at the right to answer each question.

    a. Describe the data shown in the graph. **See margin.**

4b. $1,000

    b. How large is each interval?

    c. Which interval has the greatest number of states? **$5,000–5,999**

4d. The intervals from $0 to $2,999 have been omitted.

    d. Why is there a jagged line in the horizontal axis?

    e. Make a frequency table of the data. **See margin.**

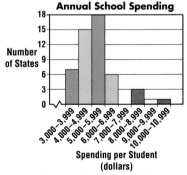

144 Chapter 4 Statistics: Analyzing Data

■ Reteaching the Lesson ■

Activity Have students search newspapers or magazines for frequency tables or information that can be used to make a frequency table. Ask them to create bar graphs or histograms with the data. Remind students that graphs make comparison of data easier.

Additional Answer

4e.

Spending per Student (thousands)	Number of States
$3,000–3,999	7
$4,000–4,999	15
$5,000–5,999	18
$6,000–6,999	6
$7,000–7,999	0
$8,000–8,999	3
$9,000–9,999	1
$10,000–10,999	1

EXERCISES

Practice

5a. Sample answer: incubation times for eggs of different birds

Family Activity

Survey at least 20 relatives and friends about their birthdays. Make a bar graph of the frequencies of birthdays in each month. Write about any patterns you see.

8a. Sample answer: numbers of threatened animals of different species

8b. See Answer Appendix.

Applications and Problem Solving

5. Use the table of incubation times to answer each question.

Egg Incubation Time

Bird	Chicken	Duck	Goose	Pigeon	Turkey
Time (days)	21	30	30	18	26

   a. Describe the data in the table.
   b. Construct a bar graph of the data.   **See Answer Appendix.**

6. Use the histogram at the right to answer each question.   **a. See margin.**

   a. Describe the data in the histogram.
   b. How large is each interval?   **3 inches**
   c. Which interval of heights has the most presidents?   **72–74 in.**
   d. Construct a frequency table of the data of presidents' heights.   **See margin.**
   e. Can you tell from the histogram or the table the exact height of the tallest president? Explain.   **See margin.**
   f. What percent of the presidents were less than 6 feet tall?   **55%**

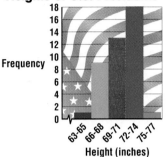

7. The suggested retail prices of several different bicycle helmets tested by *Zillions* magazine are shown below. Make a frequency table and a histogram of the prices.   **See Answer Appendix.**

   $25 $30 $30 $30 $30 $30 $40 $40
   $40 $40 $40 $40 $40 $43 $50 $50

8. **Life Science** Use the table of U.S. threatened species to answer each question.

   a. Describe the data in the table.
   b. Construct a bar graph of the data.
   c. How many species are recorded in the bar graph?   **115**
   d. *Write a Problem* using the data in the table.   **See students' work.**

U.S. Threatened Species

Species	Number
Amphibians	6
Birds	16
Clams	6
Crustaceans	3
Fishes	40
Insects	9
Mammals	9
Reptiles	19
Snails	7

Source: Fish and Wildlife Service

9. **Civics** The United Nations was formed at the end of World War II to maintain international peace and security. The year that each of the 184 member nations entered the U.N. are shown in the table below. Make a histogram of the data.   **See Answer Appendix.**

United Nations Entry

1945–1954	1955–1964	1965–1974	1975–1984	1985–1994
59	55	21	21	28

Source: *The World Almanac, 1997*

Lesson 4-1 Bar Graphs and Histograms **145**

Additional Answers

6d.

Heights of U.S. Presidents	
Height (in.)	Frequency
63–65	1
66–68	9
69–71	13
72–74	18
75–77	1

6e. No; the individual heights are not listed. You can tell only that the tallest president was between 75 and 77 inches tall.

Assignment Guide

Core: 5–9 odd, 12–14
Enriched: 6, 8–10, 12–14

Family Activity

Have students share their bar graphs with the class. Ask students to note any patterns throughout the graphs. Then have them combine their information into a single graph. When they finish, ask students if the pattern they noticed in their individual graphs is evident in the combined graph.

CHAPTER Project

Exercise 11 asks students to advance to the next stage of work on the Chapter Project. You may want to have students include their finished histograms in their portfolios.

Additional Answer

6a. Sample answer: the heights of the U.S. Presidents, divided into 3-inch intervals

Study Guide Masters, p. 26

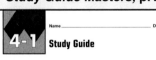

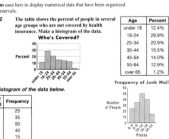

Lesson 4-1 **145**

4 ASSESS

Closing Activity
Speaking Ask students to discuss the advantages of using a bar graph to display data. Ask them to consider the types of information or statistics that could be depicted using a bar graph.

Additional Answers

12a. You cannot tell from the table how many nations joined the U.N. in 1994 because the data are organized in intervals.

12b. Sample answers: Large growth occurred in 1945–1954 because the U.N. was formed in 1945 and a large number of nations joined at its beginning. The interval 1985–1994 included the time of the break up of the Soviet Union, and a number of nations that became independent at that time joined the U.N. in the years soon after.

Practice Masters, p. 26

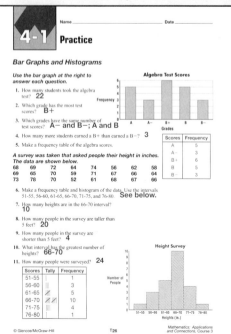

10a. 0–2,999 thousand acres

10c. See Answer Appendix.

10. **Natural Resources** This frequency table shows the numbers of states with different amounts of National Forest Service land.

   a. How much National Forest Service land do most states have?

   b. What fraction of the states have less than 9,000,000 acres of National Forest Service land? $\frac{4}{5}$

   c. Make a histogram of the data.

National Forest Service Land	
Acres (thousands)	States
0–2,999	38
3,000–5,999	1
6,000–8,999	1
9,000–11,999	4
12,000–14,999	1
15,000–17,999	2
18,000–20,999	2
21,000–23,999	1

Source: U.S. Forest Service

11. **Working on the CHAPTER Project** If you chose an art museum, construct a bar graph of the data on museum visitors on page 139. Be sure to include your chosen museum. If you chose another type of museum, find data about visitors to those types of museum and construct a bar graph. **See students' work.**

12. **Critical Thinking** Study the information in the table for Exercise 9.

    a. How many nations entered the U.N. in 1994? Explain your answer.

    b. Research world events in the times when there were large numbers of nations that joined the U.N. Explain why the growth occurred.

12a–b. See margin.

Mixed Review

13. Estimate the discount if a $58 jacket is marked down 24%. *(Lesson 3-6)*

14. **Test Practice** Evaluate $xy + |-4|$ if $x = 2$ and $y = 12$. *(Lesson 2-1)* **C**

    A 24 **B** 27 **C** 28 **D** 30

13. Sample answer: $60 ÷ 4 = $15

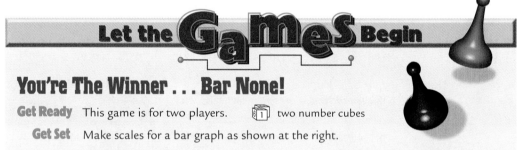

You're The Winner... Bar None!

Get Ready This game is for two players. 🎲 two number cubes

Get Set Make scales for a bar graph as shown at the right.

Go
- Player A chooses a sum. Then player B chooses one of the remaining sums.
- The players take turns rolling the number cubes. With each roll, the player finds the sum on the cubes and creates or lengthens the bar for that sum on the graph.
- The winner is the player whose bar reaches a frequency of 5 first.
- In the next round, player B chooses his or her bar first.

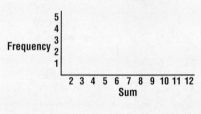

interNET CONNECTION Visit www.glencoe.com/sec/math/mac/mathnet for more games.

146 Chapter 4 Statistics: Analyzing Data

■ Extending the Lesson ■

Enrichment Masters, p. 26

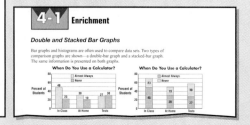

Let the Games Begin

Remind students that *frequency* means the number of times an event happens. As students play the game, ask them to look for any patterns that develop in the frequency of the numbers that come up. Ask them if this will have an effect on the numbers they select in the next round. *Additional resources for this game can be found on page 48 of the* **Classroom Games.**

GRAPHING CALCULATORS

4-1B Histograms

A Follow-Up of Lesson 4-1

graphing calculator

You can use a graphing calculator to make a histogram for a set of data.

TRY THIS

Work with a partner.

The list is the record low temperatures for each state and Washington, D.C.
-27 -80 -40 -29 -45 -61 -32 -17 -15 -2 -17 12 -60 -35
-36 -47 -40 -37 -16 -48 -40 -35 -51 -59 -19 -40 -70 -47
-50 -46 -34 -50 -52 -34 -60 -39 -27 -54 -42 -23 -19 -58
-32 -23 -69 -50 -30 -48 -37 -54 -66

Step 1 Begin by entering the data into the calculator's memory. Press [STAT] [ENTER] to see the lists. If there are numbers in the first list, press [▲] [CLEAR] [ENTER] to clear them. Then enter the data by entering each number and pressing [ENTER].

Step 2 Next choose the type of graph. Press [2nd] [STAT PLOT] to display the menu. Choose the first plot by pressing [ENTER]. Use the arrow and [ENTER] keys to highlight "on", the histogram, L1 for the Xlist, and 1 as the frequency.

Step 3 Choose the display window. Access the menu by pressing [WINDOW]. Choose appropriate range settings. For this data, use -80 to 20 with a scale of 10 on the *x*-axis and 0 to 15 with a scale of 5 on the *y*-axis.

Step 4 Display the graph by pressing [GRAPH].

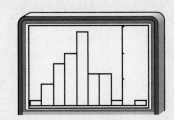

ON YOUR OWN

1. Press the [TRACE] key. The right and left arrow keys allow you to find the frequency of each interval. How many temperatures are between -60 and -50? **8**
2. Use the data on iced tea prices on page 153 to make a histogram. **See margin.**
3. How does the graphing calculator determine the size of the intervals? **The scale for the *x*-axis is the interval.**

Lesson 4-1B **TECHNOLOGY LAB** 147

Additional Answer
2.

Have students write a paragraph about how a histogram could be used in an article describing voting behavior in a local or national election.

4-2 Lesson Notes

Instructional Resources
- *Study Guide Masters*, p. 27
- *Practice Masters*, p. 27
- *Enrichment Masters*, p. 27
- Transparencies 4-2, A and B
- *Assessment and Evaluation Masters*, p. 99
- *Hands-On Lab Manual*, p. 71
- CD-ROM Program
  - Resource Lesson 4-2

Recommended Pacing	
Standard	Day 4 of 13
Honors	Day 4 of 13
Block	Day 2 of 7

1 FOCUS

5-Minute Check
(Lesson 4-1)

Construct a bar graph of the following data: 22% of trucks and vans sold in one year were white, 17% green, 12% medium red, 10% teal, 9% black, 8% bright red, and 7% medium blue.

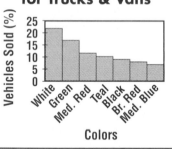

Popular Colors for Trucks & Vans

The 5-Minute Check is also available on **Transparency 4-2A** for this lesson.

Motivating the Lesson

Hands-On Activity Give each group of students a circle with the center marked. Have them cut out the circle and divide it into 5 unequal wedges. Have students measure the angle of each wedge and find the sum of the measures.

148   Chapter 4

4-2 Circle Graphs

What you'll learn
You'll learn to construct and interpret circle graphs.

When am I ever going to use this?
Circle graphs are often used in advertising to compare parts to a whole.

Word Wise
circle graph

What kind of book do you choose for a quiet afternoon? The table shows the number of adult books published in 1996 in different subjects.

A **circle graph** compares parts of a set of data to the whole set. For example, a circle graph of the book subject data could be used to compare the number of fiction books to the total number of books published.

Subject	Books Published
Art, literature, music	4,864
Business, law	9,862
Fiction, general	7,843
History	6,130
Philosophy, psychology	4,346
Science, technology	8,680
Sports, recreation	2,407

Source: *Publishers Weekly*

Example CONNECTION

Language Arts Construct a circle graph using the data above.

Step 1 Find the total number of books published.

4864 [+] 9862 [+] 7843 [+] 6130 [+] 4346 [+] 8680 [+] 2407 [=] *44132*

Step 2 Find the ratio that compares the number of books in each subject to the total number of books published. Round each ratio to the nearest hundredth.

Since you are comparing the number of books for each subject to the total, each ratio represents the percent of the whole circle.

Art, literature, music: $\frac{4,864}{44,132}$

4864 [÷] 44132 [=] *0.11021481*

$0.11021481 \approx 0.11$

Business, law: $\frac{9,862}{44,132}$

9862 [÷] 44132 [=] *0.223465966*

$0.223465966 \approx 0.22$

Fiction, general: $\frac{7,843}{44,132}$

7843 [÷] 44132 [=] *0.177716849*

$0.177716849 \approx 0.18$

History: $\frac{6,130}{44,132}$

6130 [÷] 44132 [=] *0.138901477*

$0.138901477 \approx 0.14$

Philosophy, psychology: $\frac{4,346}{44,132}$

4346 [÷] 44132 [=] *0.098477295*

$0.098477295 \approx 0.10$

Science, technology: $\frac{8,680}{44,132}$

8680 [÷] 44132 [=] *0.196682679*

$0.196682679 \approx 0.20$

Sports, recreation: $\frac{2,407}{44,132}$

2407 [÷] 44132 [=] *0.054540923*

$0.054540923 \approx 0.05$

148   Chapter 4 Statistics: Analyzing Data

Cross-Curriculum Cue

Inform the other teachers on your team that your students are studying circle graphs. Suggestions for curriculum integration are:

Geography: population distribution
Earth Science: land resources
Health: distribution of calories in a healthful diet

Study Hint

Estimation You can check your solution by adding the measures of the angles. If the sum is close to 360, you are probably correct.

Study Hint

Technology Many spreadsheets and word processing programs will construct circle graphs from data you enter.

Step 3 Since there are 360° in a circle, multiply each ratio by 360 to find the number of degrees for each section of the graph. When necessary, round to the nearest degree.

Art, literature, music: 0.11 × 360 = 39.6 $39.6 \approx 40°$

Business, law: 0.22 × 360 = 79.2 $79.2 \approx 79°$

Fiction, general: 0.18 × 360 = 64.8 $64.8 \approx 65°$

History: 0.14 × 360 = 50.4 $50.4 \approx 50°$

Philosophy, psychology: 0.10 × 360 = 36

Science, technology: 0.20 × 360 = 72

Sports, recreation: 0.05 × 360 = 18

Step 4 Use a compass to draw a circle and a radius. Then use a protractor to draw a 40° angle. *You can start with any of the angles.*

Step 5 From the new radius, draw the next angle. Repeat for each of the remaining angles. Label each section and write each ratio as a percent. Then give the graph a title.

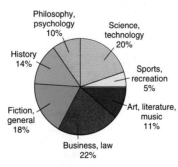

Books Published, 1996

---

 2 TEACH

Transparency 4-2B contains a teaching aid for this lesson.

Thinking Algebraically The process of determining how large each slice of the circle graph should be can be thought of as a proportion. Each category represents a part of the whole population and corresponds to each section of the graph representing an equivalent part of the whole 360 degrees.

In-Class Example

For the Example
Construct a circle graph using the data in the table.

Age Groups in Lodi	
Age (years)	Percent
0–19	34%
20–39	17%
40–59	21%
60 and over	28%

Age Groups in Lodi

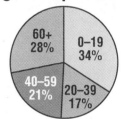

Lesson 4-2 Circle Graphs **149**

3 PRACTICE/APPLY

Check for Understanding

If students need additional practice or instruction after completing Exercises 1–3, one of these options may be helpful.
- Extra Practice, see p. 615
- Reteaching Activity
- *Transition Booklet*, pp. 7–8
- *Study Guide Masters*, p. 27
- *Practice Masters*, p. 27

Assignment Guide

Core: 5, 7, 9–11
Enriched: 4–8 even, 9–11

Additional Answer

3d.

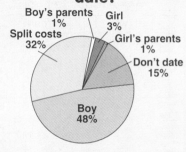

Who pays when you date?
- Boy's parents 1%
- Girl 3%
- Split costs 32%
- Girl's parents 1%
- Don't date 15%
- Boy 48%

Study Guide Masters, p. 27

CHECK FOR UNDERSTANDING

Communicating Mathematics

Read and study the lesson to answer each question.

1. **Identify** the Disney theme park that about one-fourth of the visitors choose. **Tokyo Disneyland**

2. **Describe** when a circle graph is used to show data. **Circle graphs are used to compare parts of a whole.**

1996 Disney Visitors (millions)
- MGM Studios 8.0
- Other 1.1
- Tokyo Disneyland 16.0
- Euro Disneyland 8.8
- EPCOT Center 9.7
- Disneyland 10.3
- Magic Kingdom 11.2

Source: Amusement Business

Guided Practice

3a. 100%

3. The results of a survey of high school students are shown in the table.
   a. What should the total in the ratio column be?
   b. What total would you expect in the degree column? **360°**
   c. Copy and complete the table. **See table.**
   d. Make a circle graph of the data. **See margin.**

Who Pays When You Date?

Category	Number	Ratio	Degrees in graph
Boy	1,200	0.48	172.8°
Split costs	800	0.32	115.2°
Girl	75	0.03	10.8°
Girl's parents	25	0.01	3.6°
Boy's parents	25	0.01	3.6°
Don't date	375	0.15	54°
Total	2,500	1	360°

Source: BKG Youth poll

4a. See Answer Appendix.
4b. cassette

Practice

EXERCISES

4. **Music** The table shows the sales of each type of recorded music for 1995.
   a. Make a circle graph of the data.
   b. Which format accounts for about one-fifth of the total music sales?

5. **History** The Medal of Honor is the highest military award for bravery in the United States. There were 124 medals given to veterans of World War I, 433 for World War II, 131 for the Korean War, 239 for the Vietnam War, 2 for Somalia, and 18 in peacetime. Make a circle graph of the medals of honor. **See Answer Appendix.**

Format	Sales ($ millions)
CD	9,401.7
CD Single	88.6
Cassette	2,303.6
Cassette Single	236.3
LP/EP	25.1
Vinyl Single	46.7
Video	220.3

Source: Recording Industry Assn. of America

6. **Geography** The table shows the areas of the five counties in the state of Hawaii. Make a circle graph for the areas of the Hawaiian counties.

County	Hawaii	Honolulu	Kalawao	Kauai	Maui
Area (sq mi)	4,028	600	13	623	1,159

Source: U.S. Department of Commerce

6. See Answer Appendix.

150 Chapter 4 Statistics: Analyzing Data

Reteaching the Lesson

Activity Have students choose a circle graph from a magazine or newspaper article. Have them determine whether the graph is drawn correctly and redraw the graph if it is incorrect.

Error Analysis
Watch for students who find angle measures incorrectly.
Prevent by having students add the measures of all sections before drawing their graphs. The sum should be 360°.

Applications and Problem Solving

7–8. See margin.

7. *Charities* According to the American Association of Fund-Raising Counsel, Americans gave $143.84 billion to charity in 1995. Corporations gave $7.4 billion, foundations gave $10.44 billion, $9.77 billion was left in wills, and individuals gave $116.23 billion. Make a circle graph of the contributions.

8. *History* The table shows the birthplaces of the signers of the Declaration of Independence. Make a circle graph of the data.

Location	Signers	Location	Signers	Location	Signers
Connecticut	5	Massachusetts	9	Rhode Island	2
Delaware	2	New York	3	South Carolina	4
Maine	1	New Jersey	3	United Kingdom	8
Maryland	5	Pennsylvania	5	Virginia	9

Source: *World Almanac,* 1997

9. No; the percents are not parts of a whole because the categories overlap.

9. *Critical Thinking* According to Nielsen Media Research, 98% of U.S. households owned at least one television in 1995. 38% had 2 TVs, 28% had 3 or more TVs, 81% had a VCR, and 65% received cable. Can this information be displayed in a circle graph? Explain.

Mixed Review

10. See Answer Appendix.

10. *Entertainment* One hundred students were asked how many hours they watch television each week. The results are given in the table. Make a histogram of the data. *(Lesson 4-1)*

Hours	Frequency
0–2	4
3–5	8
6–8	22
9–11	32
12–14	30
15–17	4

11. **Test Practice** Solve $h = 18 - (-4)$. *(Lesson 2-5)* **A**

   A 22   **B** 14   **C** -22 **D** -14

MATH IN THE MEDIA

Computer Comparisons

The report described below appeared on the *NBC Nightly News* on May 25, 1995.

Tom Brokaw read a report saying that 14 percent of African-Americans, 13 percent of Hispanics, and 27 percent of Caucasians use computers. He went on to say that computer use by Caucasians was equal to that of African-Americans and Hispanics combined.

1–3. See Answer Appendix.

The circle graph shows the composition of the United States population by race.

1. Find the percent of the population that each ethnic group represents.

2. Use the figures from the news report with the data from the circle graph to find the number of African-Americans, Hispanics, and Caucasians who use computers.

3. Is it true that "computer use by Caucasians was equal to that of African-Americans and Hispanics combined"? Explain.

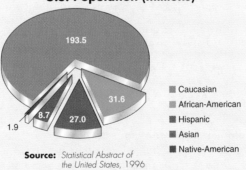

U.S. Population (millions)

193.5, 31.6, 8.7, 27.0, 1.9

- Caucasian
- African-American
- Hispanic
- Asian
- Native-American

Source: *Statistical Abstract of the United States,* 1996

Lesson 4-2 Circle Graphs **151**

---

4 ASSESS

Closing Activity

Writing Have students work in pairs to write an explanation of the steps for creating a circle graph to display a given set of data.

Chapter 4, Quiz A (Lessons 4-1 and 4-2) is available in the *Assessment and Evaluation Masters,* p. 99.

Additional Answers

7. **Giving to Charities**

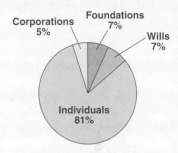

8. **Declaration Signers**

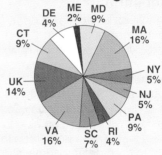

Practice Masters, p. 27

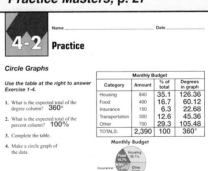

Extending the Lesson

Enrichment Masters, p. 27

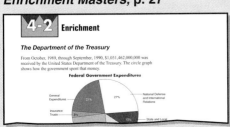

MATH IN THE MEDIA

Ask students if presenting the information in the news report as fractions would make it easier or harder to understand. Have students look for other news reports that use graphs and percents.

Lesson 4-2 **151**

Motivating Students

Students learned in the Chapter Opener on page 138 about museum collections, attendance figures, and different jobs at a museum. They may be interested in a career as a museum curator, archivist, director, conservator, or technician. To start the discussion, ask students questions about museums.

- What are the 3 main types of museums? **art, history, and science**
- Where is the Baseball Hall of Fame Museum located? **Cooperstown, New York**
- When did the U.S. Congress establish the Smithsonian? **August 10, 1846**

Making the Math Connection

When writing an annual budget, museum director Spencer Crew analyzes data to determine how much money the museum can spend the following year. Such data are effectively presented as bar graphs, histograms, and circle graphs. Crew needs to be able to interpret information presented in graphs to prepare and administer the annual budget.

Working on *Your Turn*

Students may want to work in groups to develop a list of museums in their city or state. Make sure students understand that they don't have to have a master's or doctoral degree to work at a museum.

*An additional School to Career activity is available on page 30 of the **School to Career Masters**.*

HISTORY

Spencer R. Crew
MUSEUM DIRECTOR

The National Museum of American History is one of the 14 museums of the Smithsonian Institution. Located in Washington, D.C., it contains over 17 million artifacts including the gowns of the U.S. First Ladies and the original Star-Spangled Banner that inspired the writing of the national anthem. Spencer Crew, director since 1992, leads museum staff in making decisions about research, collecting, and developing new exhibitions and public programs. He is also responsible for overseeing the resources of the museum: its funds, its people, and its spaces.

Museum staffs are skilled in many areas and enjoy sharing their expertise with the public. Curators, historians, museum specialists, museum technicians, and archivists collect, manage, and catalogue artifacts and documents. All museum staff need a good education and hands-on experience. Curators and directors usually have masters or doctoral degrees. To work in one of these museum jobs, you should have a good background in a wide variety of subjects including history, art, science, business, and math. You should also have good team skills, be well-organized, and enjoy working with the public.

For more information:
American Association of Museums
1225 I St., NW, Suite 200
Washington, DC 20005

interNET
CONNECTION
www.glencoe.com/sec/math/mac/mathnet

I think working in a museum like Spencer Crew does would be fun.

Your Turn
Compile a list of museums in your city or state. At which museums do you think it would be fun to work? Explain why, and describe the types of courses and experience you would need to work there.

152 Chapter 4 Statistics: Analyzing Data

More About Spencer Crew

- Spencer Crew is the first African American to be appointed director of the Smithsonian Institution's National Museum of American History.
- He earned a bachelor's degree from Brown University in 1971, an M.A. from Rutgers University in 1973, and a Ph.D. from Rutgers in 1979.
- Crew's expertise is in African American urban history. His interest is in the migration of African Americans to northern cities and the development of African American communities in the twentieth century.

4-3 Line Plots

What you'll learn
You'll learn to construct and interpret line plots.

When am I ever going to use this?
Line plots are useful in recording data from surveys.

Word Wise
line plot
data analysis

One of the ways that you can display data is in a **line plot**. In a line plot, data is organized using a number line.

HANDS-ON MINI-LAB

Work with a partner. tape measure tape · markers

You will make a human line plot of the heights of your class members.

Try This
- Measure and record your partner's height in inches.
- As a class, use masking tape to place a number line on the floor. The distances between consecutive numbers should be the same.
- Have each person stand above his or her height on the number line. People of the same height should form a column above the height.

Talk About It
1. Which height has the most people? How can you tell?
2. Which heights had more boys or more girls? **See students' work.**

1. See students' work. The height with the most people is the one with the longest column of people.

LOOK BACK
You can refer to Lesson 2-1 for information on number lines.

It is usually not practical to arrange people or objects to form a line plot. You can draw a line plot using a symbol to represent each data point.

Zillions magazine compared the taste and price of lemon-flavored iced teas. The teas were ranked very good (VG), good (G), or fair (F). The results are shown in the table.

Tea	Rating	Price per 8-oz serving
Arizona	G	37¢
Celestial Seasonings Uptown Express	G	35¢
Homemade iced tea	VG	8¢
Lipton Brisk	F	21¢
Lipton Iced Tea Mix	G	8¢
Lipton Original	G	34¢
Mistic	G	42¢
Nestea	G	37¢
Nestea Cool	F	21¢
Nestea Iced Tea Mix	G	7¢
Safeway Select Refreasher	G	30¢
ShopRite	G	23¢
Snapple	G	36¢
Tetley	F	21¢
Tetley Lemon Frost	VG	36¢
Twinings Ceylon Blend	G	40¢
Veryfine Chillers	G	20¢

Source: *Zillions* magazine, Aug/Sept 1995

Lesson 4-3 Line Plots **153**

4-3 Lesson Notes

Instructional Resources
- *Study Guide Masters*, p. 28
- *Practice Masters*, p. 28
- *Enrichment Masters*, p. 28
- Transparencies 4-3, A and B
- CD-ROM Program
  - Resource Lesson 4-3
  - Interactive Lesson 4-3

Recommended Pacing	
Standard	Day 5 of 13
Honors	Day 5 of 13
Block	Day 3 of 7

1 FOCUS

 5-Minute Check (Lesson 4-2)

Forty eighth graders who were asked if they exercised at least 30 minutes a day gave the following responses: 24—always, 5—sometimes, and 11—never.
1. Find the percent in each category. **60%, 12.5%, 27.5%**
2. In a circle graph, how many degrees should be used for each section? **216°, 45°, 99°**

 The 5-Minute Check is also available on **Transparency 4-3A** for this lesson.

2 TEACH

 Transparency 4-3B contains a teaching aid for this lesson.

Using the Mini-Lab After students have made the human line plot, have the girls and the boys form separate line plots. How does the appearance of the line plot change? Do the line plots support students' conclusions in Exercise 2?

Multiple Learning Styles

Verbal/Linguistic
Have students write a news article or feature magazine story to accompany the *Zillions* magazine tea survey. Students can include fictional quotes from taste testers and responses from the tea companies.

Motivating the Lesson
Communication Give students a histogram. Ask these questions.
- What is the largest value in the data? What is the least?
- Which data occurs most often?

Then discuss how the identity of individual pieces of data cannot be determined from this type of graph.

Lesson 4-3 **153**

In-Class Example

For the Example
Construct a line plot for this inventory of breathing rates.

Breathing Rates (breaths/minute)			
13-W	11-M	13-M	14-W
10-W	16-M	12-M	13-M
15-M	13-W	11-W	13-W

M = man, W = woman

3 PRACTICE/APPLY

Check for Understanding

If students need additional practice or instruction after completing Exercises 1–4, one of these options may be helpful.
- Extra Practice, see p. 615
- Reteaching Activity
- *Study Guide Masters*, p. 28
- *Practice Masters*, p. 28
- Interactive Mathematics Tools Software

Study Guide Masters, p. 28

To make a line plot of the tea prices, use a number line that includes all of the data values. In this case the least value is 7 and the greatest is 42, so a number line from 0 to 45 is a good choice. Mark an × for each price above the number line to complete the line plot.

You can see that there are two groups of prices in which most of the iced teas fall. One group is 20–23 cents per serving, and the other is 34–37 cents per serving. The lowest and highest prices are 7 and 42 cents.

Studying data and making observations is called **data analysis**. Altering the line plot can show more information about the iced tea prices.

Example APPLICATION

Money Matters Make a new line plot for the iced tea prices by replacing each × with a letter rating the quality of the iced tea. Use V for very good, G for good, and F for fair. How much is the least expensive tea with a good rating?

Refer to the original list of iced tea ratings.

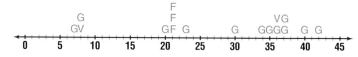

The least expensive tea that was rated good costs 7¢ per serving.
Which tea would you choose and why?

CHECK FOR UNDERSTANDING

Communicating Mathematics

Read and study the lesson to answer each question.

1. *Describe* how to make a line plot of a set of data. **See margin.**

2. In general, frozen juices are less expensive per serving than chilled juices.

2. The line plot for a *Zillions* orange juice test is shown below. *Tell* what you observe about the types and costs per serving of the juices.
   f = frozen concentrate, c = chilled

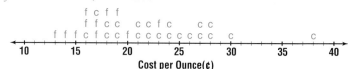

HANDS-ON MATH

3. *Make* a line plot of your classmates' heights found in the Mini Lab. **See students' work.**

Guided Practice

4. The table lists the Calories per serving of several brands of yogurt. **a. See Answer Appendix.**

4b. Most of the yogurts have 140 Calories. The extremes are 90 and 250 Calories.

   a. Make a line plot of the data.
   b. Analyze the line plot. Write all of the conclusions that you can.
   c. How many yogurts can you choose that have less than 200 Calories per serving? **18**

Yogurt Calories

90	90	90
100	110	110
140	140	140
140	140	140
150	160	170
170	180	190
220	240	250

154 Chapter 4 Statistics: Analyzing Data

■ Reteaching the Lesson ■

Activity Give each group of 4 or 5 students a snack-size bag of colored candies. Have each person count the number of each color and record the results on a line plot. Ask them to analyze their plot.

Additional Answer

1. Sample answer: Make a number line with appropriate numbers for the data points. Then mark an × for each data point above the appropriate number.

EXERCISES

Practice

5. See Answer Appendix.

5. Make a line plot of the concert ticket prices listed below.

| $20 | $24 | $31 | $31 | $35 | $21 | $33 | $43 | $35 | $30 |
| $37 | $48 | $42 | $28 | $30 | $27 | $21 | $30 | $25 | $29 |

6. Lina and Paige are playing backgammon. The results of their first twenty-six rolls of two dice are shown below.

| 8 | 6 | 5 | 10 | 6 | 9 | 7 | 4 | 6 | 11 | 9 | 2 | 10 |
| 7 | 3 | 10 | 8 | 6 | 2 | 9 | 7 | 5 | 8 | 3 | 7 | 5 |

a. Make a line plot of the data. **See Answer Appendix.**
b. What are the greatest and least sums? **11 and 2**
c. Which sum occurred most often? **6 and 7**

7. In a variation of backgammon, a player gets another turn if they roll 1 and 2. In these 26 rolls, how many times did a player get another turn? **twice**

Applications and Problem Solving

8a–b. See Answer Appendix.

9b. Sample answer: The NFC has won more Super Bowls. The margins tend to be high, with NFC teams usually winning by larger margins than AFC teams.

8. *Nutrition* The numbers of Calories per serving in selected chocolate-chip and chocolate sandwich cookies are listed at the right. *c = chocolate-chip, s = sandwich*

Cookie Calories

70-c	80-c	100-s
100-s	130-c	130-s
130-c	140-s	140-c
150-c	150-s	150-s
160-c	160-s	

a. Make a line plot of the data. Use *c* to represent chocolate chip cookies and *s* to represent chocolate sandwich cookies.
b. Analyze your graph.

9. *Sports* The National Football League began choosing its champion in the Super Bowl in 1967. The list below shows the margin of victory and the winning league for the first 31 Super Bowl games. *A = AFC, N = NFC*

25-N	19-N	9-A	16-A	3-A	21-N	7-A	17-A
10-A	4-A	18-A	17-N	4-A	12-A	17-A	5-N
10-N	29-A	22-N	36-N	19-N	32-N	4-N	45-N
1-N	13-N	35-N	17-N	10-N	23-N	14-N	

a. Make a line plot of the data. **See Answer Appendix.**
b. What do you observe about the winning margins?

10. *Critical Thinking* The list of Super Bowl margins in Exercise 9 is given in order of years: first 25-N, then 19-N, and so on. Describe any patterns you see in the margins or in the winning league over the years of the Super Bowl. **See margin.**

Mixed Review

11. *Civics* Make a circle graph of the 1996 presidential votes. *(Lesson 4-2)* **See margin.**

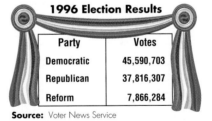

1996 Election Results

Party	Votes
Democratic	45,590,703
Republican	37,816,307
Reform	7,866,284

Source: Voter News Service

12. **Test Practice** Choose the fraction that is less than 35%. *(Lesson 3-4)* **C**

A $\frac{2}{5}$ B $\frac{3}{8}$ C $\frac{1}{6}$ D $\frac{5}{12}$

13. *Home Economics* If you were to add lace to the edge of a bedspread that is 66 by 96 inches, how much lace would you need? *(Lesson 1-8)* **324 inches**

Lesson 4-3 Line Plots **155**

Extending the Lesson

Enrichment Masters, p. 28

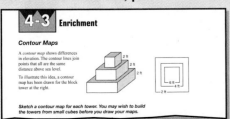

Activity Give groups of students definitions for *outlier* (an especially large or small value) and *cluster* (an isolated group of values). Have each group create a set of data containing outliers, clusters, and gaps and then draw a line plot of the data.

Assignment Guide

Core: 5–9 odd, 10–13
Enriched: 6, 8–13

4 ASSESS

Closing Activity

Writing Have students work with a partner to choose a data category, create a list of values, and draw a line plot. Then ask them to write a paragraph that includes their interpretation of the data.

Additional Answers

10. Sample answer: The AFC won most of the early Super Bowls, and the NFC has won most of the Super Bowls in more recent years.

11. **1996 Presidential Votes**

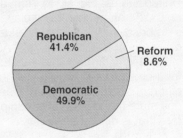

Republican 41.4%
Reform 8.6%
Democratic 49.9%

Practice Masters, p. 28

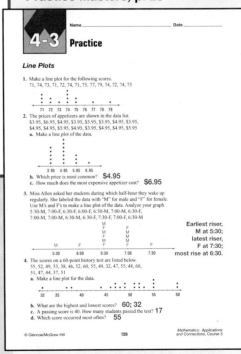

Lesson 4-3 **155**

4-3B LAB Notes

GET READY

Objective Students will use maps to display statistical data.

Optional Resources
Hands-On Lab Masters
- worksheet, pp. 43–44

MANAGEMENT TIPS

Recommended Time
45 minutes

Getting Started You may wish to ask a social studies teacher for help in obtaining an outline map. A transparency made of this map could be helpful. Have students locate their state on a map of the United States. Explain to them that maps are often used to display data because maps are visual and often easier to read than lists. Ask students to compare the statistics for their state with those for three different states, including at least one state that has an ownership rate that is different from their state's rate.

The **Activity** demonstrates how maps can be used to enhance the display of statistics. Explain to students that, although statistical intervals can be manipulated to express a point of view, they can also be used to provide more detailed information. To illustrate this, ask students how their map will change when intervals of $0–100 million, $101–200 million, and more than $200 million are used.

Teaching Tip When students make their line plots of each state's statistical information, several intervals will be stacked high with state abbreviations. Students may find the plots easier to organize and read if they draw them on graph paper, using one grid square for each abbreviation.

156 Chapter 4

HANDS-ON LAB

COOPERATIVE LEARNING

4-3B Maps and Statistics

A Follow-Up of Lesson 4-3

 colored pencils

 outline map of the United States

Have you ever seen a map of the United States with statistics shown by shading or coloring individual states? Maps like these are used often in newspapers and magazines. The map below shows how many businesses are owned and operated by African-Americans in each state. Why do you think newspapers use maps instead of lists?

Source: U.S. Census Bureau

- above-average ownership rate and above-average business growth
- above average business growth; below-average ownership rate
- above average ownership rate; below-average business growth
- below-average ownership rate and below-average business growth

TRY THIS

Work with a partner.

Sales of Bird Watching and Feeding Items

State	Sales (millions)	State	Sales (millions)	State	Sales (millions)
AL	53.6	LA	51.3	OH	123.1
AK	121.3	ME	64.8	OK	55.3
AZ	128.4	MD	83.0	OR	94.3
AR	54.4	MA	124.4	PA	256.4
CA	662.6	MI	267.6	RI	19.1
CO	179.6	MN	97.5	SC	51.6
CT	55.6	MS	34.9	SD	20.7
DE	11.5	MO	165.0	TN	76.2
FL	477.0	MT	76.3	TX	155.3
GA	49.7	NE	23.1	UT	57.0
HI	66.5	NV	56.5	VT	22.7
ID	33.3	NH	57.0	VA	108.3
IL	131.7	NJ	87.5	WA	136.3
IN	64.6	NM	80.9	WV	26.6
IA	30.4	NY	219.0	WI	224.8
KS	23.5	NC	92.4	WY	62.2
KY	57.5	ND	6.6		

Source: Southwick Associates

- The table lists each state's sales of items for watching, feeding, and photographing wild birds. Make a line plot of the data using the state abbreviations instead of ×s.

- Mapmakers usually organize data into fewer than 7 categories. Separate the data using $0-50 million, $51-100 million, $101-200 million, and more than $200 million as your categories. *Because the intervals are chosen to support a point of view, they are often not equal as they are in a histogram.*

- Choose four colored pencils. Use colors ranging from light to dark to correspond with the ranges from least to greatest. Color each state on a United States map according to its category.

156 Chapter 4 Statistics: Analyzing Data

2. Sample answer: Because the map will be neater, more concise, and more to the point.

ON YOUR OWN

Refer to your map of birdwatching and feeding item sales.

1. The intervals used to shade the states are not all equal. Why do you think the intervals were chosen as they were? **Sample answer: The data are clustered in these intervals.**

2. Explain why a mapmaker may not want to use more than seven colors to code a map that shows statistical information.

3. What areas of the country show the highest sales? Why do you think this is true? **See margin.**

4. Compare the map to the line plot that you made. Which would you use if you wanted to show the areas of the country in which bird watching is most popular? Explain. **See margin.**

5. Why do you think newspapers and magazines often use maps to show information instead of using tables? **Maps are more visual than tables.**

6. *Earth Science* The table shows the average number of tornadoes that occur in each state each year. **a. See students' work.**

Average Number of Tornadoes Each Year							
State	Average per Year	State	Average per Year	State	Average per Year	State	Average per Year
AL	22	IN	20	NE	37	SC	10
AK	0	IA	36	NV	1	SD	29
AZ	4	KS	40	NH	2	TN	12
AR	20	KY	10	NJ	3	TX	139
CA	5	LA	28	NM	9	UT	2
CO	26	ME	2	NY	6	VT	1
CT	1	MD	3	NC	15	VA	6
DE	1	MA	3	ND	21	WA	2
FL	53	MI	19	OH	15	WV	2
GA	21	MN	20	OK	47	WI	21
HI	1	MS	26	OR	1	WY	12
ID	3	MO	26	PA	10		
IL	27	MT	6	RI	0		

Source: National Severe Storm Forecast Center

a. Color a map to illustrate this data.

b. How would you change your map if you were a meteorologist trying to get a new National Weather Service tornado research center located in your state?

c. How would you change the map if you were an opponent of having a National Weather Service center located in your state? **b-c. See Answer Appendix.**

7. Compare and contrast a histogram with a map display of data. Which type of graph is easier to adjust to change the impression that is given? Explain your reasoning. **See margin.**

Lesson 4-3B HANDS-ON 157

 Have students write a paragraph that compares and contrasts the effectiveness of line plots versus maps for displaying data.

Additional Answer

7. Sample answer: Both use range numbers to display data. However, a map need not use the same size intervals and a histogram does. Answers may vary. Generally a map is easier to change because there is more flexibility in the data intervals.

ASSESS

Have students complete Exercises 1–7. Watch for students who have difficulty determining the number of states for each interval. Remind students that the total number of states must equal 50.

Use Exercise 6 to determine whether students understand how to make and interpret a statistical map.

Additional Answers

3. The north-east quarter of the country, plus California and Florida. Sample answer: There are more birds to be viewed in these areas, there are more people in California than in other states, and there are more retirees with time for leisure activities in Florida.

4. It is easier to see geographic patterns in the map. You can't see geographic patterns in the line plot because only the numbers are shown.

Hands-On Lab Masters, p. 43

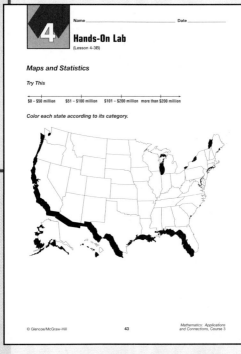

4-4 Lesson Notes

Instructional Resources
- *Study Guide Masters*, p. 29
- *Practice Masters*, p. 29
- *Enrichment Masters*, p. 29
- Transparencies 4-4, A and B
- *Assessment and Evaluation Masters*, pp. 98–99
- *Science and Math Lab Manual*, p. 73–76
- *School to Career Masters*, p. 30
- *Technology Masters*, p. 59
- CD-ROM Program
  - Resource Lesson 4-4

Recommended Pacing	
Standard	Day 7 of 13
Honors	Day 6 of 13
Block	Day 3 of 7

1 FOCUS

5-Minute Check
(Lesson 4-3)

Make a line plot for the total runs scored in 14 baseball games.

9 5 6 2 6 4 5
6 11 8 4 1 9 6

```
            ×
            ×
      × × ×
× ×   × × ×   × ×     ×
1 2 3 4 5 6 7 8 9 10 11
        Total Runs
```

The 5-Minute Check is also available on **Transparency 4-4A** for this lesson.

Motivating the Lesson

Problem Solving The annual salary of the president of a company is $90,000. The salary of each of the three employees is $16,000. The president stated proudly that the average salary at the company is $34,500. Ask students if the president's claim is correct. Then ask them if this $34,500 "average salary" gives a very clear picture of all the company salaries.

4-4 Measures of Central Tendency

What you'll learn
You'll learn to find the mean, median, and mode of a set of data.

When am I ever going to use this?
You'll use mean, median, and mode to analyze the scores on your tests.

Word Wise
measures of central tendency
mean
median
mode

Study Hint
Reading Math When you read a table or graph, be sure to note the units of the data given. For example, in the table, the units are hundreds of dollars. So a tourist from Canada spends 4 hundreds, or $400.

If you could visit any place in the world, where would you go? Many people from other countries choose to come to the United States! The table shows how much a foreign tourist spends on a trip to the United States.

Tourist Dollars	
Country	Spending (hundreds of dollars)
Canada	4
France	20
Germany	22
Italy	24
Japan	24
Mexico	4
United Kingdom	22
Venezuela	28

Source: *World Almanac*, 1997

Suppose one tourist from each of these countries visits your community. How much would each one spend on average? There are three common ways to describe a set of data. These ways are called the **measures of central tendency**. They are the mean, the mode, and the median.

Although the word *average* can be used for any of the measures of central tendency, most people use it to refer to the **mean**. For the dollar amounts above, you can find the mean as shown below.

$$\frac{4 + 20 + 22 + 24 + 24 + 4 + 22 + 28}{8} = \frac{148}{8} \text{ or } 18.5$$

The mean amount spent is 18.5 hundreds or $1,850.
Notice that the mean of a set of data may or may not be a member of the set.

Mean	The mean of a set of data is the sum of the data divided by the number of pieces of data.

A second measure of central tendency is the **mode**. The mode is the piece of data that appears most often.

4   4   20   22   22   24   24   28

There are two 4s, two 22s, and two 24s in this set of data. So there are three modes, 4, 22, and 24. If there was another 4 in the data set, 4 would be the only mode. A set of data in which no numbers appear more than once has no mode. *A mode is always a member of the data set.*

Mode	The mode of a set of data is the number or item that appears most often.

The final measure of central tendency is the **median**. The median is the middle number when the data are written in order from least to greatest.

158 Chapter 4 Statistics: Analyzing Data

Multiple Learning Styles

Logical Ask students to record the number of commercials shown during one hour of television viewing each day for five days. Ask them to determine the mean, mode, and median for the data on commercials shown per hour. Ask students to estimate how much time they spend watching commercials compared to viewing the actual TV shows.

Consider the data set {1, 4, 7, 12, 16, 21, 21, 29, 33}. There are nine numbers in the set, and they are in order from least to greatest. So the median is the fifth number, or 16.

If the number of data is even, as in the set of tourist spending, there are two middle numbers. In that case, the median is the mean of the numbers.

$$4 \quad 4 \quad 20 \quad \underbrace{22 \quad 22}_{\frac{22+22}{2} = 22} \quad 24 \quad 24 \quad 28$$

The median of the tourist spending is 22 hundreds or $2,200. *The median is not necessarily a member of the set of data. In this case, it is in the set.*

Median	The median of a set of data is the number in the middle when the data are arranged in order. When there are two middle numbers, the median is their mean.

You can learn more about different sets of data by comparing the mean, median, and mode.

Example — APPLICATION

Sports The table shows the number of home runs hit by each major league baseball team in the 1996 season.

a. Find the mean, mode, and median number of home runs for the National League teams.

Begin by writing the National League numbers in order.

National League		American League	
Team	Home Runs	Team	Home Runs
Colorado	221	Baltimore	257
Atlanta	197	Seattle	245
Cincinnati	191	Oakland	243
Chicago Cubs	175	Kansas City	243
San Francisco	153	Texas	221
Florida	150	Cleveland	218
Los Angeles	150	Boston	209
Montreal	148	Detroit	204
San Diego	147	Chi. White Sox	195
New York Mets	147	California	192
St. Louis	142	Milwaukee	178
Pittsburgh	138	Toronto	177
Philadelphia	132	NY Yankees	162
Houston	129	Minnesota	118

Source: *World Almanac, 1997*

129 132 138 142 147 147 148
150 150 153 175 191 197 221

Mean   Find their mean by dividing their sum by 14.
2220 ÷ 14 = *158.5714286*
The mean is about 159 home runs.

Mode   The mode is the most frequent number. In this data set, 147 and 150 each appear twice. So the modes are 147 and 150.

Median   The median is the middle number. There are an even number of data, so the median is the mean of the 7th and 8th numbers.

$$\frac{148 + 150}{2} = 149$$

The median is 149 home runs. *(continued on the next page)*

Lesson 4-4 Measures of Central Tendency   **159**

2 TEACH

 Transparency 4-4B contains a teaching aid for this lesson.

Reading Mathematics Help students understand the difference between the median and the mode. Remind them that the median strip of a highway is the middle of the road. Tell them that the word *mode* also means "a common style," something that is seen often.

In-Class Example

The table shows the cost to repair front and rear bumpers of some vehicles that underwent "bumper basher" tests in 1997.

a. Find the mean, mode, and median repair costs for the Ford vehicles listed. **$714.40, none, $750**

b. Find the mean, mode, and median repair costs for the GM vehicles listed. **$346.80, $0, $312**

c. Which brand has higher repair costs for the vehicles listed? **Ford**

Ford Mercury	
Vehicle	Repair Costs
Windstar	$1,373
Mercury Mystique	750
Ford Escort sedan	315
Ford Taurus wagon	687
Ford Escort wagon	447

General Motors	
Vehicle	Repair Costs
Chevrolet Venture	$ 0
Pontiac Transport	312
Oldsmobile Cutlass	0
Buick Century	519
Chevrolet Cavalier	903

Source: Consumers Union of U.S., Inc., 1997

Teaching Tip In the Example, ask students to look at the number of home runs for the teams in the American League. Point out that all of these except one are above 160.

Reteaching the Lesson

Activity Have each group of three students generate a list of 10 random numbers by rolling a number cube ten times and recording the number. Then one student should explain how to find the mean of the numbers, one how to find the median, and one how to find the mode.

Error Analysis
Watch for students who divide by the wrong number when finding a mean.
Prevent by having them circle each item of data and then count the number of items marked.

Lesson 4-4   **159**

3 PRACTICE/APPLY

Check for Understanding

If students need additional practice or instruction after completing Exercises 1–7, one of these options may be helpful.
- Extra Practice, see p. 615
- Reteaching Activity
- *Transition Booklet*, pp. 5–6
- *Study Guide Masters*, p. 29
- *Practice Masters*, p. 29

Assignment Guide

Core: 9–21 odd, 23–26
Enriched: 8–18 even, 20, 21, 23–26

Additional Answer

1. Find the sum of the data and divide by the number of data. $(1 + 3 + 4 + 6 + 6 + 6 + 8 + 9 + 10 + 10 + 14) \div 11$
   The data are written in order in the line plot. Since there are eleven pieces of data, the median is the sixth number. The median of the set is 6.

   In a line plot, you can find the mode by choosing the number that has the most *x*s. In this data set, the mode is 6.

Study Guide Masters, p. 29

160 Chapter 4

Did you know? Hank Aaron is Major League Baseball's home run king! He hit 755 homers in his career. Mr. Aaron is now a senior vice-president of the Atlanta Braves.

b. Find the mean, mode, and median number of home runs for the American League teams.

The numbers for the American League in order are:

| 118 | 162 | 177 | 178 | 192 | 195 | 204 |
| 209 | 218 | 221 | 243 | 243 | 245 | 257 |

The mean is $\frac{2,862}{14}$ or about 204.

The mode is 243 because it occurs twice.

The median is $\frac{204 + 209}{2}$ or 206.5.

c. Which league has the higher average number of home runs?

The American League has the higher average number of home runs. All of the measures of central tendency for the American League are greater than the corresponding measures for the National League.

CHECK FOR UNDERSTANDING

Communicating Mathematics

Read and study the lesson to answer each question.

1. *Explain* how to find the mean, median, and mode of the data shown in the line plot. Then find the mean, median, and mode. **See margin.**

2. *Analyze* the measures of central tendency for the National League home run data in the Example. Which measure best represents the data? Explain your choice. **See margin.**

3. *You Decide* Dolores says that all measures of central tendency must be members of the data set. Carol disagrees. Who is correct and why?

3. Carol is correct. The mean and the median may not be members of the set of data.

Guided Practice

Find the mean, median, and mode for each set of data. When necessary, round to the nearest tenth. **6. 79¢; 81.5¢; no mode**

4. 9, 8, 15, 8, 20, 23, 16, 5, 6, 14, 12, 25, 18, 22, 24   **15; 15; 8**

5. 36, 38, 33, 34, 32, 30, 34, 35   **34; 34; 34**

6. 85¢, 87¢, 69¢, 74¢, 70¢, 98¢, 54¢, 89¢, 65¢, 82¢, 81¢, 94¢

7. *Money Matters* The list shows the suggested retail prices of several different personal stereos.   **b. See margin.**

| $40 | $45 | $50 | $59 | $60 | $69 | $85 | $111 | $120 |

7c. Because the value of the new data piece changes the total of the data drastically, the mean is affected most.

a. Find the mean, median, and mode prices.   **$71; $60; no mode**

b. Which measure of central tendency best represents the prices? Explain.

c. If a price of $17 is added to the list, which measure of central tendency is affected most? Explain.

160 Chapter 4 Statistics: Analyzing Data

Additional Answers

2. Sample answer: The median best represents the data. The mean is higher than most of the data because it is affected by a number that is greater than most of the data, 221. The mode is slightly less than most of the data.

7b. Sample answer: There is a lot of variation in the prices. The median is a good representative price because half of the prices are above it and half are below.

EXERCISES

Practice

Find the mean, median, and mode for each set of data. When necessary, round to the nearest tenth.   9. 6; 4.5; 3   10. 94.4; 93; 93

8. 20, 17, 20, 14, 19   **18; 19; 20**

9. 3, 9, 14, 3, 0, 2, 6, 11

11. 79.3; 79.5; 84

10. 93, 90, 94, 99, 92, 93, 100

11. 78, 80, 75, 73, 84, 81, 84, 79

12. $82.5; $87; $87 and $100

12. $93, $71, $83, $100, $55, $87, $79, $100, $58, $95, $87

13. 8.8, 10.0, 9.3, 8.3, 9.4, 8.9, 9.5, 8.3, 9.7, 8.1, 9.7, 8.8, 8.3   **9.0; 8.9; 8.3**

14. 1.5; 1.6; no mode

14. 1.2, 1.78, 1.73, 1.9, 1.19, 1.8, 1.24, 1.92, 1.54, 1.7, 1.42, 1.0

15. 27, 19, 22, 41, 24, 28, 40, 33, 35, 33, 35, 48, 32   **32.1; 33; 33 and 35**

16.    0.5; 0.6; 0.6

17. Sample answer: {1, 2, 3, 4, 5, 6, 7}

18. See margin.

20b. See Answer Appendix.

17. Give an example of a set of data that has no mode.

18. Is it possible for a set of numerical data to have no mean? Explain.

19. Construct a set of data with at least five different numbers for which the mean, median, and mode are all the same number.   **See margin.**

Applications and Problem Solving

20. *Highway Safety* In 1995, President Clinton signed a bill that allows states to set speed limits on their highways. Montana posted its limit as "reasonable and prudent." The speed limits in miles per hour for rural highways in the other states and the District of Columbia are given.

70	65	75	65	70	75	55	65	55	70	70	55	75
65	65	65	70	65	65	65	65	65	65	65	70	70
75	75	65	55	65	65	65	70	65	75	65	65	65
65	75	65	70	75	65	65	70	65	65	75		

a. Find the mean, median, and mode speed limits.   **67; 65; 65**

b. What would you say is the "average" speed limit? Explain.

21. *Entertainment* After release of *Star Wars Special Edition*, the average American had seen *Star Wars* seven times. How was that number found?

22. *Working on the* **CHAPTER Project**   Research to find the number of items your museum has acquired in each of the last ten years.

a. Find the mean, median, and mode of items acquired.

b. If it takes 15 minutes to catalog each item, about how much employee time would be required for cataloging each year?

c. Estimate the annual budget for catalog staff in your museum. Support your answer with facts and figures.   **a-c. See students' work.**

Star Wars Special Edition

Mixed Review

21. See Answer Appendix.

24. See Answer Appendix.

23. *Critical Thinking* Of the mean, median, and mode, which is most affected by adding a very large or very small data value to a set of data? Explain.   **See Answer Appendix.**

24. *Highway Safety* Make a line plot of the data in Exercise 20. *(Lesson 4-3)*

25. *Algebra* Solve $\frac{5}{8} = \frac{m}{24}$. *(Lesson 3-3)*   **15**

26. **Test Practice** Choose the solution of $c = -22 + 5$. *(Lesson 2-3)*   **B**

A -27 B -17 C 17 D 27

Lesson 4-4 Measures of Central Tendency   **161**

Extending the Lesson

Enrichment Masters, p. 29

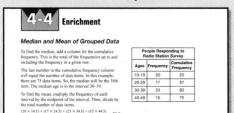

Activity Have students solve this problem: *The median of five one-digit numbers is 7. The mode is 2. What are the numbers?*   **2, 2, 7, 8, 9**

CHAPTER Project

Exercise 22 asks students to advance to the next stage of work on the Chapter Project. You may ask students to discuss the practicality of their answers to part c.

4 ASSESS

Closing Activity

Speaking Write the numbers 1, 2, 2, and 3 on the chalkboard. Have students describe how they would find the mean, median, and mode of the set of data.

Chapter 4, Quiz B (Lessons 4-3 and 4-4) is available in the *Assessment and Evaluation Masters*, p. 98.

Mid-Chapter Test (Lessons 4-1 through 4-4) is available in the *Assessment and Evaluation Masters*, p. 99.

Additional Answers

18. No, any set of numerical data can be added and divided by the number of data.

19. Sample answer: The mean, median, and mode of {1, 2, 3, 4, 4, 5, 6, 7} are all 4.

Practice Masters, p. 29

Measures of Central Tendency

Find the mean, median, and mode for each set of data. If necessary, round to the nearest tenth.

1. 2, 8, 3, 7, 5, 5, 7, 3, 5   **5; 5; 5**

2. 30, 37, 42, 44, 44, 46, 49, 52   **43; 44; 44**

3. 21, 27, 20, 29, 23, 21, 21, 27, 26, 25   **24; 24; 21**

4. 335, 428, 360, 480, 342   **389; 360; none**

5. 6.42, 7.38, 5.29, 6.37, 6.9, 7.42, 5.3, 6.6   **6.5; 6.5; none**

6. 4, 5, 5, 6, 6, 6, 6, 7   **5.6; 6; 6**

7. 2, 6, 10, 12, 8, 4   **7; 7; none**

8. 13.6, 18.24, 16.7, 14.38, 15.29   **15.6; 15.3; none**

9. 28, 32, 39, 25   **31; 30; none**

10. 3, 3, 3, 9, 9, 4, 4, 4, 5   **4.9; 4; 3 and 4**

11. 29, 42, 44, 44, 45, 50, 51, 51, 52, 54, 58, 59, 61, 62, 63, 66, 71, 72   **54.1; 53; 44 and 51**

12. **22.5; 23; 23**

Lesson 4-4   **161**

4-4B LAB Notes

GET READY

Objective Students use a spreadsheet to find the mean of a set of data.

Technology Resources
Suggested spreadsheet software
- Lotus 1·2·3
- ClarisWorks
- Microsoft Excel

Instructional Resources
CD-ROM Program
• Interactive Lesson 4-4B

MANAGEMENT TIPS

Recommended Time
30 minutes

Getting Started Review how data are entered with your spreadsheet software.

You may wish to have students work in pairs to enter the spreadsheet program into the computer. Have one student read the instructions while his or her partner enters the information on the spreadsheet.

Have students work through the exercises using the data they generated.

Using a Calculator Many graphing calculators will calculate the mean for a list of data. On the TI-83, select "mean" (from the LIST MATH menu) and enter the data for which you want to find the mean.

ASSESS

After students complete Exercises 1–4, ask them how they could use the spreadsheet to determine the mean score for each of the tests. Ask them what formula they would use to generate the answer.

162 Chapter 4

TECHNOLOGY LAB SPREADSHEETS

4-4B Finding a Mean

A Follow-Up of Lesson 4-4

- computer
- spreadsheet software

Would you like to know your grades before you get your report card? One way that you can do that is to use a spreadsheet program.

Many teachers give several tests and determine the final grades by finding the mean of the scores. A portion of a spreadsheet program that Mr. Gutierrez uses to find the mean of his students' test scores is shown.

TRY THIS

Work with a partner.

Use the spreadsheet to determine the mean test score for each student.

	A	B	C	D	E	F
1	Student	Test 1	Test 2	Test 3	Test 4	Mean
2	Alejandra	78	76	81	83	=(B2+C2+D2+E2)/4
3	Jessica	72	83	85	83	=(B3+C3+D3+E3)/4
4	Kelli	84	82	85	88	=(B4+C4+D4+E4)/4
5	Raheem	88	92	90	91	=(B5+C5+D5+E5)/4
6	Steven	90	88	87	92	=(B6+C6+D6+E6)/4

The formula (B2+C2+D2+E2)/4 adds the values in cells B2, C2, D2, and E2 and then divides the sum by 4.

Each box in the spreadsheet is called a *cell*. The formulas in the cells in column F find the mean of the scores that are entered in the cells in columns B, C, D, and E. The formula first finds the sum of the scores and then divides the sum by 4 to find the average. The printout below shows the results when the calculations are complete.

	A	B	C	D	E	F
1	Student	Test 1	Test 2	Test 3	Test 4	Mean
2	Alejandra	78	76	81	83	79.5
3	Jessica	72	83	85	83	80.75
4	Kelli	84	82	85	88	84.75
5	Raheem	88	92	90	91	90.25
6	Steven	90	88	87	92	89.25

4. Make columns F, G, and H for the three additional scores. The cells in column I would be the formulas for the mean. The formulas for the mean should find the sum of the seven scores and divide the sum by 7.

ON YOUR OWN

1. Use the spreadsheet to determine Editon's average test score if his grades were 92, 84, 89, and 95. **90**
2. If Kelli had scored 2 points higher on each test, how would her average have been affected? **It would be 2 points higher.**
3. What would Raheem have to score on a fifth test to average 90%? **89**
4. How could you change the spreadsheet to find the mean of seven quiz scores?

162 Chapter 4 Statistics: Analyzing Data

 Math Journal Ask students to write a paragraph about how they could use a spreadsheet to keep track of their grades in all courses throughout the school year.

4-5 Measures of Variation

What you'll learn
You'll learn to find the range and quartiles of a set of data.

When am I ever going to use this?
Doctors use measures of variation to determine whether new medicines are effective.

Word Wise
variation
range
quartile
interquartile range
upper quartile
lower quartile
outlier

Americans may think of Hollywood for movies. But India holds the record for producing the most feature-length films. They average 700 films per year compared to 450 from the U.S.

"And the award goes to..." Do you dream of someday accepting an Academy Award? Moviemakers have been honoring their best with Academy Awards since 1927. The ages of the best actress winners for the last 25 years are listed below.

25	26	28	30	30	30	31	32
32	32	33	35	35	36	37	40
40	41	41	44	48	48	60	73	80

Frances McDormand
Best Actress 1996

The list shows that the data extends from a low of 25 to a high of 80. The *spread* of data is called the **variation**.

One way to measure the variation of a set of data is to find the **range**.

| Range | The range of a set of data is the difference between the greatest and the least numbers in the set. |

The range of ages of the award winners is 80−25 or 55.

When you are considering large sets of data, such as thousands of college entrance exam scores, it is often helpful to separate the data into four equal parts called **quartiles**. The quartiles are used to find another measure of variation called the **interquartile range**.

| Interquartile Range | The interquartile range is the range of the middle half of the data. |

Find the interquartile range of the actresses' ages.

Step 1 Find the median. The median is the 13th number when the data are in order from least to greatest, 35. It separates the data into two halves, which are shown in brackets.

[25 26 28 30 30 30 31 32 32 32 33 35] 35
[36 37 40 40 41 41 44 48 48 60 73 80]

Step 2 Next find the median of the upper half of the data. This number is the **upper quartile**. There are 12 numbers in the upper half of the data. The median is halfway between the 6th and 7th numbers above the median.

[25 26 28 30 30 30 31 32 32 32 33 35] 35
[36 37 40 40 41 |41 44| 48 48 60 73 80]

upper quartile = $\frac{41 + 44}{2}$ or 42.5

Lesson 4-5 Measures of Variation **163**

2 TEACH

 Transparency 4-5B contains a teaching aid for this lesson.

Modeling Mathematics Make a list of 27 randomly selected numbers and write them in order in a column on notebook paper, one number per line. Cut out the strip that contains the numbers. Fold the strip in half to find the median and in half again to find the lower and upper quartiles. Compare these results with the computed results for the data.

In-Class Examples

For Example 1

a. Find the range for the data in the frequency table below. **3**

Number	Frequency
3	2
4	3
5	7
6	3

b. Find the interquartile range. **1**

For Example 2
Determine whether there are any outliers in the following set of data. **no**

6   15   18   21   23   26   32

Teaching Tip In Example 1, reinforce that quartiles show a division of the data into four parts. The median divides the data in half, and the upper and lower quartiles divide each half into two parts.

Step 3 Find the **lower quartile**. The lower quartile is the median of the lower half of the data. In this data set, the lower quartile is $\frac{30 + 31}{2}$ or 30.5.

[25   26   28   30   30 |30   31| 32   32   32   33   35]   35
[36   37   40   40   41   41   44   48   48   60   73   80]

Step 4 Find the interquartile range. This is the middle half of the data, which is between the lower quartile and the upper quartile. To find the interquartile range, subtract the lower quartile from the upper quartile. In this data set, the interquartile range is 42.5–30.5 or 12. This means that 50% of the actresses' ages are between 30.5 years and 42.5 years inclusive, a span of 12 years.

Example 1 APPLICATION

Sports Thousands of kids have strapped on in-line skates. The frequency table shows the prices of different models of in-line skates reviewed by *Zillions* magazine.

a. Find the range of this data.
b. What is the interquartile range?

Price	Tally	Frequency
279	I	1
275	IIII	4
239	II	2
229	I	1
220	I	1
200	II	2
185	I	1
159	I	1
139	II	2
116	I	1
115	II	2
109	I	1
100	I	1
79	I	1
69	I	1
59	I	1
50	I	1
49	I	1
39	I	1

a. The range of the data is the difference between the greatest and least numbers. The range of the data is 279–39 or 240.

b. Notice that the data are written in order in the table. The first step to finding the interquartile range is to find the median.

The median is $\frac{139 + 159}{2}$ or 149.

Each half of the data has 13 numbers. The upper quartile, UQ, is the 7th number in the upper half of the data.

UQ = 239

The lower quartile, LQ, is the 7th number in the lower half of the data.

LQ = 100

The interquartile range is 239 − 100 or 139.

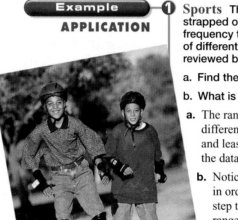

In some sets of data, one or both of the extreme values are much greater or less than the other data. Data that are more than 1.5 times the interquartile range from the quartiles are called **outliers**.

164 Chapter 4 Statistics: Analyzing Data

Reteaching the Lesson

Activity Place 19 counters side-by-side. Have a student locate the middle counter and slide it upward. Have two more students locate the middle counters of the two smaller groups and slide them downward. Explain that the interquartile range is the difference between the two lowered counters.

Error Analysis
Watch for students who find quartiles by dividing the range into four equal sections.
Prevent by stressing that *quartile* refers to quarters of the number of items, not quarters of the range.

Example 2 APPLICATION

Entertainment Refer to the beginning of the lesson. Determine whether there are any outliers in the actresses' ages.

We found that the upper quartile is 42.5 and the lower quartile is 30.5. The interquartile range is 12. So data more than 1.5 · 12, or 18, above the upper quartile or below the lower quartile are outliers.

Find the limits for the outliers.

Subtract 18 from the lower quartile.   30.5 − 18 = 12.5
Add 18 to the upper quartile.   42.5 + 18 = 60.5

So, 12.5 and 60.5 are the limits for the outliers. There are two outliers in the data, 73 and 80.

CHECK FOR UNDERSTANDING

Communicating Mathematics

Read and study the lesson to answer each question.

1. *Summarize* the differences between the measures of variation and the measures of central tendency. **See margin.**

2. *Show* how to divide the data into quartiles. **See margin.**
   {135, 170, 125, 174, 136, 145, 180, 156, 188}

Guided Practice

Find the range, median, upper and lower quartiles, interquartile range, and any outliers for each set of data.

3. 7; 16; 17, 13; 4; no outliers
4. 2.0; 1.5; 1.85, 1.15; 0.7; 3.0
5. 38; 52; 57, 48; 9; 22
6. 75; 85.5; 90, 50; 40; no outliers

3. 12, 13, 14, 16, 17, 17, 19
4. 1.4, 1.8, 1.0, 1.1, 1.6, 1.2, 3.0, 1.9
5. 43, 55, 49, 49, 53, 48, 57, 60, 57, 60, 47, 51, 59, 22

6. **Nutrition** The list below shows the numbers of Calories per tablespoon of different brands of margarine.

   100, 70, 70, 90, 50, 90, 50, 90, 100, 50, 90, 100, 90, 50, 25, 81

   Find the range, median, quartiles, interquartile range, and any outliers.

EXERCISES

Practice

Find the range, median, upper and lower quartiles, interquartile range, and any outliers for each set of data.

7. 9; 58; 61, 56; 5; no outliers
8. 19; 4; 8.5, 2; 6.5; 19
9. 22; 37; 40.5, 34; 6.5; 51
10. 20; 117; 118; 109; 9; no outliers
12–15. See margin.

7. 54, 54, 58, 58, 58, 59, 60, 62, 63
8. 9, 0, 2, 8, 19, 5, 3, 2
9. 38, 43, 36, 37, 32, 37, 29, 51
10. 117, 118, 120, 109, 117, 117, 100
11. 2.3, 2.3, 3.8, 2.6, 3.7, 2.9, 6.1, 2.3, 2.9, 2.5, 3.5   3.8; 2.9; 3.7, 2.3; 1.4; 6.1
12. 198, 166, 190, 155, 146, 184, 135, 180, 145
13. 55, 76, 104, 65, 62, 79, 63, 57, 52, 72, 57, 73, 55, 60, 80, 53
14. 349, 341, 351, 357, 353, 346, 342, 315, 336, 339, 313, 342, 347, 351, 353, 304, 356, 342, 357

15.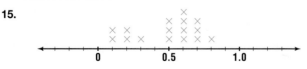

Lesson 4-5 Measures of Variation

3 PRACTICE/APPLY

Check for Understanding

If students need additional practice or instruction after completing Exercises 1–6, one of these options may be helpful.
• Extra Practice, see p. 616
• Reteaching Activity, see p. 164
• *Study Guide Masters,* p. 30
• *Practice Masters,* p. 30

Assignment Guide

Core: 7–17 odd, 19, 20
Enriched: 8–16 even, 17–20
Self Test: All 1–5

Additional Answers

1. The measures of variation describe the dispersal of data. The measures of central tendency describe the set as a whole.

2. First write the data in order and find the median. In this set, the median is 156. The upper quartile is the median of the upper half of the data. In this set, the upper quartile is $\frac{174 + 180}{2}$ or 177. The lower quartile is the median of the lower half of the data. In this set, the lower quartile is $\frac{135 + 136}{2}$ or 135.5.

Study Guide Masters, p. 30

Additional Answers

12. 63; 166; 187; 145.5; 41.5; no outliers
13. 52; 62.5; 74.5; 56; 18.5; 104
14. 53; 346; 353; 339; 14; 304, 313, and 315
15. 0.7; 0.55; 0.65; 0.25; 0.4; no outliers

4 ASSESS

Closing Activity

Speaking State a lower quartile and an upper quartile and have students state the interquartile range. Then state an interquartile range and one of the quartiles and have students state the other quartile. Conduct a similar activity involving the two extreme values and the range of a set of data.

Mid-Chapter Self Test

The Mid-Chapter Self Test reviews concepts in Lessons 4-1 through 4-5. Lesson numbers are given so students can review concepts not yet mastered.

Additional Answers

17a. Sample answer: {1, 1, 2, 2, 2, 5, 9, 9, 9, 10, 10} and {1, 4, 4, 4, 4, 5, 5, 5, 9, 10, 10}

17b. Sample answer: {1, 2, 5, 7, 9, 10, 12, 14, 15, 17, 22} and {0, 2, 5, 7, 9, 10, 12, 14, 15, 17, 27}

Practice Masters, p. 30

Applications and Problem Solving

16a. 2,710; 1,380; 2,147, 1190.5; 956.5; 3,710

16b. Ohio-Allegheny and Colorado (AZ)

Mixed Review

16. Geography The table shows the lengths of the 16 longest rivers in North America.

a. Find the range, the median, the upper and lower quartiles, the interquartile range, and any outliers of the river lengths.

b. Which river is closest to the median length?

17. Critical Thinking Create two different sets of data that meet the following conditions.

a. the same range, different interquartile ranges

b. the same median and quartiles, but different ranges

a–b. See margin.

River	Length (mi)
Arkansas	1,459
Churchill, Man.	1,000
Colorado (AZ)	1,450
Columbia	1,243
Mackenzie	1,025
Mississippi	2,340
Upper Mississippi	1,171
Mississippi-Missouri-Red Rock	3,710
Missouri	2,315
Missouri-Red Rock	2,540
Ohio-Allegheny	1,310
Peace	1,210
Red	1,290
Rio Grande (OK-TX-LA)	1,900
Snake	1,038
Yukon	1,979

Source: Geological Survey

18. Civics Are you looking forward to getting a driver's license? The list below shows the earliest ages at which a person can get a driver's license in each state and the District of Columbia.

16 16 16 16 14 16 16 16 18 16 16 15 16
15 16 16 16 14 16 15 16 16 16.5 14 15 14
16 16 13 14 14 16 15 16 16 14 14 14
16 14 16 16 15 14 14 15 16 16 16 16

Find the range, median, upper and lower quartiles, the interquartile range, and any outliers of the data.   5; 16; 16, 14; 2; no outliers

19. Life Science Find the mean, median, and mode of the plant heights 22, 4, 1, 12, 5, 22, 5, 25, 25, 19, 23, 24, 11, 16, 3, and 22. *(Lesson 4-4)*   14.9; 17.5; 22

20. Test Practice Choose the value of the expression 6^3. *(Lesson 1-2)*   **C**

A 18   B 63   C 216   D 729

CHAPTER 4 Mid-Chapter Self Test

The frequency table shows the grams of sugar per serving in 28 cereals made for adults.   1–3. See Answer Appendix.

1. Use the intervals 0-2, 3-5, 6-8, and 9-11 to make a histogram of the data. *(Lesson 4-1)*

2. Make a circle graph to show the percent of cereals with 0-2, 3-5, 6-8, and 9-11 grams of sugar. *(Lesson 4-2)*

3. Make a line plot of the data. *(Lesson 4-3)*

4. Find the mean, median, and mode amounts of sugar. *(Lesson 4-4)*   4.1; 3.5; 3

5. Find the range, upper and lower quartiles, the interquartile range, and any outliers of the data. *(Lesson 4-5)*   11; 6, 2; 4; none

Grams	Tally	Frequency
0						5
1		0
2					3
3							6
4			1
5						5
6						4
7			1
8		0
9			1
10			1
11			1

166 Chapter 4 Statistics: Analyzing Data

Extending the Lesson

Enrichment Masters, p. 30

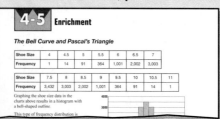

Activity Have small groups of students create a set of data and then find its mean. Next, have them find the difference between each piece of data and the mean, listing the differences separately. Finally, have them find the mean of these differences. This value is the *mean variation*.

TECHNOLOGY LAB

GRAPHING CALCULATORS

4-5B Box-and-Whisker Plots

A Follow-Up of Lesson 4-5

graphing calculator

A *box-and-whisker plot* uses a number line with the median, the quartiles, and the extreme values to represent a set of data. You can make a box-and-whisker plot with a graphing calculator.

TRY THIS

Work with a partner.

The list shows the ages of the winners of the Academy Award for Best Actor for the last 25 years. Use a graphing calculator to make a box-and-whisker plot of the data.

29	31	31	35	36	36	37
37	38	39	41	41	42	44
47	48	50	51	52	53	53
55	60	61	76

Step 1 Begin by entering the data into the calculator's memory.

Press STAT ENTER to see the lists. If there are numbers in the first list, press ▲ CLEAR ENTER to clear them. Then enter the data by entering each number and pressing ENTER.

Step 2 Next choose the type of graph. Press 2nd [STAT PLOT] to display the menu. Choose the first plot by pressing ENTER. Use the arrow and ENTER keys to highlight "on", the modified box-and-whisker plot, L1 for the Xlist, and 1 as the frequency.

Step 3 Choose the display window. Access the menu by pressing WINDOW. Choose appropriate range settings for the *x* values. The window 25 to 80 with a scale of 5 includes all of this data.

Step 4 Display the graph by pressing GRAPH.

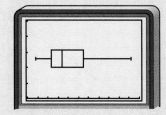

ON YOUR OWN

1. Press the TRACE key. The right and left arrow keys allow you to find the minimum, the maximum, the quartiles, and the median of the data. What are these values for the Academy Award winners' ages?
2. Use the data on the ages of the winners of the Academy Award for Best Actress on page 163 to make a box-and-whisker plot. How does the plot compare to the one above?
3. Does the graphing calculator show outliers? Explain. **1–3. See Answer Appendix.**

Lesson 4-5B **TECHNOLOGY LAB** 167

Math Journal Ask students to write a paragraph that explains how the median, quartiles, and extremes of a set of data are used to make a box-and-whisker plot.

TECHNOLOGY 4-5B LAB Notes

GET READY

Objective Students use a graphing calculator to construct and interpret box-and-whisker plots.

Technology Resources
- TI-80, TI-81, TI-82, or TI-83 graphing calculator

MANAGEMENT TIPS

Recommended Time
20 minutes

Getting Started Before students begin the lesson, briefly review the process for entering data on a graphing calculator. Stress that constructing a box-and-whisker plot builds on the skills learned in Lesson 4-5 by displaying the measures found there.

When they have finished entering their data, ask volunteers to identify the parts of the plot as students work through Exercise 1. Ask students why it may be easier to display data using a box-and-whisker plot.

In-Class Example

Use a graphing calculator to create a box-and-whisker plot for these automobile gas mileages.

| 31 | 27 | 12 | 23 | 45 | 24 | 39 | 19 |
| 48 | 24 | 29 | 22 | 29 | 17 | 34 |

Use 12 and 48 for the range settings and a scale of 2.

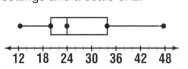

ASSESS

After students answer Exercises 1–3, ask them how box-and-whisker plots help them to interpret data using medians, quartiles, and extreme values.

Technology Lab 4-5B **167**

4-6 Lesson Notes

Instructional Resources
- *Study Guide Masters,* p. 31
- *Practice Masters,* p. 31
- *Enrichment Masters,* p. 31
- Transparencies 4-6, A and B
- *Assessment and Evaluation Masters,* p. 100
- *Science and Math Lab Manual,* p. 57–60
- *Technology Masters,* p. 60
- CD-ROM Program
  - Resource Lesson 4-6

Recommended Pacing	
Standard	Day 9 of 13
Honors	Day 9 of 13
Block	Day 5 of 7

1 FOCUS

5-Minute Check
(Lesson 4-5)

Use the following data.

22 25 33 31 28
22 26 22 30 27
31 28 33 23 29

1. Find the range. **11**
2. Find the median. **28**
3. Find the upper and lower quartiles. **31; 23**
4. Find the interquartile range. **8**
5. Identify any outliers. **none**

The 5-Minute Check is also available on **Transparency 4-6A** for this lesson.

2 TEACH

Transparency 4-6B contains a teaching aid for this lesson.

Using the Mini-Lab Explain that scatter plots are useful for spotting relationships between sets of data that may be difficult to see from the data itself. Emphasize that these relationships among data do not prove their validity. The fact that many tall people have long arms does not prove that height and arm length are directly related.

4-6 Integration: Algebra
Scatter Plots

What you'll learn
You'll learn to construct and interpret scatter plots.

When am I ever going to use this?
You can use scatter plots to observe trends in data and make predictions.

Word Wise
scatter plot

When you graph two sets of data as ordered pairs, you form a **scatter plot**. You can use scatter plots to look for trends in data.

HANDS-ON MINI-LAB

Work in a small group.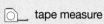

Try This 1–3. See students' work.

1. Measure each group member's height to the nearest centimeter. Then measure each person's arm from shoulder to wrist.

2. Graph the data on a coordinate graph similar to the one at the right. The point on the graph represents a person whose height is 155 centimeters and whose arm length is 50 centimeters.

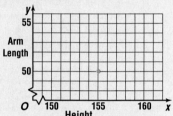

Talk About It

3. Use your graph to predict the height of a person who has an arm length of 52 centimeters. Does your graph indicate that taller people tend to have longer or shorter arms?

Scatter plots can suggest whether two sets of data are related. Imagine a line drawn so that half of the points are above it and half are below it. If the line slopes upward to the right, there is a *positive* relationship. A line that slopes downward to the right indicates a *negative* relationship.

Positive Relationship   **Negative Relationship**

Example APPLICATION

1 Demographics The scatter plot at the right compares a person's age to the expected number of years he or she has yet to live.

a. Describe the point with the box around it.

The point with the box around it is at (53, 30). It shows that a person 53 years old can expect to live 30 more years.

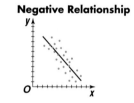

168 Chapter 4 Statistics: Analyzing Data

Motivating the Lesson

Communication Refer students to the lesson opener. Ask them to consider why it is important to look for trends when analyzing data. Explain that many business decisions such as what to produce and the sale price of goods are based on observed trends in consumer spending.

b. What type of relationship is shown by the scatter plot?

The points suggest a line that slopes downward to the right. So the scatter plot shows a negative relationship. This means that the older a person is, the fewer years he or she has yet to live.

Some scatter plots show that there is no relationship between sets of data.

Examples

Determine whether each scatter plot shows a positive, negative, or no relationship.

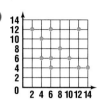

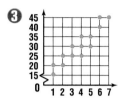

The points in the scatter plot are very spread out. There appears to be no relationship between these data.

This scatter plot suggests a line that slopes upward to the right. These data appear to have a positive relationship.

CHECK FOR UNDERSTANDING

Communicating Mathematics

Read and study the lesson to answer each question.

1. *Tell* how to draw a scatter plot for two sets of data. **See margin.**
2. *Draw* a scatter plot that shows a negative relationship. **See students' work.**

3. *Measure* the distance around the closed fist and the length of the forearm of each person in your group. Create a scatter plot of the ordered pairs. Do you see a relationship? Explain. **See students' work.**

Guided Practice

Determine whether a scatter plot of the data below would show a *positive*, *negative*, or *no* relationship. 5. negative 6. positive

4. age and height **positive**
5. age of a used car and its value
6. hours worked and earnings
7. grades and distance to school **no relationship**

8. **Health** The scatter plot shows the number of Calories in different fruits compared to the number of milligrams of calcium they offer. Does there appear to be a positive, a negative, or no relationship between Calories and calcium in fruit? **positive**

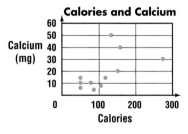

Lesson 4-6 Integration: Algebra Scatter Plots **169**

■ Reteaching the Lesson ■

Activity Draw and label a coordinate plane on the chalkboard. Name four points and have students plot them. Ask students to plot points showing a positive relationship. Erase these and have them plot points showing a negative relationship. Then have them plot points showing no relationship.

Additional Answer

1. Write the data as ordered pairs, and graph each ordered pair on a coordinate grid.

In-Class Examples

For Example 1
Refer to the scatter plot.

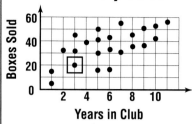

a. Describe what (3, 20) represents. **3 years membership and the sale of 20 boxes**

b. What type of relationship is shown by the scatter plot? **positive**

For Examples 2 and 3
Determine whether the scatter plot below shows a *positive*, a *negative*, or *no relationship*. **no relationship**

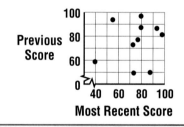

Study Guide Masters, p. 31

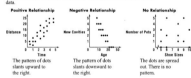

Lesson 4-6 **169**

3 PRACTICE/APPLY

Check for Understanding
If students need additional practice or instruction after completing Exercises 1–8, one of these options may be helpful.
- Extra Practice, see p. 616
- Reteaching Activity, see p. 169
- *Study Guide Masters*, p. 31
- *Practice Masters*, p. 31
- Interactive Mathematics Tools Software

Assignment Guide
Core: 9–21 odd, 23–26
Enriched: 10–20 even, 21–26

4 ASSESS

Closing Activity
Modeling Have students research how long it takes a car to stop at various speeds. Then ask them to use small round stickers and a poster to illustrate the relationship between speed and stopping distance. **positive**

Chapter 4, Quiz C (Lessons 4-5 and 4-6) is available in the *Assessment and Evaluation Masters*, p. 100.

Practice Masters, p. 31

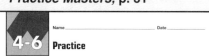

EXERCISES

Practice Determine whether a scatter plot of the data below would show a *positive, negative,* or *no* relationship. **10. positive**

12. no relationship
13. negative

9. height and test scores   **no relationship**
10. years of education and income
11. speed and distance covered   **positive**
12. height and month of birth
13. temperature and cooking time
14. bank balance and interest earned   **positive**
15. miles per gallon and weight of a car   **negative**
16. playing time and points scored in a basketball game   **positive**
17. a U.S. president's time in office and age when elected   **no relationship**
18. outside temperature and amount of heating fuel used   **negative**
19. number of people in a household and weekly grocery bill   **positive**
20. number of pages in a book and number of copies sold   **no relationship**

Applications and Problem Solving

21a. See Answer Appendix.

22a. See Answer Appendix.

*inter*NET CONNECTION
For the latest information on the states, visit:
www.glencoe.com/sec/math/mac/mathnet

21. **Aviation** The table shows the number of seats and the average airborne speed of the most commonly used aircraft.
    a. Make a scatter plot of the seats and speeds.
    b. Does the scatter plot show a positive, a negative, or no relationship? **positive**

22. **History** Use a reference book or Internet search to find the area and year of entrance of each state in the United States.
    a. Make a scatter plot of the data.
    b. Does the scatter plot show a positive, a negative, or no relationship? **Sample answer: positive**

Aircraft	Seats	Speed (mph)
B747-400	396	538
B747-100	395	521
B747-200/300	342	535
B777	291	512
DC-10-40	288	506
L-1011-100/200	285	490
DC-10-10	283	503
DC-10-30	277	523
A300-600	267	468
MD-11	261	526
L-1011-500	222	524
B757-300ER	219	496
B757-200	186	465
B767-200ER	180	487
A310-300	172	498

Source: Air Transport Assoc. of America

23. **Critical Thinking** A scatter plot of skateboard sales and swim suit sales shows a positive relationship. **a–b. See Answer Appendix.**
    a. Why might this be true?
    b. Does this mean that one factor caused the other? Explain.

Mixed Review
24. 30; 115; 122, 105; 17; no outliers

24. **Statistics** Find the range, median, upper and lower quartiles, interquartile range, and any outliers of {115, 117, 111, 121, 110, 127, 116, 126, 105, 115, 100, 103, 122, 130, 101, 100, 108, 130}. *(Lesson 4-5)*

25. **Test Practice** Which is in order from least to greatest? *(Lesson 2-2)* **A**
    A $-4, 2, 8$ **B** $4, -1, -6$ **C** $-1, 2, -4, 8$ **D** $0, -1, -4$

26. **Life Science** The United States lists 55 mammals as endangered. That is 8 less than one-fourth of those on international endangered species lists. How many species are on international lists? *(Lesson 1-7)* **252**

170   Chapter 4 Statistics: Analyzing Data

Extending the Lesson

Enrichment Masters, p. 31

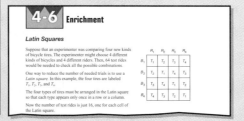

Activity Have students write down the last names of 25 people they know, recording the total number of letters and the total number of vowels in each name. Students should then use graph paper to make a scatter plot comparing the number of letters versus the number of vowels and decide what type of relationship is shown.

4-7 Choosing an Appropriate Display

What you'll learn
You'll learn to choose an appropriate display for a set of data.

When am I ever going to use this?
Knowing how to choose the correct type of graph can help you present information in an understandable way.

Alexander Tsiaras of New York, New York, is working to expand the field of medical photography. One of his breakthroughs is a new view of a human fetus. The table shows the length of an average fetus at different times of development. How would you display this information in a statistical graph? *This will be solved in Example 1.*

Time (weeks)	Length (mm)
4	7
8	30
12	75
16	180
20	250
24	300
28	350
32	410
36	450
38	500

So far in this chapter you have learned how to construct different types of statistical graphs. But when you are given a set of data to display, you must choose which type of graph would be appropriate for the data.

As you decide what type of graph to use, it helps to ask yourself these two questions:
- What type of information is this?
- What do I want my graph to show?

Consider each type of graph when choosing a display for your data. The displays you have studied are as follows.
- table
- histogram
- line plot
- bar graph
- circle graph
- scatter plot

Example 1 CONNECTION

Life Science Refer to the beginning of the lesson. What type of display is most appropriate for displaying the data on fetus size if you are looking for a relationship between time and size?

There are two sets of numerical data to be compared. So we can eliminate displays that are used for one set of data — bar graph, histogram, and circle graph.

A table allows us to see each size at each time. However, because it does not allow a visual interpretation, it is not the best choice.

A line plot is not appropriate. The two sets of data are not similar, so comparing them with a line plot using different symbols would not be appropriate.

A scatter plot is the most appropriate choice. It allows us to plot the data as ordered pairs and determine whether there is a positive, negative, or no relationship between time and size.

Lesson 4-7 Choosing an Appropriate Display **171**

Classroom Vignette

"I generate class surveys using a spreadsheet format. Students answer questions regarding their height, number of pets, siblings, shoe size, wingspan, and so on. We then learn to use a graphing calculator to enter the data and create histograms, scatter plots, and box-and-whisker plots to explore the data and examine relationships."

Robert Allen, Teacher
Explorer Middle School
Phoenix, AZ

4-7 Lesson Notes

Instructional Resources
- *Study Guide Masters*, p. 32
- *Practice Masters*, p. 32
- *Enrichment Masters*, p. 32
- Transparencies 4-7, A and B
- *Classroom Games*, p. 11–12
- *Hands-On Lab Masters*, p. 10–11
- CD-ROM Program
  - Resource Lesson 4-7

Recommended Pacing	
Standard	Day 10 of 13
Honors	Day 10 of 13
Block	Day 6 of 7

1 FOCUS

 5-Minute Check
(Lesson 4-6)

Determine whether a scatter plot of the data below would show a *positive*, a *negative*, or *no* relationship.

1. person's height and area code **no**
2. age and total number of meals eaten **positive**
3. hourly wage and number of hours needed to earn $1,000 **negative**

 The 5-Minute Check is also available on **Transparency 4-7A** for this lesson.

Motivating the Lesson
Communication Ask students to read the lesson opener. Explain that the visual presentation of statistical data is useful in many ways and that knowing how to create and interpret statistical displays will help them as they write reports and read newspapers and magazines.

Lesson 4-7 **171**

2 TEACH

 Transparency 4-7B contains a teaching aid for this lesson.

Using Logical Reasoning
Explain that eliminating possibilities is a strategy for solving problems indirectly. When uncertain about which display to use, first eliminate the displays that are inappropriate.

In-Class Examples

For Example 1
What type of graph would be appropriate to compare the drivers' ages with their driving speeds? **scatter plot**

For Example 2
What is the best way to display data to show how lumber is used? **circle graph**

---

3 PRACTICE/APPLY

Check for Understanding
If students need additional practice or instruction after completing Exercises 1–7, one of these options may be helpful.
- Extra Practice, see p. 616
- Reteaching Activity
- *Study Guide Masters,* p. 32
- *Practice Masters,* p. 32

Study Guide Masters, p. 32

172 Chapter 4

Example 2 CONNECTION

Earth Science You are given data on the amount of space in landfills used for different types of waste. What is the best way to display the data to show that a certain type of waste occupies a much larger amount of space than most others?

A table is a good choice if you want to show the individual data. However, it may be difficult to see quickly which piece of data is much larger.

The line plot is appropriate. The reader would be able to see a number that is much larger than the rest. However, this plot may not be very visually appealing and the reader would not be able to tell which type of waste takes up the greatest amount of space.

A bar graph is a good choice. It would show which type has the greatest amount of space visually. However, a circle graph is most appropriate. A category with a greater amount of space would be shown with a larger section of the graph. The distribution of the other data would be shown for comparison.

CHECK FOR UNDERSTANDING

Communicating Mathematics

Read and study the lesson to answer each question. **1. See margin.**

1. *Give an example* of a set of data that is best displayed using a histogram.
2. *Make a list* of the advantages and disadvantages of each type of display you have studied. **See students' work.**
3. *Find* a display of data in a newspaper or on the Internet. Do you think the most appropriate type of display was used? Explain. **See students' work.**

Guided Practice
4–7. Sample answers given. See Answer Appendix for explanations.

Choose the most appropriate type of display for each data set and situation. Explain your choice.

4. ages and average heart rates in a television news report **scatter plot**
5. test scores in a science class **histogram or line plot**
6. points scored by individual members of a basketball team compared to the team total **circle graph**
7. **Employment** You are given the annual salaries of fifty men and fifty women with similar work experience and education. How would you display the data to show how the salaries of men and women compare? **scatter plot**

172 Chapter 4 Statistics: Analyzing Data

■ Reteaching the Lesson ■

Activity Ask students to read newspapers and magazines to find examples of data displayed in graph form. Ask them to determine whether the data could have been better presented by using a different type of display. Have students reconstruct the data using a different graph display.

Additional Answer
1. Sample answer: data on the number of immigrants who came to the United States in each decade.

EXERCISES

Practice

8–18. Sample answers given. See Answer Appendix for explanations.

Choose the most appropriate type of display for each data set and situation. Explain your choice.

8. ages of state fair attendees in marketing information for the fair **histogram**

9. the amounts of sales by different divisions of a company in information for budget decisions **circle graph**

10. number of televisions in surveyed homes in a newspaper article **histogram**

11. scatter plot
12. scatter plot
13. line plot
14. circle graph

11. plant height measurements made every 2 days in a science fair report

12. math test scores for two classes for comparing teaching techniques

13. home prices in a neighborhood to show that a home is underpriced

14. percent of people who own a Macintosh computer compared to all computer owners in advertising

15. annual crime index in an advertisement for a self defense class **histogram or scatter plot**

16. histogram or scatter plot

16. percent of Americans who are computer owners in each year in a computer advertisement

17. table or bar graph

17. numbers of Americans whose first language is Spanish, Mandarin, or Hindi

18. prices of ice creams rated good or fair in a consumer magazine **line plot**

Applications and Problem Solving

19. *Money Matters* The table shows the first-class postal rates for different years. Would a histogram be appropriate for this data? Explain.

Year	1971	1974	1975	1978	1981	1985	1988	1991	1995
Rate	8¢	10¢	13¢	15¢	18¢	22¢	25¢	29¢	32¢

Source: USA Today research

20. *Sports* The table below shows the number of gold medals won by each team that won medals in the 1996 Summer Olympic Games. The United States won 44 gold medals. What type of display would you use if you were writing an article on how well the American athletes performed? **See margin.**

Medals	0	1	2	3	4	5	7	9	13	15	16	20	26	44
Countries	26	18	7	8	7	1	3	3	1	1	1	1	1	1

Source: World Almanac, 1997

19. See Answer Appendix.
21. See students' work.

21. *Working on the CHAPTER Project* Study the information you found about your museum. Choose an appropriate display for one of the data sets. Then construct the display and add it to your brochure or poster.

22. *Critical Thinking* Describe a situation in which someone might choose a display other than the one that is most appropriate for the data. **See margin.**

Mixed Review

23. *Life Science* Make a scatter plot of the data on fetus size on page 171. *(Lesson 4-6)* **See Answer Appendix.**

24. **Test Practice** What is the area of a rectangle whose length is 14 meters and whose perimeter is 64 meters? *(Lesson 1-8)* **C**

    A 896 square meters B 700 square meters
    C 252 square meters D 50 square meters

25. Write the product 5 • 5 • 8 • 8 • 8 using exponents. *(Lesson 1-2)* $5^2 \cdot 8^3$

Lesson 4-7 Choosing an Appropriate Display **173**

Extending the Lesson

Enrichment Masters, p. 32

Activity Ask students to conduct a two-question survey about their classmates' most and least favorite courses. When the survey is completed, separate the students into small groups. Ask each group to display the data from the survey in the type of graph that works best. Ask each group to explain the reasons for their choice.

Assignment Guide

Core: 9–19 odd, 22–25
Enriched: 8–18 even, 19, 20, 22–25

CHAPTER Project

Exercise 21 asks students to advance to the next stage of work on the Chapter Project. You may wish to have students explain why they chose a particular graph to display their information.

4 ASSESS

Closing Activity

Writing Ask students to list types of data they are examining in their other courses. Have them suggest graphs that would best display the data and explain their choices.

Additional Answers

20. A line plot is a good choice because it would show that 44 is an outlier and most of the rest of the data are concentrated around 0–4.

22. Sample answer: When someone is trying to make a claim that the data will not necessarily support.

Practice Masters, p. 32

4-7 Practice

Choosing an Appropriate Display

Choose the most appropriate type of display for each data set and situation. Sample answers are given.

1. percentages of different kinds of juice sold in the United States **circle graph**
2. number of amateurs who play various musical instruments **bar graph**
3. number of evening meals American adults cook at home in an average week **histogram**
4. market shares of all fast-food companies **circle graph**
5. year and number of cans recycled for a recycling industry report **scatter plot**
6. percentage of college seniors' plans for five years in the future **circle graph**
7. cost of making each type of money (pennies, nickels, dimes, quarters, half-dollars, dollar bills) **table**
8. sales of five best-selling albums of all time **bar graph**
9. composition of snack foods in a brand of party mix **circle graph**
10. attendance at top five theme parks for a year-end report **bar graph**
11. percentage of movie-goers who attend each month to show most popular month for movie-going **line plot**
12. age groups and allowances given **histogram**

Lesson 4-7 **173**

4-8 Lesson Notes

Instructional Resources
- *Study Guide Masters*, p. 33
- *Practice Masters*, p. 33
- *Enrichment Masters*, p. 33
- Transparencies 4-8, A and B
- *Assessment and Evaluation Masters*, p. 100
- CD-ROM Program
  - Resource Lesson 4-8
  - Extended Activity 4-8

Recommended Pacing	
Standard	Day 11 of 13
Honors	Day 11 of 13
Block	Day 6 of 7

1 FOCUS

5-Minute Check
(Lesson 4-7)

Choose the most appropriate type of display for each data set and situation. Explain.
1. survey results in a magazine article showing what types of vehicle are most popular in America **circle graph**
2. people's age compared their weight in a medical journal **scatter plot**
3. annual cookie sales of 10 Girl Scout troops in a regional scouting newsletter **bar graph**

See students' work for explanations.

 The 5-Minute Check is also available on **Transparency 4-8A** for this lesson.

Motivating the Lesson

Problem Solving A toothpaste ad shows a smiling movie star and the caption "More beautiful people brush with Glisten." Ask students how the ad misleads. Lead students to see that the ad is vague and that it appeals to vanity rather than being based on hard facts.

4-8 Misleading Graphs and Statistics

What you'll learn
You'll learn to recognize when graphs and statistics are misleading.

When am I ever going to use this?
Knowing how to recognize misleading graphs and statistics can help you read and interpret advertisements.

Word Wise
sample

Have you studied president Alf Landon in Social Studies? Probably not. Landon lost the 1936 presidential election to Franklin Roosevelt. But Landon and the readers of *Literary Digest* were confident of a victory.

Literary Digest mailed more than 10 million surveys to households listed in telephone books and car registration books. That method would work well today, but in 1936, only the rich could afford telephones and cars. The people surveyed, called the **sample**, were not representative of the general population. For this reason, the prediction was wrong, and *Literary Digest* was soon out of business.

Statisticians often use samples to represent larger groups. They must make sure that their samples are representative of the larger group.

Example APPLICATION

1 Marketing When Coca-Cola hired Michael Jordan as a spokesperson, they were looking for someone who would appeal to teenagers. Before hiring him, they considered celebrities from sports and entertainment. Would they have gotten good results from each survey?
 a. 200 teens at a Chicago Bulls basketball game
 b. 25 teens at a shopping mall
 c. 500 students at a number of middle and high schools

 a. A survey at a Chicago Bulls basketball game would probably favor Chicago Bulls players. This would not be a good choice unless they were trying to choose between several Bulls players.
 b. The teens at a mall would represent the segment of the population that spend time at the mall. However, they may not represent all of the teens in the United States.
 c. This sample is large, and if the schools are chosen in a number of different areas, this survey would probably give very good results.

Once a survey is taken, the results can be reported in a misleading way. In Lesson 4-4, you learned about the mean, median, and mode of a set of data. In the data set {$76, $76, $78, $78, $78, $80, $92, $95, $97, $97, $99, $100, $100}, the mean is $88, the median is $92, and the mode is $78.

A company whose product sells for $91 could say that the "average price" is $92 to demonstrate that their product is inexpensive. But that is misleading. It does not give a complete picture of the data.

174 Chapter 4 Statistics: Analyzing Data

Graphs can also be misleading. Here are some ways a graph may be misleading.
- Numbers are omitted on an axis, but no break is shown.
- The tick marks on the axes are not the same distance apart or do not have the same-sized intervals.

Example 2 APPLICATION

Entertainment Nielsen Media Research studies the popularity of television programs. The graphs below show the results of a Nielsen study of the top five programs for the 1995-1996 season.
a. Which graph is not misleading? Explain your reasoning.
b. How could you change the misleading graph to make it accurate?

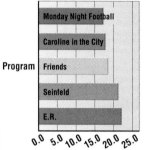
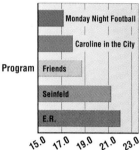

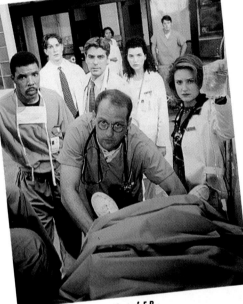
Cast of E.R.

a. Both graphs display the same information. However, the graph on the right has omitted the numbers from 0 to 15 on the horizontal axis. Since the bar representing *E.R.* is about twice as long as the bar for *Friends,* it appears that about twice as many people watch *E.R.* Since this is not true, the graph on the right is misleading.

b. Adding a break in the horizontal axis and each of the bars would point out the gap between 0 and 15.

1. Sample answers: good places - schools, telephone surveys; bad places: computer stores, libraries
2. See students' work.

CHECK FOR UNDERSTANDING

Communicating Mathematics

Read and study the lesson to answer each question.

1. *List* two good places and two bad places to conduct a survey on the number of teenagers who have access to a computer.
2. *Transform* one of the graphs in this chapter into a misleading graph.

Lesson 4-8 Misleading Graphs and Statistics **175**

 Investigations for the Special Education Student

This blackline master helps you plan for the needs of your special education students by providing long-term projects along with teacher notes. Investigation 2, *Commercial Success,* may be used with this chapter.

2 TEACH

 Transparency 4-8B contains a teaching aid for this lesson.

Using Discussion Students should understand that statistical information is often used to reinforce an idea or promote a company's product. Explain to them that they should not accept statistics cited in speeches or advertisements as the whole truth. Ask students to think of examples of misleading statistics that they have encountered and then discuss the content of the misleading information.

In-Class Examples

For Example 1
Each day, a radio station broadcasts a "question of the day" concerning the local public school system to which listeners can call in their "yes" or "no" votes. Will the results provide a true representation of the public's opinion? Explain. **Probably not; the station's listeners may not be representative of the whole community.**

For Example 2
Mayoral candidate Doneti produced this graph showing her voter approval ratings.

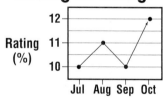

a. List the candidate's monthly ratings. **10%, 11%, 10%, 12%**

b. How does the graph mislead? **It shows only the 10–12% range on the vertical axis, exaggerating the apparent importance of the September-October gain. The graph is drawn as an arrow. The title is misleading.**

Lesson 4-8 **175**

3 PRACTICE/APPLY

Check for Understanding

If students need additional practice or instruction after completing Exercises 1–6, one of these options maybe helpful.
- Extra Practice, see p. 617
- Reteaching Activity
- *Study Guide Masters*, p. 33
- *Practice Masters*, p. 33

Assignment Guide

Core: 7–17 odd, 19–22
Enriched: 8–14 even, 16–22

Additional Answers

3. People at a Gloria Estefan concert will be more likely than the general public to choose her.
4. All ages and interests will be represented.
5. Only people who are concerned enough about the situation will call.

Study Guide Masters, p. 33

Guided Practice
3–5. See margin for explanations.

Decide whether each location is a good place to find a representative sample for the selected survey. Justify your answer.

3. favorite singer at a Gloria Estefan concert **no**
4. number of books read in a month at a shopping mall **yes**
5. passage of a school levy in a television phone-in poll **no**
6. *Fashion* Of the 95% of schools who allow students to wear jeans, many have restrictions such as no jeans with holes. Consider the graph at the right.
   a. How much greater is the percent of schools who allow jeans with restrictions than those that allow jeans with no restrictions? **44.6%**
   b. How does the difference in the percents compare with the difference in area of the two pairs of jeans? **See Answer Appendix.**
   c. Is this a misleading graph? Explain. **See Answer Appendix.**

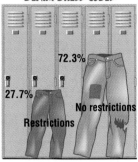

Source: *Seventeen* Magazine Market Research Council

EXERCISES

Practice
7–15. See Answer Appendix for explanations.

Decide whether each location is a good place to find a representative sample for the selected survey. Justify your answer.

7. favorite kind of entertainment at a movie theater **no**
8. whether families own pets in an apartment complex **no**
9. choice of mayoral candidate in a telephone poll **yes**
10. taste test of a snack food at a grocery store **yes**
11. Americans who are bilingual in a mall in San Antonio, Texas **no**
12. favorite teacher in a school cafeteria **yes**
13. teenagers' favorite magazine at five different high schools **yes**
14. favorite drink at a coffee house **no**
15. career choice at a community college **no**

Applications and Problem Solving

16a. Yes; the values represented by the graphs are the same.
16b–c. See Answer Appendix.

16. *Marketing* The two graphs below show the results of a taste test of Mrs. Field's cookies.

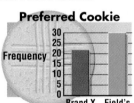
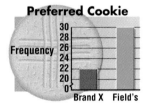

a. Do both graphs contain the same information? Explain.
b. One of the graphs is misleading. Explain why it is misleading.
c. Which graph would you choose if you were creating advertising for Mrs. Field's? Explain your reasoning.

176 Chapter 4 Statistics: Analyzing Data

Reteaching the Lesson

Activity Ask students to find examples of graphs and other statistical displays in newspapers or magazines. Ask them to interpret the information in the graph and to decide whether the graph is misleading or is a true representation of the data. Have students explain their answers.

Error Analysis
Watch for students who believe that all statistical information represents a true sample of the population.
Prevent by explaining that statistics often are accurate representations of data but they can also be used to support a particular point of view.

17. **Math History** The timeline shows some of the major events in the history of mathematics.

The Elements published.	Eratosthenes creates sieve for finding primes.	Circle divided into 360 degrees.	Hypatia publishes in geometry.	Tsu Ch'ung-chih approximates pi.	Al-Khowârizmî writes about Algebra.
300	230	180 B.C.	A.D. 410	480	820

17a. Yes; the space between the tick marks is not consistent with the time between each date.

17b–c. See students' work.

a. In what way is the timeline misleading? Explain your answer.
b. Draw a timeline. Include the events in the timeline above and the publication of the Chinese book *Arithmetic in Nine Sections* in 100 B.C., introduction of the Greek number system in 450 B.C., and the destruction of the University of Alexandria in A.D. 641.
c. Choose one of the events listed on your timeline for further research. Describe the event and how you can see that part of mathematics used today.

18. *Careers* If you could trade places with an adult, who would you choose? A survey of 5,000 students asked that question. The results are shown in the graph.

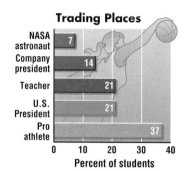

a. How does the number of students who chose an astronaut compare to the number who chose the U.S. President? **See margin.**

18b. yes

b. Are the sizes of the bars consistent with the differences in percents?
c. Is this graph misleading? Explain. **See margin.**

19. *Critical Thinking* A toothpaste commercial says "Recommended by 4 out of 5 dentists." **a–b. See margin.**
    a. Would this claim be reasonable if 10 dentists were surveyed? Explain.
    b. Dr. Glover was a part of the survey. She also recommended several other toothpastes. Explain how the commercial was written to misrepresent Dr. Glover's recommendation.

Mixed Review

20. **Test Practice** What type of graph is most appropriate for displaying the change in house prices over several years? *(Lesson 4-7)* **A**

    A scatter plot   B circle graph
    C bar graph   D line plot

21. Express 4.8% as a decimal. *(Lesson 3-4)* **0.048**

22. *Games* Hector borrowed $100 from the bank to buy Park Place. If he passes go, collects $200, and repays the bank, how much money will Hector have? *(Lesson 2-3)* **$100**

Lesson 4-8 Misleading Graphs and Statistics **177**

---

4 ASSESS

Closing Activity

Modeling Have students make a deceptive graph or create a misleading statistic from a set of data.

Chapter 4, Quiz D (Lessons 4-7 and 4-8) is available in the *Assessment and Evaluation Masters*, p. 100.

Additional Answers

18a. Three times as many students chose the U.S. President as chose astronaut.

18c. No; the representation is consistent with the data.

19a. The fewer the dentists surveyed, the less reliable the results are.

19b. The commercial makes it sound as if the dentists recommended their brand of toothpaste over all others.

Practice Masters, p. 33

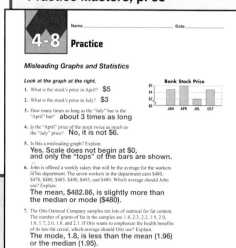

Extending the Lesson

Enrichment Masters, p. 33

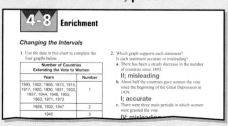

Activity Explain to students that nutritional information on food packaging should be read carefully. Ask them to record the amount of each food that they eat in one day. At the end of the day, have them compare the serving sizes that are listed on the packages with the amount of food that they ate.

Lesson 4-8 **177**

CHAPTER 4 Study Guide and Assessment

Vocabulary

After completing this chapter, you should be able to define each term, concept, or phrase and give an example or two of each.

Statistics
bar graph (p. 142)
box-and-whisker plot (p. 167)
circle graph (p. 148)
data analysis (p. 154)
frequency table (p. 142)
histogram (p. 142)
interquartile range (p. 163)
line plot (p. 153)
lower quartile (p. 164)
mean (p. 158)
measures of central tendency (p. 158)
median (pp. 158, 159)
mode (p. 158)

outlier (p. 164)
quartiles (p. 163)
range (p. 163)
sample (p. 174)
statistics (p. 142)
upper quartile (p. 163)
variation (p. 163)

Algebra
scatter plot (p. 168)

Problem Solving
make a table (p. 140)

Understanding and Using the Vocabulary

Choose the correct term or number to complete each sentence.

1. A (<u>scatter plot</u>, line plot) is a graph whose ordered pairs consist of two sets of data.
2. A (frequency table, <u>histogram</u>) is a bar graph that shows the frequency of data in intervals.
3. The (interquartile range, <u>range</u>) of a set of numbers is the difference between the least and greatest number in the set.
4. (<u>An outlier</u>, A variation) is a piece of data that is more than 1.5 times the interquartile range from one of the quartiles.
5. A small group that is representative of a larger population is called a (<u>sample</u>, mode).
6. The (<u>mean</u>, median) is the sum of the data divided by the number of pieces of data.
7. If you want to show how parts compare to a whole, the best display to use is a (line plot, <u>circle graph</u>).
8. The range is one of the (measures of central tendency, <u>measures of variation</u>).

In Your Own Words

9. *Explain* the difference between the mean and the median of a set of data. **The mean is the average of the set and the median is the middle number when the set is arranged in order.**

MindJogger Videoquizzes

MindJogger Videoquizzes provide an alternative review of concepts presented in this chapter. Students work in teams to answer questions, gaining points for correct answers. The questions are presented in three rounds.
Round 1 Concepts–5 questions
Round 2 Skills–4 questions
Round 3 Problem Solving–4 questions

Study Guide and Assessment   Chapter 4

Objectives & Examples

Upon completing this chapter, you should be able to:

● construct and interpret bar graphs and histograms *(Lesson 4-1)*

Construct a histogram for {1, 1, 2, 3, 4, 4, 4, 5, 5, 5, 6, 7, 7, 7, 8, 8, 8, 8, 9, 9, 9, 10, 10, 10, 10}.

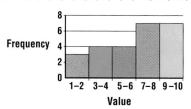

● construct and interpret circle graphs *(Lesson 4-2)*

Each day, an average of 2,749 Americans enroll in a language class. Of these, 46 choose Chinese; 64 Japanese; 93 Russian; 754 French; and 1,127 Spanish. Make a circle graph of the data.

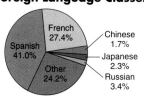

● construct and interpret line plots *(Lesson 4-3)*
Construct a line plot for the data {1, 1, 2, 3, 4, 4, 4, 4, 5, 5, 5, 6, 7, 7, 7, 8, 8, 8, 8, 9, 9, 9, 10, 10, 10, 10}.

Review Exercises

Use these exercises to review and prepare for the chapter test.

Use the histogram at the left to answer each question. 11. 7-8 and 9-10 each have 7 data.
10. How large is each interval? **2**
11. Which interval has the most data?
12. What percent of the data is below 7? **44%**
13. Construct a histogram for the following history test scores. **See Answer Appendix.**
56, 87, 67, 74, 77, 84, 94, 80, 72, 58, 87, 90, 68, 90, 60, 73, 74, 82, 68, 64

14. See Answer Appendix.
14. *Geography* Make a circle graph of the data on the area of the Great Lakes.

Lake	Area (sq mi)
Erie	9,910
Huron	23,010
Michigan	22,300
Ontario	7,540
Superior	31,700

Source: Carnegie Library of Pittsburgh

15. Make a circle graph of the time that you spend each week on studying, sleeping, eating, watching television, and other activities. **See students' work.**

16. See Answer Appendix.
The list shows ages of the first 24 people to enter an art museum one day.
30, 22, 23, 25, 22, 30, 28, 29, 26, 29, 27, 27, 24, 24, 30, 27, 24, 26, 27, 23, 24, 27, 23, 27

16. Construct a line plot for the data.
17. What is the most common age? **27**
18. What are the ages of the oldest and youngest people? **22 and 30**

Chapter 4 Study Guide and Assessment **179**

Objectives & Examples

This section reviews the skills and concepts of the chapter and shows completely worked examples.

Review Exercises

These exercises provide practice for the corresponding objectives.

Assessment and Evaluation Masters, pp. 87–88

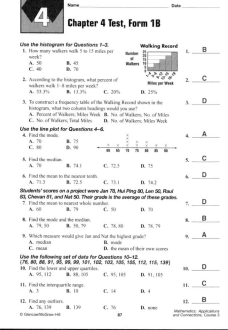

Assessment and Evaluation

Six forms of Chapter 4 Test are available in the *Assessment and Evaluation Masters* as shown in the chart.

Chapter 4 Test, Form 1B, is shown at the right. Chapter 4 Test, Form 2B, is shown on the next page.

1A	Multiple Choice	Honors
1B	Multiple Choice	Average
1C	Multiple Choice	Basic
2A	Free Response	Honors
2B	Free Response	Average
2C	Free Response	Basic

Study Guide and Assessment **179**

Chapter 4 Study Guide and Assessment

Objectives & Examples | Review Exercises

● find the mean, median, and mode of a set of data
(Lesson 4-4)

Find the mean, median, and mode for {53, 55, 58, 63, 66}.

mean: $\frac{53 + 55 + 58 + 63 + 66}{5} = 59$

median: 58   no mode

Find the mean, median, and mode for each set of data. When necessary, round to the nearest tenth. 19. 7.7; 6.8; no mode

19. 5.6, 6.5, 6.8, 9.6, 10.1
20. 0, 2, 1, 5, 3, 7, 8, 5, 9   **4.4; 5; 5**
21. $16, $20, $21, $25, $18, $19, $20, $21
    20; 20; 20 and 21

● find the range and quartiles of a set of data
(Lesson 4-5)

Find the range, median, quartiles, and interquartile range of {1, 2, 2, 3, 3, 3, 4, 4, 5, 6, 6}.

Range: 6 − 1 = 5   median: 3
LQ: 2   UQ: 5
Interquartile range: 5 − 2 = 3

22–24. See Answer Appendix.
Find the range, median, upper and lower quartiles, interquartile range, and any outliers for each set of data.

22. 8, 9, 5, 10, 7, 6, 2, 4
23. 195, 121, 135, 123, 138, 150, 122, 136, 149, 124, 149, 151, 152
24. 80, 91, 82, 83, 77, 79, 78, 75, 75, 88, 84, 82, 61, 93, 88, 85, 84, 89, 62, 79

● construct and interpret scatter plots
(Lesson 4-6)

A scatter plot is a graph of points whose ordered pairs consist of two sets of data.

Determine whether the scatter plot of the data below would show a *positive*, *negative*, or *no* relationship. 26. no relationship

25. age and income   **positive**
26. temperature and day of the week

● choose an appropriate display for a set of data
(Lesson 4-7)

Choose the most appropriate type of display for a set of data that compares a person's age and their height. The data can be written in ordered pairs, so a scatter plot is most appropriate.

Choose the most appropriate type of display for each data set and situation. Explain your choice. 27–28. **Sample answers given.**

27. number of computers in surveyed homes in magazine article   **histogram**
28. prices of different types of sandwiches at a deli   **line plot**

● recognize when graphs and statistics are misleading *(Lesson 4-8)*

A sample must be representative of the larger group. A graph must have an appropriate scale.

29–30. See Answer Appendix.
Decide whether each location is a good place to find a representative sample for the selected survey. Justify your answer.

29. favorite movie at a movie theater
30. favorite dessert at an ice cream shop

---

180   Chapter 4 Statistics: Analyzing Data

Study Guide and Assessment — Chapter 4

Applications & Problem Solving

31. Make a Table Survey your classmates about their favorite dessert. Record the results in a frequency table. *(Lesson 4-1A)*

32. Agriculture The table shows the thousands of tons of several principal crops grown in the United States in 1996. Make a bar graph of the data. *(Lesson 4-1)*

Crop	Tons (thousands)
Almonds	304.3
Hazelnuts	39.0
Pecans	134.0
Walnuts	234.0

33. Wildlife Management The list shows lengths in inches of catfish captured and released in Grand Lake, Oklahoma.
8, 6, 4, 10, 9, 14, 4, 5, 12, 13, 10, 8
Make a line plot of the data. *(Lesson 4-3)*

31. See students' work.
32–33. See Answer Appendix.
34b. 5.74; 4.285, 1.83; 2.455; no outliers

34. Earth Science The list shows the amount of rainfall in inches in north central Texas for each month of 1994.
1.27, 2.40, 1.78, 2.79, 7.01, 1.68, 3.69, 1.88, 3.14, 6.53, 4.88, 3.37 **a. 3.37; 2.97; no mode**

a. What is the mean, median, and mode rainfall amount for 1994? *(Lesson 4-4)*

b. Find the range, upper and lower quartiles, interquartile range, and any outliers of the data. *(Lesson 4-5)*

35. Sales The graphs show Guillermo's sales for January through July. Which graph might be misleading and why? *(Lesson 4-8)* **See margin.**

Alternative Assessment

Performance Task

Jacqui Franklin is trying to find the least expensive store in town from which to rent a video. The data she has collected about the costs of renting a video shows that some stores have a one-day rate and others have a two-day rate. Do you think it would help her to display the costs in a statistical graph? Why or why not? If so, what type of display should Jacqui use? **See students' work.**

What types of displays would not be appropriate for Jacqui to use? Explain. **See students' work.**

A practice test for Chapter 4 is provided on page 650.

Completing the CHAPTER Project Use the following checklist to make sure that your brochure or poster is complete.

☑ You have included a clear, well-organized description of the museum, its purpose, and its history.

☑ The bar graph of the data visitor information is clear and easy to read.

☑ The measures of central tendency are accurate, and the estimates of the budget requirements are included.

 Select a graph you drew in this chapter and place it in your portfolio. Attach a note explaining how it illustrates one of the important concepts in this chapter.

Additional Answer

35. The graph on the right is misleading because the intervals are unequal on the vertical scale.

Standardized Test Practice

The Standardized Test Practice may be used to help students prepare for standardized tests. The test items are written in the same style as those in state proficiency tests and standardized tests like CAT, CTBS, ITBS, MAT, SAT, and Terra Nova. The test items cover skills and concepts covered up to this point in the text.

The pages can be used as an overnight assessment. After students have completed the pages, discuss how each problem can be solved, or provide copies of the solutions from the *Solutions Manual*.

Assessment and Evaluation Masters, p. 103

CHAPTERS 1-4 Standardized Test Practice
Assessing Knowledge & Skills

Section One: Multiple Choice

There are twelve multiple-choice questions in this section. Choose the best answer. If a correct answer is *not here*, choose the letter for Not Here.

1. What is 33.67 rounded to the nearest tenth? **D**
   A 33
   B 34
   C 33.6
   D 33.7

2. 36.5% = **H**
   F 36.5
   G 3.65
   H 0.365
   J 0.0365

3. If $3m + 8 = 32$, what is the value of m? **C**
   A 12
   B 10
   C 8
   D 6

4. What is $5 \cdot 5 \cdot 5 \cdot 8 \cdot x \cdot x$ written in exponential notation? **F**
   F $5^3 \cdot 8 \cdot x^2$
   G $40x^2$
   H $3^5 \cdot 4 \cdot x^2$
   J $3^5 \cdot 4 \cdot 2x$

5. The line plot shows the heights in inches of the Washington Middle School girls' basketball team. **D**

   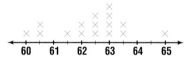

   Which sentence best describes the data?
   A The range is 6 inches.
   B Most of the players are 63 inches tall.
   C There are 8 players.
   D Most of the players are between 62 and 64 inches tall.

Please note that Questions 6-12 have five answer choices.

6. A bolt of fabric is 36 inches long. It is cut into two pieces so that one piece is 14 inches longer than the other. Which equation could be used to find the shorter piece, z? **F**
   F $z + (z + 14) = 36$
   G $z^2 + 14 = 36$

   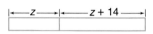

   H $2(z + 14) = 36$
   J $2z - 14 = 36$
   K $2z + (z - 14) = 36$

7. Bryanna invited 8 friends to a soccer game. The cost of tickets is $9.50 per person. Which is the best estimate for the cost of the game? **B**
   A $70
   B $85
   C $90
   D $100
   E $125

8. Stacy and Ebony went to the store to buy 3 bags of chips for a party. They had $8 and wanted to spend no more than that on the chips. Which inequality could they use to find the price, p, of each bag of chips so that the total would be at or below $8? **J**
   F $p + 3 \le 8$
   G $3p \ge 8$
   H $8 > 3p$
   J $3p \le 8$
   K $8 - p \le 3$

182 Chapters 1-4 Standardized Test Practice

◄◄◄ **Instructional Resources**
Another cumulative review is shown at the left and is available in the *Assessment and Evaluation Masters*, p. 103.

Standardized Test Practice Chapters 1–4

9. A tire filled with air has a leaking valve. The change in the amount of air is –16 cubic centimeters per hour. At this rate, in how many hours will the total change be –100 cubic centimeters? **D**
   - A 16 hours
   - B 12 hours
   - C 10.5 hours
   - D 6.25 hours
   - E Not Here

10. Scoma's Restaurant bought 58 pounds of shrimp at the market for $190.82 and 42 pounds of lobster for $186.90. What was the cost per pound of the shrimp? **K**
    - F $2.39
    - G $2.84
    - H $3.19
    - J $4.45
    - K Not Here

11. The total price for a jacket and pants was $184.62. This price included $9.97 in sales tax. What was the price of the jacket and pants before tax? **A**
    - A $174.65
    - B $176.85
    - C $177.15
    - D $182.65
    - E $194.59

12. A truck carrying lumber travels 899.1 miles on 74 gallons of diesel fuel. How many miles per gallon did the truck average? **J**
    - F 10.85 mpg
    - G 11.21 mpg
    - H 11.86 mpg
    - J 12.15 mpg
    - K 12.85 mpg

Test-Taking Tip

You can guess on standardized tests. If you know that one or more answers for a particular problem are definitely wrong, then it can be to your advantage to guess from the choices that remain. Random guessing will not increase your score, but educated guessing can make a difference.

Section Two: Free Response

This section contains four questions for which you will provide short answers. Write your answers on your paper.

13. $-3 + 8 + (-3) + 1 =$ **3**

14. Mrs. Campas teaches a second grade class of 23 students. If she wants to buy 9 stickers for each student and the stickers cost $0.07 each, what is the best estimate of the cost of the stickers? Round to the nearest dollar. **$14**

15. Write the equation to represent *three more than twice a number equals 14*.
    $2n + 3 = 14$

16. George, Justin, and Benito wanted to collect winter hats for a local shelter. Their goal is 120 hats. The table shows their progress so far.

Name	Week 1	Week 2	Week 3	Total
George	4	14	8	26
Justin	12	30	7	49
Benito	5	5	13	23

    How many more hats do they need to collect to reach their goal? **22 hats**

Test-Taking Tip

Suggest to students that they try to estimate what the answer should be before they begin calculations. Sometimes answers can be eliminated because the suggested answer is much larger or much smaller than the estimate.

Assessment and Evaluation Masters, pp. 101–102

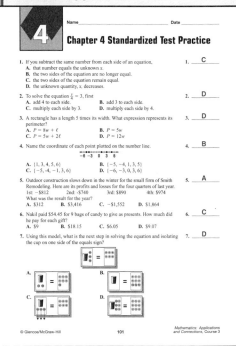

Instructional Resources ▶▶▶

Additional standardized test practice is shown at the right and is available in the *Assessment and Evaluation Masters,* pp. 101–102.

CHAPTER 5
Geometry: Investigating Patterns

Previewing the Chapter

Overview
This chapter develops basic concepts in geometry. Students learn the classic methods of construction using a compass and a straightedge and learn to determine the traceability of polygons as networks. Students study the properties of parallel lines and the classification and properties of triangles and quadrilaterals. Comparing and transforming geometric figures are introduced with concepts on congruence, similarity, and symmetry. Students solve problems by interpreting Venn diagrams.

Lesson (pages)	Lesson Objectives	NCTM Standards	Standardized Tests	State/Local Objectives
5-1A (186–187)	Measure and construct congruent line segments and angles.	1–4, 12, 13	CAT, CTBS, TN	
5-1 (188–192)	Identify lines that are parallel and types of angles formed by parallel lines and transversals.	1–4, 7, 9, 12, 13	CTBS, ITBS, SAT, TN	
5-1B (193)	Construct a line parallel to a given line.	1–3, 12, 13		
5-2A (194–195)	Use a Venn diagram to solve problems.	1–4, 7, 8, 10, 12	CAT, CTBS, TN	
5-2 (196–199)	Classify triangles by their angles and their sides, and find measures of missing angles in triangles.	1–4, 7, 9, 12, 13	CAT, CTBS, ITBS, SAT, TN	
5-3A (200)	Determine whether a network is traceable.	1–4, 8, 11, 12		
5-3 (201–204)	Classify quadrilaterals.	1–4, 7, 9, 12, 13	ITBS, SAT	
5-4A (205)	Use geoboards to reflect a figure.	1–4, 8, 12, 13		
5-4 (206–209)	Identify line symmetry and rotational symmetry.	1–4, 7, 8, 12	ITBS	
5-5 (210–212)	Verify congruent triangles by using SSS, ASA, and SAS.	1–4, 7, 9, 12	CTBS, ITBS, TN	
5-5B (213)	Construct a triangle congruent to a given triangle.	1–3, 4, 12, 13		
5-6A (214)	Draw a dilation of a figure.	1–4, 7, 12, 13		
5-6 (215–218)	Identify similar triangles.	1–4, 7, 9, 12	CTBS, ITBS, MAT, TN	
5-7 (220–223)	Create Escher-like drawings using translations and rotations.	1–4, 8, 12		

CAT = California Achievement Tests, CTBS = Comprehensive Tests of Basic Skills, ITBS = Iowa Tests of Basic Skills, MAT = Metropolitan Achievement Tests, SAT = Stanford Achievement Tests, TN = Terra Nova

Organizing the Chapter

CD-ROM
All of the blackline masters in the Teacher's Classroom Resources are available on the **Electronic Teacher's Classroom Resources** CD-ROM.

LESSON PLANNING GUIDE

Lesson	Extra Practice (Student Edition)	Study Guide	Practice	Enrichment	Assessment & Evaluation	Classroom Games	Diversity	Hands-On Lab	School to Career	Science and Math Lab Manual	Technology	Transparencies A and B
5-1A								45				
5-1	p. 617	34	34	34								5-1
5-1B								46				
5-2A	p. 617											
5-2	p. 618	35	35	35	127							5-2
5-3A								47				
5-3	p. 618	36	36	36		13–16		72			61	5-3
5-4A								48				
5-4	p. 618	37	37	37	126, 127		31					5-4
5-5	p. 619	38	38	38							62	5-5
5-5B								49				
5-6A								50				
5-6	p. 619	39	39	39	128				31			5-6
5-7	p. 619	40	40	40	128							5-7
Study Guide/ Assessment					113–125, 129–131							

OTHER CHAPTER RESOURCES

Student Edition
Chapter Project, pp. 185, 209, 223, 227
School to Career, p. 219
Let the Games Begin, p. 218

Technology
 CD-ROM Program
 Interactive Mathematics Tools Software

Teacher's Classroom Resources

Applications
Family Letters and Activities, pp. 61–62
Investigations and Projects Masters, pp. 33–36

Meeting Individual Needs
Investigations for the Special Education Student, p. 29

Teaching Aids
Answer Key Masters
Block Scheduling Booklet
Lesson Planning Guide
Solutions Manual

Professional Publications
Glencoe Mathematics Professional Series

Chapter 5 **184b**

Planning the Chapter

 MindJogger Videoquizzes provide a unique format for reviewing concepts presented in the chapter.

ASSESSMENT RESOURCES

Student Edition
Mixed Review, pp. 192, 199, 204, 209, 212, 218, 223
Mid-Chapter Self Test, p. 204
Math Journal, pp. 211
Study Guide and Assessment, pp. 224–227
Performance Task, p. 227
Portfolio Suggestion, p. 227
Standardized Test Practice, pp. 228–229
Chapter Test, p. 651

Assessment and Evaluation Masters
Multiple-Choice Tests (Forms 1A, 1B, 1C), pp. 113–118
Free-Response Tests (Forms 2A, 2B, 2C), pp. 119–124
Performance Assessment, p.125
Mid-Chapter Test, p. 126
Quizzes A–D, pp. 127–128
Standardized Test Practice, pp. 129–130
Cumulative Review, p. 131

Teacher's Wraparound Edition
5-Minute Check, pp. 188, 196, 201, 206, 210, 215, 220
Building Portfolios, p. 184
Math Journal, pp. 187, 193, 200, 205, 213, 214
Closing Activity, pp. 192, 195, 199, 204, 209, 212, 218, 223

Technology
Test and Review Software
MindJogger Videoquizzes
CD-ROM Program

MATERIALS AND MANIPULATIVES

Lesson 5-1A
compass*†
protractor*†
ruler*

Lesson 5-1
straightedge*
protractor*†
colored pencils
notebook paper

Lesson 5-1B
compass*†
straightedge*

Lesson 5-3A
straightedge*

Lesson 5-3
protractor*†
ruler*

Lesson 5-4A
geoboard*†
geobands*†

Lesson 5-4
transparencies
overhead markers†
straight pin

Lesson 5-5B
compass*†
straightedge*

Lesson 5-6A
brass fasteners
cardboard strips
map
ruler*

Lesson 5-6
scissors*
tracing paper
index cards

Lesson 5-7
paper clips
scissors*
tape

*Glencoe Manipulative Kit   †Glencoe Overhead Manipulative Resources

PACING CHART

See pages T25–T27 for the Course Planning Calendar.

COURSE	DAY 1	DAY 2	DAY 3	DAY 4	DAY 5	DAY 6	DAY 7
Standard	Chapter Project	Lessons 5-1A & 5-1		Lesson 5-2A	Lesson 5-2	Lessons 5-3A & 5-3	
Honors	Chapter Project, Lessons 5-1 & 5-1B		Lesson 5-2A	Lesson 5-2	Lessons 5-3A & 5-3		Lessons 5-4A & 5-4
Block	Chapter Project, Lessons 5-1A & 5-1	Lessons 5-2A & 5-2	Lessons 5-3A & 5-3	Lessons 5-4A & 5-4	Lesson 5-5	Lessons 5-6A & 5-6	Lesson 5-7

Interactive Mathematics:
Activities and Investigations

is an activity-based program that may be used as an enhancement for chapters in *Mathematics: Applications and Connections*.

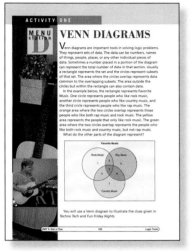

Unit 5, Activity One, Menu D
Use with Lesson 5-2A.

Summary Students work in groups to solve logic problems by using Venn diagrams. They draw appropriate diagrams to represent the information given.

Math Connection Students use Venn diagrams to solve nonroutine problems. Venn diagrams represent a set of data. Usually a rectangle is used to represent the given set and circles are used to represent the subsets of the given set.

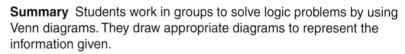

Unit 4, Activity Four
Use with Lesson 5-4.

Summary Students work in groups or with a partner to explore symmetry by cutting folded paper. They then cut folded squares in a prescribed way and predict the design being formed. Finally, they determine how a square of paper should be folded and cut to make a given design.

Math Connection By experimenting with the physical attributes of figures, students can more easily identify the properties and symmetry of figures.

DAY 8	DAY 9	DAY 10	DAY 11	DAY 12	DAY 13	DAY 14	DAY 15
Lessons 5-4A & 5-4		Lesson 5-5	Lessons 5-6A & 5-6		Lesson 5-7	Study Guide and Assessment	Chapter Test
(continue from Day 7)	Lessons 5-5 & 5-5B		Lessons 5-6A & 5-6		Lesson 5-7	Study Guide and Assessment	Chapter Test
Study Guide and Assessment, Chapter Test							

Chapter 5 **184d**

Enhancing the Chapter

APPLICATIONS

Classroom Games, pp. 13–16

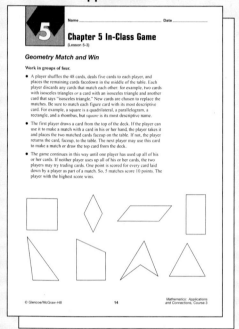

Diversity Masters, p. 31

School to Career Masters, p. 31

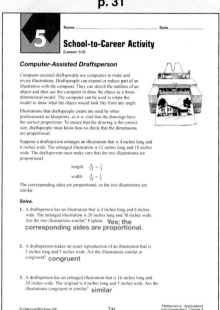

Family Letters and Activities, pp. 61–62

MANIPULATIVES/MODELING

Hands-On Lab Masters, p. 72

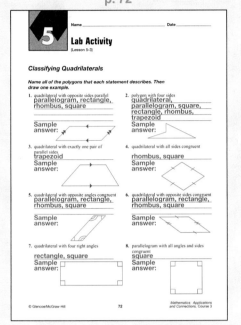

ASSESSMENT/EVALUATION

Assessment and Evaluation Masters, pp. 126–128

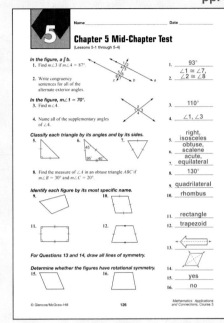

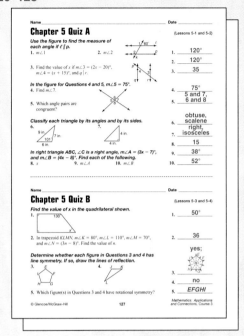

TECHNOLOGY/MULTIMEDIA

Technology Masters, pp. 61–62

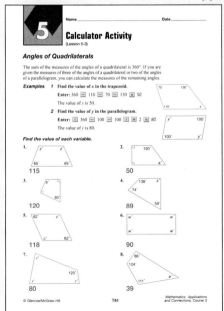

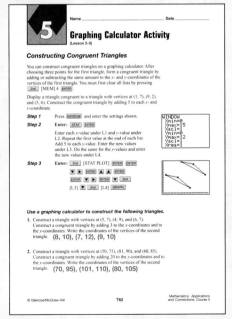

MEETING INDIVIDUAL NEEDS

Investigations for the Special Education Student, p. 29

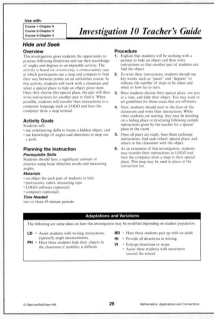

CHAPTER 5 NOTES

Theme: Advertising

Advertisers often use eye-catching patterns to bring attention to their products. Tessellations are such a pattern. Tessellations have long been used to decorate buildings in Spain and the Middle East. The word *tessellate,* which comes from a Greek word meaning "four," means "to arrange small squares or tiles in a mosaic pattern." The first tessellations were made with square tiles. Now tessellations are made with many shapes.

Question of the Day Suppose an advertiser wishes to make a tessellation for a cat food product. Ask students how they might draw a cat so that it could be reflected, rotated, and translated. **Drawings will vary.**

Assess Prerequisite Skills

Ask students to read through the list of objectives presented in "What you'll learn in Chapter 5." You may wish to ask them what each of the objectives means or if they have experienced or used any of these math concepts before.

 Building Portfolios

Encourage students to revise their portfolios as they study this chapter. They should include examples of their logos, as well as interesting rotations and reflections from their work in this chapter.

 Math and the Family

In the *Family Letters and Activities* booklet (pp. 61–62), you will find a letter to the parents explaining what students will study in Chapter 5. An activity appropriate for the whole family is also available.

184 Chapter 5

CHAPTER 5

Geometry: Investigating Patterns

 you'll learn in Chapter 5

- to identify lines that are parallel and types of angles formed by parallel lines and transversals,
- to solve problems by using Venn diagrams,
- to classify triangles and quadrilaterals,
- to identify line symmetry and rotational symmetry,
- to identify congruent triangles and similar triangles, and
- to create Escher-like drawings using translations and rotations.

184 Chapter 5 Geometry: Investigating Patterns

 CD-ROM Program

Activities for Chapter 5
- Chapter 5 Introduction
- Interactive Lessons 5-1 and 5-4
- Extended Activity 5-7
- Assessment Game
- Resource Lessons 5-1 through 5-7

CHAPTER Project

BE TRUE TO YOUR SCHOOL

In this project, you will design a school logo using a tessellation. You will begin by choosing a basic shape that will tessellate. After you determine whether your basic shape has symmetry, you will use one or more transformations to create a new shape. You will use this new shape to cover a poster board to form your school logo.

Getting Started

- A tessellation is a tile-like pattern formed by repeating shapes to fill a plane without gaps or overlaps. Some tessellations require just one shape, while others use more than one shape.
- Study various logos used by companies, sports teams, schools, and so on. List the ideas that you would like to incorporate in your logo.
- Find one or more shapes that will tessellate.
- Pick the colors that you want to use for your school logo.

Technology Tips

- Surf the **Internet** to search for ideas for your school logo.
- Use a **word processor** to write about the symmetry of your basic shape.
- Use **computer software** to design your school logo.

 interNET CONNECTION For up-to-date information on tessellations, visit:
www.glencoe.com/sec/math/mac/mathnet

Working on the Project

You can use what you'll learn in Chapter 5 to help you make your school logo.

Page	Exercise
209	20
223	16
227	Alternative Assessment

interNET CONNECTION

Glencoe has made every effort to ensure that the website links for *Mathematics: Applications and Connections* at www.glencoe.com/sec/math/mac/mathnet are current and contain appropriate content. However, these website links are not under Glencoe's control.

Instructional Resources ▶▶▶
A recording sheet to help students organize their data for the Chapter Project is shown at the right and is available in the *Investigations and Projects Masters*, p. 36.

CHAPTER Project NOTES

Objectives Students should
- identify line symmetry and rotational symmetry.
- translate, rotate, and reflect figures.
- create Escher-like tessellations.

Project Pointer You may suggest that students begin a *Project Folder* to keep their work as they complete each stage of the Chapter Project. The completed project may also be added to their portfolios.

Using Computer Software The drawing tools in software such as *HyperCard, ClarisWorks, HyperStudio,* and *Paint* can by used to make shapes that can tessellate. Directions for specific software can be found on the web. In general, students first draw a square. Then using scissors or a lasso, cut out a piece from one side of the square and move it to the opposite side. Do this again for the top or bottom. Decorate the shape. Then create an identical shape with contrasting colors. Tile the shapes.

Investigations and Projects Masters, p. 36

Chapter 5 Project

Be True to Your School

Page 185, Getting Started
- Ideas to incorporate in logo:
- Shape(s) to tessellate:
- School colors:

Page 209, Working on the Chapter Project, Exercise 20
a. Draw all lines of symmetry.

b. List angles of rotation, if any.

Page 223, Working on the Chapter Project, Exercise 16
Tessellate your shape.

5-1A HANDS-ON LAB Notes

GET READY

Objective Students measure and construct congruent line segments and angles.

Optional Resources
Hands-On Lab Masters
- protractors, p. 21
- worksheet, p. 45

Overhead Manipulative Resources
- compass
- protractor

Manipulative Kit
- compass
- protractor
- ruler

MANAGEMENT TIPS

Recommended Time
45 minutes

Getting Started Sketch this arrangement of three points on the chalkboard.

A B

C

Ask these questions.
- How many line segments are needed to connect the points in all possible ways? **3**
- How many angles will result? **3**

Sketch this arrangement of four points on the chalkboard. Repeat the questions above. **6; 16, not counting those angles measuring 180°**

A B

C D

Activities 1 and 2 demonstrate how to measure line segments and how to construct a segment congruent to a given segment.

HANDS-ON LAB

COOPERATIVE LEARNING

5-1A Measuring and Constructing Line Segments and Angles

A Preview of Lesson 5-1

- compass
- protractor
- ruler

You know two systems of measurement for determining length: customary and metric. In geometry, both systems may be used to measure line segments.

TRY THIS

Work with a partner.

① Use a ruler to measure line segments.
- Line segments are named by their endpoints. The segment below may be named $\overline{PF}$ or $\overline{FP}$.

P   F

- The sides of a polygon are formed by line segments. *ABCD* is a polygon with four sides. $\overline{AC}$ is a diagonal of polygon *ABCD*. Measure each line segment to the nearest tenth of a centimeter. Record the measurements.

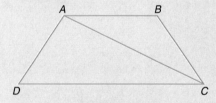

② Construct a line segment congruent to $\overline{PF}$ by following these steps.
- Draw a long line with a ruler or straightedge. Put a point on the line. Label it *Z*.

- Open your compass to the same width as the length of $\overline{PF}$.
- Put the point of the compass at *Z* and draw an arc that intersects the line. Label this intersection *Y*. Segments with the same length are said to be *congruent*. Segment *ZY* is congruent to segment *PF*, or $\overline{ZY} \cong \overline{PF}$.

ON YOUR OWN

Trace each line segment. Then construct a line segment congruent to it. 1–4. See margin.

1.
   E   F

2.

3.

4. Measure each line segment in Exercises 1–3 to the nearest tenth of a centimeter. Then measure each line segment that you constructed. How do these measures compare?

5. **Look Ahead** Suppose the line segments of the polygon above are extended. Which line segments do you think will never intersect? $\overline{AB}$ and $\overline{CD}$

186 Chapter 5 Geometry: Investigating Patterns

Additional Answers

1.
2.
3.
4. 3 cm, 4.5 cm, 1.7 cm; 3 cm, 4.5 cm, 1.7 cm; They are the same.

A circle can be divided into 360 equal sections. Each section is one degree. Degrees are used to measure angles. A protractor can be used to measure an angle in degrees or to draw an angle with a given degree measure.

TRY THIS

Work with a partner.

3 Use a protractor to measure an angle.
- Trace ∠ADC from the polygon on page 186. Extend sides $\overrightarrow{DA}$ and $\overrightarrow{DC}$. ($\overrightarrow{DA}$ means ray DA. When naming a ray, the endpoint must be given first, followed by any other point on the ray. A ray begins at the endpoint and continues in one direction from that point.)
- Place the protractor over the angle with the center point on vertex D and the horizontal line aligned with side $\overrightarrow{DC}$.
- Side $\overrightarrow{DC}$ should intersect the point on the protractor marked 0°.
- Using the scale that begins with 0°, count off the degrees until you reach side $\overrightarrow{DA}$.
- Read the measurement. To say, *the measure of angle ADC is 55 degrees*, we can write $m\angle ADC = 55°$.

4 Use a protractor to draw an angle with a measure of 115°.
- Draw a ray. Label the endpoint R and put a point labeled S on the ray.

- Align the protractor so that the center is at R and the horizontal line aligns with $\overrightarrow{RS}$.
- Find the scale containing 0° along $\overrightarrow{RS}$. Count along that scale until you reach 115°. Label this point Q.
- Draw $\overrightarrow{RQ}$. $m\angle QRS = 115°$.

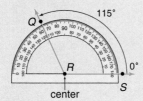

ON YOUR OWN

Use a protractor to draw an angle with the given measure.   6–8. See margin.
6. 70°   7. 90°   8. 175°

9. Trace polygon ABCD on page 186. Measure ∠BAC and ∠ACD. (*Hint:* Extend any rays that will help you to measure the angles.)   **25°, 25°**

10. **Look Ahead** In polygon ABCD on page 186, what is the relationship between ∠BAC and ∠ACD?   **They are congruent.**

Lesson 5-1A HANDS-ON LAB   **187**

 Have students write a paragraph telling how congruence with line segments is similar to congruence with angles. **To be congruent, the two figures must have the same measure.**

Additional Answers

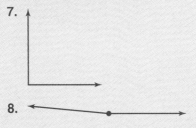

5-1 Lesson Notes

Instructional Resources
- *Study Guide Masters*, p. 34
- *Practice Masters*, p. 34
- *Enrichment Masters*, p. 34
- Transparencies 5-1, A and B
- CD-ROM Program
  - Resource Lesson 5-1
  - Interactive Lesson 5-1

Recommended Pacing	
Standard	Days 2 & 3 of 15
Honors	Days 1 & 2 of 15
Block	Day 1 of 8

1 FOCUS

5-Minute Check
(Chapter 4)

Use the following test scores for Exercises 1–3.

48 57 44 52 49 36
50 54 41 44 39 38

1. Find the mean, median, and mode for these 12 test scores. **mean: 46; median: 46; mode: 44**
2. Find the lower and upper quartiles. **40, 51**
3. Make a line plot for the test scores.

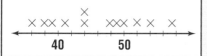

The 5-Minute Check is also available on **Transparency 5-1A** for this lesson.

Motivating the Lesson

Hands-On Activity Select a rectangular wall in the classroom. State the distance between the top and bottom edges at one end and ask students to estimate the distance at the other end. **the same** Ask why the distances are the same. **The floor and ceiling are parallel and therefore always the same distance apart.**

188 Chapter 5

5-1 Parallel Lines

What you'll learn
You'll learn to identify lines that are parallel and types of angles formed by parallel lines and transversals.

When am I ever going to use this?
Knowing how to identify types of angles formed by parallel lines can help you make proportional drawings.

Word Wise
parallel lines
transversal
alternate interior angles
alternate exterior angles
corresponding angles
vertical angles
supplementary angles

Artists must be able to draw people of various proportions. One method that artists use is to draw lines that divide a person's body or face into equal parts. Then they use line segment lengths and angle measures to draw the correct body proportions. The lines that divide the person into equal parts are called **parallel lines**.

Parallel lines are lines in a plane that will never intersect. If line p is parallel to line q, we write $p \parallel q$.

A line that intersects two or more other lines is called a **transversal**. The figure at the right shows a transversal, line ℓ, that intersects two parallel lines, so that eight angles are formed.

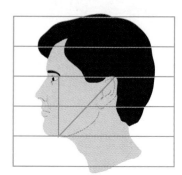

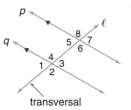

The red arrowheads on the lines p and q mean that $p \parallel q$.

HANDS-ON MINI-LAB

Work with a partner.   straightedge   protractor
colored pencils   notebook paper

Try This
- Draw two parallel lines using the lines on your notebook paper. Draw a third line that intersects these two lines.
- Label the angles formed as shown. Measure each angle and record its measurement.
- Use different colored pencils to circle the numbers of all of the angles that are congruent. *Congruent angles have the same measure.*

Talk About It
1. List the pairs of congruent angles in which one angle is formed by the transversal and one of the parallel lines and the other angle is formed by the transversal and the other parallel line.

1. ∠1, ∠5; ∠4, ∠8;
   ∠2, ∠6; ∠3, ∠7;
   ∠4, ∠6; ∠3, ∠5;
   ∠1, ∠7; ∠2, ∠8

188 Chapter 5 Geometry: Investigating Patterns

Cross-Curriculum Cue

Inform the other teachers on your team that your classes are studying geometry. Suggestions for curriculum integration are:
Art: the golden rectangle
Social Studies: maps, flags
Industrial Technology: shapes in metal working and woodworking

2. Which pairs of congruent angles are on the same side of the transversal? ∠1, ∠5; ∠4, ∠8; ∠2, ∠6; ∠3, ∠7
3. Which pairs of congruent angles are on opposite sides of the transversal? **See margin.**
4. Using your results from Exercises 1-3, describe the locations of the sets of congruent angles in relationship to the transversal and the parallel lines. **See margin.**

Congruent angles formed by parallel lines and a transversal have special names.

Study Hint
Reading Math The symbol ≅ means *is congruent to*.

Congruent Angles with Parallel Lines

If a pair of parallel lines is intersected by a transversal, these pairs of angles are congruent.

alternate interior angles:
∠4 ≅ ∠6, ∠3 ≅ ∠5

alternate exterior angles:
∠1 ≅ ∠7, ∠2 ≅ ∠8

corresponding angles:
∠1 ≅ ∠5, ∠2 ≅ ∠6, ∠3 ≅ ∠7, ∠4 ≅ ∠8

Examples

In the figure, $p \parallel q$ and $m\angle 1 = 85°$.

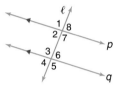

❶ Find $m\angle 3$.

∠1 and ∠3 are corresponding angles, so they are congruent.
$m\angle 3 = m\angle 1$
$m\angle 3 = 85°$

❷ Find $m\angle 7$.

∠3 and ∠7 are alternate interior angles, so they are congruent.
$m\angle 7 = m\angle 3$
$m\angle 7 = 85°$

INTEGRATION

❸ Algebra Find the value of x if $m\angle 7 = (3x - 10)°$, $m\angle 8 = 110°$, and $r \parallel s$.

∠7 and ∠8 are alternate interior angles formed by parallel lines r and s and a transversal. These angles are congruent, so their measures are equal. You can write an equation.

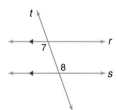

$m\angle 7 = m\angle 8$
$3x - 10 = 110$
$\quad 3x = 120$ *Add 10 to each side.*
$\quad\ \ x = 40$ *Divide each side by 3.*

The value of x is 40.

Lesson 5-1 Parallel Lines   189

2 TEACH

  Transparency 5-1B contains a teaching aid for this lesson.

Using the Mini-Lab Assist students in correctly identifying congruent angles formed by parallel lines intersected by a transversal. Point out the zigzag pattern of congruent angles in these figures.

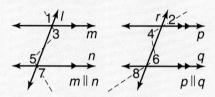

Teaching Tip In Example 1, point out that p and q are parallel even though they are not vertical or horizontal lines.

In-Class Examples

In the figure, $h \parallel k$ and $m\angle 2 = 50°$.

For Example 1
Find $m\angle 6$. **50°**

For Example 2
Find $m\angle 4$. **50°**

For Example 3
Find the value of z if $m\angle 1 = 124°$, $m\angle 5 = 3z - 14°$, and $h \parallel k$. **46**

Additional Answers for the Mini-Lab

3. ∠3, ∠5; ∠4, ∠6; ∠2, ∠8; ∠2, ∠4; ∠1, ∠3; ∠1, ∠7; ∠5, ∠7; ∠6, ∠8

4. Congruent angles are outside the parallel lines and on opposite sides of the transversal, are in the same exact location at each intersection of a parallel line and the transversal, and are between the parallel lines and on opposite sides of the transversal.

Using the Mini-Lab Guide students in adding the measure of various angles together. Remind students that the angles along a straight line add up to 180° and that all of the angles in a circle add up to 360°.

In-Class Examples

In the figure, $m\angle 1 = 45°$.

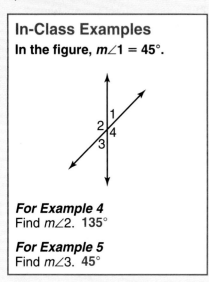

For Example 4
Find $m\angle 2$. 135°

For Example 5
Find $m\angle 3$. 45°

Additional Answers for the Mini-Lab

2b. Yes; $\angle 1, \angle 2$; $\angle 1, \angle 4$; $\angle 3, \angle 4$; $\angle 5, \angle 6$; $\angle 5, \angle 8$; $\angle 6, \angle 7$; $\angle 7, \angle 8$; the angles are found next to each other.

3. If lines intersect, then opposite angles are congruent and adjacent angles are supplementary.

HANDS-ON MINI-LAB

Work with a partner.   straightedge   protractor

Try This

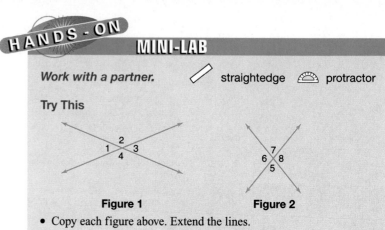

Figure 1   Figure 2

- Copy each figure above. Extend the lines.
- Measure and then record the measures of angles 1-8.

Talk About It

1a. In each figure, name the pairs of angles that are congruent.
 b. Describe their position to each other.
2a. What is the sum of the measures of $\angle 2$ and $\angle 3$? **180°**
 b. Are there other pairs of angles in the figures with the same sum? If so, list them and describe where each of the pairs of angles is found in the diagrams. See margin.
3. Using your results from Exercises 1 and 2, write a summary about the angles formed by two intersecting lines. See margin.

1a. $\angle 1, \angle 3$; $\angle 2, \angle 4$; $\angle 5, \angle 7$; $\angle 6, \angle 8$
1b. These angles are non-adjacent angles formed by intersecting lines.

Angles formed by intersecting lines have special relationships.

Straight angles have measures equal to 180°.

| **Vertical Angles and Supplementary Angles** | **Vertical angles** are opposite angles formed by the intersection of two lines. Vertical angles are congruent. ($\angle 1 \cong \angle 3$, $\angle 2 \cong \angle 4$) **Supplementary angles** are two angles whose measures have a sum of 180°. ($\angle 1$ and $\angle 2$ are an example of supplementary angles.) |

Examples

In the figure, $m\angle 1 = 65°$.

4 Find $m\angle 2$.

$\angle 1$ and $\angle 2$ are supplementary angles, so $m\angle 1 + m\angle 2 = 180°$.
$65 + m\angle 2 = 180$
$m\angle 2 = 115°$

5 Find $m\angle 3$.

$\angle 1$ and $\angle 3$ are vertical angles, so they are congruent.
$m\angle 3 = m\angle 1$
$m\angle 3 = 65°$

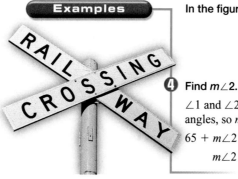

190   Chapter 5 Geometry: Investigating Patterns

💡 **Investigations for the Special Education Student**

This blackline master booklet helps you plan for the needs of your special education students by providing long-term projects along with teacher notes. Investigation 10, *Hide and Seek,* may be used with this chapter.

CHECK FOR UNDERSTANDING

Communicating Mathematics

Read and study the lesson to answer each question. 1–2. See margin.

1. *Write* a definition for parallel lines.
2. *Give* two examples of parallel lines in your surroundings.

3. *Draw* two parallel lines using the lines on your notebook paper. Draw a third line that intersects these lines.

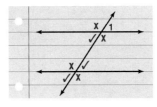

   a. Trace ∠1 on a piece of tracing paper. Use the traced angle to determine which angles are congruent to ∠1. Place a ✓ in each angle that is congruent to ∠1.
   b. Trace one of the angles that is not marked. Use the traced angle to determine which angles are congruent to this angle. Place an X in each of these angles.

Guided Practice

Use the figure at the right for Exercises 4–5.

4. Find $m\angle 5$ if $m\angle 7 = 95°$. **95°**
5. Find $m\angle 7$ if $m\angle 4 = 110°$. **110°**

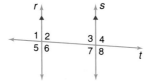

Find the measure of each angle in the figure if $p \parallel q$ and $m\angle 7 = 60°$.

6. $m\angle 1$ **60°**
7. $m\angle 8$ **120°**

8. *Algebra* Find the value of x in the figure below. **22**

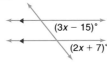

9. *Furniture* A single piece of wood is used for both the backrest of a chair and its rear legs. If the inside angle that the wood makes with the floor is 100° and the seat is parallel to the floor, what are the values of x and y? **100; 80**

EXERCISES

Practice

Use the figure at the right for Exercises 10–15.

10. Find $m\angle 6$ if $m\angle 2 = 35°$. **35°**
11. Find $m\angle 7$ if $m\angle 6 = 45°$. **135°**
12. Find $m\angle 3$ if $m\angle 5 = 77°$. **77°**
13. Find $m\angle 1$ if $m\angle 3 = 138°$. **138°**
14. Find $m\angle 4$ if $m\angle 8 = 122°$. **122°**
15. Find $m\angle 7$ if $m\angle 1 = 68°$. **68°**

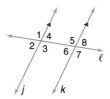

Lesson 5-1 Parallel Lines **191**

■ Reteaching the Lesson ■

Activity Each group of 3 students uses a spinner numbered 1–8 and the figure from Examples 1 and 2. One player spins, names an angle in the figure that is congruent to the angle whose number was spun, and tells why the angles are congruent. The player scores one point for each correct answer. After five rounds, the player with the highest score wins.

3 PRACTICE/APPLY

Check for Understanding

If students need additional practice or instruction after completing Exercises 1–9, one of these options may be helpful.
- Extra Practice, see p. 617
- Reteaching Activity
- *Study Guide Masters*, p. 34
- *Practice Masters*, p. 34

Assignment Guide

Core: 11–27 odd, 28–32
Enriched: 10–24 even, 26–32

Additional Answers

1. Parallel lines are lines that are the same distance apart and never meet.
2. Sample answers: top and bottom lines of a chalkboard, yard lines on a football field

Study Guide Masters, p. 34

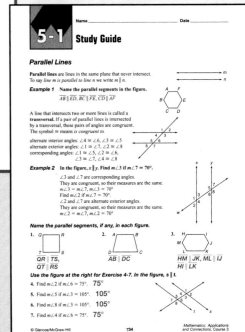

Lesson 5-1 **191**

4 ASSESS

Closing Activity

Speaking Draw the figure from Guided Practice, Exercises 4 and 5, on the chalkboard. Name an angle. Have students name an angle that is congruent to it and tell why they are congruent. Repeat using other angles.

Additional Answers

26. Yes, the two corners at that intersection have measures of 108° and 72°. Therefore, it is just within the safety limits.

31.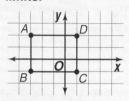

14 units

Practice Masters, p. 34

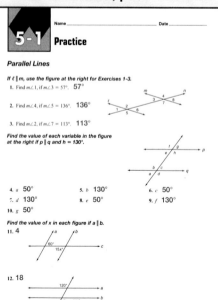

Find the measure of each angle in the figure if $s \parallel t$, $q \parallel r$, and $m\angle 8 = 75°$.

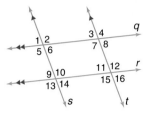

16. $m\angle 1$ 75°
17. $m\angle 3$ 75°
18. $m\angle 5$ 105°
19. $m\angle 7$ 105°
20. $m\angle 9$ 75°
21. $m\angle 11$ 75°

Find the value of x in each figure.

22. 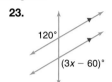 4
23. 60
24. 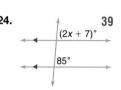 39

25. The measure of one angle formed by two intersecting lines is 90°. Find the measures of the other angles formed by these lines. 90°, 90°, 90°

Applications and Problem Solving

26. *Urban Planning* Ambulances can't safely make turns of less than 70°. The angle at the southeast corner of Delavan and Elmwood is 108°. Should the proposed site of the hospital emergency entrance at the northeast corner of Bidwell and Elmwood be approved? Explain your answer. See margin.

27. *Marching Band* During a performance, a marching band forms two intersecting lines. The measure of $\angle 1$ is 65°. If the band wants their lines to be straight, what should the measure of $\angle 2$ be? 115°

28. $m\angle S = 70°$, $m\angle T = 110°$, $m\angle U = 70°$

28. *Critical Thinking* The figure at the right is a parallelogram. It has two pairs of parallel sides. If $m\angle R = 110°$, find the measures of the other angles of the parallelogram. (*Hint:* Extend the sides of the parallelogram.)

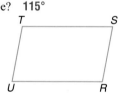

Mixed Review

29. **Test Practice** Which of the following would be the best place to find a representative sample for a survey of favorite types of music? *(Lesson 4-8)* C

   A a jazz club
   B a symphony performance
   C a mall
   D a high school dance
   E a retirement center

30. *Algebra* Solve $\frac{2}{n} = \frac{7}{98}$. *(Lesson 3-3)* 28

31. *Geometry* Graph the points $A(-3, 2)$, $B(-3, -1)$, $C(1, -1)$, and $D(1, 2)$ on the same coordinate plane. Draw $\overline{AB}$, $\overline{BC}$, $\overline{CD}$, and $\overline{AD}$. Find the perimeter of the rectangle formed. *(Lesson 2-10)* See margin.

32. Find the value of 5^3. *(Lesson 1-2)* 125

192 Chapter 5 Geometry: Investigating Patterns

Extending the Lesson

Enrichment Masters, p. 34

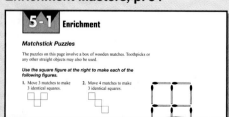

Activity Have students work in small groups to solve the following problem: *In how many points can three lines intersect? Draw sketches to illustrate your answer.*

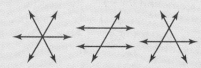

COOPERATIVE LEARNING

5-1B Constructing Parallel Lines

A Follow-Up of Lesson 5-1

- compass
- straightedge

Before computers, navigators on ships and airplanes used a compass and a straightedge to plot their course. You can use a compass and a straightedge to construct a line parallel to a given line.

TRY THIS

Work with a partner.

Step 1 Draw a line and label it ℓ.

Step 2 Choose a point P not on line ℓ.

Step 3 Draw a line through point P so that it intersects line ℓ. Label the point of intersection point Q.

Step 4 Place the point of your compass at point Q and draw a large arc. Label the point where the arc crosses line ℓ as Point R, and label where it crosses line PQ as point S.

Step 5 With the same compass opening, place the compass point at P and draw a large arc. Label the point of intersection with line PQ as point T.

Step 6 Use your compass to measure the distance between points S and R.

Step 7 With the compass opened the same amount, place the compass point at point T and draw an arc to intersect the arc already drawn. Label this point of intersection point U.

Step 8 Draw a line through points P and U. Label this line m. You have drawn $\ell \parallel m$.

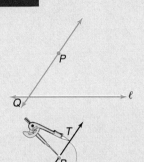

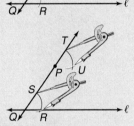

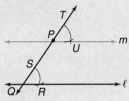

ON YOUR OWN

Trace each line. Then construct a line parallel to it. 1–2. See Answer Appendix.

1.
2.

3. **Reflect Back** Examine your constructions. What type of angles did you use to create the parallel lines? Measure the angles. Are the angles congruent? **See margin.**

Lesson 5-1B HANDS-ON LAB **193**

Have students write a paragraph describing how to construct parallel lines using alternate exterior angles.

Additional Answer
3. corresponding angles; yes

HANDS-ON 5-1B LAB Notes

GET READY

Objective Students construct a line parallel to a given line.

Optional Resources
Hands-On Lab Masters
- worksheet, p. 46

Manipulative Kit
- compass

Overhead Manipulative Resources
- compass

MANAGEMENT TIPS

Recommended Time
25 minutes

Getting Started Ask students to list ways in which they could draw a line parallel to a line you draw on the chalkboard. Then ask how they would verify that the two lines are parallel. Give students time to practice using their compasses.

The **Activity** shows students how to use a compass and straightedge to construct a line parallel to a given line. After students complete the activity, ask them to name the pair of congruent angles used to create the parallel lines. ∠*TPU*, ∠*SQR* Then ask them how they could construct other lines parallel to ℓ and m.

ASSESS

Have students complete Exercises 1 and 2. Watch for students who have difficulty following the steps. Make sure they understand which points to use when drawing arcs.

Use Exercise 3 to determine whether students understand the relationship between parallel lines and congruent angles.

Hands-On Lab 5-1B **193**

5-2A THINKING LAB Notes

Objective Students use a Venn diagram to solve problems.

Recommended Pacing	
Standard	Day 4 of 15
Honors	Day 3 of 15
Block	Day 2 of 8

1 FOCUS

Getting Started Have students act out a situation similar to the one presented at the beginning of the lesson but pertaining just to one class. Then have them work in pairs to answer Exercises 1 and 2. Discuss the results as a class.

2 TEACH

Teaching Tip In Exercise 3, have students start with the information that only 11 customers like both mushroom and pepperoni pizza. They can then fill in the non-intersecting sections by subtracting the number in the intersection from the number who liked each pizza topping. Then the students can use these numbers to determine how many liked neither topping.

In-Class Example

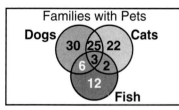

a. How many families have dogs and cats? **28 families**
b. How many families have both fish and dogs but no cats? **6 families**

THINKING LAB PROBLEM SOLVING

5-2A Use a Venn Diagram

A Preview of Lesson 5-2

The Spring Carnival committee at Barrington Middle School needs to hire a band. They decided to take a school survey to find the type of band that most of the students like. They are discussing the results. Let's listen in!

Will: Of the students who turned in surveys, 74 like country, 189 like rock, and 43 like rap. 55 students like rock and country, 12 like rock and rap, and 23 like rap and country. Only 8 students like all three types of music.

How many students turned in a survey?

Well, there are 74 + 189 + 43 + 55 + 12 + 23 + 8, or 404 responses in all.

That can't be right because there are only 325 students in our school.

But some of the surveys had more than one response. Let's make a Venn diagram to show the responses.

Dario:

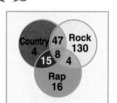

THINK ABOUT IT

Work with a partner. 1. See Answer Appendix.

1. **Explain** what the brown center region of the Venn diagram represents. How many students are represented by this region?

2. **Use** the Venn diagram to find how many students turned in a survey. **224**

3. Use a **Venn diagram** to solve the following problem. **See Answer Appendix.**

   Napoli's Pizza conducted a survey of 75 customers. The results showed that 35 people like mushroom pizza, 41 like pepperoni, and 11 like both mushroom and pepperoni pizza. How many like neither mushroom nor pepperoni pizza?

4. The Venn diagram below shows the types of books owned by Carter Middle School library. The circle for biographies inside of the nonfiction area indicates that all of the biographies are also nonfiction. Since the nonfiction and fiction areas do not intersect, there are no books that are both fiction and nonfiction. **State** the total number of nonfiction books. **675 books**

Carter Middle School Library

Fiction	Nonfiction 591
487	84 Biographies

194 Chapter 5 Geometry: Investigating Patterns

Reteaching the Lesson

Activity Out of 100 music students surveyed, 63 play piano and 49 play guitar. Have students draw and cut out two rectangles from grid paper to represent 63 and 49 units. Have them overlap the two rectangles so that only 100 grid squares show. Help students see that overlapped grid squares will tell them how many students play both instruments.

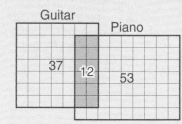

ON YOUR OWN

5. The last step of the 4-step plan for problem solving asks you to *examine* the solution. *Explain* how you can use your answer in Exercise 3 to confirm that the number of customers surveyed is accurately represented in your Venn diagram. **The numbers in the Venn diagram must have a sum of 75.**

6. *Write a Problem* that can be solved by using a Venn diagram. **See margin.**

7. *Look Ahead* Draw a Venn diagram showing the relationship of triangles that have at least two congruent sides and triangles that have all three sides congruent. **See Answer Appendix.**

MIXED PROBLEM SOLVING

Strategies
Look for a pattern.
Solve a simpler problem.
Act it out.
Guess and check.
Draw a diagram.
Make a chart.
Work backward.

Solve. Use any strategy.

8. *Marketing* The results of a supermarket survey showed that 83 customers chose wheat cereal, 83 chose rice cereal, and 20 chose corn cereal. Of those customers who bought exactly two boxes of cereal, 6 bought corn and wheat, 10 bought rice and corn, and 12 bought rice and wheat. Four customers bought all three. Make a Venn diagram of this information. **See Answer Appendix.**

9. *Advertising* The graph shows who spent money on radio advertising in 1995. **a. 4 times**

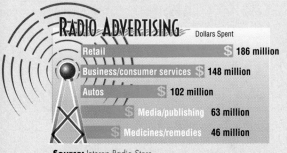

Source: Interep Radio Store

   a. About how many times more money was spent by retail companies than by the medicine/remedy companies?

   b. Sears spent 67 million dollars on radio advertising in 1995. How much more money did Sears spend than the entire medicine/remedy industry? **$21 million**

10. *Data Analysis* The school cafeteria surveyed 36 students about their dessert preference. The results are listed below. **a–c. See margin.**

Number of Students	Preference of Students
25	cake
20	ice cream
15	pie
2	all three
1	no desserts
15	cake or ice cream
8	pie or cake
3	ice cream only

   a. Draw a Venn diagram that will represent the responses.
   b. How many students prefer only pie? either pie or ice cream?
   c. What two desserts should the cafeteria order? Explain.

11. There are about 3,907,000 miles of roads in the United States. Twelve percent of these roads are in Texas and California. About how many miles of roads are located in these two states? **D**

   A less than 40,000 miles
   B between 50,000 and 100,000 miles
   C between 300,000 and 400,000 miles
   D between 400,000 and 500,000 miles
   E more than 500,000 miles

Lesson 5-2A THINKING LAB 195

Extending the Lesson

Activity Have students work in small groups to solve the following problem: *A survey of 100 tourists showed that 45 had visited France, 48 China, 26 Chile, 10 Chile and China, 6 Chile and France, 8 France and China, and 5 all three countries. How many had visited China, but not France or Chile?* **35**

3 PRACTICE/APPLY

Check for Understanding
Use the results from Exercise 4 to determine whether students comprehend how to use Venn diagrams to solve problems.

Extra Practice If students need additional practice in problem solving, extra practice is available on the following pages.
- Use a Venn Diagram, see p. 617
- Mixed Problem Solving, see pp. 645–646

Assignment Guide
All: 5–11

4 ASSESS

Closing Activity
Speaking Have students describe the best strategy for completing a Venn diagram. Ask them to explain why a Venn diagram is useful for the types of problems solved in the Thinking Lab. When items overlap, Venn diagrams can help sort them out.

Additional Answers
6. Sample answer: There are 26 students in a math class. The class takes a survey and finds that 14 students have pet dogs and 10 students have cats. Four students have both. How many students have neither a dog nor a cat?

10a.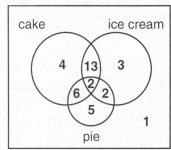

10b. 5 students; 4 students

10c. Cake and pie; there are only 4 students who do not like either pie or cake.

Thinking Lab 5-2A 195

5-2 Lesson Notes

Instructional Resources
- *Study Guide Masters*, p. 35
- *Practice Masters*, p. 35
- *Enrichment Masters*, p. 35
- Transparencies 5-2, A and B
- *Assessment and Evaluation Masters*, p. 127

 CD-ROM Program
- Resource Lesson 5-2

Recommended Pacing	
Standard	Day 5 of 15
Honors	Day 4 of 15
Block	Day 2 of 8

1 FOCUS

5-Minute Check
(Lesson 5-1)

Use the figure for Exercises 1–2. In the figure, $a \parallel b$, $m\angle 3 = 120°$, and $m\angle 8 = 60°$.

1. Find $m\angle 7$. **120°**
2. Find $m\angle 6$. **60°**
3. Find the value of x if $m\angle 1 = 120°$ and $m\angle 5 = (4x + 16)°$. **26**

The 5-Minute Check is also available on **Transparency 5-2A** for this lesson.

Motivating the Lesson

Hands-On Activity Ask students to find examples of triangles in the classroom. List these on the chalkboard. Have students describe similarities and differences among the triangles.

5-2 Classifying Triangles

What you'll learn
You'll learn to classify triangles by their angles and their sides and to find measures of missing angles in triangles.

When am I ever going to use this?
Knowing how to classify triangles can help you describe bicycle frames.

Word Wise
polygon
triangle
acute
right
obtuse
scalene
isosceles
equilateral
perpendicular
complementary angles

Cultural Kaleidoscope

Because of its strength and beauty, marble has been valued throughout the ages. The ancient Greeks used marble for sculptures and buildings.

A **polygon** is a simple closed figure in a plane formed by three or more line segments. A **triangle** is a polygon formed by three line segments. Triangles can be classified by their angles and by their sides.

Examine the Eiffel Tower at the right and locate as many different-shaped triangles as you can. You can see that all of the triangles have at least two angles that are less than 90°. The third angle and its measure determine the classification of a triangle.

Acute angles have measures less than 90°.
Right angles have measures equal to 90°.
Obtuse angles have measures greater than 90°, but less than 180°.

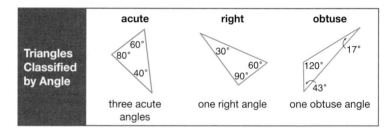

Triangles can also be classified by the number of congruent sides.
Congruent sides are often marked with a slash through the sides.

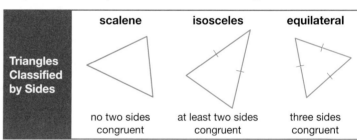

These side classifications can be organized in a Venn diagram to show their relationship.

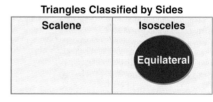

196 Chapter 5 Geometry: Investigating Patterns

Classroom Vignette

"If a computer is available, students can use software like *The Geometric Supposer: Triangles* or LOGO for more practice in classifying triangles. Working in pairs, one student creates a triangle while the other classifies it and justifies their answer. Then they switch roles."

Clare L. Polzin

Clare L. Polzin, Teacher
Red Rock Central Junior High School
Lamberton, MN

Study Hint
Reading Math △HIJ is read as triangle HIJ.

Triangles are named by letters at their vertices. The triangle at the right can be named △HIJ.
A vertex is the point where two sides meet.

In Example 1, the ⌐ symbol in △EFG indicates that ∠E is a right angle. When segments meet to form right angles, they are **perpendicular**.

Examples

Classify each triangle by its angles and by its sides.

1

△EFG is a right, isosceles triangle.

2

△XYZ is an acute, equilateral triangle.

The sum of the measures of the three angles of a triangle is 180°. You can check this by folding the angles of a triangle as shown in the diagram. Since the three angles together form a straight angle, we know that they add up to 180°.

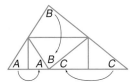

Example
INTEGRATION

3 Algebra Find the measure of each angle in △ABC if ∠C is a right angle, $m\angle A = (x + 50)°$, and $m\angle B = (x + 10)°$.

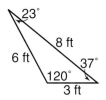

Explore   You can draw and label a triangle to represent the problem. You know the expressions for two angles and that the third angle is a right angle. You need to know the measure of each angle.

Plan   Find the value of x by using the equation $m\angle A + m\angle B + m\angle C = 180$. Then substitute the value of x into $m\angle A = (x + 50)°$ and $m\angle B = (x + 10)°$, to find $m\angle A$ and $m\angle B$.

Solve

$$m\angle A + m\angle B + m\angle C = 180$$
$$(x + 50) + (x + 10) + 90 = 180$$
$$2x + 150 = 180 \quad \text{Add like terms.}$$
$$2x = 30 \quad \text{Subtract 150.}$$
$$x = 15 \quad \text{Divide by 2.}$$

By substitution, $m\angle A = 15 + 50$ or 65° and $m\angle B = 15 + 10$ or 25°. Therefore, the measures of the three angles of the triangle are 65°, 25° and 90°.

Examine   You know that the sum of the measures of the angles of a triangle is 180°. $65 + 25 + 90 = 180$ ✓

Lesson 5-2 Classifying Triangles **197**

---

2 TEACH

 Transparency 5-2B contains a teaching aid for this lesson.

Reading Mathematics Make sure that students know that there are two ways to classify triangles—by their sides and by their angles.

In-Class Examples
Classify each triangle by its angles and by its sides.

For Example 1

[triangle with 23°, 8 ft, 6 ft, 37°, 120°, 3 ft]

obtuse, scalene

For Example 2

right, isosceles

For Example 3
Find the value of x in △ABC if $m\angle A = 90°$, $m\angle B = 32°$, and $m\angle C = x°$. **58**

Teaching Tip In Example 2, point out that because the triangle has two congruent sides, it is isosceles as well as equilateral. By agreement, such a triangle is named by its more specific classification, equilateral.

3 PRACTICE/APPLY

Check for Understanding

If students need additional practice or instruction after completing Exercises 1–8, one of these options may be helpful.
- Extra Practice, see p. 618
- Reteaching Activity
- *Study Guide Masters*, p. 35
- *Practice Masters*, p. 35
- Interactive Mathematics Tools Software

Assignment Guide

Core: 9–23 odd, 24–27
Enriched: 10–20 even, 22–27

Additional Answers

1. Both isosceles triangles and equilateral triangles have congruent sides. An isosceles triangle has at least two sides congruent, and an equilateral triangle has three sides congruent.

3. Jaali is correct. A triangle has three angles whose sum is 180°. If a triangle had two obtuse angles, then the sum of two of the three angles would already be greater than 180°.

Study Guide Masters, p. 35

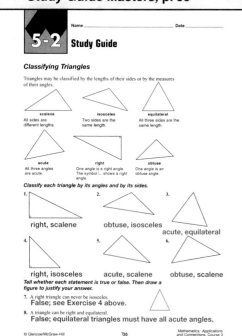

198 Chapter 5

The two acute angles in a right triangle have a special relationship. Since the sum of the measures of the angles in the triangle is 180° and the right angle is 90°, the sum of the two acute angles must be 180° − 90° or 90°. Two angles whose sum is 90° are **complementary angles**.

Complementary Angles	**Complementary angles** are two angles whose measures have a sum of 90°. ($\angle 1$ and $\angle 2$ are an example of complementary angles.)

CHECK FOR UNDERSTANDING

Communicating Mathematics

Read and study the lesson to answer each question. **1. See margin.**

1. *Compare and contrast* isosceles triangles and equilateral triangles.
2. *Describe* the types of angles that are in a right triangle.
3. *You Decide* Jaali and Adia were drawing triangles on paper and then describing them to each other. Adia told Jaali that she drew a triangle that had two obtuse angles. Jaali said that it couldn't be done. Who was correct? Explain.   See margin.

2. One angle is a 90° angle and the other two angles are each acute.

Guided Practice

Classify each triangle by its angles and by its sides.

4. right, isosceles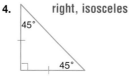

5. acute, scalene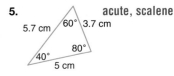

6. *Algebra* Find the measure of each angle in $\triangle RST$ if $m\angle R = x°$, $m\angle S = (x + 20)°$, and $m\angle T = 2x°$. $m\angle R = 40°$, $m\angle S = 60°$, $m\angle T = 80°$

7. *True* or *false?* A pair of angles can be both complementary and supplementary.   false

8b. $m\angle 1 = 32°$,
$m\angle 2 = 148°$,
$m\angle 3 = 64°$,
$m\angle 4 = 116°$,
$m\angle 5 = 32°$

8. In the diagram, $\overrightarrow{DE} \parallel \overrightarrow{AB}$, $m\angle DEC = 116°$, and $m\angle BAC = 32°$.
   a. Classify $\triangle ABC$ by its angles.   obtuse
   b. Find the measure of each numbered angle in the drawing.

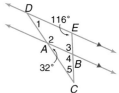

EXERCISES

Practice

Classify each triangle by its angles and by its sides.

9. obtuse, isosceles

10. obtuse, scalene

11.
acute, isosceles

198 Chapter 5 Geometry: Investigating Patterns

Reteaching the Lesson

Activity Label an index card for each of the seven possible angle/side classifications for triangles. Prepare seven more cards, each picturing one of those triangles. Shuffle the cards, arrange them facedown, and have pairs of students play a memory game. Each student in turn reveals two cards. If the cards match, the student collects the cards and gets another turn. If not, the cards are turned facedown again and another student has a turn. The student with the most cards collected at the end wins.

Classify each triangle by its angles and by its sides.

12.
acute, equilateral

13.
acute, isosceles

14.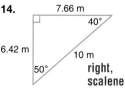
right, scalene

15. **Algebra** Find the value of x in $\triangle XYZ$ if $m\angle X = 2x°$, $m\angle Y = 64°$, and $m\angle Z = (2x - 16)°$. **33**

16. **Algebra** Find the measure of each angle in $\triangle EFG$ if $\angle E$ is a right angle, $m\angle F = (2x + 5)°$, and $m\angle G = (x + 25)°$.

16. $m\angle E = 90°$, $m\angle F = 45°$, $m\angle G = 45°$

17. **Algebra** Find the measure of each angle in $\triangle ABC$ if $m\angle A = 65°$, $m\angle B = 3x°$, and $m\angle C = (x + 15)°$. $m\angle A = 65°$, $m\angle B = 75°$, $m\angle C = 40°$

Tell whether each statement is *true* or *false*. Then draw a figure to justify your answer. 18–19. See margin for drawings.

18. A triangle can have three acute angles. **true**

19. An obtuse isosceles triangle has two acute angles. **true**

20. What type of triangle has one pair of perpendicular sides? **right triangle**

21. An equiangular triangle has three congruent angles. What is the measure of each angle of an equiangular triangle? **60°, 60°, 60°**

Applications and Problem Solving

22. **Bicycles** Bicycles with small frames have three tubes in the center that form a triangle. Classify the triangle used in the bike frame by its sides. **scalene**

23. **Architecture** Triangles are used to stabilize buildings and bridges. Find a photograph in a newspaper or magazine that shows triangles used in architecture. Classify the triangles. **See students' work.**

24. **Critical Thinking** Use toothpicks, straws, or pretzel sticks to build the figure below. Remove the stated number of sticks to get the specific number of triangles. Draw your figure. **a–d. See Answer Appendix.**

Remove	Number of Triangles
a. 2	3
b. 2	2
c. 3	1
d. 3	2

Mixed Review

25. **Geometry** Find the value of x in the figure at the right. (Lesson 5-1) **40**

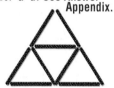

26. Solve $n = -28 + 73$. (Lesson 2-3) **45**

27. **Test Practice** If Rodolfo worked a total of 36 hours over a 4-day period and was paid $5.35 per hour, how much did he earn before deductions? (Lesson 1-5) **B**

A $770.40   B $192.60   C $182.80   D $21.40   E Not Here

Lesson 5-2 Classifying Triangles **199**

Extending the Lesson

Enrichment Masters, p. 35

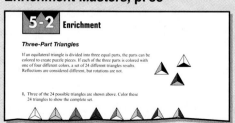

Activity Ask students to find triangles in their surroundings and to suggest other ways triangles are used. **Sample answers: access ramps, building supports, layout of walkways, escalators; in architecture, navigation, surveying land**

4 ASSESS

Closing Activity

Modeling Have students use toothpicks to illustrate each of the following:
- an acute, scalene triangle
- a right, isosceles triangle
- an obtuse, scalene triangle
- an equilateral triangle

Chapter 5, Quiz A (Lessons 5-1 and 5-2) is available in the *Assessment and Evaluation Masters*, p. 127.

Additional Answers

18.

19.

Practice Masters, p. 35

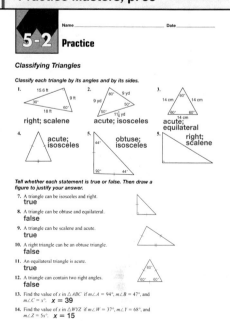

Lesson 5-2 **199**

5-3A Hands-On Lab Notes

GET READY

Objective Students determine whether a network is traceable.

Optional Resources
Hands-On Lab Masters
- worksheet, p. 47

Overhead Manipulative Resources
- regular polygon transparency

Manipulative Kit
- ruler

MANAGEMENT TIPS

Recommended Time
25 minutes

Getting Started Ask students to draw a grid of streets in a neighborhood and imagine the route a mail carrier must travel. Have them attempt to trace on the grid a route that will reach each street without covering any street more than once. They must be sure both sides of all streets are covered. Explain that networks are studied to increase efficiency in such situations.

The **Activity** explores the traceability of a figure. To trace a figure, have students copy it onto a sheet of paper and use a different color ink to trace it.

ASSESS

Have students complete Exercises 1–6. Watch for students who may have difficulty determining traceability. Advise them to try several starting points, marking off vertices as they are eliminated.

Use Exercise 7 to determine whether students have discovered a pattern for determining traceability.

Hands-On Lab

COOPERATIVE LEARNING

5-3A Polygons as Networks

A Preview of Lesson 5-3

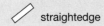 straightedge

When people talk about networking, they usually mean that they are making contact with other people and exchanging ideas and information. In mathematics, a *network* is a figure consisting of points, called *nodes*, and paths that join various nodes to one another, called *edges*.

TRY THIS

Work with a partner.

Step 1 Make a copy of each polygon. Draw the diagonals from one vertex as shown.

A diagonal is a segment that connects any two nonconsecutive vertices.

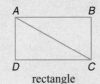 rectangle   pentagon   hexagon

Step 2 Label the vertices of each polygon as shown in the rectangle. These figures can be considered networks. The vertices are the nodes, and the line segments are the edges.

Step 3 Vertices or nodes can be called *odd* or *even*. A vertex is odd if it has an odd number of line segments meeting at the vertex. An even vertex has an even number of line segments meeting at the vertex. For each figure, examine the vertices. Which vertices are odd and which are even?

Step 4 A figure is considered *traceable* if you can trace it without covering any line segment twice, and without taking your pencil off the paper. For example, one way the rectangle can be traced is by following *ABCADC*. Therefore, it is traceable. Determine whether the pentagon and hexagon are traceable.

ON YOUR OWN

Draw a heptagon (7-sided polygon) and an octagon (8-sided polygon) and their diagonals from one vertex. Then complete the table.

	Name of Polygon	Number of Odd Vertices	Traceable or Not Traceable
1.	rectangle	2	traceable
2.	pentagon	2	traceable
3.	hexagon	4	not traceable
4.	heptagon	4	not traceable
5.	octagon	6	not traceable

6. Examine your answers for Exercises 1–5. Make a conjecture about traceability.

7. **Look Ahead** Will any four-sided figure with one diagonal have the same traceable result as a rectangle? Explain. **6–7. See margin.**

200 Chapter 5 Geometry: Investigating Patterns

Additional Answers

6. If the number of odd vertices is two, then the figure is traceable. You must begin at one of the odd vertices and end at the other.

7. Yes; any 4-sided figure with one diagonal will have 2 odd vertices.

 Math Journal Have students write a paragraph explaining what a network represents. Ask them to describe a traceable network. You may wish to have them include an example of a network in everyday life.

5-3 Classifying Quadrilaterals

What you'll learn
You'll learn to classify quadrilaterals.

When am I ever going to use this?
Knowing how to identify quadrilaterals can help you design room layouts.

Word Wise
quadrilateral
parallelogram
rectangle
rhombus
square
trapezoid

Have you ever noticed that members of the same family often have the same characteristics? Families of polygons also share certain characteristics. Each member of the **quadrilateral** family has four sides and four angles. Each member of the family has additional characteristics or attributes that distinguish it from other family members.

 A **parallelogram** is a quadrilateral with two pairs of opposite sides that are parallel.

 A **rectangle** is a parallelogram with four right angles.

 A **rhombus** is a parallelogram with all sides congruent.

 A **square** is a parallelogram with all sides congruent and four right angles.

 A **trapezoid** is a quadrilateral with exactly one pair of opposite sides that are parallel.

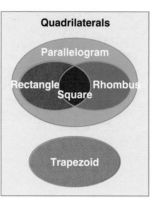

Quadrilaterals are named by the letters at their vertices, in consecutive order. The figure at the right can be called parallelogram ABCD.

Examples

Identify all of the names that describe each quadrilateral.

1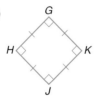

Quadrilateral GHJK is a square. It is also a rhombus, a rectangle, and a parallelogram. When a quadrilateral can fit into several types of definitions, we call it by the name that tells the most about it. In this case, the quadrilateral is called a square.

2

Quadrilateral QRST is a trapezoid. A trapezoid in which the two nonparallel sides are congruent is called an *isosceles trapezoid*.

Lesson 5-3 Classifying Quadrilaterals **201**

Multiple Learning Styles

Kinesthetic Have students work in groups to design and build simple kites using the quadrilaterals discussed in the lesson. Have the class vote for the most interesting kite and the kite most likely to fly well.

5-3 Lesson Notes

Instructional Resources
- *Study Guide Masters*, p. 36
- *Practice Masters*, p. 36
- *Enrichment Masters*, p. 36
- Transparencies 5-3, A and B
- *Classroom Games*, pp. 13–16
- *Hands-On Lab Masters*, p. 72
- *Technology Masters*, p. 61
- CD-ROM Program
  - Resource Lesson 5-3

Recommended Pacing
Standard	Days 6 & 7 of 15
Honors	Days 5 & 6 of 15
Block	Day 3 of 8

1 FOCUS

5-Minute Check
(Lesson 5-2)

Classify each triangle by its angles and by its sides.

1.   right, scalene

2. 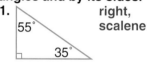 obtuse, isosceles

3. Find the value of *x* in △ABC if m∠A = 47°, m∠B = x°, and m∠C = 38°. 95

The 5-Minute Check is also available on **Transparency 5-3A** for this lesson.

Motivating the Lesson

Communication Draw these figures on the chalkboard.

Have the students compare and contrast these quadrilaterals. Guide them to consider congruence of angles and sides and to look for parallel lines and right angles.

Lesson 5-3 **201**

2 TEACH

 Transparency 5-3B contains a teaching aid for this lesson.

In-Class Examples
Identify all of the names that describe each quadrilateral.

For Example 1

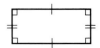

parallelogram, rectangle

For Example 2

parallelogram, rhombus, square, rectangle

Using the Mini-Lab To show that the sum of the measures of the angles in a quadrilateral is 360°, divide the figures into two triangles by drawing one of the diagonals. Point out that the sum of the measures of the angles of the quadrilateral is the same as the sum of the measures of the angles in the two triangles, which is 180° + 180°, or 360°.

Teaching Tip In Example 3, emphasize that the measurement of all four angles combined must be 360°.

In-Class Example
For Example 3
In quadrilateral $ABCD$, $m\angle A = 90°$, $m\angle B = 90°$, and $m\angle C = 120°$. Find $m\angle D$. 60°

We know about quadrilaterals and their sides. Let's explore quadrilaterals and their angles.

 **MINI-LAB**

Work in groups of four. 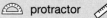 protractor ruler

Try This
- Draw a set of the six quadrilaterals described on page 201.
- Measure the angles of each figure and record them on the figure.
- Measure the sides of each figure and record them on the figure.
- Measure and record the lengths of the diagonals of each figure.

Talk About It
1. Copy and complete the table.

	Quadrilateral	Parallelogram	Trapezoid	Rectangle	Rhombus	Square
Opposite sides parallel?	no	yes	one pair	yes	yes	yes
Opposite sides congruent?	no	yes	no	yes	yes	yes
All sides congruent?	no	no	no	no	yes	yes
Diagonals congruent?	no	no	no	yes	no	yes
All right angles?	no	no	no	yes	no	yes
Opposite angles congruent?	no	yes	no	yes	yes	yes

2. What is the sum of the angle measures for each figure? **360°**

The Mini-Lab suggests that the sum of the measures of the angles of a quadrilateral is 360°.

Example 3 INTEGRATION

Algebra Find the value of x in the rhombus.

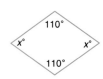

Explore You know that the figure is a rhombus and a rhombus is a quadrilateral. The angle measures are 110°, $x°$, 110°, and $x°$. You need to know the value of x.

Plan Write and solve an equation to find the value of x. The sum of the measures of the angles is 360°.

Solve
$110 + x + 110 + x = 360$
$220 + 2x = 360$ *Add like terms.*
$2x = 140$ *Subtract 220 from each side.*
$x = 70$ *Divide each side by 2.*

The value of x is 70.

202 Chapter 5 Geometry: Investigating Patterns

> *Examine* You know the sum of the measures of the angles of a quadrilateral is 360°.
>
> $110 + 70 + 110 + 70 = 360$ ✓

CHECK FOR UNDERSTANDING

Communicating Mathematics

Read and study the lesson to answer each question.

1. *Tell* what characteristics all of the members of the quadrilateral family share.
2. *Explain* why a parallelogram is not considered a trapezoid. **See margin.**

HANDS-ON MATH

1. They each have four sides and four angles.

3. *Draw* an isosceles trapezoid. Measure the angles, lengths of the sides, and lengths of the diagonals. Record them on the figure. **See students' work.**
   a. Are the diagonals congruent? **yes**
   b. Which angles are congruent? **See margin.**

Guided Practice

Sketch each figure. Let Q = quadrilateral, P = parallelogram, R = rectangle, S = square, RH = rhombus, and T = trapezoid. Write all of the letters that describe it inside the figure.

4.  Q, P, R

5.  Q, T

Tell whether each statement is *true* or *false*. Then draw a figure to justify your answer. **6–7. See margin for drawings.**

6. Every square is a parallelogram. **true**
7. The diagonals of a rectangle are perpendicular. **false**
8. *Algebra* In trapezoid WXYZ, $m\angle W = 2a°$, $m\angle X = 40°$, $m\angle Y = 110°$, and $m\angle Z = 70°$. Find the value of a. **70**

EXERCISES

Practice

Sketch each figure. Let Q = quadrilateral, P = parallelogram, R = rectangle, S = square, RH = rhombus, and T = trapezoid. Write all of the letters that describe it inside the figure.

9. Q, P, R, S, RH

10. Q, P

11. Q, P, RH

12. Q, T

13. Q

14. Q

Lesson 5-3 Classifying Quadrilaterals **203**

---

3 PRACTICE/APPLY

Check for Understanding
If students need additional practice or instruction after completing Exercises 1–8, one of these options may be helpful.
- Extra Practice, see p. 618
- Reteaching Activity
- *Study Guide Masters*, p. 36
- *Practice Masters*, p. 36

Assignment Guide
Core: 9–21 odd, 22–25
Enriched: 10–18 even, 20–25
All: Self Test 1–5

Additional Answers

2. A parallelogram has two pairs of opposite sides parallel, and a trapezoid has only one pair of opposite sides parallel.

3b. The angles at each end of either of the parallel sides are congruent to each other.

6.

7.

Study Guide Masters, p. 36

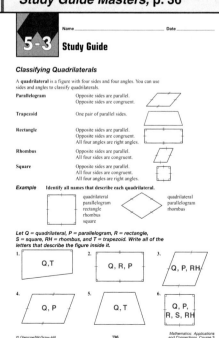

Reteaching the Lesson

Activity Have students work in pairs to find examples of quadrilaterals in the classroom. They should describe each example, measure its sides and angles, record each measurement, and identify each type.

Error Analysis
Watch for students who confuse two or more types of quadrilaterals.
Prevent by having students make flashcards in the shapes of the various types, with all of the characteristics for each type listed on the reverse.

Lesson 5-3 **203**

4 ASSESS

Closing Activity
Speaking Describe the properties of a particular type of quadrilateral. For example, *The opposite sides are parallel. There are four congruent sides and four right angles.* Have the students identify the quadrilateral.

Mid-Chapter Self Test
The Mid-Chapter Self Test reviews the concepts in Lessons 5-1 through 5-3. Lesson references are given so students can review concepts not yet mastered.

Additional Answers for the Self Test

4.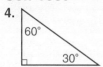

5.

Practice Masters, p. 36

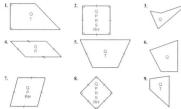

Tell whether each statement is *true* or *false*. Then draw a figure to justify your answer. 15–18. See Answer Appendix for drawings.

15. A rhombus is a square. **false**
16. A trapezoid can have only one right angle. **false**
17. The diagonals of a square cut each other in half. **true**
18. The diagonals of a square cut opposite angles in half. **true**

19. Name all of the quadrilaterals that have congruent diagonals. **rectangle, square**

Applications and Problem Solving
20. trapezoids, squares, and rhombuses
21b. $m\angle A = 70°$, $m\angle B = 110°$, $m\angle C = 70°$, $m\angle D = 110°$

20. **Design** The stained glass window at the right uses quadrilaterals in the design. Identify the quadrilaterals.

21. **Algebra** In parallelogram $ABCD$, $m\angle B = (x + 30)°$ and $m\angle D = (2x - 50)°$.
   a. Solve for x. **80**
   b. Find the measure of each angle in parallelogram $ABCD$.

22. **Critical Thinking** I am a quadrilateral. I have one pair of parallel sides. My longest side is twice as long as each of the other sides. My perimeter is 20 centimeters.
   a. What quadrilateral am I? **isosceles trapezoid**
   b. Draw me and label the lengths of my sides. **See Answer Appendix.**

Mixed Review

23. **Algebra** Find the value of x in $\triangle RST$ if $m\angle R = 54°$, $m\angle S = 90°$, and $m\angle T = 3x°$. *(Lesson 5-2)* **12**

24. Solve $d = 28(-12)$. *(Lesson 2-7)* **−336**

25. **Test Practice** Jamilah paid $42.83 for a sweater, which includes $2.78 sales tax. What was the price of the sweater before tax? *(Lesson 1-4)* **A**

   A $40.05   B $42.55   C $42.61   D $45.61   E Not Here

CHAPTER 5 Mid-Chapter Self Test

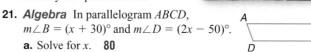

In the diagram $n \parallel p$.

1. Find the value of x. *(Lesson 5-1)* **64**
2. Find $m\angle ABC$. *(Lesson 5-1)* **116°**
3. Classify $\triangle BCD$ by its angles and by its sides. *(Lesson 5-2)* **acute, scalene**

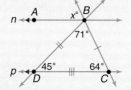

Tell whether each statement is *true* or *false*. Then draw a figure to justify your answer.

4. If the sum of the measures of two angles of a triangle is 90°, it must be an isosceles right triangle. *(Lesson 5-2)* **False; see margin for drawing.**

5. All sides of a parallelogram are congruent. *(Lesson 5-3)* **False; see margin for drawing.**

204 Chapter 5 Geometry: Investigating Patterns

Extending the Lesson

Enrichment Masters, p. 36

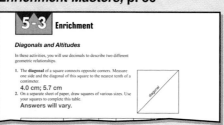

Activity The flag of Djibouti is made up of two trapezoids and an isosceles triangle. Have students find illustrations of the flags of other countries. Ask students to sketch several flags and identify the different geometric shapes in them.

COOPERATIVE LEARNING

5-4A Reflections

A Preview of Lesson 5-4

 geoboards

geobands

If you look into a mirror, you see yourself reflected in the mirror. It appears that both you and your reflected image are the same distance from the mirror. When an image is reflected, it is "flipped" over the reflection line. The mirror is the line of reflection.

TRY THIS

Work with a partner.

Step 1 Place two geoboards next to each other. Use a geoband to create the figure shown.

Step 2 The edge where the two boards meet will be the line of reflection. Vertex A is 2 pegs left of the line of reflection. Find a peg on the same row that is 2 pegs right of the line of reflection. Place a geoband around this peg.

Step 3 Vertex B is 5 pegs left of the line. Find a peg on the same row that is 5 pegs right of the line. Stretch the geoband to fit around this peg.

Step 4 Find the corresponding pegs for vertices C and D. Complete the figure with the geoband. The figure on the geoboard on the right is the reflection of the image on the geoboard on the left.

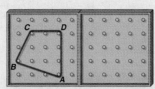
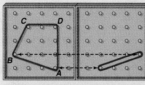
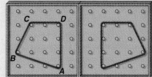

ON YOUR OWN

Use side-by-side geoboards to reflect each figure.   1–4. See Answer Appendix.

1.
2.
3.

4. Place a third geoboard to the right of the second geoboard and reflect your reflected image from Exercise 1. How does this new image compare to the original figure?

5. **Look Ahead** Compare the original figures and your reflected figures from Exercises 1–3. Are their corresponding sides congruent? Explain the relationship between these two figures. **See margin.**

Lesson 5-4A **205**

 Have students write a paragraph comparing and contrasting a given figure and its reflection.

Additional Answer

5. Yes, the corresponding sides are congruent. The figures are exactly the same except they are facing opposite directions.

HANDS-ON 5-4A LAB Notes

GET READY

Objective Students use geoboards to reflect a figure.

Optional Resources
Hands-On Lab Masters
• square dot paper, p. 13
• worksheet, p. 48

Overhead Manipulative Resources
• geoboard
• geobands

Manipulative Kit
• geoboard
• geobands

MANAGEMENT TIPS

Recommended Time
25 minutes

Getting Started Draw a geometric figure and its reflection on the chalkboard. Ask students to describe how the figures are alike and how they are different.

The **Activity** demonstrates the reflection of a geometric figure over a straight line. Ask students if the figures would be reflections of each other if the geoboard on the left were placed on the right. **yes**

Teaching Tip If some students have difficulty identifying corresponding sides on a reflection, have them use four colors of geobands to create the original figure. Then use four more geobands to create the reflection so that the corresponding sides are the same color.

ASSESS

Have students complete Exercises 1–4. Make sure that students count pegs accurately and do not jump rows.

Use Exercise 5 to determine whether students understand the concept of reflection.

Hands-On Lab 5-4A **205**

5-4 Lesson Notes

Instructional Resources
- *Study Guide Masters*, p. 37
- *Practice Masters*, p. 37
- *Enrichment Masters*, p. 37
- Transparencies 5-4, A and B
- *Assessment and Evaluation Masters*, pp. 126, 127
- *Diversity Masters*, p. 31
- CD-ROM Program
  - Resource Lesson 5-4
  - Interactive Lesson 5-4

Recommended Pacing	
Standard	Days 8 & 9 of 15
Honors	Days 7 & 8 of 15
Block	Day 4 of 8

1 FOCUS

5-Minute Check
(Lesson 5-3)

Identify all of the names that describe each quadrilateral.

1. parallelogram rhombus

2. parallelogram rectangle square rhombus

3. trapezoid

 The 5-Minute Check is also available on **Transparency 5-4A** for this lesson.

5-4 Symmetry

What you'll learn
You'll learn to identify line symmetry and rotational symmetry.

When am I ever going to use this?
Knowing how to describe and define lines of symmetry can help you design and create logos.

Word Wise
line symmetry
reflection
rotation
rotational symmetry

Chevrolet is one of many companies that use a logo to mark its products. The custom of using logos to identify manufacturers began at a time when many people could not read. Today, manufacturers want their logos to be easily identified from many viewpoints. They often design their logo to have line or rotational symmetry.

A figure has **line symmetry** if it can be folded so that one half of the figure coincides with the other half.

The Hallmark logo, at the right, has line symmetry. The right half is a **reflection** of the left half, and the center line is the *line of reflection*. The line of reflection is also called the *line of symmetry*.

Some figures have more than one line of symmetry. The Metropolitan Life Insurance Company's logo has one vertical, one horizontal, and two diagonal lines of symmetry.

Examples

① Trace each letter. Draw all lines of symmetry.

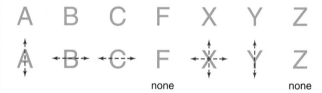

none   none

CONNECTION

② **Art** Artist Scott Kim uses reflections of words or names as part of his art. Patricia's reflected name is at the right. Create a reflection design for your name.

- Fold a piece of tracing paper in half. Write your name along the fold of the tracing paper.
- Turn the paper over. Use another color to trace everything you see through the paper.
- Open the paper to see the design. The colored part is the reflection of your original design.

206 Chapter 5 Geometry: Investigating Patterns

Classroom Vignette

"If students have difficulty tracing (paper too thick to see through), they can use small mirrors to establish the location of each line of symmetry."

Suetta Gladfelter, Teacher
Caroline Middle School
Milford, VA

Another type of symmetry uses **rotations**, or spins. When a compass is turned completely around, it forms a circle. A circle contains 360°. If a figure can be turned less than 360° about its center and still look like the original, then the figure has **rotational symmetry**.

Example — APPLICATION

3 **Advertising** Refer to the beginning of the lesson. Trace the Metropolitan Life Insurance symbol and determine whether it has rotational symmetry.

original   90° turn   180° turn   270° turn   360° turn

When you turn the symbol about its center, it looks like the original. Therefore, it has rotational symmetry.

MINI-LAB

Work with a partner.   straight pin   transparency   overhead markers

Try This
- Trace the star onto a transparency.
- Place the transparency over the star.
- Use a straight pin through the center of the star to hold the transparency in place.
- Rotate the transparency until the two stars match. Record how many times the two figures match. Do not count the original position twice.
- Divide 360° by the number of times it matched. This is the first angle of rotation.
- For additional angles of rotation, add the first angle of rotation to the previous angle. Stop when you reach 360°.

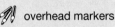

Talk About It
1. How many angles of rotation are there in the star? Give the angles of rotation. **5; 72°, 144°, 216°, 288°, 360°**
2. Repeat the lab for each of the quadrilaterals—parallelogram, rectangle, rhombus, square, and trapezoid. **See students' work.**
3. Which quadrilaterals have rotational symmetry? What are their angles of rotation?
4. Determine which quadrilaterals have line symmetry. Are there any quadrilaterals that have both rotational and line symmetry? List them.

3. parallelogram, 180°, 360°; rectangle, 180°, 360°; rhombus, 180°, 360°; square, 90°, 180°, 270°, 360°
4. rectangle, rhombus, square, isosceles trapezoid; rectangle, rhombus, square

---

Motivating the Lesson
Hands-On Activity Give each student an equilateral triangle cut from construction paper. Ask students to find all of the possible ways to fold the triangle in half so that the halves match. **3 ways**

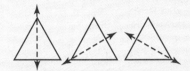

2 TEACH

Transparency 5-4B contains a teaching aid for this lesson.

In-Class Examples

For Example 1
Draw all lines of symmetry in the letters of the word GEOMETRY.

For Example 2
Create a reflection design with one side of a drawn face. **See students' drawings.**

For Example 3
Determine whether the sun symbol of New Mexico has rotational symmetry. **It does.**

Teaching Tip In Example 3, point out that the rotation can be both clockwise and counterclockwise.

Using the Mini-Lab You may wish to have students use sheets of tracing paper or patty paper instead of transparencies. For Exercises 2–4, you may wish to provide photocopies of several quadrilaterals.

3 PRACTICE/APPLY

Check for Understanding

If students need additional practice or instruction after completing Exercises 1–7, one of these options may be helpful.
- Extra Practice, see p. 618
- Reteaching Activity
- *Study Guide Masters*, p. 37
- *Practice Masters*, p. 37

Assignment Guide

Core: 9–19 odd, 21–24
Enriched: 8–16 even, 17–19, 21–24

Family Activity

Have students share with the class the examples of symmetry they found in their homes. Then have students identify similar examples in the classroom or school.

Additional Answers

1. In line symmetry, half of the figure is the reflection of the other half.
14. 8, 9, 10, 11, 12

Study Guide Masters, p. 37

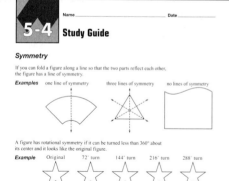
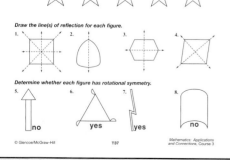

208 Chapter 5

CHECK FOR UNDERSTANDING

Communicating Mathematics

Read and study the lesson to answer each question.

1. *Explain* how a reflection and line symmetry are related. **See margin.**
2. *Draw* a letter of the alphabet that has both line symmetry and rotational symmetry. **Sample answer: H**

HANDS-ON MATH

3. *Draw* several geometric shapes on paper and then draw the same shapes on a transparency. Determine whether the shapes have rotational symmetry. Draw a geometric shape that has an infinite number of angles of rotation. **See Answer Appendix.**

Guided Practice

Trace each figure. Determine whether the figure has line symmetry. If so, draw the line(s) of reflection.

4.    5.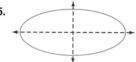

6. Which of the figures in Exercises 4-5 have rotational symmetry? **both of them**

7. *Geography* The "Union Jack", a common name for the flag of the United Kingdom, has multiple symmetries. Trace the flag. Draw the lines of symmetry. Name the angles of rotation.
180°, 360°

EXERCISES

Practice

Trace each figure. Determine whether the figure has line symmetry. If so, draw the lines of reflection.

8.   9.   10. **no lines of symmetry**

11.   12.   13. **no lines of symmetry**

Family Activity

Go on a symmetry hunt with family members. First choose a location, such as rooms in your home or an area outside your home. Then, see how many examples of line and rotational symmetry you can find in 15 minutes.

14. See margin.

14. Which of the figures in Exercises 8-13 have rotational symmetry?
15. Which types of triangles—*scalene, isosceles, equilateral*—have line symmetry? Which have rotational symmetry? **See Answer Appendix.**
16. What capital letters of the alphabet produce the same letter after being rotated 180°? **H, I, N, O, S, X, Z**

Applications and Problem Solving

17. *Life Science* The diatom is a single-celled marine animal that has multiple symmetries. Trace the diatom.
   a. Draw any lines of symmetry.
   b. Give any angles of rotation. **120°, 240°, 360°**

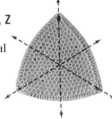

208 Chapter 5 Geometry: Investigating Patterns

Reteaching the Lesson

Activity Give students regular polygons cut from paper. Have them fold the figures to discover all possible lines of symmetry. Inform students that the point where the lines of symmetry intersect is the center point. Now have them repeat the steps of the Mini-Lab using their polygons.

Error Analysis
Watch for students who have difficulty identifying lines of symmetry.
Prevent by having students trace the shapes, fold them in half, and hold the shapes perpendicular to a mirror. Repeat until all lines of symmetry have been found.

18. **Sports** Determine whether the emblem for the American Football Conference and the emblem for the National Football Conference display symmetry. If so, what type of symmetry? **AFC, yes, line symmetry; NFC, yes rotational symmetry.**

19. **Art** Handcrafted tiles were used as decoration in the homes of wealthy merchants and nobles during the Renaissance period in Europe. Usually the artist did not know in advance how the purchaser would use the tile, so the artist made them symmetric. Find all of the lines of symmetry in the tile pattern. Does it also have rotational symmetry? If so, what are the angles of rotation? **yes, 90°, 180°, 270°, 360°**

20. **Working on the CHAPTER Project** Refer to the shape you found on page 185.
    a. Draw all lines of symmetry for your shape.
    b. Does your shape have rotational symmetry? If so, what are the angles of rotation? **a–b. See students' work.**

21. **Critical Thinking** A printer puts four pages on each side of a large sheet of paper when printing books. Decide how and where to print pages 1 through 8 so that the pages will be in the correct order after assembling. **See Answer Appendix.**

Mixed Review

22. **Algebra** In trapezoid $LMNO$, $m\angle L = 120°$, $m\angle M = 120°$, $m\angle N = 2x°$, and $m\angle O = 60°$. Find the value of x. *(Lesson 5-3)* **30**

23. **Test Practice** The bar graph shows the results of a survey on popular bagel flavors. How many flavors were chosen by fewer than 5 students? *(Lesson 4-1)* **A**

    A 2 flavors B 3 flavors
    C 4 flavors D 5 flavors
    E 6 flavors

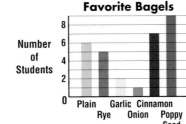

24. **Algebra** The product of 9 and r is 54. Find the number. *(Lesson 1-6)* **6**

Lesson 5-4 Symmetry 209

Extending the Lesson

Enrichment Masters, p. 37

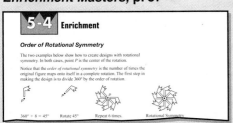

Activity The capital letter T has a vertical line of symmetry. The capital letter D has a horizontal line of symmetry. Have students find words that are vertically or horizontally symmetric. Sample answers:

-DEBBIE-

M
A
T
H

CHAPTER Project

Exercise 20 asks students to progress to the next stage of the Chapter Project. Students may wish to work in pairs to find all types of symmetry for their shapes.

4 ASSESS

Closing Activity
Writing Have students write a few sentences describing how they would decide whether a figure has line symmetry, rotational symmetry, or neither.

Chapter 5, Quiz B (Lessons 5-3 and 5-4) is available in the *Assessment and Evaluation Masters*, p. 127.

Mid-Chapter Test (Lessons 5-1 through 5-4) is available in the *Assessment and Evaluation Masters*, p. 126.

Practice Masters, p. 37

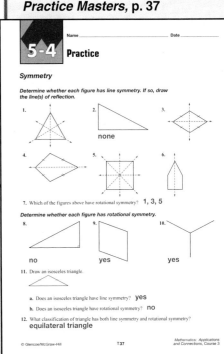

Lesson 5-4 **209**

5-5 Lesson Notes

Instructional Resources
- *Study Guide Masters*, p. 38
- *Practice Masters*, p. 38
- *Enrichment Masters*, p. 38
- Transparencies 5-5, A and B
- *Technology Masters*, p. 62
- CD-ROM Program
  - Resource Lesson 5-5

Recommended Pacing	
Standard	Day 10 of 15
Honors	Days 9 & 10 of 15
Block	Day 5 of 8

1 FOCUS

5-Minute Check
(Lesson 5-4)

Determine whether each figure has line symmetry. If so, draw the lines of reflection.

1. 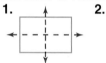   2.

Determine whether the figure has rotational symmetry. If so, find the angle of rotation.

3. 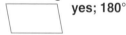 yes; 180°

 The 5-Minute Check is also available on **Transparency 5-5A** for this lesson.

2 TEACH

 Transparency 5-5B contains a teaching aid for this lesson.

Thinking Algebraically Point out to students that because the measures of the angles in a triangle always total 180°, the value of an unknown angle measure in a triangle can be calculated if the measures of the other two angles are known.

5-5 Congruent Triangles

What you'll learn
You'll learn to verify congruent triangles by using SSS, ASA, and SAS.

When am I ever going to use this?
Knowing how to verify congruence in triangles can help in the construction of buildings.

Word Wise
congruent triangles
corresponding parts

Study Hint
Reading Math Matching arcs on angles show that the angles are congruent.

Nature is composed of many shapes and patterns. If we look closely at birds, reptiles, and plants, we can identify specific shapes. The skin pattern on the head of a snake and the wings on a hummingbird are shaped like triangles.

Triangles that have the same size and shape are called **congruent triangles**.

The following triangles are congruent.

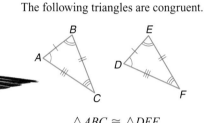

$\triangle ABC \cong \triangle DEF$

Congruent Angles	Congruent Sides
$\angle A \cong \angle D$	$\overline{AB} \cong \overline{DE}$
$\angle B \cong \angle E$	$\overline{BC} \cong \overline{EF}$
$\angle C \cong \angle F$	$\overline{AC} \cong \overline{DF}$

Parts of congruent triangles that "match" are called **corresponding parts**. For example, in the triangles above, $\angle B$ corresponds to $\angle E$, and $\overline{AC}$ corresponds to $\overline{DF}$.

Two triangles are congruent when all of the corresponding parts are congruent. However, you do not need to know that all six corresponding parts are congruent to know that the triangles are congruent. Sometimes you only need to know that three of the six corresponding parts are congruent.

Congruent Triangles	Two triangles are congruent if the following corresponding parts of the triangles are congruent.
	three sides (SSS) / two angles and the included side (ASA) / two sides and the included angle (SAS)

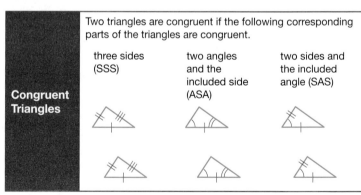

210 Chapter 5 Geometry: Investigating Patterns

Motivating the Lesson

Problem Solving Draw two congruent triangles on the chalkboard. Give the measures of the sides and the angles of one triangle. Ask students how they would find the remaining measures of the second triangle if they knew only three of them.

Multiple Learning Styles

Interpersonal Have students work in groups of three. Ask each group to draw three pairs of congruent triangles—one congruent by SSS, one by ASA, and one by SAS. Then have the students exchange sets of triangles within their groups and identify the types of congruence.

Examples

Determine whether each pair of triangles is congruent. If so, write a congruence statement and tell why the triangles are congruent.

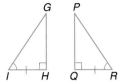 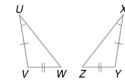

The statement △GHI ≅ △PQR tells us that G corresponds to P, H corresponds to Q, and I corresponds to R.

1 Since $m\angle H = 90°$ and $m\angle Q = 90°$, $\angle H \cong \angle Q$. Also, $\overline{IH} \cong \overline{RQ}$ and $\angle I \cong \angle R$. Therefore, $\triangle GHI \cong \triangle PQR$ by ASA.

2 $\overline{UV} \cong \overline{XY}$, $\angle VUW \cong \angle YXZ$, and $\overline{VW} \cong \overline{YZ}$. This is two sides and an angle congruent. But the angle is not included between the sides. So we cannot conclude that the triangles are congruent.

INTEGRATION **3** Algebra △ABC is congruent to △RST. Find the value of y.

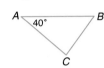 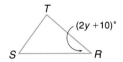

Since $\triangle ABC \cong \triangle RST$, the corresponding angles are congruent.

$m\angle A = m\angle R$

$40 = 2y + 10$ *Replace $m\angle A$ with 40 and $m\angle R$ with $2y + 10$.*

$30 = 2y$ *Subtract 10 from each side.*

$15 = y$ *Divide each side by 2.*

CHECK FOR UNDERSTANDING

Communicating Mathematics

Read and study the lesson to answer each question. 1–3. See Answer Appendix.

1. **Explain** what must be true for two figures to be congruent.
2. **Describe** an *included* angle. Draw an example.
3. **Write** a paragraph describing three ways to show that two triangles are congruent. Draw a diagram to illustrate each method.

Guided Practice

Determine whether each pair of triangles is congruent. If so, write a congruence statement and tell why the triangles are congruent.

4. yes; △MNP ≅ △XYZ; SSS

4. 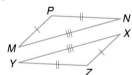   5.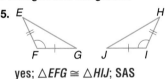

   yes; △EFG ≅ △HIJ; SAS

6. Find the value of x in the two congruent triangles. 9

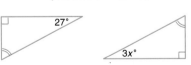

Lesson 5-5 Congruent Triangles **211**

In-Class Examples

Determine whether each pair of triangles is congruent. If so, write a congruence statement and tell why the triangles are congruent.

For Example 1

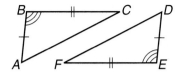

$\angle B \cong \angle E$, $\overline{AB} \cong \overline{DE}$, and $\overline{BC} \cong \overline{EF}$. Therefore $\triangle ABC \cong \triangle DEF$, by SAS.

For Example 2

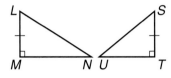

$\overline{LM} \cong \overline{ST}$, and $\angle M \cong \angle T$. However, $\overline{MN}$ is not congruent to $\overline{UT}$. Therefore, the two triangles are not congruent.

For Example 3

△ABC is congruent to △DEF. Find the value of x. 20

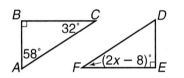

Study Guide Masters, p. 38

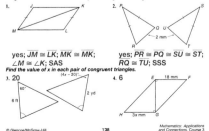

Reteaching the Lesson

Activity Cut five pairs of congruent triangles from construction paper and mix them up. Have students find the congruent pairs and explain how they know that they have correctly matched each triangle with its partner.

Error Analysis

Watch for students who write congruences incorrectly.

Prevent by pointing out that the order of letters in congruence indicates corresponding parts.

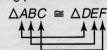

Lesson 5-5 **211**

3 PRACTICE/APPLY

Check for Understanding

If students need additional practice or instruction after completing Exercises 1–7, one of these options may be helpful.
- Extra Practice, see p. 619
- Reteaching Activity, see p. 211
- Study Guide Masters, p. 38
- Practice Masters, p. 38

Assignment Guide

Core: 9–15 odd, 16–18
Enriched: 8–14 even, 15–18

4 ASSESS

Closing Activity

Writing On the chalkboard, write the congruence statement △MGE ≅ △FRJ. Have students sketch two congruent triangles and label the vertices properly for the congruence. Then have them list the three pairs of sides and three pairs of angles that are congruent. $\overline{MG} \cong \overline{FR}$, $\overline{ME} \cong \overline{FJ}$, $\overline{GE} \cong \overline{RJ}$ ∠M ≅ ∠F, ∠G ≅ ∠R, ∠E ≅ ∠J

Practice Masters, p. 38

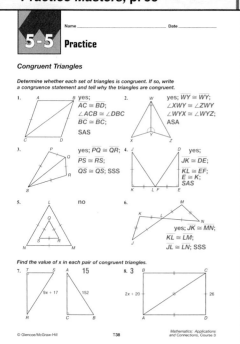

interNET CONNECTION

For the latest statistics on music sales, visit:
www.glencoe.com/sec/math/mac/mathnet

7. Statistics The circle graph shows the percent breakdown of all music sales in 1996. If the intersections of each consecutive radius with the circle are connected with a line segment, are any of the triangles formed congruent? If so, tell why the triangles are congruent. **Yes; the triangles formed by the sections for urban contemporary music and country music; SAS.**

EXERCISES

Practice Determine whether each pair of triangles is congruent. If so, write a congruence statement and tell why the triangles are congruent.

8.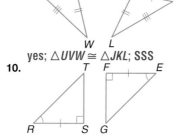
yes; △UVW ≅ △JKL; SSS

9.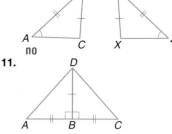
no

10.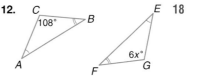
yes; △RST ≅ △EFG; ASA

11.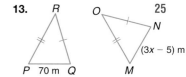
yes; △ABD ≅ △CBD; SAS

Find the value of *x* in each pair of congruent triangles.

12. C 108° B   E 18
    A   6x° F G

13. R O 25
       N
    P 70 m Q (3x − 5) m M

14. Refer to Example 1. If $m\angle I = 55°$, find $m\angle R$. **55°**

Applications and Problem Solving

15. **Design** The symbol for the National Council of Teachers of Mathematics is shown at the right. Which triangles in the symbol appear to be congruent? **See Answer Appendix.**

16. **Critical Thinking** Find at least four ways to divide a square into four congruent shapes. **See Answer Appendix.**

Mixed Review

17. **Geometry** Determine whether the figure at the right has rotational symmetry. *(Lesson 5-4)* **no**

18. **Test Practice** Juana has $79.43. She wants to buy a video game system that costs $212.16. How much does she need to save? *(Lesson 1-1)* **C**
A $132.16 B $180.42 C $132.73 D $291.59 E Not Here

212 Chapter 5 Geometry: Investigating Patterns

Extending the Lesson

Enrichment Masters, p. 38

Activity Have students work in groups to find all the ways to divide a geoboard into four congruent quadrilaterals. **Sample answer:**

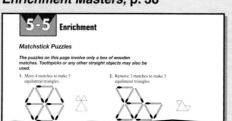

HANDS-ON LAB

COOPERATIVE LEARNING

5-5B Constructing Congruent Triangles

A Follow-Up of Lesson 5-5

- compass
- straightedge

Construction engineers and architects often need to copy figures from one part of their drawing to other places. They can use a compass and a straightedge to do this.

TRY THIS

Work with a partner.

Construct a triangle congruent to $\triangle ABC$.

Step 1 Use a straightedge to draw a line. Put a point on it labeled G.

Step 2 Open your compass to the same width as the length of $\overline{AB}$. Put the compass point at G. Draw an arc that intersects the line. Label this point of intersection H.

Step 3 Open your compass to the same width as the length of $\overline{AC}$. Place your compass point at G and draw an arc above the line.

Step 4 Open your compass to the same width as the length of $\overline{BC}$. Place the compass point at H and draw an arc above the line so that it intersects the arc that you drew in Step 3. Label this point of intersection K.

Step 5 Draw $\overline{HK}$ and $\overline{GK}$.
$\triangle ABC \cong \triangle GHK$

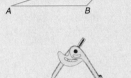
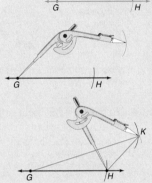

ON YOUR OWN

Trace each triangle. Then construct a triangle congruent to it.   1–3. See Answer Appendix.

1.   2.   3.

4. **Reflect Back** How do the corresponding angles of two congruent triangles compare?   **The corresponding angles are congruent.**

Lesson 5-5B HANDS-ON LAB **213**

Math Journal Have students write a paragraph explaining why this construction results in triangles that are congruent.

HANDS-ON LAB Notes

5-5B

GET READY

Objective Students construct a triangle congruent to a given triangle.

Optional Resources
Hands-On Lab Masters
- worksheet, p. 49

Overhead Manipulative Resources
- compass
- straightedge

Manipulative Kit
- compass
- straightedge

MANAGEMENT TIPS

Recommended Time
25 minutes

Getting Started Use a straightedge and chalk to draw a simple house on the chalkboard. Ask students to suggest ways that an exact copy of the house can be drawn elsewhere on the chalkboard. Then ask them how they would do this without using a ruler and a protractor.

The **Activity** demonstrates how to construct a triangle congruent to a given triangle by using SSS. When students have completed the construction, have them compare the new triangle with the original by overlaying the two sheets of paper and holding them up to a light source.

ASSESS

Have students complete Exercises 1–3. Watch for students who may have difficulty working with a compass. You may want them to practice copying segments before constructing triangles.

Use Exercise 4 to determine whether students understand the relationship between congruent angles and congruent triangles.

Hands-On Lab 5-5B **213**

HANDS-ON LAB 5-6A Notes

GET READY

Objective Students draw a dilation of a figure.

Optional Resources
Hands-On Lab Masters
- worksheet, p. 50

Manipulative Kit
- ruler
- scissors

MANAGEMENT TIPS

Recommended Time
45 minutes

Getting Started Have students read the first paragraph of the lesson. Ask them to name examples of other reduced or enlarged images. **Sample answers: photographs, scale drawings, TV screens**

The **Activity** demonstrates one way to enlarge an image. Provide students with cardboard sturdy enough to hold two pencils securely. To prevent slippage, direct students to punch holes just big enough to accommodate the brass fasteners and pencils. You may wish to use thin markers instead of pencils. Make sure that the maps are simple in detail.

ASSESS

Have students complete Exercises 1–3. Assist students who have difficulty holding their pantographs steady. Adjust the position of the second pencil or marker if necessary.

Use Exercise 4 to determine whether students understand the relationship between dilation and congruence.

214 Chapter 5

HANDS-ON LAB

COOPERATIVE LEARNING

5-6A Dilations

A Preview of Lesson 5-6

- 3 cardboard strips
- 3 brass fasteners
- map
- ruler

Before copy machines were available to enlarge and reduce maps, cartographers, or mapmakers, used pantographs to enlarge or reduce maps and charts. In mathematics, an enlargement or a reduction of an image is called a *dilation*.

TRY THIS

Work with a partner.
Draw an enlargement of a map.

Step 1 Cut one of the strips of cardboard into two congruent pieces.

Step 2 Assemble the pantograph as shown in the diagram using brass fasteners at locations *A, B,* and *C*. Make sure that $\overline{BC} \cong \overline{AP}$ and $\overline{AB} \cong \overline{PC}$.

Step 3 Punch holes in the cardboard strips at locations *P* and *Q* and insert two pencils into the holes.

Step 4 To make an enlargement of your map, hold the pantograph stationary at location *H*. Trace the original map with the pencil at location *P*. The pencil at location *Q* will draw an enlargement of your map.

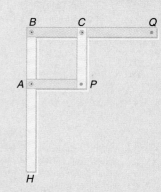

ON YOUR OWN

Trace each drawing. Then use the pantograph to draw an enlargement of it.   1–3. See Answer Appendix.

1.   2.   3.

4. **Look Ahead** Measure the angles in the drawing in Exercise 1. Then measure the corresponding angles in your enlargement. How do these measures compare? **Corresponding angles have the same measure.**

214 Chapter 5 Geometry: Investigating Patterns

Math Journal Have students write a paragraph comparing and contrasting the original figure in Exercise 1 with its enlargement. Are the angles in the two figures congruent? Why or why not?

5-6 Similar Triangles

What you'll learn
You'll learn to identify similar triangles.

When am I ever going to use this?
Knowing how to identify similar triangles can help you recognize triangular relationships in architecture.

Word Wise
similar triangles

The Alcoa Office Building in San Francisco, California, uses triangular braces to help secure the building in the event of an earthquake. The triangles marked in the photograph have the same shape, but are different sizes. Triangles that have the same shape, but may differ in size, are called **similar triangles**.

HANDS-ON MINI-LAB

Work with a partner. tracing paper, scissors

Try This
Trace each pair of triangles and cut them out.

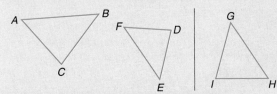

- Compare the angles in each pair of triangles.
- For each pair of triangles, see if you can place one figure inside the other with equal amount of space separating the corresponding sides in every case.

Talk About It
1. What did you find when you compared the angles?
2. If you can fit one triangle inside another with equal space separating all corresponding sides, the triangles are similar. Name the similar figures. **The triangles on the left are similar.**
3. Make a conjecture about triangles with congruent corresponding angles. **Triangles with congruent corresponding angles are similar.**

1. The triangles on the left have congruent corresponding angles. The triangles on the right do not have congruent corresponding angles.

Study Hint
Reading Math The symbol ~ means *is similar to*.

The Mini-Lab suggests that if corresponding angles of two triangles are congruent, the triangles are similar. In the figure below, $\triangle ABC$ is similar to $\triangle PQR$. This is written $\triangle ABC \sim \triangle PQR$.

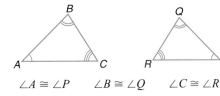

$\angle A \cong \angle P$ $\angle B \cong \angle Q$ $\angle C \cong \angle R$

Lesson 5-6 Similar Triangles   **215**

5-6 Lesson Notes

Instructional Resources
- *Study Guide Masters*, p. 39
- *Practice Masters*, p. 39
- *Enrichment Masters*, p. 39
- Transparencies 5-6, A and B
- *Assessment and Evaluation Masters*, p. 128
- *School to Career Masters*, p. 31
- CD-ROM Program
  - Resource Lesson 5-6

Recommended Pacing	
Standard	Days 11 & 12 of 15
Honors	Days 11 & 12 of 15
Block	Day 6 of 8

1 FOCUS

5-Minute Check
(Lesson 5-5)

Determine whether each pair of triangles is congruent. If so, tell why the triangles are congruent.

1.
   yes, ASA

2.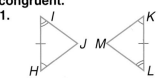
   yes, SAS

3. Find the value of *x* for $\triangle ABC \cong \triangle DEF$. **7**
   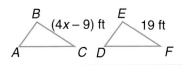

The 5-Minute Check is also available on **Transparency 5-6A** for this lesson.

Motivating the Lesson

Problem Solving Draw two similar triangles on the chalkboard. Provide measures for two pairs of corresponding angles. Ask students to find the measures of the missing angles.

Lesson 5-6   **215**

2 TEACH

 Transparency 5-6B contains a teaching aid for this lesson.

Using the Mini-Lab You may wish to use a photocopier to enlarge the set of triangles and give one copy to each pair of students. Students can then more easily cut out and manipulate the triangles.

In-Class Examples

For Example 1
Determine whether the triangles are similar. Because $\angle D \cong \angle L$, $\angle E \cong \angle M$, and $\angle F \cong \angle N$, the triangles are similar.

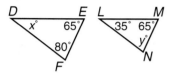

For Example 2
If $\triangle HIJ \sim \triangle PQR$, find the value of y. **28**

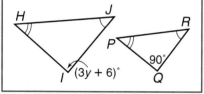

Additional Answers

1.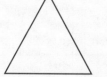

2. **Sample answer:** When triangles are similar, they are the same shape, but may be a different size. When triangles are not similar, they are not the same shape.

3. Helki; since the sum of 3 angles of a triangle is 180°, the unmarked angle in each of the triangles must be 75°. Since corresponding angles are congruent, the triangles are similar.

4e. If two figures are congruent, then the corresponding angles are congruent and the triangles are similar.

Example 1 Determine whether the triangles are similar.

The sum of the angles of a triangle is 180°.

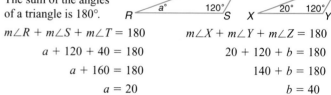

$m\angle R + m\angle S + m\angle T = 180$ $\qquad$ $m\angle X + m\angle Y + m\angle Z = 180$
$a + 120 + 40 = 180$ $\qquad\qquad\quad$ $20 + 120 + b = 180$
$a + 160 = 180$ $\qquad\qquad\qquad\qquad$ $140 + b = 180$
$a = 20$ $\qquad\qquad\qquad\qquad\qquad\;\;$ $b = 40$

Since $\angle R \cong \angle X$, $\angle S \cong \angle Y$, and $\angle T \cong \angle Z$, the triangles are similar.

You can use the relationships of similar figures to find missing measures.

Example 2 **INTEGRATION** **Algebra** If $\triangle ABC \sim \triangle UVW$, find the value of x.

Since $\triangle ABC \sim \triangle UVW$, the corresponding angles are congruent.

$m\angle A = m\angle U$
$2x - 13 = 85$
$2x = 98$ $\qquad$ *Add 13 to each side.*
$x = 49$ $\qquad$ *Divide each side by 2.*

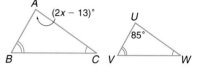

CHECK FOR UNDERSTANDING

Communicating Mathematics

Read and study the lesson to answer each question. 1–3. See margin.

1. **Draw** a pair of similar triangles.

2. **Compare and contrast** a pair of triangles that are similar and a pair that are not similar.

3. **You Decide** Helki says that the triangles are similar. Edith disagrees. She says that the unmarked angles may not be congruent. Who is correct? Explain.

HANDS-ON MATH

4. **Draw** a triangle. Trace the triangle onto tracing paper. Place the corresponding angles on top of each other. **c. yes**

   a. Are the angles congruent? **yes**
   b. Are the sides congruent? **yes**
   c. Are the two triangles congruent?
   d. Are the two triangles similar? **yes**
   e. Explain why two congruent triangles are also similar. See margin.

Guided Practice

5. Tell whether the triangles are *congruent*, *similar*, or *neither*. Justify your answer.

   5. Similar; corresponding angles are congruent.

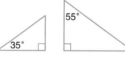

6. Find the value of x in the pair of similar triangles. **11**

216 Chapter 5 Geometry: Investigating Patterns

7. The figure at the right is known as the Sierpinski Triangle. How does this figure relate to the congruent and similar triangles? **See margin.**

EXERCISES

Practice

Tell whether each pair of triangles is *congruent*, *similar*, or *neither*. Justify your answer. 9. Similar; corresponding angles are congruent.

8.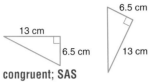
congruent; SAS

9.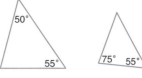

10. Neither; corresponding angles are not congruent.

10.

11.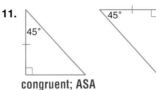
congruent; ASA

Find the value of *x* in each pair of similar triangles.

12. 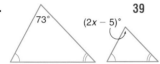 39

13. 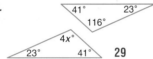 29

Applications and Problem Solving

14. **Architecture** The bell tower at Old City Hall in Toronto, Canada, is shown at the left. Describe the four triangles that form the peak of the tower as *congruent, similar,* or *neither*. **congruent**

15. **Earth Science** Two trees are standing near each other in a park. At any time of day, the rays of sun will hit the ground in the park at the same angle. If both of the trees are perpendicular to the ground, what can you say about the triangles formed by the trees and their shadows as shown below? Explain. **Similar; corresponding angles are congruent.**

Lesson 5-6 Similar Triangles **217**

■ Reteaching the Lesson ■

Activity Have students draw triangles and measure their angles. Have them enlarge or reduce the triangles using a photocopier and then measure the angles again. Ask students if the triangles will always be similar when doing this. **yes**

4 ASSESS

Closing Activity

Modeling Have students cut several lengths of paper approximately $\frac{1}{2}$-inch wide, with at least two strips of each length. Direct students to use the paper strips to assemble two identical triangles. Then tell students to fold the paper strips that form one of the triangles in half and to reassemble the triangles. Ask students if the two triangles are similar and to explain why or why not. **Yes; angles are congruent.**

Chapter 5, Quiz C (Lessons 5-5 and 5-6) is available in the *Assessment and Evaluation Masters*, p. 128.

Additional Answer

16. $\triangle RST \sim \triangle RXY$; since $\overline{XY} \parallel \overline{ST}$, $\angle RST \cong \angle RXY$ and $\angle RTS \cong \angle RYX$. $\angle SRT \cong \angle XRY$ because they are the same angle. The corresponding angles are congruent, so the triangles are similar.

Practice Masters, p. 39

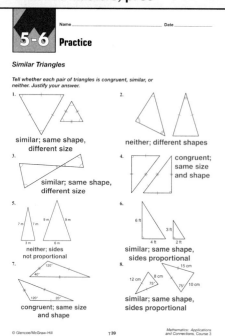

---

16. **Critical Thinking** Refer to $\triangle RST$. If $\overline{XY} \parallel \overline{ST}$, what is true about $\triangle RST$ and $\triangle RXY$? Explain. **See margin.**

Mixed Review

17. yes; $\triangle ABC \cong \triangle EFG$; SSS

17. **Geometry** Determine whether the pair of triangles is congruent. If so, write a congruence statement and tell why the triangles are congruent. *(Lesson 5-5)*

18. **Test Practice** Vanessa, Seth, Sue, and Jordan are collecting aluminum cans to recycle. Their goal is 100 pounds of cans. The table shows how their collection has progressed so far. *(Lesson 4-1A)*

Name	M	T	W	T	F	Total
Vanessa	1	2	1	2	3	9
Seth	1	3	4	2	0	10
Sue	3	1	3	3	0	10
Jordan	5	3	2	1	4	15

According to the information in the table, how many more pounds of cans do they need to collect to reach their goal? **C**

A 44 lb   **B** 54 lb   **C** 56 lb   **D** 65 lb   **E** 72 lb

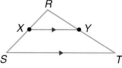

Let the Games Begin

Shape Relations

Get Ready This game is for teams of two players.   18 index cards

Get Set Draw each of the following figures on an index card: equilateral triangle, isosceles triangle, scalene triangle, acute triangle, obtuse triangle, and right triangle. Duplicate this set exactly. On the remaining set of 6 cards, draw a similar figure for each figure.

Go
- Mix the original set of 6 cards. Then mix the duplicate set and the similar set of cards together.
- Place each stack facedown. One member from each team turns over the top card on each pile. All four players write down all relationships they know between those two figures.
- Then, each team makes one list of the relationships without duplication from their two lists. Each team reads their list. Each team that has a true relationship earns one point for their team. Any team that has listed a true relationship that the other team did not list gets two points instead of one point for that particular relationship.
- Continue these steps until the larger pile has all been used once and the smaller pile has been used twice. The team with the most points wins.

interNET CONNECTION Visit www.glencoe.com/sec/math/mac/mathnet for more games.

218 Chapter 5 Geometry: Investigating Patterns

Extending the Lesson

Enrichment Masters, p. 39

To keep the pace of the game moving, have students use a timer to limit the time they have to make their lists.

*Additional resources for this game can be found on pages 49–50 of the **Classroom Games**.*

ADVERTISING

I love art class! I would like to work on advertising for big companies someday!

Elena Baca
GRAPHIC DESIGNER

Elena Baca was born in Peru and raised in Egypt. She has been a graphic designer for more than ten years. Her list of clients includes American Airlines, GTE Corporation, Hyatt Hotels, Johnson and Johnson Medical Inc., and Just My Size. She has won many awards in her field, including awards for T-shirt designs and for web page designs.

A person who is interested in becoming a graphic designer needs artistic talent, creativity, and imagination. Math will help a graphic designer see patterns, make scale models, and create designs that are pleasing to the eye. Two out of three people entering the field today have a college degree or some college coursework. Whether a person goes to college or not, a good portfolio containing samples is a necessity.

For more information:
Design International
3748 22nd Street
San Francisco, CA 94114

interNET CONNECTION
www.glencoe.com/sec/math/mac/mathnet

Your Turn
Design your own wrapping paper. Write a paragraph summarizing how you used math to create your design.

More About Elena Baca
- Ms. Baca graduated cum laude from the University of Texas at Arlington in 1992 with a Bachelor of Fine Arts with an emphasis in Visual Communication. While there, she studied drawing, metal jewelry, Etruscan art history, modern art history, printmaking, and video, in addition to her basic design and core courses.
- Ms. Baca won a Silver Addy (an advertising award) in 1994, for self promotion, and another in 1997, for four-color collateral design. She has also had some of her T-shirt, letterhead, and logo designs published in 1996 and 1997 by Rockport Publisher.

Motivating Students
As students learned in the Chapter Opener on pages 184–185, geometric shapes can be used in logos and other graphic displays, and can be tessellated. They may be interested in a career in advertising, graphic design, or a related field. To start the discussion, you may ask students questions about the history of advertising and logos.
- Where was the oldest known written advertisement found? **Thebes; it is 3,000 years old.**
- What started the modern era of advertisement and when? **the invention of movable type by Johannes Gutenberg, in Europe; in 1450**
- When did the registered trademark of Levi Strauss & Co. first appear? **1870s; in San Francisco**
- When did the Morton Salt girl first appear on a package of salt? **1914**

Making the Math Connection
Ms. Baca uses computer software to design logos and other graphics. She makes grids by drawing parallel lines and then rotating a copy of them 90°. She uses percentages to calculate dilations of shapes, as well as in applying color and estimating costs.

Working on *Your Turn*
Students may want to work in pairs to design their wrapping paper. They should then work individually to write their summaries. Make sure they understand that T-shirts, web pages, and wrapping paper are just a few of the possible design applications of geometry. In addition, graphic designer is just one of the many jobs that benefit from an ability to use geometric patterns.

*An additional School to Career activity is available on page 31 of the **School to Career Masters**.*

5-7 Lesson Notes

Instructional Resources
- *Study Guide Masters*, p. 40
- *Practice Masters*, p. 40
- *Enrichment Masters*, p. 40
- Transparencies 5-7, A and B
- *Assessment and Evaluation Masters*, p. 128

 CD-ROM Program
- Resource Lesson 5-7
- Extended Activity 5-7

Recommended Pacing	
Standard	Day 13 of 15
Honors	Day 13 of 15
Block	Day 7 of 8

1 FOCUS

5-Minute Check
(Lesson 5-6)

Tell whether each pair of triangles is *congruent*, *similar*, or *neither*.

1.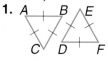

congruent by SSS and similar

2.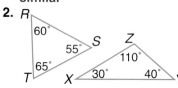

neither

3.

similar

 The 5-Minute Check is also available on **Transparency 5-7A** for this lesson.

5-7 Transformations and M. C. Escher

What you'll learn
You'll learn to create Escher-like drawings using translations and rotations.

When am I ever going to use this?
Knowing how to use translations and rotations can help you analyze how to make products without wasting materials.

Word Wise
tessellation
transformation
translation

A **tessellation** is a tiling made up of copies of the same shape or shapes that fit together without gaps and without overlapping. Brick walls and some quilts are tessellations. Maurits Cornelis Escher (1898–1972) was a Dutch artist whose work used tessellations.

Unique, interlocking, puzzle-like patterns can be made by using **transformations** to change tessellations. Transformations are movements of geometric figures. One transformation commonly used is a slide, or **translation**, of a figure.

Source: © M.C. Escher/Cordon Art—Baam—
Collection Haags Gemeentemuseum—The Hauge

HANDS-ON MINI-LAB

Work with a partner. paper clips   scissors   tape

Try This

- Stack two sheets of paper together. Fold the paper in half twice to make eight layers and then paperclip them together.
- Draw a square on the top layer. Cut out the square so that you have 8 identical squares. Paper clip them together.
- On the left side of the squares, cut out a section. Tape that section onto the right side of all eight squares in the same position. Moving the part from one side to the opposite side is a translation.
- On the top of the squares, cut out a section. Tape that section onto the bottom of all eight squares in the same position.

- Arrange the eight pieces in two rows of four pieces so that they cover a flat surface without overlapping and without leaving gaps.

Talk About It 1–2. See margin.

1. Create another tessellation by repeating the steps above, but use a parallelogram instead of a square. Describe your design.
2. What are the differences between using a square and using a parallelogram to create a tessellation?

220 Chapter 5 Geometry: Investigating Patterns

Additional Answers for the Mini-Lab

1. Sample answer:

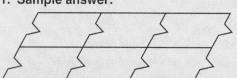

Parallelograms with triangles cut out on one side and added to the other

2. The only difference is the shape of the original figure.

220 Chapter 5

Example 1 Make an Escher-like drawing using the modification shown at the right.

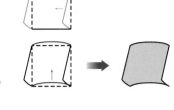

Draw a square. Copy the pattern shown. Translate the change shown on the right side to the left side.

Now translate the change shown on the bottom side to the top side to complete the pattern unit.

Repeat this pattern on a tessellation of squares. *It is sometimes helpful to complete one pattern unit, cut it out, and trace it for the other units.*

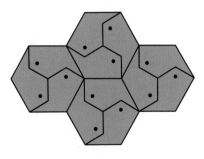

Other tessellations can be modified by using changes with a rotation. Rotations are transformations that involve a turn around a point.

Example 2 Create a tile for tessellating by using a rotation around a vertex of a parallelogram base figure.

base figure   change   change rotated around right vertex   final tile

Now repeat this pattern unit on a tessellation of parallelograms.

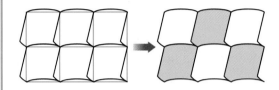

Lesson 5-7 Transformations and M. C. Escher **221**

Multiple Learning Styles

Intrapersonal Have students select their favorite tessellation out of the ones they designed. Ask them to write an entry in their journal that includes the tessellation, as well as a critique of it. What do they like best about it? What would they like to improve? Have them decorate their tessellations with colors or penciled designs.

Motivating the Lesson

Communication Sketch the rectangle below that is covered by congruent right triangles. Ask students to find other congruent figures with which a rectangle can be covered.

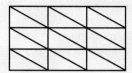

2 TEACH

Transparency 5-7B contains a teaching aid for this lesson.

Using the Mini-Lab Encourage students to be creative. They should keep their designs simple enough to cut out, yet different from others. They may wish to make animal shapes.

In-Class Examples

For Example 1
Make an Escher-like drawing using these modifications.

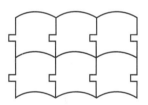

For Example 2
Create a tile for tessellating by using a rotation around a vertex of an equilateral triangle base figure, and repeat this pattern in a tessellation of triangles.

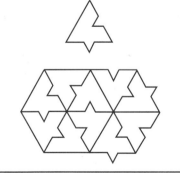

Lesson 5-7 **221**

Teaching Tip In Example 3, look closely at the picture to find the pattern that Escher used for his lithograph.

In-Class Example

For Example 3
Refer to Example 3 in the Student Edition. What transformation was used in Escher's lithograph? **rotation**

---

3 PRACTICE/APPLY

Check for Understanding
If students need additional practice or instruction after completing Exercises 1–6, one of these options may be helpful.
- Extra Practice, see p. 619
- Reteaching Activity
- *Study Guide Masters*, p. 40
- *Practice Masters*, p. 40

Additional Answers
1. a tiling of a plane made with copies of the same shape or shapes fit together without gaps or overlaps

3c.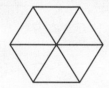

Study Guide Masters, p. 40

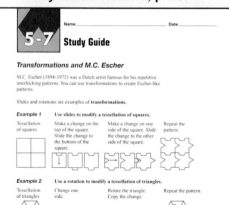

Example 3 CONNECTION

Art The Escher lithograph entitled *Study of Regular Division of the Plane with Reptiles* was created in 1939. Study the figures carefully.

a. Trace the outside edge of one pattern unit.

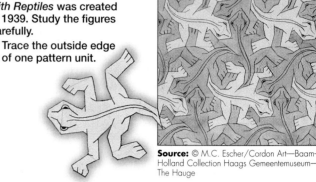

Source: © M.C. Escher/Cordon Art—Baarn—Holland Collection Haags Gemeentemuseum—The Hauge

b. What geometric shape do you think was used for this tessellation?

By looking at what was deleted and then added to make the reptiles, you can see that the shape that was used was a six-sided polygon, or a hexagon.

CHECK FOR UNDERSTANDING

Communicating Mathematics

Read and study the lesson to answer each question.

1. *Describe* a tessellation. **See margin.**
2. *Name* five shapes that will tessellate.

HANDS-ON MATH

3. *Trace* the equilateral triangle.
   a. Measure each angle of the triangle.
   b. There are 360° around a point. Divide 360° by the measure of each angle of the triangle. **6**
   c. Use the triangle to create a tessellation using rotation about one of the vertices. How many figures are needed to fill all the space around the vertex? **6 figures, see margin.**
   d. What is the relationship between the measure of a rotated angle and the number of figures needed to fill the space around the vertex? **See margin.**

2. Sample answers: isosceles trapezoid, parallelogram, right triangle, equilateral triangle, square

3a. 60°, 60°, 60°

Guided Practice

4. Make an Escher-like drawing for the pattern described at the right. Use a tessellation of two rows of three squares as the base. **See Answer Appendix.**

5. Tell whether the pattern in Exercise 4 involves *translations* or *rotations*. **translations**

6. *Hobbies* Name the polygon and the transformation used to produce a sheet of the Pacific 97 stamps at the left. **triangles; rotation**

222 Chapter 5 Geometry: Investigating Patterns

■ Reteaching the Lesson ■

Activity Most students will have an easier time making tessellations with squares than with parallelograms or triangles. Have students begin with an array of squares. Then have them replace all of the vertical segments with the same pattern. You may also choose to have them replace all of the horizontal segments.

Additional Answer

3d. The number of figures needed to fill the space equals 360 divided by the measure of the angle.

EXERCISES

Practice

7–9. See Answer Appendix.

Make an Escher-like drawing for each pattern described. For squares, a tessellation of two rows of three squares as the base. For the triangle, use a tessellation of two rows of five equilateral triangles as the base.

7.
8.
9.

Tell whether each pattern involves translations or rotations.

10. Exercise 7
    translation
11. Exercise 8
    rotation
12. Exercise 9
    rotation

13. Create an Escher-like drawing from a pattern of your own. **See students' work.**

Applications and Problem Solving

14. They would fit on a baking sheet without any gaps or waste.

14. *Food Design* Crackers are often baked into shapes that tessellate. Explain why a company would want to design tessellating crackers.

Tostitos   **Harvest Crisps**   **CHEEZ-ITS**

15. *Construction* A floor installer lays a two-by-two tile at the right. Examine each design and determine whether *translations* or *rotations* were used to create each pattern.

a.
b.
c.

  translation   rotation   rotation

16. *Working on the* **Chapter Project** Refer to the shape you found on page 185. Tessellate your shape as described in this lesson. Use your shape to cover a poster board. Color your design. **See students' work.**

17. *Critical Thinking* Refer to the tessellation.
    a. What polygon was transformed to produce the pattern unit used to create this tessellation? **square**
    b. What transformation(s) were used to create the pattern unit? Illustrate. **Translation; see Answer Appendix for drawings.**

Mixed Review

18. *Geometry* Draw two similar right triangles. *(Lesson 5-6)* **See Answer Appendix.**

19. **Test Practice** High temperatures for 12 cities on March 20 were: 40, 72, 74, 35, 58, 64, 40, 67, 40, 75, 68, 51. What is the range of this set of data? *(Lesson 4-5)* **C**

    A 75   B 51   C 40   D 11

Lesson 5-7 Transformations and M. C. Escher   **223**

Extending the Lesson

Enrichment Masters, p. 40

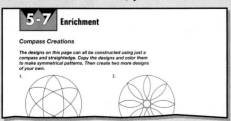

Activity Have students read more about M. C. Escher and the connections he found between mathematics and art. Escher created impossible worlds where water runs uphill and people climb stairs forever. Have students display Escher prints and attempt to explain how the artist created them.

Study Guide and Assessment

Study Guide and Assessment

Vocabulary

This section provides a listing of the new terms, properties, and phrases that were introduced in this chapter. Have students define each term and provide an example or two of it, if appropriate.

Understanding and Using the Vocabulary

These exercises check students' understanding of the terms by using a variety of verbal formats including matching, completion, and true/false.

Glossaries A complete glossary of terms appears on pages 702–710. The glossary also appears in Spanish on pages 711–721.

Additional Answer

9. If a figure can be turned less than 360° about its center and it looks like the original, then the figure has rotational symmetry.

CHAPTER 5 Study Guide and Assessment

Vocabulary

After completing this chapter, you should be able to define each term, concept, or phrase and give an example or two of each.

Geometry

acute (p. 196)
alternate exterior angles (p. 189)
alternate interior angles (p. 189)
ASA congruence (p. 210)
complementary angles (p. 198)
congruent triangles (p. 210)
corresponding angles (p. 189)
corresponding parts (p. 210)
dilation (p. 214)
equilateral (p. 196)
isosceles (p. 196)
line symmetry (p. 206)
obtuse (p. 196)
parallel lines (p. 188)
parallelogram (p. 201)
perpendicular (p. 197)
polygon (p. 196)
quadrilateral (p. 201)
rectangle (p. 201)

reflection (p. 206)
rhombus (p. 201)
right (p. 196)
rotation (p. 207)
rotational symmetry (p. 207)
SAS congruence (p. 210)
scalene (p. 196)
similar triangles (p. 215)
square (p. 201)
SSS congruence (p. 210)
supplementary angles (p. 190)
tessellation (p. 220)
transformation (p. 220)
translation (p. 220)
transversal (p. 188)
trapezoid (p. 201)
triangle (p. 196)
vertical angles (p. 190)

Problem Solving
use a Venn diagram (p. 194)

Understanding and Using the Vocabulary

Choose the correct term or symbol to complete each sentence.

1. The symbol ___?___ means *is similar to*. **i**
2. ___?___ are transformations that involve a turn about a given point. **f**
3. ___?___ are tilings formed by repeating a shape or shapes. **g**
4. A figure that can be folded exactly in half is said to have ___?___. **b**
5. ___?___ angles have measures greater than 90° and less than 180°. **e**
6. The symbol ___?___ means *is parallel to*. **j**
7. A parallelogram that has four congruent sides is a(n) ___?___. **a**
8. A(n) ___?___ is a line that intersects two other lines. **c**

In Your Own Words 9. See margin.

9. *Tell* how to determine whether a figure has rotational symmetry.

a. rhombus
b. line symmetry
c. transversal
d. acute
e. obtuse
f. rotations
g. tessellations
h. ≅
i. ~
j. ∥

224   Chapter 5 Geometry: Investigating Patterns

MindJogger Videoquizzes

MindJogger Videoquizzes provide an alternative review of concepts presented in this chapter. Students work in teams to answer questions, gaining points for correct answers. The questions are presented in three rounds.

Round 1 Concepts–5 questions
Round 2 Skills–4 questions
Round 3 Problem Solving–4 questions

Study Guide and Assessment Chapter 5

Objectives & Examples

Upon completing this chapter, you should be able to:

● identify lines that are parallel and types of angles formed by parallel lines and transversals *(Lesson 5-1)*

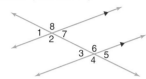

If $m\angle 6 = 135°$, then $m\angle 2 = 135°$.
If $m\angle 5 = 43°$, then $m\angle 6 = 137°$.
If $m\angle 8 = 118°$, then $m\angle 6 = 118°$.

● classify triangles by their angles and their sides and find the measures of missing angles in triangles *(Lesson 5-2)*

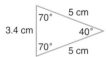

The triangle is acute and isosceles.

● classify quadrilaterals *(Lesson 5-3)*

The names that describe this quadrilateral are parallelogram and rectangle.

Review Exercises

Use these exercises to review and prepare for the chapter test.

Refer to the figure below for Exercises 10–12.

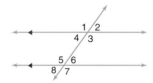

10. Find $m\angle 3$ if $m\angle 5 = 105°$. **105°**
11. Find $m\angle 1$ if $m\angle 2 = 53°$. **127°**
12. Find $m\angle 1$ if $m\angle 7 = 118°$. **118°**

Classify each triangle by its angles and by its sides. 14. obtuse, isosceles

13.

right, scalene

15. In $\triangle RST$, $m\angle R = 33°$ and $m\angle S = 72°$. What is $m\angle T$? **75°**

Sketch each figure. Let Q = quadrilateral, P = parallelogram, R = rectangle, S = square, RH = rhombus, and T = trapezoid. Write all of the letters that describe it inside the figure.

16. 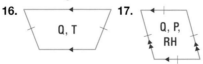   Q, T   17. Q, P, RH

Chapter 5 Study Guide and Assessment **225**

Objectives & Examples

This section reviews the skills and concepts of the chapter and shows completely worked examples.

Review Exercises

These exercises provide practice for the corresponding objectives.

Assessment and Evaluation Masters, pp. 115–116

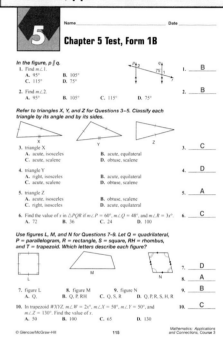

Assessment and Evaluation

Six forms of Chapter 5 Test are available in the *Assessment and Evaluation Masters* as shown in the chart.

Chapter 5 Test, Form 1B, is shown at the right. Chapter 5 Test, Form 2B, is shown on the next page.

1A	Multiple Choice	Honors
1B	Multiple Choice	Average
1C	Multiple Choice	Basic
2A	Free Response	Honors
2B	Free Response	Average
2C	Free Response	Basic

Study Guide and Assessment **225**

Study Guide and Assessment

Chapter 5 Study Guide and Assessment

Additional Answers
22. yes; △KLM ≅ △TRS; ASA
23. yes; △WXY ≅ △PQR; SSS
24. Similar; corresponding angles are congruent.

Assessment and Evaluation Masters, pp. 121–122

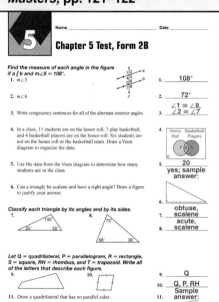
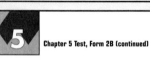

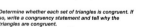

Objectives & Examples

● identify line symmetry and rotational symmetry *(Lesson 5-4)*

Draw all lines of symmetry for a square.

The figure also has rotational symmetry.

● verify congruent triangles by using SSS, ASA, and SAS *(Lesson 5-5)*

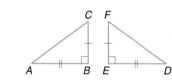

△ABC ≅ △DEF by SAS.

● identify similar triangles *(Lesson 5-6)*

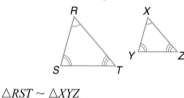

△RST ~ △XYZ

● create Escher-like drawings using translations and rotations *(Lesson 5-7)*

An example of a tessellation of squares is a floor covered by tiles. By modifying these squares, an Escher-like drawing can be made.

Review Exercises

Trace each figure. Determine whether the figure has line symmetry. If so, draw the lines of reflection.

18.   19.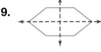

20. Does the figure in Exercise 18 have rotational symmetry? **no**
21. Does the figure in Exercise 19 have rotational symmetry? **yes**

Determine whether each pair of triangles is congruent. If so, write a congruence statement and tell why the triangles are congruent. 22–23. See margin.

22.   23.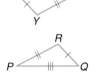

Tell whether each pair of triangles is congruent, similar, or neither. Justify your answer. 25. congruent; SAS

24.   25. ...

24. See margin.

26–27. See Answer Appendix.

Make an Escher-like drawing for each pattern described.

26.   27.

226 Chapter 5 Geometry: Investigating Patterns

Test and Review Software

You may use this software, a combination of an item generator and item bank, to create your own tests or worksheets. Types of items include free response, multiple choice, short answer, and open ended.

CD-ROM Program

The CD-ROM Program contains an Assessment Game whose questions review the concepts in this chapter.

Study Guide and Assessment   Chapter 5

Applications & Problem Solving

28. Traffic The shape of a ROAD NARROWS sign is a parallelogram. How many pairs of sides of a ROAD NARROWS sign are parallel?
(Lesson 5-1)   **2 pairs of sides**

29. Use a Venn Diagram Oliver Middle School offers different activities before and after school. Twenty-one students have signed up for aerobics classes. Thirty students have signed up for basketball. Six students have signed up for both activities. What is the total number of students who have signed up for these two activities?
(Lesson 5-2A)   **45 students**

30. Construction A 12-foot ladder is leaning against a house. The base of the ladder is 4 feet from the house. Classify the triangle formed by the house, the ground, and the ladder by its angles.
(Lesson 5-2)   **right triangle**

31. Architecture Frank Lloyd Wright was one of the most influential American architects. Identify the types of quadrilaterals Wright used in the roof of the home shown below. *(Lesson 5-3)*
trapezoids, rectangles

Alternative Assessment

● **Performance Task**

Suppose you are designing wallpaper for a friend. Your friend wants the design to contain a figure composed of any combination of two or more quadrilaterals or triangles. Your friend also wants the figure to have at least four lines of symmetry. Draw a figure to meet these requirements.   **See students' work.**

Your friend wants to know if your figure will tessellate. If it will, draw the tessellation. If it won't, explain why.   **See students' work.**

A practice test for Chapter 5 is provided on page 651.

● **Completing the CHAPTER Project**

Use the following checklist to make sure your school logo is complete.

☑ The original shape is shown with all of its lines of symmetry.

☑ Information about the rotational symmetry of the original shape is accurate.

☑ The logo is used to neatly fill the poster board.

 Select one of the words you learned in this chapter and place the word and its definition in your portfolio. Attach a note to it explaining why you selected it.

 Performance Assessment

Additional performance assessment tasks for this chapter are included in the *Assessment and Evaluation Masters* on page 125. A scoring guide is also provided on page 137.

Applications & Problem Solving

This section provides additional practice in solving real-world problems that involve the skills of this chapter.

Alternative Assessment

The ***Performance Task*** provides students with a performance assessment opportunity to evaluate their work and understanding.

 CHAPTER Project

Students should complete the final stages of their project and prepare a class demonstration of their results. A scoring guide for the project is available in the *Investigations and Projects Masters*, p. 35.

 Students should add to their portfolios at this time.

Assessment and Evaluation Masters, p. 125

Standardized Test Practice

The Standardized Test Practice may be used to help students prepare for standardized tests. The test items are written in the same style as those in state proficiency tests and standardized tests like CAT, CTBS, ITBS, MAT, SAT, and Terra Nova. The test items cover skills and concepts covered up to this point in the text.

The pages can be used as an overnight assessment. After students have completed the pages, discuss how each problem can be solved, or provide copies of the solutions from the *Solutions Manual*.

Assessment and Evaluation Masters, p. 131

CHAPTERS 1–5 Standardized Test Practice
Assessing Knowledge & Skills

Section One: Multiple Choice

There are nine multiple-choice questions in this section. Choose the best answer. If a correct answer is *not here*, choose the letter for Not Here.

1. The line plot shows the amount of allowance received by 30 students. How many students received $10 or less? **A**

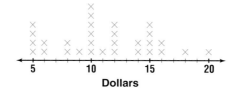

   A 15
   B 10
   C 5
   D 3

2. How many lines of symmetry does an equilateral triangle have? **H**
   F 0
   G 1
   H 3
   J 6

3. How many pairs of congruent triangles are formed by the diagonals of rectangle *ABCD*? **C**

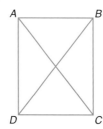

   A 2
   B 3
   C 4
   D 5

Please note that Questions 4–9 have five answer choices.

4. Mindy earns $7 each weekday for her paper route. She earns between $10 and $15 on Saturday and on Sunday. What is a reasonable amount for Mindy to earn during a seven-day week? **H**
   F less than $45
   G between $45 and $55
   H between $55 and $65
   J between $65 and $75
   K more than $75

5. The Venn diagram shows the relationship of five sets of eighth grade girls. Each set and the intersections shown have at least one member in it.

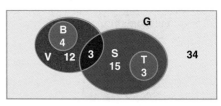

   G = set of all eighth grade girls
   B = set of all basketball players
   V = set of all volleyball players
   S = set of all softball players
   T = set of all tennis players

   Which is a valid conclusion concerning these sets? **D**
   A There is no volleyball player on the basketball team.
   B There is no tennis player on the softball team.
   C All softball players play volleyball.
   D Some girls play both softball and volleyball.
   E All tennis players play basketball.

228 Chapters 1–5 Standardized Test Practice

◀◀◀ **Instructional Resources**

Another cumulative review is shown at the left and is available in the *Assessment and Evaluation Masters*, p. 131.

Standardized Test Practice Chapters 1–5

6. The graph shows how the world's land is distributed. What percent of the world's land is *not* in the North America or South America? **F**

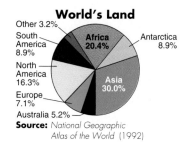

World's Land
Other 3.2%
South America 8.9%
Africa 20.4%
Antarctica 8.9%
North America 16.3%
Asia 30.0%
Europe 7.1%
Australia 5.2%
Source: National Geographic Atlas of the World (1992)

F 74.8%   G 83.7%
H 30.1%   J 25.2%
K 16.3%

7. Thomas bought a gym bag that was reduced 30%. If the bag originally cost $20, what was the final cost of the bag including 6% sales tax? **B**

A $14.00   B $14.84
C $16.48   D $20.00
E $20.84

8. Moses wants to paint a wall that measures 12 feet by 15 feet. There are two windows on the wall that each measure 2 feet by 3 feet. How much surface area will be covered with paint? **H**

F 180 ft²   G 178 ft²
H 168 ft²   J 152 ft²
K 150 ft²

9. During the day, the temperature rose 6°F every two hours. At 8 A.M., the temperature was −25°F. What was the temperature at 2 P.M.? **B**

A 11°F   B −7°F
C −43°F   D −13°F
E Not Here

Section Two: Free Response

This section contains seven questions for which you will provide short answers. Write your answers on your paper.

10. Solve $5x + 32 < 42$. Show the solution on a number line.
    $x < 2$; see margin for number line.

11. Simplify $(3a + b) + (2a + 4b)$. **$5a + 5b$**

12. Solve $a = 20 \cdot (-2)$. **−40**

13. Find the absolute value of −28. **28**

14. Find the value of x. **25**

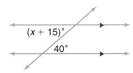
$(x + 15)°$
$40°$

15. Solve $b = -56 \div (-7)$. **8**

16. Which point on the graph below is in the third quadrant? **D**

Additional Answer

10.

Instructional Resources ▶▶▶

Additional standardized test practice is shown at the right and is available in the *Assessment and Evaluation Masters*, pp. 129–130.

Test-Taking Tip

Test anxiety or getting overly worried about taking a standardized test keeps some students from doing their best. Try to find a moment of quiet before a test in order to focus. The best way to reduce test anxiety is to prepare and relax.

Test-Taking Tip

One of the important skills in test-taking is managing stress. Tell students that taking a deep, cleansing breath occasionally as they take the test will often reduce tension.

Assessment and Evaluation Masters, pp. 129–130

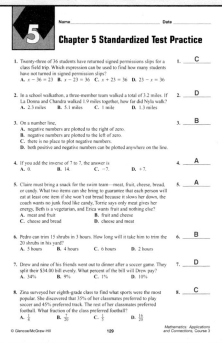

CHAPTER 6
Exploring Number Patterns

Previewing the Chapter

Overview
This chapter develops basic ideas in number theory. Patterns are stressed. Students investigate prime factorization, greatest common factor, and least common multiple and then apply these concepts to fraction simplification and probability. Rational numbers are introduced, their relationship to decimals is investigated, and methods of comparing and ordering them are developed. Students also learn to solve problems by making a list.

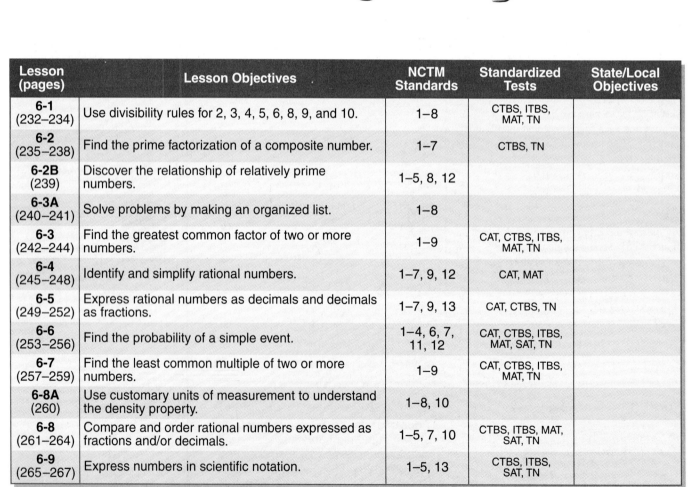

Lesson (pages)	Lesson Objectives	NCTM Standards	Standardized Tests	State/Local Objectives
6-1 (232–234)	Use divisibility rules for 2, 3, 4, 5, 6, 8, 9, and 10.	1–8	CTBS, ITBS, MAT, TN	
6-2 (235–238)	Find the prime factorization of a composite number.	1–7	CTBS, TN	
6-2B (239)	Discover the relationship of relatively prime numbers.	1–5, 8, 12		
6-3A (240–241)	Solve problems by making an organized list.	1–8		
6-3 (242–244)	Find the greatest common factor of two or more numbers.	1–9	CAT, CTBS, ITBS, MAT, TN	
6-4 (245–248)	Identify and simplify rational numbers.	1–7, 9, 12	CAT, MAT	
6-5 (249–252)	Express rational numbers as decimals and decimals as fractions.	1–7, 9, 13	CAT, CTBS, TN	
6-6 (253–256)	Find the probability of a simple event.	1–4, 6, 7, 11, 12	CAT, CTBS, ITBS, MAT, SAT, TN	
6-7 (257–259)	Find the least common multiple of two or more numbers.	1–9	CAT, CTBS, ITBS, MAT, TN	
6-8A (260)	Use customary units of measurement to understand the density property.	1–8, 10		
6-8 (261–264)	Compare and order rational numbers expressed as fractions and/or decimals.	1–5, 7, 10	CTBS, ITBS, MAT, SAT, TN	
6-9 (265–267)	Express numbers in scientific notation.	1–5, 13	CTBS, ITBS, SAT, TN	

CAT = California Achievement Tests, CTBS = Comprehensive Tests of Basic Skills, ITBS = Iowa Tests of Basic Skills, MAT = Metropolitan Achievement Tests, SAT = Stanford Achievement Tests, TN = Terra Nova

Organizing the Chapter

CD-ROM
All of the blackline masters in the Teacher's Classroom Resources are available on the **Electronic Teacher's Classroom Resources** CD-ROM.

LESSON PLANNING GUIDE

Lesson	Extra Practice (Student Edition)	Blackline Masters (page numbers)											Transparencies A and B
		Study Guide	Practice	Enrichment	Assessment & Evaluation	Classroom Games	Diversity	Hands-On Lab	School to Career	Science and Math Lab Manual	Technology	
6-1	p. 620	41	41	41							64	6-1
6-2	p. 620	42	42	42		17–18						6-2
6-2B								51				
6-3A	p. 620											
6-3	p. 621	43	43	43	155							6-3
6-4	p. 621	44	44	44								6-4
6-5	p. 621	45	45	45	154, 155				32			6-5
6-6	p. 622	46	46	46								6-6
6-7	p. 622	47	47	47	156							6-7
6-8A								52				
6-8	p. 622	48	48	48				73				6-8
6-9	p. 623	49	49	49	156		32				63	6-9
Study Guide/ Assessment					141–153, 157–159							

Other Chapter Resources

Student Edition
Chapter Project, pp. 231, 248, 252, 264, 271
Math in the Media, p. 267
Let the Games Begin, p. 238

Technology
 CD-ROM Program
 Interactive Mathematics Tools Software

Teacher's Classroom Resources

Applications
Family Letters and Activities, pp. 63–64
Investigations and Projects Masters, pp. 37–40

Meeting Individual Needs
Transition Booklet, pp. 5–6, 9–10, 13–18
Investigations for the Special Education Student, pp. 13–14

Teaching Aids
Answer Key Masters
Block Scheduling Booklet
Lesson Planning Guide
Solutions Manual

Professional Publications
Glencoe Mathematics Professional Series

Planning the Chapter

MindJogger Videoquizzes provide a unique format for reviewing concepts presented in the chapter.

Assessment Resources

Student Edition
Mixed Review, pp. 234, 238, 244, 248, 252, 256, 259, 264, 267
Mid-Chapter Self Test, p. 248
Math Journal, pp. 244, 263
Study Guide and Assessment, pp. 268–271
Performance Task, p. 271
Portfolio Suggestion, p. 271
Standardized Test Practice, pp. 272–273
Chapter Test, p. 652

Assessment and Evaluation Masters
Multiple-Choice Tests (Forms 1A, 1B, 1C), pp. 141–146
Free-Response Tests (Forms 2A, 2B, 2C), pp. 147–152
Performance Assessment, p. 153
Mid-Chapter Test, p. 154
Quizzes A–D, pp. 155–156
Standardized Test Practice, pp. 157–158
Cumulative Review, p. 159

Teacher's Wraparound Edition
5-Minute Check, pp. 232, 235, 242, 245, 249, 253, 257, 261, 265
Building Portfolios, p. 230
Math Journal, pp. 239, 260
Closing Activity, pp. 234, 238, 241, 244, 248, 252, 256, 259, 264, 267

Technology
Test and Review Software
MindJogger Videoquizzes
CD-ROM Program

Materials and Manipulatives

Lesson 6-1
calculator

Lesson 6-2
grid paper†
calculator

Lesson 6-2B
colored pencils
tracing paper

Lesson 6-5
calculator

Lesson 6-7
colored pencils

Lesson 6-8A
12-inch ruler*
calculator

Lesson 6-8
calculator

Lesson 6-9
calculator

*Glencoe Manipulative Kit †Glencoe Overhead Manipulative Resources

Pacing Chart

See pages T25–T27 for the Course Planning Calendar.

COURSE	DAY 1	DAY 2	DAY 3	DAY 4	DAY 5	DAY 6	DAY 7
Standard	Chapter Project	Lesson 6-1	Lessons 6-2 & 6-2B		Lesson 6-3A	Lesson 6-3	Lesson 6-4
Honors	Chapter Project & Lesson 6-1	Lessons 6-2 & 6-2B		Lesson 6-3A	Lesson 6-3	Lesson 6-4	Lesson 6-5
Block	Chapter Project & Lesson 6-1	Lessons 6-2 & 6-2B	Lessons 6-3A & 6-3	Lessons 6-4 & 6-5	Lessons 6-6 & 6-7	Lessons 6-8A & 6-8	Lessons 6-9

The *Transition Booklet* (Skills 1, 3, 5–7) can be used to practice comparing and ordering decimals, multiplying decimals, simplifying fractions, and working with mixed numbers, improper fractions, and equivalent fractions.

Interactive Mathematics:
Activities and Investigations

is an activity-based program that may be used as an enhancement for chapters in *Mathematics: Applications and Connections*.

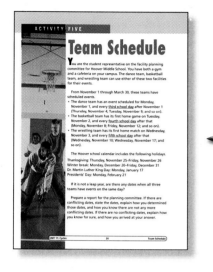

Unit 11, Activity Five,
Team Schedule
Use with Lesson 6-3A.

Summary Students work in groups to create a schedule for the school gym and cafeteria to accommodate the school's team without a facility conflict. Each group prepares a written explanation of its solution and the method(s) the students used to arrive at their conclusion.

Math Connection Students may use listings or patterns and multiples to solve the problem of scheduling several school events. This activity involves cyclical patterns.

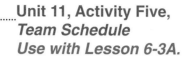

Unit 10, Activity Two,
Probable Cause—What If . . .
Use with Lesson 6-6.

Summary Students work in groups to determine the probability of five given situations. They then explain the process they used to arrive at their answers.

Math Connection Students explore probability by analyzing five situations. Probability is the ratio of the number of ways an outcome can occur to the number of possible outcomes.

DAY 8	DAY 9	DAY 10	DAY 11	DAY 12	DAY 13	DAY 14	DAY 15
Lesson 6-5	Lesson 6-6	Lesson 6-7	Lessons 6-8A & 6-8		Lesson 6-9	Study Guide and Assessment	Chapter Test
Lesson 6-6	Lesson 6-7	Lesson 6-8	Lesson 6-9	Study Guide and Assessment	Chapter Test		
Study Guide and Assessment, Chapter Test							

Chapter 6 **230d**

Enhancing the Chapter

APPLICATIONS

Classroom Games, pp. 17–18

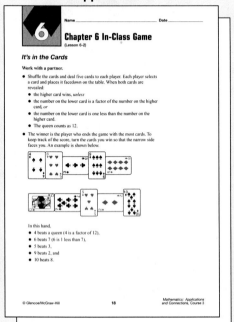

Diversity Masters, p. 32

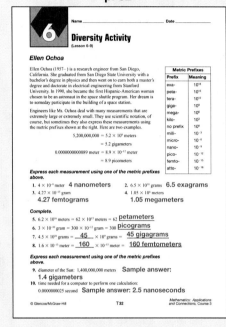

School to Career Masters, p. 32

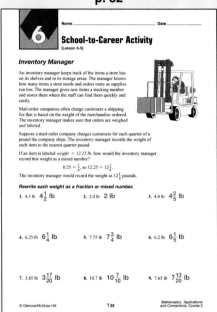

Family Letters and Activities, pp. 63–64

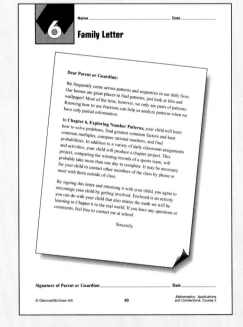

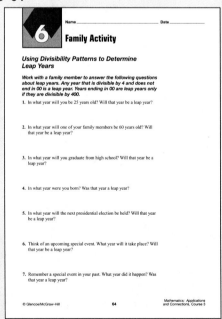

230e Chapter 6

MANIPULATIVES/MODELING

Hands-On Lab Masters, p. 73

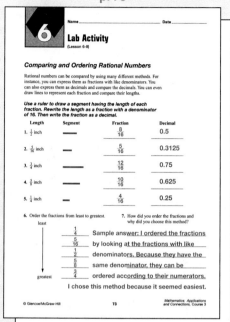

ASSESSMENT/EVALUATION

Assessment and Evaluation Masters, pp. 154–156

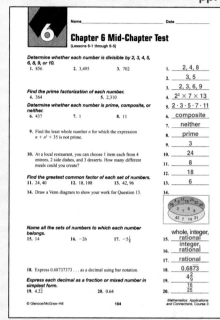

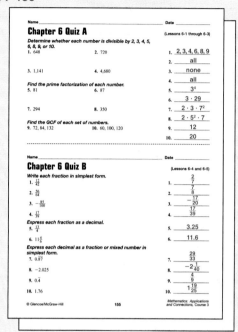

TECHNOLOGY/MULTIMEDIA

Technology Masters, pp. 63–64

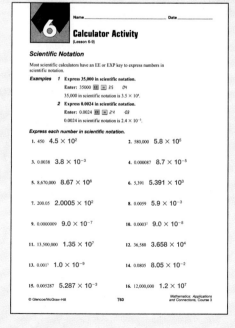

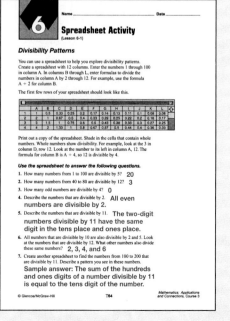

MEETING INDIVIDUAL NEEDS

Investigations for the Special Education Student, pp. 13–14

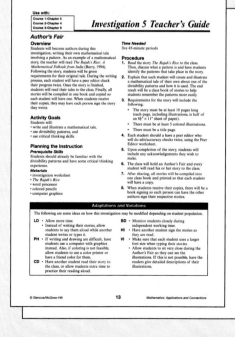

Chapter 6 **230f**

CHAPTER 6 NOTES

Theme: Sports
The WNBA is not the first women's basketball league. The All American Red Heads Team, formed in 1936, used men's rules and competed against men's teams. The women's Professional Basketball League consisted of eight teams and played three seasons, starting in December of 1978. The Liberty Basketball Association (1991) and the Women's World Basketball Association (1992), though short-lived, showed that interest in women's professional basketball persisted. The WNBA began its first season on June 21, 1997.

Question of the Day The WNBA had eight teams during its first season. Assuming all teams had equal talent, what was the probability that the Houston Comets would win the 1997 championship? $\frac{1}{8}$

Assess Prerequisite Skills
Ask students to read through the list of objectives presented in "What you'll learn in Chapter 6." You may wish to ask them what each of the objectives means or if they have experienced or used any of these math concepts before.

 Building Portfolios
Encourage students to revise their portfolios as they study this chapter. They may wish to include examples of concepts learned in this chapter as they apply to other school subjects.

 Math and the Family
In the *Family Letters and Activities* booklet (pp. 63–64), you will find a letter to the parents explaining what students will study in Chapter 6. An activity appropriate for the whole family is also available.

CHAPTER 6 Exploring Number Patterns

What you'll learn in Chapter 6
- to solve problems by making a list,
- to find the greatest common factor and the least common multiple of two or more numbers,
- to identify, simplify, and compare rational numbers,
- to find the probability of a simple event, and
- to express numbers in scientific notation.

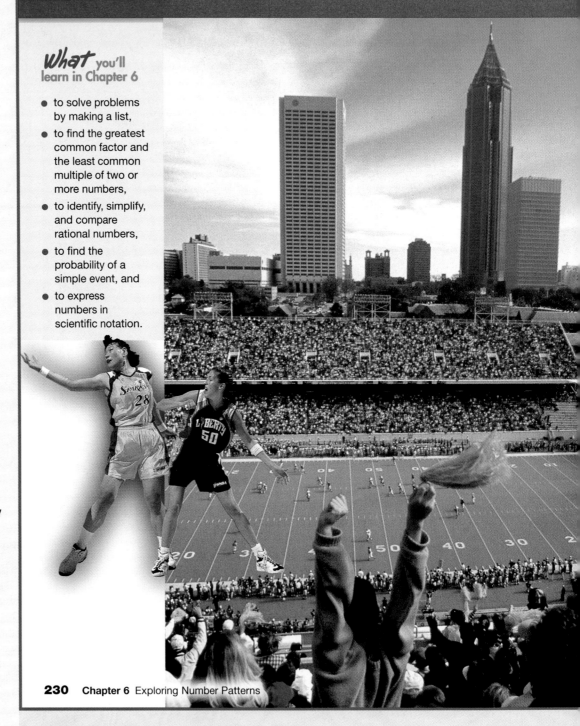

230 Chapter 6 Exploring Number Patterns

 CD-ROM Program

Activities for Chapter 6
- Chapter 6 Introduction
- Interactive Lessons 6-3, 6-7, 6-8
- Extended Activity 6-6
- Assessment Game
- Resource Lessons 6-1 through 6-9

CHAPTER Project

TAKE ME OUT TO THE BALL GAME

In this project, you will write a news article on the record of a professional sports team for the past several years. You will include a comparison of the seasons in the article.

Getting Started

- Decide which professional league you would like to study. Two possibilities are the Women's National Basketball Association (WNBA) or the National Football League (NFL).
- Choose a sports team from your chosen league.
- Research your league and your team. For each of the past several years, find the total number of regular season games played, the number of games won, and team that won the league championship.

Technology Tips

- Use an **electronic almanac** to do your research.
- Use a **spreadsheet** to organize data and make calculations.
- Use a **word processor** to write your news article.

 For up-to-date information on professional sports, visit:
www.glencoe.com/sec/math/mac/mathnet

Working on the Project

You can use what you'll learn in Chapter 6 to help you write a news article about your chosen team.

Page	Exercise
248	33
252	41
264	32
271	Alternative Assessment

Chapter 6 **231**

interNET CONNECTION

Glencoe has made every effort to ensure that the website links for *Mathematics: Applications and Connections* at www.glencoe.com/sec/math/mac/mathnet are current and contain appropriate content. However, these website links are not under Glencoe's control.

Instructional Resources ▶▶▶

A recording sheet to help students organize their data for the Chapter Project is shown at the right and is available in the *Investigations and Projects Masters*, p. 40.

CHAPTER Project NOTES

Objectives Students should
- learn how to organize data.
- learn to compare data by using their knowledge of rational numbers and probability.
- be able to express data they collect and make comparisons of the data in verbal form.

Project Pointer You may suggest that students begin a *Project Folder* to keep their work as they complete each stage of the Chapter Project. The completed project may also be added to their portfolios.

Using Spreadsheets Have students set up the framework for their spreadsheets at the beginning of the chapter. For example, have them set up column heads such as *Number of Games Played, Number of Games Won,* and *Championship Team*. Then as the students work through the chapter, they can fill in the data and add rows for the totals.

Investigations and Projects Masters, p. 40

---

6 Chapter 6 Project

Take Me Out to the Ball Game

Page 248, Working on the Chapter Project, Exercise 33

League: _____

Team name: _____

Year	Number of Games Played	Number of Games Won	Name of Championship Team	Fraction of Games Won

Page 252, Working on the Chapter Project, Exercise 41

Fraction of Games Won	Decimal Equivalent

Page 264, Working on the Chapter Project, Exercise 32

The fractions in order from least to greatest are:

The team's best season was in:

The team's worst season was in:

© Glencoe/McGraw-Hill   40   *Mathematics: Applications and Connections, Course 3*

Chapter 6 Project **231**

6-1 Lesson Notes

Instructional Resources
- *Study Guide Masters,* p. 41
- *Practice Masters,* p. 41
- *Enrichment Masters,* p. 41
- Transparencies 6-1, A and B
- *Technology Masters,* p. 64

 CD-ROM Program
- Resource Lesson 6-1

Recommended Pacing

Standard	Day 2 of 15
Honors	Day 1 of 13
Block	Day 1 of 8

1 FOCUS

5-Minute Check
(Chapter 5)

1. Name the parallel segments in the figure. $\overline{MQ} \parallel \overline{NP}$

2. Classify the triangle by its angles and by its sides. **obtuse; isosceles**

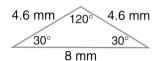

3. Draw all lines of symmetry for a rectangle.

 The 5-Minute Check is also available on **Transparency 6-1A** for this lesson.

Motivating the Lesson
Hands-On Activity Ask students to find 6 ways to separate 12 counters equally into smaller groups with no counters left over. $1 \times 12, 2 \times 6, 3 \times 4, 4 \times 3, 6 \times 2, 12 \times 1$

6-1 Divisibility Patterns

What you'll learn
You will learn to use divisibility rules for 2, 3, 4, 5, 6, 8, 9, and 10.

When am I ever going to use this?
Knowing how to use divisibility rules can help you determine which years are leap years.

Word Wise
divisible

LOOK BACK
You can refer to Lesson 1-2 to review factors.

Fina is the drum major for the Madison High School Band. The band has 144 marching members. How many ways can Fina arrange the band members so that they are marching in a rectangular formation? *This problem will be solved in Example 1.*

To answer this question, Fina must find all of the factor pairs for 144. If a number is a factor of a given number, we can also say the given number is **divisible** by the factor.

The following rules can help determine whether a number is divisible by 2, 3, 4, 5, 6, 8, 9, or 10.

A number is divisible by:
- 2 if the ones digit is divisible by 2.
- 3 if the sum of the digits is divisible by 3.
- 4 if the number formed by the last two digits is divisible by 4.
- 5 if the ones digit is 0 or 5.
- 6 if the number is divisible by 2 *and* 3.
- 8 if the number formed by the last three digits is divisible by 8.
- 9 if the sum of the digits is divisible by 9.
- 10 if the ones digit is 0.

Example 1 APPLICATION

School Refer to beginning of the lesson. How many ways can Fina arrange the band members?

2: The ones digit, 4, is divisible by 2. So 144 is divisible by 2.
3: $1 + 4 + 4$ or 9 is divisible by 3. So 144 is divisible by 3.
4: 44 is divisible by 4. So 144 is divisible by 4.
5: The last digit is 4. So 144 is *not* divisible by 5.
6: 144 is divisible by 2 and 3. So 144 is divisible by 6.
8: 144 is divisible by 8.
9: $1 + 4 + 4$ or 9 is divisible by 9. So 144 is divisible by 9.
10: The last digit is 4. So 144 is *not* divisible by 10.

Fina also knows that $1 \times 144 = 144$ and $12 \times 12 = 144$.

The possible formations are 1 by 144, 2 by 72, 3 by 48, 4 by 36, 6 by 24, 8 by 18, 9 by 16, and 12 by 12.

How do you know these are all of the formations?

Chapter 6 Exploring Number Patterns

Classroom Vignette

"I really like to have students use the Sieve of Eratosthenes to find prime numbers and discover divisibility patterns in numbers. I have students use different colors and shapes to denote divisibility patterns."

Susan Stier, Teacher
Belle Plaine Jr. High School
Belle Plaine, MN

Examples

② Determine whether 2,416 is divisible by 2, 3, 4, 5, 6, 8, 9, or 10.

2: Yes, 6 is divisible by 2.
3: No, 2 + 4 + 1 + 6 or 13 is *not* divisible by 3.
4: Yes, 16 is divisible by 4.
5: No, the ones digit is *not* a 0 or 5.
6: No, 2,416 is divisible by 2, but *not* by 3.
8: Yes, 416 is divisible by 8.
9: No, 2 + 4 + 1 + 6 or 13 is *not* divisible by 9.
10: No, the ones digit is *not* 0.

Therefore, 2,416 is divisible by 2, 4, and 8, but not by 3, 5, 6, 9 or 10.

INTEGRATION

③ **Geometry** The area of a rectangle is 4,329 square inches. The length and width of the rectangle are whole numbers. Find two possible dimensions of the rectangle.

The sum of the digits of 4,329 is 4 + 3 + 2 + 9, or 18, which is divisible by 3 and 9. So 4,329 is divisible by 3 and 9.

4329 ÷ 3 = *1443*   4329 ÷ 9 = *481*

Two possible dimensions of the rectangle are 3 inches by 1,443 inches and 9 inches by 481 inches.

CHECK FOR UNDERSTANDING

Communicating Mathematics

Read and study the lesson to answer each question.

3. Alonzo; if the sum of the digits is divisible by 9, it is also divisible by 3.

1. **Explain** what is meant by the statement "*a* is divisible by *b*." Assume that *a* and *b* are whole numbers.   **b is a factor of a.**

2. **Tell** why an odd number cannot be divisible by 6.   **See Answer Appendix.**

3. **You Decide** Alonzo says that any number that is divisible by 9 is also divisible by 3. Pamela disagrees. Who is correct? Explain.

Guided Practice

Determine whether each number is divisible by 2, 3, 4, 5, 6, 8, 9, or 10.

4. 48   **2, 3, 4, 6, 8**   5. 153   **3, 9**   6. 2,470   **2, 5, 10**

7. Is 6 a factor of 198?   **yes**   8. Is 795 divisible by 10?   **no**

9. Sample answer: 1,110

9. Use mental math to find a four-digit number that is divisible by 3 and 10.

10. Sample answers: 3 × 145, 5 × 87

10. Find two ways to write 435 as a product of two whole number factors.

11. 1 by 40, 2 by 20, 4 by 10, 5 by 8

11. **School** In preparation for a pep assembly, the students are placing 40 chairs on the stage for the football team. If the chairs are to be placed in a rectangular array, what are the possible arrangements for the chairs?

Lesson 6-1 Divisibility Patterns   **233**

Reteaching the Lesson

Activity Have each student prepare a chart with the numbers 2, 3, 4, 5, 6, 8, 9, and 10 as the column heads. In each column, have students list numbers divisible by the number heading that column. Then show that the numbers in each column satisfy the divisibility rule for the number heading the column.

Error Analysis
Watch for students who add digits to test divisibility by numbers other than 3 or 9. **Prevent by** having students multiply the factors to confirm that they produce the desired number.

---

2 TEACH

 Transparency 6-1B contains a teaching aid for this lesson.

Using Mental Math Write the number 360 on the chalkboard. Ask students if it is divisible by 2, 3, 4, 5, 6, 8, 9, or 10. **It is divisible by each number.**

In-Class Examples

For Example 1
How many ways could a dance company with 138 members be arranged in rows that form a rectangle?
1 by 138, 2 by 69, 3 by 46, and 6 by 23

For Example 2
Determine whether 456 is divisible by 2, 3, 4, 5, 6, 8, 9, or 10. **456 is divisible by 2, 3, 4, 6, and 8, but not by 5, 9, or 10.**

For Example 3
Find two possible dimensions of a rectangle with an area of 7,938 square centimeters. **Sample answers: 3 cm by 2646 cm, 6 cm by 1,323 cm, 9 cm by 882 cm**

Study Guide Masters, p. 41

Lesson 6-1   **233**

3 PRACTICE/APPLY

Check for Understanding
If students need additional practice or instruction after completing Exercises 1–11, one of these options may be helpful.
- Extra Practice, see p. 620
- Reteaching Activity, see p. 233
- *Study Guide Masters*, p. 41
- *Practice Masters*, p. 41
- Interactive Mathematics Tools Software

Assignment Guide
Core: 13–33 odd, 34–38
Enriched: 12–30 even, 32–38

4 ASSESS

Closing Activity
Modeling Use collections of books, pencils, or other items. Have students separate each group into smaller, equal-sized groups and state the factors indicated.

Additional Answer
31. Sample answer: $2 \times 8{,}592$, $3 \times 5{,}728$, $4 \times 4{,}296$, $6 \times 2{,}864$, $8 \times 2{,}148$

Practice Masters, p. 41

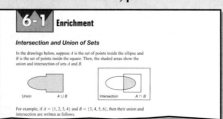

234 Chapter 6

EXERCISES

Practice Determine whether each number is divisible by 2, 3, 4, 5, 6, 8, 9, or 10.

16. 2, 4, 5, 10

12. 56 2, 4, 8
13. 165 3, 5
14. 323 none
15. 918 2, 3, 6, 9
16. 1,700
17. 2,865 3, 5
18. 12,357 3, 9
19. 16,084 2, 4
20. Is 3 a factor of 777? yes
21. Is 5 a factor of 232? no
22. Is 989 divisible by 9? no
23. Is 2,348 divisible by 4? yes
24. Is 16,454 divisible by 8? no
25. Is 2 a factor of 88,096? yes

Use mental math to find a number that satisfies the given conditions.

26–30. Sample answers given.

26. a three-digit number divisible by 3, 6, and 9 522
27. a four-digit number divisible by 2 but not by 6 1,004
28. a five-digit number *not* divisible by 2, 5 or 9 10,001

29. 2×42, 3×28, 4×21
30. $3 \times 4{,}707$, $9 \times 1{,}569$

29. Find three ways to write 84 as a product of two whole number factors.
30. Find two different pairs of whole number factors of 14,121.
31. Find five different pairs of whole number factors of 17,184. See margin.

Applications and Problem Solving

32. *Calendar* Any year that is divisible by 4 and does *not* end in 00 is a leap year. Years ending in 00 are leap years only if they are divisible by 400.
   a. Was 1492 a leap year? yes
   b. Will 2015 be a leap year? no
   c. Were you born in a leap year? See students' work.

33. *Civics* Each star in the U.S. flag represents a state. If another state joins the Union, could the stars be arranged in a rectangular array? Explain. Yes; $3 \times 17 = 51$.

34. *Critical Thinking* Write a rule to determine whether a number is divisible by 15.

Mixed Review

34. A number is divisible by 15 if the sum of its digits is divisible by 3 and the ones digit is 0 or 5.

35. **Test Practice** Which of the following shapes will *not* tessellate?
   (*Lesson 5-7*) A
   A pentagon
   B square
   C equilateral triangle
   D regular hexagon

36. *Statistics* Refer to the bar graph. How many times more students watch 0 to 5 hours than those who watch 15 to 17 hours? (*Lesson 4-1*) 3 times more

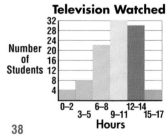

37. Express the ratio $\frac{\$2.88}{9 \text{ ounces}}$ as a unit rate. (*Lesson 3-1*) $\frac{\$0.32}{1 \text{ ounce}}$

38. Solve $p = 85 + (-47)$. (*Lesson 2-3*) 38

234 Chapter 6 Exploring Number Patterns

Extending the Lesson

Enrichment Masters, p. 41

Activity Have students work in small groups to explore divisibility patterns in numbers greater than 10, such as 11, 12, and 15. Ask them to start by charting multiples of the number in question and then try to find a rule to explain the pattern. For example, double digits under 100, such as 33, can be divided by 11.

6-2 Prime Factorization

What you'll learn
You'll learn to find the prime factorization of a composite number.

When am I ever going to use this?
Knowing how to find the prime factorization of a number can help you make secret codes.

Word Wise
prime number
composite number
prime factorization
factor tree
Fundamental Theorem of Arithmetic

For over 2,000 years, mathematicians have been fascinated by numbers with exactly two factors. Using computers, today's mathematicians have discovered very large numbers with only two factors.

In the following Mini-Lab, you will investigate the number of factors for various numbers.

MINI-LAB

Work with a partner. grid paper

Draw a rectangle with 2 square units on the grid paper. A 1-by-2 rectangle is the only way to draw such a rectangle using whole grid squares.

Try This 1–3. See Answer Appendix.
1. Draw as many rectangles as possible with an area of 3 square units.
2. Draw as many rectangles as possible with an area of 4 square units.
3. Continue to draw all possible rectangles for areas from 5 square units to 10 square units.

Talk About It
4. Which areas had only one rectangle? **2, 3, 5, 7**
5. Which areas had more than one rectangle? **4, 6, 8, 9, 10**

Did you know In 1994, Paul Gage and David Slowinski used a supercomputer to find a prime number with 258,716 digits.

When a whole number greater than 1 has *exactly* two factors, 1 and itself, it is called a **prime number**. From the results of the Mini-Lab, you can see that the numbers 2, 3, 5, and 7 are prime numbers.

When a whole number greater than 1 has more than two factors, it is called a **composite number**. From the results of the Mini-Lab, you can see that the numbers 4, 6, 8, 9, and 10 are composite numbers.

The numbers 0 and 1 are *neither* prime *nor* composite. Notice that 0 has an endless number of factors and 1 has only one factor, itself.

A composite number can be expressed as a product of prime numbers. To find the prime factors of any composite number, begin by expressing the number as a product of two factors. Then continue to factor until all the factors are prime. When the number is expressed as a product of factors that are all prime, the expression is called the **prime factorization** of the number.

Lesson 6-2 Prime Factorization **235**

Multiple Learning Styles

 Logical To determine whether a number is prime, you need test only for divisibility by prime numbers less than the square root of the number (the square root can be found using a calculator). For example, the number 97 need be tested only by 2, 3, 5, and 7 because $\sqrt{97} \approx 9.8$. Ask students which numbers they would use to test 147 for divisibility. **2, 3, 5, 7, and 11**

6-2 Lesson Notes

Instructional Resources
- *Study Guide Masters*, p. 42
- *Practice Masters*, p. 42
- *Enrichment Masters*, p. 42
- Transparencies 6-2, A and B
- *Classroom Games*, pp. 17–18
- CD-ROM Program
  - Resource Lesson 6-2

Recommended Pacing	
Standard	Days 3 & 4 of 15
Honors	Days 2 & 3 of 13
Block	Day 2 of 8

1 FOCUS

 5-Minute Check
(Lesson 6-1)
Determine whether each number is divisible by 2, 3, 4, 5, 6, 8, 9, or 10.
1. 777 **3**
2. 3,605 **5**
3. 802 **2**
4. 5,328 **2, 3, 4, 6, 8, 9**
5. 13,712 **2, 4, 8**

The 5-Minute Check is also available on **Transparency 6-2A** for this lesson.

Motivating the Lesson
Problem Solving Have students use their calculators to find all whole numbers 1 through 20 by which 1,649 is divisible. **1, 17**

2 TEACH

 Transparency 6-2B contains a teaching aid for this lesson.

Using the Mini-Lab Make sure students recognize that 1×4 and 4×1 describe the same rectangle. Make sure students include $1 \times n$ as one of the rectangles for each digit.

Lesson 6-2 **235**

In-Class Examples

For Example 1
Find the prime factorization of 210. $2 \cdot 3 \cdot 5 \cdot 7$

For Example 2
Show that the expression $n^2 - n + 41$ yields a prime number for $n = 4$ and 10.

$n = 4$: $n^2 - n + 41$
$= 4^2 - 4 + 41$
$= 53$

$n = 10$: $n^2 - n + 41$
$= 10^2 - 10 + 41$
$= 131$

Teaching Tip Part of the prime factorization of 30,030 is shown below. Without computing, what two prime numbers are missing? Explain.

$30,030 = \blacksquare \cdot \blacksquare \cdot 7 \cdot 13 \cdot 3 \cdot 11$

Because 30,030 ends with the digit 0, the primes must be 2 and 5.

3 PRACTICE/APPLY

Check for Understanding
If students need additional practice or instruction after completing Exercises 1–10, one of these options may be helpful.
- Extra Practice, see p. 620
- Reteaching Activity
- *Study Guide Masters*, p. 42
- *Practice Masters*, p. 42

Each diagram shows a different way to find the prime factorization of 60. These diagrams are called **factor trees**. Use the divisibility rules and mental math to factor 60.

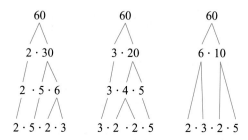

Notice that the bottom row of "branches" in each factor tree is the same except for the order in which the factors are written. Every number has a unique set of prime factors. This property of numbers is called the **Fundamental Theorem of Arithmetic**.

You can use a calculator and the divisibility rules to find the prime factorization.

Example 1 Find the prime factorization of 1,530.

Divide by prime factors until the quotient is prime.

prime factors
↓

1530 ÷ 2 = 765   *0 is divisible by 2. So, 1,530 is divisible by 2.*
765 ÷ 5 = 153   *The last digit is 5. So, 765 is divisible by 5.*
153 ÷ 3 = 51   *$1+5+3$ or 9 is divisible by 3. 153 is divisible by 3.*
51 ÷ 3 = 17   *17 is a prime number.*

$1,530 = 2 \cdot 3^2 \cdot 5 \cdot 17$

Study Hint
Technology When you use a calculator to find prime factors, it is helpful to record them on paper as you find each one.

For centuries, mathematicians have tried to find an expression for calculating every prime number. While an expression may yield some prime numbers, none has been found to yield all prime numbers.

Example 2 INTEGRATION **Algebra** Show that the expression $n^2 - n + 41$ yields prime numbers for $n = 3, 5,$ and 7.

$n = 3$: $n^2 - n + 41 = 3^2 - 3 + 41$
$= 47$ *Replace n with 3.*
47 has only two factors, 1 and 47. So 47 is a prime number.

$n = 5$: $n^2 - n + 41 = 5^2 - 5 + 41$
$= 61$ *Replace n with 5.*
61 has only two factors, 1 and 61. So 61 is a prime number.

$n = 7$: $n^2 - n + 41 = 7^2 - 7 + 41$
$= 83$ *Replace n with 7.*
83 has only two factors, 1 and 83. So 83 is a prime number.

236 Chapter 6 Exploring Number Patterns

■ Reteaching the Lesson ■

Activity Have students use blocks or tiles to form a rectangle for a given number. Point out that only one rectangle can be formed for a prime number and more than one rectangle can be formed for a composite number.

CHECK FOR UNDERSTANDING

Communicating Mathematics

Read and study the lesson to answer each question. 2. See Answer Appendix.

1. *Compare and contrast* prime and composite numbers. See margin.
2. *Draw* two different factor trees to find the prime factorization of 24.
3. *Draw* as many rectangles as possible with areas of 11 square units, 12 square units, 13 square units, 14 square units, and 15 square units.
   a. Which areas have only one rectangle? **11, 13**
   b. Which areas have more than one rectangle? **12, 14, 15**

HANDS-ON MATH

3. See Answer Appendix.

Guided Practice

Determine whether each number is *prime, composite,* or *neither.*

4. 45 **composite** 5. 23 **prime** 6. 1 **neither**

Find the prime factorization of each number.

7. 49 7^2 8. 32 2^5 9. 225 $3^2 \cdot 5^2$

10. **Algebra** Is the value of $2a + 3b$ prime or composite if $a = 11$ and $b = 7$? **prime**

EXERCISES

Practice

13. composite

Determine whether each number is *prime, composite,* or *neither.*

11. 13 **prime** 12. 27 **composite** 13. 96 14. 37 **prime**
15. 0 **neither** 16. 177 **composite** 17. 233 **prime** 18. 507 **composite**

Find the prime factorization of each number.

25. $2^2 \cdot 5^2 \cdot 7^2$
26. $3^2 \cdot 5^2 \cdot 17$

19. 25 5^2 20. 36 $2^2 \cdot 3^2$ 21. 75 $3 \cdot 5^2$ 22. 80 $2^4 \cdot 5$
23. 117 $3^2 \cdot 13$ 24. 72 $2^3 \cdot 3^2$ 25. 4,900 26. 3,825

27. What is the least prime number that is a factor of 1,365? **3**
28. What is the greatest prime number that is less than 60? **59**
29. What is the least prime number that is greater than 60? **61**

Applications and Problem Solving

30a. Sample answer: 3 and 5, 11 and 13, 17 and 19

30. **Number Theory** Primes that differ by two are called *twin primes.*
    a. Give three examples of twin primes.
    b. Are there any primes that differ by 1? Explain. See margin.

31. **Patterns** Consider the number pattern. a. 128, 256, 512
    a. What are the seventh, eighth, and ninth values in the pattern?

	1st	2nd	3rd	4th	5th	6th	7th	8th	9th
Pattern	2	4	8	16	32	64	?	?	?

    b. Write an algebraic expression you could use to find the *n*th value of the pattern. (*Hint:* Use the prime factorization.) 2^n

32. **Algebra** Find a whole number, *n*, for which the expression $n^2 - n + 41$ is not prime. **Sample answer: 41**

Lesson 6-2 Prime Factorization 237

Classroom Vignette

"To introduce prime and composite numbers, I use color tiles arranged in rectangles. If they can be arranged in more than one way to form a rectangle, then they are composite. Remind students that squares are also rectangles. You can also use grid paper to do this activity if color tiles are not available."

Janine Lopez

Janine Lopez, Teacher
Broomfield Heights Middle School
Broomfield, CO

Assignment Guide

Core: 11–31 odd, 33–37
Enriched: 12–28 even, 30–37

Additional Answers

1. Both prime and composite numbers are whole numbers. 0 and 1 are neither prime nor composite. Prime numbers have exactly 2 factors, 1 and the number itself. Composite numbers have more than 2 factors.

30b. Yes; 2 and 3 are the only primes that differ by 1. Any other pair of numbers that differ by 1 will have one even number which equals 2 times another number and therefore will not be prime.

Study Guide Masters, p. 42

6-2 Study Guide

Prime Factorization

A whole number greater than 1 with exactly two factors, 1 and itself, is called a **prime number.**

Example 1 19 is a prime number. It has only 1 and 19 as factors.

A whole number greater than 1 with more than two factors is called a **composite number.**

Example 2 18 is a composite number. It has 1, 2, 3, 6, 9, and 18 as factors.

The numbers 0 and 1 are neither prime nor composite.

A composite number may be written as the product of prime numbers. This product is the **prime factorization** of the number.

Example 3 Find the prime factorization of 660.

The prime factorization of 660 is $3 \cdot 11 \cdot 5 \cdot 2 \cdot 2$.

Determine whether each number is prime, composite, or neither.

1. 28 **composite** 2. 47 **prime** 3. 39 **composite** 4. 61 **prime**
5. 53 **prime** 6. 0 **neither** 7. 159 **composite** 8. 1 **neither**

Find the prime factorization of each number.

9. 30 $3 \cdot 2 \cdot 5$ 10. 155 $31 \cdot 5$ 11. 169 $13 \cdot 13$ 12. 100 $2 \cdot 2 \cdot 5 \cdot 5$
13. 86 $2 \cdot 43$ 14. 98 $2 \cdot 7 \cdot 7$ 15. 495 $5 \cdot 3 \cdot 3 \cdot 11$ 16. 40 $2 \cdot 2 \cdot 2 \cdot 5$

Lesson 6-2 237

4 ASSESS

Closing Activity

Writing State a range of five numbers, for example, 41 through 45. Have students list the prime numbers in the range. Then have them write the prime factorizations of the composite numbers in the range.

Additional Answer

33b. Yes; in order for a number to be a perfect square, all prime factors must be able to be arranged in pairs.

33a. All exponents are even.

35. Similar; the figures are the same shape, but are different sizes.

33. **Critical Thinking** Find the prime factorization of the perfect squares 9, 36, 100, and 144. $9 = 3^2$, $36 = 2^2 \cdot 3^2$, $100 = 2^2 \cdot 5^2$, $144 = 2^4 \cdot 3^2$
   a. What characteristics do the prime factorizations have?
   b. Is this true for all perfect squares? Explain. See margin.

Mixed Review

34. **Test Practice** Ogima must arrange the chairs in the cafeteria in a rectangular pattern. If there are 486 chairs, in which of the following ways can they be arranged? *(Lesson 6-1)* C
   A 27 rows of 19 chairs   B 18 rows of 26 chairs
   C 27 rows of 18 chairs   D 18 rows of 31 chairs

35. **Geometry** Tell whether the pair of figures are *congruent*, *similar*, or *neither*. Explain. *(Lesson 5-6)*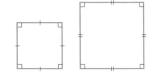

36. Express 38.5% as a decimal and as a fraction in simplest form. *(Lesson 3-4)* $0.385; \frac{77}{200}$

37. **Algebra** Write an inequality for *Four times a number is less than 20.* Then solve the inequality. *(Lesson 1-9)* $4x < 20; x < 5$

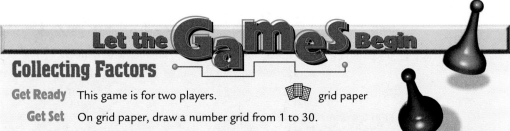

Collecting Factors

Get Ready   This game is for two players.   grid paper

Get Set   On grid paper, draw a number grid from 1 to 30.

Go
- The first player selects a number from the grid and crosses it out. The player then crosses out all of the factors of the selected number. The player is awarded the sum of the factors. For example, if the player selects 10, then he or she crosses out 1, 2, 5, and 10. The player receives $1 + 2 + 5 + 10$ or 18 points.

1	2	3	4	5
6	7	8	9	10
11	12	13	14	15
16	17	18	19	20
21	22	23	24	25
26	27	28	29	30

- The second player selects a number that has not been crossed out and crosses out all of its factors including the number itself. The player is awarded the sum of the factors that had not already been crossed out.
- If a player selects a number with no factors available other than itself, the player receives no points and loses his or her turn.
- The players continue to take turns choosing numbers until the numbers remaining have no other factors available.
- The player with the most points wins.

interNET CONNECTION Visit www.glencoe.com/sec/math/mac/mathnet for more games.

Practice Masters, p. 42

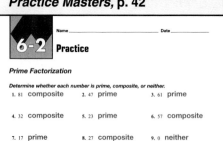

Extending the Lesson

Enrichment Masters, p. 42

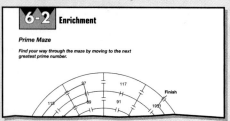

Students may wish to use a different color ink or pencil for each turn in order to avoid confusion. For a more advanced game, the grid can be expanded to a higher number.

*Additional resources for this game can be found on page 51 of the **Classroom Games**.*

HANDS-ON LAB

COOPERATIVE LEARNING

6-2B Basketball Factors

A Follow–Up of Lesson 6-2

- colored pencils
- tracing paper

Ms. Salgado coaches the eighth grade girls' basketball team. She is also a mathematics teacher. Sometimes she uses drills at basketball practice that reinforce concepts such as factors.

In one such drill, the ten players stand in a circle as indicated at the right. The player with the ball (B) calls out a number, such as 4, and passes the ball to the fourth player to her right. The person catching the ball continues the pattern by passing the ball to the fourth person to her right and so on.

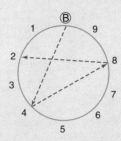

Sometimes the player who starts the drill calls out a number that results in every player around the circle touching the ball once before it returns to her. When this happens, the team is given a water break.

TRY THIS

Work with a partner.

Step 1 Use tracing paper and colored pencils to record the different paths the ball takes when each of the numbers 1 through 9 is called.

Step 2 Copy and complete the table to determine when the team gets a water break.

Number of Players	Number Called	Water Break?
10	1	yes
10	2	no
10	3	yes
10	4	no
10	5	no
10	6	no
10	7	yes
10	8	no
10	9	yes

ON YOUR OWN

1. What numbers can be called so the team can take a break? **1, 3, 7, 9**
2. What is the relationship between the factors of 10 and the factors of the numbers in the answer to Exercise 1? **See margin.**
3. What is the relationship between the factors of 10 and the factors of other numbers in the table? **See margin.**
4. *Reflect Back* Suppose there are 12 girls in the circle. Use factoring to determine which numbers will allow the girls to have a water break. **1, 5, 7, 11**

Lesson 6-2B HANDS-ON LAB **239**

Math Journal

Ask students to write a paragraph about other situations in which factors could help in problem solving.

Additional Answers

2. The only factor that is the same for 10 and each number is 1.

3. There is a factor other than 1 that is the same for 10 and each number.

HANDS-ON 6-2B LAB Notes

GET READY

Objective Students discover the relationship of relatively prime numbers.

Optional Resources
Hands-On Lab Masters
- worksheet, p. 51

MANAGEMENT TIPS

Recommended Time
30 minutes

Getting Started Ask students to solve this problem: *The girls' and boys' basketball teams both had games today. The girls play every 4 days, the boys every 5 days. When will they next play on the same day?* **in 20 days**

The **Activity** helps students see the paths that the ball takes around the circle. It doesn't matter which number starts each path. After students have drawn all nine paths, ask this question: *For what numbers is the path of the ball the same (except for direction)?* **1 and 9, 2 and 8, 3 and 7, 4 and 6**

ASSESS

Have students complete Exercises 1–3. Watch for students who have difficulty tracing the ball's path. Make sure they carefully count the appropriate number of spaces to the right for each pass.

Use Exercise 4 to determine whether students have discovered the relationship between the diagram, prime numbers, and factoring.

Hands-On Lab 6-2B **239**

6-3A THINKING LAB Notes

Objective Students solve problems by making an organized list.

Recommended Pacing	
Standard	Day 5 of 15
Honors	Day 4 of 13
Block	Day 3 of 8

1 FOCUS

Getting Started Have students act out the opening scenario. Then ask students if they consider a pepperoni, mushroom, and green pepper pizza the same as a pepperoni, green pepper, and mushroom pizza.

2 TEACH

Teaching Tip Explain that the make-a-list strategy is used to organize information into a logical framework. This allows the data to be analyzed more easily so that patterns, if they exist, can be spotted.

In-Class Example
A vending machine takes quarters, dimes, and nickels. George wants to buy a bag of pretzels that costs 55¢. How many different combinations of coins can George use to purchase the pretzels by putting exact change in the machine?
11

THINKING LAB PROBLEM SOLVING

6-3A Make a List
A Preview of Lesson 6-3

Atepa and Curtis are helping to make and sell pizzas to raise money for their class trip. The students can add pepperoni, mushrooms, and/or green peppers to the basic cheese pizza. Atepa and Curtis are making posters to advertise the sale. Let's listen in!

Atepa: We can take orders faster if we list all of the different pizzas we can make.

Curtis: Okay, let's start the list with a plain cheese pizza.

Atepa: Then we can add the pizzas with just one topping to our list. Add pizza with pepperoni, pizza with mushrooms, and pizza with green peppers.

Curtis: We must also add pizzas with two or three toppings to our list.

Atepa: Good! If that's all the different pizzas, we can tell how many choices there are on the posters.

THINK ABOUT IT

Work with a partner. 1. Atepa and Curtis have included all the types of pizzas that they can make.

1. **Analyze** Atepa's and Curtis' thinking. Have they included all of the types of pizzas they can make?

2. **Make a list** of all the types of pizzas the students can make. How many different pizzas can Atepa and Curtis advertise on the posters? **See margin.**

3. Apply the **make a list** strategy to solve the following problem.
   Trish wants to buy a cookie from a vending machine. The cookie costs 45¢. If Trish uses exact change, how many different combinations of nickels, dimes, and quarters can she use? **8 combinations**

240 Chapter 6 Exploring Number Patterns

■ Reteaching the Lesson ■

Activity A player throws three darts at this dartboard, with each striking the board. Have small groups make a list of the possible scores. **3, 6, 9, 11, 12, 14, 17, 19, 22, 27**

(Dartboard with rings labeled 1, 4, 9)

Additional Answer

2. cheese; pepperoni; mushrooms; green peppers; pepperoni and mushrooms; pepperoni and green peppers; mushrooms and green peppers; pepperoni, mushrooms, and green peppers
8 pizzas

ON YOUR OWN

4. The fourth step of the 4-step plan for problem solving tells you to *examine* the solution. *Tell* what things you should look for when examining a list used to solve a problem.

5. *Write a Problem* which you could solve using the make a list strategy.

6. *Look Ahead* Explain how the make a list strategy could be used to find what numbers are factors of both 12 and 18. **4–6. See margin.**

MIXED PROBLEM SOLVING

STRATEGIES
Look for a pattern.
Solve a simpler problem.
Act it out.
Guess and check.
Draw a diagram.
Make a chart.
Work backward.

Solve. Use any strategy.

7. *Entertainment* Paper cups come in packages of 40 or 75. Dimas needs 350 paper cups for the school party. How many packages of each size should he buy? **See margin.**

8. *Sports* In the World Series, two teams play each other until one team wins 4 games.
   a. What is the least number of games needed to determine a winner of the World Series?
   b. What is the greatest number of games needed to determine a winner? **7 games**
   c. How many different ways can a team win the World Series in six games or less? (*Hint:* The team that wins the series must win the last game.) **15 ways**

9. *Camp* The camp counselor lists 21 chores on separate pieces of paper and places them in a basket. The counselor takes one piece of paper, and each camper takes one as it is passed around the circle. There is one piece of paper left when the basket returns to the counselor. How many people could be in the circle if the basket goes around the circle more than once? **10, 5, 4, or 2 people**

10. *Critical Thinking* Copy and complete the factor tree. **See margin.**

8a. 4 games
12. Gail, 7 cards; Callie, 5 cards

11. *Technology* Seven line segments are used to make the digits 0 to 9 on a digital clock. (The number 1 is made using the line segments on the right.) **b. bottom left**
    a. In forming these digits, which line segment is used most often? **bottom right**
    b. Which line segment is used the least?

12. *Hobbies* Gail says to Callie, "If you give me one of your football cards, I will have twice as many football cards as you have." Callie answers, "If you give me one of your cards, we will have the same numbers of cards." How many cards do each of the girls have?

13. **Test Practice** How many factors does 21 have? **C**
    A exactly two factors
    B exactly three factors
    C exactly four factors
    D exactly five factors

Lesson 6-3A **THINKING LAB** 241

3 PRACTICE/APPLY

Check for Understanding
Use the results from Exercise 3 to determine whether students apply the make-a-list strategy effectively.

Extra Practice If students need additional practice in problem solving, extra practice is available on the following pages.
- Make an Organized List, see p. 620
- Mixed Problem Solving, see pp. 645–646

Assignment Guide

All: 4–13

4 ASSESS

Closing Activity
Writing A two-symbol code begins with A, B, C, or D and ends with 1, 2, or 3. Have students make a list of all of the possible codes. A1, A2, A3, B1, B2, B3, C1, C2, C3, D1, D2, D3

Additional Answers

4. Make sure all possibilities are listed and no possibility is listed twice.

5. Sample answer: A sandwich can be made with ham, turkey, American cheese and/or Swiss cheese on white bread or wheat bread. How many different sandwiches can be made?

6. List all the factors of 12 and all the factors of 18. Compare the lists and give the factors that are listed for both numbers.

7. 5 packages of 40 and 2 packages of 75

Extending the Lesson

Activity Have the class plan a strategy for a 128-page school yearbook. How many pages will be devoted to activities, candid pictures, faculty pictures, and student pictures? How many student pictures will go on each page?

Additional Answer
10.

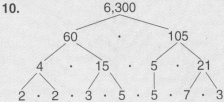

Thinking Lab 6-3A **241**

6-3 Lesson Notes

Instructional Resources
- *Study Guide Masters*, p. 43
- *Practice Masters*, p. 43
- *Enrichment Masters*, p. 43
- Transparencies 6-3, A and B
- *Assessment and Evaluation Masters*, p. 155
- CD-ROM Program
  - Resource Lesson 6-3
  - Interactive Lesson 6-3

Recommended Pacing	
Standard	Day 6 of 15
Honors	Day 5 of 13
Block	Day 3 of 8

1 FOCUS

5-Minute Check
(Lesson 6-2)

Determine whether each number is *prime, composite,* or *neither.*
1. 19 **prime**
2. 21 **composite**
3. 95 **composite**

Find the prime factorization of each number.
4. 54 $2 \cdot 3^3$
5. 2,736 $2^4 \cdot 3^2 \cdot 19$

The 5-Minute Check is also available on **Transparency 6-3A** for this lesson.

2 TEACH

Transparency 6-3B contains a teaching aid for this lesson.

Reading Mathematics This lesson presents a mathematical concept in verbal form. Point out that the words *greatest, common,* and *factor* indicate a method for finding the GCF:
- List all factors.
- Find all common factors.
- Which common factor is the *greatest?*

6-3 Greatest Common Factor

What you'll learn
You'll learn to find the greatest common factor of two or more numbers.

When am I ever going to use this?
Knowing how to find the greatest common factor can help you solve problems involving home improvements.

Word Wise
greatest common factor (GCF)

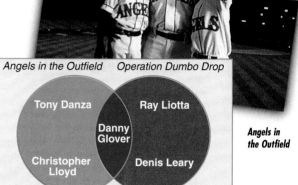

Different actors and actresses star in various movies. The Venn diagram shows the stars of *Angels in the Outfield* and *Operation Dumbo Drop.*

Angels in the Outfield

LOOK BACK
You can refer to Lesson 5-2A to review Venn diagrams.

Danny Glover, Tony Danza, and Christopher Lloyd starred in *Angels in the Outfield.* Danny Glover, Ray Liotta, and Denis Leary starred in *Operation Dumbo Drop.* Notice that Danny Glover starred in both movies.

A Venn diagram is a visual way to examine common elements. Numbers may have common factors. The greatest of the factors common to two or more numbers is called the **greatest common factor (GCF)** of the numbers.

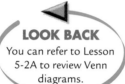

① Find the greatest common factor of 36 and 60.

First, make a list of the factors of 36 and 60.

Factors of 36: 1, 2, 3, 4, 6, 9, 12, 18, 36
Factors of 60: 1, 2, 3, 4, 5, 6, 10, 12, 15, 20, 30, 60

Then, use a Venn diagram to summarize the information.

Factors of 36: 9, 18, 36
Common: 1, 2, 3, 4, 6, 12
Factors of 60: 5, 10, 15, 20, 30, 60

Notice that 1, 2, 3, 4, 6, and 12 are common factors of 36 and 60. The GCF is 12.

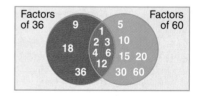
Operation Dumbo Drop

242 Chapter 6 Exploring Number Patterns

Motivating the Lesson
Communication Choose two motor vehicles. Have students list characteristics (factors) of each separately. Then have them list which characteristics they have in common. A Venn diagram could be used for this. Sample answer: Both a sport utility vehicle and a truck have 4-wheel drive.

You can also use prime factorization to find the GCF.

Examples

2 Use prime factorization to find the GCF of 90, 54, and 81.

Use a factor tree to find the prime factorization of each number.

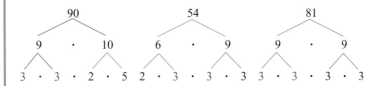

90, 54, and 81 have 3 and 3 as common prime factors. The product of these prime factors, 3 · 3 or 9, is the GCF of 90, 54, and 81.

APPLICATION

3 **Interior Design** Toshiro wants to resurface the top of a table with ceramic tiles. The table is 30 inches long and 24 inches wide. What is the largest square tile that he can use without having to use any partial squares?

Explore   You know the dimensions of the table. You want to know the largest square tile that he can use without having to use partial squares.

Plan   The length of the side of the square must be a factor of both 30 and 24. You want the largest square, so you need to find the greatest common factor of these numbers.

Solve   Write each number as a product of prime factors.

30 = 2 · 3 · 5
24 = 2 · 2 · 2 · 3

Thus, the GCF is 2 · 3 or 6. The tile should be 6 inches on each side.

Examine   Write all of the factors of 30 and 24 using a Venn diagram.

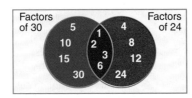

The greatest common factor of 30 and 24 is 6. The answer is correct.

CHECK FOR UNDERSTANDING

Communicating Mathematics

Read and study the lesson to answer each question.

1. **Name** the common factors of 40 and 60. What is the greatest common factor of these numbers? **1, 2, 4, 5, 10, 20; 20**

2. **Write** two numbers whose GCF is 16. **Sample answer: 32 and 48**

Lesson 6-3 Greatest Common Factor **243**

Reteaching the Lesson

Activity Provide schematic drawings like the one below to guide students to the GCF.

12 = [2] · [2] · [3]

18 =   [2] · [3] · [3]

GCF =   [2] · [3]   = [6]

Error Analysis
Watch for students who find the GCF by simply picking out the greatest factor.
Prevent by stressing the word *common*. The GCF must be a factor that is common to all the numbers.

In-Class Examples

For Example 1
Find the GCF of 24 and 40. **8**

For Example 2
Use prime factorization to find the GCF of 36, 72, and 108. **36**

For Example 3
Suzie plans to make a patio with squares of concrete. The patio area is 24 feet long and 14 feet wide. What is the largest size square of concrete that she can pour without having to pour any partial squares? **2 square feet**

Teaching Tip In Example 2, have students use a calculator to find all factors of 90. Then have them test each factor to see if it is also a factor of both 54 and 81.

3 PRACTICE/APPLY

Check for Understanding
If students need additional practice or instruction after completing Exercises 1–8, one of these options may be helpful.
- Extra Practice, see p. 621
- Reteaching Activity
- *Study Guide Masters*, p. 43
- *Practice Masters*, p. 43
- Interactive Mathematics Tools Software

Study Guide Masters, p. 43

6-3 Study Guide

Greatest Common Factor

The greatest of the factors common to two or more number is the **greatest common factor (GCF).** One way to find the GCF is to list all of the factors of each number. Then find the greatest number that is in both lists.

Example 1 Find the GCF of 32 and 48.

Factors of 32: 1, 2, 4, 8, 16, 32
Factors of 48: 1, 2, 3, 4, 6, 8, 12, 16, 24, 48

The common factors of 32 and 48 are 1, 2, 4, 8, and 16.
The greatest common factor of 32 and 48 is 16.

You can use prime factorization to find the greatest common factor.

Example 2 Find the greatest common factor of 90 and 120.

90 = 3 · 30   120 = 3 · 40
   = 3 · 15 · 2   = 3 · 10 · 4
   = 3 · 5 · 3 · 2   = 3 · 2 · 5 · 2 · 2

90 and 120 have 2, 3, and 5 as common factors.
The product of the common factors is the greatest common factor.
2 · 3 · 5 = 30
30 is the greatest common factor of 90 and 120.

Find the GCF of each set of numbers.

1. 24, 40 **8**   2. 18, 30 **6**   3. 12, 48 **12**

4. 36, 90 **18**   5. 50, 20 **10**   6. 15, 17 **1**

7. 52, 16 **4**   8. 63, 42 **21**   9. 25, 40 **5**

10. 36, 12, 72 **12**   11. 28, 49, 105 **7**   12. 30, 45, 75 **15**

Assignment Guide

Core: 9–25 odd, 26–29
Enriched: 10–22 even, 24–29

4 ASSESS

Closing Activity

Speaking Write a pair of prime factorizations on the chalkboard (they must have several common factors). Have students working in pairs describe how they would find the GCF. Then have them state the GCF.

Chapter 6, Quiz A (Lessons 6-1 through 6-3) is available in the *Assessment and Evaluation Masters,* p. 155.

Practice Masters, p. 43

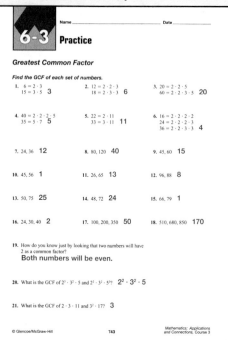

---

3. *Draw* a Venn diagram showing the factors of 30 and 42. **See Answer Appendix.**
    a. Name the GCF of 30 and 42. **6**
    b. Verify your answer by using factor trees to find the prime factorization of each number. **See Answer Appendix.**

Guided Practice Find the GCF of each set of numbers.

4. 45, 20 **5**   5. 27, 54 **27**   6. 26, 34, 64 **2**   7. 45, 105, 75 **15**

8. *Crafts* A quilt pattern calls for congruent square pieces. Some of the fabrics chosen for the quilt come in a width of 48 inches, and others come in a width of 60 inches. If no fabric is to be wasted, what is the greatest length that can be used for the sides of the squares? **12 in.**

EXERCISES

Practice Find the GCF of each set of numbers.

9. 21, 28 **7**   10. 24, 48 **24**   11. 63, 84 **21**   12. 60, 80 **20**
13. 24, 36, 42 **6**   14. 36, 15, 45 **3**   15. 35, 49, 84 **7**   16. 36, 72, 90 **18**

18. 30
19. 42
20. 102

17. 36, 60, 84 **12**   18. 210, 330, 150   19. 84, 126, 210   20. 510, 714, 306

21. What is the GCF of $2 \cdot 3^3 \cdot 5^2$ and $2^3 \cdot 3 \cdot 5^2$? **150**

22. How do you know just by looking that two numbers will have 5 as a common factor? **Both numbers will end in either 5 or 0.**

23. Write three numbers that have a GCF of 7. **Sample answer: 7, 14, 21**

Applications and Problem Solving

24. *Interior Design* Refer to Example 3. How many 6-inch squares will be needed to cover the 30-inch by 24-inch table? **20 tiles**

25. *Carpentry* Mali wants to cut two boards to make shelves. One board is 72 inches long, and the other is 54 inches long. Mali wants each shelf to be the same length and does not want to waste any of the wood.
    a. What is the longest possible length of the shelves? **18 in.**
    b. How many of these shelves will Mali have? **7 shelves**

26b. False; sample counterexample: 4 and 9 are relatively prime, but neither is prime.

26. *Critical Thinking* Numbers that have a GCF of 1 are *relatively prime*. Use this definition to determine whether each statement is true or false and tell why. **a. True; their only common factor is 1.**
    a. Any two prime numbers are relatively prime.
    b. If two numbers are relatively prime, one of them must be prime.

Mixed Review

27. Find the prime factorization of 56. *(Lesson 6-2)* $2^3 \cdot 7$

28. *Geometry* Determine whether the figure has rotational symmetry. *(Lesson 5-4)* **yes**

29. **Test Practice** Lucrecia is an experienced diver. During a dive, she descended from the surface at the rate of 6 meters per minute. What was her location at the end of 12 minutes? *(Lesson 2-7)* **D**
    A −6 m   B −12 m   C −18 m   D −72 m   E Not Here

244 Chapter 6 Exploring Number Patterns

Extending the Lesson

Enrichment Masters, p. 43

Activity Have small groups find the missing digits in this division problem.

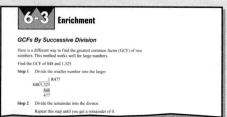

6-4 Rational Numbers

What you'll learn
You'll learn to identify and simplify rational numbers.

When am I ever going to use this?
Knowing how to simplify fractions can help you understand data such as 6 out of 30 games won.

Word Wise
rational number
simplest form

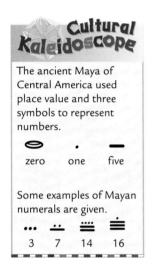

Cultural Kaleidoscope

The ancient Maya of Central America used place value and three symbols to represent numbers.

zero   one   five

Some examples of Mayan numerals are given.

3   7   14   16

About 5,000 years ago, Egyptians used hieroglyphics to represent numbers.

1   10   100   1,000

The Egyptian concept of fractions was mostly limited to fractions with numerators of 1. The hieroglyphics were placed under the symbol ⬭ to indicate the number was a denominator. Study the examples of Egyptian fractions.

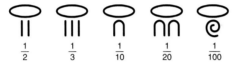
$\frac{1}{2}$ $\frac{1}{3}$ $\frac{1}{10}$ $\frac{1}{20}$ $\frac{1}{100}$

Fractions, their decimal equivalents, and integers are examples of **rational numbers**.

| Rational Numbers | Any number that can be expressed in the form $\frac{a}{b}$, where a and b are integers and $b \neq 0$, is called a rational number. |

You can use a Venn diagram to show how sets within the rational numbers are related.

In the Venn diagram, notice that whole numbers and integers are included in the set of rational numbers. This is because integers, such as 4 and -7, can be written as $\frac{4}{1}$ and $\frac{-7}{1}$.

Rational Numbers: -1.3, $\frac{1}{2}$, -0.29
Integers: -80, $2\frac{2}{3}$
Whole Numbers: $-\frac{7}{1}$, $-\frac{2}{5}$
0, 4, $\frac{3}{3}$, $\frac{6}{1}$, $-\frac{9}{3}$, 0.4

The set of rational numbers also includes decimals because they can be written as fractions with a power of ten for the denominator. For example, -0.29 can be written as $\frac{-29}{100}$.

Mixed numbers are also included in the set of rational numbers because any mixed number can be written as an improper fraction. For example, $5\frac{3}{4}$ can be written as $\frac{23}{4}$.

Lesson 6-4 Rational Numbers 245

 Investigations for the Special Education Student

This blackline master booklet helps you plan for the needs of your special education students by providing long-term projects along with teacher notes. Investigation 5, *Author's Fair*, may be used with this chapter.

6-4 Lesson Notes

Instructional Resources
- *Study Guide Masters*, p. 44
- *Practice Masters*, p. 44
- *Enrichment Masters*, p. 44
- Transparencies 6-4, A and B
- CD-ROM Program
  - Resource Lesson 6-4

Recommended Pacing	
Standard	Day 7 of 15
Honors	Day 6 of 13
Block	Day 4 of 8

1 FOCUS

 5-Minute Check
(Lesson 6-3)

Find the GCF for each set of numbers.
1. 16, 28   4
2. 10, 14   2
3. 14, 35   7
4. 12, 18, 36   6
5. 81, 18, 9   9

The 5-Minute Check is also available on **Transparency 6-4A** for this lesson.

Motivating the Lesson
Communication State that the word *ratio* can be defined as a comparison of two numbers. Ask students to guess the meaning of the word *rational*. Sample answer: capable of being expressed as a comparison of two numbers

2 TEACH

 Transparency 6-4B contains a teaching aid for this lesson.

Using Cooperative Groups
When you discuss the Venn diagram, have small groups each draw a number line from -10 to 10. Have them graph each of the rational numbers except -80 shown in the Venn diagram.

Lesson 6-4   245

Teaching Tip For Examples 1–4, draw a number line on the chalkboard. Indicate various points on the line and ask students to say whether the coordinate of each point is a whole number, an integer, and/or a rational number.

In-Class Examples

Name all the sets of numbers to which each number belongs.

For Example 1
412 whole numbers, integers, rationals

For Example 2
$-\frac{10}{2}$ integers, rationals

For Example 3
$-2\frac{3}{10}$ rationals

For Example 4
0.55 rationals

For Example 5
Write $\frac{18}{24}$ in simplest form. $\frac{3}{4}$

For Example 6
The average annual precipitation in Anchorage, Alaska, is about 16 inches. Express this amount in feet.
$1\frac{1}{3}$ feet

Examples

Name all the sets of numbers to which each number belongs.

① **25** 25 is a whole number and an integer. Since it can be written as $\frac{25}{1}$, it is also a rational number.

② **−8** −8 is an integer. Since −8 can be written as $\frac{-8}{1}$ or $\frac{8}{-1}$, it is also a rational number.

③ **$-7\frac{2}{3}$** Since $-7\frac{2}{3}$ can be written as $\frac{-23}{3}$ or $\frac{23}{-3}$, it is a rational number.

④ **0.621** Since 0.621 can be written as $\frac{621}{1,000}$, it is a rational number.

Study Hint
Reading Math $-7\frac{2}{3}$ is read as negative seven and two thirds.

When a rational number is represented as a fraction, it is often expressed in **simplest form**. A fraction is in simplest form when the GCF of the numerator and denominator is 1.

Examples

⑤ Write $\frac{30}{45}$ in simplest form.

Method 1 Divide by the GCF.

$30 = 2 \cdot 3 \cdot 5$ } The GCF of 30
$45 = 3 \cdot 3 \cdot 5$ and 45 is $3 \cdot 5$ or 15.

$\frac{30}{45} \rightarrow \frac{30 \div 15}{45 \div 15} = \frac{2}{3}$

Method 2 Use prime factorization.

$\frac{30}{45} = \frac{2 \cdot 3 \cdot 5}{3 \cdot 3 \cdot 5} = \frac{2}{3}$

The slashes indicate that the numerator and denominator are divided by $3 \cdot 5$, the GCF.

Since the GCF of 2 and 3 is 1, the fraction $\frac{2}{3}$ is in simplest form.

APPLICATION

⑥ **Space Exploration** On February 20, 1962, John Glenn became the first American to orbit Earth. He traveled around Earth 3 times. The trip took a total of 4 hours and 48 minutes. Express the length of time in terms of hours using a fraction in simplest form.

Write 48 minutes as hours and add to 4 hours.

48 minutes equals $\frac{48}{60}$ hour.

Write $\frac{48}{60}$ in simplest form.

$48 = 2 \cdot 2 \cdot 2 \cdot 2 \cdot 3$ } The GCF of 48 and
$60 = 2 \cdot 2 \cdot 3 \cdot 5$ 60 is $2 \cdot 2 \cdot 3$ or 12.

$\frac{48}{60} \rightarrow \frac{48 \div 12}{60 \div 12} = \frac{4}{5}$

John Glenn's trip took 4 plus $\frac{4}{5}$ or $4\frac{4}{5}$ hours.

CHECK FOR UNDERSTANDING

Communicating Mathematics

Read and study the lesson to answer each question. 1. Sample answer: $\frac{1}{2}$

1. *Write* an example of a rational number that is not an integer.
2. *Explain* how to express $\frac{30}{50}$ in simplest form. See margin.

Guided Practice

Name all the sets of numbers to which each number belongs.

3. $1\frac{3}{4}$ rationals
4. -45 integers, rationals

Write each fraction in simplest form.

5. $\frac{8}{72}$ $\frac{1}{9}$
6. $\frac{27}{45}$ $\frac{3}{5}$
7. $-\frac{60}{75}$ $-\frac{4}{5}$
8. $\frac{36}{54}$ $\frac{2}{3}$

9. *Theater* The Antoinette Perry, or Tony, Award is given to exceptional plays and the people who make them happen. The award weighs 1 pound 10 ounces. Write the weight in pounds using a fraction in simplest form. $1\frac{5}{8}$ lb

EXERCISES

Practice

Name all sets of numbers to which each number belongs.

10. 6.2   rationals
11. 0   See margin.
12. $-5\frac{5}{7}$ rationals
13. 77   See margin.
14. -9 See margin.
15. -0.85 rationals

Write each fraction in simplest form.

16. $-\frac{3}{9}$ $-\frac{1}{3}$
17. $\frac{15}{25}$ $\frac{3}{5}$
18. $\frac{36}{81}$ $\frac{4}{9}$
19. $-\frac{18}{54}$ $-\frac{1}{3}$
20. $\frac{14}{66}$ $\frac{7}{33}$
21. $\frac{24}{54}$ $\frac{4}{9}$
22. $-\frac{15}{24}$ $-\frac{5}{8}$
23. $\frac{48}{72}$ $\frac{2}{3}$
24. $\frac{24}{120}$ $\frac{1}{5}$
25. $-\frac{66}{88}$ $-\frac{3}{4}$
26. $\frac{72}{98}$ $\frac{36}{49}$
27. $-\frac{45}{100}$ $-\frac{9}{20}$

28. Write two other names for -20.   Sample answers: $-\frac{20}{1}, \frac{20}{-1}$
29. Write 16 out of 40 as a fraction in simplest form. $\frac{2}{5}$
30. Both numerator and denominator of a fraction are even. Can you tell whether the fraction is in simplest form? Explain.   See margin.

Applications and Problem Solving

31. *Weather* Yuma, Arizona, has an average of 175 days a year with temperatures of 90°F or above. Using 365 days in a year, write a fraction representing these hot days as a part of a year. Express the fraction in simplest form. $\frac{35}{73}$

Lesson 6-4 Rational Numbers   **247**

Reteaching the Lesson

Activity Have students model a fraction by shading the appropriate sections of a rectangle. Then have them shade sections of a congruent rectangle to model the equivalent fraction in simplest form.

Error Analysis
Watch for students whose fractions are simplified but not in simplest form.
Prevent by having them factor the numerator and denominator of each answer to look for other common factors.

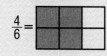   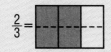

Exercise 33 asks students to advance to the next stage of work on the Chapter Project. You may wish to have students work in groups to complete their tables.

4 ASSESS

Closing Activity
Modeling Provide students with a thermometer with negative temperatures indicated. Ask students to identify whole numbers, integers, and rational numbers on the thermometer.

Mid-Chapter Self Test
The Mid-Chapter Self Test reviews the concepts in Lessons 6-1 through 6-4. Lesson references are given so students can review concepts not yet mastered.

Additional Answers
32a. $\frac{4}{15}$
32b. $\frac{11}{15}$
32c. $\frac{4}{15}$
32d. $\frac{1}{5}$

Practice Masters, p. 44

---

32. **Football** The graph shows how well the first 30 teams to win the Super Bowl did the following year. Write a fraction in simplest form to represent each of the following. **a–d. See margin.**
    a. teams that did not play in the play-offs
    b. teams that played in the play-offs
    c. teams that played in the Super Bowl
    d. teams that won the Super Bowl the next year

FOLLOWING YEAR STATUS OF THE FIRST 30 SUPER BOWL CHAMPS
- Missed play-offs 8
- Made it to play-offs 22
- Made it to AFC/NFC title game 14
- Made it to Super Bowl 8
- Won Super Bowl 6

Source: USA TODAY research, NFL

33. **Working on the CHAPTER Project** Refer to the information you gathered about a sports team on page 231. For each year, write a fraction in simplest form representing the portion of regular season games won. Organize your results in a table. **See students' work.**

34. Yes; $\frac{4}{6.4}$ can be written as $\frac{40}{64}$ or $\frac{5}{8}$.

34. **Critical Thinking** Does $\frac{4}{6.4}$ name a rational number? Explain.

Mixed Review

35. Find the GCF of 28, 126, and 56. *(Lesson 6-3)* **14**

36. No; people in a pet store would probably own at least one pet.

36. **Statistics** Would a pet store be a good location to find a representative sample for a survey of number of pets owned? Explain. *(Lesson 4-8)*

37. Solve $-42 \div (-14) = p$. *(Lesson 2-8)* **3**

38. **Test Practice** If $3r + 8 = 35$, what is the value of r? *(Lesson 1-7)* **D**
    A 52 B 43 C 14 D 9

CHAPTER 6 Mid-Chapter Self Test

Determine whether each number is divisible by 2, 3, 4, 5, 6, 8, 9, or 10. *(Lesson 6-1)*

1. 880 **2, 4, 5, 8, 10**
2. 12,321 **3, 9**

Find the prime factorization of each number. *(Lesson 6-2)*

3. 68 $2^2 \cdot 17$
4. 400 $2^4 \cdot 5^2$
5. 63 $3^2 \cdot 7$

Find the GCF for each set of numbers. *(Lesson 6-3)*

6. 28, 70 **14**
7. 12, 18, 30 **6**

Write each fraction in simplest form. *(Lesson 6-4)*

8. $\frac{12}{30}$ $\frac{2}{5}$
9. $-\frac{81}{99}$ $-\frac{9}{11}$

10. **Weather** On the day of winter solstice, Hilo, Hawaii, has 10 hours and 55 minutes of sunshine. Express this length of time in terms of hours. *(Lesson 6-4)* $10\frac{11}{12}$ h

248 Chapter 6 Exploring Number Patterns

Extending the Lesson

Enrichment Masters, p. 44

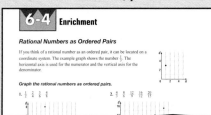

Activity The design illustrates $\frac{6}{12} = \frac{1}{2}$. Have students create their own designs and name the equivalent fractions modeled.

6-5 Rational Numbers and Decimals

What you'll learn
You'll learn to express rational numbers as decimals and decimals as fractions.

When am I ever going to use this?
Knowing how to change between fractions and decimals can help you interpret calculator readouts.

Word Wise
terminating decimal
repeating decimal
bar notation

The chart shows the recycling habits of Americans. How could you express the portion of iron and steel recycled as a decimal? How could you express the portion of paper recycled as a decimal?

Remember that a fraction is another way of writing a division problem. Any fraction can be expressed as a decimal by dividing the numerator by the denominator.

Portion of Trash Recycled
$\frac{1}{4}$ of all iron and steel
$\frac{1}{3}$ of all paper
$\frac{1}{10}$ of all wood
$\frac{1}{5}$ of all plastic

Source: *Vitality*, March, 1996

$\frac{1}{4}$ means $1 \div 4$.

```
      0.25
   4)1.00      Annex zeros to the
      8        numerator: 1 = 1.00.
      ──
      20
      20
      ──
       0
```

$\frac{1}{4} = 0.25$

$\frac{1}{3}$ means $1 \div 3$.

```
      0.333...    The three dots means
   3)1.000        the three keeps
      9           repeating.
      ──
      10
       9
      ──
       10
        9
       ──
        1
```

$\frac{1}{3} = 0.333...$

Study Hint

Mental Math It is helpful to memorize these commonly used fraction-decimal equivalencies.

$\frac{1}{2} = 0.5$ $\frac{1}{3} = 0.\overline{3}$
$\frac{1}{4} = 0.25$ $\frac{1}{5} = 0.2$
$\frac{1}{8} = 0.125$
$\frac{1}{10} = 0.1$

The fraction $\frac{1}{4}$ can be expressed as the decimal 0.25. A decimal like 0.25 is called a **terminating decimal** because the division ends, or terminates, when the remainder is zero.

The fraction $\frac{1}{3}$ can be expressed as the decimal 0.333... . A decimal like 0.333... is called a **repeating decimal**. Since it is impossible to write all the digits, you can use **bar notation** to show that the 3 repeats.

$$0.33333... = 0.\overline{3}$$

Lesson 6-5 Rational Numbers and Decimals **249**

Cross-Curriculum Cue

Inform the other teachers on your team that your students are finding decimal and fractional equivalents and learning scientific notation. Suggestions for curriculum integration are:

Economics: stock market quotations
Music: harmonic intervals
Earth Science: astronomical measurements, weather

6-5 Lesson Notes

Instructional Resources
- *Study Guide Masters*, p. 45
- *Practice Masters*, p. 45
- *Enrichment Masters*, p. 45
- Transparencies 6-5, A and B
- *Assessment and Evaluation Masters*, pp. 154, 155
- *School to Career Masters*, p. 32
- CD-ROM Program
  - Resource Lesson 6-5

Recommended Pacing

Standard	Day 8 of 15
Honors	Day 7 of 13
Block	Day 4 of 8

1 FOCUS

5-Minute Check
(Lesson 6-4)

Name all the sets of numbers to which each number belongs.

1. -5 integers, rationals
2. 0.37 rationals

Write each fraction in simplest form.

3. $-\frac{9}{18}$ $-\frac{1}{2}$
4. $\frac{6}{10}$ $\frac{3}{5}$
5. $-\frac{7}{21}$ $-\frac{1}{3}$

 The 5-Minute Check is also available on **Transparency 6-5A** for this lesson.

Motivating the Lesson

Hands-On Activity Show a collection of 10 identical objects. Have a student remove 5 of them. Ask the following questions.
- How can you write the fraction of objects removed? $\frac{1}{2}$
- How can you write the portion removed as a decimal? Write: $\frac{5}{10} = \frac{1}{2} = 0.5$

Lesson 6-5 **249**

2 TEACH

Transparency 6-5B contains a teaching aid for this lesson.

Using Mental Math Draw attention to the equivalencies listed in the Study Hint on p. 249. Ask students to use them to find the following decimal equivalents mentally.

a. $\frac{3}{4}$ 0.75 b. $\frac{2}{5}$ 0.4
c. $\frac{7}{10}$ 0.7 d. $\frac{4}{5}$ 0.8
e. $\frac{3}{8}$ 0.375 f. $\frac{3}{10}$ 0.3

In-Class Examples

For Example 1
Express 0.707070 . . . using bar notation. $0.\overline{70}$

For Example 2
Express 6.00943943 . . . using bar notation. $6.00\overline{943}$

For Example 3
One out of every nine students at Meadowlake Middle School volunteers to plant trees on Earth Day. Express this fraction as a decimal. $0.\overline{1}$

For Example 4
Express 0.42 as a fraction or mixed number in simplest form. $\frac{21}{50}$

For Example 5
Express -7.84 as a fraction or mixed number in simplest form. $-7\frac{21}{25}$

For Example 6
Express $0.\overline{8}$ as a fraction. $\frac{8}{9}$

Teaching Tip Have students use their calculators and bar notation to express each fraction as a decimal.

a. $\frac{17}{99}$ $0.\overline{17}$ b. $\frac{41}{99}$ $0.\overline{41}$
c. $\frac{65}{99}$ $0.\overline{65}$ d. $\frac{87}{99}$ $0.\overline{87}$

Ask them what pattern they observe. Then ask them to use their results to write 0.585858 . . . as a fraction. $\frac{58}{99}$

Examples

Express each decimal using bar notation.

1 0.454545...
The digits 45 repeat.
$0.454545... = 0.\overline{45}$

2 0.1345345...
The digits 345 repeat.
$0.1345345... = 0.1\overline{345}$

CONNECTION

3 **Geography** North America contains about $\frac{1}{6}$ of the dry land of Earth. Express this fraction as a decimal.

Use a calculator.
1 ÷ 6 = 0.166666667
The calculator rounded the answer 0.166666666... to 0.166666667.
$\frac{1}{6} = 0.1\overline{6}$

Study Hint
Technology Most calculators round answers, but some truncate answers. *Truncate* means to cut-off at a certain place-value position, ignoring the digits that follow.

Every terminating decimal can be expressed as a fraction with a denominator of 10, 100, 1,000, and so on. Thus, terminating decimals are rational numbers.

Examples

Express each decimal as a fraction or mixed number in simplest form.

4 0.85
$0.85 = \frac{85}{100}$
$= \frac{17}{20}$ *Simplify. The GCF of 85 and 100 is 5.*

5 -4.125
$-4.125 = -4\frac{125}{1,000}$
$= -4\frac{1}{8}$ *Simplify.*

Repeating decimals can also be expressed as fractions. Thus, repeating decimals are rational numbers.

Example

INTEGRATION

6 **Algebra** Express $0.\overline{7}$ as a fraction.

Let $N = 0.\overline{7}$ or 0.777.... Then $10N = 7.777....$ *Multiply N by 10, because 1 digit repeats.*

Subtract $N = 0.777...$ to eliminate the repeating part, 0.777....

$10N = 7.777...$
$\underline{-1N = 0.777...}$ $N = 1N$
$9N = 7$
$N = \frac{7}{9}$ *Divide each side by 9.*

So, $0.\overline{7} = \frac{7}{9}$

Check: 7 ÷ 9 = 0.777777778 ✓

250 Chapter 6 Exploring Number Patterns

Example 7 INTEGRATION

Algebra Express $3.\overline{16}$ as a fraction.

Let $N = 3.\overline{16}$. Then $100N = 316.\overline{16}$.   *Multiply N by 100, because 2 digits repeat.*

$$100N = 316.\overline{16}$$
$$-N = 3.\overline{16}$$ *Subtracting eliminates the repeating part, $0.\overline{16}$.*
$$99N = 313$$

$$N = \frac{313}{99} \text{ or } 3\frac{16}{99}$$ So, $3.\overline{16} = \frac{313}{99}$ or $3\frac{16}{99}$.

Check: 313 ÷ 99 = 3.161616162 ✓

Study Hint
Reading Math $3.\overline{16}$ is read as *three point sixteen repeating*.

In-Class Example
For Example 7
Express $0.\overline{33}$ as a fraction. $\frac{1}{3}$

3 PRACTICE/APPLY

Check for Understanding
If students need additional practice or instruction after completing Exercises 1–12, one of these options may be helpful.
- Extra Practice, see p. 621
- Reteaching Activity
- *Study Guide Masters*, p. 45
- *Practice Masters*, p. 45

Assignment Guide
Core: 13–39 odd, 42–47
Enriched: 14–36 even, 38–40, 42–47

Additional Answer
3. Nadia; $\frac{1}{4} = 0.25$ but $0.\overline{25} = 0.252525\ldots$

CHECK FOR UNDERSTANDING

Communicating Mathematics

1. 0.012; $\frac{3}{250}$
2. so the repeating part can be eliminated

Read and study the lesson to answer each question.

1. *Write* twelve thousandths as a decimal and then as a fraction in simplest form.
2. *Tell* why one of the steps in expressing a repeating decimal as a fraction involves multiplying by a power of 10.
3. *You Decide* Heather says that $0.\overline{25}$ equals $\frac{1}{4}$. Nadia disagrees. Who is correct? Explain. **See margin.**

Guided Practice

4. Express 2.353535… using bar notation. $2.\overline{35}$
5. Write the first ten decimal places of $-0.6\overline{17}$. -0.6171717171

Express each fraction or mixed number as a decimal.

6. $\frac{3}{5}$ 0.6
7. $\frac{7}{11}$ $0.\overline{63}$
8. $-4\frac{13}{25}$ -4.52

Express each decimal as a fraction or mixed number in simplest form.

9. 0.66 $\frac{33}{50}$
10. $0.\overline{2}$ $\frac{2}{9}$
11. $-1.\overline{15}$ $-1\frac{5}{33}$

12. **Measurement** A micrometer measured the thickness of the wall of a plastic pipe as 0.084 inch. What fraction of an inch is this? $\frac{84}{1,000}$ or $\frac{21}{250}$ in.

EXERCISES

Practice

Express each decimal using bar notation.   15. $7.0\overline{74}$

13. 0.255555… $0.2\overline{5}$
14. $-15.345345345\ldots$ $-15.\overline{345}$
15. 7.0747474…

17. $-0.30\overline{53053053}$

Write the first ten decimal places of each decimal.   18. 0.3818181818

16. $0.\overline{15}$ 0.1515151515
17. $-0.\overline{305}$
18. $0.3\overline{81}$

Lesson 6-5 Rational Numbers and Decimals   **251**

■ Reteaching the Lesson ■

Activity Have students work in pairs using three number cubes and a calculator. One student rolls the cubes and uses the numbers to write a fraction. The lowest number becomes the numerator, and the sum of the other two numbers becomes the denominator. The other student finds the decimal equivalent of the fraction.

Study Guide Masters, p. 45

6-5 Study Guide

Rational Numbers and Decimals

To change a fraction to a decimal, divide the numerator by the denominator.

Example 1 Express $\frac{5}{8}$ as a decimal.

Use a calculator.
5 ÷ 8 = 0.625
$\frac{5}{8} = 0.625$
The remainder is 0.
0.625 is a terminating decimal.

Use paper and pencil.
0.625
8)5.000   *Annex zeros as needed.*
$\underline{-48}$
20
$\underline{-16}$
40
$\underline{-40}$
0   $\frac{5}{8} = 0.625$

A **terminating decimal** can be written as a fraction with a denominator of 10, 100, 1000, and so on.

Examples Express each decimal as a fraction or mixed number in simplest form.

2 $0.52 = \frac{52}{100}$
$= \frac{13}{25}$

3 $-1.4375 = -1\frac{4375}{10,000}$
$= -1\frac{7}{16}$

Express each fraction as a decimal.

1. $\frac{3}{5}$ 0.6
2. $-\frac{7}{10}$ -0.7
3. $-\frac{7}{4}$ -1.75
4. $-\frac{15}{32}$ -0.46875
5. $\frac{3}{20}$ 0.15

6. $5\frac{7}{25}$ 5.28
7. $-\frac{24}{16}$ -1.5
8. $-6\frac{3}{8}$ -6.375
9. $\frac{14}{35}$ 0.4
10. $\frac{5}{40}$ 0.125

Express each decimal as a fraction or mixed number in simplest form.

11. 0.08 $\frac{2}{25}$
12. -3.75 $-3\frac{3}{4}$
13. 0.015 $\frac{3}{200}$
14. -0.6 $-\frac{3}{5}$
15. 1.05 $1\frac{1}{20}$

16. -12.34 $-12\frac{17}{50}$
17. 2.1875 $2\frac{3}{16}$
18. -0.875 $-\frac{7}{8}$
19. -8.85 $-8\frac{17}{20}$
20. 9.5 $9\frac{1}{2}$

Exercise 41 asks students to advance to the next stage of work on the Chapter Project. You may wish to have students work in pairs to complete their tables.

4 ASSESS

Closing Activity

Speaking Tell students that you are thinking of a certain repeating decimal. Then tell them the number of repeating digits, 1, 2, or 3. Ask them to tell by what power of 10 they would multiply N in the process of expressing the decimal as a fraction.

Chapter 6, Quiz B (Lessons 6-4 and 6-5) is available in the *Assessment and Evaluation Masters*, p. 155.

Mid-Chapter Test (Lessons 6-1 through 6-5) is available in the *Assessment and Evaluation Masters*, p. 154.

Practice Masters, p. 45

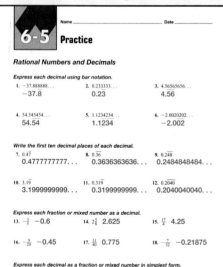

Express each fraction or mixed number as a decimal.

19. $\frac{3}{4}$ 0.75
20. $\frac{4}{9}$ $0.\overline{4}$
21. $-\frac{7}{25}$ -0.28
22. $\frac{3}{22}$ $0.1\overline{36}$
23. $-5\frac{2}{3}$ $-5.\overline{6}$
24. $7\frac{5}{33}$ $7.\overline{15}$
25. $12\frac{5}{8}$ 12.625
26. $-8\frac{15}{44}$ $-8.34\overline{09}$

Express each decimal as a fraction or mixed number in simplest form.

27. 0.88 $\frac{22}{25}$
28. $0.\overline{45}$ $\frac{5}{11}$
29. $0.8\overline{3}$ $\frac{5}{6}$
30. -0.225 $-\frac{9}{40}$
31. $-1.\overline{5}$ $-1\frac{5}{9}$
32. 7.08 $7\frac{2}{25}$
33. $-5.\overline{67}$ $-5\frac{67}{99}$
34. $2.1\overline{5}$ $2\frac{7}{45}$

35. When $\frac{131}{200}$ is expressed as a decimal, is it a *terminating* or *repeating* decimal? **terminating**

36. Suppose $N = 0.\overline{894}$. To change this number to a fraction, what power of ten should you multiply times N? 10^3 or 1,000

37. How would you use a calculator to express $5\frac{6}{11}$ as a decimal?
5 6 ÷ 11 = 5.545454545

Applications and Problem Solving

40. See Answer Appendix.

38. *History* During the fourteenth and fifteenth centuries, printing presses used type cut from wood blocks. Each block was $\frac{7}{8}$ inch thick. Write this as a decimal. **0.875 in.**

39. *Life Science* An egg of a hummingbird weighs 0.013 ounce. Write this as a fraction of an ounce. Express the fraction in simplest form. $\frac{13}{1,000}$ oz

40. *Technology* A spreadsheet's display and two calculators' displays for $6 \div 9$ are listed below. How would each display the answer to $4 \div 11$?
   a. 0.7
   b. 0.666666667
   c. 0.666666666

41. *Working on the* **CHAPTER Project** Refer to the fractions that you found in Exercise 33 on page 248. Express each fraction as a decimal. Add these results to your table.

41. See students' work.

42. *Critical Thinking* A *unit fraction* is a fraction that has 1 as its numerator.
   a. Write the four greatest unit fractions that are terminating decimals. Express each fraction as a decimal. **a–b. See Answer Appendix.**
   b. Write the four greatest unit fractions that are repeating decimals. Express each fraction as a decimal.

Mixed Review

43. integers, rationals
44. obtuse, scalene

43. Name all the sets of numbers to which -38 belongs. *(Lesson 6-4)*

44. *Geometry* Classify the triangle at the right according to its angles and its sides. *(Lesson 5-2)*
57 cm, 29°, 21°, 27 cm, 130°, 36 cm

45. *Algebra* Solve $\frac{3}{17} = \frac{n}{68}$. *(Lesson 3-3)* **12**

46. Without graphing, tell in which quadrant the point $M(6, -4)$ lies. *(Lesson 2-10)* **IV**

47. **Test Practice** Find the value of $2^3 \cdot 6^2$. *(Lesson 1-2)* **C**
   A 48   B 144   C 288   D 864

252 Chapter 6 Exploring Number Patterns

Extending the Lesson

Enrichment Masters, p. 45

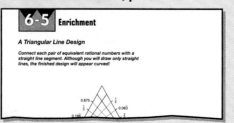

Activity Have students study the following table.

Fraction	$\frac{5}{9}$	$\frac{74}{99}$	$\frac{415}{999}$
Decimal	$0.\overline{5}$	$0.\overline{74}$	$0.\overline{415}$

Ask them to express each decimal as a fraction.
a. $0.\overline{512}$ $\frac{512}{999}$
b. $0.\overline{68}$ $\frac{68}{99}$

252 Chapter 6

6-6

Integration: Probability
Simple Events

What you'll learn
You'll learn to find the probability of a simple event.

When am I ever going to use this?
Knowing how to find the probability of a simple event can help you play some games wisely.

Word Wise
outcome
sample space
random
event
probability

An ancient Hebrew game is played with a four-sided top called a dreidel and tokens called gelt. Each side of the top has one of the following Hebrew letters.

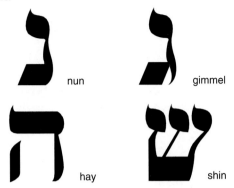

The players take turns spinning the dreidel. Depending on the spin, the player may add gelt to the center pot or take gelt from the center pot. For example, if a spin results in a hay, the player takes half of the gelt from the pot. What is the probability of spinning a hay?

When the dreidel is spun, there are four equally-likely results or **outcomes**. The list of all possible outcomes is called the **sample space**. The sample space of the dreidel is nun, gimmel, hay, and shin. Each outcome has the same chance of occurring. When all outcomes have an *equally-likely* chance of happening, we say that the outcomes happen at **random**.

An **event** is a specific outcome or type of outcome. In this case, the event is spinning a hay. **Probability** is the chance that an event will happen.

| **Probability** | Probability = $\dfrac{\text{number of ways that an event can occur}}{\text{number of possible outcomes}}$ |

The probability of spinning a hay is $\frac{1}{4}$.

Use the scale below when you consider the probability of an event. When it is *impossible* for an event to happen, its probability is 0. An outcome that is *certain* to happen has a probability of 1.

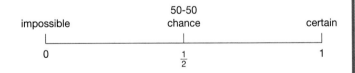

The probability of an outcome is written as a number from 0 to 1.

Lesson 6-6 Integration: Probability Simple Events **253**

6-6 Lesson Notes

Instructional Resources
- *Study Guide Masters*, p. 46
- *Practice Masters*, p. 46
- *Enrichment Masters*, p. 46
- Transparencies 6-6, A and B
- CD-ROM Program
  - Resource Lesson 6-6
  - Extended Activity 6-6

Recommended Pacing	
Standard	Day 9 of 15
Honors	Day 8 of 13
Block	Day 5 of 8

1 FOCUS

5-Minute Check
(Lesson 6-5)

Express each fraction or mixed number as a decimal.
1. $\frac{1}{8}$ 0.125
2. $\frac{9}{4}$ 2.25
3. $1\frac{5}{9}$ $1.\overline{55}$

Express each decimal as a fraction or mixed number in simplest form.
4. $0.\overline{5}$ $\frac{5}{9}$
5. -2.8 $-2\frac{4}{5}$

The 5-Minute Check is also available on **Transparency 6-6A** for this lesson.

Motivating the Lesson

Problem Solving Ask students to rate the likelihood of each of the following events on a scale of 0 to 10 (0 = impossible, 10 = certain).
- The sun will come up tomorrow.
- The next president will be a woman.
- It will literally rain cats and dogs today.

Lesson 6-6 253

2 TEACH

 Transparency 6-6B contains a teaching aid for this lesson.

Reading Mathematics
Have students discuss the everyday meanings of the words *outcome, random, event,* and *probably.* Then help students relate the everyday meanings with the terms in the lesson.

In-Class Examples

For Example 1
A 12-hour mantel clock strikes the hour. What is the probability that the hour is divisible by 3? $\frac{1}{3}$

For Example 2
A popular game at a fair requires a player to toss a table tennis ball onto saucers floating in water. Suppose there are 50 white saucers, 15 yellow saucers, and 5 red saucers. Assume a ball is thrown randomly and always lands in a saucer.
a. What is the probability that the ball will land on a white saucer? $\frac{5}{7}$
b. On a yellow saucer? $\frac{3}{14}$
c. On a red saucer? $\frac{1}{14}$

Teaching Tip For Example 2 in the Student Edition, assume that the pin will land on the square.

3 PRACTICE/APPLY

Check for Understanding
If students need additional practice or instruction after completing Exercises 1–10, one of these options may be helpful.
• Extra Practice, see p. 622
• Reteaching Activity
• *Study Guide Masters,* p. 46
• *Practice Masters,* p. 46

Additional Answer
1. Sample answer: An even number picked at random is divisible by 2.

254 **Chapter 6**

Examples

1 A bag contains 5 red marbles, 4 yellow marbles, and 3 blue marbles. Lashanda picks a marble without looking. What is the probability that she will pick a blue marble?

$$P(\text{blue}) = \frac{\text{blue marbles}}{\text{total number of marbles}}$$

$$= \frac{3}{12} \text{ or } \frac{1}{4} \text{ or } 0.25$$

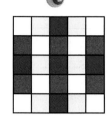

INTEGRATION **2** **Geometry** A pin is dropped at random onto the square at the right. The point of the pin lands in one of the small regions.

Study Hint
Reading Math $P(\text{blue})$ is read as *the probability of blue.*

a. What is the probability that the point lands inside a yellow region?

$$P(\text{yellow}) = \frac{\text{yellow squares}}{\text{total number of squares}}$$

$$= \frac{6}{25} \text{ or } 0.24$$

b. What is the probability that the point lands inside a blue or white region?

$$P(\text{blue or white}) = \frac{\text{blue squares} + \text{white squares}}{\text{total number of squares}}$$

$$= \frac{5 + 8}{25}$$

$$= \frac{13}{25} \text{ or } 0.52$$

c. What is the probability that the point lands inside a green region?

Since there are no green regions, the probability is 0.

CHECK FOR UNDERSTANDING

Communicating Mathematics
Read and study the lesson to answer each question. 1–2. See margin.
1. *Write* an example of an outcome with a probability of 1.
2. *Draw* a spinner where the probability of an outcome of red is $\frac{1}{5}$.

HANDS-ON MATH
3. *Toss* a coin 50 times. Record the results as heads or tails. Are the results what you expected? Explain. **See students' work.**

Guided Practice
State the probability of each outcome as a fraction and as a decimal.
4. A date picked at random is a Saturday. $\frac{1}{7}$; $0.\overline{142857}$
5. This is a history book. **0**

A number cube is rolled. Find each probability.
6. $P(3)$ $\frac{1}{6}$ 7. $P(\text{even})$ $\frac{1}{2}$ 8. $P(3 \text{ or } 5)$ $\frac{1}{3}$ 9. $P(\text{less than } 7)$ **1**

10. *Games* The game of Scrabble has 100 tiles including 2 blank tiles, 9 tiles with A, 12 tiles with E, 9 tiles with I, 8 tiles with O, and 4 tiles with U.
a. What is the probability that the first tile picked is blank? $\frac{1}{50}$
b. What is the probability that the first tile picked is not blank? $\frac{49}{50}$
c. What is the probability that the first tile picked is A, E, I, O, or U? $\frac{21}{50}$

254 Chapter 6 Exploring Number Patterns

■ **Reteaching the Lesson** ■

Activity Have pairs of students calculate the probabilities of each outcome when a number cube is tossed.
1. 6 $\frac{1}{6}$ 2. an odd number $\frac{1}{2}$

Now have students roll a number cube 50 times, record the results, and compare them to the calculated probabilities.

Additional Answer
2.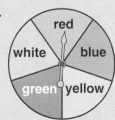

EXERCISES

Practice

State the probability of each outcome as a fraction and as a decimal.

11. A number cube is rolled and shows a number greater than 1. $\frac{5}{6}$; $0.8\overline{3}$
12. A month picked at random starts with J. $\frac{1}{4}$; 0.25
13. A date picked at random is February 30. 0
14. A two-digit number picked at random is divisible by 10. $\frac{1}{10}$; 0.1
15. An integer picked at random is a rational number. 1
16. A positive one-digit number picked at random is prime. $\frac{2}{5}$; 0.4

The spinner at the right is used for a game. Find each probability.

17. $P(7)$ $\frac{1}{8}$
18. $P(2, 3,$ or $4)$ $\frac{3}{8}$
19. $P(\text{odd})$ $\frac{1}{2}$
20. $P(\text{less than 6})$ $\frac{5}{8}$
21. $P(\text{not 8})$ $\frac{7}{8}$
22. $P(\text{greater than 2})$ $\frac{3}{4}$

The letters of the word "associative" are written one each on 11 identical slips of paper and shuffled in a hat. A blindfolded student draws one slip of paper. Find each probability.

23. $P(s)$ $\frac{2}{11}$
24. $P(\text{vowel})$ $\frac{6}{11}$
25. $P(\text{not } r)$ 1
26. $P(s \text{ or } t)$ $\frac{3}{11}$
27. $P(\text{consonant})$ $\frac{5}{11}$
28. $P(\text{not } i)$ $\frac{9}{11}$

29. Suppose a person living in your home is picked at random. What is the probability that the person is a female? **See students' work.**
30. Can a probability ever be greater than 1? Explain. **See margin.**
31. Can a probability ever be less than 0? Explain. **See margin.**
32. *Write a Problem* in which the answer will be a probability of $0.\overline{3}$. **See margin.**

Applications and Problem Solving

33. *Games* The game of UNO has 25 red cards, 25 green cards, 25 yellow cards, 25 blue cards, and 8 wild cards.

   a. What is the probability that the first card dealt is a yellow card? $\frac{25}{108}$
   b. Find the probability that the first card dealt is a wild card. $\frac{2}{27}$
   c. There are two number 7 cards for each color. What is the probability that the first card dealt is a 7? $\frac{2}{27}$
   d. There is only one number 0 card for each color. What is the probability that the first card dealt is a 0? $\frac{1}{27}$

Lesson 6-6 Integration: Probability Simple Events **255**

Assignment Guide

Core: 11–33 odd, 35–39
Enriched: 12–32 even, 33–39

Additional Answers

30. No; the number of ways an event can occur cannot exceed the number of possible outcomes.
31. No; the number of ways something can occur cannot be negative.
32. Sample answer: Find the probability that a die is rolled and it shows a 1 or a 2.

4 ASSESS

Closing Activity

Modeling Have students use a deck of ten index cards numbered 1 through 10 to make up problems involving the probability of drawing certain cards. Have students explain their solutions.

34c. The population of the West has increased and has become a larger part of the total population of the U.S.

Mixed Review

34. **History** The U.S. Bureau of the Census divides the United States into four regions. The table shows the population of these four regions in 1890 and 1990. **a. 0.050**

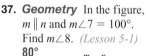

a. Suppose a person living in the United States in 1890 was picked at random. What is the probability that the person lived in the West? Write your answer as a decimal to the nearest thousandth.

b. Suppose a person living in the United States in 1990 was picked at random. What is the probability that the person lived in the West? Write your answer as a decimal to the nearest thousandth. **0.212**

c. How has the population of the West changed?

35. **Critical Thinking** If the probability that an event will occur is $\frac{3}{7}$, what is the probability that the event will not occur? $\frac{4}{7}$

36. **History** In 1864, Abraham Lincoln won the presidential election with about 0.55 of the popular vote. Write this as a fraction in simplest form. *(Lesson 6-5)* $\frac{11}{20}$

37. **Geometry** In the figure, $m \parallel n$ and $m\angle 7 = 100°$. Find $m\angle 8$. *(Lesson 5-1)* **80°**

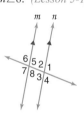

38. **Test Practice** To try to determine the most popular color for cars driven in the Minneapolis-St. Paul area, the colors of cars passing through an intersection in one hour were counted. The results are listed in the table.

Which color is the mode? *(Lesson 4-4)* **C**

A black   B white   C red   D blue

39. **Algebra** Solve $-12 + t = -8$. *(Lesson 2-9)* **4**

256   Chapter 6 Exploring Number Patterns

Extending the Lesson

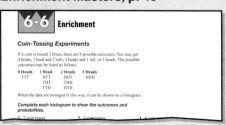

Activity The odds that an event will occur are the ratio of favorable outcomes to unfavorable outcomes. When a number cube is tossed, the probability of rolling a 2 is $\frac{1}{6}$. The odds in favor of rolling a 2 are 1:5. Have small groups create and exchange problems involving the calculation of odds.

Practice Masters, p. 46

6-7 Least Common Multiple

What you'll learn
You'll learn to find the least common multiple of two or more numbers.

When am I ever going to use this?
Knowing how to find the LCM of integers can help you find when certain events such as the arrival of buses will coincide.

Word Wise
multiple
least common multiple (LCM)

Rebecca is buying hot dogs and hot dog buns for the photography club picnic. Hot dogs come in packages of 10. Hot dog buns came in packages of 8. Rebecca wants to have the same number of hot dogs and hot dog buns. What is the least number of hot dogs and hot dog buns she can buy?

You can use multiples and the problem-solving strategy *make a list* to answer this question. A **multiple** of a number is the product of that number and any whole number.

MINI-LAB

Work with a partner.   2 colored pencils

Try This
- List the numbers from 1 to 100 on a sheet of paper.
- Use divisibility rules to cross out all of the multiples of 10.
- Using a different color, cross out all of the multiples of 8.

Talk About It   2. common multiples of 10 and 8
1. Which numbers were crossed out by both colors?   **40, 80**
2. How would you describe these numbers?
3. What is the least number crossed out by both colors?   **40**

In the Mini-Lab, you discovered some common multiples of 10 and 8. The least of the nonzero common multiples of two or more numbers is called the **least common multiple (LCM)** of the numbers. The LCM of 10 and 8 is 40. So, the least number of hot dogs and hot dog buns Rebecca can buy is 40.

Examples

1 List the first four multiples of 15.

multiples of 15
$0 \cdot 15 = 0$
$1 \cdot 15 = 15$
$2 \cdot 15 = 30$
$3 \cdot 15 = 45$

2 List the first four multiples of y.

multiples of y
$0 \cdot y = 0$
$1 \cdot y = y$
$2 \cdot y = 2y$
$3 \cdot y = 3y$

Lesson 6-7 Least Common Multiple **257**

Multiple Learning Styles

Auditory/Musical To find the LCM of three numbers, assign one of the numbers and a noise maker to each of three groups. Read a list of numbers. If the number read is a multiple of that group's number, they can make their noise. When all three noises sound at once, that number is the LCM. This can be adapted for the GCF.

6-7 Lesson Notes

Instructional Resources
- *Study Guide Masters*, p. 47
- *Practice Masters*, p. 47
- *Enrichment Masters*, p. 47
- Transparencies 6-7, A and B
- *Assessment and Evaluation Masters*, p. 156
- CD-ROM Program
  - Resource Lesson 6-7
  - Interactive Lesson 6-7

Recommended Pacing	
Standard	Day 10 of 15
Honors	Day 9 of 13
Block	Day 5 of 8

1 FOCUS

5-Minute Check
(Lesson 6-6)

A drawer contains 3 yellow socks, 4 red socks, and 5 blue socks. One sock is drawn at random. Find each probability.
1. $P(\text{red})$ $\frac{1}{3}$
2. $P(\text{red or blue})$ $\frac{3}{4}$
3. $P(\text{sock})$ 1
4. $P(\text{hat})$ 0
5. $P(\text{yellow or blue})$ $\frac{2}{3}$

The 5-Minute Check is also available on **Transparency 6-7A** for this lesson.

Motivating the Lesson
Hands-On Activity Have students cut 4 × 6 rectangles from graph paper. Ask them to find the smallest square that can be constructed entirely of rectangles of this size.   **12 × 12**

2 TEACH

Transparency 6-7B contains a teaching aid for this lesson.

Using the Mini-Lab If colored pencils are not available, students can circle multiples of 10 and draw Xs through multiples of 8.

Lesson 6-7   **257**

In-Class Examples

For Example 1
List the first four multiples of 3.
0, 3, 6, 9

For Example 2
List the first four multiples of $2n$.
0, $2n$, $4n$, $6n$

For Example 3
Find the LCM of 9, 12, and 18.
36

For Example 4
Use prime factorization to find the LCM of 21 and 27. **189**

For Example 5
Use prime factorization to find the LCM of 48, 20, and 30. **240**

Teaching Tip After working the examples, have students compare and contrast the two methods for finding the LCM, listing multiples and using prime factorization.

3 PRACTICE/APPLY

Check for Understanding

If students need additional practice or instruction after completing Exercises 1–10, one of these options may be helpful.
- Extra Practice, see p. 622
- Reteaching Activity
- *Study Guide Masters*, p. 47
- *Practice Masters*, p. 47

Study Guide Masters, p. 47

Example

3 Find the LCM of 8, 9, and 12.

multiples of 8: 0, 8, 16, 24, 32, 40, 48, 56, 64, 72, 80, …
multiples of 9: 0, 9, 18, 27, 36, 45, 54, 63, 72, 81, 90, …
multiples of 12: 0, 12, 24, 36, 48, 60, 72, 84, 96, 108, 120, …

The LCM of 8, 9, and 12 is 72. Remember that the LCM is a nonzero number.

Prime factorization can be used to find the LCM of a set of numbers. A common multiple contains *all* the prime factors of each number in the set. The LCM contains *each* factor the greatest number of times it appears in the set.

Examples

Use prime factorization to find the LCM of each set of numbers.

4 12, 18

$12 = 2 \cdot 2 \cdot 3$ *Express each common factor*
$18 = 2 \cdot 3 \cdot 3$ *and all other factors.*
$2 \cdot 2 \cdot 3 \cdot 3$ *Multiply all the factors, using the common factors only once.*

The LCM of 12 and 18 is $2^2 \cdot 3^2$ or 36.

Check: Make a list.
multiples of 12: 12, 24, 36, 48, …
multiples of 18: 18, 36, 54, … ✓

5 9, 21, 24

$9 = 3 \cdot 3$ or 3^2 *The greatest power of 2 is 2^3.*
$21 = 3 \cdot 7$ *The greatest power of 3 is 3^2.*
$24 = 2 \cdot 2 \cdot 2 \cdot 3$ or $2^3 \cdot 3$ *The greatest power of 7 is 7^1.*

The LCM of 9, 21, and 24 is $2^3 \cdot 3^2 \cdot 7$ or 504.

CHECK FOR UNDERSTANDING

Communicating Mathematics

Read and study the lesson to answer each question.

1. *Describe* two ways to find the LCM of 10 and 16. **See Answer Appendix.**
2. *Write* three numbers with the LCM of 24. **Sample answer: 6, 8, 12**

HANDS-ON MATH

3. *List* the numbers from 1 to 100. Cross out all of the multiples of 9. Using a different color, cross out all of the multiples of 12.
   a. Which numbers were crossed out by both colors? **36, 72**
   b. What is the LCM of 9 and 12? **36**

Guided Practice

List the first six multiples of each number or algebraic expression.

4. 11   **0, 11, 22, 33, 44, 55**
5. t **0, t, $2t$, $3t$, $4t$, $5t$**

258 Chapter 6 Exploring Number Patterns

Reteaching the Lesson

Activity Many students will choose to list multiples when finding the LCM. Prepare charts like the one below for students. LCM = 12

×	1	2	3	4	5	6
6	6	12	18	24	30	36
4	4	8	12	16	20	24

Error Analysis
Watch for students who confuse LCM and GCF.
Prevent by having them find both the LCM and GCF of several pairs of numbers. Point out that the LCM is a multiple, so it is *greater* than the given numbers and the GCF is a factor, so it is *less* than the given numbers.

Find the LCM of each set of numbers.

6. 8, 20 **40**   7. 15, 18 **90**   8. 2, 7, 8 **56**   9. 8, 28, 30 **840**

10. *Life Science* Cicadas emerge from the ground every 17 years. The number of one type of caterpillar peaks every 5 years. If the peak cycles of the caterpillar and the cicadas coincided in 1998, what will be the next year in which they coincide? **2083**

EXERCISES

Practice

List the first six multiples of each number or algebraic expression.

13. 0, 14, 28, 42, 56, 70
14. 0, 25, 50, 75, 100, 125
15. 0, 150, 300, 450, 600, 750
16. 0, 2x, 4x, 6x, 8x, 10x

11. 7 **0, 7, 14, 21, 28, 35**   12. 5 **0, 5, 10, 15, 20, 25**   13. 14
14. 25   15. 150   16. $2x$

Find the LCM of each set of numbers.   24. 420   25. 1,225

17. 12, 16 **48**   18. 7, 12 **84**   19. 20, 50 **100**   20. 16, 24 **48**
21. 2, 3, 5 **30**   22. 4, 8, 12 **24**   23. 7, 21, 5 **105**   24. 10, 12, 14
25. 35, 25, 49   26. 24, 12, 6 **24**   27. 68, 170, 4 **340**   28. 45, 10, 6 **90**

29. Determine whether 285 is a multiple of 15. **yes**

30. when one number is a factor of the other

30. When will the LCM of two numbers be one of the numbers?
31. When will the LCM of two numbers be the product of the numbers? **when the GCF is 1**

Applications and Problem Solving

32. *Civics* In the United States, a president is elected every four years. Members of the House of Representatives are elected every two years. Senators are elected every six years. If a voter had the opportunity to vote for a president, a representative, and a senator in 1996, what will be the next year the voter has a chance to make a choice for a president, a representative, and the same Senate seat? **2008**

33. *Bus Routes* Bus Route A reaches a particular bus stop every 12 minutes. Bus Route B reaches the same bus stop every 20 minutes. If a bus from Route A and a bus from Route B are at the bus stop at 7:00 A.M., what is the next time that a bus from each route will be at the bus stop? **8:00 A.M.**

34. *Critical Thinking* Find the LCM of $8xy$ and $6x^2yz$. **$24x^2yz$**

Mixed Review

35. 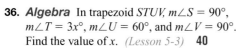 A box contains 7 orange marbles, 9 blue marbles, and 5 green marbles. If a marble is picked at random, what is the probability that the marble will be orange? *(Lesson 6-6)* **C**

   A 0   **B** $\frac{7}{20}$ **C** $\frac{1}{3}$ **D** 1

36. *Algebra* In trapezoid *STUV*, $m\angle S = 90°$, $m\angle T = 3x°$, $m\angle U = 60°$, and $m\angle V = 90°$. Find the value of *x*. *(Lesson 5-3)* **40**

37. *Algebra* Solve $r + 125 = 483$. Check your solution. *(Lesson 1-4)* **358**

38. *Manufacturing* A furniture company produces 15 rocking chairs in one hour. How long will it take to produce 45 chairs? *(Lesson 1-1)* **3 h**

Lesson 6-7 Least Common Multiple **259**

Extending the Lesson

Enrichment Masters, p. 47

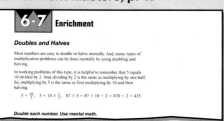

Activity Let *m* and *n* represent two numbers. Let *g* represent the GCF of the numbers. The Greek mathematician Euclid found that the LCM of *m* and *n* is given by the expression $\frac{m \cdot n}{g}$. Use his method to find the LCM of 108 and 352. **9,504**

6-8A Hands-On Lab Notes

GET READY

Objective Students learn to use customary units of measurement to understand the density property.

Optional Resources
Hands-On Lab Masters
- number line, p. 15
- worksheet, p. 52

Overhead Manipulative Resources
- number line transparency

Manipulative Kit
- 12-inch ruler

MANAGEMENT TIPS

Recommended Time
35 minutes

Getting Started Take a strip of paper, fold it in half, tear, and throw away half. Take remaining piece, fold in half, tear, and throw away half. Keep going. Ask students, theoretically, how many times could you do this? **an infinite number**

In the **Activity,** a compass can be used to find the midpoint of a line segment. Arcs of equal radius are drawn using the endpoints of the line segment as centers. The line passing through the points of intersection of the arcs passes through the midpoint of the line segment.

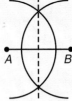

ASSESS

Have students complete Exercises 1–3. Students who have trouble locating the midpoint on a number line may benefit from making a physical number line and folding it in half to match the endpoints to find the midpoint.

Use Exercise 4 to determine whether students understand the density property.

260 Chapter 6

Hands-On Lab — COOPERATIVE LEARNING

6-8A Density Property
A Preview of Lesson 6-8

- 12-inch ruler
- calculator

How many numbers are there between $-1\frac{3}{8}$ and $-1\frac{1}{4}$? Would you guess a few, many, or none? The *density property* states that between any two rational numbers, no matter how close they may seem, there is at least one other rational number.

TRY THIS

Work with a partner.

Step 1 Near the middle of a piece of paper, draw an eight-inch line segment parallel to the long side of the paper. Mark the ends and each one-inch interval with a vertical dash. Below the line, label the left endpoint 0, the right endpoint 2, and the middle point 1.

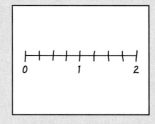

Step 2 Copy and complete the *Midpoint* column of the chart to find the midpoint between each pair of endpoints.

Step 3 Express the numbers in the *Endpoints* column as decimals. Find the mean of the two decimals. Record each result in the *Mean* column.

Endpoints	Midpoint of Endpoints	Mean of Endpoints
0 and 2	1	1
1 and 2	$1\frac{1}{2}$	1.5
1 and $1\frac{1}{2}$	$1\frac{1}{4}$	1.25
$1\frac{1}{2}$ and $1\frac{1}{4}$	$1\frac{3}{8}$	1.375
$1\frac{3}{8}$ and $1\frac{1}{4}$	$1\frac{5}{16}$	1.3125

ON YOUR OWN

1. What is the relationship between the numbers in the *Midpoint* column and the numbers in the *Endpoints* column? **The midpoint is the mean of the endpoints.**

2. What is the relationship between the numbers in the *Midpoint* column and the numbers in the *Mean* column? **same**

3. Tell how you would find the midpoint between $\frac{3}{4}$ and $\frac{7}{8}$. **Add them together and divide by 2.**

4. **Look Ahead** List two decimals with all but the last digits the same; for example, 4.26 and 4.27. Add a digit other than 0 to the lesser number. What is the relationship between the original numbers and the number formed by adding the digit? **See Answer Appendix.**

260 Chapter 6 Exploring Number Patterns

 Math Journal Have students write a paragraph using the density property of rational numbers to explain Zeno's dichotomy paradox: *Because any moving object must travel half the total distance before it travels the total distance, it will never reach the end.* **There are an infinite number of midpoints.**

6-8 Comparing and Ordering Rational Numbers

What you'll learn
You'll learn to compare and order rational numbers expressed as fractions and/or decimals.

When am I ever going to use this?
Knowing how to order rational numbers can help you determine which team has a better record.

Word Wise
least common denominator (LCD)

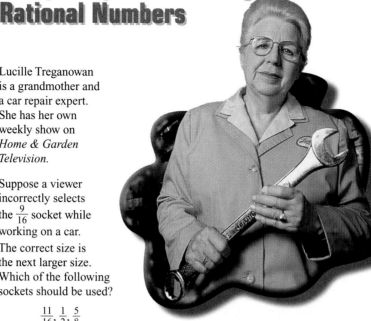

Lucille Treganowan is a grandmother and a car repair expert. She has her own weekly show on *Home & Garden Television*.

Suppose a viewer incorrectly selects the $\frac{9}{16}$ socket while working on a car. The correct size is the next larger size. Which of the following sockets should be used?

$$\frac{11}{16}, \frac{1}{2}, \frac{5}{8}$$

One way to compare two rational numbers is to express them as fractions with like denominators. Any common denominator may be used, but the computation may be easier if the **least common denominator (LCD)** is used. The least common denominator is the LCM of the denominators.

Another way to compare these fractions is to express them as decimals and then compare the decimals.

Socket Size	Fraction Method	Decimal Method
$\frac{9}{16}$	$\frac{9}{16}$	0.5625
$\frac{11}{16}$	$\frac{11}{16}$	0.6875
$\frac{1}{2}$	$\frac{8}{16}$	0.5
$\frac{5}{8}$	$\frac{10}{16}$	0.625

Now it is easy to order the fraction. The socket order from least to greatest is $\frac{1}{2}, \frac{9}{16}, \frac{5}{8}$, and $\frac{11}{16}$. The viewer should use the $\frac{5}{8}$ socket since it is the next size larger than the $\frac{9}{16}$ socket.

Lesson 6-8 Comparing and Ordering Rational Numbers **261**

6-8 Lesson Notes

Instructional Resources
- *Study Guide Masters*, p. 48
- *Practice Masters*, p. 48
- *Enrichment Masters*, p. 48
- Transparencies 6-8, A and B
- *Hands-On Lab Masters*, p. 73
- CD-ROM Program
  - Resource Lesson 6-8
  - Interactive Lesson 6-8

Recommended Pacing	
Standard	Days 11 & 12 of 15
Honors	Day 10 of 13
Block	Day 6 of 8

1 FOCUS

5-Minute Check
(Lesson 6-7)

List the first six multiples of each number.
1. 9 **0, 9, 18, 27, 36, 45**
2. k **0, k, $2k$, $3k$, $4k$, $5k$**

Find the LCM of each set of numbers.
3. 10, 15 **30**
4. 4, 8 **8**
5. 4, 6, 9 **36**

The 5-Minute Check is also available on **Transparency 6-8A** for this lesson.

Motivating the Lesson
Hands-On Activity Present a collection of six pencils or drinking straws of different lengths. While disguising the difference in the lengths, have each of six volunteers draw a pencil or straw. Have the remaining students arrange the volunteers according to the lengths of the pencils or straws, from shortest to longest. Point out that fractions and decimals can also be ordered from least to greatest.

Lesson 6-8 **261**

2 TEACH

 Transparency 6-8B contains a teaching aid for this lesson.

Using Connections Recall for students that measurements are easier to compare when they are expressed in the same units. For example, it is easy to see that 50 inches is greater than 4 feet if 4 feet is expressed as 48 inches. Similarly, rational numbers can be compared by expressing them with the same denominators or in the same decimal form.

In-Class Examples

For Example 1
An environmental inspection of 24 lakes in Blair County found 17 are polluted. Of the 18 lakes in King County, 13 are polluted. Which county has a better record for unpolluted lakes?
Blair

For Example 2
Replace ● with $<$, $>$, or $=$ to make a true sentence.
$-\frac{2}{3}$ ● -0.65 $<$

For Example 3
Order $0.\overline{16}$, $\frac{1}{6}$, $\frac{3}{20}$, and 0.16 from least to greatest. $\frac{3}{20}$, 0.16, $0.\overline{16}$, $\frac{1}{6}$

For Example 4
Find the median of 3.444, $3\frac{9}{20}$, and $3\frac{4}{9}$. $3\frac{4}{9}$

Teaching Tip Referring to Example 3, point out that $\frac{17}{9}$ is almost $\frac{18}{9}$, or 2. In addition, $\frac{3}{2}$ is $1\frac{1}{2}$. Therefore, the only numbers too close together to order using mental math are 1.8 and $\frac{17}{9}$.

Example APPLICATION

① **Sports** In the first round of the 1997 NCAA Baseball Tournament, the University of Florida played St. John's University. Before the tournament, Florida had won 38 of their 60 games. St. John's had won 35 of their 50 games. Which team had the better record?

Explore You are given the number of wins and the number of games for each team. You are asked to determine which team had a better record.

Plan Florida won $\frac{38}{60}$ of their games. St. John's won $\frac{35}{50}$ of their games. The LCD of these fractions is 300. Express each fraction as a fraction with a denominator of 300.

Solve Florida: $\frac{38}{60} = \frac{190}{300}$

St. John's: $\frac{35}{50} = \frac{210}{300}$

Since $\frac{190}{300} < \frac{210}{300}$, St. John's University had the better record.

Examine Express each record as a decimal.

Florida: 38 ÷ 60 = *0.633333333*

St. John's: 35 ÷ 50 = *0.7*

The decimals verify that St. John's had the better record.

When comparing or ordering rational numbers, it is usually easier and faster if all of the numbers are decimals.

Examples

② Replace ● with $<$, $>$, or $=$ to make -0.5 ● $-\frac{3}{7}$ a true sentence.

First, express $-\frac{3}{7}$ as a decimal.

3 +/− ÷ 7 = *−0.428571429*

Now compare it to -0.5. In the tenths place, $-5 < -4$.
Therefore $-0.5 < -\frac{3}{7}$.

Study Hint
Mental Math Since
$1.8 = 1.800$ and
$1.\overline{8} = 1.888...,$
$1.8 < 1.\overline{8}$.

③ Order 1.8, 1.07, $\frac{17}{9}$, and $\frac{3}{2}$ from least to greatest.

17 ÷ 9 = *1.888888889*   3 ÷ 2 = *1.5*

$\frac{17}{9} = 1.\overline{8}$ $\frac{3}{2} = 1.5$

The order from least to greatest is 1.07, $\frac{3}{2}$, 1.8, and $\frac{17}{9}$.

262 Chapter 6 Exploring Number Patterns

Reteaching the Lesson

Activity Have each student in the group write a fraction or decimal on a sheet of paper. Students then reveal their numbers and work together to order the numbers from least to greatest. You may wish to direct some students to choose fractions and others to choose decimals.

Error Analysis
Watch for students who have difficulty ordering negative numbers.
Prevent by having them plot the numbers on a number line and then copy the numbers in order.

Example 4

Statistics Find the median of 17.4, 15, $17\frac{1}{2}$, $17\frac{3}{8}$, and 16.9.

To find the median, the numbers must be in order.

$17\frac{1}{2} = 17.5$ $\qquad\qquad 17\frac{3}{8} = 17.375$

The order from least to greatest is 15, 16.9, $17\frac{3}{8}$, 17.4, and $17\frac{1}{2}$. The median is $17\frac{3}{8}$ since it is the number in the middle.

LOOK BACK You can refer to Lesson 4-4 to review median.

CHECK FOR UNDERSTANDING

Communicating Mathematics

Read and study the lesson to answer each question. 1–3. See margin.

1. **Draw** a 10-by-10 square on grid paper. Let the large square represent 1. Use the square to explain why $0.3 > 0.08$.

2. **Describe** two different ways to compare $\frac{5}{6}$ and $\frac{8}{9}$.

3. **Write** a fraction and a decimal that are between $\frac{2}{5}$ and $\frac{4}{7}$. Explain how you know these numbers are between the given numbers.

Guided Practice

4. Find the LCD for $\frac{7}{12}$ and $\frac{3}{8}$. **24**

Replace each ● with <, >, or = to make a true sentence.

5. $-6.2 \; ● \; -6.02$ **<**
6. $\frac{6}{11} \; ● \; 0.5\overline{3}$ **>**
7. $9.6 \; ● \; 9\frac{3}{5}$ **=**

8. Order $\frac{1}{2}, \frac{4}{5}, \frac{2}{5}, \frac{3}{4}$, and 0 from least to greatest. **0, $\frac{2}{5}, \frac{1}{2}, \frac{3}{4}, \frac{4}{5}$**

9. Order $\frac{3}{8}$, 0.376, 0.367, and $\frac{2}{5}$ from greatest to least. **$\frac{2}{5}$, 0.376, $\frac{3}{8}$, 0.367**

10. **Entertainment** Refer to the ride times for nine coasters at Paramount's Canada's Wonderland. Find the median of the ride times. **$2\frac{1}{6}$ min**

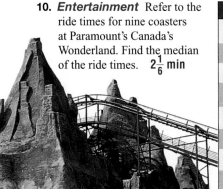

Coaster	Ride Time
Dragon Fyre	$2\frac{1}{6}$ min
Mighty Canadian Minebuster	$2.\overline{6}$ min
Wilde Beast	2.5 min
Ghoster Coaster	$1\frac{5}{6}$ min
SkyRider	$2\frac{5}{12}$ min
Thunder Run	1.75 min
The Bat	$1\frac{5}{6}$ min
Vortex	1.75 min
TOP GUN	$2\frac{5}{12}$ min

EXERCISES

Practice Find the LCD for each pair of fractions.

11. $\frac{5}{16}, \frac{3}{8}$ **16**
12. $\frac{1}{6}, \frac{2}{9}$ **18**
13. $\frac{5}{18}, \frac{4}{15}$ **90**

3 PRACTICE/APPLY

Check for Understanding

If students need additional practice or instruction after completing Exercises 1–10, one of these options may be helpful.
- Extra Practice, see p. 622
- Reteaching Activity, see p. 262
- *Transition Booklet,* pp. 5–6, 13–14
- *Study Guide Masters,* p. 48
- *Practice Masters,* p. 48

Assignment Guide

Core: 11–31 odd, 33–37
Enriched: 12–30 even, 31, 33–37

Additional Answer

1.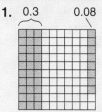

0.3 is represented by 30 small squares. 0.08 is represented by 8 small squares.

Study Guide Masters, p. 48

Additional Answers

2. (1) Express each fraction as a decimal. Compare the decimals.
(2) The LCD of the fractions is 18. Express each fraction as a fraction with a denominator of 18. Compare the numerators.

3. Sample answer: $\frac{1}{2}$, 0.53; Since $\frac{2}{5} = 0.4$, $\frac{4}{7} = 0.\overline{571428}$, and $\frac{1}{2} = 0.5$, $\frac{1}{2}$ and 0.53 are between the given numbers.

Exercise 32 asks students to advance to the next stage of work on the Chapter Project. You may have the whole class discuss how to approach the task before beginning.

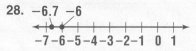

Closing Activity

Speaking Write a pair of fractions on the chalkboard. Have students tell which fraction is greater. Repeat with a pair of decimals and then with a fraction and a decimal.

Additional Answer
28.
```
  -6.7  -6
←●——————●——→
-7-6-5-4-3-2-1 0 1
```
On the number line -6.7 is to the left of -6.

Practice Masters, p. 48

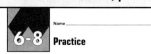

264 Chapter 6

Replace each ● with <, >, or = to make a true sentence.

14. $3\frac{3}{7}$ ● $3\frac{4}{9}$ **<**   15. 1.4 ● 1.403 **<**   16. $\frac{1}{6}$ ● $0.1\overline{6}$ **=**

17. $-\frac{7}{2}$ ● $-\frac{9}{4}$ **<**   18. $\frac{5}{7}$ ● $\frac{9}{21}$ **>**   19. $11\frac{1}{8}$ ● 11.26 **<**

20. -5.2 ● $5\frac{1}{5}$ **<**   21. 0.77 ● $0.\overline{7}$ **<**   22. 5.92 ● $5\frac{23}{25}$ **=**

Order each set of rational numbers from least to greatest.

24. $\frac{1}{2}, \frac{6}{11}, 0.6, 0.\overline{63}$

23. $\frac{1}{9}, \frac{1}{10}, -\frac{1}{3}, -\frac{1}{4}$ $-\frac{1}{3}, -\frac{1}{4}, \frac{1}{10}, \frac{1}{9}$ 24. $0.6, \frac{6}{11}, \frac{1}{2}, 0.\overline{63}$

25. $-4.75, -4\frac{2}{3},$
    $-4.5, -4\frac{2}{5},$
    $-4.1\overline{9}$

25. $-4.5, -4\frac{2}{3}, -4\frac{2}{5}, -4.1\overline{9}, -4.75$ 26. $0.182, 0.182\overline{5}, 0.18\overline{2}, 0.\overline{18}$

27. Which is least, $\frac{7}{11}, 0.6, \frac{2}{3}, 0.\overline{63},$ or $\frac{8}{13}$? **0.6**

26. $0.\overline{18}, 0.182,$ 28. Using a number line, explain why $-6.7 < -6$. **See margin.**
    $0.18\overline{2}, 0.182\overline{5}$

29. Match each number with a point on the number line.

a. 0.567 **Q**   b. $\frac{7}{9}$ **S**

c. $\frac{3}{16}$ **P**   d. 0.7 **R**

Applications and Problem Solving

30. **School** Marco scored 17 out of 20 on his history test and 21 out of 25 on his science test. Did Marco score better on his history test or his science test? **history**

31. **Photography** The shutter speed on Natasha's camera is set at $\frac{1}{250}$ second. If Natasha wants to increase the shutter time, should she set the speed at $\frac{1}{500}$ second or $\frac{1}{125}$ second? $\frac{1}{125}$

32. **Working on the** CHAPTER Project Refer to the table of fractions of games your chosen team won during the regular seasons. Order the fractions from least to greatest. Compare the seasons of the teams. **See students' work.**

33. **Critical Thinking** Are there any rational numbers between $0.\overline{4}$ and $\frac{4}{9}$? Explain. **no;** $0.\overline{4} = \frac{4}{9}$

Mixed Review

34. **Test Practice** The #5 train runs every 8 minutes. The #3 train runs every 6 minutes. If they both leave the station at 8:12 A.M., when will be the next time the trains leave the station together? *(Lesson 6-7)* **C**

A 9:36 A.M.
B 9:00 A.M.
C 8:36 A.M.
D 8:24 A.M.

35. Sample answer: 18   35. Estimate 20.5% of 89. *(Lesson 3-6)*

36. 21   36. Find the absolute value of -21. *(Lesson 2-1)*

37. 35   37. Evaluate $4^2 + 6(12 - 8) - 15 \div 3$. *(Lesson 1-3)*

264 Chapter 6 Exploring Number Patterns

Enrichment Masters, p. 48

Extending the Lesson

Activity Point out that being able to compare and order rational numbers helps when comparing probability. Have the class choose a topic and calculate the probability of various outcomes. Then have them order the outcomes from least likely to most likely to occur.

6-9 Scientific Notation

What you'll learn
You'll learn to express numbers in scientific notation.

When am I ever going to use this?
Knowing how to express numbers in scientific notation can help you in science classes.

Word Wise
scientific notation

In 1996, Disneyland in Anaheim, California, had about 15,000,000 visitors. The number 15,000,000 is in standard form. It can also be written as $1.5 \times 10,000,000$ or 1.5×10^7. The number 1.5×10^7 is in **scientific notation**. Large numbers like 15,000,000 are written in scientific notation to lessen the chance of omitting a zero or misplacing the decimal point.

When a number is expressed in scientific notation, it is written as a product of a factor and a power of 10. The factor must be greater than or equal to 1 and less than 10.

Multiplying by a positive power of 10 moves the decimal point to the right the same number of places as the exponent.

Example 1

Express 2.483×10^5 in standard form.

$2.483 \times 10^5 = 2.483 \times 100,000$
$\qquad\qquad\quad = 248,300$

2.48300
5 places

Scientific notation is also used to express very small numbers. Study the pattern of products. Notice that multiplying by a negative power of 10 moves the decimal point to the left the same number of places as the absolute value of the exponent.

$4.16 \times 10^2 = 416$
$4.16 \times 10^1 = 41.6$
$4.16 \times 10^0 = 4.16$
$4.16 \times 10^{-1} = 0.416$
$4.16 \times 10^{-2} = 0.0416$
$4.16 \times 10^{-3} = 0.00416$
$4.16 \times 10^{-4} = 0.000416$

Example 2 CONNECTION

Life Science The diameter of a red blood cell is about 7.4×10^{-4} centimeter. Write this number in standard form.

$7.4 \times 10^{-4} = 7.4 \times \frac{1}{10^4}$
$\qquad\qquad = 7.4 \times \frac{1}{10,000}$
$\qquad\qquad = 7.4 \times 0.0001$
$\qquad\qquad = 0.00074$

0007.4
4 places

The diameter of a red blood cell is about 0.00074 centimeter.

Look at Examples 1 and 2. What is the relationship between the number of places the decimal point is moved and the absolute value of the exponent? This relationship allows you to use mental math to express any number in scientific notation.

Lesson 6-9 Scientific Notation **265**

Multiple Learning Styles

Verbal/Linguistic Have students write a few sentences to complete each statement.
- A number is written in standard form. To express it in scientific notation, _?_.
- A number is written in scientific notation. To express it in standard form, _?_.

6-9 Lesson Notes

Instructional Resources
- *Study Guide Masters*, p. 49
- *Practice Masters*, p. 49
- *Enrichment Masters*, p. 49
- Transparencies 6-9, A and B
- *Assessment and Evaluation Masters*, p. 156
- *Diversity Masters*, p. 32
- *Technology Masters*, p. 63
- CD-ROM Program
  - Resource Lesson 6-9

Recommended Pacing	
Standard	Day 13 of 15
Honors	Day 11 of 13
Block	Day 7 of 8

1 FOCUS

5-Minute Check
(Lesson 6-8)

Replace each ● with <, >, or = to make a true sentence.
1. $\frac{1}{5}$ ● $\frac{1}{4}$ <
2. 5.2 ● -5.3 >
3. 0.4 ● 0.07 >
4. $\frac{3}{5}$ ● 0.6 =
5. Order $0.66, 0.\overline{6}, 0.\overline{65},$ and $0.6\overline{65}$ from least to greatest.
   $0.65, 0.66, 0.665,$ and $0.\overline{6}$

The 5-Minute Check is also available on **Transparency 6-9A** for this lesson.

Motivating the Lesson

Problem Solving On the chalkboard, write the digit 9 followed by 23 zeros. Tell students that the Sahara Desert has been estimated to contain that many grains of sand. Ask students to suggest ways to write the number more concisely.

2 TEACH

Transparency 6-9B contains a teaching aid for this lesson.

Thinking Algebraically Point out that numbers in scientific notation follow the form $a \times 10^n$, where a is greater than or equal to 1 and less than 10 and n is an integer.

Lesson 6-9 **265**

In-Class Examples

For Example 1
Express 3.7902×10^4 in standard form. **37,902**

For Example 2
A nanosecond is 1×10^{-9} of a second. Express this in standard form. **0.000000001**

For Example 3
Express 79,400.6 in scientific notation. **7.94006×10^4**

For Example 4
Express 0.0000012 in scientific notation. **1.2×10^{-6}**

For Example 5
Enter 0.00000069 into a calculator.
6.9 [EE] 7 [+/−] 6.9 -07

3 PRACTICE/APPLY

Check for Understanding
If students need additional practice or instruction after completing Exercises 1–9, one of these options may be helpful.
- Extra Practice, see p. 623
- Reteaching Activity
- *Transition Booklet*, pp. 9–10
- *Study Guide Masters*, p. 49
- *Practice Masters*, p. 49

Study Guide Masters, p. 49

266 Chapter 6

Examples

Study Hint
Technology The calculator display **5.23 06** means 5.23×10^6.

Express each number in scientific notation.

③ 243,900,000
$243,900,000 = 2.439 \times 10^8$
The decimal point moves 8 places to the left.

④ 0.000000595
$0.000000595 = 5.95 \times 10^{-7}$
The decimal point moves 7 places to the right.

Example 5 shows how to enter a number in a calculator with too many digits to fit on the display screen in standard form.

Example

⑤ Enter 0.00000000627 into a calculator.
First write the number in scientific notation.
$0.00000000627 = 6.27 \times 10^{-9}$
Then enter the number.
6.27 [EE] 9 [+/−]   6.27 -09

CHECK FOR UNDERSTANDING

Communicating Mathematics
Read and study the lesson to answer each question.

1. *Explain* why 36.2×10^3 and 0.362×10^{-4} are not written in scientific notation. **See Answer Appendix.**
2. *Write* a number in scientific notation. Then write the same number in standard form. **Sample answer: $7.1 \times 10^3 = 7,100$**

Guided Practice

Family Activity
Find four very large numbers and four very small numbers in a newspaper. Write these numbers in standard form and in scientific notation.

Express each number in standard form. 4. 0.0000119 5. 234,000,000
3. 4.882×10^5 **488,200** 4. 1.19×10^{-5} 5. 2.34×10^8

Express each number in scientific notation. 6. 2.085×10^9
6. 2,085,000,000   7. 0.054 **5.4×10^{-2}**   8. 0.000091 **9.1×10^{-5}**

9. *Life Science* The Giganotosaurus weighed about 1.4×10^4 pounds. Write this number in standard notation. **14,000 lb**

EXERCISES

Practice

10. 3,110,700
11. 8,080,000,000
12. 0.002331
13. 0.000000075
15. 0.0000252
18. 543,000,000

Express each number in standard form.
10. 3.1107×10^6 11. 8.08×10^9 12. 2.331×10^{-3}
13. 7.5×10^{-8} 14. 1.05×10^4 **10,500**   15. 2.52×10^{-5}
16. 6.81×10^{-2} **0.0681**   17. 2.021×10^2 **202.1**   18. 5.43×10^8

Express each number in scientific notation. 19–27. See margin.
19. 0.00767   20. 5,750,000,000   21. 400,400
22. 0.0000051   23. 0.000033   24. 54,800,000
25. 7,600   26. 0.00083   27. 0.000000025

266 Chapter 6 Exploring Number Patterns

■ Reteaching the Lesson ■

Activity Have pairs of students write six decimals (between 1 and 10) and six powers of 10 on separate sets of index cards. Keeping the sets separate, turn the cards facedown. Have students take turns drawing one card from each set and expressing the product in standard form.

Additional Answers
19. 7.67×10^{-3} 20. 5.75×10^9
21. 4.004×10^5 22. 5.1×10^{-6}
23. 3.3×10^{-5} 24. 5.48×10^7
25. 7.6×10^3 26. 8.3×10^{-4}
27. 2.5×10^{-8}

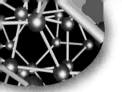

Applications and Problem Solving

28. A googol is a number written with a 1 followed by 100 zeros. Write a googol in scientific notation. 1×10^{100}

29. Write the product of 0.00004 and 0.0008 in scientific notation. 3.2×10^{-8}

30. *Physical Science* An oxygen atom has a mass of 2.66×10^{-23} grams. Explain how to enter this number into a calculator.   2.66 [EE] 23 [+/−]

31. *History* The United States purchased Alaska for $7,200,000. Write this dollar amount in scientific notation. 7.2×10^{6} dollars

32. *Critical Thinking* The galaxy NGC1232 is over 65 million light years from Earth. A light year is 9.46×10^{12} kilometers. Use scientific notation to express the distance to the galaxy in kilometers. 6.149×10^{20} km

Mixed Review

33. Order 7.35, $7\frac{2}{7}$, and $\frac{37}{5}$ from least to greatest. *(Lesson 6-8)* $7\frac{2}{7}$, 7.35, $\frac{37}{5}$

34. *Statistics* Determine whether a scatter plot of the speed of a car and the stopping distance would show a positive, negative, or no relationship. *(Lesson 4-6)*   positive

35. Solve $-319 - (-98) = w$. *(Lesson 2-5)* -221

36. **Test Practice**   A board 42 inches long is cut in 2 pieces so that 1 piece is 7 inches longer than the other. Which equation could be used to find ℓ, the length of the shorter piece? *(Lesson 1-6)*   **A**

A $\ell + (\ell + 7) = 42$
B $2(\ell + 7) = 42$
C $\ell^2 + 7 = 42$
D $2\ell - 7 = 42$
E $42 + 7 = 2\ell$

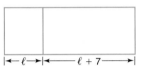

MATH IN THE MEDIA

Fox Trot

1. She multiplied 94 by 24, then multiplied the product by 60, and finally multiplied this product by 60.

1. How did she determine 2,256 hours, 135,260 minutes, and 8,121,600 seconds?
2. Write 8,121,600 seconds in scientific notation. 8.1216×10^{6} s
3. There are 1,000,000,000 nanoseconds in a second. Use scientific notation to write the number of nanoseconds in 8,121,600 seconds. 8.1216×10^{15} nanoseconds
4. Suppose there are 88 days of summer vacation. Write the number of seconds for this vacation in standard form and scientific notation.   7,603,200 s; 7.6032×10^{6} s

Lesson 6-9 Scientific Notation   **267**

Extending the Lesson

Enrichment Masters, p. 49

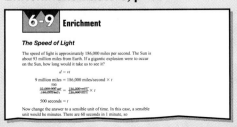

MATH IN THE MEDIA

Have students choose a span of time and calculate how long it will last in days, hours, and seconds. Then have them express the amounts of time in scientific notation. You may wish to have students challenge each other to convert scientific notation into standard form.

Study Guide and Assessment

Chapter 6

Vocabulary

After completing this chapter, you should be able to define each term, concept, or phrase and give an example or two of each.

Number and Operations
bar notation (p. 249)
composite number (p. 235)
density property (p.260)
divisible (p. 232)
factor tree (p. 236)
Fundamental Theorem of Arithmetic (p. 236)
greatest common factor (GCF) (p. 242)
least common denominator (LCD) (p. 261)
least common multiple (LCM) (p. 257)
multiple (p. 257)
prime factorization (p. 235)
prime number (p. 235)
rational number (p. 245)

repeating decimal (p. 249)
scientific notation (p. 265)
simplest form (p. 246)
terminating decimal (p. 249)

Probability
event (p. 253)
outcome (p. 253)
probability (p. 253)
random (p. 253)
sample space (p. 253)

Problem Solving
make a list (p. 240)

Understanding and Using the Vocabulary

Choose the correct term or number to complete each sentence.

1. A (factor, <u>multiple</u>) of a whole number is the product of the number and any other whole number.
2. A whole number greater than 1 that has more than two factors is called a (prime, <u>composite</u>) number.
3. A fraction is in (bar notation, <u>simplest form</u>) when the GCF of the numerator and denominator is 1.
4. The least common denominator of two fractions is the (<u>LCM</u>, GCF) of the denominators.
5. An outcome that is certain to happen has a probability of (0, <u>1</u>).
6. $0.4\overline{5}$ is an example of a (<u>repeating</u>, terminating) decimal.
7. The scientific notation for 32,000 is ($\underline{3.2 \times 10^4}$, 32×10^3).
8. A number that can be expressed as a fraction where the numerator and denominator are both integers is called a (prime, <u>rational</u>) number.

In Your Own Words

9. *Explain* how to express a terminating decimal as a fraction.

268 Chapter 6 Exploring Number Patterns

MindJogger Videoquizzes

MindJogger Videoquizzes provide an alternative review of concepts presented in this chapter. Students work in teams to answer questions, gaining points for correct answers. The questions are presented in three rounds.

Round 1 Concepts–5 questions
Round 2 Skills–4 questions
Round 3 Problem Solving–4 questions

Study Guide and Assessment   Chapter 6

Objectives & Examples

Upon completing this chapter, you should be able to:

- **use divisibility rules for 2, 3, 4, 5, 6, 8, 9, and 10** *(Lesson 6-1)*

  747 is divisible by 3 and 9 since $7 + 4 + 7 = 18$. Since 747 is not even and does not end in 5 or 0, it is *not* divisible by 2, 4, 5, 6, 8, or 10.

- **find the prime factorization of a composite number** *(Lesson 6-2)*

  Write the prime factorization of 56.
  $56 = 2 \cdot 28$
  $= 2 \cdot 2 \cdot 14$
  $= 2 \cdot 2 \cdot 2 \cdot 7$ or $2^3 \cdot 7$

- **find the greatest common factor of two or more numbers** *(Lesson 6-3)*

  Find the GCF of 24 and 32.
  factors of 24: 1, 2, 3, 4, 6, 8, 12, 24
  factors of 32: 1, 2, 4, 8, 16, 32
  The GCF of 24 and 32 is 8.

- **identify and simplify rational numbers** *(Lesson 6-4)*

  Write $\frac{60}{140}$ in simplest form.

  $\frac{60}{140} = \frac{\cancel{2} \cdot \cancel{2} \cdot 3 \cdot \cancel{5}}{\cancel{2} \cdot \cancel{2} \cdot \cancel{5} \cdot 7} = \frac{3}{7}$

- **express rational numbers as decimals and decimals as fractions** *(Lesson 6-5)*

  Write $\frac{4}{11}$ as a decimal.
  $\frac{4}{11} = 4 \div 11$
  $= 0.363636\ldots$ or $0.\overline{36}$

Review Exercises

Use these exercises to review and prepare for the chapter test.

Determine whether each number is divisible by 2, 3, 4, 5, 6, 8, 9, or 10.

10. 533 none
11. 2,435 5
12. 332 2, 4
13. 4,298 2
14. 16,548 2, 3, 4, 6
15. 111 3

Find the prime factorization of each number.

16. 24 $2^3 \cdot 3$
17. 63 $3^2 \cdot 7$
18. 150 $2 \cdot 3 \cdot 5^2$
19. 202 $2 \cdot 101$
20. 44 $2^2 \cdot 11$
21. 81 3^4

Find the GCF for each set of numbers.

22. 16, 64 16
23. 14, 42 14
24. 14, 21, 42 7
25. 22, 33, 55 11
26. 160, 320, 480 160

Write each fraction in simplest form.

27. $\frac{15}{21}$ $\frac{5}{7}$
28. $\frac{16}{20}$ $\frac{4}{5}$
29. $\frac{54}{63}$ $\frac{6}{7}$
30. $-\frac{55}{66}$ $-\frac{5}{6}$

Express each fraction or mixed number as a decimal.

31. $\frac{5}{8}$ 0.625
32. $\frac{16}{33}$ $0.\overline{48}$
33. $2\frac{3}{5}$ 2.6
34. $5\frac{4}{9}$ $5.\overline{4}$

Chapter 6 Study Guide and Assessment **269**

Objectives & Examples

This section reviews the skills and concepts of the chapter and shows completely worked examples.

Review Exercises

These exercises provide practice for the corresponding objectives.

Assessment and Evaluation Masters, pp. 143–144

Assessment and Evaluation

Six forms of Chapter 6 Test are available in the *Assessment and Evaluation Masters* as shown in the chart.

Chapter 6 Test, Form 1B, is shown at the right. Chapter 6 Test, Form 2B, is shown on the next page.

1A	Multiple Choice	Honors
1B	Multiple Choice	Average
1C	Multiple Choice	Basic
2A	Free Response	Honors
2B	Free Response	Average
2C	Free Response	Basic

Study Guide and Assessment **269**

Study Guide and Assessment

Chapter 6 Study Guide and Assessment

Objectives & Examples | Review Exercises

Express $2.\overline{18}$ as a fraction.
Let $N = 2.\overline{18}$. Then $100N = 218.\overline{18}$.
$100N = 218.\overline{18}$
$-N = 2.\overline{18}$
$99N = 216$
$N = \frac{216}{99}$ or $2\frac{2}{11}$. So, $2.\overline{18} = 2\frac{2}{11}$.

Express each decimal as a fraction or mixed number in simplest form.

35. 0.65 $\frac{13}{20}$
36. -0.075 $-\frac{3}{40}$
37. $0.\overline{16}$ $\frac{16}{99}$
38. $6.\overline{3}$ $6\frac{1}{3}$
39. 8.375 $8\frac{3}{8}$
40. $-1.\overline{15}$ $-1\frac{5}{33}$

● find the probability of a simple event
(Lesson 6-6)

$\text{Probability} = \frac{\text{number of ways an event can occur}}{\text{number of possible outcomes}}$

A bag contains six red, three blue, and nine white marbles. A blindfolded student draws a marble. Find each probability.

41. $P(\text{white})$ $\frac{1}{2}$
42. $P(\text{red})$ $\frac{1}{3}$
43. $P(\text{not red})$ $\frac{2}{3}$
44. $P(\text{red or blue})$ $\frac{1}{2}$

● find the least common multiple of two or more numbers (Lesson 6-7)

Find the LCM of 6 and 9.
  multiples of 6: 0, 6, 12, 18, 24, . . .
  multiples of 9: 0, 9, 18, 27, 36, . . .
The LCM of 6 and 9 is 18.

Find the LCM for each set of numbers.

45. 15, 30 30
46. 8, 10 40
47. 21, 24 168
48. 8, 12, 18 72
49. 5, 10, 175 350

● compare and order rational numbers expressed as fractions and/or decimals (Lesson 6-8)

Is $\frac{1}{3} > 0.33$?
$\frac{1}{3} = 0.\overline{3}$ and $0.\overline{3} > 0.33$. So, $\frac{1}{3} > 0.33$.

Replace each ● with <, >, or = to make a true sentence.

50. -8.39 ● -8.4 $>$
51. $\frac{5}{12}$ ● $\frac{3}{4}$ $<$
52. $\frac{2}{3}$ ● 0.6 $>$
53. 0.5 ● $0.\overline{5}$ $<$

● express numbers in scientific notation
(Lesson 6-9)

Express 0.000165 in scientific notation.
$0.000165 \to 1.65 \times 10^{-4}$
4 places

54. 7.41×10^6 55. 6.48×10^{-5}

Express each number in scientific notation.

54. 7,410,000
55. 0.0000648
56. 17,500 1.75×10^4
57. 0.00055
58. 87,500,000
59. 0.00476
57. 5.5×10^{-4} 58. 8.75×10^7 59. 4.76×10^{-3}

270 Chapter 6 Exploring Number Patterns

Test and Review Software

You may use this software, a combination of an item generator and item bank, to create your own tests or worksheets. Types of items include free response, multiple choice, short answer, and open ended.

CD-ROM Program

The CD-ROM Program contains an Assessment Game whose questions review the concepts in this chapter.

270 Chapter 6

Study Guide and Assessment — Chapter 6

Applications & Problem Solving

60. Make a List Laura and Emilio are saving money to attend the state fair. Two tickets cost $30.00. Laura starts with $10.15 and saves $1.50 a week. Emilio starts with $8.50 and saves $1.75 a week. When will they save enough money? *(Lesson 6-3A)*
4th week

61. Sports A batting average is stated as a decimal rounded to the nearest thousandth. Ellis got two hits in eight times at bat. What is his batting average? *(Lesson 6-5)* **0.250**

62. Games To finish the game of Trouble, the number cube must show the number equal to the number of spaces you have left to move. If you have one space left to move, what is the probability that the number cube will show a one? *(Lesson 6-6)* $\frac{1}{6}$

63. Geometry A pin is dropped onto the rectangle below. If the point of the pin lands in one of the small square regions, what is the probability that the point lands inside a shaded region? *(Lesson 6-6)* $\frac{3}{5}$

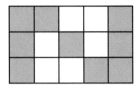

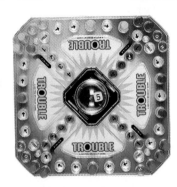

Alternative Assessment

● Performance Task

Suppose your friend was absent when you learned how to express terminating and repeating decimals as fractions. You decide to use the decimals 0.83, $0.8\overline{3}$, and $0.\overline{83}$ to show your friend what she missed. How would you explain the differences among these decimals? How would you explain how to change these decimals to fractions? **See Answer Appendix.**

Your friend was also absent when you learned about comparing and ordering rational numbers. How can you use decimals to show her how to order 0.83, $0.8\overline{3}$, and $0.\overline{83}$? How can you use the fractions to verify your answer? **See Answer Appendix.**

A practice test for Chapter 6 is provided on page 652.

● Completing the CHAPTER Project

Use the following checklist to make sure your news article is complete.

☑ The chart showing the fraction of games won is accurate.

☑ The seasons are compared in the article.

Add any finishing touches that you would like to make your article interesting.

 Select one of the words you learned in this chapter. Place the word and its definition in your portfolio. Attach a note explaining why you selected the word.

 Performance Assessment

Additional performance assessment tasks for this chapter are included in the *Assessment and Evaluation Masters* on page 153. A scoring guide is also provided on page 165.

Standardized Test Practice

The Standardized Test Practice may be used to help students prepare for standardized tests. The test items are written in the same style as those in state proficiency tests and standardized tests like CAT, CTBS, ITBS, MAT, SAT, and Terra Nova. The test items cover skills and concepts covered up to this point in the text.

The pages can be used as an overnight assessment. After students have completed the pages, discuss how each problem can be solved, or provide copies of the solutions from the *Solutions Manual*.

Assessment and Evaluation Masters, p. 159

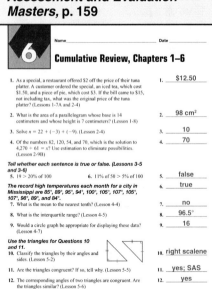

CHAPTERS 1–6
Standardized Test Practice
Assessing Knowledge & Skills

Section One: Multiple Choice

There are thirteen multiple-choice questions in this section. Choose the best answer. If a correct answer is *not here*, choose the letter for Not Here.

1. What is the least prime factor of 42? **C**
   A 4
   B 3
   C 2
   D 1

2. If $\triangle JKL$ is similar to $\triangle QRP$, what is the value of x? **F**

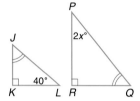

   F 20
   G 25
   H 40
   J 50

3. Express 0.0024 in scientific notation. **C**
   A 2.4×10^3
   B 2.4×10^4
   C 2.4×10^{-3}
   D 2.4×10^{-4}

4. Write $0.\overline{24}$ as a fraction. **F**
   F $\frac{8}{33}$
   G $\frac{1}{3}$
   H $\frac{24}{100}$
   J $\frac{6}{25}$

5. Which equation is equivalent to $x + 5 = 12$? **B**
   A $x = 12$
   B $x + 5 - 5 = 12 - 5$
   C $x + 10 = 24$
   D $x + 5 - 5 = 12 + 5$

6. What is the probability of the spinner landing on a prime number? **G**

   F 0.25
   G 0.5
   H 0.75
   J 1

7. If $\frac{y}{-2} = -4$, what is the value of y? **C**
   A 2
   B -2
   C 8
   D -8

Please note that Questions 8–13 have five answer choices.

8. Carolyn makes $5.65 per hour for working her regularly scheduled hours. If she works any extra hours, she will earn $9.25 per hour. If r is the number of regularly scheduled hours and x is the number of extra hours, which number sentence can be used to find the total pay, p, that Carolyn makes in one week? **G**
   F $p = 5.65 + 9.25x$
   G $p = 5.65r + 9.25x$
   H $p = 5.65r \times 9.25x$
   J $p = (5.25r + 9.25)x$
   K $p = 5.65r + 9.25$

9. Frank bought a suit for $149.99, a tie for $12.49, and a shirt for $14.99. What is the best estimate of the total amount he paid? **B**
   A $160
   B $180
   C $200
   D $220
   E $240

272 Chapters 1–6 Standardized Test Practice

◀◀◀ **Instructional Resources**
Another cumulative review is shown at the left and is available in the *Assessment and Evaluation Masters*, p. 159.

10. Which integers are graphed on the number line? H

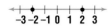

F {−2, 0, 2}
G {−3, 0, 3}
H {−2, 2}
J {−3, 3}
K {0}

11. Mirna paid $42.33 for a school jacket. If the price included $3.10 for sales tax, what was the price of the jacket before tax? D
    A $47.43
    B $45.43
    C $40.23
    D $39.23
    E $35.00

12. Rashid lost an average of two pounds a week on his diet. What was the change in his weight after 5 weeks? G
    F −5 lb
    G −10 lb
    H −15 lb
    J −2 lb
    K Not Here

13. Mr and Mrs. Dixon are having the ceiling of their rectangular living room painted. The living room is 14 feet wide and 20 feet long. A painter charges $0.24 for each square foot to be painted. How much will it cost the Dixons to have their ceiling painted? C
    A $33.60
    B $53.40
    C $67.20
    D $81.80
    E Not Here

Section Two: Free Response

This section contains three questions for which you will provide short answers. Write your answers on your paper.

14. If the perimeter of the isosceles triangle is 45 units, what are the lengths of the sides?

10 units, 10 units, 25 units

15. Which two triangles are similar? I and II

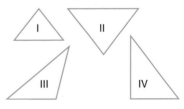

16. What is the perimeter of the rectangle? 41 m

Interdisciplinary Investigation

GET READY

This optional investigation is designed to be completed by a pair of students over several weeks.

Mathematical Overview
This investigation utilizes the concepts from Chapters 1–6.
- finding measures of variation
- choosing a display for data
- writing numbers in scientific notation

Time Management	
Gathering Data	45 minutes
Calculations	30 minutes
Creating Graphs	30 minutes
Summarizing Data	15 minutes
Presentation	15 minutes

Instructional Resources
- *Investigations and Projects Masters*, pp. 37–40

Investigations and Projects Masters, p. 8

274 Chapter 6

Interdisciplinary Investigation

WHAT IN THE WORLD IS "WYSIWYG?"

Do you spend your spare time "on line"? Do you really get "bugged" when you "crash" and must "reboot"? Do you worry that some "hacker" will invade your "database"? If these terms are familiar to you, then you are probably one of the millions of people who use computers.

WYSIWYG (wizzywig) is an acronym for "What you see is what you get." It means that what you see on your computer screen is exactly what will be printed.

What You'll Do

In this investigation, you will research data about computer sales, display the results, and make predictions.

Materials
 almanacs or other factual books
 calculator
 grid paper

Procedure

1. Work in pairs. Locate current information on yearly computer sales. Find the number of computers and the dollar sales of computers for the last ten years. Write the numbers in both standard and scientific notation.
2. Display the change in sales over the last ten years graphically.
3. Use the data to find the average cost per computer for each year.
4. Display the change in the cost of computers graphically.
5. Make some predictions about the sales of computers in the next ten years.

Technology Tips

- Surf the **Internet** to find information about computers.
- Use a **spreadsheet** to calculate the mean cost per computer.
- Use **graphing software** to display the data.

274 Interdisciplinary Investigation What in the World is "WYSIWYG?"

◀◀◀ **Instructional Resources**
A recording sheet to help students organize their data for this investigation is shown at the left and is available in the *Investigations and Projects Masters*, p. 8.

 Cooperative Learning
This investigation offers an excellent opportunity for using cooperative learning groups. For more information on cooperative learning strategies and group management, see *Cooperative Learning in the Mathematics Classroom*.

Making the Connection

Use the data collected about computers as needed to help in these investigations.

Language Arts

Write a newspaper article predicting the future of computers. Use the information that you gathered in the investigation. Be creative and describe the changes in computers that you think will occur. Explain how people will be affected.

Science

Research the impact of computers on the environment. Consider the amounts of paper and electricity used.

Music

Investigate how musicians can use computers in their work.

Go Further

- Survey 10 to 20 adults. Ask them how computers have changed their lives. Make a list of these changes.
- Investigate the binary number system. Find out how it works and why it is important to computers.

 For current information on computers, visit: www.glencoe.com/sec/math/mac/mathnet

 You may want to place your work on this investigation in your portfolio.

Interdisciplinary Investigation What in the World is "WYSIWYG?"

MANAGEMENT TIPS

Working in Pairs Students should work separately to collect data and then pool their findings. They may split calculation tasks and check each other's work. Then they should work together to display the results, make predictions, and present their findings.

Making the Connection

You may wish to alert other teachers on your team that your students may need their assistance in this investigation.
Language Arts Students may need help with expository and persuasive styles of writing.
Science Students will need information on materials used in the production of computers as well as the amount of paper used by computer operators.
Music Students will need information on music software and synthesizers. Students should know that computers are used to record, enhance, and play music.

Investigations and Projects Masters, p. 7

ASSESS

You may require an oral presentation as well as a written report from the students. Ask students to explain what they have learned about computer sales and the impact computers have had on various aspects of their lives in the conclusions to their reports.

Instructional Resources ▶▶▶

Sample solutions for this investigation are provided in the *Investigations and Projects Masters* on p. 6. The scoring guide for assessing student performance shown at the right is also available on p. 7.

Interdisciplinary Investigation

CHAPTER 7
Algebra: Using Rational Numbers

Previewing the Chapter

Overview
This chapter builds on the concepts learned in Chapter 6. Methods for applying all of the arithmetic operations to rationals are developed. Students apply these methods to solve equations and inequalities involving rational numbers. Rational numbers are used in finding the areas of triangles and trapezoids and the circumference of circles. Students learn to solve problems by finding a pattern. They then apply the strategy to sequences, including the Fibonacci sequence.

Lesson (pages)	Lesson Objectives	NCTM Standards	Standardized Tests	State/Local Objectives
7-1 (278–280)	Add and subtract fractions with like denominators.	1–5, 7–9, 13	CAT, CTBS, ITBS, MAT, SAT, TN	
7-2 (281–285)	Add and subtract fractions with unlike denominators.	1–4, 6, 7, 9	CAT, CTBS, ITBS, MAT, SAT, TN	
7-3 (286–289)	Multiply fractions.	1–5, 7, 9	CAT, CTBS, ITBS, MAT, SAT, TN	
7-4 (290–292)	Identify and use rational number properties.	1–5, 7, 9		
7-5A (294–295)	Solve problems by finding and extending a pattern.	1–5, 7–9	MAT	
7-5 (296–299)	Recognize and extend arithmetic and geometric sequences.	1–5, 7–9, 12	CTBS, ITBS, MAT, SAT, TN	
7-5B (300)	Discover the numbers that make up the Fibonacci sequence.	1–5, 8		
7-6 (301–304)	Find the areas of triangles and trapezoids.	1–4, 7, 9, 12, 13	CTBS, ITBS, MAT, SAT, TN	
7-6B (305–306)	Connect algebra and geometry to find the area of a triangle.	1–4, 7, 8, 9, 12, 13		
7-7A (307–308)	Explore the value of pi.	1–6, 8, 12, 13		
7-7 (309–311)	Find the circumference of circles.	1–5, 7, 9, 12, 13	CTBS, ITBS, MAT, SAT, TN	
7-8 (312–314)	Divide fractions.	1–5, 7, 9, 12	CAT, CTBS, ITBS, MAT, SAT, TN	
7-9 (315–317)	Solve equations involving rational numbers.	1–4, 7, 9	CTBS, ITBS, MAT, SAT, TN	
7-10 (318–321)	Solve inequalities involving rational numbers and graph their solutions.	1–9	CTBS, SAT, TN	

CAT = California Achievement Tests, CTBS = Comprehensive Tests of Basic Skills, ITBS = Iowa Tests of Basic Skills, MAT = Metropolitan Achievement Tests, SAT = Stanford Achievement Tests, TN = Terra Nova

Organizing the Chapter

CD-ROM

All of the blackline masters in the Teacher's Classroom Resources are available on the **Electronic Teacher's Classroom Resources** CD-ROM.

Lesson	Extra Practice (Student Edition)	Blackline Masters (page numbers)										
		Study Guide	Practice	Enrichment	Assessment & Evaluation	Classroom Games	Diversity	Hands-On Lab	School to Career	Science and Math Lab Manual	Technology	Transparencies A and B
7-1	p. 623	50	50	50								7-1
7-2	p. 623	51	51	51		19–20						7-2
7-3	p. 624	52	52	52	183				33			7-3
7-4	p. 624	53	53	53								7-4
7-5A	p. 624											
7-5	p. 625	54	54	54	182, 183					77–80		7-5
7-5B								53				
7-6	p. 625	55	55	55				74				7-6
7-6B								54				
7-7A								55				
7-7	p. 625	56	56	56							65	7-7
7-8	p. 626	57	57	57	184							7-8
7-9	p. 626	58	58	58			33				66	7-9
7-10	p. 626	59	59	59	184							7-10
Study Guide/ Assessment					169–181, 185–187							

Other Chapter Resources

Student Edition
Chapter Project, pp. 277, 298, 311, 325
Math in the Media, p. 289
School to Career, p. 293
Let the Games Begin, p. 285

Technology
 CD-ROM Program
 Interactive Mathematics Tools Software

Applications
Family Letters and Activities, pp. 65–66
Investigations and Projects Masters, pp. 41–44
Meeting Individual Needs
Transition Booklet, pp. 7–8, 15–28
Investigations for the Special Education Student, pp. 15, 17–18

Teacher's Classroom Resources

Teaching Aids
Answer Key Masters
Block Scheduling Booklet
Lesson Planning Guide
Solutions Manual

Professional Publications
Glencoe Mathematics Professional Series

Planning the Chapter

 MindJogger Videoquizzes provide a unique format for reviewing concepts presented in the chapter.

ASSESSMENT RESOURCES

Student Edition
Mixed Review, pp. 280, 284, 289, 292, 299, 304, 311, 314, 317, 321
Mid-Chapter Self Test, p. 299
✏ Math Journal, pp. 291, 310
Study Guide and Assessment, pp. 322–325
Performance Task, p. 325
📂 Portfolio Suggestion, p. 325
Standardized Test Practice, pp. 326–327
Chapter Test, p. 653

Assessment and Evaluation Masters
Multiple-Choice Tests (Forms 1A, 1B, 1C), pp. 169–174
Free-Response Tests (Forms 2A, 2B, 2C), pp. 175–180
Performance Assessment, p. 181
Mid-Chapter Test, p. 182
Quizzes A–D, pp. 183–184
Standardized Test Practice, pp. 185–186
Cumulative Review, p. 187

Teacher's Wraparound Edition
5-Minute Check, pp. 278, 281, 286, 290, 296, 301, 309, 312, 315, 318
📂 Building Portfolios, p. 276
✏ Math Journal, pp. 300, 306, 308
Closing Activity, pp. 280, 284, 289, 292, 295, 299, 304, 311, 314, 317, 321

Technology
💾 Test and Review Software
📼 MindJogger Videoquizzes
💿 CD-ROM Program

MATERIALS AND MANIPULATIVES

Lesson 7-1
grid paper
colored pencils

Lesson 7-2
number cubes*
counters*†

Lesson 7-3
lined paper†
ruler*
markers

Lesson 7-5B
grid paper

Lesson 7-6
grid paper

Lesson 7-6B
geoboard*†
spreadsheet software

Lesson 7-7A
grid paper
circular objects
tape

Lesson 7-7
calculator

Lesson 7-8
calculator

Lesson 7-9
calculator

*Glencoe Manipulative Kit   †Glencoe Overhead Manipulative Resources

PACING CHART

See pages T25–T27 for the Course Planning Calendar.

COURSE	DAY 1	DAY 2	DAY 3	DAY 4	DAY 5	DAY 6	DAY 7
Standard	Chapter Project	Lesson 7-1	Lesson 7-2	Lesson 7-3	Lesson 7-4	Lesson 7-5A	Lesson 7-5
Honors	Chapter Project & Lesson 7-1	Lesson 7-2	Lesson 7-3	Lesson 7-4	Lesson 7-5A	Lessons 7-5 & 7-5B	
Block	Chapter Project & Lesson 7-1	Lessons 7-2 & 7-3	Lessons 7-4 & 7-5A	Lessons 7-5 & 7-6	Lessons 7-7A & 7-7	Lesson 7-8 & 7-9	Lesson 7-10

The *Transition Booklet* (Skills 2, 6–12) can be used to practice simplifying fractions; adding, subtracting, multiplying, and dividing fractions and mixed numbers; and adding and subtracting decimals.

Interactive Mathematics:
Activities and Investigations

is an activity-based program that may be used as an enhancement for chapters in *Mathematics: Applications and Connections.*

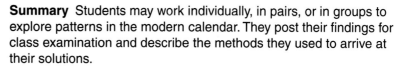

Unit 11, Activity One, Menu A
Use with Lesson 7-5A.

Summary Students may work individually, in pairs, or in groups to explore patterns in the modern calendar. They post their findings for class examination and describe the methods they used to arrive at their solutions.

Math Connection Students examine time measurement units and how the modern calendar is designed. They find and interpret information from the *Time Measurement Reference* in the Data Bank to analyze the structure and design of the modern calendar.

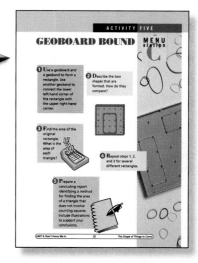

Unit 9, Activity Five, Menu C
Use before Lesson 7-6.

Summary Students work in groups using geoboards to discover how to find the area of a triangle. They prepare a report identifying the method they developed. The method used is not to involve counting squares.

Math Connection Students use a geoboard to discover that the area of a triangle formed by connecting the diagonals of a rectangle is one-half the area of the rectangle.

	DAY 8	DAY 9	DAY 10	DAY 11	DAY 12	DAY 13	DAY 14	DAY 15
	Lesson 7-6	Lessons 7-7A & 7-7		Lesson 7-8	Lesson 7-9	Lesson 7-10	Study Guide and Assessment	Chapter Test
	Lessons 7-6 & 7-6B		Lesson 7-7	Lesson 7-8	Lesson 7-9	Lesson 7-10	Study Guide and Assessment	Chapter Test
	Study Guide and Assessment, Chapter Test							

Enhancing the Chapter

APPLICATIONS

Classroom Games, pp. 19–20
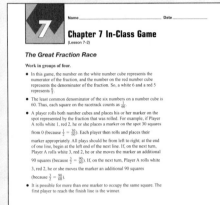

Diversity Masters, p. 33
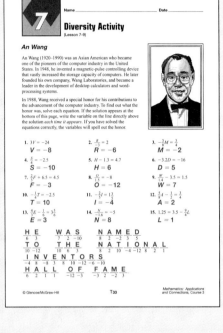

School to Career Masters, p. 33
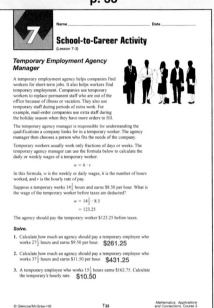

Family Letters and Activities, pp. 65–66

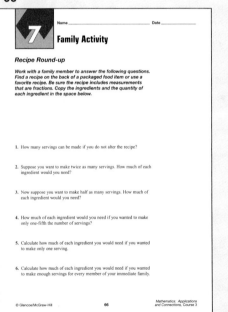

Science and Math Lab Manual, pp. 77–80
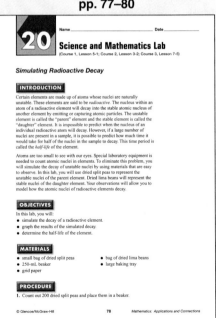

MANIPULATIVES/MODELING

Hands-On Lab Masters,
p. 74

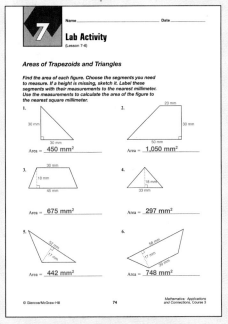

ASSESSMENT/EVALUATION

Assessment and Evaluation Masters,
pp. 182–184

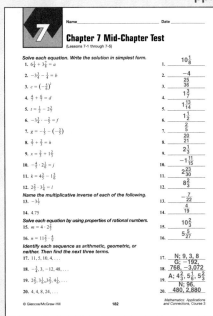

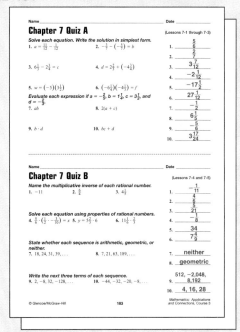

TECHNOLOGY/MULTIMEDIA

Technology Masters,
pp. 65–66

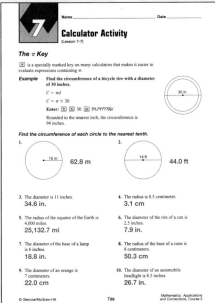

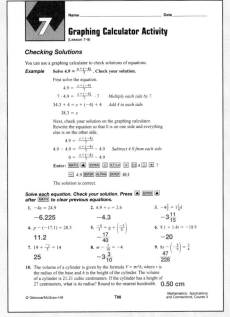

MEETING INDIVIDUAL NEEDS

*Investigations for the Special
Education Student,* pp. 15, 17–18

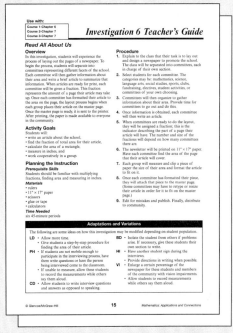

Chapter 7 **276f**

CHAPTER 7 NOTES

Theme: Nature
Many types of patterns occur in nature—such as cycles, geometric designs, and physical laws. Johann Kepler's observations of the planets' movements helped him develop the three laws of planetary motion. Isaac Newton discovered astronomical patterns and wrote the law of gravitation. Today, scientists study weather patterns to help predict storms. Lightning has fractal patterns. Tornadoes, whirlwinds, hurricanes, and typhoons have spiral patterns.

Question of the Day Growth in plants and animals occurs through cell division. One cell divides into two. Each newly formed cell divides into two new cells, which divide again. What is the pattern of the total number of cells at each stage? 2, 4, 8, 16, 32, . . .

Assess Prerequisite Skills
Ask students to read through the list of objectives presented in "What you'll learn in Chapter 7." You may wish to ask them what each of the objectives means or if they have experienced or used any of these math concepts before.

 Building Portfolios
Encourage students to revise their portfolios as they study this chapter. They may find some lessons more challenging than others and wish to include work that shows how they have improved their mathematical skills.

 Math and the Family
In the *Family Letters and Activities* booklet (pp. 65–66), you will find a letter to the parents explaining what students will study in Chapter 7. An activity appropriate for the whole family is also available.

CHAPTER 7 — Algebra: Using Rational Numbers

What you'll learn in Chapter 7
- to compute with fractions and mixed numbers,
- to solve problems by using patterns,
- to recognize and extend arithmetic and geometric sequences,
- to find the areas of triangles and trapezoids, and find the circumference of circles, and
- to solve equations and inequalities involving rational numbers.

 CD-ROM Program

Activities for Chapter 7
- Chapter 7 Introduction
- Interactive Lessons 7-2, 7-3, 7-7
- Extended Activity 7-7
- Assessment Game
- Resource Lessons 7-1 through 7-10

CHAPTER Project

PATTERNS IN NATURE

In this project, you will explore one of the mathematical patterns found in nature. You will use what you find to create a poster or collage, or to present a slide show with narration.

Getting Started

- Collect a variety of pinecones. If these are not available, check with your science teacher. Identify the clockwise and counterclockwise spirals of scales from the bottoms of the cones. Count the number of individual spirals in each direction. Record the numbers of spirals you find.
- Study the arrangements of leaves on the stem of a variety of plants. If you look down on a plant, leaves are often arranged so that each one gets water and light. Make a drawing of each plant. Number the leaves from the bottom of the stem to the top. What is the difference between the numbers of the leaves that are aligned? How many times do you go around the stem to reach aligning leaves? Record these numbers for several different types of plants.

Technology Tips

- Use an **electronic encyclopedia** to do your research.
- Write portions of your poster or write your slide show script using **word processing** or **presentation** software.
- Record the information you gather about pinecone spirals or plant leaves in a **database**.

*inter*NET CONNECTION For up-to-date information on patterns and nature, visit:
www.glencoe.com/sec/math/mac/mathnet

Working on the Project

You can use what you'll learn in Chapter 7 to find patterns in nature.

Page	Exercise
298	29
311	26
325	Alternative Assessment

*inter*NET CONNECTION

Glencoe has made every effort to ensure that the website links for *Mathematics: Applications and Connections* at www.glencoe.com/sec/math/mac/mathnet are current and contain appropriate content. However, these website links are not under Glencoe's control.

Instructional Resources ▶▶▶

A recording sheet to help students organize their data for the Chapter Project is shown at the right and is available in the *Investigations and Projects Masters*, p. 44.

CHAPTER Project NOTES

Objectives Students should
- gain an awareness of mathematical patterns in the world around them.
- develop their skills in presenting knowledge both verbally and visually.

Project Pointer You may suggest that students begin a *Project Folder* to keep their work as they complete each stage of the Chapter Project. The completed project may also be added to their portfolios.

Using the Drawings Students may have difficulty connecting two-dimensional drawings to three-dimensional images. Encourage them to think about how they would draw a pinecone or leaves on a stem. You may wish to make a transparency of an enlarged three-dimensional drawing and show it to the students. Point at parts of the drawing and model them with your hands.

Investigations and Projects Masters, p. 44

Chapter 7 Project

Patterns In Nature

Page 277, Getting Started
No. of Clockwise Spirals:
No. of Counterclockwise Spirals:

Page 298, Working on the Chapter Project, Exercise 29
a. 1, 1, 2, 3, 5, 8, 13, 21, ___, ___, ___
   The sequence is:
b.

Page 311, Working on the Chapter Project, Exercise 26
a. Use this space to draw centimeter boxes.

b. Use this space to draw spirals found in nature.

7-1 Lesson Notes

Instructional Resources
- *Study Guide Masters*, p. 50
- *Practice Masters*, p. 50
- *Enrichment Masters*, p. 50
- Transparencies 7-1, A and B
- CD-ROM Program
  - Resource Lesson 7-1

Recommended Pacing	
Standard	Day 2 of 15
Honors	Day 2 of 15
Block	Day 1 of 8

1 FOCUS

5-Minute Check
(Chapter 6)

1. Find the prime factorization of 180. $2^2 \cdot 3^2 \cdot 5$
2. Find the GCF for 36 and 84. 12
3. Write $-\frac{12}{28}$ in simplest form. $-\frac{3}{7}$
4. Express 0.65 as a fraction in simplest form. $\frac{13}{20}$
5. Find the LCM for 18 and 24. 72

The 5-Minute Check is also available on **Transparency 7-1A** for this lesson.

Motivating the Lesson

Hands-On Activity Use 1-cup and 2-cup liquid measures to illustrate fractions. Pour water from the small cup into the large one to find sums. For example, $\frac{3}{4} + \frac{3}{4} = \frac{6}{4}$ or $1\frac{1}{2}$. Show subtraction by pouring water back into the smaller cup.

2 TEACH

Transparency 7-1B contains a teaching aid for this lesson.

Using the Mini-Lab Point out that in a sum the larger number is not always the denominator. For example, $\frac{3}{5} + \frac{4}{5} = \frac{7}{5}$, not $\frac{5}{7}$.

7-1 Adding and Subtracting Like Fractions

What you'll learn
You'll learn to add and subtract fractions with like denominators.

When am I ever going to use this?
Knowing how to add and subtract fractions can help you decorate your room.

Word Wise
mixed number

1. $\frac{3}{5} + \frac{1}{5} = \frac{4}{5}$
2. Sample answer: Erase one of the shaded sections.

You can use rectangles to model adding fractions.

HANDS-ON MINI-LAB

Work with a partner. grid paper colored pencils

Use models to add $\frac{3}{5}$ and $\frac{1}{5}$.

Try This
- On grid paper, draw a rectangle like the one shown. This rectangle shows fifths.
- With a colored pencil, shade three sections of the rectangle to represent $\frac{3}{5}$.
- With a different colored pencil, shade one more section of the rectangle to represent $\frac{1}{5}$.

Talk About It
1. What sum is modeled by your shading?
2. How could you use this model to show $\frac{4}{5} - \frac{1}{5} = \frac{3}{5}$?

In the Mini-Lab, the rational numbers $\frac{3}{5}$ and $\frac{1}{5}$ have like denominators. Fractions with like denominators are also called *like fractions*. The results of the Mini-Lab suggest this rule for adding like fractions.

Adding Like Fractions	Words:	To add fractions with like denominators, add the numerators.
	Symbols:	Arithmetic $\frac{2}{5} + \frac{1}{5} = \frac{3}{5}$
		Algebra $\frac{a}{c} + \frac{b}{c} = \frac{a+b}{c}, c \neq 0$

When the sum of two fractions is greater than 1, the sum is written as a **mixed number**, usually in simplest form. A mixed number is the sum of a whole number and a fraction.

Example 1 Find $\frac{1}{2} + \frac{1}{2} + \frac{1}{2}$.

$\frac{1}{2} + \frac{1}{2} + \frac{1}{2} = \frac{3}{2}$ *Add the numerators.*

$= 1\frac{1}{2}$ *Rename $\frac{3}{2}$ as a mixed number, $1\frac{1}{2}$.*

278 Chapter 7 Algebra: Using Rational Numbers

Classroom Vignette

"Some of my students have trouble remembering which numbers are the numerator and denominator. I have them remember **N**otre **D**ame—N comes first and stands for numerator; D comes second and stands for denominator."

Dennis Iffland, Teacher
East Toledo Junior High School
Toledo, OH

Dennis Iffland

278 Chapter 7

Subtracting like fractions is similar to adding them.

Subtracting Like Fractions	**Words:**	To subtract fractions with like denominators, subtract the numerators.
	Symbols: Arithmetic	$\frac{2}{3} - \frac{1}{3} = \frac{1}{3}$
	Algebra	$\frac{a}{c} - \frac{b}{c} = \frac{a-b}{c}, c \neq 0$

Rational numbers include both positive and negative fractions. Use the rules for adding and subtracting integers to determine the sign of the sum or difference of two rational numbers.

Examples

2 Solve $x = -\frac{3}{8} - \left(-\frac{5}{8}\right)$.

$x = -\frac{3}{8} - \left(-\frac{5}{8}\right)$

$x = \frac{-3 - (-5)}{8}$ *Since the denominators are the same, subtract the numerators.*

$x = \frac{2}{8}$ or $\frac{1}{4}$ *Simplify.*

APPLICATION

3 **Home Decorating** Suppose you bought two storage blocks for your bedroom. One is $14\frac{3}{4}$ inches high. The other is $9\frac{3}{4}$ inches high. You want to stack them and fit them under the window, which is 24 inches from the floor. Will they fit?

Find the sum of the heights of the blocks.

$14\frac{3}{4} + 9\frac{3}{4} = 23\frac{6}{4}$ *Add the whole numbers and then the fractions.*

$= 24\frac{2}{4}$ or $24\frac{1}{2}$ *Rename the improper fraction.*

The sum of the blocks is higher than 24 inches. They will not fit under the window.

CHECK FOR UNDERSTANDING

Communicating Mathematics

Read and study the lesson to answer each question.

1. *Write* the subtraction sentence shown by the following model. $\frac{2}{3} - \frac{1}{3} = \frac{1}{3}$

2. *Explain* how adding and subtracting rational numbers is similar to adding and subtracting integers. **See margin.**

HANDS-ON MATH

3. *Model* $\frac{3}{4} + \frac{3}{4}$ using a diagram and find the sum. **See Answer Appendix.**

Lesson 7-1 Adding and Subtracting Like Fractions **279**

In-Class Examples

For Example 1
Find $\frac{2}{5} + \frac{2}{5} + \frac{2}{5}$. $1\frac{1}{5}$

For Example 2
Solve $n = \frac{4}{9} - \frac{7}{9}$. $-\frac{1}{3}$

For Example 3
Suppose you and your brother want to put your desks side-by-side along an 8-foot wall in your den. Your desk is $3\frac{3}{4}$ feet wide. Your brother's desk is $3\frac{1}{4}$ feet wide. Will they fit?
Yes; $3\frac{3}{4} + 3\frac{1}{4} = 7$.

3 PRACTICE/APPLY

Check for Understanding

If students need additional practice or instruction after completing Exercises 1–8, one of these options may be helpful.
- Extra Practice, see p. 623
- Reteaching Activity
- *Transition Booklet,* pp. 15–20, 23–24
- *Study Guide Masters,* p. 50
- *Practice Masters,* p. 50
- Interactive Mathematics Tools Software

Study Guide Masters, p. 50

■ **Reteaching the Lesson** ■

Activity Draw a number line from -1 to 1. Divide each unit into eighths. Have students use the line to illustrate addition and subtraction of fractions with denominators of 8.

Additional Answer

2. Sample answer: Adding and subtracting rational numbers is similar to adding and subtracting integers because you use the same rules to determine the sign of the sum or difference.

Lesson 7-1 **279**

Assignment Guide

Core: 9–23 odd, 25–28
Enriched: 10–22 even, 23–28

4 ASSESS

Closing Activity

Modeling Have students use fraction circles or fraction bars to model addition and subtraction equations.

Additional Answers

24c. 1; Sample answer: The sum of the probabilities equals one because you can roll either a 2 or not a 2.

28. Sample answer:

Practice Masters, p. 50

280 Chapter 7

Guided Practice Solve each equation. Write the solution in simplest form.

4. $\frac{3}{8} + \frac{5}{8} = a$ 1
5. $\frac{3}{4} + \left(-\frac{1}{4}\right) = d$ $\frac{1}{2}$
6. $k = -\frac{2}{3} - \frac{1}{3}$ -1
7. $y = 2\frac{3}{5} - \left(-1\frac{4}{5}\right)$ $4\frac{2}{5}$
8. **Algebra** Evaluate $a + b$ if $a = -\frac{3}{8}$ and $b = -2\frac{7}{8}$. $-3\frac{1}{4}$

EXERCISES

Practice Solve each equation. Write the solution in simplest form.

9. $\frac{3}{8} - \frac{1}{8} = b$ $\frac{1}{4}$
10. $-\frac{5}{6} + \frac{1}{6} = f$ $-\frac{2}{3}$
11. $m = -\frac{4}{5} - \frac{3}{5}$ $-1\frac{2}{5}$
12. $r = \frac{5}{6} - \left(-\frac{7}{6}\right)$ 2
13. $-2\frac{1}{2} - \frac{1}{2} = d$ -3
14. $\frac{1}{4} - \frac{3}{4} = z$ $-\frac{1}{2}$
15. $c = -\frac{7}{12} - \frac{5}{12}$ -1
16. $5\frac{2}{3} - \left(-2\frac{2}{3}\right) = x$ $8\frac{1}{3}$
17. $-2\frac{5}{9} - \frac{5}{9} = y$ $-3\frac{1}{9}$

19. $-\frac{1}{2}$

18. Subtract $-2\frac{1}{3}$ from $5\frac{1}{3}$. $7\frac{2}{3}$
19. What is the sum of $-\frac{5}{12}$ and $-\frac{1}{12}$?

Evaluate each expression if $a = -\frac{2}{5}$ and $b = \frac{3}{5}$.

20. $a + b$ $\frac{1}{5}$
21. $b - a$ 1
22. $a - (a + b)$ $-\frac{3}{5}$

Applications and Problem Solving

23. **Manufacturing** Plastic straps are often wound around large cardboard boxes to reinforce them during shipping. Suppose the ends of the strap must overlap $\frac{7}{16}$ inch to fasten. How long is the plastic strap around the box? $106\frac{3}{16}$ in. or 8 ft $10\frac{3}{16}$ in.

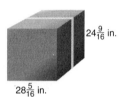

24. **Probability** A number cube is rolled.
    a. What is the probability of rolling a 2 or a 5? $\frac{1}{3}$
    b. What is the probability of *not* rolling a 2 or a 5? $\frac{2}{3}$
    c. Find the sum of the answers to parts a and b. Why is this true?   **See margin.**

25. **Critical Thinking** A bookworm chewed its way from the first page of Volume I to the last page of Volume III. The books are on a bookshelf standing next to each other. If each volume is 2 inches thick and the covers are $\frac{3}{16}$-inch thick, how many inches did the bookworm chew? $2\frac{3}{4}$ in.

Mixed Review

26. Express 9.2×10^{-4} in standard form. *(Lesson 6-9)*   **0.00092**

27. **Test Practice** One centimeter is about 0.392 inch. What fraction of an inch is this? *(Lesson 6-5)*   **C**

   A $\frac{49}{500}$ in.   **B** $\frac{98}{125}$ in.
   C $\frac{49}{125}$ in.   **D** $\frac{392}{100}$ in.

28. **Geometry** Draw an isosceles right triangle. *(Lesson 5-2)*   **See margin.**

280   Chapter 7 Algebra: Using Rational Numbers

Enrichment Masters, p. 50

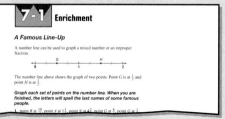

Extending the Lesson

Activity Have students study the following equations, $\frac{1}{6} + \frac{1}{6} = \frac{1}{3}$; $\frac{3}{8} + \frac{3}{8} = \frac{3}{4}$; $\frac{5}{12} + \frac{5}{12} = \frac{5}{6}$. Explain the relationship between the answers and the fractions being added. **The numerators and denominators of the addends are the same, and the denominator of the sum is half that of the denominators of the addends.**

7-2 Adding and Subtracting Unlike Fractions

What you'll learn
You'll learn to add and subtract fractions with unlike denominators.

When am I ever going to use this?
Knowing how to add and subtract fractions with unlike denominators is helpful when mixing paint.

Quoits (pronounced kóits) was one of the five original games included in the ancient Greek pentathlon. It is similar to horseshoes, but is played with metal rings. The object of the game is to throw the quoit onto or as close as possible to a vertical metal pole.

A quoit weighs about $3\frac{1}{2}$ pounds and is 6 inches across with a hole in the middle. If each section of metal is $1\frac{5}{8}$ inches wide, how large is the hole in the center of the quoit? *This problem will be solved in Example 2.*

In order to find the width of the hole in the center, you must subtract the total width of the metal, $1\frac{5}{8} + 1\frac{5}{8}$, from the width of the quoit, 6. The whole number 6 does not contain a fraction. To subtract, we must rename 6.

Adding and Subtracting Unlike Fractions	To find the sum or difference of two fractions with unlike denominators, rename the fractions with a common denominator. Then add or subtract and simplify.

Example **Algebra** Evaluate $m - n$ if $m = \frac{5}{6}$ and $n = -\frac{2}{3}$.

INTEGRATION

The least common denominator (LCD) is the least common multiple (LCM) of the denominators. The LCD is helpful in renaming fractions for adding and subtracting.

$\begin{aligned} m - n &= \frac{5}{6} - \left(-\frac{2}{3}\right) && \text{Replace } m \text{ with } \frac{5}{6} \text{ and } n \text{ with } -\frac{2}{3}. \\ &= \frac{5}{6} - \left(-\frac{4}{6}\right) && \text{Use the LCM of 6 and 3 to rename } -\frac{2}{3} \text{ as } -\frac{4}{6}. \\ &= \frac{5}{6} + \left(\frac{4}{6}\right) && \text{Subtract } -\frac{4}{6} \text{ by adding its inverse, } \frac{4}{6}. \\ &= \frac{9}{6} && \text{Add the numerators.} \\ &= 1\frac{3}{6} \text{ or } 1\frac{1}{2} && \text{Simplify.} \end{aligned}$

Lesson 7-2 Adding and Subtracting Unlike Fractions **281**

7-2 Lesson Notes

Instructional Resources
- *Study Guide Masters*, p. 51
- *Practice Masters*, p. 51
- *Enrichment Masters*, p. 51
- Transparencies 7-2, A and B
- *Classroom Games*, pp. 19–20
- CD-ROM Program
  - Resource Lesson 7-2
  - Interactive Lesson 7-2

Recommended Pacing	
Standard	Day 3 of 15
Honors	Day 2 of 15
Block	Day 2 of 8

1 FOCUS

5-Minute Check
(Lesson 7-1)
Solve each equation. Write the solution in simplest form.
1. $\frac{5}{9} + \frac{1}{9} = k$ $\frac{2}{3}$
2. $-\frac{7}{8} + \left(-\frac{5}{8}\right) = p$ $-1\frac{1}{2}$
3. $x = \frac{7}{10} - \frac{17}{10}$ -1
4. $\ell = \frac{5}{8} - \left(-\frac{2}{8}\right)$ $\frac{7}{8}$
5. Evaluate $b + a$ if $a = \frac{1}{3}$ and $b = -\frac{2}{3}$. $-\frac{1}{3}$

The 5-Minute Check is also available on **Transparency 7-2A** for this lesson.

Motivating the Lesson
Communication Ask students to name real-life situations that require adding and subtracting fractions. **Sample answer: calculating the quantities of construction materials needed to build a house**

Classroom Vignette

"I use an activity relating music and time signatures as it relates to fractions. Then I bring in the various string instruments I play to demonstrate what these fractions mean."

Jerry W. Murkerson, Teacher
West Bainbridge Middle School
Bainbridge, GA

Jerry W. Murkerson

Lesson 7-2 **281**

2 TEACH

 Transparency 7-2B contains a teaching aid for this lesson.

Using Connections Remind students that in Lesson 7-1 they added and subtracted fractions with like denominators. This lesson shows how to rename fractions so they have like denominators. Then the fractions can be added and subtracted using the methods students know.

Teaching Tip In Example 1, point out that the LCD of the fractions is the same as the LCM of their denominators.

In-Class Examples

For Example 1
Evaluate $j - k$ if $j = \frac{7}{8}$ and $k = -\frac{1}{4}$. $1\frac{1}{8}$

For Example 2
A door with a window is $2\frac{1}{2}$ feet wide. If there are $6\frac{1}{4}$ inches of wood on each side of the glass, what is the width of the window? $17\frac{1}{2}$ in.

For Example 3
Solve $2\frac{3}{4} + 3\frac{7}{8} = w$. $6\frac{5}{8}$

To subtract a fraction or mixed number from a whole number, you must rename the whole number with an improper fraction. To add or subtract mixed numbers with unlike denominators, first rename the fractions with a common denominator.

Examples APPLICATION

② Recreation Refer to the beginning of the lesson. Find the width of the hole in a quoit.

Explore What do you know? The width of the quoit is 6 inches. Each section of metal is $1\frac{5}{8}$ inches wide.

Plan Add the two metal widths and subtract from 6.

Solve Find the total width of the metal.

$$1\frac{5}{8} + 1\frac{5}{8} = 2\frac{10}{8}$$
$$= 3\frac{2}{8} \text{ or } 3\frac{1}{4}$$

Now subtract $3\frac{1}{4}$ from 6.

$$6 - 3\frac{1}{4} = 5\frac{4}{4} - 3\frac{1}{4} \quad \textit{Rename 6 as } 5\frac{4}{4}.$$
$$= 2\frac{3}{4} \quad \textit{Simplify.}$$

The hole is $2\frac{3}{4}$ inches across.

Examine See if the sum of the parts is equal to 6.

$$1\frac{5}{8} + 1\frac{5}{8} + 2\frac{3}{4} \stackrel{?}{=} 6$$
$$1\frac{5}{8} + 1\frac{5}{8} + 2\frac{6}{8} \stackrel{?}{=} 6 \quad \textit{Rename } \frac{3}{4} \textit{ as } \frac{6}{8}.$$
$$4\frac{16}{8} \stackrel{?}{=} 6$$
$$6 = 6 \checkmark$$

③ Solve $6\frac{3}{8} - 2\frac{5}{6} = x$.

Estimate: $6 - 3 = 3$

The LCM of 8 and 6 is 24.

$$\begin{array}{ccccc} 6\frac{3}{8} & \rightarrow & 6\frac{9}{24} & \rightarrow & 5\frac{33}{24} \\ -2\frac{5}{6} & & -2\frac{20}{24} & & -2\frac{20}{24} \\ \hline & & & & 3\frac{13}{24} \end{array}$$

Rename $6\frac{9}{24}$ as $5\frac{33}{24}$. Why?

So, $3\frac{13}{24} = x$. *Compare to the estimate.*

282 Chapter 7 Algebra: Using Rational Numbers

CHECK FOR UNDERSTANDING

Communicating Mathematics

1. $\frac{2}{3} + \frac{1}{4} = \frac{11}{12}$

Read and study the lesson to answer each question.

1. *Write* the addition sentence shown by the model.

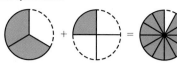

2. *Explain* why you might rename $5\frac{3}{8}$ as $4\frac{11}{8}$ in a subtraction problem. **See margin.**

HANDS-ON MATH

3. *Model* $\frac{7}{12} - \frac{1}{3}$ using the diagram. Then find the difference. $\frac{1}{4}$; See Answer Appendix for model.

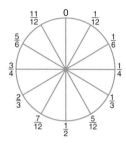

Guided Practice

Complete.

4. $5\frac{1}{3} = 4\frac{\blacksquare}{3}$ **4**
5. $3\frac{2}{5} = 2\frac{\blacksquare}{5}$ **7**
6. $4\frac{5}{8} = 3\frac{\blacksquare}{8}$ **13**

Solve each equation. Write the solution in simplest form.

7. $\frac{3}{4} + \frac{1}{2} = z$ $1\frac{1}{4}$
8. $4\frac{3}{4} + \left(-\frac{1}{3}\right) = x$ $4\frac{5}{12}$
9. $c = \frac{1}{4} - \left(-\frac{5}{8}\right)$ $\frac{7}{8}$
10. $h = -\frac{7}{10} + \frac{1}{5}$ $-\frac{1}{2}$
11. $3\frac{7}{8} + 5\frac{5}{24} = d$ $9\frac{1}{12}$
12. $5\frac{1}{3} - 3\frac{3}{4} = n$ $1\frac{7}{12}$

13. **Algebra** Evaluate $c - d$ if $c = -\frac{3}{4}$ and $d = -12\frac{7}{8}$. $12\frac{1}{8}$

14. **Algebra** Simplify $1\frac{7}{8}a - \frac{1}{4}a$. $1\frac{5}{8}a$

EXERCISES

Practice

Solve each equation. Write the solution in simplest form.

15. $\frac{3}{8} + \frac{2}{3} = k$ $1\frac{1}{24}$
16. $-\frac{5}{6} + \frac{1}{2} = g$ $-\frac{1}{3}$
17. $m = -1\frac{4}{5} - \frac{3}{10}$ $-2\frac{1}{10}$
18. $y = -\frac{3}{4} + \left(-\frac{1}{3}\right)$ $-1\frac{1}{12}$
19. $9\frac{1}{3} - 2\frac{1}{2} = h$ $6\frac{5}{6}$
20. $5 - 3\frac{1}{3} = f$ $1\frac{2}{3}$
21. $s = 5 - 3\frac{3}{4}$ $1\frac{1}{4}$
22. $j = -4\frac{1}{2} - \frac{3}{5}$ $-5\frac{1}{10}$
23. $b = -4\frac{1}{4} - 5\frac{5}{8}$ $-9\frac{7}{8}$
24. $d = 3\frac{2}{3} + \left(-5\frac{3}{4}\right)$ $-2\frac{1}{12}$
25. $7\frac{3}{4} - \left(-1\frac{1}{8}\right) = x$ $8\frac{7}{8}$
26. $-6\frac{5}{6} - 4\frac{7}{8} = w$ $-11\frac{17}{24}$

28. $-1\frac{1}{8}$

27. Subtract $-6\frac{1}{4}$ from 9. $15\frac{1}{4}$
28. What is the sum of $-\frac{5}{8}$ and $-\frac{1}{2}$?
29. What is $2\frac{3}{8}$ less than $-8\frac{1}{5}$? $-10\frac{23}{40}$
30. Find the sum of $\frac{1}{2}$, $\frac{1}{3}$, and $\frac{1}{5}$. $1\frac{1}{30}$

Lesson 7-2 Adding and Subtracting Unlike Fractions **283**

Reteaching the Lesson

Activity Have students use a ruler to model addition and subtraction of fractions with denominators of 2, 4, or 8. For example, to add $1\frac{1}{2} + \frac{5}{8}$, have students draw a segment $1\frac{1}{2}$ inches long, then extend the segment $\frac{5}{8}$ inch farther to the right (to a length of $2\frac{1}{8}$ inches).

Error Analysis
Watch for students who rename the denominator but not the numerator.
Prevent by having students estimate answers. Calculated and estimated answers will generally differ markedly if students do not rename the numerators.

---

3 PRACTICE/APPLY

Check for Understanding

If students need additional practice or instruction after completing Exercises 1–14, one of these options may be helpful.
- Extra Practice, see p. 623
- Reteaching Activity
- *Transition Booklet,* pp. 21–24
- *Study Guide Masters,* p. 51
- *Practice Masters,* p. 51
- Interactive Mathematics Tools Software

Assignment Guide

Core: 15–35 odd, 37–42
Enriched: 16–32 even, 34–42

Additional Answer

2. Sample answer: If you were subtracting $2\frac{5}{8}$ from $5\frac{3}{8}$, you would need to rename $5\frac{3}{8}$ to $4\frac{11}{8}$ so you would have enough eighths to subtract from.

Study Guide Masters, p. 51

Lesson 7-2 **283**

4 ASSESS

Closing Activity

Writing Have students work in pairs to create problems involving addition and subtraction of fractions with unlike denominators.

Evaluate each expression if $r = -\frac{5}{8}$, $s = 2\frac{5}{6}$, and $t = \frac{5}{9}$.

31. $r - s$ $-3\frac{11}{24}$
32. $r + s + t$ $2\frac{55}{72}$
33. $s - (-r)$ $2\frac{5}{24}$

Applications and Problem Solving

34. **Music** A waltz is written in $\frac{3}{4}$ (read as three-four) time. This means there are three beats in a measure and the quarter note gets one beat. The value of the notes can be expressed as fractions, and the total value of each measure is $\frac{3}{4}$. The first few measures of *The Laughing Song* are shown with the values of each note. What type of note must be used to finish the last measure? $\frac{1}{4}$

35. **Geometry** Find the perimeter of the rectangle if the width is $9\frac{3}{8}$ inches shorter than the length. $66\frac{1}{4}$ in. ($21\frac{1}{4}$ in.)

36. **Life Science** In frog jumping contests, the distance recorded is the combined lengths of three consecutive jumps. The longest jump in the United States was 21 feet $5\frac{3}{4}$ inches in 1986. However, the longest jump in the world was in South Africa in 1977 for a length of 33 feet $5\frac{1}{2}$ inches. How much farther is the world record than the U.S. record? (*Hint*: Change the lengths to inches.)

36. $143\frac{3}{4}$ inches or 11 ft $11\frac{3}{4}$ in.

37a. $a = \frac{4}{5}$, $b = \frac{1}{2}$

37. **Critical Thinking** The figure at the right is a magic square.
   a. Find the values of a and b so that each row, column, and diagonal has the same sum.
   b. Create your own magic square. Explain how you did it. **See students' work.**

a	$\frac{7}{10}$	$1\frac{1}{5}$
$1\frac{3}{10}$	$\frac{9}{10}$	b
$\frac{3}{5}$	$1\frac{1}{10}$	1

Mixed Review

38. Find the sum of $-5\frac{1}{6}$ and $2\frac{5}{6}$. (*Lesson 7-1*) $-2\frac{1}{3}$

39. **Test Practice** José correctly answered 35 out of 50 questions on his math test. How would you describe his success as a fraction in simplest form? (*Lesson 6-4*) **A**

A $\frac{7}{10}$ B $\frac{18}{25}$ C $\frac{50}{35}$ D $\frac{35}{50}$

40. **Statistics** Determine whether a scatter plot of the number of base hits to the number of times at bat would show a *positive, negative,* or *no* relationship. (*Lesson 4-6*) **positive**

41. Solve $j = -7(15)(-10)$. (*Lesson 2-7*) **1,050**

42. Write $4 \cdot 4 \cdot 5 \cdot 5 \cdot 4$ using exponents. (*Lesson 1-2*) $4^3 \cdot 5^2$

284 Chapter 7 Algebra: Using Rational Numbers

Extending the Lesson

Activity The ancient Egyptians expressed every fraction, except $\frac{2}{3}$, as a sum of unit fractions (fractions with a numerator of 1).

$$\frac{3}{4} = \frac{1}{2} + \frac{1}{4} \qquad \frac{33}{40} = \frac{1}{2} + \frac{1}{5} + \frac{1}{8}$$

Have small groups find ways to express other fractions as the sum of two or more different unit fractions.

Let the Games Begin

Fraction Track

Get Ready This game is for two to four players.

 two number cubes   counters

Get Set
- Make the game board shown.
- Each person starts with a counter on each box at the left on the "fraction track."

[Game board showing a fraction track with rows:
0/1 — 1/1
0/2 — 1/2 — 2/2
0/3 — 1/3 — 2/3 — 3/3
0/4 — 1/4 — 2/4 — 3/4 — 4/4
0/5 — 1/5 — 2/5 — 3/5 — 4/5 — 5/5
0/6 — 1/6 — 2/6 — 3/6 — 4/6 — 5/6 — 6/6]

Go
- The first player rolls two number cubes and creates a fraction. The smaller number is always the numerator. For example, suppose a player rolls a 1 and a 6. The player's fraction is $\frac{1}{6}$.
- A player may move one counter the full value of the fraction or move any number of counters as long as they total exactly the value of the fraction rolled. For example, $\frac{1}{2}$ can be played as a $\frac{1}{3}$ and a $\frac{1}{6}$ since $\frac{1}{3} + \frac{1}{6} = \frac{1}{2}$.
- If the player rolls a fraction and cannot move, play moves to the next player.
- The winner is the first player to have all of the counters at the right.

 Visit www.glencoe.com/sec/math/mac/mathnet for more games.

Let the Games Begin

When the players approach the end of the game, they may not always be able to move their counters the total value of the fraction rolled. In such situations, have the students do one of the following for consistency. Have a player lose a turn if the total for value of the fraction rolled cannot be moved. Or, have a player move counters for as many portions of the total value as possible.

*Additional resources for this game can be found on page 52 of the **Classroom Games**.*

7-3 Lesson Notes

Instructional Resources
- *Study Guide Masters*, p. 52
- *Practice Masters*, p. 52
- *Enrichment Masters*, p. 52
- Transparencies 7-3, A and B
- *Assessment and Evaluation Masters*, p. 183
- *School to Career Masters*, p. 33
- CD-ROM Program
  - Resource Lesson 7-3
  - Interactive Lesson 7-3

Recommended Pacing

Standard	Day 4 of 15
Honors	Day 3 of 15
Block	Day 2 of 8

1 FOCUS

5-Minute Check
(Lesson 7-2)

Solve each equation. Write the solution in simplest form.

1. $n = \frac{2}{3} + \frac{5}{6}$ $1\frac{1}{2}$
2. $-\frac{5}{8} - \frac{1}{4} = k$ $-\frac{7}{8}$
3. $p = 2\frac{1}{5} - \left(-1\frac{3}{10}\right)$ $3\frac{1}{2}$
4. $w = 4\frac{1}{6} + \left(-5\frac{4}{9}\right)$ $-1\frac{5}{18}$
5. Evaluate $a + b$ if $a = \frac{5}{8}$ and $b = \frac{1}{6}$. $\frac{19}{24}$

The 5-Minute Check is also available on **Transparency 7-3A** for this lesson.

Motivating the Lesson

Hands-On Activity Have students fold a sheet of paper in thirds, unfold it, and shade one third using a colored pencil. Then have them fold the sheet in half in the other direction and shade one half using a different colored pencil. Ask them to write the multiplication problem illustrated by the shading; point out that the product is represented by the region shaded with both colors. $\frac{1}{3} \times \frac{1}{2} = \frac{1}{6}$

286 Chapter 7

7-3 Multiplying Fractions

What you'll learn
You'll learn to multiply fractions.

When am I ever going to use this?
You will multiply fractions when you cut your brownie recipe in half.

Since the Hershey Kiss was introduced on July 1, 1907, its size and shape has not changed. Each Kiss weighs $\frac{1}{6}$ ounce. If about $\frac{1}{4}$ of a Kiss is composed of cocoa butter, how much cocoa butter is in each Kiss?

You need to find $\frac{1}{4}$ of $\frac{1}{6}$. This means $\frac{1}{4} \times \frac{1}{6}$. You can use models to solve the problem.

HANDS-ON MINI-LAB

Work with a partner. lined paper · ruler · markers

Try This
- Draw a rectangle that is 6 rows long. Shade 1 row to represent $\frac{1}{6}$.
- Draw vertical lines that separate the rectangle into four equal columns. Using a different color, shade 1 column to represent $\frac{1}{4}$.

Talk About It
1. How many small rectangles are inside of the large rectangle? 24
2. How many of the small rectangles are shaded by two colors? 1
3. If the large rectangle has an area of 1, what fraction of the large rectangle is shaded by two colors? $\frac{1}{24}$ square unit

In the Mini-Lab, the portion of the large rectangle where the colors overlap shows the product. A Hershey Kiss contains $\frac{1}{24}$ ounce cocoa butter. The Mini-Lab suggests the following rule.

Multiplying Fractions	Words:	To multiply fractions, multiply the numerators and multiply the denominators.
	Symbols: Arithmetic	$\frac{2}{5} \cdot \frac{2}{3} = \frac{4}{15}$
	Algebra	$\frac{a}{b} \cdot \frac{c}{d} = \frac{ac}{bd}, b \neq 0, d \neq 0$

286 Chapter 7 Algebra: Using Rational Numbers

Cross-Curriculum Cue

Inform other teachers on your team that your students are studying fractions. Suggestions for curriculum integration are:

Geography: relative size of states and countries

Earth Science: relative size and distance of planets and stars

Industrial Technology: measuring lengths, calculating perimeter and area

Family and Consumer Sciences: adapting pattern sizes and recipes

Use the rules of signs for multiplying integers when you multiply rational numbers.

Example 1

Solve $t = -\dfrac{3}{8} \cdot \dfrac{4}{5}$.

Method 1 Multiply first. Then simplify.

$t = -\dfrac{3}{8} \cdot \dfrac{4}{5}$

$t = -\dfrac{3 \cdot 4}{8 \cdot 5}$ *Multiply.*

$t = -\dfrac{12}{40}$ or $-\dfrac{3}{10}$ *Simplify.*

Method 2 Divide common factors. Then multiply.

$t = -\dfrac{3}{8} \cdot \dfrac{4}{5}$

$t = -\dfrac{3}{\underset{2}{8}} \cdot \dfrac{\overset{1}{4}}{5}$ *The GCF of 4 and 8 is 4. Divide 4 and 8 by 4.*

$t = -\dfrac{3 \cdot 1}{2 \cdot 5}$ or $-\dfrac{3}{10}$ *Multiply.*

To multiply mixed numbers, first rename them as improper fractions.

Example 2 APPLICATION

Food The product label of a can of green beans states that there are approximately $3\dfrac{1}{2}$ servings per can and each serving is $\dfrac{1}{2}$ cup. How many cups does the can contain?

Multiply the number of servings by the size of each serving.

$3\dfrac{1}{2} \times \dfrac{1}{2} = \dfrac{7}{2} \times \dfrac{1}{2}$ *Rename $3\dfrac{1}{2}$ as $\dfrac{7}{2}$.*

$= \dfrac{7}{4}$

$= 1\dfrac{3}{4}$

The can contains about $1\dfrac{3}{4}$ cups.

LOOK BACK
You can refer to Lesson 1-2 to review exponents.

The rules for exponents also hold for rational numbers. For example, the expression $\left(\dfrac{2}{5}\right)^3$ means $\dfrac{2}{5} \cdot \dfrac{2}{5} \cdot \dfrac{2}{5}$.

Example 3 INTEGRATION

Algebra Evaluate $(xy)^2$ if $x = -\dfrac{2}{5}$ and $y = \dfrac{2}{3}$.

$(xy)^2 = \left(-\dfrac{2}{5} \cdot \dfrac{2}{3}\right)^2$ *Replace x with $-\dfrac{2}{5}$ and y with $\dfrac{2}{3}$.*

$= \left(-\dfrac{4}{15}\right)^2$ *Multiply inside the parentheses.*

$= \left(-\dfrac{4}{15}\right)\left(-\dfrac{4}{15}\right)$

$= \dfrac{16}{225}$

Lesson 7-3 Multiplying Fractions **287**

---

2 TEACH

 Transparency 7-3B contains a teaching aid for this lesson.

Using the Mini-Lab Point out to students that to take $\dfrac{1}{3}$ of something means that you are dividing something into 3 parts and taking 1 part. Therefore, $\dfrac{1}{6} \cdot \dfrac{1}{4}$ is something divided into 6 parts and then each of those sixths is divided into 4 parts, resulting in 24 sections that are each $\dfrac{1}{24}$ of the whole.

In-Class Examples

For Example 1
Solve $x = \dfrac{5}{8} \cdot \dfrac{6}{25} \cdot \dfrac{3}{20}$

For Example 2
A box of pancake mix makes $25\dfrac{1}{2}$ servings. If each serving uses $\dfrac{1}{3}$ cup of pancake mix, how many cups of mix does the box contain? $8\dfrac{1}{2}$ **cups**

For Example 3
Evaluate $(mn)^2$ if $m = -\dfrac{5}{8}$ and $n = \dfrac{2}{3}$. $\dfrac{25}{144}$

Teaching Tip In the Examples, point out that unlike addition and subtraction, multiplication of fractions does not require finding a common denominator.

Multiple Learning Styles

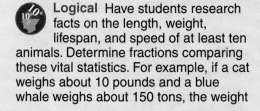

Logical Have students research facts on the length, weight, lifespan, and speed of at least ten animals. Determine fractions comparing these vital statistics. For example, if a cat weighs about 10 pounds and a blue whale weighs about 150 tons, the weight of one cat equals $\dfrac{1}{30,000}$ of the weight of a blue whale. Students can display photos or drawings of the animals, the data about them, and the fractions they calculated.

Lesson 7-3 **287**

3 PRACTICE/APPLY

Check for Understanding

If students need additional practice or instruction after completing Exercises 1–9, one of these options may be helpful.
- Extra Practice, see p. 624
- Reteaching Activity
- *Transition Booklet*, pp. 25–26
- *Study Guide Masters*, p. 52
- *Practice Masters*, p. 52
- Interactive Mathematics Tools Software

Assignment Guide
Core: 11–29 odd, 30–34
Enriched: 10–26 even, 28–34

Additional Answers

1. Sample answer: Three out of four rows are shaded to represent $\frac{3}{4}$. Three out of eight columns are shaded to represent $\frac{3}{8}$.

2. Sample answer: Substitute $-\frac{1}{2}$ for a and $-\frac{2}{3}$ for b. Multiply $-\frac{1}{2} \cdot -\frac{2}{3} \cdot -\frac{2}{3}$.

3. $\frac{3}{5}$

Study Guide Masters, p. 52

CHECK FOR UNDERSTANDING

Communicating Mathematics

Read and study the lesson to answer each question. 1–3. See margin.

1. **Tell** how the model shows the product of $\frac{3}{8}$ and $\frac{3}{4}$.

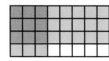

2. **Explain** how to evaluate ab^2 if $a = -\frac{1}{2}$ and $b = -\frac{2}{3}$.

3. **Draw** a model that shows the product of $\frac{4}{5}$ and $\frac{3}{4}$. Then find the product.

Guided Practice Solve each equation. Write the solution in simplest form.

4. $\frac{1}{3} \cdot \frac{3}{4} = z$ $\frac{1}{4}$
5. $\frac{4}{5} \cdot 3 = b$ $2\frac{2}{5}$
6. $d = -2\left(-\frac{5}{8}\right)$ $1\frac{1}{4}$
7. $x = \left(2\frac{2}{5}\right)^2$ $5\frac{19}{25}$
8. **Algebra** Evaluate c^2d if $c = -\frac{2}{3}$ and $d = 2\frac{5}{8}$. $1\frac{1}{6}$
9. **Geometry** Find the perimeter and area of the parallelogram. $4\frac{1}{4}$ in.; $\frac{5}{8}$ in²

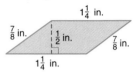

EXERCISES

Practice Solve each equation. Write the solution in simplest form.

10. $k = \frac{3}{8} \cdot \frac{4}{9}$ $\frac{1}{6}$
11. $f = -\frac{7}{8} \cdot \frac{5}{6}$ $-\frac{35}{48}$
12. $\frac{3}{4} \cdot \left(-\frac{1}{3}\right) = y$ $-\frac{1}{4}$
13. $-3\frac{3}{8} \cdot -\frac{2}{3} = j$ $2\frac{1}{4}$
14. $g = -\frac{5}{6} \cdot 1\frac{4}{5}$ $-1\frac{1}{2}$
15. $n = 2\left(-\frac{3}{10}\right)$ $-\frac{3}{5}$
16. $c = -4\frac{1}{4} \cdot \left(-3\frac{1}{3}\right)$ $14\frac{1}{6}$
17. $p = 1\frac{1}{3} \cdot 5\frac{1}{2}$ $7\frac{1}{3}$
18. $q = 9\left(-3\frac{2}{3}\right)$ -33
19. $s = \left(-\frac{3}{4}\right)^2$ $\frac{9}{16}$
20. $t = \left(1\frac{1}{8}\right)^2$ $1\frac{17}{64}$
21. $r = \left(-\frac{2}{3}\right)^3$ $-\frac{8}{27}$

22. What is one-half of $-\frac{5}{8}$ times $-3\frac{1}{5}$? 1

23. Find the product of $\frac{1}{2}, \frac{2}{3}$, and $-\frac{3}{4}$. $-\frac{1}{4}$

Evaluate each expression if $w = 1\frac{1}{3}, x = -\frac{1}{2}, y = 3\frac{3}{4}$, and $z = -\frac{4}{5}$.

24. xz $\frac{2}{5}$
25. $2y$ $7\frac{1}{2}$
26. w^2 $1\frac{7}{9}$
27. $x^3(-y)$ $\frac{15}{32}$

Applications and Problem Solving

28. **Food** Refer to Example 2 on page 287. Suppose another label on the can states that the can contained $9\frac{3}{4}$ ounces of beans before liquid was added for processing. If there are 8 ounces in a cup, does a serving of beans contain liquid, according to the information on the label? Explain.   See margin.

288 Chapter 7 Algebra: Using Rational Numbers

■ Reteaching the Lesson ■

Activity To model $\frac{1}{3} \cdot \frac{3}{4} = \frac{1}{4}$, draw a rectangle on a transparency and shade $\frac{3}{4}$ of it. Place another transparency on the first, and shade $\frac{1}{3}$ of the previously shaped region. Stress that this second shading is $\frac{1}{4}$ of the rectangle.

Additional Answer

28. Yes; according to the serving information, there are $1\frac{3}{4} \times 8$ or 14 ounces in the can, but only $9\frac{3}{4}$ ounces are beans so each serving must contain some liquid.

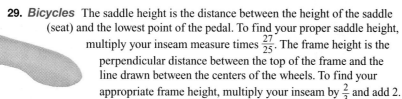

29. **Bicycles** The saddle height is the distance between the height of the saddle (seat) and the lowest point of the pedal. To find your proper saddle height, multiply your inseam measure times $\frac{27}{25}$. The frame height is the perpendicular distance between the top of the frame and the line drawn between the centers of the wheels. To find your appropriate frame height, multiply your inseam by $\frac{2}{3}$ and add 2.
    a. Antonio's inseam is 25 inches. What saddle height should he use? **27 in**.
    b. What frame height should a person with a 30-inch inseam use? **22 in**.

30. **Critical Thinking** What fraction is halfway between $\frac{1}{4}$ and $\frac{1}{3}$? $\frac{7}{24}$

Mixed Review

31. **Test Practice** Yana spent $\frac{2}{3}$ of an hour doing his homework on Monday. On Tuesday, he spent $1\frac{1}{2}$ hours doing homework. How much more time did he spend doing homework on Tuesday than on Monday? *(Lesson 7-2)* **C**
    A $2\frac{1}{6}$ h   B $\frac{13}{6}$ h   C $\frac{5}{6}$ h   D $\frac{1}{4}$ h   E $\frac{1}{6}$ h

32. Find the LCM of 9 and 30. *(Lesson 6-7)* **90**

33. **Algebra** Find the value of x in $\triangle DEF$ if $m\angle D = 4x°$, $m\angle E = 60°$, and $m\angle F = 72°$. *(Lesson 5-2)* **12**

34. **Communication** The cost of a long distance phone call is $0.15 for the first 5 minutes and $0.10 for each additional minute. If Lenora makes a 30-minute long distance phone call, how much will she be charged? *(Lesson 1-1)* **$2.65**

MATH IN THE MEDIA

"I'll take a large pizza with half-onion, two-thirds olives, nine-fifteenths mushrooms, five-eighths pepperoni, one-eighth anchovies, and extra cheese on five-ninths of the onion half."

1. What fraction of the pizza should have extra cheese? $\frac{5}{18}$

2. Is it possible to make the pizza ordered? Explain.

3. *Write a Problem* using the information in the comic. **See students' work.**

2. Sample answer: yes; but there will have to be overlapping.

7-4 Lesson Notes

Instructional Resources
- *Study Guide Masters*, p. 53
- *Practice Masters*, p. 53
- *Enrichment Masters*, p. 53
- Transparencies 7-4, A and B
- CD-ROM Program
  - Resource Lesson 7-4

Recommended Pacing	
Standard	Day 5 of 15
Honors	Day 4 of 15
Block	Day 3 of 8

1 FOCUS

5-Minute Check
(Lesson 7-3)

Solve each equation. Write the solution in simplest form.
1. $\frac{3}{5} \cdot \frac{7}{8} = k$ $\frac{21}{40}$
2. $m = -\frac{2}{3} \cdot \frac{5}{8}$ $-\frac{5}{12}$
3. $w = \left(-2\frac{1}{4}\right)\left(-1\frac{1}{3}\right)$ 3
4. $c = \left(-1\frac{1}{2}\right)^2$ $2\frac{1}{4}$
5. Evaluate $w^2(-2y)$ if $w = 2\frac{1}{2}$ and $y = 1\frac{1}{3}$. $-16\frac{2}{3}$

The 5-Minute Check is also available on **Transparency 7-4A** for this lesson.

Motivating the Lesson

Communication Have students read the beginning of the lesson. Then ask the following questions.
- Which weigh more, objects on Earth or objects on Mars? **objects on Earth**
- By what fraction could you multiply Sojourner's weight on Mars to find its weight on Earth? $\frac{13}{5}$

7-4 Properties of Rational Numbers

What you'll learn
You'll learn to identify and use rational number properties.

When am I ever going to use this?
Understanding properties of rational numbers can help you determine weights on other planets.

Word Wise
inverse property of multiplication
multiplicative inverse
reciprocal

Did you know The Sojourner rover was named after 19th-century African-American reformer Sojourner Truth.

On July 4, 1997, the Pathfinder landed on Mars. The next day, the Sojourner rover rolled out of Pathfinder.

Objects on Earth are about $2\frac{3}{5}$ times heavier than objects on Mars. If the Sojourner weighs about 9 pounds on Mars, how much did it weigh on Earth? *This problem will be solved in Example 2.*

Sojourner rover

All of the properties that were true for addition and multiplication of integers are also true for addition and multiplication of rational numbers.

Property	Arithmetic	Algebra
Commutative	$\frac{3}{8} + \frac{1}{3} = \frac{1}{3} + \frac{3}{8}$	$a + b = b + a$
	$\frac{1}{3} \cdot \frac{2}{5} = \frac{2}{5} \cdot \frac{1}{3}$	$a \cdot b = b \cdot a$
Associative	$\left(-\frac{1}{2} + \frac{2}{3}\right) + \frac{4}{5} = -\frac{1}{2} + \left(\frac{2}{3} + \frac{4}{5}\right)$	$(a + b) + c = a + (b + c)$
	$\left(-\frac{1}{2} \cdot \frac{2}{3}\right) \cdot \frac{4}{5} = -\frac{1}{2} \cdot \left(\frac{2}{3} \cdot \frac{4}{5}\right)$	$(a \cdot b) \cdot c = a \cdot (b \cdot c)$
Identity	$\frac{3}{4} + 0 = \frac{3}{4}$	$a + 0 = a$
	$-\frac{1}{2} \cdot 1 = -\frac{1}{2}$	$a \cdot 1 = a$

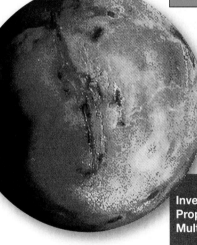

In Chapter 2, you learned about the inverse property of addition. A similar property that applies to the multiplication of rational numbers is called the **inverse property of multiplication**. Two numbers whose product is 1 are **multiplicative inverses**, or **reciprocals**, of each other. For example, $-\frac{5}{6}$ and $-\frac{6}{5}$ are multiplicative inverses because $-\frac{5}{6} \cdot \left(-\frac{6}{5}\right) = 1$.

Inverse Property of Multiplication	Words:	The product of a rational number and its multiplicative inverse is 1.
	Symbols: Arithmetic	$\frac{2}{5} \cdot \frac{5}{2} = 1$
	Algebra	$\frac{a}{b} \cdot \frac{b}{a} = 1$, where $a, b \neq 0$.

290 Chapter 7 Algebra: Using Rational Numbers

Multiple Learning Styles

Visual/Spatial Have students use grid paper and colored pencils as they did in the Mini-Labs in Lessons 7-1 and 7-3 to test properties of rational numbers. For example, for the associative property, have students color $\frac{1}{3} \cdot \frac{1}{4}$ on one grid, and then have them color $\frac{1}{2}$ of that same grid. Then have them color $\frac{1}{4} \cdot \frac{1}{2}$ on a second grid and color in $\frac{1}{3}$ of the second grid. Have them compare grids.

Example 1 Name the multiplicative inverse of $2\frac{3}{4}$.

$2\frac{3}{4} = \frac{11}{4}$ *Rename the mixed number as an improper fraction.*

$\frac{11}{4} \times \square = 1$ *What number can you multiply by $\frac{11}{4}$ to get 1?*

$\frac{11}{4} \times \frac{4}{11} = 1$

The multiplicative inverse of $2\frac{3}{4}$ is $\frac{4}{11}$.

You can use the distributive property to find products involving mixed numbers.

Example 2 CONNECTION

Physical Science Refer to the beginning of the lesson. How much did the Sojourner rover weigh on Earth?

Method 1 Use the distributive property.

$w = 2\frac{3}{5} \cdot 9$

$= 9 \cdot 2\frac{3}{5}$ *Commutative property*

$= 9\left(2 + \frac{3}{5}\right)$

$= (9 \cdot 2) + \left(9 \cdot \frac{3}{5}\right)$ *Distributive property*

$= 18 + \frac{27}{5}$

$= 18 + 5\frac{2}{5}$ or $23\frac{2}{5}$

Method 2 Multiply first. Then simplify.

$w = 2\frac{3}{5} \cdot 9$

$= \frac{13}{5} \cdot \frac{9}{1}$

$= \frac{117}{5}$

$= 23\frac{2}{5}$

The Sojourner rover weighed about 23 pounds on Earth.

LOOK BACK
You can refer to Lesson 1-7 to review the distributive property.

CHECK FOR UNDERSTANDING

Communicating Mathematics

Read and study the lesson to answer each question.

1. *Write* the reciprocal of $\frac{5}{9}$. $\frac{9}{5}$

2. *Identify* the property that allows you to compute $1\frac{1}{8} \cdot 8$ as $8 \cdot 1\frac{1}{8}$. Then state the product.   **commutative property of multiplication; 9**

3. *Write* a sentence or two explaining how to find the multiplicative inverse of $-3\frac{7}{8}$.   **Change $-3\frac{7}{8}$ to an improper fraction and then find the reciprocal; $-\frac{8}{31}$.**

Guided Practice   Name the multiplicative inverse of each of the following.

4. $-\frac{1}{5}$ -5

5. $\frac{7}{8}$ $\frac{8}{7}$ or $1\frac{1}{7}$

6. $\frac{3}{4}$ $1\frac{1}{3}$

Solve each equation using properties of rational numbers.

7. $\frac{3}{4} \cdot \left(-\frac{2}{3} \cdot -\frac{1}{2}\right) = x$ $\frac{1}{4}$

8. $y = -3\frac{3}{5} \cdot \frac{1}{3}$ $-1\frac{1}{5}$

9. **Algebra**   Evaluate ab if $a = \frac{1}{2}$ and $b = -2\frac{4}{5}$. $-1\frac{2}{5}$

Lesson 7-4 Properties of Rational Numbers   **291**

2 TEACH

 Transparency 7-4B contains a teaching aid for this lesson.

Reading Mathematics Emphasize that *commutative* refers to the order of the numbers being added or multiplied and *associative* refers to the way the numbers are grouped. Show how these meanings apply as you discuss the properties.

In-Class Examples

For Example 1
Name the multiplicative inverse of $7\frac{5}{8}$. $\frac{8}{61}$

For Example 2
A Chihuahua is 6 inches tall. A German shepherd is $3\frac{2}{3}$ times the height of the Chihuahua. Find the height of the German shepherd.   **22 inches**

3 PRACTICE/APPLY

Check for Understanding
If students need additional practice or instruction after completing Exercises 1–9, one of these options may be helpful.
- Extra Practice, see p. 624
- Reteaching Activity
- *Study Guide Masters*, p. 53
- *Practice Masters*, p. 53

Study Guide Masters, p. 53

Reteaching the Lesson

Activity Write the names of the eight properties discussed in this lesson on eight index cards. Write examples of the properties on eight other index cards. Shuffle the cards and arrange them facedown. Each student in turn draws two cards. If they match, the student keeps the pair and draws again. If they don't match, the cards are returned to their places facedown. The player with the most cards at the end wins.

Lesson 7-4   **291**

Assignment Guide

Core: 11–31 odd, 32–36
Enriched: 10–28 even, 30–36

4 ASSESS

Closing Activity

Speaking Write examples of the properties of rational numbers on the chalkboard, using both arithmetic and algebraic expressions. Have students identify the properties.

Practice Masters, p. 53

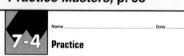

EXERCISES

Practice Name the multiplicative inverse of each rational number.

10. $8 \quad \frac{1}{8}$
11. $-\frac{2}{3} \quad -\frac{3}{2}$
12. $2\frac{3}{5} \quad \frac{5}{13}$
13. $-0.2 \quad -5$
14. $-1 \quad -1$
15. $2.25 \quad \frac{4}{9}$
16. $\frac{a}{b} \quad \frac{b}{a}$
17. $-x \quad -\frac{1}{x}$

Solve each equation using properties of rational numbers.

18. $p = 6 \cdot 3\frac{5}{6} \quad 23$
19. $\left(\frac{2}{3} \cdot -\frac{1}{2}\right) \cdot \frac{3}{8} = q \quad -\frac{1}{8}$
20. $r = 16\frac{2}{9} \cdot \frac{3}{4} \quad 12\frac{1}{6}$
21. $x = -7 \cdot 8\frac{1}{2} \quad -59\frac{1}{2}$
22. $\frac{3}{5} \cdot 10\frac{1}{3} = t \quad 6\frac{1}{5}$
23. $w = -\frac{4}{5}\left(\frac{5}{8} \cdot \frac{10}{11}\right) \quad -\frac{5}{11}$

24. Is $\frac{5}{6}$ the reciprocal of $1\frac{1}{5}$? Explain. **yes; $\frac{5}{6} \cdot \frac{6}{5} = 1$**

25. Are $-2\frac{1}{2}$ and $\frac{2}{5}$ reciprocals? Explain. **No; the reciprocal of $-2\frac{1}{2}$ is $-\frac{2}{5}$.**

Evaluate each expression if $u = -1\frac{1}{6}$, $v = \frac{2}{3}$, and $w = -\frac{3}{4}$.

26. $v^3 \quad \frac{8}{27}$
27. $4w + 3v \quad -1$
28. $w^2(6-v) \quad 3$
29. $uvw \quad \frac{7}{12}$

Applications and Problem Solving

30. *Life Science* The average Alaskan brown bear is about $1\frac{1}{8}$ times as long as a grizzly bear. If the average grizzly bear is 8 feet long, how long is an Alaskan brown bear? **9 feet**

31. *Geometry* Find the area of the rectangle. **$36\frac{7}{8}$ in²**

    $14\frac{3}{4}$ in.
    $2\frac{1}{2}$ in.

32. *Critical Thinking*
    a. Copy and complete the table.

n	1	2	3	4
n^2	1	4	9	16
$\frac{1}{n^2}$	1	$\frac{1}{4}$	$\frac{1}{9}$	$\frac{1}{16}$

    32b. Sample answer: The numbers in the n^2 row are getting larger. The numbers in the $\frac{1}{n^2}$ row are getting smaller.

    b. What pattern do you notice as you move across each row to the right?
    c. What relationship do you see between n^2 and $\frac{1}{n^2}$? **The numbers in the rows are reciprocals of each other.**

Mixed Review

33. Solve $-3\frac{5}{8}\left(-5\frac{1}{4}\right) = s$. *(Lesson 7-3)* $19\frac{1}{32}$

34. Find the GCF of 24 and 64. *(Lesson 6-3)* **8**

35. Estimate 48% of 178. *(Lesson 3-6)* $\frac{1}{2} \times 180 = 90$

36. **Test Practice** Brianna's Girl Scout troop sold 3,275 boxes of cookies last year. This year they sold 5,163 boxes. How many more boxes did they sell this year than last year? *(Lesson 2-5)* **A**

    A 1,888   B 1,988   C 2,998   D 3,008   E Not Here

292 Chapter 7 Algebra: Using Rational Numbers

Extending the Lesson

Enrichment Masters, p. 53

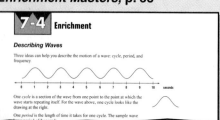

Activity Have students work together to research the reflexive, symmetric, and transitive properties of equality. Have them prepare a table like the one on the first page of the lesson with arithmetic and algebraic statements of each property.

HEALTH

Dr. Vivian Pinn
PATHOLOGIST

Vivian Pinn is a pathologist and the director of the Office of Research on Women's Health at the National Institutes of Health. When Dr. Pinn entered the University of Virginia Medical School in 1963, she was the only African-American and the only woman.

Pathology is the study of the nature, cause, progression, and effects of diseases. A person who is interested in becoming a pathologist should consider an undergraduate degree in biology or chemistry with a strong emphasis in mathematics. Then the person must attend a medical school to complete a medical degree.

For more information:
American Medical Association
515 North State Street
Chicago, IL 60610

www.glencoe.com/sec/math/mac/mathnet

I really enjoy my science class! I think that I would like to be a doctor someday!

Your Turn
The life of a doctor is very different from that shown on television dramas. Interview a doctor to find out about their experiences in completing medical training. Ask him or her about the importance of mathematics in their career. Write a paragraph about your findings.

More About Dr. Vivian Pinn

- Dr. Pinn has received The American Medical Women's Association President's Citation and The African American Women's Achievement Award.
- Dr. Pinn attended public schools in Lynchburg, Virginia. She received a bachelor's degree from Wellesley College in Massachusetts and a medical degree from the University of Virginia in 1967. She completed postgraduate training in pathology at Massachusetts General Hospital.

Motivating Students
As students learned in the chapter opener on pages 276–277, there are many mathematical patterns in nature. Pathologists look for ways a diseased body varies from the normal body. Students may be interested in a career in pathology or a related field. To start the discussion, you may ask students who is credited with these landmarks in medical history.
- smallpox vaccine **Edward Jenner, 1796**
- inoculations against rabies **Louis Pasteur, 1885**
- properties of penicillium **Alexander Fleming, 1928**

Making the Math Connection
Dr. Pinn helps lead research studies such as the Women's Health Initiative, in which about 160,000 U.S. women will participate over many years. In this study, data will be collected and studied to find patterns and trends in women's health.

Working on *Your Turn*
Students may want to work in pairs for their interview. Have them research jobs in medicine and prepare a set of interview questions.

*An additional School to Career activity is available on page 33 of the **School to Career Masters**.*

7-5A LAB Notes

Objective Students solve problems by finding and extending a pattern.

Recommended Pacing	
Standard	Day 6 of 15
Honors	Day 5 of 15
Block	Day 3 of 15

1 FOCUS

Getting Started Have a student stand at a point halfway between the front and back walls of the classroom. Then have another student stand halfway between the first student and the front wall. Continue this pattern for two more students. Ask students to make a rule describing the pattern. Then ask them to predict the next few steps of the pattern.

2 TEACH

Teaching Tip Point out to students that the chart shows to what extent radiocarbon atoms decay every 5,700 years, so each age increases by 5,700 from the age before it. Another way to find out the number of years for $\frac{1}{64}$ of the original amount is to add 5,700 to 22,800.

PROBLEM SOLVING

7-5A Look for a Pattern

A Preview of Lesson 7-5

Santos and Becky are doing research on radiocarbon dating for a science report. Let's listen in!

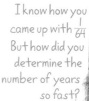

Santos

This chart shows that half of the radiocarbon atoms decay every 5,700 years.

I see that each fraction is $\frac{1}{2}$ of the previous one.

So, if there is only $\frac{1}{64}$ of the original amount left, it is 34,200 years old.

I know how you came up with $\frac{1}{64}$. But how did you determine the number of years so fast?

Radiocarbon Atoms	
Original Amount Remaining	Age (yr)
$\frac{1}{2}$	5,700
$\frac{1}{4}$	11,400
$\frac{1}{8}$	17,100
$\frac{1}{16}$	22,800
$\frac{1}{32}$	28,500

I saw that the second number of years was double the first number. Then I noticed that the fourth number was double the second number. So, to get the sixth number, I doubled the third number.

Becky

THINK ABOUT IT

3–4. See margin.

1. *Deductive reasoning* uses a rule to make a conclusion, and *inductive reasoning* makes a rule after seeing several examples. **Tell** whether Becky or Santos used inductive reasoning. Explain. **See margin.**

2. **Explain** how Becky and Santos knew the next fraction was $\frac{1}{64}$. $\frac{1}{2} \cdot \frac{1}{32} = \frac{1}{64}$

3. **Determine** the next two rows of the chart.

4. **Apply** the *look for a pattern* strategy to draw the next two figures in the pattern.

294 Chapter 7 Algebra: Using Rational Numbers

■ Reteaching the Lesson ■

Activity Take a large square and cut it in half. Set one half aside, and cut the other half in half again. Continue cutting successive pieces in half, until all of the halves are rather small. Point out that, theoretically, the remaining piece can always be cut in half. Thus, the pieces cut off never equal the whole square.

Additional Answers

1. Santos; He made a rule after seeing several examples.

3. $\frac{1}{64}$, 34,200; $\frac{1}{128}$, 39,900

4.

ON YOUR OWN

5. See margin.
5. The first step of the 4-step plan for problem solving asks you to *explore*. **Describe** how looking for a pattern relates to this step.

6. *Write a Problem* that can be solved by finding a pattern. Then ask a classmate to describe the pattern and solve the problem. **See students' work.**

7. *Look Ahead* Look for a pattern to determine the next two numbers in each list.
   a. 5, 10, 15, _?_, _?_ **20, 25**
   b. 39, 33, 27, _?_, _?_ **21, 15**

MIXED PROBLEM SOLVING

Strategies
Look for a pattern.
Solve a simpler problem.
Act it out.
Guess and check.
Draw a diagram.
Make a chart.
Work backward.

Solve. Use any strategy.

8. *Pets* Mrs. Vallez loves cats and canaries. Altogether her pets have thirty heads and eighty legs. How many cats does she have? **10 cats**

9. *Physical Science* The Italian scientist Galileo discovered that there was a relationship between the time of the swing back and forth of a pendulum and its length.

Time of Swing	Length of Pendulum
1 second	1 unit
2 seconds	4 units
3 seconds	9 units
4 seconds	16 units

   a. How long is a pendulum with a swing of 5 seconds? **25 units**
   b. How long is a pendulum with a swing of 1.5 seconds? **2.25 units**
   c. Use the time as the *x*-coordinate and the length as the *y*-coordinate to graph the data in the table. Describe the graph. **See margin.**

10. *School* Judi was assigned some math exercises for homework. She answered half of them in a study period. After school, she answered 7 more exercises. If she still has 11 exercises to do, how many exercises were assigned? **36 exercises**

11. *Geometry* What is the total number of rectangles, of any size, in the figure below? **21**

12. *Money Matters* After a shopping trip, Abbi and Sophia counted their money to see how much they had left. Abbi said to Sophia, "If I had $4 more, I would have as much as you." Sophia replied, "If I had $4 more, I would have twice as much as you." How much does each girl have? **Abbi has $8, and Sophia has $12.**

13. *Spreadsheets* The spreadsheet can be used to investigate bounce height when a ball is dropped from a height of 25 meters. The value in cell B2 tells you that each time the ball bounces it returns to a height that is $\frac{2}{5}$ or 0.4 times the previous height.

	A	B
1	Initial Height (m)	25
2	Bounce Factor	0.4
3	Height Number	Return Height
4	0	= B1
5	= A4 + 1	= B4*B2
6	= A5 + 1	= B4*B2

   a. What value will be in cell B5? **10**
   b. A different ball will return to a height that is 0.5 times the previous height. How would you modify the spreadsheet to find heights for this ball? **Change cell B2 to 0.5.**

14. **Test Practice** Which number is missing from this pattern? **C**
   . . . , _?_, 29, 22, 15, 8, . . .
   A 30   B 35   C 36   D 37

Lesson 7-5A **THINKING LAB** 295

■ Extending the Lesson ■

Activity Have students work in pairs to find several patterns that begin with the numbers 1, 2, ... They should give the rule governing each pattern and the next three numbers in the pattern. **Sample answers: 1, 2, 3, 4, 5, add 1; 1, 2, 4, 8, 16, multiply by 2.**

3 PRACTICE/APPLY

Check for Understanding
Use the results from Exercise 4 to determine whether students comprehend how to find a pattern.

Extra Practice If students need additional practice in problem solving, extra practice is available on the following pages.
- Look for a Pattern, see p. 624
- Mixed Problem Solving, see pp. 645–646

Assignment Guide

All: 5–14

4 ASSESS

Closing Activity
Modeling Have groups of students work together to model patterns. Have each student devise a pattern. Then have the group build or draw models of all of the patterns. You may wish to build a classroom display of the most interesting models.

Additional Answers
5. Sample answer: Looking for a pattern involves determining what you know.

9c.

Thinking Lab 7-5A **295**

7-5 Lesson Notes

Instructional Resources
- *Study Guide Masters*, p. 54
- *Practice Masters*, p. 54
- *Enrichment Masters*, p. 54
- Transparencies 7-5, A and B
- *Assessment and Evaluation Masters*, pp. 182, 183
- *Science and Math Lab Manual*, pp. 77–80
- CD-ROM Program
  - Resource Lesson 7-5

Recommended Pacing
Standard	Day 7 of 15
Honors	Days 6 & 7 of 15
Block	Day 4 of 8

1 FOCUS

5-Minute Check
(Lesson 7-4)

Name the multiplicative inverse of each rational number.
1. $-3\frac{1}{2}$ $-\frac{2}{7}$
2. $\frac{3}{4}$ $\frac{4}{3}$

Evaluate each expression if $x = -1\frac{1}{3}$ and $y = 2\frac{1}{6}$.
3. xy $-2\frac{8}{9}$
4. $2x - 4y$ $11\frac{1}{3}$
5. $x^2(y - 3)$ $-1\frac{13}{27}$

The 5-Minute Check is also available on **Transparency 7-5A** for this lesson.

Motivating the Lesson
Problem Solving Suppose Lenore's score on her weekly math quizzes increased by the same amount each week. If her first two quiz scores were 75 and 78, what was her score on the seventh quiz? 93

7-5

Integration: Patterns and Functions
Sequences

What you'll learn
You'll learn to recognize and extend arithmetic and geometric sequences.

When am I ever going to use this?
Sequences can be used to find the cost of a long-distance telephone call.

Word Wise
sequence
term
arithmetic sequence
common difference
geometric sequence
common ratio

Have you ever made yourself a hot bowl of instant oatmeal on a cold winter morning? If so, you know that it is made with a mix and boiling water. A box of Quaker Oats has the following table to determine the right amounts of water and oatmeal mix.

Servings	1	2	3
Water	1 c	$1\frac{3}{4}$ c	$2\frac{1}{2}$ c
Oats	$\frac{1}{2}$ c	1 c	$1\frac{1}{2}$ c

If you look at the amounts of water in order, the numbers form a **sequence**. A sequence is a list of numbers in a certain order, such as 1, $1\frac{3}{4}$, $2\frac{1}{2}$, ... or $\frac{1}{2}$, 1, $1\frac{1}{2}$, ... Each number is called a **term** of the sequence. When the difference between any two consecutive terms is the same, the sequence is called an **arithmetic sequence**. The difference is called the **common difference**.

Examples / APPLICATION

1 **Cooking** Refer to the beginning of the lesson. How many cups of water are needed for the next three numbers of servings of oatmeal?

Find the difference between consecutive terms.

The difference between any two consecutive terms is $\frac{3}{4}$. So, this is an arithmetic sequence. Add $\frac{3}{4}$ to the last term of the sequence, and continue adding until the next three terms are found. The next three serving sizes will need $3\frac{1}{4}$ cups, 4 cups, and $4\frac{3}{4}$ cups of water.

2 State whether the sequence 15, 7, 0, −6, . . . is arithmetic. Then find the next three terms.

Since there is no common difference, the sequence is *not* arithmetic. But we can still identify a pattern to this sequence. Add −5 to the last term of the sequence, and continue adding the next greater integer until the next three terms are found. The next three terms are −11, −15, and −18.

Study Hint
Reading Math The three dots following a list of numbers are read as *and so on.*

296 Chapter 7 Algebra: Using Rational Numbers

Investigations for the Special Education Student

This blackline master booklet helps you plan for the needs of your special education students by providing long-term projects along with teacher notes. Investigations 6 and 7, *Read All About Us* and *Wall Street Week*, may be used with this chapter.

The first term of a sequence can be represented by a_1, the second term a_2, the third term a_3, and so on up to the nth term, a_n. To find the nth term of an arithmetic sequence, you can use the formula $a_n = a_1 + (n-1)d$, where n is the number of the term you want to find and d is the common difference.

Example **INTEGRATION**

3 **Algebra** Find the fifteenth term, a_{15}, in the sequence 12.3, 13.8, 15.3, . . .

This sequence is arithmetic, and the common difference is 1.5.

$a_n = a_1 + (n - 1)d$
$a_{15} = 12.3 + (15 - 1)1.5$ *Replace n with 15, a_1 with 12.3, and d with 1.5.*
$a_{15} = 12.3 + (14)(1.5)$
$a_{15} = 12.3 + 21$
$a_{15} = 33.3$

So, the fifteenth term is 33.3.

Sometimes, the consecutive terms of a sequence are formed by multiplying by a constant factor. This type of sequence is called a **geometric sequence**. The factor is called the **common ratio**.

Examples

State whether each sequence is geometric. Then find the next three terms.

4 3, −6, 12, −24, . . .

Since there is a common ratio, −2, the sequence is a geometric sequence. Multiply the last term of the sequence by −2, and continue multiplying until the next three terms are found. The next three terms are 48, −96, and 192.

5 96, 48, 12, 2, . . .

Since there is no common ratio, the sequence is *not* geometric. However, the sequence does have a pattern. Multiply the last term by $\frac{1}{8}$, and continue multiplying by $\frac{1}{10}$ and then $\frac{1}{12}$ to find the next three terms. The next three terms are $\frac{1}{4}$, $\frac{1}{40}$, and $\frac{1}{480}$.

Study Hint
Problem Solving Look for a pattern when deciding whether a sequence is geometric. Does each consecutive term result from multiplying the previous term by the same number?

Lesson 7-5 Integration: Patterns and Functions Sequences 297

---

2 TEACH

 Transparency 7-5B contains a teaching aid for this lesson.

Thinking Algebraically Show students that they can represent sequences algebraically. For example, one arithmetic sequence is $a_n + 3 = a_{(n+1)}$, resulting in a_1, a_2, a_3, ... If $a_1 = 2$, then the sequence is 2, 5, 8, 11, 14, ...

In-Class Examples

For Example 1
Refer to the beginning of the lesson. How many cups of oats are needed for the next three serving sizes of oatmeal?
2 cups, $2\frac{1}{2}$ cups, 3 cups

For Example 2
State whether the sequence −8, −15, −22, −29, . . . is arithmetic. Then write the next three terms. **yes; −36, −43, −50**

For Example 3
Find the thirtieth term in the sequence −32, −21, −10, 1, ... **287**

State whether each sequence is geometric. Then find the next three terms.

For Example 4
$-\frac{1}{3}$, 1, −3, 9, . . . **yes; −27, 81, −243**

For Example 5
1, 2, 6, 24, . . . **no; 120, 720, 5,040**

Teaching Tip In Example 3, point out the logic of the formula: To reach the last term, a_n, start with the first term, a_1, and add the common difference, d, once for each term except the last, or $(n - 1)$ times.

Study this pattern to derive the formula.

$a_1 = a_1$
$a_2 = a_1 + (1)d$
$a_3 = a_1 + (2)d$
$a_4 = a_1 + (3)d$
$a_5 = a_1 + (4)d$
$\vdots$
$a_n = a_1 + (n - 1)d$

Lesson 7-5 297

3 PRACTICE/APPLY

Check for Understanding

If students need additional practice or instruction after completing Exercises 1–10, one of these options may be helpful.
- Extra Practice, see p. 625
- Reteaching Activity
- *Transition Booklet*, pp. 7–8
- *Study Guide Masters*, p. 54
- *Practice Masters*, p. 54
-  Interactive Mathematics Tools Software

Assignment Guide

Core: 11–27 odd, 31–36
Enriched: 12–28 even, 30–36
All: Self Test 1–10

CHAPTER Project

Exercise 29 asks students to advance to the next stage of work on the Chapter Project. You may wish to organize a class trip to a nature park to look for Fibonacci numbers in plant patterns.

Additional Answer

1. When the difference between any two consecutive terms is the same, the sequence is arithmetic.

Study Guide Masters, p. 54

Communicating Mathematics

2. Sample answer: 2, 6, 18, 54
3. Joanne is correct. The common difference is −4.

Guided Practice

Practice
4. A; 10, 12, 14
5. N; −11, −14, −16
6. G; 160, 320, 640
7. G; 48, −96, 192
8. N; 120, 720, 5,040
9. A; 14, $16\frac{1}{2}$, 19
13. G; $-\frac{1}{64}, -\frac{1}{256}, -\frac{1}{1,024}$
19. G; $\frac{7}{9}, \frac{7}{27}, \frac{7}{81}$
22. A; 107.35, 95.05, 82.75
23. N; 111, 136, 124
29a. 34, 55, 89; Neither; there is no common difference or common ratio.

Applications and Problem Solving

29b. Sample answer: The numbers are all Fibonacci numbers.

CHECK FOR UNDERSTANDING

Read and study the lesson to answer each question.
1. *Explain* how to determine whether a sequence is arithmetic. **See margin.**
2. *Write* a geometric sequence with a common ratio of 3.
3. *You Decide* Joanne says that the next term in the sequence 12, 8, 4, 0, ... is found by adding −4 to the last term. Marcie says that the next term is found by multiplying the last term by $\frac{1}{2}$. Who is correct? Explain.

Identify each sequence as *arithmetic*, *geometric*, or *neither*. Then find the next three terms.

4. 2, 4, 6, 8, ...
5. 11, 4, −2, −7, ...
6. 10, 20, 40, 80, ...
7. 3, −6, 12, −24, ...
8. 1, 1, 2, 6, 24, ...
9. 4, $6\frac{1}{2}$, 9, $11\frac{1}{2}$, ...

10. *Money Matters* Marianne has $5 in her bank. Suppose she puts $1.50 into her bank each week. If she does not take any money out, how much will be in the bank after 20 weeks? **$35**

EXERCISES

Identify each sequence as *arithmetic*, *geometric*, or *neither*. Then find the next three terms. 16. G; 1.25, 0.625, 0.3125 17. G; 162, −486, 1,458

11. 1, 3, 9, 27, ... **G; 81, 243, 729**
12. −6, −4, −2, 0, ... **A; 2, 4, 6**
13. $-4, -1, -\frac{1}{4}, -\frac{1}{16}, ...$
14. −5, 0, 4, 7, ... **N; 9, 10, 10**
15. 20, 24, 28, 32, ... **A; 36, 40, 44**
16. 20, 10, 5, 2.5, ...
17. 2, −6, 18, −54, ...
18. $1, \frac{1}{2}, \frac{1}{6}, \frac{1}{24}, ...$ **N; $\frac{1}{120}, \frac{1}{720}, \frac{1}{5,040}$**
19. 63, 21, 7, $2\frac{1}{3}, ...$
20. 1, 2, 5, 10, 17, ... **N; 26, 37, 50**
21. $4\frac{1}{2}, 4\frac{1}{6}, 3\frac{5}{6}, 3\frac{1}{2}, ...$ **A; $3\frac{1}{6}, 2\frac{5}{6}, 2\frac{1}{2}$**
22. 156.55, 144.25, 131.95, 119.65, ...
23. 84, 72, 97, 85, 110, 98, 123, ...
24. −1, 1, −1, 1, ... **G; −1, 1, −1**
25. What are the first four terms in an arithmetic sequence with a common difference of $3\frac{1}{3}$ if the first term is 4? **4, $7\frac{1}{3}, 10\frac{2}{3}$, 14**
26. Find the twentieth term in the sequence 18, 14, 10, 6, ... **−58**
27. What is the thirtieth term in the sequence 1.5, 4, 6.5, 9, ... ? **74**
28. The sixth term of a geometric sequence is 81. The common ratio is $-\frac{1}{3}$. Find the first five terms. **−19,683, 6,561, −2,187, 729, −243**
29. *Working on the* **CHAPTER Project** The sequence 1, 1, 2, 3, 5, 8, 13, 21, ... is called the *Fibonacci sequence*.
   a. Find the next three terms in the sequence. Is the sequence arithmetic, geometric, or neither? Explain.
   b. Compare the numbers in the Fibonacci sequence to the numbers you found in the pinecone spirals and plant leaves. Describe your findings.
   c. Find other plant materials and look for Fibonacci numbers. Add pictures and descriptions to your display or slide show. **See students' work.**

298 Chapter 7 Algebra: Using Rational Numbers

Reteaching the Lesson

Activity Prepare five pairs of index cards with the first terms of an arithmetic sequence on one index card of a pair and the next three on the other. Mix the cards and deal them faceup. Have one student at a time choose a pair and explain why the six terms form an arithmetic sequence. Repeat the activity using geometric sequences.

Error Analysis
Watch for students who have difficulty identifying types of sequences. Some sequences may initially appear to be both; for example, 2, 4, 6, 8, . . . and 2, 4, 8, 16, ...
Prevent by having students compare consecutive terms for several terms in the sequence.

298 Chapter 7

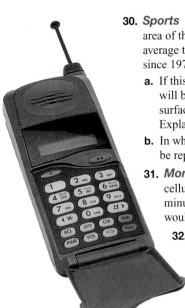

30. **Sports** The graph shows how the area of the hitting surface of an average tennis racket has increased since 1976.

    a. If this pattern continues, what will be the next area of the hitting surface of an average racket? Explain.

    b. In what year would this increase be reported? Explain.

31. **Money Matters** Suppose a cellular phone service charges a base rate of $20 per month and $0.25 per minute for each call. If you talked for 55 minutes last month, how much would you be charged before taxes? **$33.75**

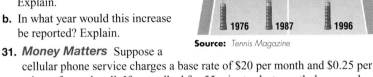

Source: *Tennis Magazine*

32. **Critical Thinking** The second term of an arithmetic sequence is 3, and the sixth term is 9. **a. 1.5; (9 − 3) ÷ 4 = 1.5**

    a. What is the common difference? How did you find it?

    b. Explain how to find the common difference of an arithmetic sequence if you know any two terms. **Find the difference between the two known terms and divide by the difference between the terms' numbers.**

Mixed Review

30a. 153 in²; Sample answer: The amount of increase is 11 in² less.

30b. 2003; Sample answer: The amount of the increase is 2 years less.

33. **Test Practice** What number will make $\frac{4}{5} + \frac{2}{3} = \frac{2}{3} + g$ true? *(Lesson 7-4)* **B**

    A $\frac{6}{8}$ B $\frac{4}{5}$ C $\frac{3}{5}$ D $\frac{2}{3}$

34. Find the LCD of $-\frac{5}{6}$ and $\frac{3}{8}$. *(Lesson 6-8)* **24**

35. **Probability** A number cube is rolled. What is the probability it shows an odd number? *(Lesson 6-6)* $\frac{1}{2}$

36. **Statistics** Find the interquartile range of the set {239, 226, 232, 212, 243, 216, 250}. *(Lesson 4-5)* **27**

Mid-Chapter Self Test

Solve each equation. Write the solution in simplest form.
(Lessons 7-1, 7-2, and 7-3)

1. $\frac{3}{8} + \left(-\frac{1}{8}\right) = d$ $\frac{1}{4}$
2. $z = \frac{1}{8} - \frac{7}{8}$ $-\frac{3}{4}$
3. $12 - 5\frac{3}{5} = g$ $6\frac{2}{5}$
4. $x = -7\frac{3}{4} + \left(-3\frac{4}{5}\right)$ $-11\frac{11}{20}$
5. $8\frac{3}{4}(-11) = b$ $-96\frac{1}{4}$
6. $k = \left(-\frac{5}{6}\right)^2$ $\frac{25}{36}$

7. Find the product of $\frac{1}{2}$ and $12\frac{4}{5}$ using the distributive property. *(Lesson 7-4)* **See margin.**

8. **Construction** Lavar is tiling the wall over a kitchen counter. The space requires 7 rows of tiles with $15\frac{1}{2}$ tiles in each row. How many tiles will be in the finished space? *(Lesson 7-4)* **$108\frac{1}{2}$ tiles**

Identify each sequence as *arithmetic*, *geometric*, or *neither*. Then find the next three terms. *(Lesson 7-5)*

9. 768, 192, 48, ... **G; 12, 3, $\frac{3}{4}$**

10. 20, 23, 26, 29, ... **A; 32, 35, 38**

Lesson 7-5 Integration: Patterns and Functions Sequences **299**

7-5B Hands-On Lab Notes

GET READY

Objective Students discover the numbers that make up the Fibonacci sequence.

Optional Resources
Hands-On Lab Masters
- grid paper, p. 10
- worksheet, p. 53

MANAGEMENT TIPS

Recommended Time
20 minutes

Getting Started Ask students to give examples of sequences that are neither arithmetic nor geometric. **Sample answer: 1, 4, 9, 16, ... (the sequence of perfect squares)**

The **Activity** helps students discover the numbers in the Fibonacci sequence. Before students begin creating a road, emphasize that the width of the road remains constant. Explain that they should consider both the orientation of each brick and the pattern the bricks make.

ASSESS

Have students complete Exercises 1–4. Watch for students who have trouble drawing all of the possible road combinations. You may wish to have students remain in groups while completing the Exercises.

Use Exercise 5 to determine whether students understand that the Fibonacci sequence is an example of a sequence that is neither arithmetic nor geometric.

Hands-On Lab — Cooperative Learning

7-5B The Fibonacci Sequence

A Follow-Up of Lesson 7-5

📋 grid paper

Leonardo was born about A.D. 1170 in the city of Pisa, Italy. This famous mathematician was also known by the name Fibonacci. Leonardo created story problems centered on a series of numbers that became known as the *Fibonacci sequence*. In this lab, you will discover this sequence.

TRY THIS

Work in groups of three.

Step 1 Using grid paper or dot paper, draw a "brick" that is 2 units long and 1 unit wide. If you build a "road" of grid paper bricks, there is only one way to build a road that is 1 unit long.

Step 2 Using two bricks from Step 1, draw all of the different roads possible that are 2 units long.

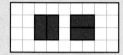

Step 3 Using three bricks, draw all of the different roads possible that are 3 units long.

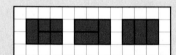

ON YOUR OWN

1. Draw all of the different roads possible that are 4 units long, 5 units long, and 6 units long using the indicated number of bricks. **See students' work.**
2. Copy and complete the chart.

Length of Road	0	1	2	3	4	5	6
Number of Ways to Build the Road	1	1	2	3	5	8	13

3. Look for a pattern in the numbers in the second row. How is each number related to the previous numbers? **Each number is the sum of the previous two numbers.**
4. Without making a drawing, how many ways are there to build a road that is 7 units long? **21 ways**
5. *Reflect Back* The numbers in the second row of the chart are the first numbers in the Fibonacci sequence. Write the ratios of consecutive terms of the Fibonacci sequence, $\frac{1}{1}, \frac{2}{1}, \frac{3}{2}, \frac{5}{3}, \frac{8}{5}, ..., \frac{89}{55}$, as decimals rounded to the nearest thousandth. Describe the pattern. **Sample answer: All the decimals are about 1.62.**

300 Chapter 7 Algebra: Using Rational Numbers

 Have students write a paragraph describing the Fibonacci sequence. Ask them to write a story problem using the Fibonacci sequence, just as Leonardo did.

7-6 Integration: Geometry
Area of Triangles and Trapezoids

What you'll learn
You'll learn to find the areas of triangles and trapezoids.

When am I ever going to use this?
Many shapes can be subdivided into triangles or trapezoids to make it easier to determine their areas.

Word Wise
base
altitude
height
trapezoid

Sailboat races were conducted in 8 different classes of competition in the 1996 Olympics. In the Star Class, the mainsail area cannot exceed 20.6 square meters. The dimensions of a mainsail are shown in the diagram. Does it qualify for the Star Class? *This problem will be solved in Example 1.*

Any side of a triangle can be used as a **base**. A line segment perpendicular to the base from the opposite vertex is called the **altitude**. The length of the altitude is called the **height**.

In the triangle at the right, side $\overline{CD}$ is the base, and $\overline{EF}$ is the altitude.

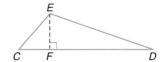

Did you know?
The Star Class has been part of 13 of the last 14 Olympics, making it the oldest Olympic sailing class.

HANDS-ON MINI-LAB

Work with a partner. grid paper

Try This
- Copy the triangles shown onto a piece of grid paper. Label as shown.

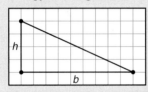 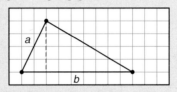

- Sketch the figure that is a result of moving each triangle as shown.

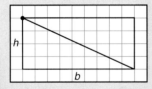 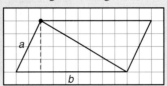

Talk About It   3. See margin.
1. What is the length of the base and height of each parallelogram or rectangle? $b = 9, h = 4$
2. Find the area of each parallelogram or rectangle. 36 units^2
3. Explain how the area of each parallelogram or rectangle is related to the product of the base and height of the corresponding triangle.

Lesson 7-6 Integration: Geometry Area of Triangles and Trapezoids **301**

Additional Answer for the Mini-Lab
3. The area of the parallelogram and the rectangle is twice the product of the height and base of each triangle.

2 TEACH

 Transparency 7-6B contains a teaching aid for this lesson.

Using the Mini-Lab Remind students that each triangle is congruent to its repositioned figure. Therefore, the base and altitude are the same in both. Have students turn their sketches upside down to illustrate this.

In-Class Examples

For Example 1
Find the area of the triangle.

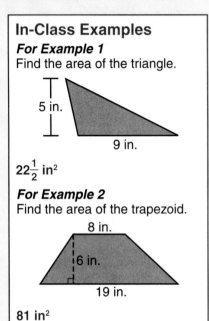

$22\frac{1}{2}$ in²

For Example 2
Find the area of the trapezoid.

81 in²

The Mini-Lab suggests the following rule.

Area of a Triangle
Words: The area of a triangle is equal to half the product of its base and height.
Symbols: $A = \frac{1}{2}bh$
Model:

Example APPLICATION

1 Sports Refer to the beginning of the lesson. Does the mainsail qualify for the Star Class?

$A = \frac{1}{2}bh$

$A = 0.5 \;[\times]\; 3.5 \;[\times]\; 6.55$ *Replace b with 3.5 and h with 6.55.*

$A = 11.4625$

The area of the mainsail is 11.4625 square meters. Since 11.4625 is less than 20.6, the mainsail qualifies for the Star Class.

A **trapezoid** is a quadrilateral with exactly one pair of parallel sides. These parallel sides are its bases. When you draw a diagonal in a trapezoid, two triangles are formed. The area of the trapezoid is the sum of the areas of the triangles.

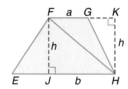

The triangles are $\triangle EFH$ and $\triangle FGH$. The altitudes of these triangles, $\overline{FJ}$ and $\overline{HK}$, are congruent. Both are h units long. The base of $\triangle FGH$ is a units long. The base of $\triangle EFH$ is b units long.

area of trapezoid $EFGH$ = area of $\triangle FGH$ + area of $\triangle EFH$

$\phantom{\text{area of trapezoid } EFGH} = \frac{1}{2}ah + \frac{1}{2}bh$

$\phantom{\text{area of trapezoid } EFGH} = \frac{1}{2}h(a + b)$ *Distributive property*

Area of a Trapezoid
Words: The area of a trapezoid is equal to the product of half the height and the sum of the bases.
Symbols: $A = \frac{1}{2}h(a + b)$
Model:

Example 2 Find the area of the trapezoid.

$A = \frac{1}{2}h(a + b)$

$A = \frac{1}{2} \cdot 3(12 + 6)$ *Replace h with 3, a with 12, and b with 6.*

$A = \frac{1}{2} \cdot 3 \cdot 18$ or 27 The area is 27 square feet.

302 Chapter 7 Algebra: Using Rational Numbers

CHECK FOR UNDERSTANDING

Communicating Mathematics

Read and study the lesson to answer each question.   1. See Answer Appendix.

1. *Draw* an isosceles triangle. Sketch and label the three possible altitudes.
2. *Explain* how to tell which two sides of a trapezoid are the bases.
3. *Copy* the trapezoid on a piece of grid paper. Cut it out and then cut along the dashed line. Move the parts so they form a parallelogram. Explain how the height and length of the base of the parallelogram are related to the height and the sum of the lengths of the bases of the original trapezoid.
   2-3. See margin.

Guided Practice

State the measure of the base(s) and the height of each triangle or trapezoid. Then find the area of each figure.

4. 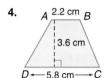 $a = 2.2$ cm, $b = 5.8$ cm, $h = 3.6$ cm, $A = 14.4$ cm^2

5. $b = 2\frac{2}{3}$ ft, $h = 3\frac{3}{4}$ ft, $A = 5$ ft^2

Find the area of each figure.

6. triangle: base, 7.5 cm; height, 3.75 cm   14.0625 cm^2
7. triangle: base, $3\frac{3}{4}$ in.; height, $2\frac{1}{8}$ in. $3\frac{63}{64}$ in^2
8. trapezoid: bases $7\frac{1}{6}$ yd and $5\frac{2}{3}$ yd; height, 8 yd $51\frac{1}{3}$ yd^2
9. trapezoid: bases 30 ft and 22 ft; height, 9 ft   234 ft^2

10. **Geography** The shape of the state of Virginia is close to a triangle.
    a. Estimate the area of Virginia. $A \approx 49{,}248$ mi^2
    b. Research to find the actual area. Compare to your estimate.   Sample answer: 42,777 mi^2

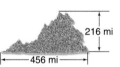

EXERCISES

Practice

State the measure of the base(s) and the height of each triangle or trapezoid. Then find the area of each figure.   14–16. See margin.

12. $b = 30$ cm, $h = 12$ cm, $A = 180$ cm^2
13. $a = 12$ yd, $b = 18$ yd, $h = 10$ yd, $A = 150$ yd^2

11.
12. $b = 3$ ft, $h = 4$ ft, $A = 6$ ft^2
    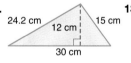
13.

14.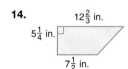
15.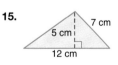
16.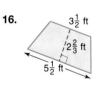

Lesson 7-6 Integration: Geometry Area of Triangles and Trapezoids **303**

Have each student bring one example to class, either the object itself or a model of the object with dimensions included. Prepare a class display of the examples, and have small groups or the class as a whole calculate the total area of all of the shapes.

4 ASSESS

Closing Activity

Writing Have students work in pairs to create problems that involve finding the area of trapezoids and triangles. Have them trade problems with another pair and solve the problems they receive.

Additional Answers

29. Sample answer:

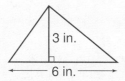

32. The area to be fertilized is 8,736 ft². He needs to buy 5 bags of fertilizer.

Practice Masters, p. 55

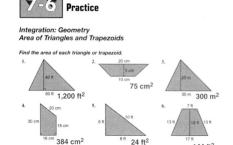

Find the area of each triangle.

	base	height	
17.	5 in.	9 in.	$22\frac{1}{2}$ in²
19.	12 cm	5.5 cm	33 cm²
21.	$6\frac{1}{2}$ yd	2 yd	$6\frac{1}{2}$ yd²

	base	height	
18.	8 ft	5 ft	20 ft²
20.	3.75 m	2.2 m	4.125 m²
22.	$9\frac{3}{4}$ in.	$12\frac{1}{8}$ in.	$59\frac{7}{64}$ in²

Find the area of each trapezoid.

23. $37\frac{1}{2}$ yd²
24. 20.25 km²
25. 29.49 m²
26. 14.4 cm²
27. 9 ft²
28. $11\frac{61}{64}$ in²

	base (a)	base (b)	height
23.	6 yd	9 yd	5 yd
25.	4.5 m	5.33 m	6 m
27.	$3\frac{1}{2}$ ft	$5\frac{1}{2}$ ft	2 ft

	base (a)	base (b)	height
24.	4 km	9.5 km	3 km
26.	2.2 cm	5.8 cm	3.6 cm
28.	$4\frac{3}{4}$ in.	$5\frac{7}{8}$ in.	$2\frac{1}{4}$ in.

29. Draw and label a triangle that has an area of 9 square inches. **See margin.**

30. A trapezoid has an area of 81 square meters, and the lengths of its bases are 8 meters and 19 meters. Find the height of this trapezoid. **6 m**

Applications and Problem Solving

31. **Algebra** The shorter base of a trapezoid is x units long. The longer base of the trapezoid is one and one-half times as long as the shorter base. Find the length of the bases if the area of the trapezoid is 150 square yards and the height is 10 yards. $a = 12, b = 18$

32. **Landscaping** A diagram of Mr. Stone's yard is shown. The rectangles represent the house and driveway. He wants to fertilize his lawn. The label on a bag of fertilizer indicates that one bag will cover 2,000 square feet. How much fertilizer should he buy? **See margin.**

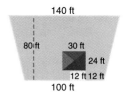

33. **Critical Thinking** What is the effect on the area of a trapezoid if the length of each base is doubled and the height is doubled? **The area is quadrupled.**

Mixed Review

34. **Patterns** Write the next three terms in the sequence 80, 76, 72, 68, ... *(Lesson 7-5)* **64, 60, 56**

35. Express 0.000084 in scientific notation. *(Lesson 6-9)* 8.4×10^{-5}

36. **Test Practice** If $\triangle EFG$ is congruent to $\triangle ABC$, then — *(Lesson 5-5)* **D**

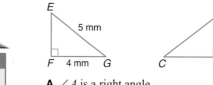

A $\angle A$ is a right angle.
B the perimeter of $\triangle EFG$ is 11 millimeters.
C $\overline{FG} \cong \overline{AB}$.
D the measure of side $\overline{BC}$ is 4 millimeters.
E $m\angle E = m\angle C$.

37. **Algebra** Evaluate $5a^2 - (b + c)$ if $a = 3, b = 8,$ and $c = 10$. *(Lesson 1-3)* **27**

304 Chapter 7 Algebra: Using Rational Numbers

Extending the Lesson

Enrichment Masters, p. 55

Activity Have students work in small groups to find the amount of material needed to cover the trapezoidal sides and rectangular base of this wastebasket. **876 in²**

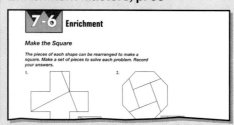

HANDS-ON LAB

COOPERATIVE LEARNING

7-6B Area and Pick's Theorem

A Follow-Up of Lesson 7-6

- geoboard or dot paper
- spreadsheet software

Pick's Theorem is a formula that you can use to find the area of a shape formed on a geoboard. You can investigate this formula by using a spreadsheet.

Pick's Theorem is: $\frac{\text{Boundary Points}}{2} + \text{Interior Points} - 1$.

TRY THIS

Work in groups of four.

1. Make these shapes on your geoboard. *If you don't have a geoboard, draw them on dot paper.*
   - a triangle with an area of 1.5 square units
   - a rectangle with an area of 2 square units
   - a triangle with an area of 3 square units
   - a rectangle with an area of 4 square units
   - a triangle with an area of 4.5 square units
   - a rectangle with an area of 6 square units

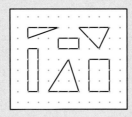

2. Copy the spreadsheet.

	A	B	C	D
1	Area	Boundary Points	Interior Points	Pick's Theorem
2	1.5	5	0	= B2/2 + C2 − 1
3	2	6	0	= B3/2 + C3 − 1
4	3	6	1	= B4/2 + C4 − 1
5	4	10	0	= B5/2 + C5 − 1
6	4.5	5	3	= B6/2 + C6 − 1
7	6	10	2	= B7/2 + C7 − 1

- Count the number of dots on the boundary of each shape. Then count the number of dots in the interior of each shape. Record these numbers in columns B and C of the spreadsheet.
- Use the spreadsheet formula in column D to find the results of using Pick's Theorem.

Lesson 7-6B HANDS-ON LAB

HANDS-ON 7-6B LAB Notes

GET READY

Objective Students connect algebra and geometry to find the area of a triangle.

Optional Resources
Hands-On Lab Masters
- square dot paper, p. 12
- worksheet, p. 54

Manipulative Kit
- geoboard
- geobands

Overhead Manipulative Resources
- geoboard and geobands
- rectangular dot paper transparency

Suggested spreadsheet software
- *Lotus 1·2·3*
- *ClarisWorks*
- *Microsoft Excel*

MANAGEMENT TIPS

Recommended Time
50 minutes

Getting Started Have students describe methods for finding the areas of geometric figures.
Sample answers: Use a formula; trace the figure on grid paper and count the squares inside it.

Using Dot Paper If students are using dot paper, they must use a straightedge and accurately draw segments, because their success in this lab will require them to distinguish between dots lying on a triangle and those lying inside the triangle.

Hands-On Lab 7-6B

ASSESS

After students answer Exercises 1–5, ask them to determine how they would use Pick's Theorem to find area when a spreadsheet is not available.

ON YOUR OWN

1. Compare your results in column D to the area of each shape. What do you notice? **They are the same.**

2. If b represents the number of dots on the boundary of a figure, i represents the number of dots in its interior, and A represents its area, write a formula for Pick's Theorem. $A = \dfrac{b}{2} + i - 1$.

3. Do you think Pick's Theorem would work for a square? Show several squares on your geoboard to support your answer. **Yes; see students' work.**

4. Do you think Pick's Theorem would work for trapezoids? Show examples to support your answer. **Yes; see students' work.**

5. Make each of the figures below on your geoboard or dot paper.

I.
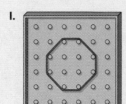

II.
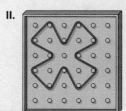

III.
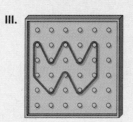

IV.
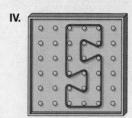

a. How could you find the area of each figure without using Pick's Theorem? (*Hint:* Look for ways to divide each figure into rectangles, triangles, or trapezoids.) **Count the squares.**

b. Investigate Pick's Theorem with each figure. With which types of figures can you use Pick's Theorem? **Pick's Theorem works for all of the figures.**

306 Chapter 7 Algebra: Using Rational Numbers

Math Journal Have students write a sentence or two in their own words explaining how they could find the area of a triangle drawn on dot paper. **Sample answer: Divide the number of dots on the triangle by 2, add the number of dots inside the triangle, and subtract 1.**

Hands-On Masters, p. 54

COOPERATIVE LEARNING

7-7A Graphing Pi
A Preview of Lesson 7-7

- grid paper
- circular objects
- tape

Press $\boxed{\pi}$ on your calculator. Have you ever wondered why π is such a special number? Before you investigate π, let's study some properties related to circles.

A *circle* is a set of points in a plane that are the same distance from a given point in the plane, called the *center*. The distance across the circle through its center is its *diameter*. The *circumference* of a circle is the distance around the circle.

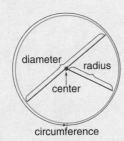

TRY THIS

Work in groups of three or four.

Step 1 Tape several pieces of grid paper together to form a large coordinate grid. Draw the horizontal and vertical axes so that the origin is 1 or 2 units from the lower left-hand corner.

Step 2 Place the bottom of a can or other circular object so that it touches the origin and the *x*-axis cuts the circle in half. Make a mark on the can where it crosses the *x*-axis. Then trace the circumference.

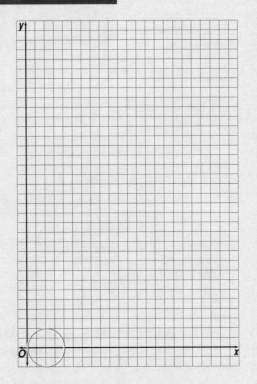

Lesson 7-7A HANDS-ON LAB **307**

HANDS-ON 7-7A LAB Notes

GET READY

Objective Students explore the value of pi.

Optional Resources
Hands-On Lab Masters
- grid paper, p. 10
- worksheet, p. 55

MANAGEMENT TIPS

Recommended Time
30 minutes

Getting Started Have students press $\boxed{\pi}$ on their calculators. Point out that during the fifth century, a Chinese mathematician found the rational approximation $\frac{355}{113}$ for π. Have students divide 355 by 113 and compare the quotient with π. **The quotient is identical for the first six decimal places.**

The **Activity** leads students through an exploration of the value of π. After completing the activity, ask students to estimate how the circumference of each circular object compares to its diameter. **Each circumference is about 3 times the diameter.**

Hands-On Lab 7-7A **307**

ASSESS

Have students complete Exercise 1. Watch for students who have difficulty distinguishing between the *x*- and *y*-coordinates and between the circumference and diameter. Review the terms with the students.

Use Exercise 2 to determine whether students understand the relationship among π, the circumference, and the diameter.

Additional Answers

1c. Answers will vary; the ratios should be close to $\frac{22}{7}$ or 3.14.

1d. Answers will vary; the ratios should be closer to $\frac{22}{7}$ or 3.14.

2. Sample answer: Since the *x*-coordinates represent circumference and the *y*-coordinates represent diameter, the ratio of *x*-coordinates to *y*-coordinates is π.

Hands-On Lab Masters, p. 55

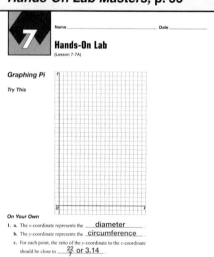

Step 3 Roll the can up one rotation, keeping the path of its edge on the *y*-axis. Draw a point on the graph where the can completes one rotation.

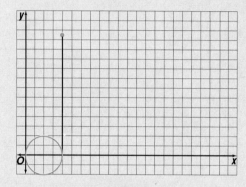

Step 4 Repeat Steps 2 and 3 for three to five more circular objects.

Step 5 From the origin, draw a curve that connects the points. Describe the curve.

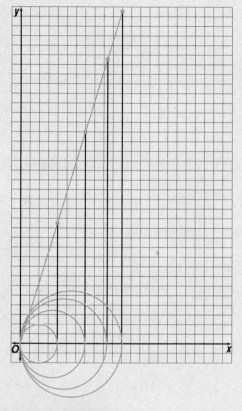

ON YOUR OWN

1. Find the *x*-coordinate and *y*-coordinate of each of the points you graphed.
   a. What does the *x*-coordinate represent? **diameter**
   b. What does the *y*-coordinate represent? **circumference**
   c. Determine the ratio of the *y*-coordinate to the *x*-coordinate for each point. **See margin.**
   d. What is the mean of the ratios? **See margin.**

2. **Look Ahead** The Greek letter π (pi) represents the ratio of the circumference of a circle to its diameter. What do you think would be a good approximation for π? Explain. **See margin.**

308 Chapter 7 Algebra: Using Rational Numbers

 Have students write a paragraph describing π and how they might use π outside of math class.

7-7 Integration: Geometry
Circles and Circumference

What you'll learn
You'll learn to find the circumference of circles.

When am I ever going to use this?
Knowing how to find the circumference of a circle can help you determine the amount of fencing needed to enclose a circular garden.

Word Wise
circle
center
radius
diameter
circumference

Is the Ferris wheel one of your favorite amusement park rides? The largest Ferris wheel was built in Yokohama City, Japan in 1985. The wheel is 328 feet in diameter. What distance would you travel in one revolution if you rode on the largest Ferris wheel? *This problem will be solved in Example 2.*

A **circle** is a set of points in a plane that are the same distance from a given point in the plane. The given point is called the **center**. The distance from the center to any point on the the circle is called the **radius (r)**. The distance across the circle through the center is its **diameter (d)**. The **circumference (C)** of a circle is the distance around the circle. The diameter of a circle is twice its radius, or $d = 2r$.

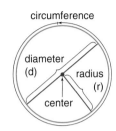

The relationship you discovered in Lesson 7-7A is true for all circles. The circumference of a circle divided by its diameter is always 3.1415926.... The Greek letter π (pi) represents this number. The numbers 3.14 and $\frac{22}{7}$ are often used as approximations for π.

Circumference	**Words:** The circumference of a circle is equal to its diameter times π, or 2 times its radius times π.	**Model:** 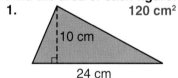
	Symbols: $C = \pi d$ or $C = 2\pi r$	

Example

1 Find the circumference of a circle with a radius of 14 meters.

Method 1 Use $\frac{22}{7}$ for π.

$C = 2\pi r$

$C = 2 \cdot \frac{22}{7} \cdot 14$

$C = 2 \cdot \frac{22}{\underset{1}{7}} \cdot \frac{\overset{2}{14}}{1}$

$C \approx 88$ m

Method 2 Use 3.14 for π.

$C = 2\pi r$

$C = 2 \cdot 3.14 \cdot 14$

$C \approx 87.92$ m

The circumference is about 88 meters.

Lesson 7-7 Integration: Geometry Circles and Circumference **309**

In-Class Examples

For Example 1
Find the circumference of a circle with a diameter of 25 inches. for $\pi \approx \frac{22}{7}$: ≈ 79 in.; for $\pi \approx 3.14$: ≈ 78.5 in.

For Example 2
The tip of the minute hand of a clock is 14 centimeters from the center of the clock's face. How far will the tip of the hand travel in 1 hour? Use a calculator. ≈ 88 cm

3 PRACTICE/APPLY

Check for Understanding
If students need additional practice or instruction after completing Exercises 1–9, one of these options may be helpful.
- Extra Practice, see p. 625
- Reteaching Activity
- *Study Guide Masters*, p. 56
- *Practice Masters*, p. 56

Additional Answer
3. Sample answer: Divide 22 by 3.14. Then round to the nearest tenth.

Study Guide Masters, p. 56

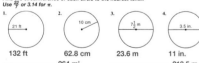

Most calculators have a π key that gives a more precise approximation of pi.

Example APPLICATION — **Entertainment** Refer to the beginning of the lesson. What distance would you travel in one revolution if you rode on the Ferris wheel?

$C = \pi d$

$C = \pi \cdot 328$ *Replace d with 328.*

[π] [×] 328 [=] *1030.44239*

You would travel about 1,030 feet in one revolution.

CHECK FOR UNDERSTANDING

Communicating Mathematics

Read and study the lesson to answer each question. 2. $C \approx 6r$

1. **Identify** two numbers that are approximations for π. $\frac{22}{7}$ and 3.14
2. **Tell** how to estimate the circumference of a circle if you know the radius.
3. **Write** a few sentences explaining how you can find the diameter of a circle that has a circumference of 22 inches. **See margin.**

Answers are calculated using $\frac{22}{7}$ or 3.14 and then rounded.

Guided Practice

Find the circumference of each circle to the nearest tenth. Use $\frac{22}{7}$ or 3.14 for π.

4. 14 in. — 44 in.   5. 9 in. — 56.5 in.   6. 7.5 cm — 23.6 cm

7. 106.8 cm   7. The radius is 17 centimeters.   8. The diameter is $\frac{1}{2}$ foot. $1\frac{4}{7}$ ft

9. **Archaeology** Stonehenge is a circular array of giant stones in England. The diameter of Stonehenge is 30.5 meters. Find the approximate distance it is around Stonehenge. **95.8 meters**

EXERCISES

Practice Find the circumference of each circle to the nearest tenth. Use $\frac{22}{7}$ or 3.14 for π.

10. $4\frac{3}{8}$ in. — 13.8 in.   11. 10.5 mm — 66 mm   12. 12.56 m — 39.4 m   13. 29.5 cm — 185.3 cm

14. The radius is 30 yards. **188.4 yd**   15. The diameter is 18 feet. **56.5 ft**
16. The diameter is 13.3 meters. **41.8 m**   17. The radius is $5\frac{1}{4}$ inches. **33 in.**
18. 28.3 cm   18. The radius is 4.5 centimeters.   19. The diameter is 19.33 meters.
19. 60.7 m   20. The diameter is $6\frac{1}{2}$ feet. $20\frac{3}{7}$ ft   21. The radius is $6\frac{3}{4}$ inches. $42\frac{3}{7}$ in.

310 Chapter 7 Algebra: Using Rational Numbers

Reteaching the Lesson

Activity The formula $C = \pi d$ indicates that the circumference of a circle is slightly more than three times the diameter. Have students use $C = 3d$ to estimate the circumference. Then assist them in choosing a reasonable value for π in Exercises 4–8.

Error Analysis
Watch for students who confuse radius and diameter when calculating circumference.
Prevent by having students underline the word *radius* or *diameter* in each problem and then match problems with radius to $C = 2\pi r$ and those with diameter to $C = \pi d$.

22. 1 unit

22. The circumference of a circle is π units. What is the diameter of the circle?
23. The circumference of a circle is 2.041 meters. Find its radius.   **0.325 m**

Applications and Problem Solving

24. *Forestry* Loggers use a rule to estimate the amount of board feet in a live tree. They find the circumference of the tree in inches and divide by π. They square the quotient, multiply by the height of the tree in feet, and divide by 144. To allow for bark, they multiply by 0.9. Finally, the number of board feet is found by multiplying by 6. How many board feet are in a 40-foot tree with a circumference of 63 inches?   **603.83 board feet**

25. *Manufacturing* How many peanut butter sandwiches have you eaten in your lifetime? The label that goes around a jar of Jif peanut butter overlaps itself by $\frac{3}{8}$ inch to allow for sealing the label on the jar. If the diameter of the jar is 3 inches, what is the length of the Jif label? $9\frac{45}{56}$ **or about 9.8 inches**

26. Working on the **CHAPTER Project**

Nautilus shells, hurricanes, and galaxies all form spirals. You can model a spiral using the Fibonacci numbers and circles.

a. Use a ruler to draw two squares with sides 1 centimeter long as shown. Then add a square with sides of 2 centimeters, 3 centimeters, 5 centimeters, and so on.

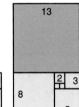

Use a compass to draw quarter circles in each square so that a continuous curve results. The result is a *Fibonacci spiral*.   **See margin.**

b. Look for spirals in nature. Add their descriptions and pictures to your project.   **See students' work.**

27. *Critical Thinking* Three tennis balls are packaged one on top of the other in a can. Which measure is greater, the height of the can or the circumference of the can? Explain.   **circumference; height = 3d, C ≈ 3.14d**

Mixed Review

28. *Earth Science* The sketch of the cross section of a stream was made by taking a depth reading every 5 feet. Estimate the area of the cross section. *(Lesson 7-6)*   **570 ft²**

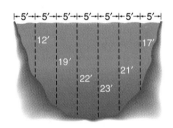

29. **Test Practice** What number will make $\frac{3}{4} \cdot \frac{7}{8} = n \cdot \frac{7}{8}$ true? *(Lesson 7-4)*   **B**

A $\frac{4}{8}$ B $\frac{3}{4}$ C $\frac{10}{12}$ D $\frac{7}{8}$

30. *Number Sense* Use divisibility rules to determine whether 284 is divisible by 6. Explain why or why not. *(Lesson 6-1)*

30. **No; 284 is divisible by 2 but not divisible by 3.**

Lesson 7-7 Integration: Geometry   Circles and Circumference   **311**

Extending the Lesson

Enrichment Masters, p. 56

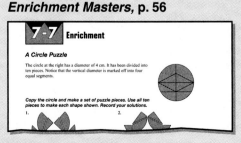

Activity The circumference of Earth at the equator is about 24,900 miles. Ask students what the approximate distance is from the center of Earth to its crust.   **3,963 mi** Have students research the circumference of planets, stars, and moons and calculate their radii and diameters.

Assignment Guide

Core: 11–25 odd, 27–30
Enriched: 10–24 even, 25, 27–30

CHAPTER Project

Exercise 26 asks students to advance to the next stage of work on the Chapter Project. You may wish to have students use library and Internet research as well as field experience when looking for spirals in nature.

4 ASSESS

Closing Activity

Speaking Draw the circle on the chalkboard. Ask students to identify a radius, the diameter, and the center.

$\overline{AK}$ or $\overline{KT}$; $\overline{AT}$; K

Additional Answer

26a.

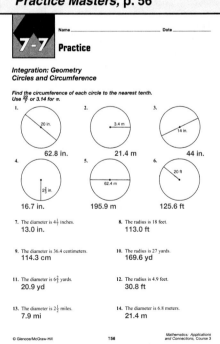

Practice Masters, p. 56

7-8 Lesson Notes

Instructional Resources
- *Study Guide Masters*, p. 57
- *Practice Masters*, p. 57
- *Enrichment Masters*, p. 57
- Transparencies 7-8, A and B
- *Assessment and Evaluation Masters*, p. 184

CD-ROM Program
- Resource Lesson 7-8

Recommended Pacing	
Standard	Day 11 of 15
Honors	Day 11 of 15
Block	Day 6 of 8

1 FOCUS

5-Minute Check
(Lesson 7-7)

Find the circumference of each circle to the nearest tenth. Use $\frac{22}{7}$ or 3.14 for π.

1.
22.0 m

2.
75.4 in.

3. The radius is 1.5 meters.
9.4 m

 The 5-Minute Check is also available on **Transparency 7-8A** for this lesson.

2 TEACH

 Transparency 7-8B contains a teaching aid for this lesson.

Thinking Algebraically Use the identity property of multiplication to justify the division algorithm:

$$\frac{\frac{3}{3}}{\frac{3}{4}} = \frac{3}{\frac{3}{4}} \cdot \frac{\frac{4}{3}}{\frac{4}{3}} = \frac{3 \cdot \frac{4}{3}}{1} = 3 \cdot \frac{4}{3}$$
↑
1

312 Chapter 7

7-8 Dividing Fractions

What you'll learn
You'll learn to divide fractions.

When am I ever going to use this?
Knowing how to divide fractions can help you find the diameter of a circle when the circumference is known.

Have you thought of a career as a teacher? If so, you will need to earn a bachelor's degree in education. In 1993, about 108,000 people earned bachelor's degrees in education in the United States. Men earned a little more than $\frac{1}{5}$ of those degrees. Approximately how many men earned a bachelor's degree in education in 1993?

You could find this number in two ways.

Method 1 Divide by 5.

$$108000 \;\boxed{\div}\; 5 \;\boxed{=}\; 21600$$

Method 2 Multiply by $\frac{1}{5}$.

$$108000 \;\boxed{\times}\; 0.2 \;\boxed{=}\; 21600 \quad \frac{1}{5} = 0.2$$

About 21,600 men earned bachelor's degrees in education in 1993.

In the example above, notice that dividing by 5 and multiplying by $\frac{1}{5}$ gave you the same result. Also note that 5 and $\frac{1}{5}$ are multiplicative inverses.

Dividing Fractions	Words:	To divide by a fraction, multiply by its multiplicative inverse.
	Symbols:	Arithmetic $\quad \frac{5}{6} \div \frac{2}{3} = \frac{5}{6} \cdot \frac{3}{2}$
		Algebra $\quad \frac{a}{b} \div \frac{c}{d} = \frac{a}{b} \cdot \frac{d}{c}$, where $b, c, d \neq 0$.

Example

1 Solve $a = \frac{-6}{-1\frac{1}{2}}$.

$a = \frac{-6}{-1\frac{1}{2}}$ *Estimate:* $-6 \div (-2) = 3$

$a = -6 \div \left(-1\frac{1}{2}\right)$ *Remember that the fraction bar means division.*

$a = -\frac{6}{1} \cdot \left(-\frac{2}{3}\right)$ *Dividing by $-1\frac{1}{2}$ or $-\frac{3}{2}$ is the same as multiplying by $-\frac{2}{3}$.*

$a = -\frac{\overset{2}{6}}{1} \cdot \left(-\frac{2}{\underset{1}{3}}\right)$

$a = 4$

312 Chapter 7 Algebra: Using Rational Numbers

Motivating the Lesson
Hands-On Activity Have students use rulers to find the number of $\frac{1}{4}$-inch sections in $2\frac{3}{4}$ inches. **11**

Write $2\frac{3}{4} \div \frac{1}{4} = 11$. Show that the same result is obtained by multiplying $2\frac{3}{4}$ by 4.

Multiple Learning Styles

 Interpersonal Give each student in a group a sheet of paper with problems involving the four operations on fractions. All of the problems must be solved by having each student take turns verbalizing a step in the process while everyone else in the group performs that step. Repeat until all of the problems are solved.

Examples

Solve each equation.

2 $\frac{3}{4} \div \frac{5}{6} = y$

$\frac{3}{4} \div \frac{5}{6} = y$

$\frac{3}{4} \cdot \frac{6}{5} = y$

$\frac{3}{\underset{2}{4}} \cdot \frac{\overset{3}{6}}{5} = y$

$\frac{9}{10} = y$

3 $f = -3\frac{1}{4} \div 2\frac{1}{2}$

$f = -3\frac{1}{4} \div 2\frac{1}{2}$

$f = -\frac{13}{4} \div \frac{5}{2}$

$f = -\frac{13}{\underset{2}{4}} \cdot \frac{\overset{1}{2}}{5}$

$f = -\frac{13}{10}$ or $-1\frac{3}{10}$

APPLICATION

4 **Food** The first major hamburger chain in the United States was White Castle. They are famous for their square burgers. Their burgers are $2\frac{1}{2}$ inches on a side. How many White Castle hamburgers can fit across a grill that is 30 inches wide?

$30 \div 2\frac{1}{2} = \frac{30}{1} \div \frac{5}{2}$ *Rename 30 as $\frac{30}{1}$ and $2\frac{1}{2}$ as $\frac{5}{2}$.*

$= \frac{\overset{6}{30}}{1} \cdot \frac{2}{\underset{1}{5}}$

$= \frac{12}{1}$ or 12

You can fit 12 White Castle hamburgers across a grill that is 30 inches wide.

 Did you know? The ground beef hamburger was first made in Hamburg, Germany. Hamburgers were introduced in the United States at the 1904 World's Fair in St. Louis, Missouri.

CHECK FOR UNDERSTANDING

Communicating Mathematics

Read and study the lesson to answer each question.

1. **Explain** how the model shows the quotient of 2 and $\frac{2}{3}$.

   1. There are 3 sets of $\frac{2}{3}$ to equal 2.

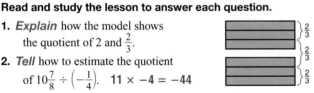

2. **Tell** how to estimate the quotient of $10\frac{7}{8} \div \left(-\frac{1}{4}\right)$. $11 \times -4 = -44$

Guided Practice

Solve each equation. Write the solution in simplest form.

3. $\frac{2}{3} \div \frac{3}{4} = a$ $\frac{8}{9}$

4. $\frac{1}{2} \div 6 = d$ $\frac{1}{12}$

5. $r = -\frac{3}{8} \div \frac{9}{10}$ $-\frac{5}{12}$

6. $q = 2\frac{1}{2} \div 1\frac{1}{4}$ 2

7. What is the value of $\frac{2\frac{2}{3}}{4}$? $\frac{2}{3}$

8. **Cooking** How many slices of pepperoni, each $\frac{1}{16}$ inch thick, can be cut from a stick 8 inches long?   **128 slices**

Lesson 7-8 Dividing Fractions **313**

Assignment Guide

Core: 9–21 odd, 23–26
Enriched: 10–20 even, 21–26

4 ASSESS

Closing Activity

Writing Have each student write a paragraph describing how to solve $a = \frac{3}{4} \div 1\frac{5}{16}$. $a = \frac{4}{7}$

Chapter 7, Quiz C (Lessons 7-6 through 7-8) is available in the *Assessment and Evaluation Masters*, p. 184.

Additional Answer
26. Sample answer:

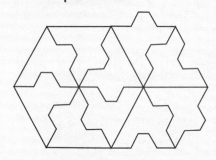

Practice Masters, p. 57

EXERCISES

Practice Solve each equation. Write the solution in simplest form.

9. $12 \div -3\frac{1}{3} = j$ $-3\frac{3}{5}$
10. $\frac{2}{3} \div \frac{5}{6} = x$ $\frac{4}{5}$
11. $r = \frac{9}{10} \div (-6)$ $-\frac{3}{20}$
12. $y = 1\frac{1}{2} \div \frac{1}{2}$ 3
13. $-2\frac{1}{4} \div \left(-\frac{3}{8}\right) = t$ 6
14. $\frac{7}{8} \div 2\frac{4}{5} = b$ $\frac{5}{16}$
15. $4\frac{1}{2} \div \frac{3}{4} = h$ 6
16. $f = -3\frac{3}{4} \div \left(-2\frac{1}{2}\right)$ $1\frac{1}{2}$
17. $c = -12\frac{1}{4} \div 4\frac{2}{3}$ $-2\frac{5}{8}$

18. **Algebra** Evaluate $\frac{d}{q}$ if $d = -3\frac{3}{4}$ and $q = 6\frac{2}{3}$. $-\frac{9}{16}$

19. What is the quotient of $-3\frac{3}{5}$ and -10? $\frac{9}{25}$

20. **Algebra** Evaluate $\frac{a^2}{b^2}$ if $a = -\frac{4}{5}$ and $b = \frac{2}{3}$. $1\frac{11}{25}$

Applications and Problem Solving

21. **Geometry** The circumference of the circle is $93\frac{1}{2}$ feet. Find the diameter of the circle. Use $\frac{22}{7}$ for π. $29\frac{3}{4}$ feet

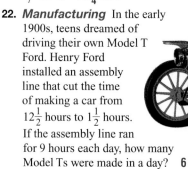

22. **Manufacturing** In the early 1900s, teens dreamed of driving their own Model T Ford. Henry Ford installed an assembly line that cut the time of making a car from $12\frac{1}{2}$ hours to $1\frac{1}{2}$ hours. If the assembly line ran for 9 hours each day, how many Model Ts were made in a day? 6

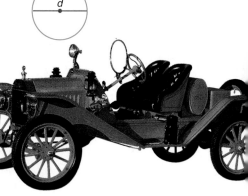

23. **Critical Thinking** Which is greater, mn or $\frac{m}{n}$, if $m > 0$ and $0 < n < 1$? Explain. $\frac{m}{n}$ is greater; Dividing by a proper fraction is actually multiplying by an improper fraction.

Mixed Review

24. **Geometry** Find the circumference of a circle with a radius of 2.7 millimeters. *(Lesson 7-7)* **16.96 mm**

25. **Test Practice** Paige has been jogging every day for the past four weeks. At the beginning of each week, she has been increasing the length of her daily jog. If she continues jogging according to the pattern shown in the table, how many minutes will she spend jogging each day during her fifth week of jogging? *(Lesson 7-5)* **B**

Week	Time Jogging (min)
1	8
2	16
3	24
4	32
5	?

A 32 min   B 40 min   C 48 min   D 56 min

26. **Geometry** Make an Escher-like drawing for the pattern shown. Use a tessellation of two rows of five equilateral triangles as your base. *(Lesson 5-7)* **See margin.**

314 Chapter 7 Algebra: Using Rational Numbers

Extending the Lesson

Enrichment Masters, p. 57

Activity Ask students to solve this problem: *Use the numbers 2, 3, 4, and 5 to write a division problem with a quotient of $\frac{3}{10}$.* $\frac{3}{4} \div \frac{5}{2} = \frac{3}{10}$ Have students work in small groups to create similar problems.

7-9 Solving Equations

What you'll learn
You'll learn to solve equations involving rational numbers.

When am I ever going to use this?
Knowing how to solve equations can help you find the length of rectangle when you know the area and width.

Would you pay $121,000 for an ink pen? The Meisterstück Solitaire Royal fountain pen is 24K or 24-karat gold and is encased with 4,810 diamonds. It can be ordered and takes 6 months to make.

The amount of gold in jewelry is usually expressed using the karat system. One karat means that the metal is $\frac{1}{24}$ pure gold. Thus, pure gold is 24-karat. If there are $2\frac{1}{2}$ ounces of gold in a gold bracelet that weighs 6 ounces, how should the bracelet be advertised using karats?

You can apply the skills you have learned for rational numbers to solve equations containing rational numbers. You can find the amount of gold in jewelry, G, using the formula $G = \frac{k}{24} \cdot w$, where k is the number of karats and w is the weight.

$$G = \frac{k}{24} \cdot w$$

$$2\frac{1}{2} = \frac{k}{24} \cdot 6 \quad \text{Replace } G \text{ with } 2\frac{1}{2} \text{ and } w \text{ with } 6.$$

$$\frac{5}{2} \cdot \frac{1}{6} = \frac{k}{24} \cdot 6 \cdot \frac{1}{6} \quad \text{Multiply each side by } \frac{1}{6}, \text{ the multiplicative inverse of 6.}$$

$$\frac{5}{12} = \frac{k}{24}$$

$$24\left(\frac{5}{12}\right) = 24 \cdot \frac{k}{24} \quad \text{Multiply each side by 24.}$$

$$10 = k$$

The bracelet should be advertised as 10-karat gold.

Example

1 Solve $-\frac{5}{8}x = \frac{3}{4}$. Check your solution.

$$-\frac{5}{8}x = \frac{3}{4}$$

$$-\frac{8}{5} \cdot -\frac{5}{8}x = -\frac{8}{5} \cdot \frac{3}{4} \quad \text{Multiply each side by } -\frac{8}{5}. \text{ Why?}$$

$$x = -\frac{6}{5} \text{ or } -1\frac{1}{5}$$

Check: $-\frac{5}{8}x = \frac{3}{4}$

 $-\frac{5}{8}\left(-\frac{6}{5}\right) \stackrel{?}{=} \frac{3}{4}$ Replace x with $-\frac{6}{5}$.

$$\frac{3}{4} = \frac{3}{4} \checkmark$$

Lesson 7-9 Solving Equations **315**

In-Class Examples

For Example 1
Solve $\frac{8}{9}k = -\frac{4}{15}$. $-\frac{3}{10}$

For Example 2
Bobby Bonilla helped lead the Florida Marlins to victory in the 1997 World Series. He played in 153 regular season games, and his batting average was 0.297. He had 562 at bats. How may hits did he have? **167**

For Example 3
Solve $\frac{k}{2.6} + 18 = 18.5$. **1.3**

Teaching Tip When working the examples, remind students to isolate the fraction with the variable on one side before multiplying by the denominator.

3 PRACTICE/APPLY

Check for Understanding
If students need additional practice or instruction after completing Exercises 1–10, one of these options may be helpful.
- Extra Practice, see p. 626
- Reteaching Activity
- *Study Guide Masters,* p. 58
- *Practice Masters,* p. 58

Study Guide Masters, p. 58

Examples — APPLICATION

For the latest baseball statistics, visit:
www.glencoe.com/sec/math/mac/mathnet

2 Sports In baseball, a player's batting average is determined by dividing the number of hits by the number of at-bats. On August 22, 1997, Edgar Martinez' batting average was 0.327, and he had 447 at-bats. How many hits did he have?

$$0.327 = \frac{h}{447}$$
$$447 \cdot 0.327 = \frac{447}{1} \cdot \frac{h}{447} \quad \textit{Multiply each side by 447.}$$
$$146.169 = h$$

Mr. Martinez had 146 hits.

Check: 146 ÷ 447 = 0.326621924

Since batting averages are rounded to the thousandths place, the answer is correct.

3 Solve $\frac{b}{1.5} - 13 = -2.2$. Check your solution.

$$\frac{b}{1.5} - 13 = -2.2$$
$$\frac{b}{1.5} - 13 + 13 = -2.2 + 13 \quad \textit{Add 13 to each side.}$$
$$\frac{b}{1.5} = 10.8$$
$$1.5 \cdot \frac{b}{1.5} = 1.5 \cdot 10.8 \quad \textit{Multiply each side by 1.5.}$$
$$b = 16.2$$

Check: $\frac{b}{1.5} - 13 = -2.2 \quad \textit{Replace b with 16.2.}$

16.2 ÷ 1.5 − 13 = −2.2 ✓

CHECK FOR UNDERSTANDING

Communicating Mathematics

2. Subtract $\frac{1}{4}$ from each side and then divide each side by 3.

Read and study the lesson to answer each question.

1. *Write* an equation that is solved by using the multiplicative inverse of $1\frac{1}{4}$. **Sample answer:** $-1\frac{1}{4}a + 4 = 8$

2. *Explain* how to solve $3n + \frac{1}{4} = -\frac{7}{8}$.

3. *You Decide* Esohe says that to solve the equation $7m - \frac{2}{3} = 1\frac{2}{3}$, the first step is to add $\frac{2}{3}$ to each side. Priya says the first step is to divide each side by 7. Whose method would you use and why? **See margin.**

Guided Practice

Solve each equation. Check your solution.

4. $\frac{3}{4}v = -\frac{7}{8}$ $-1\frac{1}{6}$
5. $2.3y = 6.9$ **3**
6. $-4.2 = \frac{c}{7}$ -29.4
7. $t + 0.25 = -4.125$ **−4.375**
8. $\frac{1}{3}x - \frac{3}{8} = -\frac{1}{2}$ $-\frac{3}{8}$
9. $-1.6w + 3.5 = 0.48$ **1.8875**

10. *Physical Science* Use the formula $F = \frac{9}{5}C + 32$ to find the Celsius temperature C for the temperature of 68°F. **20°C**

316 Chapter 7 Algebra: Using Rational Numbers

Reteaching the Lesson

Activity Have students identify if an equation is a one-step or two-step equation. Then have them identify the operations involved. Have them verbally express how they are going to solve each equation in Exercises 4–9.

Additional Answer

3. Sample answer: Esohe's method uses the order of operations in reverse and is probably easier then Priya's method because it requires dividing two fractions.

EXERCISES

Practice

Solve each equation. Check your solution.

11. $-12 = -\frac{z}{5}$ **60**
12. $1.1 + d = -4.4$ **−5.5**
13. $2\frac{1}{2}x = 5\frac{3}{4}$ **$2\frac{3}{10}$**
14. $u - (-0.03) = 3.2$ **3.17**
15. $\frac{n}{2} = -1.6$ **−3.2**
16. $2a = -12$ **−6**
17. $\frac{t}{3.2} = -4.5$ **−14.4**
18. $1.8 = 0.6 - g$ **−1.2**
19. $-c + \frac{1}{6} = \frac{3}{5}$ **$-\frac{13}{30}$**
20. $\frac{2}{3h} - (-3) = 6$ **$\frac{2}{9}$**
21. $\frac{x-10}{5} = 2.5$ **22.5**
22. $-12 + \frac{w}{4} = 9$ **84**
23. $4.7 = 1.2 - 7m$ **−0.5**
24. $\frac{c}{2} + (-3) = 5$ **16**
25. $1.6 = \frac{-8+k}{-7}$ **−3.2**

26. Find the solution of $-2 = j - \frac{2}{5}$. **$-1\frac{3}{5}$**
27. What is the solution of $7p - \left(-\frac{2}{9}\right) = \frac{5}{9}$? **$\frac{1}{21}$**

Applications and Problem Solving

28. **Geometry** The volume of a cone is given by the formula $V = \frac{1}{3}\pi r^2 h$ where r is the radius and h is the height of the cone. The volume of a cone is 346.5 cubic inches. If the cone has a radius of 3.5 inches, what is its height? **27 in.**

29. **Air-Conditioning** An estimate of the size of a room air conditioner needed is given by the formula, $B = 27\ell w$, where B is the number of British Thermal Units per hour, ℓ is the length of the room, and w is the width of the room. A room air conditioner is rated at 4,000 BTUs. What is the maximum length room it can cool if the room is $11\frac{1}{2}$ feet wide? **about 12.9 ft**

30. **Safety** The time that a traffic light remains yellow is given by the formula $t = \frac{1}{2}s + 1$, where t is the time in seconds and s is the speed limit. What is the yellow time for a traffic light on a street with a speed limit of 25 miles per hour? **$13\frac{1}{2}$ seconds**

31. **Critical Thinking** Using the formula in Exercise 10, find the temperature that is the same on both the Fahrenheit and Celsius scales. **−40°F = −40°C**

Mixed Review

32. Solve $a = \frac{7}{12} \div \frac{3}{4}$. *(Lesson 7-8)* **$\frac{7}{9}$**
33. **Number Sense** Find the prime factorization of 48. *(Lesson 6-2)* **$2^4 \times 3$**
34. **Test Practice** The graph shows how Melissa spends her day. What percent of Melissa's day is not spent in school or studying? *(Lesson 3-4)* **B**
    A 70%
    B 62%
    C 38%
    D 30%
    E Not Here

Melissa's Day
Sleeping 33%, School 30%, Eating 12%, Work 17%, Studying 8%

35. Find the value of $4^2 - 2^3$. *(Lesson 1-2)* **8**

Lesson 7-9 Solving Equations **317**

Extending the Lesson

Enrichment Masters, p. 58

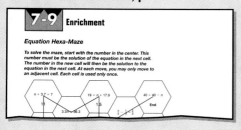

Activity Scientists often use the Kelvin temperature scale. Kelvin and Celsius temperatures are related by the formula $K = C + 273.2$. Find the Kelvin temperature for each Celsius temperature.
a. −147.7°C   **125.5 K**
b. 56°C   **329.2 K**
c. 95°C   **368.2 K**
d. 365°C   **638.2 K**

Assignment Guide

Core: 11–29 odd, 31–35
Enriched: 12–26 even, 28–35

4 ASSESS

Closing Activity

Speaking Write $\frac{3}{4}m = 2\frac{5}{8} \cdot 2\frac{2}{7}$ on the chalkboard. Have students work in pairs describing each step in solving the equation to each other. Finish by asking all the pairs to report their solution for m to the class. $1\frac{1}{7}$ Discuss the steps with the class to resolve any points of difficulty.

Practice Masters, p. 58

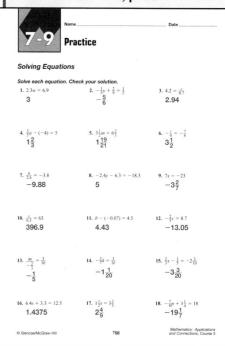

Lesson 7-9 **317**

7-10 Lesson Notes

Instructional Resources
- *Study Guide Masters*, p. 59
- *Practice Masters*, p. 59
- *Enrichment Masters*, p. 59
- *Transparencies 7-10, A and B*
- *Assessment and Evaluation Masters*, p. 184
- CD-ROM Program
  - Resource Lesson 7-10

Recommended Pacing	
Standard	Day 13 of 15
Honors	Day 13 of 15
Block	Day 7 of 8

1 FOCUS

5-Minute Check
(Lesson 7-9)

Solve each equation. Check your solution.
1. $\frac{a}{25} = -3$ -75
2. $7.2 + b = -2.5$ -9.7
3. $3\frac{3}{4}c = 5\frac{5}{8}$ $1\frac{1}{2}$
4. $d - (-5.6) = 4.5$ -1.1
5. $\frac{11+y}{6} = 4.3$ 14.8

The 5-Minute Check is also available on **Transparency 7-10A** for this lesson.

Motivating the Lesson

Problem Solving Ask students how they would solve this problem. *Neil has $36.75 to spend on a gift for his parents for their twentieth anniversary. If he has to pay a 5% sales tax, what is the highest priced item he can afford?* (Hint: The multiplicative inverse of 5% is 20.) **$35.00**

7-10 Solving Inequalities

What you'll learn

You'll learn to solve inequalities involving rational numbers and graph their solutions.

When am I ever going to use this?

You can use inequalities to find the minimum number of signatures needed for a petition.

Have you ever ordered something from a catalog? Then you know the price listed in the catalog is not the total cost of your order! Usually, taxes and shipping and handling fees are added.

Shipping and handling fees are often determined by the total price of the items you order. The table is from the Harriet Carter catalog. Regardless of how much you order, the shipping and handling charges, s, can be expressed as the inequality $s \leq \$11.98$.

In Lesson 1-9, you learned to solve inequalities by solving the related equations. You can solve inequalities involving rational numbers in the same way.

Example 1 Solve $\frac{1}{2}x - 6 > 5\frac{1}{4}$. Graph the solution on a number line.

Solve the related equation $\frac{1}{2}x - 6 = 5\frac{1}{4}$.

$$\frac{1}{2}x - 6 = 5\frac{1}{4}$$

$\frac{1}{2}x - 6 + 6 = 5\frac{1}{4} + 6$ *Add 6 to each side.*

$\frac{1}{2}x = 11\frac{1}{4}$ *Simplify.*

$2\left(\frac{1}{2}x\right) = 2\left(11\frac{1}{4}\right)$ *Multiply each side by 2.*

$x = 22\frac{2}{4}$ or $22\frac{1}{2}$

The solution must be either numbers greater than $22\frac{1}{2}$ or numbers less than $22\frac{1}{2}$. Let's test two numbers, one greater than $22\frac{1}{2}$ and one less than $22\frac{1}{2}$, to see which is correct.

Try 24.

$\frac{1}{2}(24) - 6 \stackrel{?}{>} 5\frac{1}{4}$ *Replace x with 24.*

$12 - 6 \stackrel{?}{>} 5\frac{1}{4}$

$6 > 5\frac{1}{4}$ *true*

Try 10.

$\frac{1}{2}(10) - 6 \stackrel{?}{>} 5\frac{1}{4}$ *Replace x with 10.*

$5 - 6 \stackrel{?}{>} 5\frac{1}{4}$

$-1 > 5\frac{1}{4}$ *false*

Numbers greater than $22\frac{1}{2}$ make up the solution set. So, $x > 22\frac{1}{2}$. Graph the solution on a number line.

Since $22\frac{1}{2}$ is not included, an open circle is used.

318 Chapter 7 Algebra: Using Rational Numbers

Examples

2 Solve $\frac{d-2.5}{1.25} \leq -3.8$. Graph the solution on a number line.

Solve the related equation $\frac{d-2.5}{1.25} = -3.8$.

$$\frac{d-2.5}{1.25} = -3.8$$
$$(1.25)\left(\frac{d-2.5}{1.25}\right) = (1.25)(-3.8) \quad \textit{Multiply each side by 1.25.}$$
$$d - 2.5 = -4.75 \quad \textit{Simplify.}$$
$$d - 2.5 + 2.5 = -4.75 + 2.5 \quad \textit{Add 2.5 to each side.}$$
$$d = -2.25$$

The solution includes −2.25. *Why?* Now test two numbers to determine whether the solution also includes numbers greater than −2.25 or numbers less than −2.25.

Try 0.   Try −2.5.
$\frac{0-2.5}{1.25} \stackrel{?}{<} -3.8$ $\frac{-2.5-2.5}{1.25} \stackrel{?}{<} -3.8$
$\frac{-2.5}{1.25} \stackrel{?}{<} -3.8$ $\frac{-5}{1.25} \stackrel{?}{<} -3.8$
$-2 < -3.8$ *false* $-4 < -3.8$ *true*

Numbers less than 2.25 are included in the solution. So, $d \leq -2.25$.
The number line shows $d \leq -2.25$.

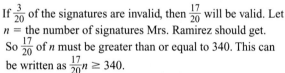

Since −2.25 is included, a solid dot is used.

CONNECTION

3 **Civics** Mrs. Ramirez needs 340 valid signatures on a petition before she can enter the race for a city council position. The elections board told her that about $\frac{3}{20}$ of the signatures usually prove to be invalid. How many signatures should she get to ensure that she will qualify for the ballot?

If $\frac{3}{20}$ of the signatures are invalid, then $\frac{17}{20}$ will be valid. Let $n =$ the number of signatures Mrs. Ramirez should get. So $\frac{17}{20}$ of n must be greater than or equal to 340. This can be written as $\frac{17}{20}n \geq 340$.

Solve the related equation $\frac{17}{20}n = 340$.

$$\frac{17}{20}n = 340$$
$$\frac{20}{17} \cdot \frac{17}{20}n = \frac{20}{17} \cdot 340 \quad \textit{Multiply each side by } \frac{20}{17}.$$
$$n = 400 \quad \textit{Simplify.}$$

If you test numbers greater than and less than 400, you find the solution set includes numbers greater than 400. So, $n \geq 400$.

Mrs. Ramirez needs at least 400 signatures to ensure her petition.

In Examples 1, 2, and 3, notice that the inequality symbol in the solution is the same as the one in the original inequality. Is this always true?

Lesson 7-10 Solving Inequalities **319**

---

2 TEACH

 Transparency 7-10B contains a teaching aid for this lesson.

Using Logical Reasoning
Encourage students to check the reasonableness of a solution by using logical reasoning. For example, in the inequality $\frac{1}{3}x - 8 > 4\frac{1}{3}$, x must be fairly large for one third of it to have eight subtracted from it and still be greater than $4\frac{1}{3}$.

In-Class Examples

For Example 1
Solve $\frac{2}{3}y + 5 < 6\frac{1}{3}$. Graph the solution on a number line. **$y < 2$**

For Example 2
Solve $\frac{f-3.25}{2.5} \geq -1.8$. Graph the solution on a number line. **$f \geq -1.25$**

For Example 3
The Rin Tin Tin company manufactures tin cans. About $\frac{1}{26}$ of the tin in each production cycle is scrapped and recycled. 1,000 cans are produced in each cycle, and each can weighs $\frac{1}{8}$ pound. How much tin is needed per cycle? **at least 130 lb**

Teaching Tip In Example 3, remind students that $\frac{20}{17}$ is the multiplicative inverse of $\frac{17}{20}$.

Lesson 7-10 **319**

In-Class Example

For Example 4
Solve $-7a - 6 > 15$. Graph the solution on a number line.
$a < -3$

3 PRACTICE/APPLY

Check for Understanding

If students need additional practice or instruction after completing Exercises 1–9, one of these options may be helpful.
- Extra Practice, see p. 626
- Reteaching Activity
- *Study Guide Masters*, p. 59
- *Practice Masters*, p. 59

Additional Answer

1. The process for solving inequalities is the same. If the process includes dividing or multiplying by a negative number, the solution will have the opposite inequality sign.

Study Guide Masters, p. 59

320   Chapter 7

Example 4 Solve $-4x + 5 < 11$. Graph the solution on a number line.

Solve the related equation $-4x + 5 = 11$.

$$-4x + 5 = 11$$
$$-4x + 5 - 5 = 11 - 5 \quad \text{Subtract 5 from each side.}$$
$$-4x = 6$$
$$\left(-\frac{1}{4}\right)(-4x) = \left(-\frac{1}{4}\right)6 \quad \text{Multiply each side by the reciprocal of } -4, -\frac{1}{4}.$$
$$x = -\frac{3}{2} \text{ or } -1\frac{1}{2}$$

The solution includes numbers greater than $-1\frac{1}{2}$ or less than $-1\frac{1}{2}$. Test two numbers to see which is correct.

Try 2.
$-4(2) + 5 \stackrel{?}{<} 11$
$-8 + 5 \stackrel{?}{<} 11$
$-3 < 11$ true

Try -2.
$-4(-2) + 5 \stackrel{?}{<} 11$
$8 + 5 \stackrel{?}{<} 11$
$13 < 11$ false

Numbers greater than $-1\frac{1}{2}$ are included in the solution.
So, $x > -1\frac{1}{2}$. The number line shows the solution set.

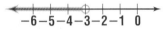

Notice that the inequality sign of the solution in Example 4 is the opposite of that in the original inequality. Also notice that in solving the related equation, you multiplied by a negative number. Whenever you multiply or divide by a negative number in solving the related equation, the solution will have the opposite inequality sign from that given in the original inequality.

CHECK FOR UNDERSTANDING

Communicating Mathematics

Read and study the lesson to answer each question.

1. *Compare* solving inequalities involving rational numbers with solving inequalities involving whole numbers.   **See margin.**
2. *Tell* whether each statement is *true* or *false*. If false, explain how to change the inequality to make it true.
    a. If $\frac{1}{2}x > -7$, then $x < -14$.   false; $x > -14$
    b. If $3x > -15$, then $x > -5$.   true

Guided Practice

3–8. See Answer Appendix for graphs.
8. $x > -2\frac{1}{4}$

Solve each inequality. Graph the solution on a number line. 5. $y \geq -16.5$

3. $-6x > 18$ $x < -3$
4. $\frac{3}{8}a < -\frac{5}{8}$ $a < -1\frac{2}{3}$
5. $3.3 + y \geq -13.2$
6. $n + 3\frac{7}{8} \leq 6\frac{1}{4}$ $n \leq 2\frac{3}{8}$
7. $\frac{2}{3}h - (-3) > 6$ $h > 4\frac{1}{2}$
8. $-2x - 1\frac{1}{2} < 3$

9. **Money Matters** Theo wants to earn at least $20 this week to go to the fair. His father said he would pay him $9 for mowing the lawn and $2.75 an hour to weed the flower bed. Theo mows the lawn. Use the inequality $9 + \$2.75h \geq \$20$ to find the number of hours he needs to weed to earn at least $20. **at least 4 hours**

320   Chapter 7 Algebra: Using Rational Numbers

Reteaching the Lesson

Activity Have students use red and black licorice sticks to model inequalities. Have black represent positive numbers and red represent negative numbers. Long sticks can be cut into equal portions to represent fractions. Have students pay particular attention to what happens when an inequality is multiplied or divided by a negative number.

Error Analysis
Watch for students who use the wrong inequality symbol when solving inequalities that involve dividing or multiplying by a negative.
Prevent by reminding students to check their solutions.

EXERCISES

Practice

10–27. See Answer Appendix for graphs.

11. $p < 3.17$
16. $z \geq -1\frac{3}{5}$
21. $h \leq \frac{5}{22}$
22. $c \geq -\frac{52}{75}$
23. $b \geq 11.4$
25. $g > -9\frac{1}{5}$
26. $q > 1.8875$
27. $n > -8.97$

Solve each inequality. Graph the solution on a number line.

10. $2w > -12$ $w > -6$
11. $p + 0.03 < 3.2$
12. $\frac{8}{5}r \geq 4$ $r \geq 2\frac{1}{2}$
13. $-12 \leq -\frac{1}{7}y$ $y \leq 84$
14. $3.6 > \frac{f}{0.9}$ $f < 3.24$
15. $-\frac{3}{5}k > \frac{2}{3}$ $k < -1\frac{1}{9}$
16. $z - \frac{2}{5} \geq -2$
17. $t + \frac{1}{4} < -4\frac{1}{8}$ $t < -4\frac{3}{8}$
18. $-6 > \frac{1}{3}m$ $m < -18$
19. $2\frac{1}{2}d > 5\frac{3}{4}$ $d > 2\frac{3}{10}$
20. $5 + 8x > 59$ $x > 6\frac{3}{4}$
21. $-11h + 15 \geq 12\frac{1}{2}$
22. $-\frac{5}{8}c + \frac{1}{6} \leq \frac{3}{5}$
23. $\frac{2}{1.5}b - 13 \geq 2.2$
24. $\frac{2x - 15}{3} > 3$ $x > 12$
25. $\frac{1}{2}g + 3 > -1\frac{3}{5}$
26. $-3.2q + 7 < 0.96$
27. $\frac{n}{2.3} - 1.3 > -5.2$

28. Find the solution of $4p - \frac{1}{5} > -7\frac{2}{5}$. $p > -1\frac{4}{5}$

29. Choose the inequality for which $h \leq -5\frac{1}{2}$ is a solution.   **b**

   a. $-4h + 7 > 15$
   b. $-\frac{4}{5}h - \frac{7}{5} \geq 3$
   c. $-\frac{4}{5}h - \frac{7}{5} \leq 3$

Applications and Problem Solving

30. **Food** Do you ever grab some cookies for an afternoon snack? Nutritionists recommend that you select cookies that contains 3 grams or less of fat per serving and no more than 8 to 12 grams of sugar. Which cookies meet these standards?   **Famous Amos Iced Gingersnaps and Keebler Elfin Delights**

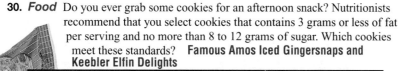

Cookie (cookies per serving)	Grams per Serving	
	Fat	Sugar
Famous Amos Iced Gingersnaps (6)	1	8
Keebler Elfin Delights 50% Reduced Fat Crème Sandwich Cookies (2)	2.5	10
Nabisco Reduced Fat Oreos (3)	3.5	13
Nabisco Cinnamon Teddy Grahams (25)	4	8

Source: *Vitality*

31. **Transportation** Car Safe offers a special rate for long term parking. It charges $5.50 for the first day and $4 for each additional day not to exceed $50 for a 15-day period. Use the inequality $\$5.50 + \$4d \leq \$50$ to find out when you will owe $50 for your stay.   **on days 12–15**

32. the numbers between $-1\frac{5}{6}$ and $2\frac{1}{3}$; sample answers: $-1, 0, 1\frac{5}{6}$

32. **Critical Thinking** The expression *the numbers between $-2\frac{1}{2}$ and $3\frac{1}{3}$* can be written as $-2\frac{1}{2} < n < 3\frac{1}{3}$. What does the algebraic sentence $-1\frac{1}{3} < x + \frac{1}{2} < 2\frac{5}{6}$ represent? Give examples of these numbers.

Mixed Review

33. **Algebra** Solve $\frac{2}{3}x + 4 = -6$. *(Lesson 7-9)*   **-15**

34. **Geometry** Find the area of a triangle whose base is 2.5 inches and whose height is 3.5 inches. *(Lesson 7-6)*   **4.375 in²**

35. **Probability** Two number cubes are rolled. Find the probability that the sum of the numbers showing is 7. *(Lesson 6-6)* $\frac{1}{6}$

36. **Test Practice**   Rick took his father out to dinner for his birthday. When the bill came, Rick's father reminded him that it is customary to tip the server 15% of the bill. If the bill was for $19.60, a good estimate for the tip is — *(Lesson 3-6)*   **D**

   A $6.   **B** $5.   **C** $4.   **D** $3.   **E** $2.

Lesson 7-10 Solving Inequalities   **321**

Extending the Lesson

Enrichment Masters, p. 59

Activity Write the inequality $-7a + \frac{1}{2} \leq b$ on the chalkboard. Ask students the following questions.
- If *a* is negative, can *b* be negative?   **no**
- If *b* is negative, can *a* be positive?   **yes**

Have students substitute a series of values for *a* to see the range of values for *b*.

Assignment Guide

Core: 11–31 odd, 32–36
Enriched: 10–28 even, 30–36

4 ASSESS

Closing Activity

Writing Have students work in pairs to write problems involving inequalities. Then use the problems for a class-wide review. Sample problem: *A furnace repair person charges $50 to come to your house and $25 for each half hour. Without including parts, what is the longest period of time this person can work without having the bill exceed $125?* $1\frac{1}{2}$ **hours**

Practice Masters, p. 59

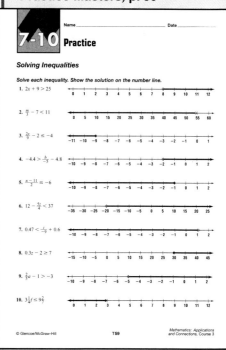

Study Guide and Assessment

Vocabulary

This section provides a listing of the new terms, properties, and phrases that were introduced in this chapter. Have students define each term and provide an example or two of it, if appropriate.

Understanding and Using the Vocabulary

These exercises check students' understanding of the terms by using a variety of verbal formats including matching, completion, and true/false.

Glossaries A complete glossary of terms appears on pages 702–710. The glossary also appears in Spanish on pages 711–721.

Additional Answer

9. Sample answer: Rewrite the mixed numbers as fractions. Multiply the numerators and multiply the denominators. Then simplify.

CHAPTER 7 Study Guide and Assessment

Vocabulary

After completing this chapter, you should be able to define each term, concept, or phrase and give an example or two of each.

Algebra
arithmetic sequence (p. 296)
common difference (p. 296)
common ratio (p. 297)
Fibonacci sequence (p. 300)
geometric sequence (p. 297)
sequence (p. 296)
term (p. 296)

Numbers and Operations
inverse property of multiplication (p. 290)
mixed number (p. 278)
multiplicative inverse (p. 290)
reciprocal (p. 290)

Geometry
altitude (p. 301)
base (pp. 301, 302)
center (pp. 307, 309)
circle (pp. 307, 309)
circumference (pp. 307, 309)
diameter (pp. 307, 309)
height (p. 301)
Pick's Theorem (p. 305)
radius (p. 309)
trapezoid (p. 302)

Problem Solving
deductive reasoning (p. 294)
inductive reasoning (p. 294)
look for a pattern (pp. 294–295)

Understanding and Using the Vocabulary

Choose the letter of the term or number that best matches each phrase.

1. the sum of a whole number and a fraction **g**
2. the LCM of 3 and 9 **c**
3. a sequence whose terms increase or decrease by a constant factor **h**
4. a sequence having the same difference between any two consecutive terms **i**
5. a quadrilateral with exactly one pair of parallel sides **j**
6. the distance across a circle through the center **e**
7. the distance around a circle **f**
8. the number you would multiply each side of the equation $\frac{2}{3}t = -\frac{4}{9}$ by to solve it **b**

a. $\frac{2}{3}$
b. $\frac{3}{2}$
c. 9
d. 18
e. diameter
f. circumference
g. mixed number
h. geometric sequence
i. arithmetic sequence
j. trapezoid

In Your Own Words

9. *Describe* how to multiply two mixed numbers. See margin.

322 Chapter 7 Algebra: Using Rational Numbers

 MindJogger Videoquizzes

MindJogger Videoquizzes provide an alternative review of concepts presented in this chapter. Students work in teams to answer questions, gaining points for correct answers. The questions are presented in three rounds.
Round 1 Concepts–5 questions
Round 2 Skills–4 questions
Round 3 Problem Solving–4 questions

Study Guide and Assessment Chapter 7

Objectives & Examples

Upon completing this chapter, you should be able to:

● add and subtract fractions with like denominators *(Lesson 7-1)*

Solve $b = \frac{1}{5} - \frac{3}{5}$.

$b = \frac{1-3}{5}$ *Subtract the numerators.*

$b = \frac{-2}{5}$ or $-\frac{2}{5}$

● add and subtract fractions with unlike denominators *(Lesson 7-2)*

Solve $h = \frac{2}{3} + \frac{3}{4}$.

$h = \frac{8}{12} + \frac{9}{12}$ *Rename each fraction.*

$h = \frac{17}{12}$ or $1\frac{5}{12}$ *Simplify.*

● multiply fractions *(Lesson 7-3)*

Solve $f = -\frac{5}{8} \cdot \frac{1}{3}$.

$f = \frac{-5 \cdot 1}{8 \cdot 3}$ *Multiply the numerators and the denominators.*

$f = -\frac{5}{24}$

● identify and use rational number properties *(Lesson 7-4)*

Commutative: $a + b = b + a$
$a \cdot b = b \cdot a$

Associative: $(a + b) + c = a + (b + c)$
$(a \cdot b) \cdot c = a \cdot (b \cdot c)$

Identity: $a + 0 = a$
$a \cdot 1 = a$

Multiplicative Inverse: $\frac{a}{b} \cdot \frac{b}{a} = 1$, where $a, b \neq 0$.

Review Exercises

Use these exercises to review and prepare for the chapter test.

Solve each equation. Write the solution in simplest form.

10. $\frac{2}{9} - \frac{5}{9} = n$ $-\frac{1}{3}$

11. $w = -\frac{1}{8} - \frac{3}{8}$ $-\frac{1}{2}$

12. $x = \frac{5}{11} + \frac{6}{11}$ 1

Solve each equation. Write the solution in simplest form.

13. $m = -\frac{3}{5} + \frac{2}{3}$ $\frac{1}{15}$

14. $z = 5 - 1\frac{2}{5}$ $3\frac{3}{5}$

15. $-4\frac{1}{2} + (-6\frac{2}{3}) = t$ $-11\frac{1}{6}$

Solve each equation. Write the solution in simplest form.

16. $p = \left(-\frac{1}{6}\right)\left(-\frac{4}{5}\right)$ $\frac{2}{15}$

17. $2\frac{2}{3}\left(-3\frac{1}{8}\right) = s$ $-8\frac{1}{3}$

18. $-\frac{3}{7} \cdot \frac{5}{7} = k$ $-\frac{15}{49}$

19. $g = \frac{8}{9} \cdot 3\frac{1}{4}$ $2\frac{8}{9}$

Name the multiplicative inverse of each rational number.

20. $\frac{3}{4}$ $\frac{4}{3}$

21. $-5\frac{1}{3}$ $-\frac{3}{16}$

Solve each equation using properties of rational numbers.

22. $p = 4 \cdot 6\frac{1}{4}$ 25

23. $\left(\frac{3}{5}\right)\left(-2\frac{1}{6}\right) = b$ $-1\frac{3}{10}$

Objectives & Examples

This section reviews the skills and concepts of the chapter and shows completely worked examples.

Review Exercises

These exercises provide practice for the corresponding objectives.

Assessment and Evaluation Masters, pp. 171–172

Chapter 7 Study Guide and Assessment **323**

Assessment and Evaluation

Six forms of Chapter 7 Test are available in the *Assessment and Evaluation Masters* as shown in the chart.

Chapter 7 Test, Form 1B, is shown at the right. Chapter 7 Test, Form 2B, is shown on the next page.

1A	Multiple Choice	Honors
1B	Multiple Choice	Average
1C	Multiple Choice	Basic
2A	Free Response	Honors
2B	Free Response	Average
2C	Free Response	Basic

Study Guide and Assessment **323**

Study Guide and Assessment

Chapter 7 Study Guide and Assessment

Objectives & Examples

● recognize and extend arithmetic and geometric sequences (Lesson 7-5)

State whether the sequence 2, 4, 8, 16, . . . is geometric. Then write the next three terms.

Since there is a common ratio, 2, the sequence is geometric. The next three terms are 32, 64, and 128.

● find the areas of triangles and trapezoids (Lesson 7-6)

Find the area of the triangle.

$A = \frac{1}{2}bh$

$A = \frac{1}{2}(3.4)(2.6)$

$A = 4.42$ cm^2

2.6 cm
3.4 cm

● find the circumference of circles (Lesson 7-7)

$C = \pi d$

$C \approx 3.14(6)$

$C \approx 18.84$ in^2

6 in.

● divide fractions (Lesson 7-8)

Solve $r = \frac{7}{12} \div \left(-\frac{2}{3}\right)$.

$r = \frac{7}{12} \cdot -\frac{3}{2}$ Use the multiplicative inverse.

$r = -\frac{7}{8}$ Simplify.

● solve equations involving rational numbers (Lesson 7-9)

Solve $\frac{3}{4}n = \frac{7}{8}$.

$\frac{4}{3} \cdot \frac{3}{4}n = \frac{4}{3} \cdot \frac{7}{8}$ Multiply each side by $\frac{4}{3}$.

$n = \frac{7}{6}$ or $1\frac{1}{6}$

Review Exercises

Identify each sequence as *arithmetic, geometric,* or *neither.* Then find the next three terms.

24. $-7, -4, -1, 2, \ldots$ A; 5, 8, 11
25. $32, 16, 8, 4, \ldots$ G; 2, 1, $\frac{1}{2}$
26. $6, 14, 21, 27, \ldots$ N; 32, 36, 39
27. $80, 68, 56, 44, \ldots$ A; 32, 20, 8

Find the area of each figure described below.

28. triangle: base, $2\frac{1}{4}$ in.; height, $3\frac{1}{2}$ in. $3\frac{15}{16}$ in^2
29. trapezoid: bases, 10.5 m and 7 m; height, 6.4 m   56 m^2

Find the circumference of each circle described to the nearest tenth.

30. The diameter is $4\frac{1}{3}$ feet.   13.6 ft
31. The radius is 2.6 meters.   16.3 m
32. The radius is 18 yards.   113.0 yd
33. The diameter is 6.5 inches.   20.4 in.

Solve each equation. Write the solution in simplest form.

34. $j = -\frac{7}{9} \div \left(-\frac{1}{3}\right)$ $2\frac{1}{3}$
35. $4\frac{2}{5} \div 2 = f$ $2\frac{1}{5}$
36. $2\frac{1}{7} \div \left(-3\frac{3}{5}\right) = d$ $-\frac{25}{42}$
37. $q = \frac{3}{10} \div 2\frac{2}{5}$ $\frac{1}{8}$

Solve each equation. Check your solution.

38. $\frac{x}{4} = -2.2$ -8.8 39. $-7.2 = \frac{r}{1.6}$ -11.52
40. $2b - \frac{6}{7} = 5\frac{2}{7}$ $3\frac{1}{14}$ 41. $t - (-0.8) = 4$ 3.2

324   Chapter 7 Algebra: Using Rational Numbers

Test and Review Software

You may use this software, a combination of an item generator and item bank, to create your own tests or worksheets. Types of items include free response, multiple choice, short answer, and open ended.

CD-ROM Program

The CD-ROM Program contains an Assessment Game whose questions review the concepts in this chapter.

Assessment and Evaluation Masters, pp. 177–178

Study Guide and Assessment — Chapter 7

Objectives & Examples

• solve inequalities involving rational numbers and graph their solutions. *(Lesson 7-10)*

Solve $-3.7x < 11.4$.

$-3.8x = 11.4$ *Solve the related equation.*
$\dfrac{-3.8x}{-3.8} = \dfrac{11.4}{-3.8}$ *Divide each side by -3.8.*
$x = -3$

Test numbers greater than -3 and less than -3 to find that the solution is $x > -3$.

Review Exercises

Solve each inequality. Graph the solution on a number line. 42–45. See Answer Appendix for graphs.

42. $-5 < \dfrac{x}{4}$ $-20 < x$

43. $-\dfrac{d}{7} \le 3$ $d \ge -21$

44. $36 + \dfrac{k}{5} \ge 51$ $k \ge 75$

45. $2.6v + 9 < 28.76$ $v < 7.6$

Applications & Problem Solving

This section provides additional practice in solving real-world problems that involve the skills of this chapter.

Applications & Problem Solving

46. **Farming** A farmer needs to fence a rectangular pasture. The length of the pasture is $20\tfrac{3}{4}$ yards, and the width is $20\tfrac{1}{3}$ yards. How many feet of fence does he need to buy? *(Lesson 7-2)* $246\tfrac{1}{2}$ ft

47. **Look for a Pattern** Bo rides in the May bike marathon. To train for the event, he rides 4 miles the first day, 8 miles the second day, and 12 miles the third day. If he continues this pattern, how many miles will Bo have ridden in all by the end of the fourth day? *(Lesson 7-5A)* **40 miles**

48. **Flooring** Ms. Herr owns a carpet and tile store. She needs to special order some tile for the floor area shown below. What is the area of the tile needed? *(Lesson 7-6)*
$82\tfrac{1}{6}$ ft^2

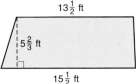

Alternative Assessment

The **Performance Task** provides students with a performance assessment opportunity to evaluate their work and understanding.

CHAPTER Project

Students should complete the final stages of their project and prepare a class demonstration of their results. A scoring guide for the project is available in the *Investigations and Projects Masters*, p. 181.

PORTFOLIO Students should add to their portfolios at this time.

Alternative Assessment

Performance Task

You are designing a patio that is shaped like a trapezoid. You want the area of the patio to be between 130 and 150 square feet. One of the bases of the trapezoid must be 8 feet long. Make a sketch of a possible trapezoid. Label your sketch with the dimensions. What is the area of your patio?

You want to place a round table with a circumference of about 18.84 feet in the middle of your patio. Will the table fit on your patio? Explain. **See students' work.**

A practice test for Chapter 7 is provided on page 653.

Completing the CHAPTER Project

Use the following checklist to make sure your display or slide show is complete.

☑ You have included a clear, well-organized description of the Fibonacci sequence and the Fibonacci spiral.

☑ Include pictures of examples of Fibonacci numbers in nature.

PORTFOLIO Select one of the equations you solved in this chapter and place it in your portfolio. Attach a note to it explaining why you selected it.

Assessment and Evaluation Masters, p. 181

Chapter 7 Study Guide and Assessment **325**

Performance Assessment

Additional performance assessment tasks for this chapter are included in the *Assessment and Evaluation Masters* on page 181. A scoring guide is also provided on page 193.

Standardized Test Practice

The Standardized Test Practice may be used to help students prepare for standardized tests. The test items are written in the same style as those in state proficiency tests and standardized tests like CAT, CTBS, ITBS, MAT, SAT, and Terra Nova. The test items cover skills and concepts covered up to this point in the text.

The pages can be used as an overnight assessment. After students have completed the pages, discuss how each problem can be solved, or provide copies of the solutions from the *Solutions Manual*.

Assessment and Evaluation Masters, p. 187

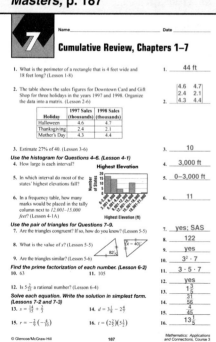

CHAPTERS 1–7 Standardized Test Practice
Assessing Knowledge & Skills

Section One: Multiple Choice

There are eleven multiple-choice questions in this section. Choose the best answer. If a correct answer is *not here,* mark the letter for Not Here.

1. A four-sided figure with exactly one pair of parallel sides is a — **D**
   A rectangle.
   B rhombus.
   C parallelogram.
   D trapezoid.

2. Express 42.8% as a decimal. **H**
   F 42.8
   G 4.28
   H 0.428
   J 0.0428

3. If $\frac{x}{5} + 4 = 32$, what is the value of x? **D**
   A 5.6
   B 7.2
   C 28
   D 140

4. What is the greatest common factor of 35 and 42? **G**
   F 6
   G 7
   H 14
   J 1,470

5. What is the probability that Tashima will draw a blue jelly bean from a jar containing 3 black jelly beans, 2 red jelly beans, and 5 yellow jelly beans? **A**
   A 0
   B $\frac{1}{10}$
   C $\frac{1}{3}$
   D 1

6. The histogram shows the quiz scores of the students in Mr. Guerrero's class.

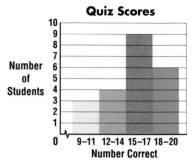

   How many students had scores less than 15? **G**
   F 4
   G 7
   H 16
   J 22

Please note that Questions 7–11 have five answer choices.

7. Carla averages 3 miles per hour walking. Jorge averages 7 miles per hour on his bicycle. They start at the same point and go in opposite directions on a straight road. Which equation could be used to find the number of hours h, when they would be 12 miles apart? **C**

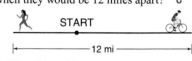

   A $7h - 3h = 15$
   B $(7h)(3h) = 12$
   C $3h + 7h = 12$
   D $15 + 7h = 12$
   E $12 + 3h = 7h$

326   Chapters 1–7 Standardized Test Practice

◀◀◀ **Instructional Resources**

Another cumulative review is shown at the left and is available in the *Assessment and Evaluation Masters*, p. 187.

Standardized Test Practice   Chapters 1–7

8. Luther works 20 hours each week. So far this week, he has worked 6.5 hours on Tuesday and 4.75 hours on Thursday. Which sentence could be used to find h, the number of hours remaining for Luther to work this week?   **H**
   F $6.5 + 4.75 = h + 20$
   G $6.5 + 4.75 = h \times 20$
   H $h = 20 - (6.5 + 4.75)$
   J $h - 6.5 + 4.75 = 20$
   K $h = 20 - 6.5 + 4.75$

9. Jaya works 21 hours a week and earns $122.85. How much money does Jaya earn per hour?   **B**
   A $6.50
   B $5.85
   C $5.05
   D $4.11
   E Not Here

10. Nora's classmates collected 4,656 aluminum cans last year. This year they collected 6,024 cans. How many more cans did they collect this year than last year?   **F**
    F 1,368
    G 1,468
    H 2,478
    J 2,632
    K Not Here

11. If an airplane travels 365 miles per hour, how many miles will it travel in 4 hours?   **D**
    A 1,240 mi
    B 1,260 mi
    C 1,440 mi
    D 1,460 mi
    E Not Here

Test-Taking Tip
Most standardized tests have a time limit, so you must use your time wisely. Some questions will be easier than others. If you cannot answer a question quickly, go on to another. If there is time remaining when you are finished, go back to the questions you skipped.

Section Two: Free Response

This section contains eight questions for which you will provide short answers. Write your answers on your paper.

12. Which point is in the first quadrant?   **point L**

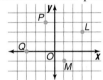

13. State the measure of the bases and the height of the trapezoid. Then find the area.

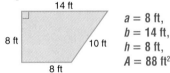

$a = 8$ ft,
$b = 14$ ft,
$h = 8$ ft,
$A = 88$ ft^2

14. What is the value of $35 \cdot (-2)$?   **−70**

15. Evaluate the expression ab if $a = -44$ and $b = 11$.   **−484**

16. What is the multiplicative inverse of $-8\frac{1}{4}$? $-\frac{4}{33}$

17. Express 3.49×10^{-3} in standard form.
    0.00349

18. State the greater of the numbers, 7.089 or 7.09.   **7.09**

19. Does an isosceles triangle have line symmetry? If so, draw the lines of reflection.   **Yes; see margin.**

Additional Answer
19.

CHAPTER 8
Applying Proportional Reasoning

Previewing the Chapter

Overview
This chapter explores proportion and percents. The percent proportion, the percent equation, large and small percents, percent of change, and simple interest are explored. Students identify corresponding parts of similar polygons and solve problems involving similar triangles and scale drawings. They learn to graph dilations on a coordinate plane. Students also learn to solve more difficult problems by first solving a simpler problem.

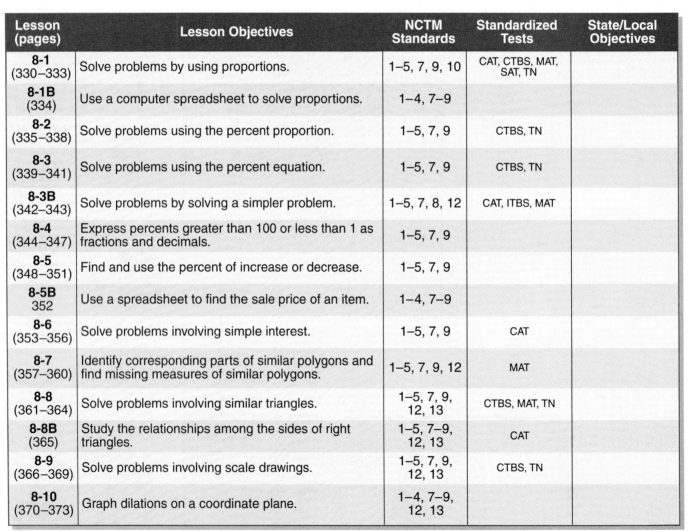

Lesson (pages)	Lesson Objectives	NCTM Standards	Standardized Tests	State/Local Objectives
8-1 (330–333)	Solve problems by using proportions.	1–5, 7, 9, 10	CAT, CTBS, MAT, SAT, TN	
8-1B (334)	Use a computer spreadsheet to solve proportions.	1–4, 7–9		
8-2 (335–338)	Solve problems using the percent proportion.	1–5, 7, 9	CTBS, TN	
8-3 (339–341)	Solve problems using the percent equation.	1–5, 7, 9	CTBS, TN	
8-3B (342–343)	Solve problems by solving a simpler problem.	1–5, 7, 8, 12	CAT, ITBS, MAT	
8-4 (344–347)	Express percents greater than 100 or less than 1 as fractions and decimals.	1–5, 7, 9		
8-5 (348–351)	Find and use the percent of increase or decrease.	1–5, 7, 9		
8-5B 352	Use a spreadsheet to find the sale price of an item.	1–4, 7–9		
8-6 (353–356)	Solve problems involving simple interest.	1–5, 7, 9	CAT	
8-7 (357–360)	Identify corresponding parts of similar polygons and find missing measures of similar polygons.	1–5, 7, 9, 12	MAT	
8-8 (361–364)	Solve problems involving similar triangles.	1–5, 7, 9, 12, 13	CTBS, MAT, TN	
8-8B (365)	Study the relationships among the sides of right triangles.	1–5, 7–9, 12, 13	CAT	
8-9 (366–369)	Solve problems involving scale drawings.	1–5, 7, 9, 12, 13	CTBS, TN	
8-10 (370–373)	Graph dilations on a coordinate plane.	1–4, 7–9, 12, 13		

CAT = California Achievement Tests, CTBS = Comprehensive Tests of Basic Skills, ITBS = Iowa Tests of Basic Skills, MAT = Metropolitan Achievement Tests, SAT = Stanford Achievement Tests, TN = Terra Nova

Organizing the Chapter

CD-ROM
All of the blackline masters in the Teacher's Classroom Resources are available on the **Electronic Teacher's Classroom Resources** CD-ROM.

LESSON PLANNING GUIDE

Lesson	Extra Practice (Student Edition)	Blackline Masters (page numbers)										
		Study Guide	Practice	Enrichment	Assessment & Evaluation	Classroom Games	Diversity	Hands-On Lab	School to Career	Science and Math Lab Manual	Technology	Transparencies A and B
8-1	p. 627	60	60	60						61–64		8-1
8-1B												
8-2	p. 627	61	61	61								8-2
8-3	p. 627	62	62	62	211							8-3
8-3B	p. 628											
8-4	p. 628	63	63	63			34			81–84		8-4
8-5	p. 628	64	64	64	210, 211	21–23						8-5
8-5B												
8-6	p. 629	65	65	65							67	8-6
8-7	p. 629	66	66	66								8-7
8-8	p. 629	67	67	67	212							8-8
8-8B								56				
8-9	p. 630	68	68	68				75				8-9
8-10	p. 630	69	69	69	212				34		68	8-10
Study Guide/ Assessment					197–209, 213–215							

Other Chapter Resources

Student Edition
Chapter Project, pp. 329, 338, 351, 377
Math in the Media, p. 364
Let the Games Begin, pp. 333, 369

Technology
CD-ROM Program
Interactive Mathematics Tools Software

Teacher's Classroom Resources

Applications
Family Letters and Activities, pp. 67–68
Investigations and Projects Masters, pp. 45–48

Meeting Individual Needs
Transition Booklet, pp. 5–6, 17–18, 29–34
Investigations for the Special Education Student, pp. 31–34

Teaching Aids
Answer Key Masters
Block Scheduling Booklet
Lesson Planning Guide
Solutions Manual

Professional Publications
Glencoe Mathematics Professional Series

Planning the Chapter

 MindJogger Videoquizzes provide a unique format for reviewing concepts presented in the chapter.

ASSESSMENT RESOURCES

Student Edition
Mixed Review, pp. 333, 338, 341, 347, 351, 356, 360, 364, 369, 373
Mid-Chapter Self Test, p. 356
Math Journal, pp. 337, 362
Study Guide and Assessment, pp. 374–377
Performance Task, p. 377
Portfolio Suggestion, p. 377
Standardized Test Practice, pp. 378–379
Chapter Test, p. 654

Assessment and Evaluation Masters
Multiple-Choice Tests (Forms 1A, 1B, 1C), pp. 197–202
Free-Response Tests (Forms 2A, 2B, 2C), pp. 203–208
Performance Assessment, p. 209
Mid-Chapter Test, p. 210
Quizzes A–D, pp. 211–212
Standardized Test Practice, pp. 213–214
Cumulative Review, p. 215

Teacher's Wraparound Edition
5-Minute Check, pp. 330, 335, 339, 344, 348, 353, 357, 361, 366, 370
Building Portfolios, p. 328
Math Journal, pp. 334, 352, 365
Closing Activity, pp. 333, 338, 341, 343, 347, 351, 356, 360, 364, 369, 373

Technology
Test and Review Software
MindJogger Videoquizzes
CD-ROM Program

MATERIALS AND MANIPULATIVES

Lesson 8-1
index cards

Lesson 8-1B
computer spreadsheet software

Lesson 8-3
calculator

Lesson 8-4
grid paper
markers

Lesson 8-5
grid paper
scissors*

Lesson 8-5B
computer spreadsheet software

Lesson 8-6
calculator

Lesson 8-7
overhead transparency
flashlight
ruler*
protractor*†

Lesson 8-8B
protractor*†
metric ruler*
calculator

Lesson 8-9
map with a distance scale

Lesson 8-10
grid paper†
straightedge*

*Glencoe Manipulative Kit   †Glencoe Overhead Manipulative Resources

PACING CHART

See pages T25–T27 for the Course Planning Calendar.

COURSE	DAY 1	DAY 2	DAY 3	DAY 4	DAY 5	DAY 6	DAY 7
Standard	Chapter Project	Lesson 8-1	Lesson 8-2	Lesson 8-3	Lesson 8-3B	Lesson 8-4	Lesson 8-5
Honors	Chapter Project, Lessons 8-1 & 8-1B		Lesson 8-2	Lesson 8-3	Lesson 8-3B	Lesson 8-4	Lessons 8-5 & 8-5B
Block	Chapter Project & Lesson 8-1	Lessons 8-2 & 8-3	Lessons 8-3B & 8-4	Lessons 8-5 & 8-6	Lessons 8-7 & 8-8	Lessons 8-9 & 8-10	Study Guide and Assessment, Chapter Test

The *Transition Booklet* (Skills 1, 7, 13–15) can be used to practice comparing and ordering decimals, mixed numbers, and improper fractions and to use the customary and metric systems of measurement.

Interactive Mathematics:
Activities and Investigations

is an activity-based program that may be used as an enhancement for chapters in *Mathematics: Applications and Connections.*

Unit 3, Activity Four
Use with Lesson 8-8.

Summary Students indirectly measure the height of a tall object using shadows and proportions. They repeat the procedure with another tall object and write a report on their findings.

Math Connection Students use the shadows cast by two objects and the height of one of the objects to estimate the height of the taller object. Similar right triangles and proportions can be used to find the unknown measure.

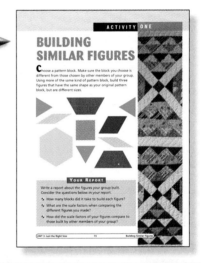

Unit 3, Activity One
Use with Lesson 8-9.

Summary Students work in groups as individuals and pairs to complete this activity. Using pattern blocks, students explore similar figures, scale factors, and the relationships among the scale factor, perimeter, and area of each figure.

Math Connection Students choose a pattern block. Next, they use a tessellation and more of the chosen pattern block to build a larger geometric figure of the same shape. Then, they use ratios to find the relationship between the scale factor and the perimeter and area of the figure.

DAY 8	DAY 9	DAY 10	DAY 11	DAY 12	DAY 13	DAY 14	DAY 15
Lesson 8-6	Lesson 8-7	Lessons 8-8 & 8-8B		Lesson 8-9	Lesson 8-10	Study Guide and Assessment	Chapter Test
Lesson 8-6	Lesson 8-7	Lessons 8-8 & 8-8B		Lesson 8-9	Lesson 8-10	Study Guide and Assessment	Chapter Test

Enhancing the Chapter

APPLICATIONS

Classroom Games, pp. 21–23
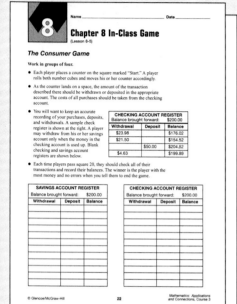

Diversity Masters, p. 34
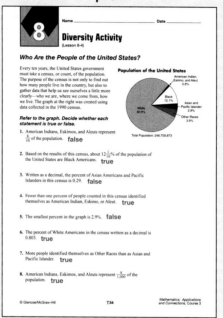

School to Career Masters, p. 34
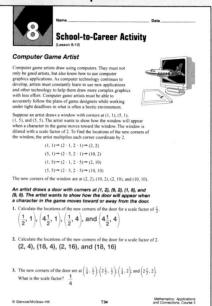

Family Letters and Activities, p. 67–68
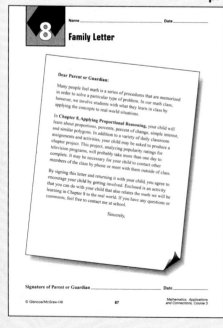

Science and Math Lab Manual, pp. 61–64, 81–84

MANIPULATIVES/MODELING

Hands-On Lab Masters,
p. 75

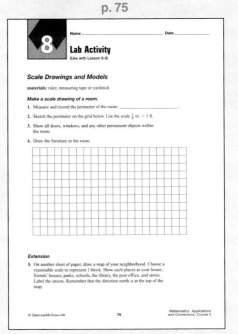

ASSESSMENT/EVALUATION

Assessment and Evaluation Masters,
p. 210–212

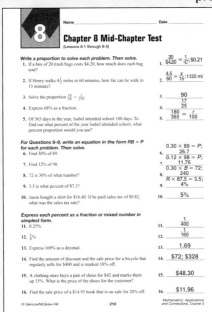

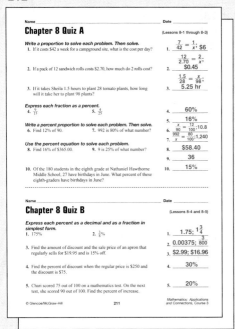

TECHNOLOGY/MULTIMEDIA

Technology Masters,
pp. 67–68

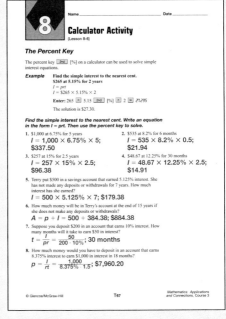

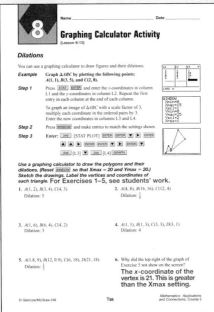

MEETING INDIVIDUAL NEEDS

Investigations for the Special Education Student, pp. 31–34

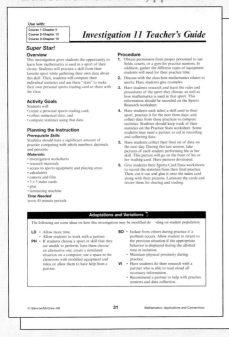

Chapter 8 **328f**

CHAPTER 8 NOTES

Theme: Television

Nielsen Media Research now measures more than just television viewership. In July 1996, Nielsen announced a new study called the Home Technology Report. This report measures how many U.S. homes own video-game systems, home computers, cellular phones, satellite dishes, and CD-ROMs and tracks subscription levels for on-line services. To more accurately reflect the U.S. population, researchers scientifically select by phone number the persons to be surveyed. Researchers first calculate the proportion of telephones in each geographic area and then select numbers from each area.

Question of the Day According to the first Home Technology Report, 8 percent of U.S. residents (12 years and older) live in households that access an on-line service. If there are about 212,500,000 people 12 and older in the U.S., how many people live in households with on-line service? **about 17 million** Express this as a ratio. **about 1:10**

Assess Prerequisite Skills

Ask students to read through the list of objectives presented in "What you'll learn in Chapter 8." You may wish to ask them what each of the objectives means or if they have experienced or used any of these math concepts before.

 Building Portfolios

Encourage students to revise their portfolios as they study this chapter. They may find some lessons more challenging than others and wish to include work that shows how they have improved their mathematical skills.

 Math and the Family

In the *Family Letters and Activities* booklet (pp. 67–68), you will find a letter to the parents explaining what students will study in Chapter 8. An activity appropriate for the whole family is also available.

CHAPTER 8 — Applying Proportional Reasoning

What you'll learn in Chapter 8

- to solve problems by using proportions,
- to solve problems involving percents,
- to solve problems by first solving a simpler problem,
- to find missing measures of similar polygons, and
- to graph dilations on a coordinate plane.

328 Chapter 8 Applying Proportional Reasoning

 CD-ROM Program

Activities for Chapter 8
- Chapter 8 Introduction
- Interactive Lessons 8-3, 8-5, 8-10
- Extended Activity 8-9
- Assessment Game
- Resource Lessons 8-1 through 8-10

CHAPTER Project

STAY TUNED!

In this project, you will write a newspaper article discussing the popularity of television programs. You will poll your classmates to determine how your peers rate several programs. You will also analyze the Nielsen ratings for one of these programs.

Getting Started

- Each person in the class should pick a television program to follow in the Nielsen ratings.
- As a class, make a list of all of the shows picked.
- Poll the class members to determine how many watch each program. Record the results of the poll in the form of a fraction. For example, if there are 25 students in your class and 6 of them watch *Home Improvement*, record $\frac{6}{25}$ for that program.
- Nielsen ratings are expressed in percents. For example, a rating of 10 indicates that 10% of the 97 million households with a television watched the program. Record the Nielsen ratings for your program for three consecutive weeks.

Technology Tips

- Use the **Internet** to search for the Nielsen ratings for your television program.
- Use a **spreadsheet** to organize the class ratings and the Nielsen ratings.
- Use a **word processor** to write your article.

interNET CONNECTION For up-to-date information on the Nielsen ratings, visit:
www.glencoe.com/sec/math/mac/mathnet

Working on the Project

You can use what you'll learn in Chapter 8 to change fractions to percents and to compare the ratings.

Page	Exercise
338	45
351	37
377	Alternative Assessment

interNET CONNECTION
Glencoe has made every effort to ensure that the website links for *Mathematics: Applications and Connections* at www.glencoe.com/sec/math/mac/mathnet are current and contain appropriate content. However, these website links are not under Glencoe's control.

Instructional Resources ▶▶▶
A recording sheet to help students organize their data for the Chapter Project is shown at the right and is available in the *Investigations and Projects Masters*, p. 48.

Chapter 8 **329**

CHAPTER Project NOTES

Objectives Students should
- gain an understanding of survey sampling.
- learn to apply proportions and percents to surveys.
- develop their ability to organize information verbally and to form generalizations from it.

Project Pointer You may suggest that students begin a *Project Folder* to keep their work as they complete each stage of the Chapter Project. The completed project may also be added to their portfolios.

Familiarize students with newspaper writing style. Because readers often skim newspapers, journalists pack the basic information in the beginning, or *lead*, of an article. Have students answer *who? what? when? where? why?* and *how?* first. Have them supply any details after the lead. Finally, have them write an interesting but short ending. Newspaper articles that are too long often get chopped from the ending.

Investigations and Projects Masters, p. 48

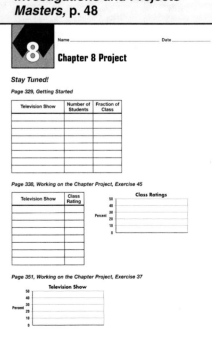

Chapter 8 Project **329**

8-1 Lesson Notes

Instructional Resources
- *Study Guide Masters*, p. 60
- *Practice Masters*, p. 60
- *Enrichment Masters*, p. 60
- Transparencies 8-1, A and B
- *Science and Math Lab Manual*, pp. 61–64

 CD-ROM Program
- Resource Lesson 8-1

Recommended Pacing	
Standard	Day 2 of 15
Honors	Days 1 & 2 of 15
Block	Day 1 of 7

1 FOCUS

 5-Minute Check
(Chapter 7)

Solve each equation. Write the solution in simplest form.

1. $q = \frac{1}{2} + \frac{5}{6}$ $1\frac{1}{3}$
2. $\left(-1\frac{1}{3}\right)\left(1\frac{1}{8}\right) = p$ $-1\frac{1}{2}$
3. State whether the sequence 18, 25, 32, 39, ... is *arithmetic, geometric,* or *neither.* Then write the next three terms. **arithmetic; 46, 53, 60**
4. The base of a triangle is 12 inches, and the height is 3 inches. Find the area. **18 in²**
5. Solve $-3.2h - 3.2 \leq 25.6$. $h \geq -9$ Show the solution on a number line.

 The 5-Minute Check is also available on **Transparency 8-1A** for this lesson.

Motivating the Lesson
Problem Solving Tell students that a local bus has a 3-hour route and makes 224 stops that last 15 seconds each. Ask them what the total time spent at stops on a 5-hour route. **93 min 20 s**

8-1 Using Proportions

What you'll learn
You'll learn to solve problems by using proportions.

When am I ever going to use this?
Knowing how to use proportions can help you find an actual length represented on a scale drawing.

 LOOK BACK
You can refer to Lesson 3-3 for information on proportions and cross products.

Are you a "channel-surfer" during commercials? A typical 30-minute TV program in the United States has about 8 minutes of commercials. At that rate, how many commercial minutes are shown during a 3-hour TV movie?

This problem can be solved using a proportion. A 3-hour TV movie is 3×60 or 180 minutes long. In the proportion below, c represents the number of commercial minutes during a 3-hour TV movie.

commercial minutes $\rightarrow$ $\dfrac{8}{30} = \dfrac{c}{180}$ $\leftarrow$ commercial minutes
program length $\rightarrow$ $\phantom{\dfrac{8}{30} = \dfrac{c}{180}}$ $\leftarrow$ program length

Cross products can be used to solve this proportion. *You will solve this problem in Exercise 5.*

Example 1
CONNECTION

Health To burn off the Calories from a 12-ounce can of regular cola, you would have to cycle 24 minutes at 13 miles per hour. How long would you have to cycle at the same speed to burn off the Calories from a 16-ounce bottle of cola?

Explore You must cycle for 24 minutes to burn off the Calories from a 12-ounce can of cola. You want to know how long you would need to cycle to burn off the Calories from a 16-ounce bottle of cola.

Study Hint
Mental Math Sometimes you can solve a proportion mentally by using equivalent fractions.
$\dfrac{3}{11} = \dfrac{12}{x}$
THINK: $3 \times 4 = 12$
$11 \times 4 = 44$
So, $x = 44$.

Plan Let m represent the number of minutes needed to burn the Calories from a 16-ounce bottle of cola. Write a proportion.

minutes of cycling $\rightarrow$ $\dfrac{24}{12} = \dfrac{m}{16}$ $\leftarrow$ minutes of cycling
ounces of cola $\rightarrow$ $\phantom{\dfrac{24}{12} = \dfrac{m}{16}}$ $\leftarrow$ ounces of cola

Solve Solve the proportion.

$\dfrac{24}{12} = \dfrac{m}{16}$

$24 \cdot 16 = 12 \cdot m$ *Find the cross products.*

$384 = 12m$

$\dfrac{384}{12} = \dfrac{12m}{12}$ *Divide each side by 12.*

$32 = m$

You will need to cycle 32 minutes to work off the Calories from a 16-ounce bottle of regular cola.

330 Chapter 8 Applying Proportional Reasoning

 Cross-Curriculum Cue

Inform the other teachers on your team that your students are studying proportions. Suggestions for curriculum integration are:

Geography: land to water
Earth Science: weather (rainy days to sunny days per month and per year)
Physical Education: long-distance running statistics

Examine Check your solution by substituting 32 for m in your original ratios. Simplify each ratio.

$$\overset{\div 12}{\frac{24}{12}} = \frac{2}{1} \qquad \overset{\div 16}{\frac{32}{16}} = \frac{2}{1}$$
$$\underset{\div 12}{} \qquad \underset{\div 16}{}$$

Since both ratios are equivalent to the same fraction, $\frac{2}{1}$, they form a proportion. So, 32 is correct.

Sometimes you need to round your answer so it makes sense.

Example 2

CONNECTION

Life Science About 1 out of every 5 people is left handed. If there are 28 students in a class, how many would you expect to be left handed?

Let s represent the number of left-handed students.

left handed → $\frac{1}{5} = \frac{s}{28}$ ← left handed
size of group → ← size of group

$1 \cdot 28 = 5 \cdot s$ *Find the cross products.*

$28 = 5s$

$\frac{28}{5} = \frac{5s}{5}$ *Divide each side by 5.*

$5.6 = s$

You would expect to find 5 or 6 left-handed students out of 28 students.

CHECK FOR UNDERSTANDING

Communicating Mathematics

Read and study the lesson to answer each question.

1. *Tell* which proportions from the list below could be used to solve the problem. Be prepared to defend your answer.
   If a 15-minute long-distance call costs $2.10, how much does a similar 12-minute call cost?

1a–c. See margin for reasons.

a. $\frac{15}{2.10} = \frac{12}{c}$ **yes**   b. $\frac{15}{2.10} = \frac{c}{12}$ **no**   c. $\frac{c}{12} = \frac{2.10}{15}$ **yes**

2. *Write a Problem* that could be solved by using a proportion. Then write a proportion that could be used to solve the problem. **See margin.**

3–4. See margin for sample proportions.

Guided Practice

Write a proportion to solve each problem. Then solve.

3. If 16 ounces equals a pound, then 56 ounces equals p pounds. $3\frac{1}{2}$ **lb**

4. If $\frac{1}{4}$ inch on a scale drawing represents 1 foot, then 3 inches on the drawing represent f feet. **12 ft**

Lesson 8-1 Using Proportions **331**

■ **Reteaching the Lesson** ■

Activity Use a newspaper or magazine to find information presented as a ratio. An example might be: "12 out of 50 people surveyed said they enjoyed the movie." Then ask students how many out of 200 would have enjoyed the movie. **48**

Additional Answers

2. Sample answer: If 5 peaches cost $2, how much do 15 peaches cost?
   $\frac{5}{2} = \frac{15}{c}$

3. $\frac{16}{1} = \frac{56}{p}$

4. $\frac{\frac{1}{4}}{1} = \frac{3}{f}$

---

2 TEACH

 Transparency 8-1B contains a teaching aid for this lesson.

Using Calculators Show that students can solve a proportion by cross-multiplying two of the known terms and then dividing by the remaining known term.

In-Class Examples

For Example 1
An examination of 150 elm trees found 18 suffering from Dutch elm disease. How many of 3,800 elms can be expected to have the disease? **456 elms**

For Example 2
In 26 times at bat, Baxter hit 3 home runs. At that rate, how many home runs would he be expected to hit in 475 times at bat? **about 55 home runs**

Teaching Tip In Example 2, it may be helpful to remind students of the rules for rounding. Remind them to examine their answers to see whether they are reasonable.

Teaching Tip Survey the class and possibly other classes to see if "1 out of 5 is left-handed" applies.

3 PRACTICE/APPLY

Check for Understanding
If students need additional practice or instruction after completing Exercises 1–7, one of these options may be helpful.
- Extra Practice, see p. 627
- Reteaching Activity
- *Transition Booklet,* pp. 29–34
- *Study Guide Masters,* p. 60
- *Practice Masters,* p. 60

Additional Answers

1a. Each ratio represents the time to the cost.

1b. One ratio represents the time to the cost, and the other represents the cost to the time.

1c. Each ratio represents the cost to the time.

Lesson 8-1 **331**

Assignment Guide

Core: 9–21 odd, 23–26
Enriched: 8–18 even, 20–26

Teaching Tip You can use Exercise 20 as a science/math lab activity. Have students record their heartbeat for 10 seconds and then multiply by 6. Find the class average.

Additional Answers

8. $\frac{1}{12} = \frac{8}{x}$
9. $\frac{100}{1} = \frac{170}{m}$
10. $\frac{12}{4.20} = \frac{1}{x}$
11. $\frac{12}{96} = \frac{4}{x}$
12. $\frac{2}{15} = \frac{35}{m}$
13. $\frac{35}{2.5} = \frac{420}{m}$

Study Guide Masters, p. 60

Solve each problem.

5. **Advertising** Refer to the beginning of the lesson. How many commercial minutes are shown during a 3-hour TV movie? **48 min**

6. **Photography** A photograph is 3 inches wide and 5 inches long. If the width of an enlargement of this photograph is 9 inches, find the length of the enlargement. **15 in.**

7. **Life Science** A microscope slide shows 35 red blood cells out of 60 blood cells. How many red blood cells would be expected in a blood sample of 75,000 blood cells? **43,750 red blood cells**

EXERCISES

Practice

8–13. See margin for sample proportions.

Write a proportion to solve each problem. Then solve. 9. 1.7 m

8. If 1 foot equals 12 inches, 8 feet equals x inches. **96 in.**
9. If 100 centimeters equals a meter, then 170 centimeters equals m meters.
10. If a roll of 12-exposure film costs $4.20 to develop, then each picture costs x cents to develop. **35¢**
11. If 12 eggs cost 96¢, then 4 eggs cost x cents. **32¢**
12. If Trevor can type 2 pages in 15 minutes, then he can type 35 pages in m minutes. **$262\frac{1}{2}$ min**
13. If 35 copies are made in 2.5 minutes, then 420 copies will take m minutes. **30 min**

Solve each problem. 14. $37\frac{1}{2}$ mi

14. If Maka can bicycle 25 miles in 2 hours, how far can she bicycle in 3 hours?
15. How many feet are in $2\frac{3}{4}$ miles? (*Hint:* 1 mile = 5,280 feet) **14,520 ft**
16. **Money Matters** If it costs $90 to feed a family of 3 for one week, how much will it cost to feed a family of 5 for a week? **$150**
17. If a car can travel 536 miles on 16 gallons of gasoline, how far can it travel on 10 gallons of gasoline? **335 mi**
18. **Fund-raising** The student council at Maple Springs Middle School decided to sell school sweatshirts as a fund-raiser. They surveyed 75 students. Of those students, 53 said they would buy a sweatshirt. If there are 1,023 students in the school, about how many students can you expect to buy a sweatshirt? **about 723 students**
19. **Earth Science** The ratio of a person's weight on the moon to his or her weight on Earth is 1 to 6.
    a. If a person weighs 126 pounds on Earth, how much would he or she weigh on the moon? **21 lb**
    b. Find your weight on the moon. **See students' work.**

Applications and Problem Solving

20. **Health** The average human heart beats 72 times in 60 seconds.
    a. How many times does it beat in 15 seconds? **18 beats**
    b. Take your pulse. Record the number of beats in 30 seconds.
    c. Use a proportion to determine how many times your heart beats in 45 seconds. **b–c. See students' work.**

332 Chapter 8 Applying Proportional Reasoning

21. **Technology** Each 10K of memory in a microcomputer can store 10,240 bytes of information. Find the number of bytes that can be stored in 128K memory. **131,072 bytes**

22. **Life Science** An area of 2,500 square feet of grass produces enough oxygen for a family of 4. **a. 3,125 ft²**
    a. Find the area of grass needed to supply a family of 5 with oxygen.
    b. Find the number of people that could be supplied with oxygen by a lawn that covers 3,800 square feet. **about 6 people**

23. Sample answer:
$\frac{15}{120} = \frac{25}{p}$,
$\frac{120}{15} = \frac{p}{25}$,
$\frac{15}{25} = \frac{120}{p}$,
$\frac{25}{15} = \frac{p}{120}$

23. **Critical Thinking** Write four different proportions that could be used to solve this problem.
    If 15 gallons of punch are needed for 120 people, how many people will 25 gallons serve?

Mixed Review

24. **Test Practice** What is the solution of $2r + 3 > 11$? *(Lesson 7-10)* **A**
    A $r > 4$ B $r < 3$ C $r > 7$ D $r < 7$

25. Express $2.\overline{44}$ as a mixed number in simplest form. *(Lesson 6-5)* $2\frac{4}{9}$

26. **Inventory** A restaurant serves 300 cups of coffee each day. If one pound of coffee is used to make 100 cups, how many pounds will the restaurant need for a week? *(Lesson 1-1)* **21 lb**

4 ASSESS

Closing Activity

Writing A motorist drove 110 miles in 2 hours. Have students write and solve a problem involving the same unit rate but a different time.

Let the Games Begin

What's Missing?

Get Ready This game is for two players. — 10 index cards

Get Set Write one of the following numbers on one side of each card.
2, 3, 4, 6, 9, 12, 15, 18, 24, 30

Go
- Shuffle the cards and arrange them facedown.
- One player chooses three cards and writes a proportion using the three numbers drawn.
- Both players solve for the missing number in the proportion.
- The first person to correctly solve the proportion earns one point. Both players check the solution using cross products.
- The players take turns picking the cards. The player with the most points after a given time period wins.

interNET CONNECTION Visit www.glencoe.com/sec/math/mac/mathnet for more games.

Lesson 8-1 Using Proportions 333

Extending the Lesson

Enrichment Masters, p. 60

Let the Games Begin

As students play the game, ask them to look for patterns in arranging the numbers so that they might gain an advantage over their opponents.

Practice Masters, p. 60

TECHNOLOGY LAB 8-1B Notes

GET READY

Objective Students use a computer spreadsheet to solve proportions.

Technology Resources
Suggested spreadsheet software
- Lotus 1·2·3
- ClarisWorks
- Microsoft Excel

MANAGEMENT TIPS

Recommended Time
30 minutes

Getting Started Review the process for entering formulas into a spreadsheet. Make sure students understand how cells are used and named.

When they have finished entering their data, ask students to estimate the answers. After the spreadsheet program generates the values for each cell, ask students whether their estimations were close to the actual value. Have students work through the exercises.

Using Calculators If spreadsheet software is unavailable, students can use calculators to determine their answers. Have students work in groups to make tables matching the spreadsheet example. Then have them use calculators to solve each equation.

ASSESS

After students answer Exercises 1–3, ask them how they could use mental math to determine their answers when a computer is unavailable.

TECHNOLOGY LAB SPREADSHEETS

8-1B Proportions
A Follow-Up of Lesson 8-1

- computer
- spreadsheet software

Chili Dip
1 c. cottage cheese $\frac{1}{4}$ c. chili sauce
$\frac{1}{4}$ t. onion powder $\frac{1}{4}$ c. skim milk
3 T. grated Parmesan cheese
Mix all ingredients in blender until smooth. Chill. Serves 20.

The Elegant Eatery caters food for various functions. Their chili dip recipe is shown at the right. A spreadsheet can be used to determine how much of each ingredient to use for various sized groups. The proportion below will give the number of batches needed.

$$\frac{\text{number of people expected}}{\text{number of servings per batch}} = \frac{\text{number of batches needed}}{1 \text{ batch}}$$

Since B1 is the number of people and the number of servings per batch is 20, we can rewrite the proportion $\frac{B1}{20} = \frac{x}{1}$. The formula in cell B2 uses B1 ÷ 20 to find the number of batches that need to be made.

TRY THIS

Work with a partner.

Use the spreadsheet to determine the recipe for a function with 120 people by making the following substitution.

B1 = 120

	CHILI DIP RECIPE		
	A	B	C
1	People To Serve	B1	
2	Batches Needed	= B1/20	
3	INGREDIENT	NUMBER	
4	cottage cheese	= B2	cups
5	chili sauce	= B2/4	cups
6	onion powder	= B2/4	t.
7	skim milk	= B2/4	cups
8	parmesan cheese	= B2 * 3	T.

ON YOUR OWN

1. Use the spreadsheet to find the amount of each ingredient needed to make enough chili dip for 180 people. **See margin.**
2. How could you change the spreadsheet if one batch of chili dip served 12 people? **Change the formula in cell B2 to B1/12.**
3. How could you change the spreadsheet if the recipe included $\frac{1}{8}$ teaspoon of Tabasco sauce? **See margin.**

334 Chapter 8 Applying Proportional Reasoning

Additional Answers

1. 9 cups cottage cheese, $2\frac{1}{4}$ cups chili sauce, $2\frac{1}{4}$ teaspoons onion powder, $2\frac{1}{4}$ cups skim milk, 27 tablespoons parmesan cheese

3.

Math Journal Have students write a paragraph about how a spreadsheet could be used to determine gas mileage or sales tax.

8-2 The Percent Proportion

What you'll learn
You'll learn to solve problems using the percent proportion.

When am I ever going to use this?
Knowing how to use the percent proportion can help you determine the amount of tax you must pay.

Word Wise
percentage
base
rate
percent proportion

About 29% of the United States is forest or woodland. The symbol % is a common symbol for the word percent. A *percent* is a ratio that compares a number to 100. Percent also means *hundredths*, or *per hundred*. So 29% of the United States means that 29 square miles out of every 100 square miles or $\frac{29}{100}$ of the land is forest.

The models below represent 50%, 75%, and 10%.

$\frac{1}{2} = \frac{50}{100} = 50\%$ $\frac{3}{4} = \frac{75}{100} = 75\%$ $\frac{1}{10} = \frac{10}{100} = 10\%$

You can use a proportion to find a percent.

Example 1

What percent of the circle is shaded?

Let *x* represent the percent of the circle that is shaded.

parts shaded → $\frac{5}{8} = \frac{x}{100}$ ← percent shaded
parts in whole →   ← percent in whole

$5 \cdot 100 = 8 \cdot x$ *Find the cross products.*

$500 = 8x$

$\frac{500}{8} = \frac{8x}{8}$ *Divide each side by 8.*

$62.5 = x$

So, 62.5% of the circle is shaded.

Did you know?

The tallest tree in the world is a coastal redwood in Redwood National Park, California. In 1995, it was measured to be $365\frac{1}{2}$ feet tall, which is 60 feet taller than the Statue of Liberty.

In Example 1, the number of parts shaded, 5, is called the **percentage (P)**. The total number of parts, 8, is called the **base (B)**. The ratio $\frac{62.5}{100}$ is called the **rate (r)**.

$\frac{5}{8} = \frac{62.5}{100}$ → $\frac{\text{Percentage}}{\text{Base}} = \text{Rate}$

If *r* represents the percent, the proportion can be written as $\frac{P}{B} = \frac{r}{100}$. This proportion is called the **percent proportion**.

Lesson 8-2 The Percent Proportion 335

Investigations for the Special Education Student

This blackline master booklet helps you plan for the needs of your special education students by providing long-term projects along with teacher notes. Investigation 11, *Super Star!*, may be used with this chapter.

8-2 Lesson Notes

Instructional Resources
- *Study Guide Masters*, p. 61
- *Practice Masters*, p. 61
- *Enrichment Masters*, p. 61
- Transparencies 8-2, A and B
- CD-ROM Program
  - Resource Lesson 8-2

Recommended Pacing
Standard	Day 3 of 15
Honors	Day 3 of 15
Block	Day 2 of 7

1 FOCUS

5-Minute Check
(Lesson 8-1)

Solve each problem.

1. If Cynthia can walk 3 miles in one hour, how far can she walk at the same rate in $2\frac{1}{2}$ hours? **7.5 miles**

2. There are eight pints in a gallon. How many pints are there in $6\frac{3}{4}$ gallons? **54 pints**

3. There are 5,280 feet in a mile. How many feet are in $5\frac{1}{4}$ miles? **27,720 ft**

The 5-Minute Check is also available on **Transparency 8-2A** for this lesson.

Motivating the Lesson

Hands-On Activity Assign students to groups of three or four. Have each student make a 10 × 10 square grid to use as a percent model. Ask students to shade a portion of their grids and to quiz each other on what percent each shaded area represents.

Lesson 8-2 335

2 TEACH

 Transparency 8-2B contains a teaching aid for this lesson.

Reading Mathematics Point out to students that in *percentage* and *proportion,* the prefixes *per-* and *pro-* both mean "for each": *Percentage* means "for each hundred." *Proportion* comes from a Latin word meaning "for each share." In addition, *rate* comes from a Latin word meaning "proportion" and refers to the measurement of one thing in regard to another. For example, a mileage rate is often expressed as miles per hour.

In-Class Examples

For Example 1
What percent of the circle is shaded?
75%

For Example 2
Refer to the graph in Example 2 of the Student Edition to find the total number of teams in the NFL, the NBA, and major league baseball. **87**

For Example 3
What number is 45% of 198?
89.1

Teaching Tip Before discussing Example 3, ask students to estimate an answer without using the percent proportion to solve the problem. Encourage them to make estimation a regular part of their problem-solving routine when working with percents.

Additional Answer
2. 2 out of 5 parts are shaded.
$\frac{2}{5} = \frac{40}{100} = 40\%$

Example 2 APPLICATION

Sports Major league sports use playoffs to determine the league champion. Use the information in the graph to find the number of teams in the National Basketball Association (NBA).

Making the Playoffs
Sport (number of playoff teams)

NBA (16) — 55%
NFL (12) — 40%
Major league baseball (8) — 29%

Percent of Teams

Source: USA TODAY research, 1996

$\frac{P}{B} = \frac{r}{100}$ *Percent proportion*

$\frac{16}{B} = \frac{55}{100}$ $P = 16, r = 55$

$16 \cdot 100 = B \cdot 55$ *Find the cross products.*

$1{,}600 = 55B$

$\frac{1{,}600}{55} = \frac{55B}{55}$ *Divide each side by 55.*

1,600 ÷ 55 = 29.09090909

$29.09 \approx B$

Since there cannot be part of a team, there are 29 teams in the National Basketball Association.

The percent proportion can be used to find a percentage.

Example 3 What number is 35% of 450?

$\frac{P}{B} = \frac{r}{100}$ *Percent proportion*

$\frac{P}{450} = \frac{35}{100}$ $B = 450, r = 35$

$P \cdot 100 = 450 \cdot 35$ *Find the cross products.*

$100P = 15{,}750$

$\frac{100P}{100} = \frac{15{,}750}{100}$ *Divide each side by 100.*

$P = 157.5$

157.5 is 35% of 450.

CHECK FOR UNDERSTANDING

Communicating Mathematics
1. Percent means per one hundred and *r* is the number out of one hundred.

Read and study the lesson to answer each question.

1. *Explain* why the value of *r* in $\frac{P}{B} = \frac{r}{100}$ represents a percent.
2. *Tell* why the following statement is true. **See margin.**

40% of the figure is shaded.

336 Chapter 8 Applying Proportional Reasoning

Reteaching the Lesson

Activity Have students use decimal models to express several fractions whose denominators are factors of 100. Have them write the percent represented. Then have them write the percent proportion that represents how the percent can be determined.

Error Analysis
Watch for students who confuse the base and the percentage when working with the percent proportion.
Prevent by reminding students that base is equal to the total number of pieces represented.

3. The percent proportion can be used to find the percentage, the base, or the percent. *Write* an example of each type of problem. Include the percent proportion needed to solve each problem. **See margin.**

Guided Practice

4. Write a percent to represent the shaded portion of the figure. **13%**

Express each fraction as a percent.

5. $\frac{1}{20}$ **5%**   6. $\frac{3}{10}$ **30%**   7. $\frac{19}{50}$ **38%**   8. $\frac{7}{40}$ **17.5%**

Write a percent proportion to solve each problem. Then solve. Round to the nearest tenth.

9. $\frac{P}{60} = \frac{15}{100}$; 9
10. $\frac{70}{280} = \frac{r}{100}$; 25%
11. $\frac{15}{45} = \frac{r}{100}$; 33.3%
12. $\frac{2}{B} = \frac{16}{100}$; 12.5

9. Find 15% of 60.
10. 70 is what percent of 280?
11. 15 is what percent of 45?
12. 2 is 16% of what number?

13. **Taxes** Mrs. Reeder stayed at a hotel during a mathematics teachers' conference. Her bill listed the room tax as $19.46 and the room rate as $139.00. Write the tax as a percent of the room rate. **14%**

EXERCISES

Practice

Write a percent to represent the shaded portion of each figure.

14.  **25%**   15. 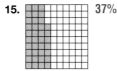 **37%**   16.   **95%**

Express each fraction as a percent. 24. 82.5% 25. 18.75% 28. 27.5%

17. $\frac{1}{5}$ **20%**   18. $\frac{7}{10}$ **70%**   19. $\frac{13}{25}$ **52%**   20. $\frac{9}{20}$ **45%**   21. $\frac{41}{50}$ **82%**   22. $\frac{7}{8}$ **87.5%**

23. $\frac{1}{2}$ **50%**   24. $\frac{33}{40}$ 25. $\frac{3}{16}$ 26. $\frac{24}{25}$ **96%**   27. $\frac{7}{28}$ **25%**   28. $\frac{11}{40}$

29. $\frac{P}{66} = \frac{25}{100}$; 16.5
30. $\frac{P}{80} = \frac{13}{100}$; 10.4
31. $\frac{16}{48} = \frac{r}{100}$; 33.3%
32. $\frac{22}{110} = \frac{r}{100}$; 20%
33. $\frac{5}{B} = \frac{26}{100}$; 19.2
34. $\frac{8}{B} = \frac{13}{100}$; 61.5
35. $\frac{7}{20} = \frac{r}{100}$; 35%
36. $\frac{P}{70} = \frac{27}{100}$; 18.9
37. $\frac{60}{B} = \frac{15}{100}$; 400
38. $\frac{P}{65} = \frac{47}{100}$; 30.6
39. $\frac{45}{B} = \frac{35}{100}$; 128.6
40. $\frac{9}{15} = \frac{r}{100}$; 60%

Write a percent proportion to solve each problem. Then solve. Round to the nearest tenth.

29. Find 25% of 66.
30. Find 13% of 80.
31. 16 is what percent of 48?
32. 22 is what percent of 110?
33. 5 is 26% of what number?
34. 8 is 13% of what number?
35. 7 is what percent of 20?
36. What is 27% of 70?
37. 60 is 15% of what number?
38. Find 47% of 65.
39. 45 is 35% of what number?
40. 9 is what percent of 15?
41. **Geometry** What percent of the area of the square is shaded? **10%**

Lesson 8-2 The Percent Proportion **337**

3 PRACTICE/APPLY

Check for Understanding

If students need additional practice or instruction after completing Exercises 1–13, one of these options may be helpful.
* Extra Practice, see p. 627
* Reteaching Activity, see p. 336
* *Study Guide Masters*, p. 61
* *Practice Masters*, p. 61

Assignment Guide

Core: 15–43 odd, 46–52
Enriched: 14–40 even, 42–44, 46–52

Additional Answer

3. Sample answer: Percentage: Find 20% of 40.
$\frac{P}{40} = \frac{20}{100}$
Base: 3 is 40% of what number? $\frac{3}{B} = \frac{40}{100}$
Percent: 5 is what percent of 25? $\frac{5}{25} = \frac{r}{100}$

Study Guide Masters, p. 61

Chapter Project

Exercise 45 asks students to advance to the next stage of work on the Chapter Project. Students may elect to use graphing software to make their graphs. You may want to have students include their graphs in their class portfolios.

4 ASSESS

Closing Activity

Modeling Have students use visual representation of percents to illustrate examples of percents. They may use square grids and circles, or they may devise other shapes for their models.

Practice Masters, p. 61

Applications and Problem Solving

42. *Jewelry* Jewelers measure gold in karats. Pure gold is 24 karat gold.
    a. In the United States, most jewelry is 18 karat gold. What percent of the 18-karat jewelry is gold? **75%**
    b. In the Netherlands, most jewelry is 22 karat gold. What percent of the 22-karat jewelry is gold? **about 91.7%**

43. *Royalties* Consuela Reyes writes rap songs and receives a royalty of 8% of the sales of her songs. Last month she received a royalty check for $3,896.00. How much money was spent on her songs? **$48,700**

44. *Education* Jamal scored a 92% on his science test. If there were 25 questions on the test, how many questions did Jamal answer correctly? **23 questions**

45. *Working on the CHAPTER Project* Refer to the class poll that you took on television programs on page 329. **a–b. See students' work.**
    a. Change each fraction to a percent, and identify which television program is the most popular in your class.
    b. Make a graph to display your data.

46. Hurt; 7 out of 13 is about 53.8% which is less than 56%.

46. *Critical Thinking* Hao made 56% of his free throws in the first half of the basketball season. If he makes 7 shots out of the next 13 attempts, will it help or hurt his average? Explain.

Mixed Review

47. *Baking* A recipe calls for 4 cups of flour for 64 cookies. How much flour is needed for 96 cookies? *(Lesson 8-1)* **6 cups**

48. Solve $g = \frac{5}{6} \div \frac{4}{3}$. *(Lesson 7-8)* $\frac{5}{8}$

49. Write $\frac{54}{81}$ in simplest form. *(Lesson 6-4)* $\frac{2}{3}$

50. **Test Practice** What is the value of x in the figure? *(Lesson 5-1)* **B**
    A 14°
    B 42°
    C 98°
    D 140°

51. Express *240 shrimp in 6 pounds* as a unit rate. *(Lesson 3-1)* **40 shrimp/1 pound**

52. Find $|-20|$. *(Lesson 2-1)* **20**

338 Chapter 8 Applying Proportional Reasoning

Extending the Lesson

Enrichment Masters, p. 61

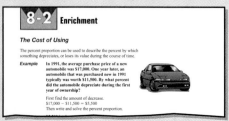

Activity Ask students to make a chart that tallies how they spend their time in school each day. They should track how much time they spend in homeroom, walking to class, at lunch, and in each class. Ask students to use their charts to determine the percent of time spent for each activity.

8-3 Integration: Algebra
The Percent Equation

What you'll learn
You'll learn to solve problems using the percent equation.

When am I ever going to use this?
Knowing how to use the percent equation can help you understand grades given in percents.

The M&M/Mars candy company gave consumers the opportunity to vote for a new color of M&M to replace the tan-colored ones. Blue won with over 54% of the vote. The table shows the new mix of colors. In a 16-ounce bag of M&M's, there are about 525 pieces of candy. How many would you expect to be red?

You have learned to find 20% of 525 by using the percent proportion. Another way to solve this problem is to express 20% as a decimal and then multiply.

$0.20 \times 525 = 105$ *Estimate: $\frac{1}{5}$ of 500 is 100.*

In a 16-ounce bag of M&M's, there should be about 105 red ones. In this example, 20% is the rate, 525 is the base, and 105 is the percentage.

M&M's Plain Chocolate Candies

Color	Percent
brown	30%
yellow	20%
red	20%
orange	10%
green	10%
blue	10%

Source: M & M/Mars

You can use an equation to describe how the quantities are related.

Percent Equation	Words:	The rate (R) times the base (B) is equal to the percentage (P).
	Symbols:	$RB = P$

In the percent equation, the rate is usually expressed as a decimal.

Examples

1 Find 4% of $625.   *Estimate: 1% of 600 is 6; $4 \times 6 = 24$*

$RB = P$ *Percent equation*
$0.04 \times 625 = P$ *Replace R with 4% or 0.04 and B with 625.*
$25 = P$

So, 4% of $625 is $25.   *Compare to the estimate.*

Study Hint
Technology To find 4% of $625 with your calculator, press 4 [2nd] [%] [×] 625 [=] 25.

2 12 is 60% of what number?   *Estimate: $60\% \approx \frac{1}{2}$; 12 is $\frac{1}{2}$ of 24.*

$RB = P$ *Percent equation*
$0.6B = 12$ *Replace R with 60% or 0.6 and P with 12.*
$\frac{0.6B}{0.6} = \frac{12}{0.6}$ *Divide each side by 0.6.*

12 [÷] 0.6 [=] 20

So, 12 is 60% of 20.   *Check your answer by finding 60% of 20.*

Lesson 8-3 Integration: Algebra The Percent Equation **339**

8-3 Lesson Notes

Instructional Resources
- *Study Guide Masters*, p. 62
- *Practice Masters*, p. 62
- *Enrichment Masters*, p. 62
- Transparencies 8-3, A and B
- *Assessment and Evaluation Masters*, p. 72
- CD-ROM Program
  - Resource Lesson 8-3
  - Interactive Lesson 8-3

Recommended Pacing	
Standard	Day 4 of 15
Honors	Day 4 of 15
Block	Day 2 of 7

1 FOCUS

5-Minute Check
(Lesson 8-2)

Express each fraction as a percent.

1. $\frac{19}{100}$ **19%**

2. $\frac{2}{5}$ **40%**

Write a percent proportion to solve each problem. Then solve.

3. What is 20% of 25?
   $\frac{P}{25} = \frac{20}{100}$; **5**

4. 18 is 45% of what number?
   $\frac{18}{B} = \frac{45}{100}$; **40**

5. 9 is what percent of 12?
   $\frac{9}{12} = \frac{r}{100}$; **75%**

The 5-Minute Check is also available on **Transparency 8-3A** for this lesson.

Motivating the Lesson

Problem Solving The cost of traveling to the Citrus Bowl with the band next January is $500 per person. Imagine that you earned $450 mowing lawns last summer, which you invest this January at 5.5% interest. How much interest would you earn on this investment in 12 months? **$24.75** How much more money would you need to save to go on this trip? **$25.25**

Lesson 8-3 **339**

2 TEACH

Transparency 8-3B contains a teaching aid for this lesson.

Thinking Algebraically Reassure students that the equation $RB = P$ is merely a different form of the percent proportion, not a whole new approach to percents.

In-Class Examples

For Example 1
Find 28% of 370. **103.6**

For Example 2
39 is 75% of what number? **52**

For Example 3
A picture frame sells for $29. The state sales tax is $1.74. What is the sales tax rate? **6%**

---

3 PRACTICE/APPLY

Check for Understanding
If students need additional practice or instruction after completing Exercises 1–9, one of these options may be helpful.
- Extra Practice, see p. 627
- Reteaching Activity
- *Study Guide Masters*, p. 62
- *Practice Masters*, p. 62

Study Guide Masters, p. 62

Example APPLICATION

3 Money Matters Betty bought supplies at the hardware store that cost $28.40. If she paid sales tax of $1.42, what was the sales tax rate?

Explore   You know the amount of the purchases and the sales tax. You need to find the rate.

Plan   To find the rate, use the percent equation. Let B represent the base, $28.40, and let P represent the percentage, $1.42.

Solve
$$RB = P \quad \text{Percent equation}$$
$$R \cdot 28.40 = 1.42 \quad \text{Replace B with 28.40 and P with 1.42.}$$
$$\frac{R \cdot 28.40}{28.40} = \frac{1.42}{28.40} \quad \text{Divide each side by 28.40.}$$

1.42 ÷ 28.40 = 0.05

So, the sales tax rate is 0.05 or 5%.

Examine   5% of $28.40 is 1.42. The answer checks.

CHECK FOR UNDERSTANDING

Communicating Mathematics

Read and study the lesson to answer each question.

1. *Write* an equation to find the percent of questions answered correctly if you correctly answered 28 out of 34 questions on a history test. $R \cdot 34 = 28$

2. *Tell* what number you would use for R in the equation $RB = P$ if the rate is 35%.   **0.35**

Guided Practice

3. $0.47 \cdot 52 = P$; 24.44
4. $0.56 \cdot 80 = P$; 44.8
5. $R \cdot 90 = 36$; 40%
6. $R \cdot 51 = 17$; $33\frac{1}{3}$%

Write an equation in the form $RB = P$ for each problem. Then solve.

3. What number is 47% of 52?
4. What number is 56% of 80?
5. What percent of 90 is 36?
6. $17 is what percent of $51?
7. $48 is 30% of what amount? $0.30 \cdot B = 48$; $160
8. Fifty is 10% of what number? $0.10 \cdot B = 50$; 500
9. **Money Matters** The price of a home video game system is $198. If the sales tax on the system is $8.91, what is the sales tax rate?   **4.5%**

EXERCISES

Practice

10–27. See Answer Appendix for equations.

11. $88\frac{1}{3}$%

Write an equation in the form $RB = P$ for each problem. Then solve.

10. Find 16.5% of 60.   **9.9**
11. Fifty-five is what percent of 66?
12. 75 is 50% of what number?   **150**
13. 15% of what number is 30?   **200**
14. 18 is what percent of 60?   **30%**
15. Find 24% of 72.   **17.28**
16. Find $33\frac{1}{3}$% of 420.   **140**
17. 16 is $66\frac{2}{3}$% of what number?   **24**
18. 45 is what percent of 150?   **30%**
19. 6 is what percent of 300?   **2%**

340   Chapter 8 Applying Proportional Reasoning

Reteaching the Lesson

Activity Separate students into groups of three. Ask one student to write a percent problem in sentence form, a second student to write the percent proportion for the problem, and a third to write the percent equation. Have students compare the proportion and the equation. Repeat several times.

Error Analysis
Watch for students who fail to write percents as decimals in the percent equation, yielding an answer too large.
Prevent by having students estimate answers before using the equation.

20. Find 5% of 3,200.   **160**
21. What percent of 80 is 25?   **31.25%**
22. Find 28% of $231.   **$64.68**
23. $6 is what percent of $50?   **12%**
24. 15 is 75% of what number?   **20**
25. What percent of 80 is 70?   **87.5%**
26. $1.47 is 7% of what amount?   **$21**
27. $54 is 8% of what amount?   **$675**

28. There are 35 students. Eight of them are wearing green. What percent are wearing green?   **about 22.9%**

29. Twenty percent of what amount is $5,000?   **$25,000**

Applications and Problem Solving

30. *Life Science* Earth's age is estimated at 4.6 billion years. If life began 3.5 billion years ago, for what percent of Earth's existence has life been present?

30. about 76.1%

31. *Music* In a six-month period, Americans ages 12 to 54 buy an average of 11 CDs, tapes, or records. The graph below shows the reasons they give for making a purchase. Suppose you survey 25 people. Predict how many of those surveyed bought music because they saw the video.   **about 11 people**

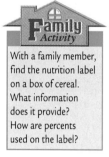

With a family member, find the nutrition label on a box of cereal. What information does it provide? How are percents used on the label?

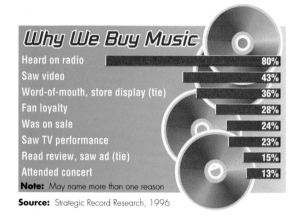

32. No; sample answer: 100 + 0.1(100) = 110, 110 − 0.1(110) = 99

Mixed Review

32. *Critical Thinking* Todd adds 10% of a number to the number. Then he subtracts 10% of the total. Is the result equal to his original number? Explain.

33. *Recycling* Americans recycled 35.4% of the aluminum products discarded in 1993. If Americans discarded 3,000,000 tons of aluminum products, how many tons were recycled? *(Lesson 8-2)*   **1,062,000 tons**

34. Express $\frac{7}{8}$ as a percent. *(Lesson 3-4)*   **87.5%**

35. **Test Practice**   Order the integers 7, 12, −61, 23, −22, −6, and 0 from greatest to least. *(Lesson 2-2)*   **D**

  A −61, 23, −22, 12, 7, −6, 0
  B 0, −6, 7, 12, −22, 23, −61
  C −61, −22, −6, 0, 7, 12, 23
  D 23, 12, 7, 0, −6, −22, −61

36. *Algebra* Solve $q + 6.8 = 15.2$. Check your solution. *(Lesson 1-4)*   **8.4**

Lesson 8-3 Integration: Algebra The Percent Equation   **341**

Extending the Lesson

Enrichment Masters, p. 62

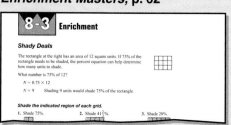

Activity Have students survey the class to determine the birth month of each student. Then ask each student to calculate the percent of students born in each month. As a class, create a table on the chalkboard to display the results.

Assignment Guide

Core: 11–31 odd, 32–36
Enriched: 10–28 even, 30–36

Have students note the grams of sugar, the grams of carbohydrates, and the Calories in one serving of several cereals. Ask them to calculate the percent of carbohydrates that is sugar in each cereal. As a class, create a table that relates the percent of carbohydrates that is sugar and the Calories per serving.

4 ASSESS

Closing Activity

Speaking Ask students to compare and contrast the percent equation and the percent proportion as a means of solving percent problems.

Chapter 8, Quiz A (Lessons 8-1 through 8-3) is available in the *Assessment and Evaluation Masters*, p. 211.

Practice Masters, p. 62

8-3B Thinking Lab Notes

Objective Students solve problems by solving a simpler problem.

Recommended Pacing	
Standard	Day 5 of 15
Honors	Day 5 of 15
Block	Day 3 of 7

1 FOCUS

Getting Started Have students act out the situation presented at the beginning of the lesson. Then have them work in pairs to answer Exercises 1–3.

2 TEACH

Teaching Tip As students work through the opening problem, ask them to think about how this method may be used to solve other real-life problems or situations that they have encountered.

In-Class Example

Emilio has invited 120 people to his grandparents' anniversary party. He needs to tell the restaurant how many beef and chicken dinners to prepare. The ten responses he's received indicate that 4 people would like beef and 6 people would like chicken. Based on this, how many of each dinner should Emilio order? **48 beef, 72 chicken**

Additional Answer

1. Both 100 and 300 are divisible by 20, but they are not divisible by 18 or 22. Therefore, the computations are easier if 20 is used instead of 18 or 22. Sample answer: 25

PROBLEM SOLVING

8-3B Solve a Simpler Problem

A Follow-Up of Lesson 8-3

Shaniqua and Michael are on the holiday social committee. They need to order cups of ice cream for the 300 students who will be attending the holiday dance. Let's listen in!

> It would take too much time to ask everyone what flavor of ice cream he or she wants. How will we know how many of each flavor to order?
> — *Michael*

> I took this survey of 20 of our classmates. We could use the results of the survey to determine how many of each flavor to order.
> — *Shaniqua*

> Since 45% prefer chocolate chip ice cream, let's write the percent equation $RB = P$ and solve for P.

> In this case, R is 45% or 0.45, and B is 300. Since $0.45 \times 300 = 135$, we should order about 135 cups of chocolate chip ice cream.

Flavor	Vanilla	Chocolate	Strawberry	Chocolate Chip	Peanut Butter
Number	1	5	3	9	2
Percent	5%	25%	15%	45%	10%

THINK ABOUT IT

Work with a partner. 1–3. See margin.

1. **Explain** why Michael chose to survey 20 students instead of 18 or 22 students. What other number might have been a good number to survey?

2. **Describe** how Shaniqua and Michael could use proportions to decide how many cups of chocolate chip ice cream to order.

3. **Apply** the **solve a simpler problem** strategy to determine how many cups of each flavor of ice cream to order.

342 Chapter 8 Applying Proportional Reasoning

■ Reteaching the Lesson ■

Activity Ask students to review the fraction-percent equivalencies chart in Lesson 3-5. Have students work together to write problems using percents that can be solved by writing a simpler problem that uses a fraction.

Additional Answers

2. Since 9 out of 20 students want chocolate chip ice cream, they could use the proportion $\frac{9}{20} = \frac{c}{300}$.

3. vanilla, 15 cups; chocolate, 75 cups; strawberry, 45 cups; chocolate chip, 135 cups; peanut butter, 30 cups

ON YOUR OWN

4. The second step of the 4-step plan for problem solving asks you to *plan*. *Explain* why this step is important when using the strategy of solving a simpler problem.   **See margin.**

5. *Write a Problem* that can be solved by solving a simpler problem.   **Sample answer: Predict who will win a presidential election. Instead of surveying all voters, you would survey a smaller group that is representative of all voters.**

6. *Refer* to the beginning of Lesson 8-3. How many of each color of M&M's would you expect in a 16-ounce bag of the candy? **about 158 brown, 105 yellow, 105 red, 53 orange, 53 green, 53 blue**

MIXED PROBLEM SOLVING

Solve. Use any strategy.

Strategies
Look for a pattern.
Solve a simpler problem.
Act it out.
Guess and check.
Draw a diagram.
Make a chart.
Work backward.

7. *Life Science* A paramecium, which is 0.23 millimeter long, is pictured in a science book. If the paramecium in the picture is 6 millimeters long, about how many times has the paramecium been magnified for the book? **about 26 times**

8. *Geography* Minnesota is known for its lakes. In fact, 5.75% of its area is covered with water. If Minnesota has 4,854 square miles of inland water, find the total area of Minnesota. **about 84,417 mi²**

9. *Laundry* Two clothespins are needed to hang one towel on a clothesline. One clothespin can be used on a corner of one towel and a corner of the towel next to it. What is the least number of clothespins you need to hang 8 towels? **9 clothespins**

10. *Geometry* The perimeter of a rectangle is 22 meters. Its area is 24 square meters. What are the dimensions of the rectangle? **8 m by 3 m**

11. *Health* A frozen lean dinner entree contains 200 Calories. About 25% of the Calories are from fat. How many Calories are from fat? **50 Calories**

12. *Statistics* The graph shows three of the major concerns of adults in the United States today.

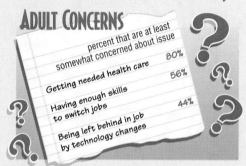

Source: KRC Research & Consulting for *U.S News*

a. If there are about 142,146,000 people between 20 and 59 years of age in the United States, how many people in this age group have each concern? **See margin.**

b. What do you think are the major concerns of teenagers today? How do these concerns compare with those of adults? **See students' work.**

13. **Test Practice** If the probability of a certain event is 100%, which statement is true?   **B**

A The probability is more than 1.
B The probability is equal to 1.
C The probability is less than 1.
D The probability is equal to 0.

Lesson 8-3B THINKING   **343**

3 PRACTICE/APPLY

Check for Understanding
Use the results from Exercise 3 to determine whether students understand how to solve a difficult problem by first solving a simpler problem.

Extra Practice If students need additional practice in problem solving, extra practice is available on the following pages.
• Solve a Simpler Problem, see p. 628
• Mixed Problem Solving, see pp. 645–646

Assignment Guide

All: 4–13

4 ASSESS

Closing Activity
Writing Show students a photo of a large crowd of people. Ask them to write a paragraph telling how they would determine the number of people in the photo.

Additional Answers
4. **Sample answer: You must plan to solve a simpler problem that will help you to find the answer to the more complex problem.**

12a. **health care, 113,716,800; switching jobs, 79,601,760; technology, 62,544,240**

Extending the Lesson

Activity Students can work in groups to write difficult problems that other students can solve by solving a simpler problem first. For example, one group could prepare an opinion poll and another group survey the class and tally the results. Other groups could research how many students are in their grade or school. They can use the results of their survey to determine the probable outcomes of polling their whole grade or school.

8-4 Lesson Notes

Instructional Resources
- *Study Guide Masters*, p. 63
- *Practice Masters*, p. 63
- *Enrichment Masters*, p. 63
- Transparencies 8-4, A and B
- *Diversity Masters*, p. 34
- *Hands-On Lab Masters*, p. 10
- *Science and Math Lab Manual*, pp. 81–84

CD-ROM Program
- Resource Lesson 8-4

Recommended Pacing	
Standard	Day 6 of 15
Honors	Day 6 of 15
Block	Day 3 of 7

1 FOCUS

5-Minute Check
(Lesson 8-3)

Write an equation in the form $RB = P$ for each problem. Then solve.

1. Find 18% of 72.
   $0.18 \times 72 = P$; 12.96
2. $36 is 25% of what amount?
   $0.25B = 36$; $144
3. What number is 38% of 154?
   $0.38 \times 154 = P$; 58.52
4. Find 40% of 860.
   $0.40 \times 860 = P$; 344
5. Jenny bought a baseball glove for $26.00. If she paid $2.08 in sales tax, what was the sales tax rate?
   $R \cdot 26.00 = 2.08$; 8%

 The 5-Minute Check is also available on **Transparency 8-4A** for this lesson.

Motivating the Lesson

Communication A coach said that in order for the team to win, each player would have to give a 110% effort. Ask students to explain what this statement means.

344 Chapter 8

8-4 Large and Small Percents

What you'll learn
You'll learn to express percents greater than 100 or less than 1 as fractions and decimals.

When am I ever going to use this?
Knowing how to express percents greater than 100 can help you understand the meaning of a sales increase of 110%.

Farmers from the United States produce 50% of the world's soybeans, 40% of the world's corn, 25% of the world's beef, and 15% of the world's cotton. However, farmers who live in the United States represent only 0.3% of the total number of farmers in the world. What does 0.3% mean? *You will answer this question in Exercise 1.*

You can use grid paper to investigate percents less than 1 such as 0.3% and percents greater than 100 such as 125%.

Study Hint
Reading Math 0.3% is read as *three tenths percent*. 125% is read as *one hundred twenty-five percent*. $\frac{1}{2}$% is read as *one-half percent*.

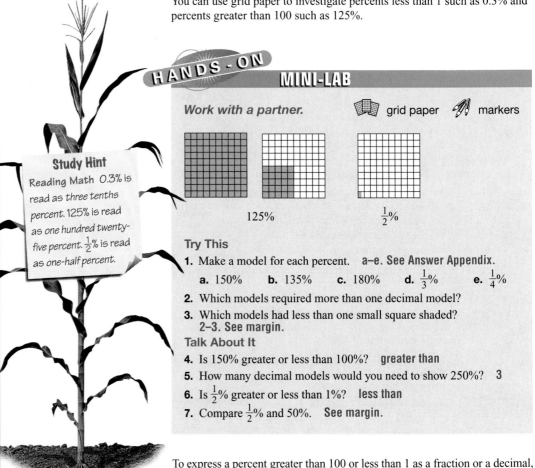

HANDS-ON MINI-LAB

Work with a partner.   grid paper   markers

125% $\frac{1}{2}$%

Try This
1. Make a model for each percent.   a–e. See Answer Appendix.
   a. 150%   b. 135%   c. 180%   d. $\frac{1}{3}$%   e. $\frac{1}{4}$%
2. Which models required more than one decimal model?
3. Which models had less than one small square shaded?
   2–3. See margin.

Talk About It
4. Is 150% greater or less than 100%?   **greater than**
5. How many decimal models would you need to show 250%?   **3**
6. Is $\frac{1}{2}$% greater or less than 1%?   **less than**
7. Compare $\frac{1}{2}$% and 50%.   **See margin.**

To express a percent greater than 100 or less than 1 as a fraction or a decimal, you can use the same procedure you used for percents from 1 to 100.

344   Chapter 8 Applying Proportional Reasoning

Additional Answers for the Mini-Lab
2. 125%, 150%, 135%, 180%
3. $\frac{1}{2}$%, $\frac{1}{3}$%, $\frac{1}{4}$%
7. $\frac{1}{2}$% means $\frac{1}{2}$ out of 100 and 50% means 50 out of 100.

Examples

Express each percent as a fraction or mixed number in simplest form.

1 0.8%

$0.8\% = \frac{0.8}{100}$ *Multiply the numerator and denominator by 10.*

$= \frac{8}{1,000}$ or $\frac{1}{125}$

LOOK BACK
You can refer to Lesson 7-8 for information on dividing fractions.

2 145%

$145\% = \frac{145}{100}$ *Change the improper fraction to a mixed number.*

$= 1\frac{45}{100}$ or $1\frac{9}{20}$

3 $\frac{1}{5}$%

$\frac{1}{5}\% = \frac{\frac{1}{5}}{100}$

$= \frac{1}{5} \div 100$ *The fraction bar indicates division.*

$= \frac{1}{5} \times \frac{1}{100}$ *To divide by 100, multiply by its multiplicative inverse, $\frac{1}{100}$.*

$= \frac{1}{500}$

Express each percent as a decimal.

4 137%

$137\% = \frac{137}{100}$

$= 1.37$

5 0.7%

$0.7\% = \frac{0.7}{100}$

$= \frac{7}{1,000}$ or 0.007

APPLICATION

6 Business The five states with the greatest increases in female-owned companies from 1987 to 1996 are shown in the graph. Order these states according to their percent increase starting with the state with the greatest increase.

Increases in Female-Owned Companies
Florida +106.3%
Nevada +130.0%
Georgia +112.4%
Idaho +104.1%
New Mexico +108.0%
Source: National Foundation for Women Business Owners

130.0% > 112.4% > 108.0% > 106.3% > 104.1%
Nevada   Georgia   New Mexico   Florida   Idaho

In order, the states are Nevada, Georgia, New Mexico, Florida, and Idaho.

Lesson 8-4 Large and Small Percents   **345**

---

2 TEACH

Transparency 8-4B contains a teaching aid for this lesson.

Using the Mini-Lab Remind students that one 10 × 10 decimal model completely shaded represents 100%. If a percent is greater than 100%, an additional 10 × 10 decimal model must be used to represent it. When students have completed the Mini-Lab, ask them the following questions: How many squares would you shade to represent 140%? 260%? 370%? **140; 260; 370**

In-Class Examples

For Example 1
Express 0.6% as a fraction. $\frac{3}{500}$

For Example 2
Express 138% as a mixed number. $1\frac{19}{50}$

For Example 3
Express $\frac{1}{4}$% as a fraction. $\frac{1}{400}$

For Example 4
Express 158% as a decimal. **1.58**

For Example 5
Express 0.9% as a decimal. **0.009**

For Example 6
Refer to the graph in Example 6 of the Student Edition. Determine how adding 25% to the percent of female-owned companies in Idaho would change the order of the states. **Idaho would be second.**

Teaching Tip In Example 4, remind students that if the percent exceeds 100, the decimal will be greater than 1.

3 PRACTICE/APPLY

Check for Understanding

If students need additional practice or instruction after completing Exercises 1–11, one of these options may be helpful.
- Extra Practice, see p. 628
- Reteaching Activity
- *Transition Booklet*, pp. 5–6
- *Study Guide Masters*, p. 63
- *Practice Masters*, p. 63

Assignment Guide
Core: 13–35 odd, 37–42
Enriched: 12–34 even, 35–42

Additional Answers

3a.

3b.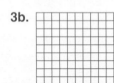

Study Guide Masters, p. 63

CHECK FOR UNDERSTANDING

Communicating Mathematics

1. 3 out of every 1000 farmers in the world live in the U.S.

HANDS-ON MATH

Guided Practice

Read and study the lesson to answer each question.

1. **Tell** what is meant by the following statement.
   Farmers who live in the United States represent only 0.3% of the total number of farmers in the world.
2. **Compare** 125% and 1 using >, <, or =.   **125% > 1**
3. **Make a model** for each percent.   **a–b. See margin.**
   a. 110%   b. 0.25%

Express each percent as a fraction or mixed number in simplest form.

4. 0.04% $\frac{1}{2,500}$ 5. 175% $1\frac{3}{4}$ 6. $\frac{1}{4}$% $\frac{1}{400}$

Express each percent as a decimal.

7. 155% **1.55**   8. 0.25% **0.0025**   9. 123.5% **1.235**

10. Order 25%, $\frac{1}{4}$%, 125%, and 1 from least to greatest. $\frac{1}{4}$%, 25%, 1, 125%

11. **Taxes** The percent of tax returns audited by the Internal Revenue Service varies from year to year. The percent of returns audited for five years are given in the table.

Year	1990	1991	1992	1993	1994
Percent of Audits	1.17%	1.06%	0.92%	1.08%	2.21%

a. Write each percent as a decimal.
b. Which year had the greatest percent of audits?   **1994**
c. Which year had the least percent of audits?   **1992**

11a.
1.17% = 0.0117;
1.06% = 0.0106;
0.92% = 0.0092;
1.08% = 0.0108;
2.21% = 0.0221

EXERCISES

Practice

Express each percent as a fraction or mixed number in simplest form.

15. $\frac{3}{5,000}$
16. $1\frac{73}{200}$
19. $134\frac{47}{100}$

12. 0.5% $\frac{1}{200}$ 13. 115% $1\frac{3}{20}$ 14. $\frac{1}{8}$% $\frac{1}{800}$ 15. 0.06%   16. 136.5%

17. 245% $2\frac{9}{20}$ 18. $\frac{3}{4}$% $\frac{3}{400}$ 19. 13,447%   20. $\frac{4}{25}$% $\frac{1}{625}$ 21. $33\frac{1}{3}$% $\frac{1}{3}$

Express each percent as a decimal.   **33. 104%, 1, 1.04%, 0.4%, 0.04%**

23. 0.0003
24. 2.25
25. 0.1025
26. 0.00009
27. 0.00079
28. 0.003
29. 0.004
30. 23.552
31. 1.104

22. 118% **1.18**   23. 0.03%   24. 225%   25. 10.25%   26. 0.009%

27. 0.079%   28. $\frac{3}{10}$%   29. $\frac{2}{5}$%   30. 2,355.2%   31. $110\frac{2}{5}$%

32. Order $\frac{1}{5}$%, 20%, 200%, and 1 from least to greatest. $\frac{1}{5}$%, 20%, 1, 200%

33. Write 1.04%, 0.4%, 104%, 0.04%, and 1 in order from greatest to least.

34. Which of the following numbers is the greatest? the least?   **7; $\frac{3}{8}$%**
    0.8, 67%, 7, $\frac{3}{8}$%

346 Chapter 8 Applying Proportional Reasoning

Reteaching the Lesson

Activity Prepare a deck of cards in which half of the cards have fractional percents and the other half have the decimal equivalents. Each player receives 6 cards, and the game is played like rummy. In turn, players draw from the remaining cards or pick the card discarded by the previous player, and then discard. The winner is the first person to make three matching pairs.

Error Analysis
Watch for students who express fractional percents as decimals by simply changing the fraction to a decimal.
Prevent by emphasizing that the first step is to change the fraction to a decimal and then to move the decimal point two places to the left.

Applications and Problem Solving

35. Movies The average cost of a ticket at a movie theater in the United States increased about 181% from 1970 to 1995. Write this percent as a decimal. **1.81**

36. Education The graph shows the percent of high school students studying various languages.

36a. See margin.

a. Write each percent as a fraction in simplest form.

b. Which language do the greatest percent of students study? **Spanish**

36c. Russian

c. Which language do the least percent of students study?

Percent of High School Students Taking Languages

French	9.8%
German	2.7%
Italian	0.4%
Japanese	0.23%
Latin	1.5%
Russian	0.15%
Spanish	23.5%

Source: USA TODAY

37. Critical Thinking The average cost of four years of college is expected to double in about 20 years. Explain whether this means that the cost is expected to be 200% of the current cost. $200\% = \frac{200}{100} = \frac{2}{1} = 2$; therefore, 200% means to double.

Mixed Review

38. Sports Eleven of the 48 members of the football team are on the field. What percent of the team members are playing? *(Lesson 8-3)* **about 22.9%**

39. Geometry Find the circumference of the circle. *(Lesson 7-7)* **about 51.5 cm**

8.2 cm

40. Find the GCF of 36, 108, and 180. *(Lesson 6-3)* **36**

41. Test Practice Two angles of a triangle have the same measure, and the third angle measures 68°. Let m represent the measure in degrees of each of the congruent angles. Which equation could be used to find m? *(Lesson 5-2)* **C**

A $m + 68 = 180$
B $m + 68 = 90$
C $2m + 68 = 180$
D $2m - 68 = 180$
E $2m + 68 = 360$

42. Solve $u = -8(10)(-12)$. *(Lesson 2-7)* **960**

Lesson 8-4 Large and Small Percents **347**

Extending the Lesson

Enrichment Masters, p. 63

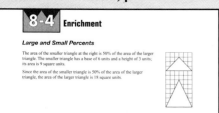

Activity Have students work together to make a list of situations in which a small percent increase might be more significant than a large percent increase. Example: An increase of 200% for a group of 2 is only 4 more people. An increase of 0.2% for a group of 1 million is 2,000 people.

4 ASSESS

Closing Activity

Writing Have students work in pairs to write sensible and non-sensible statements involving percents greater than 100%. For example, "William ate 115% of the pizza" is not sensible, but "ticket prices went up 150%" is sensible.

Additional Answer

36a. $9.8\% = \frac{49}{500}$;

$2.7\% = \frac{27}{1,000}$;

$0.4\% = \frac{1}{250}$;

$0.23\% = \frac{23}{10,000}$;

$0.15\% = \frac{3}{2,000}$;

$1.5\% = \frac{3}{200}$;

$23.5\% = \frac{47}{200}$

Practice Masters, p. 63

8-5 Lesson Notes

Instructional Resources
- *Study Guide Masters*, p. 64
- *Practice Masters*, p. 64
- *Enrichment Masters*, p. 64
- Transparencies 8-5, A and B
- *Assessment and Evaluation Masters*, p. 211
- *Classroom Games*, pp. 21–23
- *Hands-On Lab Masters*, p. 10
- CD-ROM Program
  - Resource Lesson 8-5
  - Interactive Lesson 8-5

Recommended Pacing	
Standard	Day 7 of 15
Honors	Day 7 of 15
Block	Day 4 of 7

1 FOCUS

5-Minute Check
(Lesson 8-4)

Express each percent as a fraction or mixed number in simplest form.

1. 0.4% $\frac{1}{250}$
2. 168% $1\frac{17}{25}$
3. $\frac{3}{8}$% $\frac{3}{800}$

Express each percent as a decimal.

4. 243% 2.43
5. 0.08% 0.0008

The 5-Minute Check is also available on **Transparency 8-5A** for this lesson.

Motivating the Lesson

Hands-On Activity Display a set of ten pencils on a desk where students can see them. Then ask a volunteer to remove some of the pencils from the set. Ask the class to express this decrease as a percent. You may also ask students to add pencils to the set and to express the increase as a percent.

8-5 Percent of Change

What you'll learn
You'll learn to find and use the percent of increase or decrease.

When am I ever going to use this?
Knowing how to use percent of decrease can help you determine discount prices.

Word Wise
percent of change
percent of increase
percent of decrease
discount
markup
selling price

You can express an increase or a decrease in terms of a percent.

HANDS-ON MINI-LAB

Work with a partner.   grid paper   scissors

Try This 1–2. See margin.

1. Draw a rectangle 10 units long and 1 unit wide. Then draw another rectangle 7 units long and 1 unit wide.

2. Cut out the second rectangle and place it over the first rectangle. Align one end of the rectangles. Shade the squares that are not covered.

3. Write a percent that represents the shaded squares in the first rectangle using the first rectangle as the base for the percent. **30%**

Talk About It 5. 20%; increase

4. In Step 3, you found the percent of change in the area from the first rectangle to the second rectangle. Does this percent of change represent an increase or a decrease? **decrease**

5. Cut out a third rectangle that is 12 units long and 1 unit wide. Find the percent of change in the area from the first to the third rectangle. Does this change represent an increase or a decrease?

When an increase or decrease is expressed as a percent, the percent is called the **percent of change**.

Examples

Find each percent of change. Round to the nearest percent.

❶ original: 25
new: 28

First, find the amount of change.

$28 - 25 = 3$

Then, find the percent using the original number, 25, as the base.

$\frac{3}{25} = \frac{r}{100}$

$3 \cdot 100 = 25 \cdot r$

$300 = 25r$

$\frac{300}{25} = \frac{25r}{25}$

$12 = r$

The percent of change is 12%.

❷ original: 45
new: 30

First, find the amount of change.

$45 - 30 = 15$

Then, find the percent using the original number, 45, as the base.

$\frac{15}{45} = \frac{r}{100}$

$15 \cdot 100 = 45 \cdot r$

$1{,}500 = 45r$

$\frac{1{,}500}{45} = \frac{45r}{45}$ *Use a calculator.*

$33.3 \approx r$

The percent of change is 33.3%.

348 Chapter 8 Applying Proportional Reasoning

Additional Answers for the Mini-Lab

1.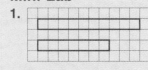

2.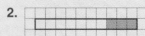

In Example 1, the new number is greater than the original number, so the percent of change is a **percent of increase**. In Example 2, the new number is less than the original number, so the percent of change is a **percent of decrease**.

Do you shop at sales? The amount by which the regular price is reduced is called the **discount**. The percent of the discount is a percent of decrease. You can find the sale price by subtracting the discount.

Example 3 APPLICATION

Shopping The Just for Fun Sporting Goods store advertises that all in-line skates are on sale for 20% off the regular price. Find the sale price of a pair of in-line skates that cost $135.

Method 1
First, use the percent equation to find the amount of the discount.
$$RB = P$$
$$0.20 \cdot 135 = P$$
$$27 = P$$
Then, subtract the $27 discount from the regular price, $135.
$$135 - 27 = 108$$

Method 2
First, find the percent paid.
$$100\% - 20\% = 80\%$$
Then, use the percent equation to find the sale price of the skates.
$$RB = P$$
$$0.80 \cdot 135 = P$$
$$108 = P$$

The sale price of the in-line skates is $108.

Did you know? The first world championships of in-line skating were held in 1992. Derek Parra holds two overall world championship titles. Kimberly Ames holds the world records in 12- and 24-hour road racing.

A store sells an item for more than it paid for that item. The extra money is used to cover the expenses of the store and to make a profit. The increase in the price is called the **markup**. The percent of a markup is a percent of increase. The amount the customer pays for an item is called the **selling price**.

Example 4 APPLICATION

Business The Just for Fun Sporting Goods store prices items 30% over the price paid by the store. If the store purchases a tennis racket for $65, find the selling price of the racket.

Method 1
Use the percent proportion to find the markup.
$$\frac{m}{65} = \frac{30}{100}$$
$$m \cdot 100 = 65 \cdot 30$$
$$100m = 1{,}950$$
$$\frac{100m}{100} = \frac{1{,}950}{100}$$
$$m = 19.50$$
Add the markup to the price paid by the store.
$$65.00 + 19.50 = 84.50$$

Method 2
The customer will pay 100% of the price plus an extra 30% of the price. Find 100% + 30% or 130% of the price paid by the store.
$$\frac{s}{65} = \frac{130}{100}$$
$$s \cdot 100 = 65 \cdot 130$$
$$100s = 8{,}450$$
$$\frac{100s}{100} = \frac{8{,}450}{100}$$
$$s = 84.50$$

The selling price of the racket for the customer is $84.50.

2 TEACH

 Transparency 8-5B contains a teaching aid for this lesson.

Using the Mini-Lab Have students think of the short rectangle as a long one that has lost some squares. The *amount* of change in area is 3 squares. The *percent* of change in area, 30%, is a comparison of the amount of change, 3, to the original area, 10.

In-Class Examples

For Example 1
Find the percent of change. Round to the nearest percent.
original: 53; new: 61 **15%**

For Example 2
Find the percent of change. Round to the nearest percent.
original: 65; new: 40 **38%**

For Example 3
Find the price of an item that usually costs $175 but is on sale for 40% off. **$105**

For Example 4
If a store purchases a pair of sneakers for $20 and marks them up 75%, how much will the consumer pay? **$35**

Teaching Tip In Example 3, point out that the discount of 20% is also the "percent of change" that students worked to find in Examples 1 and 2.

Multiple Learning Styles

 Interpersonal Separate students into small groups and ask them to research population growth or decline between 1980 and 1990 for five cities in their state. Possible sources for such information include almanacs and the Internet. Ask them to make percent models that represent the percent of change of the population.

3 PRACTICE/APPLY

Check for Understanding

If students need additional practice or instruction after completing Exercises 1–10, one of these options may be helpful.
- Extra Practice, see p. 628
- Reteaching Activity
- *Study Guide Masters*, p. 64
- *Practice Masters*, p. 64

Assignment Guide

Core: 11–35 odd, 38–43
Enriched: 12–32 even, 34–36, 38–43

Additional Answers

2. If the second number is greater than the first number, it is a percent of increase. If the second number is less than the first number, it is a percent of decrease.

3.

Study Guide Masters, p. 64

CHECK FOR UNDERSTANDING

Communicating Mathematics

1. Find the amount of the change.

HANDS-ON MATH

Guided Practice

Read and study the lesson to answer each question.
1. *State* the first step in finding the percent of change.
2. *Explain* how you know whether a percent of change is a percent of increase or a percent of decrease. **See margin.**
3. *Make a model* to show a decrease of 20%. **See margin.**

Find each percent of change. Round to the nearest percent.

4. original: 5   **20%**
   new: 4
5. original: 325   **23%**
   new: 400

Find the sale price of each item to the nearest cent.

6. $29.95 jeans, 25% off   **$22.46**
7. $300.00 stereo, $33\frac{1}{3}$% off   **$200.00**

Find the selling price for each item given the amount paid by the store and the markup. Round to the nearest cent.   8. **$202.50**

8. $150.00 skis, 35% markup
9. $12.00 shirt, 40% markup   **$16.80**

10. *Advertising* In a 1997 television commercial, Rally's claimed that their Big Buford was 75% larger than the Big Mac. The Big Buford contains $\frac{1}{3}$ pound of ground beef, while the Big Mac contains $\frac{1}{4}$ pound ground beef. Does the Big Buford contain 75% more beef by weight than the Big Mac? Explain.   **No; the percent of increase is only about 33.3%.**

EXERCISES

Practice   Find each percent of change. Round to the nearest percent.

11. original: 10   **40%**
    new: 6
12. original: 50   **34%**
    new: 67
13. original: 12   **67%**
    new: 20
14. original: 80   **31%**
    new: 55
15. original: 775   **6%**
    new: 825
16. original: 835   **8%**
    new: 900

Find the sale price of each item to the nearest cent.

17. $14.50 CD, 10% off   **$13.05**
18. $39.95 sweater, 25% off   **$29.96**
19. $3.59 tennis balls, $\frac{1}{4}$ off   **$2.69**
20. $119.50 lamp, $\frac{1}{3}$ off   **$79.67**
21. $13.00 book, 20% off   **$10.40**
22. $18.00 shirt, 30% off   **$12.60**

Find the selling price for each item given the amount paid by the store and the markup. Round to the nearest cent.

23. **$1,440.00**
25. **$19.60**
26. **$116.00**
27. **$73.75**
28. **$65.25**

23. $1,200.00 computer, 20% markup
24. $15.00 belt, 35% markup   **$20.25**
25. $14.00 video tapes, 40% markup
26. $87.00 coat, $33\frac{1}{3}$% markup
27. $59.00 camera, 25% markup
28. $45.00 dress, 45% markup

350   Chapter 8 Applying Proportional Reasoning

Reteaching the Lesson

Activity Have groups of three students spin a spinner twice, recording both numbers spun. The first student states the amount of change from the first number to the second number. The second student states the fraction comparing the amount of change with the first number. The third student determines the percent of change.

Error Analysis
Watch for students who use the new or second number as the comparison with the amount of change.
Prevent by reinforcing the idea that the amount of change is compared to the first amount because that is the number that has changed.

30. about 16.7%

29. A math class had 25 students. If 2 more students enrolled in the class, what is the percent of change?   **8%**

30. Find the discount rate on a $25 watch that regularly sells for $30.

31. What is the sale price of a $250 bicycle on sale at 10% off?   **$225**

32. Find the markup rate on a $60 jacket that sells for $75.   **25%**

33. A store has a markup of 30%. If a store buys a sweatshirt for $15, what is the selling price?   **$19.50**

Applications and Problem Solving

34. *Technology* In 1992, there were 87 cable television networks. In 1996, there were 162 cable television networks.   **a. about 86.2%**
    a. Find the percent of change in the number of cable networks.
    b. Is this a percent of *increase* or a percent of *decrease*?   **increase**

35. *Publishing* In 1994, the circulation of the magazine *Boy's Life* was 1,242,722. In 1995, its circulation was 1,229,052.
    a. Find the percent of change in the circulation of *Boy's Life*.   **about 1.1%**
    b. Is this a percent of *increase* or a percent of *decrease*?   **decrease**

36. *Population* Between the 1980 census and the 1990 census, the United States population increased by approximately 22,164,873 or 9.8%.
    a. What was the census population in 1980?   **about 226,172,174 people**
    b. What was the census population in 1990?   **about 248,337,047 people**
    c. Assume the percent of change remains the same. Predict the population for the 2000 census.   **about 272,674,078 people**

37. *Working on the* **CHAPTER Project** Refer to the Nielsen ratings you collected on page 329.   **a–c. See students' work.**
    a. Compute the percent of change from the first week to the second week.
    b. Compute the percent of change from the second week to the third week.
    c. How do the Nielsen ratings compare with your class ratings? Draw a double-line graph to show these results.

38. The store lost money because 50% of the listed price is less than the original number.

38. *Critical Thinking* A store has a 50% markup on sweaters. When the sweaters do not sell, they are put on sale at 50% off the listed price. All the sweaters are sold. Did the store break even, make a profit, or lose money? Explain.

Mixed Review

39. Express $\frac{2}{5}$% as a fraction in simplest form. *(Lesson 8-4)* $\frac{1}{250}$

40. **Test Practice** A deck is constructed in the shape of a trapezoid. The parallel sides of the deck are 24 feet and 18 feet, and the height of the trapezoid is 47 feet. What is the area of the deck? *(Lesson 7-6)* **A**
    A 987 sq ft   **B** 432 sq ft   **C** 164 sq ft   **D** 89 sq ft

41. 6 + 7 + 2 = 15, 15 ÷ 3 = 5; yes

41. Use divisibility rules to determine whether 672 is divisible by 3. *(Lesson 6-1)*

42. *Statistics* Refer to the line plot. Find the median and mode of the data. *(Lesson 4-4)* **21; 21**

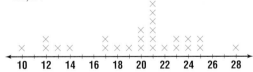

43. Find the value of $9^2 - 2^3$. *(Lesson 1-2)*   **73**

Lesson 8-5 Percent of Change   **351**

TECHNOLOGY 8-5B LAB Notes

GET READY

Objective Students learn to use a spreadsheet to find the sale price of an item.

Technology Resources
Suggested spreadsheet software:
- *Lotus 1·2·3*
- *ClarisWorks*
- *Microsoft Excel*

MANAGEMENT TIPS

Recommended Time
30 minutes

Getting Started Remind students that the *fill down* command can be used to find the values of an entire column when the same formula is being used for each cell.

When they have finished entering their data and the program has generated values for each item, ask students to think about how the values would change if the discount was 40% and how the equation could be rewritten to reflect this change.

Have students work through the exercises using the data they generated. If spreadsheet software is not available, students can use calculators to find the values in the table.

In-Class Example
Have students find the sale price of a $42.99 pair of shoes discounted at a rate of 20%.
$34.39

ASSESS

After students complete Exercises 1–5, ask them how they would determine the sale price of an item when they are shopping and neither a spreadsheet nor calculator is available.

TECHNOLOGY LAB — SPREADSHEETS

8-5B Discounts

A Follow-Up of Lesson 8-5

 computer

 spreadsheet software

Suppose you are the manager of the casual clothes department of a department store. The store has frequent sales when many of the items are discounted by the same percent. You use the spreadsheet below to generate sale signs for each item. The discount rate is entered into cell B1. Then the formulas in the cells in column C determine the sale prices.

TRY THIS

Work with a partner.

Use the spreadsheet to determine the price of each item during the Midnight Madness Sale when the prices are discounted 25%. To find these prices, substitute 25 for B1.

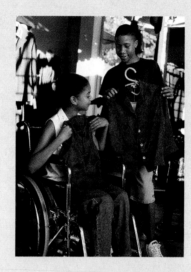

	A	B	C
1	Discount Rate	B1	
2	Item	Original Price	Sale Price
3	Cotton Sweaters	29.99	= (100-B1)/100 * B3
4	Denim Jackets	36.29	= (100-B1)/100 * B4
5	Team Sweatshirts	24.89	= (100-B1)/100 * B5
6	Sport Socks 3-Pack	6.59	= (100-B1)/100 * B6
7	T-Shirts	7.99	= (100-B1)/100 * B7

5. | 8 | suede jackets | $99.59 | (100-B1)/100*B8 |

ON YOUR OWN

1. *List* the prices of the items during the Midnight Madness Sale. **1–2. See margin.**
2. *Explain* why the formula in cell C3 correctly finds the sale price of cotton sweaters.
3. Use the spreadsheet to find the sale price of a denim jacket if the discount rate is 35%. **$23.59**
4. What is the discount on a T-shirt if the discount rate is 40%? **$3.20**
5. Suppose you wanted to add a row 8 to the spreadsheet for a $99.59 suede jacket. List each of the cell entries (A8, B8, and C8) that you would enter.

352 Chapter 8 Applying Proportional Reasoning

Additional Answers

1. sweaters, $22.49; jackets, $27.22; sweatshirts, $18.67; socks, $4.94; T-shirts, $5.99
2. $(100 - B1)$ is the percent paid. Since percent means hundredths, the sale price is $\frac{100 - B1}{100}$ times the original price (B3).

 Math Journal Have students write a paragraph about how a family could use a spreadsheet to plan for saving 20% of their monthly budget.

8-6 Simple Interest

What you'll learn
You'll learn to solve problems involving simple interest.

When am I ever going to use this?
Knowing how to solve problems involving simple interest can help you determine the interest on your savings account.

Word Wise
interest
principal

Darius wants to buy a car after high school graduation. Through his paper route, gifts, and odd jobs, he now has $750. He would like to deposit his money in a bank savings account. The newspaper listed several interest rates for certificates of deposit (CDs) as shown at the right. If he puts his money in a five-year CD, how much will he have at the end of the 5 years?

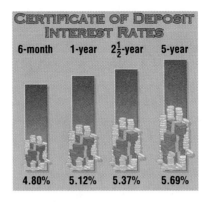

Interest is the amount paid or earned for the use of money. The formula $I = prt$ is used to solve problems involving simple interest. In the formula, I represents the interest, p represents the amount of money invested, called the **principal**, r represents the annual interest *rate*, and t represents *time* in years.

$$I = prt$$
$$I = 750 \cdot 0.0569 \cdot 5 \quad p = 750, r = 5.69\% \text{ or } 0.0569, t = 5$$

750 [×] .0569 [×] 5 [=] 213.375

$$I = 213.375$$

To the nearest cent, Darius will earn $213.38 in interest. He will have $750.00 + $213.38 or $963.38 in five years.

Study Hint
Reading Math $I = prt$ means interest equals principal times rate times time.

Example APPLICATION

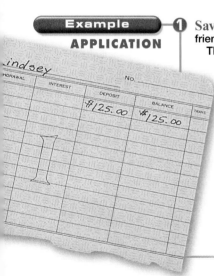

Savings Lindsey sold stationery to her family and her mother's friends. She deposited the $125 she earned in a savings account. The account earns 5.18% interest annually. If she does not deposit or withdraw any money for 18 months, how much will she have in her account?

Since there are 12 months in a year, 18 months equals $\frac{18}{12}$ or $\frac{3}{2}$ years. Use the simple interest formula to find the amount of interest earned during this time.

$$I = prt$$
$$I = 125 \cdot 0.0518 \cdot \frac{3}{2} \quad p = 125, r = 5.18\% \text{ or } 0.0518, t = \frac{3}{2}$$

125 [×] .0518 [×] 3 [÷] 2 [=] 9.7125

$$I = 9.7125$$

To the nearest cent, the interest will be $9.71. Lindsey will have $125.00 + $9.71 or $134.71 in her account in 18 months.

Lesson 8-6 Simple Interest

8-6 Lesson Notes

Instructional Resources
- *Study Guide Masters*, p. 65
- *Practice Masters*, p. 65
- *Enrichment Masters*, p. 65
- Transparencies 8-6, A and B
- *Technology Masters*, p. 67
- CD-ROM Program
  - Resource Lesson 8-6

Recommended Pacing	
Standard	Day 8 of 15
Honors	Day 8 of 15
Block	Day 4 of 7

1 FOCUS

5-Minute Check
(Lesson 8-5)

Find each percent of change. Round to the nearest percent.

1. original: 25; new: 19 **24%**
2. original: 70; new: 96 **37%**
3. original: 123; new: 214 **74%**
4. Find the sale price of a $34.50 blouse at 30% off. **$24.15**
5. Mega Electronics purchased a VCR for $215.00. Find its selling price with a 25% markup. **$268.75**

The 5-Minute Check is also available on **Transparency 8-6A** for this lesson.

Motivating the Lesson
Problem Solving Ask students how much money Christopher Columbus would have in his account in 1999 if he had invested $100 at 6% simple interest on his first trip to America in 1492. **$3,142**

Lesson 8-6

2 TEACH

Transparency 8-6B contains a teaching aid for this lesson.

Thinking Algebraically Show students that they can also use the simple interest formula to solve for principal, rate, or time. Ask them what the simple interest rate is on the money in a savings account that earns $15.75 interest from a $375 deposit after 12 months.
$r = \frac{I}{pt}$; $r = \frac{15.60}{375 \cdot 1}$; $r = 4.16\%$

In-Class Examples

For Example 1
Willis invested $2,800 in a savings account at $6\frac{1}{2}\%$ simple interest. Find the amount in the account after 42 months.
$3,437

For Example 2
Renatta borrowed $4,000 for college. He will be paying $85 per month for five years. Find the simple interest rate on the loan. **5.5%**

For Example 3
A credit card company charges 18.9% annual interest on the unpaid balance. Find the monthly interest on an $863 unpaid balance. **$13.59**

Teaching Tip Guide students in converting interest rates. Remind them that when converting percent to decimals, they can simply move the decimal point two places to the left.

Interest is charged to you when you borrow money. When this happens, the principal is the amount borrowed.

Examples
APPLICATION

② Money Matters Raul Franco borrowed $2,400 to purchase a multimedia computer, printer, monitor, and software for his family. He will be paying $125 each month for the next 24 months. Find the simple interest rate on his loan.

Explore You know the amount of the loan, the amount of each payment, and the number of payments. You are asked to find the simple interest rate.

Plan Before you can determine the rate of the interest, you must find the amount of the interest. The amount of interest equals the total amount paid minus the principal. Then, use the simple interest formula to find the rate of interest. Remember that 24 months equals 2 years.

Solve Total amount paid = $125 × 24 or $3,000
Interest (I) = $3,000 − $2,400 or $600

$I = prt$ *Simple interest formula*
$600 = 2,400 \cdot r \cdot 2$ *$I = 600, p = 2,400, t = 2$*
$600 = 4,800r$
$\frac{600}{4,800} = \frac{4,800r}{4,800}$ *Divide each side by 4,800.*
$0.125 = r$ 600 ÷ 4800 = 0.125

The rate of interest is 0.125 or 12.5%.

Examine Use the simple interest formula to determine the amount of interest on $2,400 at 12.5% over 2 years.

$I = prt$
$I = 2,400 \cdot 0.125 \cdot 2$
$I = 600$ 2400 × .125 × 2 = 600

Since this is the same amount you calculated as the interest paid by Mr. Franco, the answer seems reasonable.

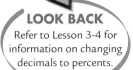
LOOK BACK
Refer to Lesson 3-4 for information on changing decimals to percents.

APPLICATION

③ Money Matters Each month, a credit card company charges an annual interest rate of 18.9% on the unpaid balance of an account. Mrs. Hill pays $100 of her $695 balance. If she does not charge any other items, find the amount of interest she will pay next month.

$I = prt$
$I = 595 \cdot 0.189 \cdot \frac{1}{12}$ $p = 695 − 100$ or 595, $r = 18.9\%$ or 0.189, $t = \frac{1}{12}$
$I = 9.37125$ 595 × .189 ÷ 12 = 9.37125

To the nearest cent, the amount of interest for next month is $9.37.

354 Chapter 8 Applying Proportional Reasoning

CHECK FOR UNDERSTANDING

Communicating Mathematics

1. $I = prt$; I represents the amount of interest, p represents the amount of money that is initially deposited in a savings account or the amount of money borrowed, r represents the percent of interest, and t represents the time in years.

Read and study the lesson to answer each question.

1. *Write* the simple interest formula. Explain in your own words the meaning of each value.
2. *Explain* the advantages and disadvantages of using credit to buy the computer items in Example 2.   **See margin.**
3. *You Decide* Janay and Sonia are trying to find the interest on a bank account of $360 for 8 months if the rate of interest is 5.5%. Each student's work is shown below. Who is correct? Explain your reasoning.   **See margin.**

Janay	Sonia
$I = 360 \cdot 0.055 \cdot \frac{2}{3}$	$I = 360 \cdot 0.055 \cdot 8$
$I = 13.20$	$I = 158.40$

Guided Practice

Find the simple interest to the nearest cent.

4. $500 at 6.5% for 3 years   **$97.50**
5. $230 at 12% for 8 months   **$18.40**

Find the total amount in each account to the nearest cent.

6. $660 at 5.25% for 2 years   **$729.30**
7. $685 at 6.3% for 21 months   **$760.52**
8. *History* In 1626, Peter Minuit paid the natives of Manhattan Island $24 in trade goods for their island. If they had taken the money and invested it at a simple interest rate of 9.8%, what would the balance be today?   **See students' work.**

EXERCISES

Practice

9. $112.50
11. $77.24
13. $133.20
14. $23.84
15. $840.00
17. $294.93
18. $275.56
19. $421.38

Find the simple interest to the nearest cent.

9. $300 at 7.5% for 5 years
10. $770 at 16% for 6 months   **$61.60**
11. $668 at 9.25% for 15 months
12. $285 at 8.5% for $2\frac{1}{2}$ years   **$60.56**
13. $360 at 18.5% for 2 years
14. $175 at 5.45% for 30 months

Find the total amount in each account to the nearest cent.

15. $800 at 7.5% for 8 months
16. $460 at 4.5% for 2 years   **$501.40**
17. $235 at 8.5% for 3 years
18. $240 at 6.35% for 28 months
19. $385 at 12.6% for 9 months
20. $190 at 5.45% for $1\frac{1}{2}$ years   **$205.53**

21. Suppose $1,250 is placed in a savings account for 2 years. Find the simple interest if the interest rate is 4.5%.   **$112.50**

22. A savings account starts with $980. If the simple interest rate is 5%, find the total amount in the account after 9 months.   **$1,016.75**

Lesson 8-6 Simple Interest **355**

Reteaching the Lesson

Activity Have pairs of students role play "banker and saver." The banker announces the rate, and the saver deposits money in the account and states how long the money will stay in the account. The banker calculates the annual interest and pays that amount. Players then switch roles.

Error Analysis

Watch for students who write time incorrectly in the formula $I = prt$.
Prevent by having students identify each variable, including the proper units. Have them identify the time period in the interest rate and the time units to verify that they correspond.
$I = 5\%$ **per year** $t = 3\frac{1}{2}$ **years**

---

3 PRACTICE/APPLY

Check for Understanding

If students need additional practice or instruction after completing Exercises 1–8, one of these options may be helpful.
- Extra Practice, see p. 629
- Reteaching Activity
- *Study Guide Masters*, p. 65
- *Practice Masters*, p. 65

Assignment Guide

Core: 9–25 odd, 26–29
Enriched: 10–22 even, 24–29
All: Self Test 1–10

Additional Answers

2. The advantages of using credit are you can have the items right away and you can budget $125 each month for the items. The disadvantages are you pay $600 more for the items and you must pay $125 each month even if you have emergencies and need the money for other things.

3. Janay; 8 months equal $\frac{8}{12}$ or $\frac{2}{3}$ year. Sonia forgot to change the time to years.

Study Guide Masters, p. 65

Lesson 8-6 **355**

4 ASSESS

Closing Activity
Speaking Have students tell the number they would write in the formula $I = prt$ for each of the following:
- $p = \$5,000$ **5,000**
- $r = 9\frac{1}{2}\%$ **0.095**
- $t = 3$ months **0.25**

Have students use $I = prt$ and the values above to compute the interest. **$118.75**

Mid-Chapter Self Test
The Mid-Chapter Self Test reviews concepts in Lessons 8-1 through 8-6. Lesson references are given so students can review concepts not yet mastered.

Additional Answer
26b. If the interest rate for the $2\frac{1}{2}$-year certificate of deposit decreases in $2\frac{1}{2}$ years, Darius would get less money. If it increases, he will get even more money.

Practice Masters, p. 65

Applications and Problem Solving

24. Credit union at 12%; the monthly rate is 1% which is less than 1.25%.

interNET CONNECTION
For the latest CD information, visit: www.glencoe.com/sec/math/mac/mathnet

Mixed Review

---

23. **Estrella's** bank statement listed a balance of $328.80. She originally opened the account with a $200 deposit and a simple interest rate of 4.6%. If there were no deposits or withdrawals, how long ago was the account opened? **14 years**

24. *Money Matters* Antonio Lopez can get a small loan from his credit union at 12% simple interest. He can also use his credit card to get a cash advance at a 1.25% per month interest rate. If he intends to pay back the loan within a month, which is the better rate? Explain.

25. *Credit Cards* In 1994, the worldwide charges on MasterCard totaled $390,000,000,000. Assume that these charges remained unpaid for one year. Using 18% as the interest rate, find the amount of interest due on these charges. **$70,200,000,000**

26. *Critical Thinking* Refer to the beginning of the lesson. Suppose Darius had put his money in a 30-month CD. After 30 months or $2\frac{1}{2}$ years, he redeposited the original $750 plus interest into another 30-month CD with the same interest rate. **a. more by $1.52**
   a. Would he have more or less money than with the 5-year CD?
   b. How would future changes in interest rates affect which investment would be better? **See margin.**

27. *Fund-raising* The Roosevelt Middle School Pep Club is selling sweatshirts. They purchase the sweatshirts for $19.00 and sell them for $25.00. What is the percent of markup? *(Lesson 8-5)* **about 31.6%**

28. **Test Practice** What is the prime factorization of 756? *(Lesson 6-2)* **B**
   A $2 \cdot 3 \cdot 7$ B $2^2 \cdot 3^3 \cdot 7$ C $2 \cdot 3^5 \cdot 7$ D $2 \cdot 3 \cdot 7^5$

29. Solve $n = -282 + 41$. *(Lesson 2-3)* **−241**

CHAPTER 8 Mid-Chapter Self Test

1. *Agriculture* A 4-acre field has a yield of 112 bushels of wheat. What yield can be expected from a 42-acre field? *(Lesson 8-1)* **1,176 bushels**

Write a percent proportion to solve each problem. Then solve. *(Lesson 8-2)*

2. Find 35% of 700. $\frac{P}{700} = \frac{35}{100}$; **245**
3. 63 is what percent of 84? $\frac{63}{84} = \frac{r}{100}$; **75%**

Write an equation in the form of $RB = P$ for each problem. Then solve. *(Lesson 8-3)*

4. 9 is 45% of what number? $0.45 \cdot B = 9$; **20**
5. Find 55% of 86. $0.55 \cdot 86 = R$; **47.3**

Express each percent as a fraction or mixed number in simplest form. *(Lesson 8-4)*

6. 0.6% $\frac{3}{500}$
7. 285% $2\frac{17}{20}$

8. *Shopping* What is the sale price of a $79 pair of boots on sale at 25% off? *(Lesson 8-5)* **$59.25**

Find the simple interest to the nearest cent. *(Lesson 8-6)*

9. $750 at 6.25% for 3 years **$140.63**
10. $430 at 13.5% for 6 months **$29.03**

356 Chapter 8 Applying Proportional Reasoning

Extending the Lesson

Enrichment Masters, p. 65

Activity Have students work in small groups to solve this problem: *How many years will it take the accumulated interest to exceed the principal at these simple interest rates: 6%, 8%, 10%?*

16.7 years; $12\frac{1}{2}$ years; 10 years

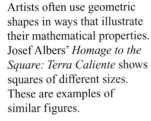

Integration: Geometry
Similar Polygons

What you'll learn
You'll learn to identify corresponding parts of similar polygons and to find missing measures of similar polygons.

When am I ever going to use this?
Knowing how to find missing measures of similar polygons can help you find actual sizes of figures shown in a scale drawing.

Word Wise
similar polygons
pentagon

LOOK BACK
You can refer to Lesson 5-6 for information on similar triangles.

Artists often use geometric shapes in ways that illustrate their mathematical properties. Josef Albers' *Homage to the Square: Terra Caliente* shows squares of different sizes. These are examples of similar figures.

In mathematics, a polygon is a simple, closed figure in a plane formed by three or more sides. A quadrilateral is a polygon with four sides. The quadrilaterals shown below have the same shape, but differ in size. These quadrilaterals are **similar polygons**.

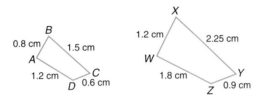

quadrilateral $ABCD \sim$ quadrilateral $WXYZ$
Recall that the symbol $\sim$ means "is similar to."

Corresponding angles of similar polygons are congruent. *You can use a protractor to make sure the angles have the same measure.*

Using corresponding angles as a guide, you can easily identify the corresponding sides.

$\angle A \cong \angle W$ $\angle B \cong \angle X$ $\overline{AB} \leftrightarrow \overline{WX}$ $\overline{BC} \leftrightarrow \overline{XY}$

$\angle C \cong \angle Y$ $\angle D \cong \angle Z$ $\overline{CD} \leftrightarrow \overline{YZ}$ $\overline{DA} \leftrightarrow \overline{ZW}$

A special relationship also exists among the corresponding sides of the polygons. Compare the ratios of the lengths of the corresponding sides.

$\frac{AB}{WX} = \frac{0.8}{1.2}$ or $\frac{2}{3}$ $\frac{BC}{XY} = \frac{1.5}{2.25}$ or $\frac{2}{3}$ $\frac{CD}{YZ} = \frac{0.6}{0.9}$ or $\frac{2}{3}$ $\frac{DA}{ZW} = \frac{1.2}{1.8}$ or $\frac{2}{3}$

As you can see, the ratios of the lengths of the corresponding sides all equal $\frac{2}{3}$. Since the ratios are all equivalent, you can form proportions using corresponding sides.

Lesson 8-7 Integration: Geometry   Similar Polygons   **357**

8-7 Lesson Notes

Instructional Resources
- *Study Guide Masters*, p. 66
- *Practice Masters*, p. 66
- *Enrichment Masters*, p. 66
- Transparencies 8-7, A and B
- CD-ROM Program
  - Resource Lesson 8-7

Recommended Pacing

Standard	Day 9 of 15
Honors	Day 9 of 15
Block	Day 5 of 7

1 FOCUS

5-Minute Check
(Lesson 8-6)

Find the simple interest to the nearest cent.

1. $427 at 13.2% for 48 months
   $225.46
2. $516 at 12.7% for 4 years
   $262.13
3. $830 at 7.3% for $2\frac{1}{2}$ years
   $151.48
4. $707 at 8.3% for 39 months
   $190.71
5. Find the total amount in an account with $750 invested at 6.1% for 9 months.
   $784.31

The 5-Minute Check is also available on **Transparency 8-7A** for this lesson.

Motivating the Lesson

Communication Ask students to give examples of pairs of objects that have the same shape but different sizes. **Sample answers: a photo and its enlargement; different sizes of the same T-shirt**

Classroom Vignette

"I have students create a unique shape using pattern blocks. They draw a scale figure of the shape on an index card and include step-by-step instructions on how to make the shape on the back of the card. Students exchange cards and try to build the figure from the instructions."

Cindy W. Hansard

Cindy Hansard, Teacher
North Forsyth School
Cumming, GA

Lesson 8-7   **357**

2 TEACH

 Transparency 8-7B contains a teaching aid for this lesson.

Using the Mini-Lab Ask students whether the orientation of the transparency is important. Help them see that the transparency must be parallel to the wall and the flashlight parallel to the floor for the projected image to be similar to the triangle on the transparency.

In-Class Example

For Example 1
If quadrilaterals EFGH and PQRS are similar, find the length of side $\overline{QR}$. **36.25 m**

Teaching Tip Have students complete this statement: *Similar polygons have the same* **shape**; *but different* **sizes**; *corresponding angles are* **congruent**; *and corresponding sides are* **in proportion**.

| Similar Polygons | Two polygons are similar if their corresponding angles are congruent and their corresponding sides are in proportion. |

 **MINI-LAB**

Work in groups of 3 or 4.   ☐ transparency   flashlight

Try This   ruler   protractor

- Draw a scalene triangle on a transparency.
- Hold a flashlight 6 inches directly behind the transparency and project the triangle's image onto a paper held against a wall 12 inches in front of the transparency. Trace the projected image of the triangle onto the paper.
- Measure the angles of each triangle. Record your results.
- Measure the lengths of the sides of the new triangle. Find the ratio of the length of each side of the new triangle to the corresponding side of the original triangle.

Talk About It

1. Are the two triangles similar? Explain your answer.
2. If the flashlight were held at an angle below the triangle on the overhead transparency, would the projected triangle be similar to the original triangle? Test your answer. **No; see students' work.**

1. Yes; the corresponding angles have the same measure and the corresponding sides are in proportion.

Proportions are useful in finding the missing length of a side in any pair of similar polygons.

Example 1 A *pentagon* is a polygon with five sides. If pentagons ABCDE and HIJKL are similar, find the length of side $\overline{IJ}$.

$\overline{AB}$ corresponds to $\overline{HI}$, and $\overline{BC}$ corresponds to $\overline{IJ}$. So you can write a proportion.

Study Hint
Estimation You can often tell which angles of similar polygons are corresponding by estimating their angle measure instead of measuring each angle.

$\dfrac{AB}{HI} = \dfrac{BC}{IJ}$

$\dfrac{5}{4} = \dfrac{7}{x}$ $AB = 5, HI = 4, BC = 7, IJ = x$

$5 \cdot x = 4 \cdot 7$ *Find the cross products.*

$5x = 28$

$\dfrac{5x}{5} = \dfrac{28}{5}$ *Divide each side by 5.*

$x = 5.6$

The length of side $\overline{IJ}$ is 5.6 meters.

358 Chapter 8 Applying Proportional Reasoning

Reteaching the Lesson

Activity Have students draw a triangle ABC and measure its sides. Have them draw $\overline{DE} \parallel \overline{BC}$ (D on $\overline{AB}$ and E on $\overline{AC}$) and measure the sides of △ADE. Have them calculate $\dfrac{AD}{AB}$, $\dfrac{AE}{AC}$, and $\dfrac{DE}{BC}$, and explain their findings. **The ratios are equal; △ADE and △ABC are similar.**

Error Analysis
Watch for students who incorrectly match corresponding sides of similar polygons.
Prevent by reminding students to align polygons and clearly label each side.

Example 2
APPLICATION

Design Ashley designed a logo for her school. The logo, which is 5 inches wide and 8 inches long, will be enlarged and used on a school sweatshirt. If the logo will be $12\frac{1}{2}$ inches wide on the sweatshirt, find its length.

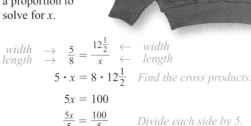

Let x represent the length. Use a proportion to solve for x.

$$\begin{array}{r} \text{width} \rightarrow \\ \text{length} \rightarrow \end{array} \frac{5}{8} = \frac{12\frac{1}{2}}{x} \begin{array}{l} \leftarrow \text{width} \\ \leftarrow \text{length} \end{array}$$

$5 \cdot x = 8 \cdot 12\frac{1}{2}$ *Find the cross products.*

$5x = 100$

$\frac{5x}{5} = \frac{100}{5}$ *Divide each side by 5.*

$x = 20$

The logo will be 20 inches long.

In-Class Example
For Example 2
Noriko drew a detailed map for her class's website. The map is 6 inches wide and 9 inches long on a monitor. If it is 4 inches wide on a smaller monitor, what is its length? **6 in.**

3 PRACTICE/APPLY

Check for Understanding
If students need additional practice or instruction after completing Exercises 1–7, one of these options may be helpful.
- Extra Practice, see p. 629
- Reteaching Activity, see p. 358
- *Study Guide Masters*, p. 66
- *Practice Masters*, p. 66

CHECK FOR UNDERSTANDING

Communicating Mathematics
Read and study the lesson to answer each question. 1–2. See margin.

1. **Tell** how you can determine whether two polygons are similar.

2. **Compare and contrast** similar polygons and congruent polygons.

HANDS-ON MATH

3. **Draw** a parallelogram on an overhead transparency. With the help of some of your classmates, use a flashlight to draw a similar parallelogram.
   See students' work.

Guided Practice

4. No; the corresponding angles are congruent, but $\frac{2}{4} \ne \frac{8}{14}$.

5. Sample answer: $\frac{10}{5} = \frac{6}{x}$; 3

7. $2\frac{1}{2}$ in. by $3\frac{1}{8}$ in.

4. Tell whether the rectangles are similar. Explain your reasoning.

5. In the figure at the right, $\triangle ABC \sim \triangle EBD$. Write a proportion to find the missing measure x. Then find the value of x.

6. Refer to Example 1 on page 358. Find the length of $\overline{HL}$. **9.6 m**

7. **Publishing** Portrait proofs for the yearbook are 4 inches by 5 inches. The yearbook staff must reduce the proofs so that they can fit 3 photographs across the page. To do this, the ratio of the original to the reduced print is 8:5. Find the dimensions of the pictures as they will appear in the yearbook.

Lesson 8-7 Integration: Geometry Similar Polygons **359**

Study Guide Masters, p. 66

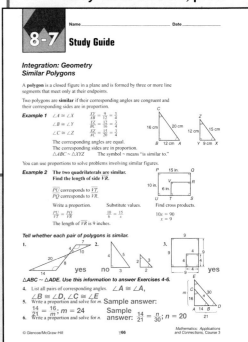

Additional Answers

1. If 2 polygons have corresponding congruent angles and have corresponding sides that are in proportion, then the polygons are similar.

2. Both similar polygons and congruent polygons have corresponding congruent angles. The corresponding sides of similar polygons are in proportion. The corresponding sides of congruent polygons are congruent. Congruent polygons are always similar polygons, but similar polygons are not necessarily congruent polygons.

Lesson 8-7 **359**

Assignment Guide

Core: 9–17 odd, 18–22
Enriched: 8–14 even, 16–22

Additional Answers

11. $\frac{5}{3} = \frac{x}{4}$

12. $\frac{6}{3} = \frac{8}{x}$

13. $\frac{10}{6} = \frac{5}{x}$

17b. Sample answer:

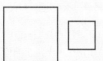

17c. All angles are right angles, so all angles of one square are congruent to all angles of any other square. Since all 4 sides of a square are the same length, corresponding sides are always in proportion.

4 ASSESS

Closing Activity

Writing These triangles are similar. Have students write two ratios equal to $\frac{k}{c}$. $\frac{m}{a}, \frac{n}{b}$

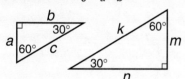

Practice Masters, p. 66

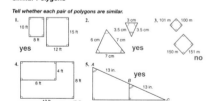

EXERCISES

Practice Tell whether each pair of polygons is similar. Explain your reasoning.

8. Yes; the corresponding angles are congruent, and $\frac{15}{45} = \frac{12}{36} = \frac{8}{24}$.

9. Yes; the corresponding angles are congruent and $\frac{3}{2} = \frac{3}{2} = \frac{3}{2} = \frac{3}{2}$.

10. No; the corresponding angles are congruent, but $\frac{4}{5} \neq \frac{5}{7}$.

8.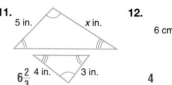

9.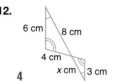

10.

Each pair of polygons is similar. Write a proportion to find the missing measure x. Then find the value of x. 11–13. See margin for sample proportions.

11. $6\frac{2}{3}$
12. 4
13. 3

Refer to Example 1 on page 358.

14. Find the length of $\overline{CD}$. **10 m**
15. Find the length of $\overline{DE}$. **8.75 m**

Applications and Problem Solving

16. *Hobbies* A world's smallest miniature model railroad with a scale of 1:1,400 was made by Bob Henderson of Gravenhurst, Ontario. If the engine measures $\frac{5}{16}$ inch, how long is the real engine? $437\frac{1}{2}$ in. or about $36\frac{1}{2}$ feet

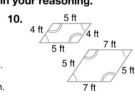

17. *Art* Refer to the picture at the beginning of the lesson. *Regular polygons* have all sides equal and all angles equal.
    a. Are the squares in the picture regular polygons? **yes**
    b. Draw two squares of different sizes than the ones in the picture. Are they similar to the ones in the picture? See margin; yes.
    c. Are all squares similar polygons? Explain. Yes; see margin.

18. *Critical Thinking* A rectangle is 5 centimeters by 7 centimeters. The sides of a similar rectangle are twice as long.
    a. What is the ratio of the perimeter of the first rectangle to the perimeter of the second rectangle? **1:2**
    b. What is the ratio of the area of the first rectangle to the area of the second rectangle? **1:4**

Mixed Review

19. *Money Matters* Shala's savings account earned $4.56 in 6 months at a simple interest rate of 4.75%. How much was in her account? *(Lesson 8-6)* **$192**

20. **Test Practice** What is the multiplicative inverse of $-\frac{1}{a}$? *(Lesson 7-4)* **B**
    A $\frac{1}{a}$ **B** $-a$ **C** a **D** -1

21. Express $9\frac{3}{5}$ as a decimal. *(Lesson 6-5)* **9.6**

22. Solve $\frac{-108}{12} = a$. *(Lesson 2-8)* **−9**

360 Chapter 8 Applying Proportional Reasoning

Extending the Lesson

Enrichment Masters, p. 66

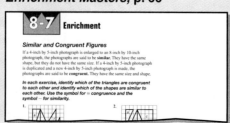

Activity Have pairs of students find the height of a tall object (CD) using a yardstick (EB) and measuring distances AB and AC. $\frac{DC}{EB} = \frac{AC}{AB}$ or $DC = \frac{36 \cdot AC}{AB}$

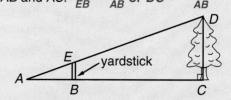

8-8 Indirect Measurement

What you'll learn
You'll learn to solve problems involving similar triangles.

When am I ever going to use this?
Knowing how to solve problems involving similar triangles can help you find the height of trees or other tall structures.

Word Wise
indirect measurement

The rope-stretchers, or surveyors, of ancient Egypt used a technique called shadow reckoning to determine the heights of tall objects. The height of a staff and the length of its shadow are proportional to the height of an object and its shadow. The objects and their shadows form two sides of similar triangles from which a proportion can be written.

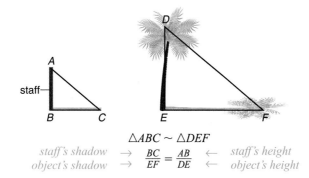

$\triangle ABC \sim \triangle DEF$

staff's shadow → $\dfrac{BC}{EF} = \dfrac{AB}{DE}$ ← staff's height
object's shadow ← object's height

Using proportions to find a measurement is called **indirect measurement**.

Example 1 — APPLICATION

Tourism While touring Egypt, Isabel visited the Great Pyramid of Cheops. At the same time of day that Isabel's shadow was 0.6 meter, the pyramid's shadow was 56 meters. If Isabel is 1.5 meters tall, how high is the Great Pyramid of Cheops?

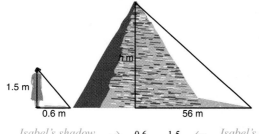

Isabel's shadow → $\dfrac{0.6}{56} = \dfrac{1.5}{h}$ ← Isabel's height
pyramid's shadow ← pyramid's height

$0.6 \cdot h = 56 \cdot 1.5$ *Find the cross products.*

$0.6h = 84$

$\dfrac{0.6h}{0.6} = \dfrac{84}{0.6}$ *Divide each side by 0.6.*

$h = 140$

The Great Pyramid of Cheops is 140 meters tall.

You can also use similar triangles that do not involve shadows to find missing measurements.

Cultural Kaleidoscope
There are ten pyramids and a huge statue of a sphinx at Giza, just southwest of Cairo, Egypt. The Great Pyramid of Cheops is the largest of these pyramids.

Lesson 8-8 Indirect Measurement **361**

8-8 Lesson Notes

Instructional Resources
- *Study Guide Masters*, p. 67
- *Practice Masters*, p. 67
- *Enrichment Masters*, p. 67
- Transparencies 8-8, A and B
- *Assessment and Evaluation Masters*, p. 212
- CD-ROM Program
  - Resource Lesson 8-8

Recommended Pacing
Standard	Days 10 & 11 of 15
Honors	Days 10 & 11 of 15
Block	Day 5 of 7

1 FOCUS

5-Minute Check
(Lesson 8-7)

1. Tell whether the rectangles are similar. **no**

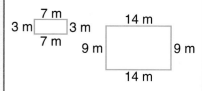

2. The triangles are similar. Find the value of x. **18 in.**

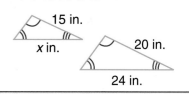

The 5-Minute Check is also available on **Transparency 8-8A** for this lesson.

Motivating the Lesson
Problem Solving On a sunny day, you are looking up at your kite, which has gotten caught at the very top of a tall tree. How could you determine the height of the tree? Measure your shadow and the tree's shadow. Solve the following proportion.

$$\dfrac{\text{your height}}{\text{your shadow's length}} = \dfrac{\text{tree's height}}{\text{tree's shadow's length}}$$

Multiple Learning Styles
Visual/Spatial Have students find pictures or drawings that show a situation in which it would be nearly impossible to measure the distance directly. Have them state why each image fits in this category. Then have them devise a plan for using indirect measurement.

Lesson 8-8 **361**

2 TEACH

 Transparency 8-8B contains a teaching aid for this lesson.

Using Discussion Explain that methods of indirect measurements use similar triangles and their proportional sides. The ratio of two known measurements can be applied to two other similarly related measures, one of which is unknown.

In-Class Examples

For Example 1
The boat sail reaches 8 feet above the water. How tall is the cliff? **192 ft**

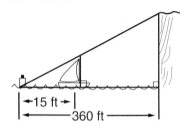

For Example 2
The triangles are similar. Find AB, the distance across the canyon. **175 ft**

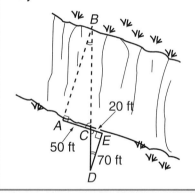

3 PRACTICE/APPLY

Check for Understanding

If students need additional practice or instruction after completing Exercises 1–6, one of these options may be helpful.
- Extra Practice, see p. 629
- Reteaching Activity
- *Study Guide Masters*, p. 67
- *Practice Masters*, p. 67

Example **APPLICATION**

② Engineering Delaware and Elmwood Avenues are parallel. Elmwood now ends at Forest Avenue, but the city is planning to extend Elmwood to Military Avenue. The triangles are similar triangles. Find the length of the Elmwood extension.

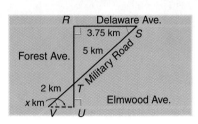

$\triangle RST \sim \triangle UVT$

$\dfrac{RS}{UV} = \dfrac{ST}{VT}$

$\dfrac{3.75}{x} = \dfrac{5}{2}$ *RS = 3.75, UV = x, ST = 5, VT = 2*

$3.75 \cdot 2 = x \cdot 5$ *Find the cross products.*

$7.5 = 5x$

$\dfrac{7.5}{5} = \dfrac{5x}{5}$ *Divide each side by 5.*

$1.5 = x$

The part of Elmwood Avenue that is under construction is 1.5 kilometers long.

CHECK FOR UNDERSTANDING

Communicating Mathematics

Read and study the lesson to answer each question.

1. **Tell** what is meant by indirect measurement. **See Answer Appendix.**
2. **Draw and label** a diagram for the following problem. Then write an appropriate proportion. **See margin.**

   A staff's shadow is 9 feet and a tree's shadow is 15 feet. If the staff is 6 feet tall, how tall is the tree?

3. **Write a Problem** that requires indirect measurement. Explain how to solve the problem. **See Answer Appendix.**

Guided Practice

Write a proportion for each problem and then solve it. Assume the triangles are similar.

4. Find the distance between Lewiston and Buffalo.
   Sample answer: $\dfrac{2}{7.5} = \dfrac{x}{30}$; 38 mi

5. A guy wire is attached to the top of a telephone pole and goes to the ground 9 feet from its base. When Honovi stands under the guy wire so that his head touches it, he is 2 feet 3 inches from where the wire goes into the ground. If Honovi is 5 feet tall, how tall is the telephone pole? (*Hint:* Make a drawing and identify two similar triangles.) Sample answer: $\dfrac{h}{5} = \dfrac{9}{\text{---}}$; 20 ft

362 Chapter 8 Applying Proportional Reasoning

■ **Reteaching the Lesson** ■

Activity Have students stretch a string from a point on a wall (P) to a point on the floor (A) and then position a 12-inch ruler as shown in the figure at the right. Students should measure $\overline{AC}$ and $\overline{PF}$, find PF using $\dfrac{PF}{BC} = \dfrac{AF}{AC}$, and check by measuring $\overline{PF}$.

Additional Answer
2. Sample answer:

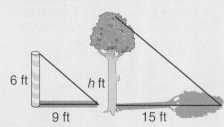

Some people like to collect scale models of cars or trains. Scale models are frequently used in the movie industry.

Example APPLICATION

② Movies In the movie *Honey, I Blew Up the Kid*, a child who was 28 inches tall was shown to be 112 feet tall. Find the scale needed for the props in the movie.

Write the ratio of the height of the child to the height he appeared to be in the movie. Then simplify the ratio.

$$\frac{28 \text{ inches}}{112 \text{ feet}} = \frac{1 \text{ inch}}{4 \text{ feet}}$$

The scale is 1 inch = 4 feet.

Some designers prefer that both values in the scale be in the same units.

$$\frac{1 \text{ inch}}{4 \text{ feet}} = \frac{1 \text{ inch}}{(4 \times 12) \text{ inches}} = \frac{1 \text{ inch}}{48 \text{ inches}} = \frac{1}{48}$$

The scale is 1:48. *Note that if the scale uses the same units, it is not necessary to include them.*

CHECK FOR UNDERSTANDING

Communicating Mathematics

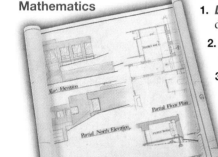

Read and study the lesson to answer each question.

1. **Describe** why it is sometimes necessary to use scale drawings and models.

2. **Name** three occupations or careers that use scale drawings or models. Tell how they use scale drawings or models.

3. **You Decide** On a blueprint, 1 inch represents 3 feet. Renee says the scale is 1:3, but Sierra disagrees. Sierra says the scale is 1:36. Who is correct? Explain. **See margin.**

Guided Practice

1. Some objects are too small or too large to show the actual item in the space available.

2. See margin.

4. The figure at the right is a scale drawing of the west side of a cabin. In the drawing, the side of each square represents 2 feet. Find the actual size of each of the following.
   a. the width of the cabin **24 ft**
   b. the height of the cabin **18 ft**
   c. the width of the door **4 ft**
   d. the height of the door **7 ft**

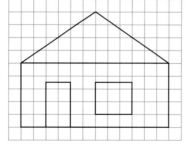

5. If the actual distance between two airports is 500 kilometers and the distance on a map is 4 centimeters, what is the scale for the map? **1 cm = 125 km or 1:12,500,000**

Lesson 8-9 Scale Drawings and Models **367**

■ **Reteaching the Lesson** ■

Activity Have students use this diagram to set up proportions in scale drawings.

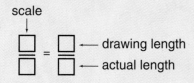

Additional Answers

2. Sample answer: Architects use blueprints and scale models to show their ideas to their clients. Interior designers use scale drawings to show placement of furniture. Movie producers use miniature models for some scenes.

3. Sierra; the scale is 1 in. = 3 ft or 1 in. = 36 in., which is 1:36.

> **Assignment Guide**
> **Core:** 7–15 odd, 16–18
> **Enriched:** 8–12 even, 13–18

6. *Movies* One of the models of King Kong used in the filming of the 1933 movie was only 18 inches tall. In the movie, King Kong was seen as 24 feet high. What was the scale used? **3 in. = 4 ft or 1:16**

EXERCISES

Practice

7. The figure at the right is a scale drawing of a house plan. In the drawing, the side of each square represents 4 feet. Find the length and width of each room.
   a. living room   **16 ft by 18 ft**
   b. bedroom 1   **8 ft by 12 ft**
   c. kitchen   **8 ft by 8 ft**
   d. bedroom 3   **8 ft by 10 ft**

8. On a map, the scale is 1 inch = 90 miles. Find the actual distance for each map distance.   a. **1,305 mi**   b. **360 mi**   c. **315 mi**

	From	To	Map Distance
a.	Tampa Bay, Florida	Buffalo, New York	$14\frac{1}{2}$ inches
b.	Columbus, Ohio	Chicago, Illinois	4 inches
c.	Montgomery, Alabama	Savannah, Georgia	$3\frac{1}{2}$ inches

9. A scale model of the Statue of Liberty is 10 inches high. If the Statue of Liberty is 305 feet tall, find the scale of the model.   **10 in. = 305 ft or 1:366**

10. A model of a 1997 Ferrari F355 Spider is 10 inches long. If the actual car is 14 feet, find the scale of the model.   **5 in. = 7 ft or 5:84**

11. The distance between Little Rock, Arkansas, and Memphis, Tennessee, is 140 miles. If the scale on a map is 1 inch = 50 miles, find the distance between the two cities on the map. $2\frac{4}{5}$ **in.**

12. *Life Science* An amoeba is pictured in a science book. If it is 2.4 centimeters long in the picture and the scale is 200:1, find the actual size of the amoeba.   **0.012 cm**

Applications and Problem Solving
13. **2.6 cm**

13. *Geography* Madurodam is a tourist attraction in the Netherlands. It is a miniature town representing the country. It includes typical houses, historic buildings, windmills, oil rigs, and tulips. The scale of Madurodam is 1:25. If a typical tulip is 65 centimeters tall, how tall are the tulips in Madurodam?

368   Chapter 8 Applying Proportional Reasoning

Study Guide Masters, p. 68

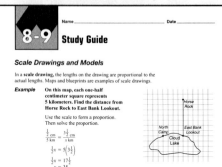

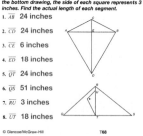

14. **Earth Science** On a weather map, a hurricane is shown $2\frac{1}{4}$ inches southeast of Puerto Rico. The hurricane is traveling northwest at approximately 20 miles per hour. If the scale of the map is $\frac{3}{5}$ inch = 300 miles, how long before the hurricane should reach Puerto Rico? **$56\frac{1}{4}$ h or 56 h 15 min**

15. **Hobbies** A doll house is being designed as a replica of a townhouse. The scale is 1 inch = 12 inches. If the actual townhouse has the outside dimensions of 25 feet by 35 feet, find the outside dimensions of the doll house. **25 in. by 35 in.**

16. **Critical Thinking** Kansas is approximately the shape of a rectangle measuring 480 miles by 267 miles. As a social studies project, you must make a scale drawing of the state on a piece of $8\frac{1}{2}$-inch by 11-inch paper. What should 1 inch on the drawing represent so that the drawing is as large as possible? **$43\frac{2}{3}$ mi**

Mixed Review

17. **Geography** The Pyramid of the Sun at Teotihuacán near Mexico City casts a shadow 13.3 meters long. At the same time, a 1.83-meter tall tourist casts a shadow 0.4 meters long. How tall is the Pyramid of the Sun? *(Lesson 8-8)* **60.8475 m**

18. **Test Practice** Solve $t = 5\frac{1}{6} + 4\frac{2}{9}$. Write the solution in simplest form. *(Lesson 7-2)* **B**

    A $9\frac{1}{3}$ B $9\frac{7}{18}$ C $10\frac{1}{3}$ D $10\frac{7}{18}$

Let the Games Begin

Map Sense

Get Ready This game is for two players.   map with a distance scale

Get Set Choose a map with a distance scale.

Go
- One player chooses two locations on the map.
- The second player estimates the actual distance.
- Both players write and solve a proportion using the map scale and the map distance between the two chosen locations. If the difference between the actual distance and the estimated distance is less than 10% of the actual distance, the player who made the estimate earns a point.
- Players exchange roles.
- The player with the most points wins the game.

interNET CONNECTION Visit www.glencoe.com/sec/math/mac/mathnet for more games.

Extending the Lesson

Enrichment Masters, p. 68

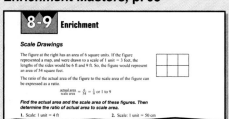

Let the Games Begin

If the students master estimating distance quickly, you can expand the game to include stops. For example, how far is it from Sioux City, IA, to Little Rock, AK, if you stop in Fort Wayne, IN, and Saint Louis, MO? **about 1,900 km**

---

4 ASSESS

Closing Activity

Modeling Have students work together to plan a scale model of their school. What information would they need? Where would they find the answers? **They could consult blueprints available in city hall. As a team they could measure outside walls.** How large would they make the model? What materials would they use?

Practice Masters, p. 68

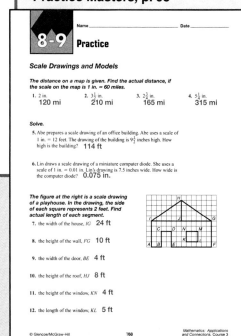

8-10 Lesson Notes

Instructional Resources
- *Study Guide Masters*, p. 69
- *Practice Masters*, p. 69
- *Enrichment Masters*, p. 69
- Transparencies 8-10, A and B
- *Assessment and Evaluation Masters*, p. 112
- *Hands-On Lab Masters*, p. 10
- *School to Career Masters*, p. 34
- *Technology Masters*, p. 68
- CD-ROM Program
  - Resource Lesson 8-10
  - Interactive Lesson 8-10

Recommended Pacing	
Standard	Day 13 of 15
Honors	Day 13 of 15
Block	Day 6 of 7

1 FOCUS

5-Minute Check
(Lesson 8-9)

A drawing of a train uses the scale 1 in. = 12 ft.
1. How long does a 60-foot locomotive appear in the drawing? **5 in.**
2. In the drawing, the caboose is $2\frac{1}{4}$ inches long. Find its actual length. **27 ft**
3. On a map, the scale is 7 cm = 850 km. How far is it from Amarillo, TX, to Albuquerque, NM, if the cities are 3.5 centimeters apart on the map? **425 km**

The 5-Minute Check is also available on **Transparency 8-10A** for this lesson.

Motivating the Lesson

Communication Ask students to explain how the zoom feature on a camera works.

8-10

Integration: Geometry
Dilations

What you'll learn
You'll learn to graph dilations on a coordinate plane.

When am I ever going to use this?
Knowing about dilations can help you show depth in two-dimensional drawings.

Word Wise
dilation
scale factor

The Renaissance architect Sebastian Serlio designed a stage set in 1545 that created the illusion of distance and depth using **dilations**. The arched doorways on the right are the same shape, but vary in size. Can you find other parts of the set design that show similar shapes?

In mathematics, the process of enlarging or reducing an image is called a dilation. Since the dilated image has the same shape as the original, the two images are similar. The ratio of the dilated image to the original is called the **scale factor**.

Example

A dilated image is usually named using the same letters as the original figure, but with primes, as in $\triangle ABC \sim \triangle A'B'C'$.

1 Graph $\triangle ABC$ with vertices $A(2, 4)$, $B(6, 2)$, and $C(8, 6)$. Graph the image of $\triangle ABC$ with a scale factor of $\frac{3}{2}$.

To find the vertices of the dilation image, multiply each coordinate in the ordered pairs by $\frac{3}{2}$.

$A(2, 4) \rightarrow \left(2 \cdot \frac{3}{2}, 4 \cdot \frac{3}{2}\right) \rightarrow A'(3, 6)$

$B(6, 2) \rightarrow \left(6 \cdot \frac{3}{2}, 2 \cdot \frac{3}{2}\right) \rightarrow B'(9, 3)$

$C(8, 6) \rightarrow \left(8 \cdot \frac{3}{2}, 6 \cdot \frac{3}{2}\right) \rightarrow C'(12, 9)$

Now graph A', B', and C'. Connect them to form $\triangle A'B'C'$. $\triangle A'B'C'$ is the dilation of $\triangle ABC$ with a scale factor of $\frac{3}{2}$.

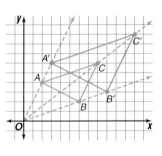

To check your graph, draw lines through the origin (0, 0) and each of the vertices of the original figure. If the vertices of the dilated figure do not lie on those same lines, you have made a mistake.

370 Chapter 8 Applying Proportional Reasoning

In art, the point where the lines connecting corresponding points intersect is called the *vanishing point*. In Example 1, the origin (0, 0) is the vanishing point. Can you find the vanishing point in Sebastian Serlio's drawing?

Sometimes you need to find the scale factor of a dilation.

Example 2 APPLICATION

Movies A projector focuses an image from the 16-millimeter film onto a screen that is 1.5 meters wide. What is the scale factor?

Write a ratio of the width of the image to the width of the film. Change meters to millimeters. Then write the scale factor in simplest form.

$$\frac{1.5 \text{ meters}}{16 \text{ millimeters}} = \frac{(1.5 \times 1{,}000) \text{ millimeters}}{16 \text{ millimeters}}$$

$$= \frac{1{,}500 \text{ millimeters}}{16 \text{ millimeters}}$$

$$= \frac{375}{4}$$

The scale factor is $\frac{375}{4}$ or 93.75.

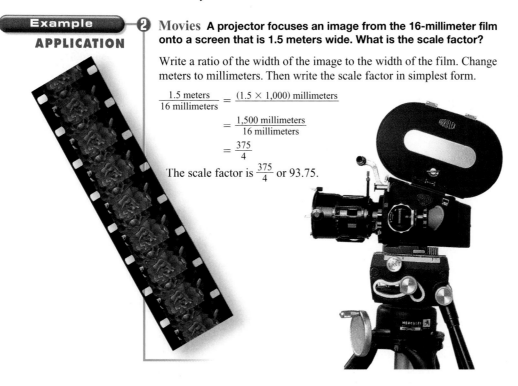

What happens if the scale factor of the dilation is less than 1, or perhaps equal to 1?

HANDS-ON MINI-LAB

Work with a partner.   grid paper   straightedge

1. Graph $\triangle ABC$ with vertices $A(2, 4)$, $B(8, 6)$, and $C(10, 12)$.
2. Find the coordinates for a dilation with a scale factor of $\frac{1}{2}$.
3. On the same coordinate plane, graph $\triangle A'B'C'$.   See margin.

Talk About It   4. The dilated image is smaller than the original.
4. How does the dilated image compare to the original?
5. Graph a dilation of $\triangle ABC$ with a scale factor of 1. How do the two triangles compare?   See margin.

1. See margin.
2. $A'(1, 2)$, $B'(4, 3)$, $C'(5, 6)$

Lesson 8-10 Integration: Geometry   Dilations   **371**

Additional Answer for the Mini-Lab
5. The graph of a dilation of $\triangle ABC$ with a scale factor of 1 is the same as the graph of $\triangle ABC$ as shown in Exercise 1. The dilated figure is the same size and shape as the original.

---

2 TEACH

Transparency 8-10B contains a teaching aid for this lesson.

In-Class Examples

For Example 1
Graph $\triangle ABC$ with vertices $A(2, 0)$, $B(0, 1)$, and $C(3, 2)$. Graph the image of $\triangle ABC$ with a scale factor of 3.

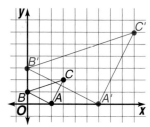

For Example 2
One way to draw the silhouette of a person or thing is to place the object between a light and a wall and then trace the shadow on paper. What is the scale factor of a 15-inch-high image traced from a 6-inch-high object?   **2.5**

Teaching Tip   Point out that dilations can be performed on all shapes. The shape in Example 2 is a rectangle. Have the students try graphing dilations of rectangles or parallelograms as in Example 1.

Using the Mini-Lab   Point out that the vertices of the dilated triangle lie on the lines drawn from the origin through the vertices of the original triangle, just as in the Example 1.

Additional Answer for the Mini-Lab
1 and 3.

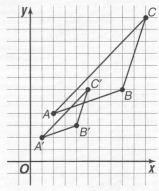

Lesson 8-10   **371**

3 PRACTICE/APPLY

Check for Understanding

If students need additional practice or instruction after completing Exercises 1–7, one of these options may be helpful.
- Extra Practice, see p. 630
- Reteaching Activity
- *Study Guide Masters*, p. 69
- *Practice Masters*, p. 69

Assignment Guide

Core: 9–19 odd, 20–23
Enriched: 8–16 even, 18–23

Additional Answers

1. A dilation is an enlargement or reduction of an image.

2. Multiply each member of the ordered pair (x, x) by the scale factor k. The new coordinates are (kx, ky).

3a and 3c.

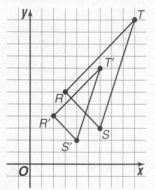

Study Guide Masters, p. 69

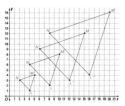

CHECK FOR UNDERSTANDING

Communicating Mathematics

Read and study the lesson to answer each question. 1–2. See margin.

1. *Describe* in your own words the meaning of dilation.
2. *Write* a general rule for finding the new coordinates of any ordered pair (x, y) for a dilation with a scale factor of k.

HANDS-ON MATH

3a. *Graph* $\triangle RST$ with vertices $R(3, 6)$, $S(6, 3)$, and $T(9, 12)$. See margin.
   b. *Write* the coordinates of a dilation with scale factor of $\frac{2}{3}$.
   c. *Graph* $\triangle R'S'T'$ on the same coordinate plane. See margin.
   d. *Compare* the two triangles. See margin.

3b. $R'(2, 4)$, $S'(4, 2)$, $T'(6, 8)$

Guided Practice

4. Find the coordinates of the image of $A(4, 12)$ for a dilation with a scale factor of 2.5. $A'(10, 30)$

5. In the figure at the right, the green triangle is a dilation of the blue triangle. Find the scale factor. **1.5**

6. Triangle XYZ has vertices $X(3, 6)$, $Y(6, 9)$ and $Z(12, 6)$. Find the coordinates of its image for a dilation with a scale factor of $\frac{4}{3}$. Graph $\triangle XYZ$ and its dilation. See Answer Appendix.

7. **Photocopying** Some photocopiers have various settings to produce copies that are larger or smaller than the original.

7a. 200%; 50% = $\frac{1}{2}$ so the image would be half the size, but 200% = 2.00 so the image would be twice as large.

   a. Suppose an enlargement is required. Should the machine be set on 50% or 200%? Explain.
   b. If the machine is set on 75%, what is the scale factor of the dilation? $\frac{3}{4}$

EXERCISES

Practice

Find the coordinates of the image of each point for a dilation with a scale factor of $\frac{3}{4}$.

8. $A(8, 4)$ $A'(6, 3)$
9. $B(10, 16)$ $B'\left(7\frac{1}{2}, 12\right)$
10. $C(6, 4)$ $C'\left(4\frac{1}{2}, 3\right)$

Triangle DRT has vertices $D(-4, 12)$, $R(-2, -4)$, and $T(8, 6)$. Find the coordinates of its image for a dilation with each given scale factor. Graph $\triangle DRT$ and each dilation. 11–13. See Answer Appendix.

11. 2
12. $\frac{1}{4}$
13. $\frac{5}{4}$

372 Chapter 8 Applying Proportional Reasoning

■ Reteaching the Lesson ■

Activity Use a number line to show dilations in one dimension.

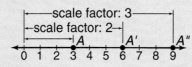

After several one-dimensional dilations, extend the idea to two dimensions.

Additional Answer

3d. The 2 triangles are the same shape. The dilated figure is smaller than the original.

In each figure, the green figure is a dilation of the blue figure. Find each scale factor.

14. $\frac{1}{3}$ 15. 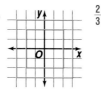 2   16. $\frac{2}{3}$

17. Draw a rectangle and its dilated image if the scale factor is 1.5. **See Answer Appendix.**

Applications and Problem Solving

18. **Art** Examine Van Gogh's painting of Montmartre, Paris. How did he create the illusion of depth in the two-dimensional painting? **See margin.**

19. **Geometry** Graph polygon $ABCD$ with vertices $A(2, 3)$, $B(8, 3)$, $C(8, 7)$, and $D(2, 7)$.

19a. 20 units; 24 units²

a. Find the perimeter and area of polygon $ABCD$.

19b. See Answer Appendix.

b. Graph the image of polygon $ABCD$ for a dilation with a scale of 3.

19c. 60 units; 216 units²

c. Find the perimeter and area of polygon $A'B'C'D'$.

d. Write a ratio in simplest form comparing the perimeter of $A'B'C'D'$ to the perimeter of $ABCD$. **3:1**

e. Write a ratio in simplest form comparing the area of $A'B'C'D'$ to the area of $ABCD$. **9:1**

f. Make a conjecture about your findings. **See margin.**

20. **Critical Thinking** Examine the triangle.

a. How many different sizes of triangles do you see? **3 sizes**

b. What is the scale factor of each size as compared to the large triangle? $\frac{1}{3}, \frac{2}{3}, 1$

c. How many triangles are there altogether? **13 triangles**

Mixed Review

21. **Scale Drawings** On a scale drawing of a house with a scale of $\frac{1}{4}$ inch = 1 foot, the length of the house is 14 inches. Find the length of the actual house. *(Lesson 8-9)* **56 ft**

22. **Test Practice** Express 52,380,000 in scientific notation. *(Lesson 6-9)* **D**

A 52.38×10^6 B 5.238×10^8 C 5.238×10^6 D 5.238×10^7

23. **Geometry** *True* or *false*? Every square is a rhombus. *(Lesson 5-3)* **true**

Lesson 8-10 Integration: Geometry Dilations **373**

Extending the Lesson

Enrichment Masters, p. 69

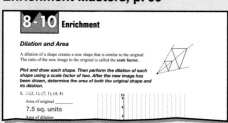

Activity Which scale for this building yields the largest drawing possible on a 24 in. × 36 in. sheet of paper?

$\frac{1}{8}$ in. = 1 ft,

$\frac{3}{16}$ in. = 1 ft, or $\frac{1}{4}$ in. = 1 ft

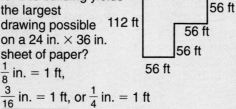

---

4 ASSESS

Closing Activity

Speaking On the chalkboard, write an ordered pair and the image of the ordered pair after applying a scale factor. Have students name the scale factor.

Chapter 8, Quiz D (Lessons 8-9 and 8-10) is available in the *Assessment and Evaluation Masters,* p. 212.

Additional Answers

18. He used dilations with a focal point in the bottom right quadrant of the picture.

19f. The perimeter of the dilation of a polygon is equal to the perimeter of the original polygon times the scale factor. The area of the dilation of a polygon is equal to the area of the original polygon times the square of the scale factor.

Practice Masters, p. 69

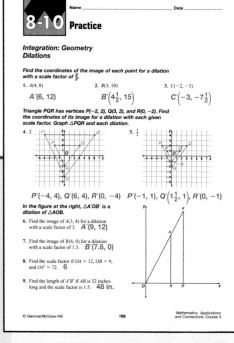

Lesson 8-10 **373**

Study Guide and Assessment

CHAPTER 8 Study Guide and Assessment

Vocabulary (teacher notes)

This section provides a listing of the new terms, properties, and phrases that were introduced in this chapter. Have students define each term and provide an example or two of it, if appropriate.

Understanding and Using the Vocabulary (teacher notes)

These exercises check students' understanding of the terms by using a variety of verbal formats including matching, completion, and true/false.

Glossaries A complete glossary of terms appears on pages 702–710. The glossary also appears in Spanish on pages 711–721.

Additional Answers

8. Multiply the principal ($500) times the rate (4.5%) times the time in years (1).
9. Sample answer: $\frac{80}{10} = \frac{x}{15}$; $120
10. Sample answer: $\frac{5}{4} = \frac{x}{10}$; $12\frac{1}{2}$ in.
11. $\frac{P}{18} = \frac{45}{100}$
12. $\frac{18}{27} = \frac{r}{100}$
13. $\frac{P}{80} = \frac{86}{100}$
14. $\frac{21}{B} = \frac{14}{100}$
15. $0.66 \cdot 7{,}000 = P$
16. $R \cdot 56 = 7$
17. $0.30 \cdot B = 15$
18. $R \cdot 500 = 60$

Vocabulary

After completing this chapter, you should be able to define each term, concept, or phrase and give an example or two of each.

Number and Operations
- base (p. 335)
- discount (p. 349)
- interest (p. 353)
- markup (p. 349)
- percent of change (p. 348)
- percent of decrease (p. 349)
- percent of increase (p. 349)
- percent proportion (p. 335)
- percentage (p. 335)

principal (p. 353)
rate (p. 335)
selling price (p. 349)

Measurement
- indirect measurement (p. 361)

Problem Solving
- solve a simpler problem (p. 342)

Geometry
- dilation (p. 370)
- pentagon (p. 358)
- scale (p. 366)
- scale drawing (p. 366)
- scale factor (p. 370)
- scale model (p. 366)
- similar polygons (p. 357)
- trigonometry (p. 365)

Understanding and Using the Vocabulary

Choose the letter of the term that best matches each phrase.

1. value represented by P in the equation $RB = P$ **d**
2. polygons that have the same shape but may differ in size   **b**
3. process of enlarging or reducing an image in mathematics   **f**
4. percent of increase or decrease   **e**
5. finding a measurement using proportions   **a**
6. drawing that represents a larger or smaller object   **c**
7. ratio of a dilated image to the original   **g**

a. indirect measurement
b. similar polygons
c. scale drawing
d. percentage
e. percent of change
f. dilation
g. scale factor

In Your Own Words

8. *Explain* how you would find the simple interest on $500 invested at 4.5% for 1 year. **See margin.**

Objectives & Examples

Upon completing this chapter, you should be able to:

- solve problems by using proportions *(Lesson 8-1)*

3 tapes cost $29.97. How much do 5 tapes cost?

$$\frac{3}{29.97} = \frac{5}{x} \rightarrow 3x = 29.97(5) \rightarrow x = 49.95$$

5 tapes cost $49.95.

Review Exercises

Use these exercises to review and prepare for the chapter test.

Write a proportion to solve each problem. Then solve. 9–10. See margin.

9. Luis earns $80 in 10 hours. How much will he earn in 15 hours?
10. A turtle can move 5 inches in 4 minutes. How far can the turtle move in 10 minutes?

374   Chapter 8 Applying Proportional Reasoning

MindJogger Videoquizzes

MindJogger Videoquizzes provide an alternative review of concepts presented in this chapter. Students work in teams to answer questions, gaining points for correct answers. The questions are presented in three rounds.

Round 1 Concepts–5 questions
Round 2 Skills–4 questions
Round 3 Problem Solving–4 questions

Study Guide and Assessment — Chapter 8

Objectives & Examples

- **solve problems using the percent proportion** *(Lesson 8-2)*

  75 is what percent of 250?

  $\frac{75}{250} = \frac{r}{100}$ Percent proportion

  $7{,}500 = 250r$ Find the cross products.

  $30 = r$ Divide each side by 250.

  So, 75 is 30% of 250.

- **solve problems using the percent equation** *(Lesson 8-3)*

  Find 15% of 82.

  $RB = P$ Percent equation

  $0.15 \cdot 82 = P$ $R = 0.15$ and $B = 82$.

  $12.3 = P$

  So, 15% of 82 is 12.3.

- **express percents greater than 100 or less than 1 as fractions and decimals** *(Lesson 8-4)*

  Express 130% as a mixed number.

  $130\% = \frac{130}{100} = 1\frac{30}{100} = 1\frac{3}{10}$

- **find and use the percent of increase or decrease** *(Lesson 8-5)*

  Find the percent of change of the price of a sweater that was $37.50 and is now on sale for $30.00.

  $37.50 - 30 = 7.50$

  $\frac{7.50}{37.50} = \frac{r}{100}$ Percent proportion

  $750 = 37.50r$ Find the cross products.

  $20 = r$ Divide each side by 37.50.

  The percent of change is 20%.

Review Exercises

Write a percent proportion to solve each problem. Then solve. Round to the nearest tenth. **11–14. See margin for proportions.**

11. Find 45% of 18.   **8.1**
12. 18 is what percent of 27?   **66.7%**
13. What is 86% of 80?   **68.8**
14. 21 is 14% of what number?   **150**

Write an equation in the form $RB = P$ for each problem. Then solve.

15. What is 66% of 7,000?   **4,620**
16. 7 is what percent of 56?   **12.5%**
17. 15 is 30% of what number?   **50**
18. 60 is what percent of 500?   **12%**

15–18. See margin for equations.

Express each percent as a fraction or mixed number in simplest form.

19. 215% $2\frac{3}{20}$
20. $\frac{1}{7}\%$ $\frac{1}{700}$

Express each percent as a decimal.

21. 0.6%   **0.006**
22. 126.5%   **1.265**

Find each percent of change. Round to the nearest percent.

23. original: 8, new: 10   **25%**
24. original: 10, new: 15   **50%**
25. original: 18, new: 12   **33%**
26. original: 900, new: 725   **19%**

27. Find the percent of change if the original number of members in the club was 35 and now there are 49 members.   **40%**

Chapter 8 Study Guide and Assessment **375**

Objectives & Examples

This section reviews the skills and concepts of the chapter and shows completely worked examples.

Review Exercises

These exercises provide practice for the corresponding objectives.

Assessment and Evaluation Masters, pp. 199–200

Assessment and Evaluation

Six forms of Chapter 8 Test are available in the *Assessment and Evaluation Masters* as shown in the chart.

Chapter 8 Test, Form 1B, is shown at the right. Chapter 8 Test, Form 2B, is shown on the next page.

Form	Type	Level
1A	Multiple Choice	Honors
1B	Multiple Choice	Average
1C	Multiple Choice	Basic
2A	Free Response	Honors
2B	Free Response	Average
2C	Free Response	Basic

Chapter 8 Test, Form 1B

1. A car uses 8 gallons of gasoline to travel 192 miles. How many gallons are used to go 144 miles?
   A. 10   B. 4   C. 12   D. 6   **1. D**

2. A recipe calls for 2 pounds, 8 ounces of apples for 8 servings. How many apples, in pounds and ounces, are needed for 12 servings?
   A. 3 pounds   B. 3 pounds, 12 ounces
   C. 2 pounds, 6 ounces   D. 4 pounds   **2. B**

3. 250 is what percent of 375?
   A. $33\frac{1}{3}\%$ B. 70%   C. $62\frac{1}{2}\%$ D. $66\frac{2}{3}\%$ **3. D**

4. Express $\frac{14}{25}$ as a percent.
   A. 56%   B. 5.6%   C. 28%   D. 50%   **4. A**

5. Fifty dogs were enrolled in an obedience school. Eighteen were German shepherds, 7 were sheepdogs, and 25 were mixed breeds. To find the combined percent of German shepherds and sheepdogs, which percent proportion would you use?
   A. $\frac{25}{50} = \frac{r}{100}$ B. $\frac{50}{25} = \frac{P}{100}$
   C. $\frac{25}{50} = \frac{100}{P}$ D. $\frac{25}{100} = \frac{P}{50}$ **5. A**

6. What is 45% of 260?
   A. 117   B. 1.17   C. 11.7   D. 11,700   **6. A**

7. 56 is 20% of what number?
   A. 70   B. 280   C. 40   D. 90   **7. B**

8. Express 12.75% as a fraction in simplest form.
   A. $\frac{51}{400}$ B. $\frac{51}{200}$ C. $\frac{51}{40}$ D. $\frac{1{,}275}{200}$ **8. A**

9. Express 0.08% as a decimal.
   A. 0.8   B. 0.008   C. 0.08   D. 0.0008   **9. D**

10. Find the percent of change in taxes if the old taxes were $38.00 and the new taxes are $41.61.
    A. 10.5%   B. 9.5%   C. 11%   D. 12%   **10. B**

11. Find the amount of discount and the sale price of a $6.99 cassette that is on sale for 20% off. Round to the nearest cent.
    A. $1.40; $5.59   B. $1.39; $5.49
    C. $1.38; $5.61   D. $1.45; $5.44   **11. A**

Chapter 8 Test, Form 1B (continued)

12. Find the simple interest to the nearest cent on $250 at 7% annually for 6 months.
    A. $9.00   B. $105.00   C. $29.17   D. $8.75   **12. D**

13. Find the annual rate of simple interest if the principal is $2,000, the interest is $375, and the time is 1 year.
    A. 19%   B. 18.75%   C. 20%   D. 1.875%   **13. B**

14. If the polygons $ABCD$ and $HIJK$ are similar, find the measure of side $\overline{HI}$.
    A. 13.5   B. 6   C. 12   D. 9   **14. B**

15. In the polygons above, which angle in polygon $ABCD$ corresponds to $\angle KHI$?
    A. $\angle DAB$ B. $\angle ABD$ C. $\angle BCD$ D. $\angle CDA$ **15. A**

16. A tree casts a shadow 18 feet long at the same time that a building that is 198 feet tall casts a 72-foot shadow. How tall is the tree?
    A. 50 feet   B. 16 feet   C. 49.5 feet   D. 11 feet   **16. C**

17. The distance on a map is 2.75 centimeters. Find the actual distance if the scale on the map is 1 centimeters = 60 kilometers.
    A. 165 kilometers   B. 21 kilometers
    C. 180 kilometers   D. 150 kilometers   **17. A**

18. The distance on a map is $1\frac{3}{8}$ inches. Find the actual distance if the scale is 1 inch:40 miles.
    A. 60 miles   B. 45 miles   C. 55 miles   D. 50 miles   **18. C**

19. Triangle ABC has vertices $A(-2, -4)$, $B(-3, 1)$, and $C(2, 2)$. Find the vertices for a dilation with a scale factor of 0.5.
    A. $A'(1, -2)$, $B'(1.5, -0.5)$, $C'(-1, -1)$
    B. $A'(-1, -3)$, $B'(-2, 2)$, $C'(1, 1)$
    C. $A'(-1, -2)$, $B'(-1.5, 0.5)$, $C'(1, 1)$
    D. $A'(-3, -5)$, $B'(-4, 0)$, $C'(1, 1)$ **19. C**

20. An image that is $12\frac{1}{2}$ inches wide on the transparency is 35 inches wide on a screen. What is the scale factor?
    A. 28   B. 2.8   C. 3.6   D. 3   **20. B**

Study Guide and Assessment **375**

Study Guide and Assessment

Chapter 8 Study Guide and Assessment

Objectives & Examples | Review Exercises

● solve problems involving simple interest
(Lesson 8-6)

Find the simple interest for $350 at 5% for 3 years.

$I = 350 \cdot 0.05 \cdot 3$

$= 52.50$ The interest is $52.50.

Find the simple interest to the nearest cent.
28. $100 at 8.5% for 2 years   **$17.00**
29. $780 at 6% for 8 months   **$31.20**
30. $125 at 5.55% for 2.5 years   **$17.34**
31. $260 at 17.5% for 18 months   **$68.25**

● identify corresponding parts of similar polygons and find missing measures of similar polygons
(Lesson 8-7)

Two polygons are similar if their corresponding angles are congruent and their corresponding sides are in proportion.

In the figure, $\triangle ABC \sim \triangle RST$. Find the length of each side.
32. $\overline{BC}$ $7\frac{1}{2}$ in.
33. $\overline{RS}$ $5\frac{1}{3}$ in.

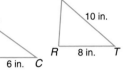

● solve problems involving similar triangles
(Lesson 8-8)

A mailbox casts a shadow 6.5 meters long, while a nearby tree casts a shadow 52 meters long. If the mailbox is 2 meters high, how tall is the tree?

mailbox shadow → $\frac{6.5}{2} = \frac{52}{x}$ ← tree shadow
mailbox height →   ← tree height

$6.5x = 104$

$x = 16$

The tree is 16 meters tall.

Solve.
34. A house casts a shadow $5\frac{1}{2}$ feet long. At the same time, a light post casts a $2\frac{1}{4}$-foot shadow. If the light post is 6 feet high, how tall is the house? $14\frac{2}{3}$ ft
35. A flagpole casts a shadow 9 feet long at the same time that an office building casts a 15-foot shadow. If the office building is 60 feet tall, how tall is the flagpole?   **36 ft**

● solve problems involving scale drawings
(Lesson 8-9)

In a drawing, 3 cm = 45 m. Find the length of a line representing 75 meters.

$\frac{3 \text{ centimeters}}{45 \text{ meters}} = \frac{x \text{ centimeters}}{75 \text{ meters}}$

$225 = 45x$

$5 = x$

The line is 5 centimeters long.

On a map, 2 inches = 5 miles. Find the actual distance for each map distance.
36. 4 inches   **10 mi**
37. 5.5 inches $13\frac{3}{4}$ mi
38. 12 inches   **30 mi**
39. 9 inches $22\frac{1}{2}$ mi

376   Chapter 8 Applying Proportional Reasoning

Assessment and Evaluation Masters, pp. 205–206

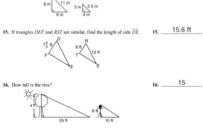

📀 Test and Review Software

You may use this software, a combination of an item generator and item bank, to create your own tests or worksheets. Types of items include free response, multiple choice, short answer, and open ended.

💿 CD-ROM Program

The CD-ROM Program contains an Assessment Game whose questions review the concepts in this chapter.

376   Chapter 8

Study Guide and Assessment Chapter 8

- graph dilations on a coordinate plane
(Lesson 8-10)

In a dilation with a scale factor of k, the ordered pair (x, y) becomes the ordered pair (kx, ky).

Find the coordinates of the image of each point for a dilation with a scale factor of 4.

40. $A(6, 9)$ **(24, 36)**   41. $B(15, 3)$ **(60, 12)**
42. $C(2, 5)$ **(8, 20)**   43. $D(1, 4)$ **(4, 16)**

Applications & Problem Solving

44. Entertainment Domingo hired a rock band to play at the school dance. The band charges $3,000 per performance. The manager required a 20% deposit. How much did Domingo pay for the deposit? *(Lesson 8-2)* **$600**

45. Solve a Simpler Problem How many cuts are needed to separate a long board into 19 shorter pieces with the same width as the original board? *(Lesson 8-3B)* **18 cuts**

46. Money Matters Emily wants to buy a CD player that costs $189.95. If she waits to buy the CD player until it is on sale for $132.97, what is the percent of savings? *(Lesson 8-5)* **about 30.0%**

47. Travel A map of the roads between four cities is shown below. How far is it from Main City to Carlton to the nearest mile? *(Lesson 8-8)* **107 mi**

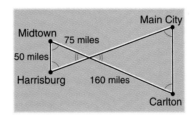

Alternative Assessment

Performance Task

Suppose you earn $42 for working 8 hours. A friend says that you can use the following proportion to find out how much you would earn for 24 hours.

$$\frac{42}{8} = \frac{24}{h}$$

Is your friend correct? Why or why not? If not, write the correct proportion. Use the correct proportion to solve the problem.

Determine how to find out how much you would earn for 24 hours of work without using proportions. **See Answer Appendix.**

A practice test for Chapter 8 is provided on page 654.

Completing the CHAPTER Project

Use the following checklist to make sure your newspaper article is complete.

☑ The class rating for each program is given both as a fraction and as a percent.

☑ The Nielsen ratings for your program are recorded for three consecutive weeks and the percent of change is computed.

☑ The class ratings and the Nielsen ratings are compared.

 Select one of the problems you solved in this chapter and place it in your portfolio. Attach a note explaining how your problem illustrates one of the important concepts covered in this chapter.

 Performance Assessment

Additional performance assessment tasks for this chapter are included in the *Assessment and Evaluation Masters* on page 209. A scoring guide is also provided on page 221.

Applications & Problem Solving

This section provides additional practice in solving real-world problems that involve the skills of this chapter.

Alternative Assessment

The *Performance Task* provides students with a performance assessment opportunity to evaluate their work and understanding.

CHAPTER Project

Students should complete the final stages of their project and prepare a class demonstration of their results. A scoring guide for the project is available in the *Investigations and Projects Masters*, p. 47.

 Students should add to their portfolios at this time.

Assessment and Evaluation Masters, p. 209

Standardized Test Practice

The Standardized Test Practice may be used to help students prepare for standardized tests. The test items are written in the same style as those in state proficiency tests and standardized tests like CAT, CTBS, ITBS, MAT, SAT, and Terra Nova. The test items cover skills and concepts covered up to this point in the text.

The pages can be used as an overnight assessment. After students have completed the pages, discuss how each problem can be solved, or provide copies of the solutions from the *Solutions Manual*.

Assessment and Evaluation Masters, p. 215

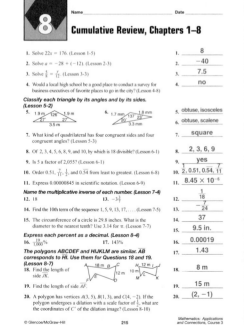

CHAPTERS 1–8 Standardized Test Practice
Assessing Knowledge & Skills

Section One: Multiple Choice

There are eleven multiple-choice questions in this section. Choose the best answer. If a correct answer is *not here,* choose the letter for Not Here.

1. What is the probability that Kinsey will draw a green marble from a jar containing 3 white marbles, 2 blue marbles, 5 orange marbles, and no other marbles?   **A**
   A 0
   B $\frac{1}{10}$
   C $\frac{1}{3}$
   D 1

2. If $2x - 5 = -3$, what is the value of x?   **H**
   F -4
   G -1
   H 1
   J 4

3. $6a + 5a - 2a =$ **A**
   A $9a$
   B $13a$
   C $11a - 2$
   D $9a^3$

4. $0.3\% =$ **J**
   F 3
   G 0.3
   H 0.03
   J 0.003

5. How many lines of symmetry does a square have?   **D**
   A none
   B 1
   C 2
   D 4

Please note that Questions 6–11 have five answer choices.

6. The Venn diagram shows the relationship of four sets of students in the music program at Jones Middle School.

   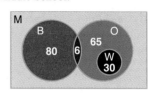

   M = the set of all students in the school music program.
   B = the set of all students in the band.
   O = the set of all students in the orchestra.
   W = the set of all students in the woodwinds.

   Which is a valid conclusion concerning these sets?   **H**
   F There is no orchestra student in the band.
   G All woodwind students are in the band.
   H All woodwind students are in the orchestra.
   J There are students in the band that are not in the music program.
   K All orchestra students play woodwinds.

7. The student senate surveyed the 8th grade class to determine what kind of decorations to have for the winter dance. Of the 320 students surveyed, 100 preferred streamers. If the same survey were conducted among the entire middle school population of 1,280 students, what is a reasonable number of students in the entire middle school that would prefer streamers?   **B**
   A 300
   B 400
   C 500
   D 600
   E 700

378 Chapters 1–8 Standardized Test Practice

◀◀◀ **Instructional Resources**

Another cumulative review is shown at the left and is available in the *Assessment and Evaluation Masters,* p. 215.

Standardized Test Practice Chapters 1–8

8. If it costs approximately $150 per week to feed a family of 5, how much should it cost to feed a family of 7? **J**
   F $30
   G $150
   H $180
   J $210
   K $250

9. Bananas are on sale for $0.32 per pound. If Iggy's Ice Cream Parlor pays $30.08 for a shipment of bananas, how many pounds were purchased? **A**
   A 94 lb
   B 99 lb
   C 104 lb
   D 127 lb
   E Not Here

10. Suppose $840 is deposited in a savings account that earned 4.5% interest. If there are no deposits or withdrawals, how much was in the account at the end of 6 months? **G**
    F $1,066.80
    G $858.90
    H $226.80
    J $18.90
    K Not Here

11. Including tax, the cost of renting a movie on videotape for 2 days is $2.50 and $2.00 for each day after that. What is the cost for renting a movie for 7 days? **B**
    A $10.00
    B $12.50
    C $14.00
    D $16.50
    E Not Here

Test-Taking Tip

When you prepare for a standardized test, review basic definitions such as the ones below.
- The least common multiple is the least of the common multiples of two or more numbers.
- A number is prime if it has exactly two factors, 1 and itself.

You can use a basic definition when answering Question 12.

Section Two: Free Response

This section contains five questions for which you will provide short answers. Write your answers on your paper.

12. What is the least prime factor of 54? **2**

13. What is the greatest common factor of 14 and 50? **2**

For Questions 14 and 15, use the following information.

These are the daily low temperatures (°F) for two weeks in April.

36, 42, 38, 50, 48, 44, 46,
50, 52, 49, 48, 45, 46, 48

14. What is the median temperature? **47°**

15. What is the interquartile range of the data? **5**

16. An architectural drawing measuring 9 inches wide by 15 inches long is enlarged. The new enlargement is 36 inches wide. How long is the enlargement? **60 in.**

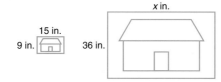

Chapters 1–8 Standardized Test Practice **379**

CHAPTER 9
Algebra: Exploring Real Numbers

Previewing the Chapter

Overview

With the inclusion of irrational numbers, this chapter fills in the number lines begun in Chapter 2 (Algebra: Using Integers) and continued in Chapter 7 (Algebra: Using Rational Numbers). Students work extensively with square roots before learning the definition of irrational numbers. The Pythagorean Theorem is developed and applied in a variety of ways, especially in finding distances, graphing irrational numbers, and discovering relationships in 30°-60° and 45°-45° right triangles. Students will learn how to solve problems by using a formula and by drawing a diagram.

Lesson (pages)	Lesson Objectives	NCTM Standards	Standardized Tests	State/Local Objectives
9-1 (382–384)	Find square roots of perfect squares.	1–7, 12	CTBS, SAT, TN	
9-2A (385)	Use models to estimate square roots.	1–5, 7, 8		
9-2 (386–389)	Estimate square roots.	1–5, 7, 8, 12		
9-3 (390–394)	Identify and classify numbers in the real number system.	1–4, 6, 7, 9	CAT, ITBS, SAT	
9-4A (396–397)	Explore the relationships in a right triangle.	1–5, 7, 8, 12, 13		
9-4 (398–401)	Use the Pythagorean Theorem.	1–5, 7–9, 12, 13		
9-5A (402–403)	Solve problems by drawing a diagram.	1–5, 7, 8	CAT, MAT	
9-5 (404–407)	Solve problems using the Pythagorean Theorem.	1–5, 7, 8, 12, 13		
9-5B (408–409)	Use the Pythagorean Theorem to graph irrational numbers on a number line.	1–5, 7, 12, 13		
9-6 (410–413)	Find the distance between points in the coordinate plane.	1–5, 7, 12, 13	TN	
9-7 (414–417)	Find missing measures in 30°-60° right triangles and 45°-45° right triangles.	1–5, 7–9, 12, 13		

CAT = California Achievement Tests, CTBS = Comprehensive Tests of Basic Skills, ITBS = Iowa Tests of Basic Skills, MAT = Metropolitan Achievement Tests, SAT = Stanford Achievement Tests, TN = Terra Nova

Organizing the Chapter

CD-ROM — All of the blackline masters in the Teacher's Classroom Resources are available on the **Electronic Teacher's Classroom Resources** CD-ROM.

LESSON PLANNING GUIDE

Lesson	Extra Practice (Student Edition)	Blackline Masters (page numbers)										
		Study Guide	Practice	Enrichment	Assessment & Evaluation	Classroom Games	Diversity	Hands-On Lab	School to Career	Science and Math Lab Manual	Technology	Transparencies A and B
9-1	p. 630	70	70	70		35					69	9-1
9-2A								57				
9-2	p. 631	71	71	71	239							9-2
9-3	p. 631	72	72	72								9-3
9-4A								58				
9-4	p. 631	73	73	73	238, 239							9-4
9-5A	p. 632											
9-5	p. 632	74	74	74							70	9-5
9-5B								59				
9-6	p. 632	75	75	75	240				35			9-6
9-7	p. 633	76	76	76	240	25–29		76				9-7
Study Guide/ Assessment					225–237, 241–243							

Other Chapter Resources

Student Edition
Chapter Project, pp. 388, 393, 413, 421
School to Career, p. 395
Let the Games Begin, p. 389

Technology
 CD-ROM Program
 Interactive Mathematics Tools Software

Teacher's Classroom Resources

Applications
Family Letters and Activities, pp. 69–70
Investigations and Projects Masters, pp. 49–52

Meeting Individual Needs
Investigations for the Special Education Student, pp. 19–20

Teaching Aids
Answer Key Masters
Block Scheduling Booklet
Lesson Planning Guide
Solutions Manual

Professional Publications
Glencoe Mathematics Professional Series

Planning the Chapter

 MindJogger Videoquizzes provide a unique format for reviewing concepts presented in the chapter.

Assessment Resources

Student Edition
Mixed Review, pp. 384, 388, 394, 401, 407, 413, 417
Mid-Chapter Self Test, p. 394
Math Journal, pp. 383, 399
Study Guide and Assessment, pp. 418–421
Performance Task, p. 421
Portfolio Suggestion, p. 421
Standardized Test Practice, pp. 422–423
Chapter Test, p. 655

Assessment and Evaluation Masters
Multiple-Choice Tests (Forms 1A, 1B, 1C), pp. 225–230
Free-Response Tests (Forms 2A, 2B, 2C), pp. 231–236
Performance Assessment, p. 237
Mid-Chapter Test, p. 238
Quizzes A–D, pp. 239–240
Standardized Test Practice, pp. 241–242
Cumulative Review, p. 243

Teacher's Wraparound Edition
5-Minute Check, pp. 382, 386, 390, 398, 404, 410, 414
Building Portfolios, p. 380
Math Journal, pp. 385, 397, 409
Closing Activity, pp. 384, 389, 394, 401, 403, 407, 413, 417

Technology
Test and Review Software
MindJogger Videoquizzes
CD-ROM Program

Materials and Manipulatives

Lesson 9-1
calculator

Lesson 9-2A
algebra tiles*† or base-ten blocks*
calculator

Lesson 9-2
index cards
markers

Lesson 9-3
calculator

Lesson 9-4A
dot paper*†
geoboard*†
geobands*†

Lesson 9-4
calculator

Lesson 9-5
calculator

Lesson 9-5B
compass*†
straightedge*†

Lesson 9-6
calculator
grid paper†
ruler*†

Lesson 9-7
calculator
compass*†
protractor*†
ruler*†
scissors*

*Glencoe Manipulative Kit †Glencoe Overhead Manipulative Resources

Pacing Chart

See pages T25–T27 for the Course Planning Calendar.

COURSE	DAY 1	DAY 2	DAY 3	DAY 4	DAY 5	DAY 6	DAY 7
Standard	Chapter Project	Lesson 9-1	Lessons 9-2A & 9-2		Lesson 9-3	Lessons 9-4A & 9-4	
Honors	Chapter Project & Lesson 9-1	Lesson 9-2	Lesson 9-3	Lesson 9-4	Lesson 9-5A	Lessons 9-5 & 9-5B	
Block	Chapter Project & Lesson 9-1	Lessons 9-2A & 9-2	Lessons 9-3 & 9-4A	Lessons 9-4 & 9-5A	Lessons 9-5 & 9-6	Lesson 9-7	Study Guide and Assessment, Chapter Test

Interactive Mathematics:
Activities and Investigations

is an activity-based program that may be used as an enhancement for chapters in *Mathematics: Applications and Connections.*

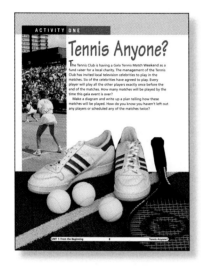

Unit 1, Activity One
Use with Lesson 9-5A.

Summary Students work in groups to make a diagram showing which tennis players in a charity fund raiser play each other and in what order. Diagrams, charts, or written instructions allow students different choices in presenting their conclusions.

Math Connection Students draw a diagram to solve a problem. They need to determine how many games each tennis player must play and then support and communicate their reasoning in writing.

Unit 15, Activity One
Use with Lesson 9-6.

Summary Students work in groups at various work stations to perform activities that explore the mathematical ideas of motion. The activities involve distance, time, and speed. At the end of each activity, they graph the results and discuss their graphs.

Math Connection Students graph results in a coordinate plane. Then they look at the graphs to determine a pattern if one exists.

DAY 8	DAY 9	DAY 10	DAY 11	DAY 12	DAY 13	DAY 14	DAY 15
Lesson 9-5A	Lesson 9-5	Lesson 9-6	Lesson 9-7	Study Guide and Assessment	Chapter Test		
Lesson 9-6	Lesson 9-7	Study Guide and Assessment	Chapter Test				

Enhancing the Chapter

APPLICATIONS

Classroom Games, pp. 25–29

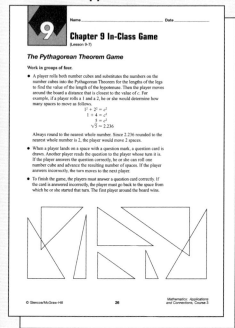

Diversity Masters, p. 35

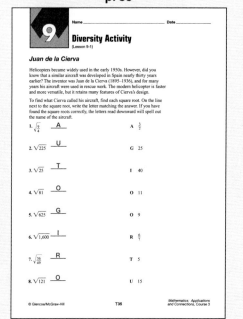

School to Career Masters, p. 35

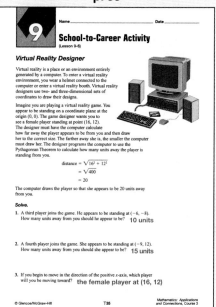

Family Letters and Activities, pp. 69–70

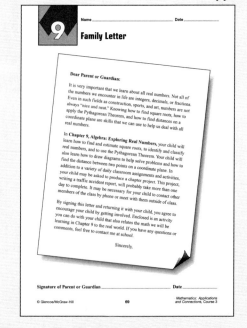

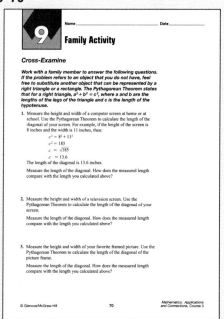

MANIPULATIVES/MODELING

Hands-On Lab Masters, p. 76

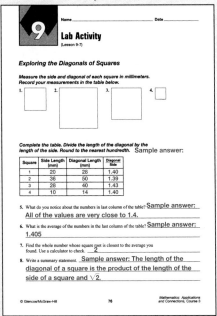

ASSESSMENT/EVALUATION

Assessment and Evaluation Masters, pp. 238–240

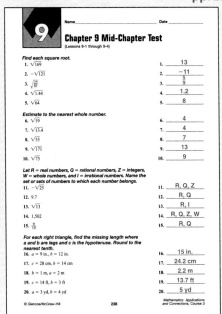

TECHNOLOGY/MULTIMEDIA

Technology Masters, pp. 69–70

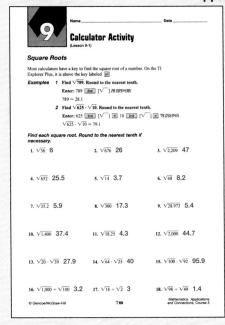

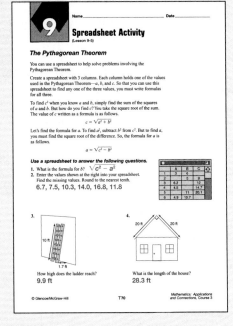

MEETING INDIVIDUAL NEEDS

Investigations for the Special Education Student, pp. 19–20

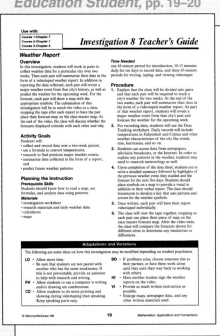

CHAPTER 9 NOTES

Theme: Traffic Safety
Information from accident reports helps the police and insurance agencies reconstruct accidents. Officers measure and record the final position of collided vehicles, the condition of the road, and the weather. They use measurements of skid marks and the drag factor of the road to calculate the speed of the vehicles at the time of the crash. They take photos of the scene and make scale drawings.

Question of the Day Suppose traffic officers arrive at the scene of a four-car collision at a three-way intersection. Have students draw a possible diagram reconstructing directions, angles, and locations of the vehicles. Ask students how they would plot the accident on a coordinate plane and calculate distances. **Answers will vary.**

Assess Prerequisite Skills
Ask students to read through the list of objectives presented in "What you'll learn in Chapter 9." You may wish to ask them what each of the objectives means or if they have experienced or used any of these math concepts before.

 Building Portfolios
Encourage students to revise their portfolios as they study this chapter. Students may decide to include diagrams to show their understanding of the concepts of this chapter.

 Math and the Family
In the *Family Letters and Activities* booklet (pp. 69–70), you will find a letter to the parents explaining what students will study in Chapter 9. An activity appropriate for the whole family is also available.

CHAPTER 9 — Algebra: Exploring Real Numbers

What you'll learn in Chapter 9
- to estimate and find square roots,
- to identify and classify numbers,
- to use the Pythagorean Theorem,
- to solve problems by drawing a diagram, and
- to find the distance between points in the coordinate plane.

 CD-ROM Program

Activities for Chapter 9
- Chapter 9 Introduction
- Interactive Lessons 9-2, 9-4
- Assessment Game
- Resource Lessons 9-1 through 9-7

CHAPTER Project

PROCEED WITH CAUTION

Traffic accidents can happen at any time and in any weather condition. Police officers often reconstruct the details of the accident. If an officer doesn't actually see the accident, how can he or she know what happened? Officers look for certain clues at the accident scene. In this project, you will use square roots, real numbers, and distance on a coordinate plane to write a traffic accident report for the accident scene shown in the diagram.

Getting Started

- Research how to investigate an accident scene and how to write an accident report. What types of information should go into an accident report?

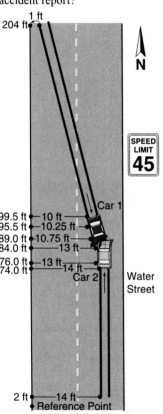

Technology Tips

- Use a **word processor** to write your accident report.
- Surf the **Internet** for more information.

interNET CONNECTION For up-to-date information on traffic, visit:
www.glencoe.com/sec/math/mac/mathnet

Working on the Project

You can use what you'll learn in Chapter 9 to help you write a traffic report.

Page	Exercise
388	25
393	32
413	21
421	Alternative Assessment

CHAPTER Project NOTES

Objectives Students should
- plot real numbers on a coordinate plane.
- find distance by using the Pythagorean Theorem.

Project Pointer You may suggest that students begin a *Project Folder* to keep their work as they complete each stage of the Chapter Project. The completed project may also be added to their portfolios.

Using the Diagram Discuss the diagram with your students. What can they tell about how the accident occurred? Car 1 swerved out of its lane into Car 2, which was traveling in the opposite direction.

Investigations and Projects Masters, p. 52

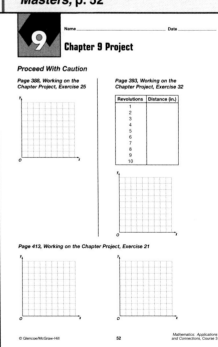

Instructional Resources ▶▶▶
A recording sheet to help students organize their data for the Chapter Project is shown at the right and is available in the *Investigations and Projects Masters*, p. 52.

interNET CONNECTION
Glencoe has made every effort to ensure that the website links for *Mathematics: Applications and Connections* at www.glencoe.com/sec/math/mac/mathnet are current and contain appropriate content. However, these website links are not under Glencoe's control.

9-1 Lesson Notes

Instructional Resources
- *Study Guide Masters*, p. 70
- *Practice Masters*, p. 70
- *Enrichment Masters*, p. 70
- Transparencies 9-1, A and B
- *Diversity Masters*, p. 35
- *Technology Masters*, p. 69

CD-ROM Program
- Resource Lesson 9-1

Recommended Pacing	
Standard	Day 2 of 13
Honors	Day 1 of 11
Block	Day 1 of 7

1 FOCUS

5-Minute Check
(Chapter 8)

1. On a map, 2 inches equal 31 miles. City A is 3 inches from City B. What is the actual distance from City A to City B? Write a proportion to solve the problem. Then solve. $\frac{2}{31} = \frac{3}{m}$; $46\frac{1}{2}$ miles

2. The Good Food Grocery Store advertises that its store brand of cereal is on sale for 35% off the regular price. Find the sale price of a box of cornflakes that costs $2.25. Round to the nearest cent. **$1.46**

3. A tower is 125 meters tall. A souvenir of it is 25 centimeters high. What is the scale factor? **0.002**

The 5-Minute Check is also available on **Transparency 9-1A** for this lesson.

Motivating the Lesson
Problem Solving Ask students to find the length of a side of a square garden if the area of the garden is 400 square feet. **20 feet**

382 Chapter 9

9-1 Square Roots

What you'll learn
You'll learn to find square roots of perfect squares.

When am I ever going to use this?
Knowing how to find square roots can help you find the dimensions of a square work space.

Word Wise
perfect square
square root
principal square root
radical sign

Did you know? The square root symbol we use today evolved from a symbol that Christoff Rudolff used in 1525. It resembled a small r for the initial of the word *radix*.

The Pythagorean school of ancient Greece focused on the study of philosophy, mathematics, and natural science. The students, called Pythagoreans, made many advances in these fields. One of their studies was *figurate numbers*. By drawing pictures of various numbers, patterns can be discovered. For example, some whole numbers can be represented by drawing dots arranged in squares.

1	4	9	16	64
1 by 1	2 by 2	3 by 3	4 by 4	8 by 8

Numbers that can be pictured in squares of dots are called **perfect squares**. The number of dots in each row or column in the square is called a **square root** of the perfect square. The perfect square 9 has a square root of 3 because there are 3 rows and 3 columns. We say that 8 is a square root of 64 because $64 = 8 \times 8$ or 8^2.

Square Root	If $x^2 = y$, then x is a square root of y.

Perfect squares also include decimals and fractions like 0.09 and $\frac{4}{9}$ since $(0.3)^2 = 0.09$ and $\left(\frac{2}{3}\right)^2 = \frac{4}{9}$.

It is also true that $(-8)^2 = 64$ and $(-12)^2 = 144$. So we can say that -8 is also a square root of 64 and -12 is a square root of 144.

The positive square root of a number is called the **principal square root**. The symbol $\sqrt{}$, called a **radical sign**, is used to indicate the principal square root.

$\sqrt{25} = 5$ *principal square root*

$-\sqrt{25} = -5$ *negative square root*

Examples

Study Hint
Reading Math $\sqrt{225}$ is read as *the square root of 225*.

❶ Find $\sqrt{225}$.

The symbol $\sqrt{}$ indicates the principal square root.

Since $15^2 = 225$, $\sqrt{225} = 15$.

❷ Find $-\sqrt{49}$.

The $-\sqrt{}$ symbol indicates the negative square root.

Since $(-7)^2 = 49$, $-\sqrt{49} = -7$.

382 Chapter 9 Algebra: Exploring Real Numbers

Classroom Vignette

"I like to have my students use their calculators to calculate expressions like $(-8)^2$ and -8^2 for *negative eight squared* so they can see that when they square a negative number, the parentheses are necessary.

Beth Murphy Anderson, Teacher
Brownell Talbot School
Omaha, NE

Some equations that involve squares can be solved by taking the square root of each side of the equation.

Examples
INTEGRATION

3 **Algebra** Find a number that when squared equals 361.

Let n be the number.

Then $n^2 = 361$.

$\sqrt{n^2} = \sqrt{361}$ *Take the square root of each side.*

361 [2nd] [√] *19*   *Use a calculator.*

$n = 19$ or $n = -19$

The number could be either 19 or -19.

APPLICATION

4 **Design** Ms. Lewis is creating several cubicle work spaces for her consulting business. One square workspace needs to have a floor area of 576 square feet. Draw a diagram and describe how she could find the length of each side for the cubicle.

Use the formula for the area of a square.

$A = s^2$

$576 = s^2$ *Replace A with 576.*

$\sqrt{576} = s$ *Take the square root of each side.*

576 [2nd] [√] *24*   *Use a calculator.*

$s = 24$ or $s = -24$

Therefore, each side of the workspace must be 24 feet.

-24 does not make sense because length is never negative.

s ft

576 square feet   s ft

CHECK FOR UNDERSTANDING

Communicating Mathematics

Read and study the lesson to answer each question.

1. *Write* the symbol for the negative square root of 100. $-\sqrt{100}$

2. *Tell* why $\frac{36}{81}$ is a perfect square.   **See Answer Appendix.**

3. *Write* a description of the relationship between a perfect square and its two square roots. Give a specific example and include a drawing of dots arranged in a square.   **See Answer Appendix.**

Guided Practice

Find each square root.

4. $\sqrt{100}$ **10**   5. $\sqrt{81}$ **9**   6. $\sqrt{121}$ **11**   7. $-\sqrt{64}$ **−8**

8. *Algebra* Solve $x^2 = 144$.   **12, −12**

9. *History* The Great Pyramid at Giza, built around 2600 B.C., is one of the "Seven Wonders of the Ancient World." It has a square base with an area of 602,176 square feet. How long is each side of the base?   **776 feet**

Lesson 9-1 Square Roots **383**

Reteaching the Lesson

Activity Have students use a calculator to find the first fifteen perfect squares that are whole numbers. **1, 4, 9, 16, 25, 36, 49, 64, 81, 100, 121, 144, 169, 196, 225**

Error Analysis
Watch for students who confuse finding $\sqrt{9}$ and solving $x^2 = 9$.
Prevent by stressing that $\sqrt{9}$ indicates the principal square root, but $x^2 = 9$ asks for all possible roots, both positive and negative.

---

2 TEACH

 Transparency 9-1B contains a teaching aid for this lesson.

Reading Mathematics Inform students that *radical* in the context of square roots does not mean "extreme." Rather, it comes from the Latin word *radix,* meaning "root."

In-Class Examples

For Example 1
Find $\sqrt{324}$. **18**

For Example 2
Find $-\sqrt{\dfrac{144}{225}}$. $-\dfrac{4}{5}$

For Example 3
Find a number that when squared equals 676. **26 or −26**

For Example 4
An architect is asked to design a square backyard deck with an area of 169 square feet. Draw a diagram and describe how she could find the length of each side for the deck. **13 ft**

s ft

169 sq ft   s ft

Study Guide Masters, p. 70

Lesson 9-1 **383**

3 PRACTICE/APPLY

Check for Understanding
If students need additional practice or instruction after completing Exercises 1–9, one of these options may be helpful.
- Extra Practice, see p. 630
- Reteaching Activity
- *Study Guide Masters*, p. 70
- *Practice Masters*, p. 70
- Interactive Mathematics Tools Software

Assignment Guide
Core: 11–29 odd, 30–33
Enriched: 10–26 even, 27–33

4 ASSESS

Closing Activity
Modeling Have students use square tiles to construct squares and then to use the lengths of their sides to find the square roots of the areas.

Additional Answer
25. 13, −13; 13 is the principal square root because it is the positive square root.

Practice Masters, p. 70

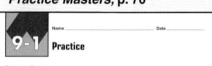

EXERCISES

Practice Find each square root.

10. $\sqrt{25}$ 5
11. $-\sqrt{9}$ −3
12. $\sqrt{256}$ 16
13. $-\sqrt{36}$ −6
14. $\sqrt{400}$ 20
15. $\sqrt{\frac{4}{9}}$ $\frac{2}{3}$
16. $\sqrt{0.16}$ 0.4
17. $-\sqrt{\frac{64}{100}}$ $-\frac{4}{5}$
18. $\sqrt{441}$ 21
19. $-\sqrt{196}$ −14
20. $\sqrt{\frac{121}{625}}$ $\frac{11}{25}$
21. $-\sqrt{2.89}$ −1.7

Solve each equation.

22. $x^2 = 81$ 9, −9
23. $0.25 = x^2$ 0.5, −0.5
24. $x^2 = \frac{16}{25}$ $\frac{4}{5}, -\frac{4}{5}$

25. Find two square roots of 169. Identify the principal square root. **See margin.**

26. *Geometry* If the area of a square is 1.69 square meters, what is the length of each of its sides? **1.3 meters**

Applications and Problem Solving

27. *Agriculture* Pecan trees are planted in square patterns to take advantage of land space and for ease of harvesting. If you wanted to plant 289 trees, how many rows and how many trees in each row would you plant?

28. *Design* The designs of tile floors are often planned on a piece of grid paper. Suppose you have 100 1-inch square blue tiles.
   a. What is the largest square you could make with these tiles? **10 in. × 10 in.**
   b. If you wanted to make two square designs that were the same size, what is the largest that each square could be? How many tiles would be left over?
   c. Describe three different-sized squares that you could make at the same time out of your 100 tiles. How many tiles are left over?

27. 17 rows of 17 trees in each row
28b. 7 in. × 7 in.; 2 tiles
28c. Sample answers: 8 in. × 8 in., 5 in. × 5 in., 3 in. × 3 in.; 11 tiles.

29. *Critical Thinking* Is the product of two perfect squares always a perfect square? Explain why or why not. **See Answer Appendix.**

Mixed Review

30. See Answer Appendix.

30. *Geometry* Graph $\overline{EF}$ with endpoints $E(2, 6)$ and $F(4, -4)$. Then graph its image for a dilation with a scale factor of 2. *(Lesson 8-10)*

31. *Geometry* Find the area of a trapezoid with bases 9 and 12 centimeters long and a height of 11 centimeters. *(Lesson 7-6)* **115.5 cm²**

32. **Test Practice** A ladder leaning against the side of a building forms a 75° angle with the ground. Approximately what is the measure of the angle between the ground and the other side of the ladder? *(Lesson 5-1)* **C**
   A 60° B 75° C 105° D 255°

33. *Statistics* Find the mean, median, and mode of the data 6, 4, 6, 12, 10, 8, 7, 12, 11, 9. *(Lesson 4-4)* **8.5; 8.5; 6, 12**

384 Chapter 9 Algebra: Exploring Real Numbers

Extending the Lesson

Enrichment Masters, p. 70

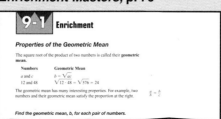

Activity The Pythagoreans also studied triangular numbers. These can be arranged as dots in triangles. Any two consecutive triangular numbers add up to a square. Have students work in pairs to discover several triangular numbers. **1, 3, 6, 10, ...** Then have them test whether any two adjacent triangles add up to a square. **yes: 1 + 3 = 4, 6 + 10 = 16, and so on**

COOPERATIVE LEARNING

9-2A Estimating Square Roots
A Preview of Lesson 9-2

- algebra tiles or base-ten blocks
- calculator

Suppose you try to arrange 50 tiles into a square. You discover that it's impossible. This suggests that 50 is not a perfect square. In this lab, you will estimate the square roots of numbers that are not perfect squares.

TRY THIS

Work with a partner.
Use tiles to estimate the square root of 50.

- Arrange 50 tiles into the largest square possible.
  The largest possible square has 49 tiles, with one left over.

- Add tiles until you have the next larger square.
  You need to add 7 tiles on top and 7 tiles on the side, and then the leftover tile from above can be placed in the upper right corner. Therefore, you added 14 new tiles to make a square that has 64 tiles.

- The square root of 49 is 7, and the square root of 64 is 8. Therefore, the square root of 50 is between the whole numbers 7 and 8. $\sqrt{50}$ is closer to 7 because 50 is closer to 49 than 64.

$\sqrt{49} = 7$

$\sqrt{64} = 8$

- Verify the estimate with a calculator. 50 [2nd] [√] 7.071067812

ON YOUR OWN

For each number, arrange tiles or base-ten blocks into the largest square possible. Then add tiles until you have the next larger square. To the nearest whole number, estimate the square root of each number.

1. 20   **4**
2. 76   **9**
3. 133 **12**
4. 150 **12**

5. **Look Ahead** Describe another method that could be used to find the square root of a number by squaring numbers to estimate, rather than by taking square roots to estimate. **See margin.**

Lesson 9-2A HANDS-ON LAB   **385**

Additional Answer

5. You can square numbers by guess and check until you find a number that when squared is less than the square root you are finding and then the next greater number that when squared is just greater than the square root.

 Have students write a paragraph explaining how to estimate $\sqrt{67}$ using tiles.

9-2A LAB Notes

GET READY

Objective Students use models to estimate square roots.

Optional Resources
Hands-On Lab Masters
- algebra tiles, p. 29
- centimeter grid, p. 11
- worksheet, p. 57

Overhead Manipulative Resources
- area/algebra tiles

Manipulative Kit
- algebra tiles or base-ten blocks

MANAGEMENT TIPS

Recommended Time
20 minutes

Getting Started Before beginning the Activity, have students make a list of perfect squares and their principal square roots to use as a reference in building models.

Teaching Tip Centimeter grid paper can be cut into squares to use for this activity. Sugar cubes also work well.

In the **Activity,** students add tiles to make the next largest perfect square. Remind them that a square must have the same number of tiles on each of the four sides. If they add 6 tiles to finish the column, there will be 7 tiles in the columns and 8 tiles in the rows, which is not a square.

Ask students to demonstrate how they would estimate $\sqrt{10}$ with tiles. Then have them use a number line to indicate where $\sqrt{10}$ lies.

ASSESS

Have students complete Exercises 1–4. Watch for students who may have difficulty modeling larger numbers. Point out that the square root of 100 is 10. Therefore, they will need more than 10 squares on each side.

Use Exercise 5 to determine whether students have discovered how to estimate square roots.

Hands-On Lab 9-2A   **385**

9-2 Lesson Notes

Instructional Resources
- *Study Guide Masters*, p. 71
- *Practice Masters*, p. 71
- *Enrichment Masters*, p. 71
- Transparencies 9-2, A and B
- *Assessment and Evaluation Masters*, p. 239

 CD-ROM Program
- Resource Lesson 9-2
- Interactive Lesson 9-2

Recommended Pacing	
Standard	Days 3 & 4 of 13
Honors	Day 2 of 11
Block	Day 2 of 7

1 FOCUS

5-Minute Check
(Lesson 9-1)

Find each square root.
1. $\sqrt{289}$ **17**
2. $-\sqrt{484}$ **−22**
3. $\sqrt{\frac{25}{64}}$ **$\frac{5}{8}$**
4. Solve $x^2 = 100$. **10, −10**
5. The area of a town square is 961 square meters. Find the length of a side. **31 m**

 The 5-Minute Check is also available on **Transparency 9-2A** for this lesson.

Motivating the Lesson
Hands-On Activity Have students use calculators to find the square root of each whole number from 9 to 16. Write the square roots on the board. Ask students to draw a conclusion. **Sample answer: As the number increases, so does its square root.**

9-2 Estimating Square Roots

Deep down inside, Coach Knott had always wanted to be a math teacher.

What you'll learn
You'll learn to estimate square roots.

When am I ever going to use this?
Knowing how to estimate square roots can help you approximate how far you can see to the horizon.

Could you count off by $\sqrt{7}$? The class is stumped since 7 is not a perfect square.

We can estimate the square roots of numbers that are not perfect squares. For example, the number 130 is not a perfect square. However, we know that 130 is between two perfect squares, 121 and 144. So the square root of 130 is between 11 and 12.

Study the tiled squares at the right. Notice that the 11 × 11 square contains 121 squares, and the 12 × 12 square contains 144 squares. If you shade 130 squares, you can see that you are closer to 121 squares than you are to 144 squares.

Since 130 is closer to 121 than 144, the best whole number estimate for the square root of 130 is 11.

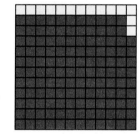

Examples

1 Estimate $\sqrt{78}$.

$64 < 78 < 81$ *64 and 81 are perfect squares.*
$8^2 < 78 < 9^2$
$8 < \sqrt{78} < 9$

Since 78 is closer to 81 than 64, the best whole number estimate for $\sqrt{78}$ is 9.

INTEGRATION

2 **Geometry** Heron, an Egyptian mathematician, created a formula that finds the area of a triangle if the length of each side is known.

If the measures of the sides of a triangle are a, b, and c, then the area of the triangle equals $\sqrt{s(s-a)(s-b)(s-c)}$ where s equals $\frac{1}{2}$ of the perimeter of the triangle.

Find the area of the triangular piece of land shown at the right.

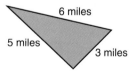

386 Chapter 9 Algebra: Exploring Real Numbers

Multiple Learning Styles

 Logical Have students write the squares of the numbers 1–10. Then have them find the difference between each consecutive pair of perfect squares. What pattern occurs? **The amount increases by two each time.** Ask students to use this information and estimation techniques to establish a rule for estimating square roots. **If you round to whole numbers only, the first half of the whole numbers between perfect squares is closer to the lower number; the second half is closer to the higher number.** Does this rule test true for the numbers between 19^2 and 20^2? **yes**

$a = 3$, $b = 5$, $c = 6$, and $s = \frac{1}{2}(3 + 5 + 6)$ or 7.

$A = \sqrt{s(s - a)(s - b)(s - c)}$ *Substitute the values for a, b, c,*
$= \sqrt{7(7 - 3)(7 - 5)(7 - 6)}$ *and s into the formula.*
$= \sqrt{7(4)(2)(1)}$
$= \sqrt{56}$

The area is $\sqrt{56}$. Now, estimate $\sqrt{56}$.

$49 < 56 < 64$
$7^2 < 56 < 8^2$
$7 < \sqrt{56} < 8$

Since $49 < 56 < 64$, we know that the best whole number estimate for $\sqrt{56}$ is between 7 and 8. Because 56 is closer to 49 than to 64, $\sqrt{56}$ is closer to 7 than 8. Therefore, the area of the triangular piece of land is approximately 7 square miles.

CHECK FOR UNDERSTANDING

Communicating Mathematics

Read and study the lesson to answer each question.

1. *Tell* how the drawing at the right can be used to estimate $\sqrt{12}$. **1–2. See margin.**

2. *Draw* a figure that can be used to explain why the square root of 75 is between 8 and 9.

Guided Practice

Estimate to the nearest whole number.

3. $\sqrt{68}$ **8**
4. $\sqrt{29}$ **5**
5. $\sqrt{135}$ **12**
6. $\sqrt{11}$ **3**

7. *Physical Science* The distance you can see to the horizon can be estimated by using the formula $d = 1.22 \times \sqrt{h}$. In the formula, d represents the distance you can see, in miles, and h represents the height your eyes are from the ground, in feet. If you are standing on the ground, suppose your eyes are about 5 feet above ground level. About how far can you see to the horizon? **about 2.7 miles**

EXERCISES

Practice

Estimate to the nearest whole number.

8. $\sqrt{23}$ **5**
9. $\sqrt{15}$ **4**
10. $\sqrt{44}$ **7**
11. $\sqrt{13.5}$ **4**
12. $\sqrt{113}$ **11**
13. $\sqrt{200}$ **14**
14. $\sqrt{408}$ **20**
15. $\sqrt{31}$ **6**
16. $\sqrt{23.5}$ **5**
17. $\sqrt{5}$ **2**
18. $\sqrt{250}$ **16**
19. $\sqrt{136}$ **12**

20. Find the nearest whole number estimate for $\sqrt{65.25}$. **8**

21. *Algebra* Estimate the solution of $y = \sqrt{78}$ to the nearest whole number. **9**

22. *Algebra* If $t = \sqrt{221}$, what is an estimate for the value of t? **14.9**

Lesson 9-2 Estimating Square Roots **387**

---

2 TEACH

 Transparency 9-2B contains a teaching aid for this lesson.

Thinking Algebraically To estimate square roots, we use a process similar to solving algebraic equations mentally. Point out that the process involves trying reasonable values in the inequality $\sqrt{a} < \sqrt{x} < \sqrt{b}$ where x is the known number and a and b are its nearest perfect squares.

In-Class Examples

For Example 1
Estimate $\sqrt{34}$. **6**

For Example 2
Use Heron's formula to find the area of a triangular wall hanging with sides measuring 4 meters, 5 meters, and 7 meters. **about 10 square meters**

3 PRACTICE/APPLY

Check for Understanding

If students need additional practice or instruction after completing Exercises 1–7, one of these options may be helpful.
- Extra Practice, see p. 631
- Reteaching Activity
- *Study Guide Masters*, p. 71
- *Practice Masters*, p. 71

Assignment Guide

Core: 9–23 odd, 26–29
Enriched: 8–22 even, 23, 24, 26–29

Additional Answers

1. If three more tiles are shaded, then 12 tiles are shaded with four unshaded. Therefore, $\sqrt{12}$ is closer to $\sqrt{9}$ than to $\sqrt{16}$; $\sqrt{12} \approx 3$.

2. [grid figure]

■ **Reteaching the Lesson** ■

Activity Name a whole number. Have students write the nearest perfect squares greater than and less than the number and circle the closer number. Then have them write the square root of the circled number. Example: ㊱ 39 49 **6** Students can continue this exercise with a partner.

Lesson 9-2 **387**

Exercise 25 asks students to advance to the next stage of work on the Chapter Project. Students need to use the formula $s = \sqrt{30df}$. Point out that $\sqrt{a \cdot b} = \sqrt{a} \cdot \sqrt{b}$. Use $a = 4$ and $b = 16$ to demonstrate this on the chalkboard.

Additional Answer

25b. Sample answer:

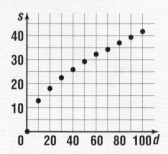

Study Guide Masters, p. 71

Applications and Problem Solving

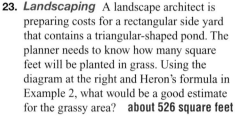

23. **Landscaping** A landscape architect is preparing costs for a rectangular side yard that contains a triangular-shaped pond. The planner needs to know how many square feet will be planted in grass. Using the diagram at the right and Heron's formula in Example 2, what would be a good estimate for the grassy area? **about 526 square feet**

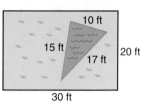

24. **Physical Science** Suppose you are visiting Washington, D.C., and you climb the 898 steps to the top of the Washington Monument. Your eyes are approximately 520 feet above the ground. Using the formula in Exercise 7, about how far can you see to the horizon? **about 28 miles**

25. **Working on the CHAPTER Project**
When investigating a car accident, police officers can estimate how fast a car was going by measuring the length of the skid marks. The formula $s = \sqrt{30df}$ gives the minimum speed s of the car in miles per hour, where d is the distance in feet the car skidded after its brakes were applied and f is a drag factor that depends on the road surface. Some typical drag factors are listed in the table.

Road Surface	Drag Factor (f)
dry concrete	0.82
packed gravel	0.66
wet concrete	0.57

   a. A car left skid marks for 60 feet on a dry, concrete road surface. If the speed limit on the road were 35 mph, was the car speeding when the brakes were applied? If so, by how much? **yes, about 3 mph**

   b. Choose ten different values for d and evaluate the formula for a wet, concrete road surface. Graph the resulting ordered pairs on a coordinate grid and describe the graph. **See margin.**

26. **Critical Thinking** Name all of the whole numbers whose square roots are between 5 and 6. **26, 27, 28, 29, 30, 31, 32, 33, 34, 35**

Mixed Review

27. Find the principal square root of 900. *(Lesson 9-1)* **30**

28. **Money Matters** Eleven-year old Currito regularly performs some of the jobs listed in the chart below. *(Lesson 8-5)*

Pay Rates			
Job	Unit of Pay	Ages 9-11	Ages 12-14
Baby-sitting	per hour	$3	$3.50
Mowing lawns	per lawn	$5	$6
Other yard work	per job	$2.88	$5
Pet care	per hour	$1	$3
Washing cars	per car	$4.50	$5

   a. If in the course of one week, Currito mows 3 lawns, takes care of the neighbor's dog each day for 1 hour, and washes 2 cars, how much money will he earn for the week? **$31**

   b. What is the percent of change in earnings if he were 12 years old? **about 58% increase**

388 Chapter 9 Algebra: Exploring Real Numbers

29. **Test Practice** The Venn diagram shows the relationship of four sets of students at Natalie's school. Which is a valid conclusion? *(Lesson 5-2)* **E**

U = set of all students at Natalie's middle school
M = set of students in Natalie's music class
G = set of all eighth-grade girls
B = set of all eighth-grade basketball players

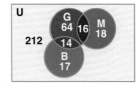

A No eighth-grade girl is in Natalie's music class.
B All eighth-grade girls are in Natalie's music class.
C All students in Natalie's music class are eighth-grade basketball players.
D All students in Natalie's music class are eighth-grade girls.
E No eighth-grade basketball players are in Natalie's music class.

4 ASSESS

Closing Activity
Speaking Write $\sqrt{8}$, $\sqrt{32}$, $\sqrt{95}$, and $\sqrt{172}$ on the chalkboard. Have students in pairs explain to each other how to estimate each square root.

Chapter 9, Quiz A (Lessons 9-1 and 9-2) is available in the *Assessment and Evaluation Masters*, p. 239.

Let the Games Begin

Estimate and Eliminate

40 index cards   4 markers

Get Ready This game is for four players.

Get Set Each player is given 10 index cards. Player 1 writes one of each of the whole numbers 1 – 10 on each card. Player 2 writes the square of one of each of the whole numbers 1 – 10 on each card. Player 3 writes a different whole number between 11-50, that is not a perfect square, on each card. Player 4 writes a different whole number between 51-99, that is not a perfect square, on each card.

Go
- Mix all 40 cards together. The dealer deals all 40 cards to the four players, one at a time.

- In turn, moving clockwise, each player lays down any pair(s) of a perfect square and its square root that they have in their hand. The two cards should be laid perpendicular to each other.

- After the first round, any player, during their turn: (1) lays down a square and a square root pair, or (2) lays down a card on the card that is perpendicular to the best approximate square root if it is already on the table in front of a player. For example, a 21 should be laid on a 25 if the 5 and 25 pair is already on the table. A player makes as many plays as possible during their turn.

- After each complete round, each player passes one card to their left, and the play resumes.

- The winner is the first person who places all of their 10 cards on the table.

 Visit www.glencoe.com/sec/math/mac/mathnet for more games.

Lesson 9-2 Estimating Square Roots **389**

Extending the Lesson
Enrichment Masters, p. 71

When students lay their cards on the table, they should be faceup so other students can play on them. Placing a card on the best approximate square root will help students learn to estimate square roots quickly.

Practice Masters, p. 71

9-2 Practice

Estimating Square Roots

Estimate to the nearest whole number.

1. $\sqrt{84}$ 9	2. $\sqrt{10}$ 3	3. $\sqrt{69}$ 8
4. $\sqrt{99}$ 10	5. $\sqrt{120}$ 11	6. $\sqrt{78}$ 9
7. $\sqrt{250}$ 16	8. $\sqrt{444}$ 21	9. $\sqrt{51}$ 7
10. $\sqrt{78}$ 9	11. $\sqrt{300}$ 17	12. $\sqrt{123}$ 11
13. $\sqrt{199}$ 14	14. $\sqrt{171}$ 13	15. $\sqrt{286}$ 17
16. $\sqrt{730}$ 27	17. $\sqrt{17.8}$ 4	18. $\sqrt{630}$ 25
19. $\sqrt{1,230}$ 35	20. $\sqrt{8.42}$ 3	21. $\sqrt{0.09}$ 0
22. $\sqrt{80.95}$ 9	23. $\sqrt{1.05}$ 1	24. $\sqrt{47.25}$ 7

Lesson 9-2 389

9-3 Lesson Notes

Instructional Resources
- *Study Guide Masters*, p. 72
- *Practice Masters*, p. 72
- *Enrichment Masters*, p. 72
- Transparencies 9-3, A and B

CD-ROM Program
- Resource Lesson 9-3

Recommended Pacing	
Standard	Day 5 of 13
Honors	Day 3 of 11
Block	Day 3 of 7

1 FOCUS

5-Minute Check
(Lesson 9-2)

Estimate to the nearest whole number.
1. $\sqrt{50}$ **7**
2. $\sqrt{94}$ **10**
3. $\sqrt{13}$ **4**
4. $\sqrt{31}$ **6**
5. Estimate the solution of $y = \sqrt{449}$ to the nearest whole number. **21**

The 5-Minute Check is also available on **Transparency 9-3A** for this lesson.

Motivating the Lesson

Communication Ask students to give two examples for each of the following: (Sample answers are given.)
- whole number **7, 1**
- integer **−5, 8**
- rational number $\frac{7}{8}$, **−5.82, 10**
- terminating decimal **0.641**
- repeating decimal $0.\overline{81}$

9-3 The Real Number System

What you'll learn
You'll learn to identify and classify numbers in the real number system.

When am I ever going to use this?
Knowing how to use real numbers can help you estimate distance and time.

Word Wise
irrational number
real number

LOOK BACK
You can refer to Lesson 6-4 to review rational numbers.

Have you ever gotten caught out in a thunderstorm and wished you knew how long it would be before it's over? Meteorologists use the formula $t^2 = \frac{d^3}{216}$ to estimate the amount of time that a thunderstorm will last. In this formula, t is the time in hours, and d is the diameter of the storm in miles.

Numbers such as 6, 8.4, and $10\frac{1}{2}$ are common replacements for d in $t^2 = \frac{d^3}{216}$.

These numbers are from sets of numbers you've studied.

Whole Numbers	$\{0, 1, 2, 3, 4, \ldots\}$
Integers	$\{\ldots, -2, -1, 0, 1, 2, \ldots\}$
Rational Numbers	{all numbers that can be expressed in the form $\frac{a}{b}$, where a and b are integers and $b \neq 0$}

The Venn diagram represents a summary of the classification of rational numbers.

Remember, terminating or repeating decimals are rational numbers since they can be expressed as fractions. The square roots of perfect squares are also rational numbers. For example, $\sqrt{0.09}$ is a rational number because $\sqrt{0.09} = 0.3$, a rational number.

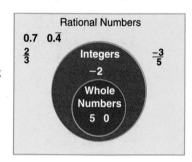

Where do numbers like $\sqrt{2}$ and $\sqrt{5}$ fit into the Venn diagram? Notice what happens when you find $\sqrt{2}$ and $\sqrt{5}$ with your calculator.

$\sqrt{2}$: 2 [2nd] [$\sqrt{\ }$] *1.414213562*
$\sqrt{5}$: 5 [2nd] [$\sqrt{\ }$] *2.236067978*

390 Chapter 9 Algebra: Exploring Real Numbers

Cross-Curriculum Cue

Inform the other teachers on your team that your students are studying the classification of numbers. Suggestions for curriculum integration are:
Biology: Linnaean classification
Social Studies: political units
Language: parts of speech

The numbers do not terminate nor is there a pattern of repeating digits. These numbers are not rational numbers. Numbers like $\sqrt{2}$ and $\sqrt{5}$ are called **irrational numbers**.

Irrational Number	An irrational number is a number that cannot be expressed as $\frac{a}{b}$, where a and b are integers and b does not equal 0.

Examples

Determine whether each number is rational or irrational.

1 0.16666666 . . .

This decimal ends in a repeating pattern. It is a rational number and can be expressed as $\frac{1}{6}$.

2 0.161161116111161111116 . . .

This decimal ends in a pattern that will never repeat or terminate. It is an irrational number.

3 π

$\pi = 3.1415926 \ldots$
This decimal does not repeat or terminate. It is an irrational number.

You have graphed rational numbers on a number line. If you graph all of the rational numbers, you would still have some "holes" in the number line. The irrational numbers "fill in" the number line. The set of rational and irrational numbers combine to form the set of **real numbers**. The graph of all real numbers is the entire number line without any "holes."

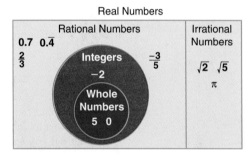

To graph irrational numbers, you can use a calculator or a table of squares and square roots to find approximate square roots in decimal form.

Example

4 Graph $\sqrt{5}$, π, and $-\sqrt{2}$ on a number line.

Lesson 9-3 The Real Number System **391**

2 TEACH

Transparency 9-3B contains a teaching aid for this lesson.

Reading Mathematics The word *rational* applied to numbers comes from the word *ratio*. It does not mean logical or reasonable numbers, but numbers that can be represented by a ratio, such as $\frac{a}{b}$ where b is any integer but zero. Every rational number can be represented by a repeating or terminating decimal; an irrational number cannot.

In-Class Examples
Determine whether each number is rational or irrational.

For Example 1
0.333333 . . . rational

For Example 2
2.449489743 . . . irrational

For Example 3
$\sqrt{3}$ irrational

For Example 4
Graph $\sqrt{19}$, $-\sqrt{70}$, and 2π on a number line.

Teaching Tip Point out that in Example 2, the number 0.161161116111161111116 has a predictable pattern but the number is not rational because the exact pattern does not repeat.

Lesson 9-3 **391**

In-Class Example

For Example 5
Use the formula $A = \pi r^2$ to calculate the area of a circle with a radius of 2 feet. Is the answer an exact representation of the area? **12.566; no, because π is an irrational number.**

3 PRACTICE/APPLY

Check for Understanding

If students need additional practice or instruction after completing Exercises 1–10, one of these options may be helpful.
- Extra Practice, see p. 631
- Reteaching Activity
- *Study Guide Masters*, p. 72
- *Practice Masters*, p. 72

Additional Answers

1.

   √15 graphed between 3 and 4
   0 1 2 3 4

2. $\sqrt{\frac{25}{36}}$ is a rational number because it can be expressed as a fraction. 25 and 36 are both perfect squares so, $\sqrt{\frac{25}{36}} = \frac{5}{6}$, and $\frac{5}{6}$ is rational.

3. Their teacher is correct because it depends on what number of which you are taking the square root. Examples: $\sqrt{9} = 3$ and $3 > 9$, but $\sqrt{0.09} = 0.3$ and $0.3 > 0.09$.

6.
   √7 graphed between 2 and 3
   0 1 2 3

7.
   √20 graphed between 4 and 5
   2 3 4 5

So far in this textbook, you have solved equations that have rational number solutions. However, some equations have solutions that are irrational numbers.

Example 5 — APPLICATION

Meteorology Refer to the beginning of the lesson. Use the formula to determine how long a thunderstorm will last if it is 7.8 miles wide.

$$t = \frac{d^3}{216}$$
$$t^2 = \frac{7.8^3}{216} \quad \text{Replace } d \text{ with } 7.8.$$
$$\sqrt{t^2} = \sqrt{2.197} \quad \text{Take the square root of each side.}$$
2.197 [2nd] [√] *1.4822280553* Use a calculator.
$$t \approx 1.5$$

The thunderstorm will last about 1.5 hours.

CHECK FOR UNDERSTANDING

Communicating Mathematics

Read and study the lesson to answer each question. 1–3. See margin.

1. **Draw** a number line and graph $\sqrt{15}$.

2. **Explain** why $\sqrt{\frac{25}{36}}$ is a rational number.

3. **You Decide** Lan and Sharon disagree about whether a square root of a number is greater or less than the original number. Lan claims that the square root is less than the number, and Sharon says that the square root is greater. Their teacher told them that they were both correct. Who is really correct? Explain and give some examples.

8. −12.2, 12.2

Guided Practice

Let R = real numbers, Q = rational numbers, Z = integers, W = whole numbers, and I = irrational numbers. Name the set or sets of numbers to which each real number belongs.

4. $-\sqrt{9}$ **Z, Q, R**
5. $0.\overline{27}$ **Q, R**

Find an estimate for each square root. Then graph the square root on a number line. 6–7. See margin for number lines.

6. $\sqrt{7}$ **2.6**
7. $\sqrt{20}$ **4.5**

Solve each equation. Round solutions to the nearest tenth.

8. $x^2 = 148$
9. $y^2 = 59$ **−7.7, 7.7**

10. **Carpentry** Hotah and his aunt are building a deck. In one space, they need a board that measures $\sqrt{110}$ feet. About how long is the board? **about 10.5 feet**

392 Chapter 9 Algebra: Exploring Real Numbers

Reteaching the Lesson

Activity Have students work in small groups to plan and write a paragraph explaining the Venn diagram on page 390. Have them give a definition of each set shown in the diagram and several additional examples of each set.

Error Analysis
Watch for students who identify all negative numbers as irrational.
Prevent by pointing out that the sign of a number has no bearing on whether the number is rational or irrational. A number is irrational only if it cannot be expressed as a fraction and its decimal doesn't repeat.

EXERCISES

Practice

Let R = real numbers, Q = rational numbers, Z = integers, W = whole numbers, and I = irrational numbers. Name the set or sets of numbers to which each real number belongs.

11. 36 **W, Z, Q, R**
12. $\sqrt{31}$ **I, R**
13. 0.121121112 **I, R**
14. $\frac{5}{8}$ **Q, R**
15. $-\sqrt{121}$ **Z, Q, R**
16. $\frac{\sqrt{10}}{\sqrt{49}}$ **I, R**

Find an estimate for each square root. Then graph the square root on a number line. 17–22. See Answer Appendix for number lines.

17. $\sqrt{6}$ **2.4**
18. $-\sqrt{27}$ **−5.2**
19. $\sqrt{50}$ **7.1**
20. $-\sqrt{99}$ **−9.9**
21. $\sqrt{108}$ **10.4**
22. $\sqrt{300}$ **17.3**

Solve each equation. Round solutions to the nearest tenth.

23. $x^2 = 69$ **−8.3, 8.3**
24. $y^2 = 12$ **−3.5, 3.5**
25. $x^2 = 75$ **−8.7, 8.7**
26. $n^2 = 41$ **−6.4, 6.4**
27. $x^2 = 925$ **−30.4, 30.4**
28. $t^2 = 0.28$ **−0.5, 0.5**

29. Which is greater, $\sqrt{\frac{25}{4}}$ or $\frac{8}{3}$? **$\frac{8}{3}$**

Applications and Problem Solving

30. *Algebra* In the geometric sequence 4, 12, __?__, 108, 324, the missing number is called the *geometric mean* of 12 and 108. It can be found by simplifying $\sqrt{ab}$ where a and b are the numbers on either side of the geometric mean. Find the missing number. **36**

31. *Physical Science* The formula $d = 16t^2$ represents the distance, d, in feet, that an object falls in t seconds. Copy and complete the table of a ball falling from different distances. Do you need the negative square root? What pattern can you find? **See margin.**

Distance (ft)	Time (s)
128	
64	
32	
16	
8	

32. *Working on the* Police officers often use a rolling device known as a trundle wheel to measure distances at an accident scene. **a–b. See margin.**

   a. If a trundle wheel has a radius of 8 inches, what distance on the ground has been measured when the wheel has made one complete revolution? Name the set of numbers to which this distance belongs.

   b. Make a table of the distances measured in 1, 2, 3, …, 10 revolutions. Then graph the ordered pairs and describe the graph.

33. *Critical Thinking* The area of each of the smallest squares at the right is 10 square units. **a. $\sqrt{10} \approx 3.2$ units**

   a. What is the length of each side of the smallest squares to the nearest tenth?

   b. Find the length of a diagonal of the largest square to the nearest tenth. **$4\sqrt{10} \approx 12.6$ units**

Lesson 9-3 The Real Number System **393**

Assignment Guide

Core: 11–31 odd, 33–37
Enriched: 12–30 even, 31, 33–37
All: Self Test 1–10

CHAPTER Project

Exercise 32 asks students to advance to the next stage of work on the Chapter Project. You may wish to have students draw a diagram of the trundle wheel, marking the radius of the wheel on the diagram. Have students review Lessons 7-7A and 7-7 regarding the relationship between radius and circumference.

Additional Answer

31.

Distance	Time
128 ft	$\sqrt{8} \approx 2.8$
64 ft	$\sqrt{4} = 2$
32 ft	$\sqrt{2} \approx 1.4$
16 ft	$\sqrt{1} = 1$
8 ft	$\sqrt{\frac{1}{2}} \approx 0.7$

Study Guide Masters, p. 72

Additional Answers

32a. 16π inches, or about 50.22 inches; I, R

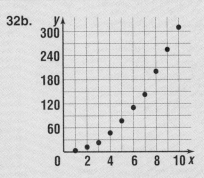

32b.

4 ASSESS

Closing Activity

Writing Have students give examples of rational and irrational numbers. Then have them compare and contrast the numbers as to whether they can be expressed as ratios or fractions. Does the decimal representation of the number go on without repeating?

Mid-Chapter Self Test

The Mid-Chapter Self Test reviews concepts in Lessons 9-1 through 9-3. Lesson references are given so students can review concepts not yet mastered.

Practice Masters, p. 72

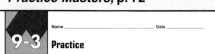

Enrichment Masters, p. 72

Mixed Review

interNET CONNECTION
For the latest color survey, visit:
www.glencoe.com/sec/math/mac/mathnet

34. Find the nearest whole number estimate for $\sqrt{60}$. Then check your estimate by using a calculator. *(Lesson 9-2)* **8**

35. **Statistics** What is your favorite color? Pantone Color Institute surveyed people about color preferences. Of the 1,000 women surveyed, 280 said blue was their favorite. If next year's survey was increased to include 2,500 women, how many would you predict say blue is their favorite color? *(Lesson 8-1)* **about 700**

36. **Forestry** The ranger at Crosswoods Nature Preserve recorded the heights of several trees in an area of the preserve. Make a histogram of this data. *(Lesson 4-1)* **See Answer Appendix.**

Height (ft)	Number of Trees
11-20	4
21-30	10
31-40	14
41-50	22
51-60	28
61-70	18

37. **Test Practice** Eighth-grade students who were randomly sampled gave the following responses when they were asked how many extracurricular activities in which they were involved: 1, 3, 2, 0, 4, 3, 1, 5, 2, 3, 4, 2, 0, 1, 2. Which expression shows how to determine what percent of the sampled students are involved in 3 or more extracurricular activities? *(Lesson 3-4)* **E**

A $\frac{3}{6} \times 100$ **B** $\frac{3}{12} \times 100$ **C** $\frac{3}{15} \times 100$

D $\frac{6}{9} \times 100$ **E** $\frac{6}{15} \times 100$

CHAPTER 9 — Mid-Chapter Self Test

Find each square root. *(Lesson 9-1)*

1. $\sqrt{1}$ **1**
2. $-\sqrt{0.0016}$ **−0.04**

3. **Geometry** If the area of a square is 2.25 square meters, what is the length of its side? *(Lesson 9-1)* **1.5 m**

Estimate to the nearest whole number. *(Lesson 9-2)*

4. $\sqrt{90}$ **9**
5. $\sqrt{28}$ **5**
6. $\sqrt{226}$ **15**

Let R = real numbers, Q = rational numbers, Z = integers, W = whole numbers, and I = irrational numbers. Name the set or sets of numbers to which each real number belongs. *(Lesson 9-3)*

7. 0.12121212... **Q, R**
8. $\sqrt{4}$ **W, Z, Q, R**
9. $\sqrt{21}$ **I, R**

10. **Algebra** Solve $y^2 = 75$. Round the solution to the nearest tenth. *(Lesson 9-3)* **−8.7, 8.7**

394 Chapter 9 Algebra: Exploring Real Numbers

Extending the Lesson

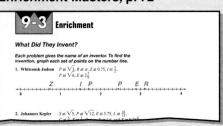

Activity Have students answer and explain the following problem. *The number k is a perfect square. Is the square root of k rational or irrational? Why?* **Rational; the square root of a perfect square is a whole number, which is also a rational number.**

SCHOOL to CAREER

LAW ENFORCEMENT

Evelio H. Quinonez
POLICE OFFICER

Someday, I'd like to be a highway patrol trooper like Trooper Quinonez.

Across the nation, highway patrol troopers like Evelio Quinonez enforce traffic laws, respond to traffic accidents, and assist motorists with car trouble. Troopers also enforce criminal laws and assist local police and sheriffs in crime investigation. Trooper Quinonez was a co-recipient of the Florida Highway Patrol Trooper of the Month award in January 1997.

To be a highway patrol trooper, you should be a U.S. citizen, have a high school education, and be in good physical condition. You should also enjoy working with people. For highway patrol work, you should have a solid background in mathematics and have good communication skills. Some departments prefer candidates with some college. Many colleges offer degrees in law enforcement and criminal justice.

For more information:
Qualifications for highway patrols vary from state to state. To find out how to become a trooper in your state, contact your state patrol agency. For example,

Florida Highway Patrol
1011 NW 111th Ave.
Miami, FL 33172

interNET CONNECTION
www.glencoe.com/sec/math/mac/mathnet

Your Turn
Traffic safety is a primary concern for state highway patrol troopers. Investigate a safety issue such as the number of accidents that occur at different speeds. Use a statistical graph to design a poster about the need for safe driving speeds that could be used in a traffic safety campaign.

School to Career: Law Enforcement **395**

More About Trooper Evelio H. Quinonez
- Trooper Quinonez was a 12-year veteran of the Florida Highway Patrol when he received this award.
- Like other applicants to the Florida Highway Patrol, Evelio Quinonez had to be a high school graduate. He also had to take training at the Florida Highway Patrol Academy in Tallahassee.
- Trooper Quinonez received his citation of Trooper of the Month for tackling a suspected carjacker and grabbing his gun before the suspect could shoot another Florida trooper.

Motivating Students
As students learned in the Chapter Project on page 381, police officers use angle and distance measurements and scale drawings to investigate traffic accidents. Students may be interested in a career in law enforcement or a related field. To begin the discussion, you may wish to ask the following questions about traffic accidents.
- In 1996, how many people on average died per day in the United States in traffic accidents? **approximately 115 people—one every 13 minutes**
- What factor contributed more often to fatal vehicle accidents—alcohol or speeding? **In 1996, alcohol was a contributing factor in 40.9% of all fatal accidents; speeding was a contributing factor in 30%.**

Making the Math Connection
The National Highway Traffic Safety Administration collects and analyzes traffic accident statistics. In cities and states, safety experts analyze accident reports using geometric formulas to find dangerous intersections and other roadway hazards that contribute to increased vehicular accidents, deaths, and injuries. In addition police use instruments such as speedometer clocks, radar detectors, average speed computers, and LIDAR (laser infrared light) monitors to measure traffic speed.

Working on *Your Turn*
If possible, students should peruse the Internet to find traffic statistics collected by state departments of transportation. Other students could call insurance companies to ask for information. You may wish to have students work in pairs to design their posters.

*An additional School to Career activity is available on page 35 of the **School to Career Masters**.*

9-4A Hands-On Lab Notes

GET READY

Objective Students explore the relationships in a right triangle.

Optional Resources

Hands-On Lab Masters
- square dot paper, p. 12
- worksheet, p. 58

Overhead Manipulative Resources
- geoboards
- geobands
- rectangular dot paper transparency

Manipulative Kit
- geoboards
- geobands

MANAGEMENT TIPS

Recommended Time
40 minutes

Getting Started Point out to students that this lab discusses only right triangles and the conclusions drawn from this lab relate only to right triangles.

The **Activity** introduces the Pythagorean Theorem by having students explore models of right triangles. You may wish to have students work in pairs, taking turns with each part of all the steps.

Hands-On Lab — Cooperative Learning

9-4A The Pythagorean Theorem

A Preview of Lesson 9-4

- geoboard
- dot paper
- geobands

In the previous lessons, you learned about the relationship between the area of a square and the length of its side. You will use this relationship when you investigate a famous rule about right triangles. This rule is called the Pythagorean Theorem.

TRY THIS

Work with a partner.

Step 1 Build a triangle like the one shown on the geoboard. Remember, because it has one right angle, it is called a *right triangle*.

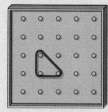

Step 2 Using the longest side of the triangle, build a square. Each side of the square has the same length as the longest side of the triangle. As you can see below, the area of this square is 2 square units.

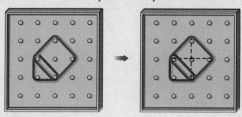

Step 3 Now build squares on the two shorter sides. Each square has an area of 1 square unit.

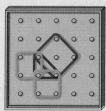

Step 4 Copy the table below and enter the results from this lab in the first row.

Area of Square on Short Side 1	Length of Short Side 1	Area of Square on Short Side 2	Length of Short Side 2	Area of Square on Longest Side	Length of Longest Side
1	1	1	1	2	$\sqrt{2}$

ON YOUR OWN

On dot paper, draw each triangle below and then build squares on each side. Record your results in your table. **1–3. See Answer Appendix.**

1.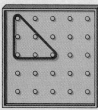
2.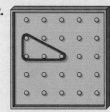
3.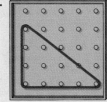

4. In a certain right triangle, the two shorter sides are 2 units and 3 units long. Without making a drawing, what is the area of the square built on the longest side? **13 units**

5. In another right triangle, the longest side is 6 units long, and one of the other sides is 2 units long. What is the area of the square built on the remaining side? **32 units**

6. Can you build a square on the longest side of a right triangle with an area of 10 square meters? If you think of 10 as the sum of 1 and 9, you can build squares like the ones shown. The square that is built on the longest side has an area of 10 square units. On dot paper, build squares on the longest side of a right triangle with areas of 13, 17, 40, and 50 square units **See Answer Appendix.**

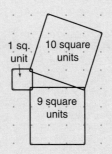

1 sq. unit
10 square units
9 square units

7. **Look Ahead** Use the results in your table to look for a relationship between the area of the two smaller squares and the area of the larger square. Write a statement that describes this relationship.
(one small side)² + (another small side)² = (longest side)²

Lesson 9-4A HANDS-ON LAB **397**

9-4 Lesson Notes

Instructional Resources
- *Study Guide Masters*, p. 73
- *Practice Masters*, p. 73
- *Enrichment Masters*, p. 73
- *Transparencies 9-4, A and B*
- *Assessment and Evaluation Masters*, pp. 238, 239

 CD-ROM Program
- Resource Lesson 9-4
- Interactive Lesson 9-4

Recommended Pacing	
Standard	Days 6 & 7 of 13
Honors	Day 4 of 11
Block	Day 4 of 7

1 FOCUS

5-Minute Check
(Lesson 9-3)

Let R = real numbers, Q = rational numbers, Z = integers, W = whole numbers, and I = irrational numbers. Name the set or sets of numbers to which each real number belongs.

1. 38 **W, Z, Q, R**
2. $\sqrt{32}$ **I, R**
3. 0.151151115 **I, R**
4. Estimate $\sqrt{8}$. Then graph the square root on a number line. **3**

5. Solve $x^2 = 70$. Round to the nearest tenth. **8.4 or −8.4**

 The 5-Minute Check is also available on **Transparency 9-4A** for this lesson.

9-4 The Pythagorean Theorem

What you'll learn
You'll learn to use the Pythagorean Theorem.

When am I ever going to use this?
You can use the Pythagorean Theorem to find how far baseball players throw a ball to reach certain positions.

Word Wise
hypotenuse
legs
Pythagorean Theorem
converse

Cultural Kaleidoscope

Theano, the wife of Pythagoras, and her two daughters carried on the Pythagorean School after Pythagoras' death.

Chou-pei Suan-king is an ancient Chinese book that was written sometime during 1200–600 B.C. It contains a diagram named *hsuan-thu*, which offers an interesting puzzle showing the relationship among the sides of a right triangle.

The sides of each right triangle have lengths of 3, 4, and 5 units. The longest side of each triangle, called the **hypotenuse**, is opposite the right angle. The sides that form the right angles are called the **legs**.

Many years after the Chinese, in the fifth century B.C., a Greek mathematician, Pythagoras, and the students in his school studied the 3-4-5 triangle. They noticed that if they built a square on each of the three sides, the total area of the two smaller squares was equal to the area of the large square.

Today, we call this relationship the **Pythagorean Theorem**. It is true for *any* right triangle.

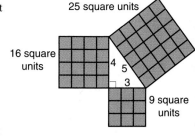

Pythagorean Theorem	Words:	In a right triangle, the square of the length of the hypotenuse is equal to the sum of the squares of the lengths of the legs.
	Symbols:	$c^2 = a^2 + b^2$ Model:

Example 1 Find the length of the hypotenuse in the triangle.

$c^2 = a^2 + b^2$ *Pythagorean Theorem*
$c^2 = 9^2 + 12^2$ *Replace a with 9 and b with 12.*
$c^2 = 81 + 144$
$c^2 = 225$
$c = \sqrt{225}$ *You can ignore $-\sqrt{225}$ because it is*
$c = 15$ *not reasonable to have a negative length.*

The length of the hypotenuse is 15 feet.

398   Chapter 9 Algebra: Exploring Real Numbers

Classroom Vignette

"I use an overhead projector to separate a square into four right triangles, then into two squares and two rectangles. I compare the areas and then derive the Pythagorean Theorem."

Denis J. Fogaroli, Teacher
Eisenhower Middle School
Succasunna, NJ

398   Chapter 9

Example 2

The hypotenuse of a right triangle is 40 meters long, and one of its legs is 18 meters long. Find the length of the other leg.

$$c^2 = a^2 + b^2$$
$$40^2 = a^2 + 18^2$$
$$1{,}600 = a^2 + 324$$
$$1{,}600 - 324 = a^2 + 324 - 324$$
$$1{,}276 = a^2$$
$$\sqrt{1{,}276} = a^2$$

1,276 [2nd] [√] 35.7211422

The length of the leg is about 35.7 meters.

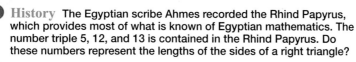

Study Hint

Estimation THINK:
30 × 30 = 900
40 × 40 = 1,600
√1,276 is between 30 and 40.

Some triangles may look like right triangles, but in reality they are not. The **converse** of the Pythagorean Theorem can be used to test whether a triangle is a right triangle.

| Converse of Pythagorean Theorem | If the sides of a triangle have lengths a, b, and c units such that $c^2 = a^2 + b^2$, then the triangle is a right triangle. |

Example 3 — CONNECTION

History The Egyptian scribe Ahmes recorded the Rhind Papyrus, which provides most of what is known of Egyptian mathematics. The number triple 5, 12, and 13 is contained in the Rhind Papyrus. Do these numbers represent the lengths of the sides of a right triangle?

$$c^2 = a^2 + b^2$$
$$13^2 \stackrel{?}{=} 5^2 + 12^2 \quad \textit{Remember, the hypotenuse is the longest side.}$$
$$169 \stackrel{?}{=} 25 + 144$$
$$169 = 169 \checkmark$$

The numbers represent the sides of a right triangle because the Pythagorean Theorem holds true for the lengths of the sides.

CHECK FOR UNDERSTANDING

Communicating Mathematics

2–3. See Answer Appendix.

Math Journal

Read and study the lesson to answer each question.

1. *Tell* the area of the shaded square shown at the right. **100 square units**

2. *Draw* a right triangle and label the right angle, the hypotenuse, and the legs.

3. Use centimeter grid paper to determine, and then *write* why a triangle with sides 6, 12, and 14 cannot be a right triangle.

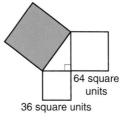

64 square units
36 square units

Lesson 9-4 The Pythagorean Theorem **399**

Motivating the Lesson

Hands-On Activity On graph paper, have students draw a right triangle with sides of 5 units, 12 units, and 13 units. Then have them draw squares on all three sides of the triangle. Ask them to find the areas of the squares. **25 units², 144 units², 169 units²** Have the students repeat the process for a right triangle with sides of 9 units, 12 units, and 15 units. What pattern do they see? **The sum of the areas of the squares on the two shorter sides of the triangles equals the area of the square on the longest side.**

2 TEACH

Transparency 9-4B contains a teaching aid for this lesson.

Reading Mathematics The word *hypotenuse* derives from the Greek word meaning "to stretch under." As the hypotenuse is always the longest side, thinking of it as the stretched side may help students remember which one it is.

In-Class Examples

For Example 1
Find the length of the hypotenuse of a right triangle with legs 10 inches and 24 inches long. **26 in.**

For Example 2
The hypotenuse of a right triangle is 16 meters long, and one of its legs is 9 meters long. Find the length of the other leg. **about 13.2 m**

For Example 3
Determine whether a triangle with sides of 13 cm, 15 cm, and 20 cm is a right triangle. **no**

3 PRACTICE/APPLY

Check for Understanding

If students need additional practice or instruction after completing Exercises 1–10, one of these options may be helpful.
- Extra Practice, see p. 631
- Reteaching Activity
- *Study Guide Masters*, p. 73
- *Practice Masters*, p. 73

Reteaching the Lesson

Activity Have students draw a right triangle on graph paper, using sides of any length, and measure the hypotenuse. Have them confirm their measurement using the Pythagorean Theorem.

Error Analysis
Watch for students who mistakenly assume that the unknown side of a triangle is always the hypotenuse.
Prevent by having students find the right angle in the triangle and identify the side opposite it.

Lesson 9-4 **399**

Assignment Guide

Core: 11–31 odd, 33–37
Enriched: 12–28 even, 30–37

Additional Answers

11. $12^2 + 9^2 = c^2$
12. $10^2 + 10^2 = x^2$
13. $a^2 + 12^2 = 30^2$
14. $4^2 + x^2 = 8^2$
15. $1^2 + 1^2 = h^2;\ h^2 + 1^2 = x^2$
16. $x^2 + 5^2 = (\sqrt{29})^2$
17. $3^2 + b^2 = 8^2$
18. $a^2 + 12^2 = 22^2$
19. $40^2 + b^2 = 41^2$
20. $a^2 + 99^2 = 101^2$
21. $48^2 + 55^2 = c^2$
22. $3.5^2 + 12.5^2 = c^2$

Study Guide Masters, p. 73

Guided Practice Write an equation you could use to find the length of the missing side of each right triangle. Then find the missing length. Round to the nearest tenth.

4. 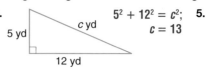 $5^2 + 12^2 = c^2;\ c = 13$

5. 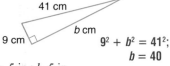 $9^2 + b^2 = 41^2;\ b = 40$

6. a, 9 ft; c, 12 ft $9^2 + b^2 = 12^2;\ b \approx 7.9$
7. a, 5 in.; b, 5 in. $5^2 + 5^2 = c^2;\ c \approx 7.1$

Determine whether each triangle with sides of given lengths is a right triangle.

8. 5 in., 10 in., 12 in.   **no**
9. 9 m, 40 m, 41 m   **yes**

10. **Sports** A baseball diamond is actually a square. How far does a catcher have to throw when he throws the ball from home plate to second base?   **about 127.3 feet**

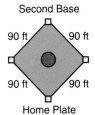

EXERCISES

Practice Write an equation you could use to find the length of the missing side of each right triangle. Then find the missing length. Round to the nearest tenth.

11–22. See margin for equations.

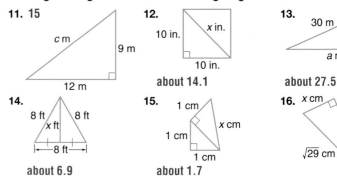

11. about 15
12. about 14.1
13. about 27.5
14. about 6.9
15. about 1.7
16. 2

17. a, 3 m; c, 8 m   about 7.4
18. b, 12 cm; c, 22 cm   about 18.4
19. a, 40 in.; c, 41 in.   9
20. b, 99 mm; c, 101 mm   20
21. a, 48 yd; b, 55 yd   73
22. a, 3.5 m; b, 12.5 m   about 13.0

Determine whether each triangle with sides of given lengths is a right triangle.

23. 18 mi, 24 mi, 30 mi   yes
24. 4 ft, 7 ft, 5 ft   no
25. 10 cm, 12 cm, 15 cm   no
26. 28 yd, 197 yd, 195 yd   yes
27. 1 ft, 1 ft, $\sqrt{2}$ ft   yes
28. 24 m, 143 m, 145 m   yes

29. What is the length of the other leg of a right triangle if the hypotenuse is 26 inches and one leg is 10 inches?   **24 inches**

400 Chapter 9 Algebra: Exploring Real Numbers

Applications and Problem Solving

30. **Geometry** A circle with a diameter of 10 centimeters is inscribed in a square. What is the length of the diagonal of the square? $10\sqrt{2} \approx 14.1$ cm

31. **Design** Sauder Woodworking Company makes entertainment centers that include a space for a television. Televisions are sized by the length of the diagonal of the rectangular screen. The designer wants to accommodate a television with a 27-inch screen, but TVs vary in width and length. The screen widths for a 27-inch TV range from 18 to 20 inches. What is the range of the heights? **between 18 and 20 inches**

32. **Write a Problem** about the photograph at the right that requires using the Pythagorean Theorem in the solution. **See students' work.**

33. **Critical Thinking** About 2000 B.C., Egyptian engineers discovered a way to make a right triangle using a rope with 12 evenly spaced knots tied in it. They attached one end of the rope to a stake in the ground. At what knot locations should the other two stakes be placed in order to form a right triangle? Draw a diagram. **See margin.**

Mixed Review

34. **Algebra** Solve $x^2 = 1.70$. Round the solution to the nearest tenth. *(Lesson 9-3)* **−1.3, 1.3**

35. At 3:00 P.M., Vinita's shadow was 2.5 feet long. The shadow of a tree was 21 feet long. If Vinita is 5.25 feet tall, estimate the height of the tree. *(Lesson 8-7)* **44 feet**

36. **Geometry** Roberto became lost when driving from his house to a restaurant. From his house, he drove west 3 blocks, north 2 blocks, east 5 blocks, south 4 blocks, east 3 blocks, and north 6 blocks before arriving at the restaurant. If Roberto's house has coordinates (0, 0), what are the coordinates of the restaurant? *(Lesson 2-10)* **(5, 4)**

37. **Test Practice** Alisa works at the Dairy Dream 20 hours each week. So far this week, she has worked $4\frac{1}{2}$ hours on Monday and $7\frac{1}{4}$ hours on Wednesday. Which sentence could be used to find h, the number of hours remaining for Alisa to work this week? *(Lesson 1-6)* **C**

A $4\frac{1}{2} + 7\frac{1}{4} = h + 20$ **B** $4\frac{1}{2} + 7\frac{1}{4} = h \times 20$

C $h = 20 - \left(4\frac{1}{2} + 7\frac{1}{4}\right)$ **D** $h - 4\frac{1}{2} - 7\frac{1}{4} = 20$

E $h = 20 - 4\frac{1}{2} + 7\frac{1}{4}$

Lesson 9-4 The Pythagorean Theorem **401**

═══ **Extending the Lesson** ═══

Enrichment Masters, p. 73

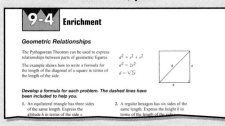

Activity Have students work in small groups to solve this problem: *A softball diamond is a square 60 feet on each side. How long will it take a ball thrown from home plate at 75 feet per second to reach second base?* **about 1.1 s**

4 ASSESS

Closing Activity

Writing Have students write the Pythagorean Theorem using the figure. $h^2 = p^2 + q^2$

Chapter 9, Quiz B (Lessons 9-3 and 9-4) is available in the *Assessment and Evaluation Masters*, p. 239.

Mid-Chapter Test (Lessons 9-1 through 9-4) is available in the *Assessment and Evaluation Masters*, p. 238.

Additional Answer

33.

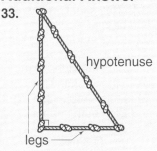

Practice Masters, p. 73

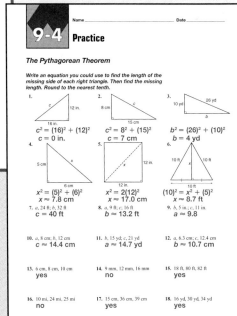

Lesson 9-4 **401**

9-5A Notes

Objective Students solve problems by drawing a diagram.

Instructional Resources
CD-ROM Program
- Resource Lesson 9-5A

Recommended Pacing	
Standard	Day 8 of 13
Honors	Day 5 of 11
Block	Day 4 of 7

1 FOCUS

Getting Started To help students visualize the concept of this lab, have them draw two unfolded cube diagrams like the one below. The sides of the cubes should measure 2 inches and be marked like the diagram. Have the students cut out one pattern and fold it into a cube and tape it. Match the sides of the folded cube to those on the diagram. Students should also mark points A and B on both cubes.

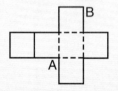

2 TEACH

Teaching Tip If some students have difficulty visualizing spatial relationships, have them use the two cubes to visualize the problem in the Thinking Lab. They should draw a line from point A to point B on the flat cube diagram and compare this line to the most direct route an invisible ant could negotiate across the folded cube.

PROBLEM SOLVING

9-5A Draw a Diagram
A Preview of Lesson 9-5

Lonzo zoomed up on his in-line skates yelling, "I've got a great puzzle for you. You'll never figure it out!" Paul rolled his eyes, but he was ready to take Lonzo's challenge. Let's listen in!

My invisible pet ant must travel from one corner to the opposite corner of a 2-inch cube. My ant knows the shortest way, do you?

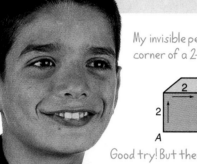

That's pretty easy. It would need to go 2 inches up, 2 inches across, and 2 inches back. That's 6 inches.

Good try! But there's a shorter way.

OK, use the diagonal across the face of the cube, which is the hypotenuse of a right triangle. This length can be found by solving $h^2 = 2^2 + 2^2$ or $h^2 = 8$. That means that $h = \sqrt{8}$, which is a little less than 3. So, now you have a little less than 3 inches plus the length of the edge to the back that is 2 inches. This makes the path length slightly less than 5 inches.

That's better, but it's still not the shortest way. Remember, the shortest distance between two points is a straight line.

Lonzo

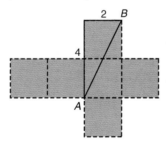

Maybe a picture would help. A cube has 6 faces and they all fit together like this. You need to use a hypotenuse, but the right triangle you use has a 2 inch leg and a 4 inch leg. Therefore, $h^2 = 2^2 + 4^2$ or $h^2 = 20$. That means that $h = \sqrt{20}$, which is about 4.5 inches. This must be the shortest path.

Paul

402 Chapter 9 Algebra: Exploring Real Numbers

■ Reteaching the Lesson ■

Activity Have students work in small groups to draw diagrams to solve the problem. *What is the maximum number of pieces that can be cut from circular pizzas using the given number of straight cuts? (They need not go through the center.)*

a. 1 2   **b.** 2 4   **c.** 3 7   **d.** 4 11

THINK ABOUT IT

Work with a partner.

1. *Compare and contrast* the three solutions. Are all three paths feasible for an ant to travel?

2. *Draw* a different diagram that could be used to solve this problem. How is it different? What path would the ant travel and how long is that path? See Answer Appendix.

1. Yes each path is along the faces of the cube, so they are feasible.

3. *Draw a diagram* to solve the following problem.

   A section of a theater is arranged so that each row has the same number of seats. Annie is seated in the fifth row from the front and the third row from the back. Her seat is sixth from the left and second from the right. How many seats are in this section?
   See Answer Appendix.

ON YOUR OWN

5. See students' work.

4. The first step of the 4-step plan for problem solving asks you to *explore* the problem. *Explain* how you determined what information was given, what was needed, and if there was too much information given for Exercise 3.
See students' work.

5. *Write a Problem* that can be solved by drawing a diagram. Exchange problems with another student and solve each others' problem.

6. *Look Ahead* Draw a diagram that could be used for solving Exercise 10 on page 407.
See Answer Appendix.

MIXED PROBLEM SOLVING

12a. See margin.

Solve. Use any strategy.   11. 11 quarters and 19 nickels 12b. See Answer Appendix.

Strategies
Look for a pattern.
Solve a simpler problem.
Act it out.
Guess and check.
Draw a diagram.
Make a chart.
Work backward.

7. *Recreation* Ms. Llarena's dance class is evenly spaced in a circle. If the sixth person is directly opposite the sixteenth person, how many people are in the circle? **20 people**

8. *Number Theory* When a number is decreased by 4, the result is −23. What is the number? **−19**

9. *Geometry* Two sides of a triangle have the same length. The third side is 2 meters long. If the perimeter of the triangle is 20 meters, find the lengths of the sides. **9 meters**

10. *Food* Of the 30 members in a cooking club, 20 like to mix salads, 17 prefer baking desserts, and 8 like to do both.
    a. How many like to mix salads, but not bake desserts? **12 people**
    b. How many do not like either baking desserts or mixing salads? **1 person**

11. *Money* Matsuko has some quarters and nickels in her pocket. She has 8 more nickels than quarters. The total value is $3.70. How many of each coin does Matsuko have?

12. *Entertainment* The graph shows the numbers of types of outdoor grills sold.
    a. How does the number of charcoal grills compare to the number of gas grills?
    b. Make a circle graph of the information.

fire it up! Millions of Grills Sold
Charcoal 7.9
Gas 4.3
Electric 0.16
Source: Barbecue Industry Association

13. **Test Practice** There are eight people at a business meeting. Each person shakes hands with everyone else exactly once. How many handshakes occur? **C**
    A 8   B 16   C 28   D 64

Lesson 9-5A THINKING LAB 403

In-Class Example

Ask students to draw a diagram to solve this problem. There are four sticks of one length and four more exactly half the length of the first four. Show how to enclose exactly three squares of equal area with the eight sticks.

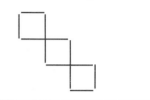

3 PRACTICE/APPLY

Check for Understanding
Use the results of Exercise 3 to determine whether students understand how to solve a problem using a diagram.

Extra Practice If students need additional practice in problem solving, extra practice is available on the following pages.
- Drawing a Diagram, see p. 632
- Mixed Problem Solving, see pp. 645–646

Assignment Guide
All: 4–13

4 ASSESS

Closing Activity
Writing Have students write a paragraph describing how using a diagram helped them solve Exercise 3.

Additional Answer
12a. Sample answer: There were almost twice as many charcoal grills sold as gas grills.

Extending the Lesson

Activity Have students create problems that other students can solve with diagrams. Sample problem: *A table can seat 3 people on each side and 1 at each end. How many people can be seated if three of these tables are placed end-to-end?*

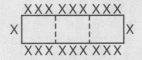

20 people

Thinking Lab 9-5A   403

9-5 Lesson Notes

Instructional Resources
- Study Guide Masters, p. 74
- Practice Masters, p. 74
- Enrichment Masters, p. 74
- Transparencies 9-5, A and B
- Technology Masters, p. 70
- CD-ROM Program
  - Resource Lesson 9-5

Recommended Pacing	
Standard	Day 9 of 13
Honors	Days 5 & 9 of 11
Block	Day 5 of 7

1 FOCUS

5-Minute Check
(Lesson 9-4)

1. Write an equation to find the length of the missing side of the triangle. Then find the missing length.
   $c = \sqrt{8^2 + 15^2}$; 17 in.

2. Is a triangle measuring 9 meters, 12 meters, and 15 meters a right triangle? **yes**

3. What is the length of the other leg of a right triangle if the hypotenuse is 13 meters and one leg is 12 meters? **5 meters**

 The 5-Minute Check is also available on **Transparency 9-5A** for this lesson.

Motivating the Lesson

Communication Point out that the Pythagorean Theorem has many everyday applications. Ask this question: *What part of a 25-inch TV set measures 25 inches?* **the diagonal; In the figure, $a^2 + b^2 = 25$.**

9-5 Using the Pythagorean Theorem

What you'll learn
You'll learn to solve problems using the Pythagorean Theorem.

When am I ever going to use this?
You can use the Pythagorean Theorem to find the distance a gymnast will run on a floor mat.

Word Wise
Pythagorean triple

The Anasazi Indians built pueblos in the southwest United States around A.D. 900, long before the European Americans reached that area. Bandelier National Monument in New Mexico has preserved some of their ancient ruins including the amazing pueblos and cave dwellings. Visitors can still climb ladders, sit in the ancient caves, and admire the spectacular view.

The stone walls are rugged, making it difficult to measure the height of caves from the ground. However, using the Pythagorean Theorem, archeologists can easily calculate the heights.

 **APPLICATION**

Archeology A 10-foot ladder that is placed 4 feet from the base of the vertical rock wall just reaches the entrance of the cave. How high is the cave entrance from the ground?

Explore Draw a diagram. The wall, the ground, and the ladder form a right triangle and that the ladder is the hypotenuse. You need to know the height of the cave.

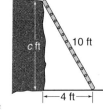

Plan Let c = the height of the cave, g = the length on the ground from the wall to the bottom of the ladder, and ℓ = the length of the ladder. Find the value of c by using the Pythagorean Theorem.

Solve
$$c^2 + g^2 = \ell^2$$
$$c^2 + 4^2 = 10^2 \quad \text{Substitute the values for } g \text{ and } \ell \text{ into the Pythagorean Theorem.}$$
$$c^2 + 16 = 100$$
$$c^2 + 16 - 16 = 100 - 16 \quad \text{Subtract 16 from each side.}$$
$$c^2 = 84$$
$$c = \sqrt{84}$$

84 [2nd] [√] 9.1651513 9   *Use a calculator.*

The cave is about 9.2 feet above the ground.

Examine Check to see if the measures satisfy the Pythagorean Theorem.
$$4^2 + 9.2^2 \stackrel{?}{=} 10^2$$
$$16 + 84.64 \stackrel{?}{=} 100.64$$
$$100.64 \approx 100 \quad \checkmark \quad \text{The numbers aren't exactly the same because 9.2 was rounded.}$$

404 Chapter 9 Algebra: Exploring Real Numbers

Multiple Learning Styles

 Auditory/Musical Have students split into groups to create jingles to help commit triples and the Pythagorean Theorem to memory. Sample jingles: *a* times *a*, plus *b* times *b*, will always come to *c* times *c*. Sides 3, 4, and long side 5, make the Theorem come alive.

Example 2 — APPLICATION

Sports Drew Bledsoe and Terry Glenn play for the New England Patriots. On a play, the announcer said, "Bledsoe drops back to the 10-yard line and fires the ball to Glenn. Glenn catches it at the 50-yard sideline. What a catch!" If Bledsoe was 25 yards from the sideline when he dropped back to the 10-yard line, how far did he throw the ball?

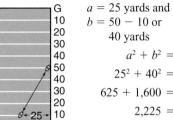
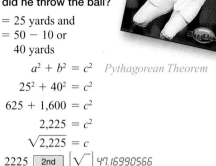

$a = 25$ yards and
$b = 50 - 10$ or
40 yards

$$a^2 + b^2 = c^2 \quad \text{Pythagorean Theorem}$$
$$25^2 + 40^2 = c^2$$
$$625 + 1{,}600 = c^2$$
$$2{,}225 = c^2$$
$$\sqrt{2{,}225} = c$$

2225 [2nd] [√] 47.16990566

Drew Bledsoe's pass was just over 47 yards.

You know that a triangle with sides 3, 4, and 5 units is a right triangle because these numbers satisfy the Pythagorean Theorem. Such numbers are called **Pythagorean triples**. By using multiples of a primitive set of Pythagorean triples, you can create additional triples. *The triple 3-4-5 is called a primitive Pythagorean triple because the numbers are relatively prime.*

Example 3 — INTEGRATION

Number Patterns

a. Multiply the triple 3-4-5 by the numbers 2, 3, 4, and 10 to find more Pythagorean triples.

b. Compare the ratio of the areas of a 3-4-5 triangle and a 30-40-50 triangle to the ratio of the sides of the two triangles.

a. You can organize your answers in a table. Multiply each Pythagorean triple entry by the same number and then check the Pythagorean relationship.

	a	b	c	Check: $a^2 + b^2 = c^2$
original	3	4	5	9 + 16 = 25 ✓
× 2	6	8	10	36 + 64 = 100 ✓
× 3	9	12	15	81 + 144 = 225 ✓
× 4	12	16	20	144 + 256 = 400 ✓
× 10	30	40	50	900 + 1,600 = 2,500 ✓

b. Find the area of each triangle.

$A = \frac{1}{2}bh$ *Formula for area of a triangle*

3-4-5 triangle:
$A = \frac{1}{2}(3)(4)$
$A = 6$

30-40-50 triangle:
$A = \frac{1}{2}(30)(40)$
$A = 600$

(continued on the next page)

Study Hint
Reading Math ×2 means multiply by 2, ×3 means multiply by 3, and so on.

Lesson 9-5 Using the Pythagorean Theorem

---

2 TEACH

Transparency 9-5B contains a teaching aid for this lesson.

Using Connections Students can find the primitive Pythagorean triple for any Pythagorean triple by dividing each member of the triple by the GCF of the three numbers. Challenge students to find the width and height of the TV set in the Motivating problem by informing them that the dimensions are based on a primitive Pythagorean triple. **15 in. × 20 in.**

In-Class Examples

For Example 1
Find h, the height of the tent. **about 5.2 ft**

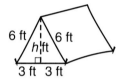

For Example 2
To save energy in a cold climate, a house has a roof that slants from the middle of the house to the ground on the north side. The house is 48 feet wide and 18 feet tall. How long is the roof? **30 ft**

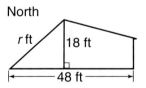

For Example 3
A right triangle can have sides with lengths of 5, 12, and 13 units. Does this pattern work if multiplied by 2 and 3? **yes** Organize data into a table to find out.

	a	b	c	$a^2 + b^2 = c^2$
original	5	12	13	25 + 144 = 169
× 2	10	24	26	100 + 576 = 676
× 3	30	72	78	900 + 5,184 = 6,084

3 PRACTICE/APPLY

Check for Understanding

If students need additional practice or instruction after completing Exercises 1–4, one of these options may be helpful.
- Extra Practice, see p. 632
- Reteaching Activity
- *Study Guide Masters*, p. 74
- *Practice Masters*, p. 74

Assignment Guide

Core: 5–13 odd, 14–16
Enriched: 6, 8, 9–16

Additional Answer

2. The Pythagorean Theorem is an equation and therefore, by substituting two of the three unknowns, you can solve for the third one.

Study Guide Masters, p. 74

---

ratio of areas = $\frac{6}{600}$ or $\frac{1}{100}$

ratio of sides = $\frac{3}{30}$ or $\frac{1}{10}$, $\frac{4}{40}$ or $\frac{1}{10}$, and $\frac{5}{50}$ or $\frac{1}{10}$

Each pair of sides has a ratio of $\frac{1}{10}$, and the ratio of the areas is $\frac{1}{100}$.

CHECK FOR UNDERSTANDING

Communicating Mathematics

Read and study the lesson to answer each question.

1. *Write* two Pythagorean triples other than 5-12-13 that are in the 5-12-13 family. **Sample answer: 10-24-26 and 15-36-39**

2. *Explain* why you can use any two sides of a right triangle to find the third side. **See margin.**

Guided Practice

Write an equation that can be used to answer the question. Then solve. Round to the nearest tenth.

3. How long is the lake?
$x^2 + 21^2 = 30^2$;
$x = \sqrt{459} \approx$
21.4 miles

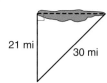

4. **Geography** The state of Wyoming is shaped like a rectangle that is about 360 miles by 280 miles.
   a. What is the longest distance you could go in a straight line and still remain in Wyoming?
   b. Estimate the longest distance in your state. **See students' work.**
   a. $360^2 + 280^2 = x^2$; $x = \sqrt{208,000} \approx 456.1$ miles

EXERCISES

Practice

5. $5^2 + 8^2 = x^2$; $x = \sqrt{89} \approx 9.4$ miles

6. $15^2 = 3^2 + x^2$; $x = \sqrt{216} \approx 14.7$ feet

7. $x^2 = 9^2 + 18^2$; $x = \sqrt{405} \approx 20.1$ feet; $x^2 = 9^2 + 12^2$; $x = \sqrt{225} = 15$ feet

Write an equation that can be used to answer each question. Then solve. Round to the nearest tenth.

5. How far apart are the planes?

6. How high does the ladder reach?

7. How long is each rafter?

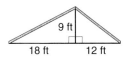

8. Name the primitive Pythagorean triple to which 16-30-34 belongs. **8-15-17**

406 Chapter 9 Algebra: Exploring Real Numbers

Reteaching the Lesson

Activity Separate students into groups of three. Have each group prepare sets of 10 cards numbered 1 to 10 for one pile and three cards labeled "leg," "leg," and "hypotenuse" for another. One student shuffles each set of cards separately and draws two cards from each pile, another sketches a triangle relating to the numbers, and the third student finds the unknown length. The answers will usually be irrational numbers.

Applications and Problem Solving

9. **Safety** For safety reasons, the base of a 24-foot ladder should be at least 8 feet from the wall. How high can a 24-foot ladder safely reach? $\sqrt{512} \approx 22.6$ **feet**

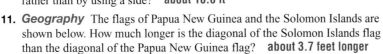

10. **Sports** At 13 years old, Dominique Moceanu was the youngest gymnast ever to win the U.S. Senior National Championship. Using the diagonal of the floor mat for her run, flip, and finish, gives her more distance. The routine is performed on a 40-foot by 40-foot mat. How much more room does she get by using the diagonal rather than by using a side? **about 16.6 ft**

11. **Geography** The flags of Papua New Guinea and the Solomon Islands are shown below. How much longer is the diagonal of the Solomon Islands flag than the diagonal of the Papua New Guinea flag? **about 3.7 feet longer**

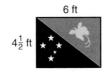
Papua New Guinea

Solomon Islands

12. **Recreation** The acceleration ramp for a skateboard competition is 20 meters long, and it extends 15 meters from the base of the starting point. How high is the ramp? **about 13.2 meters**

13. **Geography** Jewel likes to hike in the San Miguel Mountains at Bandelier National Monument. The park boundaries are two legs of a right triangle. One leg is 6 kilometers long, and the other leg is 11.8 kilometers long. The trail from the Rio Grande River to Jewel's campsite is slightly longer than the hypotenuse of the triangle. Approximate the length of her hike. **about 13.2 km**

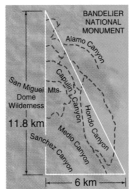

14. **Critical Thinking** If you only know the length of one side of a right triangle, can you find the lengths of the other two sides? Draw diagrams to support your answer. **See Answer Appendix.**

Mixed Review

15. **Test Practice** Triangle DEF is a right triangle. Its hypotenuse is 15 inches long, and one leg is 12 inches long. How long is the other leg? (Lesson 9-4) **B**

    A 11 in. B 9 in. C 4 in. D 3 in.

16. Factor 162 completely. (Lesson 6-2) $2 \cdot 3^4$

In *The Wizard of Oz*, Scarecrow recites a formula, supposedly the Pythagorean Theorem. Find the tape at home or in the library. Does the scarecrow recite the formula correctly? Explain.

Lesson 9-5 Using the Pythagorean Theorem **407**

Extending the Lesson

Enrichment Masters, p. 74

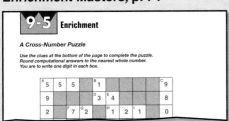

Activity Have students prepare a table with five columns labeled m, n, $m^2 - n^2$, $2mn$, and $m^2 + n^2$. Let them choose several values for m and n so that one is odd, the other even, and $m > n$. Find what they observe when they evaluate the expressions using the values of m and n. **They are Pythagorean triples.**

Family Activity

Have students write down what they hear the scarecrow say and discuss it with their families. The scarecrow says: "The sum of the square roots of any two sides of an isosceles triangle is equal to the square root of the remaining side." The Pythagorean Theorem, in contrast, pertains to right triangles, not isosceles triangles; to the sum of the squares, not the square roots; and to specific sides, not any two sides.

4 ASSESS

Closing Activity

Writing Have students work in pairs to create problems that can be solved using the Pythagorean Theorem.

Practice Masters, p. 74

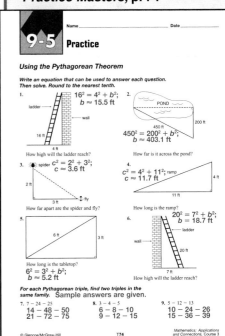

9-5B LAB Notes

GET READY

Objective Students use the Pythagorean Theorem to graph irrational numbers on a number line.

Optional Resources
Hands-On Lab Masters
- number lines, p. 15
- worksheet, p. 59

Overhead Manipulative Resources
- number lines

Manipulative Kit
- compass
- ruler

MANAGEMENT TIPS

Recommended Time
45 minutes

Getting Started To introduce the technique of this lab, ask students to express the following numbers as the sum or difference of two squares—20, 11, and 17. **20:** $2^2 + 4^2$ or $6^2 - 4^2$; **11:** $6^2 - 5^2$; **17:** $1^2 + 4^2$

Activity 1 shows students how to use a compass and a number line to find the length of the hypotenuse of a right triangle. Side $C = OB$. Although students can use rulers to mark off number lines, you may prefer to have them use compasses kept a constant distance between units.

COOPERATIVE LEARNING

9-5B Graphing Irrational Numbers

A Follow-Up of Lesson 9-5

 compass

 straightedge

Lewis Howard Latimer (1848-1928), an African-American engineer and drafter, made the drawings needed to get a patent for the telephone, which was invented by Alexander Graham Bell.

Drafters, inventors, and engineers used to use a compass and straightedge to copy measurements. Today, they use computers to generate duplicate measurements.

You already know how to graph integers and rational numbers on a number line. How would you graph an irrational number like $\sqrt{5}$?

TRY THIS

Work with a partner.

1 Graph $\sqrt{5}$.

- Find two numbers whose squares have a sum or difference of 5. One pair that works is 1 and 2, since $1^2 + 2^2 = 5$.

- Draw a number line. At 2 on the number line, construct a perpendicular line segment 1 unit long. *You can construct a perpendicular segment by folding the number line at 2, making sure the two parts of the number line align. The fold is a segment perpendicular to the number line.*

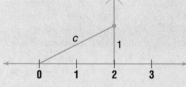

- Draw the line segment shown from 0 to the top of the 1 unit segment. Label it c.

- The Pythagorean Theorem can be used to show that c is $\sqrt{5}$ units long.

  $c^2 = a^2 + b^2$
  $c^2 = 1^2 + 2^2$ *Replace a with 1 and b with 2.*
  $c^2 = 5$
  $c = \sqrt{5}$

- Open the compass to the length of c. With the tip of the compass at 0, draw an arc that intersects the number line at B. The distance from 0 to B is $\sqrt{5}$ units.

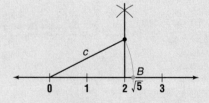

408   Chapter 9 Algebra: Exploring Real Numbers

2 Graph $\sqrt{7}$.

- Find two numbers whose squares have a sum or difference of 7. One pair that works is 4 and 3, since $4^2 - 3^2 = 7$.

- Draw a number line. At 3 on the number line, construct a perpendicular line segment. Put the tip of the compass at 0. With the compass at 4 units, construct an arc that intersects the perpendicular line segment. Label the perpendicular leg a.

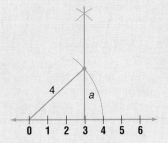

- The Pythagorean Theorem can be used to show that a is $\sqrt{7}$ units long.
  $c^2 = a^2 + b^2$
  $4^2 = a^2 + 3^2$ *Replace c with 4 and b with 3.*
  $16 = a^2 + 9$
  $7 = a^2$
  $\sqrt{7} = a$

- Open the compass to the length of a. With the tip of the compass at 0, draw an arc that intersects the number line at D. The distance from 0 to D is $\sqrt{7}$ units.

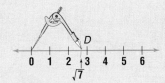

ON YOUR OWN

Graph each number on a number line. 1–9. See Answer Appendix.

1. $\sqrt{2}$ 2. $\sqrt{3}$ 3. $\sqrt{8}$ 4. $\sqrt{10}$

5. Explain how you graphed $\sqrt{2}$.
6. Describe two different ways to graph $\sqrt{8}$.
7. Explain how to graph $-\sqrt{5}$.
8. Explain how the graph of $\sqrt{2}$ can be used to locate the point that represents $\sqrt{3}$.
9. **Reflect Back** For each number from 11-20 that is not a perfect square, write the equation that is the sum or difference of two squares that could be used in order to graph the square root of that number.

Lesson 9-5B HANDS-ON LAB **409**

 Have students write a paragraph that compares Activity 1 to Activity 2. Students should explain why Activity 2 has an extra step.

9-6 Lesson Notes

Instructional Resources
- *Study Guide Masters*, p. 75
- *Practice Masters*, p. 75
- *Enrichment Masters*, p. 75
- Transparencies 9-6, A and B
- *Assessment and Evaluation Masters*, p. 240
- *School to Career Masters*, p. 35

CD-ROM Program
- Resource Lesson 9-6

Recommended Pacing	
Standard	Day 10 of 13
Honors	Day 9 of 11
Block	Day 5 of 7

1 FOCUS

5-Minute Check
(Lesson 9-5)

1. Tony drove 12 miles south and then 16 miles west. How far is he from his starting point? **20 miles**

2. How far is the helicopter from the school? Write an equation to answer the question. Then solve. Round to the nearest tenth.

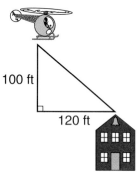

$100^2 + 120^2 = x^2$;
$x = 156.2$ ft

3. Name the primitive Pythagorean triple to which 24-45-51 belongs. **8-15-17**

The 5-Minute Check is also available on **Transparency 9-6A** for this lesson.

9-6

Integration: Geometry
Distance on the Coordinate Plane

What you'll learn
You'll learn to find the distance between points in the coordinate plane.

When am I ever going to use this?
Knowing how to find the distance between points in the coordinate plane can help you estimate the distance between places on a map.

In 1791, Pierre L'Enfant was hired to create a plan for the nation's capital city, Washington, D.C. He placed the Capitol at the center of a grid-style layout for the city. Below is a map of Capitol Hill and its surrounding area with a centimeter grid laid over top.

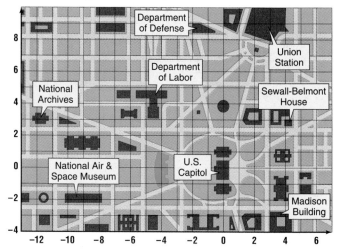

When points are graphed on a coordinate grid, you can use ordered pairs to determine the distance between two points.

LOOK BACK
You can refer to Lesson 2-10 to review the coordinate plane.

HANDS-ON MINI-LAB

Work with a partner. grid paper   ruler

Try This
- On your grid paper, draw a coordinate system and label the *x*- and *y*-axes.
- Plot points $A(2, -1)$ and $B(5, 6)$ on your graph and connect them with a line segment.
- Draw a horizontal line through A and a vertical line through B. Call the point of intersection C.

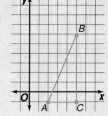

Talk About It
1. Find the length of $\overline{AC}$. **3 units**
2. What is the length of $\overline{BC}$? **7 units**
3. Explain how you can find the length of $\overline{AB}$. Then find AB.

3. Use the Pythagorean Theorem; $AB = \sqrt{58} \approx 7.6$ units

410   Chapter 9 Algebra: Exploring Real Numbers

Multiple Learning Styles

 Kinesthetic Use paper plates labeled *X*, *O*, and *Y* to locate the vertices of a right triangle on the gymnasium floor. Let *O* be the vertex at the right angle and assigned the ordered pair (0, 0). Have students walk from each vertex to another and record the number of steps. Then have them write the ordered pair that represents points *X* and *Y*. Have them use the vertices to compute *XY* (in terms of steps) and compare it to the number of steps they recorded.

Examples

① Graph the ordered pairs (2, 0) and (5, −4). Then find the distance between the points.

Let c = the distance between the two points, $a = 3$, and $b = 4$.

Use the Pythagorean Theorem.

$c^2 = a^2 + b^2$
$c^2 = 3^2 + 4^2$
$c^2 = 9 + 16$
$c^2 = 25$
$c = \sqrt{25}$
$c = 5$

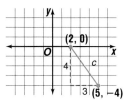

The points are 5 units apart.

APPLICATION

② **Travel** Refer to the map at the beginning of the lesson. Suppose that the Capitol is located at the origin. How many miles is it from the National Air and Space Museum to the Department of Defense if a unit on the grid is about 0.05 mile?

The coordinates of the National Air and Space Museum are (−8, −2) and the coordinates of the Department of Defense are (−2, 9).

Let c = the distance between the National Air and Space Museum and the Department of Defense. Then $a = 6$ and $b = 11$.

$c^2 = a^2 + b^2$ *Pythagorean Theorem*
$c^2 = 6^2 + 11^2$
$c^2 = 36 + 121$
$c^2 = 157$
$c = \sqrt{157}$

157 [2nd] [√] *12.52996409* *Use a calculator.*

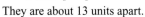

They are about 13 units apart.

Multiply 13 units by 0.05 mile per unit to find the number of miles.

$13 \times 0.05 = 0.65$

The National Air and Space Museum is about 0.65 mile from the Department of Defense.

Did you know? Suppose all of the Smithsonian treasures were lined up in one long exhibit. If a person spent 1 second looking at each item nonstop for 24 hours a day, it would take that person more than 2.5 years to see all of them.

CHECK FOR UNDERSTANDING

Communicating Mathematics

Read and study the lesson to answer each question. 1–2. See margin.

1. *Draw* the triangle that you can use to find the distance between points at (−2, 1) and (−5, −3) on the coordinate plane.

2. *Write* the steps you could use to find the distance between points at (5, 5) and (2, 2).

Lesson 9-6 Integration: Geometry Distance on the Coordinate Plane **411**

Additional Answers

1.

2. 1) Graph the points (5, 5) and (2, 2) on a coordinate plane and connect them with a line segment.
   2) Draw a horizontal line through (2, 2) and a vertical line through (5, 5).
   3) Use the Pythagorean Theorem.

Motivating the Lesson

Hands-On Activity Have students plot the point P(4, 6) on a coordinate plane. Students should set a compass for a radius of 5 based on units of the graph paper and draw a circle with center at P. Then have them name as many points as possible located on the circle. This shows how a coordinate plane can be used to plot points for any figure. **Sample answers:** (4, 1); (4, 11); (−1, 6); (9, 6); (0, 3); (0, 9); (8, 9); (8, 3)

2 TEACH

 Transparency 9-6B contains a teaching aid for this lesson.

Using the Mini-Lab Point out that the length of $\overline{AC}$ equals the difference between the x-coordinates of A and C: $5 − 2 = 3$. The length of $\overline{BC}$ equals the difference between the y-coordinates of B and C: $6 − (−1) = 7$.

In-Class Examples

For Example 1
Graph the ordered pairs (−1, 2) and (7, −4) on a coordinate plane. Then find the distance between the points. **10 units**

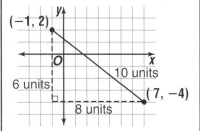

For Example 2
Use the map and grid on page 410 to find how far it is from the National Archives to the Madison Building if you could walk directly from one point to the other. Each unit of the grid equals 0.05 mile. **using grid points (−12, 3) and (4, −3): about 0.9 miles**

Teaching Tip When working Examples 2 in the Student Edition and the Teachers' Wraparound Edition, first establish which grid point you are using for the buildings, as they may lie on more than one intersection.

Lesson 9-6 **411**

3 PRACTICE/APPLY

Check for Understanding

If students need additional practice or instruction after completing Exercises 1–7, one of these options may be helpful.
- Extra Practice, see p. 632
- Reteaching Activity
- *Study Guide Masters*, p. 75
- *Practice Masters*, p. 75

Assignment Guide

Core: 9–19 odd, 22–25
Enriched: 8–16 even, 18–20, 22–25

Additional Answer

3. Both distances are 5 units or about 0.25 mile. Because the Madison Building is at (4, −3) and the Sewall-Belmont House is at (4, 3), the distances are the same.

Study Guide Masters, p. 75

9-6 Study Guide

Distance on the Coordinate Plane

8. 5.8 units
9. 4.1 units
10. 5.1 units
11. 7.1 units
12. 7.6 units
13. 4.5 units
14. 11.3 units
15. 7 units
16. 13.5 units

HANDS-ON MATH

3. Find the distance from the Capitol to the Sewall-Belmont House and from the Capitol to the Madison Building. Are the distances the same or different? Explain. **See margin.**

Guided Practice

4. Find the distance between the pair of points whose coordinates are given. Round to the nearest tenth. **5**

Graph each pair of ordered pairs. Then find the distance between the points. Round to the nearest tenth. **5–6. See Answer Appendix for graphs.**

5. (−2, 4), (3, −1)   **7.1 units**
6. (1, 4), (6, 2)   **5.4 units**

7. **Maps** On a street map, the side of each square represents 1 mile. The Wright Ball Park is located at (1, 2), and the Weaver Rollerdome is located at (6, 10). A diagonal street runs directly between the two locations. Approximately how far is it from Wright Ball Park to Weaver Rollerdome? **about 9.4 miles**

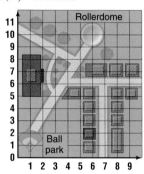

EXERCISES

Practice

Find the distance between each pair of points whose coordinates are given. Round to the nearest tenth.

8. 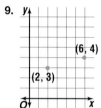   9. (graph with (6, 4) and (2, 3))   10. (graph with (2, 4) and (−3, 3))

Graph each pair of ordered pairs. Then find the distance between the points. Round to the nearest tenth. **11–16. See Answer Appendix for graphs.**

11. (3, −2), (2, 5)
12. (−1, 0), (2, 7)
13. (1, 5), (3, 1)
14. (−6, 3), (2, −5)
15. (−5, 0), (2, 0)
16. (7, −4), (−3, 5)

17. The coordinates of points R and S are (4, 3) and (1, 6). What is the distance between the points? **4.2 units**

Applications and Problem Solving

18. **Geometry** A triangle on the coordinate plane has vertices $A(2, -1)$, $B(-2, 2)$, and $C(-6, 14)$.
    a. Draw the triangle. **See Answer Appendix.**
    b. Find the perimeter of the triangle. **about 34.6 units**

412   Chapter 9 Algebra: Exploring Real Numbers

Reteaching the Lesson

Activity Have students toss a pair of number cubes to form an ordered pair. Toss them again to name replacements for the x- and y-coordinates. For example, if the original point is A(4, 1), replace x to from B(6, 1) and replace y to form C(4, 3). Have students plot the points and find the length of $\overline{BC}$.

Error Analysis
Watch for students who count across the diagonal lines on the grid to find the length of hypotenuses.
Prevent by having students use the Pythagorean Theorem.

412   Chapter 9

19. **Geometry** A circle goes through the points $A(3, 0)$, $B(6, 3)$, $C(9, 0)$, and $D(6, -3)$. Desiree connected those points and determined that they formed a square inside the circle. What is the length of a side of the square? $\sqrt{18}$

20. **Graphing Calculator** The graphing calculator program will find the distance between two points in the coordinate plane. To use the program, first enter it into the calculator's memory. To access the program memory, press PRGM ▶ ▶ ENTER. Enter all the program instructions at the right. Then press 2nd [QUIT] to return to the Home screen. To run the program, press PRGM and choose the program from the list by pressing the number next to its name then ENTER. Enter the coordinates of the points as the program asks for them. *X1 refers to the x-coordinate of the first point.* The calculator automatically rounds the distance to the nearest tenth of a unit.

```
Prgm 1: DISTANCE
: Fix 1
: Disp "ENTER X1"
: Input A
: Disp "ENTER Y1"
: Input B
: Disp "ENTER X2"
: Input C
: Disp "ENTER Y2"
: Input D
: √((A-C)²+(B-D)²)→E
: Disp "THE DISTANCE IS"
: Disp E
```

Use the program to find the distance between each pair of points.

20a. 10.8 units
20b. 27.7 units
20c. 11.0 units
20d. 9.2 units
20e. 179.4 units
20f. 6.2 units ≈ 4.2 units

a. $(3, 3), (-7, -1)$ b. $(-12, 1), (15, -5)$ c. $(9, 1), (-2, 1)$
d. $(-8, 3), (-2, -4)$ e. $(-47, 21), (125, 72)$ f. $(2.4, 6.1), (0.2, 0.3)$

21. **Working on the CHAPTER Project**
Police officers may use a technique called the *coordinate method* to diagram an accident scene. An officer chooses a reference point such as a mile post or a telephone pole. For each item to be diagramed, the officer places a marker along the side of the road at a point that is directly across from the item.

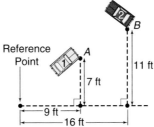

a. Place the accident scene diagram above on a coordinate plane and determine how far point A is from point B. **about 8 feet**

b. Simulate an accident scene in the school parking lot. Make a diagram on a coordinate grid and write an accident report. **See students' work.**

22. **Critical Thinking** The distance between points A and B is 17 units. Find the value of x if the coordinates of A and B are $(-3, x)$ and $(5, 2)$. **17 or −13**

Mixed Review

23. **Test Practice** What is the height of the tower? *(Lesson 9-5)* **B**

A 8 feet   B 31.5 feet
C 992 feet   D 49.9 feet

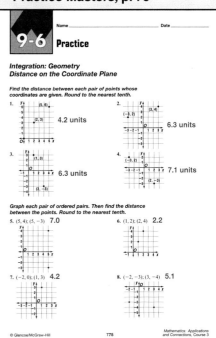

24. Express $8.\overline{72}$ as a mixed number. *(Lesson 6-5)* $8\frac{8}{11}$

25. Solve $f = 8(-3)(-4)^2$. *(Lesson 2-7)* **−384**

Lesson 9-6 Integration: Geometry Distance on the Coordinate Plane **413**

9-7 Lesson Notes

Instructional Resources
- *Study Guide Masters*, p. 76
- *Practice Masters*, p. 76
- *Enrichment Masters*, p. 76
- Transparencies 9-7, A and B
- *Assessment and Evaluation Masters*, p. 240
- *Classroom Games*, pp. 25–29
- *Hands-On Lab Masters*, p. 76
- CD-ROM Program
  - Resource Lesson 9-7

Recommended Pacing

Standard	Day 11 of 13
Honors	Day 9 of 11
Block	Day 6 of 7

1 FOCUS

5-Minute Check
(Lesson 9-6)

Graph each pair of ordered pairs. Then find the distance between the points. Round to the nearest tenth.

1. (1, 2), (5, 5) **5 units**

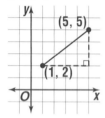

2. (−2, −2), (3, 0) **5.4 units**

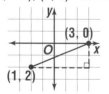

3. (2, 6), (−3, 1) **7 units**

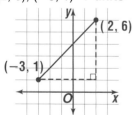

 The 5-Minute Check is also available on **Transparency 9-7A** for this lesson.

9-7 Integration: Geometry
Special Right Triangles

What you'll learn
You'll learn to find missing measures in 30°-60° right triangles and 45°-45° right triangles.

When am I ever going to use this?
Knowing how to find missing measures in special right triangles can help you determine roof span length.

The sides of right triangles with 30° and 60° angles have a special relationship.

MINI-LAB

Work with a partner. compass   protractor
 ruler   scissors

Try This
- Construct and cut out an equilateral triangle.
- Fold the triangle in half and cut along the fold line.
- Measure each side and each angle. Measure the length to the nearest tenth of a centimeter.
- Repeat the above steps for at least two other equilateral triangles.

Talk About It The hypotenuse is twice as long.
What is the relationship between the length of the hypotenuse and the length of the side opposite the 30° angle?

LOOK BACK
You can refer to Lesson 5-1A to review how to measure and construct line segments and angles.

The relationship you discovered in the Mini-Lab is always true in a 30°-60° right triangle. The length of the hypotenuse is twice the length of the side opposite the 30° angle.

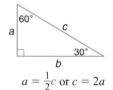

$a = \frac{1}{2}c$ or $c = 2a$

Therefore, the ratio $\frac{\text{length of hypotenuse}}{\text{length of side opposite the 30° angle}}$ is $\frac{2}{1}$.

Example 1 Find the lengths *a* and *b*.

Step 1 Find *a*.
$a = \frac{1}{2}c$
$a = \frac{1}{2}(18)$ *Replace c with 18.*
$a = 9$

414 Chapter 9 Algebra: Exploring Real Numbers

Motivating the Lesson
Problem Solving Have students help you draw a baseball diamond and label the angle measures on the chalkboard. **four 90° angles** Then ask students to calculate the angles from third base to first base. **four 45° angles** Ask: *What does the dividing line form?* **the hypotenuse of two right triangles**

Step 2 Find b.

$$a^2 + b^2 = c^2 \quad \text{\textit{Pythagorean Theorem}}$$
$$9^2 + b^2 = 18^2 \quad \text{\textit{Replace a with 9 and c with 18.}}$$
$$81 + b^2 = 324$$
$$b^2 = 243$$
$$b = \sqrt{243}$$

243 [2nd] [√] *15.58845727*

$$b \approx 15.6$$

The length a is 9 feet, and the length b is about 15.6 feet.

There is also a special relationship between the sides of a triangle with two 45° angles.

HANDS-ON MINI-LAB

Work with a partner.   protractor   scissors

Try This
- Construct and cut out a square.
- Fold the square in half and cut along the diagonal.
- Measure each side and each angle. Measure the length to the nearest tenth of a centimeter.
- Repeat the above steps for at least two other squares.

Talk About It

What is the relationship between the length of the legs in the 45°-45° right triangles?   **The lengths of the legs are equal.**

In a 45°-45° right triangle, the lengths of the legs are equal. $a = b$

Example 2 APPLICATION

Construction Find the width that a roof will span if the roof forms a 45° angle with the horizontal and one side of the roof is 10 feet. The top of the roof forms a 90° angle.

Step 1 Find a.
Since the roof is a 45°-45° right triangle, $a = b$.
Therefore, $a = 10$.

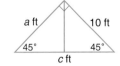

(continued on the next page)

Lesson 9-7 Integration: Geometry   Special Right Triangles   **415**

Investigations for the Special Education Student

This blackline master booklet helps you plan for the needs of your special education students by providing long-term projects along with teacher notes. Investigation 8, *Weather Report,* may be used with this chapter.

---

2 TEACH

Transparency 9-7B contains a teaching aid for this lesson.

Using the Mini-Lab In the Mini-Lab on page 414, students should draw a segment of arbitrary length for the base of the triangle. Setting the compass radius to the length of this segment and placing the point at each end of the segment, they should draw intersecting arcs above the segment. The point of intersection is the third vertex of the equilateral triangle.

In-Class Example

For Example 1
Find the lengths a and b.
$a = 13$, $b \approx 22.5$

Using the Mini-Lab One easy way to construct a square is to fold one corner of a rectangular piece of paper to the opposite edge and cut off the part that does not lie in the folded part.

In-Class Example

For Example 2
Find the length of the sides of a stained-glass attic window if the lower corners are both 45° and its height is 60 centimeters. The top of the window forms a 90° angle.

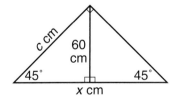

The sides of the window are about 84.9 centimeters each; the bottom side is 120 centimeters.

Lesson 9-7   **415**

3 PRACTICE/APPLY

Check for Understanding

If students need additional practice or instruction after completing Exercises 1–6, one of these options may be helpful.
- Extra Practice, see p. 633
- Reteaching Activity
- *Study Guide Masters*, p. 76
- *Practice Masters*, p. 76

Assignment Guide

Core: 7–17 odd, 18–20
Enriched: 8–12 even, 14–20

Additional Answers

1. $c = 2a$ where a is the leg opposite the 30° angle
2. two legs have equal length and there is a right angle
3. See students' work.

Study Guide Masters, p. 76

Step 2 Find c.

$c^2 = a^2 + b^2$ *Pythagorean Theorem*
$c^2 = 10^2 + 10^2$ *Replace a and b with 10.*
$c^2 = 100 + 100$
$c^2 = 200$
200 [2nd] [√] *14.14213562*
$c \approx 14.1$

The roof spans about 14.1 feet.

CHECK FOR UNDERSTANDING

Communicating Mathematics

Read and study the lesson to answer each question.   1–3. See margin.

1. *Write* a sentence describing the relationship between the hypotenuse of a 30°-60° right triangle and the leg opposite the 30° angle.
2. *Tell* why a triangle with two 45° angles is an isosceles right triangle.
3. *Draw* an acute triangle and an altitude. Describe the triangles formed by the sides and the altitude of the original triangle. Describe what you know about the two new triangles.

Guided Practice

Find the missing lengths. Round decimal answers to the nearest tenth.

4. $a = 8$ ft; $c \approx 11.3$ ft
5. $a = 12$ mm; $b \approx 20.8$ mm

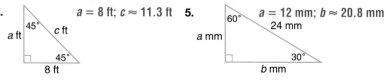

6. **Signs** A stop sign is cut from a square piece of sheet metal. If each side of the sign is 1 foot long and all of the angles are equal, how long were the sides of the original square piece of metal?   **about 2.4 feet**

7. $b = 3.2$ m; $c \approx 4.5$ m
8. $b \approx 8.7$ cm; $c = 10$ cm;

EXERCISES

Practice

Find the missing lengths. Round decimal answers to the nearest tenth.

9. $a = 6$ in.; $b \approx 10.4$ in.
10. $a = 9.5$ in.; $b \approx 16.5$ in.
11. $b \approx 34.6$ ft; $c = 40$ ft
12. $a = 6$ cm; $c \approx 8.5$ cm

7.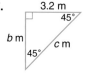
8.
9.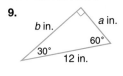
10.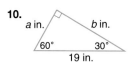
11.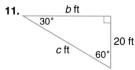
12.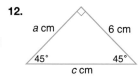

416 Chapter 9 Algebra: Exploring Real Numbers

■ Reteaching the Lesson ■

Activity Have students work in pairs drawing 30°-60° triangles and 45°-45° triangles with sides of varying lengths. Then have them calculate the lengths of the hypotenuses. Ask them to use all of their examples to test the two rules for special triangles given in the lesson.

Applications and Problem Solving

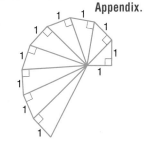

15a. $\sqrt{2}$, $\sqrt{3}$, $\sqrt{4}$, or 2, $\sqrt{5}$

15b. yes, the smallest triangle; yes, fourth triangle.

13. The length of the hypotenuse of a 30°-60° right triangle is 7.5 inches. Find the length of the side opposite the 30° angle. **3.75 inches**

14. *Geometry* Draw a 30°-60° right triangle whose hypotenuse is 16 units. Draw the height to the hypotenuse. In each of the two new triangles, draw the height to the hypotenuse. In each of the four new triangles, draw the height to the hypotenuse. Find and label the measure of each angle and find the length of as many sides as you can without using the Pythagorean Theorem. **See Answer Appendix.**

15. *Geometry* Use the diagram at the right.
    a. Find the length of the hypotenuse in each of the four smallest triangles.
    b. Are there any 45°-45° right triangles? Are there any 30°-60° right triangles?
    c. Look at the pattern. What do you think is the length of the hypotenuse of the next triangle? Verify your answer by calculating the length. $\sqrt{6}$

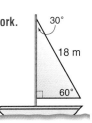

16. *Sports* Did you have your first taste of team sports playing Tee-ball? Tee-ball is similar to softball, but is played without a pitcher and on a smaller field. The baselines on a Tee-ball diamond are 40 feet long. How far is it from home plate to second base on a Tee-ball diamond? **about 56.6 feet**

17. *Write a Problem* that involves finding the length of a side of a 30°-60° right triangle. **See students' work.**

18. *Critical Thinking* Determine how much material is needed to make the sail shown at the right. **about 70 square meters**

Mixed Review

19. **Test Practice** What is the distance between points $Q(1, -3)$ and $R(6, 9)$? *(Lesson 9-6)* **D**

    A 9 units   B 5 units   C 12 units   D 13 units

20. *Money* The cost of making a $1 bill and each coin is listed in the chart. What percent of the $1 bill's value and each coin's value is used to make it? *(Lesson 8-3)*
    $1, 3%;
    half dollar, 15.6%;
    quarter, 14.8%;
    dime, 17%;
    nickel, 58%;
    penny, 80%

Extending the Lesson

Enrichment Masters, p. 76

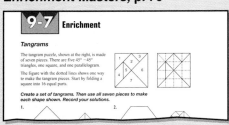

Activity For each of several 30°-60° right triangles, have students divide the length of the longer leg by that of the shorter leg. Have them compare their results with the values in a square root table and write a sentence summarizing the results. **Sample answer: In a 30°-60° right triangle, the length of the longer leg is $\sqrt{3}$ times the length of the shorter leg.**

---

4 ASSESS

Closing Activity

Speaking Draw several special right triangles on the chalkboard, indicating one length on each triangle. Have students describe how they would find the missing lengths.

Chapter 9, Quiz D (Lesson 9-7) is available in the *Assessment and Evaluation Masters*, p. 240.

Practice Masters, p. 76

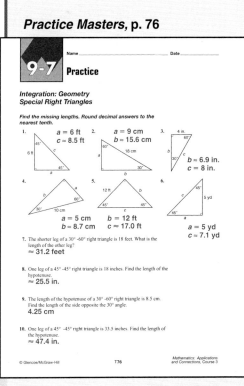

Study Guide and Assessment

Chapter 9

Vocabulary

This section provides a listing of the new terms, properties, and phrases that were introduced in this chapter. Have students define each term and provide an example or two of it, if appropriate.

Understanding and Using the Vocabulary

These exercises check students' understanding of the terms by using a variety of verbal formats including matching, completion, and true/false.

Glossaries A complete glossary of terms appears on pages 702–710. The glossary also appears in Spanish on pages 711–721.

Additional Answer

9. Answers will vary. Sample answer: Rational numbers can be expressed in the form $\frac{a}{b}$, where a and b are integers and $b > 0$, but irrational numbers cannot be expressed in that form; rational: 0.75, irrational: $\sqrt{3}$

Vocabulary

After completing this chapter, you should be able to define each term, concept, or phrase and give an example or two of each.

Number and Operations
irrational number (p. 391)
perfect square (p. 382)
principal square root (p. 382)
radical sign (p. 382)
real numbers (p. 391)
square root (p. 382)

Geometry
converse (p. 399)
hypotenuse (p. 398)
legs (p. 398)
Pythagorean Theorem (pp. 396-398)
Pythagorean triple (p. 405)

Problem Solving
draw a diagram (p. 402)

Understanding and Using the Vocabulary

Choose the letter of the term that best matches each statement or phrase.

1. the symbol $\sqrt{}$ **b**
2. the square of a rational number   **g**
3. the combined set of rational numbers and irrational numbers   **c**
4. can always be expressed as a terminating or repeating decimal   **d**
5. one of the sides that forms the right angle in a right triangle   **j**
6. The Pythagorean Theorem is true for this type of triangle.   **h**
7. In a 30°-60° right triangle, the length of the side opposite the 30° angle is one-half the length of this side.   **i**
8. the name of x in relation to y if $x^2 = y$ **f**

a. triangle
b. radical sign
c. real numbers
d. rational number
e. irrational number
f. square root
g. perfect square
h. right triangle
i. hypotenuse
j. leg

In Your Own Words

9. *Explain* the difference between a rational number and an irrational number. Give an example of each. **See margin.**

418 Chapter 9 Algorithm: Exploring Real Numbers

MindJogger Videoquizzes

MindJogger Videoquizzes provide an alternative review of concepts presented in this chapter. Students work in teams to answer questions, gaining points for correct answers. The questions are presented in three rounds.

Round 1 Concepts–5 questions
Round 2 Skills–4 questions
Round 3 Problem Solving–4 questions

Study Guide and Assessment Chapter 9

Objectives & Examples

Upon completing this chapter, you should be able to:

- **find square roots of perfect squares** *(Lesson 9-1)*
  Find $\sqrt{64}$.
  Since $8^2 = 64$, $\sqrt{64} = 8$.

- **estimate square roots** *(Lesson 9-2)*
  Estimate $\sqrt{42}$.
  $36 < 42 < 49$
  $6^2 < 42 < 7^2$
  $6 < \sqrt{42} < 7$
  Since 42 is closer to 36 than to 49, the best whole number estimate is 6.

- **identify and classify numbers in the real number system** *(Lesson 9-3)*
  Determine whether 7.43 is a rational or irrational number.
  7.43 is a terminating decimal.
  Therefore, 7.43 is a rational number.

- **use the Pythagorean Theorem** *(Lesson 9-4)*
  Find the missing length of the right triangle.
  $c^2 = a^2 + b^2$
  $c^2 = 5^2 + 8^2$
  $c^2 = 25 + 64$
  $c^2 = 89$
  $c = \sqrt{89}$

  89 [2nd] [√] [=] 9.433981132

  The hypotenuse is about 9.4 meters long.

Review Exercises

Use these exercises to review and prepare for the chapter test.

Find each square root.

10. $\sqrt{225}$ 15
11. $\sqrt{6.25}$ 2.5
12. $-\sqrt{\dfrac{4}{9}}$ $-\dfrac{2}{3}$
13. $\sqrt{\dfrac{36}{81}}$ $\dfrac{2}{3}$
14. $-\sqrt{100}$ -10
15. $\sqrt{3.24}$ 1.8

Estimate to the nearest whole number.

16. $\sqrt{135}$ 12
17. $\sqrt{50.1}$ 7
18. $\sqrt{696}$ 26
19. $\sqrt{320}$ 18
20. $\sqrt{19.25}$ 4
21. $\sqrt{230}$ 15

Let R = real numbers, Q = rational numbers, Z = integers, W = whole numbers, and I = irrational numbers. Name the set or sets of numbers to which each real number belongs. 24. I, R

22. $\sqrt{32}$ I, R
23. -12 Z, Q, R
24. $-0.171171117...$
25. $0.333333...$ Q, R

Write an equation you could use to find the length of the missing side of each right triangle. Then find the missing length. Round to the nearest tenth.

26.
    $18^2 + 24^2 = c^2$;
    $c = 30$ ft

27.
    $4^2 + b^2 = 9.5^2$;
    $b \approx 8.6$ m

Objectives & Examples

This section reviews the skills and concepts of the chapter and shows completely worked examples.

Review Exercises

These exercises provide practice for the corresponding objectives.

Assessment and Evaluation Masters, pp. 227–228

Assessment and Evaluation

Six forms of Chapter 9 Test are available in the *Assessment and Evaluation Masters* as shown in the chart.

Chapter 9 Test, Form 1B, is shown at the right. Chapter 9 Test, Form 2B, is shown on the next page.

1A	Multiple Choice	Honors
1B	Multiple Choice	Average
1C	Multiple Choice	Basic
2A	Free Response	Honors
2B	Free Response	Average
2C	Free Response	Basic

Study Guide and Assessment **419**

Study Guide and Assessment

Additional Answers
34. $a = 12.5$ in.; $b \approx 21.7$ in.
35. $b \approx 10.4$ km; $c = 12$ km
36. $a = 10$ mm; $c \approx 14.1$ mm
37. $b = 7$ m; $c \approx 9.9$ m

Assessment and Evaluation Masters, pp. 233–234

Chapter 9 Study Guide and Assessment

Objectives & Examples

• solve problems using the Pythagorean Theorem *(Lesson 9-5)*

How tall is the tree? Let h represent the height of the tree.

$$h = a^2 + b^2$$
$$62^2 = 40^2 + h^2$$
$$3{,}844 = 1{,}600 + h^2$$
$$2{,}244 = h^2$$
$$\sqrt{2{,}244} = h$$
$$47.4 \approx h$$

The tree is about 47.4 feet tall.

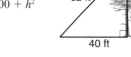

• find the distance between points in the coordinate plane *(Lesson 9-6)*

Find c.
$$c^2 = a^2 + b^2$$
$$c^2 = 4^2 + 2^2$$
$$c^2 = 16 + 4$$
$$c^2 = 20$$
$$c = \sqrt{20}$$
$$c \approx 4.5$$

The distance is about 4.5 units.

• find the missing measures in 30°-60° right triangles and 45°-45° right triangles *(Lesson 9-7)*

In a 30°-60° right triangle, $a = \frac{1}{2}c$ or $c = 2a$.

In a 45°-45° right triangle, $a = b$.

Review Exercises

Write an equation that can be used to answer each question. Then solve. Round to the nearest tenth.

28. How wide is the window?

29. How far is the airplane from the airport?

Graph each pair of ordered pairs. Then find the distance between each pair of points. Round to the nearest tenth.

30. $(3, 4), (2, 7)$ 3.2 units
31. $(-1, 2), (4, 8)$ 7.8 units
32. $(0, -3), (5, 5)$ 9.4 units
33. $(-6, 2), (-4, 5)$ 3.6 units

30–33. See Answer Appendix for graphs.

Find the missing lengths. Round decimal answers to the nearest tenth. 34–37. See margin.

34.

35.

36.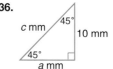

37.

420   Chapter 9 Algebra: Exploring Real Numbers

Test and Review Software

You may use this software, a combination of an item generator and item bank, to create your own tests or worksheets. Types of items include free response, multiple choice, short answer, and open ended.

CD-ROM Program

The CD-ROM Program contains an Assessment Game whose questions review the concepts in this chapter.

Study Guide and Assessment Chapter 9

Applications & Problem Solving

38. *Design* Albert is an interior designer. His client wants him to decorate a square room that has an area of 484 square feet. What is the length of each side of the room? *(Lesson 9-1)* **22 feet**

39. *Gardening* A diagram of Nina's garden is shown below. How long is the walkway to the nearest tenth of a foot? *(Lesson 9-4)*
9.4 feet

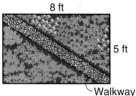

40. *Draw a Diagram* Ashland Middle School's band director arranges the band so that the same number of students are in each row. Charmaine is in the third row from the front and the fourth row from the back. Her place in the row is fifth from the left and third from the right. How many students are in the band? *(Lesson 9-5A)* **42 students**

41. *Construction* Meda needs to reach a window that is 8 feet above ground level. When she placed a ladder so that it reached the bottom of the window, the base of the ladder was 6 feet from the house. How long is the ladder? *(Lesson 9-5)* **10 feet**

Alternative Assessment

● **Performance Task**
You have four landscape timbers. They are 13 feet long, 10 feet long, 9 feet long, and 12 feet long. You want to use three of the timbers to make a flower garden shaped like a right triangle. You must use all of the timbers, but you can cut one of the timbers in half and use one or both halves. Is it possible to build the garden? Explain why or why not. **See margin.**

Suppose you decide that one of the legs of the right triangle must be 9 feet long. Can you create your garden using the landscape timbers you have? Explain why or why not.

No; the sides would need to be at least 9 feet, 12 feet, and 15 feet long.

A practice test for Chapter 9 is provided on page 655.

● **Completing the**
Now that you have studied the concepts in this chapter, you should have the skills to complete your chapter project.

☑ Summarize your research of traffic accident investigations and reports. Explain why it is important to carefully measure and record the location of each item at the accident scene and pertinent landmarks.

☑ Refer to the traffic accident diagram on page 381. Write an accident report for the accident scene. Be sure to include all appropriate information, including the estimated speed of each car.

 Select one of the assignments you completed in this chapter and place it in your portfolio. Attach a note explaining why you selected it.

Chapter 9 Study Guide and Assessment **421**

Additional Answer for the Performance Task
Yes; cut the 10-foot timber in half. The garden will have legs of 5 feet and 12 feet and a hypotenuse of 13 feet.

Performance Assessment
Additional performance assessment tasks for this chapter are included in the *Assessment and Evaluation Masters* on page 237. A scoring guide is also provided on page 249.

Applications & Problem Solving
This section provides additional practice in solving real-world problems that involve the skills of this chapter.

Alternative Assessment
The *Performance Task* provides students with a performance assessment opportunity to evaluate their work and understanding.

Students should complete the final stages of their project and prepare a class demonstration of their results. A scoring guide for the project is available in the *Investigations and Projects Masters*, p. 51.

 Students should add to their portfolios at this time.

Assessment and Evaluation Masters, p. 237

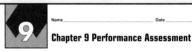

Study Guide and Assessment **421**

Standardized Test Practice

CHAPTERS 1–9

Assessing Knowledge & Skills

Section One: Multiple Choice

There are eleven multiple-choice questions in this section. Choose the best answer. If a correct answer is *not here*, choose the letter for Not Here.

1. A red blood cell is about 0.00075 centimeter long. How is this measure expressed in scientific notation? **B**
   - A 0.75×10^{-3}
   - B 7.5×10^{-4}
   - C 7.5×10^{-5}
   - D 75×10^{-5}

2. $\triangle RST$ is a right triangle. What is the length of $\overline{RT}$? **G**

   - F 4 cm
   - G 6 cm
   - H 8 cm
   - J 10 cm

3. $-\sqrt{0.81} =$ **B**
   - A -0.09
   - B -0.9
   - C 0.09
   - D 0.9

4. What ordered pair represents the intersection of line t and line m? **H**
   - F $(2, 3)$
   - G $(-2, -3)$
   - H $(2, -3)$
   - J $(-2, 3)$

   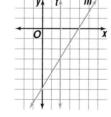

Please note that Questions 5-11 have five answer choices.

5. To the nearest inch, how long is the diagonal of a 20-inch by 16-inch window? **D**
   - A 12 in.
   - B 24 in.
   - C 25 in.
   - D 26 in.
   - E 27 in.

6. Arturo wants to wallpaper a wall that measures 10 feet by 15 feet. A window that is 3 feet by 4 feet is in the center of the wall. What is the wallpapered area? **H**

   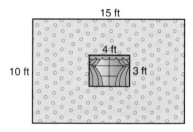

   - F 110 square feet
   - G 120 square feet
   - H 138 square feet
   - J 144 square feet
   - K 150 square feet

7. A drawing measures 4.5 centimeters long and 5.5 centimeters wide. An enlargement is similar to the original and its length is 36 centimeters. What is the width of the enlargement? **C**
   - A 10 cm
   - B 24 cm
   - C 44 cm
   - D 64 cm
   - E 124 cm

422 Chapters 1–9 Standardized Test Practice

Standardized Test Practice   Chapters 1–9

8. Margi worked for a total of 33 hours in a 4-day period. How much did she earn before deductions, if she was paid $5.25 per hour?   **J**
   - F $132.00
   - G $135.00
   - H $171.60
   - J $173.25
   - K Not Here

9. Felica's bedroom is 15 feet by 15 feet. If the ceiling is 8 feet high and a can of paint will cover 120 square feet, how many cans will be needed to cover only the walls?   **A**
   - A 4
   - B 5
   - C 6
   - D 7
   - E 8

10. Jill runs on a 400-meter track. Last week she ran 4.5 kilometers on Monday, 3.5 kilometers on Wednesday, and 4.75 kilometers on Friday. What was the average length of Jill's runs on those 3 days?   **K**
    - F 3.25 km
    - G 4 km
    - H 4.75 km
    - J 12.75 km
    - K Not Here

11. Foothill Middle School sold 2,053 rolls of wrapping paper last year. This year the school sold 1,837 rolls. The profit that the school makes on each roll of paper sold is $2.25. How much more profit did they have last year than this year?   **C**
    - A $216
    - B $405
    - C $486
    - D $4133.25
    - E Not Here

Test-Taking Tip

Most of the basic formulas that you need in order to answer the questions on a standardized test are usually given to you in the test booklet. However, it is a good idea to review the formulas before the test. For example, if your test includes the Pythagorean Theorem, familiarize yourself with the formula and its uses.

Section Two: Free Response

This section contains six questions for which you will provide short answers. Write your answers on your paper.

12. To use a calculator to divide 8 by $3\frac{2}{5}$, what decimal number should you enter for $3\frac{2}{5}$?   **3.4**

13. If $x + 2\frac{2}{3} = -4\frac{1}{3}$, what is the value of x?   **−7**

14. What is $0.\overline{39}$ written as a fraction? $\frac{13}{33}$

15. Which figure could contain exactly one right angle: a rectangle, an isosceles triangle, an acute triangle, or an obtuse triangle?   **isosceles triangle**

16. $-(8)^3 =$ **−512**

17. To the nearest tenth, what is the distance between points A and B?   **4.5 units**

Test-Taking Tip

In addition to reviewing the Pythagorean Theorem— $a^2 + b^2 = c^2$ —suggest that students review other useful geometric formulas such as those for area, perimeter, and circumference.

Assessment and Evaluation Masters, pp. 241–242

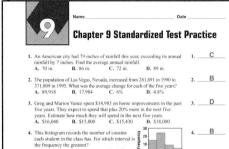

Instructional Resources ▶▶▶

Additional standardized test practice is shown at the right and is available in the *Assessment and Evaluation Masters*, pp. 241–242.

Interdisciplinary Investigation

GET READY

This optional investigation is designed to be completed by a group of 4 students over several days or several weeks.

Mathematical Overview

This investigation utilizes the concepts from Chapters 7–9.
- applying proportional reasoning
- drawing a diagram

Time Management	
Gathering Data	30 minutes
Calculations	40 minutes
Creating Diagrams	40 minutes
Summarizing Data	30 minutes
Presentation	10 minutes

Instructional Resources

- *Investigations and Projects Masters*, pp. 9–12
- Manipulative Kit
  - rulers or tape measure

Investigations and Projects Masters, p. 12

Interdisciplinary Investigation

MATH AT THE MALL

What is your favorite store at the mall? Maybe it's the bakery, the music store, or the photo shop. Did you know that they use math at all of these stores?

When you go shopping, you have dozens of choices. How do you decide which is the best use of your hard-earned money?

What You'll Do

In this investigation, you will calculate and compare the ratio of areas of similar polygons to help you choose which picture frame or cake to buy.

Materials grid paper   tape measure

 unlined paper   calculator

Procedure

1. Work in groups of four. Your first stop at the mall is the photo shop. Suppose you have a photograph you took on your family vacation that you would like to enlarge and frame. The original photograph is 6 inches by 8 inches. There are frames available in the following standard sizes.

8" by 10"	9" by 12"	10" by 12"
11" by 14"	12" by 16"	14" by 18"
16" by 20"	18" by 24"	20" by 24"
22" by 28"	24" by 36"	30" by 40"

   Use proportions and similar figures to determine which standard size frames are suitable for an enlargement of your photograph if you do not wish to crop any of the image.

   The photo shop also offers custom-made frames in any size you specify. List three possible custom frame sizes you could use.

2. Your next stop is at the bakery to check out the prices for different sizes of cakes. Draw a diagram of each cake. How do the sizes and the areas of the cakes compare? Write the ratio of each cake's area to its price. Is one of the cakes less expensive per square inch? What other things are there to consider when choosing which cake to order?

Schmidt's Bakery
Custom cakes for every occasion. Choose chocolate, white, or marble.

Size		Price
Small	9" x 13"	$7.99
Medium	13" x 18"	$15.99
Large	18" x 26"	$29.99

424 Interdisciplinary Investigation Math at the Mall

◄◄◄ **Instructional Resources**
A recording sheet to help students organize their data for this investigation is shown at the left and is available in the *Investigations and Projects Masters*, p. 12.

 Cooperative Learning

This investigation offers an excellent opportunity for using cooperative learning groups. For more information on cooperative learning strategies and group management, see *Cooperative Learning in the Mathematics Classroom*.

Making the Connection

Use the information about the frame and cake sizes as needed to help in these investigations.

Language Arts

Use what you learned about the area of similar polygons to decide whether different-sized enlargements at a local photo shop are priced fairly. Write a paragraph supporting your opinion.

Art

Find a small drawing or cartoon. Make an enlargement that has 9 times the area. Write a paragraph explaining how you know the ratio of the areas is 1 to 9.

Social Studies

Choose a lake on a map. Approximate the actual area of the lake by using the scale provided on the map.

Go Further

- Draw a polygon such as a rectangle. Multiply the measures of the sides of the polygon by 2, 3, 4, and 5, and draw the enlargements. Investigate how the area of the enlarged polygons compare to the area of the original polygon.

- Most copiers do not enlarge by 300%. How could you enlarge a photograph by 300%?

 For current information on math at the mall, visit:
www.glencoe.com/sec/math/mac/mathnet

 You may want to place your work on this investigation in your portfolio.

Technology Tips

- Use **geometry software** to draw your figures and calculate the areas.
- Use a **spreadsheet** to make a table and calculate the ratios and areas.

Interdisciplinary Investigation Math at the Mall **425**

MANAGEMENT TIPS

Working in Teams Each person on the team should work on drawing diagrams when working on Steps 1 and 2. After everyone draws and discusses the diagrams, have students answer the questions as a group.

Making the Connection

You may wish to alert other teachers on your team that your students may need their assistance in this investigation.
Language Arts Students may need help learning to write and edit their work more effectively.
Art Ask the art teacher to show the class how enlargements are made in drawing classes by using grids.
Social Studies Students may benefit from learning about map projections, scales, and enlargements.

Investigations and Projects Masters, p. 11

SCORING GUIDE

Interdisciplinary Investigation
(Student Edition, Pages 424–425)

Math at the Mall

Level	Specific Criteria
3 Superior	• Shows a thorough understanding of the concepts of *calculating the area of different polygons and understanding how to apply the data to different areas of concern such as the ratios of cost to size.* • Uses appropriate strategies to solve problems. • Computations are correct. • Written explanations are exemplary. • Charts, model, and any statements included are appropriate and sensible. • Goes beyond the requirements of some or all problems.
2 Satisfactory, with minor flaws	• Shows understanding of the concepts of *calculating the area of different polygons and understanding how to apply the data to different areas of concern such as the ratios of cost to size.* • Uses appropriate strategies to solve problems. • Computations are mostly correct. • Written explanations are effective. • Charts, model, and any statements included are appropriate and sensible. • Satisfies the requirements of problems.
1 Nearly Satisfactory, with obvious flaws	• Shows understanding of most of the concepts of *calculating the area of different polygons and understanding how to apply the data to different areas of concern such as the ratios of cost to size.* • May not use appropriate strategies to solve problems. • Computations are mostly correct. • Written explanations are satisfactory. • Charts, model, and any statements included are appropriate and sensible. • Satisfies the requirements of problems.
0 Unsatisfactory	• Shows little or no understanding of the concepts of *calculating the area of different polygons and understanding how to apply the data to different areas of concern such as the ratios of cost to size.* • Does not use appropriate strategies to solve problems. • Computations are incorrect. • Written explanations are not satisfactory. • Charts, model, and any statements included are not appropriate or sensible. • Does not satisfy the requirements of the problems.

ASSESS

Ask students to prepare an oral presentation of the investigation. The presentation should include explanations of visuals—diagrams of proportional sizes, diagrams showing sizes and areas, and, if technology was used, a representation of spreadsheets and geometry printouts.

Instructional Resources ▶▶▶

Sample solutions for this investigation are provided in the *Investigations and Projects Masters* on p. 10. The scoring guide for assessing student performance shown at the right is also available on p. 11.

CHAPTER 10
Algebra: Graphing Functions

Previewing the Chapter

Overview
This chapter introduces the relationship that has been described as most important in all of mathematics, the function. Students learn to complete function tables, graph ordered pairs, and draw the lines or curves suggested by the plotted points. Linear and quadratic functions are addressed, and students learn to solve systems of equations graphically. Students use graphs to solve problems. They also graph translations, reflections, and rotations on coordinate planes.

Lesson (pages)	Lesson Objectives	NCTM Standards	Standardized Tests	State/Local Objectives
10-1 (428–431)	Complete function tables.	1–4, 7–9		
10-1B (432)	Create function tables by using a graphing calculator.	7, 9		
10-2 (433–435)	Graph functions by using function tables.	1–5, 7–9, 12		
10-3 (437–440)	Find solutions of equations with two variables.	1–5, 7–9	CAT, MAT	
10-4A (441)	Determine a linear function to describe data.	1–5, 8, 10		
10-4 (442–444)	Graph linear functions by plotting points.	1–5, 7–9, 12		
10-4B (445)	Graph linear equations on a graphing calculator.	1–4, 8–10		
10-5 (446–449)	Solve systems of linear equations by graphing.	1–4, 7–9, 12		
10-6A (450–451)	Solve problems by using a graph.	1–4, 7, 8, 10	CTBS, TN	
10-6 (452–455)	Graph quadratic functions.	1–4, 7–9, 12		
10-7 (456–459)	Graph translations on a coordinate plane.	1–5, 7–9, 12		
10-8 (460–463)	Graph reflections on a coordinate plane.	1–5, 7–9, 12		
10-9 (464–467)	Graph rotations on a coordinate plane.	1–5, 7–9, 12		

CAT = California Achievement Tests, CTBS = Comprehensive Tests of Basic Skills, ITBS = Iowa Tests of Basic Skills, MAT = Metropolitan Achievement Tests, SAT = Stanford Achievement Tests, TN = Terra Nova

Organizing the Chapter

CD-ROM
All of the blackline masters in the Teacher's Classroom Resources are available on the **Electronic Teacher's Classroom Resources** CD-ROM.

LESSON PLANNING GUIDE

Lesson	Extra Practice (Student Edition)	Blackline Masters (page numbers)											
		Study Guide	Practice	Enrichment	Assessment & Evaluation	Classroom Games	Diversity	Hands-On Lab	School to Career	Science and Math Lab Manual	Technology	Transparencies A and B
10-1	p. 633	77	77	77							71	10-1
10-1B												
10-2	p. 633	78	78	78								10-2
10-3	p. 634	79	79	79	267							10-3
10-4A								60				
10-4	p. 634	80	80	80								10-4
10-4B												
10-5	p. 634	81	81	81	266, 267						72	10-5
10-6A	p. 635											
10-6	p. 635	82	82	82						65–68		10-6
10-7	p. 635	83	83	83	268							10-7
10-8	p. 636	84	84	84				77	36			10-8
10-9	p. 636	85	85	85	268	31–36	36					10-9
Study Guide/ Assessment					253–265, 269–271							

OTHER CHAPTER RESOURCES

Student Edition
Chapter Project, pp. 427, 444, 449, 467, 471
Math in the Media, p. 440
School to Career, p. 436
Let the Games Begin, pp. 431, 455

Technology
 CD-ROM Program
 Interactive Mathematics Tools Software

Teacher's Classroom Resources

Applications
Family Letters and Activities, pp. 71–72
Investigations and Projects Masters, pp. 53–56

Meeting Individual Needs
Investigations for the Special Education Student, pp. 39–42

Teaching Aids
Answer Key Masters
Block Scheduling Booklet
Lesson Planning Guide
Solutions Manual

Professional Publications
Glencoe Mathematics Professional Series

Planning the Chapter

MindJogger Videoquizzes provide a unique format for reviewing concepts presented in the chapter.

Assessment Resources

Student Edition
Mixed Review, pp. 431, 435, 440, 444, 449, 455, 459, 463, 467
Mid-Chapter Self Test, p. 449
Math Journal, pp. 439, 466
Study Guide and Assessment, pp. 468–471
Performance Task, p. 471
Portfolio Suggestion, p. 471
Standardized Test Practice, pp. 472–473
Chapter Test, p. 656

Assessment and Evaluation Masters
Multiple-Choice Tests (Forms 1A, 1B, 1C), pp. 253–258
Free-Response Tests (Forms 2A, 2B, 2C), pp. 259–264
Performance Assessment, p. 265
Mid-Chapter Test, p. 266
Quizzes A–D, pp. 267–268
Standardized Test Practice, pp. 269–270
Cumulative Review, p. 271

Teacher's Wraparound Edition
5-Minute Check, pp. 428, 433, 437, 442, 446, 452, 456, 460, 464
Building Portfolios, p. 426
Math Journal, pp. 432, 441, 445
Closing Activity, pp. 431, 435, 440, 444, 449, 451, 455, 459, 463, 467

Technology
Test and Review Software
MindJogger Videoquizzes
CD-ROM Program

Materials and Manipulatives

Lesson 10-1B
graphing calculator

Lesson 10-4A
large rubber band*†
paper clips
ruler*
masking tape
paper cup
washers

Lesson 10-4
grid paper†
ruler*

Lesson 10-4B
graphing calculator

Lesson 10-5
grid paper†
straightedge

Lesson 10-6
spinners*†
index cards
grid paper†

Lesson 10-7
grid paper†
unlined paper
straightedge
scissors*
colored pencils

Lesson 10-8
grid paper†
straightedge
geomirror*

Lesson 10-9
protractor*†

*Glencoe Manipulative Kit   †Glencoe Overhead Manipulative Resources

Pacing Chart

See pages T25–T27 for the Course Planning Calendar.

COURSE	DAY 1	DAY 2	DAY 3	DAY 4	DAY 5	DAY 6	DAY 7
Standard	Chapter Project	Lesson 10-1	Lesson 10-2	Lesson 10-3	Lessons 10-4A & 10-4		Lesson 10-5
Honors	Chapter Project & Lessons 10-1 & 10-1B		Lesson 10-2	Lesson 10-3	Lessons 10-4 & 10-4B		Lesson 10-5
Block	Chapter Project & Lesson 10-1	Lessons 10-2 & 10-3	Lessons 10-4A & 10-4	Lessons 10-5 & 10-6A	Lessons 10-6 & 10-7	Lessons 10-8 & 10-9	Study Guide and Assessment, Chapter Test

Interactive Mathematics:
Activities and Investigations

is an activity-based program that may be used as an enhancement for chapters in *Mathematics: Applications and Connections.*

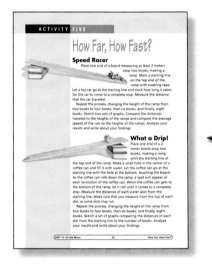

Unit 15, Activity Five
Use with Lesson 10-2.

Summary Students complete tasks that illustrate velocity and acceleration. They experiment with rolling toy cars down ramps and compare the distance, speed, and time of the cars. Then they sketch graphs from the data that they collect.

Math Connection Students compare the speed and acceleration of a toy car when the ramp is at different heights. They also analyze the speed and acceleration of a toy car going down one ramp and up another.

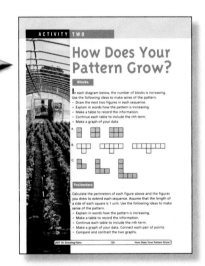

Unit 16, Activity Two
Use with Lesson 10-4.

Summary Students look for patterns in sequences of blocks, perimeters, and shapes that will help them extend the sequences to the *n*th term. For each sequence, they make a table, graph their data, and explain how the pattern is changing.

Math Connection Students look for patterns in the area, perimeter, and shape of each figure in a sequence in order to extend the sequence. A sequence is a set of elements that are in a specific order.

DAY 8	DAY 9	DAY 10	DAY 11	DAY 12	DAY 13	DAY 14	DAY 15
Lesson 10-6A	Lesson 10-6	Lesson 10-7	Lesson 10-8	Lesson 10-9	Study Guide and Assessment	Chapter Test	
Lesson 10-6A	Lesson 10-6	Lesson 10-7	Lesson 10-8	Lesson 10-9	Study Guide and Assessment	Chapter Test	

Enhancing the Chapter

APPLICATIONS

Classroom Games, p. 31–36

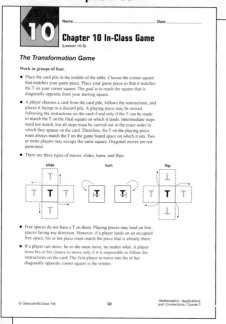

Diversity Masters, p. 36

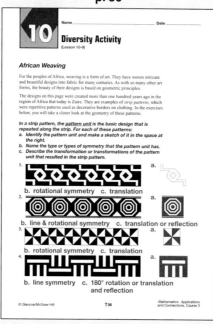

School to Career Masters, p. 36

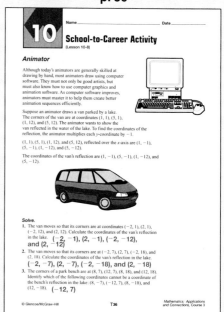

Family Letters and Activities, pp. 71–72

Family Letters and Activities, pp. 71–72

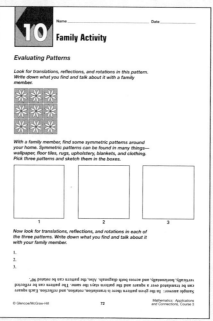

Science and Math Lab Manual, pp. 65–68

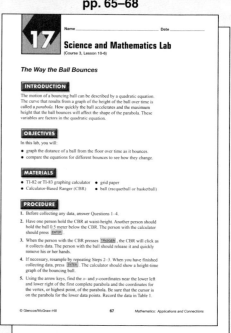

MANIPULATIVES/MODELING

Hands-On Lab Masters,
p. 77

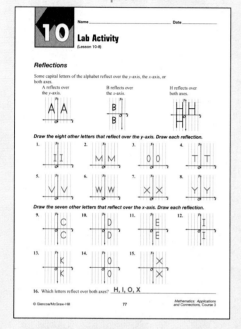

ASSESSMENT/EVALUATION

Assessment and Evaluation Masters,
pp. 266–268

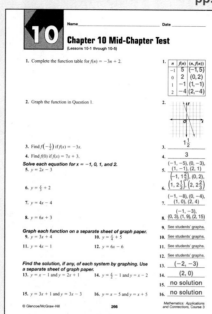
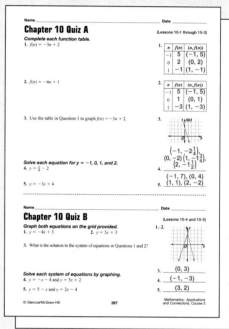

TECHNOLOGY/MULTIMEDIA

Technology Masters,
pp. 71–72

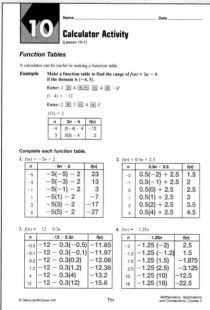
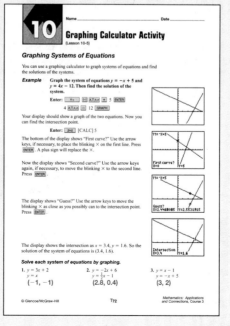

MEETING INDIVIDUAL NEEDS

Investigations for the Special Education Student, pp. 39–42

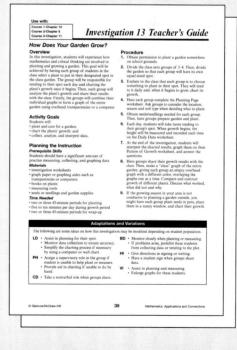

Chapter 10   **426f**

CHAPTER 10 NOTES

Theme: Computer Games
Computer game designers are constantly looking for ways to improve their products. Developments such as virtual reality and advanced 3-D graphics make games more interactive and dynamic than ever before.

Question of the Day Many computer games operate using linear functions. How would you instruct a computer to move a game piece three spaces to the right and four spaces up? This translation can be communicated by the ordered pair (3, 4).

Assess Prerequisite Skills
Ask students to read through the list of objectives presented in "What you'll learn in Chapter 10." You may wish to ask them what each of the objectives means or if they have experienced or used any of these math concepts before.

 Building Portfolios
Encourage students to revise their portfolios as they study this chapter. They may wish to include graphs of a variety of different functions.

 Math and the Family
In the *Family Letters and Activities* booklet (pp. 71–72), you will find a letter to the parents explaining what students will study in Chapter 10. An activity appropriate for the whole family is also available.

426 Chapter 10

CHAPTER 10 Algebra: Graphing Functions

What you'll learn in Chapter 10

- to complete function tables and graph linear and quadratic functions,
- to find solutions of equations with two variables,
- to solve systems of linear equations by graphing,
- to solve problems by using a graph, and
- to graph translations, reflections, and rotations on a coordinate plane.

426 Chapter 10 Algebra: Graphing Functions

 CD-ROM Program

Activities for Chapter 10
- Chapter 10 Introduction
- Interactive Lessons 10-2, 10-7, 10-8, 10-9
- Extended Activity 10-1
- Assessment Game
- Resource Lessons 10-1 through 10-9

CHAPTER Project

GAMES PEOPLE PLAY

In this project, you will use functions to design a computer game. After word processing, playing games is the most popular use for computers. A good computer game has a storyline and characters, a goal that the player is trying to achieve, obstacles that the player must face, and rewards for overcoming the obstacles. Many computer games use sophisticated graphics and animation. Movement of everything on the screen is controlled by mathematical formulas.

Getting Started

- Research computer graphics to find out how a grid like the coordinate plane is used to position pictures on a computer screen.
- Research computer animation to find out how objects appear to move on a computer screen.
- Brainstorm a theme or story for your game.

Technology Tips

- Use an **electronic encyclopedia** to do your research.
- Use a **word processor** to write out your game plan.
- Use a **software drawing program** to sketch a screen.

 For up-to-date information on designing computer games, visit:
www.glencoe.com/sec/math/mac/mathnet

Working on the Project

You can use what you'll learn in Chapter 10 to help you design your computer game.

Page	Exercise
444	21
449	22
467	18
471	Alternative Assessment

interNET CONNECTION

Glencoe has made every effort to ensure that the website links for *Mathematics: Applications and Connections* at **www.glencoe.com/sec/math/mac/mathnet** are current and contain appropriate content. However, these website links are not under Glencoe's control.

Instructional Resources ▶▶▶

A recording sheet to help students organize their data for the Chapter Project is shown at the right and is available in the *Investigations and Projects Masters,* p. 56.

CHAPTER Project NOTES

Objectives Students should
- gain an understanding of how computer games are developed.
- be able to explain the relationship between math functions and the creation of computer games.

Project Pointer You may suggest that students begin a *Project Folder* to keep their work as they complete each stage of the Chapter Project. The completed project may also be added to their portfolios.

Students may want to work in groups to complete the Chapter Project. Some students may feel more comfortable creating the storyline for the game while others will be more at ease doing computer research. By working with their peers, students will develop a better understanding of their own abilities and will have the opportunity to learn from other students.

Investigations and Projects Masters, p. 56

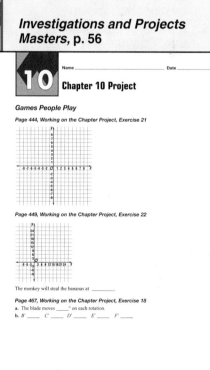

10-1 Lesson Notes

Instructional Resources
- *Study Guide Masters*, p. 77
- *Practice Masters*, p. 77
- *Enrichment Masters*, p. 77
- *Transparencies 10-1, A and B*
- *Technology Masters*, p. 71
- CD-ROM Program
  - Resource Lesson 10-1
  - Extended Activity 10-1

Recommended Pacing

Standard	Day 2 of 14
Honors	Day 1 of 14
Block	Day 1 of 7

1 FOCUS

5-Minute Check
(Chapter 9)

1. Find $-\sqrt{81}$. **−9**
2. Estimate $\sqrt{52}$ to the nearest whole number. **7**
3. Give an example of an irrational number. **Sample answers:** π, $\sqrt{11}$
4. A right triangle has a hypotenuse measuring 15 inches and a leg measuring 12 inches. Find the length of the other leg. **9 in.**
5. The coordinates of points C and D are (6, −3) and (2, 5). What is the distance between the points? Round to the nearest tenth. **8.9 units**

 The 5-Minute Check is also available on **Transparency 10-1A** for this lesson.

Motivating the Lesson

Communication Have students read the paragraphs at the beginning of the lesson. Ask them whether age or loss of height is a function of the other. **Loss of height is a function of age.** Then have them suggest other things that are related functions. **Sample answers: rain and crop growth, sunshine and warmth**

10-1 Functions

What you'll learn
You'll learn to complete function tables.

When am I ever going to use this?
You will evaluate functions when you study motion in physical science.

Word Wise
function
function table
domain
range

LOOK BACK
Refer to Lesson 1-7 for information on solving two-step equations.

In the 1997 movie *HONEY, WE SHRUNK OURSELVES*, Wayne Szalinski's invention shrinks four adults to a height of just $\frac{3}{4}$ inch! Of course, the movie is make believe. But did you know that after a person turns 30, he or she shrinks about 0.06 centimeter per year? At that rate, how many centimeters shorter would a person be when he or she is 35, 42, 57, 61, and 83 years old? *This problem will be solved in Example 3.*

The amount of height loss depends upon a person's age. A relationship like this where one thing depends upon another is called a **function**. In a function, one or more operations are performed on one number to get another. So, the second number depends on, or is a function of, the first.

Functions are often written as equations. Suppose the variable n is used for the first number, or the *input*. The notation for the second number, or the *output*, is $f(n)$. This is read *the function of n*, or more simply *f of n*. The operations performed in the function are sometimes called the *rule*. You can organize the input, rule, and output of a function into a **function table**.

Copy and complete the function table to find the function values of $\{-1, 0, 1, 2, 3\}$ for $f(n) = n - 3$.

Substitute each value of n, or input, into the function rule. Then simplify to find the output.

$f(n) = n - 3$
$f(-1) = (-1) - 3$ or -4 *Replace n with −1.*
$f(0) = (0) - 3$ or -3
$f(1) = (1) - 3$ or -2
$f(2) = (2) - 3$ or -1
$f(3) = (3) - 3$ or 0

Input n	Rule $n - 3$	Output $f(n)$
−1		
0		
1		
2		
3		

Input n	Rule $n - 3$	Output $f(n)$
−1	−1 − 3	−4
0	0 − 3	−3
1	1 − 3	−2
2	2 − 3	−1
3	3 − 3	0

The set of input values in a function is called the **domain**. The set of output values is called the **range**. If the domain contains all values of n, then the range contains all values of $f(n)$.

428 Chapter 10 Algebra: Graphing Functions

Multiple Learning Styles

 Naturalist Have students think about how science is used to explain the laws of nature. For example, botanists study plants to determine how much water and sunlight different plants need to grow. Scientists use this information to develop functions, or equations, about how much water and light a specific species of plant needs.

Examples

2 Make a function table to find the range of $f(n) = 2n - 6$ if the domain is $\{-2, -1, 0, \frac{1}{2}, 14\}$.

First make a function table. List the domain, or input values.

Next, substitute each member of the domain into the function to find the range, or output values, of $f(n)$.

Domain n	$2n - 6$	Range $f(n)$
-2	$2(-2)-6$	-10
-1	$2(-1)-6$	-8
0	$2(0)-6$	-6
$\frac{1}{2}$	$2(\frac{1}{2})-6$	-5
14	$2(14)-6$	22

The range is $\{-10, -8, -6, -5, 22\}$.

CONNECTION **3** **Life Science** Refer to the beginning of the lesson. If a person shrinks 0.06 centimeter per year, how many centimeters shorter would the person be when he or she is 35, 42, 57, 61, and 83 years old?

Each year, the person shrinks 0.06 cm. To find the total height loss, multiply the number of years after age 30 by 0.06.

$f(n) = 0.06(n - 30)$

The person would be
0.3 cm shorter at 35,
0.72 cm shorter at 42,
1.62 cm shorter at 57,
1.86 cm shorter at 61,
and 3.18 cm shorter at 83 years old.

n	$0.06(n - 30)$	$f(n)$
35	$0.06(35 - 30)$	0.3
42	$0.06(42 - 30)$	0.72
57	$0.06(57 - 30)$	1.62
61	$0.06(61 - 30)$	1.86
83	$0.06(83 - 30)$	3.18

CHECK FOR UNDERSTANDING

Communicating Mathematics

Read and study the lesson to answer each question.   1. domain, range

1. *State* the mathematical names for the input values and the output values.
2. *Define* *function* in your own words.   See Answer Appendix.
3. *You Decide* Raul says that $f(6)$ for $f(n) = 2n^2 - 18$ is 54. Candace says that $f(6) = 126$. Who is correct and why?   Raul is correct. Candace's solution did not follow the order of operations.

Guided Practice

4. Copy and complete the function table.

n	$3n$	$f(n)$
-2	$3(-2)$	-6
0	$3(0)$	0
2	$3(2)$	6
$3\frac{1}{2}$	$3(3\frac{1}{2})$	$10\frac{1}{2}$
4.3	$3(4.3)$	12.9

Find each function value.

5. $f(6)$ if $f(n) = 2n - 8$ 4
6. $f(-3)$ if $f(n) = \frac{1}{2}n + 4$ $2\frac{1}{2}$

Lesson 10-1 Functions **429**

Reteaching the Lesson

Activity Have students work in groups to define the function for each table.

n	f(n)
3	-3
6	0
8	2

$f(n) = n - 6$;

n	f(n)
-1	-7
0	-2
1	3

$f(n) = 5n - 2$;

n	f(n)
1	2
2	5
3	10

$f(n) = n^2 + 1$

Error Analysis
Watch for students who confuse n and $f(n)$ and their values.
Prevent by having students review function machines in Lesson 1-7B. Then have them make a machine for each table and follow each value for n through step-by-step to find its corresponding value for $f(n)$.

---

2 TEACH

 Transparency 10-1B contains a teaching aid for this lesson.

Reading Mathematics The introduction to this lesson includes many mathematical concepts presented in verbal form. Make sure students understand the meanings of *function, function table, domain,* and *range.*

In-Class Examples

For Example 1
Copy and complete the function table to find the function values of $\{12, 13, 14, 15\}$ for $f(n) = \frac{1}{3}n$.

Input n	Rule $\frac{1}{3}n$	Output $f(n)$
12	$\frac{1}{3}(12)$	4
13	$\frac{1}{3}(13)$	$4\frac{1}{3}$
14	$\frac{1}{3}(14)$	$4\frac{2}{3}$
15	$\frac{1}{3}(15)$	5

For Example 2
Make a function table to find the range of $f(n) = -2x - 4$ if the domain is $\{-2, 0, 2, 4\}$.
$\{0, -4, -8, -12\}$

For Example 3
Refer to Example 3 in the Student Edition. Determine how many centimeters shorter a person would be when he or she is 32, 46, 53, 66, and 75 years old. **0.12, 0.96, 1.38, 2.16, 2.70**

Teaching Tip In Example 3, students may find it helpful to replace n with a for *age*.

3 PRACTICE/APPLY

Check for Understanding
If students need additional practice or instruction after completing Exercises 1–7, one of these options may be helpful.
- Extra Practice, see p. 633
- Reteaching Activity
- *Study Guide Masters,* p. 77
- *Practice Masters,* p. 77
- Interactive Mathematics Tools Software

Lesson 10-1 **429**

Assignment Guide

Core: 9–21 odd, 22–25
Enriched: 8–18 even, 19–25

Additional Answers

8.
n	2n − 3
−4	2(−4) − 3
−1	2(−1) − 3
0	2(0) − 3
1.5	2(1.5) − 3
$4\frac{1}{2}$	$2\left(4\frac{1}{2}\right) - 3$

9.
n	−5n
−4	−5(−4)
−2	−5(−2)
0	−5(0)
$2\frac{1}{4}$	$-5\left(2\frac{1}{4}\right)$
5.7	−5(5.7)

10.
n	−0.5n + 1
−2	−0.5(−2) + 1
0	−0.5(0) + 1
2.5	−0.5(2.5) + 1
8	−0.5(8) + 1
14.9	−0.5(14.9) + 1

Study Guide Masters, p. 77

7. Life Science Veterinarians have used the rule that one year of a dog's life is equivalent to seven years of human life. This can be written as $f(n) = 7n$, where n represents the dog's age and $f(n)$ represents the equivalent human age. If Wendy's dog Goldie is 6 years old, what is Goldie's human equivalent age? **42**

EXERCISES

Practice Copy and complete each function table. 8–10. See margin for expressions.

8.
n	2n − 3	f(n)
−4		−11
−1		−5
0		−3
1.5		0
$4\frac{1}{2}$		6

9.
n	−5n	f(n)
−4		20
−2		10
0		0
$2\frac{1}{4}$		$-11\frac{1}{4}$
5.7		−28.5

10.
n	−0.5n + 1	f(n)
−2		2
0		1
2.5		−0.25
8		−3
14.9		−6.45

Find each function value.

11. $f(4)$ if $f(n) = n - 6$ **−2**
12. $f(-2)$ if $f(n) = 4n + 1$ **−7**
13. $f(-8)$ if $f(n) = 3n + 24$ **0**
14. $f\left(\frac{5}{8}\right)$ if $f(n) = n + \frac{1}{6}$ **$\frac{19}{24}$**
15. $f(2.2)$ if $f(n) = n^2 + 1$ **5.84**
16. $f(-3.4)$ if $f(n) = 2n^2 - 5$ **18.12**
17. If $f(n) = -5n - 4$, find $f\left(\frac{4}{5}\right)$. **−8**
18. Find $f(-3.3)$ if $f(n) = 8.5 - 2n$. **15.1**

Applications and Problem Solving

19. **Earth Science** Have you ever noticed that you see the lightning before you hear the thunder? You can use that time difference to estimate the distance in feet between you and a lightning strike. If n is the number of seconds between the lightning and the thunder and $f(n)$ is the number of feet between you and the lightning, then $f(n) = 1,100n$.

   interNET CONNECTION
   For more information on math and the weather, visit:
   www.glencoe.com/sec/math/mac/mathnet

   a. Copy and complete the function table.

Time (seconds) n	Distance (feet) f(n)
2	2,200
5	5,500
11	12,100
18	19,800

   b. If you are two miles from a lightning strike, in how many seconds would you hear thunder? Round to the nearest tenth of a second. **9.6 s**

430 Chapter 10 Algebra: Graphing Functions

20. **Sports** Each team in a bowling league has an equal chance of winning because team members are given handicaps based on their bowling averages. Enrico's handicap is 30, so the function used to determine his handicapped score is $f(n) = n + 30$, where n is the game score. Make a function table to show Enrico's handicapped scores if he bowled 153, 144, 161, 163, and 166 in the first five matches of the season. **See margin.**

21. **Write a Problem** that can be solved using a function. **See students' work.**

22. **Critical Thinking** A diagonal of a polygon connects vertices that are not adjacent. A triangle has no diagonals and a rectangle has 2.
    a. Sketch polygons with 3, 4, 5, 6, 7, and 8 sides. Include the diagonals.
    b. If there are n sides in a polygon, write an expression for the number of diagonals that can be drawn from one vertex. **$n - 3$**
    c. Use $f(n)$ to write an equation for the relationship between n, the number of sides in a polygon, and $f(n)$, the number of diagonals. $f(n) = \frac{n(n-3)}{2}$

    22a. See Answer Appendix.

Mixed Review

23. **Geometry** The length of a leg of a 45°-45° right triangle is 6.4 feet. Find the length of the hypotenuse to the nearest tenth. *(Lesson 9-7)* **9.1 feet**

24. Find $\sqrt{\frac{9}{16}}$. *(Lesson 9-1)* $\frac{3}{4}$

25. **Test Practice** Fifteen out of the 60 eighth-graders at Seabring Junior High are on the track team. What percent of the eighth-graders are on the track team? *(Lesson 8-2)* **B**

    A 15%   **B** 25%   **C** 45%   **D** 60%   **E** Not Here

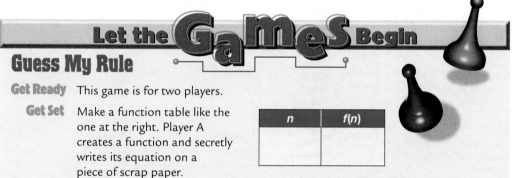

Guess My Rule

Get Ready This game is for two players.

Get Set Make a function table like the one at the right. Player A creates a function and secretly writes its equation on a piece of scrap paper.

n	f(n)

Go
- Player B writes a value for n in the function table. Then Player A writes the corresponding value for $f(n)$ in the table.
- Player B may guess the function rule at any time. If the guess is correct, the round ends and Player B receives 10 points. If the guess is incorrect, Player A receives 5 points and Player B may guess again or continue writing values of n.
- After Player B guesses the function, Player A receives one point for each value of n Player B wrote.
- Exchange positions for the next round.
- The winner is the player with the most points after four rounds.

 Visit www.glencoe.com/sec/math/mac/mathnet for more games.

Lesson 10-1 Functions **431**

■ Extending the Lesson ■

Enrichment Masters, p. 77

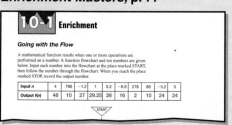

Students may find it helpful to use calculators while playing the game. At the end of each round, be sure that Player B understands the equation and knows why each answer was either right or wrong.

---

4 ASSESS

Closing Activity

Speaking Write a function on the chalkboard. Name values from the domain and have students name the corresponding values in the range.

Additional Answer

20.

Game Score	Handicapped Score
n	f(n)
153	183
144	174
161	191
163	193
166	196

Practice Masters, p. 77

Lesson 10-1 **431**

TECHNOLOGY LAB

10-1B LAB Notes

GET READY

Objective Students create function tables by using a graphing calculator.

Technology Resources
- graphing calculator (TI-80, TI-81, TI-82, or TI-83)

Optional Resources
Hands-On Lab Masters
- calculator layout (TI-83), p. 33

MANAGEMENT TIPS

Recommended Time
30 minutes

Getting Started Allow students to experiment with the table functions of their calculators before beginning the lab.

After they have entered the domain values and the calculator has displayed the range values, have students look for a pattern. Are the range values increasing or decreasing as the domain values increase? **increasing**

Have students work through the exercises using the same technique.

Using Other Calculators
If graphing calculators are not available, regular calculators can be used to find the range values. Have students fill in the charts with the functions and the domain values first, and then calculate each range value.

ASSESS

After students answer Exercises 1–4, ask them to compare and contrast all of the function tables. Is there a pattern to the range values? **yes** Is the pattern the same for all of the tables? **Yes; all of the range values increase or decrease in a linear fashion.**

GRAPHING CALCULATORS

10-1B Function Tables

A Follow-Up of Lesson 10-1

 graphing calculator

You can use a graphing calculator to create function tables. When you enter the function and the domain values, the calculator will find the corresponding range values.

TRY THIS

Work with a partner.

Make a function table to find the range of $f(n) = 4n + 2$ if the domain is $\{-5, -3, 0, 0.5, 12\}$.

The graphing calculator uses X for domain values and Y for range values. So the function can be represented by $Y = 4X + 2$.

Step 1 Enter the function into the Y= list.
Press [Y=] to access the function list. Then enter the function by pressing 4 [X,T,θ,n] [+] 2.

Step 2 Next, set up your table. Press [2nd] [TBLSET] to display the table setup screen. Use the arrow and [ENTER] keys to highlight Indpnt: ASK and Depend: Auto. *These selections allow you to enter values of X and have the calculator find the values for Y.*

Step 3 Access the table by pressing [2nd] [TABLE]. The calculator will display an empty function table.

Step 4 Now input the domain values. Enter each value and press [ENTER]. The calculator will display each range value.

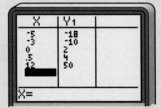

4. less than −16; as the x values become greater, the y values are becoming smaller

ON YOUR OWN

Use a graphing calculator to complete each function table.

1.
X	Y = 15X
−3	−45
−2	−30
0	0
1.5	22.5
4.8	72

2.
X	Y = 5X − 4
−8	−44
−1.7	−12.5
0.3	−2.5
3	11
5.4	23

3.
X	Y = 3 − 2X
−14	31
−3.3	9.6
1.3	0.4
4.5	−6
9.4	−15.8

4. Study the table you created for Exercise 3. If you entered a value of X that is greater than 10, would the value of Y be greater or less than −16? Explain.

432 Chapter 10 Algebra: Graphing Functions

 Math Journal Have students write a paragraph about how the table functions on a graphing calculator could be helpful in building a staircase.

10-2 Using Tables to Graph Functions

What you'll learn
You'll learn to graph functions by using function tables.

When am I ever going to use this?
You can compare prices by graphing functions.

You may have heard of the *Titanic*, the "unsinkable" luxury ship that sank April 15, 1912. But have you heard of her sister ships, the *Britannic* and the *Olympic*? These three ships were nearly identical even though one was a cruise ship, one was a hospital ship, and the other a troopship.

Robert Ballad visited the wrecks of the *Titanic* and *Britannic*. He located the shipwrecks using sonar. Sonar units find distances to objects using the time it takes to reflect sound. The function used is $f(n) = 727n$, where $f(n)$ is the distance to the object in meters and n is the time in seconds.

You have made function tables to represent functions. Another way to represent a function is by graphing. When a function is graphed, the domain and range values are written as ordered pairs.

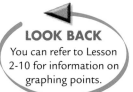

LOOK BACK
You can refer to Lesson 2-10 for information on graphing points.

To form an ordered pair, let the x-coordinate be the value of n. The y-coordinate is the value of $f(n)$. The table shows ordered pairs for the function $f(n) = 2n + 1$. Each ordered pair is graphed on the coordinate grid. Notice that the x-axis is labeled n and the y-axis is labeled $f(n)$.

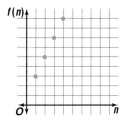

What pattern do the points suggest?

n	$2n + 1$	$f(n)$	$(n, f(n))$
1	2(1) + 1	3	(1, 3)
2	2(2) + 1	5	(2, 5)
3	2(3) + 1	7	(3, 7)
4	2(4) + 1	9	(4, 9)

Example 1 CONNECTION

History Refer to the beginning of the lesson. Draw a graph of the sonar function $f(n) = 727n$.

Step 1 Make a function table.

n	$727n$	$f(n)$	$(n, f(n))$
0	727(0)	0	(0, 0)
5	727(5)	3,635	(5, 3,635)
10	727(10)	7,270	(10, 7,270)
15	727(15)	10,905	(15, 10,905)

Step 2 Choose several values of n and find the value of $f(n)$ for each.

Step 3 Write the ordered pairs.

Step 4 Graph the ordered pairs. Draw the line that the points suggest.

Lesson 10-2 Using Tables to Graph Functions **433**

10-2 Lesson Notes

Instructional Resources
- *Study Guide Masters*, p. 78
- *Practice Masters*, p. 78
- *Enrichment Masters*, p. 78
- Transparencies 10-2, A and B
- CD-ROM Program
  - Resource Lesson 10-2
  - Interactive Lesson 10-2

Recommended Pacing	
Standard	Day 3 of 14
Honors	Day 3 of 14
Block	Day 2 of 7

1 FOCUS

5-Minute Check
(Lesson 10-1)

1. Copy and complete the function table.

n	$4n - 3$	$f(n)$
-1	4(-1) - 3	-7
0	4(0) - 3	-3
1	4(1) - 3	1
2	4(2) - 3	5
3	4(3) - 3	9

2. Find $f(5)$ if $f(n) = n^2 - 3$. **22**

The 5-Minute Check is also available on **Transparency 10-2A** for this lesson.

Motivating the Lesson
Problem Solving A redwood tree grows at three times the rate of a maple tree. Ask students to suggest ways that the two growth rates could be modeled. **Sample answers: animation; different colored stick lengths; graphing**

2 TEACH

Transparency 10-2B contains a teaching aid for this lesson.

Thinking Algebraically As students graph ordered pairs in which the domain values are equally spaced, ask them to look for a pattern on the graph and predict the next three values in the pattern.

Lesson 10-2 **433**

In-Class Examples

For Example 1
A blizzard at the Slippery Ski Area deposited $\frac{1}{2}$ foot of snow per hour atop a 3-foot snow base. If n represents the number of hours, $f(n) = 3 + \frac{1}{2}n$ represents the depth of snow after each hour since the blizzard began. Draw a graph of this function.

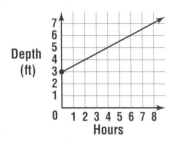

For Example 2
Graph the function $f(n) = n^2 - 5$.

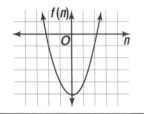

Teaching Tip In Example 2, you may wish to remind students that the possible values of n are infinite, or endless, because the domain is infinite.

Study Guide Masters, p. 78

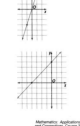

Not all functions have graphs that are straight lines. The graphs of many functions also lie in more than the first quadrant.

Example 2 Graph the function $f(n) = 2n^2 + 1$.

Make a function table for different values of n.

Then graph the ordered pairs and connect the points to complete the graph. There are infinitely many values possible for n. If you graphed all of them, the points would form a smooth curve. Draw a curve suggested by these points.

n	$f(n)$	$(n, f(n))$
-5	51	(-5, 51)
-3	19	(-3, 19)
-1	3	(-1, 3)
0	1	(0, 1)
1	3	(1, 3)
3	19	(3, 19)
5	51	(5, 51)

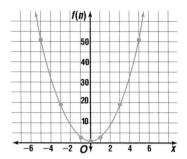

1. Use the n values for x and the $f(n)$ values for y and graph the coordinates (x, y).
2. $f(0)$ is the location of the sonar unit.

CHECK FOR UNDERSTANDING

Communicating Mathematics

Read and study the lesson to answer each question.

1. Write in your own words how you can use a function table to graph a function.
2. Tell what $f(0)$ means in terms of the sonar described in Example 1.

Guided Practice
3–4. See Answer Appendix for graphs.

Copy and complete each function table. Then graph the function.

3. $f(n) = n + 5$

n	$f(n)$	$(n, f(n))$
-2	3	(-2, 3)
-1	4	(-1, 4)
0	5	(0, 5)
1	6	(1, 6)
2	7	(2, 7)

4. $f(n) = \frac{4}{n}$

n	$f(n)$	$(n, f(n))$
$\frac{1}{2}$	8	$(\frac{1}{2}, 8)$
1	4	(1, 4)
2	2	(2, 2)
4	1	(4, 1)
16	$\frac{1}{4}$	$(16, \frac{1}{4})$

5. **Driver Safety** A "tailgater" risks an accident by following cars too closely. The function $f(n) = 3.48n - 20$ estimates the stopping distance $f(n)$ in feet for a car traveling n miles per hour.

n	$f(n)$	$(n, f(n))$
10	14.8	(10, 14.8)
20	49.6	(20, 49.6)
30	84.4	(30, 84.4)
40	119.2	(40, 119.2)
50	154	(50, 154)

a. Copy and complete the function table.
b. Graph the function. See Answer Appendix.

434 Chapter 10 Algebra: Graphing Functions

Reteaching the Lesson

Activity Students may become confused about which values in their tables to use to write the ordered pair as in Example 1. Use colored pencils or highlighters to call out the columns representing n and $f(n)$.

Error Analysis
Watch for students who confuse the x- and y-axes when graphing coordinate pairs.
Prevent by reminding them that the value of n is always found on the x-axis.

EXERCISES

Practice

6–11. See Answer Appendix for graphs.

Copy and complete each function table. Then graph the function.

6. $f(n) = n + 4$

n	f(n)	(n, f(n))
−3	1	(−3, 1)
−1	3	(−1, 3)
1	5	(1, 5)
3	7	(3, 7)
5	9	(5, 9)

7. $f(n) = -8n$

n	f(n)	(n, f(n))
−2	16	(−2, 16)
0	0	(0, 0)
0.5	−4	(0.5, −4)
1	−8	(1, −8)
2	−16	(2, −16)

8. $f(n) = 2n + 3$

n	f(n)	(n, f(n))
−4	−5	(−4, −5)
−1	1	(−1, 1)
0	3	(0, 3)
2	7	(2, 7)
5	13	(5, 13)

9. $f(n) = 8 - n$

n	f(n)	(n, f(n))
−4	12	(−4, 12)
−1	9	(−1, 9)
0	8	(0, 8)
6	2	(6, 2)
10	−2	(10, −2)

10. $f(n) = \dfrac{8}{n}$

n	f(n)	(n, f(n))
$\frac{1}{2}$	16	$(\frac{1}{2}, 16)$
2	4	(2, 4)
4	2	(4, 2)
8	1	(8, 1)
16	$\frac{1}{2}$	$(16, \frac{1}{2})$

11. $f(n) = n^2 + 3$

n	f(n)	(n, f(n))
−3	12	(−3, 12)
−2	7	(−2, 7)
−1	4	(−1, 4)
0	3	(0, 3)
1	4	(1, 4)
2	7	(2, 7)
3	12	(3, 12)

12. Choose values for n and graph $f(n) = 0.4n + 5$.

13. Choose values for n and graph $f(n) = -(n^2)$.

12–13. See Answer Appendix.

Applications and Problem Solving

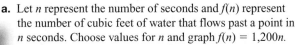

14. *Sports* In October 1995, Joe Dengler and his team attempted the first rafting expedition on China's Shuiluo river. The 150-mile river cuts through the mountains descending 3,000 feet. At the start of the trip, Joe estimated that the river was flowing at 1,200 cubic feet per second. **a. See Answer Appendix.**

  a. Let n represent the number of seconds and $f(n)$ represent the number of cubic feet of water that flows past a point in n seconds. Choose values for n and graph $f(n) = 1,200n$.

  b. Team member Beth Rypins hiked ahead to check the rapids. How much water passed by in the 8 minutes Beth was gone? **576,000 cubic feet**

15. *Life Science* A person's recommended weight is a function of his or her height. The function is $w = \dfrac{11(h - 40)}{2}$ for a weight of w pounds and a height of h inches.

  a. Choose values of h and graph the function using the h values as the x-coordinates and the values of w as the y-coordinates.

  b. Monifa is 5 feet 4 inches tall. According to this formula, what is Monifa's recommended weight? **132 pounds   a. See Answer Appendix.**

16. *Critical Thinking* Graph the absolute value function $f(n) = |n| + 2$. Describe the shape and range of this graph.

16. The graph forms a "v" and its range is a set of positive numbers greater than or equal to 2.

Mixed Review

17. 3,809.48 m

17. *History* Refer to the beginning of the lesson. How far beneath the surface is the *Titanic* if it takes 5.24 seconds for sound to reflect? *(Lesson 10-1)*

18. **Test Practice** What is the GCF of 72, 132, and 24? *(Lesson 6-3)* **D**

  A 2   **B** 3   **C** 6   **D** 12

Lesson 10-2 Using Tables to Graph Functions   **435**

Extending the Lesson

Enrichment Masters, p. 78

Activity Have students work in small groups to graph $f(n) = \sqrt{25 - n^2}$.

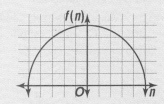

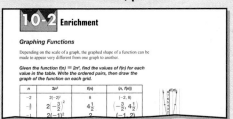

---

3 PRACTICE/APPLY

Check for Understanding

If students need additional practice or instruction after completing Exercises 1–5, one of these options may be helpful.
- Extra Practice, see p. 633
- Reteaching Activity, see p. 434
- *Study Guide Masters,* p. 78
- *Practice Masters,* p. 78
- Interactive Mathematics Tools Software

Assignment Guide

Core: 7–11 odd, 12, 14, 16–18
Enriched: 6–12 even, 13–18

4 ASSESS

Closing Activity

Modeling Have students graph the ordered pairs (−4, 2), (−2, 4), (−1, 8), (1, −8), (2, −4), (4, −2) on centimeter grid paper using small dot stickers. Ask students to suggest the shape of the graph and add additional dots of a different color to the graph. Have them verify their guesses by graphing more ordered pairs that satisfy $f(n) = \dfrac{-8}{n}$.

Practice Masters, p. 78

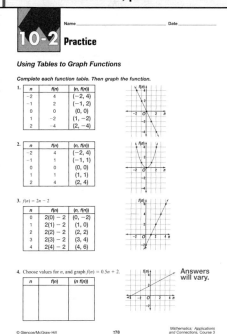

Lesson 10-2   **435**

Motivating Students
As students learned in the Chapter Opener on page 427, there are many opportunities in the field of computer game design. Students may be interested in that or a related field. To start the discussion, you may ask students about the history of computers and computer games.
- When was the first mechanical computing device invented? **500 B.C.**
- When was the first digital computer invented? **1946**
- What was the first video game and when did it first appear in arcades? ***Pong*, 1972**

Making the Math Connection
In her job as a computer game designer, Roberta Williams uses math everyday. Mathematical functions enable computer programmers to create visual screen effects such as movement, which is one of the most important features of computer games.

Working on *Your Turn*
Students can arrange an interview with a local computer programmer or the designer of their favorite game. Prior to the interview, they should research their subject and prepare a list of questions. Possible questions include: how the interviewee became interested in computers and what mathematics courses he or she studied in school that are useful at work.

*An additional School to Career activity is available on page 36 of the **School to Career Masters**.*

COMPUTER PROGRAMMING

Roberta Williams
COMPUTER GAME DESIGNER

Roberta Williams co-founded Sierra On-Line with her husband in their kitchen in 1979. Today, Sierra is one of the world's top computer game sellers. Ms. Williams' popular *King's Quest* adventure game series has sold over 7 million copies! As a game designer, she invents a plot and characters for a game and then works closely with a team of computer programmers and artists to create the game.

To be a computer programmer, you should have a solid background in mathematics and science and take courses in keyboarding, computer skills, and business. Although some computer programmers can get jobs with a degree from a junior or technical college, most programmers have a bachelor's degree.

For more information:
The Association for
   Computing Machinery
1515 Broadway
New York, NY 10036

www.glencoe.com/sec/math/mac/mathnet

How can I get to be a successful computer game designer like Roberta Williams?

Your Turn
Interview a computer programmer about the wide variety of work programmers do. Be sure to ask them how they use mathematics in their job.

More About Roberta Williams
- Williams was 26 when she and her husband founded Sierra On-Line.
- Since 1980, Roberta Williams has created more than 20 computer games that have sold a total of almost seven million copies worldwide.
- Among the Williams-created titles: *Mystery House, Mixed-Up Mother Goose, Dark Crystal,* the *King's Quest* series, and *Phantasmagoria*.

10-3 Equations with Two Variables

What you'll learn
You'll learn to find solutions of equations with two variables.

When am I ever going to use this?
Knowing how to solve equations with two variables can help you to find how long a car trip will take.

What are your career plans? For many people, a college degree is a big part of their future plans. One way to estimate future college costs is to multiply the current cost of the college of your choice by the inflation factor. An estimated inflation factor for 4 years is 1.34. So if the cost of a college is n dollars, the cost in four years will be $f(n) = 1.34n$.

Harvard University, Cambridge, MA

Sometimes functions do not use the $f(n)$ notation. Instead, they use two variables — one to represent the input and one to represent the output. For example, you could write the college cost function as $y = 1.34c$ if y is your cost in four years and c is the current cost. Other functions you know are written in this way.

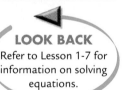

Finding the perimeter of a square	$P = 4s$
The distance you travel at 65 miles per hour	$d = 65t$
Changing Celsius to Fahrenheit	$F = \frac{9}{5}C + 32$

Values for the two variables that make the equation true form an ordered pair that is a solution of the equation. Most equations with two variables have an infinite number of solutions.

One variable in an equation with two variables represents the input. The other variable represents the output of the function. When x and y are used, x usually represents the input.

Example

LOOK BACK
Refer to Lesson 1-7 for information on solving equations.

 Find four solutions of $2x + 3 = y$. Write the solutions as a set of ordered pairs.

First, make a function table to find the ordered pairs. Use x for the input and y for the output.

Choose any four values for x. Then complete the table.

x	$2x + 3$	y	(x, y)
−3	2(−3) + 3	−3	(−3, −3)
0	2(0) + 3	3	(0, 3)
1	2(1) + 3	5	(1, 5)
3	2(3) + 3	9	(3, 9)

Four solutions of the equation $2x + 3 = y$ are $\{(-3, -3), (0, 3), (1, 5), (3, 9)\}$.

Lesson 10-3 Equations with Two Variables **437**

10-3 Lesson Notes

Instructional Resources
- *Study Guide Masters*, p. 79
- *Practice Masters*, p. 79
- *Enrichment Masters*, p. 79
- Transparencies 10-3, A and B
- *Assessment and Evaluation Masters*, p. 267

 CD-ROM Program
- Resource Lesson 10-3

Recommended Pacing	
Standard	Day 4 of 14
Honors	Day 4 of 14
Block	Day 2 of 7

1 FOCUS

5-Minute Check
(Lesson 10-2)
Copy and complete the function table for $f(n) = 2n - 1$. Then graph the function.

n	$f(n)$	$(n, f(n))$
−1	−3	(−1, −3)
0	−1	(0, −1)
1	1	(1, 1)
2	3	(2, 3)

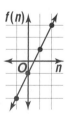

 The 5-Minute Check is also available on **Transparency 10-3A** for this lesson.

Motivating the Lesson
Problem Solving Write the formula $C = \pi d$ on the chalkboard. Ask students to identify the two variables. *C and d* Then ask them which variable represents the input and which one represents the output. *d; C* Have them find the circumference of circles with diameters of 1, 2, 3, and 4 centimeters. Use 3.14 for π. **3.14 cm, 6.28 cm, 9.42 cm, 12.56 cm**

Cross-Curriculum Cue

Inform the other teachers on your team that your students are studying equations with two variables to show the relationship between two quantities. Suggestions for curriculum integration are:

Economics: interest rates for loans
Physical Education: maximum aerobic heart rate
Physical Science: rate of speed

Lesson 10-3 **437**

2 TEACH

 Transparency 10-3B contains a teaching aid for this lesson.

Using Connections Continue the relationship between functions and formulas suggested in the first paragraph of the lesson by writing the equation $y = 4x + 2$ on the chalkboard. Name a value for x and ask students to name the corresponding value for y and the ordered pair.

In-Class Examples

For Example 1
Find four solutions of $-2x + 5 = y$. Write the solutions as a set of ordered pairs.
Sample answer: $\{(-1, 7), (0, 5), (1, 3), (2, 1)\}$

For Example 2
Tourists visiting Toronto from the United States must be able to convert prices in Canadian dollars to the equivalent amount in U.S. dollars. Assume one Canadian dollar is worth 0.73 U.S. dollar. Determine which ordered pairs in the set $\{(0.73, 1.00), (5.50, 6.50), (14.60, 20.00), (25.00, 26.95)\}$ represent solutions of this function. **(0.73, 1.00), (14.60, 20.00)**

Teaching Tip Have students read Example 2 and then identify what x and y represent. Students can then reread the example, substituting x for number of chirps and y for temperature before translating the verbal statement into an equation.

In Example 1, we found four ordered pairs that are solutions of $2x + 3 = y$. There are many more solutions. *Can you name another?*

Example 2
CONNECTION

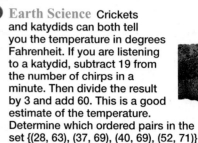

Earth Science Crickets and katydids can both tell you the temperature in degrees Fahrenheit. If you are listening to a katydid, subtract 19 from the number of chirps in a minute. Then divide the result by 3 and add 60. This is a good estimate of the temperature. Determine which ordered pairs in the set {(28, 63), (37, 69), (40, 69), (52, 71)} represent solutions of this function.

First, translate the verbal sentence into an equation.
Let $x =$ the number of chirps and $y =$ the temperature.

temperature	equals	19 less than the number of chirps divided by 3	plus	60
y	$=$	$\dfrac{x - 19}{3}$	$+$	60

Did you know? The hot humid days of summer, when the crickets and katydids are chirping, are called the "dog days" because Sirius, the dog star, is visible then.

Try each ordered pair in the equation.

Try (28, 63).
$y = \dfrac{x - 19}{3} + 60$
$63 \stackrel{?}{=} \dfrac{28 - 19}{3} + 60$
$63 \stackrel{?}{=} \dfrac{9}{3} + 60$
$63 \stackrel{?}{=} 3 + 60$
$63 = 63$ ✓
(28, 63) is a solution.

Try (37, 69).
$69 = \dfrac{x - 19}{3} + 60$
$69 \stackrel{?}{=} \dfrac{37 - 19}{3} + 60$
$69 \stackrel{?}{=} \dfrac{18}{3} + 60$
$69 \stackrel{?}{=} 6 + 60$
$69 \neq 66$
(37, 69) is not a solution.

Try (40, 69).
$y = \dfrac{x - 19}{3} + 60$
$69 \stackrel{?}{=} \dfrac{40 - 19}{3} + 60$
$69 \stackrel{?}{=} \dfrac{21}{3} + 60$
$69 \stackrel{?}{=} 7 + 60$
$69 \neq 67$
(40, 69) is not a solution.

Try (52, 71).
$y = \dfrac{x - 19}{3} + 60$
$71 \stackrel{?}{=} \dfrac{52 - 19}{3} + 60$
$71 \stackrel{?}{=} \dfrac{33}{3} + 60$
$71 \stackrel{?}{=} 11 + 60$
$71 = 71$ ✓
(52, 71) is a solution.

The ordered pairs (28, 63) and (52, 71) are solutions. The ordered pair (28, 63) represents 28 chirps and a temperature of 63° F.

Check by graphing the function and the points.

CHECK FOR UNDERSTANDING

Communicating Mathematics

Read and study the lesson to answer each question.

1. *Find* four ordered pairs that are solutions of $2x + 3 = y$ other than those found in Example 1. **Sample answers: $(-2, -1), (-1, 1), (2, 7), (4, 11)$**

2. *Show* how you could use the method in Example 2 to check the solutions in Example 1. **See margin.**

3. *Write* two examples of equations with two variables that you have used in your math or science classes. **See students' work.**

Guided Practice

Copy and complete the table for each equation.

4. $y = 4x - 1$

x	y
-1	-5
0	-1
2	7
4	15

5. $0.5x = y$

x	y
-4	-2
0	0
5	2.5
16	8

6. Find four solutions of $y = \frac{1}{4}x + 5$. **Sample answers: $(0, 5), (4, 6), \left(6, 6\frac{1}{2}\right), (8, 7)$**

7. **Geometry** The equation $s = 180(n - 2)$ relates the sum of the measures of the angles s formed by the sides of a polygon to the number of sides n. Find four ordered pairs (n, s) that are solutions of the equation.
Sample answers: $(3, 180), (4, 360), (5, 540), (6, 720)$

EXERCISES

Practice

Copy and complete the table for each equation.

8. $y = x + 1$

x	y
-3	-2
-1	0
1	2
2	3

9. $y = -1.5x$

x	y
-6	9
-1	1.5
3	-4.5
16	-24

10. $y = 5x - 2$

x	y
$-\frac{4}{5}$	-6
2	8
$\frac{7}{15}$	$\frac{1}{3}$

11. $y = 7 + 0.2x$

x	y
-5	6
0	7
5	8
10	9

12. $1 - 5x = y$

x	y
0	1
-2	11
4	-19
10	-49

13. $y = \frac{x}{3} + 4$

x	y
-3	3
3	5
6	6
8	$6\frac{2}{3}$

14. $(-1, -1), (0, 3),$
    $(1, 7), (2, 11)$

15. $(-1, 2), (0, 1),$
    $(1, 0), (2, -1)$

16. $(0, 3), (2, -1),$
    $(4, -5), (6, -9)$

Find four solutions of each equation. **14–17. Sample answers given.**

14. $y = 4x + 3$

15. $y = -x + 1$

16. $y = -2x + 3$

17. List four solutions of $y = 15x - 57$. **$(0, -57), (2, -27), (4, 3), (6, 33)$**

Lesson 10-3 Equations with Two Variables **439**

---

3 PRACTICE/APPLY

Check for Understanding

If students need additional practice or instruction after completing Exercises 1–7, one of these options may be helpful.
- Extra Practice, see p. 634
- Reteaching Activity
- *Study Guide Masters*, p. 79
- *Practice Masters*, p. 79

Assignment Guide

Core: 9–19 odd, 20–23
Enriched: 8–16 even, 18–23

Additional Answer

2. Substitute each ordered pair into the equation and verify that the equation is true.

Study Guide Masters, p. 79

Study Guide

Equations with Two Variables

The set of ordered pairs that make an equation true is the solution set for the equation.

Example Find four solutions of $y = 1.5x + 2$. Write the solutions as a set of ordered pairs.

Choose values for x.   Calculate y values.   Write ordered pairs.
Let $x = -3$. $y = 1.5(-3) + 2$ or -2.5 $(-3, -2.5)$
Let $x = 0$. $y = 1.5(0) + 2$ or 2 $(0, 2)$
Let $x = 2$. $y = 1.5(2) + 2$ or 5 $(2, 5)$
Let $x = 5$. $y = 1.5(5) + 2$ or 9.5 $(5, 9.5)$

Four solutions of the equation $y = 1.5x + 2$, are $\{(-3, -2.5), (0, 2), (2, 5), (5, 9.5)\}$.

Complete the table for each equation.

1. $y = 0.5x + 3$

x	y
-4	1
0	3
2	4
10	8

2. $y = -3x + 3$

x	y
-3	12
0	3
5	-12
10	-27

3. $y = \frac{x}{2} + 4$

x	y
-6	1
-2	3
4	6
12	10

Find four solutions of each equation. **Answers may vary.**

4. $y = 4x - 2$
Sample solutions: $(0, -2), (-1, -6), (1, 2), (2, 6)$

5. $y = -2x + 1$
Sample solutions: $(0, 1), (-1, 3), (1, -1), (2, -3)$

6. $y = -2x + 5$
Sample solutions: $(0, 5), (-1, 7), (1, 3), (2, 1)$

7. $y = -\frac{1}{2}x - 4$
Sample solutions: $(0, -4), (-2, -3), \left(1, -4\frac{1}{2}\right), (2, -5)$

8. $y = 2x + 3$
Sample solutions: $(0, 3), (-1, 1), (1, 5), (2, 7)$

9. $y = 2.5x - 2$
Sample solutions: $(0, -2), (-1, -4.5), (1, 0.5), (2, 3)$

Reteaching the Lesson

Activity Give each group of students an equation and a range of values from which to choose a value of x. Then have them find four solutions.

Error Analysis
Watch for students who evaluate expressions incorrectly.
Prevent by having students review evaluating algebraic expressions in Lesson 1-3.

Lesson 10-3 **439**

4 ASSESS

Closing Activity
Writing Have students choose an equation with two variables and write three solutions of the equation as a set of ordered pairs.

Chapter 10, Quiz A (Lessons 10-1 through 10-3) is available in the *Assessment and Evaluation Masters*, p. 267.

Additional Answers
18a. If Latanya's cost $= \ell$ and the current costs $= c$, $\ell = 1.34c - 1,000$.

18b. Texas A&M: $13,997.28; Purdue University: $19,309.04; The University of Virginia: $23,710.94

20a. Sample answers: $(0, -2)$, $(1, 2)$, $(2, 6)$, $(3, 10)$

20b. Sample answer: $y = 4x - 2$ is easier to solve because the variable y is on one side by itself.

21. See students' tables.

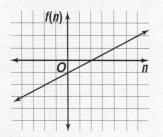

Practice Masters, p. 79

440 Chapter 10

Applications and Problem Solving
18a–b. See margin.

19. Sample answers: $(-459.67, 0)$, $(32, 273.15)$, $(98.6, 310.15)$
21. See margin.

Mixed Review

18. **Finances** Refer to the beginning of the lesson. In the regional science fair, Latanya won a scholarship that will pay $1,000 per year for college.

College	Current Annual Cost
Texas A & M	$11,192
Purdue University	$15,156
The University of Virginia	$18,441

a. Write an equation with two variables for Latanya's college costs in four years.
b. Use the current costs listed in the table to find Latanya's annual cost for each of the schools she is considering.

19. **Physical Science** Chemists often use the Kelvin scale to measure temperature. The equation $1.8K = F + 459.67$ relates degrees Fahrenheit F to kelvins K. Write three ordered pairs (F, K) that represent equivalent Fahrenheit and Kelvin temperatures.

20. **Critical Thinking** The equations $y = 4x - 2$ and $8x - 2y = 4$ have the same solution set. a–b. See margin.
   a. Find four solutions of the first equation. Then verify that the ordered pairs are solutions of the second equation.
   b. For which equation do you think it is easier to find solutions? Explain.

21. **Algebra** Choose values for n and graph $f(n) = \frac{n}{2} - 1$. *(Lesson 10-2)*

22. **Test Practice** Which of the following does not describe the figure? *(Lesson 5-3)* **C**
   A parallelogram   B square
   C trapezoid   D rhombus

23. **Music** The number of pages in *Rock Star* magazine in the last nine issues is 196, 188, 184, 200, 168, 176, 192, 160, and 180. Find the median, upper quartile, and lower quartile of this data. *(Lesson 4-5)* **184; 194; 172**

MATH IN THE MEDIA

1. If the small candles are 1 year, how old is Herman? **50**

2. Let x represent the number of large candles and y represent the number of small candles. Write an equation for the numbers of candles that can be used to show an age of 60 years. $15x + y = 60$

3. Use your equation to find all of the different ways to show an age of 60 years.
   $(4, 0), (3, 15), (2, 30), (1, 45), (0, 60)$

"The big ones are 15 years."

440 Chapter 10 Algebra: Graphing Functions

■ Extending the Lesson ■
Enrichment Masters, p. 79

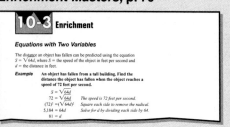

MATH IN THE MEDIA
Ask students to determine how many candles would be on their birthday cake if the large candles represent 3 years and the small candles 1 year.

COOPERATIVE LEARNING

10-4A Graphing Relationships

A Preview of Lesson 10-4

- large rubber band
- 2 paper clips
- ruler
- paper cup
- 10 washers
- tape
- grid paper

Scientists and engineers often graph relationships to look for patterns in the points they graph.

TRY THIS

Work in groups of four.

Step 1 Use a pencil to punch a small hole in the bottom of the cup. Place one paper clip onto the rubber band. Push the other end of the rubber band through the hole in the cup. Attach the second paper clip to the other end of the rubber band to keep it from coming back out of the hole.

Step 2 Copy the table at the right.

Step 3 Tape the top paper clip to the edge of a desk. Measure and record the distance from the bottom of the desk to the bottom of the cup. Drop one washer into the cup. Measure and record the new distance.

Step 4 Keep dropping washers in the cup and recording distances until you have ten sets of distances.

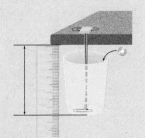

Number of Washers x	Distance y

ON YOUR OWN

1. Refer to your table of recorded data. Let x represent the number of washers and y the distance from the paper clip to the bottom of the cup. Do you think that the distance is a function of the number of washers? Explain.   **See margin.**

2. On a coordinate grid, graph the ordered pairs formed by your data. What pattern do the points suggest?

3. **Look Ahead** Combine your data with the rest of the class. Graph the ordered pairs. A line that is drawn to represent the data closely is called a *line of best fit*. Draw a line of best fit for the data.   **2–3. See Answer Appendix.**

Lesson 10-4A   **441**

Have students read about Robert Hooke (1635–1703), who studied the behavior of elastic materials. Ask them to write a short paragraph about how their experiment relates to Hooke's findings.

Additional Answer

1. Sample answer: Yes; each additional washer stretches the rubber band about the same additional distance.

HANDS-ON 10-4A LAB Notes

GET READY

Objective Students determine a linear function to describe data.

Optional Resources
Hands-On Lab Masters
- grid paper, p. 11
- worksheet, p. 60

Overhead Manipulative Resources
- coordinate grid transparency

Manipulative Kit
- ruler
- coordinate grid stamp

MANAGEMENT TIPS

Recommended Time
30 minutes

Getting Started Ask students to describe what happens to a grocery store scale when another orange of the same weight is added to an orange already on the scale. Explain that the weight readout approximately doubles.

The **Activity** demonstrates a relationship between two variables, the weight held by the rubber band and the distance the rubber band stretches.

Students should measure the distance the rubber band stretches as accurately as possible. Ask them to consider what would happen if they continued to drop washers into the cup. **The pattern would repeat until the rubber band reached its elastic limit.**

ASSESS

Have students complete Exercises 1–3. Watch for students who may have trouble graphing their ordered pairs. Make sure they realize that the number of washers is the x-coordinate and the distance is the y-coordinate.

Hands-On Lab 10-4A   **441**

10-4 Lesson Notes

Instructional Resources
- *Study Guide Masters*, p. 80
- *Practice Masters*, p. 80
- *Enrichment Masters*, p. 80
- Transparencies 10-4, A and B
- *Hands-On Lab Masters*, p. 59
- CD-ROM Program
  - Resource Lesson 10-4

Recommended Pacing	
Standard	Days 5 and 6 of 14
Honors	Days 5 and 6 of 14
Block	Day 3 of 7

1 FOCUS

5-Minute Check
(Lesson 10-3)

1. Copy and complete the table.

$y = 3x - 4$	
x	y
−4	−16
0	−4
3	5
6	14

2. Find four solutions of $y = -x + 2$. Write the solutions as a set of ordered pairs. Sample answer: {(−1, 3), (0, 2), (1, 1), (2, 0)}

The 5-Minute Check is also available on **Transparency 10-4A** for this lesson.

2 TEACH

Transparency 10-4B contains a teaching aid for this lesson.

Using the Mini-Lab Ask students to explain why the graph appears only in the first quadrant. Explain that in this situation neither the number of packs of cigarettes nor the number of hours lost to smoking makes sense as a negative number.

10-4 Graphing Linear Functions

What you'll learn
You'll learn to graph linear functions by plotting points.

When am I ever going to use this?
Accountants graph linear functions to show costs of manufactured items.

Word Wise
linear function

Experts estimate that smoking a pack of cigarettes takes a half hour off the average person's life. That may not sound like much, but heavy smokers can take years off their lives. A graph of the equation $y = 0.5x$ can show how much time is lost to smoking.

HANDS-ON MINI-LAB

Work with a partner. grid paper   ruler

Try This
- Make a function table to find ten solutions of the equation $y = 0.5x$.
- The solutions include the ordered pairs (0, 0), (1, 0.5), (2, 1), and (4, 2). *(4, 2) means 4 packs of cigarettes take 2 hours off a person's life.*
- Graph all of the ordered pairs and look for a pattern in the points. *Since all of the coordinates are positive, you need only show the first quadrant of the coordinate plane.*

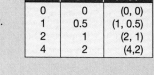

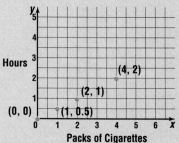

Talk About It
1. What shape is suggested by the graphs of the ordered pairs? **straight line**
2. Sketch the figure. **See students' work.**
3. Find four more solutions of the equation. Do the graphs of these ordered pairs lie on the figure you sketched? **yes**
4. What conclusion can you make about other solutions of $y = 0.5x$? **They are on the line too.**

In the Mini-Lab, you saw that all solutions of the equation suggested a straight line. An equation in which the graphs of the solutions form a line is called a **linear function**.

The graphs of linear functions that interpret real-life activities often lie in the first quadrant. However, most linear functions lie in at least two quadrants of the coordinate plane.

442 Chapter 10 Algebra: Graphing Functions

Motivating the Lesson
Hands-On Activity Place 20 items of the same weight on each side of a pan balance. Ask students to record what happens to the balance when two objects are removed from the left side and only one is removed from the right. Continue removing items at this rate until no objects remain on the left. Explain to students that this experiment demonstrates a linear relationship.

Examples

Study Hint

Estimation Estimate how long the axes should be before you graph a function. The graph may be more manageable if each unit on the axis represents 2, 3, 5, or some other number greater or less than 1.

1 Graph $y = \frac{x}{2} + 1$.

Begin by making a function table. List at least three values for x.

Graph each ordered pair. Connect the points with a line. Add arrows to the ends of the line to show that the line continues indefinitely.

x	y	(x, y)
−4	−1	(−4, −1)
0	1	(0, 1)
4	3	(4, 3)

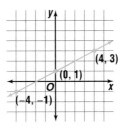

APPLICATION

2 *Sports* In 1989, Mark Wellman became the first paraplegic to climb Yosemite's 3,595-foot *El Capitan*. As he and his partner climbed, they had to prepare for lower temperatures at the top. If the temperature at sea level is 78° F, the function $t = 78 - 5.4h$ describes the temperature t at a height of h thousand feet above sea level. Graph the temperature function.

Make a function table. Then graph the points and complete the line.

h	t	(h, t)
0	78.0	(0, 78.0)
2	67.2	(2, 67.2)
5	51.0	(5, 51.0)

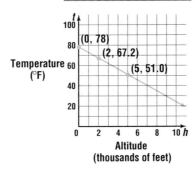

In-Class Examples

For Example 1

Graph $y = \frac{x}{2} - 2$.

For Example 2

The Ups-n-Downs Company has monthly operating expenses of $10,000. The function $P = 140b - 10,000$ describes the net profit P when b batches of their product are sold at a profit of $140 per batch. Graph this function.

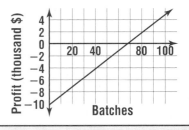

3 PRACTICE/APPLY

Check for Understanding

If students need additional practice or instruction after completing Exercises 1–7, one of these options may be helpful.
- Extra Practice, see p. 634
- Reteaching Activity
- *Study Guide Masters*, p. 80
- *Practice Masters*, p. 80

Study Guide Masters, p. 80

CHECK FOR UNDERSTANDING

Communicating Mathematics

Read and study the lesson to answer each question.

1. *Explain* why you should use at least three ordered pairs to graph a linear function. **The third point is for a check.**

2. *Tell* why only points in the first quadrant make sense for the graph in the Mini-Lab. **You cannot smoke a negative number of cigarettes.**

HANDS-ON MATH

3. *Extend* the graph in the Mini-Lab to find the amount of time taken off the life of a person who smokes ten packs of cigarettes. **5 hours**

Lesson 10-4 Graphing Linear Functions **443**

Reteaching the Lesson

Activity Have students work in pairs to make a function table showing the number of inches in whole numbers of feet from 1 to 10. Have them list the ordered pairs (feet, inches) and graph the function.

Error Analysis
Watch for students who incorrectly graph positive and negative coordinates. **Prevent by** naming each quadrant and telling them the signs of points graphed there: I (+, +), II (−, +), III (−, −), IV (+, −).

Lesson 10-4 **443**

Assignment Guide
Core: 9–19 odd, 22–25
Enriched: 8–18 even, 19, 20, 22–25

CHAPTER Project
Exercise 21 asks students to advance to the next stage of work on the Chapter Project. You may want students to work in groups as they graph the equation.

4 ASSESS

Closing Activity
Modeling Separate students into small groups and ask them to think of a machine, such as a scale or pulley, that can be used to show a linear relationship between two variables. Ask students to describe the relationship or to give a demonstration using objects commonly found in the classroom.

Additional Answers
20c. The ordered pair (x, y) represents the time in minutes and the tank level.

23. Sample answer: $\{(1, 13), (2, 16), (3, 19), (4, 22)\}$

Practice Masters, p. 80

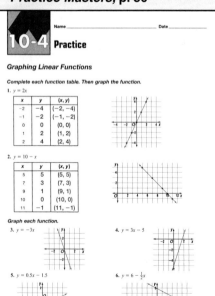

Guided Practice Graph each function. 4–7. See Answer Appendix.
4. $y = 3x$
5. $y = 10 - x$
6. $y = 6x + 4$

7. *History* American Indians used wampum beads for jewelry and money when English settlers came to New England. Wampum beads were made of polished clam shell. Three black beads could be traded for one English penny. Graph the function $p = 3w$.

EXERCISES

Practice Graph each function. 8–18. See Answer Appendix.
8. $y = 5x$
9. $y = 25 - x$
10. $y = 3x + 1$
11. $y = 15 - 3x$
12. $y = -4x$
13. $y = 1.5x + 2.5$
14. $y = \frac{1}{3}x + 5$
15. $y = 10 - \frac{x}{4}$
16. $y = 60 - 5x$
17. Draw a graph of $y = 75 - 3x$.
18. Graph the linear function $y = -\frac{1}{2}x + 6$.

19. See Answer Appendix.

Applications and Problem Solving

19. *Food* Maple syrup is made by boiling the sap of maple trees. The function $y = 40x$ describes the relationship between the number of gallons of sap y, used to make x gallons of maple syrup. Graph the function.

20. *Transportation* Because of their tolerance of hot weather, camels have been used for transportation in Africa and in the Middle East for centuries. A thirsty camel can drink up to 25 gallons of water in 10 minutes. **b. See Answer Appendix.**
   a. Suppose a camel is drinking from a 25-gallon tank. Let x be the number of minutes he has been drinking and y be the number of gallons of water remaining in the tank. Write two ordered pairs that relate the gallons of water in the tank y to the number of minutes x. **(0, 25), (10, 0)**
   b. Graph the linear function that contains these two points.
   c. How could you use the graph to determine the number of gallons in the tank at any time? **See margin.**

21. *Working on the* CHAPTER Project A computer game includes an animated insect that moves across the screen. The insect is at (x, y). Graph the insect's path as it moves along $y = 2x + 5$. **See Answer Appendix.**

22. *Critical Thinking*
   a. Graph $y = 2x$ and $y = -\frac{1}{2}x$ on the same coordinate plane. **See Answer Appendix.**
   b. At what point do the graphs intersect? **(0, 0)**
   c. Describe how these lines are related. **perpendicular lines**

Mixed Review
23. *Algebra* Find four solutions of $y = 3x + 10$. *(Lesson 10-3)* **See margin.**
24. *Patterns* Name the eighth term in the sequence 81, 27, 9, ... *(Lesson 7-5)* $\frac{1}{27}$
25. **Test Practice** There are 5,280 feet in one mile. Jun-Ko and Lorie rode their bicycles 3 miles. How many feet is this? *(Lesson 2-7)* **D**
   A 1,760 ft   B 5,280 ft   C 10,560 ft   D 15,840 ft

444 Chapter 10 Algebra: Graphing Functions

Extending the Lesson

Enrichment Masters, p. 80

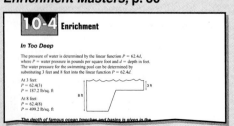

Activity Line *a* is steeper than line *b*. Have students work in groups to find a way to measure the steepness (or slope) of a line.

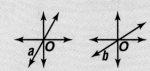

TECHNOLOGY LAB

GRAPHING CALCULATORS

10-4B Linear Functions
A Follow-Up of Lesson 10-4

graphing calculator

A graphing calculator can be a valuable tool to study functions. You need to set an appropriate range before you can create a graph. For example, a viewing window of $[-10, 10]$ by $[-10, 10]$ with a scale factor of 1 on both axes means that the graph will show x-coordinates from -10 to 10 and y-coordinates from -10 to 10. The scale factor of 1 indicates that the tick marks on both axes are one unit apart. This is called the *standard viewing window*. *You can choose a viewing window other than the standard by pressing* WINDOW *and entering your values.*

TRY THIS

Work with a partner.

Graph $y = 3x + 1$ in the standard viewing window.

Step 1 Choose the standard viewing window by pressing ZOOM 6.

Step 2 Enter and graph the equation. Press Y= to access the function list. Then enter the equation by pressing 3 X,T,θ,n + 1. Complete the graph by pressing GRAPH.

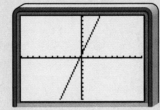

You will need to clear the graphics screen before you can graph another function. To do this, press Y= and clear any equations in the list.

ON YOUR OWN

Graph each linear function on a graphing calculator. Then sketch the graph on a piece of paper. 1–6. See Answer Appendix.

1. $y = x - 4$
2. $y = 4 - x$
3. $y = 5$
4. $y = 3x$
5. $y = 3x + 5$
6. $y = -3x$

7. Describe the graphs of $y = x - 4$ and $y = 4 - x$. **perpendicular lines**
8. What will the graph of $y = c$ look like if c is any number? **See margin.**

Lesson 10-4B TECHNOLOGY 445

Math Journal
Have students write a paragraph about how a graphing calculator could be helpful in designing a wheelchair ramp.

Additional Answer
8. Horizontal line c units above the x-axis if c is positive and c units below the x-axis if c is negative.

TECHNOLOGY
10-4B LAB Notes

GET READY

Objective Students learn to graph linear equations on a graphing calculator.

Technology Resources
- graphing calculator (TI-80, TI-81, TI-82, or TI-83)

MANAGEMENT TIPS

Recommended Time
30 minutes

Getting Started Have students press GRAPH to see if any graphs are visible on the screen. Then have them press Y= and delete any information found there. If a statistical graph appears, press 2nd [STAT PLOT] 4 to clear the screen.

After students have entered the data and the graph appears on the screen, ask students to compare it with the one in their books. Ask them to graph the equation $y = 3x - 1$. How has the view in the window changed?

Have students work through the exercises using the outlined steps.

Using Computers If graphing calculators are not available, students can use specialized computer software to graph their equations. Students may want to work in groups having one student say the equation aloud while another enters it into the computer. Another member of the group can sketch the results on graph paper.

ASSESS

Have students complete Exercises 1–8. Watch for students who incorrectly enter negative numbers on the calculator. Remind them that the calculator key for a negative sign is (−).

10-5 Lesson Notes

Instructional Resources
- *Study Guide Masters*, p. 81
- *Practice Masters*, p. 81
- *Enrichment Masters*, p. 81
- Transparencies 10-5, A and B
- *Assessment and Evaluation Masters*, pp. 266, 267
- *Technology Masters*, p. 72

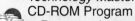 CD-ROM Program
- Resource Lesson 10-5

Recommended Pacing	
Standard	Day 7 of 14
Honors	Day 7 of 14
Block	Day 4 of 7

1 FOCUS

5-Minute Check
(Lesson 10-4)

Graph each function.
1. $y = 2x + 3$

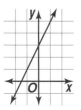

2. $y = 3x - 1$

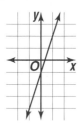

 The 5-Minute Check is also available on **Transparency 10-5A** for this lesson.

Motivating the Lesson
Problem Solving A car rental agency charges $24 per day to rent a car plus $0.09 per mile. Ask students to write an algebraic expression for the one-day charge if a renter drives m miles. $24 + 0.09m$

10-5 Graphing Systems of Equations

What you'll learn
You'll learn to solve systems of linear equations by graphing.

When am I ever going to use this?
Business owners solve systems of linear equations to determine whether a product will be profitable.

Word Wise
system of equations

Working students may find their mailboxes stuffed with offers for credit cards. But how do you know which one is the best choice? One way is by comparing your spending habits with each offer.

Ana Torres is comparing a MasterCard and a Visa. The MasterCard offer has no annual fee and an annual interest rate of 18%. The Visa has a $25 annual fee and an annual interest rate of 16%. Which offer is better for Ana? *This problem will be solved in Example 2.*

A set of two or more equations is called a **system of equations**. When you find an ordered pair that is a solution of all of the equations in the system, you have solved the system. The ordered pair for the point where the graphs of the equations intersect is the solution of the system of equations.

Example 1 Graph the system of equations $y = 2x - 5$ and $y = 1 - x$. Then find the solution of the system.

Step 1 Make a function table for each equation.

$y = 2x - 5$		$y = 1 - x$	
x	y	x	y
0	−5	−1	2
1	−3	1	0
3	1	4	−3

Step 2 Graph each equation on one coordinate grid. Then locate the point where the graphs intersect.

Step 3 The lines intersect at the point whose coordinates are $(2, -1)$. So, the solution of the system of equations is $(2, -1)$.

Step 4 Check the solution in *both* equations.

$y = 2x - 5$ $y = 1 - x$
$-1 \stackrel{?}{=} 2(2) - 5$ *Substitute 2 for x* $-1 \stackrel{?}{=} 1 - 2$
$-1 \stackrel{?}{=} 4 - 5$ *and −1 for y.* $-1 = -1$ ✓
$-1 = -1$ ✓

The solution checks in both equations.

446 Chapter 10 Algebra: Graphing Functions

Multiple Learning Styles

Verbal/Linguistic Have students work in pairs with a divider between them so that they cannot see each others' work. Give one student in each pair a system of equations to graph. The other student graphs the system by performing only the tasks that he or she hears from the first student. When complete, have them check the work for accuracy. Have students switch roles for another system.

Example APPLICATION **2** **Money Matters** Refer to the beginning of the lesson. Which card is the better choice for Ana?

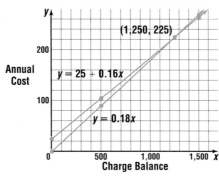

Let x represent the balance and y represent the annual cost. The total paid for a MasterCard can be represented by $y = 0.18x$. For a Visa, the cost is represented by $y = 25 + 0.16x$.

Graph both equations on the same coordinate plane. You can use a calculator to find the y-values for each value of x quickly.

MasterCard $y = 0.18x$		Visa $y = 25 + 0.16x$	
x	y	x	y
0	0	0	25
500	90	500	105
1,500	270	1,500	265

Graph each function. The graphs intersect at (1,250, 225).

If Ana carries a balance of less than $1,250, the MasterCard is the better choice. If her balance is more than $1,250, Ana should choose the Visa.

Not all systems of equations can be solved by graphing.

 **MINI-LAB**

Work with a partner. grid paper straightedge

Try This
Graph the system of equations $y = x + 2$ and $y = x - 5$.
• Make a function table for each equation.
• Graph the ordered pairs and draw the lines.

Talk About It
1. What type of lines are these?   **parallel**
2. Where do the lines meet?   **The lines do not meet.**
3. What conclusion can you make about the solution of this system of equations?   **There is no solution.**

Lesson 10-5 Graphing Systems of Equations **447**

---

2 TEACH

 Transparency 10-5B contains a teaching aid for this lesson.

In-Class Examples

For Example 1
Graph the system of equations $y = -2x + 1$ and $y = 3x - 4$. Then find the solution of the system. **(1, −1)**

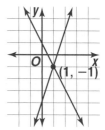

For Example 2
Baby Sharam weighed 6 pounds at birth and gained 2 pounds per month. Baby Beth weighed 9 pounds at birth and gained $\frac{1}{2}$ pound per month. Use a graph to determine when the babies' weights were the same. **3 mo**

Teaching Tip In Example 2, students should check their answers by replacing x and y in both of the equations with the values of their solutions.

Using the Mini-Lab Have students study the two equations, propose a third equation that might graph as a line parallel to the first two graphs, and test their proposal. Any equation of the form $y = x + C$ (where C represents a constant) will work.

Lesson 10-5 **447**

3 PRACTICE/APPLY

Check for Understanding

If students need additional practice or instruction after completing Exercises 1–7, one of these options may be helpful.
- Extra Practice, see p. 634
- Reteaching Activity
- *Study Guide Masters*, p. 81
- *Practice Masters*, p. 81
- Interactive Mathematics Tools Software

Assignment Guide

Core: 9–21 odd, 23–26
Enriched: 8–20 even, 21, 23–26
All: Self Test, 1–5

Exercise 22 asks students to advance to the next stage of work on the Chapter Project. Before they graph the equation, ask students to estimate whether or not the monkey will steal the bananas. You may wish to have students work in groups to complete this exercise.

Study Guide Masters, p. 81

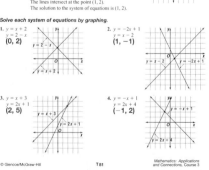

CHECK FOR UNDERSTANDING

Communicating Mathematics

Read and study the lesson to answer each question.

1. *Write* in your own words what is meant by a system of equations. **two or more equations**

2. *Identify* the solution of the system of equations graphed at the right. **(−1, 3)**

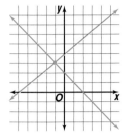

HANDS-ON MATH

3. *Sketch* a graph for a system of equations that has no solution. **See Answer Appendix.**

Guided Practice
4–6. See Answer Appendix for graphs.

Solve each system of equations by graphing. **6. no solution**

4. $y = 3 + x$
   $y = 2x + 5$ **(−2, 1)**

5. $y = x$
   $y = 4 - x$ **(2, 2)**

6. $y = -2x - 6$
   $y = -2x - 3$

7. **Economics** In March 1997, American exports were $71.4 billion, down $1.1 billion from February. Imports were $80.5 billion, up $1.2 billion from February. If the trend continues, exports for a month could be represented by $y = 71.4 - 1.1x$ and imports by $y = 80.5 + 1.2x$ where x is the number of months since March. Would there be a time after March when exports and imports are equal? If so, when? **See Answer Appendix.**

EXERCISES

Practice
8–18. See Answer Appendix for graphs.

Solve each system of equations by graphing.

8. $y = 2x + 1$
   $y = 7 - x$ **(2, 5)**

9. $y = 4 - 2x$
   $y = 2x$ **(1, 2)**

10. $y = x - 4$ **(4, 0)**
    $y = -4x + 16$

11. $y = 3x + 1$ **no solution**
    $y = 3x - 1$

12. $x + y = 7$
    $3x - y = 5$ **(3, 4)**

13. $y = 4x - 15$
    $y = x + 3$ **(6, 9)**

14. $y = -x - 1$
    $y = -2x + 4$ **(5, −6)**

15. $y = 4x - 13$
    $y = -\frac{1}{2}x + 5$ **(4, 3)**

16. $2x + y = 5$
    $x - y = 1$ **(2, 1)**

17. Graph the system $y = 4x - 1$ and $x + 2y = 16$. Find the solution of the system. **(2, 7)**

18. Find the solution of the system $y = 3.4x - 5$ and $11x - 5y = -5$. **(5, 12)**

19. *Write a Problem* involving a system of equations whose solution is (2, 8). **See students' work.**

Applications and Problem Solving

20. **Business** Part of making a business plan is finding the *break-even point*. The break-even point is the point where the cost of doing business and the revenue is the same. The break-even point is the intersection of the functions that describe the cost and the revenue. Suppose the cost function for a company is $y = 1,365 + 2.5x$ and the revenue function is $y = 6x$.
    a. What is the break-even point? **(390, 2,340)**
    b. If profit is the revenue minus the cost, what is the profit at the break-even point? **$0**

448 Chapter 10 Algebra: Graphing Functions

Reteaching the Lesson

Activity Have students work in pairs, each using a sheet of tracing paper. Students draw and number coordinate axes identically, then each graphs one equation of the system. Students then hold their sheets together and read the solution of their system of equations.

Error Analysis
Watch for students who solve systems of equations incorrectly because they draw graphs carelessly.
Prevent by insisting they use sharpened pencils and straightedges and that they check the solution in both equations.

21. **Aviation** The first nonstop coast to coast flight across the U.S. was made by a Fokker T-2 in 1923. It took about 27 hours. In 1987, the same flight took just over 4 hours in a Boeing 747. The Fokker averaged 93 miles per hour, and the Boeing plane averaged about 600 miles per hour. Suppose the flights had been made on the same day with a 5-hour lead for the Fokker. **a. See Answer Appendix.**

   a. Let y represent miles and x represent hours. The Fokker could be represented by $y = 93x$. The Boeing could be represented by $y = 600(x - 5)$. Graph the system of equations.

   b. Would the Boeing have passed the Fokker? If so, approximately when? **Yes; about 6 hours after the Fokker leaves.**

22. **Working on the CHAPTER Project** In a computer game, if you cross the path of the monkey, it will steal the bananas you have collected. The monkey is moving along $y = -3x + 23$. If the player is traveling along $y = 2x + 3$, will the monkey steal the bananas? If so, where? **yes; (4, 11)**

23. See Answer Appendix for graph.

23. **Critical Thinking** Graph $5x + y = 4$ and $10x + 2y = 8$ on the same grid.

   a. Describe the graphs. **They are the same line.**

   b. Does this system of equations have a solution? If so, what is the solution? **Yes; all of the points on the graph represent solutions.**

Mixed Review

24. **Test Practice** Which is the graph of $y = -3x$? *(Lesson 10-4)* **C**

 A B C D

25. Compute $\frac{5}{6} \cdot 3\frac{3}{7}$ mentally. *(Lesson 7-3)* $2\frac{6}{7}$

26. **Geography** There are 5,828,987 people who live in El Salvador. The area of the country is 8,124 square miles. Find the average number of people per square mile in El Salvador. *(Lesson 3-1)* **about 718 people per square mile**

CHAPTER 10 Mid-Chapter Self Test

1. Copy and complete the function table for $f(n) = 3n + 5$. *(Lesson 10-1)*

2. Use the table you made in Exercise 1 to graph $f(n) = 3n + 5$. *(Lesson 10-2)* **2–4. See Answer Appendix.**

3. Find four solutions of $y = 2.5x + 1$. *(Lesson 10-3)*

4. Graph $y = 6x + 5$. *(Lesson 10-4)*

5. Find the solution of the system $y = 14x - 23$ and $4x + 2y = 18$. *(Lesson 10-5)* **(2, 5)**

n	3n + 5	f(n)
-2	3(-2) + 5	-1
-1	3(-1) + 5	2
0	3(0) + 5	5
3	3(3) + 5	14
5	3(5) + 5	20

Extending the Lesson

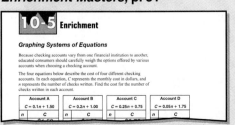

Enrichment Masters, p. 81

Activity Felicia is six years older than James but eight years younger than twice his age. Let x represent James' age and y represent Felicia's age. Write and graph equations to find James' and Felicia's ages. $y = x + 6$; $y = 2x - 8$; **James, 14; Felicia, 20**

4 ASSESS

Closing Activity

Writing State that the graphs of two equations intersect at the point whose coordinates are (4, 6). Ask students to write what they can conclude. **The values $x = 4$ and $y = 6$ satisfy both equations. The point (4, 6) is the solution for the system of equations.**

Chapter 10, Quiz B (Lessons 10-4 and 10-5) is available in the *Assessment and Evaluation Masters,* p. 267.

Mid-Chapter Test (Lessons 10-1 through 10-5) is available in the *Assessment and Evaluation Masters,* p. 266.

Mid-Chapter Self Test

The Mid-Chapter Self Test reviews the concepts in Lessons 10-1 through 10-5. Lesson references are given so students can review concepts not yet mastered.

Practice Masters, p. 81

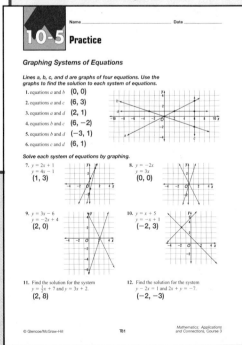

10-6A THINKING LAB Notes

Objective Solve problems by using a graph.

Recommended Pacing	
Standard	Day 8 of 14
Honors	Day 8 of 14
Block	Day 4 of 7

1 FOCUS

Getting Started Have students act out the situation presented at the beginning of the lesson. Then have them work in pairs to solve Exercises 1–3. Direct students to *USA Today, Newsweek,* or other newspapers and magazines as possible sources for graphs in Exercise 3.

2 TEACH

Teaching Tip You may want to use the graph to reinforce what students learned in Lesson 10-5. Ask them to analyze what the intersection of the graphs for Latin America/Caribbean and Africa means.

In-Class Example
Use the graph to estimate the lumber production in the Latin American/Caribbean forests in 1962 and 1990. Determine the percent increase or decrease in the lumber production rate for Latin America/Caribbean from 1962 to 1990. **about 22 million cubic meters in 1962 and about 63 million cubic meters in 1992; about 186% increase**

THINKING LAB PROBLEM SOLVING
10-6A Use a Graph
A Preview of Lesson 10-6

Wesley

Wesley and Kate are lab partners in their earth science class. Their assignment is to write a paper about the effects of cutting down the rainforests. Let's listen in!

Kate

This article I found says that the rain forests are being cleared for lumber production, cattle ranches, farms, dams, and highways.

We should have something about each of those topics in our paper. Mr. Hatano likes it when we back up what we say with numbers or graphs. Let's start with the lumber.

Okay. There's a graph in here that shows the lumber production for different parts of the world.

Let's put the graph in the paper and point out how much the annual lumber production has grown.

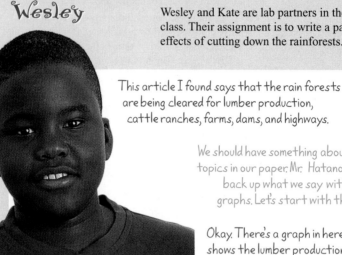

The graph shows that in 1962 it was about 70 million cubic meters. In 1991, it was up to about 160 million cubic meters.

So in 29 years, the production increased 90 million cubic meters. If we write that as a percent, it's 90 divided by 70 or a 129% increase!

THINK ABOUT IT

Work with a partner. 1–2. See margin.

1. **Tell** why the graphs of lumber production are not straight lines.
2. **Write** a sentence about the change of lumber production in Africa from 1962 to 1991.
3. **Find** a graph in a newspaper, a magazine, or on the Internet. Write a sentence explaining the information contained in the graph. **See students' work.**

450 Chapter 10 Algebra: Graphing Functions

■ Reteaching the Lesson ■

Activity Have students work in pairs to find additional examples of graphs in newspapers and magazines. For each example, ask students to describe the purpose of the graph and the meaning of each axis. Then have them mark one point on the graph and explain its significance.

Additional Answers

1. The change in the lumber production from one year to the next is not constant.
2. The lumber production increased slowly from 1962 to 1991.

ON YOUR OWN

4. The first step of the 4-step plan for problem solving is to *explore* the problem. **Describe** two things you could observe about lumber production by examining the graph. **See margin.**

5. **Look Ahead** Refer to the graph in Example 1 on page 452. Which grows faster as you move along the graph, the *x* values or the *y* values? **y values**

6. **Write a Problem** that can be answered using the graph on teen credit. **See students' work**

Source: Aragon Consulting Group

MIXED PROBLEM SOLVING

STRATEGIES
Look for a pattern.
Solve a simpler problem.
Act it out.
Guess and check.
Draw a diagram.
Make a chart.
Work backward.

Solve. Use any strategy.

7. **Money Matters** A 1997 auction of items owned by members of the *Beatles* was held in Tokyo, Japan. Paul McCartney's birth certificate was sold for $14,613 and sold again for $73,064. How much more did the second purchaser pay than the first? **$58,451**

8. **Earth Science** Graciela and Lisa's science paper is on chlorofluorocarbon (CFC) production. They found the graph below.

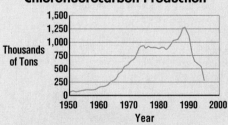

Source: Du Pont, Worldwatch estimates

 a. About how much greater was the CFC production in 1995 than in 1950?
 b. How much was the CFC production reduced from 1988 to 1995? **a-b. See margin.**

9. **Patterns** A number is doubled and then -19 is added to it. If the result is 11, what was the original number? **15**

10. **Health** The rate of death among liver cancer patients has dropped gradually. Choose the graph that shows a gradual decrease over time. **b**

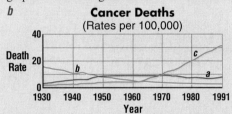

Source: National Center for Health Statistics

11. **Fashion** Eric, Julian, and Louisa each chose different stones for their class rings. Their stones are black onyx, blue sapphire, and purple amethyst. Julian's stone is not amethyst. Eric wished he had chosen onyx. Louisa's stone is blue. What stone does each student have?

12. **History** The Lighthouse of Pharos was one of the Seven Wonders of the World. It was 500 feet tall. The tallest structure in the United States is a television tower in Blanchard, North Dakota, which is 2,063 feet tall. About how many times taller is the television tower than the lighthouse? **11-12. See margin.**

13. **Test Practice** Approximately 36,460,000 people attended college football games in 1994. What is this number rounded to the nearest million? **A**
 A 36,000,000   B 1,000,000
 C 40,000,000   D 36,500,000

Lesson 10-6A **THINKING LAB** 451

3 PRACTICE/APPLY

Check for Understanding
Use the results from Exercise 3 to determine whether students comprehend how to read and interpret information presented in a graph.

Extra Practice If students need additional practice in problem solving, extra practice is available on the following pages.
- Use a Graph, see p. 635
- Mixed Problem Solving, see pp. 645–646

Assignment Guide
All: 4–13

ASSESS

Closing Activity
Writing Have students create word problems based on the graph showing Lumber Production in the Tropics.

Additional Answers
4. There is more production in Asia than in Africa or Latin America. The production increased more from 1962 to 1991 in Asia than it did in Africa or Latin America.
8a. about 250 thousand tons
8b. about 950 thousand tons
11. Eric—amethyst; Julian—onyx; Louisa—sapphire
12. about 4 times as tall

■ Extending the Lesson ■

Activity Ask students to list topics from current events or from their other classes that could be illustrated well with graphs. You may then wish to choose one topic for the class or small groups to present in graph form.

Thinking Lab 10-6A **451**

10-6 Lesson Notes

Instructional Resources
- *Study Guide Masters*, p. 82
- *Practice Masters*, p. 82
- *Enrichment Masters*, p. 82
- Transparencies 10-6, A and B
- *Science and Math Lab Masters*, pp. 65–68

 CD-ROM Program
- Resource Lesson 10-6

Recommended Pacing	
Standard	Day 9 of 14
Honors	Day 9 of 14
Block	Day 5 of 7

1 FOCUS

5-Minute Check
(Lesson 10-5)

Solve each system of equations by graphing.

1. $y = 3x - 1$
   $y = x + 5$
   $(3, 8)$

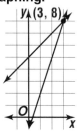

2. $y = -4x + 1$
   $y = 2x - 5$
   $(1, -3)$

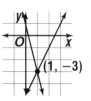

 The 5-Minute Check is also available on **Transparency 10-6A** for this lesson.

2 TEACH

 Transparency 10-6B contains a teaching aid for this lesson.

Reading Mathematics The formulas and equations in this lesson are more complex than students have seen thus far. Make sure they can interpret the operations expressed in each formula or equation. You may wish to have them write in words how they would evaluate each formula.

10-6 Graphing Quadratic Functions

What you'll learn
You'll learn to graph quadratic functions.

When am I ever going to use this?
You will use quadratic functions when you study gravity in physical science.

Word Wise
quadratic function

Do you dream of hitting the track in a Hot Rod? You may not have to wait. The National Hot Rod Association sponsors Junior Drag Racing Championships for up and coming drivers. Cars half the size of the pros' compete on a one-eighth mile track.

In 1997, the *Texas Raceway* team took the championship. The distance their car traveled can be described by the function $d(t) = \frac{1}{2}at^2$, where a is the rate of acceleration and t is time. This is a **quadratic function**. In a quadratic function, the greatest power is 2.

If you knew the acceleration and wanted to know how far the team's car had traveled at each second, you could evaluate $d(t)$ for several values of t. A better way is to graph the function.

Example — APPLICATION

Study Hint

Reading Math Notation other than $f(n)$ can be used for functions. The first letters of the quantities represented are often used. For example, in $d(t)$, d may represent distance.

① Sports Suppose the *Texas Raceway* car accelerated at a rate of 4.5 feet per second per second. Graph $d(t) = \frac{1}{2}at^2$ to find the number of feet that the car traveled after 6 seconds.

Substitute 4.5 for a in the function.

$d(t) = \frac{1}{2}at^2$

$d(t) = \frac{1}{2}(4.5)t^2$

$d(t) = 2.25t^2$

Now make a function table.

t	$d(t)$	$(t, d(t))$
0	$2.25(0)^2 = 0$	$(0, 0)$
2	$2.25(2)^2 = 9$	$(2, 9)$
4	$2.25(4)^2 = 36$	$(4, 36)$
6	$2.25(6)^2 = 81$	$(6, 81)$

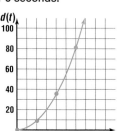

Now graph the ordered pairs. Notice that the points suggest a curve. Sketch a smooth curve to connect the points. At that rate of acceleration, the car will have traveled 81 feet after 6 seconds.

452 Chapter 10 Algebra: Graphing Functions

Motivating the Lesson

Hands-On Activity Explain to students that the *quad-* in *quadratic* means square. Then draw a large coordinate grid on the chalkboard or place one on the bulletin board. Ask students to use an 8-ft length of string or yarn to graph $y = -x^2$. Have them use tape on the chalkboard or thumbtacks on the bulletin board to affix the string and start with $x = 0$, $x = 1$, $x = -1$, $x = 2$, $x = -2$, and proceed.

Often the graphs of quadratic functions that represent real-life situations are only in the first quadrant because negative numbers are unreasonable. But the graph of a quadratic function or equation can be in any quadrant.

Example 2 Graph $y = -3x^2 + 6$.

Make a table and graph the ordered pairs.

x	y	(x, y)
−2	−3(−2)² + 6 = −6	(−2, −6)
−1.5	−3(−1.5)² + 6 = −0.75	(−1.5, −0.75)
−1	−3(−1)² + 6 = 3	(−1, 3)
0	−3(0)² + 6 = 6	(0, 6)
1	−3(1)² + 6 = 3	(1, 3)
1.5	−3(1.5)² + 6 = −0.75	(1.5, −0.75)
2	−3(2)² + 6 = −6	(2, −6)

The graphs of the ordered pairs suggest a downward curve. Connect the points with a smooth curve.

A graph with this shape is called a parabola. A parabola can also curve up, to the left, or to the right.

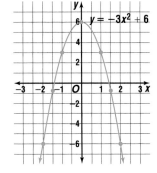

Study Hint

Technology You can use the $\boxed{x^2}$ key on your calculator to evaluate expressions with square numbers quickly. For example, for $x = 1.5$, press $1.5 \boxed{x^2}$.

In-Class Examples

For Example 1
The number of feet that an object falls in t seconds is given by the function $f(t) = 16t^2$. Graph the function to find how far an object falls in 4 seconds.

t	(t, f(t))
0	(0, 0)
1	(1, 16)
2	(2, 64)
3	(3, 144)
4	(4, 256)

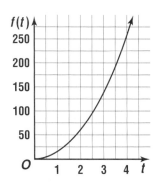

For Example 2
Graph $y = x^2 - 4$.

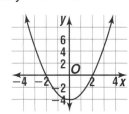

Teaching Tip After working Example 2, point out that a parabola that opens to the left or right would solve for x. For example, $x = y^2$ would open to the right, and $x = -y^2$ would open to the left.

CHECK FOR UNDERSTANDING

Communicating Mathematics

Read and study the lesson to answer each question.

1. **State** the highest power in a quadratic function. **2**

2. **Explain** why the graphs of some real-life functions such as in Example 1 use only the first quadrant of the graph.

3. **You Decide** Nicole says that $y = \frac{x^2}{3}$ is a quadratic function. Eduardo disagrees. Who is correct and why? **See Answer Appendix.**

2. There are no negative numbers in the domain or the range.

Guided Practice

4. Copy and complete the function table for $y = x^2 - 1$. Then graph the function. See Answer Appendix for graph.

x	y	(x, y)
−2	3	(−2, 3)
−1	0	(−1, 0)
0	−1	(0, −1)
1	0	(1, 0)
2	3	(2, 3)

Graph each quadratic function. **5–7. See Answer Appendix.**

5. $y = x^2$ 6. $f(x) = -2x^2$ 7. $y = 2x^2 + 1$

Lesson 10-6 Graphing Quadratic Functions **453**

3 PRACTICE/APPLY

Check for Understanding
If students need additional practice or instruction after completing Exercises 1–8, one of these options may be helpful.
- Extra Practice, see p. 635
- Reteaching Activity
- *Study Guide Masters*, p. 82
- *Practice Masters*, p. 82
- Interactive Mathematics Tools Software

Reteaching the Lesson

Activity Use the equation $y = x^2 + 2$. Ask: "If x is 0, what is y?" **2** "If x is 1, what is y?" **3** Continue in this manner using positive values for x, having students graph each ordered pair. Now have students fold their graphs along the y-axis and use symmetry to complete the graph.

Error Analysis
Watch for students who evaluate expressions like $3a^2$ as $(3a)^2$.
Prevent by reviewing the order of operations and writing expressions in expanded form ($3 \cdot a \cdot a$).

Lesson 10-6 **453**

Assignment Guide

Core: 9–23 odd, 25–28
Enriched: 10–22 even, 23–28

Additional Answers

9. (−3, −9), (−1, −1), (0, 0), (1.5, −2.25), (2, −4)
10. (−3, −13.5), (−1.5, −3.375), (0, 0), (2, −6), (4, −24)
11. (−2, 4), (−1.5, 0.5), (0, −4), (3, 14), (4, 28)

Study Guide Masters, p. 82

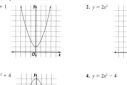

---

8. **Physical Science** If you have stood on a beach and watched a boat become tiny as it approaches the horizon, you may have wondered how far you could see. Standing on the shore, you can see d miles if your eyes are $f(d)$ feet above the ground. This is described by the function $f(d) = \frac{2}{3}d^2$.
   a. Graph the function $f(d) = \frac{2}{3}d^2$. **See Answer Appendix.**
   b. Find how many feet above the ground your eyes would need to be in order to see a friend waving from a small boat 4 miles from shore. **10 feet 8 inches**

EXERCISES

Practice

9–11. See margin for ordered pairs and see Answer Appendix for graphs.

Copy and complete each function table. Then graph the function.

9. $y = -x^2$

x	y	(x, y)
−3	−9	
−1	−1	
0	0	
1.5	−2.25	
2	−4	

10. $f(x) = -1.5x^2$

x	f(x)	(x, f(x))
−3	−13.5	
−1.5	−3.375	
0	0	
2	−6	
4	−24	

11. $f(n) = 2n^2 - 4$

n	f(n)	(n, f(n))
−2	4	
−1.5	0.5	
0	−4	
3	14	
4	28	

Graph each quadratic function. **12–20. See Answer Appendix.**

12. $y = 3x^2$
13. $f(x) = -2x^2 + 10$
14. $y = 5x^2 + 1$
15. $f(n) = 1.5n^2 - 1$
16. $f(x) = \frac{1}{2}x^2 + 2$
17. $y = -1.5x^2 - 1$
18. $y = -2x^2 + 6$
19. $y = 10 - x^2$
20. $f(n) = n^2 + n$

21. Consider the function $y = 3x^2 - 4$. Which ordered pairs from the set {(−2, 8), (−1, −7), (0, −4), (1, −1), (2, −8)} are solutions? **(−2, 8), (0, −4), (1, −1)**

22. Determine which ordered pairs from the set {(−2, 3), (−1.5, −2.75), (2, 1), (3.5, −7.75), (0, 1)} are solutions of $f(n) = -n^2 + n + 1$. **(−1.5, −2.75), (3.5, −7.75), (0, 1)**

Applications and Problem Solving

23. **Fireworks** Is going to see fireworks one of your family's favorite Independence Day activities? You can estimate the height of a fireworks rocket using a quadratic function. If the rocket is fired from the ground with an initial velocity of 40 meters per second, then the height of the rocket after t seconds is $h = 40t - 4.9t^2$.
    a. Graph the height function. **See margin.**
    b. Estimate the height of the rocket after 3.5 seconds. **about 80 m**
    c. What is the maximum height the rocket reaches? **about 81 m**

24. **Aviation** The formula $a = 0.66d^2$ represents the number of miles d that can be seen when flying at a height of a feet.
    a. Graph the function. **See margin.**
    b. In 1988, the passengers and crew of the *Nashua Number One* set a record for the most people on an untethered balloon flight. If the balloon was 328 feet above the ground, use your graph to estimate how far the passengers and crew could see. **about 22 miles**

454 Chapter 10 Algebra: Graphing Functions

Additional Answers

23a.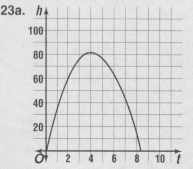

24a.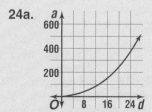

25. Critical Thinking Graph $y = x^2$, $y = x^2 + 2$, and $y = x^2 + (-2)$ on the same coordinate plane. *See margin.*

   a. How do the graphs of $y = x^2 + 2$ and $y = x^2 + (-2)$ differ from the graph of $y = x^2$? *a–b. See margin.*

   b. Write a statement comparing the graph of $y = x^2 + c$ to the graph of $y = x^2$ depending on the value of c.

Mixed Review

26. Which ordered pair best represents the intersection of lines ℓ and k? *(Lesson 10-5)* **B**

   A $(-6, 5)$ **B** $(-5, 6)$
   C $(6, -5)$ **D** $(5, -6)$

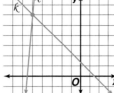

27. Money Matters When Joaquín filled his car with gasoline on Monday, the price was $1.15 per gallon. The next Monday when he went to fill the tank the price was $1.19 per gallon. Find the percent of change in gasoline prices. *(Lesson 8-5)* **3.5%**

28. Algebra Solve $158 = s - 36$. Check your solution. *(Lesson 1-4)* **194**

Let the Games Begin

Parabola Hit or Miss

Get Ready This game is for two players.

- two spinners
- 10 index cards
- grid paper

Get Set Copy the equations below onto index cards, one per card. Shuffle the cards and place them facedown in a pile.

$y = -x^2 + 5$ $y = 1 - 2x^2$ $y = -\frac{1}{2}x^2 + 14$ $y = x^2 + 3$

$y = 4x^2 - 12$ $y = 1.5x^2$ $y = x^2$ $y = 3x^2 + 20$

$y = \frac{1}{2}x^2 - 10$ $y = -2 - x^2$

Label equal sections of two spinners each with -6, -4, -3, -1, 0, 2, 5, and 6. Decide which spinner names x-coordinates and which names y-coordinates.

Go
- One player spins both spinners and plots the ordered pair on a coordinate grid.
- The second player selects the top card from the pile and graphs the quadratic function.
- If the graphed point lies within the parabola, it's a hit, and the second player gets a point. If the graphed point lies outside of the parabola, it's a miss, and the first player gets a point.
- Exchange roles and continue. The player with the most points after all of the cards have been selected wins.

interNET CONNECTION Visit www.glencoe.com/sec/math/mac/mathnet for more games.

10-7 Lesson Notes

Instructional Resources

- *Study Guide Masters*, p. 83
- *Practice Masters*, p. 83
- *Enrichment Masters*, p. 83
- *Transparencies 10-7, A and B*
- *Assessment and Evaluation Masters*, p. 268
- *Hands-On Lab Masters*, p. 17
- CD-ROM Program
  - Resource Lesson 10-7
  - Interactive Lesson 10-7

Recommended Pacing	
Standard	Day 10 of 14
Honors	Day 10 of 14
Block	Day 5 of 7

1 FOCUS

5-Minute Check
(Lesson 10-6)

Graph each quadratic function.

1. $y = 2 - x^2$

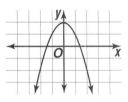

2. $y = x^2 - 3$

 The 5-Minute Check is also available on **Transparency 10-7A** for this lesson.

Motivating the Lesson

Communication Ask students to read the opening paragraph of the lesson. Explain that translations are used in other professions also; for example, amusement park designers and car manufacturers use motion geometry.

10-7 Integration: Geometry
Translations

What you'll learn
You'll learn to graph translations on a coordinate plane.

When am I ever going to use this?
Animators use translations to create moving images for movies.

Word Wise
translation

LOOK BACK
Refer to Lesson 5-7 for more information on translations.

Have you conquered the castle in *Super Mario 64*? If you play video games like *Mario*, you have seen an application of *motion geometry*. The game designers use computers to move points on the screen, or pixels, to create the illusion of movement. In Chapter 5, you learned that a sliding motion is called a **translation**.

HANDS-ON MINI-LAB

Work with a partner. 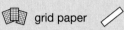 grid paper   straightedge
 colored pencils   unlined paper   scissors

Try This

- Graph $\triangle RST$ with vertices $R(-6, 1)$, $S(-1, -1)$, $T(-3, 3)$ on a coordinate plane.
- Trace $\triangle RST$ and cut it out.
- Place the cutout over $\triangle RST$. Translate the cutout 3 units right. Then translate the cutout 2 units down. Trace the cutout with a colored pencil.

Talk About It

1. Write the ordered pairs that describe the locations of the three vertices of $\triangle RST$ after the translations. $R'(-3, -1)$, $S'(2, -3)$, $T'(0, 1)$
2. How do the coordinates of the vertices of the translation compare to the coordinates of the original vertices?

2. The *x*-coordinates are 3 more, and the *y*-coordinates are 2 less.

In a coordinate plane, a translation down or to the left is negative. A translation up or to the right is positive. The translation in the Mini-Lab could be written as $(3, -2)$. If we add 3 to the *x*-coordinate and -2 to the *y*-coordinate of each vertex of the original triangle, the results are the coordinates of the vertices of the translated triangle.

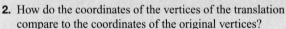

Translation	**Words:**	To translate a point as described by an ordered pair, add the coordinates of the ordered pair to the coordinates of the point.
	Symbols:	**Arithmetic:** $(4, -3)$ translated by $(-1, 2)$ becomes $(3, -1)$.
		Algebra: (x, y) translated by (a, b) becomes $(x + a, y + b)$.

456 Chapter 10 Algebra: Graphing Functions

Example 1

The vertices of △ABC are A(4, −3), B(0, 2), and C(5, 1). Graph △ABC. Then graph the triangle after a translation 4 units left and 3 units up.

The location of a point after being moved is often written using a prime. So the new coordinates of A are written as A′, B as B′, and C as C′.

vertex		4 left, 3 up	translation	
A(4, −3)	+	(−4, 3)	→	A′(0, 0)
B(0, 2)	+	(−4, 3)	→	B′(−4, 5)
C(5, 1)	+	(−4, 3)	→	C′(1, 4)

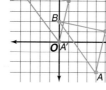

The coordinates of the vertices of the translated triangle are A′(0, 0), B′(−4, 5), and C′(1, 4).

Graph A′, B′, and C′ and draw the translated triangle.

Study Hint

Reading Math A′ is read as A prime. If point A has been moved twice, the new point is labeled A″, which is read as A double prime.

Sometimes you may need to know how a figure was moved. You can use your knowledge of translations and algebra to find the ordered pair that describes the translation.

Example 2

INTEGRATION

Algebra Parallelogram MNPQ has vertices M(−3, 0), N(−1, −1), P(0, 2), and Q(−2, 3). Use an equation to find the ordered pair that describes the translation if M′ has coordinates (2, −1). Then graph parallelogram M′N′P′Q′.

Explore You know the coordinates of the vertices of MNPQ and the coordinates of M′. You need to find the ordered pair for the translation and graph M′N′P′Q′.

Plan Use an equation to compare each coordinate of M and M′. Find the ordered pair for the translation. Then use the ordered pair to find the coordinates of N′, P′, and Q′.

Solve Let (a, b) represent the ordered pair for the translation.

M has coordinates (−3, 0) and M′ has coordinates (2, −1), so we can say M(−3, 0) + (a, b) → M′(2, −1). Write an equation for each coordinate.

x-coordinate	y-coordinate
−3 + a = 2	0 + b = −1
a = 5	b = −1

The ordered pair for the translation is (5, −1). a is positive, so the first move is 5 units right. b is negative, so the next move is 1 unit down.

(continued on the next page)

Lesson 10-7 Integration: Geometry Translations **457**

Multiple Learning Styles

Visual/Spatial Ask students to draw a small picture of something that moves, such as a car, train, person, or animal. Have them cut out the drawing and place it on a coordinate plane. Students can then choose a direction for the object and plot its movement on the plane. Have them write one or two sentences to explain the movement of their drawing and the coordinates used.

2 TEACH

Transparency 10-7B contains a teaching aid for this lesson.

Using the Mini-Lab Tell students that while every point on the triangle must be translated, concentrating on the vertices ensures that each side of the triangle moves in the appropriate direction. Have them mark the horizontal and vertical path of each translated vertex.

Teaching Tip In Example 1, remind students that a vertex is the point where two sides of a figure come together. A triangle has three vertices.

In-Class Examples

For Example 1
The vertices of △ABC are A(3, −2), B(0, 0), and C(−5, −6). Graph △ABC. Then graph the triangle after a translation 2 units right and 6 units up.

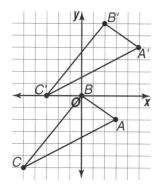

For Example 2
Triangle RST has vertices R(−1, 2), S(0, −2), and T(8, 1). Use an equation to find the ordered pair that describes the translation if R′ has coordinates (−2, 5). Then graph △R′S′T′.
(−1, 3)

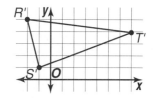

Lesson 10-7 **457**

3 PRACTICE/APPLY

Check for Understanding

If students need additional practice or instruction after completing Exercises 1–7, one of these options may be helpful.
- Extra Practice, see p. 635
- Reteaching Activity
- *Study Guide Masters*, p. 83
- *Practice Masters*, p. 83

Assignment Guide

Core: 9–17 odd, 19–22
Enriched: 8–16 even, 17–22

Additional Answers

3.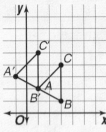

4.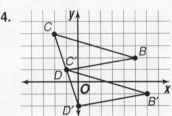

Study Guide Masters, p. 83

Find the coordinates of N', P', and Q'.

$N(-1, -1) + (5, -1) \rightarrow N'(4, -2)$
$P(0, 2) + (5, -1) \rightarrow P'(5, 1)$
$Q(-2, 3) + (5, -1) \rightarrow Q'(3, 2)$

Now graph $M'N'P'Q'$.

Examine Check your work by graphing $MNPQ$ and translating each vertex by $(5, -1)$. Do the points correspond to $M'N'P'Q'$?

CHECK FOR UNDERSTANDING

Communicating Mathematics

Read and study the lesson to answer each question.

1. **Describe** the type of movement that results from a translation. **slide**
2. **Write** a sentence to describe the translation named by $(-3, 4)$.
3. **Graph** triangle ABC with vertices $A(1, 2)$, $B(3, 1)$, and $C(3, 4)$. Show the position of the triangle after a translation of $(-2, 1)$. **See margin.**

   2. Move 3 units left and 4 units up.

Guided Practice Find the coordinates of the vertices of each figure after the translation described. Then graph the figure and its translation. **4–5. See margin for graphs.**

4. $B'(6, -1)$
   $C'(-1, 1)$
   $D'(0, -2)$

4. $\triangle BCD$ with vertices $B(5, 2)$, $C(-2, 4)$, and $D(-1, 1)$, translated by $(1, -3)$

5. rectangle $HJKL$ with vertices $H(1, 3)$, $J(4, 0)$, $K(3, -1)$, and $L(0, 2)$, translated by $(3, -4)$ $H'(4, -1)$, $J'(7, -4)$, $K'(6, -5)$, $L'(3, -2)$

6. Triangle RST has vertices $R(-2, 3)$, $S(-3, 0)$, and $T(1, 1)$. When translated, R' has coordinates $(3, 5)$.
   a. Describe the translation using an ordered pair. **(5, 2)**
   b. Find the coordinates of S' and T'. **$S'(2, 2)$, $T'(6, 3)$**

7. **Geometry** If a figure can be translated onto another so that all of the vertices correspond, the figures are congruent. To determine whether two figures are congruent, find the ordered pair that translates each vertex of one figure to the corresponding vertex of the other. If all of these ordered pairs are the same, the figures are congruent. Is $\triangle ABC$ with vertices $A(-2, 5)$, $B(9, 0)$, and $C(-2, -2)$ congruent to $\triangle XYZ$ with vertices $X(4, 4)$, $Y(15, -1)$, and $Z(4, -3)$? **yes**

EXERCISES

Practice Find the coordinates of the vertices of each figure after the translation described. Then graph the figure and its translation. **8–11. See Answer Appendix.**

8. $\triangle EFG$ with vertices $E(-5, -2)$, $F(-2, 3)$, and $G(2, -3)$, translated by $(6, 3)$
9. $\triangle PQR$ with vertices $P(0, 0)$, $Q(-3, -4)$, and $R(1, 3)$, translated by $(-6, 3)$
10. square $SQAR$ with vertices $S(2, 1)$, $Q(4, 3)$, $A(2, 5)$, and $R(0, 3)$, translated by $(-1, 3)$
11. rectangle $WXYZ$ with vertices $W(-4, 1)$, $X(2, 4)$, $Y(3, 2)$, and $Z(-3, -1)$, translated by $(-1, 4)$

458 Chapter 10 Algebra: Graphing Functions

■ **Reteaching the Lesson** ■

Activity Have students write a problem that states the coordinates of the vertices of a figure and a translation. Then have them exchange problems and graph the translated figure.

Additional Answer

5.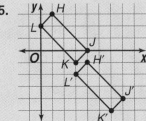

12–13. See Answer Appendix.

12. parallelogram *FGHJ* with vertices *F*(7, 5), *G*(5, 2), *H*(7, 0), and *J*(9, 3), translated by (−4, −2)

13. pentagon *ABCDE* with vertices *A*(−2, −1), *B*(0, −1), *C*(1, 1), *D*(−1, 3), and *E*(−3, 1), translated by (−2, 1)

14. Triangle *NMP* has vertices *M*(4, 2), *N*(−8, 0), and *P*(6, 7). When translated, *M′* has coordinates (−2, 4).
   a. Describe the translation using an ordered pair. **(−6, 2)**
   b. Find the coordinates of *N′* and *P′*. **N′(−14, 2), P′(0, 9)**

15. The vertices of parallelogram *QRST* are *Q*(−10, 2), *R*(−4, 0), *S*(6, 2), and *T*(0, 4). *S′* has coordinates (8, −3).
   a. Describe the translation using an ordered pair. **(2, −5)**
   b. Find the coordinates of *Q′*, *R′*, and *T′*. **Q′(−8, −3), R′(−2, −5), T′(2, −1)**

16b. A′(2, −1), B′(−3, −3), C′(−2, −7), D′(2, −7), F′(5, −2)

16. Hexagon *ABCDEF* has vertices *A*(0, 2), *B*(−5, 0), *C*(−4, −4), *D*(0, −4), *E*(6, −2), and *F*(3, 1). When translated, *E′* has coordinates (8, −5).
   a. Describe the translation using an ordered pair. **(2, −3)**
   b. Find the coordinates of *A′*, *B′*, *C′*, *D′* and *F′*.

Applications and Problem Solving

17. Sample answer: In early years the center moved west, then it moved south and west.

17. **Demographics** The center of population is the point around which a nation's population is equally distributed. That is, half of the population lives north of this point and half lives south; half lives east of it and half lives west. Describe the translations of the median center of the U.S. population.

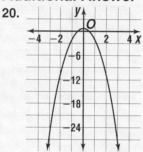
Source: Statistical Abstract of the United States, 1994

18. **Music** Composers refer to *transposing* music from one key to another. When music is transposed, each note is translated the same distance up or down the musical scale.
   a. The music at the top is from Mozart's *Clarinet Quintet in A Major* in the key of C major. Copy the music at the bottom and translate the music to the key of A major. The first note is done for you.
   b. How would you describe the transformation in part a mathematically? **−2**

19. The final position of the figure is the same as the original position of the figure.

19. **Critical Thinking** A figure is translated by (4, −1). Then the result is translated by (−4, 1). Without graphing, what is the final position of the figure?

Mixed Review

20. **Algebra** Graph $y = -3x^2 + 1$. *(Lesson 10-6)* **See margin.**

21. **Test Practice** Which is the best estimate for $\sqrt{317}$? *(Lesson 9-2)* **D**
    A 15   B 16   C 17   D 18   E 19

22. Express $35.70 for 6 pounds as a unit rate. *(Lesson 3-1)* **$5.95/1 pound**

Lesson 10-7 Integration: Geometry Translations 459

4 ASSESS

Closing Activity
Modeling Have students place congruent triangles at different positions on a coordinate axis and name ordered pairs that describe the translation from one position to the other.

Chapter 10, Quiz C (Lessons 10-6 and 10-7) is available in the *Assessment and Evaluation Masters*, p. 268.

Additional Answer
20.

Practice Masters, p. 83

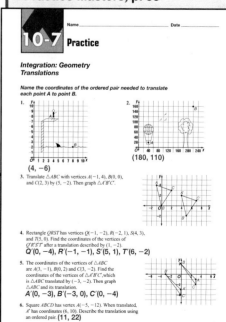

Extending the Lesson

Enrichment Masters, p. 83

Activity Have students brainstorm in small groups to make a list of applications for the translation of figures. **Sample answer: A single shape is repeated through translation to create a wallpaper pattern.**

10-8 Lesson Notes

Instructional Resources
- *Study Guide Masters*, p. 84
- *Practice Masters*, p. 84
- *Enrichment Masters*, p. 84
- Transparencies 10-8, A and B
- *Hands-On Lab Masters*, p. 77
- *School to Career Masters*, p. 36
- CD-ROM Program
  - Resource Lesson 10-8
  - Interactive Lesson 10-8

Recommended Pacing	
Standard	Day 11 of 14
Honors	Day 11 of 14
Block	Day 6 of 7

1 FOCUS

5-Minute Check
(Lesson 10-7)

△EFG is translated by (−3, 2). Find the coordinates of the vertices of △E'F'G'.
1. E(3, 7)   (0, 9)
2. F(−1, 3) (−4, 5)
3. G(1, −2) (−2, 0)
4. Point N(−1, 4) is translated to point N'(−2, 0). Describe the translation using an ordered pair. (−1, −4)
5. Point S(5, −3) is translated to point S'(−1, 5). Describe the translation using an ordered pair. (−6, 8)

The 5-Minute Check is also available on **Transparency 10-8A** for this lesson.

Motivating the Lesson

Hands-On Activity Give students various symmetrical shapes. Have them place a mirror perpendicular to the shape and move it to a position where the image on the paper plus the image in the mirror look like the original image. Explain that the edge of the mirror represents a line of symmetry.

10-8 Integration: Geometry
Reflections

What you'll learn
You'll learn to graph reflections on a coordinate plane.

When am I ever going to use this?
You will study reflections when you study light and sound in physical science.

LOOK BACK
Refer to Lesson 5-4 for more information on reflections.

The techniques for making Hopi pottery have been passed down from mother to daughter for centuries. Many of the patterns are made by repeated transformations. Dextra Quotskuyva designed the seed jar at the right to be *symmetric*.

Something is symmetric if you can fold the design so that the halves coincide. In Chapter 5, you learned that this fold line is called the *line of symmetry*. One half of the design is the *reflection* of the other.

Every point of the original figure has a corresponding point on the other side of the line of symmetry.

HANDS-ON MINI-LAB

Work with a partner.   grid paper   straightedge   geomirror

Try This

Step 1 Graph △PQR with vertices P(2, 1), Q(7, 3), and R(3, 5).

Step 2 Place a geomirror on the x-axis. Use the reflection of △PQR to draw △P'Q'R'.

Step 3 Now place the geomirror on the y-axis and draw the reflection of △PQR as △P"Q"R".

Talk About It

1. Fold the grid along the x-axis. What do you notice about △PQR and △P'Q'R'? **They are the same size and shape.**
2. Compare the coordinates of P, Q, and R to the coordinates of P', Q' and R'. What pattern do you notice?
3. Make a conjecture about the result for △PQR and △P"Q"R" if you fold the grid again, this time along the y-axis. Fold to verify your conjecture.
4. Write a sentence comparing the coordinates of P, Q, and R to the coordinates of P"Q" and R".

2. They have the same x value, but opposite y values.
3. Sample answer: They will be the same size and shape and have the same y value, but opposite x values.
4. They have the same y value, but opposite x values.

460   Chapter 10   Algebra: Graphing Functions

In the Mini-Lab, you reflected △PQR over the x-axis to draw △P'Q'R'. For these figures, the x-axis is the line of symmetry. The coordinates of the vertices have a special relationship.

Reflection over the x-axis		
Words:	To reflect a point over the x-axis, use the same x-coordinate and multiply the y-coordinate by −1.	
Symbols:	Arithmetic:	(4, −3) becomes (4, 3).
	Algebra:	(x, y) becomes (x, −y).

Example 1

Reflect parallelogram ABCD with vertices A(1, 2), B(0, 0), C(−5, 0), and D(−4, 2) over the x-axis.

Find the coordinates of each vertex after the reflection.

$A(1, 2) \rightarrow (1, 2 \cdot -1) \rightarrow A'(1, -2)$
$B(0, 0) \rightarrow (0, 0 \cdot -1) \rightarrow B'(0, 0)$
$C(-5, 0) \rightarrow (-5, 0 \cdot -1) \rightarrow C'(-5, 0)$
$D(-4, 2) \rightarrow (-4, 2 \cdot -1) \rightarrow D'(-4, -2)$

Graph ABCD and A'B'C'D'.

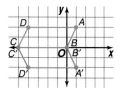

In the Mini-Lab, you also investigated the relationship for the coordinates of points reflected over the y-axis.

Reflection over the y-axis		
Words:	To reflect a point over the y-axis, multiply the x-coordinate by −1 and use the same y-coordinate.	
Symbols:	Arithmetic:	(4, −3) becomes (−4, −3).
	Algebra:	(x, y) becomes (−x, y).

You can use a reflection over the y-axis to help graph some quadratic functions.

Example 2 INTEGRATION

Algebra The graph of $y = x^2 + 4$ is symmetric about the y-axis. Complete a function table and reflect the points over the y-axis to complete the graph.

Find each function value.

x	y	(x, y)
0	$(0)^2 + 4$	(0, 4)
1	$(1)^2 + 4$	(1, 5)
2	$(2)^2 + 4$	(2, 8)
3	$(3)^2 + 4$	(3, 13)
4	$(4)^2 + 4$	(4, 20)

(continued on the next page)

Lesson 10-8 Integration: Geometry Reflections **461**

---

2 TEACH

 Transparency 10-8B contains a teaching aid for this lesson.

Using the Mini-Lab Geomirrors are also known as MIRAs. Point out that the x-coordinates of vertices P, Q, and R do not change when they are reflected over the x-axis. However, their y-coordinates do change.

Teaching Tip In Example 1, remind students that the term *reflection* refers to the image after it has been flipped over a line of symmetry.

In-Class Examples

For Example 1
Reflect △ABC with vertices A(4, −3), B(−3, −2) and C(2, −5) over the x-axis.

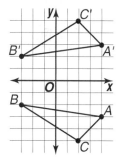

For Example 2
The graph of $y = x^2 - 3$ is symmetric about the y-axis. Complete the function table and reflect the points over the y-axis to complete the graph.

x	y
0	−3
1	−2
2	1
3	6
4	13

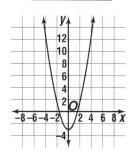

3 PRACTICE/APPLY

Check for Understanding

If students need additional practice or instruction after completing Exercises 1–6, one of these options may be helpful.
- Extra Practice, see p. 636
- Reteaching Activity
- *Study Guide Masters*, p. 84
- *Practice Masters*, p. 84

Additional Answers

1. Answers may vary. Sample answer: In a mirror, the object seen is the flip of the real object. In a geometric reflection, what is on one side of a line is flipped from what is on the other side of the line.

2. Sample answer: Reflection over the *x*-axis: same *x*, opposite *y*; Reflection over the *y*-axis: same *y*, opposite *x*

Study Guide Masters, p. 84

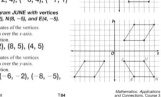

Now reflect each point over the *y*-axis.

$(0, 4) \rightarrow (0 \cdot -1, 4) \rightarrow (0, 4)$
$(1, 5) \rightarrow (1 \cdot -1, 5) \rightarrow (-1, 5)$
$(2, 8) \rightarrow (2 \cdot -1, 8) \rightarrow (-2, 8)$
$(3, 13) \rightarrow (3 \cdot -1, 13) \rightarrow (-3, 13)$
$(4, 20) \rightarrow (4 \cdot -1, 20) \rightarrow (-4, 20)$

Finally graph each point and its reflection. Complete by drawing a smooth curve.

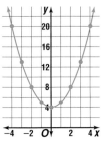

CHECK FOR UNDERSTANDING

Communicating Mathematics

Read and study the lesson to answer each question. 1–2. See margin.

1. *Write* a sentence explaining how a mirror reflection is related to a geometric reflection.

2. *Develop* a way to remember how to reflect a point over one of the axes.

HANDS-ON LAB

3. *Sketch* the figure at the right. Then use a geomirror to draw all of the lines of symmetry.

Guided Practice

4. Name the line of symmetry for the pair of figures. *y*-axis

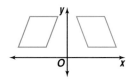

5a–c. See Answer Appendix.

5. Graph $\triangle ABC$ with vertices $A(-2, 3)$, $B(5, 1)$, and $C(4, 5)$.
   a. Reflect $\triangle ABC$ over the *x*-axis
   b. Reflect $\triangle ABC$ over the *y* axis.
   c. Is $\triangle ABC$ congruent to its reflections? That is, are the lengths of corresponding sides in the ratio 1 to 1? Justify your answer.

6. Yes; see students' work.

6. **Architecture** Georgian architecture was the major architectural style in England and the American colonies during the 1700s. Study the photo of the Georgian style Blenheim Palace at the right. Does it exhibit symmetry? If so, sketch the building and draw the line of symmetry.

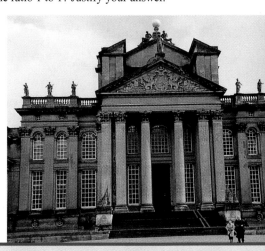

462 Chapter 10 Algebra: Graphing Functions

Reteaching the Lesson

Activity Have students work in pairs. One student graphs a point and writes its coordinates. The other rolls a number cube to determine an axis (even = *x*-axis, odd = *y*-axis.), calls out the axis, names the coordinates of the point's reflection over that axis, and then graphs the reflection.

Error Analysis
Watch for students who find coordinates of transformations incorrectly.
Prevent by having students graph transformed images of figures to be sure that they are congruent to the original figures.

EXERCISES

Practice Name the line of symmetry for each pair of figures.

7.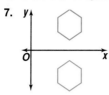
   x-axis

8.
   y-axis

9.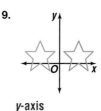
   y-axis

Graph each figure. Then draw its reflections over the *x*-axis and over the *y*-axis. 10–12. **See Answer Appendix.**

10. $\triangle MNP$ with vertices $M(-8, 4)$, $N(-3, 8)$, and $P(-2, 2)$

11. square $WXYZ$ with vertices $W(-3, 2)$, $X(1, -2)$, $Y(-3, -6)$, and $Z(-7, -2)$

12. trapezoid $PQRS$ with vertices $P(4, -2)$, $Q(8, -2)$, $R(8, 5)$, and $S(2, 5)$

13. Graph rectangle $RECT$ with vertices $R(-3, 3)$, $E(3, 3)$, $C(3, -3)$, and $T(-3, -3)$.
    a. Reflect $RECT$ over the *x*-axis. **a–c. See Answer Appendix.**
    b. Reflect $RECT$ over the *y*-axis on the same coordinate plane.
    c. What do you observe about the three graphs?

14. Graph $A(2, 6)$ and $B(5, 5)$.
    a. Reflect A and B over the *y*-axis and graph A' and B'. **See Answer Appendix.**
    b. Complete $A'ABB'$. What type of figure is formed? **trapezoid**

Applications and Problem Solving

15. **Art** Refer to the beginning of the lesson. Draw your own pottery design using one or more reflections. **See students' work.**

16. **Life Science** If you look at your face in the mirror and imagine a vertical line drawn between your eyes, you will find that most of your features are symmetric. Studies show that people of all different cultures find faces that are symmetric to be attractive. Find a photograph of a friend or a famous person. What features of his or her face are symmetric? Are there features that are not symmetric? **See students' work.**

17. **Critical Thinking** The point $D(4, 0)$ is its own reflection with respect to the *x*-axis. Write the coordinates of a point that is its own reflection with respect to the *y*-axis. **See margin.**

Mixed Review

18. **Test Practice** What are the coordinates of $W(-6, 3)$ if it is translated by $(2, -1)$? *(Lesson 10-7)* **A**
    A $(-4, 2)$ B $(-8, 4)$ C $(-4, 4)$ D $(-7, 5)$

19. **Advertising** How many times do you see commercials before you buy the product? The results of a survey on how many times people watch infomercials before they buy are shown. Make a circle graph of this data. *(Lesson 4-2)* **See Answer Appendix.**

Times Seen Before We Buy	Percent
Once	27%
Twice	31%
Three times	18%
Four times	9%
Five or more times	15%

Source: National Infomercial Marketing Assn.

Lesson 10-8 Integration: Geometry Reflections **463**

Extending the Lesson

Enrichment Masters, p. 84

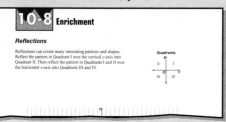

Activity Ask students to find examples of company logos with one or more lines of symmetry. Have students copy the logos, draw the lines of symmetry, and display the drawings on the bulletin board.

Assignment Guide

Core: 7–15 odd, 17–19
Enriched: 8–14 even, 15–19

4 ASSESS

Closing Activity

Speaking Name a point and say "*x*-axis" or "*y*-axis." Have students name the coordinates of the reflection of the point over the axis named.

Additional Answer

17. Any ordered pair with 0 as its *x*-coordinate is its own reflection over the *y*-axis.

Practice Masters, p. 84

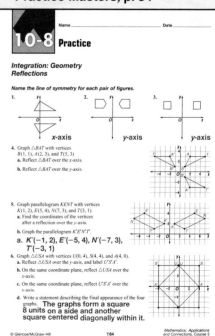

Lesson 10-8 **463**

10-9 Lesson Notes

Instructional Resources

- *Study Guide Masters,* p. 85
- *Practice Masters,* p. 85
- *Enrichment Masters,* p. 85
- Transparencies 10-9, A and B
- *Assessment and Evaluation Masters,* p. 268
- *Classroom Games,* pp. 31–36
- *Diversity Masters,* p. 36
- CD-ROM Program
  - Resource Lesson 10-9
  - Interactive Lesson 10-9

Recommended Pacing	
Standard	Day 12 of 14
Honors	Day 12 of 14
Block	Day 6 of 7

1 FOCUS

5-Minute Check
(Lesson 10-8)

△*ABC* has vertices *A*(−3, 1), *B*(0, −2), and *C*(5, −1). Find the coordinates of the vertices after a reflection over the given axis.

1. *x*-axis *A*′(−3, −1), *B*′(0, 2), *C*′(5, 1)
2. *y*-axis *A*′(3, 1), *B*′(0, −2), *C*′(−5, −1)

The 5-Minute Check is also available on **Transparency 10-9A** for this lesson.

Motivating the Lesson

Communication Explain to students that rotations are used in logos, as well as on flags, to give the illusion of a circle, which suggests completeness.

2 TEACH

Transparency 10-9B contains a teaching aid for this lesson.

Using the Mini-Lab This lab generates a great deal of information. Stress the importance of a carefully planned and organized approach to data recording.

464 Chapter 10

10-9

What you'll learn
You'll learn to graph rotations on a coordinate plane.

When am I ever going to use this?
You can use rotations to create quilt patterns.

Word Wise
rotation

Cultural Kaleidoscope

Nations that have a common history often share the same colors in their flag. For example, the flags of the Arab nations contain black, green, red, and white.

3. They are switched and then the *x*-coordinate is multiplied by −1.
5. Both coordinates are multiplied by −1.

Integration: Geometry
Rotations

Have you ever studied vexillology *(vek suh LAHL uh jee)*? Vexillology is the study of the history and symbolism of flags. Egyptian art shows us that the first flags were symbols attached to the tops of poles and carried into battle. Cloth flags were first made in China about 3000 B.C. The colors and symbols of national flags are often chosen to represent heritage or an event in history.

The symbols on the flags of Anguilla and the Isle of Man can be made using a **rotation**. A rotation moves a figure about a central point.

Anguilla   Isle of Man

HANDS-ON MINI-LAB

Work in groups of two or three.   protractor

Try This

- The graph models the blades of a fan. Record the coordinates of each lettered point.
- Measure ∠*MON*, ∠*NOP*, ∠*POQ*, and ∠*QOM*. Record your measurements.
- As the fan blades turn, each point of one blade will occupy the previous location of the corresponding point on another blade. Make a list of the corresponding points as the blade rotates.

Talk About It

1. Which direction is the fan turning? **counterclockwise**
2. How many degrees did the blade rotate to move from point *M* to point *N*? **90°**
3. How do the coordinates of *M* compare to the coordinates of *N*?
4. How many degrees did the blade rotate to move from point *M* to point *P*? **180°**
5. Compare the coordinates of *M* and *P*.

464 Chapter 10 Algebra: Graphing Functions

Investigations for the Special Education Student

This blackline-master booklet helps you plan for the needs of your special education students by providing long-term projects along with teacher notes. Investigation 13, *How Does Your Garden Grow?*, may be used with this chapter.

As you discovered in the Mini-Lab, the coordinates of points rotated 90° and 180° are related.

Rotation of 90° counterclockwise	Words:	To rotate a figure 90° counterclockwise about the origin, switch the coordinates of each point and then multiply the new first coordinate by −1.
	Symbols:	Arithmetic: $P(6, 2) \rightarrow P'(-2, 6)$ Algebra: $P(x, y) \rightarrow P'(-y, x)$
Rotation of 180°	Words:	To rotate a figure 180° about the origin, multiply both coordinates of each point by −1.
	Symbols:	Arithmetic: $P(6, 2) \rightarrow P'(-6, -2)$ Algebra: $P(x, y) \rightarrow P'(-x, -y)$

Examples

1 Triangle *TVW* has vertices *T*(3, 2), *V*(6, 3), and *W*(4, 6). Graph △*TVW* and rotate it 90° counterclockwise about the origin. Then graph △*T'V'W'*.

To rotate △*TVW* 90°, switch the coordinates of each vertex and multiply the first by −1.

$T(3, 2) \rightarrow T'(-2, 3)$
$V(6, 3) \rightarrow V'(-3, 6)$
$W(4, 6) \rightarrow W'(-6, 4)$

Graph △*T'V'W'*.

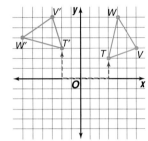

Study Hint

Estimation You can verify a 90° rotation using a corner of a piece of paper. If you can place the paper so that the corner is on the origin and *T* and *T'* are on the edges, you have probably rotated correctly.

2 Use △*TVW* from Example 1. Rotate it 180° about the origin. Then graph △*T"V"W"*.

To rotate △*TVW* 180°, multiply the coordinates of each vertex by −1.

$T(3, 2) \rightarrow T''(-3, -2)$
$V(6, 3) \rightarrow V''(-6, -3)$
$W(4, 6) \rightarrow W''(-4, -6)$

Graph △*T"V"W"*.

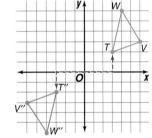

You know that some figures have line symmetry. Other figures have *rotational* or *point symmetry*. If a figure has rotational symmetry, when you turn it around its center point, there is at least one other position in which the figure looks the same as it did originally. *Any figure will look the same after a 360° turn, so a figure must look the same after a rotation of less than 360° to have rotational symmetry.*

Lesson 10-9 Integration: Geometry Rotations **465**

In-Class Examples

For Example 1
Triangle *MNP* has vertices of *M*(−2, −3), *N*(−3, 1), and *P*(−1, −1). Graph △*MNP* and rotate it 90° counterclockwise about the origin. Then graph △*M'N'P'*.

For Example 2
Use △*MNP* above. Rotate it 180° about the origin. Then graph △*M"N"P"*.

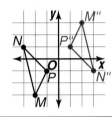

Teaching Tip For Examples 1 and 2 in the Student Edition, point out that multiplying a vertex by −1 moves it from a positive position to a negative one or vice versa. When rotating by 90°, only one coordinate moves from a positive position to a negative one, or vice versa. When rotating by 180°, both coordinates (*x* and *y*) move from a positive position to a negative position, or vice versa.

Classroom Vignette

"To help students visualize rotations, have them use patty paper to trace the figure to be rotated. Then use a pin or pencil point to maintain the center of rotation and turn the patty paper to actually do the rotation."

Cindy J. Boyd, Teacher
Abilene High School
Abilene, TX

In-Class Example

For Example 3
Refer students to the flag for the Isle of Man. Ask them if the symbol would look different if it were rotated 180°. **Yes, it would be upside down.**

3 PRACTICE/APPLY

Check for Understanding
If students need additional practice or instruction after completing Exercises 1–6, one of these options may be helpful.
- Extra Practice, see p. 636
- Reteaching Activity
- *Study Guide Masters*, p. 85
- *Practice Masters*, p. 85

Assignment Guide
Core: 7–17 odd, 19–22
Enriched: 8–14 even, 15–17, 19–22

Additional Answer
2. The corresponding sides are all the same length; they are just moved to another place in the coordinate plane.

Study Guide Masters, p. 85

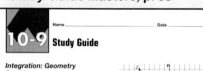
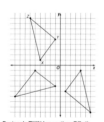

Example 3 APPLICATION

Flags Refer to the beginning of the lesson. Find the measures of the counterclockwise turns that make the symbol on the flag of the Isle of Man look the same.

Label the symbol to keep track of the rotations.

Original   120° turn   240° turn   360° turn

The parts of the symbol are in different positions, but it looks the same after each turn. A 360° turn returns the figure to its original position. So divide 360° into thirds to find that the counterclockwise turns are 120° and 240°.

CHECK FOR UNDERSTANDING

Communicating Mathematics

Read and study the lesson to answer each question. **2. See margin.**

1. *State* the quadrant in which a triangle will lie if it is in the first quadrant and is rotated 90° counterclockwise. **second quadrant**

2. *Explain* why a figure and its rotation image are congruent figures.

3. *Write* three examples of rotating objects you see every day. **See students' work.**

Guided Practice

4. Determine whether the figures represent a rotation. **no**

5. Graph $\triangle ABC$ with vertices $A(1, 3)$, $B(6, 7)$, and $C(9, 1)$.
   a. Rotate the triangle 90° counterclockwise and graph $\triangle A'B'C'$. **a–b. See Answer Appendix.**
   b. Rotate the triangle 180° and graph $\triangle A''B''C''$.

6. *Geometry* An equilateral triangle has rotational symmetry. Draw three other figures that have rotational symmetry. **See students' work.**

EXERCISES

Practice

Determine whether each pair of figures represents a rotation. Write *yes* or *no*.

7. **yes** 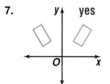   8. **no** 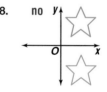   9. **yes**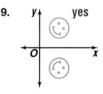

10. Graph $\triangle XYZ$ with vertices $X(-6, -9)$, $Y(-1, -3)$, and $Z(-8, -5)$.
    a. Rotate the triangle 90° counterclockwise and graph $\triangle X'Y'Z'$.
    b. Rotate the triangle 180° and graph $\triangle X''Y''Z''$. **a–b. See Answer Appendix.**

466 Chapter 10 Algebra: Graphing Functions

Reteaching the Lesson

Activity Fold a square piece of paper along the diagonal to form an isosceles right triangle. Label the right angle *A*, the second angle *B*, and the third angle *C*. Attach vertex *A* of the triangle (the right angle) to the origin of a coordinate system. Have students spin the triangle around the origin and read the coordinates of vertex *B*. Then have them read the coordinates of vertex *C*, which will be the 90° rotation image of *B*.

11. Rectangle RECT has vertices $R(-5, 4)$, $E(-5, -2)$, $C(-3, -2)$, and $T(-3, 4)$. Graph RECT.   **a–b. See Answer Appendix.**
    a. Graph $R'E'C'T'$, the rotation of RECT 90° counterclockwise.
    b. Rotate the rectangle 180° and graph $R''E''C''T''$.

12. Graph trapezoid ABCD with vertices $A(3, -2)$, $B(7, -2)$, $C(9, -7)$, and $D(-1, -7)$.   **a–b. See Answer Appendix.**
    a. Rotate the trapezoid 90° counterclockwise and graph $A'B'C'D'$.
    b. Rotate the trapezoid 180° and graph $A''B''C''D''$.

13. $(4, -1), (1, -4),$ and $(5, 8)$

13. After a triangle is rotated 180°, its vertices are at $(-4, 1)$, $(-1, 4)$, and $(-5, -8)$. What were the coordinates of the vertices before the rotation?

14. the original figure

14. A figure is rotated 180° and then its image is rotated 180°, what is the result?

Applications and Problem Solving

15. **Marketing** Marketers create corporate logos to identify products from their company. Some corporate logos have rotational symmetry.

American Automobile Association

Mitsubishi

Chevrolet

   a. Which of these logos exhibit rotational symmetry?
   b. For each logo that has rotational symmetry, find the degree turns that show the symmetry.   **Mitsubishi: 120°, 240°; Chevrolet: 180°**

15a. Mitsubishi and Chevrolet

16. **Art** How many degrees is hedgehog A rotated to create hedgehog B in this piece of African art?   **180°**

17. See Answer Appendix.

17. **Entertainment** In a deck of cards, the eight of diamonds has rotational symmetry. What other cards have rotational symmetry?

18. **Working on the CHAPTER Project**
    A computer game has a spinning windmill. One blade has vertices $A(-6, 4)$, $B(-11, 2)$, $C(-10, -2)$, $D(-6, -2)$, $E(0, 0)$, and $F(-3, 3)$. After 6 equal rotations, A' has coordinates $(-4, -6)$.
    a. How many degrees does the blade move on each rotation?   **15°**
    b. Find the coordinates of the other vertices of the blade after 6 rotations.

18b. $B'(-2, -11)$, $C'(2, -10)$, $D'(2, -6)$, $E'(0, 0)$, $F'(-3, -3)$

19. **Critical Thinking** If the point at (x, y) is rotated 90° *clockwise*, what are the new coordinates?   **$(y, -x)$**

Mixed Review

20. **Geometry** Graph $\triangle RST$ with vertices $R(3, 3)$, $S(0, 0)$, and $T(6, -1)$. (Lesson 10-8)   **a–b. See Answer Appendix.**
    a. Reflect $\triangle RST$ over the x-axis.
    b. Reflect $\triangle RST$ over the y-axis.

21. **Test Practice** It is 326 kilometers from Milford to Loveland. If there are 1,000 meters in a kilometer, use scientific notation to write the distance from Milford to Loveland in meters. (Lesson 6-9)   **D**
    A 3.26×10^6 B 32.6×10^5 C 326×10^5 D 3.26×10^5

22. Solve $q = 360 \div (-40)$. (Lesson 2-8)   **–9**

Lesson 10-9 Integration: Geometry Rotations   467

Extending the Lesson

Enrichment Masters, p. 85

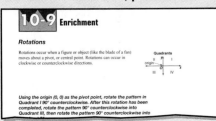

Activity Ask students to find out why a time exposure of the night sky taken through a telescope pointed directly at the North Star produces a photo showing rotation arcs of stars, centered at the North Star. **The arcs are caused by the rotation of Earth about an axis pointed at the North Star.**

Study Guide and Assessment

Vocabulary

This section provides a listing of the new terms, properties, and phrases that were introduced in this chapter. Have students define each term and provide an example or two of it, if appropriate.

Understanding and Using the Vocabulary

These exercises check students' understanding of the terms by using a variety of verbal formats including matching, completion, and true/false.

Glossaries A complete glossary of terms appears on pages 702–710. The glossary also appears in Spanish on pages 711–721.

CHAPTER 10 Study Guide and Assessment

Vocabulary

After completing this chapter, you should be able to define each term, concept, or phrase and give an example or two of each.

Patterns and Functions
domain (p. 428)
function (p. 428)
function table (p. 428)
linear function (p. 442)
quadratic function (p. 452)
range (p. 428)
standard viewing window (p. 445)
system of equations (p. 446)

Geometry
line of symmetry (p. 460)
reflection (p. 460)
rotation (p. 464)
symmetric (p. 460)
translation (p. 456)

Problem Solving
use a graph (p. 450)

Understanding and Using the Vocabulary

Choose the correct term or number to complete each sentence.

1. The (<u>domain</u>, range) is the set of input values of a function.
2. The range is the set of (input, <u>output</u>) values of a function.
3. When you find a common solution of two or more equations, you are solving a (function, <u>system of equations</u>).
4. A function in which the graphs of the solutions form a line is called a (<u>linear</u>, quadratic) function.
5. A function in which the greatest power is (<u>two</u>, three) is called a quadratic function.
6. The movement of a figure 2 units right and 4 units down is a (rotation, <u>translation</u>).
7. $A(2, 1) \rightarrow A'(-2, 1)$ describes a (<u>reflection</u>, translation) over the y-axis.
8. In a (<u>function</u>, rotation), one or more operations are performed on one number to get another.
9. Substitute the values into the original equation and check to see that a true sentence results.

In Your Own Words

9. *Tell* how to determine whether an ordered pair is a solution of a function.

468 Chapter 10 Algebra: Graphing Functions

MindJogger Videoquizzes

MindJogger Videoquizzes provide an alternative review of concepts presented in this chapter. Students work in teams to answer questions, gaining points for correct answers. The questions are presented in three rounds.
Round 1 Concepts–5 questions
Round 2 Skills–4 questions
Round 3 Problem Solving–4 questions

Study Guide and Assessment Chapter 10

Objectives & Examples

Upon completing this chapter, you should be able to:

● complete function tables *(Lesson 10-1)*

Find $f(5)$ if $f(n) = 2n + 6$.
$f(5) = 2(5) + 6$ *Replace n with 5.*
$f(5) = 10 + 6$
$f(5) = 16$

● graph functions by using function tables *(Lesson 10-2)*

Graph $f(n) = n - 1$.

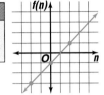

n	f(n)	(n, f(n))
-2	-3	(-2, -3)
0	-1	(0, -1)
2	1	(2, 1)

● find solutions of equations with two variables *(Lesson 10-3)*

Find a solution of $y = 5x + 2$.
Let $x = 3$.
$y = 5(3) + 2$
$y = 15 + 2$
$y = 17$

A solution of $y = 5x + 2$ is (3, 17).

● graph linear functions by plotting points *(Lesson 10-4)*

Make a function table and choose at least three values for x. Then graph the ordered pairs and connect the points with a line.

Review Exercises

Use these exercises to review and prepare for the chapter test.

10. Copy and complete the function table for $f(n) = 1 - 3n$.

n	1 − 3n	f(n)
-2	1 − 3(−2)	7
0	1 − 3(0)	1
3	1 − 3(3)	−8

11. Copy and complete the function table for $f(n) = 3n + 1$. Then graph the function. **See Answer Appendix for graph.**

n	f(n)	(n, f(n))
-3	-8	(-3, -8)
0	1	(0, 1)
1	4	(1, 4)

Choose values for n and graph each function. 12–13. **See Answer Appendix.**

12. $f(n) = 0.5n - 4$

13. $f(n) = -2n$

Copy and complete the table for each equation.

14. $y = 1 - 1.5x$

x	y
-4	7
0	1
2	-2

15. $y = x - 4$

x	y
-2	-6
4	0
6	2

16. Find four solutions of $y = 5x - 10$. **Sample answers: {(−1, −15), (0, −10), (2, 0), (4, 10)}**

Graph each function. 17–20. **See Answer Appendix.**

17. $y = -5x$

18. $y = 4 + 2x$

19. $y = 1\frac{1}{2} - 2\frac{1}{2}x$

20. $y = \frac{x}{2} - 2$

Chapter 10 Study Guide and Assessment **469**

Objectives & Examples

This section reviews the skills and concepts of the chapter and shows completely worked examples.

Review Exercises

These exercises provide practice for the corresponding objectives.

Assessment and Evaluation Masters, pp. 255–256

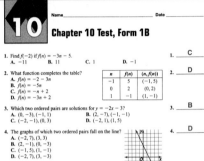

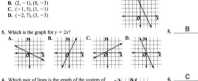

 Assessment and Evaluation

Six forms of Chapter 10 Test are available in the *Assessment and Evaluation Masters* as shown in the chart.

Chapter 10 Test, Form 1B, is shown at the right. Chapter 10 Test, Form 2B, is shown on the next page.

1A	Multiple Choice	Honors
1B	Multiple Choice	Average
1C	Multiple Choice	Basic
2A	Free Response	Honors
2B	Free Response	Average
2C	Free Response	Basic

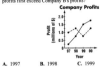

Study Guide and Assessment **469**

Study Guide and Assessment

Chapter 10 Study Guide and Assessment

Objectives & Examples | Review Exercises

- **solve systems of linear equations by graphing** *(Lesson 10-5)*

  The coordinates of the point where the graphs of two linear equations intersect is the solution of the system of equations.

  Solve each system of equations by graphing.
  21. $y = 5x$ **(1, 5)**   22. $y = 3x - 6$ **(1, −3)**
      $y = x + 4$ $y = x - 4$

  23. Graph the system $y = 6x$ and $y = x + 5$. Find the solution of the system. **(1, 6)**

  21–23. See Answer Appendix for graphs.
  24–27. See Answer Appendix.

- **graph quadratic functions** *(Lesson 10-6)*

  Make a function table. Then graph the ordered pairs and draw a smooth curve connecting the points.

  Graph each quadratic function.
  24. $y = 0.5x^2 + 4$
  25. $y = x^2 - 2$
  26. $f(n) = 5 - n^2$
  27. $y = -\frac{1}{2}x^2 - 3$

  28–29. See Answer Appendix.

- **graph translations on a coordinate plane** *(Lesson 10-7)*

  To translate a point by (a, b), add a to the x-coordinate and add b to the y-coordinate.

  Graph each figure and its translation.
  28. $\triangle XYZ$ with vertices $X(2, 2)$, $Y(3, 5)$, and $Z(5, 3)$, translated by $(-2, -4)$
  29. rectangle $ABCD$ with vertices $A(-3, -1)$, $B(0, -1)$, $C(0, 1)$, and $D(-3, 1)$, translated by $(3, 2)$

- **graph reflections on a coordinate plane** *(Lesson 10-8)*

  reflection over the x-axis
  (x, y) becomes $(x, -y)$.

  reflection over the y-axis
  (x, y) becomes $(-x, y)$.

  30. Graph square $EFGH$ with vertices $E(2, 5)$, $F(4, 5)$, $G(4, 3)$, and $H(2, 3)$ and its reflection over the x-axis.
  31. Graph $\triangle TUV$ with vertices $T(-4, -5)$, $U(-3, -3)$, and $V(-5, -2)$ and its reflection over the y-axis.
  30–31. See Answer Appendix.

  32–33. See Answer Appendix.

- **graph rotations on a coordinate plane** *(Lesson 10-9)*

  rotation 90° counterclockwise
  (x, y) becomes $(-y, x)$.

  rotation 180°
  (x, y) becomes $(-x, -y)$.

  32. Graph rectangle $IJKL$ with vertices $I(1, 4)$, $J(3, 6)$, $K(6, 3)$, and $L(4, 1)$ and its rotation of 90° counterclockwise.
  33. Graph $\triangle QRS$ with vertices $Q(2, 2)$, $R(4, -1)$, and $S(2, -3)$ and its 180° rotation.

Test and Review Software

You may use this software, a combination of an item generator and item bank, to create your own tests or worksheets. Types of items include free response, multiple choice, short answer, and open ended.

CD-ROM Program

The CD-ROM Program contains an Assessment Game whose questions review the concepts in this chapter.

Study Guide and Assessment Chapter 10

Applications & Problem Solving

34–36. See Answer Appendix.

34. Sports Golfers are given a handicap based on their average. Tatanka's handicap is 15, so $f(n) = n - 15$, where n is his actual score, is used to determine his handicapped score. Make a function table of his handicapped scores if Tatanka had actual scores of 85, 87, and 89. *(Lesson 10-1)*

35. Communication Code flags can send messages between ships at sea. Tell whether each code flag displays symmetry. If a flag has symmetry, describe it as line or rotational symmetry. *(Lessons 10-8 and 10-9)*

A   E   N   T   Z

36. Business One company's profit is described by the equation $p = 100x - 200$. Another company's profit is described by $p = 100x + 200$. They plan to merge when their profit is the same. At what point will that occur? Explain your answer. *(Lesson 10-5)*

37. Use a Graph The graph below shows the number of farms in the United States from 1940 to 1995. *(Lesson 10-6A)*

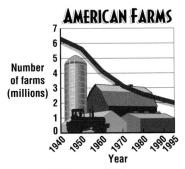

Source: *World Almanac*, 1997

a. About how many more farms were there in 1950 than in 1980?

b. During which ten-year period was the decrease in the number of farms the greatest? **from 1950 to 1960**

37a. Sample answer: about 3.3 million

Alternative Assessment

● **Performance Task**
Ofelia drew a graph of the cost of renting videos at two stores near her home using x as the number of videos and y as the cost. One line passes through (0, 5) and (5, 15). The other line passes through the origin and (5, 15). Graph the lines. Describe the situations shown by the lines.
See Answer Appendix.
One of the stores decided to change its prices. Ofelia drew a new graph and found that the two lines she drew did not intersect. Describe how the prices could have changed.
See Answer Appendix.

● **Completing the CHAPTER Project**
Use the following checklist to make sure that your project is complete.
☑ Outline your computer game plan.
☑ Use a grid to design the background for a scene from your game.
☑ Design the characters in your game.
☑ Use functions to describe how the characters move on the screen.

 Select one of the assignments from this chapter and place it in your portfolio. Attach a note to it explaining why you selected it.

A practice test for Chapter 10 is provided on page 656.

Performance Assessment
Additional performance assessment tasks for this chapter are included in the *Assessment and Evaluation Masters* on page 265. A scoring guide is also provided on page 277.

Standardized Test Practice

The Standardized Test Practice may be used to help students prepare for standardized tests. The test items are written in the same style as those in state proficiency tests and standardized tests like CAT, CTBS, ITBS, MAT, SAT, and Terra Nova. The test items cover skills and concepts covered up to this point in the text.

The pages can be used as an overnight assessment. After students have completed the pages, discuss how each problem can be solved, or provide copies of the solutions from the *Solutions Manual*.

Assessment and Evaluation Masters, p. 271

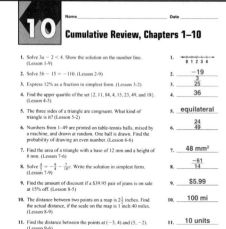

◀◀◀ **Instructional Resources**
Another cumulative review is shown at the left and is available in the *Assessment and Evaluation Masters*, p. 271.

472   Chapter 10

CHAPTERS 1–10 Standardized Test Practice
Assessing Knowledge & Skills

Section One: Multiple Choice

There are eleven multiple-choice questions in this section. Choose the best answer. If the correct answer is *not here,* choose the letter for Not Here.

1. What number will make the number sentence true? **B**
   $$\frac{2}{3} + \frac{5}{9} = \frac{5}{9} + h$$
   A $\frac{7}{12}$
   B $\frac{2}{3}$
   C $\frac{3}{9}$
   D $\frac{4}{3}$

2. $0.9 =$ **J**
   F 9%
   G 0.9%
   H 0.09%
   J 90%

3. An electronics store is having a going out of business sale. The store advertises 35% off the price of a $210 VCR. To find the amount saved on the cost of the VCR, multiply $210 by — **B**
   A 0.035.
   B 0.35.
   C 0.65.
   D 0.065.

4. If $\frac{3}{8} = \frac{c}{12}$, which statement is true? **H**
   F $8c > 3(12)$
   G $8c < 3(12)$
   H $8c = 3(12)$
   J Not enough information

5. Hakeem has been increasing the length of his walks at the beginning of each week for three weeks.

Week	Number of Miles
1	1.4 miles
2	2.8 miles
3	4.2 miles
4	

   If Hakeem continues at the same rate, how many miles will he walk during the fourth week of training? **D**
   A 5.0 miles
   B 5.2 miles
   C 5.4 miles
   D 5.6 miles

Please note that Questions 6–11 have five answer choices.

6. A store sold $22,057 worth of merchandise in the first three weeks of the month. In the fourth week, they sold $6,900 worth of merchandise. A reasonable conclusion would be that the store sold — **H**
   F less than $2,500 per week.
   G between $2,500 and $5,000 per week.
   H between $5,000 and $7,500 per week.
   J between $7,500 and $10,000 per week.
   K more than $10,000 per week.

7. Which set are solutions of the equation $y = 2x - 4$? **A**
   A $\{(-1, -6), (1, -2)\}$
   B $\{(-6, -1), (-2, 1)\}$
   C $\{(1, -2), (3, -2)\}$
   D $\{(-2, 1), (2, 3)\}$
   E $\{(9, 14), (-2, 0)\}$

472   Chapters 1–10 Standardized Test Practice

Standardized Test Practice   Chapters 1–10

8. In a scale drawing of a room, 1 unit = 6 inches. What are the scale dimensions of a 40 in. by 60 in. table? **G**

   F 4 units by 6 units
   G $6\frac{2}{3}$ units by 10 units
   H 8 units by 10 units
   J 24 units by 36 units
   K 240 units by 360 units

9. Tasha bought three bags of apples for apple bobbing. The bags weighed 5.5 pounds, 8.4 pounds, and 7.35 pounds. What was the total weight of the apples? **B**

   A 20.45 lb
   B 21.25 lb
   C 23.60 lb
   D 25.55 lb
   E Not Here

10. If Mike worked a total of 31 hours over a four-day period and is paid $7.25 per hour, how much did he earn before deductions? **H**

    F $29.00
    G $124.00
    H $224.75
    J $899.00
    K Not Here

11. Four friends are saving money to buy a computer game that costs $64.95. Lydia has saved $17.94, Owen $12.55, Tomás $16.85, and Alma $15.90. What is the total that the four friends have saved? **E**

    A $61.24
    B $62.84
    C $64.86
    D $64.94
    E Not Here

Test-Taking Tip

When taking a test, wear a watch. If you know how much time you have left, you can pace yourself more accurately. Most tests do not allow calculator watches, so leave those watches at home.

12. See margin.

Section Two: Free Response

This section contains six questions for which you will provide short answers. Write your answers on your paper.

12. Which figure has rotational symmetry?

13. What is the least prime factor of 24? **2**

14. What is the solution of the system of equations? **(−1, 1)**

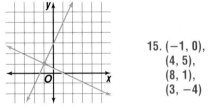

15. (−1, 0), (4, 5), (8, 1), (3, −4)

15. What are the coordinates of the rectangle MATH translated by (2, −1) if M(−3, 1), A(2, 6), T(6, 2), and H(1, 3)?

16. To the nearest inch, how long is the diagonal of a 20 in. by 50 in. window? **54 in.**

17. Find the value of x in the square. **10**

Chapters 1–10 Standardized Test Practice **473**

Additional Answer
12.
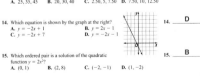

Instructional Resources ▶▶▶
Additional standardized test practice is shown at the right and is available in the *Assessment and Evaluation Masters*, pp. 269–270.

Test-Taking Tip
Time management is an important skill in test taking. This tip can help students judge how much time they can spend on each question.

Assessment and Evaluation Masters, pp. 269–270

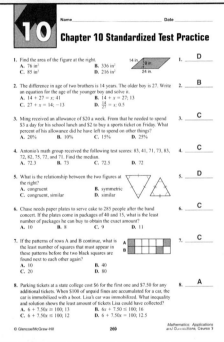

CHAPTER 11
Geometry: Using Area and Volume

Previewing the Chapter

Overview
This chapter combines algebra and geometry to explore volume and area. Students solve problems by making models of three-dimensional figures. They also draw various views of solid figures and their nets. The formulas for the areas of circles, surface areas of cylinders, and volumes of prisms, cylinders, pyramids, and cones are presented. Students also study measurement precision and the use of significant digits.

Lesson (pages)	Lesson Objectives	NCTM Standards	Standardized Tests	State/Local Objectives
11-1 (476–479)	Find the areas of circles.	1–5, 7, 9, 11, 12		
11-2A (480–481)	Solve problems by making a model.	1–4, 7, 12	CTBS, MAT, TN	
11-2 (482–485)	Identify and sketch three-dimensional figures.	1–4, 12, 13	MAT	
11-3 (486–489)	Find the volumes of prisms and cylinders.	1–4, 7, 12	MAT, SAT	
11-4 (490–493)	Find the volumes of pyramids and cones.	1–4, 7, 9, 12	SAT	
11-5A (494)	Recognize a solid from its net and sketch it, and vice versa.	1–4, 12		
11-5 (495–498)	Find the surface areas of rectangular and triangular prisms.	1–4, 7, 12		
11-6 (499–502)	Find the surface areas of cylinders.	1–4, 7, 12, 13		
11-6B (503)	Investigate how surface area and volume are related.	1–4, 7, 12, 13		
11-7 (504–507)	Analyze measurements.	1–5, 7, 13		

CAT = California Achievement Tests, CTBS = Comprehensive Tests of Basic Skills, ITBS = Iowa Tests of Basic Skills, MAT = Metropolitan Achievement Tests, SAT = Stanford Achievement Tests, TN = Terra Nova

Organizing the Chapter

CD-ROM
All of the blackline masters in the Teacher's Classroom Resources are available on the **Electronic Teacher's Classroom Resources** CD-ROM.

LESSON PLANNING GUIDE

Lesson	Extra Practice (Student Edition)	Blackline Masters (page numbers)									
		Study Guide	Practice	Enrichment	Assessment & Evaluation	Classroom Games	Diversity	Hands-On Lab	School to Career	Science and Math Lab Manual	Technology	Transparencies A and B
11-1	p. 636	86	86	86			37					11-1
11-2A	p. 637											
11-2	p. 637	87	87	87	295	37–39						11-2
11-3	p. 637	88	88	88						69–72	73	11-3
11-4	p. 638	89	89	89	294, 295			78				11-4
11-5A								61				
11-5	p. 638	90	90	90							74	11-5
11-6	p. 638	91	91	91	296				37			11-6
11-6B								62				
11-7	p. 639	92	92	92	296							11-7
Study Guide/ Assessment					281–293, 297–299							

Other Chapter Resources

Student Edition
Chapter Project, pp. 485, 502, 507, 511
Math in the Media, p. 507
Let the Games Begin, p. 498

Technology
 CD-ROM Program
 Interactive Mathematics Tools Software

Teacher's Classroom Resources
Applications
Family Letters and Activities, pp. 73–74
Investigations and Projects Masters, pp. 57–60
Meeting Individual Needs
Transition Booklet, pp. 29–36
Investigations for the Special Education Student, pp. 43–44

Teaching Aids
Answer Key Masters
Block Scheduling Booklet
Lesson Planning Guide
Solutions Manual

Professional Publications
Glencoe Mathematics Professional Series

Planning the Chapter

MindJogger Videoquizzes provide a unique format for reviewing concepts presented in the chapter.

Assessment Resources

Student Edition
- Mixed Review, pp. 479, 485, 489, 493, 498, 502, 507
- Mid-Chapter Self Test, p. 493
- Math Journal, pp. 484, 506
- Study Guide and Assessment, pp. 508–511
- Performance Task, p. 511
- Portfolio Suggestion, p. 511
- Standardized Test Practice, pp. 512–513
- Chapter Test, p. 657

Assessment and Evaluation Masters
- Multiple-Choice Tests (Forms 1A, 1B, 1C), pp. 281–286
- Free-Response Tests (Forms 2A, 2B, 2C), pp. 287–292
- Performance Assessment, p. 293
- Mid-Chapter Test, p. 294
- Quizzes A–D, pp. 295–296
- Standardized Test Practice, pp. 297–298
- Cumulative Review, p. 299

Teacher's Wraparound Edition
- 5-Minute Check, pp. 476, 482, 486, 490, 495, 499, 504
- Building Portfolios, p. 474
- Math Journal, pp. 494, 503
- Closing Activity, pp. 479, 481, 485, 489, 493, 498, 502, 507

Technology
- Test and Review Software
- MindJogger Videoquizzes
- CD-ROM Program

Materials and Manipulatives

Lesson 11-1
calculator

Lesson 11-2A
centimeter cubes*

Lesson 11-2
isometric dot paper*†

Lesson 11-3
calculator
centimeter cubes*

Lesson 11-4
calculator
compass*†
construction paper
rice
ruler*†
scissors*
tape

Lesson 11-5A
boxes of different sizes and shapes
ruler*†
scissors*

Lesson 11-5
grid paper†
scissors*
tape
centimeter cubes*

Lesson 11-6
cardboard tube
ruler*†
scissors*
calculator

Lesson 11-6B
calculator

*Glencoe Manipulative Kit †Glencoe Overhead Manipulative Resources

Pacing Chart

See pages T25–T27 for the Course Planning Calendar.

COURSE	DAY 1	DAY 2	DAY 3	DAY 4	DAY 5	DAY 6	DAY 7
Standard	Chapter Project	Lesson 11-1	Lesson 11-2A	Lesson 11-2	Lesson 11-3	Lesson 11-4	Lessons 11-5A & 11-5
Honors	Chapter Project & Lesson 11-1	Lesson 11-2A	Lesson 11-2	Lesson 11-3	Lesson 11-4	Lesson 11-5	Lessons 11-6 & 11-6B
Block	Chapter Project & Lesson 11-1	Lessons 11-2A & 11-2	Lessons 11-3 & 11-4	Lessons 11-5A & 11-5	Lessons 11-6 & 11-7	Study Guide and Assessment, Chapter Test	

The *Transition Booklet* (Skills 13–16) can be used for practice with measurement in the customary and metric systems.

Interactive Mathematics:
Activities and Investigations

is an activity-based program that may be used as an enhancement for chapters in *Mathematics: Applications and Connections.*

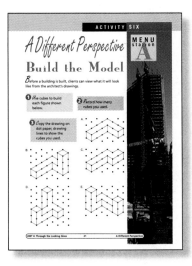

Unit 4, Activity Six, Menu A
Use with Lesson 11-2.

Summary Students work in groups using blocks to build three-dimensional structures from isometric drawings. They then record how many cubes they used to build each structure and copy the isometric drawing, adding lines to show the number of cubes used.

Math Connection Students interpret isometric drawings, build models of their interpretations, and use isometric drawings to depict other three-dimensional structures adding greater detail. This involves length and use of perspective.

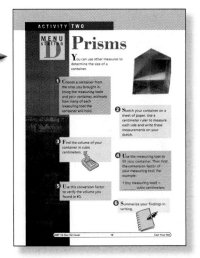

Unit 14, Activity Two, Menu D
Use with Lesson 11-4.

Summary Students work in groups using various measuring tools and materials to find the volumes of right rectangular prisms. Then they find the actual volume of the prisms and develop an equivalence factor so they can convert between their measuring tool and cubic centimeters.

Math Connection Students measure the volumes of right rectangular prisms using various units and calculate the actual volume using $V = Bh$. They use proportional reasoning to determine how much of each nonstandard unit equals 1 cubic centimeter.

DAY 8	DAY 9	DAY 10	DAY 11	DAY 12	DAY 13	DAY 14	DAY 15
(continue from Day 7)	Lessons 11-6 & 11-6B		Lesson 11-7	Study Guide and Assessment	Chapter Test		
(continue from Day 7)	Lesson 11-7	Study Guide and Assessment	Chapter Test				

Enhancing the Chapter

APPLICATIONS

Classroom Games, pp. 37–39
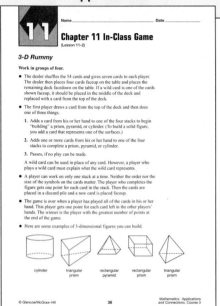

Diversity Masters, p. 37
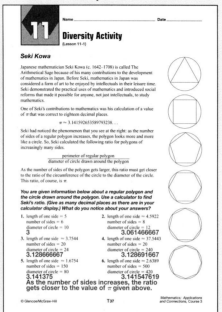

School to Career Masters, p. 37
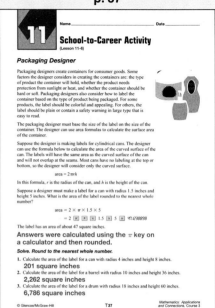

Family Letters and Activities, pp. 73–74
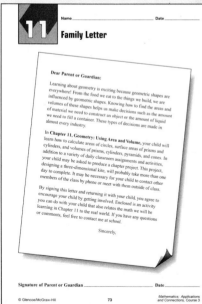
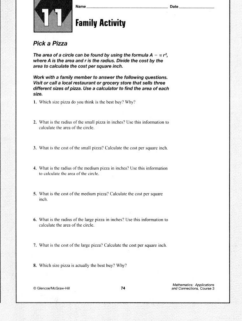

Science and Math Lab Manual, pp. 69–72
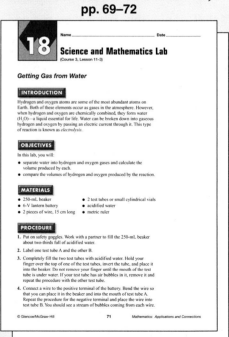

MANIPULATIVES/MODELING

Hands-On Lab Masters, p. 78

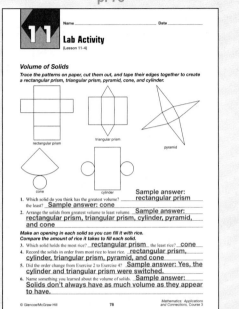

ASSESSMENT/EVALUATION

Assessment and Evaluation Masters, pp. 294–296

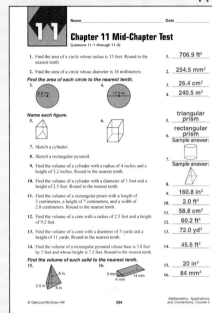

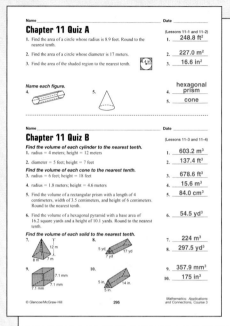

TECHNOLOGY/MULTIMEDIA

Technology Masters, pp. 73–74

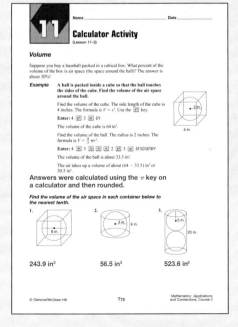

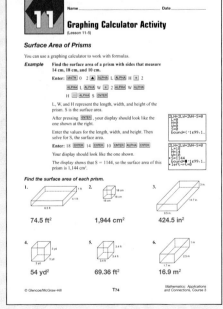

MEETING INDIVIDUAL NEEDS

Investigations for the Special Education Student, pp. 43–44

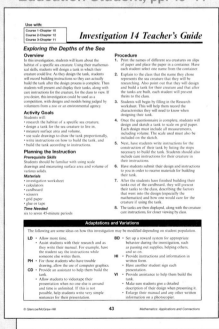

Chapter 11   **474f**

CHAPTER 11 NOTES

Theme: Kites

Kites probably originated more than 2,000 years ago in China. People all over the world have used kites for ceremonies, fighting and spying, and scientific experimentation. In 1749 in Scotland, Alexander Wilson measured temperature at different altitudes with kites. At the beginning of the twentieth century, Orville and Wilbur Wright and Alexander Graham Bell, working separately, built kites that could carry people. They proved that objects heavier than air could fly.

Question of the Day Alexander Graham Bell invented a tetrahedral kite in 1902. A tetrahedron is the strongest structure known. How much surface area would one tetrahedron measuring 40 centimeters on each edge have? **about 2,771.3 cm²**

Assess Prerequisite Skills

Ask students to read through the list of objectives presented in "What you'll learn in Chapter 11." You may wish to ask them what each of the objectives means or if they have experienced or used any of these math concepts before.

Building Portfolios

Encourage students to revise their portfolios as they study this chapter. Diagrams of three-dimensional figures and calculations of volume and surface areas of these figures should be kept and revised as students plan and design their three-dimensional kites.

Math and the Family

In the *Family Letters and Activities* booklet (pp. 73–74), you will find a letter to the parents explaining what students will study in Chapter 11. An activity appropriate for the whole family is also available.

CHAPTER 11

Geometry: Using Area and Volume

What you'll learn in Chapter 11

- to find the areas of circles,
- to solve problems by making a model,
- to find the volumes of prisms, cylinders, pyramids, and cones,
- to find the surface areas of prisms and cylinders, and
- to analyze measurements.

CD-ROM Program

Activities for Chapter 11
- Chapter 11 Introduction
- Interactive Lessons 11-3, 11-5, 11-6
- Extended Activity 11-2
- Assessment Game
- Resource Lessons 11-1 through 11-7

CHAPTER Project

UP, UP, AND AWAY

In this project, you will design a three-dimensional kite. You will make a three-dimensional scale drawing of the kite, compute the amount and cost of materials you will need to make the kite, and determine the instrument(s) you will need to make the appropriate measurements to build the kite. You will place your plans and information on your kite in a folder.

Getting Started

- Research different styles and shapes of three-dimensional kites. Choose a type to design.
- Research the cost of the materials you would need to make the kite.

Technology Tips

- Use an **electronic encyclopedia** to do your research.
- Use a **drawing program** to sketch your design.
- Use a **spreadsheet** to record the cost of materials.
- Use a **word processor** to write the information about your kite.

 interNET CONNECTION For up-to-date information on kites, visit:
www.glencoe.com/sec/math/mac/mathnet

Working on the Project

You can use what you learn in Chapter 11 to draw your kite and determine what you will need to make the kite.

Page	Exercise
485	16
502	16
507	18
511	Alternative Assessment

CHAPTER Project NOTES

Objectives Students should
- be able to draw three-dimensional figures on grid or dot paper.
- calculate surface area.
- determine appropriate measuring instruments.

Project Pointer You may suggest that students begin a *Project Folder* to keep their work as they complete each stage of the Chapter Project. The completed project may also be added to their portfolios.

At the end of the project, students may want to make models of their kites to see how they fly. Remind students that the weight and the durability of the materials used for the kites is as important as the amount of material used.

Investigations and Projects Masters, p. 60

interNET CONNECTION

Glencoe has made every effort to ensure that the website links for *Mathematics: Applications and Connections* at www.glencoe.com/sec/math/mac/mathnet are current and contain appropriate content. However, these website links are not under Glencoe's control.

Instructional Resources ▶▶▶

A recording sheet to help students organize their data for the Chapter Project is shown at the right and is available in the *Investigations and Projects Masters*, p. 60.

11-1 Lesson Notes

Instructional Resources
- Study Guide Masters, p. 86
- Practice Masters, p. 86
- Enrichment Masters, p. 86
- Transparencies 11-1, A and B
- Diversity Masters, p. 37
- CD-ROM Program
  - Resource Lesson 11-1

Recommended Pacing

Standard	Day 2 of 13
Honors	Day 1 of 11
Block	Day 1 of 6

1 FOCUS

5-Minute Check
(Chapter 10)

1. Complete the table and graph the linear function $y = 2x - 1$.

x	y
2	3
1	1
0	-1
-1	-3

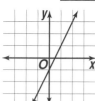

2. Graph the quadratic function $y = x^2 + 1$.

3. Graph parallelogram ABCD with vertices A(2, 2), B(5, 6), C(3, 8), D(0, 4) and its reflection over the y-axis.

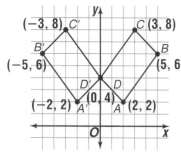

11-1 Area of Circles

What you'll learn
You'll learn to find the areas of circles.

When am I ever going to use this?
Knowing how to find the areas of circles can help you compare the size of pizzas and choose the better buy.

LOOK BACK
You can refer to Lesson 7-7 to review circles.

The Floral Clock located in Niagara Falls, Canada, is a favorite tourist attraction. Each year over 19,000 plants are arranged to form the circular face of the working clock which has a diameter of 40 feet. The School of Horticulture fertilizes the area before planting. If $1\frac{1}{2}$ ounces of fertilizer are needed for each square foot of planting area, how many pounds of fertilizer should be used for the clock? *You will solve this problem in Example 2.*

In order to solve this problem, you need to find the area of the clock. Finding the area of a circle can be related to finding the area of a parallelogram. Draw several equally-spaced radii of a circle. Cut the circle along the radii to form wedge-like pieces. Rearrange the pieces to form a parallelogram-shaped figure.

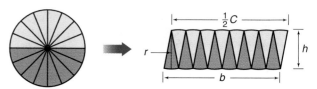

The length of each wedge from the point to its curved side is the same as the radius (r) of the circle. So the height of the parallelogram is r. The curved sides of the wedges form the circumference (C) of the circle. The base of the parallelogram is made of half of these curves, or half of the circumference of the circle.

$A = bh$ *Formula for the area of a parallelogram*
$A = \frac{1}{2}C \cdot r$ $b = \frac{1}{2}C, h = r$
$A = \frac{1}{2} \cdot 2\pi r \cdot r$ $C = 2\pi r$
$A = \pi r^2$

So, the formula for the area of a circle is $A = \pi r^2$.

Area of a Circle		
Words:	The area (A) of a circle equals π times the radius (r) squared.	
Symbols:	$A = \pi r^2$	**Model:**

The 5-Minute Check is also available on **Transparency 11-1A** for this lesson.

Examples

1 Find the area of a circle with a radius of 6 inches to the nearest square inch.

Method 1 Use paper and pencil.

$A = \pi r^2$

$A \approx 3.14 \cdot 6^2$ *Use 3.14 for π.*

$A \approx 3.14 \cdot 36$

$A \approx 113.04$

Method 2 Use a calculator.

$A = \pi r^2$

$A = \pi \cdot 6^2$

[π] [×] 6 [x²] [=] *113.0973355*

$A \approx 113$

The area of the circle is about 113 square inches.

APPLICATION

Did you know?

The largest clock is located in Beauvais, France. It is 40 feet high, 20 feet long, and 9 feet wide and contains 90,000 parts.

2 **Horticulture** Refer to the beginning of the lesson. How many pounds of fertilizer should be used for the clock?

Explore You know the diameter of the circular clock and the amount of fertilizer needed for each square foot. You must find how many pounds of fertilizer are needed for the clock.

Plan Since the diameter of the clock is 40 feet, the radius is half of 40 or 20 feet. Use the radius to find the area of the clock. Then calculate the number of ounces of fertilizer needed and change this answer to pounds.

Solve $A = \pi r^2$ *Formula for the area of a circle*

$A = \pi \cdot 20^2$ *r = 20*

$A \approx 1{,}257$ [π] [×] 20 [x²] [=] *1256.637061*

Multiply 1,257 by $1\frac{1}{2}$ to find how many ounces of fertilizer are needed.

$1{,}257 \times 1\frac{1}{2} = \frac{3{,}771}{2}$ or $1{,}885\frac{1}{2}$

Divide by 16 to find the number of pounds of fertilizer needed.

$1{,}885\frac{1}{2} \div 16 = \frac{3{,}771}{32}$ or about 118   *1 lb = 16 oz*

About 118 pounds of fertilizer are needed.

Examine Estimate the area of the clock.

$3 \times 20^2 = 3 \times 400$ or 1,200 square feet

Since $1{,}200 \times 1\frac{1}{2} = 1{,}800$, the number of ounces needed seems reasonable.

Remember that the probability of an event is defined as the ratio of the number of ways something can happen to the total possible outcomes. Probability can also be related to the area of a figure.

Lesson 11-1 Area of Circles   **477**

Investigations for the Special Education Student

This blackline master booklet helps you plan for the needs of your special education students by providing long-term projects along with teacher notes. Investigation 14, *Exploring the Depths of the Sea,* may be used with this chapter.

Motivating the Lesson

Problem Solving Ask students which is the better buy, a 16-inch square pizza for $9 or two 12-inch square pizzas for $11. **a 16-inch for $9** How would they calculate the better buy if the pizzas were round? **Calculate the area of each size of pizza. Divide the first pizza's area by $9. Multiply the second pizza's area by 2, then divide by $11.**

2 TEACH

 Transparency 11-1B contains a teaching aid for this lesson.

Reading Mathematics *The Nine Chapters on the Mathematical Procedures,* compiled in China in the first century A.D., gives elementary formulas for the area of the circle—"multiply the diameter by itself, triple this, divide by four." This involves using 3 for π. The book also gives the formula for the area of a circle as "multiplying half the diameter and half the circumference, one gets the area."

In-Class Examples

For Example 1
Find the area of a circle with a radius of 28 centimeters to the nearest square centimeter. **2,463 cm²**

For Example 2
If this circle is painted on a playground and the paint costs $0.09 a square foot, how much did the paint cost? **about $7.07**

Teaching Tip Students may not be convinced by the "parallelogram" on page 476 because it has curved edges. Point out that if the circle was cut into smaller and smaller pieces, the sides of the figure would more closely approximate line segments.

Lesson 11-1   **477**

In-Class Example

For Example 3
If you give your dog a bone and let it wander anywhere in your yard, what are the chances that the dog will chew it on the circle of freshly planted grass? **about $\frac{13}{100}$ or 13%**

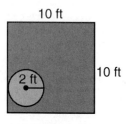

3 PRACTICE/APPLY

Check for Understanding
If students need additional practice or instruction after completing Exercises 1–7, one of these options may be helpful.
- Extra Practice, see p. 636
- Reteaching Activity
- *Study Guide Masters*, p. 86
- *Practice Masters*, p. 86
- Interactive Mathematics Tools Software

Study Guide Masters, p. 86

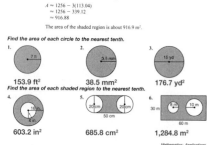

Example 3 INTEGRATION

Probability To win a penny toss game at the carnival, you must toss a penny in the red section of the square board. What is the probability that the center of a penny thrown onto the square at random will land in the red section?

To find the probability of landing in the red section, you need to know the area of the red section and the area of the entire board.

area of red section = area of large circle − area of inner white circle

$$= \pi \cdot 6^2 - \pi \cdot 2^2$$
$$= 36\pi - 4\pi$$
$$= 32\pi$$
$$\approx 101 \quad \boxed{32} \times \boxed{\pi} = \text{100.5309649}$$

area of square board = s^2
$$= 16^2 \text{ or } 256$$

$P(\text{landing in red section}) = \frac{\text{area of red section}}{\text{total area}}$
$$\approx \frac{101}{256}$$

The probability of landing in the red section is about $\frac{101}{256}$ or about 39.5%.

Study Hint
Mental Math Use the distributive property to combine numbers containing π.
$36\pi - 4\pi$
$= (36 - 4)\pi$
$= 32\pi$

CHECK FOR UNDERSTANDING

Communicating Mathematics

Read and study the lesson to answer each question.

1. *Tell* how you can estimate the area of a circle whose radius is 10 meters.
2. *Explain* whether a person is likely to win the penny toss game in Example 3.
3. *You Decide* Alisa says that the area of the semicircle equals $\frac{1}{2}\pi \cdot 5^2$. Megan says that it equals $\frac{1}{2}\pi \cdot 10^2$. Cheryl says that it equals $\pi \cdot 5^2$. Who is correct? Explain.

1. Sample answer: Multiply 3 times 10^2. The area is about 300 square meters.
2–3. See margin.

Guided Practice

Answers are calculated using the π key on a calculator and then rounded.

Find the area of each circle to the nearest tenth.

4. diameter, 24 millimeters **452.4 mm²**
5. radius, 13 feet **530.9 ft²**
6. Find the area of the shaded region to the nearest tenth. **54.9 ft²**

7. **Art** Sculptor Selma Burke sculpted the profile of former president Franklin D. Roosevelt used on the dime first minted in 1946. The diameter of a dime is 17.91 millimeters. Find the area of one side of a dime. **about 251.93 mm²**

478 Chapter 11 Geometry: Using Area and Volume

■ Reteaching the Lesson ■

Activity To help students visualize what they are measuring when they calculate the area of a circle, have them draw three different circles on graph paper using a compass set to multiples of graph-paper units. Have them count squares to estimate each area. Then have them calculate each area and compare the results with their estimates.

Additional Answers

2. A person is not likely to win the penny toss game half of the time, but he or she should win about 2 out of 5 times.

3. Alisa; since the diameter is 10 m, the radius is 5 m. Therefore, the area of the whole circle equals $\pi \cdot 5^2$ and the area of the semicircle is $\frac{1}{2}\pi \cdot 5^2$.

EXERCISES

Practice Find the area of each circle to the nearest tenth.

8. 1,963.5 cm²
9. 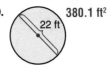 380.1 ft²
10.  706.9 in²

11. radius, 19 meters 1,134.1 m²
12. diameter, 8 yards 50.3 yd²
13. radius, 9 centimeters 254.5 cm²

Find the area of each shaded region to the nearest tenth.

14. 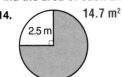 14.7 m²
15. 10 cm 514.2 cm²
16. 25.1 in²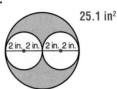

17. Find the radius of a circle whose area is 64π square inches. **8 in.**

18. 3 cm²; 13 cm²; 28 cm²; 50 cm²

18. Determine the areas of circles with radii of 1 centimeter, 2 centimeters, 3 centimeters, and 4 centimeters to the nearest whole number.

   a. Complete the table.

x	Radius	1	2	3	4
y	Area	3	13	28	50

   b. Use the data in the table to plot points on a coordinate grid. Let the radius be the x value and the area be the y value. **See margin.**

   c. Connect the points. Does the graph form a straight line or a curve? **curve**

   d. Use your graph to predict the area of a circle with a radius of 5 centimeters. Use the area formula to check your prediction.
      See students' work; π · 5² ≈ 79 cm².

Applications and Problem Solving

19. **Probability** Suppose you throw a dart at random at the dartboard and you hit the board. What is the probability that the dart lands in the red section? $\frac{8}{9}$

20. **Food** In 1987, a circular pizza with a diameter of 100 feet established a world record. In 1991, a square pizza with sides of 100 feet attempted to break the record. Which pizza had the greater area? how much greater? **the square pizza; 2,146 ft²**

21. **Critical Thinking** If the ratio of the radii of two circles is 3 to 1, what is the ratio of the areas? **9:1**

Mixed Review

22. See margin.

22. **Geometry** Graph △HIJ with vertices H(2, 2), I(4, −1), and J(1, −2). Then graph △H'I'J' after a rotation of 90° counterclockwise. (Lesson 10-9)

23. **Test Practice** Write 1,276% as a decimal. (Lesson 8-4) **C**
   A 1,276
   B 127.6
   C 12.76
   D 1.276

24. Express 15.363636... using bar notation. (Lesson 6-5) $15.\overline{36}$

Lesson 11-1 Area of Circles **479**

Assignment Guide

Core: 9–21 odd, 22–24
Enriched: 8–18 even, 19–24

4 ASSESS

Closing Activity

Modeling Have students choose circular objects in the room, measure their radii, and calculate their area.

Additional Answers

18b.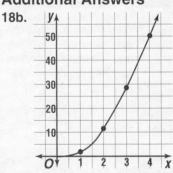

22.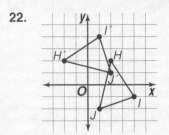

Practice Masters, p. 86

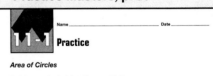

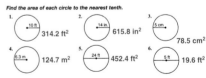

7. Find the area of a circle that has a diameter of 60 feet. **2,827.4 ft²**

8. Find the area of a circle that has a radius of 22 feet. **1,520.5 ft²**

9. Find the diameter of a circle that has an area of 36π square inches. **12 in.**

Find the area of each shaded region to the nearest tenth.

10. 24.8 ft²
11. 13.7 ft²
12. 7.9 ft²

13. A circular flower garden has a diameter of 16 feet. At the center of the garden is a circular pool 5 feet in diameter. If a coin is tossed at random into the garden, what is the probability that the coin will land in the pool? about $\frac{1}{10}$

Extending the Lesson

Enrichment Masters, p. 86

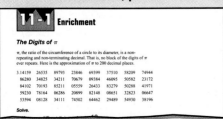

Activity Have students use compasses to draw this figure. Then have them use the formula A ≈ 0.54r² to find the total area of the shaded regions. **Answers depend on the radius.**

Lesson 11-1 **479**

THINKING LAB 11-2A Notes

Objective Students solve problems by making a model.

Recommended Pacing	
Standard	Day 3 of 13
Honors	Day 2 of 11
Block	Day 2 of 6

1 FOCUS

Getting Started Ask students to describe situations where building a model might help clarify or solve a problem. *Sample answers:* a model of a new building; a three-dimensional representation of machinery to see how parts fit together

2 TEACH

Teaching Tip In the drawings, show students that every dark line indicates an edge where a cube is not touched by another cube. Ask students if they could construct the model correctly from only one view shown in the example. **no** from two? **yes**

In-Class Example
Arrange 12 cubes so that the least possible number of cube faces are showing.

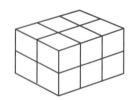

THINKING LAB PROBLEM SOLVING

11-2A Make a Model
A Preview of Lesson 11-2

For their community living project, Lorena and Jeffrey suggest building a children's play area using pre-fabricated cubical units. Lorena and Jeffrey study the plans showing each side of the building. Let's listen in!

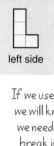

left side

front

right side

back

top

If we use blocks and our plans to build a model, we will know how many pre-fabricated units we need. Remember, the dark lines indicate a break in the surface.

From the top view, I can see we need 6 blocks to form the base of our building.

From the left and right views, I can see we need to build a three-story tower in each back corner.

There is just one story along the front, but there are two stories in the middle of the back of the building.

THINK ABOUT IT

Work with a partner.

1. **Tell** what each block in the model represents.
2. **Make a model** of the building. How many pre-fabricated cubical units are needed?
3. **Explain** how to build the model by starting with one of the side views. **See margin.**

1. a pre-fabricated unit
2. See students' work; 11 units.

4. **Apply** the **make a model** strategy to determine how many pre-fabricated cubical units are needed for the following plans. **9 units**

left side

front

right side

back

top

480 Chapter 11 Geometry: Using Area and Volume

▪ Reteaching the Lesson ▪

Activity Have students use centimeter cubes to construct a different model and then draw top, side, and front views of the model. After drawings are finished, students may exchange with classmates and attempt to construct a model from the diagrams received.

Additional Answer

3. Sample answer: Start with a tower of 3 blocks and 1 block in front of it. Then add a tower of 2 blocks to the right of the original tower, and place 1 block in front of it. Finally, make a tower of 3 blocks to the right of the tower of 2, and place 1 block in front of it.

ON YOUR OWN

5. The last step of the 4-step plan for problem solving asks you to *examine* your solution. *Explain* what you need to check if you are using a model to solve a problem. **See margin.**

6. *Write a Problem* that can be solved by making a model. **See margin.**

7. **Look Ahead** The figure is called a rectangular prism. Use cubes to make a model of a rectangular prism that is 3 units long, 2 units wide, and 4 units high. **See students' work.**

MIXED PROBLEM SOLVING

STRATEGIES
Look for a pattern.
Solve a simpler problem.
Act it out.
Guess and check.
Draw a diagram.
Make a chart.
Work backward.

Solve. Use any strategy.

8. **Education** Ricardo is reading a 216-page book for his book report. He needs to read twice as many pages as he has already read to finish the book. How many pages has he read so far? **72 pages**

9. **Test Practice** Four cubes are glued together. If you could pick them up and look at them from all sides, which of these arrangements shows the fewest squares? **A**

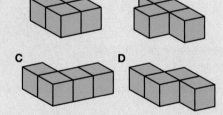

10. **Design** Edu-Toys is designing a new package to hold a set of 30 alphabet blocks. Each block is a cube with each edge of the cube being 2 inches long. Give two possible dimensions for the box. **Sample answer: 10 in. by 6 in. by 4 in. or 10 in. by 12 in. by 2 in**

11. **Architecture** The Sears Tower in Chicago is one of the tallest structures in the world. It is actually a building of varying heights. The diagram shows a top view of the building with the number of stories for each section indicated. Use grid paper to draw the view of the Sears Tower from each side.
See Answer Appendix.

50 stories	89 stories	66 stories
110 stories	110 stories	89 stories
66 stories	89 stories	50 stories

12. **Money Matters** Dion paid $45 for a jacket that was on sale at 40% off. What was the original price of the jacket? **$75**

13. **Test Practice** Which of the following is *not* a top or side view of the figure? **C**

A   B

C   D

Lesson 11-2A THINKING LAB **481**

Extending the Lesson

Activity Ask students to solve this problem with a model and diagram. *How high and wide can you build a staircase made of cubes if you must use exactly 40 cubes and each step is one cube deep?*

4 cubes wide and 4 cubes high

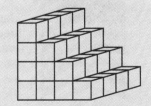

---

3 PRACTICE/APPLY

Check for Understanding
Use the results from Exercise 4 to determine whether students understand how to apply the make-a-model strategy.

Extra Practice If students need additional practice in problem solving, extra practice is available on the following pages.
- Make a Model, see p. 637
- Mixed Problem Solving, see pp. 645–646

Assignment Guide

All: 5–13

4 ASSESS

Closing Activity
Modeling Have students work in groups to design a simple childrens' play area. After they agree on a plan, ask them to model it with blocks and describe it to the class.

Additional Answers
5. Check to make sure the model satisfies all of the information.
6. Sample answer: Arrange 8 cubes so that the least number of squares is showing.

Thinking Lab 11-2A **481**

11-2 Lesson Notes

Instructional Resources
- *Study Guide Masters*, p. 87
- *Practice Masters*, p. 87
- *Enrichment Masters*, p. 87
- Transparencies 11-2, A and B
- *Assessment and Evaluation Masters*, p. 295
- *Classroom Games*, pp. 37–39
- CD-ROM Program
  - Resource Lesson 11-2
  - Extended Activity 11-2

Recommended Pacing	
Standard	Day 4 of 13
Honors	Day 3 of 11
Block	Day 2 of 6

1 FOCUS

5-Minute Check
(Lesson 11-1)

1. Find the area of a circle with radius of 4 meters to the nearest tenth. **50.3 m²**
2. Find the area of the shaded region to the nearest tenth. **75.4 cm²**

3. What is the probability that a coin dropped on the circle in Exercise 2 will land on the shaded area? **about $\frac{1}{2}$**

The 5-Minute Check is also available on **Transparency 11-2A** for this lesson.

Motivating the Lesson

Hands-On Activity Display a cardboard box in various positions. Ask students the following questions.
- How many sides, including the top and bottom, are there? **6**
- What is the greatest number of sides you can see at once? **3**

482 Chapter 11

11-2 Three-Dimensional Figures

What you'll learn
You'll learn to identify and sketch three-dimensional figures.

When am I ever going to use this?
Knowing how to sketch three-dimensional figures can help you make drawings in art class.

Word Wise
solid
prism
face
edge
vertex
base
pyramid

Geologists classify crystals by the number of flat surfaces and the angles at which the surfaces meet.

 isometric (cubic)    orthorhombic

 hexagonal

monoclinic   tetragonal   triclinic

In geometry, these three-dimensional figures are called **solids**. Some common geometric solids are shown below.

rectangular prism   triangular prism   pyramid   cone   cylinder

You can use isometric dot paper to draw geometric solids.

Example **1**

Use isometric dot paper to sketch a rectangular prism that is 3 units high, 1 unit long, and 2 units wide.

Step 1 Lightly draw the bottom of the prism 1 unit by 2 units.

Step 2 Lightly draw the vertical segments at the vertices of the base. Each segment is 3 units high.

Step 3 Complete the top of the prism.

Step 4 Go over your lines. Use dashed lines for the edges of the prism you cannot see from your perspective and solid lines for the edges you can see.

482 Chapter 11 Geometry: Using Area and Volume

Classroom Vignette

"I have my students use straws and trash bag ties to create three-dimensional geometric figures. They also use magazine pictures to create a pictorial geometry handbook that gives real-world examples of the geometry terms."

Steve Werges, Teacher
LaSalle Springs Middle School
Glencoe, MD

You could draw several other perspectives of the same prism.

Prisms, like the ones above, have flat surfaces. The surfaces of a prism are called **faces**. The faces meet to form the **edges** of the prism. The edges meet at corners called **vertices**.

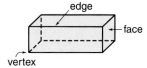

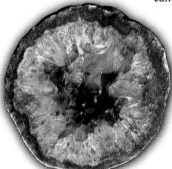

All prisms have at least one pair of faces that are parallel and congruent. These are called **bases** and are used to name the prism. The prisms shown above are *rectangular prisms*.

A **pyramid** is a solid that has a polygon for a base and triangles for sides. It is named according to its base.

Examples

2 Describe the solid in the drawing. Include its name and the number of faces, edges, and vertices.

The base of the figure is a hexagon. The other faces are triangles. The figure is a hexagonal pyramid. It has a total of 7 faces. It has 12 edges and 7 vertices.

APPLICATION **3** **Architecture** An architect's sketch shows the plans for a pedestal for a sculpture in front of a library.

a. What are the dimensions of the bottom of the base?

The base is 8 units by 3 units.

b. What are the dimensions of the top of the center section?

The top is 2 units by 3 units.

c. What is the area of the top of the center section?

The area is 2 × 3 or 6 square units.

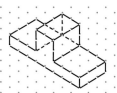

Lesson 11-2 Three-Dimensional Figures

Teaching Tip A blackline master for isometric dot paper can be found on page 13 of the *Hands-On Lab Masters*.

---

2 TEACH

Transparency 11-2B contains a teaching aid for this lesson.

Reading Mathematics The word *prism* comes from a Greek word meaning "something sawed or bitten." Students might identify these figures with the way some rocks, such as columnar basalt, break in an exact pattern that looks sawed.

In-Class Examples

For Example 1
Use isometric dot paper to sketch a rectangular prism that is 2 units high, 4 units long, and 1 unit deep.

For Example 2
Describe the solid in the drawing. Include its name and the number of faces, edges, and vertices. **a triangular prism with 5 faces, 9 edges, and 6 vertices**

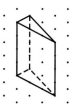

For Example 3
The figure shows a pedestal for the conductor of a band.

a. What are the dimensions of the bottom of the base? **3 units by 2 units**

b. What are the dimensions of the top of the center section? **1 unit by 2 units**

c. What is the area of the top of the center section? **2 square units**

3 PRACTICE/APPLY

Check for Understanding

If students need additional practice or instruction after completing Exercises 1–6, one of these options may be helpful.
- Extra Practice, see p. 637
- Reteaching Activity
- *Study Guide Masters*, p. 87
- *Practice Masters*, p. 87

Assignment Guide

Core: 7–15 odd, 17–19
Enriched: 8–14 even, 15, 17–19

Additional Answers

1. Solid lines are used for the edges that you can see and dashed lines are used for the edges you cannot see.

4.

6.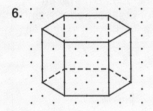

Study Guide Masters, p. 87

CHECK FOR UNDERSTANDING

Communicating Mathematics

Read and study the lesson to answer each question.

1. *Explain* which lines should be solid and which lines should be dashed when drawing a solid figure. **See margin.**

2. *Name* each figure.

   a.   b.   c. 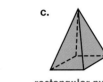 d.

   triangular prism   cone   rectangular pyramid   cylinder

3. *Write* a paragraph comparing a pentagonal prism and a pentagonal pyramid. Include a drawing of each solid. **3. See Answer Appendix.**

Guided Practice

4. See margin.

5b. 2 units by 5 units by 3 units

4. Use isometric dot paper to draw a triangular pyramid with an isosceles triangle as its base.

5a. Name the solid at the right. **rectangular prism**

   b. Give the dimensions of the solid.
   c. How many faces does the solid have? **6**
   d. How many edges does the solid have? **12**
   e. How many vertices does the solid have? **8**
   f. Draw the solid from a different perspective. **See Answer Appendix.**

6. *Pets* The Roosevelt Middle School science classes have a pet lizard that lives in an aquarium with a hexagonal base and height of 4 units. Sketch the aquarium. **See margin.**

EXERCISES

Practice

Use isometric dot paper to draw each solid. **7–9. See Answer Appendix.**

7. a rectangular prism that is 3 units high, 4 units long, and 5 units deep
8. a triangular prism that is 5 units high
9. a pyramid with a square base
10a. Name the solid at the right. **hexagonal prism**
   b. What is the height of the solid? **3 units**
   c. How many faces does the solid have? **8**
   d. How many edges does the solid have? **18**
   e. How many vertices does the solid have? **12**
   f. Draw the solid from a different perspective. **See Answer Appendix.**

11. Write a few sentences to tell how you would draw a cylinder. Then draw one. **See Answer Appendix.**

12. Write a few sentences to tell how you would draw a cone. Then draw one. **See Answer Appendix.**

484 Chapter 11 Geometry: Using Area and Volume

Reteaching the Lesson

Activity To help students draw prisms, begin with three edges joined at the vertex representing length, height, and depth. To complete the figure, draw segments parallel to those already drawn.

Error Analysis
Watch for students who cannot distinguish hidden from visible edges.
Prevent by using a model of a rectangular prism to show which parts of the prism are visible and which are not.

484 Chapter 11

13. On grid paper, draw the front view, back view, two side views, and top view of the figure. **See Answer Appendix.**

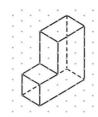

Applications and Problem Solving

14. *History* Solids with faces that are polygons are called *polyhedrons*. During the 1700s, Swiss mathematician Leonard Euler discovered a relationship among the vertices (V), faces (F), and edges (E), of polyhedrons.

    a. Copy and complete the table.

Solid	Vertices (V)	Faces (F)	Edges (E)	V + F − E
triangular prism	6	5	9	6 + 5 − 9 = 2
triangular pyramid	4	4	6	4 + 4 − 6 = 2
rectangular prism	8	6	12	8 + 6 − 12 = 2
rectangular pyramid	5	5	8	5 + 5 − 8 = 2
pentagonal prism	10	7	15	10 + 7 − 15 = 2

   b. What is the relationship among the vertices, faces, and edges of polyhedrons? **See margin.**

   c. If a prism has 10 faces and 24 edges, how many vertices does it have? Name the prism. **16; octagonal prism**

15. *Interior Design* Toshiko is arranging storage cubes to form a wall unit in her bedroom. Sketch two designs that can be made using 6 cubes. **See Answer Appendix.**

16. *Working on the* **CHAPTER Project** Refer to the kite you chose on page 475. Make a three-dimensional scale drawing of your kite. Label the measurements. **See students' work.**

17. *Critical Thinking* What is the least number of faces that a prism can have? What is the least number of faces that a pyramid can have? Draw examples to support your answers. **5 faces; 4 faces; See margin for drawings.**

Mixed Review

18. *Sports* During an attempted foul shot, the shooter stands behind the foul line. All other players must stay out of the shaded region. What is the area of this region? *(Lesson 11-1)* **about 236.5 ft²**

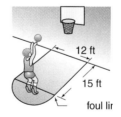

19. **Test Practice** The formula $A = s^2$ is used to find the area of a square. If the area of a square is 40 mm², what is the approximate length of one side of the square? *(Lesson 9-3)* **A**

   A 6.3 mm   **B** 7.5 mm   **C** 18.6 mm   **D** 25.2 mm

Lesson 11-2 Three-Dimensional Figures   **485**

Extending the Lesson

Enrichment Masters, p. 87

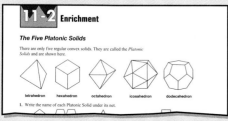

Activity Have groups of students combine 2 to 4 three-dimensional models of the figures in the lesson to create more elaborate shapes. Then have them work together to sketch the new figures on isometric dot paper.

11-3 Lesson Notes

Instructional Resources
- *Study Guide Masters*, p. 88
- *Practice Masters*, p. 88
- *Enrichment Masters*, p. 88
- Transparencies 11-3, A and B
- *Classroom Games*, p. 37–39
- *Science and Math Lab Manual*, pp. 69–72
- *Technology Masters*, p. 73
- CD-ROM Program
  - Resource Lesson 11-3
  - Interactive Lesson 11-3

Recommended Pacing	
Standard	Day 5 of 13
Honors	Day 4 of 11
Block	Day 3 of 6

1 FOCUS

5-Minute Check
(Lesson 11-2)

1. Draw a rectangular prism that is 5 units high, 3 units long, and 2 units deep.

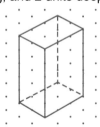

2. Describe the solid in the drawing. Include its name and the number of faces, edges, and vertices.
triangular prism, 5 faces, 9 edges, and 6 vertices

3. Draw the front, back, side, and top views of the solid.

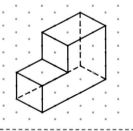

11-3 Volume of Prisms and Cylinders

What you'll learn
You'll learn to find the volumes of prisms and cylinders.

When am I ever going to use this?
Knowing how to find volume can help you determine how much ice a cooler can hold.

Word Wise
volume
circular cylinder

SHOE

In the cartoon, the word **volume** has two meanings. Volume can mean one book in a set. However in geometry, volume is the measure of the space occupied by a solid. It is measured in cubic units. Two common units of measure for volume are cubic centimeter (cm^3) and cubic inch (in^3).

HANDS-ON MINI-LAB

Work with a partner. 12 cubes

Try This
- Model the prism at the right.
- Model at least three other rectangular prisms.
- Assume each edge of the cubes represents one unit and each cube represents one cubic unit. Copy and complete the following table.

Prism	Area of Base	Height	Volume
1	4 square units	2 units	8 cubic units
2			
3			
4			

Talk About It
1. Describe how the area of the base and the height of a prism are related to its volume.
2. Write a formula for computing the volume of a prism.

1. The area of the base times the height equals the volume.
2. $V = Bh$ where V is the volume, B is the area of the base, and h is the height.

Study Hint
Reading Math
Remember that the B in the formula $V = Bh$ represents the area of the base.

Volume of a Prism	**Words:** The volume (V) of a prism is the area of the base (B) times the height (h). **Symbols:** $V = Bh$ **Model:**

486 Chapter 11 Geometry: Using Area and Volume

Sample answer:

front   back   side   top

The 5-Minute Check is also available on **Transparency 11-3A** for this lesson.

In the case of a rectangular prism, the area of the base (B) equals the length (ℓ) times the width (w). For a rectangular prism, the formula $V = Bh$ becomes $V = \ell wh$.

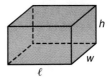

Examples

Find the volume of each prism.

1

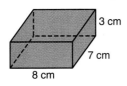

This is a rectangular prism. Use the formula $V = \ell wh$.

$V = \ell wh$

$V = 8 \cdot 7 \cdot 3$ $\ell = 8, w = 7, h = 3$

$V = 168$ The volume is 168 cubic centimeters.

2

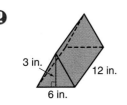

This is a triangular prism. First, find the area of the triangular base.

$A = \frac{1}{2}bh$

$A = \frac{1}{2} \cdot 6 \cdot 3$ $b = 6, h = 3$

$A = 9$

Now find the volume of the prism.

$V = Bh$

$V = 9 \cdot 12$ $B = 9, h = 12$

$V = 108$ The volume is 108 cubic inches.

You can also use the formula $V = Bh$ to find the volume of a cylinder. Most cylinders have circular bases. These cylinders are called **circular cylinders**. For a circular cylinder, the area of the base is the area of a circle (πr^2). So, the formula for finding the volume of a cylinder is $V = \pi r^2 h$.

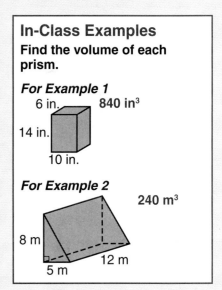

Volume of a Cylinder	Words:	The volume (V) of a cylinder is the area of the base (B) times the height (h).
	Symbols:	$V = Bh$ or $V = \pi r^2 h$ Model:

Lesson 11-3 Volume of Prisms and Cylinders   **487**

Multiple Learning Styles

Auditory/Musical Let students experiment with five identical glass or ceramic cylinders to find how the volume of water in each cylinder affects its tone when struck. Identical coffee mugs will work for this. Students should be able to develop a partial musical scale based on the volume of water. Benjamin Franklin developed a musical instrument called the "glass harmonica" played by rubbing the edges of a set of glasses filled with different volumes of water.

Motivating the Lesson

Hands-On Activity Bring in two cylindrical drinking glasses or vases. Have students measure their heights and diameters in centimeters. Have students guess which holds more. Then have them use measuring cups to fill each with water to determine their volumes. Make a chart including the cylinders' measures and estimated volumes.

2 TEACH

Transparency 11-3B contains a teaching aid for this lesson.

Using the Mini-Lab After students finish the Mini-Lab, ask them to use all 12 cubes to construct rectangular prisms of all possible dimensions and then fill in a table like the one in the Mini-Lab. Students should conclude that if they use 12 cubes to construct a prism, it will contain 12 cubic units no matter what its dimensions— 12 by 1 by 1, 4 by 3 by 1, or another variation.

In-Class Examples

Find the volume of each prism.

For Example 1

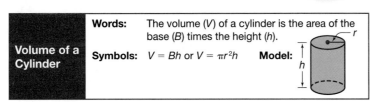

Teaching Tip In this textbook, the word *cylinder* means a circular cylinder.

Teaching Tip Make sure students can differentiate between a radius measurement and a diameter measurement in drawings of cylinders.

Lesson 11-3   **487**

In-Class Example

For Example 3
The figure is a vase that a decorator intends to fill with multicolored sand. How much sand will the vase hold? **about 565.5 cm³**

3 PRACTICE/APPLY

Check for Understanding
If students need additional practice or instruction after completing Exercises 1–6, one of these options may be helpful.
• Extra Practice, see p. 637
• Reteaching Activity
• *Study Guide Masters*, p. 88
• *Practice Masters*, p. 88

Assignment Guide

Core: 7–19 odd, 20–22
Enriched: 8–16 even, 17–22

Study Guide Masters, p. 88

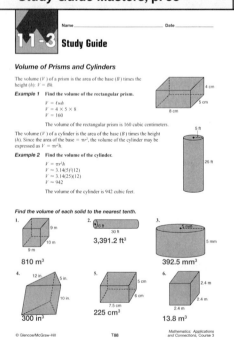

Example 3 APPLICATION

Agriculture The diagram shows the dimensions of a silo. What is the volume of the grain that can be stored in the silo?

The diameter of the base of the cylinder is 16 feet. So, the radius is $\frac{1}{2} \cdot 16$ or 8 feet.

$V = \pi r^2 h \qquad r = 8, h = 40$
$V = \pi \cdot 8^2 \cdot 40 \quad$ *Estimate:* $3 \times 8^2 \times 40 = 7,680$
$V \approx 8,042 \quad$ [π] [×] 8 [x²] [×] 40 [=] *8042.477193*

The volume of the grain is about 8,042 cubic feet.

CHECK FOR UNDERSTANDING

Communicating Mathematics

Read and study the lesson to answer each question.

1. *Compare and contrast* the formula used for finding the volume of a rectangular prism, a triangular prism, and a cylinder. **See Answer Appendix.**

2. *Draw* a cube. Let *s* represent the length of each side of the cube. Write a formula for the volume of a cube. **See margin.**

HANDS-ON MATH

3. *Model* a rectangular prism using 18 cubes. **See students' work.**
   a. Find the area of the base of the prism. **See students' work.**
   b. Find the height of the prism. **See students' work.**
   c. Find the product of the area of the base times the height of the prism. How does this product relate to the volume of the prism? **18; they are the same.**

Guided Practice
Answers are calculated using the π key on a calculator and then rounded.

Find the volume of each solid to the nearest tenth.

4. 18 in³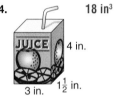

5. 1,125.7 cm³

6. *Energy* Strategic petroleum reserves at Bryan Mound, Texas, are connected to port and terminal facilities at Freeport, Texas, by two pipelines. Each pipeline is $2\frac{1}{2}$ feet in diameter and 4 miles long. Find the maximum volume of the petroleum that can be in two pipelines. **about 207,345 ft³**

EXERCISES

Practice Find the volume of each solid to the nearest tenth.

7. 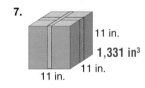 1,331 in³

8. 3,185.6 cm³

9. 2,880 cm³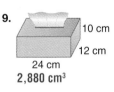

488 Chapter 11 Geometry: Using Area and Volume

■ Reteaching the Lesson ■

Activity Have students build cylinders out of stacks of pennies and calculate the volumes of the cylinders. The thickness of a newly minted penny is 1.55 mm. The diameter is about 19 mm. Students should find that each penny increases the volume by approximately 439 mm³.

Additional Answer

2.

$V = s^3$

10.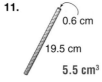
4 ft
$5\frac{1}{2}$ ft $7\frac{1}{2}$ ft
82.5 ft³

11. 0.6 cm
19.5 cm
5.5 cm³

12.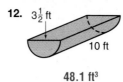
$3\frac{1}{2}$ ft
10 ft
48.1 ft³

13. Estimate the volume of a cylinder that is 10 feet tall and whose diameter is 14 feet.   **Sample answer: 1,500 ft³**

14. For a rectangular prism, $\ell = 3$, $w = 2$, and $h = 5$.
    a. Find the volume of the prism.   **30 units³**
    b. Suppose the height of the prism is doubled. What is the volume of the new prism?   **60 units³**
    c. What is the ratio of the volume of the new prism to the volume of the original prism?   **2:1**
    d. What happens to the volume of a prism if the height is doubled?

14d. The volume is doubled.

15d. It is multiplied by 8.

15. For a rectangular prism, $\ell = 4$, $w = 1$, and $h = 6$.
    a. Find the volume of the prism.   **24 units³**
    b. Suppose all three of the dimensions of the prism are doubled. What is the volume of the new prism?   **192 units³**
    c. What is the ratio of the volume of the new prism to the volume of the original prism?   **8:1**
    d. What happens to the volume of a prism if all three dimensions are doubled?

16. What happens to the volume of a cylinder if the radius is doubled?
    It is quadrupled.

17a. See margin.

Applications and Problem Solving

17. *Health* The inside of a refrigerator in a medical laboratory measures 17 inches by 18 inches by 42 inches.
    a. Tanisha Adams estimates she needs at least 8 cubic feet to refrigerate the samples from the lab. Is the refrigerator large enough for the samples? Explain.
    b. Tell how you would change the dimensions of the refrigerator to double its volume.   **Double one of the dimensions.**

18. *Environment* The cylindrical flue of an incinerator has a diameter of 30 inches and a height of 120 feet. If it takes 15 minutes for the contents of the flue to be expelled into the air, what is the volume of the substances being expelled each hour?   **about 2,356.2 ft³**

19. *Critical Thinking* The diagram shows the dimensions of a swimming pool. If the water line is one foot below the top of the pool, what is the volume of the water in the pool? (*Hint:* Think of the pool as a prism with height of 20 feet.)   **7,200 ft³**

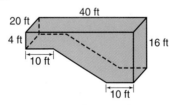
40 ft, 20 ft, 4 ft, 10 ft, 16 ft, 10 ft

Mixed Review

20. Draw a hexagonal prism that is 5 units tall. (*Lesson 11-2*)   **See margin.**

21. **Test Practice**   Doralina is enclosing a circular area of her yard that measures 31 feet in diameter. How much fencing will she need to buy? (*Lesson 7-7*)   **B**
    A 48.69 ft   B 97.39 ft   C 754.77 ft   D 3,019.07 ft

22. *Algebra* Solve $2x - 8 = 12$. (*Lesson 1-7*)   **10**

Lesson 11-3 Volume of Prisms and Cylinders   **489**

Extending the Lesson

Enrichment Masters, p. 88

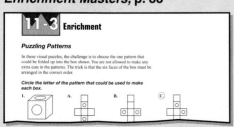

Activity A rectangular prism has a volume of 36 cm³. Have students make a list showing all the possible whole-number dimensions of the prism.
$1 \times 1 \times 36$; $1 \times 2 \times 18$; $1 \times 3 \times 12$;
$1 \times 4 \times 9$; $1 \times 6 \times 6$; $2 \times 2 \times 9$;
$2 \times 3 \times 6$; $3 \times 3 \times 4$

---

4 ASSESS

Closing Activity

Speaking Have students work in pairs to find the volumes of prisms and cylinders. One students says the measure of the sides of a solid; the other says the correct formula. Together they calculate the volume. Students should alternate roles.

Additional Answers

17a. No, the inside dimensions of the refrigerator are $1\frac{5}{12}$ feet by $1\frac{1}{2}$ feet by $3\frac{1}{2}$ feet. Therefore the volume is $1\frac{5}{12} \times 1\frac{1}{2} \times 3\frac{1}{2}$ or $7\frac{7}{16}$ ft³ which is less than 8 ft³.

20.

Practice Masters, p. 88

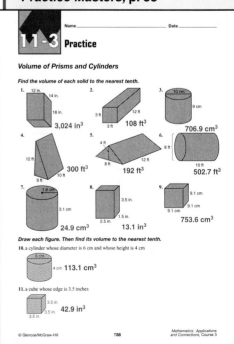

Lesson 11-3   **489**

11-4 Lesson Notes

Instructional Resources
- *Study Guide Masters*, p. 89
- *Practice Masters*, p. 89
- *Enrichment Masters*, p. 89
- Transparencies 11-4, A and B
- *Assessment and Evaluation Masters*, pp. 294, 295
- *Hands-On Lab Masters*, p. 78
- CD-ROM Program
  - Resource Lesson 11-4

Recommended Pacing	
Standard	Day 6 of 13
Honors	Day 5 of 11
Block	Day 3 of 6

1 FOCUS

5-Minute Check
(Lesson 11-3)

1. Find the volume of a rectangular prism that is 5 centimeters long, 4 centimeters wide, and 8 centimeters tall. **160 cm³**

2. Find the volume of a triangular prism that has a base that is 33 square inches and a height of 7 inches. **231 in³**

3. Sandy fences in a circular area 6 feet in diameter. If the fence is 3 feet high, what is the maximum volume of compost she can throw into her fenced enclosure? **about 84.8 ft³**

 The 5-Minute Check is also available on **Transparency 11-4A** for this lesson.

Motivating the Lesson

Problem Solving On the chalkboard, draw a cylinder and a cone, both the same height and base. Ask students *If the cone and cylinder were containers full of ice cream, which would hold more?* **the cylinder** Then, have students guess how much more ice cream would be in the cylinder container than in the cone container. **3 times**

11-4 Volume of Pyramids and Cones

What you'll learn
You'll learn to find the volumes of pyramids and cones.

When am I ever going to use this?
Knowing how to find the volume of cones can help you determine the volume of a mound of sand.

Word Wise
circular cone
altitude

Do you like to buy ice cream in a cone? A sugar cone used for ice cream is an example of the geometric solid called a **circular cone**.

A segment that goes from the vertex of the cone to its base and is perpendicular to the base is called the **altitude**. The height of a cone is measured along the altitude.

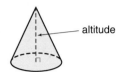

HANDS-ON MINI-LAB

Work with a partner. construction paper, rice, compass, ruler, scissors, tape

Try This

- Draw and cut out a circle with a radius of $1\frac{1}{2}$ inches. Draw and cut out a rectangle that is $9\frac{3}{4}$ inches by $2\frac{5}{8}$ inches. Wrap and tape the rectangle around the circle to form a cylinder with an open top.

- Draw and cut out a circle with a radius of 3 inches. Fold the circle into quarters and open it to form a cone.

- Set the cylinder and cone on your desk with the base of each figure down. Are the solids about the same height? **yes**

- Place the base of the cone on top of the cylinder. Are the circles that form the bases of the cylinder and cone the same size? **yes**

- Fill the cone with rice. Slide the ruler across the top of the rice to make sure the cone is full. Pour the rice into the cylinder. Repeat until the cylinder is filled.

Talk About It

1. How many times did you fill the cone in order to fill the cylinder?
2. How is the area of the base of the cone related to the area of the base of the cylinder? **They are the same.**
3. A cone and a cylinder have the same base and height. What is the ratio of the volume of the cone to the volume of the cylinder? **1:3**
4. Write a formula for finding the volume of a cone.

1. about 3
4. $V = \frac{1}{3}Bh$ where V is the volume, B is the area of the base, and h is the height.

490 Chapter 11 Geometry: Using Area and Volume

The results of the Mini-Lab suggest the formula for finding the volume of a cone.

Volume of a Cone	Words:	The volume (V) of a cone equals one-third the area of the base (B) times the height (h).
	Symbols:	$V = \frac{1}{3}Bh$ or $V = \frac{1}{3}\pi r^2 h$
	Model:	

Example 1 — APPLICATION

Business Manuel fills parfait glasses with vanilla custard and strawberry whip at the Harvest Restaurant. The glasses are in the shape of a cone. If each glass is 7 centimeters across and 15 centimeters tall, what is the volume of each dessert?

If the glass is 7 centimeters across, the radius of the base of the cone is $\frac{1}{2} \times 7$ or 3.5 centimeters.

$V = \frac{1}{3}(\pi r^2)h$ *Replace r with 3.5 and h with 15.*

$V = \frac{1}{3}(\pi \cdot 3.5^2) \cdot 15$ *Estimate: $\frac{1}{3} \cdot 3 \cdot 4^2 \cdot 15 = 240$*

$V \approx 192.4$ 1 ÷ 3 × π × 3.5 x² × 15 = 192.42255

The volume of each dessert is about 192.4 cubic centimeters.

The relationship between a pyramid and a prism is similar to the relationship between a cylinder and a cone. If you have a pyramid and a prism with the same base and height, the ratio of the volume of the pyramid to the volume of the prism is 1 to 3.

Volume of a Pyramid	Words:	The volume (V) of a pyramid equals one-third the area of the base (B) times the height (h).
	Symbols:	$V = \frac{1}{3}Bh$
	Model:	

Example 2

A tetrahedron is a pyramid with a triangular base. Find the volume of the tetrahedron.

15 in.
5 in.   12 in.

First find the area of the triangular base.

$A = \frac{1}{2}bh$ *Formula for area of a triangle*

$A = \frac{1}{2} \times 5 \times 12$ $b = 5, h = 12$

$A = 30$

Then, find the volume of the pyramid.

$V = \frac{1}{3}Bh$ *Formula for area of a pyramid*

$V = \frac{1}{3} \cdot 30 \cdot 15$ $B = 30, h = 15$

$V = 150$

The volume of the tetrahedron is 150 cubic inches.

Lesson 11-4 Volume of Pyramids and Cones

---

2 TEACH

 Transparency 11-4B contains a teaching aid for this lesson.

Using the Mini-Lab Ask students to solve each problem mentally, using the formula they found in the Mini-Lab.
- Find the volume of a cone with a base of 3 square inches and a height of 3 inches. **3 in³**
- A base of 27 square inches and a height of 9 inches. **81 in³**

Teaching Tip In this textbook, the word *cone* means a circular cone.

In-Class Examples
For Example 1
How much shaved ice would it take to fill a snow cone level with the top? Round to the nearest tenth. **11.8 cm³**

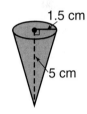
1.5 cm
5 cm

For Example 2
Find the volume of the tetrahedron. **80 m³**

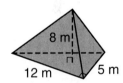
8 m
12 m   5 m

Teaching Tip In Example 2, students calculate the area of a triangle and then the area of a pyramid. Make certain that they remember that the volume of a triangle is one-half the base times the height.

3 PRACTICE/APPLY

Check for Understanding

If students need additional practice or instruction after completing Exercises 1–6, one of these options may be helpful.
- Extra Practice, see p. 638
- Reteaching Activity
- *Study Guide Masters*, p. 89
- *Practice Masters*, p. 89

Assignment Guide

Core: 7–19 odd, 20–22
Enriched: 8–16 even, 17–22
All: Self Test 1–5

Additional Answers

1. Since the area of the base of a cone is πr^2, $V = \frac{1}{3}Bh$ becomes $V = \frac{1}{3}(\pi r^2)h$ or $V = \frac{1}{3}\pi r^2 h$.
2. To find the volume of a prism or a pyramid, you multiply the area of the base times the height. This product is the volume of a prism. To find the volume of a pyramid, you must multiply the product by $\frac{1}{3}$.

Study Guide Masters, p. 89

492 Chapter 11

CHECK FOR UNDERSTANDING

Communicating Mathematics

Read and study the lesson to answer each question. 1–2. See margin.

1. *Explain* why $V = \frac{1}{3}Bh$ and $V = \frac{1}{3}\pi r^2 h$ can both be used to find the volume of a cone.
2. *Compare and contrast* the formulas for the volume of a prism and the volume of a pyramid.

HANDS-ON MATH

3a. Draw and cut out five squares that are 2 inches on each side. Tape the squares together to form a cube with an open top. **See students' work.**
b. Draw and cut out four isosceles triangles with a base of 2 inches and a height of $2\frac{1}{4}$ inches. Tape the triangles together to form a square pyramid without the base. **See students' work.** c. **They are the same.**
c. Compare the bases and heights of the prism and the pyramid.
d. Fill the pyramid with rice. Slide a ruler across the top to make sure the pyramid is full. Pour the rice into the prism. Repeat until the prism is filled. How many times did you need to fill the pyramid to fill the prism? **3 times**
e. Suppose you have a pyramid and a prism with the same base and height. What do you think would be the ratio of the volume of the pyramid to the volume of the prism? **1:3**

Answers are calculated using the π key on the calculator and then rounded.

Guided Practice Find the volume of each solid to the nearest tenth.

4. 4,241.2 cm³ (cone, 18 cm height, 15 cm radius)
5. 121.8 m³ (pyramid, 7.1 m, 6.2 m, 8.3 m)

6. **Physical Science** A model of a volcano constructed for science class has a height of 10 inches and a diameter of 9 inches. What is the volume of the volcano? **about 212.1 in³**

EXERCISES

Practice Find the volume of each solid to the nearest tenth.

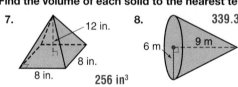

7. 256 in³ (pyramid, 12 in., 8 in., 8 in.)
8. 339.3 m³ (cone, 6 m, 9 m)
9. 56 yd³ (pyramid, 7 yd, 6 yd, 8 yd)

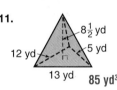
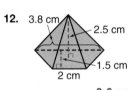

10. 2,565.6 in³ (cone, $24\frac{1}{2}$ in., 20 in.)
11. 85 yd³ (pyramid, $8\frac{1}{2}$ yd, 5 yd, 12 yd, 13 yd)
12. 3.6 cm³ (pyramid, 3.8 cm, 2.5 cm, 1.5 cm, 2 cm)

13. The area of the base of a hexagonal pyramid is 45.3 square meters. Its height is 19.7 meters. Estimate the volume of the pyramid. **Sample answer: 300 m³**

492 Chapter 11 Geometry: Using Area and Volume

Reteaching the Lesson

Activity Have students use reference materials to find the dimensions of the Great Pyramid at Giza, Egypt, and then calculate its volume. **The pyramid has a square base 756 feet on a side, and it was originally 481 feet tall. Therefore, its volume is about 91.6 million cubic feet.**

Error Analysis
Watch for students who confuse the *b* (base) in the formula for the area of a triangle with *B* (base) in the formula for volume of a pyramid.
Prevent by suggesting that students think of *b* as representing a smaller value—the base of the triangle, in units. Likewise, *B* represents a larger value—the base of the pyramid, in units².

14. The radius of a cone is 5 inches. The height is 8 inches.
    a. Find the volume of the cone.   about 209.44 in³
    b. Suppose the radius of the cone is doubled and the height remains the same. What is the volume of the new cone?   about 837.76 in³
    c. What is the ratio of the volume of the new cone to the volume of the original cone?   4:1
    d. What happens to the volume of a cone if the radius is doubled?

14d. It is quadrupled.
15. It is doubled.

15. What happens to the volume of a pyramid if the height is doubled?

16. A cone and a cylinder each have a diameter of 10 centimeters. A prism and a pyramid each have square bases that are 10 centimeters on a side. All of the solids have a height of 8 centimeters. Without calculating the volume of the solids, order the solids according to their volume starting with the solid with the least volume.   **cone, pyramid, cylinder, prism**

17. about 31,672,000 ft³
Applications and Problem Solving

17. *Architecture* The Great American Pyramid in Memphis, Tennessee, was built as a memorial to American music. It is 321 feet high and has a base that is about 296,000 square feet. Find the approximate volume of this pyramid.

18. *Highway Maintenance* A mixture of salt and sand is used to melt snow and ice on highways. A mound of this mixture is in a conical shape. It has a diameter of 18 feet and a height of 12 feet.
    a. Find the volume of the salt and sand.   about 1,017.9 ft³
    b. If 1 cubic foot of this mixture is needed for each 500 square feet of highway, how many square feet of highway can be treated with the salt and sand in this mound?   about 508,950 ft²

19. The new height would be $\frac{1}{4}$ of the original height.

19. *Critical Thinking* Suppose the radius of a cone is doubled. How could you change the height so that the volume will remain the same?

Mixed Review

20. **Test Practice**   What is the volume of a rectangular solid that measures 4 feet by 7 feet by 3 feet? *(Lesson 11-3)*   **C**
    A 14 ft³   **B** 28 ft³   **C** 84 ft³   **D** 128 ft³

21. *Money Matters* Find the percent of change if last month's electric bill was $52.50 and this month's bill is $57.20. *(Lesson 8-5)*   about 9%

22. Order 23, −8, 0, −16, 51, −51, and −30 from greatest to least. *(Lesson 2-2)*
    51, 23, 0, −8, −16, −30, −51

CHAPTER 11 Mid-Chapter Self Test

1. *Sports* In track, a shot-putter must stay inside a circle with a diameter of 7 feet. What is the area in which the athlete has to move in this competition? *(Lesson 11-1)*   about 38.5 ft²

2. Use isometric dot paper to draw a rectangular prism that is 2 units high, 5 units long, and 3 units deep. *(Lesson 11-2)*   **See margin.**

Find the volume of each solid to the nearest tenth. *(Lessons 11-3 and 11-4)*

3. 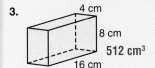 512 cm³

4.   923.6 in³

5. 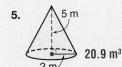 20.9 m³

Lesson 11-4 Volume of Pyramids and Cones   **493**

Extending the Lesson

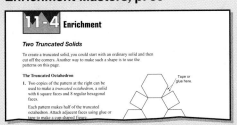

Activity Have students work in small groups to find the volume of the lower section of the cone.   about 980.2 in³

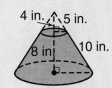

11-5A LAB Notes

GET READY

Objective Students recognize a solid from its net and sketch it, and vice versa.

Optional Resources
Hands-On Lab Masters
• worksheet, p. 61
Manipulative Kit
• rulers
• scissors

Recommended Time
30 minutes

Getting Started Ask students to give examples of objects that can be formed from patterns. Bring patterns for making clothing, furniture, or holiday decorations to show the class.

The **Activity** directs students to explore the surface area of solids by drawing nets. Students will use the connection between solids and nets when studying surface area of various solids in the lessons to come. Before students cut open their boxes, ask them to make a rough sketch predicting how the flattened box will look.

ASSESS

Have students complete Exercises 1–4. Watch for students that may have difficulty identifying shapes and congruence.

Use Exercise 5 to determine whether students have discovered the connections among surface area, nets, and solids.

HANDS-ON LAB
COOPERATIVE LEARNING

11-5A Nets

 scissors

 boxes in different sizes and shapes

 ruler

A Preview of Lesson 11-5

Every solid with at least one flat surface can be formed from a two-dimensional pattern called a *net*.

TRY THIS

Work with a partner.

Step 1 Each person should pick at least one empty box. You can choose a cereal box, a pasta box, or any other type of box. Tape the box shut.

Step 2 Flatten each box by cutting along some of the edges. You should have only one piece for each box when it is flattened. Try to cut each box in a different way.

Step 3 Trace or sketch each flattened box showing each of the shapes used for the sides of the box. Your drawing is called a net.

Step 4 Cut out each net. Fold the net along the interior lines to form a box.

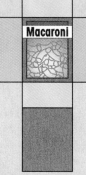

1. See students' work. Nets of boxes that are rectangular prisms will be made up of rectangles.
2. See students' work. Nets of boxes that are rectangular prisms will have 6 shapes.
3. See students' work. Nets of boxes that are rectangular prisms will have 3 sets of congruent shapes.

ON YOUR OWN

1. **Describe** each shape that makes up each net.
2. **Tell** how many shapes are needed for each net.
3. **Identify** any congruent shapes in each net.
4. **Describe** the solid formed by the net shown at the right. Sketch the solid. **See margin.**
5. **Look Ahead** Use one of your boxes. **a–d. See students' work.**
   a. Measure the length, width, and height of the box.
   b. Estimate the total area of the sides of the box.
   c. Compute the area of each side of the box. Record the area of each surface on your net.
   d. Find the sum of the areas of the sides of the box.

494 Chapter 11 Geometry: Using Area and Volume

Math Journal Have students write a brief paragraph describing how to make a net of a building.

Additional Answer
4. triangular prism;

11-5 Surface Area of Prisms

What you'll learn
You'll learn to find the surface areas of rectangular and triangular prisms.

When am I ever going to use this?
Knowing how to find surface area can help you determine the amount of canvas needed to construct a tent.

Word Wise
surface area

The faces of the three-dimensional figures that form Tony Smith's modern sculpture *Wandering Rocks* are constructed of stainless steel. The artist first constructed the sculpture in plywood and then ordered the stainless steel for the final product. Mr. Smith had to know the **surface area** of the sculpture to determine the amount of stainless steel to order. Surface area is the sum of the areas of all faces or surfaces of a solid.

Study Hint
Reading Math Remember that volume is given in cubic units and surface area is given in square units.

1. 3 units, 2 units, 6 units
2. 12 units², 18 units², 12 units², 18 units², 6 units², 6 units²

HANDS-ON MINI-LAB

Work with a partner. grid paper scissors tape

Try This
- Draw the pattern on grid paper and cut it out. *This pattern is an example of a net.*
- Fold the pattern along the red lines and tape the edges to form a rectangular prism.

Talk About It
1. What are the length, width, and height of the prism?
2. What is the area of each face of the prism?
3. What is the total surface area? **72 units²**

To find the surface area of a rectangular prism, you must first determine the area of each of its six faces.

Example 1 Find the surface area of the rectangular prism.

Faces	Area
top and bottom	8 × 5 or 40
front and back	8 × 10 or 80
right and left sides	5 × 10 or 50

Add the areas.
2(40) + 2(80) + 2(50) = 340
The surface area of the rectangular prism is 340 square centimeters.

Lesson 11-5 Surface Area of Prisms **495**

11-5 Lesson Notes

Instructional Resources
- *Study Guide Masters*, p. 90
- *Practice Masters*, p. 90
- *Enrichment Masters*, p. 90
- Transparencies 11-5, A and B
- *Technology Masters*, p. 74
- CD-ROM Program
  - Resource Lesson 11-5
  - Interactive Lesson 11-5

Recommended Pacing	
Standard	Days 7 & 8 of 13
Honors	Day 6 of 11
Block	Day 4 of 6

1 FOCUS

5-Minute Check
(Lesson 11-4)
Find the volume of each solid to the nearest tenth.

1.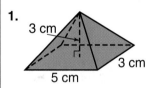
   15 cm³

2.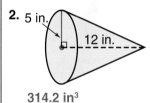
   314.2 in³

3.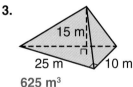
   625 m³

The 5-Minute Check is also available on **Transparency 11-5A** for this lesson.

Motivating the Lesson
Communication Ask, *Which will evaporate faster, a wide shallow pool or a deep narrow pool of the same length and containing the same amount of water? Explain.*
the shallow pool because more surface area is exposed for evaporation

Lesson 11-5 **495**

2 TEACH

 Transparency 11-5B contains a teaching aid for this lesson.

Using the Mini-Lab Ask students to calculate the increase in total surface area if the prism measured $4 \times 5 \times 12$ units. **184 units²**

In-Class Examples

For Example 1
Find the surface area of the rectangular prism. **142 cm²**

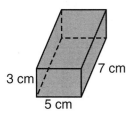

For Example 2
A European candy company sells its chocolate in triangular prisms. How much cardboard would they need to make a box of the following dimensions? **206 cm²**

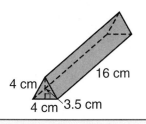

Teaching Tip In Example 2 of the Student Edition, point out that the ends of the tent are equilateral triangles.

Additional Answer

1. Face   Area
   front ℓh
   back ℓh
   top ℓw
   bottom ℓw
   side wh
   side wh
   The sum of the 6 faces is $\ell h + \ell h + \ell w + \ell w + wh + wh$ or $2\ell h + 2\ell w + 2wh$.

Example 2 — APPLICATION

Camping The Kennedy Middle School Camping Club has designed a pup tent. The tent has a canvas floor as well as canvas sides. How much canvas will the students need to construct the tent?

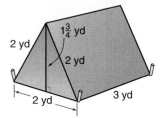

Explore You know the dimensions of a tent that is in the shape of a triangular prism. You need to find the amount of canvas needed to construct the tent.

Plan To find the amount of canvas needed, you must find the surface area of a triangular prism. To find the surface area, you must find the area of each face of the prism.

Solve Two faces of the tent are triangles, and the area of each triangle is $\frac{1}{2} \times 2 \times 1\frac{3}{4}$ or $1\frac{3}{4}$ square yards. Three faces of the tent are rectangles, and the area of each rectangle is 3×2 or 6 square yards.

area of triangular faces $2 \times 1\frac{3}{4}$ or $3\frac{1}{2}$
area of rectangular faces 3×6 or 18
                                                   Total $21\frac{1}{2}$

The students will need $21\frac{1}{2}$ square yards of canvas.

Examine

Faces		Area
front	$\frac{1}{2} \times 2 \times 1\frac{3}{4}$ or	$1\frac{3}{4}$
back	$\frac{1}{2} \times 2 \times 1\frac{3}{4}$ or	$1\frac{3}{4}$
floor	3×2 or	6
right side	3×2 or	6
left side	3×2 or	6
	Total	$21\frac{1}{2}$ ✓

CHECK FOR UNDERSTANDING

Communicating Mathematics

Read and study the lesson to answer each question.

1. *Explain* why the expression $2\ell h + 2\ell w + 2wh$ could be used to find the surface area of the rectangular prism. **See margin.**

2. *Write* an expression for the surface area of a cube if each edge is s units long. **$6s^2$**

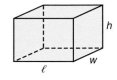

496 Chapter 11 Geometry: Using Area and Volume

Reteaching the Lesson

Activity Have students draw a net of the figure in Example 2 of the Student Edition on grid paper. Then have them cut it out, fold it, and tape it as they did for the figure in the Mini-Lab. You may wish to have students color each face on both models a different color or to number each face. Have them refer to both models when working the exercises.

Error Analysis
Watch for students who omit one or more faces when calculating the total surface area.
Prevent by having students systematically list all faces of the prism.

HANDS-ON MATH

3. *Draw* a net for a rectangular prism that is 2 units by 3 units by 4 units on grid paper. Draw a net for a rectangular prism that is 4 units by 6 units by 8 units on grid paper. **See Answer Appendix.**
   a. Find the surface area of each prism. **52 units²; 208 units²**
   b. Find the ratio of the length of each side of the first prism to the length of the corresponding side of the second prism. **1:2**
   c. Write a ratio comparing the surface area of the first prism to the surface area of the second prism. **1:4**
   d. Do the ratios from part b and c form a proportion? Explain. **No; $\frac{1}{2} \neq \frac{1}{4}$.**

Guided Practice

Family Activity
Look for rectangular containers in your home. Choose one and work with a family member to design and cut out all the possible nets that will completely cover the solid without overlapping.

Find the surface area of each prism to the nearest tenth.

4. 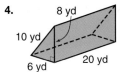 **528 yd²**
5. 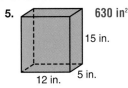 **630 in²**

6. Estimate the surface area of a rectangular prism with a length of 2.06 cm, a width of 1.89 cm, and a height of 0.91 cm. **Sample answer: 16 cm²**

7. *Sports* A water skiing ramp is built in the shape of a triangular prism. Salali is planning to paint it with water repellent paint. Find the surface area to be painted. **17.04 m²**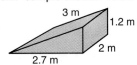

EXERCISES

Practice

Find the surface area of each prism to the nearest tenth.

8. 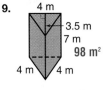 **294 ft²**
9. 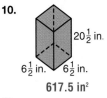 **98 m²**
10.  **617.5 in²**
11. 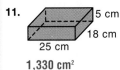 **1,330 cm²**
12. 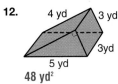 **48 yd²**
13. 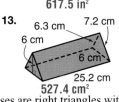 **527.4 cm²**

14. Find the surface area of a triangular prism whose bases are right triangles with legs 6 feet and 8 feet and whose height is 10 feet. **288 ft²**

15. Estimate the surface area of a cube with each edge 9.7 meters long.

16. Suppose the length of each edge of a cube is doubled. Find the ratio of the surface area of the original cube to the surface area of the new cube. **1:4**

17. *Write a Problem* that requires computing the surface area of a prism. **See margin.**

15. Sample answer: 600 m²

Applications and Problem Solving

18. *Horticulture* Find the area of the glass needed to cover the roof and sides of the greenhouse. **412 ft²**

Lesson 11-5 Surface Area of Prisms **497**

Additional Answer

17. Sample answer: Mary is redecorating her room. She wants to cover the 5 sides of an open storage cube with contact paper. If each edge of the cube is 2 ft long, what is the area that needs to be covered with contact paper?

---

3 PRACTICE/APPLY

Check for Understanding
If students need additional practice or instruction after completing Exercises 1–7, one of these options may be helpful.
- Extra Practice, see p. 638
- Reteaching Activity
- *Study Guide Masters*, p. 90
- *Practice Masters*, p. 90
- Interactive Mathematics Tools Software

Family Activity

Have each student bring in one net that has been cut out, folded, and taped together. Have them decorate the net to represent the original container. Have the students use these models to explain how to find the surface area of a solid.

Assignment Guide

Core: 9–19 odd, 20–23
Enriched: 8–16, 18–23

Study Guide Masters, p. 90

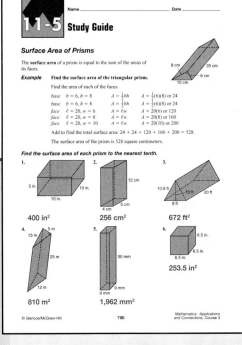

Lesson 11-5 **497**

4 ASSESS

Closing Activity
Speaking Bring various rectangular and triangular prisms to class. Have small groups of students calculate the surface areas of the shapes and then show the class how they did so.

Additional Answer
22.

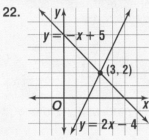

19. Block #4; its surface area is 96 in² which is less than the surface areas of the other blocks (196 in², 112 in², and 168 in²).

19. **Physical Science** The rate that ice melts depends on the amount of surface area. The greater the surface area, the faster it will melt. Which of the following blocks of ice will be the last to melt? Explain.

   block 1:   1 inch by 2 inches by 32 inches
   block 2:   4 inches by 8 inches by 2 inches
   block 3:   16 inches by 4 inches by 1 inch
   block 4:   4 inches by 4 inches by 4 inches

20. **Critical Thinking** The length of each side of a cube is 3 inches. Suppose the cube is painted and then it is cut into 27 smaller congruent cubes.
   a. How many of the smaller cubes will be painted on three sides? **8 cubes**
   b. How many of the smaller cubes will be painted on two sides? **12 cubes**
   c. How many of the smaller cubes will be painted on one side? **6 cubes**
   d. How many of the smaller cubes will not be painted at all? **1 cube**

Mixed Review

21. **Geometry** Find the volume of a cone with a radius of 2 meters and a height of 12 meters. *(Lesson 11-4)* **about 50.3 m³**

22. **Algebra** Solve the system $y = -x + 5$ and $y = 2x - 4$ by graphing. *(Lesson 10-5)* **See margin.**

23. **Test Practice** What is the GCF of 84 and 36? *(Lesson 6-3)* **D**
    A 2   **B** 3   **C** 6   **D** 12

Architest

Get Ready This game is for two players.   cubes

Get Set Each player receives 15 cubes.

Go
- Each player designs a structure with some of his or her cubes. The player then draws the top, front, back, and side views of the structure and computes its volume and surface area. Assume each cube represents one cubic unit.

- Player A tries to guess Player B's structure. Player A does this by asking Player B for information about the structure. Pieces of information include the volume, surface area, or drawing of one view of the structure. Player A receives 7 points for correctly guessing Players B's structure after receiving one piece of information, 6 points for correctly guessing after two pieces of information, and so on. If Player A cannot guess Player B's structure after receiving all 7 pieces of information, then Player B receives 3 points.

- Player B now tries to guess Player A's structure, according to the same rules.

- Play continues with different structures. The player with the most points at the end of the game wins.

interNET CONNECTION Visit www.glencoe.com/sec/math/mac/mathnet for more games.

498 Chapter 11 Geometry: Using Area and Volume

Practice Masters, p. 90

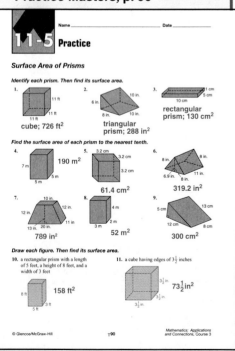

Extending the Lesson
Enrichment Masters, p. 90

When guessing each other's structures, students should consider the attributes of various styles of structures. Ask them which attributes help them the most to guess correctly quicker.

11-6 Surface Area of Cylinders

What you'll learn
You'll learn to find the surface areas of cylinders.

When am I ever going to use this?
Knowing how to find the surface area of cylinders can help you determine the amount of sheet metal needed to make a can.

People in the United States consume more soft drinks than any other type of beverage. Soft drinks are usually sold in containers in the shape of cylinders because they are easy to hold, relatively easy to manufacture, and can be easily sealed. The amount of material needed to manufacture a beverage container depends on the surface area of the cylinder.

As with prisms, you find the surface area of a cylinder by finding the area of the two bases and adding the area of the side. However, the "side" of a cylinder is one curved surface.

HANDS-ON MINI-LAB

Work with a partner.   cardboard tube   ruler   scissors

Try This
- Measure the length of the cardboard tube and its diameter in centimeters.
- Draw a line down the tube so that it is perpendicular to the bases of the cylindrical shape.
- Cut along the line and flatten the tube.

Talk About It
1. Describe the shape of the flattened tube. What part of the cylinder does the flattened tube represent? **a rectangle; the curved surface**
2. Measure the dimensions of the flattened tube in centimeters.
3. Find the circumference of either base of the original tube. How is this measure related to the dimensions of the flattened tube?
4. How is the height of the tube related to the dimensions of the flattened tube?
5. What is the area of the flattened tube? **See students' work.**
6. Write an expression for the area of the curved surface of a cylinder using d for the diameter of the cylinder and h for the height of the cylinder. πdh

2. See students' work.
3. See students' work; the circumference is the same as the measure of one side of the rectangle.
4. The height of the tube is the same as the measure of one side of the rectangle.

LOOK BACK
You can refer to Lesson 7-7 for information on circumference.

In the Mini-Lab, you see that the curved side of the cylinder can be flattened into a rectangle. The rectangle has the same height (h) as the cylinder. Its length is the same measure as the circumference (C) of the base. So, the area of the curved surface is Ch or πdh or $2\pi rh$.

The area of each base of a cylinder is πr^2. So the sum of the areas of the bases is $\pi r^2 + \pi r^2$, or $2\pi r^2$.

Lesson 11-6 Surface Area of Cylinders **499**

11-6 Lesson Notes

Instructional Resources
- *Study Guide Masters*, p. 91
- *Practice Masters*, p. 91
- *Enrichment Masters*, p. 91
- Transparencies 11-6, A and B
- *Assessment and Evaluation Masters*, p. 296
- *School to Career Masters*, p. 37
- CD-ROM Program
  - Resource Lesson 11-6
  - Interactive Lesson 11-6

Recommended Pacing
Standard	Days 9 & 10 of 13
Honors	Days 7 & 8 of 11
Block	Day 5 of 6

1 FOCUS

5-Minute Check
(Lesson 11-5)

Find the surface area of each prism to the nearest tenth.

1.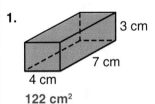

   3 cm, 7 cm, 4 cm

   122 cm²

2.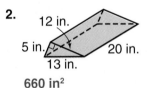

   12 in., 5 in., 20 in., 13 in.

   660 in²

3. Find the surface area of this structure before a door is cut to make it into a doghouse. Assume the doghouse has no floor. **228 ft²**

   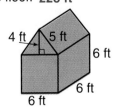

   4 ft, 5 ft, 6 ft, 6 ft, 6 ft

The 5-Minute Check is also available on **Transparency 11-6A** for this lesson.

Motivating the Lesson
Hands-On Activity Pass out examples of rectangular prisms and cylinders. Have students in groups describe similarities and differences between the two types of objects.

Lesson 11-6 **499**

2 TEACH

 Transparency 11-6B contains a teaching aid for this lesson.

Using the Mini-Lab Show students a can label to convince them that the side of a circular can is actually a rectangle.

In-Class Examples

For Example 1
Find the surface area of the cylinder to the nearest tenth.
1,105.8 cm²

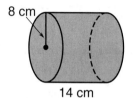

For Example 2
Stuffing mix is available in boxes and canisters. The two figures hold about the same amount of stuffing. Which container requires less material to make? how much less? **the canister; about 19.3 in² less**

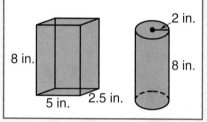

Teaching Tip While working the examples, remind students of the difference in formulas for the area of a circle and the circumference of a circle.

The surface area of a cylinder is the sum of the areas of the two circular bases and the area of the curved surface.

Surface Area of a Cylinder		
	Words:	The surface area of a cylinder equals two times the area of the circular bases ($2\pi r^2$) plus the area of the curved surface ($2\pi rh$).
	Symbols:	surface area = $2\pi r^2 + 2\pi rh$
	Model:	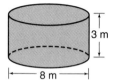

Examples

1 Find the surface area of the cylinder.

The circular base of the cylinder has a diameter of 8 meters. This means the radius is $\frac{1}{2} \times 8$ or 4 meters. The height of the cylinder is 3 meters. Use this information to find the area of each surface.

surface area = $2\pi(4^2) + 2\pi(4)(3)$

Estimate: $2 \times 3 \times 4^2 + 2 \times 3 \times 4 \times 3 = 168$

2 × π × 4 x² + 2 × π × 4 × 3 = *175.9291886*

The surface area of the cylinder is about 175.9 square meters.

APPLICATION

2 Construction An airport has changed the carrels used for public telephones. The new carrels are half of a cylinder with an open top. The old carrels consisted of four sides of a rectangular prism. How much less material is needed to construct a new carrel than an old carrel?

Did you know Although the United States has more telephones than any other nation, the small country of Monaco has the most telephones per person. It has 96 telephones for every 100 people.

First, find the area of each surface of a new carrel.

bottom $\quad \frac{1}{2}\pi \cdot 12^2 \quad \rightarrow \quad$ 1 ÷ 2 × π × 12 x² +

curved surface $\quad \frac{1}{2}(\pi \cdot 24) \cdot 42 \quad \rightarrow \quad$ 1 ÷ 2 × π × 24 × 42

$\quad\quad\quad\quad\quad\quad$ Total $\rightarrow$ = *1809.557368*

Multiple Learning Styles

 Visual/Spatial Have students make two-dimensional drawings of a cube, a cylinder, and a triangular pyramid. Use these drawings to find the surface area of each solid. Also, find the volume of each solid.

Then, find the area of each surface of an old carrel.

bottom	24 × 12	→	24 × 12 +
back	24 × 42	→	24 × 42 +
right side	12 × 42	→	12 × 42 +
left side	12 × 42	→	12 × 42
	Total	→	= 2304

About 2,304 − 1,810 or 494 square inches of material are saved.

CHECK FOR UNDERSTANDING

Communicating Mathematics

Read and study the lesson to answer each question.

1. *Explain* why the expression $2\pi r^2 + 2\pi rh$ could be used to find the surface area of the cylinder. **See margin.**

2. *Write a Problem* where you need to find the surface area of a cylinder.

   2. Sample answer: A can has a radius of 8 cm and a height of 12 cm. Find the area of the material needed to make the can.

3. *You Decide* The two solids have about the same volume. Geraldo says that the surface area of the cylinder is less than the surface area of the rectangular solid. Katie disagrees. Who is correct? Explain. **See margin.**

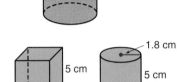

HANDS-ON MATH

4. *Trace* the bases of a can on grid paper. Cut out the circles. Measure the height of the can. Cut a long strip of grid paper so that its width is the height of the can. Wrap the strip around the can. Cut the excess paper off so that the strip just fits around the can.   **a. two circles and a rectangle**

   a. Describe the shape of each of the three pieces you have created
   b. Assume each square on the grid represents one square unit. Use the three pieces to estimate the surface area of the can.   **See students' work.**

Answers are calculated using the π key on the calculator and then rounded.

Guided Practice

5. Find the surface area of the cylinder to the nearest tenth.   **766.9 cm²**

6. A cylinder has a diameter of 7.88 centimeters and a height of 6.04 centimeters. Estimate the area of the curved surface of the cylinder.   **Sample answer: 144 cm²**

7. *Community Services* A community is building a cylindrical tank to store water. The diameter of the tank will be 20 feet, and its height will be 24 feet. The tank will be lined with a material to prevent corrosion and contamination. How much of this material will be needed to line the tank?   **about 2,136.3 ft²**

Lesson 11-6 Surface Area of Cylinders **501**

Reteaching the Lesson

Activity Have students measure a circular salt container. Then have them cut the container into two circular bases and a rectangle. Show that the length of the rectangle equals the circumference of the base. Have students use their measurements to calculate the surface area.

Error Analysis
Watch for students who confuse surface area and volume.
Prevent by teaching the mnemonic, "Surface skin, volume in."

3 PRACTICE/APPLY

Check for Understanding
If students need additional practice or instruction after completing Exercises 1–7, one of these options may be helpful.
- Extra Practice, see p. 638
- Reteaching Activity
- *Study Guide Masters,* p. 91
- *Practice Masters,* p. 91

Additional Answers
1. area of top πr^2
   area of bottom πr^2
   area of curved surface πdh or $\pi(2r)h$ or $2\pi rh$

   The surface area of the cylinder is $\pi r^2 + \pi r^2 + 2\pi rh$ or $2\pi r^2 + 2\pi rh$.

3. Geraldo
   Surface Area of Rectangular Prism
top	10 cm²
bottom	10 cm²
front	25 cm²
back	25 cm²
side	10 cm²
side	10 cm²
Total	90 cm²

   Surface Area of Cylinder
top	about 10.2 cm²
bottom	about 10.2 cm²
curved surface	about 56.5 cm²
Total	about 76.9 cm²

Study Guide Masters, p. 91

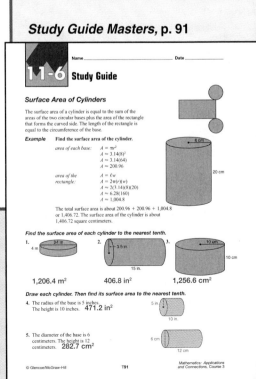

Lesson 11-6 **501**

Assignment Guide
Core: 9–15 odd, 17–19
Enriched: 8–14 even, 15, 17–19

CHAPTER Project

Exercise 16 asks students to advance to the next stage of work on the Chapter Project. You may wish to have students work in groups to determine the cost and amount of materials needed. Such information could be gathered by visiting or calling stores, or searching through newspaper ads, catalogs, or the Internet.

4 ASSESS

Closing Activity
Modeling Provide a selection of cylinders (cans, potato chip containers, and so on). Have students measure the cylinders and find their surface areas.

Chapter 11, Quiz C (Lessons 11-5 and 11-6) is available in the *Assessment and Evaluation Masters*, p. 296.

Practice Masters, p. 91

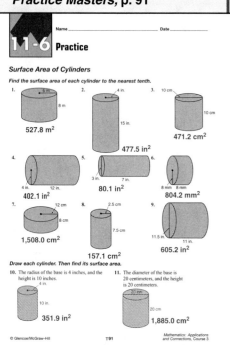

EXERCISES

Practice Find the surface area of each cylinder to the nearest tenth.

8. 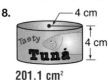 4 cm, 4 cm
201.1 cm²

9. 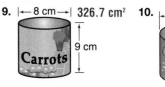 8 cm, 326.7 cm², 9 cm

10. $3\frac{1}{4}$ in., $4\frac{1}{2}$ in.
62.5 in²

11. Find the area of the curved surface of a cylinder with a radius of 3 feet and a height of 9 feet. **about 169.6 ft²**

12. Find the surface area of a cylinder with a diameter of 4.6 meters and a height of 4.6 meters. **about 99.7 m²**

13. Estimate the surface area of a cylinder with a diameter of 10 inches and a height of 10 inches. **Sample answer: 450 in²**

Applications and Problem Solving

14. *Maintenance* The toddler pool at the community center is cylindrical. It has a radius of 12 feet and is 2 feet deep. If one gallon of paint covers 24 square feet, how many gallons of paint need to be purchased to paint the toddler pool? **26 gal**

15. *Mail* Find the amount of metal needed to construct the mailbox.
about 227.4 in² 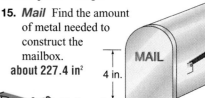 4 in., 10 in., 4 in.

16. *Working on the* **CHAPTER Project** Refer to the kite you designed on page 485.
   a. Determine the amount of materials you will need to make your kite. Some of your materials are sold by the length, and some of the materials are sold by the square unit. **a–b. See students' work.**
   b. Determine the cost of these materials.

17. *Critical Thinking* Will the surface area of a cylinder increase more if you double the height or double the radius of the base? Explain. **See Answer Appendix.**

Mixed Review

18. *Geometry* Find the surface area of a rectangular prism that is 5 inches long, 3 inches wide, and 2 inches tall. *(Lesson 11-5)* **18. 62 in²**

19. **Test Practice** Haloke invited 20 people to a skating party. The cost of food and skating is $5.17 per person. Which is the best estimate of the cost of the party? *(Lesson 3-6)* **D**
   A less than $40
   B between $50 and $60
   C between $70 and $80
   D between $100 and $110
   E more than $130

502 Chapter 11 Geometry: Using Area and Volume

Extending the Lesson

Enrichment Masters, p. 91

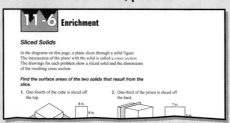

Activity Have students compare the surface area for a cylinder found by using the formula $2\pi r(r + h)$ with the surface area found by using the formula in this lesson. Have them write about their observations.

COOPERATIVE LEARNING

11-6B Surface Area and Volume

A Follow-Up of Lesson 11-6

calculator

The average cylindrical soft drink can holds 355 milliliters of beverage. In the metric system, 1 milliliter = 1 cubic centimeter. Therefore, the volume of one of these cans is about 355 cubic centimeters. A manufacturer uses a can that has a radius of 3.3 centimeters and a height of 10.4 centimeters, but is considering changing the shape of the can. The manufacturer would like the cans to have the least surface area possible, unless it would hurt sales.

TRY THIS

Work with a partner.

Step 1 Design three more cylinders that have a volume between 354 and 356 cubic centimeters. Use a calculator to help determine appropriate measurements. Record the dimensions and volume of each cylinder in a chart. Include at least one cylinder whose radius is less than 3.3 centimeters and at least one whose radius is more than 3.3 centimeters.

Cylinder	Height	Radius	Volume	Surface Area
A	10.4 cm	3.3 cm	355.8 cm^3	284.1 cm^2
B				
C				
D				

Step 2 Compute the surface area of each cylinder.

Step 3 Design three rectangular prisms that have a volume between 354 and 356 cubic centimeters. Record the dimensions and volume of each prism in a chart.

Prism	Length	Width	Height	Volume	Surface Area
E					
F					
G					

Step 4 Compute the surface area of each prism.

ON YOUR OWN

1. Which container resulted in the greatest surface area?   **1–3. See students' work.**
2. Which container resulted in the least surface area?
3. *Write* a paragraph recommending a container for the manufacturer to use. Give your reasons. Consider the appearance as well as the surface area.
4. *Reflect Back* In general, what type of container would hold the most liquid for the least amount of surface area?   **a cylinder**

 Have students write a paragraph explaining why a manufacturer would minimize surface area rather than volume and how she would do so.

HANDS-ON 11-6B LAB Notes

GET READY

Objective Students investigate how surface area and volume are related.

Optional Resources
Hands-On Lab Masters
- cylinder pattern, p. 24
- worksheet, p. 62

MANAGEMENT TIPS

Recommended Time
45 minutes

Getting Started Blow up a balloon. Ask students to describe the difference between the surface area and the volume of the balloon. Ask students to list several factors that a manufacturer might take into consideration when choosing a product container.

The **Activity** asks students to find the surface area of various-sized cylinders and prisms. By allowing students to select dimensions for the shapes, they get a better opportunity to discover the relationship between surface area and volume. Before students finish Step 1, point out that because the radius is *squared* in the formula for the volume of a cylinder, students must alter the height more than the radius in their search for equal volumes.

ASSESS

Have students complete Exercises 1–3. Watch for students who may have difficulty calculating volume and surface area. Have them review earlier lessons in the chapter.

Use Exercise 4 to determine whether students have discovered one of the relationships between surface area and volume.

11-7 Lesson Notes

Instructional Resources
- *Study Guide Masters*, p. 92
- *Practice Masters*, p. 92
- *Enrichment Masters*, p. 92
- Transparencies 11-7, A and B
- *Assessment and Evaluation Masters*, p. 296

 CD-ROM Program
- Resource Lesson 11-7

Recommended Pacing	
Standard	Day 11 of 13
Honors	Day 9 of 11
Block	Day 5 of 6

1 FOCUS

 5-Minute Check
(Lesson 11-6)

1. Find the surface area of the cylinder to the nearest tenth. **1,131.0 in²**

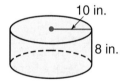

2. Find how much material is required to cover the surface of this cheese. **about 113.1 in²** What is the volume of the cheese? **about 84.8 in³**

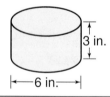

 The 5-Minute Check is also available on **Transparency 11-7A** for this lesson.

Motivating the Lesson
Communication Display a postage scale and a bathroom scale. Ask students, *Which scale would be best for weighing a pencil? Justify your answer.* **Postage scale; it gives more precise readings for small items.**

11-7 Integration: Measurement
Precision and Significant Digits

What you'll learn
You'll learn to analyze measurements.

When am I ever going to use this?
Knowing how to analyze measurements can help you know how precise of an instrument you need for a particular measurement.

Word Wise
precision
significant digits
greatest possible error
relative error

In the 1996 Olympics, Beth Botsford won the gold medal in the 100-meter backstroke. Her time was recorded as 61.19 seconds.

No measurement can be any more exact than the scale being used to make the measurement. The **precision** of a measurement depends on the unit of measure being used. The timing device used in the Olympics had a precision of 0.01 second. *How does the precision of this instrument compare with the precision of your watch?*

The digits you record when you measure are **significant**. These digits indicate the precision of the measurement. In Ms. Botsford's Olympic time, there are four significant digits.

Ms. Botsford may have taken 0.005 second more or less than the 61.19 seconds. The instrument used to record her time could not distinguish a more precise time. The **greatest possible error** is half the smallest unit used to make the measurement. For her time, the greatest possible error is 0.005 second.

The **relative error** of measurement is found by comparing the greatest possible error with the measurement itself.

$$\text{relative error} = \frac{\text{greatest possible error}}{\text{measurement}}$$

The relative error of Ms. Botsford's time can be determined by dividing 0.005 by 61.19.

$$\text{relative error} = \frac{0.005}{61.19} \text{ or about } 0.00008$$

Compare the two measurements of a paper clip.

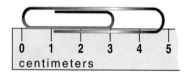 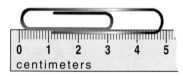

measurement: 5 cm
precision: 1 cm
significant digits: 1
greatest possible error: 0.5 cm
relative error: $\frac{0.5}{5}$ or 0.1

measurement: 4.9 cm
precision: 0.1 cm
significant digits: 2
greatest possible error: 0.05 cm
relative error: $\frac{0.05}{4.9}$ or about 0.01

504 Chapter 11 Geometry: Using Area and Volume

 Cross-Curriculum Cue

Inform the other teachers on your team that your students are studying how to determine the precision of given measurements. Suggestions for curriculum integration are:

Earth Science: atomic weights, light years, carbon dating
Geography: population statistics, distances
History: election results

Example 1 CONNECTION

Physical Science The mass of a substance is determined to be 0.0045 kilogram. Analyze this measurement.

- The mass of the substance has been measured to the nearest 0.0001 kilogram.
- There are two significant digits. The zeros in 0.0045 are used to show only the place value of the decimal and are not counted as significant digits.
- The greatest possible error is 0.00005 kilogram.
- The relative error is $\frac{0.00005}{0.0045}$ or about 0.011.

Significant digits and greatest possible error also apply to rounded data.

Example 2 APPLICATION

Entertainment The graph shows the number of people who attended the largest state fairs.

a. Are the numbers exact?

No, the measurements are to the nearest 0.1 million or 100,000 people.

b. How many significant digits are used in these numbers?

The numbers have two significant digits.

Source: International Association of Fairs and Expositions, 1995

c. What is the greatest possible error?

The greatest possible error is 0.05 million or 50,000 people.

d. Can you be sure that 100,000 more people attended the Ohio State Fair than the State Fair of Oklahoma?

No, 1,760,000 could have attended the Ohio State Fair and 1,740,000 could have attended the State Fair of Oklahoma. In this case, the difference would only be 20,000 people.

Cultural Kaleidoscope

A world's fair is an international exposition. The first world's fair was held in London in 1851. The Eiffel Tower was built for the world's fair in Paris in 1889.

CHECK FOR UNDERSTANDING

Communicating Mathematics

Read and study the lesson to answer each question.

1. *Explain* why a six-inch ruler would not be a good instrument to use in measuring the length of a horse. What would be a better instrument to use to measure the horse? Name a measurement where a six-inch ruler would be an appropriate instrument to use. **See margin.**

2. *Describe* a situation where two measurements have the same greatest possible error, but different relative errors. **See Answer Appendix.**

Lesson 11-7 Integration: Measurement Precision and Significant Digits **505**

Reteaching the Lesson

Activity Have students measure the length of the classroom using several different measuring instruments (for example, yardstick, 6-foot tape, metric tape with millimeter markings). Have them analyze their measurements and explain which is the most precise and why.

Additional Answer

1. No, the length of a horse would be greater than 6 inches. A tape measure would be a better choice for measuring a horse. A 6-inch ruler could be used to measure a paperclip.

---

2 TEACH

 Transparency 11-7B contains a teaching aid for this lesson.

Thinking Algebraically Point out that if an object's mass is measured to be 7.1 grams with a relative error of about 0.007, students can find the greatest possible error (GPE) by using GPE = *relative error · measurement*.
0.0497 ≈ 0.05 Thus, the smallest unit used to make this measurement is 0.05 × 2 or 0.1 gram.

In-Class Examples

For Example 1
A piston is measured as 135.07 millimeters high. Analyze this measurement. **There are five significant digits. The greatest possible error is 0.005 mm. The relative error is $\frac{0.005}{135.07}$ or about 0.000037**

For Example 2
The graph shows the number of popular votes for each of the 1996 presidential candidates.

1996 Popular Election Results
Votes (millions)

Bill Clinton	45.6
Bob Dole	37.8
H. Ross Perot	7.8

Source: *The World Almanac and Book of Facts, 1997*

a. Are the numbers exact? **No, they are to the nearest 0.1 million people.**

b. How many significant digits are used in these numbers? **3 for Clinton and Dole and 2 for Perot**

c. What is the greatest possible error? **0.05 million or 50,000 people**

d. Can you be sure that 7,800,000 more people voted for Clinton than for Dole? **No; 45,560,000 could have voted for Clinton and 37,840,000 could have voted for Dole. In this case, the difference would be only 7,720,000 people.**

Lesson 11-7 **505**

3 PRACTICE/APPLY

Check for Understanding

If students need additional practice or instruction after completing Exercises 1–6, one of these options may be helpful.
- Extra Practice, see p. 639
- Reteaching Activity
- Transition Booklet, pp. 29–36
- Study Guide Masters, p. 92
- Practice Masters, p. 92

Assignment Guide

Core: 7–17 odd, 19–22
Enriched: 8–14 even, 16, 17, 19–22

Additional Answers

4. The measurement is to the nearest $\frac{1}{4}$ in. The greatest possible error is $\frac{1}{8}$ in., and the relative error is $\frac{1}{50}$ or 0.020.

5. The measurement is to the nearest 0.1 km. There are 3 significant digits. The greatest possible error is 0.05 km, and the relative error is $\frac{0.05}{76.4}$ or about 0.00065.

Study Guide Masters, p. 92

---

3. **Write** a paragraph describing an instrument used for measuring length. How precise is the instrument? Give two examples of measurements for which the instrument would be used. **See Answer Appendix.**

Guided Practice

Analyze each measurement. Give the precision, significant digits if appropriate, greatest possible error, and relative error to two significant digits.

4. $6\frac{1}{4}$ in.   **See margin.**

5. 76.4 km   **See margin.**

6. **Clothing** The sizes of men's hats are given as the diameter of a circle before the hat is made more oval. Hat sizes start at $6\frac{1}{4}$ and go up by $\frac{1}{8}$ inch increments. How precise are the hat sizes?   **Hat sizes are to the nearest $\frac{1}{8}$ in.**

EXERCISES

Practice

Analyze each measurement. Give the precision, significant digits if appropriate, greatest possible error, and relative error to two significant digits.

7–15. See Answer Appendix.

7. 7 mm
8. 0.052 kg
9. $9\frac{1}{2}$ lb
10. 87.23 m
11. 47 min
12. 42.8 cm

13. Which would be the most precise measurement for a can of tomatoes: 2 pounds, 34 ounces, or 34.3 ounces? Explain.

14. Which is the more precise measurement for the height of the Statue of Liberty: 100 yards or 305 feet? Explain.

15. The lengths of two poles are 17 inches and 78 inches. Which measurement has the least relative error? Explain.

16a. No; the numbers are to the nearest 0.01 billion or 10,000,000 dollars

16b. $5,000,000

Applications and Problem Solving

interNET CONNECTION
For information on the sale of logo merchandise, visit: www.glencoe.com/sec/math/mac/mathnet

16. **Entertainment** The graph shows the amount of money spent on professional team logo merchandise.
   a. Are the numbers exact? Explain.
   b. What is the greatest possible error?
   c. What is the total amount of money spent on NFL, NBA, MLB, and NHL logos?   **$8,700,000,000**

Source: Sporting Goods Manufacturers Association, 1996

506 Chapter 11 Geometry: Using Area and Volume

Multiple Learning Styles

Intrapersonal Have students write a paragraph describing what sorts of measurements they might need in their projected careers and why precision would be necessary.

17. **Technology** A calculator gives the value of π as 3.141592654. How many significant digits does this value for π have?   **10 significant digits**

18. **Working on the CHAPTER Project** Refer to the kite you designed on page 485. Determine the appropriate instruments you will need to make the appropriate measurements to build the kite.   **See students' work.**

19. **Critical Thinking** To be precise when multiplying measurements, your rounded answer should have the same number of significant digits as the least precise measurement. With that in mind, find the area of a circle whose radius is 4.2 centimeters.   **55 cm²**

Mixed Review

20. **Geometry** Find the surface area of a cylinder 9 centimeters tall with a radius of 4 centimeters. *(Lesson 11-6)*   **about 326.7 cm²**

21. **Test Practice** Which equation describes the function represented by the table? *(Lesson 10-1)*   **A**

n	f(n)
-2	-7
0	-3
2	1
4	5

  A $f(n) = 2n - 3$ B $f(n) = n + 4$ C $f(n) = n - 3$
  D $f(n) = 2n + 3$ E $f(n) = n + 3$

22. **Probability** A cereal box stated that it may contain a prize. A consumer protection agency bought 50 boxes of the cereal. Five of them contained a prize. Estimate the probability of winning a prize when you buy a box of that brand of cereal. *(Lesson 6-6)* $\frac{1}{10}$

MATH IN THE MEDIA

Hi and Lois

1. What precision is the daughter, Dot, using? $\frac{1}{32}$ in.   2. Sample answer: $\frac{1}{4}$ in.
2. What precision might Dot's mother use to show the pieces are the same size?
3. Can you cut two pieces exactly the same size? Explain.   **No; no measurement is exact.**

Study Guide and Assessment

CHAPTER 11

Vocabulary

This section provides a listing of the new terms, properties, and phrases that were introduced in this chapter. Have students define each term and provide an example or two of it, if appropriate.

Understanding and Using the Vocabulary

These exercises check students' understanding of the terms by using a variety of verbal formats including matching, completion, and true/false.

Glossaries A complete glossary of terms appears on pages 702–710. The glossary also appears in Spanish on pages 711–721.

Additional Answers

13.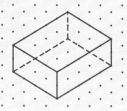

14.

15.

Vocabulary

After completing this chapter, you should be able to define each term, concept, or phrase and give an example or two of each.

Geometry
altitude (p. 490)
base (p. 483)
circular cone (p. 490)
circular cylinder (p. 487)
edge (p. 483)
face (p. 483)
net (p. 494)
prism (p. 483)
pyramid (p. 483)
solid (p. 482)
surface area (p. 495)
vertex (p. 483)
volume (p. 486)

Measurement
greatest possible error (p. 504)
precision (p. 504)
relative error (p. 504)
significant digits (p. 504)

Problem Solving
make a model (p. 480)

Understanding and Using the Vocabulary

Choose the letter of the term that best matches each statement or phrase.

1. any three-dimensional figure **f**
2. the measure of the space occupied by a solid **i**
3. a flat surface of a prism **g**
4. a figure that has an area of π times the radius squared **d**
5. the greatest possible error divided by the measurement **l**
6. a figure that has two parallel, congruent circular bases **b**
7. a figure that has a volume that is one-third the area of the base times the height **a**
8. the segment used to measure the height of a pyramid and a cone **k**
9. half the smallest unit used to make a measurement **j**

a. pyramid
b. cylinder
c. rectangular prism
d. circle
e. precision
f. solid
g. face
h. surface area
i. volume
j. greatest possible error
k. altitude
l. relative error

In Your Own Words

10. *Explain* how to find the surface area of a rectangular prism. **Find the sum of the areas of the six faces.**

 MindJogger Videoquizzes

MindJogger Videoquizzes provide an alternative review of concepts presented in this chapter. Students work in teams to answer questions, gaining points for correct answers. The questions are presented in three rounds.
Round 1 Concepts–5 questions
Round 2 Skills–4 questions
Round 3 Problem Solving–4 questions

Study Guide and Assessment Chapter 11

Objectives & Examples

Upon completing this chapter, you should be able to:

● find the areas of circles *(Lesson 11-1)*

Find the area of a circle with a radius of 5 feet.

$A = \pi r^2$
$A = \pi \cdot 5^2$ [π] [×] 5 [x²] [=] 78.53981634
$A \approx 78.5$

The area of the circle is about 78.5 square feet.

● identify and sketch three-dimensional figures *(Lesson 11-2)*

Make the edges you can see solid lines and the edges you can't see dashed lines.

● find the volumes of prisms and cylinders *(Lesson 11-3)*

Find the volume of the rectangular prism.

$V = \ell wh$
$V = 3 \cdot 4 \cdot 6$ or 72

The prism has a volume of 72 cubic meters.

Review Exercises

Use these exercises to review and prepare for the chapter test.

Find the area of each circle to the nearest tenth.

11.
153.9 m²

12.
113.1 in²

Use isometric dot paper to draw each solid. 13–15. See margin.

13. a rectangular prism that is 4 units long, 5 units wide, and 2 units high

14. a hexagonal pyramid

15. a rectangular prism with all edges 3 units long

Answers are calculated using the π key on a calculator and then rounded.

Find the volume of each solid to the nearest tenth.

16. 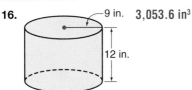 3,053.6 in³

17. 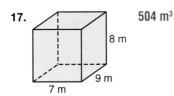 504 m³

Chapter 11 Study Guide and Assessment **509**

Objectives & Examples

This section reviews the skills and concepts of the chapter and shows completely worked examples.

Review Exercises

These exercises provide practice for the corresponding objectives.

Assessment and Evaluation Masters, pp. 283–284

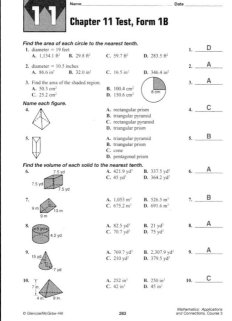

Assessment and Evaluation

Six forms of Chapter 11 Test are available in the *Assessment and Evaluation Masters* as shown in the chart.

Chapter 11 Test, Form 1B, is shown at the right. Chapter 11 Test, Form 2B, is shown on the next page.

1A	Multiple Choice	Honors
1B	Multiple Choice	Average
1C	Multiple Choice	Basic
2A	Free Response	Honors
2B	Free Response	Average
2C	Free Response	Basic

Study Guide and Assessment **509**

Chapter 11 Study Guide and Assessment

Objectives & Examples

● find the volumes of pyramids and cones
(Lesson 11-4)

Find the volume of the square pyramid.

$V = \frac{1}{3}Bh$

$V = \frac{1}{3} \cdot 4^2 \cdot 9$

$V = 48$

The volume of the square pyramid is 48 cubic feet.

● find the surface areas of rectangular and triangular prisms (Lesson 11-5)

To find the surface area of a prism, find the area of each face. Then add all the areas.

● find the surface areas of cylinders
(Lesson 11-6)

Find the surface area of a cylinder with a radius of 2 feet and a height of 5 feet.

surface area $= 2\pi r^2 + 2\pi rh$

surface area $= 2\pi(2^2) + 2\pi(2)(5)$

surface area ≈ 88.0

The surface area of cylinder is about 88.0 square feet.

● analyze measurements (Lesson 11-7)

Analyze 3.08 meters.

- The measurement is to the nearest 0.01 meter.
- There are three significant digits.
- The greatest possible error is 0.005 meter.
- The relative error is $\frac{0.005}{3.08}$ or about 0.00162.

Review Exercises

Find the volume of each solid to the nearest tenth.

18. 737.2 in³

19. 10 mm³

Find the surface area of each prism to the nearest tenth.

20. 112 ft²

21. 21.4 m²

Find the surface area of each cylinder to the nearest tenth.

22. 301.6 cm²

23. 1,633.6 yd²

Analyze each measurement. Give the precision, significant digits if appropriate, greatest possible error, and relative error to two significant digits.

24. 0.06 millimeter   24–27. See Answer Appendix.

25. 12 feet

26. $2\frac{1}{3}$ yards

27. 24.2 meters

510 Chapter 11 Geometry: Using Area and Volume

Test and Review Software

You may use this software, a combination of an item generator and item bank, to create your own tests or worksheets. Types of items include free response, multiple choice, short answer, and open ended.

CD-ROM Program

The CD-ROM Program contains an Assessment Game whose questions review the concepts in this chapter.

Study Guide and Assessment Chapter 11

Applications & Problem Solving

28. Music Find the area of the top of a compact disc if its diameter is 12 centimeters and the diameter of the hole is 1.5 centimeters. *(Lesson 11-1)*

29. Make a Model In the game of pool, a rack is used to organize the balls. There is one ball in the first row, two balls in the next row, three balls in the next row, and so on, until 5 rows are completed. How many balls can be placed in a pool rack? *(Lesson 11-2A)* **15 balls**

30. Food A soup can is in the shape of a circular cylinder. The top has a diameter of 6.5 centimeters, and the can is 9.5 centimeters tall. Find the volume of the soup can. *(Lesson 11-3)* **about 315.2 cm³**

28. about 111.3 cm²   31a. about 30.5 ft²

31. Maintenance The diagram shows the design of the trash cans in the school cafeteria. The cans need to be painted.
  a. Find the surface area of each trash can.
  b. The paint covers 200 square feet per gallon. Approximately how many trash cans can be covered with 1 gallon of paint? *(Lesson 11-6)* **about $6\frac{1}{2}$ trash cans**

Alternative Assessment

Performance Task
You have a circular swimming pool with a radius of 9 meters and a volume of about 1,017.36 cubic meters. Your friend has a circular swimming pool with a diameter that is the same as the radius of your pool. The depth of your friend's pool is the same as the depth of your pool. Explain how to find the volume of your friend's pool. Then find the volume. Compare the volume of the two pools.
See Answer Appendix.
You and your friend decide to drain the water from your pools and paint the inside surface. Your friend says that he needs 3 cans of paint for his pool. How can you use this information to determine the number of cans you will need to paint your pool? **See Answer Appendix.**

A practice test for Chapter 11 is provided on page 657.

Completing the CHAPTER Project
Use the following checklist to make sure your kite design is complete.
- ☑ The three-dimensional drawing of the kite is accurate.
- ☑ The computations involving the amount and cost of materials needed for the kite are accurate.
- ☑ A statement about the appropriate measuring instruments needed to construct the kite is included.

 Select one of the assignments from this chapter and place it in your portfolio. Attach a note to it explaining why you selected it.

Chapter 11 Study Guide and Assessment **511**

Performance Assessment
Additional performance assessment tasks for this chapter are included in the *Assessment and Evaluation Masters* on page 293. A scoring guide is also provided on page 305.

Applications & Problem Solving
This section provides additional practice in solving real-world problems that involve the skills of this chapter.

Alternative Assessment
The *Performance Task* provides students with a performance assessment opportunity to evaluate their work and understanding.

CHAPTER Project
Students should complete the final stages of their project and prepare a class demonstration of their results. A scoring guide for the project is available in the *Investigations and Projects Masters*, p. 59.

 Students should add to their portfolios at this time.

Assessment and Evaluation Masters, p. 293

Standardized Test Practice

The Standardized Test Practice may be used to help students prepare for standardized tests. The test items are written in the same style as those in state proficiency tests and standardized tests like CAT, CTBS, ITBS, MAT, SAT, and Terra Nova. The test items cover skills and concepts covered up to this point in the text.

The pages can be used as an overnight assessment. After students have completed the pages, discuss how each problem can be solved, or provide copies of the solutions from the *Solutions Manual*.

Assessment and Evaluation Masters, p. 299

CHAPTERS 1–11 Standardized Test Practice

Assessing Knowledge & Skills

Section One: Multiple Choice

There are ten multiple-choice questions in this section. Choose the best answer. If a correct answer is *not here*, choose the letter for Not Here.

1. A ladder was leaning up against a building. The ladder formed a 60° angle with the ground.

   What is the measure of the angle between the ground and the other side of the ladder?

   A 30°
   B 60°
   C 100°
   D 120°   **D**

2. If an architectural drawing measures 6 inches wide by 8 inches long, how wide is an enlargement that is 32 inches long? **H**

   F 14 inches
   G 48 inches
   H 24 inches
   J 18 inches

3. Find the volume of a 5-meter by 6-meter by 2-meter rectangular prism. **D**

   A 10 cubic meters
   B 12 cubic meters
   C 30 cubic meters
   D 60 cubic meters

4. △LMN is a right triangle. What is the length of $\overline{LM}$? **H**

   F 2 inches
   G 3 inches
   H 4 inches
   J 5 inches

Please note that Questions 5–10 have five answer choices.

5. A circular clock has a diameter of 14 inches. Which expression represents the area of the face of the clock? **A**

   A $\pi \cdot 7 \cdot 7$
   B $\pi \cdot 28$
   C $\pi \cdot 14$
   D $\pi \cdot 14 \cdot 14$
   E $2 \cdot \pi \cdot 14$

6. Which is the graph of $y = 3x - 3$? **J**

   F 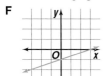   G

   H 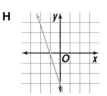   J

   K

512   Chapters 1–11 Standardized Test Practice

◀◀◀ **Instructional Resources**

Another cumulative review is shown at the left and is available in the *Assessment and Evaluation Masters*, p. 299.

Standardized Test Practice Chapters 1–11

7. A flower bed is 10 feet by 3 feet. To fill it with soil 6 inches deep, how much topsoil is needed? **B**
   A 30 cubic feet
   B 15 cubic feet
   C 120 cubic feet
   D 150 cubic feet
   E 180 cubic feet

8. On the first play of the Cowboys' game, the quarterback was sacked and lost 18 yards. On the second and third plays, the offense advanced the football a total of 12 yards. What is the total number of yards gained or lost? **F**
   F 6 yards lost
   G 6 yards gained
   H 30 yards lost
   J 30 yards gained
   K Not Here

9. Kwan-Yong bought $6\frac{1}{2}$ pounds of ground beef for $10.27 and 2 dozen cookies for $3.78. What was the cost per pound of the ground beef? **B**
   A $1.89
   B $1.58
   C $1.08
   D $2.16
   E Not Here

10. Consuela is participating in a walk-a-thon. Khalid is sponsoring her at $0.10 per mile, and Debbie is sponsoring her at $0.11 per mile. If Consuela walks 9 miles, what is the total amount she will collect from Khalid and Debbie? **H**
    F $0.90
    G $0.99
    H $1.89
    J $1.90
    K $1.99

Test-Taking Tip

Remember, on most tests, you get as much credit for correctly answering the easy questions as you do for correctly answering the difficult ones. Answer the easy questions first and then spend time on the more challenging questions.

Section Two: Free Response

This section contains three questions for which you will provide short answers. Write your answers on your paper.

11. The bases of a triangular prism are right triangles with sides of 3 feet, 4 feet, and 5 feet. The height of the prism is 10 feet. What is the surface area? **132 ft²**

12. The curved part of a can will be covered by a label. What is this area to the nearest whole number? **377 cm²**

13. Rectangle $ABCD$ has vertices $A(1, 2)$, $B(1, -3)$, $C(-5, -3)$, and $D(-5, 2)$. What are the coordinates of the rectangle after a translation 1 unit right and 2 units down? **See margin.**

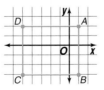

Additional Answer

13. $A'(2, 0), B'(2, -5), C'(-4, -5), D'(-4, 0)$

CHAPTER 12
Investigating Discrete Math and Probability

Previewing the Chapter

Overview
In this chapter, students examine situations that are probable, not certain, and develop methods for expressing their probability of occurring. Students study the Counting Principle, explore permutations, combinations, and compound events, and learn how Pascal's Triangle applies to the study of probability. In the Thinking Lab, students develop a technique for solving problems by acting them out. Students then apply this strategy to the calculation of experimental probabilities, using it with Punnett squares and random samples.

Lesson (pages)	Lesson Objectives	NCTM Standards	Standardized Tests	State/Local Objectives
12-1A (516–517)	Discover how to determine whether a game is fair or unfair.	1–4, 11		
12-1 (518–520)	Count outcomes by using a tree diagram or the Counting Principle.	1–4, 7, 11		
12-2 (521–523)	Find the number of permutations of objects.	1–4, 7, 11	CTBS, ITBS, MAT, SAT, TN	
12-3 (524–527)	Find the number of combinations of objects.	1–4, 7, 11	CTBS, ITBS, MAT, SAT, TN	
12-4 (528–531)	Identify patterns in Pascal's Triangle.	1–4, 7, 8, 11		
12-4B (532–533)	Discover numerical and visual patterns in Pascal's Triangle.	1–5, 7, 8		
12-5 (534–537)	Find the probability of independent and dependent events.	1–4, 7, 11	CAT, CTBS, ITBS, MAT, SAT, TN	
12-6A (538–539)	Solve problems by acting them out.	1–4, 7, 11		
12-6 (540–543)	Find experimental probability.	1–4, 7, 11	CTBS, ITBS, TN	
12-7A (544–545)	Discover how experimental probability is used in biology.	1–4, 11		
12-7 (546–548)	Predict the actions of a larger group by using a sample.	1–4, 7, 10	CTBS, ITBS, SAT, TN	

CAT = California Achievement Tests, CTBS = Comprehensive Tests of Basic Skills, ITBS = Iowa Tests of Basic Skills, MAT = Metropolitan Achievement Tests, SAT = Stanford Achievement Tests, TN = Terra Nova

Organizing the Chapter

CD-ROM
All of the blackline masters in the Teacher's Classroom Resources are available on the **Electronic Teacher's Classroom Resources** CD-ROM.

LESSON PLANNING GUIDE

Lesson	Extra Practice (Student Edition)	Blackline Masters (page numbers)											Transparencies A and B
		Study Guide	Practice	Enrichment	Assessment & Evaluation	Classroom Games	Diversity	Hands-On Lab	School to Career	Science and Math Lab Manual	Technology	
12-1A								63				
12-1	p. 639	93	93	93								12-1
12-2	p. 639	94	94	94	323						76	12-2
12-3	p. 640	95	95	95			38				75	12-3
12-4	p. 640	96	96	96	322, 323							12-4
12-4B								64				
12-5	p. 640	97	97	97								12-5
12-6A	p. 641											
12-6	p. 641	98	98	98	324			79		93–96		12-6
12-7A								65				
12-7	p. 641	99	99	99	324	41–42			38			12-7
Study Guide/ Assessment					309–321, 325–327							

Other Chapter Resources

Student Edition
Chapter Project, pp. 515, 543, 548, 553
School to Career, p. 549
Let the Games Begin, p. 543

Technology
 CD-ROM Program
 Interactive Mathematics Tools Software

Teacher's Classroom Resources

Applications
Family Letters and Activities, pp. 75–76
Investigations and Projects Masters, pp. 61–64

Meeting Individual Needs
Investigations for the Special Education Student, pp. 45–46

Teaching Aids
Answer Key Masters
Block Scheduling Booklet
Lesson Planning Guide
Solutions Manual

Professional Publications
Glencoe Mathematics Professional Series

Planning the Chapter

 MindJogger Videoquizzes provide a unique format for reviewing concepts presented in the chapter.

ASSESSMENT RESOURCES

Student Edition
Mixed Review, pp. 520, 523, 527, 531, 537, 543, 548
Mid-Chapter Self Test, p. 531
Math Journal, pp. 519, 547
Study Guide and Assessment, pp. 550–553
Performance Task, p. 553
Portfolio Suggestion, p. 553
Standardized Test Practice, pp. 554–555
Chapter Test, p. 658

Assessment and Evaluation Masters
Multiple-Choice Tests (Forms 1A, 1B, 1C), pp. 309–314
Free-Response Tests (Forms 2A, 2B, 2C), pp. 315–320
Performance Assessment, p. 321
Mid-Chapter Test, p. 322
Quizzes A–D, pp. 323–324
Standardized Test Practice, pp. 325–326
Cumulative Review, p. 327

Teacher's Wraparound Edition
5-Minute Check, pp. 518, 521, 524, 528, 534, 540, 546
Building Portfolios, p. 514
Math Journal, pp. 517, 533, 545
Closing Activity, pp. 520, 523, 527, 531, 537, 539, 543, 548

Technology
Test and Review Software
MindJogger Videoquizzes
CD-ROM Program

MATERIALS AND MANIPULATIVES

Lesson 12-1A
number cubes*

Lesson 12-2
colored pencils or markers
calculator

Lesson 12-3
index cards

Lesson 12-4B
hexagonal grid
highlighter
calculator

Lesson 12-5
counters*†
cups*†

Lesson 12-6
colored marbles
bag
number cubes*†
cups*†
coins
graphing calculator

Lesson 12-7A
counters*†
bags

Lesson 12-7
calculator

*Glencoe Manipulative Kit †Glencoe Overhead Manipulative Resources

PACING CHART

See pages T25–T27 for the Course Planning Calendar.

COURSE	DAY 1	DAY 2	DAY 3	DAY 4	DAY 5	DAY 6	DAY 7
Standard	Chapter Project	Lessons 12-1A & 12-1		Lesson 12-2	Lesson 12-3	Lesson 12-4	Lesson 12-5
Honors	Chapter Project & Lesson 12-1	Lesson 12-2	Lesson 12-3	Lessons 12-4 & 12-4B		Lesson 12-5	Lesson 12-6A
Block	Chapter Project & Lesson 12-1A	Lessons 12-1 & 12-2	Lessons 12-3 & 12-4	Lessons 12-5 & 12-6A	Lesson 12-6 & 12-7	Study Guide and Assessment, Chapter Test	

Interactive Mathematics:
Activities and Investigations

is an activity-based program that may be used as an enhancement for chapters in *Mathematics: Applications and Connections*.

Unit 17, Activity Six
Use with Lesson 12-4.

Summary Students work in groups to complete 20 rows of Pascal's Triangle and then color the triangle using multiples and remainders of numbers that they get when they roll a number cube. They make a group poster with the triangles.

Math Connection Student explore Pascal's Triangle and create different versions of Sierpinski's Triangle. Pascal's Triangle exhibits recursion, since each element is the sum of two elements in the row above it.

Unit 1, Activity Six
Use with Lessons 12-6A and 12-6.

Summary Students work in pairs to experiment with some basic properties of probability. They determine the probability of randomly selecting a given color marble from a bag and the probabilities of rolling various sums with a pair of number cubes.

Math Connection Students are involved in experimental probability activities. They then write the theoretical probabilities describing the results of their experiments. They also use counting, adding whole numbers, making bar graphs, and looking for patterns.

DAY 8	DAY 9	DAY 10	DAY 11	DAY 12	DAY 13	DAY 14	DAY 15
Lesson 12-6A	Lesson 12-6	Lessons 12-7A & 12-7		Study Guide and Assessment	Chapter Test		
Lesson 12-6	Lesson 12-7	Study Guide and Assessment	Chapter Test				

Enhancing the Chapter

APPLICATIONS

Classroom Games, pp. 41–42

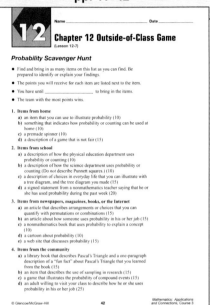

Diversity Masters, p. 38

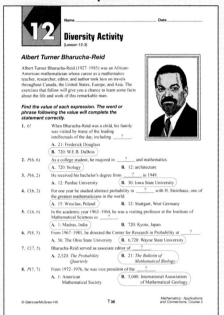

School to Career Masters, p. 38

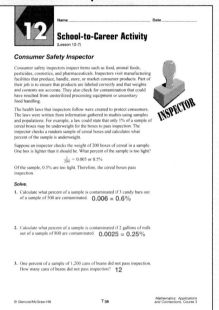

Family Letters and Activities, pp. 75–76

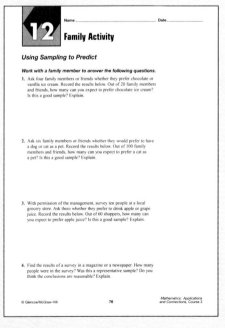

Science and Math Lab Manual, pp. 93–96

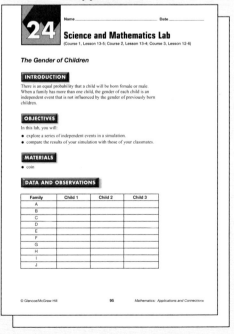

MANIPULATIVES/MODELING

Hands-On Lab Masters,
p. 79

ASSESSMENT/EVALUATION

Assessment and Evaluation Masters,
pp. 322–324

TECHNOLOGY/MULTIMEDIA

Technology Masters,
pp. 75–76

MEETING INDIVIDUAL NEEDS

Investigations for the Special Education Student, pp. 45–46

CHAPTER 12 NOTES

Theme: Athletes
Some high school students dream of playing professional sports and focus their school careers around an athletic path. This can be an unrealistic goal. There are about 12,000 public high schools in the United States. In 1995, there were about 300 colleges that had Division I basketball, about 280 with Division II basketball, and about 360 with Division III basketball. The National Basketball Association consists of 29 teams, each with 15 players on the roster.

Question of the Day What is the probability that any one boy or girl is on a sports team at your school? Students can sample their math class and try to predict from that sample. **See students' work.**

Assess Prerequisite Skills
Ask students to read through the list of objectives presented in "What you'll learn in Chapter 12." You may wish to ask them what each of the objectives means or if they have used any of these math concepts before.

 Building Portfolios
Encourage students to revise their portfolios as they study this chapter. Probability is a challenging subject, and students may find that the more work they do in probability, the greater their understanding.

 Math and the Family
In the *Family Letters and Activities* booklet (pp. 75–76), you will find a letter to the parents explaining what students will study in Chapter 12. An activity appropriate for the whole family is also available.

CHAPTER 12 — Investigating Discrete Math and Probability

 **What you'll learn in Chapter 12**

- to count outcomes by using a tree diagram or the Counting Principle,
- to find the number of permutations and combinations of objects,
- to find theoretical and experimental probability,
- to solve problems by acting them out, and
- to predict the actions of a large group by using a sample.

 CD-ROM Program

Activities for Chapter 12
- Chapter 12 Introduction
- Interactive Lessons 12-6A, 12-6
- Assessment Game
- Resource Lessons 12-1 through 12-7

CHAPTER Project

CONSIDER THE PROBABILITIES

In this project, you will write a news article about what it takes to compete at a high level in a sport. You will compute the probability that an athlete will reach the higher levels of competition in a sport of your choice. You will make a graph showing how many players out of 100,000 would be expected to play at the higher levels.

Getting Started

- Choose a sport that interests you.
- Find out how many players in your community are involved in the sport at the little league level or some other level.
- Interview a middle school or high school coach in the sport. Find out how many students play the sport for the school. Find out how many of his or her players have gone on to play in college. Ask if any of his or her players have competed in the Olympics or have played on professional teams.

Technology Tips

- Use a **spreadsheet** to organize data and make calculations.
- Use **computer software** to make the graph.
- Use a **word processor** to write your article.

 For up-to-date information on sports participation, visit:
www.glencoe.com/sec/math/mac/mathnet

Working on the Project

You can use what you'll learn in Chapter 12 to help you calculate probabilities and make predictions.

Page	Exercise
543	11
548	11
553	Alternative Assessment

interNET CONNECTION

Glencoe has made every effort to ensure that the website links for *Mathematics: Applications and Connections* at www.glencoe.com/sec/math/mac/mathnet are current and contain appropriate content. However, these website links are not under Glencoe's control.

Instructional Resources ▶▶▶

A recording sheet to help students organize their data for the Chapter Project is shown at the right and is available in the *Investigations and Projects Masters*, p. 64.

CHAPTER Project NOTES

Objectives Students should
- gain an understanding of how to make decisions based on probable outcomes.
- learn how to predict outcomes based on math principles.

Project Pointer You may suggest that students begin a *Project Folder* to keep their work as they complete each stage of the Chapter Project. The completed project may also be added to their portfolios.

Students should collect stories from magazines and newspapers about successful athletes and highlight any statistics relevant to the Chapter Project.

Investigations and Projects Masters, p. 64

Chapter 12 Project

Consider the Probabilities
Page 515, Getting Started

Name of Organization	Number of Players

Coach:
School name:
Number of players:
Number going on to high school sports:
Number going on to college sports:
Number competing in Olympics:
Number playing professionally:

Page 543, Working on the Chapter Project, Exercise 11
Probabilities:

Page 548, Working on the Chapter Project, Exercise 11

12-1A HANDS-ON LAB Notes

GET READY

Objective Students discover how to determine whether a game is fair or unfair.

Optional Resources
Hands-On Lab Masters
- worksheet, p. 63

Manipulative Kit
- number cubes

MANAGEMENT TIPS

Recommended Time
30 minutes

Getting Started Ask students what makes a game fair or unfair. Help them see that in a fair game, each player plays by the rules and has an equal chance of winning. Ask students to describe ways in which one player can have an unfair advantage in a game. Use a basketball game in which the hoops are set at different heights as an example. Then ask students how fair games can be played unfairly and how rules of a game can be changed to eliminate one player's unfair advantage.

Activity 1 uses *Scissors, Paper, Stone* as an example of a fair game. After the students play the game, ask them if the timing of each player's display of digits affects the fairness of the game.

The number game in **Activity 2** on page 517 is unfair because there are more possible even than odd outcomes. Help students understand that this is true because in multiplication the product is odd only when two odd numbers are multiplied but the product is even when at least one multiplier is even. Then ask students if the fairness of the game changes if the numbers are added rather than multiplied.

HANDS-ON LAB

COOPERATIVE LEARNING

12-1A Fair and Unfair Games

A Preview of Lesson 12-1

 2 number cubes

Many people enjoy playing games they believe are fair. A *fair game* is defined as one in which each player has an equal chance of winning. In an *unfair game,* players do *not* have an equal chance of winning.

Have you ever played Scissors, Paper, Stone? This ancient game, also known as Hic, Hacec, Hoc, is played all over the world. On the count of three, two players simultaneously display one hand with either two fingers forming a V (scissors), an open hand (paper), or a fist (stone).

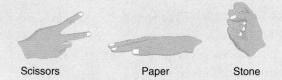

Scissors   Paper   Stone

The winner of the game is decided by the following rules.
- Scissors cut paper.
- Paper wraps stone.
- Stone breaks scissors.

If both players pick the same object, the round is a draw.

TRY THIS

Work with a partner.

1. Play 20 rounds of Scissors, Paper, Stone. Tally the number of times each player wins in a table like the one below.

Winner		
Player A	Player B	Draw

ON YOUR OWN

1. To analyze the game, make a list of all of the possible outcomes.   **See margin.**
2. How many different outcomes are possible?   **9 different outcomes**
3. How many ways can Player A win?   **3 ways**
4. How many ways can Player B win?   **3 ways**
5. How many outcomes are a draw?   **3 outcomes**
6. Is each outcome equally likely?   **yes**
7. Is Scissors, Paper, Stone a fair game? Explain.   **Yes; each player's chance of winning is $\frac{1}{3}$.**

Additional Answer

1.
Player A	Player B
scissors	scissors
scissors	paper
scissors	stone
paper	scissors
paper	paper
paper	stone
stone	scissors
stone	paper
stone	stone

TRY THIS

Work with a partner.

2. Roll two number cubes. If the product of the numbers is even, Player 1 wins. If the product of the numbers is odd, Player 2 wins. Play 20 rounds of this game. Tally the number of times each player wins in a table like the one below.

Winner	
Player 1	Player 2

ON YOUR OWN

8. To analyze the game, make a list of all of the possible outcomes. **See Answer Appendix.**
9. How many different outcomes are possible? **36 different outcomes**
10. How many ways can Player 1 win? **27 ways**
11. How many ways can Player 2 win? **9 ways**
12. Is it possible to have a draw in this game? **no**
13. Is each outcome equally likely? **no**
14. Is this game *fair* or *unfair*? Explain. **See margin.**
15. **Look Ahead** Two players each spin one of the spinners. Player X wins if the spinners show the same letter. Player Y wins if the spinners show different letters. Is this game fair or unfair? Be prepared to defend your answer. **See margin.**

Lesson 12-1A HANDS-ON LAB **517**

 Direct students to select a game and write two sets of specific directions for it. The first set of directions should make the game fair. The second set should change something to make the game unfair. **Sample answer: In soccer, it is unfair if one team plays uphill the entire game.**

Additional Answers

14. Unfair; Player 1's chance of winning is $\frac{3}{4}$, and Player 2's chance of winning is $\frac{1}{4}$.

15. Unfair; Player X's chance of winning is $\frac{4}{9}$, and Player Y's chance of winning is $\frac{5}{9}$.

ASSESS

Have students complete Exercises 1–14. Watch for students who don't understand that all possible outcomes can be predicted. Have those students calculate all possible outcomes by multiplying a value and creating a table of the results.

Product of Values						
	1	2	3	4	5	6
1	1	2	3	4	5	6
2	2	4	6	8	10	12
3	3	6	9	12	15	18
4	4	8	12	16	20	24
5	5	10	15	20	25	30
6	6	12	18	24	30	36

Use Exercise 15 to determine whether students understand that fair games offer all participants an equal chance of winning.

Hands-On Lab Masters, p. 63

Hands-On Lab 12-1A **517**

12-1 Lesson Notes

Instructional Resources
- *Study Guide Masters*, p. 93
- *Practice Masters*, p. 93
- *Enrichment Masters*, p. 93
- Transparencies 12-1, A and B
- CD-ROM Program
  - Resource Lesson 12-1

Recommended Pacing

Standard	Days 2 & 3 of 13
Honors	Day 1 of 11
Block	Day 2 of 6

1 FOCUS

5-Minute Check
(Chapter 11)

1. Find the area of the circle. **about 113.1 in²**

2. Find the surface area and volume of the rectangular prism. **136 cm²; 96 cm³**

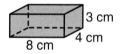

3. How many significant digits are there in the measurement 0.073 cm? **2 significant digits**

The 5-Minute Check is also available on **Transparency 12-1A** for this lesson.

Motivating the Lesson

Problem Solving Tell students that they are managers of a baseball team with four pitchers (Adams, Beck, Chi, and Dunn) and two catchers (Eck and Finn). Ask them to list all possible pitcher-catcher combinations.

12-1 Counting Outcomes

What you'll learn
You'll learn to count outcomes by using a tree diagram or the Counting Principle.

When am I ever going to use this?
Knowing how to count outcomes can help you determine the number of sandwich choices.

Word Wise
tree diagram
outcome
Counting Principle

Cultural Kaleidoscope
The Chinese wrote about the mathematics of combinations 4,000 years ago.

The games you played in Lesson 12-1A involved counting outcomes by listing them. In this lesson, you will learn two other ways to count outcomes.

Ernesto is making his lunch for school. He has a choice of wheat or rye bread. He can use ham, turkey, or beef. He can use Swiss cheese or American cheese. How many different sandwiches can he make using one bread, one meat, and one cheese?

You can draw a **tree diagram** to find the number of possible combinations or **outcomes**.

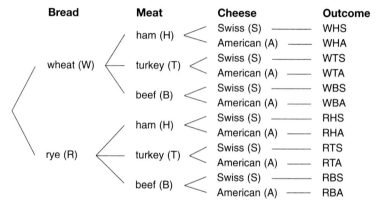

There are twelve possible sandwiches, or outcomes.

Example 1 — APPLICATION

Sports Members of a soccer team have red shorts and black shorts. They also have white jerseys, red jerseys, and black jerseys. How many different uniforms can they wear?

Use a tree diagram to find all of the possible uniforms.

Shorts	Jerseys	Outcome
red	white	red and white
	red	red and red
	black	red and black
black	white	black and white
	red	black and red
	black	black and black

There are six different uniforms.

518 Chapter 12 Investigating Discrete Math and Probability

 Cross-Curriculum Cue

Inform the other teachers on your team that your students are studying probability. Suggestions for curriculum integration are:
Civics: voting behavior, polls
Life Science: inheritance of genetic traits

You can also find the total number of outcomes by multiplying. This principle is known as the **Counting Principle**.

Counting Principle	If event M can occur in m ways and is followed by event N that can occur in n ways, then the event M followed by the event N can occur in $m \cdot n$ ways.

The number of possible sandwiches that Ernesto can make can be determined using this principle.

$$\begin{array}{ccccccc} \text{number of} & & \text{number of} & & \text{number of} & & \text{number of} \\ \text{choices for} & \times & \text{choices for} & \times & \text{choices for} & = & \text{possible} \\ \text{bread} & & \text{meat} & & \text{cheese} & & \text{sandwiches} \\ 2 & \times & 3 & \times & 2 & = & 12 \end{array}$$

Example 2 There are six different numbers on a number cube. Suppose you roll a red number cube, a green number cube, and a black number cube. What is the probability of rolling a 1 on the red number cube, a 2 on the green number cube, and a 3 on the black number cube?

Use the Counting Principle to determine the number of possible outcomes.

LOOK BACK
You can refer to Lesson 6-6 to review probability of simple events.

$$\begin{array}{ccccccc} \text{number of} & & \text{number of} & & \text{number of} & & \text{total number} \\ \text{outcomes on} & \times & \text{outcomes on} & \times & \text{outcomes on} & = & \text{of possible} \\ \text{red number} & & \text{green number} & & \text{black number} & & \text{outcomes} \\ \text{cube} & & \text{cube} & & \text{cube} & & \\ 6 & \times & 6 & \times & 6 & = & 216 \end{array}$$

There are 216 possible outcomes. Each outcome is equally likely. Only one of these outcomes shows a 1 on the red number cube, a 2 on the green number cube, and a 3 on the black number cube. The probability of this roll is $\frac{1}{216}$.

CHECK FOR UNDERSTANDING

Communicating Mathematics
1. See Answer Appendix.

Read and study the lesson to answer each question.

1. **Write a Problem** that corresponds to the tree diagram.

2. **Use** the Counting Principle to write a mathematical equation that corresponds to the tree diagram. $2 \times 4 = 8$

3. **Write** a paragraph comparing the use of a tree diagram and the use of the Counting Principle to determine the number of outcomes. **See margin.**

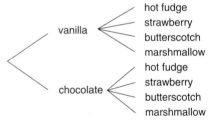

Lesson 12-1 Counting Outcomes **519**

■ Reteaching the Lesson ■

Activity When many choices present themselves, students may find an array like this one more manageable than a tree diagram.

	1	2	3	4	5
A	A1	A2	A3	A4	A5
B	B1	B2	B3	B4	B5
C	C1	C2	C3	C4	C5

Additional Answer
3. Both a tree diagram and the Fundamental Principle of Counting can be used to find the number of outcomes. A tree diagram actually shows the different outcomes. The Fundamental Principle of Counting does not show the actual outcomes, but it requires less time to solve the problem.

---

2 TEACH

 Transparency 12-1B contains a teaching aid for this lesson.

Thinking Algebraically Before introducing the Counting Principle, suggest that students draw a tree diagram for the situation in Motivating the Lesson on page 518. Discuss how to think through the tree diagram to find the number of choices. Then introduce the Counting Principle.

Teaching Tip In Example 1, point out that the number of outcomes, 6, can be found by multiplying the number of shorts, 2, by the number of jerseys, 3.

In-Class Examples

For Example 1
A red and a white number cube, each with the numbers 1, 2, 3, 4, 5, and 6 are rolled. How many outcomes are possible?
36 outcomes

For Example 2
Three coins are tossed. What is the probability that all three coins will land on heads? $\frac{1}{8}$

Study Guide Masters, p. 93

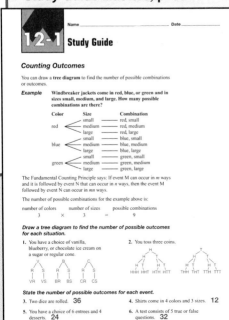

Lesson 12-1 **519**

3 PRACTICE/APPLY

Check for Understanding
If students need additional practice or instruction after completing Exercises 1–6, one of these options may be helpful.
- Extra Practice, see p. 639
- Reteaching Activity, see p. 519
- *Study Guide Masters*, p. 93
- *Practice Masters*, p. 93
- Interactive Mathematics Tools Software

Assignment Guide
Core: 7–15 odd, 16–18
Enriched: 8–12 even, 13–18

4 ASSESS

Closing Activity
Writing Have students compare and contrast tree diagrams and the Counting Principle as methods for finding the number of outcomes of an event.

Practice Masters, p. 93

Guided Practice
4. 12 outcomes; See Answer Appendix for tree diagram.

4. Suppose a number cube is rolled and a coin is tossed. Draw a tree diagram to find the number of possible outcomes.

5. Suppose you can buy a souvenir T-shirt in long or short sleeve. Each type of T-shirt comes in 5 colors and 4 sizes. How many different T-shirts are available?   **40 different T-shirts**

6. *Games* In the game of Clue, there are 6 suspects, 6 weapons, and 9 rooms. Find the number of possible solutions to the game's question of who, how, and where.   **324 possible solutions**

EXERCISES

Practice Draw a tree diagram to find the number of possible outcomes for each situation.   7–9. See Answer Appendix for tree diagrams.

7. The spinner is spun three times.   **27 outcomes**
8. Four coins are tossed.   **16 outcomes**
9. Kana has a choice of a floral, plaid, or striped blouse with a choice of tan, black, navy, or white skirt.   **12 outcomes**

State the number of possible outcomes for each event.

10. A restaurant offers a choice of three types of pasta with five types of sauce. Each pasta entrée comes with or without a meatball.   **30 outcomes**

11. There are four choices for each of five multiple-choice questions on a science quiz.   **1,024 outcomes**

12. A car comes with two or four doors, a four or six-cylinder engine, and eight exterior colors.   **32 outcomes**

Applications and Problem Solving

13. *Games* The Hopi Indians invented a game of chance called Totolospi. Players use three cane dice and a counting board. Each die can land either round side up (R) or flat side up (F).
   a. Draw a tree diagram to show the possible outcomes for the three cane dice.   **See Answer Appendix.**
   b. How many outcomes are possible?   **8 outcomes**

14. *Sports* The Silvercreek Ski Resort has four ski lifts up the mountain and eleven trails down the mountain. How many different ways can a skier go up and down the mountain?   **44 different ways**

15. *Communication* In the United States, radio and television stations use call letters that start with K or W. How many different call letters with four letters are possible?   **35,152 call letters**

16. *Critical Thinking* If x coins are tossed, write an algebraic expression for the number of possible outcomes. 2^x

Mixed Review

17. *Measurement* How many significant digits are in the measurement 14.4 centimeters? *(Lesson 11-7)*   **3 significant digits**

18. **Test Practice** The area of a square playground is 361 square feet. What is the perimeter of the playground? *(Lesson 9-1)*   **B**
   A 19 ft   B 76 ft   C 127 ft   D 361 ft

520 Chapter 12 Investigating Discrete Math and Probability

Extending the Lesson

Enrichment Masters, p. 93

Activity Have students work in small groups to solve this problem: *29,160 outcomes are possible when spinning 4 spinners. Three of the spinners have 12, 15, and 9 possible outcomes, respectively. Find the number of possible outcomes on the fourth spinner.*
18 outcomes

12-2 Permutations

What you'll learn
You'll learn to find the number of permutations of objects.

When am I ever going to use this?
Knowing how to find the number of permutations can help you find the number of ways you can order floats in a parade.

Word Wise
permutation
factorial

The Environmental Club at Greenway Junior High School has decided to complete the following projects during the school year.

- Clean up the creek that runs behind the school.
- Collect and recycle old phone books.
- Plant trees around the school on Arbor Day.

The club president plans to appoint a different chairperson for each of the projects. Seven students have volunteered to chair the projects. How many possible ways can the president appoint the chairpersons?

1st Project   Any of the 7 students can chair the first project.
2nd Project   One of the 6 remaining students can chair the second project.
3rd Project   One of the 5 remaining students can chair the third project.

Using the Counting Principle, there will be $7 \times 6 \times 5$ or 210 possible ways to pick the chairpersons from just 7 volunteers.

An arrangement or listing in which order is important is called a **permutation**. In the above example, the symbol $P(7, 3)$ represents the number of permutations of 7 volunteers taken 3 at a time.

$P(n, r)$	
Words:	$P(n, r)$ means the number of permutations of n things taken r at a time.
Symbols:	
Arithmetic	$P(7, 3) = 7 \cdot 6 \cdot 5$
Algebra	$P(n, r) = n \cdot (n - 1) \cdot (n - 2) \cdot \ldots \cdot (n - r + 1)$

HANDS-ON MINI-LAB

Work in groups of three or four.

 4 different colored pencils

Try This
- Draw four columns on notebook paper. Use the lines of the paper to complete a grid so that each row has four units.
- On the first row, color each unit a different color.
- On the next row, use the same four colors to color the units, but do not repeat the pattern you used in the first row.
- Continue coloring the units in each row until you have created all possible arrangements of the four colors.

Lesson 12-2 Permutations   **521**

Multiple Learning Styles

Kinesthetic Have three students act out all of the different orders in which they could stand one behind the other. Have other students keep track of the orders to determine whether all six ways have been acted out. Repeat with four students. Ask them what would happen with five students.

12-2 Lesson Notes

Instructional Resources
- *Study Guide Masters*, p. 94
- *Practice Masters*, p. 94
- *Enrichment Masters*, p. 94
- Transparencies 12-2, A and B
- *Assessment and Evaluation Masters*, p. 323
- *Technology Masters*, p. 76
- CD-ROM Program
  - Resource Lesson 12-2

Recommended Pacing	
Standard	Day 4 of 13
Honors	Day 2 of 11
Block	Day 2 of 6

1 FOCUS

5-Minute Check (Lesson 12-1)

1. A certain wallpaper is available in 8 colors and 6 patterns. How many combinations of color and pattern are possible? **48 combinations**

2. Draw a tree diagram to find the number of possible outcomes when both spinners are spun. **6 outcomes**

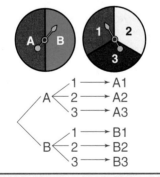

The 5-Minute Check is also available on **Transparency 12-2A** for this lesson.

2 TEACH

Transparency 12-2B contains a teaching aid for this lesson.

Using the Mini-Lab Point out that there are only three colors to choose from for the second square, two for the third, and one for the fourth.

Lesson 12-2   **521**

In-Class Examples

For Example 1
In how many different orders can six presidential candidates present their opening statements in a debate?
720 orders

For Example 2
Eight trained parakeets fly onto the stage but find only five perches. How many parakeet permutations are possible?
6,720 permutations

3 PRACTICE/APPLY

Check for Understanding

If students need additional practice or instruction after completing Exercises 1–9, one of these options may be helpful.
- Extra Practice, see p. 639
- Reteaching Activity
- *Study Guide Masters*, p. 94
- *Practice Masters*, p. 94

Additional Answer

1. Sample answer: How many 2-digit whole numbers can you write using the digits 1, 2, 3, 4, and 5 if no digit can be used twice?

Talk About It   2. Yes; order is important.
1. How many rows did you color?   24 rows
2. Is this an example of a permutation? Explain.
3. How could you determine how many rows are needed without coloring?   Find $4 \times 3 \times 2 \times 1$.

Study Hint
Technology You can use a calculator to compute factorials. To find 4!, press 4 [2nd] [x!].

The number of permutations in the Mini-Lab can be expressed as $P(4, 4)$. Notice that $P(4, 4)$ means the number of permutations of 4 things taken 4 at a time.

$$P(4, 4) = 4 \cdot 3 \cdot 2 \cdot 1$$

The mathematical notation 4! also means $4 \cdot 3 \cdot 2 \cdot 1$. The symbol 4! is read *four factorial*. $n!$ means the product of all counting numbers beginning with n and counting backward to 1. We define 0! as 1.

Examples

CONNECTION

① **Geography**   Each summer, Ms. McGatha likes to visit a different state in the United States. Over the next five years, she wants to visit the five largest states; Alaska, Texas, California, Montana, and New Mexico. How many different ways can she arrange the trips?

You must find the number of permutations of 5 states taken 5 at a time.
$$P(5, 5) = 5!$$
$$= 5 \cdot 4 \cdot 3 \cdot 2 \cdot 1$$
$$= 120$$

She can arrange the trips in 120 ways.

APPLICATION

② **Pets**   The local Humane Society needs to select 2 out of 6 dogs to feature in a newspaper advertisement. One will appear in the Saturday paper, and the other will appear in the Sunday paper. How many ways can the dogs be selected?

You must find the number of permutations of 6 dogs taken 2 at a time.
$$P(6, 2) = 6 \cdot 5$$
$$= 30$$

There are 30 different ways for the dogs to be selected. *You can draw a tree diagram to check the answer.*

CHECK FOR UNDERSTANDING

Communicating Mathematics

Read and study the lesson to answer each question.   1. See margin.
1. *Write a Problem* that could be solved by finding the value of $P(5, 2)$.
2. *Compare* 8! and $P(8, 3)$. $8! = 8 \cdot 7 \cdot 6 \cdot 5 \cdot 4 \cdot 3 \cdot 2 \cdot 1$ and $P(8, 3) = 8 \cdot 7 \cdot 6$.

522   Chapter 12 Investigating Discrete Math and Probability

Reteaching the Lesson

Activity   Have four students act out the number of different ways they can arrange themselves in four chairs. Then remove one chair and have them act out the number of different ways the four can be seated in the three chairs. Remove another chair and repeat.

Error Analysis
Watch for students using the factorial key to solve all of the permutation problems.
Prevent by reminding students that $n!$ finds the total number of possibilities. In some problems, there are limits to the number of ways items can be arranged.

Study Guide Masters, p. 94

Study Guide

Permutations

An arrangement or listing in which order is important is called a permutation.

Example 1   5 people are running a race. How many arrangements of winner, second place, and third place are possible?

There are 5 choices for winner.
Then, there are 4 choices for second place.
Finally, there are 3 possible choices for third place.
$5 \times 4 \times 3 = 60$
There are 60 permutations.

For the example above, the permutation of 5 runners taken 3 at a time may be written: $P(5, 3) = 5 \times 4 \times 3$.

The product of counting numbers beginning at n and counting backward to 1 is called n factorial. n factorial is written $n!$.

Example 2   In how many different ways can 6 people stand in a row?
$P(6, 6) = 6!$
$= 6 \times 5 \times 4 \times 3 \times 2 \times 1$
$= 720$
6 people can stand in a row 720 different ways.

Find each value.

1. $P(4, 2)$	2. $P(7, 3)$	3. $P(10, 4)$	4. $P(5, 5)$
12	210	5,040	120
5. 3!	6. 8!	7. 1!	8. 5!
6	40,320	1	120

In how many different ways can the letters of each word be arranged?

9. SUN	10. RAIN	11. CLOUDS	12. WINDY
6	24	720	120

© Glencoe/McGraw-Hill   T94   *Mathematics: Applications and Connections, Course 3*

HANDS-ON MATH

3. *Draw* three columns on notebook paper. Use the lines of the paper to complete a grid so that each row has three units. On the first row, color each unit a different color. On the next row, use the same three colors to color the units, but do not repeat the pattern you used in the first row. Continue coloring the units in each row until you have created all possible arrangements of the three colors.
   a. How many different arrangements are possible? **6 ways**
   b. How does your number of arrangements compare with 3!?
      They are the same.

Guided Practice Find each value.

4. $P(5, 3)$ **60** 5. $P(9, 2)$ **72** 6. 6! **720** 7. 2! **2**

8. How many different ways can you arrange the letters in the word *math*? **24 ways**

9. *Games* In the game Tic Tac Toe, players place an X or an O in any of the 9 locations that is empty. How many different ways can the first 3 moves of the game occur? **504 ways**

EXERCISES

Practice Find each value.

10. $P(7, 4)$ **840** 11. $P(8, 3)$ **336** 12. $P(10, 5)$ **30,240** 13. $P(6, 6)$ **720**
14. 3! **6** 15. 7! **5,040** 16. 0! **1** 17. 10! **3,628,800**
18. $P(25, 3)$ 19. $P(14, 4)$ 20. $P(7, 5)$ **2,520** 21. 8! **40,320**

18. 13,800
19. 24,024

22. How many 3-digit whole numbers can you write using the digits 1, 3, 5, 7, and 9 if no digit can be used twice? **60 whole numbers**

23. 720 ways

23. How many different ways can you arrange the letters in the word *equals*?

24. A family with 5 members is going to the mall in a car. Two people can sit in the front seat and 3 can sit in the back. If the mother is driving, how many ways can the members of the family be seated in the car? **24 ways**

Applications and Problem Solving

25. *Recreation* There were 17 floats in the 1996 Macy's Thanksgiving Day Parade in New York City. If the Santa float must be last, how many ways could the parade organizer pick the first two floats in the parade? **240 ways**

26. *Sports* There are 9 players on a baseball team. How many ways can the coach pick the first 4 batters? **3,024 ways**

27. *Critical Thinking* Compare $P(n, n)$ and $P(n, n-1)$, where n is any whole number greater than one. Explain. **See margin.**

Mixed Review

28. **Test Practice** A pizza shop has 3 different crusts, 3 different meat toppings, and 5 different vegetables. If Shalema wants a pizza with one meat and one vegetable, how many different pizzas can she order? *(Lesson 12-1)*

28. C

   A 11   B 15   C 45   D 90

29. *Geometry* State the measure of the bases and the height of the trapezoid. Then find the area of the trapezoid. *(Lesson 7-6)* **15 yd, 20 yd; 10 yd; 175 yd²**

30. *Algebra* Solve $\frac{b}{2} + 7 > 3$. Show the solution on a number line. *(Lesson 1-9)* **$b > -8$; See margin for number line.**

Lesson 12-2 Permutations **523**

Extending the Lesson

Enrichment Masters, p. 94

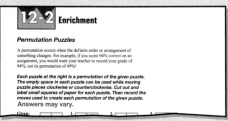

Activity Have students work in pairs to find how many ways they can arrange the letters in their first names. For each repeated letter, divide by the factorial of the number of times that letter appears. For example, the name *Tabatha* has 2 t's, 3 a's, and 7 letters. So find $\frac{P(7, 7)}{3! \cdot 2!}$. *Tabatha* can be arranged **420 ways.**

Assignment Guide

Core: 11–27 odd, 28–30
Enriched: 10–24 even, 25–30

4 ASSESS

Closing Activity

Modeling As a group project, direct students to cut out many two-inch by one-half-inch strips of paper in four different colors. Then have them form as many squares as possible using four strips in every possible combination. **24 squares**

Chapter 12, Quiz A (Lessons 12-1 and 12-2) is available in the *Assessment and Evaluation Masters*, p. 323.

Additional Answers

27. They are the same. $P(n, n) = n \cdot (n-1) \cdot (n-2) \cdot \ldots \cdot 2 \cdot 1$ and $P(n, n-1) = n \cdot (n-1) \cdot (n-2) \cdot \ldots \cdot 2$, which are equal.

30.

Practice Masters, p. 94

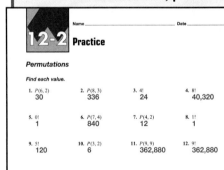

Lesson 12-2 **523**

12-3 Lesson Notes

Instructional Resources
- *Study Guide Masters*, p. 95
- *Practice Masters*, p. 95
- *Enrichment Masters*, p. 95
- Transparencies 12-3, A and B
- *Diversity Masters*, p. 38
- *Technology Masters*, p. 75
- CD-ROM Program
  - Resource Lesson 12-3

Recommended Pacing	
Standard	Day 5 of 13
Honors	Day 3 of 11
Block	Day 3 of 6

1 FOCUS

5-Minute Check
(Lesson 12-2)

Find each value.
1. $P(6, 2)$ **30**
2. $4!$ **24**
3. $P(5, 3)$ **60**
4. $6!$ **720**
5. In how many different ways can a coach name the first three batters in a ten-batter softball lineup? **720 ways**

The 5-Minute Check is also available on **Transparency 12-3A** for this lesson.

Motivating the Lesson
Communication Ask students the following questions.
- There are five digits in your ZIP code. Is the order of the digits important? **yes**
- There are five students on the planning committee for the spring picnic. Is the order of the students important? **no**

12-3 Combinations

What you'll learn
You'll learn to find the number of combinations of objects.

When am I ever going to use this?
Knowing how to find the number of combinations can help you determine the number of 4-topping pizzas you could order.

Word Wise
combination

Ms. Manzueta has been collecting books and information on butterflies for her science students to use to complete their projects. She currently has information on the Tiger Swallowtail, Giant Swallowtail, Clouded Sulfur, Viceroy, Monarch, Meadow Fritillary, and Dog Face. Each student must pick three butterflies to study. How many different groups of butterflies can the students study if they use Ms. Manzueta's resources? *This problem will be solved in Example 1.*

An arrangement or listing, like the one above, where order is not important is called a **combination**.

HANDS-ON MINI-LAB

Work with a partner. 6 index cards

Try This
- Mark each of six index cards with one of the letters A, B, C, D, E, and F.
- Select any three cards. Record your selection.
- Select another three cards, but not the same group you chose before. Record your selection.
- Continue selecting groups of three cards until all possible groups are recorded.

Talk About It
1. How many groups did you record? **20 groups**
2. In how many different orders can three letters be arranged? **6 orders**
3. Find $P(6, 3)$. **120**
4. What is the relationship between the number of groups, the number of arrangements of three cards, and $P(6, 3)$?

4. The number of groups equals $P(6, 3)$ divided by the number of ways 3 things can be arranged.

Study Hint
Reading Math $C(6, 3)$ is read as the number of combinations of 6 things taken 3 at a time.

A quick way to find the number of groupings, or combinations, of 6 cards taken 3 at a time, $C(6, 3)$, is to divide the number of permutations, $P(6, 3)$, by the number of orders 3 cards can be arranged, which is $3!$.

$$C(6, 3) = \frac{P(6, 3)}{3!}$$ *Dividing by $3!$ eliminates the combinations that are the same.*
$$= \frac{6 \cdot 5 \cdot 4}{3 \cdot 2 \cdot 1}$$
$$= \frac{120}{6} \text{ or } 20$$

There are 20 groups.

524 Chapter 12 Investigating Discrete Math and Probability

Multiple Learning Styles

Interpersonal Instruct students to read the Mini-Labs in Lessons 12-2 and 12-3 and then explain the difference between a permutation and combination to a partner. The Mini-Lab in Lesson 12-2 involves a pattern where placement is important; therefore, the example is a permutation. The Mini-Lab in Lesson 12-3 involves the selection of items where order is not important; therefore, the example is a combination.

$C(n, r)$	Words:	$C(n, r)$ means the number of combinations of n things taken r at a time.
	Symbols:	Arithmetic $C(6, 3) = \dfrac{6 \cdot 5 \cdot 4}{3!}$ or 20
		Algebra $C(n, r) = \dfrac{P(n, r)}{r!}$

Examples

CONNECTION

1 **Life Science** Refer to the beginning of the lesson. How many different groups of butterflies can the students study if they use Ms. Manzueta's resources?

Ms. Manzueta has information about 7 different butterflies. You need to find the number of combinations of 7 butterflies taken 3 at a time.

$$C(7, 3) = \dfrac{P(7, 3)}{3!}$$
$$= \dfrac{7 \cdot 6 \cdot 5}{3 \cdot 2 \cdot 1}$$
$$= \dfrac{210}{6} \text{ or } 35$$

There are 35 different groups of butterflies that the students can study.

Study Hint

Mental Math You can simplify $\dfrac{7 \cdot 6 \cdot 5}{3 \cdot 2 \cdot 1}$ before multiplying.

INTEGRATION

2 **Geometry** Six points are located on a circle. How many line segments can be drawn with these points as endpoints?

Explore There are 6 points on the circle. You want to know how many line segments can be drawn using these points as endpoints.

Plan You need to find the number of combinations of 6 things taken 2 at a time.

Solve $C(6, 2) = \dfrac{P(6, 2)}{2!}$
$$= \dfrac{6 \cdot 5}{2 \cdot 1}$$
$$= \dfrac{30}{2} \text{ or } 15$$

There are 15 possible line segments that can be drawn.

Examine Draw all of the possible line segments. There are 15 possible line segments. The answer checks.

---

2 TEACH

Transparency 12-3B contains a teaching aid for this lesson.

Using the Mini-Lab Students should search for distinct groups of three cards regardless of the order. To clarify this, point out that ABC, ACB, BAC, BCA, CAB, and CBA all count as a single group of cards, just as △ABC, △ACB, △BAC, △BCA, △CAB, and △CBA all name the same triangle.

In-Class Examples

For Example 1
From a selection of 9 pencils, Briana took 3 to math class. How many groups of 3 did she have to choose from?
84 groups

For Example 2
There are four points on a square. Each point is on a different side, none at the corners. How many line segments can be drawn between these four points?
6 segments

Teaching Tip Point out that the study hint involves simplifying. Factorials can also be simplified. $\dfrac{4!}{3!} = \dfrac{4 \cdot 3!}{3!}$, or 4

3 PRACTICE/APPLY

Check for Understanding

If students need additional practice or instruction after completing Exercises 1–11, one of these options may be helpful.
- Extra Practice, see p. 640
- Reteaching Activity
- *Study Guide Masters*, p. 95
- *Practice Masters*, p. 95

Assignment Guide

Core: 13–31 odd, 32–36
Enriched: 12–28 even, 30–36

Communicating Mathematics

2. $P(8, 3)$; To find $C(8, 3)$, you divide $P(8, 3)$ by $3!$.
3. Montega; the three positions are different, so order is important.

HANDS-ON MATH

Guided Practice

CHECK FOR UNDERSTANDING

Read and study the lesson to answer each question.

1. *Write* an expression to represent the number of six-person committees that could be formed from a group of 20 people. $C(20, 6)$
2. *Tell*, without calculating, which is greater, $P(8, 3)$ or $C(8, 3)$. Explain.
3. *You Decide* Bernice says that picking a class president, a class vice president, and a secretary from all of the members of the class is an example of a combination. Montega says it is an example of a permutation. Who is correct? Explain.
4. *Mark* each of five index cards with one of the letters A, B, C, D, and E. Select any two cards. Record your selection. Continue selecting groups of two cards until all possible groups are recorded.
   a. How many groups did you record? **10 groups**
   b. How does your number of groups compare with the value of $\frac{P(5, 2)}{2!}$? **They are the same.**

Find each value.

5. $C(8, 3)$ **56**   6. $C(9, 2)$ **36**   7. $C(5, 2)$ **10**

Determine whether each situation is a *permutation* or a *combination*.

8. placing 9 model cars in a line **permutation**
9. choosing a team of 5 players from 11 students **combination**
10. How many different combinations of 3 flowers can you choose from 12 different flowers? **220 combinations**
11. **Business** A pizza shop has 12 different toppings from which to choose. This week, if you buy a 2-topping pizza, you get 2 more toppings free. How many different ways can the special pizza be ordered? **495 ways**

EXERCISES

Practice   **Find each value.**

12. $C(5, 4)$ **5**   13. $C(7, 5)$ **21**   14. $C(14, 2)$ **91**   15. $C(10, 4)$ **210**
16. $C(8, 4)$ **70**   17. $C(9, 5)$ **126**   18. $C(20, 3)$ **1,140** 19. $C(13, 3)$ **286**

Determine whether each situation is a *permutation* or a *combination*.

20. choosing a committee of 7 from 25 members **combination**
21. five students forming a line **permutation**
22. writing a 3-digit number using no digit more than once **permutation**
23. packing 5 out of 15 outfits for a trip **combination**
24. buying 3 CDs from a selection of 18 CDs **combination**
25. placing 3 out of 18 CDs in the CD player **permutation**

526 Chapter 12 Investigating Discrete Math and Probability

Study Guide Masters, p. 95

Reteaching the Lesson

Activity Have students work in groups of four or more. Have them count to find how many groups of three they can form from the members of their group. Have them confirm their results using the formula for combinations.

Error Analysis
Watch for students who confuse permutations and combinations.
Prevent by having students decide whether order is important (permutations) or not (combinations).

26. How many different starting squads of 6 players can be picked from 10 volleyball players? **210 squads**

27. How many different 5-card hands can be dealt from a standard deck of 52 cards? **2,598,960 hands**

28. There are 20 runners in a race. In how many ways can the runners take first, second, and third place? **6,840 ways**

29. In a group of 23 people, there is about a 50% chance that two people have the same birthday. How many different combinations of 2 people could have the same birthday in the group? **253 combinations**

Applications and Problem Solving

30. *Business* The Baskin-Robbins 31 Ice Cream Stores are known for 31 flavors of ice cream.
    a. Lamel wants to buy three pints of ice cream. If each pint of ice cream is a different flavor, how many different purchases can he make? **4,495 purchases**
    b. Maria wants to buy a three-scoop ice cream cone. If each scoop is a different flavor and the order of scoops in the cone is important to her, how many different cones can she order? **26,970 cones**

31. *Games* In the game of Cribbage, a player gets two points for each combination of cards that totals 15. How many points could a player get with the hand at the right? **16 points**

32. *Critical Thinking* How many ways can a committee of 2 boys and 3 girls be formed from a class of 18 boys and 12 girls? **33,660 ways**

Mixed Review

33. How many 4-digit whole numbers can you write using the digits 1, 2, 3, 4, 5, 6, and 7 if no digit can be used twice? *(Lesson 12-2)* **840 numbers**

34. **Test Practice** What is the area of the shaded region? *(Lesson 11-1)* **D**
    A 107.5 cm²
    B 77.7 cm²
    C 59.7 cm²
    D 27.5 cm²

35. *Geography* The distance between two cities on a map is $2\frac{3}{8}$ inches. Find the actual distance between the cities if the scale on the map is 1 inch:300 miles. *(Lesson 8-9)* **$712\frac{1}{2}$ miles**

36. *Statistics* Find the mean, median, and mode for the data set. *(Lesson 4-4)*
    14, 3, 6, 8, 11, 9, 3, 2, 7   **7; 7; 3**

Lesson 12-3 Combinations   **527**

Extending the Lesson

Enrichment Masters, p. 95

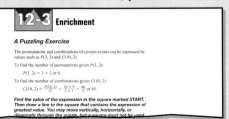

Activity Have students simplify each of the following expressions.

a. $\frac{8!}{7!}$ **8**   b. $\frac{5!}{4!}$ **5**

c. $\frac{6!}{5!}$ **6**   d. $\frac{10!}{9!}$ **10**

Then ask them to make a conjecture about the value of $\frac{247!}{246!}$. **247**

12-4 Lesson Notes

Instructional Resources
- *Study Guide Masters*, p. 96
- *Practice Masters*, p. 96
- *Enrichment Masters*, p. 96
- Transparencies 12-4, A and B
- *Assessment and Evaluation Masters*, pp. 322, 323
- CD-ROM Program
  - Resource Lesson 12-4

Recommended Pacing	
Standard	Day 6 of 13
Honors	Days 4 & 5 of 11
Block	Day 3 of 6

1 FOCUS

5-Minute Check
(Lesson 12-3)

Find each value.
1. C(4, 3) **4**
2. C(5, 2) **10**
3. C(7, 4) **35**
4. How many ways can you choose two tiles from a rack of nine? **36 ways**
5. How many different basketball teams of 5 can be formed from a squad of 10 players? **252 teams**

The 5-Minute Check is also available on **Transparency 12-4A** for this lesson.

Motivating the Lesson
Hands-On Activity Sketch the figure below (without the numbers).

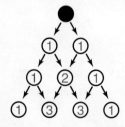

Have students find the number of ways to reach each circle, starting from the top circle and following the direction of the arrows.

12-4 Pascal's Triangle

What you'll learn
You'll learn to identify patterns in Pascal's Triangle.

When am I ever going to use this?
Knowing how to identify patterns in Pascal's Triangle can help you determine the number of combinations.

Word Wise
Pascal's Triangle

For many years, mathematicians have been interested in a special pattern called **Pascal's Triangle**. This triangle is named for a French mathematician Blaise Pascal (1623–1662). However, knowledge of the triangle is believed to have existed in China and Persia as early as the twelfth century. Published references to the triangle were made by Chinese mathematician Chu Shih-Chieh in 1303. Study the first five rows of Pascal's Triangle.

```
            1
          1   1
        1   2   1
      1   3   3   1
    1   4   6   4   1
```

Example 1 Find the pattern in Pascal's Triangle and use the pattern to complete the next two rows.

The right and left sides of the triangle are formed by ones. Notice the relationship of each inside number and the two numbers above it.

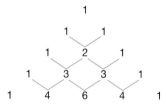

To find each inside number, add the two numbers above it.

The rows in Pascal's Triangle are numbered beginning with the top row, which is row 0.

```
Row 0                    1
Row 1                  1   1
Row 2                1   2   1
Row 3              1   3   3   1
Row 4            1   4   6   4   1
Row 5          1   5  10  10   5   1
Row 6        1   6  15  20  15   6   1
```

528 Chapter 12 Investigating Discrete Math and Probability

Multiple Learning Styles

Visual/Spatial Separate the class into 4- and 5-person groups. Form as many different 3-person teams within these groups as possible. Show that the number of possible teams can be found in Pascal's Triangle as the fourth (3 + 1) entry of the row numbered the same as the number of students in the group. For 4-person groups, the fourth entry of row 4 is 4. For 5-person groups, the fourth entry of row 5 is 10. Ask students how to use Pascal's Triangle to determine how many different 3-person teams could be formed from the entire class. **Sample answer: For 27 students, find the fourth entry in row 27.**

MINI-LAB

Work with a partner.

Try This
- Copy Pascal's Triangle formed in Example 1 and add two rows on your copy.
- Now, draw four tree diagrams to represent the possible outcomes for tossing 2, 3, 4, and 5 coins.

Talk About It
1. Which row of Pascal's Triangle has a sum equal to the number of outcomes for tossing 2 coins? 3 coins? 4 coins? 5 coins?
2. When tossing 4 coins, how many ways can the outcome of 2 heads and 2 tails result? Where is the answer located in Pascal's Triangle?
3. How many possible outcomes are there when you toss 8 coins? How many ways can you get 4 heads and 4 tails?

1. row 2; row 3; row 4; row 5
2. 6 ways; the third number in row 4
3. 256 outcomes; 70 ways

Pascal's Triangle can be used to answer questions involving combinations. To find the number of combinations of 6 things taken 4 at a time, refer to row 6 of Pascal's Triangle.

Number Taken at a Time	0	1	2	3	4	5	6
Row 6	1	6	15	20	15	6	1

There are 15 combinations of 6 things taken 4 at a time.

Example 2

Life Science Mr. Santiago is cross breeding 8 different types of tomato plants in order to find a plant that produces larger tomatoes. If he wants to cross every possible pair, how many pairs will he have?

Explore   Mr. Santiago has 8 different plants. He wants to know how many pairs he can form from these plants.

Plan   Use Pascal's Triangle to find the number of combinations of 8 things taken 2 at a time.

Solve

Number Taken at a Time	0	1	2	3	4	5	6	7	8
Row 8	1	8	28	56	70	56	28	8	1

There are 28 pairs.

Examine   Find the number of combinations of 8 things taken 2 at a time.

$$C(8, 2) = \frac{P(8, 2)}{2!}$$
$$= \frac{8 \cdot 7}{2 \cdot 1} \text{ or } 28 \checkmark$$

Lesson 12-4 Pascal's Triangle   **529**

2 TEACH

 Transparency 12-4B contains a teaching aid for this lesson.

Using the Mini-Lab Use a table like the first one on page 529 to show students why there are the same number of combinations of 6 index cards labeled A, B, C, D, E, and F taken 2 and 4 at a time.

In-Class Examples

For Example 1
Use Pascal's Triangle to find the number of combinations of 4 things taken 2 at a time.
6 combinations

For Example 2
How many different combinations of 5 letters can be chosen from the letters in the word PROBLEM?
21 combinations

Teaching Tip Explain to students that Pascal's Triangle can be used to find all combinations taken a certain number at a time. Emphasize that it does not work for permutations.

3 PRACTICE/APPLY

Check for Understanding

If students need additional practice or instruction after completing Exercises 1–8, one of these options may be helpful.
- Extra Practice, see p. 640
- Reteaching Activity
- *Study Guide Masters*, p. 96
- *Practice Masters*, p. 96

Assignment Guide

Core: 9–23 odd, 24–26
Enriched: 10–20 even, 21–26
All: Self Test 1–5

Additional Answer

1. Start with a 1. Then add pairs of numbers from row 8 to get the next 8 numbers: $1 + 8 = 9$; $8 + 28 = 36$; $28 + 56 = 84$; $56 + 70 = 126$; $70 + 56 = 126$; $56 + 28 = 84$; $28 + 8 = 36$; and $8 + 1 = 9$. The last number is a 1.

Study Guide Masters, p. 96

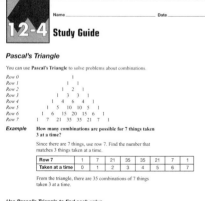

CHECK FOR UNDERSTANDING

Communicating Mathematics

Read and study the lesson to answer each question.

1. *Show* how to find row 9 of Pascal's Triangle. **See margin.**
2. *Tell* how the sum of the numbers in row 5 compares to the total number of possible ways to answer a true-or-false quiz with 5 questions. **Both equal 32.**

3. *Draw* a tree diagram to represent the possible outcomes of tossing 6 coins.
   a. How many ways can you toss 6 heads? 5 heads? 4 heads? 3 heads? 2 heads? 1 head? 0 heads? **1 way; 6 ways; 15 ways; 20 ways; 15 ways; 6 ways; 1 way**
   b. What row of Pascal's Triangle corresponds with the results of tossing 6 coins? **row 6**

Guided Practice

Use Pascal's Triangle to find each value.

4. $C(6, 2)$ **15**   5. $C(4, 3)$ **4**

Use Pascal's Triangle to answer each question.

6. **56 committees**
6. How many different committees of 3 members can be chosen from 8 people?

7. $\frac{1}{32}$
7. Five coins are tossed. What is the probability that all 5 coins will show heads?

8. **Fund-raising** The Howard Middle School Band is making pizzas to raise money for new percussion instruments. The members can add pepperoni, mushrooms, onions, and/or green peppers to the basic cheese pizzas.
   a. How many 2-item pizzas can the band prepare? **6 pizzas**
   b. How many 3-item pizzas can the band prepare? **4 pizzas**
   c. How many different pizzas can the band prepare? **16 pizzas**

EXERCISES

Practice

Use Pascal's Triangle to find each value.

9. $C(8, 3)$ **56**   10. $C(3, 2)$ **3**   11. $C(5, 2)$ **10**
12. $C(6, 5)$ **6**   13. $C(9, 5)$ **126**   14. $C(7, 5)$ **21**

Use Pascal's Triangle to answer each question.

15. **20 combinations**
15. How many combinations of 6 things taken 3 at a time are possible?
16. How many different 4-question quizzes can be formed from 7 possible questions? **35 quizzes**
17. How many branches will there be in the last column of a tree diagram showing the outcomes of tossing 6 coins? **64 branches**
18. Six coins are tossed. What is the probability that 3 will show heads and 3 will show tails? $\frac{5}{16}$

530 Chapter 12 Investigating Discrete Math and Probability

Reteaching the Lesson

Activity Have students explain where they would find $C(8, 5)$ in Pascal's Triangle. **row 8, 6th number**

Error Analysis
Watch for students who number the rows in Pascal's Triangle incorrectly.
Prevent by stressing that the first row, containing a single 1, is row 0.

19. combination of 4 things taken 2 at a time

19. What does the 6 in row 4 of Pascal's Triangle mean?

20. In row 6 of Pascal's Triangle, which combinations have the same number of possibilities? $C(6, 0)$ and $C(6, 6)$, $C(6, 1)$ and $C(6, 5)$, $C(6, 2)$ and $C(6, 4)$

Applications and Problem Solving

21b. The number of ways is equal to the number in the corresponding position in Pascal's Triangle.

21. Life Science A bee is in cell A and wants to go to cell B. The bee must travel downward from one cell directly into another.

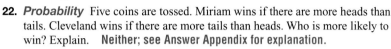

  a. How many ways can a bee get from cell A to cell B? **4**

  b. Explain how this problem relates to Pascal's Triangle.

22. Probability Five coins are tossed. Miriam wins if there are more heads than tails. Cleveland wins if there are more tails than heads. Who is more likely to win? Explain. **Neither; see Answer Appendix for explanation.**

23. Critical Thinking Compare $C(9, n)$ and $C(9, 9 - n)$, where n is a whole number and $0 \leq n \leq 9$. Explain. **See margin.**

Mixed Review

24. Find $C(9, 3)$. *(Lesson 12-3)* **84**

25. Algebra Graph the quadratic function $y = 2x^2 - 8$. *(Lesson 10-6)* **See Answer Appendix.**

26. **Test Practice** Joe Marsh and three of his co-workers ate lunch at Old Town Café. They plan to leave a 20% tip for the waiter. Two of his co-workers had turkey sandwiches, one had soup and salad, and Mr. Marsh had pasta. What other information is necessary to determine how much to leave for a tip? *(Lesson 3-5)* **B**

  A the cost of the pasta
  B the cost of the 4 meals
  C what day they had lunch
  D the soup of the day
  E the name of the waiter

CHAPTER 12 Mid-Chapter Self Test

1. Sweatshirts are available in 3 colors (black, white, and gray) and in 4 sizes (small, medium, large, and extra large). Draw a tree diagram that illustrates the outcomes. *(Lesson 12-1)* **See Answer Appendix.**

2. The school cafeteria offers a lunch special. A person can pick one of six different sandwiches, one of three beverages, and one of three types of fruit. How many different lunch specials are possible? *(Lesson 12-1)* **54 lunch specials**

3. **Music** Five band members play the trumpet. How many ways can these members be chosen for the first, second, and third chairs of the trumpet section? *(Lesson 12-2)* **60 ways**

4. How many ways can 2 student council members be elected from 7 candidates? *(Lesson 12-3)* **21 ways**

5. Use Pascal's Triangle to find the number of combinations that is possible for 7 things taken 3 at a time. *(Lesson 12-4)* **35 combinations**

Lesson 12-4 Pascal's Triangle **531**

Extending the Lesson

Enrichment Masters, p. 96

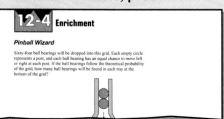

Activity Direct students to read about Blaisé Pascal (1623–1662) and the many contributions he made to the development of mathematics.

---

4 ASSESS

Closing Activity

Speaking Point to an entry in Pascal's Triangle. Have students state the number of items in a combination represented by the entry and the number taken at a time.

Chapter 12, Quiz B (Lessons 12-3 and 12-4) is available in the *Assessment and Evaluation Masters*, p. 323.

Mid-Chapter Test (Lessons 12-1 through 12-4) is available in the *Assessment and Evaluation Masters*, p. 322.

Mid-Chapter Self Test

The Mid-Chapter Self Test reviews the concepts in Lessons 12-1 through 12-4. Lesson references are given so students can review concepts not yet mastered.

Additional Answer

23. They are the same. Pascal's Triangle is symmetric. Therefore, $C(9, 0) = C(9, 9)$, $C(9, 1) = C(9, 8)$, $C(9, 2) = C(9, 7)$, and so on.

Practice Masters, p. 96

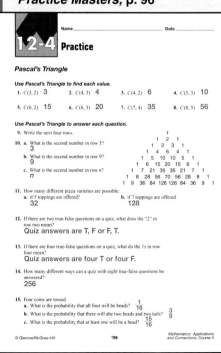

Lesson 12-4 **531**

12-4B LAB Notes

GET READY

Objective Students discover numerical and visual patterns in Pascal's Triangle.

Optional Resources
Hands-On Lab Masters
• worksheet, p. 64

MANAGEMENT TIPS

Recommended Time
45 minutes

Getting Started Have students find the sum of the numbers in each of the first six rows of Pascal's Triangle. Ask them to describe the pattern. **Each successive sum is twice the previous sum.**

Activities 1 and 2 analyze patterns in Pascal's Triangle. Make certain that all students understand that the triangle with the diagonal in Activity 1 is the same as the triangle built with hexagons in Activity 2.

In Activity 2, have students write the prime factors of the numbers in the shaded ring or any other similar ring. Ask students to describe the pattern of prime factors. **Each prime factor occurs an even number of times.**

Activity 3 on page 533 has a grid triangle that is a continuation of Pascal's Triangle. Make copies of this grid triangle and provide students with at least four copies for Activity 3. On the grids, have students number the rows starting with row 0 and note at which row four-digit numbers first appear. **row 13**

COOPERATIVE LEARNING

12-4B Patterns in Pascal's Triangle
A Follow-Up of Lesson 12-4

 hexagonal grid
 highlighter
 calculator

A famous sequence of numbers is the Fibonacci sequence, 1, 1, 2, 3, 5, 8, ... The sequence begins with 1, 1, and each number that follows is the sum of the previous two numbers. *Refer to Lesson 7-5B to review the Fibonacci sequence.*

The Fibonacci numbers and numerous number patterns can be found in Pascal's Triangle.

TRY THIS

Work with a partner.

1 • Copy Pascal's Triangle at the right and add another two rows to your copy.
• Draw diagonals beginning at each 1 along the left-hand side (passing under the 1 just above it) and extend it to the far side of the triangle as shown.
• Find the sum of the numbers through which each diagonal passes.

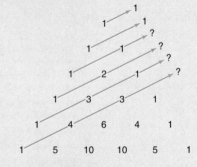

ON YOUR OWN

1. What sequence do the sums form?   **Fibonacci sequence**
2. What should the sum along the next diagonal be?   **13**

TRY THIS

Work with a partner.

2 • Find the product of the shaded ring of numbers in the Pascal's Triangle at the right.
• Find the product of any two other rings of the same shape and size in Pascal's Triangle.

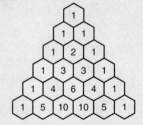

ON YOUR OWN

3. Determine the square root of the product for each ring.   **30**
4. What conclusion can you make about the product of rings in Pascal's Triangle?   **They are perfect squares.**

532   Chapter 12 Investigating Discrete Math and Probability

TRY THIS

Work with a partner.

3. • Using a hexagonal grid triangle like the one shown below, highlight all of the numbers that are multiples of 2. Highlight as many rows as necessary to determine a geometric pattern.

- On another triangle, highlight all of the numbers that are multiples of 3. Determine a geometric pattern.
- On a third triangle, highlight all of the numbers that are multiples of 4. Determine a geometric pattern.
- On a fourth triangle, highlight all of the numbers that are multiples of 7. Determine a geometric pattern.

ON YOUR OWN

5. What general visual pattern is common in all four triangles? Is this true for multiples of all numbers? Explain. **See margin.**
6. Which multiples of the numbers 2, 3, 4, and 7 have patterns that are symmetrical at each of the three vertices of the triangle? **3**
7. *Reflect Back* Write a paragraph about some of the patterns found in Pascal's Triangle. **See margin.**

Lesson 12-4B HANDS-ON **LAB** 533

Math Journal — Have students write a paragraph about any pattern found in Pascal's Triangle. The paragraph should explain why the pattern exists.

ASSESS

Have students complete Exercises 1–6. Watch for students who have difficulty working spatially and seeing patterns. To help them see patterns, have them highlight the multiples and outline the triangles.

Use Exercise 7 to determine whether students have discovered the patterns in the triangle and understand why the patterns exist.

Additional Answers

5. Inverted triangles; yes, because when two numbers next to each other have the same factor, the number below them will also have the same factor.

7. The right and left sides of Pascal's Triangle are ones. The inside numbers are the sum of the two numbers above it. If the rows are numbered from the top starting with 0, each row represents the number of combinations of items corresponding to the number of the row. The sums of the diagonals form the Fibonacci sequence. The product of the 6 numbers surrounding any number is always a perfect square. The multiples of a number form inverted triangles.

Hands-On Lab Masters, p. 64

Hands-On Lab
(Lesson 12-4B)

Patterns in Pascal's Triangle

On Your Own

1. What sequence do the sums form? The sequence of the sums is 1, 1, 2, 3, 5, and 8; the Fibonacci sequence.

2. The sum along the next diagonal should be __13__.

3. The square root of the product for each ring is __30__.

4. The products of rings in Pascal's Triangle __are perfect squares__.

5. What general visual pattern is common in all four triangles? Is this true for multiples of all numbers? Inverted triangles are visible in all four triangles. Yes; this is true for all numbers because when two numbers next to each other have the same factor, the number below them will also have the same factor.

6. The multiples of __3__ have patterns that are symmetrical at each of the three vertices.

7. *Reflect Back* Write a paragraph about some of the patterns found in Pascal's Triangle. The right left and sides of Pascal's Triangle are ones. Each inside number is the sum of the two numbers above it. If the rows are numbered from the top starting with 0, each row represents the number of combinations of items corresponding to the number of the row. The sums of the diagonals form the Fibonacci sequence. The product of the 6 numbers surrounding any number is always a perfect square. The multiples of a number form inverted triangles.

Hands-On Lab 12-4B **533**

12-5 Lesson Notes

Instructional Resources
- *Study Guide Masters*, p. 97
- *Practice Masters*, p. 97
- *Enrichment Masters*, p. 97
- Transparencies 12-5, A and B
- CD-ROM Program
  - Resource Lesson 12-5

Recommended Pacing	
Standard	Day 7 of 13
Honors	Day 6 of 11
Block	Day 4 of 6

1 FOCUS

5-Minute Check
(Lesson 12-4)

1. Write the fourth and fifth rows of Pascal's Triangle.
   Row 4: 1 4 6 4 1
   Row 5: 1 5 10 10 5 1
2. Use Pascal's Triangle to find $C(5, 2)$. **10**
3. How many combinations are possible when 4 things are taken 3 at a time? **4**

 The 5-Minute Check is also available on **Transparency 12-5A** for this lesson.

Motivating the Lesson

Problem Solving Ask students what the chances are of a coin landing heads on the tenth toss if Meg has tossed the coin 9 times and it has landed heads each time. Are the chances likely, unlikely, or even? **Even; one toss of a coin is unaffected by previous tosses so the probability of landing heads each time is $\frac{1}{2}$.**

12-5 Probability of Compound Events

What you'll learn
You'll learn to find the probability of independent and dependent events.

When am I ever going to use this?
Knowing how to find probability can help you determine whether something is likely to happen in a game.

Word Wise
compound events
independent events
dependent events

LOOK BACK
You can refer to Lesson 3-4 to review writing fractions as percents.

The United States Bureau of the Census keeps statistical information about each of the states.

Study the information about the population of Texas. What is the probability that a person from Texas picked at random will be under 18 and living in an urban area? This is an example of finding the probability of **compound events**. Compound events consist of two or more events.

Texas
$\frac{7}{25}$ of the people are under the age of 18.
$\frac{1}{10}$ of the people are 65 years old or older.
$\frac{4}{5}$ of the people live in an urban area.
$\frac{1}{5}$ of the people live in a rural area.

Source: Bureau of the Census

You can find the probability of being under the age of 18 and living in an urban area by multiplying the probability of being under the age of 18 by the probability of living in an urban area.

$$\text{number of ways it can occur} \rightarrow \frac{7}{25} \cdot \frac{4}{5} = \frac{28}{125}$$
$$\text{total numbers of possible outcomes} \rightarrow$$

The probability of a person in Texas being under the age of 18 and living in an urban area is $\frac{28}{125}$ or 22.4%.

Selecting a person under 18 years old does not depend on the selection of a person living in an urban area. We call these compound events **independent**. The outcome of one event does not affect the outcome of the other event.

Probability of Two Independent Events	Words:	The probability of two independent events can be found by multiplying the probability of the first event by the probability of the second event.
	Symbols:	$P(A \text{ and } B) = P(A) \cdot P(B)$

Example APPLICATION

1 Census Refer to the beginning of the lesson. Find the probability that a person from Texas picked at random is 65 years old or older and lives in an urban area.

$P(65 \text{ or older}) = \frac{1}{10}$ $P(\text{lives in urban area}) = \frac{4}{5}$

$P(65 \text{ or older and lives in urban area}) = \frac{1}{10} \cdot \frac{4}{5} = \frac{4}{50}$ or $\frac{2}{25}$

The probability that the two events will occur is $\frac{2}{25}$ or 8%.

534 Chapter 12 Investigating Discrete Math and Probability

 Investigations for the Special Education Student

This blackline master booklet helps you plan for the needs of your special education students by providing long-term projects along with teacher notes. Investigation 15, *The Game Show*, may be used with this chapter.

Example 2 Suppose a number cube is rolled twice. What is the probability that an even number will show both times?

$P(\text{even}) = \frac{1}{2}$

$P(\text{even and even}) = \frac{1}{2} \cdot \frac{1}{2}$ or $\frac{1}{4}$

The probability of rolling two even numbers is $\frac{1}{4}$ or 25%.

Sometimes the outcome of one event affects the outcome of another.

HANDS-ON MINI-LAB

Work with a partner.   cup   3 red counters   2 yellow counters

Try This
- Place all of the counters in the cup. Without looking, draw one counter from the cup. Do *not* put the counter back in the cup. Without looking, draw another counter from the cup.
- Copy and complete the tree diagram for this activity.

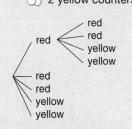

Talk About It
1. What is the probability of drawing a red counter on the first draw? $\frac{3}{5}$
2. If you drew a red counter on the first draw, what is the probability of drawing a red counter on the second draw? $\frac{1}{2}$
3. How many equally likely outcomes are there? **20 outcomes**
4. How many outcomes show 2 red counters? **6 outcomes**
5. If the first counter is replaced, is the number of outcomes affected? If so, how?

5. Yes; there are 20 outcomes if you do not replace the counter, and 25 outcomes if you replace the counter.

If the outcome of one event affects the outcome of another event, these compound events are called **dependent**. In the Mini-Lab, a counter was removed from the cup on the first draw. For the second draw, there were fewer counters from which to choose. These events are dependent.

Like independent events, the probability of two dependent events is found by multiplying probabilities. In this case, you multiply the probability that the first event occurs times the probability that the second event follows the first one.

Probability of Two Dependent Events	**Words:**	If two events, A and B, are dependent, then the probability of both events occurring is the product of the probability of A and the probability of B after A occurs.
	Symbols:	$P(A \text{ and } B) = P(A) \cdot P(B \text{ following } A)$

Lesson 12-5 Probability of Compound Events **535**

In-Class Example

For Example 3
A drawer contains 3 pairs of blue socks, 10 pairs of red socks, and 7 pairs of green socks. What is the probability that Kaitlin will randomly choose a pair of red socks followed by a pair of blue socks if the first pair is not replaced? $\frac{3}{38}$ or about 8%

3 PRACTICE/APPLY

Check for Understanding
If students need additional practice or instruction after completing Exercises 1–9, one of these options may be helpful.
- Extra Practice, see p. 640
- Reteaching Activity
- *Study Guide Masters*, p. 97
- *Practice Masters*, p. 97

Additional Answers
1. The person is not likely to be 65 years old or older and living in an urban area. Only about 8 out of 100 people would satisfy these conditions.
3. Tonya; the first spin does not affect the second spin. Therefore the probability is $\frac{2}{5} \cdot \frac{2}{5}$ or $\frac{4}{25}$.

Study Guide Masters, p. 97

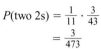

Study Hint
Reading Math In dependent events, you always assume you were successful on the first event.

Games In the game of Sorry!, there are 44 cards. Each set of cards consists of four of each denomination (1, 2, 3, 4, 5, 7, 8, 10, 11, and 12) and four Sorry cards. Cards are placed faceup after being drawn. A card with a 2 allows a player to draw again. What is the probability that the first player will draw a 2 followed by another 2?

This is an example of dependent events since the first draw affects the second draw.

First draw: $P(2) = \frac{4}{44}$ or $\frac{1}{11}$

Second draw: $P(\text{second } 2) = \frac{\text{number of 2s left}}{\text{number of cards left}}$
$= \frac{4 - 1}{44 - 1}$ or $\frac{3}{43}$

$P(\text{two 2s}) = \frac{1}{11} \cdot \frac{3}{43}$
$= \frac{3}{473}$

The probability of a 2 followed by another 2 is $\frac{3}{473}$ or about 0.6%.

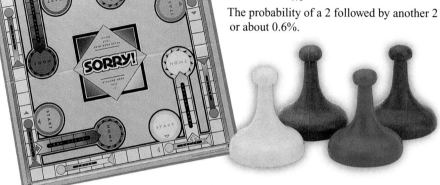

CHECK FOR UNDERSTANDING

Communicating Mathematics
Read and study the lesson to answer each question.

1. *Explain* whether the probability of $\frac{2}{25}$ or 8% found in Example 1 indicates the person picked at random is likely to be 65 years old or older and living in an urban area. **See margin.**

2. See Answer Appendix.

2. *Describe* how to find the probability of the second of two dependent events.

3. *You Decide* Nancy says that the probability of spinning a vowel on each of two consecutive spins is $\frac{2}{5} \cdot \frac{1}{4}$ or $\frac{1}{10}$. Tonya says the probability is $\frac{2}{5} \cdot \frac{2}{5}$ or $\frac{4}{25}$. Who is correct? Explain. **See margin.**

HANDS-ON MATH

4. Suppose there are 4 red and 2 yellow counters in a cup. One counter is drawn at random and is *not* returned to the cup. A second counter is drawn at random from the cup.
   a. Draw a tree diagram showing all of the possible outcomes. **See Answer Appendix.**
   b. What is the probability of drawing 2 red counters? $\frac{2}{5}$
   c. What is the probability of drawing 2 yellow counters? $\frac{1}{15}$

Guided Practice
A coin is tossed and a number cube is rolled. Find each probability.

5. P(heads and 5) $\frac{1}{12}$

6. P(tails and even number) $\frac{1}{4}$

536 Chapter 12 Investigating Discrete Math and Probability

Reteaching the Lesson

Activity What is the probability that when the cards [M] [A] [T] [H] are shuffled and dealt, they will spell MATH? $\frac{1}{24}$ Is each drawing an independent or dependent event? **dependent**

Error Analysis
Watch for students who cannot distinguish between independent and dependent events.
Prevent by asking whether the results of the second event depend on the results of the first.

There are 3 white marbles, 7 red marbles, and 5 blue marbles in a bag. Once a marble is selected, it is not replaced. Find the probability of each outcome.

7. two red marbles $\frac{1}{5}$

8. a white marble and then a blue marble $\frac{1}{14}$

9. **Games** A standard Western domino set has 28 tiles. Seven of these tiles have the same number of dots on each side and are called doubles. To start the game, each player draws a tile. What is the probability that the first and second players each draw a double if the tiles are not replaced? $\frac{1}{18}$

EXERCISES

Practice Each spinner is spun once. Find each probability.

10. $P(1 \text{ and } A)$ $\frac{1}{20}$
11. $P(2 \text{ and } B)$ $\frac{3}{20}$
12. $P(\text{even and } C)$ $\frac{1}{10}$
13. $P(\text{odd and } B)$ $\frac{3}{10}$
14. $P(\text{even and vowel})$ $\frac{1}{10}$
15. $P(\text{odd and consonant})$ $\frac{2}{5}$

Two cards are drawn from a deck of ten cards numbered 1 though 10. Once a card is selected, it is not replaced. Find the probability of each outcome.

16. a 5 and then a 3 $\frac{1}{90}$
17. two even numbers $\frac{2}{9}$
18. two numbers greater than 4 $\frac{1}{3}$
19. a 6 and then an odd number $\frac{1}{18}$
20. an odd number and then an even number $\frac{5}{18}$
21. a number greater than 7 and then a number less than 6 $\frac{1}{6}$
22. What is the probability of tossing a coin three times and getting heads each time? $\frac{1}{8}$
23. What is the probability of rolling a number cube three times and getting numbers greater than 4 each time? $\frac{1}{27}$

Applications and Problem Solving

24. **Statistics** In the United States, $\frac{3}{5}$ of all households have some kind of pet, and $\frac{1}{3}$ of all households have at least one child. What is the probability that a household picked at random will have a pet and one or more children? $\frac{1}{5}$

25. **Games** In Monopoly, there are 16 Community Chest cards. One is a Get Out of Jail card, and one is a Go to Jail card. A player keeps the Get Out of Jail card. What is the probability that the first card picked will say Get Out of Jail and the next card drawn will say Go to Jail? $\frac{1}{240}$

26. **Critical Thinking** During his career, the probability that Babe Ruth would get a hit was 0.342. What is the probability that Babe Ruth would *not* get a hit in two consecutive times at bat? **0.432964**

Mixed Review

27. Use Pascal's Triangle to find $C(6, 2)$. *(Lesson 12-4)* **15**

28. **Test Practice** Find the amount of discount for a pair of $89 shoes that are on sale at 30% off. *(Lesson 8-5)* **B**

    A $17.80   B $26.70   C $35.60   D $62.30   E Not Here

29. Find the LCM of 8, 15, and 12. *(Lesson 6-7)* **120**

12-6A THINKING LAB Notes

Objective Students solve problems by acting them out.

Instructional Resources
CD-ROM Program
- Interactive Lesson 12-6A

Recommended Pacing	
Standard	Day 8 of 13
Honors	Day 7 of 11
Block	Day 4 of 6

1 FOCUS

Getting Started Ask students to devise an experiment to find the probability of obtaining two heads when tossing two coins. Have them carry out their experiment.
Sample answer: Toss two coins ten times.
$P(HH) = \dfrac{\text{number of sucesses}}{10}$

2 TEACH

Teaching Tip Explain that the mathematical, or theoretical, probability calculated in Exercise 2 may differ from the experimental probability. Also point out that the results of another set of simulations might be entirely different.

In-Class Example
There are three traffic lights along the route from school to Lori's home. The three lights operate independently. The probability that any one of the lights is green is 0.4, and the probability that it is not green is 0.6. Estimate the probability that these lights will all be green on Lori's way home. $P(\text{green light}) = 0.4 \times 0.4 \times 0.4 = 0.064$

THINKING LAB PROBLEM SOLVING

12-6A Act It Out
A Preview of Lesson 12-6

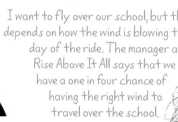

Miko and Kelsey are going on a hot air balloon ride for Miko's birthday. The girls have checked with Rise Above It All, the local air balloon company, and found that there are many considerations involved in taking the ride. Let's listen in!

Kelsey: I really want to ride in the one that is painted to look like the planet Earth. I think it is so cool! But the company says they have 6 balloons, and I can't pick the one I want.

I want to fly over our school, but that depends on how the wind is blowing the day of the ride. The manager at Rise Above It All says that we have a one in four chance of having the right wind to travel over the school.

Miko: What do you think about the chances of both of us getting our wishes?

I don't know. Let's conduct an experiment to answer that question.

That's a good idea. We can use a number cube to simulate which balloon we will get. The number one will represent the one I want. We can use a spinner with four equal sections to simulate the direction of the wind. One section will represent the wind we need to go over the school.

THINK ABOUT IT

Work with a partner.   1. See Answer Appendix.

1. **Explain** why you could use Miko's experiment, or simulation, to estimate the probability that both events will occur. Try the simulation.
2. **Compute** the probability that the girls will get the balloon that Miko wants and will travel over the school. $\dfrac{1}{24}$
3. **Tell** whether the results of the simulation tried 100 times will be exactly the same as computing the results. Explain. **See margin.**
4. **Describe** another way to simulate the possible outcomes of the balloon ride. **See Answer Appendix.**

538 Chapter 12 Investigating Discrete Math and Probability

■ Reteaching the Lesson ■

Activity Have students plan a simulation to estimate the probability that from a group of four people, one will have been born on a Monday. **Sample answer:** Spin a spinner with 7 numbers 4 times. Record each 1 as a success. Repeat many times.
$P(\text{Monday}) = \dfrac{\text{number of successes}}{\text{number of spins}}$

Additional Answer
3. No; the simulation will probably be similar to the computed results. However, simulations will vary.

ON YOUR OWN

5. The second step of the 4-step plan for problem solving asks you to *plan*. **Explain** why this step is important when using the **act it out** strategy. **You must plan a simulation that will correctly represent the situation.**

6. *Write a Problem* that can be solved by acting it out. **See margin.**

7. *Look Ahead* Explain why the act it out strategy might sometimes be called experimental probability. **See margin.**

MIXED PROBLEM SOLVING

Strategies
Look for a pattern.
Solve a simpler problem.
Act it out.
Guess and check.
Draw a diagram.
Make a chart.
Work backward.

Solve. Use any strategy.

8. **Sports** There are 16 tennis players in a tournament. If each losing player is eliminated from the tournament, how many tennis matches will be played during the tournament? **15 matches**

9. **Money Matters** Devin bought a jacket and a shirt. The total cost, not including tax, was $62.50. The jacket cost four times as much as the shirt. How much did Devin spend on each item? **jacket, $50.00; shirt, $12.50**

10. **Fashion** Namel has 6 ties. On Thursday through Sunday, Namel works at the mall. Each work day, he chooses a tie at random to wear for his job. Act it out to find the probability that Namel wears the same tie more than once in his four-day week. **Sample answer: $\frac{13}{18}$**

11. **Statistics** In a recent survey of 120 students, 50 students said they play baseball, and 60 said they play soccer. If 20 students play both sports, how many students do not play either baseball or soccer? **30 students**

12. **Physical Science** The light in the circuit will turn on if one or more switches are closed. How many combinations of open and closed switches will result in the light being on? **31 combinations**

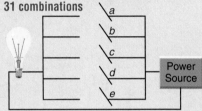

13. **Business** At a certain restaurant, toys are given with each child's meal. During the spring promotional, three different toys are given at random. **b. See margin.**
   a. Act it out to determine how many meals must be purchased in order to get all three toys. **See students' work.**
   b. What number would guarantee that you would receive all three toys? Explain.

14. **Test Practice** There are 5 red marbles, 2 yellow marbles, and 1 blue marble in a bag. One marble is selected at random, and it is not replaced. Then, a second marble is selected. Find the probability that both marbles are red. **A**

   A $\frac{5}{14}$ B $\frac{25}{64}$ C $\frac{5}{16}$ D $\frac{25}{56}$

Lesson 12-6A THINKING LAB 539

3 PRACTICE/APPLY

Check for Understanding
Use the answers in Exercise 4 to determine whether students comprehend the concept of simulation in problem solving.

Extra Practice If students need additional practice in problem solving, extra practice is available on the following pages.
- Act It Out, see p. 641
- Mixed Problem Solving, see pp. 645–646

Assignment Guide

All: 5–14

4 ASSESS

Closing Activity
Writing Have students compare and contrast mathematical, or theoretical, probability with probability found by conducting a simulation, which is experimental.

Additional Answers
6. Sample answer: Bill averages 1 basket out of 3 tries from a certain location on a basketball court. What is the probability that Bill will make 2 baskets in a row from the given location?

7. The act-it-out strategy can find the probability of an event by experimenting.

13b. No number would guarantee all 3 toys. You could be very unlucky and never get one of the toys.

Extending the Lesson

Activity A cereal maker packs a photo of one of six different sports stars in each cereal box. To model purchasing three cereal packages to obtain a certain photo, students should roll a number cube three times and see if a 1 turns up. Repeat 10 times and estimate the probability of getting the photo in three packages. Then ask students to make up problems using this problem and the one at the beginning of the lab as examples.

Thinking Lab 12-6A **539**

12-6 Lesson Notes

Instructional Resources
- *Study Guide Masters*, p. 98
- *Practice Masters*, p. 98
- *Enrichment Masters*, p. 98
- Transparencies 12-6, A and B
- *Assessment and Evaluation Masters*, p. 324
- *Hands-On Lab Masters*, p. 79
- *Science and Math Lab Manual*, pp. 93–96

CD-ROM Program
- Resource Lesson 12-6
- Interactive Lesson 12-6

Recommended Pacing	
Standard	Day 9 of 13
Honors	Day 8 of 11
Block	Day 5 of 6

1 FOCUS

5-Minute Check
(Lesson 12-5)

1. The spinner is spun twice. Find the probability of spinning a 4 followed by an odd number. $\frac{1}{8}$

2. Four lions and four tigers escaped from a cage. Find the probability that the first two out of the cage were lions. $\frac{3}{14}$

3. If 90% of a store's customers are women and 75% of the customers have a credit card, what is the probability of any one woman customer having a credit card? **67.5%**

The 5-Minute Check is also available on **Transparency 12-6A** for this lesson.

Motivating the Lesson
Communication Ask students to give examples of theories later proved untrue. **Sample answer: Earth is flat.**

540 Chapter 12

12-6 Experimental Probability

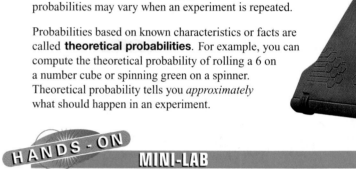

What you'll learn
You'll learn to find experimental probability.

When am I ever going to use this?
Knowing how to find experimental probability can help you find the probability that you will make a free throw.

Word Wise
simulation
experimental probability
theoretical probability

Do you like to play basketball? What is the probability that you will make a basket from the free-throw line?

Sometimes you can't tell what the probability of an event is until you conduct an experiment. For example, suppose you try 50 free throws and count the number of times you make a basket. The results of this experiment allow you to estimate the probability that you will make a basket from the free-throw line.

Probabilities that are based on frequencies obtained by conducting an experiment, or doing a **simulation** as you did in Lesson 12-6A, are called **experimental probabilities**. Experimental probabilities may vary when an experiment is repeated.

Probabilities based on known characteristics or facts are called **theoretical probabilities**. For example, you can compute the theoretical probability of rolling a 6 on a number cube or spinning green on a spinner. Theoretical probability tells you *approximately* what should happen in an experiment.

Did you know?
Around 1900, the English statistician Karl Pearson tossed a coin 24,000 times and got 12,012 heads.

HANDS-ON MINI-LAB

Work with a partner.

□ paper bag containing 10 colored marbles

Try This
- Draw one marble from the bag, record its color, and replace it in the bag. Repeat this 10 times.
- Find the experimental probability for each color of marble.
$$\text{experimental probability} = \frac{\text{number of times color was drawn}}{\text{total number of draws}}$$
- Repeat both steps described above for 20, 30, 40, and 50 draws.

Talk About It 2–3. See students' work.
1. Is it possible to have a certain color marble in the bag and never draw that color? **yes**
2. Open the bag and compute the theoretical probability of drawing each color of marble.
3. Compare the experimental probability and theoretical probability.

540 Chapter 12 Investigating Discrete Math and Probability

Classroom Vignette

"In the Mini-Lab, have students pool their data with all of the students in the class. Ask them to determine if the experimental probability of the whole class data approached the theoretical probability since the number of events increased."

Berchie Holliday
Berchie Gordon-Holliday, Supervisor
Northwest Local School District
Cincinnati, OH

Examples

APPLICATION

1 **Economics** In 1982, the United States Mint began using a different material in minting pennies to decrease their cost. The graph shows the results of a sample of 1,000 pennies. Based on this information, what is the probability that a penny you select will have a date of 1982 or later?

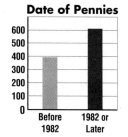

Date of Pennies

Since about 600 of the pennies selected were minted in 1982 or later, the probability of selecting a newer penny is $\frac{600}{1,000}$ or $\frac{3}{5}$. *The penny is more likely to have a date of 1982 or later than a date before 1982.*

APPLICATION

2 **Games** Nidawi and Leopoldo are playing backgammon. In this game, the combined roll of two number cubes determines each play. The graph shows the results of each roll for the game so far.

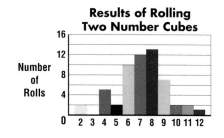

Results of Rolling Two Number Cubes

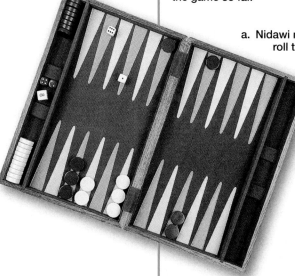

a. Nidawi needs a double six, or twelve, on the next roll to win. Based on past results, do you think she is likely to win on the next roll?

Based on the results of the rolls so far, a twelve is not very likely.

b. How many possible outcomes are there for a pair of number cubes?

There are 6 · 6 or 36 possible outcomes.

c. What is the theoretical probability of rolling a double six?

The theoretical probability is $\frac{1}{6} \cdot \frac{1}{6}$ or $\frac{1}{36}$. *The experimental probability and the theoretical probability seem to be consistent.*

CHECK FOR UNDERSTANDING

Communicating Mathematics

Read and study the lesson to answer each question.   1–2. See margin.

1. *Tell* why you might not be able to determine the theoretical probability of an event.

2. *Explain* why you would not expect the theoretical probability and the experimental probability to always be the same.

HANDS-ON MATH

3. In conducting your experiment for the Mini-Lab, could you accurately predict the probability of each color after 10 draws? What suggestion might you make to others who would conduct a similar experiment?   **See Answer Appendix.**

Lesson 12-6 Experimental Probability **541**

■ Reteaching the Lesson ■

Activity Provide a sample of 100 pennies. Have students find the experimental probability that a penny
a. was minted in 1989.
b. is more than 20 years old.
c. was minted in an even-numbered year.

Additional Answers

1. If the outcomes cannot be broken down into equally likely events, theoretical probability cannot be calculated.

2. Each experimental probability will be different. Theoretical probability tells you approximately what should happen.

---

2 TEACH

 Transparency 12-6B contains a teaching aid for this lesson.

Using the Mini-Lab Explain that as the number of times an experiment is conducted increases, the experimental and theoretical probabilities become closer.

In-Class Examples

For Example 1
Based on this data in the graph, do you think that the coin was tossed in a fair manner?

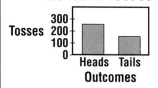

400 Coin Tosses

Theoretical probability of heads = $\frac{1}{2}$, experimental probability of heads = $\frac{5}{8}$; the coin probably was not tossed fairly.

For Example 2
The table shows the results of rolling a number cube 30 times.

Number	1	2	3	4	5	6
Frequency	2	7	5	6	4	6

a. Suppose you need to roll a 1 to win the game. Based on past results, do you think you are likely to win on the next roll? **no**

b. What is the experimental probability of rolling a 4? $\frac{1}{5}$

c. What is the theoretical probability of rolling a 4? $\frac{1}{6}$

3 PRACTICE/APPLY

Check for Understanding

If students need additional practice or instruction after completing Exercises 1–5, one of these options may be helpful.
- Extra Practice, see p. 641
- Reteaching Activity
- *Study Guide Masters*, p. 98
- *Practice Masters*, p. 98
- Interactive Mathematics Tools Software

Lesson 12-6 **541**

Assignment Guide

Core: 7, 9, 12–14
Enriched: 6, 8–10, 12–14

Have students share their game probabilities and find out whether other students calculated similar probabilities for the same game.

Additional Answers

6a. See students' work.

6b. No; tossing a paper cup does not produce equally likely outcomes.

7d. See students' diagrams; outcomes are heads, heads; heads, tails; tails, heads; tails, tails.

Study Guide Masters, p. 98

Guided Practice

4d. Experimental probabilities; they are based on experiments.

4. Roll a number cube 50 times. **a–c. See students' work.**
  a. Record the results.
  b. Based on your results, what is the probability of rolling a 2?
  c. Based on your results, what is the probability of rolling an even number?
  d. Are the probabilities in parts b and c examples of *experimental* or *theoretical* probabilities? Explain.

5. *Entertainment* A local video store has advertised that one out of every four customers will receive a free box of popcorn with their video rental. So far, 15 of the first 75 customers have won the popcorn.
  a. Based on the results so far, what is the experimental probability that a customer will win? $\frac{1}{5}$
  b. What is the theoretical probability that a customer will win? $\frac{1}{4}$

EXERCISES

Practice

Find a game in your house that uses a spinner, number cubes, or cards. Then design and carry out an experiment to find the experimental probability for one particular outcome. Compare the experimental probability to the theoretical probability.

6. Toss a paper cup 50 times and count the number of times it lands up.
  a. What is the experimental probability that the paper cup will land up?
  b. Can you determine the theoretical probability that the paper cup will land up? Explain. **a–b. See margin.**

7. Toss two coins 50 times. **d. See margin.**
  a. Record the results. **a–c. See students' work.**
  b. Based on your results, what is the probability that both coins will show heads?
  c. Based on your results, what is the probability that one coin will show heads and the other will show tails?
  d. Draw a tree diagram to show all of the possible outcomes of tossing two coins.
  e. What is the theoretical probability that both coins will show heads? $\frac{1}{4}$
  f. What is the theoretical probability that one coin will show heads and the other will show tails? $\frac{1}{2}$

8. *Graphing Calculator* You can use the random number generator function on a graphing calculator to simulate a probability experiment. The graphing calculator program will generate 30 random numbers between 1 and the number you enter for T. In order to use the program, you must first enter the program into the calculator's memory. To access the program memory, use the following keystrokes.

Enter: [PRGM] [▶] [▶] [ENTER]

Run the program. Press PRGM and choose the program from the list by pressing the number next to its name. Enter a number for T. Press enter thirty times. The calculator will display a number each time you press enter. **a–b. See students' work.**

```
PROGRAM:RANDOM
: Input T
: For (N, 1, 30)
: int((T*rand) + 1) → D
: Disp D
: Pause
: End
```

  a. Run the program to simulate 30 rolls of a number cube.

8c. Sample answer: They are consistent.

  b. Run the program to simulate 30 spins of a spinner that has 7 equal sections.
  c. How do the results compare to the theoretical probabilities of these events?

542 Chapter 12 Investigating Discrete Math and Probability

Applications and Problem Solving

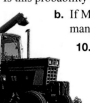

9. **Agriculture** Over the last eight years, the probability that corn seeds planted by Mr. Simon produce corn is $\frac{5}{6}$. **a. See margin.**
   a. Is this probability an experimental or a theoretical probability? Explain.
   b. If Mr. Simon wants to have 10,000 corn bearing plants, how many seeds should he plant? **12,000 seeds**

10. **Contests** The PTO at Grassland Middle School announced that "One out of every 25 students will win a movie pass." They will draw a name from each homeroom. Ambrosia complained that she does not have a $\frac{1}{25}$ chance of winning since there are 32 students in her homeroom. Is she correct? Explain. **See margin.**

11. See students' work. 11. **Working on the CHAPTER Project** Refer to the information about sports you gathered on page 515. Find the theoretical probabilities that a little league player picked at random will go on to play in higher level sports.

12. **Critical Thinking** Alfonso believes a coin is weighted to give an advantage to one side. He asked ten of his friends to toss the coin 40 times and record their results in the table. Do you think the coin is unfair? Explain. **See margin.**

Student	1	2	3	4	5	6	7	8	9	10
Number of Heads	21	22	18	26	21	21	19	22	18	29
Number of Tails	19	18	22	14	19	19	21	18	22	11

Mixed Review

13. **Test Practice** Marcus tossed a coin and rolled a number cube. What is the probability that he will toss tails and roll a multiple of 3? *(Lesson 12-5)* **D**
   A $\frac{1}{2}$ B $\frac{1}{3}$ C $\frac{1}{4}$ D $\frac{1}{6}$

14. **Algebra** Solve the system $y = 2x - 7$ and $y = -2x + 9$. *(Lesson 10-5)* **(4, 1)**

Win the Lottery

Get Ready This game is for four players.   paper bag

Get Set Use 10 pieces of paper. On each piece, write one number from 0 to 9. Place the pieces of paper in a brown bag.

Go
- Each person writes down four different lottery numbers from 0 to 9.
- Each person draws a piece of paper from the brown bag without replacement. These are the winning numbers.
- Score 2 points if you have only one number that matches the winning numbers. Score 16 points if you have two numbers that match. Score 32 points if you have three numbers that match. Score 64 points if you have all four numbers.
- Repeat the process until one player has 100 points.

interNET CONNECTION Visit www.glencoe.com/sec/math/mac/mathnet for more games.

Lesson 12-6 Experimental Probability **543**

■ **Extending the Lesson** ■

Enrichment Masters, p. 98

As a follow-up to this game, ask the students whether the game is based on combinations or permutations. **combinations**

CHAPTER Project

Exercise 11 asks students to advance to the next stage of work on the Chapter Project. To check for accuracy, you may wish to have students share their probability calculations with a partner.

4 ASSESS

Closing Activity

Writing Have students write a few sentences describing the difference between experimental probability and theoretical probability, using examples.

Chapter 12, Quiz C (Lessons 12-5 and 12-6) is available in the *Assessment and Evaluation Masters*, p. 324.

Additional Answers

9a. Experimental probability; it is based on what happened in the past.

10. Yes; they may give enough prizes so that $\frac{1}{25}$ of all of the students in the school win, but Ambrosia has a $\frac{1}{32}$ chance of winning.

12. No; there were a total of 217 heads out of 400 tries which seems reasonable.

Practice Masters, p. 98

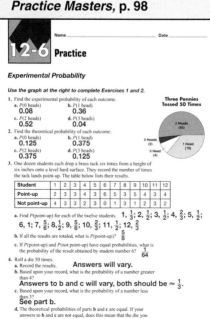

Lesson 12-6 **543**

HANDS-ON 12-7A LAB Notes

GET READY

Objective Students discover how experimental probability is used in biology.

Optional Resources
Hands-On Lab Masters
• worksheet, p. 65
Overhead Manipulative Resources
• counters
Manipulative Kit
• counters

MANAGEMENT TIPS

Recommended Time
25 minutes

Getting Started Ask students to give examples of genetic traits that appear to run in their families. Suggest traits they may not think about, such as tongue curling, taste, and longest toe.

The **Activity** requires students to understand that a dominant gene will always dominate a recessive one. Thus, all hybrid pea plants with Tt genes will be tall. Only pea plants with pure recessive tt genes will be short.

544 Chapter 12

HANDS-ON LAB

COOPERATIVE LEARNING

12-7A Punnett Squares

A Preview of Lesson 12-7

- 40 yellow counters
- 40 red counters
- 2 paper bags

In the early 1900s, Reginald Punnett developed a model to show the possible ways genes can combine at fertilization. You can use a *Punnett square* to simulate how random fertilization works.

In a Punnett square, dominant genes are shown with capital letters. Recessive genes are shown with lowercase letters. Letters representing the parent's genes are placed on the left and upper sides of the Punnett square. Letters inside the boxes of the square show the possible gene combinations for their offspring.

Suppose pea plants are being crossed. Let **T** represent the dominant gene for tallness. Let **t** represent the recessive gene for shortness. Offspring receive one gene from each parent. A pea plant with **TT** genes is pure dominant and is tall. A pea plant with **tt** genes is pure recessive and is short. A pea plant with **Tt** genes is hybrid and is tall because it has a dominant gene. Notice that the capital letter goes first in hybrid genes.

The Punnett square at the right represents a cross between two hybrid tall pea plants.

Notice that 1 out of 4 offspring is pure dominant, 1 is pure recessive, and 2 are hybrid.

	T	t
T	TT	Tt
t	Tt	tt

TRY THIS

Work with a partner.

You will simulate the offspring of two hybrid parents.

Step 1 Place 20 yellow counters and 20 red counters in a paper bag. Label the bag *female parent*. This represents a parent with one dominant gene and one recessive gene.

544 Chapter 12 Investigating Discrete Math and Probability

Step 2 Place 20 yellow counters and 20 red counters in a second paper bag. Label this bag *male parent*. This represents a parent with one dominant gene and one recessive gene.

Step 3 Make a Punnett square to show the expected offspring of these parents. Use **R** for the dominant gene, red. Use **r** for the recessive gene, yellow.

Step 4 Copy the table.

Trial	Color of Female Parent Counter	Color of Male Parent Counter	Gene Pairs
1			
2			
3			
⋮			
40			

Step 5 Shake the bags. Reach into each bag and, without looking, remove one counter. Record the colors of the counters in your table. Put each counter back into its original bag.

Step 6 Repeat Step 5 thirty-nine more times.

ON YOUR OWN

1. Refer to Step 3. Out of 40 offspring, how many would you expect to be pure dominant (**RR**)? hybrid (**Rr**)? pure recessive (**rr**)? **10, 20, 10**
2. Which of the three gene combinations did you expect to occur most often? **Rr**
3. What is the theoretical probability of an offspring having pure dominant genes? hybrid genes? pure recessive genes? $\frac{1}{4}, \frac{1}{2}, \frac{1}{4}$
4. Find each experimental probability. Describe how it compares to the corresponding theoretical probability. **See students' work.**
5. In humans, free earlobes is a dominant trait over attached earlobes. Let **F** represent free earlobes and **f** represent attached earlobes. Draw a Punnett square for each parent combination below. Find the theoretical probability for each possible offspring. **a–d. See margin.**

   a. FF, Ff   b. FF, FF   c. ff, ff   d. Ff, ff

6. **Look Ahead** Suppose 100 babies have two parents who both have the hybrid earlobe trait. How many babies would you expect to have free earlobes? attached earlobes? **75 babies; 25 babies**

Lesson 12-7A HANDS-ON LAB **545**

Have students draw a Punnett square predicting the probability that an offspring will be male (XY) or female (XX). Have students write a paragraph about why neither maleness nor femaleness is dominant and how Punnett squares predict this.

ASSESS

Have students complete Exercises 1–6. Use Exercise 4 to determine whether students understand the difference between experimental and theoretical probability.

Additional Answers

5a.
	F	F
F	FF	FF
f	Ff	Ff

$P(FF) = \frac{1}{2}, P(Ff) = \frac{1}{2}$

5b.
	F	F
F	FF	FF
F	FF	FF

$P(FF) = 1$

5c.
	f	f
f	ff	ff
f	ff	ff

$P(ff) = 1$

5d.
	F	f
f	Ff	ff
f	Ff	ff

$P(Ff) = \frac{1}{2}, P(ff) = \frac{1}{2}$

Hands-On Lab Masters, p. 65

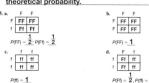

Hands-On Lab 12-7A **545**

12-7 Lesson Notes

Instructional Resources
- *Study Guide Masters,* p. 99
- *Practice Masters,* p. 99
- *Enrichment Masters,* p. 99
- *Transparencies 12-7, A and B*
- *Assessment and Evaluation Masters,* p. 324
- *Classroom Games,* pp. 41–42
- *School to Career Masters,* p. 38
- CD-ROM Program
  - Resource Lesson 12-7

Recommended Pacing

Standard	Days 10 & 11 of 13
Honors	Day 9 of 11
Block	Day 5 of 6

1 FOCUS

5-Minute Check
(Lesson 12-6)

Two coins are tossed 50 times.

Heads	Number
0	8
1	28
2	14

1. What was the experimental probability of tossing 2 heads? $\frac{7}{25}$ or 0.28
2. List the possible outcomes for tossing 2 coins. **HH, HT, TH, TT**
3. What is the theoretical probability of tossing 2 heads? $\frac{1}{4}$ or 0.25

 The 5-Minute Check is also available on **Transparency 12-7A** for this lesson.

2 TEACH

 Transparency 12-7B contains a teaching aid for this lesson.

Reading Mathematics Tell students that the word *population* means the number of people in a region or the number making up a whole. In science and math, the word *population* is used the same way but always includes a description of the exact population being sampled or discussed.

12-7   Integration: Statistics
Using Sampling to Predict

What you'll learn
You'll learn to predict the actions of a larger group by using a sample.

When am I ever going to use this?
Knowing how to predict using a sample can help you plan the type of refreshments needed for a large party.

Word Wise
sample
population

In the spring of 1997, water in the Red River in Minnesota overflowed its banks by as much as 14 miles. More than 60,000 people had to be evacuated from their homes. There was over 1 billion dollars in property damage from the flooding. Unfortunately, only about 1 out of every 15 homes had flood insurance.

As insurance companies began to determine how much money they might eventually pay in claims, they took a survey of a small group of residents in the area. This group is called a **sample**. A sample is representative of a larger group called the **population**. A sample is selected at random.

Example **APPLICATION**

LOOK BACK
Refer to Lessons 3-2 and 8-2 to review percents.

1 **Marketing** A video store is thinking about opening a new section in the store that offers only animated movies. They decide to take a survey to determine whether there is enough interest in this type of movie. For two Saturdays, they ask people at the mall if they would rent an animated movie at least every four months. The table gives the results.

a. How many people were in the survey?

$98 + 75 + 187 + 116 = 476$

476 people were surveyed.

	Yes	No
First Saturday	98	75
Second Saturday	187	116

b. What percent said they would rent an animated movie?

$98 + 187 = 285$ *285 people would rent an animated movie.*

$285 \div 476 = 0.598739496$

About 60% of the people said they would rent an animated movie at least every four months.

c. Out of 2,000 customers, how many should the video store expect to rent animated movies?

$0.60 \times 2,000 = 1,200$

1,200 customers would be expected to rent an animated movie at least every four months.

d. Would the survey have a representative sample if they asked people coming out of a movie theater that is showing an animated movie?

No, because they would tend to like animated movies.

546   Chapter 12 Investigating Discrete Math and Probability

Motivating the Lesson

Communication Ask students what population to interview to:
- determine what American teenagers prefer to eat. **all American teenagers**
- determine how many teenagers wear helmets when they in-line skate. **teenage in-line skaters**

Example — APPLICATION

2 **Weather** The experimental probability that various cities will have snow on December 25 is given in the chart. Suppose Cecilia lives to be 80 years old and spends every December in Cheyenne, Wyoming. How many times should she enjoy snow on December 25?

Find 35% of 80.

$0.35 \times 80 = 28$

Cecilia should enjoy snow 28 times.

Snow on December 25?

Boston, MA	23%
Cheyenne, WY	35%
Denver, CO	42%
Detroit, MI	50%
Minneapolis, MN	73%
Seattle, WA	5%

Source: Northeast Regional Climate Center at Cornell University

---

CHECK FOR UNDERSTANDING

Communicating Mathematics

Read and study the lesson to answer each question.

1. *Tell* how a sample group survey can be used to predict the actions of a whole population. **by setting up and solving a proportion**

2. *Compare* taking a survey and finding an experimental probability.

3. *Write* a paragraph about the importance of using a random sample to make predictions. Tell how you could find an appropriate sample to determine the favorite sport of the students at your school. Give an example of an inappropriate sample for the same question. **See Answer Appendix.**

Guided Practice

2. Taking a survey is one way to determine experimental probability.

4c. 225 red, 90 green, 18 yellow, 117 blue

4. The school bookstore sells 3-ring binders. All incoming sixth-grade students will need a binder. The binders come in 4 colors; red, green, blue, or yellow. The students who run the store decide to survey 50 sixth graders to find out their favorite color. They will use this information to order 450 binders to sell in the fall.

   a. From an alphabetical list, the students survey every ninth student. Is this a good sample? Explain. **Yes; it is randomly taken.**

   b. Of the students surveyed, 25 chose red, 10 chose green, and 2 chose yellow. How many chose blue? **13 students**

   c. For the 450-binder order, how many of each color should be ordered?

5. *Entertainment* As people leave a concert, 50 people are surveyed at random. Six people say they bought a concert T-shirt.

   a. If 6,330 people attended the concert, how many people would you predict bought T-shirts at the concert? **about 760 people**

   b. What could T-shirt sellers learn from this survey? **Sample answer: how many T-shirts to stock at future concerts**

EXERCISES

Practice

6. A survey is being taken to determine how much money the average family in the United States spends to heat their home. A sample of people from Arizona is used for the survey. Is this a good sample? Explain. **See margin.**

Lesson 12-7 Integration: Statistics Using Sampling to Predict **547**

■ Reteaching the Lesson ■

Activity Use simple fractions and percents in questions like the following to clarify sampling. *One-tenth of a representative sample of people were left-handed. How many of one million people would you expect to be left-handed?* **100,000**

Additional Answer

6. No; the winters in Arizona are very different from those in other parts of the country.

Assignment Guide

Core: 7, 9, 12–14
Enriched: 6, 8–10, 12–14

Exercise 11 asks students to advance to the next stage of work on the Chapter Project. If students make graphs with similar scales, the information will be easier for the class to understand.

4 ASSESS

Closing Activity

Speaking Have students discuss how they could use the opinions of a few dozen classmates to predict the opinions of all students their age in the United States. Then have them discuss problems with this sampling and the reliability of their predictions.

Chapter 12, Quiz D (Lesson 12-7) is available in the *Assessment and Evaluation Masters*, p. 324.

Practice Masters, p. 99

Enrichment Masters, p. 99

---

7. Three students are running for class president. Marcy surveyed some of her classmates during lunch. Seven students said they were voting for Pedro, twelve said they were voting for Jasmine, and six said they were voting for Tim.
   a. What is the size of the sample? **25 students**
   b. What percent said they were voting for Jasmine? **48%**
   c. What fraction of the students said they were voting for Tim? $\frac{6}{25}$
   d. There are 180 students in the class. Based on this survey, how many votes do you think Pedro will receive? **about 50 votes**

8. A music store surveyed some of their customers. Use the results to answer each question.

   8a. 540 customers
   8b. $\frac{38}{135}$
   8c. about 8.1%

   a. What is the size of the sample?
   b. What fraction chose country?
   c. What percent chose oldies?
   d. If the management is going to order 1,000 new CDs, how many of each type would you recommend be purchased? **See Answer Appendix.**

Favorite Type of Music	
rock	168
jazz	42
rhythm and blues	78
country	152
alternative	23
oldies	44
other	33

Applications and Problem Solving

9. *Advertising* The Sunshine Orange Juice Company conducted a taste test. Out of 600 people who compared Sunshine Orange Juice to the competitor's brand, 327 said they preferred Sunshine Orange Juice. Which of the following statements can the company correctly make in their advertising? Explain.
   a. "Consumers prefer Sunshine Orange Juice 2 to 1." **a–c. See Answer Appendix**
   b. "Over 50% of the people surveyed prefer Sunshine Orange Juice."
   c. "More people always choose Sunshine Orange Juice over our competitor's brand."

   10a. Sample answer: $\frac{15}{50} = \frac{50}{x}$

10. *Life Science* Miss Thompson is a park ranger. She wants to estimate the number of fish in a small lake. She captures and marks 50 fish. She returns them to the lake. Several weeks later, she captures another 50 fish. She notices that 15 of the fish are some of the ones she marked.
    a. Write a proportion that Miss Thompson could use to predict the number of fish in the lake.
    b. Estimate the number of fish in the lake. **about 167 fish**

11. *Working on the CHAPTER Project* Refer to the probabilities of playing higher level sports that you calculated on page 543. Use these probabilities to predict how many out of 100,000 little league players will play at these levels. Make a graph showing your predictions. **See students' work.**

12. *Critical Thinking* How could things such as the wording of a question or the tone of voice of the interviewer affect a survey? Explain. **See Answer Appendix.**

Mixed Review

13. *Probability* Define experimental probability. *(Lesson 12-6)* **See Answer Appen**

14. **Test Practice** A hiker walked 22 miles north and then walked 17 miles west. How far is the hiker from the starting point? *(Lesson 9-5)* **D**
    A 374 mi   B 112.6 mi   C 39 mi   D 27.8 mi

548 Chapter 12 Investigating Discrete Math and Probability

Extending the Lesson

Activity Have students collect examples of probabilities that are used to predict something. **Sample answers: medical statistics, accident rates, distribution of test scores**

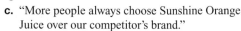
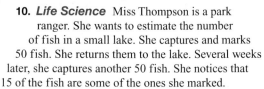

HEALTH

Debra Wein
NUTRITIONIST

Debra Wein is the founder of a nutrition consulting firm in Boston called The Sensible Nutrition Connection (SnaC). She has made presentations for several national sports and fitness organizations including the USA Track & Field Association, the American College of Sports Medicine (ACSM), and the International Dance Exercise Association (IDEA). She is a member of the Massachusetts Governor's Council on Physical Fitness & Sports. Ms. Wein has degrees in nutrition and applied physiology.

To be a nutritionist, you'll need a bachelor's degree in dietetics, food and nutrition, or food service systems management. Courses in biology, mathematics, statistics, psychology, and sociology are required for these degrees.

For more information:
American Dietetic Association
216 West Jackson Boulevard
Chicago, IL 60606

interNET CONNECTION
www.glencoe.com/sec/math/mac/mathnet

Someday, I'd like to work with athletes to help them reach their full potential.

Your Turn
Find out what types of foods an athlete should eat in the 24 hours before competition. List some examples of each of these types of foods. How many combinations of the foods listed could an athlete eat before competition? List these combinations.

School to Career: Health

More About Debra Wein
- Debra Wein received degrees in nutritional science from Cornell and Columbia Universities.
- Besides counseling athletes at her consulting firm, Ms. Wein advises athletes about nutrition through writing articles for magazines, such as *Women's Sports + Fitness* and *Environmental Nutrition*.

SCHOOL to CAREER

Motivating Students
As students learned in the Chapter Project on page 515, math can be used to analyze sports. Mathematicians, scientists, and other professionals have roles to play in amateur and professional sports. To start students thinking about careers in the sports industry ask students these questions.
- What are the main types of nutrients in food? **carbohydrates, proteins, fats, minerals, and vitamins**
- How do we measure the energy-producing value of a food? **in Calories**
- What professionals can help you recover from a sports-related injury? **orthopedist, physical therapist, chiropractor, or sports trainer**

Making the Math Connection
In nutritional counseling, Ms. Wein might use sample data to predict nutritional needs of individuals. Nutritional information from samples—such as female athletes, weight lifters, or long distance runners—is used to predict nutritional needs of the whole group.

Working on *Your Turn*
Students should know that athletes doing strenuous aerobic exercise may benefit from eating a different balance of foods than an athlete engaging in weight lifting. Students should work in groups on this project.

*An additional School to Career activity is available on page 38 of the **School to Career Masters**.*

Study Guide and Assessment

Vocabulary

This section provides a listing of the new terms, properties, and phrases that were introduced in this chapter. Have students define each term and provide an example or two of it, if appropriate.

Understanding and Using the Vocabulary

These exercises check students' understanding of the terms by using a variety of verbal formats including matching, completion, and true/false.

Glossaries A complete glossary of terms appears on pages 702–710. The glossary also appears in Spanish on pages 711–721.

CHAPTER 12 Study Guide and Assessment

Vocabulary

After completing this chapter, you should be able to define each term, concept, or phrase and give an example or two of each.

Statistics and Probability
combination (p. 524)
compound events (p. 534)
Counting Principle (p. 519)
dependent events (p. 535)
experimental probability (p. 540)
factorial (p. 522)
independent events (p. 534)
outcome (p. 518)
Pascal's Triangle (p. 528)
permutation (p.521)
population (p. 546)
sample (p. 546)
simulation (p. 540)
theoretical probability (p. 540)
tree diagram (p. 518)

Problem Solving
act it out (p. 538)

Understanding and Using the Vocabulary

Choose the letter of the term that best matches each phrase.

1. an arrangement or listing in which order is important **f**
2. an arrangement or listing in which order is not important **e**
3. the product of all counting numbers beginning with n and counting back to 1 **h**
4. a diagram used to find the total number of outcomes of an event **g**
5. a smaller group representative of a larger group **d**
6. probabilities based on frequencies obtained in an experiment **a**
7. probabilities based on known facts **b**
8. two or more events for which the outcome of one event affects the outcome of another event **i**
9. two or more events for which the outcome of one event does not affect the outcome of the other event **j**

a. experimental probabilities
b. theoretical probabilities
c. population
d. sample
e. combination
f. permutation
g. tree diagram
h. $n!$
i. dependent events
j. independent events
k. simulation

In Your Own Words 10. If one event can occur in x ways and another event can occur in y ways, the number of ways the two events can occur is xy.

10. *Write* a definition of the Counting Principle.

550 Chapter 12 Investigating Discrete Math and Probability

MindJogger Videoquizzes

MindJogger Videoquizzes provide an alternative review of concepts presented in this chapter. Students work in teams to answer questions, gaining points for correct answers. The questions are presented in three rounds.
Round 1 Concepts–5 questions
Round 2 Skills–4 questions
Round 3 Problem Solving–4 questions

Study Guide and Assessment   Chapter 12

Objectives & Examples

Upon completing this chapter, you should be able to:

● count outcomes by using a tree diagram or the Counting Principle *(Lesson 12-1)*

Two coins are tossed. How many outcomes are possible?

```
1st Coin   2nd Coin   Outcome
            H ——————— HH
   H <
            T ——————— HT
            H ——————— TH
   T <
            T ——————— TT
```

There are 4 outcomes.

Review Exercises

Use these exercises to review and prepare for the chapter test.

State the number of possible outcomes for each event.

11. A restaurant offers 4 types of soft drinks in two different sizes.   **8 outcomes**

12. A car comes in 3 models and 4 colors. There is a choice of a standard or automatic transmission.   **24 outcomes**

13. Phones come in wall or desk models with straight or coiled cords. They come in three colors—black, neutral, and white.   **12 outcomes**

Objectives & Examples

This section reviews the skills and concepts of the chapter and shows completely worked examples.

Review Exercises

These exercises provide practice for the corresponding objectives.

Assessment and Evaluation Masters, pp. 311–312

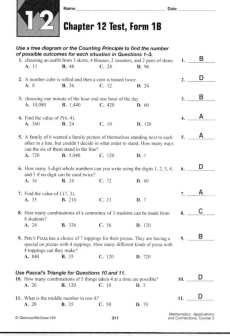

● find the number of permutations of objects *(Lesson 12-2)*

The symbol $P(n, r)$ represents the number of permutations of n things taken r at a time.

$P(n, r) = n \cdot (n - 1) \cdot (n - 2) \cdot \ldots \cdot (n - r + 1)$

Find each value.

14. $P(7, 2)$ **42**

15. $P(5, 5)$ **120**

16. How many different ways can 4 different books be arranged on a shelf?   **24 ways**

17. How many 3-digit whole numbers can you write using the digits 1, 2, 3, 4, 5, and 6 if no digit can be used twice?   **120 numbers**

● find the number of combinations of objects *(Lesson 12-3)*

The symbol $C(n, r)$ represents the number of combinations of n things taken r at a time.

$C(n, r) = \dfrac{P(n, r)}{r!}$

Find each value.

18. $C(9, 5)$ **126**

19. $C(12, 2)$ **66**

20. How many different pairs of puppies can be selected from a litter of 8?   **28 pairs**

21. How many different groups of 3 marbles can be chosen from a box containing 20 marbles?   **1,140 groups**

Chapter 12 Study Guide and Assessment   **551**

 ## Assessment and Evaluation

Six forms of Chapter 12 Test are available in the *Assessment and Evaluation Masters* as shown in the chart.

Chapter 12 Test, Form 1B, is shown at the right. Chapter 12 Test, Form 2B, is shown on the next page.

1A	Multiple Choice	Honors
1B	Multiple Choice	Average
1C	Multiple Choice	Basic
2A	Free Response	Honors
2B	Free Response	Average
2C	Free Response	Basic

Study Guide and Assessment   **551**

Chapter 12 Study Guide and Assessment

Objectives & Examples | Review Exercises

• identify patterns in Pascal's Triangle *(Lesson 12-4)*

How many different groups of 2 people can be taken from a group of 5 people?

Number Taken at a Time	0	1	2	3	4	5
Row 5	1	5	10	10	5	1

There are 10 different groups.

Use Pascal's Triangle to answer each question. 23. 70 combinations

22. Find the value of $C(9, 3)$. **84**

23. How many combinations are possible when 8 objects are chosen 4 at a time?

24. How many different kinds of 3-topping pizzas can be made when 7 toppings are offered? **35 pizzas**

• find the probability of independent and dependent events *(Lesson 12-5)*

The probability of two events can be found by multiplying the probability of the first event by the probability of the second event.

25. A number cube is rolled twice. What is the probability that an odd number will show both times? $\frac{1}{4}$

26. A bag contains 7 white marbles and 3 blue marbles. Once a marble is selected, it is not replaced. What is the probability of picking 2 blue marbles? $\frac{1}{15}$

• find experimental probability *(Lesson 12-6)*

Experimental probabilities are probabilities based on frequencies obtained by conducting an experiment.

27. Toss two coins 20 times. **a–b. See students' work.**
   a. Record the results.
   b. What is the experimental probability of no tails? of one head and one tail?
   c. What are the theoretical probabilities for part b? $\frac{1}{4}; \frac{1}{2}$

• predict the actions of a larger group by using a sample *(Lesson 12-7)*

If 30 out of a sample of 90 students prefer grape juice over orange juice, how many cartons of grape juice does the cafeteria need for 750 students?

$$\begin{array}{l}\text{grape sample} \to \\ \text{total sample} \to\end{array} \frac{30}{90} = \frac{g}{750} \begin{array}{l} \leftarrow \text{grape order} \\ \leftarrow \text{total order}\end{array}$$

30 750 90 = 250

The cafeteria needs to order 250 cartons of grape juice.

The Superfast Computer Company conducted a survey. Participants in the survey were asked to choose which computer they would buy. Of those surveyed, 48 said they would buy a Superfast computer, 34 said they would buy computer X, and 18 said they would buy computer Y.

28. What is the size of the sample? **100 people**

29. What fraction chose computer Y? $\frac{9}{50}$

30. For 8,000 people, how many would buy a Superfast computer? **3,840 people**

552 Chapter 12 Investigating Discrete Math and Probability

Study Guide and Assessment   Chapter 12

Applications & Problem Solving

31. Food County Line Restaurant has a Saturday breakfast special that offers the choices shown in the menu below. Draw a tree diagram to find the number of possible outcomes for a meal consisting of one choice from each column. *(Lesson 12-1)*

See Answer Appendix.

32. Sports There are 10 batters on a softball team. How many ways can the coach pick the first 4 batters? *(Lesson 12-2)*
5,040 ways

33. Business Midtown Music Store has 20 different CDs on a sale table. These CDs are priced 3 for $30. How many combinations of 3 CDs can be purchased? *(Lesson 12-3)*   **1,140 combinations**

34. Act It Out Each morning, Allison walks past 4 park benches. Based on her observation, she estimates that any one bench is occupied 0.75 of the time. Find the probability that all 4 benches will be occupied as Allison walks past this morning. *(Lesson 12-6A)* $\frac{81}{256}$ **or about 0.3**

Alternative Assessment

● **Performance Task**
Your friend says that he is playing a game in which the probability of event A occurring is 30% and the probability of event B occurring is 40%. Your friend says that events A and B are independent. Give an example of a game in which these conditions are met. What is P(A and then B)?   **See Answer Appendix.**

Suppose events A and B are dependent. Explain how to change your game so this is true.   **See Answer Appendix.**

A practice test for Chapter 12 is provided on page 658.

● **Completing the**
Use the following checklist to make sure your news article is complete.

☑ The probabilities that a player will play at higher levels are based on information you have collected.

☑ The predictions about how many out of 100,000 little league players will play sports at a higher level are accurate.

☑ The graph showing how many players you expect to play at higher levels is clear and concise.

 Select one of the assignments you completed in this chapter and place it in your portfolio. Attach a note explaining why you selected it.

Chapter 12 Study Guide and Assessment **553**

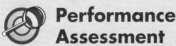 **Performance Assessment**

Additional performance assessment tasks for this chapter are included in the *Assessment and Evaluation Masters* on page 321. A scoring guide is also provided on page 333.

Applications & Problem Solving

This section provides additional practice in solving real-world problems that involve the skills of this chapter.

Alternative Assessment

The *Performance Task* provides students with a performance assessment opportunity to evaluate their work and understanding.

Students should complete the final stages of their project and prepare a class demonstration of their results. A scoring guide for the project is available in the *Investigations and Projects Masters*, p. 63.

 Students should add to their portfolios at this time.

Assessment and Evaluation Masters, p. 321

Study Guide and Assessment **553**

Standardized Test Practice

The Standardized Test Practice may be used to help students prepare for standardized tests. The test items are written in the same style as those in state proficiency tests and standardized tests like CAT, CTBS, ITBS, MAT, SAT, and Terra Nova. The test items cover skills and concepts covered up to this point in the text.

The pages can be used as an overnight assessment. After students have completed the pages, discuss how each problem can be solved, or provide copies of the solutions from the *Solutions Manual*.

Assessment and Evaluation Masters, p. 327

CHAPTERS 1–12 Standardized Test Practice
Assessing Knowledge & Skills

Section One: Multiple Choice

There are eleven multiple choice questions in this section. Choose the best answer. If a correct answer is *not here,* choose the letter for Not Here.

1. What is the probability of choosing a yellow marble from a bag containing 4 blue marbles, 3 red marbles, and 5 black marbles? **A**
   A 0
   B $\frac{1}{7}$
   C $\frac{4}{15}$
   D 1

2. Samantha is redecorating her room. She has a choice of 4 colors of paint, 3 colors of carpet, and 2 colors of curtains. How many combinations of paint, carpet, and curtain colors can she use? **H**
   F 9
   G 12
   H 24
   J 48

3. The back of the couch forms a 115° angle with its seat. Which type of angle is this? **C**
   A right
   B acute
   C obtuse
   D isosceles

4. Write $-2.\overline{8}$ as a fraction or mixed number. **G**
   F $-2\frac{4}{5}$
   G $-2\frac{8}{9}$
   H $-\frac{28}{10}$
   J $-\frac{28}{9}$

Please note that Questions 5–11 have five answer choices.

5. Which statement is correct concerning the probabilities of reaching into the jars without looking and pulling out a blue marble? **B**

   Jar 1 (100 marbles, 20 blue)   Jar 2 (200 marbles, 50 blue)

   A greater for Jar 1 than Jar 2
   B greater for Jar 2 than Jar 1
   C equal for both jars
   D both are greater than 1
   E cannot be determined

6. Which mathematical expression represents the number of ways the four shapes can be arranged in a row? **G**

   F $P(4, 1)$
   G $P(4, 4)$
   H $C(4, 1)$
   J $C(4, 4)$
   K $\frac{4!}{C(4, 4)}$

7. One centimeter is about 0.3937 inch. If Henry is 68 inches tall, estimate his height in centimeters. **E**
   A less than 140 centimeters
   B between 140 and 150 centimeters
   C between 150 and 160 centimeters
   D between 160 and 170 centimeters
   E more than 170 centimeters

554 Chapters 1–12 Standardized Test Practice

◀◀◀ **Instructional Resources**

Another cumulative review is shown at the left and is available in the *Assessment and Evaluation Masters,* p. 327.

Standardized Test Practice Chapters 1–12

8. A swimming pool filled with water is losing water through a hole at a rate of 3 gallons per minute. At this rate, in how many minutes will the pool lose 78 gallons? **H**

   F 18 minutes
   G 24 minutes
   H 26 minutes
   J 30 minutes
   K Not Here

9. One drinking cup holds 8.8 ounces. What is the largest number of drinking cups that can be filled with a 80 ounce bottle? **C**

   A 7 cups
   B 8 cups
   C 9 cups
   D 10 cups
   E 11 cups

10. Jackson Auto Rental charges $20 per day and $0.15 per mile to rent a car. Find the cost of renting a car for two days and driving 100 miles. **H**

    F $35
    G $45
    H $55
    J $65
    K Not Here

11. A paycheck has 23% deducted for federal income tax and 4.5% deducted for state income tax. What is the total percent deducted from the paycheck for these taxes? **C**

    A 18.5%
    B 22.5%
    C 27.5%
    D 100%
    E Not Here

Test-Taking Tip

When taking a standardized test, you may be required to fill in an answer sheet. Your answers may be machine scored. It is important to make your marks completely and fill in circles darkly. Answer each question only once, and make sure you check frequently that you are matching the correct problem number with the correct answer blank.

15. 1, 2, 3, 4, 6, 8, 9, 12, 18, 24, 36, 72
16. $\dfrac{AC}{DF}, \dfrac{AB}{DE}, \dfrac{BC}{EF}$

Section Two: Free Response

This section contains five questions for which you will provide short answers. Write your answers on your paper.

12. What is the least common multiple of 33 and 36? **396**

13. What is the greatest common factor of 24 and 36? **12**

14. What are the prime factors of 65? **5 and 13**

15. List the factors of 72.

16. If $\triangle ABC \sim \triangle DEF$, write three ratios relating the corresponding sides.

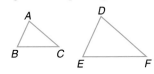

Test-Taking Tip

For students who have difficulty looking back and forth from the test to the answer booklet, suggest that they use the edge of their test booklets or scrap paper to keep track of the appropriate answer number, if permitted.

Assessment and Evaluation Masters, pp. 325–326

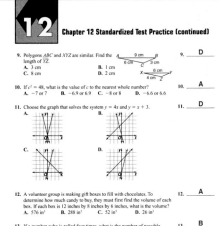

Instructional Resources ▶▶▶

Additional standardized test practice is shown at the right and is available in the *Assessment and Evaluation Masters,* pp. 325–326.

Interdisciplinary Investigation

GET READY

This optional investigation is designed to be completed by a group of 3 or 4 students over several days or several weeks.

Mathematical Overview
This investigation utilizes the concepts from Chapters 10–12.
- measuring volume
- graphing functions
- using sampling to predict

Time Management	
Gathering Data	50 minutes
Calculations	30 minutes
Creating Graphs	30 minutes
Summarizing Data	15 minutes
Presentation	10 minutes

Instructional Resources
- *Investigations and Projects Masters*, pp. 13–16
- Manipulative Kit
  - rulers

Investigations and Projects Masters, p. 16

Interdisciplinary Investigation

EXTRA! EXTRA! NEWSPAPERS MAY TAKE OVER THE PLANET!

Does your community have a recycling center? Do you recycle unwanted items? Did you know that over 40% of what we throw away in the United States is paper and paperboard? Many trees are wasted each week as a result of not recycling newspapers.

What You'll Do

In this investigation, you will collect data about a daily newspaper. You will investigate the volume of newspapers discarded by just one person and discover how that volume can accumulate over many years.

Materials   a week's supply of newspapers   ruler

 calculator   grid paper

Procedure

1. Work in a group. Collect daily editions of a newspaper for a week. Make a stack containing the newspapers. Find the volume of the newspapers.

2. Suppose you bought and saved this newspaper every day for a year. What would be the total volume of the newspapers? Mark off a space in your classroom that has approximately the same volume.

3. Under certain conditions, newspapers in a landfill can be intact after 20 or 30 years. Make a function table showing the accumulation of one person's newspapers from 1 to 20 years. Write an equation in two variables describing the function. Graph the function.

Technology Tips
- Use a **spreadsheet** to calculate the accumulation of newspapers.
- Use **graphing software** to graph your function.
- Surf the **Internet** to find information on recycling.

556 Interdisciplinary Investigation Extra! Extra! Newspapers May Take Over the Planet!

◄◄◄ **Instructional Resources**
A recording sheet to help students organize their data for this investigation is shown at the left and is available in the *Investigations and Projects Masters*, p. 16.

 Cooperative Learning

This investigation offers an excellent opportunity for using cooperative learning groups. For more information on cooperative learning strategies and group management, see *Cooperative Learning in the Mathematics Classroom*.

Making the Connection

Use the data collected about newspapers as needed to help in these investigations.

Language Arts

Write an article to encourage the recycling of newspapers. Give specific information and data about recycling newspapers.

Science

Research the conditions needed for newspapers to decompose in a landfill. Investigate the advantages and disadvantages of burning paper instead of dumping it in landfills.

Art

Find a recipe for making recycled paper. Make some recycled paper and use it for an art project.

Go Further

- Estimate the volume of another type of garbage that you discard. What would be the accumulated volume of this garbage in one year?
- Collect facts about the types and amounts of discarded items in the United States. Prepare a graph, table, or informational pamphlet displaying your findings.

interNET CONNECTION For current information on recycling, visit: www.glencoe.com/sec/math/mac/mathnet

 You may want to place your work on this investigation in your portfolio.

Interdisciplinary Investigation Extra! Extra! Newspapers May Take Over the Planet! **557**

Management Tips

Working in Teams As a team, the students should be responsible for recycling the newspapers at the end of the project.

Making the Connection

You may wish to alert other teachers on your team that your students may need their assistance in this investigation.
Language Arts Students may need help using effective words for marketing recycling ideas.
Science Students may want to investigate the affects burning the paper has on the atmosphere.
Art Students may need help finding methods for making paper.

Assess

You may require a written report or an oral presentation from the students. Ask students to explain what they have learned about the volume of trash produced and about recycling programs.

Investigations and Projects Masters, p. 15

Instructional Resources ▶▶▶

Sample solutions for this investigation are provided in the *Investigations and Projects Masters* on p. 14. The scoring guide for assessing student performance shown at the right is also available on p. 15.

SCORING GUIDE

Interdisciplinary Investigation
(Student Edition, Pages 556–557)

Extra! Extra! Newspapers May Take Over the Planet!

Level	Specific Criteria
3 Superior	• Shows a thorough understanding of the concepts of *calculating volume, writing equations with two variables,* and *graphing functions.* • Uses appropriate strategies to solve problems. • Computations are correct. • Written explanations are exemplary. • Charts, model, and any statements included are appropriate and sensible. • Goes beyond the requirements of some or all problems.
2 Satisfactory, with minor flaws	• Shows understanding of the concepts of *calculating volume, writing equations with two variables,* and *graphing functions.* • Uses appropriate strategies to solve problems. • Computations are mostly correct. • Written explanations are effective. • Charts, model, and any statements included are appropriate and sensible. • Satisfies the requirements of problems.
1 Nearly Satisfactory, with obvious flaws	• Shows understanding of most of the concepts of *calculating volume, writing equations with two variables,* and *graphing functions.* • May not use appropriate strategies to solve problems. • Computations are mostly correct. • Written explanations are satisfactory. • Charts, model, and any statements included are appropriate and sensible. • Satisfies the requirements of problems.
0 Unsatisfactory	• Shows little or no understanding of the concepts of *calculating volume, writing equations with two variables,* and *graphing functions.* • Does not use appropriate strategies to solve problems. • Computations are incorrect. • Written explanations are not satisfactory. • Charts, model, and any statements included are not appropriate or sensible. • Does not satisfy the requirements of the problems.

© Glencoe/McGraw-Hill 15 Mathematics: Applications and Connections, Course 3

CHAPTER 13
Algebra: Exploring Polynomials

Previewing the Chapter

Overview
This chapter on polynomials, heavily discovery-oriented, provides an ideal introduction to middle-level algebra. Students use algebra tiles to discover methods of operating on polynomials. Only then are methods generalized and algorithms stated. Students learn to simplify, add, subtract, multiply, and factor polynomials. Models and the distributive property are used to introduce methods of multiplying polynomials and monomials and multiplying two binomials. Students also learn to solve problems by using guess and check.

Lesson (pages)	Lesson Objectives	NCTM Standards	Standardized Tests	State/Local Objectives
13-1A (560)	Make algebra tiles for algebraic expressions.	1–4, 9, 12		
13-1 (561–564)	Represent polynomials with algebra tiles.	1–4, 7, 9		
13-2 (565–569)	Simplify polynomials by using algebra tiles.	1–4, 7, 9		
13-3 (570–572)	Add polynomials by using algebra tiles.	1–4, 7, 9		
13-4 (573–576)	Subtract polynomials by using algebra tiles.	1–4, 7, 9		
13-5A (577)	Model products by using algebra tiles.	1–4, 7, 9		
13-5 (578–581)	Multiply monomials and polynomials.	1–4, 7, 9		
13-6 (583–585)	Multiply binomials by using algebra tiles.	1–4, 7, 9		
13-7A (586–587)	Solve problems by using guess and check.			
13-7 (588–591)	Factor polynomials by using algebra tiles.	1–4, 7, 9		

CAT = California Achievement Tests, CTBS = Comprehensive Tests of Basic Skills, ITBS = Iowa Tests of Basic Skills, MAT = Metropolitan Achievement Tests, SAT = Stanford Achievement Tests, TN = Terra Nova

Organizing the Chapter

CD-ROM

All of the blackline masters in the Teacher's Classroom Resources are available on the **Electronic Teacher's Classroom Resources** CD-ROM.

LESSON PLANNING GUIDE

Lesson	Extra Practice (Student Edition)	Blackline Masters (page numbers)											
		Study Guide	Practice	Enrichment	Assessment & Evaluation	Classroom Games	Diversity	Hands-On Lab	School to Career	Science and Math Lab Manual	Technology	Transparencies A and B
13-1A								66				
13-1	p. 642	100	100	100					39			13-1
13-2	p. 642	101	101	101	351						77	13-2
13-3	p. 642	102	102	102								13-3
13-4	p. 643	103	103	103	350, 351			80				13-4
13-5A								67				
13-5	p. 643	104	104	104								13-5
13-6	p. 643	105	105	105	352		39					13-6
13-7A	p. 644											
13-7	p. 644	106	106	106	352	43–45					78	13-7
Study Guide/ Assessment					337–349, 353–355							

Other Chapter Resources

Student Edition
Chapter Project, pp. 559, 564, 585, 591, 595
School to Career, p. 582
Let the Games Begin, p. 591

Technology
 CD-ROM Program
Interactive Mathematics Tools Software

Applications
Family Letters and Activities, pp. 77–78
Investigations and Projects Masters, pp. 65–68
Meeting Individual Needs
Investigations for the Special Education Student, pp. 35–38

Teacher's Classroom Resources

Teaching Aids
Answer Key Masters
Block Scheduling Booklet
Lesson Planning Guide
Solutions Manual

Professional Publications
Glencoe Mathematics Professional Series

Chapter 13 **558b**

Planning the Chapter

 MindJogger Videoquizzes provide a unique format for reviewing concepts presented in the chapter.

Assessment Resources

Student Edition
Mixed Review, pp. 564, 569, 572, 576, 581, 585, 591
Mid-Chapter Self Test, p. 576
Math Journal, p. 562
Study Guide and Assessment, pp. 592–595
Performance Task, p. 595
Portfolio Suggestion, p. 595
Standardized Test Practice, pp. 596–597
Chapter Test, p. 659

Assessment and Evaluation Masters
Multiple-Choice Tests (Forms 1A, 1B, 1C), pp. 337–342
Free-Response Tests (Forms 2A, 2B, 2C), pp. 343–348
Performance Assessment, p. 349
Mid-Chapter Test, p. 350
Quizzes A–D, pp. 351–352
Standardized Test Practice, pp. 353–354
Cumulative Review, p. 355

Teacher's Wraparound Edition
5-Minute Check, pp. 561, 565, 570, 573, 578, 583, 588
Building Portfolios, p. 558
Math Journal, pp. 560, 577
Closing Activity, pp. 564, 569, 572, 576, 581, 585, 587, 591

Technology
Test and Review Software
MindJogger Videoquizzes
CD-ROM Program

Materials and Manipulatives

Lesson 13-1A
yellow, red, blue, and green contruction paper
scissors*

Lesson 13-1
algebra tiles*†

Lesson 13-2
algebra tiles*†

Lesson 13-3
algebra tiles*†

Lesson 13-4
algebra tiles*†

Lesson 13-5A
algebra tiles*†
product mat*†

Lesson 13-5
scientific calculator
algebra tiles*†

Lesson 13-6
algebra tiles*†
product mat*†

Lesson 13-7
algebra tiles*†

*Glencoe Manipulative Kit †Glencoe Overhead Manipulative Resources

Pacing Chart

See pages T25–T27 for the Course Planning Calendar.

COURSE	DAY 1	DAY 2	DAY 3	DAY 4	DAY 5	DAY 6	DAY 7
Honors	Chapter Project & Lesson 13-1	Lesson 13-2	Lesson 13-3	Lesson 13-4	Lesson 13-5	Lesson 13-6	Lesson 13-7A

Interactive Mathematics:
Activities and Investigations

is an activity-based program that may be used as an enhancement for chapters in **Mathematics: Applications and Connections.**

Unit 9, Activity Four
Use before Lesson 13-1.

Summary Students work in groups to discover the maximum area that can be enclosed by 170 feet of fencing. They then apply this to finding the maximum area that can be enclosed with 170 feet of fencing and one side of a barn.

Math Connection Students explore how changing the dimensions of a figure affects its area. This activity can be done as a computer investigation where students use a spreadsheet to calculate the patterns in their problem-solving process.

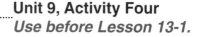

Unit 13, Activity One
Use with Lesson 13-7A.

Summary Students work in groups using guess and check and other problem-solving strategies to determine how many people ordered each type of shirt. They use their writing skills to write a one-page narrative that explains their conclusions and discusses their methods.

Math Connection Students use various problem-solving strategies. They justify their reasoning by explaining the processes used to come to a reasonable solution.

DAY 8	DAY 9	DAY 10	DAY 11	DAY 12	DAY 13	DAY 14	DAY 15
Lesson 13-7	Study Guide and Assessment	Chapter Test					

Chapter 13

Enhancing the Chapter

APPLICATIONS

Classroom Games, pp. 43–45

Diversity Masters, p. 39

School to Career Masters, p. 39

Family Letters and Activities, pp. 77–78

MANIPULATIVES/MODELING

Hands-On Lab Masters, p. 80

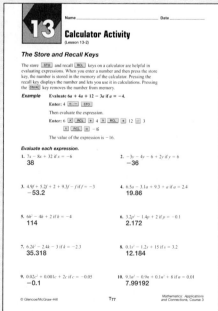

ASSESSMENT/EVALUATION

Assessment and Evaluation Masters, pp. 350–352

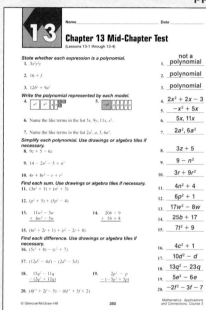

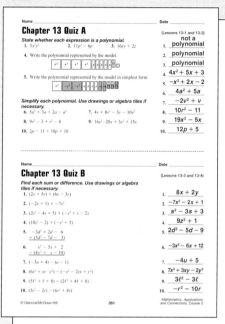

TECHNOLOGY/MULTIMEDIA

Technology Masters, pp. 77–78

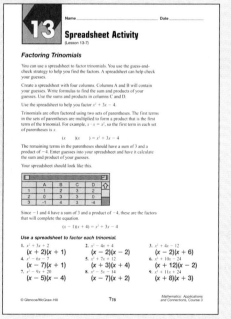

MEETING INDIVIDUAL NEEDS

Investigations for the Special Education Student, pp. 35–38

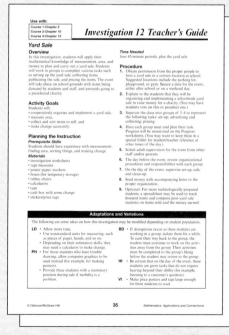

Chapter 13   558f

CHAPTER 13 NOTES

Theme: Basketball
The free throw or foul shot was created in 1894 as a penalty for rough play. A successful free throw has always been worth 1 point. If a player commits a foul against an opponent who is attempting to score, the fouled opponent receives two free throws and the possibility of scoring two points. If a team has exceeded a given number of fouls and commits another foul, the opponent gets a second free throw if he or she successfully completes the first.

Question of the Day
Possible points in a basketball game are 3 points for each goal made beyond the 3-point line, 2 points for each regular goal, and 1 point for each successful free throw. Ask students how they would model a team's final game score with an algebraic expression. $3x + 2y + z$

Assess Prerequisite Skills
Ask students to read through the list of objectives presented in "What you'll learn in Chapter 13." You may wish to ask them what each of the objectives means or if they have experienced or used any of these math concepts before.

 ## Building Portfolios
Encourage students to revise their portfolios as they study this chapter. When they complete this chapter, their portfolios should reflect their year's work.

 ## Math and the Family
In the *Family Letters and Activities* booklet (pp. 77–78), you will find a letter to the parents explaining what students will study in Chapter 13. An activity appropriate for the whole family is also available.

558 Chapter 13

CHAPTER 13
Algebra: Exploring Polynomials

What you'll learn in Chapter 13

- to represent and simplify polynomials using algebra tiles,
- to add, subtract, and factor polynomials using algebra tiles,
- to multiply monomials and polynomials using algebra tiles, and
- to solve problems by guess and check.

558 Chapter 13 Algebra: Exploring Polynomials

 ## CD-ROM Program

Activities for Chapter 13
- Chapter 13 Introduction
- Interactive Lessons 13-1, 13-3, 13-4
- Assessment Game
- Resource Lessons 13-1 through 13-7

CHAPTER Project

HIT OR MISS

In this project, you will create geometric and algebraic models that describe the number of baskets a player can expect to make when he or she shoots free throws. You will write a report explaining your models and how they can be used.

Getting Started

- Find the free throw percent of at least two of your favorite college or professional basketball players.
- Convert these percents to probabilities. For example, if the percent is 90%, then the probability is 0.9.
- What is the probability of each of these players hitting a free throw? What is the probability of their missing a free throw?
- Suppose the players are to shoot two free throws in a row. What is the probability of their hitting both free throws? missing both free throws? hitting the first and missing the second? missing the first and hitting the second?
- As you work through the lessons in the chapter, you will make area models to describe the probabilities you have just found. You will use these models to write a polynomial that you can use to find the number of points each player can expect to score if he or she attempts two free throws.

Technology Tips

- Use a **spreadsheet** to keep track of the probabilities.
- Use **computer software** to draw your geometric models.
- Use a **word processor** to write your report.

interNET CONNECTION For up-to-date information on sports statistics, visit:
www.glencoe.com/sec/math/mac/mathnet

Working on the Project

You can use what you'll learn in Chapter 13 to help you create your models.

Page	Exercise
564	34
585	21
591	25
595	Alternative Assessment

Chapter 13 **559**

interNET CONNECTION

Glencoe has made every effort to ensure that the website links for *Mathematics: Applications and Connections* at **www.glencoe.com/sec/math/mac/mathnet** are current and contain appropriate content. However, these website links are not under Glencoe's control.

Instructional Resources ▶▶▶

A recording sheet to help students organize their data for the Chapter Project is shown at the right and is available in the *Investigations and Projects Masters*, p. 68.

CHAPTER Project NOTES

Objectives Students should
- convert percents to probabilities.
- create area models to describe probabilities.
- create algebraic models to describe probabilities.

Project Pointer You may suggest that students begin a *Project Folder* to keep their work as they complete each stage of the Chapter Project. The completed project may also be added to their portfolios.

Students may need guidance entering the correct formulas for their algebraic models into their spreadsheets. Likewise, they may need help integrating their models into their reports.

Investigations and Projects Masters, p. 68

Chapter 13 Project **559**

HANDS-ON LAB 13-1A Notes

GET READY

Objective Students make algebra tiles for algebraic expressions.

Optional Resources
Hands-On Lab Masters
- algebra tiles, p. 29
- worksheet, p. 66

Overhead Manipulative Resources
- algebra tiles

Manipulative Kit
- scissors

MANAGEMENT TIPS

Recommended Time
20 minutes

Getting Started Ask students to find the areas of each figure.
36 cm²; mn units²

The **Activity** guides students in making algebra tiles for use in the rest of the chapter. Instead of using a ruler to measure their tiles, students should choose arbitrary measures of 1 unit and x units. Have students save the algebra tiles they make for use throughout this chapter.

ASSESS

Have students complete Exercises 1–4. Watch for students who may have difficulty relating width and length to area. Have them review area in Lesson 1-8.

Use Exercise 5 to determine whether students have discovered how to model monomials.

560 Chapter 13

HANDS-ON LAB

COOPERATIVE LEARNING

13-1A Algebra Tiles

A Preview of Lesson 13-1

- yellow, red, blue, and green construction paper
- scissors

Throughout this text, you have used rectangles to show multiplication problems. For example, the figure at the right shows 4×5 as a rectangle that is 4 units wide and 5 units long. Its area is 20 square units.

$4 \times 5 = 20$

In this lab, you will make models called *algebra tiles*.

TRY THIS

Work with a partner.

Step 1 From yellow construction paper, cut out a square that is 1 unit long and 1 unit wide.

Step 2 Draw a line segment that is longer than 1 unit. Since the line segment can be any length, we will say that it is x units long. From green construction paper, cut out a rectangle that is 1 unit wide and x units long.

Step 3 Using the same measure for x, cut out a square that is x units long and x units wide from blue construction paper.

ON YOUR OWN

1. Find the area of each shape and write the area on each tile.   **1, x, x^2**
2. For each tile, write a sentence that tells the relationship among the length, width, and area.   **The area is the product of the length and width.**
3. From construction paper, cut out four more sets of tiles of the same sizes and colors as the first set. Label each with its area.   **3–5. See margin.**
4. From red construction paper, cut out five sets of tiles with the same dimensions as the yellow, green, and blue tiles. Label the areas -1, $-x$, and $-x^2$.
5. *Look Ahead* Use your tiles to model $4x$.

560 Chapter 13 Algebra: Exploring Polynomials

Additional Answers

3. Sample answer: Each set of tiles will have 3 tiles with areas 1, x, and x^2.

4. Sample answer: Students' red tiles should have the same dimensions as their yellow, green, and blue tiles.

5.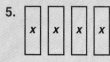

 Math Journal Have students write a paragraph discussing how algebra tiles could help them visualize $3x^2$.

13-1 Modeling Polynomials

What you'll learn
You'll learn to represent polynomials with algebra tiles.

When am I ever going to use this?
Knowing how to represent polynomials with algebra tiles will help you recognize like terms.

Word Wise
monomial
polynomial

For the latest statistics on allowances, visit: www.glencoe.com/sec/math/mac/mathnet

Do you get an allowance? A survey of 784 kids ages 9 through 14 was taken. Of those surveyed, 361 receive an allowance. Kids 13 and older were less likely to get an allowance than those younger.

Source: *Zillions,* Jan./Feb., 1997

Models are often used to help us visualize numbers. The graph models the amount of kids' allowances for ages 13 and 14.

The number 361 can be modeled as follows.

$$361 = (3 \times 10^2) + (6 \times 10) + 1$$

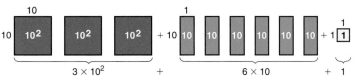

In the previous Hands-On Lab, you made algebra tiles. Algebra tiles can be used to model **monomials**.

The expressions 1, x, and x^2 are examples of monomials. A monomial is a number, a variable, or a product of a number and one or more variables.

Examples

Model each monomial using drawings or algebra tiles.

1 5

To model this expression, you need 5 yellow 1-tiles.

2 $-2x$

Use red tiles to model negative values. To model this expression, you need 2 red x-tiles.

3 $3x^2$

To model this expression, you need 3 blue x^2-tiles.

Lesson 13-1 Modeling Polynomials **561**

Cross-Curriculum Cue

Models of theories have been developed in many disciplines, so that they can be studied. Ask the other teachers on your team to discuss possible examples with their students. Suggestions for curriculum integration are:

Life Science: atomic model of matter, DNA double helix
Social Studies: Malthus' model of population growth
History: domino theory

13-1 Lesson Notes

Instructional Resources
- *Study Guide Masters,* p. 100
- *Practice Masters,* p. 100
- *Enrichment Masters,* p. 100
- Transparencies 13-1, A and B
- *School to Career Masters,* p. 39
- CD-ROM Program
  - Resource Lesson 13-1
  - Interactive Lesson 13-1

Recommended Pacing	
Honors	Day 1 of 10

1 FOCUS

5-Minute Check
(Chapter 12)

1. Cakes are sold in 5 flavors, 4 shapes, and 3 frostings. How many types are sold? **60 types**
2. In how many orders can 5 people stand in line? **120 orders**
3. Find $P(8, 4)$. **1,680**
4. Find $C(7, 2)$. **21**
5. Two coins are tossed. Find P(two heads). $\frac{1}{4}$

The 5-Minute Check is also available on **Transparency 13-1A** for this lesson.

Motivating the Lesson

Communication Display a counter and say that it represents 1 unit. Then ask the following questions.
- How could you represent 8 units? **Use 8 counters.**
- If the counter represents x units, how can you represent $5x$ units? **Use 5 counters.**

Lesson 13-1 **561**

2 TEACH

 Transparency 13-1B contains a teaching aid for this lesson.

Thinking Algebraically Students may think that $x + x = x^2$. Use the distributive property to show that the sum is $2x$.

$$x + x = 1x + 1x$$
$$= (1 + 1)x$$
$$= 2x$$

In-Class Examples

Model each monomial using drawings or algebra tiles.

For Example 1
-2

For Example 2
$5x$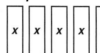

For Example 3
$-2x^2$

Model each polynomial using drawings or algebra tiles.

For Example 4
$-x^2 + x + 6$

For Example 5
$2x^2 - x - 1$

For Example 6
The expression $-x^2 + 16x$ describes the speed of a certain car after x seconds if it accelerates to 60 miles per hour in 6 seconds. Find the speed of the car after 3 seconds by evaluating the expression for $x = 3$. **39 mph**

Algebraic expressions that contain more than one monomial are called **polynomials**. A polynomial is the sum or difference of two or more monomials. You can also model polynomials with algebra tiles.

Examples Model each polynomial using drawings or algebra tiles.

④ $3x^2 + 2x + 7$

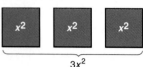

⑤ $-x^2 + 4x - 5$

Study Hint
Reading Math The prefix *poly-* means more than 1. So, a polynomial has more than 1 monomial, a polygon has more than 1 side, and so on.

Polynomial expressions can be evaluated by replacing variables with numbers and then finding the value of the numerical expression.

Example CONNECTION

LOOK BACK Refer to Lesson 1-3 to review evaluating expressions.

⑥ **Physical Science** The expression $-16x^2 + 56x$ describes the height of an arrow after x seconds if it is shot upward at 56 feet per second. Find the height of the arrow after 2 seconds by evaluating the expression for $x = 2$.

$$-16x^2 + 56x = -16(2)^2 + 56(2) \quad \text{Replace } x \text{ with 2.}$$
$$= -16(4) + 56(2)$$
$$= -64 + 112$$
$$= 48$$

After 2 seconds, the arrow is 48 feet high.

CHECK FOR UNDERSTANDING

Communicating Mathematics

Read and study the lesson to answer each question.

1. **Tell** whether each expression is a monomial or a polynomial. **a–d. See margin.**
   a. $x + 2$ b. $3x$ c. $7xy$ d. $2x^2 + x - 5$

2. **Tell** how to model a monomial like $-3x$. **Use 3 red x-tiles.**

3. **Write** a polynomial and use a drawing or algebra tiles to represent it. **See students' work.**

562 Chapter 13 Algebra: Exploring Polynomials

■ Reteaching the Lesson ■

Activity Have students model a monomial using any number of x^2-tiles only, then a second monomial using only x-tiles, and finally a third monomial using only 1-tiles. Then have them place the tiles together and name the polynomial modeled by the tiles.

Additional Answers
1a. polynomial
1b. monomial
1c. monomial
1d. polynomial

Guided Practice Write a monomial or polynomial for each model.

4.
4x

5.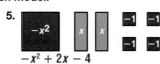
$-x^2 + 2x - 4$

Model each monomial or polynomial using drawings or algebra tiles.

6. $2x^2$ 6-7. See margin.
7. $3x - 1$

Evaluate each expression. 10. 0

8. $5x - 2$, if $x = 6$ 28
9. $x^2 + 4x$, if $x = -1$ -3
10. $x^2 - 9x + 14$, if $x = 7$

11. **Life Science** Stories about sea monsters may have come from sightings of giant squid. The expression $x - 50$ represents the length of a giant squid in feet. Evaluate the expression for $x = 120$ to find the length of a giant squid.   **70 ft**

EXERCISES

Practice Write a monomial or polynomial for each model.

12.
$-2x$

13.
$x^2 + 3x$

14.
$-x^2 + 3x - 3$

15.
$-4x + 2$

16.
$x^2 - 2x - 2$

17.
$2x^2 - x + 4$

Model each monomial or polynomial using drawings or algebra tiles.

18–23. See Answer Appendix.

18. $-3x^2$
19. $5x$
20. $-4x + 2$
21. $2x^2 - 3x$
22. $x^2 - 2x + 1$
23. $-2x^2 + x - 4$

Evaluate each expression.   26. -14 28. -5

24. $12x + 4$, if $x = 3$ 40
25. $6x^2 + 3x$, if $x = 2$ 30
26. $x^2 - 7x - 8$, if $x = 1$
27. $x^2 - 5x$, if $x = -2$ 14
28. $-x^2 + 4x - 5$, if $x = 4$
29. $2x^2 - x - 7$, if $x = -1$ -4

30. Find the value of $9x^2 + x$ if $x = 5$.   230
31. What is the sum of $3x^2$, $4x$, and -10 if $x = -3$?   5

Lesson 13-1 Modeling Polynomials   **563**

Additional Answers

6.

7.

Exercise 34 asks students to advance to the next stage of work on the Chapter Project. You may wish to have students work in pairs to make their models.

4 ASSESS

Closing Activity
Modeling Have students draw a handful of area tiles from a box, group them, and name the polynomial modeled.

Additional Answer
35.

Practice Masters, p. 100

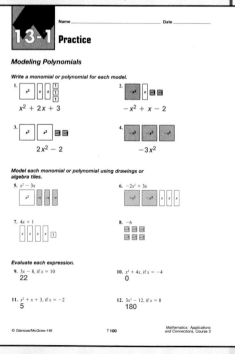

Applications and Problem Solving

32a. Use 16 red x^2-tiles and 150 yellow x-tiles.

32c. See Answer Appendix.

33b. 32 sq units

33c. rectangular prism if all sides do not have the same measure; cube if $x = 3$

Mixed Review

32. **Sports** If Juan Gonzalez of the Texas Rangers hits a baseball straight up with a speed of 150 feet per second, the height of the ball after x seconds is given by $-16x^2 + 150x$.
   a. Tell how you would use algebra tiles to model the expression $-16x^2 + 150x$.
   b. Copy and complete the table.
   c. Graph the ordered pairs and connect the points with a smooth curve.
   d. Use the graph to estimate the maximum height of the ball. **about 350 ft**

Height of Ball	
Time (s)	Height (ft)
0	0
2	236
4	344
6	324
8	176
9	54

33. **Geometry** Refer to the figure at the right.
   a. Write an expression for the surface area of the box. $2x^2 + 12x$
   b. Find the surface area when $x = 2$.
   c. Is the box a cube or a rectangular prism? Explain.

34. **Working on the CHAPTER Project**
   Refer to the probabilities you found on page 559.
   a. Use a product mat to represent the probabilities of hitting (h) or missing (m) a first free throw on the horizontal axis. Then model the probabilities of hitting or missing a second free throw on the vertical axis. **See Answer Appendix.**
   b. Use the percents of each of your players to complete the area model for the product of the probabilities of making both, missing the first then hitting the second, hitting the first then missing the second, or missing both free throws. **See students' work.**

35. **Critical Thinking** Suppose  represents an area of a and represents an area of b. Draw a model that represents an area of ab. **See margin.**

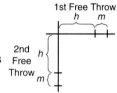

36. **Statistics** Refer to the favorite soft drink survey. For 6,300 people, how much ginger ale should the Band Boosters order? *(Lesson 12-7)* **700 ginger ales**

37. **Geometry** Find the distance between $P(-3, 4)$ and $Q(5, -2)$. *(Lesson 9-6)* **10 units**

Favorite Soft Drink	Number of Responses
Lemon Lime	17
Cola	25
Root Beer	10
Fruit	12
Ginger Ale	8

38. **Test Practice** The back of a chair forms an angle of 112° with its seat. What type of angle is this? *(Lesson 5-2)* **A**
   A obtuse   B right   C straight   D acute

39. What percent of 70 is 42? *(Lesson 3-6)* **60%**

564 Chapter 13 Algebra: Exploring Polynomials

Extending the Lesson

Enrichment Masters, p. 100

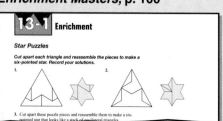

Activity Ask students to make a drawing of the polynomial $x^3 + x^2 + x + 1$.

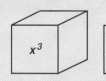

13-2 Simplifying Polynomials

What you'll learn
You'll learn to simplify polynomials by using algebra tiles.

When am I ever going to use this?
The skills used in simplifying polynomials are also used in banking.

Word Wise
term
like term
simplest form

Cultural Kaleidoscope
In 1993, Pearline Motley was the first African American honored as American Business Woman of the Year.

From 1987 to 1996, the number of businesses owned by minority women increased 153%. How many businesses did minority women own in 1996? This number can be found by adding the numbers in the minority groups. The total is $382,400 + 305,700 + 405,200$ or $1,093,300$.

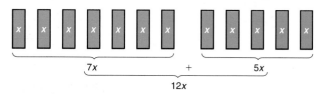

Minority Women Mean Business

1996 Businesses	
Hispanic	382,400
Asian, American Indian, Alaska Native	305,700
African American	405,200

Source: National Foundation for Women Business Owners

In the same way, the polynomial $7x + 5x$ can be simplified by adding the number of x's. We can model $7x + 5x$ with algebra tiles.

The polynomial $7x + 5x$ contains two monomials. Each monomial in a polynomial is called a **term**. The monomials $7x$ and $5x$ are called **like terms** because they have the same variable to the same power. When you use algebra tiles, you can recognize like terms because they have the same size and shape.

The model shown above suggests that you can simplify polynomials that have like terms. An expression that has no like terms is in **simplest form**. In simplest form, $7x + 5x$ is $12x$.

Example 1 Simplify $x^2 + 4x^2 + 3x$.

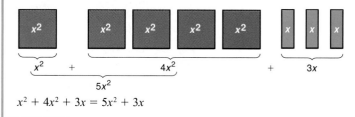

$x^2 + 4x^2 + 3x = 5x^2 + 3x$

Lesson 13-2 Simplifying Polynomials **565**

Multiple Learning Styles

Intrapersonal Ask each student to write whether they prefer to use models or another method to simplify polynomials. Encourage them to explain why they prefer a certain method.

13-2 Lesson Notes

Instructional Resources
- *Study Guide Masters*, p. 101
- *Practice Masters*, p. 101
- *Enrichment Masters*, p. 101
- Transparencies 13-2, A and B
- *Assessment and Evaluation Masters*, p. 351
- *Technology Masters*, p. 77
- CD-ROM Program
- Resource Lesson 13-2

Recommended Pacing	
Honors	Day 2 of 10

1 FOCUS

5-Minute Check (Lesson 13-1)

1. Write a polynomial for this model. $x^2 - 2x + 4$

2. Model $-x^2 + 4x$ using drawings or algebra tiles.

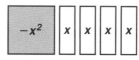

3. Evaluate $x^2 + 3x - 5$ if $x = 2$. 5

The 5-Minute Check is also available on **Transparency 13-2A** for this lesson.

Motivating the Lesson

Problem Solving Ask students to find two methods for solving this problem: *Midori bought 6 dozen eggs at one store and 5 dozen at another. How many eggs did she buy altogether?* Sample answer:
Method 1:
$6 \cdot 12 + 5 \cdot 12 = 72 + 60$, or 132;
Method 2:
$6 + 5 = 11, 11 \cdot 12 = 132$

Lesson 13-2 **565**

2 TEACH

 Transparency 13-2B contains a teaching aid for this lesson.

Using the Mini-Lab Ask students these questions.
- What property justifies the step of adding or removing a zero pair without changing the value of the set? **identity property of addition**
- What property does the zero pair model? **inverse property of addition**

In-Class Examples

For Example 1
Simplify $5x^2 + 3x + 2x^2 + 4x$.
$7x^2 + 7x$

Simplify each polynomial.

For Example 2
$2n^2 - 3n^2 + 2$ $-n^2 + 2$

For Example 3
$x^2 + y^2$ It is in simplest form.

Teaching Tip After working Example 2 in the Student Edition, point out that the variable has to be of the same power to be simplified. For example,
$3r^2 + 2r^3 + r^2 = 4r^2 + 3r^3$.

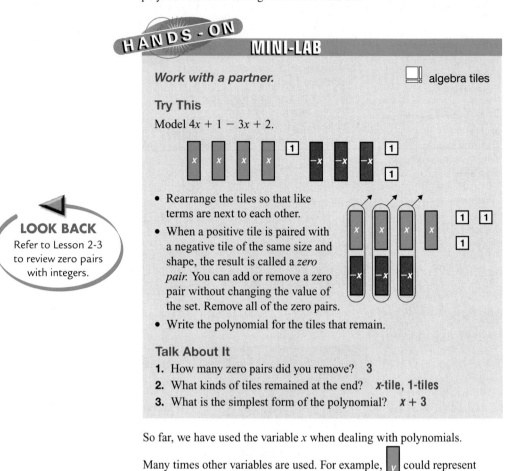

What happens when you have both positive and negative terms in a polynomial? Let's use algebra tiles to find out.

HANDS-ON MINI-LAB

Work with a partner. algebra tiles

Try This
Model $4x + 1 - 3x + 2$.

- Rearrange the tiles so that like terms are next to each other.
- When a positive tile is paired with a negative tile of the same size and shape, the result is called a *zero pair*. You can add or remove a zero pair without changing the value of the set. Remove all of the zero pairs.
- Write the polynomial for the tiles that remain.

LOOK BACK Refer to Lesson 2-3 to review zero pairs with integers.

Talk About It
1. How many zero pairs did you remove? **3**
2. What kinds of tiles remained at the end? **x-tile, 1-tiles**
3. What is the simplest form of the polynomial? **x + 3**

So far, we have used the variable x when dealing with polynomials. Many times other variables are used. For example, could represent the variable y. Then could represent $-y$.

 Simplify each polynomial.

② $2a^2 + a^2$

These are like terms because they have the same variable to the same power.

$2a^2 + a^2 = 3a^2$

③ $4x - y$

These are not like terms because the variables are different.

$4x - y$ is in simplest form.

566 Chapter 13 Algebra: Exploring Polynomials

Classroom Vignette

"I like to remind students of their work with integer counters before starting to work with polynomial models. Make sure they understand that zero pairs exist with polynomial models in the same sense that they exist with integers."

Cindy J. Boyd, Teacher
Abilene High School
Abilene, TX

Example 4 — APPLICATION

Food Matt brought 2 boxes of chocolate cookies and 1 box of peanut butter cookies to the class picnic. Tara brought 2 boxes of chocolate cookies and 3 boxes of peanut butter cookies. If c represents the number of chocolate cookies in a box, and p represents the number of peanut butter cookies in a box, then the total number of cookies is $2c + p + 2c + 3p$. Simplify the polynomial to find the total number of cookies.

$$2c + p + 2c + 3p = 4c + 4p$$

The total number of cookies is $4c + 4p$.

In-Class Example

For Example 4
During one school week, Josh wears 2 brown shirts and 3 white shirts. Alicia wears 3 brown shirts and 2 white shirts. Alby wears 4 white shirts and 1 brown shirt. If b represents a brown shirt and w represents a white shirt, $2b + 3w + 3b + 2w + 4w + 1b$ represents the total number of shirts worn. Simplify the polynomial to find the total number of shirts. **$6b + 9w$**

---

3 PRACTICE/APPLY

Check for Understanding
If students need additional practice or instruction after completing Exercises 1–11, one of these options may be helpful.
- Extra Practice, see p. 642
- Reteaching Activity
- *Study Guide Masters,* p. 101
- *Practice Masters,* p. 101

CHECK FOR UNDERSTANDING

Communicating Mathematics

Read and study the lesson to answer each question.

1. **Tell** the like terms in $2x^2 - y - 3x^2 + 2y^2 + 5y$. **$2x^2, -3x^2; -y, 5y$**

2. **Write** a polynomial in simplest form to represent the model at the right. **$x^2 + x$**

HANDS-ON MATH

3. **Draw** a model of $-2x^2 + x^2 - 2x + x - 3$. Then write a polynomial in simplest form to represent the model. **See Answer Appendix for model; $-x^2 - x - 3$.**

Guided Practice

Name the like terms in each list of terms.

4. $6x, -4x^2, -11x$ **$6x, -11x$**
5. $7x, 8y, 9z$ **none**

6. Simplify $-x^2 + 4x + 2x + x^2$ using the model. **$6x$**

Simplify each polynomial. Use drawings or algebra tiles if necessary.

7. $7x + 1 - 3x$ **$4x + 1$**
8. $-3y^2 - 2y^2 - 4y + 3$ **$-5y^2 - 4y + 3$**

Simplify each expression. Then evaluate if $x = -1$ and $y = 6$.

9. $5x - 2y + 3x + y$ **$8x - y; -14$**
10. $x + 2x - 7y + 4x$ **$7x - 7y; -49$**

11. **Money** Miyoko found three quarters, five dimes, and two nickels in her backpack. In her pocket, she had one quarter, three dimes, and three nickels.
    a. Let q, d, and n represent the value of a quarter, a dime, and a nickel, respectively. Represent all of the coins Miyoko has as a polynomial expression in simplest form. **$4q + 8d + 5n$**
    b. Evaluate the simplified expression to determine how much money Miyoko has. **$2.05**

Lesson 13-2 Simplifying Polynomials **567**

Reteaching the Lesson

Activity Have students work in groups of three. Two students each use algebra tiles to model a two-term polynomial and name it. The third student then combines the tiles, removes any zero pairs, and names the simplified polynomial modeled by the remaining tiles. Students should switch roles and repeat the activity several times.

Error Analysis
Watch for students who have difficulty simplifying polynomials mentally.
Prevent by having students record each term in tabular form so each column contains like terms.

Lesson 13-2 **567**

Assignment Guide

Core: 13–35 odd, 36–40
Enriched: 12–32 even, 33–40

EXERCISES

Practice

Name the like terms in each list of terms. 13. $3x^2, -2x^2$

12. $4, 3m^2, -5, 2m$ $4, -5$
13. $3x^2, 4x, 10, -2x^2$
14. $15y^2, 2y, 8$ none
15. $7a, 6b, 10a, 14b$
    $7a, 10a; 6b, 14b$
16. $-a^2, 3a^2, -4x^2$
    $-a^2, 3a^2$
17. $4y, 8, 9, -2y, -3y$
    $4y, -2y, -3y; 8, 9$

Simplify each polynomial using the model.

18. $3a^2 - 2a^2 + 3a$ $a^2 + 3a$

19. $2x + 3y - 4x - y$ $-2x + 2y$

20. $2x + 3 - x + x^2 - 4$ $x^2 + x - 1$

Simplify each polynomial. Use drawings or algebra tiles if necessary.

21. $2a + 1 + 3a + 4$ $5a + 5$
22. $2x^2 + 3 + 4x - 7$ $2x^2 + 4x - 4$
23. $y^2 - 5y - y^2 - 2y$ $-7y$
24. $10x + 3y - 8x + 5y$ $2x + 8y$
25. $4x^2 + 6 - x^2$ $3x^2 + 6$
26. $-a^2 + a + 10a + 9a^2$ $8a^2 + 11a$

Simplify each expression. Then evaluate if $a = 4$ and $b = -2$.

27. $17a + 11b; 46$

27. $3a + 9b + 14a + 2b$
28. $7a + 9b + 1a + 5b$ $8a + 14b; 4$
29. $a + 5b - 2a + 4b$ $-a + 9b; -22$
30. $-4a + 7a + 10b - 7b$ $3a + 3b; 6$

31. Find the value of $4x^2 + x - x^2$ if $x = 4$.   52
32. Find the sum of $8a, 7b, -2a,$ and b if $a = 3$ and $b = 2$.   34

Applications and Problem Solving

33. **Money Matters** Shana receives $50 each birthday from her uncle. Her parents put this money in a savings account with an interest rate of r. The table gives the account balance after each birthday.

    a. Write the amount Shana has in her account on her second birthday in simplest form. If r is 6% or 0.06, how much is in her account? $50r + 100, \$103$
    b. Write the balance of Shana's account after her third birthday in simplest form. $50r^2 + 150r + 150$

Birthday	Balance ($)
1	50
2	$50 + 50r + 50$
3	$50 + 100r + 50r^2$ $+ 50 + 50r + 50$

568 Chapter 13 Algebra: Exploring Polynomials

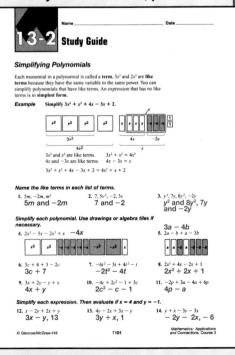

Study Guide Masters, p. 101

34. **Geometry** A new playground is to be constructed as shown at the right.
   a. Write the area of the playground in simplest form. $x^2 + 2x + 1$
   b. Find the area if x is 60. 3,721 sq yd

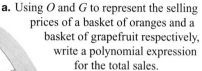

35. **School** As a fund-raiser, the school band is selling fruit. The sales for the first two weeks are shown in the table.
   a. Using O and G to represent the selling prices of a basket of oranges and a basket of grapefruit respectively, write a polynomial expression for the total sales. $114O + 96G$

Baskets of Fruit Sold		
Week	Oranges	Grapefruit
1	52	39
2	62	57

   b. If oranges cost $12 a basket and grapefruit cost $9 a basket, what was the total amount of sales? $2,232

36. **Critical Thinking** Write the polynomial modeled at the right in simplest form. $x^2 + y^2 - 1$

Mixed Review

37. **Algebra** Evaluate $5x^2 - 3x + 1$ if $x = 2$. *(Lesson 13-1)* 15

38. **Functions** Find the value of $f(3)$ if $f(n) = 2n^3 - 6$. *(Lesson 10-1)* 48

39. **Test Practice** A photograph measures 4 inches wide by 6 inches long. An enlargement is made that is 18 inches long. How wide is the enlargement? *(Lesson 8-9)* B

   A 24 in.
   B 12 in.
   C 28 in.
   D 36 in.

40. Express 8.99×10^4 in standard form. *(Lesson 6-9)* 89,900

Lesson 13-2 Simplifying Polynomials **569**

Extending the Lesson

Enrichment Masters, p. 101

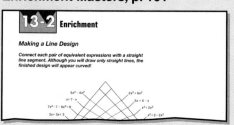

Activity Take a class poll of each student's favorite food. Have the students work together first to find a variable to represent each food and then to simplify a polynomial to represent the total number of favorite foods.

4 ASSESS

Closing Activity

Modeling Have students use algebra tiles to show how zero pairs can be used to simplify polynomials with positive and negative terms.

Chapter 13, Quiz A (Lessons 13-1 and 13-2) is available in the *Assessment and Evaluation Masters,* p. 351.

Practice Masters, p. 101

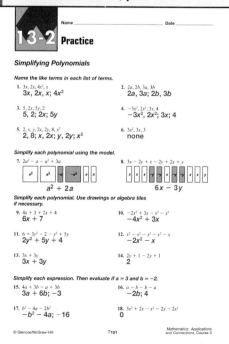

Lesson 13-2 **569**

13-3 Lesson Notes

Instructional Resources
- *Study Guide Masters*, p. 102
- *Practice Masters*, p. 102
- *Enrichment Masters*, p. 102
- Transparencies 13-3, A and B
- CD-ROM Program
  - Resource Lesson 13-3
  - Interactive Lesson 13-3

Recommended Pacing	
Honors	Day 3 of 10

1 FOCUS

5-Minute Check
(Lesson 13-2)

Simplify each polynomial. Use drawings or algebra tiles if necessary.
1. $6x + 2 + 5 + 3x$ $9x + 7$
2. $n^2 + 8n - 5n - n^2$ $3n$
3. $4a + 7b - 5b + a$ $5a + 2b$
4. Simplify $5x - 3y + 5y - 3x$. Then evaluate if $x = 3$ and $y = -3$. $2x + 2y$; 0
5. Molly has 3 quarters, 2 pennies, and 5 nickels. Esteban has 2 nickels, 4 quarters, and 1 penny. Write a polynomial to represent the total number of coins in simplest form. $7q + 3p + 7n$

The 5-Minute Check is also available on **Transparency 13-3A** for this lesson.

Motivating the Lesson

Problem Solving Ask students to solve this problem: *Find the sum of 8 hours 5 minutes 9 seconds and 3 hours 6 minutes 2 seconds.* 11 h 11 min 11 s

2 TEACH

Transparency 13-3B contains a teaching aid for this lesson.

Using the Mini-Lab Ask students to tell how the following problem is like the problem in the *Motivating the Lesson* activity above. The x^2-, x-, and 1-tiles are like hours, minutes, and seconds, respectively.

13-3 Adding Polynomials

What you'll learn
You'll learn to add polynomials by using algebra tiles.

When am I ever going to use this?
Knowing how to add polynomials is useful in carpentry.

Did you know? If you deposited $1,200 into an account with a 6% interest rate, in 30 years you would have almost $7,000. If you deposited $1,200 every year into the same account, in 30 years you would have over $107,000.

Ups and Downs of Investing

Where is the best place for your savings? A bank Certificate of Deposit (CD) is very safe, but the earnings may be low. Investing money in a mutual fund is more risky, but can produce better profits. The graph shows the earnings of a $500 investment in a CD and of a mutual fund for a recent year.

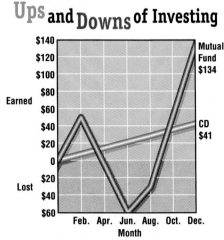

Source: *Zillions,* Aug./Sept., 1991

The mutual fund had lost $60 in June but finished the year earning a total of $134. It was worth $500 plus $134 growth.

$$\begin{array}{rl} \$500 \rightarrow & (5 \times 10^2) + (0 \times 10) + 0 \\ +134 \rightarrow & (1 \times 10^2) + (3 \times 10) + 4 \\ \hline \$634 \leftarrow & (6 \times 10^2) + (3 \times 10) + 4 \end{array}$$

The total value of the mutual fund was $634.

To find the total value, "like terms"—1s, 10s, and 100s—were added. Two or more polynomials can be added in a similar way.

HANDS-ON MINI-LAB

Work with a partner. algebra tiles

Try This

Add $2x^2 + 3x - 3$ and $x^2 - 4x + 2$.
- Model each polynomial.

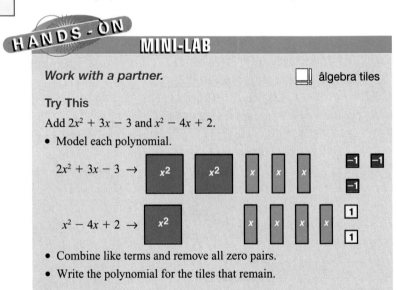

- Combine like terms and remove all zero pairs.
- Write the polynomial for the tiles that remain.

Talk About It

1. How is this method like the method for simplifying polynomials?
2. What is $(2x^2 + 3x - 3) + (x^2 - 4x + 2)$? $3x^2 - x - 1$

1. It is the same.

Multiple Learning Styles

Interpersonal Working in pairs, have each student write 4 polynomials in the form $ax^2 + bx + c$, where a, b, and c can be positive or negative. Then have students add the polynomials, each student contributing one polynomial for each sum.

Example 1 Write the two polynomials represented below. Then find their sum.

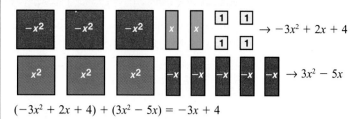

$(-3x^2 + 2x + 4) + (3x^2 - 5x) = -3x + 4$

One way to add polynomials is to write them vertically and line up like terms.

Example 2 Find $(x^2 - 4x + 3) + (2x^2 - 2x - 4)$. Then evaluate the sum if $x = -2$.

$x^2 - 4x + 3$ *Arrange like terms in columns.*
$+ 2x^2 - 2x - 4$ *Then add.*
$\overline{3x^2 - 6x - 1}$

Now evaluate for $x = -2$.
$3x^2 - 6x - 1 = 3(-2)^2 - 6(-2) - 1$ *Replace x with -2.*
$= 3(4) + 12 - 1$
$= 12 + 12 - 1$ or 23

CHECK FOR UNDERSTANDING

Communicating Mathematics
Read and study the lesson to answer each question.

1. **Tell** how to add two polynomials. **Add like terms.**

2. $2x^2 + 3x - 4$ and $x^2 - 2x - 2$; $3x^2 + x - 6$

2. **Write** the two polynomials represented at the right. Then find their sum.

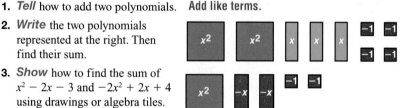

HANDS-ON MATH

3. **Show** how to find the sum of $x^2 - 2x - 3$ and $-2x^2 + 2x + 4$ using drawings or algebra tiles. **See Answer Appendix.**

Guided Practice
Find each sum. Use drawings or algebra tiles if necessary.

4. $2x^2 + x + 4$
   $+ 3x^2 + 3x + 1$
   $\overline{5x^2 + 4x + 5}$

5. $5a + 3b$
   $+ 4a + 2b$
   $\overline{9a + 5b}$

6. $4y^2 + 4y + 10$

6. $(-2y^2 + y + 8) + (6y^2 + 3y + 2)$

7. $(-3x + 1) + (2x + 5)$ $-x + 6$

8. $9r + 2s$; 20

8. Find the sum $(8r - 3s) + (r + 5s)$. Then evaluate the sum if $r = 2$ and $s = 1$.

9a. $(x + 4) + (2x - 3) + (3x + 1) = (6x + 2)$cm

9. **Geometry** Refer to the figure.
   a. Write and simplify an expression for its perimeter.
   b. Find the perimeter if $x = 4$. **26 cm**

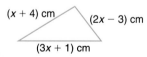

Lesson 13-3 Adding Polynomials **571**

In-Class Examples

For Example 1 Write the two polynomials represented below. Then find their sum.

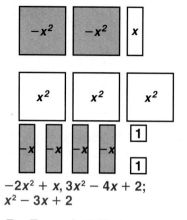

$-2x^2 + x$, $3x^2 - 4x + 2$;
$x^2 - 3x + 2$

For Example 2 Find $(x^2 - x - 1) + (-2x^2 + 4x + 2)$. Then evaluate the sum if $x = -3$.
$-x^2 + 3x + 1$; -17

Teaching Tip For Example 2, remind students that it is helpful to write the sums neatly with the terms lined up to avoid confusion when adding.

Study Guide Masters, p. 102

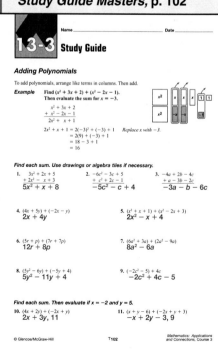

Reteaching the Lesson

Activity Write a different polynomial on each of 12 index cards and shuffle the cards. The first of two players draws two cards and writes the sum. The second player draws a card and adds the polynomial to the first player's sum. Play continues, with each player adding to the previous sum.

Error Analysis
Watch for students who add exponents when adding like terms.
Prevent by drawing a parallel between adding like terms and adding real objects: 2 pears + 3 pears = 5 pears, not 5 pears^2; $2x + 3x = 5x$, not $5x^2$.

Lesson 13-3 **571**

3 PRACTICE/APPLY

Check for Understanding

If students need additional practice or instruction after completing Exercises 1–9, one of these options may be helpful.
- Extra Practice, see p. 642
- Reteaching Activity, see p. 571
- *Study Guide Masters*, p. 102
- *Practice Masters*, p. 102
- Interactive Mathematics Tools Software

Assignment Guide

Core: 11–29 odd, 30–34
Enriched: 10–26 even, 28–34

4 ASSESS

Closing Activity

Writing Have students evaluate the sum of the polynomials $-x^2 + 3x - 2$ and $2x^2 - 3x + 5$ if $x = 3$. **12**

Practice Masters, p. 102

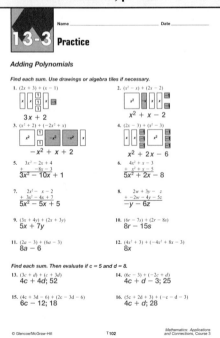

EXERCISES

Practice Find each sum. Use drawings or algebra tiles if necessary.

10. $x^2 + 2x + 1$
    $+ x^2 - 3x - 2$
    $\overline{2x^2 - x - 1}$

11. $5y^2 - 2y + 5$
    $+ y^2 + 4y - 3$
    $\overline{6y^2 + 2y + 2}$

12. $-4a^2 + 3a - 2$
    $+ 6a^2 - 5a - 7$
    $\overline{2a^2 - 2a - 9}$

13. $3x^2 - 3x + 5$
    $+ -2x^2 + 3x$
    $\overline{x^2 + 5}$

14. $4a + 6b + c$
    $+ 5a - 3b - 2c$
    $\overline{9a + 3b - c}$

15. $15q + 2r - 1$
    $+ q - 3r + 2$
    $\overline{16q - r + 1}$

16. $(2x^2 - 6x) + (x^2 - 4x)$ $3x^2 - 10x$
17. $(9a + 5b) + (3a + 7b)$ $12a + 12b$

19. $10a^2 + 2a - 3$
20. $-7x^2 + x + 1$
21. $2x^2 - 3x - 10$
22. $8g + 2h; 48$
23. $7g + 3h + 4; 51$
24. $3g - h; 11$
25. $15g + h + 11; 90$

18. $(5r - 7s) + (3r + 8s)$ $8r + s$
19. $(3a^2 + 2a) + (7a^2 - 3)$
20. $(-2x^2 + 4x - 6) + (-5x^2 - 3x + 7)$
21. $(3x^2 - 5x - 7) + (-x^2 + 2x - 3)$

Find each sum. Then evaluate if $g = 5$ and $h = 4$.

22. $(6g - 3h) + (2g + 5h)$
23. $(3g + 2h - 1) + (4g + h + 5)$
24. $(-3g + 5h) + (6g - 6h)$
25. $(14g + 3h + 2) + (g - 2h + 9)$

26. Find the sum of $-4x + 2$ and $6x^2 - 3$. $6x^2 - 4x - 1$
27. What is the value of $(p^2 - 4p) + (2p^2 + 3p)$ if $p = 3$? **24**

Applications and Problem Solving

29. $3x^2 + 21x + 6$

28. **School** Lloyd's total number of grade points for the first semester was $2A + 2B + C$. His total for the second semester was $A + 3B + D$.
    a. Add the polynomials to find his total grade points for the year. $3A + 5B + C + D$
    b. Evaluate the sum by substituting the grade point value for each variable. **30**

Grade	Grade Points
A	4
B	3
C	2
D	1
F	0

29. **Carpentry** To build the top cupboard at the right, $9x + 6$ square feet of wood is required. The bottom cupboard will requires $3x^2 + 12x$ square feet of wood. Find the total square feet of wood required for both.

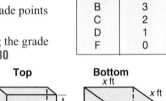

30. **Critical Thinking** If $(6x - 10y) + (9x + 5y) = 15x - 5y$, what is $(15x - 5y) - (6x - 10y)$? $9x + 5y$

Mixed Review

33. $3\frac{7}{20}$

31. **Algebra** Simplify $4a^2 + 6a - 14a^2 + 9a$. *(Lesson 13-2)* $-10a^2 + 15a$
32. **Geometry** Find the volume of a cone to the nearest cubic meter if the radius is 3.5 meters and the height is 12 meters. *(Lesson 11-4)* **154 m³**
33. **Algebra** Evaluate $a + b - c$ if $a = \frac{1}{2}$, $b = 4\frac{3}{5}$, and $c = 1\frac{3}{4}$. *(Lesson 7-2)*

34. **Test Practice** Cindy and three of her friends planned to leave a 20% tip for the waiter when they had lunch. Two of her friends had hamburgers, one had soup and salad, and Cindy had pasta. What other information is necessary to determine how much to leave for the tip? *(Lesson 3-6)* **C**

A the cost of the hamburger
B where they had lunch
C the cost of the four meals
D what day they had lunch
E the salary of the waiter

572 Chapter 13 Algebra: Exploring Polynomials

Extending the Lesson

Enrichment Masters, p. 102

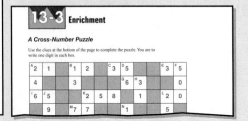

Activity Provide students with the box scores of a basketball game. Have them write each team's score as a polynomial sum of 3-point field goals (t), 2-point field goals (g), and 1-point free throws (f). Then have them add the polynomials and evaluate the sum for $t = 3$, $g = 2$, and $f = 1$.

13-4 Subtracting Polynomials

What you'll learn
You'll learn to subtract polynomials by using algebra tiles.

When am I ever going to use this?
Knowing how to subtract polynomials can help you solve problems involving interest rates and finance.

You can use algebra tiles to subtract polynomials.

HANDS-ON MINI-LAB

Work with a partner. algebra tiles

Try This

Find $(3x + 6) 2 (-1x + 3)$.

- Model the polynomial $3x + 6$.
- To subtract $(-1x + 3)$, remove 1 negative x-tile and 3 1-tiles. You can remove the 1-tiles, but there are no negative x-tiles, so you can't remove $-1x$.
- Add a zero pair, and then remove the negative x-tile.

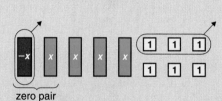

zero pair

Talk About It   3. It doesn't change the value of the expression.
1. What kinds of tiles remain? **4 x-tiles and 3 1-tiles**
2. What is $(3x + 6) - (-1x + 3)$ in simplest form? **$4x + 3$**
3. Why can you add or remove a zero pair?

To subtract an integer, it is usually easier to add its opposite. For example, $3 - 8 = 3 + (-8) = -5$. In a similar manner, to subtract a polynomial, it may be easier to add the opposite of each term of the polynomial.

Examples

Find each difference.

1
$$\begin{array}{r} 6x - 4 \\ -(2x - 1) \end{array} \quad \left\{\begin{array}{l}\text{The opposite of } 2x \text{ is } -2x.\\ \text{The opposite of } -1 \text{ is } 1.\end{array}\right\} \quad \begin{array}{r} 6x - 4 \\ +(-2x + 1) \\ \hline 4x - 3 \end{array}$$

2 $(3x^2 - 2x - 1) - (5x^2 - 3x + 4)$

$$\begin{array}{r} 3x^2 - 2x - 1 \\ -(5x^2 - 3x + 4) \end{array} \rightarrow \begin{array}{r} 3x^2 - 2x - 1 \\ +(-5x^2 + 3x - 4) \\ \hline -2x^2 + x - 5 \end{array}$$

Lesson 13-4 Subtracting Polynomials **573**

13-4 Lesson Notes

Instructional Resources
- *Study Guide Masters*, p. 643
- *Practice Masters*, p. 103
- *Enrichment Masters*, p. 103
- Transparencies 13-4, A and B
- *Assessment and Evaluation Masters*, pp. 350, 351
- *Hands-On Lab Masters*, p. 80
- CD-ROM Program
  - Resource Lesson 13-4
  - Interactive Lesson 13-4

Recommended Pacing
| Honors | Day 4 of 10 |

1 FOCUS

5-Minute Check
(Lesson 13-3)

Find each sum. Use drawings or algebra tiles if necessary.

1. $\begin{array}{r}4x^2 + 5x - 7 \\ + x^2 - 3x + 5 \\ \hline 5x^2 + 2x - 2\end{array}$

2. $(-2n^2 + 5n) + (n^2 - 2)$
   $-n^2 + 5n - 2$

3. Find the sum of $2x - 4y$ and $-5x + 2y$. Then evaluate if $x = 2$ and $y = 3$.
   $-3x - 2y$; -12

The 5-Minute Check is also available on **Transparency 13-4A** for this lesson.

Motivating the Lesson

Hands-On Activity Have students model $4x$ using algebra tiles. Ask the following questions.
- How could you subtract $2x$? **Remove two x-tiles.**
- What would you add to the model of $4x$ so that you could take out $-3x$? **Add three zero pairs.**

Investigations for the Special Education Student

This blackline master booklet helps you plan for the needs of your special education students by providing long-term projects along with teacher notes. Investigation 12, *Yard Sale*, may be used with this chapter.

Lesson 13-4 **573**

2 TEACH

 Transparency 13-4B contains a teaching aid for this lesson.

Using the Mini-Lab Ask students to explain why they must add a zero pair to solve the problem. Adding a zero pair provides one negative *x*-tile, which can then be removed. Remind students that $-1x$ is usually written as $-x$.

In-Class Examples
Find each difference.

For Example 1
$$\begin{array}{r} 7x + 6 \\ -(x + 2) \\ \hline 6x + 4 \end{array}$$

For Example 2
$(3x^2 + 4x - 5) - (-2x^2 - x - 3)$
$5x^2 + 5x - 2$

For Example 3
Two model tugboats flow down a river at the same time. The distance in meters that each boat travels is given by $4t^2 + 14t$ and $3t^2 + 32t$, where *t* equals the time in seconds. How far apart are the boats after 20 seconds? **40 meters**

---

3 PRACTICE/APPLY

Check for Understanding
If students need additional practice or instruction after completing Exercises 1–9, one of these options may be helpful.
- Extra Practice, see p. 643
- Reteaching Activity
- *Study Guide Masters*, p. 103
- *Practice Masters*, p. 103
- Interactive Mathematics Tools Software

---

574 Chapter 13

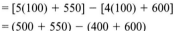

Example 3 **Physical Science** Two soapbox derby cars travel down a hill at the same time. The distance in feet that each car travels is given by $5t^2 + 55t$ and $4t^2 + 60t$. How far apart are the cars after 10 seconds?

Explore You know how far each car travels in *t* seconds. You need to find the difference between the distances and evaluate for $t = 10$.

Plan Find $(5t^2 + 55t) - (4t^2 + 60t)$. Then evaluate for $t = 10$.

Solve
$$\begin{array}{r} 5t^2 + 55t \\ -(4t^2 + 60t) \\ \hline t^2 - 5t \end{array}$$

Now evaluate for $t = 10$.

$t^2 - 5t = (10)^2 - 5(10)$ *Replace t with 10.*
$ = 100 - 50$
$ = 50$

After 10 seconds, the cars are 50 feet apart.

Examine Check by substituting 10 for *t* in the original expressions and then subtracting.

$(5t^2 + 55t) - (4t^2 + 60t) = [5(10)^2 + 55(10)] - [4(10)^2 + 60(10)]$
$= [5(100) + 550] - [4(100) + 600]$
$= (500 + 550) - (400 + 600)$
$= 1{,}050 - 1{,}000$
$= 50$ ✓

CHECK FOR UNDERSTANDING

Communicating Mathematics

Read and study the lesson to answer each question.

1. *Tell* the opposite of $2x$. $-2x$

2. $(x^2 + 4x + 6) - (3x + 2)$; $x^2 + x + 4$

2. *Write* the subtraction problem modeled at the right. What is the difference?

HANDS-ON MATH

3. *Show* how to find $(2x^2 - 3x + 3) - (x^2 + 1)$ using algebra tiles or drawings. See Answer Appendix for model; $x^2 - 3x + 2$.

Guided Practice

Find each difference. Use drawings or algebra tiles if necessary.

4. $\begin{array}{r} 9s - 1 \\ -(7s + 2) \\ \hline 2s - 3 \end{array}$

5. $\begin{array}{r} 4p^2 - 3p + 1 \\ -(2p^2 - 2p) \\ \hline 2p^2 - p + 1 \end{array}$

6. $(a^2 + 2a) - (2a^2 + 2a)$ $-a^2$

7. $(5x^2 - 3x + 2) - (3x^2 - 3)$ $2x^2 - 3x + 5$

8. Find $(7r + 5s) - (r + 4s)$. Then evaluate the difference if $r = 1$ and $s = -2$.
$6r + s$; 4

574 Chapter 13 Algebra: Exploring Polynomials

Reteaching the Lesson

Activity Write a polynomial, such as $-4x^2 - 5x + 2$, on the chalkboard. Have students state the opposite of each term. Next write a subtraction problem with the previous polynomial being subtracted. Have students complete the subtraction.

Error Analysis
Watch for students who find the additive inverse for only the first term of a polynomial when subtracting.
Prevent by reviewing the distributive property.

9. **Food** Valerie ordered 6 tacos and 3 drinks from a fast-food drive-thru. When she got home, she discovered that the bag of food contained 4 tacos and 4 drinks. Find $(6t + 3d) - (4t + 4d)$ and evaluate for the values shown in the table to find how much she was overcharged. **$1.03**

Item	Cost ($)
taco (t)	0.89
drink (d)	0.75

EXERCISES

Practice

Find each difference. Use drawings or algebra tiles if necessary.

10. $7x + 5$
    $- (3x + 4)$
    $\overline{4x + 1}$

11. $5y^2 + 9$
    $- (4y^2 + 9)$
    $\overline{y^2}$

12. $-4a^2 - 3a - 2$
    $-(2a^2 + 2a + 7)$
    $\overline{-6a^2 - 5a - 9}$

13. $3r^2 - 3rt + t^2$
    $- (2r^2 + 5rt - t^2)$
    $\overline{r^2 - 8rt + 2t^2}$

14. $(5x + 3) - (2x + 1)$ $3x + 2$

15. $(-4a + 5) - (a - 1)$ $-5a + 6$

17. $2a^2 - 5a - 9$

16. $(10c - 2d) - (6c + 3d)$ $4c - 5d$

17. $(4a^2 - 3a - 2) - (2a^2 + 2a + 7)$

18. $(-3x^2 + 2x + 1) - (x^2 + 3x - 1)$
    $-4x^2 - x + 2$

19. $(6x^2 + 2x + 9) - (3x^2 + 5x + 9)$
    $3x^2 - 3x$

Find each difference. Then evaluate if $c = -2$ and $d = 5$.

21. $2c - 11d - 4;$
    -63

20. $(3c + 4d) - (2c + 5d)$ $c - d; -7$

21. $(4c - 6d - 1) - (2c + 5d + 3)$

22. $-17c - 2d; 24$

22. $(-2c + 7d) - (15c + 9d)$

23. $9c + 6d - 10; 2$

23. $(10c + 4d - 2) - (c - 2d + 8)$

24. Find $(7a + 3b - c)$ minus $(4a - 2b + 2c)$. $3a + 5b - 3c$

25. What is $(3p^2 + 5pq - q^2)$ decreased by $(p^2 + 3pq - 2q^2)$? $2p^2 + 2pq + q^2$

27a. $-2A + 4B + 3C$

Applications and Problem Solving

26. **Money Matters** Alan borrowed $200 from his father each year for college expenses. The amount he owes his father at the beginning of his second and third years is $(400 + 200r)$ and $(600 + 600r + 200r^2)$ respectively, where r is the interest rate. **a. $200r^2 + 400r + 200$**

    a. Find how much his debt increased between his second and third years.
    b. Evaluate the increase for $r = 8\%$. **$233.28**

27. **Sports** Katrina's scoring total for Game 1 was $5A + 8B + 7C$. Her scoring total for Game 2 was $3A + 12B + 10C$.

Game	A 3 points	B 2 points	C 1 point
1	5	8	7
2	3	12	10

a. Subtract to find how many more points she scored in Game 2 than in Game 1.
b. Evaluate the difference for A = 3, B = 2, and C = 1. **5**

28. **Critical Thinking** Mentally find $(3x^2 + 7x + 9) - (5x^2 - 3x + 2)$ if $x = 0$. Explain. **7; since x is 0, the remaining expression is $9 - 2$.**

Lesson 13-4 Subtracting Polynomials **575**

Mid-Chapter Self Test

The Mid-Chapter Self Test reviews concepts in Lessons 13-1 through 13-4. Lesson references are given so students can review concepts not yet mastered.

Additional Answers for the Self Test

1.

2.

Practice Masters, p. 103

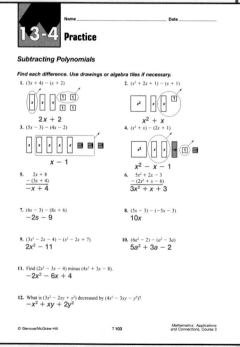

576 Chapter 13

Mixed Review

29. $6y^2 + 15y + 3$

29. **Algebra** Find the sum $(6y^2 + 7y - 5) + (8y + 8)$. *(Lesson 13-3)*

30. **Probability** Two number cubes are rolled. Find the probability that a prime number is rolled on one number cube and an odd number is rolled on the other number cube. *(Lesson 12-5)* $\frac{1}{4}$

31. **Algebra** Solve $3x - 5 = -6$. *(Lesson 7-9)* $-\frac{1}{3}$

32. **Test Practice** Miguel wants to wallpaper a wall that measures 9 feet by 12 feet. A window that is 5 feet by 3 feet is in the center of the wall. How much surface area will be covered with wallpaper? *(Lesson 1-8)* **B**

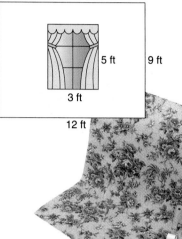

A 36 sq ft
B 93 sq ft
C 108 sq ft
D 15 sq ft
E Not Here

CHAPTER 13 Mid-Chapter Self Test

Model each monomial or polynomial using drawings or algebra tiles. *(Lesson 13-1)* **1–2. See margin.**

1. $-5x$
2. $x^2 - 2x + 4$

3. Simplify $x^2 + x - 2x + 4 - 1$ using the model. *(Lesson 13-2)* $x^2 - x + 3$

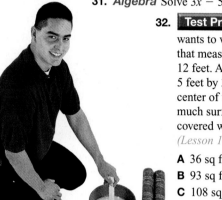

Simplify each polynomial. Use drawings or algebra tiles if necessary. *(Lesson 13-2)*

4. $6x - 4 + x$ $7x - 4$
5. $y^2 - 5y + 3 + 2y^2$ $3y^2 - 5y + 3$

Find each sum or difference. Use drawings or algebra tiles if necessary. *(Lessons 13-3 and 13-4)*

6. $(-3y^2 + 4y) + (-1y^2 - 3y)$ $-4y^2 + y$
7. $(5x^2 + 6x + 7) - (x^2 + 3x - 4)$ $4x^2 + 3x + 11$

Find each sum or difference. Then evaluate if $a = 1$ and $b = 3$. *(Lessons 13-3 and 13-4)*

8. $(-2a + 7b) + (8a - 8b)$ $6a - b$; 3
9. $(3a - 5b + 6) - (a + 2b - 1)$ $2a - 7b + 7$; -12

10. **Geometry** Write the simplified expression for the perimeter of the figure. *(Lesson 13-3)* $8x + 34$

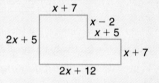

576 Chapter 13 Algebra: Exploring Polynomials

Extending the Lesson

Enrichment Masters, p. 103

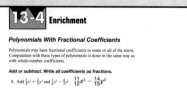

Activity Solve this problem: *By how much does the perimeter of the quadrilateral exceed that of the triangle?* $(6x + 10)$ ft

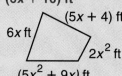

COOPERATIVE LEARNING

13-5A Modeling Products

A Preview of Lesson 13-5

- algebra tiles
- product mat

The algebra tiles that you have been using are based on the fact that the area of a rectangle is the product of the width and length.

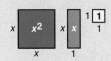

In this lab, you will use these algebra tiles to build more complex rectangles. These rectangles will help you understand how to find the product of simple polynomials. The width and length each represent a polynomial being multiplied; the area of the rectangle represents their product.

TRY THIS

Work with a partner.

Step 1 Make a rectangle with a width of 2 units and a length of $x + 1$ units. This rectangle has an area of $2(x + 1)$. Use your algebra tiles to mark off the dimensions on a product mat.

Step 2 Using the marks as a guide, fill in the rectangle with algebra tiles.

Step 3 The area of the rectangle is $x + x + 1 + 1$. In simplest form, the area is $2x + 2$. Therefore, $2(x + 1) = 2x + 2$.

ON YOUR OWN

Find each product using algebra tiles. 1–3. See Answer Appendix for models.

1. $2(x + 2)$ $2x + 4$
2. $x(x + 3)$ $x^2 + 3x$
3. $3(2x + 1)$ $6x + 3$

4. **Look Ahead** Find $3(n + 4)$.
   $3n + 12$

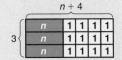

Lesson 13-5A HANDS-ON LAB **577**

Have students write a paragraph describing how they would change the first step in the *Try This* section of the lab in order to find $4(x + 2)$. **Use a length of $x + 1 + 1$ and a width of $1 + 1 + 1 + 1$.**

HANDS-ON
13-5A LAB Notes

GET READY

Objective Students model products by using algebra tiles.

Optional Resources
Hands-On Lab Masters
- algebra tiles, p. 29
- worksheet, p. 67

Overhead Manipulative Resources
- algebra tiles
- product mat

Manipulative Kit
- algebra tiles
- product mat

MANAGEMENT TIPS

Recommended Time
30 minutes

Getting Started Ask students to explain how they could use the distributive property to calculate the product 9×32 mentally.
$$9 \times 32 = 9(30 + 2)$$
$$= 9 \times 30 + 9 \times 2$$
$$= 270 + 18$$
$$= 288$$

The **Activity** helps students understand the multiplication of simple polynomials visually. Students can use frame corners to keep their tiles aligned.

ASSESS

Have students complete Exercises 1–3. Watch for students who may have difficulty deciding how to model each factor. Stress that $(x + 2)$ in Exercise 1 is one factor. Students can use a different color highlighter to mark each factor.

Use Exercise 4 to determine whether students have discovered how to multiply simple polynomials.

Hands-On Lab 13-5A **577**

13-5 Lesson Notes

Instructional Resources
- *Study Guide Masters*, p. 104
- *Practice Masters*, p. 104
- *Enrichment Masters*, p. 104
- Transparencies 13-5, A and B
- CD-ROM Program
  - Resource Lesson 13-5

Recommended Pacing	
Honors	Day 5 of 10

1 FOCUS

5-Minute Check
(Lesson 13-4)

Find each difference. Use drawings or algebra tiles if necessary.

1. $6x + 7$
   $\underline{-\ (4x + 2)}$
   $2x + 5$

2. $3n^2 - 5n + 6$
   $\underline{-\ (n^2 - 4n + 7)}$
   $2n^2 - n - 1$

3. $(-x^2 + 3x) - (2x^2 + 1)$
   $-3x^2 + 3x - 1$

The 5-Minute Check is also available on **Transparency 13-5A** for this lesson.

Motivating the Lesson
Communication Ask students what 5^2 equals. **25** What does 5^4 equal? **625** What is $25 \cdot 625$? **15,625** What does 5^6 equal? **15,625** Then ask students to compare these last two questions. **They have the same answer.** Point out that $5^2 \cdot 5^4 = 5^6$.

2 TEACH

Transparency 13-5B contains a teaching aid for this lesson.

Using the Mini-Lab Make sure students know how to find powers of numbers by using calculators. You may wish to have students check their answers by pencil and paper.

578 Chapter 13

13-5 Multiplying Monomials and Polynomials

What you'll learn
You'll learn to multiply monomials and polynomials.

When am I ever going to use this?
Knowing how to multiply polynomials can help you solve problems involving surface area.

You can use a calculator to find patterns and see how rules are developed for exponents.

TECHNOLOGY MINI-LAB

Work with a partner.   scientific calculator

Try This
Copy the table at the right. Then use a calculator to find each product and complete the table.

Factors	Product	Product Written as a Power
$10^1 \cdot 10^1$	100	10^2
$10^1 \cdot 10^2$	1,000	10^3
$10^1 \cdot 10^3$	10,000	10^4
$10^1 \cdot 10^4$	100,000	10^5
$10^1 \cdot 10^5$	1,000,000	10^6

Talk About It
1. Compare the exponents of the factors to the exponent in the products. What do you observe? **The exponents are added.**
2. Write a rule for determining the exponent of the product when you multiply powers. Test your rule by multiplying $2^2 \cdot 2^4$ using a calculator. **See students' work.**

The pattern you observed in the Mini-Lab leads to the following rule for multiplying powers that have the same base.

Product of Powers		
	Words:	You can multiply powers that have the same base by adding their exponents.
	Symbols:	For any number a and integers m and n, $a^m \cdot a^n = a^{m+n}$.

Monomials that are powers with the same base can be multiplied using the rule for the product of powers.

Examples

1 Find $6^3 \cdot 6^4$.

$6^3 \cdot 6^4 = 6^{3+4}$ or 6^7

Check: $6^3 \cdot 6^4 = (6 \cdot 6 \cdot 6)(6 \cdot 6 \cdot 6 \cdot 6)$
$= 6 \cdot 6 \cdot 6 \cdot 6 \cdot 6 \cdot 6 \cdot 6$ or 6^7 ✓

2 Find $a^2 \cdot a^3$.

$a^2 \cdot a^3 = a^{2+3}$ or a^5

578 Chapter 13 Algebra: Exploring Polynomials

You can use models or the distributive property to multiply a polynomial by a monomial. Recall that the distributive property allows you to multiply the factor *outside* the parentheses by each term *inside* the parentheses.

Examples

APPLICATION

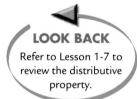

LOOK BACK
Refer to Lesson 1-7 to review the distributive property.

③ **Recreation** *Totolospi* is a game that was popular among the Hopi Indians of Oraibi, Arizona. If the rectangular game board had a width that was 5 inches less than the length, find a simplified expression for its area.

Explore   You know that the width of the game board is 5 inches less than the length. You need to use this information to find its area.

Plan   Let y represent the length.

Then $y - 5$ represents the width.

The formula for area is $A = \ell \times w$.

$A = \ell \times w$

$A = y(y - 5)$

Solve   **Method 1** Use models.

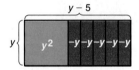

Method 2 Use the distributive property.

$y(y - 5) = y \cdot y - y \cdot 5$
$ = y^2 - 5y$

Examine   Suppose the length of the game board is 12 inches. Then the width would be $12 - 5$ or 7 inches. Its area would be $12(7)$ or 84 square inches. Check by using the expression.

$y^2 - 5y = 12^2 - 5(12)$ *Substitute 12 for y.*
$ = 144 - 60 \text{ or } 84$ ✓

④ Find $x^3(x^4 + 5)$.

$x^3(x^4 + 5) = x^3 \cdot x^4 + x^3 \cdot 5$ *Apply the distributive property.*
$ = x^7 + 5x^3$

In-Class Examples

For Example 1
Find $8^2 \cdot 8^3$. Express the answer in exponential form. 8^5

For Example 2
Find $a^5 \cdot a^7$. a^{12}

For Example 3
If a living room's width is 10 feet less than its length, find a simplified expression for its area. $A = \ell^2 - 10\ell$

For Example 4
Find $a^3(a^7 - 2)$. $a^{10} - 2a^3$

Teaching Tip For Example 4, emphasize that although you are multiplying x^3 by x^4, you *add* the exponents—3 + 4—to get x^7.

Lesson 13-5 Multiplying Monomials and Polynomials

3 PRACTICE/APPLY

Check for Understanding

If students need additional practice or instruction after completing Exercises 1–9, one of these options may be helpful.
- Extra Practice, see p. 643
- Reteaching Activity
- *Study Guide Masters*, p. 104
- *Practice Masters*, p. 104

Assignment Guide

Core: 11–27 odd, 28–32
Enriched: 10–24 even, 25–32

Additional Answers

1. Multiply $3x$ by $5x$ and multiply $3x$ by 2. Add the results.

3.
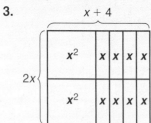

9a.
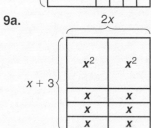

Study Guide Masters, p. 104

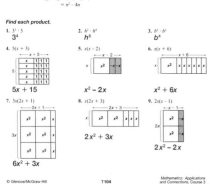

580 Chapter 13

CHECK FOR UNDERSTANDING

Communicating Mathematics

Read and study the lesson to answer each question.

1. *Tell* how to use the distributive property to find $3x(5x + 2)$. See margin.
2. *Use* a calculator to multiply $5^4 \cdot 5^6$. **9,765,625**
3. *Draw* a rectangle to model $2x(x + 4)$. Then write the product.
   See margin for model; $2x^2 + 8x$.

HANDS-ON MATH

Guided Practice   Find each product. Express the answer in exponential form.

4. $3^4 \cdot 3^2$ 3^6
5. $n^3 \cdot n^5$ n^8
6. Find $x(x + 6)$. $x^2 + 6x$

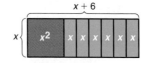

Find each product.

7. $b(b - 2)$ $b^2 - 2b$
8. $j^4(j^2 + 7)$ $j^6 + 7j^4$

9. *Gardening* A square garden plot measures x feet on each side. Suppose you double the length of the plot and increase the width by 3 feet.
   a. Draw the new garden.   See margin.
   b. Write two expressions for the area of the new plot. $2x(x + 3)$; $2x^2 + 6x$
   c. If the original plot was 10 feet on a side, what is the area of the new plot?   **260 ft²**

EXERCISES

Practice   Find each product. Express the answer in exponential form.

10. $4 \cdot 4^3$ 4^4
11. $2^6 \cdot 2^4$ 2^{10}
12. $x^2 \cdot x^7$ x^9
13. $p^5 \cdot p^9$ p^{14}

Find each product.

14. $x(x + 3)$ $x^2 + 3x$
15. $z(z - 1)$ $z^2 - z$
16. $2x(x + 2)$ $2x^2 + 4x$

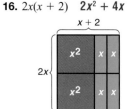

19. $2m^3 + 4m^2$ 21. $d^8 + 15d^3$

17. $b(b + 3)$ $b^2 + 3b$
18. $6t(t - 1)$ $6t^2 - 6t$
19. $m^2(2m + 4)$
20. $3y(2 + y)$ $6y + 3y^2$
21. $d^3(d^5 + 15)$
22. $2x^4(2x - 1)$ $4x^5 - 2x^4$

580 Chapter 13 Algebra: Exploring Polynomials

■ Reteaching the Lesson ■

Activity Point out to students that $x^2 \cdot x^2$ is equal to $x \cdot x \cdot x \cdot x$ which is x^4. Have students work in pairs, each writing multiplication problems involving a monomial and a polynomial for the others to solve.

23. Find the product of a^8 and a^3. a^{11}
24. Multiply $(z - 4)$ by $3z^2$. $3z^3 - 12z^2$

Applications and Problem Solving

25. **Geometry** Find the measure of the area of the shaded region in simplest form. $15d^2 + 28d$

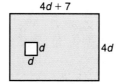

26. **Geometry** A rectangle is 5 feet longer than it is wide. Find the area of the rectangle. $(x^2 + 5x)$ ft²

27. **Money Matters** Tanisha's neighbors pay her $2 an hour to baby-sit their baby, but have promised her a raise. If her raise is x dollars, how much will she earn for 3 hours of baby-sitting? $(3x + 6)$ dollars

28. **Critical Thinking** Copy the table. Use a calculator to find each quotient and complete the table.

Division	Quotient	Quotient Written as a Power
$10^5 \div 10^1$	10,000	10^4
$10^4 \div 10^1$	1,000	10^3
$10^3 \div 10^1$	100	10^2
$10^2 \div 10^1$	10	10^1
$10^1 \div 10^1$	1	10^0

    a. Compare the exponents of the division expressions to the exponents in the quotients. What pattern do you observe?   **The exponents are subtracted.**

    b. Write a rule for determining the exponent in the quotient when you divide powers. Test your rule by dividing 7^5 by 7^3 on a calculator.
    Sample answer: $a^m \div a^n = a^{m-n}$; see students' work.

Mixed Review

29. **Algebra** Find $(8c + d) - (3c - 2d)$. Then evaluate if $c = 3$ and $d = 2$. (Lesson 13-4) $5c + 3d; 21$

30. **Probability** A test has five multiple-choice questions. Each question has three choices. How many outcomes for giving answers to the five questions are possible? (Lesson 12-1)   **243 outcomes**

31. **Test Practice** A can of green beans is 11 centimeters tall and has a diameter of 7.5 centimeters. How many square centimeters of paper will it take to make a label for the can? (Remember: The label does not cover the top or bottom of the can.) (Lesson 11-6)   **D**

    A 347.6 cm²
    B 82.5 cm²
    C 164.3 cm²
    D 259.1 cm²

32. **Architecture** In the afternoon, a building casts a shadow 137 feet long. At the same time, a nearby street sign that is 10.4 feet high casts a shadow 1.3 feet long. How tall is the building? (Lesson 8-8)   **1,096 ft**

Lesson 13-5 Multiplying Monomials and Polynomials   **581**

Extending the Lesson

Activity Have students solve $8(x + 5) - 6(x + 3) = 5(x + 2)$.   **4**

4 ASSESS

Closing Activity

Writing Have students make up a multiplication problem involving monomials and polynomials with exponents and write a paragraph detailing the steps involved in finding the product.

Practice Masters, p. 104

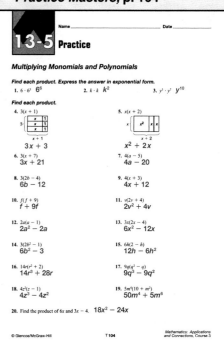

Enrichment Masters, p. 104

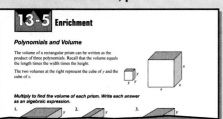

Motivating Students

Students learned in the Chapter Project on page 559 that sports statistics such as basketball free throws can be described as percents and probabilities and modeled geometrically and algebraically. They may be interested in pursuing a career that uses statistics, such as economist, psychologist, stock broker, or athletic scout. To start the discussion, you may wish to ask the following questions about basketball.

- When do scouts start looking for players? **Some colleges run summer camps for players as young as 11 years old.**
- Which sports use professional scouts? **major league football, basketball, and baseball, among others**
- Are there sources for statistics on players? **Magazines and websites offer statistics on players across the country and around the world.**

Making the Math Connection

Long before scouts watch individual athletes play, they use statistical data to narrow their choices. They also use their mathematical skills in planning scouting trips to see the most players in an economical fashion.

Working on *Your Turn*

You may wish to have students work in pairs to research basketball statistics. For instance, they could research one or two of the same teams, but select different players. Be sure to encourage students to consider researching female players as well.

*An additional School to Career activity is available on page 39 of the **School to Career Masters**.*

SPORTS MANAGEMENT

Jerry Colangelo
FORMER NBA SCOUT

Jerry Colangelo began his career in sports management as the head scout and director of merchandising for the Chicago Bulls in 1966. Then, in 1987 he bought the Phoenix Suns and has since served as president and CEO of the franchise.

A professional sports scout evaluates the athletic skills of athletes to determine their potential. A person who is interested in becoming a professional sports scout should consider obtaining a college degree or attending a special school for the chosen sport. A good understanding of mathematics, especially statistics, is necessary in analyzing data on specific players.

For more information:
Athletic Institute
200 Castlewood Drive
North Palm Beach, FL 33408

www.glencoe.com/sec/math/mac/mathnet

Your Turn
- Choose three college basketball players who play the same position. Find and list their current statistics. If you were an NBA scout, which of these players would you encourage your team to draft? Write a paragraph describing why the player you have chosen would be the best for the team. Be sure to discuss the comparison of the statistics that you found.

I love basketball! I think it would be great to be a scout for a professional team!

More About Jerry Colangelo
- Mr. Colangelo had 66 scholarship offers for college basketball and seven for professional baseball contracts.
- He graduated with a Bachelor of Arts degree in 1962 from the University of Illinois, where he earned All-Big Ten honors, averaging 15 points per game.
- Mr. Colangelo is also president and CEO of the Phoenix Mercury (WNBA) and managing general partner of the Arizona Diamondbacks (major league baseball). He helped relocate the Winnipeg Jets (NHL) to Phoenix, where they were renamed the Phoenix Coyotes.

13-6 Multiplying Binomials

What you'll learn
You'll learn to multiply binomials by using algebra tiles.

When am I ever going to use this?
Knowing how to multiply binomials is useful in landscaping and building design.

Word Wise
binomial

Did you know? The first modern bicycle appeared in England in 1885. It was called the Rover safety bicycle.

What do the words *bicycle* and *binomial* have in common? They both begin with the prefix *bi-*, meaning two. A bicycle is a cycle with two wheels. A **binomial** is a polynomial with two terms. Some examples of binomials are $x + 2$, $3y - 4$, and $r + s$. You can find the product of simple binomials by using algebra tiles.

HANDS-ON MINI-LAB

Work with a partner. □ algebra tiles □ product mat

Try This
Find $(x + 1)(x + 3)$.

- Make a rectangle with a width of $x + 1$ and a length of $x + 3$. Use algebra tiles to mark off the dimensions on a product mat.
- Using the marks as a guide, fill in the rectangle with algebra tiles.

Talk About It
1. How is this method like the method you used when multiplying a monomial by a polynomial? **It is the same.**
2. What is $(x + 1)(x + 3)$? $x^2 + 4x + 3$

You can also use the distributive property to find the product of two binomials. The figure at the right shows the rectangle from the Mini-Lab, separated into four parts. Notice that each term from the first parentheses $(x + 1)$ is multiplied by each term from the second parentheses $(x + 3)$.

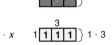

Example

1 Find $(y + 2)(2y + 1)$.

Method 1 Use models.

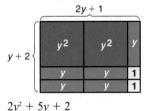

$2y^2 + 5y + 2$

Method 2 Use the distributive property.

$(y + 2)(2y + 1) = y(2y + 1) + 2(2y + 1)$
$= 2y^2 + 1y + 4y + 2$
$= 2y^2 + 5y + 2$ *Simplify.*

Lesson 13-6 Multiplying Binomials **583**

Multiple Learning Styles

Verbal/ Linguistic Ask students to write a problem for each condition.
- A polynomial plus a polynomial is a polynomial.
- A polynomial minus a polynomial is a polynomial.
- A monomial times a monomial is a monomial.
- A binomial times a binomial is a polynomial.

13-6 Lesson Notes

Instructional Resources
- *Study Guide Masters*, p. 105
- *Practice Masters*, p. 105
- *Enrichment Masters*, p. 105
- Transparencies 13-6, A and B
- *Diversity Masters*, p. 39
- *Assessment and Evaluation Masters*, p. 352
- CD-ROM Program
  - Resource Lesson 13-6

Recommended Pacing	
Honors	Day 6 of 10

1 FOCUS

5-Minute Check
(Lesson 13-5)

Find each product. Express the answer in exponential form.
1. $5^3 \cdot 5^6$ 5^9
2. $z^5 \cdot z^2$ z^7

Find each product.
3. $y(y^2 + 5)$ $y^3 + 5y$
4. $b^2(3b - 6)$ $3b^3 - 6b^2$
5. $a^5(a^2 + 13)$ $a^7 + 13a^5$

The 5-Minute Check is also available on **Transparency 13-6A** for this lesson.

Motivating the Lesson

Hands-On Activity On the chalkboard, show how 2-digit numbers, such as 43 and 12, are multiplied. That is, stress that each digit in the first number is multiplied by each digit in the second.

2 TEACH

Transparency 13-6B contains a teaching aid for this lesson.

Using the Mini-Lab Some students may have difficulty visualizing which tiles to use to complete the product. Suggest that students draw vertical and horizontal dashed lines at increment marks to more clearly define which tiles are needed.

Lesson 13-6 **583**

In-Class Examples

For Example 1
Find $(3x + 2)(x + 1)$.
$3x^2 + 5x + 2$

For Example 2
Find $(n + 6)(n + 3)$.
$n^2 + 9n + 18$

For Example 3
Find $(3a - 4)(2a + 1)$.
$6a^2 - 5a - 4$

For Example 4
The average rate of depreciation for a car is d. For every dollar of purchase price, the value of a car after 2 years is given by the formula $V = (1 - d)(1 - d)$. Write the formula in simplest form. $V = 1 - 2d - d^2$

3 PRACTICE/APPLY

Check for Understanding

If students need additional practice or instruction after completing Exercises 1–7, one of these options may be helpful.
- Extra Practice, see p. 643
- Reteaching Activity
- *Study Guide Masters*, p. 105
- *Practice Masters*, p. 105

Study Guide Masters, p. 105

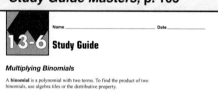

584 Chapter 13

Example 2 Find $(x + 3)(x + 2)$.

$(x + 3)(x + 2) = x(x + 2) + 3(x + 2)$
$= x^2 + 2x + 3x + 6$
$= x^2 + 5x + 6$ *Simplify.*

You can also multiply binomials vertically.

Examples 3 Find $(3t + 1)(2t + 3)$.

$(3t + 1)(2t + 3) \rightarrow$
$$\begin{array}{r} 2t + 3 \\ \times\ 3t + 1 \\ \hline 2t + 3 \\ 6t^2 + 9t \\ \hline 6t^2 + 11t + 3 \end{array}$$
Multiply as with whole numbers.
Multiply by 1.
Multiply by 3t.
Add like terms.

APPLICATION

4 Money Matters Corey opened a savings account paying an interest rate of r. For each dollar deposited, the amount in the account after two years is given by the formula $A = (1 + r)(1 + r)$. Write the formula in simplest form.

$A = (1 + r)(1 + r)$
$A = 1 \cdot (1 + r) + r \cdot (1 + r)$
$A = 1 + r + r + r^2$
$A = 1 + 2r + r^2$ *Simplify.*

CHECK FOR UNDERSTANDING

Communicating Mathematics

Read and study the lesson to answer each question.

1. **Tell** which of the following are binomials. **b, c**
   a. $2x$ b. $x + 5$ c. $3y^2 + 4x$ d. $x^2 + 2x + 8$

2. **Write** the product shown at the right.
   $2x^2 + 7x + 6$

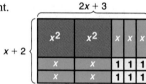

HANDS-ON MATH

3. **Draw** a rectangle to model $(2x + 2)(2x + 3)$. Then write the product.
   See Answer Appendix for model; $4x^2 + 10x + 6$.

Guided Practice

4. Find $(x + 1)(x + 2)$. $x^2 + 3x + 2$

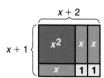

Find each product. Use drawings or algebra tiles if necessary.

5. $(x + 3)(x - 1)$ $x^2 + 2x - 3$

6. $(3a + 1)(2a + 3)$ $6a^2 + 11a + 3$

584 Chapter 13 Algebra: Exploring Polynomials

■ Reteaching the Lesson ■

Activity Write a binomial on each of 12 index cards. Have students work in pairs. One student draws two cards and models the product using algebra tiles. The second student finds the product using the distributive property. Students then compare answers.

Have each student bring in a cardboard model of the tile used to measure the area. You may wish to have the students color their tiles with colorful paper and assemble a mosaic on the floor or a wall.

11. $b^2 + 10b + 24$
12. $2x^2 - x - 3$
13. $4m^2 + 6m + 2$

Practice
14. $2x^2 + 11x + 5$
15. $d^2 + 6d + 9$
16. $3z^2 + 5z - 2$

Family Activity
Find a rectangular area in your home or a friend's home that has square tiles. Sketch the area. Then record the dimensions of the area using x to represent the length of a side of one tile.

Applications and Problem Solving

21. See Answer Appendix.

Mixed Review

7. **Geometry** A square has dimensions of x feet by x feet. A rectangle is 4 feet longer and 3 feet wider than the square. Find the area of the rectangle. $x^2 + 7x + 12$

EXERCISES

Find each product.

8. $(x + 3)(x + 3)$

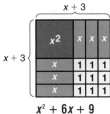

$x^2 + 6x + 9$

9. $(2x + 1)(x + 3)$

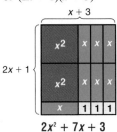

$2x^2 + 7x + 3$

10. $(x + 4)(2x + 1)$

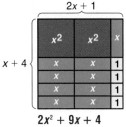

$2x^2 + 9x + 4$

Find each product. Use drawings or algebra tiles if necessary.

11. $(b + 6)(b + 4)$
12. $(x + 1)(2x - 3)$
13. $(2m + 2)(2m + 1)$
14. $(2x + 1)(x + 5)$
15. $(d + 3)(d + 3)$
16. $(3z - 1)(z + 2)$

17. Find the product of $(3a + 1)$ and $(a - 3)$. $3a^2 - 8a - 3$
18. Multiply $(z - 5)$ by $(2z + 4)$. $2z^2 - 6z - 20$

19. **Recreation** Refer to the diagram of the pool area.
    a. Find the area of the pool. $x^2 + 14x + 40$
    b. Find the area of the cement path around the pool. $9x^2 + 2x - 48$

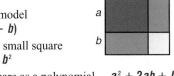

20. **Geometry** The model represents the square of a binomial.
    a. What product does this model represent? $(a + b)(a + b)$
    b. What is the area of each small square and rectangle? a^2, ab, b^2
    c. Write the area of the square as a polynomial. $a^2 + 2ab + b^2$

21. **Working on the CHAPTER Project** Refer to the model you drew in Exercise 34 on page 564. Write a polynomial represented by the model. (*Hint*: hm and mh are like terms.) Does this model work for all players? Explain.

22. **Critical Thinking** The length and width of a rectangle are $3x + 1$ and $x + 4$. Give the dimensions of a rectangle that has twice the area. Check your answer by finding each area. See students' work; $(6x + 2)(x + 4)$ or $(3x + 1)(2x + 8)$; $6x^2 + 26x + 8$

23. **Algebra** Find $t(t - 4)$. (*Lesson 13-5*) $t^2 - 4t$

24. **Test Practice** Which of the following letters does not have rotational symmetry? (*Lesson 10-9*) **C**

    A H B I C M D O

25. Find the LCM of 18, 34, and 6. (*Lesson 6-7*) 306

Lesson 13-6 Multiplying Binomials **585**

Extending the Lesson

Enrichment Masters, p. 105

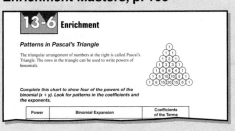

Activity Have students work in small groups to find the volume of this rectangular prism.

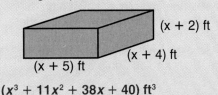

$(x^3 + 11x^2 + 38x + 40)$ ft³

Assignment Guide

Core: 9–19 odd, 22–25
Enriched: 8–18 even, 19, 20, 22–25

CHAPTER Project

Exercise 21 asks students to advance to the next stage of work on the Chapter Project. You may wish to have students work in pairs to write their polynomials.

4 ASSESS

Closing Activity

Speaking Have students identify the product represented below.

x^2	x	x	x
x^2	x	x	x
x	1	1	1

$(2x + 1)(x + 3) = 2x^2 + 7x + 3$

Chapter 13, Quiz C (Lessons 13-5 and 13-6) is available in the *Assessment and Evaluation Masters*, p. 352.

Practice Masters, p. 105

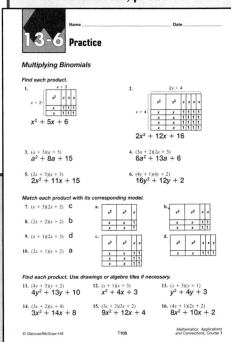

Lesson 13-6 **585**

13-7A Thinking LAB Notes

Objective Students solve problems by using guess and check.

Recommended Pacing	
Honors	Day 7 of 10

1 FOCUS

Getting Started Ask students to solve this problem: *Austin wants a baseball mitt that is available in a promotion for 275 points. If a 2-liter bottle of soda is worth 5 points, how many bottles will Austin need to collect?* 55

2 TEACH

Teaching Tip Use empty bottles and 6-pack rings to illustrate how to check the first guess. Have each bottle represent 10 bottles and each set of rings represent 10 six-packs. Then show how a diagram can be used to represent the bottles and six-packs, and use it to check the second guess. Have students make a third guess and check it mentally.

In-Class Example
A football team scored 36 points on a total of 8 touchdowns and field goals. A touchdown (plus the extra point) counts for 7 points and a field goal counts for 3 points. How many touchdowns and how many field goals did the team score?
3 touchdowns, 5 field goals

THINKING LAB PROBLEM SOLVING

13-7A Guess and Check
A Preview of Lesson 13-7

A soft drink company is offering free bikes to people who collect enough points by buying bottles or cans of soda. Ramón and Sheila are figuring out how they can get a free bike. Let's listen in!

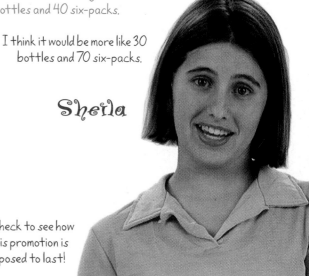

Ramón: To get a free bike, all you have to do is save up 915 points.

Sheila: A two-liter bottle is worth 5 points and a six-pack of soda is worth 10 points. So how much soda do we need to buy?

Ramón: It will be a lot — I would guess around 40 bottles and 40 six-packs.

Sheila: I think it would be more like 30 bottles and 70 six-packs.

Ramón: Let's see who's guess is closer. Forty times 5 is 200 and 40 times 10 is 400. That's only 600 points total!

Ramón: Your guess would give us 30 times 5, which is 150, plus 70 times 10, which is 700. That's still only 850 points. You're closer, but I don't know if we'll ever be able to drink all that soda.

Sheila: We better check to see how long this promotion is supposed to last!

THINK ABOUT IT

Work with a partner. 1–2. See margin.

1. **Compare and contrast** Sheila's and Ramón's guesses. How would you have guessed differently?

2. **Use** the **guess–and–check** strategy to find a better estimate of how much soda you would need to drink to win a bike.

3. **Apply** what you have learned to solve the following problem.
   The product of two consecutive odd integers is 899. What are the numbers? **29, 31**

586 Chapter 13 Algebra: Exploring Polynomials

■ Reteaching the Lesson ■

Activity One of three players secretly chooses a number from 1 to 100. A second player makes successive guesses (the first player says "high" or "low" after each guess) until finding the number. The third player counts the guesses, the total being the guesser's score. Switch roles and play again.

Additional Answers

1. Sample answer: Ramón guessed an equal number of bottles and six-packs; Sheila guessed more than twice as many six-packs than bottles.

2. Sample answer: 43 bottles and 70 six-packs would make 915 points.

ON YOUR OWN

4. The third step of the 4-step plan for problem solving asks you to *solve* the problem. *Explain* how you can use guess and check to help you find factors of a polynomial. **See students' work.**

5. *Write a Problem* that you can solve by using the guess-and-check method. Explain your answer. **See students' work.**

6. *Look Ahead* Use guess and check to find the factors of $x^2 + 5x + 4$. **$(x + 1)(x + 4)$**

MIXED PROBLEM SOLVING

STRATEGIES
Look for a pattern.
Solve a simpler problem.
Act it out.
Guess and check.
Draw a diagram.
Make a chart.
Work backward.

Solve. Use any strategy.

7. *Framing* Forty-two inches of molding are needed to frame a picture. If molding sells for $5.80 a foot, how much will the molding cost? **$20.30**

8. *Critical Thinking* Edgar's mother is five times as old as Edgar. Five years from now she will be just three times as old as he is. How old is Edgar now? **5**

9. *Business* Nami volunteers to get lunch for members of the Ecology Club. Their order of 13 burgers, 7 shakes, and 10 French fries is represented by the expression $13b + 7s + 10f$. Determine the cost of the order if burgers are $1.99 each, shakes are $1.49 each, and fries are $0.99. **$46.20**

10. *Number Theory* The product of a number and its next two consecutive whole numbers is 60. Use the guess-and-check strategy to find the number. **3**

11. *Money Matters* Stamps for postcards cost $0.20, and stamps for first-class letters cost $0.32. Louise wants to send postcards and letters to 11 friends. If she has $2.80 for stamps, how many postcards and how many letters can she send? **6 postcards, 5 letters**

12. *Retail Sales* In 1970, a desktop calculator sold for about $100. Today the same type of calculator sells for as little as $25. What is the percent of decrease of the cost of a desktop calculator? **75%**

13. **Test Practice** Teenagers were asked which they spent more time using, their computer, their video game system, or both equally. The graph shows the results of the survey.

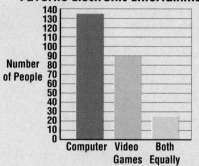

Favorite Electronic Entertainment

Source: Consumer Electronics Manufacturers Association

What was the total number of people who were surveyed? **D**

A 135   B 140   C 200
D 250   E 420

14. **Test Practice** Which word, written as shown, does not have a vertical line of symmetry? **B**

A M   B V   C W   D H
  O   O   H   A
  T   T   Y   M
  H   E

Lesson 13-7A THINKING LAB 587

■ Extending the Lesson ■

Activity An *automorphic* number is a number whose square ends with the number itself. The number 25 is automorphic because $25 \times 25 = 6\underline{25}$. Have small groups use their calculators and the guess-and-check strategy to find all the automorphic numbers less than 25.

3 PRACTICE/APPLY

Check for Understanding
Use the results from Exercise 3 to determine whether students comprehend how to use the guess-and-check strategy to solve problems.

Extra Practice If students need additional practice in problem solving, extra practice is available on the following pages.
- Guess and Check, see p. 644
- Mixed Problem Solving, see pp. 645–646

Assignment Guide

All: 4–14

4 ASSESS

Closing Activity
Modeling Give students a collection of play dimes and quarters. Ask them to use the guess-and-check strategy to find 7 coins whose total worth is $1.00. **5 dimes, 2 quarters**

Thinking Lab 13-7A **587**

13-7 Lesson Notes

Instructional Resources
- *Study Guide Masters*, p. 106
- *Practice Masters*, p. 106
- *Enrichment Masters*, p. 106
- Transparencies 13-7, A and B
- *Assessment and Evaluation Masters*, p. 352
- *Classroom Games*, pp. 43–45
- *Technology Masters*, p. 78
- CD-ROM Program
  - Resource Lesson 13-7

Recommended Pacing	
Honors	Day 8 of 10

1 FOCUS

5-Minute Check (Lesson 13-6)

Find each product. Use drawings or algebra tiles if necessary.

1. $(x + 2)(x + 3)$ $x^2 + 5x + 6$
2. $(x + 2)(x + 6)$ $x^2 + 8x + 12$
3. $(x + 3)(2x + 3)$
   $2x^2 + 9x + 9$
4. $(2x + 1)(3x + 2)$
   $6x^2 + 7x + 2$
5. $(t + 4)(t + 4)$ $t^2 + 8t + 16$

 The 5-Minute Check is also available on **Transparency 13-7A** for this lesson.

Motivating the Lesson

Problem Solving Review common factors from Lesson 6-3. Then assist students in factoring the following polynomials, looking for the common factor of the terms.
a. $4x + 12$ $4(x + 3)$
b. $x^2 + 5x$ $x(x + 5)$

2 TEACH

 Transparency 13-7B contains a teaching aid for this lesson.

Using the Mini-Lab Encourage students to keep an orderly arrangement of tiles. This will permit them to find the proper rectangle more easily. Have students suggest methods for forming the rectangle logically rather then by trial and error.

13-7 Factoring Polynomials

What you'll learn
You'll learn to factor polynomials by using algebra tiles.

When am I ever going to use this?
Knowing how to factor polynomials will help you find possible dimensions of a rectangle if you know its area.

Word Wise
factoring
trinomial

1. 2 and $x + 3$; they are the length and width of the rectangle.
2. Build a rectangle with an area of $x^2 + 4x$.

Sometimes you know a product and want to find its factors. This is called **factoring**. You can use algebra tiles to factor polynomials. You know the area of a rectangle and need to find the length and width.

HANDS-ON MINI-LAB

Work with a partner.   algebra tiles

Try This
Factor $2x + 6$.
- Model the polynomial $2x + 6$.
- Try to form a rectangle with the tiles.
- Write an expression for the length and width.

Talk About It 3. No; see students' work for model.
1. What are the factors of $2x + 6$? How are these factors related to the length and width?
2. Explain how to factor $x^2 + 4x$.
3. Model the polynomial $x^2 + 3x + 5$. Can you form a rectangle?
4. What are the factors of $3x + 5$? 1 and $3x + 5$

You can also use algebra tiles to factor **trinomials**. A trinomial is a polynomial with three terms.

Examples

1 Factor $x^2 + 3x + 2$.

Model the polynomial.   Try to form a rectangle with the tiles.

The rectangle has a width of $x + 1$ and a length of $x + 2$. Therefore, $x^2 + 3x + 2 = (x + 1)(x + 2)$.

2 Factor $x^2 + 2x + 2$.

Model the polynomial.

Try to form a rectangle with the tiles. You cannot form a rectangle with the tiles. Therefore, $x^2 + 2x + 2$ cannot be factored.

588 Algebra: Exploring Polynomials

You can also use the guess-and-check strategy to find the factors of a trinomial. Look at $(x + 3)(x + 5) = x^2 + 8x + 15$. Notice that the product of the last terms of each polynomial, 3 and 5, equals the last term in the trinomial, 15. You can use this pattern to factor polynomials.

Examples

3 Factor $x^2 + 12x + 20$.

$x^2 + 12x + 20 = (x + \underline{?})(x + \underline{?})$ *Find two integers whose product is 20.*

1st Guess: $(x + 1)(x + 20)$ *Use 1 and 20 since $1 \times 20 = 20$.*

Check: $(x + 1)(x + 20) = x(x + 20) + 1(x + 20)$
$= x^2 + 20x + x + 20$
$= x^2 + 21x + 20$ This is not correct.

2nd Guess: $(x + 2)(x + 10)$ *Use 2 and 10 since $2 \times 10 = 20$.*

Check: $(x + 2)(x + 10) = x(x + 10) + 2(x + 10)$
$= x^2 + 10x + 2x + 20$
$= x^2 + 12x + 20$ ✓ This is correct.

Therefore, $x^2 + 12x + 20 = (x + 2)(x + 10)$.

INTEGRATION

4 **Geometry** The area of a rectangle is $2x^2 + 6x + 4$. Find polynomials that might represent the dimensions of the rectangle.

Area = $2x^2 + 6x + 4$

Explore You know the area of the rectangle. You need to find the length and the width.

Plan Use the formula $\ell \times w = A$. Substitute $2x^2 + 6x + 4$ for A and find the factors.

Solve $\ell \times w = A$
$\ell \times w = 2x^2 + 6x + 4$
$\ell \times w = (2x + 4)(x + 1)$ *Use guess and check to factor the polynomial.*

So, the dimensions of the rectangle could be $(2x + 4)$ and $(x + 1)$.

Examine $(2x + 4)(x + 1) = 2x(x + 1) + 4(x + 1)$
$= 2x^2 + 2x + 4x + 4$
$= 2x^2 + 6x + 4$ ✓

The dimensions could also be $(x + 2)$ and $(2x + 2)$.

In-Class Examples

For Example 1
Factor $x^2 + 6x + 8$.
$(x + 2)(x + 4)$

For Example 2
Factor $4x^2 + 3x + 5$.
not factorable

For Example 3
Factor $3x^2 + 22x + 24$.
$(x + 6)(3x + 4)$

For Example 4
The area of a rectangle is $12x^2 + 22x + 6$. Find the dimensions of the rectangle.
$(6x + 2)(2x + 3)$ or $(3x + 1)(4x + 6)$

Teaching Tip Encourage students to check the factoring process by multiplying the factors.

3 PRACTICE/APPLY

Check for Understanding
If students need additional practice or instruction after completing Exercises 1–8, one of these options may be helpful.
- Extra Practice, see p. 644
- Reteaching Activity
- *Study Guide Masters*, p. 106
- *Practice Masters*, p. 106

CHECK FOR UNDERSTANDING

Communicating Mathematics

Read and study the lesson to answer each question. 1–2. See margin.

1. **Tell** how you know when a polynomial can be factored.
2. **You Decide** Tamika said that the factors of $x^2 + 3x + 9$ are $x + 3$ and $x + 3$. Is she correct? Explain.

Lesson 13-7 Factoring Polynomials **589**

■ Reteaching the Lesson ■

Activity Draw a parallel between the polynomial $x^2 + 5x + 6$ and composite numbers, both of which can be factored. Ask students to name the type of number to which the polynomial $x^2 + x + 1$ is similarly related. **prime number**

Additional Answers

1. Sample answer: You can form a rectangle with the algebra tiles.
2. No, $x + 3$ and $x + 3$ are the factors of $x^2 + 6x + 9$.

Assignment Guide

Core: 9–23 odd, 26–29
Enriched: 10–22 even, 23, 24, 26–29

Additional Answers

3a.

x^2	x^2	x
x	x	1
x	x	1

3b. $2x + 1$ and $x + 2$; These are the length and the width of the rectangle.

Study Guide Masters, p. 106

590 Chapter 13

Guided Practice

3a. *Model* the polynomial $2x^2 + 5x + 2$ and try to form a rectangle.
b. What are the factors? Explain how you know.
a–b. See margin.

4. Factor $x^2 + 5x + 6$. $(x + 3)(x + 2)$

If possible, factor each polynomial. Use drawings or algebra tiles if necessary.   **6.** $(x + 2)(x + 4)$ **7.** not factorable

5. $x^2 + 4x$ $x(x + 4)$ **6.** $x^2 + 6x + 8$ **7.** $x^2 + x + 3$

8. *Interior Design* When placing a rug under a dining room table, the rug should be 6 feet longer and 6 feet wider than the table to allow for pulling out chairs. A square table measures x feet on each side. Will a rug with an area of $x^2 + 12x + 36$ be suitable for underneath the table? Explain. (*Hint:* Factor the polynomial.)   **Yes, the factors are $(x + 6)(x + 6)$, so the rug's length and width are each 6 feet longer than the table's length and width.**

EXERCISES

Practice   Factor each polynomial.

9. $3x + 6$

$3(x + 2)$

10. $x^2 + 6x + 5$

$(x + 5)(x + 1)$

11. $2x^2 + 5x + 2$

$(2x + 1)(x + 2)$

If possible, factor each polynomial. Use drawings or algebra tiles if necessary.

15. $(x + 4)(x + 1)$
16. not factorable
17. $(x + 4)(x + 4)$
18. not factorable
19. $(2x + 1)(x + 3)$

12. $5x + 5$ $5(x + 1)$
13. $18x + 6$ $6(3x + 1)$
14. $3x^2 + 6x$ $3x(x + 2)$
15. $x^2 + 5x + 4$
16. $x^2 + 2x + 3$
17. $x^2 + 8x + 16$
18. $2x^2 + 2x + 3$
19. $2x^2 + 7x + 3$
20. $2x^2 + 7x + 6$
  $(2x + 3)(x + 2)$
21. Find the factors of $3x^2 + 4x$. $x(3x + 4)$
22. Factor the polynomial $x^2 + 7x + 12$. $(x + 3)(x + 4)$

Applications and Problem Solving

23. *Algebra* Simplify $\dfrac{2x^2 + 11x + 5}{x^2 + 7x + 10}$ by factoring the numerator and the denominator. $\dfrac{2x + 1}{x + 2}$

24. *Geometry* Find the length and width for a rectangle of area $3x^2 + 8x + 5$. $(3x + 5)(x + 1)$

590 Algebra: Exploring Polynomials

25a. $h^2 + 2mh + m^2 = (h + m)(h + m)$; $h + m = 1$

25b. $2 \cdot h^2 + 1 \cdot 2mh + 0 \cdot m^2 = 2h^2 + 2mh$

25c. See students' work.

Mixed Review

25. **Working on the CHAPTER Project** Refer to Exercise 21 on page 585.
   a. Factor the polynomial. What must be true of $h + m$?
   b. A player scores 2 points if 2 shots are made, 1 point if 1 is made, and 0 points if both are missed. Find the number of expected points for a player. To do this, multiply the term for 2 shots made by 2, the term for 1 made by 1, and the term for both missed by 0. Write a polynomial for the number of points a player can be expected to score.
   c. Evaluate the polynomial from part b for each of your players.

26. **Critical Thinking** For the polynomial $2x^2 + 7x + \underline{\ ?\ }$, write a positive integer in the blank that makes the polynomial factorable.
   Sample answer: 3

27. **Algebra** Find $(2k - 1)(3k + 4)$. *(Lesson 13-6)* $6k^2 + 5k - 4$

28. **Test Practice** Which would be the most precise measurement for a box of cereal? *(Lesson 11-7)* **D**
   A 1 pound   B 53 pounds
   C 510 grams D 538.5 grams

29. **Food** Three-fourths of a pan of peach cobbler is to be divided equally among 6 people. What part of the cobbler will each person receive? *(Lesson 7-8)* $\frac{1}{8}$

Let the Games Begin
Factor Challenge

Get Ready This game is for four players. • stopwatch • scissors
   • algebra tiles   • product mat   • 10 index cards

Get Set Cut each index card in half to make 20 playing cards. Each player should write a polynomial on each of 5 cards that can be modeled with algebra tiles. Four of the polynomials should be factorable, and 1 should be not factorable.

Go
• Mix the cards and place the stack facedown on the table.
• The first player turns over the top card and lays it on the table. The player then has two minutes to model the polynomial with algebra tiles, form a rectangle, and name the factors or declare the polynomial unfactorable. If the player is correct, he or she scores one point. The card is placed on the bottom of the stack and it becomes the next player's turn.
• The first player to score five points wins.

interNET CONNECTION Visit www.glencoe.com/sec/math/mac/mathnet for more games.

Lesson 13-7 Factoring Polynomials **591**

■ **Extending the Lesson** ■

Enrichment Masters, p. 106

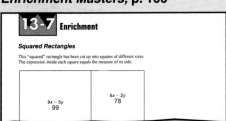

Let the Games Begin

Point out that students can write their factorable polynomials by multiplying binomials with other binomials or monomials. Students may wish to use the guess-and-check strategy to write polynomials that are not factorable.

CHAPTER Project

Exercise 25 asks students to advance to the next stage of work on the Chapter Project. You may wish to have students organize a table with column headings h^2, mh, and m^2, and players' names as row headings before writing their polynomials.

4 ASSESS

Closing Activity

Speaking Have students name the polynomial modeled below and its factors. $x^2 + 2x + 1$; $(x + 1)(x + 1)$

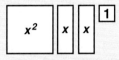

Chapter 13, Quiz D (Lesson 13-7) is available in the *Assessment and Evaluation Masters*, p. 352.

Practice Masters, p. 106

Lesson 13-7 **591**

Study Guide and Assessment

Vocabulary

This section provides a listing of the new terms, properties, and phrases that were introduced in this chapter. Have students define each term and provide an example or two of it, if appropriate.

Understanding and Using the Vocabulary

These exercises check students' understanding of the terms by using a variety of verbal formats including matching, completion, and true/false.

Glossaries A complete glossary of terms appears on pages 702–710. The glossary also appears in Spanish on pages 711–721.

CHAPTER 13 Study Guide and Assessment

Vocabulary

After completing this chapter, you should be able to define each term, concept, or phrase and give an example or two of each.

Algebra
binomial (p. 583)
factoring (p. 588)
like terms (p. 565)
monomial (p. 561)
polynomial (p. 562)
simplest form (p. 565)

term (p. 565)
trinomial (p. 588)
zero pair (p. 566)

Problem Solving
guess and check (p. 586)

Understanding and Using the Vocabulary

State whether each sentence is *true* or *false*. If false, replace the underlined word to make a true sentence.

1. The expression $b^2 - 3b$ is an example of a <u>monomial</u>.   **false; binomial**
2. A polynomial is the sum or <u>difference</u> of two or more monomials.   **true**
3. The <u>additive</u> inverse of $9y^2 - 5y + 2$ is $-9y^2 + 5y - 2$.   **true**
4. A polynomial with two terms is called a <u>binomial</u>.   **true**
5. The product of $2m$ and $m^2 + 8m$ will have <u>three</u> terms.   **false; two**
6. When <u>factoring</u>, you combine like terms to find the simplest form.   **false; simplifying**

In Your Own Words

7. *Write* the definition of a monomial in your own words.   **Sample answer: A monomial is a number, a variable, or a product of a number and one or more variables.**

Objectives & Examples

Upon completing this chapter, you should be able to:

• represent polynomials with algebra tiles
  (Lesson 13-1)
  Model $x^2 + 3x + 4$.

Review Exercises

Use these exercises to review and prepare for the chapter test.

Model each monomial or polynomial using drawings or algebra tiles. 8–11. See Answer Appendix.
8. $2x$
9. $3x^2 - 4$
10. $-2x^2 + 2x - 6$
11. $x^2 - 7$

592 Chapter 13 Algebra: Exploring Polynomials

MindJogger Videoquizzes

MindJogger Videoquizzes provide an alternative review of concepts presented in this chapter. Students work in teams to answer questions, gaining points for correct answers. The questions are presented in three rounds.
Round 1 Concepts–5 questions
Round 2 Skills–4 questions
Round 3 Problem Solving–4 questions

Study Guide and Assessment — Chapter 13

Objectives & Examples

- **simplify polynomials by using algebra tiles**
  (Lesson 13-2)
  Simplify $2x^2 + 6 + x^2$.

  $2x^2 + 6 + x^2 = 3x^2 + 6$

Review Exercises

Simplify each polynomial. Use drawings or algebra tiles if necessary.

12. $3m^2 + 3m + 10m^2 + 2m$ $13m^2 + 5m$
13. $8p - 3p - 12$ $5p - 12$
14. $12x - 2x^2 - 9x + 6x^2$ $4x^2 + 3x$
15. $5a + 7b - a - 8b$ $4a - b$

Objectives & Examples

- **add polynomials by using algebra tiles**
  (Lesson 13-3)
  Find $(2x^2 + 6x) + (3x^2 - 2x)$.

  zero pairs

  $2x^2 + 6x + 3x^2 - 2x = 5x^2 + 4x$

Find each sum. Use drawings or algebra tiles if necessary.

16. $(10m^2 - 3m) + (3m^2 + 2m)$ $13m^2 - m$
17. $(4d + 1) + (8d + 6)$ $12d + 7$
18. $(3a^2 + 6a) + (2a^2 - 5a)$ $5a^2 + a$
19. $(b^2 - 2b + 4) + (2b^2 + b - 8)$ $3b^2 - b - 4$
20. $(2x^2 - 7x) + (x^2 + 5x)$ $3x^2 - 2x$

Objectives & Examples

- **subtract polynomials by using algebra tiles**
  (Lesson 13-4)
  Find $(6x^2 + 3) - (2x^2 - 1)$.

  zero pair

  $(6x^2 + 3) - (2x^2 - 1) = 4x^2 + 4$

Find each difference. Use drawings or algebra tiles if necessary.

21. $(7g + 2) - (5g + 1)$ $2g + 1$
22. $(3c - 7) - (-3c + 4)$ $6c - 11$
23. $(2s^2 + 8) - (s^2 + 3)$ $s^2 + 5$
24. $(6k^2 - 3) - (k^2 - 5k - 2)$ $5k^2 + 5k - 1$
25. $(7p^2 + 2p - 5) - (4p^2 + 6p - 2)$
    $3p^2 - 4p - 3$

Objectives & Examples

This section reviews the skills and concepts of the chapter and shows completely worked examples.

Review Exercises

These exercises provide practice for the corresponding objectives.

Assessment and Evaluation Masters, pp. 339–340

Assessment and Evaluation

Six forms of Chapter 13 Test are available in the *Assessment and Evaluation Masters* as shown in the chart.

Chapter 13 Test, Form 1B, is shown at the right. Chapter 13 Test, Form 2B, is shown on the next page.

1A	Multiple Choice	Honors
1B	Multiple Choice	Average
1C	Multiple Choice	Basic
2A	Free Response	Honors
2B	Free Response	Average
2C	Free Response	Basic

Chapter 13 Study Guide and Assessment

Objectives & Examples | Review Exercises

● multiply monomials and polynomials
(Lesson 13-5)

Find $b^4 \cdot b^7$.

$b^4 \cdot b^7 = b^{4+7}$ or b^{11}

Find $2x(2x + 3)$.

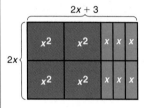

$2x(2x + 3) = 4x^2 + 6x$

Find each product. Express the answer in exponential form.

26. $7 \cdot 7^8$ 7^9
27. $12^4 \cdot 12^9$ 12^{13}
28. $x^5 \cdot x^2$ x^7
29. $d^3 \cdot d^7$ d^{10}

Find each product.
30. $y(2y + 3)$ $2y^2 + 3y$
31. $3z(z - 4)$ $3z^2 - 12z$
32. $c^3(4c^2 + 1)$ $4c^5 + c^3$
33. $2t^4(t^3 + 4)$ $2t^7 + 8t^4$

● multiply binomials by using algebra tiles
(Lesson 13-6)

Find $(x + 3)(x + 3)$.

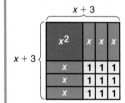

$(x + 3)(x + 3) = x^2 + 6x + 9$

Find each product. Use drawings or algebra tiles if necessary.

34. $(x + 2)(x + 3)$ $x^2 + 5x + 6$
35. $(x - 5)(2x + 1)$ $2x^2 - 9x - 5$
36. $(2x + 3)(3x + 1)$ $6x^2 + 11x + 3$
37. $(x + 3)(x - 4)$ $x^2 - x - 12$
38. $(4x + 1)(x + 1)$ $4x^2 + 5x + 1$

● factor polynomials by using algebra tiles
(Lesson 13-7)

Factor $x^2 + 4x + 4$.

Try to form a rectangle with algebra tiles.

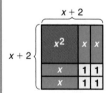

Therefore, $x^2 + 4x + 4 = (x + 2)(x + 2)$.

If possible, factor each polynomial. Use drawings or algebra tiles if necessary.

39. $14x + 7$ $7(2x + 1)$
40. $x^2 + 6x$ $x(x + 6)$
41. $x^2 + 7x + 10$ $(x + 5)(x + 2)$
42. $x^2 + 5x + 2$ not factorable
43. $2x^2 + 3x + 1$ $(2x + 1)(x + 1)$

Test and Review Software

You may use this software, a combination of an item generator and item bank, to create your own tests or worksheets. Types of items include free response, multiple choice, short answer, and open ended.

CD-ROM Program

The CD-ROM Program contains an Assessment Game whose questions review the concepts in this chapter.

Study Guide and Assessment — Chapter 13

Applications & Problem Solving

44. Gardening A diagram of Mr. Keith's flowerbed is shown below. Find its perimeter. *(Lesson 13-3)* **$(10x + 4)$ ft**

$(2x + 4)$ ft

$(3x - 2)$ ft

45. Interior Design Each side of a living room is $2x + 1$ yards long. Find the number of square yards of carpeting needed to cover the floor of the living room. *(Lesson 13-5)* **$(4x^2 + 4x + 1)$ yd^2**

46. Guess and Check Lou's house number consists of three digits. The digits in the number are consecutive and have a sum of 21. The first digit is smaller than the last digit. What is Lou's house number? *(Lesson 13-7A)* **678**

47. Money Matters Andrea borrowed $500 each year for technical school expenses. The amount she owes for the second year is $1{,}000 + 500r$, where r is the interest rate. The amount she owes for the third year is $1{,}500 + 1{,}500r + 500r^2$. How much did her debt increase between the second and third year? *(Lesson 13-4)* **$500 + 1{,}000r + 500r^2$**

Alternative Assessment

Performance Task

You are remodeling a room in your house. The original room was square. The remodeled room is a rectangle with an area of $x^2 + 8x + 15$ square feet. By how much did each dimension of the room increase? **3 ft and 5 ft**

Suppose the area of the remodeled room is 143 square feet. Can you find the actual dimensions of the remodeled room? If so, state the dimensions. Can you find the area of the original room? If so, state the area. **yes, 11 ft by 13 ft; yes, 64 sq ft**

A practice test for Chapter 13 is provided on page 659.

Completing the CHAPTER Project

Use the following checklist to make sure your poster is complete.

☑ The data for each player is clear and well organized.

☑ The drawings of your models are accurately drawn and labeled correctly.

☑ The geometric and algebraic models are accurate and correct.

 Select one of the problems you solved in this chapter and place it in your portfolio. Attach a note explaining how your problem illustrates one or more of the important concepts covered in this chapter.

Performance Assessment

Additional performance assessment tasks for this chapter are included in the *Assessment and Evaluation Masters* on page 349. A scoring guide is also provided on page 361.

Standardized Test Practice

The Standardized Test Practice may be used to help students prepare for standardized tests. The test items are written in the same style as those in state proficiency tests and standardized tests like CAT, CTBS, ITBS, MAT, SAT, and Terra Nova. The test items cover skills and concepts covered up to this point in the text.

The pages can be used as an overnight assessment. After students have completed the pages, discuss how each problem can be solved, or provide copies of the solutions from the *Solutions Manual*.

Assessment and Evaluation Masters, p. 355

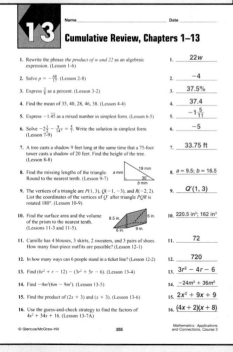

CHAPTERS 1-13

Standardized Test Practice
Assessing Knowledge & Skills

Section One: Multiple Choice

There are nine multiple choice questions in this section. Choose the best answer. If a correct answer is *not here*, choose the letter for Not Here.

1. An electronic store advertised 20% off the price of a $14 CD. To find the amount saved on the CD, multiply $14 by — **A**
   A 0.20.
   B 0.80.
   C 0.14.
   D 2.0.

2. A bagel shop offers 3 different cheeses and 3 different meats on 5 different types of bagels. If John chooses one bagel, one meat, and one cheese, how many different combinations can he order? **H**
   F 11
   G 35
   H 45
   J 55

3. Find the value of x for the given triangle. **C**

   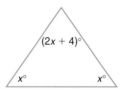

   A 24
   B 34
   C 44
   D 54

Please note that Questions 4–9 have five answer choices.

4. Jeremy was given an extra dollar on his paper route every time he sold 2 more subscriptions. Lori also earned an extra dollar every time she sold 2 subscriptions. The results of their sales are shown. How many dollars did Lori receive Monday through Wednesday? **F**

   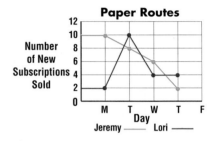

   F $8
   G $6
   H $4
   J $3
   K $2

5. Courtney owns a tent in the shape of a pyramid. The base of the tent is 8 feet by 10 feet. The tent is 7 feet tall at the center. How can you find the volume of the air inside the tent? **A**
   A Find $8 \times 10 \times 7$ and divide by 3.
   B Add $2\left(\frac{1}{2} \cdot 7 \cdot 8\right)$ and $2\left(\frac{1}{2} \cdot 7 \cdot 10\right)$.
   C Find $7 \cdot 8 \cdot 10$.
   D Find $\frac{1}{2} \cdot 7 \cdot 8 \cdot 10$.
   E Find $8 \times 10 \times 7$ and divide by 2.

596 Chapters 1–13 Standardized Test Practice

◀◀◀ **Instructional Resources**

Another cumulative review is shown at the left and is available in the *Assessment and Evaluation Masters*, p. 355.

Standardized Test Practice Chapters 1–13

6. During an overnight camping trip, the temperature dropped 4°F every hour. If the temperature was 56°F at midnight, what was the temperature at 5 A.M.? **G**
   - F 46°
   - G 36°
   - H −36°
   - J −46°
   - K Not Here

7. Kiar bought $12.48 worth of oranges at a fruit stand. If the oranges were $2.08 per pound, how many pounds were purchased? **C**
   - A 4 lb
   - B 5 lb
   - C 6 lb
   - D 7 lb
   - E Not Here

8. On Fernando's last paycheck, 28% was deducted for income tax, and 7.5% was deducted for social security tax. What is the total percent deducted from his pay for these taxes? **J**
   - F 20.5%
   - G 25.5%
   - H 100%
   - J 35.5%
   - K Not Here

9. Sandra bought $2\frac{1}{2}$ pounds of chocolate candy for $6.75. She also bought 2 pounds of popcorn for $3.24. What was the cost per pound of the chocolate candy? **D**
   - A $1.62
   - B $1.89
   - C $3.37
   - D $2.70
   - E Not Here

Section Two: Free Response

This section contains three questions for which you will provide short answers. Write your answers on your paper.

10. Which point is in the third quadrant? **K**

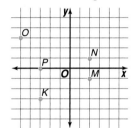

11. What is the probability of landing on an odd number? **0.5**

12. Identify the solution of the system of equations graphed below. **(3, 1)**

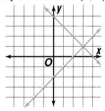

Student Handbook
Table of Contents

Extra Practice
    Basic Skills 600–604
    Lesson by Lesson 605–644
    Mixed Problem Solving 645–646

Chapter Tests 647–659

Getting Acquainted with the Graphing Calculator 660–661

Getting Acquainted with Spreadsheets 662–663

Selected Answers 664–699

Photo Credits 700–701

Glossary 702–710

Spanish Glossary 711–721

Index 722–731

Symbols, Formulas, and Measurement Conversions .... inside back cover

Basic Skills

Adding and Subtracting Decimals

1. 0.346
    -0.034
    $\overline{\mathbf{0.312}}$

2. 4.26
    $+0.72$
    $\overline{\mathbf{4.98}}$

3. 23.4
    -0.526
    $\overline{\mathbf{22.874}}$

4. 7.49
    -2.35
    $\overline{\mathbf{5.14}}$

5. 133.06
    -48.114
    $\overline{\mathbf{84.946}}$

6. 21.8
    $+434.64$
    $\overline{\mathbf{456.44}}$

7. 38.4
    $+0.77$
    $\overline{\mathbf{39.17}}$

8. 34.7
    $+31.3$
    $\overline{\mathbf{66.0}}$

9. 471.8
    -0.27
    $\overline{\mathbf{471.53}}$

10. 365.57
    -222.4
    $\overline{\mathbf{143.17}}$

11. 90
    -0.26
    $\overline{\mathbf{89.74}}$

12. 31.666
    -5.885
    $\overline{\mathbf{25.781}}$

13. 51.05
    -43.68
    $\overline{\mathbf{7.37}}$

14. 70.02
    -59.08
    $\overline{\mathbf{10.94}}$

15. 206.78
    -0.61
    $\overline{\mathbf{206.17}}$

16. 208.888
    $+2.2$
    $\overline{\mathbf{211.088}}$

17. 61.42
    $+0.77$
    $\overline{\mathbf{62.19}}$

18. 549.558
    -0.65
    $\overline{\mathbf{548.908}}$

19. 21
    -0.154
    $\overline{\mathbf{20.846}}$

20. 0.67
    $+0.63$
    $\overline{\mathbf{1.3}}$

21. $7.2 - 4.563$ **2.637**

22. $0.67 + 3.804$ **4.474**

23. $6.7 + 4.53$ **11.23**

24. $41.68 + 41.68$ **83.36**

25. $7 - 3.26$ **3.74**

26. $26.483 - 0.3$ **26.183**

27. $0.421 + 0.39$ **0.811**

28. $7.1 - 3.4$ **3.7**

29. $2.365 + 4.453$ **6.818**

30. $4.1 + 3.35$ **7.45**

31. $0.774 - 0.5$ **0.274**

32. $42.35 + 7.118$ **49.468**

33. $10 - 0.45$ **9.55**

34. $52.26 - 0.09$ **52.17**

35. $5.361 + 0.8$ **6.161**

36. $55.741 - 15.902$ **39.839**

37. $8 - 0.002$ **7.998**

38. $543.26 + 0.055$ **543.315**

39. $200.431 + 5.006$ **205.437**

40. $73.182 + 54.591$ **127.773**

41. $7.8 + 2.2$ **10**

42. $42 - 0.71$ **41.29**

43. $943.007 + 0.16$ **943.167**

44. $73.68 + 0.327$ **74.007**

45. $5.2 - 3.8$ **1.4**

46. $38.005 + 4.2$ **42.205**

47. $700.421 + 0.9$ **701.321**

48. $136.024 - 5.878$ **130.146**

49. $0.7 + 5.3$ **6**

50. $0.5 + 0.4$ **0.9**

Basic Skills

Multiplying and Dividing Decimals

1. 4.02
 × 0.16
 0.6432

2. $0.4\overline{)10.32}$ **25.8**

3. 6.3
 × 6.3
 39.69

4. $4.7\overline{)13.63}$ **2.9**

5. $4.32\overline{)2.592}$ **0.6**

6. 0.097
 × 8
 0.776

7. 22
 × 8.1
 178.2

8. $53.6\overline{)804}$ **15**

9. 2.16
 × 9.3
 20.088

10. 6.2
 × 9.4
 58.28

11. 400
 × 0.05
 20

12. $37.15\overline{)549.82}$ **14.8**

13. $10.72\overline{)269.072}$ **25.1**

14. $0.7\overline{)49}$ **70**

15. 23.6
 × 0.74
 17.464

16. $200\overline{)8.464}$ **0.04232**

17. 2.19
 × 3.55
 7.7745

18. $16.03\overline{)32.06}$ **2**

19. $6.43\overline{)0.02572}$ **0.004**

20. 6.73
 × 8.91
 59.9643

21. 4 × 0.4 **1.6**
22. 618 ÷ 0.4 **1,545**
23. 0.6 × 8 **4.8**
24. 2,276.485 ÷ 77.3 **29.45**

25. 60.125 ÷ 1.25 **48.1**
26. 2.44 × 2.22 **5.4168**
27. 0.1 × 17 **1.7**
28. 44.4 ÷ 222 **0.2**

29. 1.68 × 7 **11.76**
30. 525 ÷ 0.10 **5,250**
31. 0.01495 ÷ 2.99 **0.005**
32. 0.9 × 1 **0.9**

33. 0.7 × 23 **16.1**
34. 14.5 × 14.7 **213.15**
35. 2.209 ÷ 0.047 **47**
36. 18.8 ÷ 0.2 **94**

37. 1.375 ÷ 0.05 **27.5**
38. 65 ÷ 0.008 **8,125**
39. 900 × 0.9 **810**
40. 360 ÷ 1.8 **200**

41. 12 × 0.2 **2.4**
42. 0.39 × 0.5 **0.195**
43. 36 ÷ 0.036 **1,000**
44. 0.008 × 9 **0.072**

45. 324.52 ÷ 7.6 **42.7**
46. 29.1 × 82.7 **2,406.57**
47. 768.57 ÷ 41.1 **18.7**
48. 0.64 × 0.3 **0.192**

Basic Skills

Adding and Subtracting Fractions

1. $\frac{13}{6} - \frac{1}{6}$ 2
2. $\frac{1}{8} + \frac{1}{10}$ $\frac{9}{40}$
3. $\frac{3}{16} + \frac{1}{3}$ $\frac{25}{48}$
4. $\frac{3}{5} - \frac{1}{2}$ $\frac{1}{10}$

5. $\frac{3}{8} + \frac{1}{3}$ $\frac{17}{24}$
6. $\frac{4}{25} + \frac{7}{20}$ $\frac{51}{100}$
7. $\frac{29}{30} - \frac{7}{10}$ $\frac{4}{15}$
8. $\frac{6}{7} + \frac{2}{3}$ $1\frac{11}{21}$

9. $\frac{9}{14} - \frac{3}{14}$ $\frac{3}{7}$
10. $\frac{7}{8} - \frac{3}{4}$ $\frac{1}{8}$
11. $\frac{6}{7} - \frac{2}{3}$ $\frac{4}{21}$
12. $\frac{1}{3} - \frac{1}{4}$ $\frac{1}{12}$

13. $\frac{7}{12} + \frac{1}{3}$ $\frac{11}{12}$
14. $\frac{5}{7} - \frac{1}{7}$ $\frac{4}{7}$
15. $\frac{13}{18} - \frac{5}{9}$ $\frac{1}{6}$
16. $\frac{8}{9} - \frac{1}{9}$ $\frac{7}{9}$

17. $\frac{9}{100} + \frac{11}{100}$ $\frac{1}{5}$
18. $\frac{9}{10} - \frac{2}{5}$ $\frac{1}{2}$
19. $\frac{5}{12} + \frac{1}{3}$ $\frac{3}{4}$
20. $\frac{2}{5} + \frac{5}{14}$ $\frac{53}{70}$

21. $\frac{7}{12} - \frac{1}{3}$ $\frac{1}{4}$
22. $\frac{4}{5} - \frac{3}{5}$ $\frac{1}{5}$
23. $\frac{31}{36} - \frac{3}{4}$ $\frac{1}{9}$
24. $\frac{8}{21} + \frac{10}{21}$ $\frac{6}{7}$

25. $\frac{14}{15} - \frac{1}{15}$ $\frac{13}{15}$
26. $\frac{1}{7} + \frac{9}{28}$ $\frac{13}{28}$
27. $\frac{2}{5} + \frac{1}{10}$ $\frac{1}{2}$
28. $\frac{4}{11} + \frac{6}{11}$ $\frac{10}{11}$

29. $\frac{11}{60} - \frac{2}{15}$ $\frac{1}{20}$
30. $\frac{5}{21} + \frac{2}{21}$ $\frac{1}{3}$
31. $\frac{5}{6} - \frac{1}{2}$ $\frac{1}{3}$
32. $\frac{3}{8} + \frac{3}{8}$ $\frac{3}{4}$

33. $\frac{3}{8} - \frac{1}{4}$ $\frac{1}{8}$
34. $\frac{8}{9} + \frac{1}{9}$ 1
35. $\frac{5}{12} + \frac{1}{4}$ $\frac{2}{3}$
36. $\frac{5}{16} + \frac{1}{4}$ $\frac{9}{16}$

37. $\frac{7}{12} + \frac{1}{4}$ $\frac{5}{6}$
38. $\frac{3}{25} - \frac{1}{50}$ $\frac{1}{10}$
39. $\frac{1}{4} + \frac{1}{6}$ $\frac{5}{12}$
40. $\frac{89}{96} - \frac{11}{12}$ $\frac{1}{96}$

Basic Skills

Multiplying Fractions

1. $\frac{1}{4} \times \frac{1}{5}$ $\frac{1}{20}$
2. $\frac{2}{3} \times \frac{1}{2}$ $\frac{1}{3}$
3. $\frac{6}{7} \times \frac{7}{8}$ $\frac{3}{4}$
4. $\frac{1}{3} \times \frac{9}{10}$ $\frac{3}{10}$
5. $\frac{4}{5} \times \frac{3}{4}$ $\frac{3}{5}$
6. $80 \times \frac{7}{10}$ 56
7. $36 \times \frac{2}{9}$ 8
8. $2\frac{1}{3} \times \frac{3}{5}$ $1\frac{2}{5}$
9. $\frac{1}{8} \times 20$ $2\frac{1}{2}$
10. $\frac{1}{6} \times \frac{1}{4}$ $\frac{1}{24}$
11. $\frac{4}{5} \times \frac{12}{28}$ $\frac{12}{35}$
12. $\frac{16}{3} \times \frac{1}{8}$ $\frac{2}{3}$
13. $\frac{18}{8} \times \frac{4}{3}$ 3
14. $1\frac{3}{5} \times 15$ 24
15. $\frac{3}{50} \times \frac{10}{12}$ $\frac{1}{20}$
16. $\frac{9}{11} \times \frac{1}{4}$ $\frac{9}{44}$
17. $\frac{14}{15} \times \frac{45}{7}$ 6
18. $\frac{12}{30} \times \frac{10}{18}$ $\frac{2}{9}$
19. $\frac{6}{17} \times \frac{17}{29}$ $\frac{6}{29}$
20. $16 \times \frac{1}{10}$ $1\frac{3}{5}$
21. $3\frac{2}{5} \times \frac{10}{34}$ 1
22. $9 \times 3\frac{1}{2}$ $31\frac{1}{2}$
23. $\frac{5}{2} \times \frac{4}{10}$ 1
24. $\frac{110}{121} \times \frac{11}{55}$ $\frac{2}{11}$
25. $\frac{1}{7} \times 49$ 7
26. $\frac{81}{2} \times \frac{5}{6}$ $33\frac{3}{4}$
27. $\frac{7}{4} \times 1\frac{3}{8}$ $2\frac{13}{32}$
28. $8\frac{1}{2} \times 6\frac{2}{3}$ $56\frac{2}{3}$
29. $16 \times \frac{9}{10}$ $14\frac{2}{5}$
30. $\frac{52}{100} \times \frac{3}{4}$ $\frac{39}{100}$
31. $\frac{22}{50} \times \frac{18}{42}$ $\frac{33}{175}$
32. $\frac{5}{8} \times 36$ $22\frac{1}{2}$
33. $\frac{70}{76} \times \frac{14}{28}$ $\frac{35}{76}$
34. $\frac{60}{3} \times \frac{3}{24}$ $2\frac{1}{2}$
35. $\frac{5}{9} \times \frac{90}{100}$ $\frac{1}{2}$
36. $3\frac{6}{7} \times 2\frac{1}{3}$ 9
37. $\frac{17}{60} \times \frac{100}{12}$ $2\frac{13}{36}$
38. $\frac{42}{45} \times \frac{1}{6}$ $\frac{7}{45}$
39. $7\frac{1}{4} \times \frac{2}{3}$ $4\frac{5}{6}$
40. $\frac{32}{38} \times \frac{78}{16}$ $4\frac{2}{19}$

Basic Skills

Dividing Fractions

1. $\frac{1}{2} \div \frac{2}{3}$ $\frac{3}{4}$
2. $\frac{3}{4} \div \frac{1}{2}$ $1\frac{1}{2}$
3. $3 \div \frac{6}{7}$ $3\frac{1}{2}$
4. $\frac{1}{8} \div \frac{1}{3}$ $\frac{3}{8}$

5. $\frac{2}{3} \div \frac{1}{2}$ $1\frac{1}{3}$
6. $\frac{3}{5} \div \frac{1}{4}$ $2\frac{2}{5}$
7. $\frac{5}{6} \div \frac{2}{3}$ $1\frac{1}{4}$
8. $\frac{1}{6} \div \frac{1}{4}$ $\frac{2}{3}$

9. $8 \div \frac{1}{2}$ 16
10. $\frac{3}{8} \div \frac{6}{7}$ $\frac{7}{16}$
11. $\frac{4}{9} \div 2$ $\frac{2}{9}$
12. $\frac{5}{9} \div \frac{5}{6}$ $\frac{2}{3}$

13. $\frac{3}{4} \div \frac{3}{8}$ 2
14. $\frac{1}{6} \div 15$ $\frac{1}{90}$
15. $\frac{10}{5} \div \frac{6}{5}$ $1\frac{2}{3}$
16. $\frac{8}{9} \div \frac{4}{81}$ 18

17. $\frac{1}{8} \div \frac{1}{16}$ 2
18. $9 \div \frac{3}{4}$ 12
19. $\frac{1}{20} \div 4$ $\frac{1}{80}$
20. $5 \div \frac{4}{3}$ $3\frac{3}{4}$

21. $\frac{3}{4} \div 4\frac{1}{2}$ $\frac{1}{6}$
22. $2\frac{2}{3} \div 4$ $\frac{2}{3}$
23. $1\frac{1}{4} \div 3\frac{1}{2}$ $\frac{5}{14}$
24. $4\frac{2}{3} \div \frac{7}{8}$ $5\frac{1}{3}$

25. $\frac{9}{10} \div 2\frac{1}{4}$ $\frac{2}{5}$
26. $\frac{2}{3} \div 2\frac{1}{2}$ $\frac{4}{15}$
27. $5 \div 1\frac{1}{3}$ $3\frac{3}{4}$
28. $2\frac{1}{4} \div \frac{2}{3}$ $3\frac{3}{8}$

29. $2\frac{2}{3} \div 5\frac{1}{3}$ $\frac{1}{2}$
30. $1\frac{1}{9} \div 1\frac{2}{3}$ $\frac{2}{3}$
31. $5\frac{1}{4} \div 3$ $1\frac{3}{4}$
32. $4\frac{1}{2} \div 6\frac{3}{4}$ $\frac{2}{3}$

33. $3\frac{3}{4} \div \frac{5}{6}$ $4\frac{1}{2}$
34. $2\frac{3}{4} \div 1\frac{3}{8}$ 2
35. $6\frac{1}{2} \div 4\frac{1}{6}$ $1\frac{14}{25}$
36. $2\frac{1}{7} \div 10$ $\frac{3}{14}$

37. $2\frac{4}{5} \div 5\frac{3}{5}$ $\frac{1}{2}$
38. $1\frac{5}{9} \div 2\frac{1}{3}$ $\frac{2}{3}$
39. $5\frac{3}{4} \div 2\frac{1}{2}$ $2\frac{3}{10}$
40. $3\frac{1}{2} \div 1\frac{1}{2}$ $2\frac{1}{3}$

Extra Practice: Basic Skills

Extra Practice

Lesson 1-1 (Pages 4–7)
Use the four-step plan to solve each problem.

1. Joseph is planting bushes around the perimeter of his lawn. If the bushes must be planted 4 feet apart and Joseph's lawn is 64 feet wide and 124 feet long, how many bushes will Joseph need to purchase? **94 bushes**

2. The cost of a long distance phone call is $1.50 for the first two minutes and $0.60 for each additional minute. How much will Maria pay for a 24 minute phone call? **$14.70**

3. Find the next three numbers in the pattern. 1, 3, 7, 15, 31, ___, ___, ___
63, 127, 255

Lesson 1-2 (Pages 8–10)
Write each expression using exponents.

1. $4 \cdot 4 \cdot 4 \cdot 4$ **4^4**
2. $3 \cdot 3$ **3^2**
3. $7 \cdot 7 \cdot 7 \cdot 7 \cdot 7 \cdot 7$ **7^6**

Evaluate each expression.

4. 4^3 **64**
5. 6^2 **36**
6. 2^6 **64**

7. $5^2 \times 6^2$ **900**
8. 3×2^4 **48**
9. $10^4 \times 3^2$ **90,000**

10. $5^3 \times 1^9$ **125**
11. $2^2 \times 2^4$ **64**
12. $2 \times 3^2 \times 4^2$ **288**

13. 7^3 **343**
14. $9^2 + 3^2$ **90**
15. 0.5^2 **0.25**

Lesson 1-3 (Pages 11–15)
Evaluate each expression.

1. $15 - 5 + 9 - 2$ **17**
2. $6 \times 6 + 3.6$ **39.6**
3. $12 + 20 \div 4 - 5$ **12**

4. $6 \times 3 \div 9 - 1$ **1**
5. $(4^2 + 2^3) \times 5$ **120**
6. $24 \div 8 - 2$ **1**

7. $3 \times (4 + 5) - 7$ **20**
8. $4.3 + 24 \div 6$ **8.3**
9. $(5^2 + 2) \div 3$ **9**

Evaluate each expression if $a = 3$, $b = 6$, and $c = 5$.

10. $2a + bc$ **36**
11. ba^3 **162**
12. $\frac{bc}{a}$ **10**

13. $3a - 2b + c$ **2**
14. $(2c + b) \cdot a$ **48**
15. $\frac{2(ac)^2}{b}$ **75**

Find the solution for each equation from the given replacement set.

16. $g + 16 = 47$ {29, 31, 33, 35}   **31**
17. $y - 14 = 26$ {50, 45, 40, 35}   **40**

18. $7m = 161$ {21, 23, 25, 27}   **23**
19. $\frac{k}{16} = 19$ {300, 302, 304, 306}   **304**

Lesson 1-4 (Pages 17–20)
Solve each equation. Check your solution.

1. $g - 3 = 10$ **13**
2. $b + 7 = 12$ **5**
3. $a + 3 = 15$ **12**
4. $r - 3 = 4$ **7**
5. $t + 3 = 21$ **18**
6. $s + 10 = 23$ **13**
7. $9 + n = 13$ **4**
8. $13 + v = 31$ **18**
9. $s - 0.4 = 6$ **6.4**
10. $x - 1.3 = 12$ **13.3**
11. $18 = y + 3.4$ **14.6**
12. $7 + g = 91$ **84**
13. $63 + f = 71$ **8**
14. $0.32 = w - 0.1$ **0.42**
15. $c - 18 = 13$ **31**
16. $23 = n - 5$ **28**
17. $j - 3 = 7$ **10**
18. $18 = p + 3$ **15**
19. $12 + p = 16$ **4**
20. $25 = y - 50$ **75**
21. $x + 2 = 4$ **2**

Lesson 1-5 (Pages 21–25)
Solve each equation. Check your solution.

1. $4x = 36$ **9**
2. $39 = 3y$ **13**
3. $4z = 16$ **4**
4. $t \div 5 = 6$ **30**
5. $100 = 20b$ **5**
6. $8 = w \div 8$ **64**
7. $10a = 40$ **4**
8. $s \div 9 = 8$ **72**
9. $420 = 5s$ **84**
10. $8k = 72$ **9**
11. $2m = 18$ **9**
12. $\frac{m}{8} = 5$ **40**
13. $0.12 = 3h$ **0.04**
14. $\frac{w}{7} = 8$ **56**
15. $18q = 36$ **2**
16. $9w = 54$ **6**
17. $4 = p \div 4$ **16**
18. $14 = 2p$ **7**
19. $12 = 3t$ **4**
20. $\frac{m}{4} = 12$ **48**
21. $6h = 12$ **2**

Lesson 1-6 (Pages 27–29)
Write each phrase or sentence as an algebraic expression or equation.

1. 12 more than a number **$12 + n$**
2. 3 less than a number **$n - 3$**
3. a number divided by 4 **$\frac{n}{4}$ or $n \div 4$**
4. a number increased by 7 **$n + 7$**
5. a number decreased by 12 **$n - 12$**
6. 8 times a number **$8n$**
7. 28 multiplied by m **$28m$**
8. 15 divided by a number **$\frac{15}{n}$ or $15 \div n$**
9. 54 divided by n **$\frac{54}{n}$ or $54 \div n$**
10. 18 increased by y **$18 + y$**
11. q decreased by 20 **$q - 20$**
12. n times 41 **$41n$**
13. The difference between a number and 12 is 37. **$n - 12 = 37$**
14. The product of a number and 7 is 42. **$7n = 42$**
15. 6 less than the product of q and 4 is 18. **$4q - 6 = 18$**

Lesson 1-7A (Pages 30–31)
Solve.

1. Dwayne's weight is twice Beth's weight minus 24 pounds. Dwayne weighs 120 pounds. How much does Beth weigh? **72 pounds**

2. A store doubled its cost to determine the price for a sweater. Two months later the sweater was marked down $10.00. A week later the price was cut in half and the sweater sold for $12.52. How much did the store pay for the sweater? Did the store make or lose money on the sweater? **$17.52; lost $5.00**

3. Jonathon Michaels just bought a new car. He received a $1,500 credit for his trade-in. He gave the car dealer a $2,500 down payment and is making 24 monthly payments of $279.00 each. What is the total cost of the car? **$10,696**

4. If a number is divided by 6 and then 14 is added to it, the result is 21. What is the number? **42**

Lesson 1-7 (Pages 32–36)
Solve each equation. Check your solution.

1. $2x + 4 = 14$ **5**
2. $5p - 10 = 0$ **2**
3. $5 + 6a = 41$ **6**
4. $\frac{x}{3} - 7 = 2$ **27**
5. $18 = 6(q - 4)$ **7**
6. $18 = 4m - 6$ **6**
7. $3(r - 1) = 9$ **4**
8. $2x + 3 = 5$ **1**
9. $0 = 4x - 28$ **7**
10. $3x - 1 = 5$ **2**
11. $3z + 5 = 14$ **3**
12. $3(x - 5) = 12$ **9**
13. $9a - 8 = 73$ **9**
14. $2x - 3 = 7$ **5**
15. $3t + 6 = 9$ **1**
16. $2y + 10 = 22$ **6**
17. $15 = 2y - 5$ **10**
18. $3c - 4 = 2$ **2**
19. $6 + 2p = 16$ **5**
20. $8 = 2 + 3x$ **2**
21. $4(b + 6) = 24$ **0**

Lesson 1-8 (Pages 38–42)
Find the perimeter and area of each figure.

1.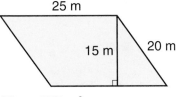
90 m; 375 m²

2.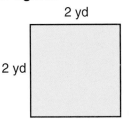
8 yd; 4 yd²

3.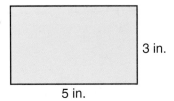
16 in.; 15 in²

4.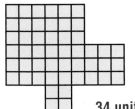
34 units; 49 units²

5.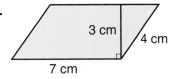
22 cm; 21 cm²

6.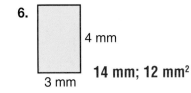
14 mm; 12 mm²

Extra Practice **607**

Lesson 1-9 (Pages 43–47)   1–21. See Answer Appendix for number lines.
Solve each inequality. Show the solution on a number line.

1. $y + 3 > 7$ $y > 4$
2. $c - 9 < 5$ $c < 14$
3. $x + 4 > 9$ $x > 5$
4. $y - 3 < 15$ $y < 18$
5. $t - 13 > 5$ $t > 18$
6. $5p < 25$ $p < 5$
7. $4x < 12$ $x < 3$
8. $15 < 3m$ $m > 5$
9. $\frac{d}{3} > 15$ $d > 45$
10. $8 < r \div 7$ $r > 56$
11. $2y + 5 > 15$ $y > 5$
12. $16 < 5d + 6$ $d > 2$
13. $3x + 2 > 11$ $x > 3$
14. $\frac{a}{3} - 2 > 1$ $a > 9$
15. $9g < 27$ $g < 3$
16. $14 < 2x + 4$ $x > 5$
17. $\frac{x}{2} - 6 > 0$ $x > 12$
18. $k + 5 < 6$ $k < 1$
19. $15 > c - 2$ $c < 17$
20. $4p > 24$ $p > 6$
21. $24 < 14 + k$ $k > 10$

Lesson 2-1 (Pages 56–58)
Graph each set of points on a number line.   1–3. See Answer Appendix for graphs.

1. $\{-8, -9, -6, -10\}$
2. $\{-3, 2, 0, -1\}$
3. $\{5, 6, 8, 7, 9\}$

Find each absolute value.

4. $|-1|$ 1
5. $|-92|$ 92
6. $|3|$ 3
7. $|160 + 32|$ 192
8. $|80 + 100|$ 180
9. $|0|$ 0
10. $|7 - 3|$ 4
11. $|3| + |-7|$ 10
12. $|-161|$ 161
13. $|150|$ 150
14. $|102| - |-2|$ 100
15. $|-116|$ 116

Lesson 2-2 (Pages 59–61)
Replace each ● with >, <, or = to make a true sentence.

1. -3 ● 0 <
2. -1 ● -2 >
3. -5 ● -4 <
4. 6 ● -7 >
5. 8 ● 10 <
6. -6 ● 6 <
7. -11 ● -20 >
8. -8 ● 2 <
9. -13 ● -12 <
10. 5 ● 2 >
11. 9 ● -8 >
12. 19 ● -19 >
13. $|-2|$ ● $|5|$ <
14. $|13|$ ● $|-19|$ <
15. $|-6|$ ● $|2|$ >
16. $|14|$ ● $|-14|$ =
17. $|0|$ ● $|-4|$ <
18. $|23|$ ● $|-20|$ >
19. $|-75|$ ● $|75|$ =
20. -71 ● 72 <
21. -15 ● -35 >

Lesson 2-3 (Pages 62–65)
Solve each equation.

1. $-7 + (-7) = h$ **−14**
2. $k = -36 + 40$ **4**
3. $m = 18 + (-32)$ **−14**
4. $47 + 12 = y$ **59**
5. $y = -69 + (-32)$ **−101**
6. $-120 + (-2) = c$ **−122**
7. $x = -56 + (-4)$ **−60**
8. $14 + 16 = k$ **30**
9. $-18 + 11 = d$ **−7**
10. $-42 + 29 = r$ **−13**
11. $h = -13 + (-11)$ **−24**
12. $x = 95 + (-5)$ **90**
13. $-120 + 2 = b$ **−118**
14. $w = 25 + (-25)$ **0**
15. $a = -4 + 8$ **4**
16. $g = -9 + (-6)$ **−15**
17. $42 + (-18) = f$ **24**
18. $-33 + (-12) = w$ **−45**
19. $-96 + (-18) = g$ **−114**
20. $-100 + 98 = a$ **−2**
21. $5 + (-7) = y$ **−2**

Lesson 2-4 (Pages 66–68)
Solve each equation. Check by solving another way.

1. $a = 7 + (-13) + 6 + (-7)$ **−7**
2. $x = -6 + 12 + (-20)$ **−14**
3. $4 + 9 + (-14) = k$ **−1**
4. $c = -20 + 0 + (-9) + 25$ **−4**
5. $b = 5 + 9 + 3 + (-17)$ **0**
6. $-36 + 40 + (-10) = y$ **−6**
7. $(-2) + 2 + (-2) + 2 = m$ **0**
8. $6 + (-4) + 9 + (-2) = d$ **9**
9. $9 + (-7) + 2 = n$ **4**
10. $b = 100 + (-75) + (-20)$ **5**
11. $x = -12 + 24 + (-12) + 2$ **2**
12. $9 + (-18) + 6 + (-3) = c$ **−6**
13. $(-10) + 4 + 6 = k$ **0**
14. $c = 4 + (-8) + 12$ **8**

Lesson 2-5 (Pages 69–72)
Solve each equation.

1. $3 - 7 = y$ **−4**
2. $-5 - 4 = w$ **−9**
3. $a = -6 - 2$ **−8**
4. $12 - 9 = x$ **3**
5. $a = 0 - (-14)$ **14**
6. $a = 58 - (-10)$ **68**
7. $n = -41 - 15$ **−56**
8. $c = -81 - 21$ **−102**
9. $26 - (-14) = y$ **40**
10. $6 - (-4) = b$ **10**
11. $z = 63 - 78$ **−15**
12. $-5 - (-9) = h$ **4**
13. $m = 72 - (-19)$ **91**
14. $-51 - 47 = x$ **−98**
15. $-99 - 1 = p$ **−100**
16. $r = 8 - 13$ **−5**
17. $-2 - 23 = c$ **−25**
18. $-20 - 0 = d$ **−20**
19. $55 - 33 = k$ **22**
20. $84 - (-61) = a$ **145**
21. $z = -4 - (-4)$ **0**

Lesson 2-6 (Pages 73–76)
Find each sum or difference. If there is no sum or difference, write *impossible*.

1. $\begin{bmatrix} 2 & 0 \\ -6 & 4 \end{bmatrix} + \begin{bmatrix} 8 & 6 \\ 1 & -3 \end{bmatrix}$ $\begin{bmatrix} 10 & 6 \\ -5 & 1 \end{bmatrix}$

2. $\begin{bmatrix} 8 & 5 & 2 \\ -1 & 3 & 11 \end{bmatrix} + \begin{bmatrix} -4 & 12 \\ 1 & 8 \\ 6 & -3 \end{bmatrix}$ **impossible**

3. $\begin{bmatrix} 1 & 2 \\ 6 & -3 \\ -5 & 7 \end{bmatrix} - \begin{bmatrix} -2 & 7 \\ 5 & 4 \\ -3 & 9 \end{bmatrix}$ $\begin{bmatrix} 3 & -5 \\ 1 & -7 \\ -2 & -2 \end{bmatrix}$

4. $\begin{bmatrix} 0 & 8 & 6 \\ 7 & -2 & 4 \\ 0 & 0 & 9 \end{bmatrix} - \begin{bmatrix} -6 & 0 & -1 \\ 2 & 4 & 0 \\ -1 & 5 & 8 \end{bmatrix}$ $\begin{bmatrix} 6 & 8 & 7 \\ 5 & -6 & 4 \\ 1 & -5 & 1 \end{bmatrix}$

5. $\begin{bmatrix} 13 & 2 \\ 7 & -1 \end{bmatrix} + \begin{bmatrix} 6 & -5 & 0 \\ 3 & 1 & -2 \\ 2 & 7 & 8 \end{bmatrix}$ **impossible**

6. $\begin{bmatrix} 13 & 5 & 7 \\ -6 & 12 & -1 \end{bmatrix} - \begin{bmatrix} 8 & -6 & 0 \\ -2 & 15 & 2 \end{bmatrix}$ $\begin{bmatrix} 5 & 11 & 7 \\ -4 & -3 & -3 \end{bmatrix}$

Lesson 2-7 (Pages 78–80)
Solve each equation.

1. $5(-2) = d$ **−10**
2. $-11(-5) = c$ **55**
3. $-5(-5) = z$ **25**
4. $x = -12(6)$ **−72**
5. $b = 2(-2)$ **−4**
6. $-3(2)(-4) = j$ **24**
7. $a = (-4)(-4)$ **16**
8. $4(21) = y$ **84**
9. $a = -50(0)$ **0**
10. $b = 3(-13)$ **−39**
11. $a = 2(2)$ **4**
12. $d = -2(-2)$ **4**
13. $x = 5(-12)$ **−60**
14. $2(2)(-2) = b$ **−8**
15. $a = 6(-4)$ **−24**
16. $x = -6(5)$ **−30**
17. $-4(8) = a$ **−32**
18. $3(-16) = y$ **−48**
19. $c = -2(2)$ **−4**
20. $6(3)(-2) = k$ **−36**
21. $y = -3(12)$ **−36**

Lesson 2-8 (Pages 81–83)
Solve each equation.

1. $a = 4 \div (-2)$ **−2**
2. $16 \div (-8) = x$ **−2**
3. $-14 \div (-2) = c$ **7**
4. $h = -18 \div 3$ **−6**
5. $-25 \div 5 = k$ **−5**
6. $n = -56 \div (-8)$ **7**
7. $x = 81 \div 9$ **9**
8. $-55 \div 11 = c$ **−5**
9. $-42 \div (-7) = y$ **6**
10. $g = 18 \div (-3)$ **−6**
11. $t = 0 \div (-1)$ **0**
12. $-32 \div 8 = m$ **−4**
13. $81 \div (-9) = w$ **−9**
14. $18 \div (-2) = a$ **−9**
15. $x = -21 \div 3$ **−7**
16. $d = 32 \div 8$ **4**
17. $8 \div (-8) = y$ **−1**
18. $c = -14 \div (-7)$ **2**
19. $-81 \div 9 = y$ **−9**
20. $q = -81 \div (-9)$ **9**
21. $-49 \div (-7) = y$ **7**

Lesson 2-9 (Pages 86–87)
Solve each equation. Check your work.

1. $-4 + b = 12$ **16**
2. $z - 10 = -8$ **2**
3. $-7 = x + 12$ **−19**
4. $a + 6 = -9$ **−15**
5. $r \div 7 = -8$ **−56**
6. $-2a = -8$ **4**
7. $r - (-8) = 14$ **6**
8. $0 = 6r$ **0**
9. $\frac{y}{12} = -6$ **−72**
10. $m + (-2) = 6$ **8**
11. $3m = -15$ **−5**
12. $c \div (-4) = 10$ **−40**
13. $5 + q = 12$ **7**
14. $\frac{16}{x} = -4$ **−4**
15. $-6f = -36$ **6**
16. $81 = -9w$ **−9**
17. $t + 12 = 6$ **−6**
18. $8 + p = 0$ **−8**
19. $0.12 = -3h$ **−0.04**
20. $12 - x = 8$ **4**
21. $14 + t = 10$ **−4**

Lesson 2-9B (Pages 90–91)
Solve.

1. At a rock concert, 2,143 sweatshirts were sold at $18.75 each. The receipts for the sale of the sweatshirts were:
   a. $4,108
   b. $40,181.25
   c. $401,812.50   **b. $40,181.25**

2. Daisuke takes about 12 breaths per minute. In a week he will take about how many breaths?   **120,960**

3. Minnesota has at least 10,000 lakes. Its total area is 84,402 square miles. On the average, Minnesota has 1 lake for how many square miles of area?
   a. 840
   b. 84
   c. 8.4   **c. 8.4**

4. If Daniel has 400 quarters, how many dollars worth of quarters does he have?
   a. $10
   b. $100
   c. $1,000   **b. $100**

5. 600 raffle tickets are sold. If there are 3 prizes and you buy 2 tickets, what is your chance of winning a prize?
   a. 1 in 10
   b. 1 in 100
   c. 1 in 1,000   **b. 1 in 100**

Lesson 2-10 (Pages 92–95)
Name the ordered pair for the coordinates of each point graphed on the coordinate plane.

1. A **(3, 1)**
2. B **(−2, −1)**
3. C **(2, −2)**
4. D **(0, −1)**
5. E **(−3, 2)**
6. F **(4, −1)**
7. G **(−5, 0)**
8. H **(−2, 1)**
9. I **(−4, −2)**

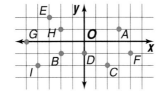

Graph each point on the same coordinate plane.   10–19. See Answer Appendix for graph.

10. (3, −2)
11. (2, 4)
12. (−1, 6)
13. (0, 5)
14. (−2, −3)
15. (−4, 0)
16. (4, −4)
17. (0, 0)
18. (3, 1)
19. (−4, −1)

Lesson 3-1 *(Pages 104–106)*
Express each ratio or rate in simplest form.

1. 27 to 9 $\frac{3}{1}$
2. 4 inches per foot $\frac{1}{3}$
3. 16 out of 48 $\frac{1}{3}$
4. 10:50 $\frac{1}{5}$
5. 40 min. per hour $\frac{2}{3}$
6. 35 is to 15 $\frac{7}{3}$
7. 16 wins, 16 losses $\frac{1}{1}$
8. 7 out of 13 $\frac{7}{13}$
9. 5 out of 50 $\frac{1}{10}$

Express each rate as a unit rate.

10. $24 per dozen **$2 each**
11. 600 students to 30 teachers $\frac{20 \text{ students}}{1 \text{ teacher}}$
12. 6 pounds gained in 12 weeks $\frac{0.5 \text{ lb}}{1 \text{ week}}$
13. $800 for 40 tickets $\frac{\$20}{1 \text{ ticket}}$
14. $6.50 for 5 pounds $\frac{\$1.30}{1 \text{ lb}}$
15. 6 inches of rain in 3 weeks $\frac{2 \text{ in.}}{1 \text{ week}}$

Lesson 3-2 *(Pages 107–110)*
Express each ratio or fraction as a percent.

1. 3 out of 5 **60%**
2. $\frac{1}{4}$ **25%**
3. $\frac{7}{10}$ **70%**
4. 39:100 **39%**
5. 11 out of 25 **44%**
6. 72.5:100 **72.5%**
7. 3 out of 4 **75%**
8. $\frac{1}{2}$ **50%**
9. $\frac{7}{20}$ **35%**
10. 93:100 **93%**

Express each percent as a fraction in simplest form.

11. 30% $\frac{3}{10}$
12. 4% $\frac{1}{25}$
13. 20% $\frac{1}{5}$
14. 85% $\frac{17}{20}$
15. 3% $\frac{3}{100}$
16. 80% $\frac{4}{5}$
17. 17% $\frac{17}{100}$
18. 55% $\frac{11}{20}$
19. 82% $\frac{41}{50}$
20. 48% $\frac{12}{25}$

Lesson 3-3 *(Pages 111–113)*
Determine whether each pair of ratios forms a proportion.

1. $\frac{3}{5}, \frac{5}{10}$ **no**
2. $\frac{8}{4}, \frac{6}{3}$ **yes**
3. $\frac{10}{15}, \frac{5}{3}$ **no**
4. $\frac{2}{8}, \frac{1}{4}$ **yes**
5. $\frac{6}{18}, \frac{3}{9}$ **yes**
6. $\frac{14}{21}, \frac{12}{18}$ **yes**
7. $\frac{4}{20}, \frac{5}{25}$ **yes**
8. $\frac{9}{27}, \frac{1}{3}$ **yes**

Solve each proportion.

9. $\frac{2}{3} = \frac{a}{12}$ **8**
10. $\frac{7}{8} = \frac{c}{16}$ **14**
11. $\frac{3}{7} = \frac{21}{d}$ **49**
12. $\frac{2}{5} = \frac{18}{x}$ **45**
13. $\frac{3}{5} = \frac{n}{21}$ **12$\frac{3}{5}$**
14. $\frac{5}{12} = \frac{b}{5}$ **2$\frac{1}{12}$**
15. $\frac{4}{36} = \frac{2}{y}$ **18**
16. $\frac{3}{10} = \frac{z}{36}$ **10$\frac{4}{5}$**
17. $\frac{2}{3} = \frac{t}{4}$ **2$\frac{2}{3}$**
18. $\frac{9}{10} = \frac{r}{25}$ **22$\frac{1}{2}$**
19. $\frac{16}{8} = \frac{y}{12}$ **24**
20. $\frac{7}{8} = \frac{a}{12}$ **10$\frac{1}{2}$**

612 Extra Practice

Lesson 3-4 (Pages 114–117)
Express each percent as a decimal.

1. 2% **0.02**
2. 25% **0.25**
3. 29% **0.29**
4. 6.2% **0.062**
5. 16.8% **0.168**
6. 14% **0.14**
7. 23.7% **0.237**
8. 42% **0.42**
9. 25.4% **0.254**
10. 98% **0.98**

Express each decimal as a percent.

11. 0.35 **35%**
12. 14.23 **1,423%**
13. 0.9 **90%**
14. 0.13 **13%**
15. 6.21 **621%**
16. 0.23 **23%**
17. 0.08 **8%**
18. 0.036 **3.6%**
19. 2.34 **234%**
20. 0.39 **39%**

Express each fraction as a percent.

21. $\frac{2}{5}$ **40%**
22. $\frac{49}{50}$ **98%**
23. $\frac{21}{50}$ **42%**
24. $\frac{1}{3}$ **$33\frac{1}{3}$%**
25. $\frac{81}{100}$ **81%**
26. $\frac{2}{25}$ **8%**
27. $\frac{11}{20}$ **55%**
28. $\frac{9}{75}$ **12%**
29. $\frac{33}{40}$ **$82\frac{1}{2}$%**
30. $\frac{1}{50}$ **2%**

Lesson 3-5 (Pages 120–123)
Compute mentally.

1. 10% of $206 **$20.6**
2. 1% of 19.3 **0.193**
3. 20% of 15 **3**
4. 87.5% of 80 **70**
5. 50% of 46 **23**
6. 12.5% of 56 **7**
7. $33\frac{1}{3}$% of 93 **31**
8. 90% of 2,000 **1,800**
9. 30% of 70 **21**
10. 40% of 95 **38**
11. $66\frac{2}{3}$% of 48 **32**
12. 80% of 25 **20**
13. 25% of 400 **100**
14. 75% of 72 **54**
15. 37.5% of 96 **36**
16. 40% of 35 **14**
17. 60% of 85 **51**
18. 62.5% of 160 **100**
19. 90% of 205 **184.5**
20. 1% of 2,364 **23.64**

Lesson 3-6A (Pages 124–125)
Solve.

1. Paige wants to leave a 15% tip for the server at a restaurant. About how much should Paige leave if the meal cost $29.63? **about $4.50**

2. In a survey of 1,138 college students, 4% said they would be willing to pay more for improved parking conditions. Is 4.5, 45 or 455 a reasonable estimate for the number of students willing to pay more? **45**

3. You spend $15.03 plus $0.90 tax at the store for a new backpack and pay with a $20 bill. Would it be more reasonable to expect $3.00 or $4.00 in change? **$4.00**

4. Paddy wants to buy a pair of in-line skates that cost $89.95. The sales tax is 6%. What is a reasonable amount for the total cost of the skates plus tax? **about $95**

5. An apple grower harvested 2,637 pounds of apples from one orchard and 1,826 pounds from another. What is a reasonable number of crates to have on hand if each create holds 15 pounds of apples? **300 crates**

Lesson 3-6 (Pages 126–129)
Estimate the percent of the area shaded.

1. 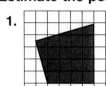 40%
2. 40%
3. 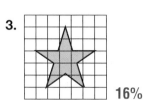 16%

Estimate.

4. 33% of 12
$\frac{1}{3} \cdot 12 = 4$

5. 24% of 84
$\frac{1}{4} \cdot 80 = 20$

6. 39% of 50
$\frac{2}{5} \cdot 50 = 20$

7. 1.5% of 135
$\frac{1}{50} \cdot 100 = 2$

Estimate the percent.

8. 11 out of 99
$\frac{10}{100} = 10\%$

9. 28 out of 89
$\frac{30}{90} = 33\frac{1}{3}\%$

10. 9 out of 20
$\frac{10}{20} = 50\%$

11. 25 out of 270
$\frac{25}{250} = 10\%$

Lesson 4-1A (Pages 140–141)

1. The list shows prices of popular video games.

   $24.99 $16.99 $44.99 $50.50 $35.99 $32.99 $10.99 $29.99 $29.99
   $43.99 $45.99 $37.99 $14.99 $18.00 $34.89 $55.80 $37.90 $40.44

   a. Make a frequency table of the prices using $10 intervals. **See Answer Appendix for sample answers.**
   b. What is the most common interval of prices? **$30.01-$40.00**

2. Use the frequency table. Which grade level has the greatest percent of students who play at least one sport? **11**

Grade	Percent of Students Who Play at Least One Sport
9	46%
10	53%
11	71%
12	63%

3. Make a frequency table for the data.

   12 21 4 15 22 5 9 6 19 11
   6   0   2 7   6   17 13 1 24 12

   See Answer Appendix for sample answers.

Lesson 4-1 (Pages 142–146)
4. About $\frac{1}{4}$ are over 70 feet tall.

Use the histogram to answer each question.

1. How large is each interval? **10 feet**
2. Which interval has the most buildings? **31–40**
3. Which interval has the least buildings? **81–90**
4. Compared to the total, how would you describe the number of buildings over 70 feet tall?
5. How does the number of buildings between 61 and 80 feet tall compare to the number of buildings between 31 and 50 feet tall? **There are 2 more buildings between 31 and 50 feet tall.**

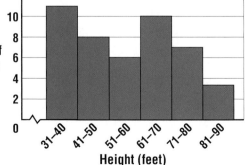
Heights of 45 Buildings

614 Extra Practice

Lesson 4-2 (Pages 148–151)

Make a circle graph for each set of data. 1–4. See Answer Appendix for graphs.

1.
Sporting Goods Sales	
Shoes	44%
Apparel	30%
Equipment	26%

2.
Energy Use in Home	
Heating/cooling	51%
Appliances	28%
Lights	21%

3.
Household income	
Primary job	82%
Secondary job	9%
Investments	5%
Other	4%

4.
Students in North High School	
White	30%
Black	28%
Hispanic	24%
Asian	18%

Lesson 4-3 (Pages 153–155)

Make a line plot for each set of data. 1–8. See Answer Appendix for line plots.

1. 8, 12, 10, 15, 11, 9, 12, 7, 14, 13, 8, 15, 17, 14, 11, 9, 8, 12, 15

2. 32, 41, 46, 38, 34, 51, 55, 49, 37, 42, 55, 46, 39, 58, 40, 35, 34, 52, 46

3. 161, 158, 163, 162, 165, 157, 159, 160, 163, 162, 158, 164, 161, 157, 166, 164

4. 78, 82, 83, 90, 58, 67, 95, 87, 97, 88, 75, 82, 78, 89, 86, 88, 79, 80, 91, 77

5. 49¢, 55¢, 77¢, 65¢, 51¢, 74¢, 68¢, 56¢, 49¢, 73¢, 62¢, 71¢, 54¢, 50¢, 70¢, 65¢

6. 2, 4, 12, 10, 2, 5, 7, 11, 7, 6, 3, 9, 12, 7, 5, 3, 7, 11, 2, 8, 7, 10, 8, 3, 7

7. 303, 298, 289, 309, 300, 294, 299, 301, 296, 308, 302, 289, 306, 308, 298, 299

8. 67, 73, 78, 61, 63, 77, 66, 75, 79, 66, 72, 69, 70, 74, 61, 63, 76, 64, 65, 78, 66

Lesson 4-4 (Pages 158–161)

Find the mean, median, and mode for each set of data. When necessary, round to the nearest tenth.

1. 2, 7, 9, 12, 5, 14, 4, 8, 3, 10
   7.4, 7.5, no mode

2. 58, 52, 49, 60, 61, 56, 50, 61 **55.9, 57, 61**

3. 122, 134, 129, 140, 125, 134, 137
   131.6, 134, 134

4. 25.5, 26.7, 20.9, 23.4, 26.8, 24.0, 25.7
   24.7, 25.5, no mode

5. 36, 41, 43, 45, 48, 52, 54, 56, 56, 57, 60, 64, 65 **52.1, 54, 56**

6.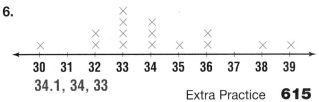
   34.1, 34, 33

Lesson 4-5 (Pages 163–166)

Find the range, median, upper and lower quartiles, interquartile range and any outliers for each set of data.

1. 15, 12, 21, 18, 25, 11, 17, 19, 20
   14; 18; 20.5, 13.5; 7; none
2. 2, 24, 6, 13, 8, 6, 11, 4
   22; 7; 12; 5; 7; 24
3. 189, 149, 155, 290, 141, 152
   149; 153.5; 189, 149; 40; 290
4. 451, 501, 388, 428, 510, 480, 390
   122; 451; 501, 390; 111; none
5. 22, 18, 9, 26, 14, 15, 6, 19, 28
   22; 18; 24, 11.5; 12.5; none
6. 245, 218, 251, 255, 248, 241, 250
   37; 248; 251; 241; 10; 218
7. 46, 45, 50, 40, 49, 42, 64
   24; 46; 50, 42; 8; 64
8. 128, 148, 130, 142, 164, 120, 152, 202
   82; 145; 158; 129; 29; 202
9. 2, 3, 2, 6, 4, 14, 13, 2, 6, 3
   12; 3.5; 6, 2; 4; 13, 14
10. 88, 84, 92, 93, 90, 96, 87, 97
    13; 91; 94.5, 87.5; 7; none
11. 378, 480, 370, 236, 361, 394, 345, 328, 388, 339
    244; 365.5; 388, 339; 49; 236, 480

Lesson 4-6 (Pages 168–170)

Determine whether a scatter plot of the data below would show a *positive*, *negative*, or *no* relationship.

1. height and hair color   **no relationship**
2. hours spent studying and test scores   **positive**
3. income and month of birth   **no relationship**
4. price of oranges and number available   **negative**
5. size of roof and number of shingles   **positive**
6. number of clouds and number of stars seen   **negative**

7.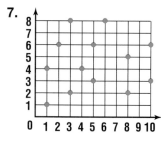
no relationship

8.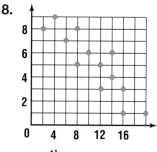
negative

9.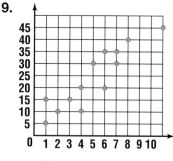
positive

Lesson 4-7 (Pages 171–173)

Choose the most appropriate type of display for each data set and situation. Explain your choice.   1–6. Sample answers given. See Answer Appendix for explanations.

1. number of fax machines in surveyed homes in newspaper article   **histogram**
2. the amount of sales by different people in a company for promotion decisions   **circle graph**
3. ages of amusement park attendees in marketing information for the park   **circle graph**
4. proficiency test scores for five consecutive years for comparing teaching techniques   **scatter plot**
5. numbers of Americans who own motorcycles, boats, and recreational vehicles in newspaper article   **table or bar graph**
6. percent of people who own Pontiacs compared to all car owners in advertising   **circle graph**

Lesson 4-8 (Pages 174–177)

Decide whether each location is a good place to find a representative sample for the selected survey. Justify your answer.

1. favorite kind of music at a jazz club   **no, probably prefer jazz**
2. the distance students travel to school on a school bus   **no, probably live far from school**
3. career choice in the student union of a large university   **yes, many choices would be represented**
4. favorite fast food restaurant at McDonald's   **no, probably prefer McDonald's**
5. whether people own cats in a telephone poll   **yes, a representative sample is possible**
6. teenagers' favorite compact disc at three different middle schools   **yes, different schools allow a diverse sample**
7. The two graphs at the right show the number of students in each grade who completed an obstacle course within the required time limit.
   a. Do both graphs contain the same information? Explain.
   b. One of the graphs is misleading. Explain why it is misleading.

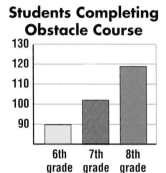
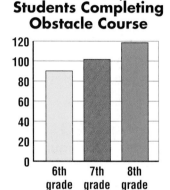

7a. **Yes, the values represented by the graph are the same.**
7b. **The first graph is misleading. It looks like twice as many 8th graders completed the obstacle course.**

Lesson 5-1 (Pages 188–192)

Use the figure at the right for Exercises 1–4.

1. Find $m\angle 6$, if $m\angle 3 = 42°$.   **42°**
2. Find $m\angle 4$, if $m\angle 3 = 71°$.   **109°**
3. Find $m\angle 1$, if $m\angle 5 = 128°$.   **128°**
4. Find $m\angle 7$, if $m\angle 2 = 83°$.   **83°**

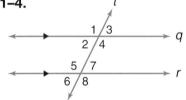

Find the value of x in each figure.

5. 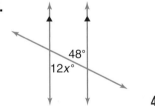   **4**

6.   **27**

Lesson 5-2A (Pages 194–195)

1. The Venn diagram shows the number of eighth grade students at Franklin Middle School who participate in sports and music activities.
   a. How many eighth graders participate in sports?   **82**
   b. How many participate in musical events?   **76**
   c. How many do not participate in sports or music?   **28**

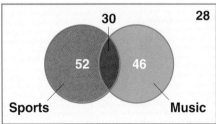

2. A tour bus is stopping at Calera City and Watertown. Fifteen people on the bus are only going to Calera City and seventeen are going only to Watertown. Twenty-two people are visiting both cities.
   a. Draw a Venn diagram that will represent this information.   **See Answer Appendix for diagram.**
   b. How many people are on the tour bus?   **54 people**

Extra Practice   **617**

Lesson 5-2 (Pages 196–199)
Classify each triangle by its angles and by its sides.

1.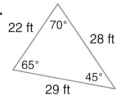
acute, scalene

2.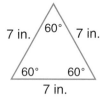
acute, equilateral

3.
right, scalene

4.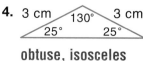
obtuse, isosceles

5.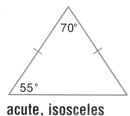
acute, isosceles

6.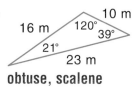
obtuse, scalene

7.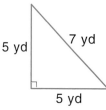
right, isosceles

8.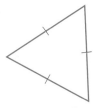
acute, equilateral

Lesson 5-3 (Pages 201–204)
Sketch each figure. Let Q = quadrilateral, P = parallelogram, R = rectangle, S = square, RH = rhombus, and T = trapezoid. Write all of the letters that describe it inside the figure.

1.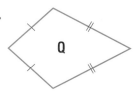
Q

2.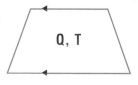
Q, T

3.
Q, P, RH

4.
Q, P, R

5.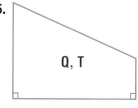
Q, T

6.
Q, P, RH

7.
Q, T

8.
Q, P, R, S, RH

Lesson 5-4 (Pages 206–209)
Trace each figure. Determine if the figure has line symmetry. If so, draw the lines of reflection.

1.

2. no

3.

4.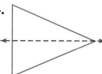

5.

6.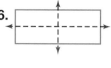

7. no

8.

9. Which of the figures in Exercises 1–8 have rotational symmetry?   **1, 3, 5, 6, 8**

Lesson 5-5 (Pages 210–212)

Determine whether each pair of triangles is congruent. If so, write a congruence statement and tell why the triangles are congruent.

1.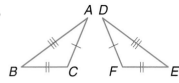
yes; △ABC ≅ △DEF; SSS

2.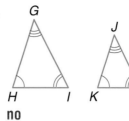
no

3.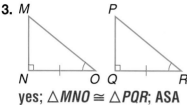
yes; △MNO ≅ △PQR; ASA

Find the value of x in each pair of congruent triangles.

4.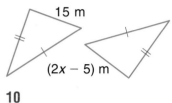
10

5.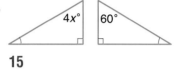
15

6.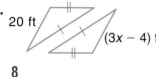
8

Lesson 5-6 (Pages 215–218)

Tell whether each pair of triangles is *congruent*, *similar*, or *neither*. Justify your answer.

1.
Similar; corresponding angles are congruent.

2.
congruent; SAS

3.
Neither; corresponding angles are not congruent.

Find the value of x in each pair of similar triangles.

4.
11

5.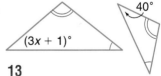
13

6.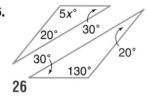
26

Lesson 5-7 (Pages 220–223)

Make an Escher-like drawing for each pattern described. For squares, use a tessellation of two rows of three squares as the base. For triangles, use a tessellation of two rows of five equilateral triangles as your base.   1–3. See Answer Appendix.

1.

2.

3.

Tell whether each pattern involves *translations* or *rotations*.

4. Exercise 1   translation

5. Exercise 2   translation

6. Exercise 3   rotation

Lesson 6-1 *(Pages 232-234)*
Determine whether each number is divisible by 2, 3, 4, 5, 6, 8, 9, or 10.

1. 210 **2, 3, 5, 6, 10**
2. 614 **2**
3. 985 **5**
4. 756 **2, 3, 4, 6, 9**
5. 432 **2, 3, 4, 6, 8, 9**
6. 96 **2, 3, 4, 6, 8**
7. 87 **3**
8. 113 **none**

9. Is 6 a factor of 936? **yes**
10. Is 9 a factor of 216? **yes**
11. Is 1,346 divisible by 2? **yes**
12. Is 1,249 divisible by 8? **no**
13. Is 752 divisible by 6? **no**
14. Is 3 a factor of 1,687? **no**
15. Is 4 a factor of 208? **yes**
16. Is 448 divisible by 8? **yes**

Lesson 6-2 *(Pages 235-238)*
Determine whether each number is *prime*, *composite*, or *neither*.

1. 17 **prime**
2. 1,258 **composite**
3. 37 **prime**
4. 483 **composite**
5. 97 **prime**
6. 0 **neither**
7. 25 **composite**
8. 61 **prime**
9. 45 **composite**
10. 419 **prime**

Find the prime factorization of each number.

11. 20 $2^2 \times 5$
12. 65 5×13
13. 52 $2^2 \times 13$
14. 30 $2 \times 3 \times 5$
15. 28 $2^2 \times 7$
16. 72 $2^3 \times 3^2$
17. 155 5×31
18. 50 2×5^2
19. 96 $2^5 \times 3$
20. 201 3×67
21. 1,250 2×5^4
22. 2,648 $2^3 \times 331$

Lesson 6-3A *(Pages 240-241)*
Solve.

1. Joshua and Martin are helping to make and sell submarine sandwiches to raise money for the marching band. They can add cheese, mushrooms, and/or green peppers to the basic submarine sandwich. How many different sandwiches can Joshua and Martin advertise? **8 sandwiches**

2. Katrina wants to buy a soda from a vending machine. The soda costs $0.65. If Katrina uses exact change, how many different combinations of nickels, dimes, and quarters can she use? **14 combinations**

3. How many three digit numbers can be formed from the digits 6, 7, 8, and 9 if no digit is repeated? **24 numbers**

4. Shaney has a basket containing 21 muffins to share with friends. She takes the first one and each friend takes one as they continue around the table. There is one muffin left when the basket returns to Shaney. How many people could have shared the muffins if the basket can go around the table more than once? **20, 10, 5, 4, or 2 people**

Lesson 6-3 (Pages 242-244)
Find the GCF for each set of numbers.

1. 8, 18 **2**
2. 6, 9 **3**
3. 4, 12 **4**
4. 18, 24 **6**
5. 8, 24 **8**
6. 17, 51 **17**
7. 65, 95 **5**
8. 42, 48 **6**
9. 64, 32 **32**
10. 72, 144 **72**
11. 54, 72 **18**
12. 60, 75 **15**
13. 16, 24 **8**
14. 12, 27 **3**
15. 25, 30 **5**
16. 48, 60 **12**
17. 16, 20, 36 **4**
18. 12, 18, 42 **6**
19. 30, 45, 15 **15**
20. 20, 30, 40 **10**
21. 81, 27, 108 **27**
22. 9, 18, 12 **3**

Lesson 6-4 (Pages 245-248)
Name all sets of numbers to which each number belongs.

1. -7 **IR**
2. $-6\frac{5}{9}$ **R**
3. 7.4 **R**
4. 86 **WIR**
5. -0.23 **R**

Write each fraction in simplest form.

6. $\frac{12}{16}$ $\frac{3}{4}$
7. $\frac{28}{32}$ $\frac{7}{8}$
8. $\frac{75}{100}$ $\frac{3}{4}$
9. $\frac{8}{16}$ $\frac{1}{2}$
10. $\frac{6}{18}$ $\frac{1}{3}$
11. $\frac{27}{36}$ $\frac{3}{4}$
12. $\frac{16}{64}$ $\frac{1}{4}$
13. $\frac{8}{32}$ $\frac{1}{4}$
14. $\frac{50}{100}$ $\frac{1}{2}$
15. $\frac{24}{40}$ $\frac{3}{5}$
16. $\frac{32}{80}$ $\frac{2}{5}$
17. $\frac{8}{24}$ $\frac{1}{3}$
18. $\frac{20}{25}$ $\frac{4}{5}$
19. $\frac{4}{10}$ $\frac{2}{5}$
20. $\frac{3}{5}$ $\frac{3}{5}$
21. $\frac{14}{19}$ $\frac{14}{19}$
22. $\frac{9}{12}$ $\frac{3}{4}$
23. $\frac{6}{8}$ $\frac{3}{4}$
24. $\frac{15}{18}$ $\frac{5}{6}$
25. $\frac{9}{20}$ $\frac{9}{20}$

Lesson 6-5 (Pages 249–252)
Express each decimal using bar notation.

1. $0.161616...$ $\mathbf{0.\overline{16}}$
2. $0.12351235...$ $\mathbf{0.\overline{1235}}$
3. $0.6666...$ $\mathbf{0.\overline{6}}$
4. $0.15151...$ $\mathbf{0.\overline{15}}$

Write the first ten decimal places of each decimal.

5. $0.\overline{09}$ **0.0909090909**
6. $0.\overline{076923}$ **0.0769230769**
7. $0.8\overline{4563}$ **0.8456345634**
8. $0.98\overline{745}$ **0.9874545454**

Express each fraction or mixed number as a decimal.

9. $\frac{2}{5}$ **0.4**
10. $2\frac{3}{11}$ **$2.\overline{27}$**
11. $-\frac{3}{4}$ **-0.75**
12. $\frac{5}{7}$ **$0.\overline{714285}$**
13. $\frac{3}{4}$ **0.75**
14. $-\frac{2}{3}$ **$-0.\overline{6}$**
15. $\frac{17}{20}$ **0.85**
16. $\frac{14}{25}$ **0.56**
17. $-6\frac{7}{10}$ **-6.7**
18. $\frac{7}{11}$ **$0.\overline{63}$**

Express each decimal as a fraction or mixed number in simplest form.

19. 0.5 $\frac{1}{2}$
20. $0.\overline{8}$ $\frac{8}{9}$
21. 0.32 $\frac{8}{25}$
22. -0.75 $-\frac{3}{4}$
23. $2.\overline{2}$ $2\frac{2}{9}$
24. $0.\overline{38}$ $\frac{38}{99}$
25. -0.486 $\frac{-243}{500}$
26. 20.08 $20\frac{2}{25}$
27. -9.36 $-9\frac{9}{25}$
28. $10.1\overline{8}$ $10\frac{17}{90}$

Lesson 6-6 (Pages 253–256)

A date is chosen at random from the calendar below. Find the probability of choosing each date.

1. The date is the thirteenth. $\frac{1}{30}$
2. The date is Friday. $\frac{2}{15}$
3. It is after the twenty-fifth. $\frac{1}{6}$
4. It is before the seventh. $\frac{1}{5}$
5. It is an odd-numbered date. $\frac{1}{2}$
6. The date is divisible by 3. $\frac{1}{3}$

November						
S	M	T	W	T	F	S
		1	2	3	4	5
6	7	8	9	10	11	12
13	14	15	16	17	18	19
20	21	22	23	24	25	26
27	28	29	30			

Lesson 6-7 (Pages 257–259)

List the first six multiples of each number or algebraic expression.

1. 6   6, 12, 18, 24, 30, 36
2. 9   9, 18, 27, 36, 45, 54
3. 15   15, 30, 45, 60, 75, 90
4. $3n$ $3n, 6n, 9n, 12n, 15n, 18n$

Find the LCM of each set of numbers.

5. 5, 15 **15**
6. 13, 39 **39**
7. 16, 24 **48**
8. 18, 20 **180**
9. 21, 14 **42**
10. 25, 30 **150**
11. 28, 42 **84**
12. 7, 13 **91**
13. 6, 30 **30**
14. 12, 42 **84**
15. 8, 10 **40**
16. 30, 10 **30**
17. 12, 18, 6 **36**
18. 15, 75, 25 **75**
19. 6, 10, 15 **30**
20. 3, 6, 9 **18**
21. 21, 14, 6 **42**
22. 12, 35, 10 **420**

Lesson 6-8 (Pages 261–264)

Replace each ● with a <, >, or = to make a true sentence.

1. -5.6 ● 4.2 **<**
2. 4.256 ● 4.25 **>**
3. 0.233 ● $0.\overline{23}$ **>**
4. $\frac{5}{7}$ ● $\frac{2}{5}$ **>**
5. $\frac{6}{7}$ ● $\frac{7}{9}$ **>**
6. $\frac{2}{3}$ ● $\frac{2}{5}$ **>**
7. $\frac{3}{8}$ ● 0.375 **=**
8. $-\frac{1}{2}$ ● 0.5 **<**
9. 12.56 ● $12\frac{3}{8}$ **>**

Order each set of rational numbers from least to greatest.

10. $0.24, 0.2, 0.245, 2.24, 0.25$ $0.2, 0.24, 0.245, 0.25, 2.24$
11. $0.\overline{3}, 0.3, 0.3\overline{4}, 0.\overline{34}, 0.33$ $0.3, 0.33, 0.\overline{3}, 0.\overline{34}, 0.3\overline{4}$
12. $\frac{2}{5}, \frac{2}{3}, \frac{2}{7}, \frac{2}{9}, \frac{2}{1}$ $\frac{2}{9}, \frac{2}{7}, \frac{2}{5}, \frac{2}{3}, \frac{2}{1}$
13. $\frac{1}{2}, \frac{5}{7}, \frac{2}{9}, \frac{8}{9}, \frac{6}{6}$ $\frac{2}{9}, \frac{1}{2}, \frac{5}{7}, \frac{8}{9}, \frac{6}{6}$

Lesson 6-9

Express each number in standard form.

1. 4.5×10^3 **4,500**
2. 2×10^4 **20,000**
3. 1.725896×10^6 **1,725,896**
4. 9.61×10^2 **961**
5. 1×10^7 **10,000,000**
6. 8.256×10^8 **825,600,000**
7. 5.26×10^4 **52,600**
8. 3.25×10^2 **325**
9. 6.79×10^5 **679,000**

Express each number in scientific notation.   11. 7.56×10^3 14. 9.1256×10^4

10. 720 7.2×10^2
11. 7,560
12. 892 8.92×10^2
13. 1,400 1.4×10^3
14. 91,256
15. 51,000 5.1×10^4
16. 145,600 1.456×10^5
17. 90,100 9.01×10^4
18. 123,568,000,000 1.23568×10^{11}

Lesson 7-1 (Pages 278–280)

Solve each equation. Write the solution in simplest form.

1. $\frac{17}{21} + \left(-\frac{13}{21}\right) = m$ $\frac{4}{21}$
2. $t = \frac{5}{11} + \frac{6}{11}$ 1
3. $k = -\frac{8}{13} + \left(-\frac{11}{13}\right)$ $-1\frac{6}{13}$
4. $-\frac{7}{12} + \frac{5}{12} = a$ $-\frac{1}{6}$
5. $\frac{13}{28} - \frac{9}{28} = g$ $\frac{1}{7}$
6. $b = -1\frac{2}{9} - \frac{7}{9}$ -2
7. $r = \frac{15}{16} + \frac{13}{16}$ $1\frac{3}{4}$
8. $2\frac{1}{3} - \frac{2}{3} = n$ $1\frac{2}{3}$
9. $-\frac{4}{35} - \left(-\frac{17}{35}\right) = c$ $\frac{13}{35}$
10. $\frac{3}{8} + \left(-\frac{5}{8}\right) = w$ $-\frac{1}{4}$
11. $s = \frac{8}{15} - \frac{2}{15}$ $\frac{2}{5}$
12. $d = -2\frac{4}{7} - \frac{3}{7}$ -3
13. $-\frac{29}{9} - \left(-\frac{26}{9}\right) = y$ $-\frac{1}{3}$
14. $2\frac{3}{5} + 7\frac{3}{5} = z$ $10\frac{1}{5}$
15. $x = \frac{5}{18} - \frac{13}{18}$ $-\frac{4}{9}$
16. $j = -2\frac{2}{7} + \left(-1\frac{6}{7}\right)$ $-4\frac{1}{7}$
17. $p = -\frac{3}{10} + \frac{7}{10}$ $\frac{2}{5}$
18. $\frac{4}{11} + \frac{9}{11} = f$ $1\frac{2}{11}$

Lesson 7-2 (Pages 281–284)

Solve each equation. Write the solution in simplest form.

1. $r = \frac{7}{12} + \frac{7}{24}$ $\frac{7}{8}$
2. $-\frac{3}{4} + \frac{7}{8} = z$ $\frac{1}{8}$
3. $\frac{2}{5} + \left(-\frac{2}{7}\right) = q$ $\frac{4}{35}$
4. $d = -\frac{3}{5} - \left(-\frac{5}{6}\right)$ $\frac{7}{30}$
5. $\frac{5}{24} - \frac{3}{8} = j$ $-\frac{1}{6}$
6. $g = -\frac{7}{12} - \frac{3}{4}$ $-1\frac{1}{3}$
7. $-\frac{3}{8} + \left(-\frac{4}{5}\right) = x$ $-1\frac{7}{40}$
8. $t = \frac{2}{15} + \left(-\frac{3}{10}\right)$ $-\frac{1}{6}$
9. $r = -\frac{2}{9} - \left(-\frac{2}{3}\right)$ $\frac{4}{9}$
10. $a = -\frac{7}{15} - \frac{5}{12}$ $-\frac{53}{60}$
11. $\frac{3}{8} + \frac{7}{12} = s$ $\frac{23}{24}$
12. $-2\frac{1}{4} + \left(-1\frac{1}{3}\right) = m$ $-3\frac{7}{12}$
13. $3\frac{2}{5} - 3\frac{1}{4} = v$ $\frac{3}{20}$
14. $b = \frac{3}{4} + \left(-\frac{4}{15}\right)$ $\frac{29}{60}$
15. $f = -1\frac{2}{3} + 4\frac{3}{4}$ $3\frac{1}{12}$
16. $-\frac{1}{8} - 2\frac{1}{2} = n$ $-2\frac{5}{8}$
17. $p = 3\frac{2}{5} - 1\frac{1}{3}$ $2\frac{1}{15}$
18. $y = 5\frac{1}{3} + \left(-8\frac{3}{7}\right)$ $-3\frac{2}{21}$

Lesson 7-3 (Pages 285–289)
Solve each equation. Write the solution in simplest form.

1. $\frac{2}{11} \cdot \frac{3}{4} = m$ $\frac{3}{22}$
2. $4\left(-\frac{7}{8}\right) = r$ $-3\frac{1}{2}$
3. $d = -\frac{4}{7} \cdot \frac{3}{5}$ $-\frac{12}{35}$
4. $g = \frac{6}{7}\left(-\frac{7}{12}\right)$ $-\frac{1}{2}$
5. $b = \frac{7}{8} \cdot \frac{1}{3}$ $\frac{7}{24}$
6. $\frac{3}{4} \cdot \frac{4}{5} = t$ $\frac{3}{5}$
7. $-1\frac{1}{2} \cdot \frac{2}{3} = k$ -1
8. $x = \frac{5}{6} \cdot \frac{6}{7}$ $\frac{5}{7}$
9. $c = 8\left(-2\frac{1}{4}\right)$ -18
10. $-3\frac{3}{4} \cdot \frac{8}{9} = q$ $-3\frac{1}{3}$
11. $\frac{10}{21} \cdot -\frac{7}{8} = n$ $-\frac{5}{12}$
12. $w = -1\frac{4}{5}\left(-\frac{5}{6}\right)$ $1\frac{1}{2}$
13. $a = 5\frac{1}{4} \cdot 6\frac{2}{3}$ 35
14. $-8\frac{3}{4} \cdot 4\frac{2}{5} = p$ $-38\frac{1}{2}$
15. $y = 6 \cdot 8\frac{2}{3}$ 52
16. $n = \left(\frac{3}{5}\right)^2$ $\frac{9}{25}$
17. $-4\frac{1}{5}\left(-3\frac{1}{3}\right) = h$ 14
18. $-8 \cdot \left(\frac{3}{4}\right)^2 = v$ $-4\frac{1}{2}$

Lesson 7-4 (Pages 290–292)
Name the multiplicative inverse of each of the following.

1. 3 $\frac{1}{3}$
2. -5 $-\frac{1}{5}$
3. $\frac{2}{3}$ $\frac{3}{2}$
4. $2\frac{1}{8}$ $\frac{8}{17}$
5. $\frac{a}{b}$ $\frac{b}{a}$
6. -8 $-\frac{1}{8}$
7. $\frac{1}{15}$ 15
8. 0.75 $\frac{4}{3}$
9. c $\frac{1}{c}$
10. $-\frac{3}{5}$ $-\frac{5}{3}$

Solve each equation using properties of rational numbers.

11. $g = 9 \cdot 2\frac{2}{3}$ 24
12. $-\frac{4}{5}\left(\frac{5}{6} \cdot \frac{3}{7}\right) = m$ $-\frac{2}{7}$
13. $k = 12\frac{1}{2} \cdot \frac{3}{4}$ $9\frac{3}{8}$
14. $w = -5 \cdot 4\frac{1}{4}$ $-21\frac{1}{4}$
15. $\frac{3}{8} \cdot 16\frac{1}{3} = d$ $6\frac{1}{8}$
16. $q = \left(\frac{3}{7} \cdot 2\frac{1}{3}\right) \cdot \frac{7}{9}$ $\frac{7}{9}$

Evaluate each expression if $a = \frac{1}{2}$, $b = -\frac{3}{4}$, and $c = -2\frac{1}{6}$.

17. a^2 $\frac{1}{4}$
18. abc $\frac{13}{16}$
19. $4a + 4b$ -1
20. ac $-1\frac{1}{12}$

Lesson 7-5A (Pages 294–295)
Solve.

1. The sum of two consecutive prime numbers is 966. What are the two numbers?   **479, 487**

2. Ms. Flores started a job making $2,000 per month. At the same time, Mr. Williams started a job making $2,400 per month. If Ms. Flores gets a $400 per month raise each year and Mr. Williams gets a $300 per month raise each year, in how many years will they both be earning the same amount of money per month?   **4 years**

3. Daisuke started a savings account in January. He puts $100 in his savings account each month. In February, Liza started a savings account. She puts $120 in her savings account each month. In what month will Daisuke and Liza have the same amount of money in their savings accounts?   **June**

4. Sergio deposits $1 into his savings account the first week, $3 the second week, $5 the third week, $7 the fourth week, and so on. How much money will he have in his savings account on the 20th week?   **$400**

Lesson 7-5 (Pages 296–299)   See Answer Appendix for terms.
Identify each sequence as *arithmetic*, *geometric* or *neither*. Then find the next three terms.

1. 1, 5, 9, 13, ...   **A**
2. 2, 6, 18, 54, ...   **G**
3. 1, 4, 9, 16, 25, ...   **N**
4. 729, 243, 81, ...   **G**
5. 2, −3, −8, −13, ...   **A**
6. 5, −5, 5, −5, ...   **G**
7. 810, −270, 90, −30, ...   **G**
8. 11, 14, 17, 20, 23, ...   **A**
9. 33, 27, 21, ...   **A**
10. 21, 15, 9, 3, ...   **A**
11. $\frac{1}{8}, -\frac{1}{4}, \frac{1}{2}, -1, ...$ **G**
12. $\frac{1}{81}, \frac{1}{27}, \frac{1}{9}, \frac{1}{3}, ...$ **G**
13. $\frac{3}{4}, 1\frac{1}{2}, 3, ...$ **G**
14. 2, 5, 9, 14, ...   **N**
15. $-1\frac{1}{4}, -1\frac{3}{4}, -2\frac{1}{4}, -2\frac{3}{4}, ...$ **A**
16. 9.9, 13.7, 17.5, ...   **A**
17. $\frac{1}{2}, 1\frac{1}{2}, 2\frac{1}{2}, 3\frac{1}{2}, ...$ **A**
18. 2, 12, 32, 62, ...   **N**
19. 3, −6, 12, −24, ...   **G**
20. 5, 7, 9, 11, 13, ...   **A**
21. −0.06, 2.24, 4.54, ...   **A**
22. 7, 14, 28, ...   **G**
23. −5.4, −1.4, 2.6, ...   **A**
24. −96, 48, −24, 12, ...   **G**
25. 4, 12, 36, ...   **G**
26. 20, 19, 18, 17, ...   **A**
27. 768, 192, 48, ...   **G**

Lesson 7-6 (Pages 301–304)
State the measures of the base(s) and the height of each triangle or trapezoid. Then find the area.

1.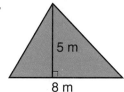
b = 8 m, h = 5 m
A = 20 m²

2.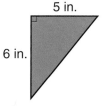
b = 5 in., h = 6 in.
A = 15 in²

3.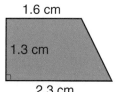
a = 1.6 cm, b = 2.3 cm,
h = 1.3 cm, A = 2.535 cm²

4.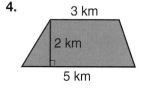
a = 3 km, b = 5 km
h = 2 km, A = 8 km²

Find the area of each figure.

5. triangle: base, $2\frac{1}{2}$ in.; height, 7 in. $8\frac{3}{4}$ **in²**
6. triangle: base, 12 cm; height, 3.2 cm   **19.2 cm²**
7. trapezoid: bases, 5 ft and 7 ft; height, 11 ft   **66 ft²**
8. trapezoid: bases, $4\frac{1}{4}$ yd and $3\frac{1}{2}$ yd; height, 5 yd $19\frac{3}{8}$ **yd²**

Lesson 7-7 (Pages 309–311)
Find the circumference of each circle to the nearest tenth. Use $\frac{22}{7}$ or 3.14 for π.

1. 14 mm, diameter
44.0 mm
2. 18 cm, diameter
56.5 cm
3. 24 in., radius
150.8 in.
4. 42 m, diameter
131.9 m

5.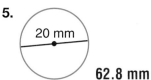
62.8 mm
6.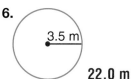
22.0 m
7.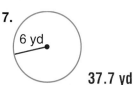
37.7 yd
8.
25.1 in.

9.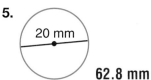
50.3 ft
10.
15.1 cm
11.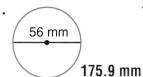
175.9 mm
12.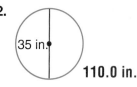
110.0 in.

Extra Practice **625**

Lesson 7-8 (Pages 312–314)
Solve each equation. Write the solution in simplest form.

1. $x = \frac{2}{3} \div \frac{3}{4}$ $\frac{8}{9}$
2. $-\frac{4}{9} \div \frac{5}{6} = c$ $-\frac{8}{15}$
3. $\frac{7}{12} \div \frac{3}{8} = q$ $1\frac{5}{9}$
4. $m = \frac{5}{18} \div \frac{2}{9}$ $1\frac{1}{4}$
5. $a = \frac{1}{3} \div 4$ $\frac{1}{12}$
6. $5\frac{1}{4} \div \left(-2\frac{1}{2}\right) = g$ $-2\frac{1}{10}$
7. $-6 \div \left(-\frac{4}{7}\right) = d$ $10\frac{1}{2}$
8. $n = -6\frac{3}{8} \div \frac{1}{4}$ $-25\frac{1}{2}$
9. $p = \frac{6}{7} \div \frac{3}{5}$ $1\frac{3}{7}$
10. $g = 3\frac{1}{3} \div (-4)$ $-\frac{5}{6}$
11. $2\frac{5}{12} \div 7\frac{1}{3} = r$ $\frac{29}{88}$
12. $v = \frac{5}{6} \div 1\frac{1}{9}$ $\frac{3}{4}$
13. $\frac{3}{8} \div (-6) = b$ $-\frac{1}{16}$
14. $k = \frac{5}{8} \div \frac{1}{6}$ $3\frac{3}{4}$
15. $4\frac{1}{4} \div 6\frac{3}{4} = w$ $\frac{17}{27}$
16. $f = 4\frac{1}{6} \div 3\frac{1}{8}$ $1\frac{1}{3}$
17. $t = 8 \div \left(-1\frac{4}{5}\right)$ $-4\frac{4}{9}$
18. $j = -5 \div \frac{2}{7}$ $-17\frac{1}{2}$
19. $\frac{3}{5} \div \frac{6}{7} = y$ $\frac{7}{10}$
20. $4\frac{8}{9} \div \left(-2\frac{2}{3}\right) = h$ $-1\frac{5}{6}$
21. $f = 8\frac{1}{6} \div 3$ $2\frac{13}{18}$
22. $k = -\frac{3}{4} \div 9$ $-\frac{1}{12}$
23. $s = 1\frac{11}{14} \div 2\frac{1}{2}$ $\frac{5}{7}$
24. $-2\frac{1}{4} \div \frac{4}{5} = z$ $-2\frac{13}{16}$

Lesson 7-9 (Pages 315–317)
Solve each equation. Check your solution.

1. $434 = -31y$ -14
2. $6x = -4.2$ -0.7
3. $\frac{3}{4}a = -12$ -16
4. $-10 = \frac{b}{-7}$ 70
5. $7.2 = \frac{3}{4}c$ 9.6
6. $2r + 4 = 14$ 5
7. $-2.4n = 7.2$ -3
8. $7 = \frac{1}{2}d - 3$ 20
9. $3.2n - 0.64 = -5.44$ -1.5
10. $\frac{t}{3} - 7 = 2$ 27
11. $\frac{3}{8} = \frac{1}{2}x$ $\frac{3}{4}$
12. $\frac{1}{2}h - 3 = -14$ -22
13. $-0.46k - 1.18 = 1.58$ -6
14. $4\frac{1}{2}s = -30$ $-6\frac{2}{3}$
15. $\frac{2}{3}f = \frac{8}{15}$ $\frac{4}{5}$
16. $\frac{2}{3}m + 10 = 22$ 18
17. $\frac{2}{3}g + 4 = 4\frac{5}{6}$ $1\frac{1}{4}$
18. $7 = \frac{1}{2}v + 3$ 8
19. $\frac{g}{1.2} = -6$ -7.2
20. $\frac{4}{7}z - 4\frac{5}{8} = 15\frac{3}{8}$ 35
21. $-12 = \frac{1}{5}j$ -60

Lesson 7-10 (Pages 318–321)
Solve each inequality. Graph the solution on a number line. See Answer Appendix for graphs.

1. $5p < 25$ $p < 5$
2. $c - 9 < 4$ $c < 13$
3. $y + 4 \geq 6$ $y \geq 2$
4. $-4 > -\frac{1}{3}k$ $k > 12$
5. $2.5 \leq \frac{h}{0.4}$ $h \geq 1$
6. $-\frac{4}{5}z > \frac{2}{5}$ $z < -\frac{1}{2}$
7. $f + \frac{1}{5} \geq 4\frac{3}{5}$ $f \geq 4\frac{2}{5}$
8. $6 + 5x > 16$ $x > 2$
9. $14 \leq 2x + 4$ $x \geq 5$
10. $-2t + 28 < 5\frac{3}{5}$ $t > 11\frac{1}{5}$
11. $\frac{7}{8}n - \frac{1}{6} > \frac{3}{4}$ $n > 1\frac{1}{21}$
12. $\frac{4}{2.6}h - 11 \leq 6.8$ $h \leq 11.57$
13. $\frac{2x + 4}{5} \geq 3$ $x \geq 5\frac{1}{2}$
14. $\frac{1}{3}j + 7 > -2\frac{1}{3}$ $j > -28$
15. $2.4 < -1.8q + 5$ $q < 1\frac{4}{9}$

Lesson 8-1 *(Pages 330–333)*
Write a proportion to solve each problem. Then solve.

1. On a radar screen the distance between two planes is $3\frac{1}{2}$ inches. If the scale is 1 inch on the screen to 2 miles in the air, what is the actual distance between the two planes. $\frac{1}{2} = \frac{3\frac{1}{2}}{x}$; **7 miles**

2. A car travels 144 miles on 4 gallons of gasoline. At this rate, how many gallons are needed to drive 450 miles? $\frac{144}{4} = \frac{450}{x}$; **12.5 gallons**

3. A park ranger stocks a pond with 4 sunfish for every three perch. Suppose 296 sunfish are put in the pond, how many perch should be stocked? $\frac{4}{3} = \frac{296}{x}$; **222 perch**

4. A furniture store bought 8 identical sofas for $4,000. How much did each sofa cost? $\frac{8}{4,000} = \frac{1}{x}$; **$500**

Lesson 8-2 *(Pages 335–338)*
Express each fraction as a percent.

1. $\frac{2}{100}$ **2%**
2. $\frac{3}{25}$ **12%**
3. $\frac{20}{25}$ **80%**
4. $\frac{10}{16}$ **62.5%**
5. $\frac{4}{6}$ **$66\frac{2}{3}$%**
6. $\frac{1}{4}$ **25%**
7. $\frac{26}{100}$ **26%**
8. $\frac{3}{10}$ **30%**
9. $\frac{21}{50}$ **42%**
10. $\frac{7}{8}$ **87.5%**
11. $\frac{1}{3}$ **$33\frac{1}{3}$%**
12. $\frac{2}{3}$ **$66\frac{2}{3}$%**
13. $\frac{2}{5}$ **40%**
14. $\frac{2}{50}$ **4%**
15. $\frac{8}{10}$ **80%**
16. $\frac{5}{12}$ **$41\frac{2}{3}$%**
17. $\frac{7}{10}$ **70%**
18. $\frac{9}{20}$ **45%**
19. $\frac{1}{2}$ **50%**
20. $\frac{3}{20}$ **15%**
21. $\frac{10}{25}$ **40%**
22. $\frac{3}{8}$ **37.5%**
23. $\frac{4}{20}$ **20%**
24. $\frac{19}{25}$ **76%**

Write a percent proportion to solve each problem. Then solve. Round answers to the nearest tenth.

25. 39 is 5% of what number? $\frac{39}{n} = \frac{5}{100}$; **780**
26. What is 19% of 200? $\frac{19}{100} = \frac{n}{200}$; **38**
27. 28 is what percent of 7? $\frac{28}{7} = \frac{n}{100}$; **400%**
28. 24 is what percent of 72? $\frac{24}{72} = \frac{n}{100}$; **33.3%**
29. 9 is $33\frac{1}{3}$% of what number? $\frac{33\frac{1}{3}}{100} = \frac{9}{n}$; **27**
30. Find 55% of 134. $\frac{55}{100} = \frac{n}{134}$; **73.7**

Lesson 8-3 *(Pages 339–341)*
Write an equation in the form P = RB for each problem. Then solve.

1. Find 5% of $73.
 $P = 0.05 \cdot 73$; **$3.65**
2. What is 15% of 15?
 $P = 0.15 \cdot 15$; **2.25**
3. Find 80% of $12.
 $P = 0.80 \cdot 12$; **$9.60**
4. What is 7.3% of 500?
 $P = 0.073 \cdot 500$; **36.5**
5. Find 21% of $720.
 $P = 0.21 \cdot 720$; **$151.20**
6. What is 12% of $62.50?
 $P = 0.12 \cdot 62.50$; **$7.50**
7. Find 0.3% of 155.
 $P = 0.003 \cdot 155$; **0.465**
8. What is 75% of $450?
 $P = 0.75 \cdot 450$; **$337.50**
9. Find 7.2% of 10.
 $P = 0.072 \cdot 10$; **0.72**
10. What is 10.1% of $60?
 $P = 0.101 \cdot 60$; **$6.06**
11. Find 23% of 47.
 $P = 0.23 \cdot 47$; **10.81**
12. What is 89% of 654?
 $P = 0.89 \cdot 654$; **582.06**
13. $20 is what percent of $64?
 $20 = R \cdot 64$; **31.25%**
14. Sixty-nine is what percent of 200?
 $69 = R \cdot 200$; **34.5%**
15. Seventy is what percent of 150?
 $70 = R \cdot 150$; **$46\frac{2}{3}$%**
16. 26 is 30% of what number?
 $26 = 0.30 \cdot B$; **$86\frac{2}{3}$**
17. 7 is 14% of what number?
 $7 = 0.14 \cdot B$; **50**
18. $35.50 is what percent of $150?
 $35.50 = R \cdot 150$; **$23\frac{2}{3}$%**
19. $17 is what percent of $25?
 $17 = R \cdot 25$; **68%**
20. 152 is 2% of what number?
 $152 = 0.02 \cdot B$; **7,600**

Lesson 8-3B (Pages 342–343)
Solve.

1. Out of 25 eighth-graders surveyed, 16 said that science was their favorite subject. If there are 173 eighth-graders at Monroe Middle School, about how many will choose science as their favorite subject? **about 110 students**

2. What is 40% of 725? **290**

3. What is the sum of the whole numbers from 1 to 100? **5,050**

4. Thirty percent of the 320 people who visited the Museum of Natural History today were students. How many of the visitors were students? **96 visitors**

5. How many cuts are needed to separate a long board into 25 smaller boards? **24 cuts**

Lesson 8-4 (Pages 344–347)
Express each percent as a fraction or mixed number in simplest form.

1. 540% $5\frac{2}{5}$
2. $\frac{25}{50}$% $\frac{1}{200}$
3. 0.02% $\frac{1}{5,000}$
4. 620% $6\frac{1}{5}$
5. 0.7% $\frac{7}{1,000}$
6. 111.5% $1\frac{23}{200}$
7. $\frac{7}{35}$% $\frac{1}{500}$
8. 0.72% $\frac{9}{1,250}$
9. 0.004% $\frac{1}{25,000}$
10. 364% $3\frac{16}{25}$
11. 0.15% $\frac{3}{2,000}$
12. 1,250% $12\frac{1}{2}$
13. $\frac{9}{10}$% $\frac{9}{1,000}$
14. 730% $7\frac{3}{10}$
15. 100.01% $1\frac{1}{10,000}$

Express each percent as a decimal. Round to the nearest ten-thousandth.

16. 0.07% **0.0007**
17. $\frac{2}{3}$% **0.0067**
18. 310% **3.1**
19. 6.05% **0.0605**
20. 7,652% **76.52**
21. $\frac{12}{50}$% **0.0024**
22. 0.93% **0.0093**
23. 200% **2**
24. 197.6% **1.976**
25. 10.75% **0.1075**
26. 0.66% **0.0066**
27. 417% **4.17**
28. 7.76% **0.0776**
29. 390% **3.9**
30. $\frac{7}{10}$% **0.007**

Lesson 8-5 (Pages 348–351)
Find each percent of change. Round to the nearest percent.

1. original: $35
   new: $29 **17%**
2. original: $550
   new: $425 **23%**
3. original: $72
   new: $88 **22%**
4. original: $25
   new: $35 **40%**
5. original: $28
   new: $19 **32%**
6. original: $46
   new: $55 **20%**
7. original: $78
   new: $44 **44%**
8. original: $120
   new: $75 **38%**

Find the sale price of each item to the nearest cent. 9. **$2,743** 10. **$12.60** 11. **$17.40**

9. $4,220 piano, 35% off
10. $14 scissors, 10% off
11. $29 book, 40% off
12. $38 sweater, 25% off **$28.50**
13. $45 pants, 50% off **$22.50**
14. $280 VCR, 25% off **$210**
15. $3,540 motorcycle, 30% off **$2,478**
16. $15.95 compact disc, 20% off **$12.76**

Find the selling price for each item given the amount paid by the store and markup. Round to the nearest cent.

17. $250.00 golf clubs, 30% markup **$325**
18. $17.00 compact disc, 15% markup **$19.55**
19. $57.00 shoes, 45% markup **$82.65**
20. $26.00 book, 20% markup **$31.20**

Lesson 8-6 (Pages 353–356)

Find the simple interest to the nearest cent.

1. $500 at 7% for 2 years   **$70**
2. $2,500 at 6.5% for 36 months   **$487.50**
3. $8,000 at 6% for 1 year   **$480**
4. $1,890 at 9% for 42 months   **$595.35**
5. $760 at 4.5% for $2\frac{1}{2}$ years   **$85.50**
6. $12,340 at 5% for 6 months   **$308.50**

Find the total amount in each account to the nearest cent.

7. $300 at 10% for 3 years   **$390**
8. $3,200 at 8% for 6 months   **$3,328**
9. $20,000 at 14% for 20 years   **$76,000**
10. $4,000 at 12.5% for 4 years   **$6,000**
11. $450 at 11% for 5 years   **$697.50**
12. $17,000 at 15% for $9\frac{1}{2}$ years   **$41,225**

Lesson 8-7 (Pages 357–360)

Tell whether each pair of polygons is similar. Explain your reasoning.

1.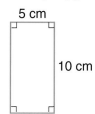

   Yes; corresponding angles have same measure, corresponding sides in proportion.

2.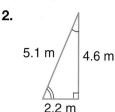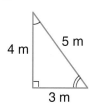

   No; corresponding sides are not in proportion.

Each pair of polygons is similar. Write a proportion to find the missing measure x. Then find the value of x.

3. 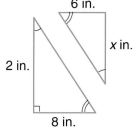 $\frac{6}{8} = \frac{x}{12}$; **9**

4. 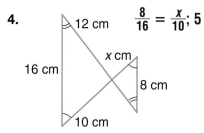 $\frac{8}{16} = \frac{x}{10}$; **5**

Lesson 8-8 (Pages 361–364)

Write a proportion for each problem and then solve it. Assume the triangles are similar.

1. A road sign casts a shadow 14 meters long, while a tree nearby casts a shadow 27.8 meters long. If the road sign is 3.5 meters high, how tall is the tree? $\frac{3.5}{x} = \frac{14}{27.8}$; **6.95 m**

2. Find the distance across Catfish Lake. $\frac{1.2}{4.5} = \frac{0.8}{x}$; **3 km**

3. A 7-foot tall flag stick on a golf course casts a shadow 21 feet long. A golfer standing nearby casts a shadow 16.5 feet long. How tall is the golfer? $\frac{7}{x} = \frac{21}{16.5}$; **5.5 ft**

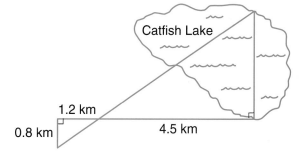

Extra Practice **629**

Lesson 8-9 (Pages 366–369)
Solve.

1. The distance between two cities on a map is 3.2 centimeters. If the scale on the map is 1 centimeter = 50 miles, find the actual distance between the two cities. **160 miles**

2. A scale model of the Empire State Building is 10 inches tall. If the Empire State Building is 1,250 feet tall, find the scale of this model. **1 inch: 125 feet**

3. On a scale drawing of a house, the dimensions of the living room are 4 inches by 3 inches. If the scale of the drawing is 1 inch = 6 feet, find the actual dimensions of the living room. **24 feet by 18 feet**

4. The trip from Columbus to Dayton is approximately 70 miles. If a scale on an Ohio map is 1 inch = 11 miles, about how far apart are the cities on the map? **6.4 inches**

Lesson 8-10 (Pages 370–373)
Find the coordinates of the image of each point for a dilation with a scale factor $\frac{5}{4}$.

1. $x(6, 8)$ $x'\left(\frac{15}{2}, 10\right)$
2. $y(4, 12)$ $y'(5, 15)$
3. $z(10, 2)$ $z'\left(\frac{25}{2}, \frac{5}{2}\right)$

Triangle ABC has vertices A(2, 2), B(−1, 4), and C(−3, −5). Find the coordinates of its image for a dilation with each given scale factor. Graph △ABC and its dilation. **See Answer Appendix.**

4. 1
5. 0.5
6. 2
7. 3
8. $\frac{1}{4}$

In each figure, the green figure is a dilation of the blue figure. Find each scale factor.

9. 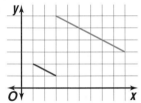 3

10. 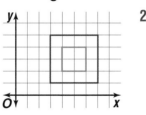 2

Lesson 9-1 (Pages 382–384)
Find each square root.

1. $\sqrt{9}$ 3
2. $\sqrt{0.16}$ 0.4
3. $\sqrt{81}$ 9
4. $\sqrt{0.04}$ 0.2
5. $-\sqrt{625}$ −25
6. $\sqrt{36}$ 6
7. $-\sqrt{169}$ −13
8. $\sqrt{144}$ 12
9. $\sqrt{2.25}$ 1.5
10. $\sqrt{961}$ 31
11. $\sqrt{324}$ 18
12. $-\sqrt{225}$ −15
13. $\sqrt{0.01}$ 0.1
14. $-\sqrt{4}$ −2
15. $-\sqrt{0.09}$ −0.3
16. $\sqrt{529}$ 23
17. $-\sqrt{484}$ −22
18. $\sqrt{196}$ 14
19. $\sqrt{0.49}$ 0.7
20. $\sqrt{1.69}$ 1.3
21. $\sqrt{729}$ 27
22. $\sqrt{0.36}$ 0.6
23. $\sqrt{289}$ 17
24. $-\sqrt{16}$ −4
25. $\sqrt{1,024}$ 32
26. $\sqrt{\frac{289}{10,000}}$ $\frac{17}{100}$
27. $\sqrt{\frac{169}{121}}$ $\frac{13}{11}$
28. $-\sqrt{\frac{4}{9}}$ $-\frac{2}{3}$
29. $-\sqrt{\frac{81}{64}}$ $-\frac{9}{8}$
30. $\sqrt{\frac{25}{81}}$ $\frac{5}{9}$

Lesson 9-2 (Pages 386–389)
Estimate to the nearest whole number.

1. $\sqrt{229}$ $\sqrt{225} = 15$
2. $\sqrt{63}$ $\sqrt{64} = 8$
3. $\sqrt{290}$ $\sqrt{289} = 17$
4. $\sqrt{27}$ $\sqrt{25} = 5$
5. $\sqrt{1.30}$ $\sqrt{1} = 1$
6. $\sqrt{8.4}$ $\sqrt{9} = 3$
7. $\sqrt{96}$ $\sqrt{100} = 10$
8. $\sqrt{19}$ $\sqrt{16} = 4$
9. $\sqrt{200}$ $\sqrt{196} = 14$
10. $\sqrt{76}$ $\sqrt{81} = 9$
11. $\sqrt{17}$ $\sqrt{16} = 4$
12. $\sqrt{34}$ $\sqrt{36} = 6$
13. $\sqrt{137}$ $\sqrt{144} = 12$
14. $\sqrt{540}$ $\sqrt{529} = 23$
15. $\sqrt{165}$ $\sqrt{169} = 13$
16. $\sqrt{326}$ $\sqrt{324} = 18$
17. $\sqrt{52}$ $\sqrt{49} = 7$
18. $\sqrt{37}$ $\sqrt{36} = 6$
19. $\sqrt{79}$ $\sqrt{81} = 9$
20. $\sqrt{18.35}$ $\sqrt{16} = 4$
21. $\sqrt{71}$ $\sqrt{64} = 8$
22. $\sqrt{117}$ $\sqrt{121} = 11$
23. $\sqrt{410}$ $\sqrt{400} = 20$
24. $\sqrt{25.70}$ $\sqrt{25} = 5$
25. $\sqrt{333}$ $\sqrt{324} = 18$
26. $\sqrt{23}$ $\sqrt{25} = 5$
27. $\sqrt{89}$ $\sqrt{81} = 9$
28. $\sqrt{47}$ $\sqrt{49} = 7$

Lesson 9-3 (Pages 390–394)
Name the set or sets of numbers to which each real number belongs. Let R = real numbers, Q = rational numbers, Z = integers, W = whole numbers, and I = irrational numbers.

1. 6.5 Q, R
2. $\sqrt{25}$ W, Z, Q, R
3. $\sqrt{3}$ I, R
4. -7.2 Q, R
5. $-0.\overline{61}$ Q, R

Find an estimate for each square root. Then graph the square root on a number line. See Answer Appendix for number lines.

6. $-\sqrt{12}$ -3.5
7. $\sqrt{23}$ 4.8
8. $\sqrt{2}$ 1.4
9. $\sqrt{10}$ 3.2
10. $-\sqrt{30}$ -5.5

Solve each equation. Round solutions to the nearest tenth.

11. $y^2 = 49$ $7, -7$
12. $x^2 = 225$ $15, -15$
13. $x^2 = 64$ $8, -8$
14. $y^2 = 79$ $8.9, -8.9$
15. $x^2 = 16$ $4, -4$
16. $y^2 = 24$ $4.9, -4.9$
17. $y^2 = 625$ $25, -25$
18. $x^2 = 81$ $9, -9$

4. $6^2 + 5^2 = c^2$; 7.8 cm
5. $12^2 + 12^2 = c^2$; 17.0 ft
6. $8^2 + 6^2 = c^2$; 10 in.
7. $20^2 + b^2 = 25^2$; 15 m
8. $9^2 + b^2 = 14^2$; 10.7 mm
9. $a^2 + 15^2 = 20^2$; 13.2 m
10. $5^2 + 50^2 = c^2$; 50.2 ft
11. $4.5^2 + b^2 = 8.5^2$; 7.2 yd

Lesson 9-4 (Pages 398–401) 2. $6^2 + 2^2 = x^2$; 6.3 cm

Write an equation you could use to find the length of the missing side of each right triangle. Then find the missing length. Round to the nearest tenth.

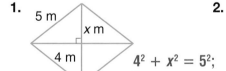
1. 5 m, x m, 4 m; $4^2 + x^2 = 5^2$; 3 m

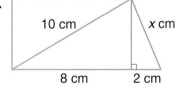
2. 10 cm, x cm, 8 cm, 2 cm

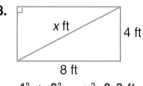
3. x ft, 4 ft, 8 ft; $4^2 + 8^2 = x^2$; 8.9 ft

4. a, 6 cm; b, 5 cm
5. a, 12 ft; b, 12 ft
6. a, 8 in.; b, 6 in.
7. a, 20 m; c, 25 m
8. a, 9 mm; c, 14 mm
9. b, 15 m; c, 20 m
10. a, 5 ft; b, 50 ft
11. a, 4.5 yd; c, 8.5 yd

Determine whether each triangle with sides of given length is a right triangle.

12. 15 m, 8 m, 17 m yes
13. 7 yd, 5 yd, 9 yd no
14. 5 in., 12 in., 13 in. yes

Lesson 9-5A (Pages 402–403)
Solve.

1. The streets in Sachi's city are arranged in square blocks. Sachi left her house and walked 5 blocks east and 2 blocks north to Jim's house. She and Jim walked 1 block north, 3 blocks west and 1 block north to school. After school Sachi walked 2 blocks west and 2 blocks south to Bella's house. How far from Bella's house is Sachi's house?   **2 blocks north**

2. If there are 3 different ticket prices and 5 different days for attending a performance of a play, how many possible choices are there for date and ticket price?   **15 choices**

3. Alice has 3 logs. The first is twice as long as the second. The third is 3 times as long as the second. Together they are 24 feet long. How long is each log?   **8 feet, 4 feet, 12 feet**

4. The streets in a town are arranged in square blocks. Brad leaves his house and rollerblades 4 blocks south and 5 blocks west to the athletic field. He then rollerblades 2 blocks south and 5 blocks east to the library. How far is the library from Brad's home?   **6 blocks south**

Lesson 9-5 (Pages 404–407)
State an equation that can be used to answer each question. Then solve. Round to the nearest tenth.

1. How far apart are the boats?

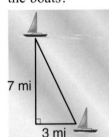

$7^2 + 3^2 = c^2$; $c = 7.6$ mi

2. How high does the ladder reach?

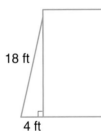

$4^2 + b^2 = 18^2$; 17.5 ft

3. How long is each rafter?

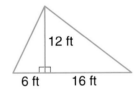

$6^2 + 12^2 = c^2$; 13.4 ft
$16^2 + 12^2 = c^2$; 20 ft

4. Name the primitive Pythagorean triple to which 24-45-51 belongs.   **8-15-17**

Lesson 9-6 (Pages 410–413)
Find the distance between each pair of points whose coordinates are given. Round to the nearest tenth.

1. (1, 2); (−3, −3)   **6.4**

2. (−1, 4); (4, 1)   **5.8**

3. (0, 4); (7, 1)   **7.6**

Graph each pair of ordered pairs. Then find the distance between the points. Round to the nearest tenth.   See Answer Appendix for graphs.

4. (−4, 2); (4, 17)   **17**
5. (5, −1); (11, 7)   **10**
6. (−3, 5); (2, 7)   **5.4**
7. (7, −9); (4, 3)   **12.4**
8. (5, 4); (−3, 8)   **8.9**
9. (−8, −4); (−3, 8)   **13**
10. (2, 7); (10, −4)   **13.6**
11. (9, −2); (3, 6)   **10**

632 Extra Practice

Lesson 9-7 (Pages 414–417)
Find the missing lengths. Round to the nearest tenth.

1.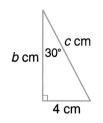
$b \approx 6.9, c = 8$

2.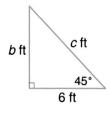
$b = 6, c \approx 8.5$

3.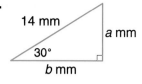
$a = 7, b \approx 12.1$

4.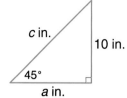
$a = 10, c \approx 14.1$

5.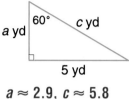
$a \approx 2.9, c \approx 5.8$

6.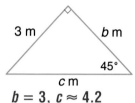
$b = 3, c \approx 4.2$

7.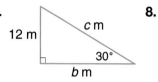
$b \approx 20.8, c = 24$

8.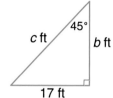
$b = 17, c \approx 24.0$

Lesson 10-1 (Pages 428–431)
Copy and complete each function table.

1. $f(n) = -4n$

n	−4n	f(n)
−2	−4(−2)	8
−1	−4(−1)	4
0	−4(0)	0
1	−4(1)	−4
2	−4(2)	−8

2. $f(n) = n + 6$

n	n + 6	f(n)
−6	−6 + 6	0
−4	−4 + 6	2
−2	−2 + 6	4
0	0 + 6	6
2	2 + 6	8

3. $f(n) = 3n + 2$

n	3n + 2	f(n)
−3.5	3(−3.5)+2	−8.5
−2.5	3(−2.5)+2	−5.5
−1.5	3(−1.5)+2	−2.5
0	3(0)+2	2
1.5	3(1.5)+2	6.5

Find each function value.

4. $f\left(\frac{1}{2}\right)$ if $f(n) = 2n - 6$ −5
5. $f(-4)$ if $f(n) = -\frac{1}{2}n + 4$ 6
6. $f(1)$ if $f(n) = -5n + 1$ −4
7. $f(6)$ if $f(n) = \frac{2}{3}n - 5$ −1
8. $f(0)$ if $f(n) = 1.6n + 4$ 4
9. $f(2)$ if $f(n) = 2n^2 - 8$ 0

Lesson 10-2 (Pages 433–435) 1–9. See Answer Appendix for graphs.
Copy and complete each function table. Then graph the function.

1. $f(n) = 6n + 2$

n	f(n)	(n, f(n))
−3	−16	(−3, −16)
−1	−4	(−1, −4)
1	8	(1, 8)
$\frac{7}{3}$	16	$\left(\frac{7}{3}, 16\right)$

2. $f(n) = -2n + 3$

n	f(n)	(n, f(n))
−2	7	(−2, 7)
−1	5	(−1, 5)
0	3	(0, 3)
1	1	(1, 1)
2	−1	(2, −1)

3. $f(n) = 4.5n$

n	f(n)	(n, f(n))
−4	−18	(−4, −18)
−2	−9	(−2, −9)
0	0	(0, 0)
1	4.5	(1, 4.5)
6	27	(6, 27)

Choose values for n and graph each function.

4. $f(n) = \frac{8}{n}$
5. $f(n) = \frac{2}{3}n + 1$
6. $f(n) = n^2 - 1$
7. $f(n) = 3.5n$
8. $f(n) = 4n - 1$
9. $f(n) = \frac{3}{5}n + \left(-\frac{1}{5}\right)$

Lesson 10-3 (Pages 437–440)
Copy and complete the table for each equation.

1. $y = 3x - 1$

x	y
−5	−16
−3	−10
−1	−4
0	−1
1	2

2. $y = \frac{x}{4} + 2$

x	y
−8	0
−4	1
0	2
4	3

3. $y = -1.5x - 3$

x	y
−4	3
−2	0
2	−6
6	−12
10	−18

4. $y = 4x - 3$

x	y
−2	−11
$\frac{1}{2}$	−1
0	−3
$2\frac{1}{4}$	6

Find four solutions of each equation. 5–12. See Answer Appendix.

5. $y = -3x + 5$
6. $y = 2x - 1$
7. $y = \frac{2}{3}x + 4$
8. $y = -0.4x$
9. $y = 12x - 8$
10. $y = \frac{3}{4}x + 2$
11. $y = -2.4x - 3$
12. $y = 5x + 7$

Lesson 10-4 (Pages 442–444)
Graph each function. 1–20. See Answer Appendix.

1. $y = -5x$
2. $y = 10x - 2$
3. $y = -2.5x - 1.5$
4. $y = 7x + 3$
5. $y = \frac{x}{4} - 8$
6. $y = 3x + 1$
7. $y = 25 - 2x$
8. $y = \frac{x}{6}$
9. $y = -2x + 11$
10. $y = 7x - 3$
11. $y = \frac{x}{2} + 5$
12. $y = 4 - 6x$
13. $y = -3.5x - 1$
14. $y = 4x + 10$
15. $y = 8x$
16. $y = -5x + \frac{1}{2}$
17. $y = \frac{x}{3} + 9$
18. $y = -7x + 15$
19. $y = 10x - 2$
20. $y = 1.5x - 7.5$

Lesson 10-5 (Pages 446–449)
Solve each system of equations by graphing. 1–15. See Answer Appendix for graphs.

1. $y = x$
   $y = -x + 4$ (2, 2)
2. $y = -x + 8$
   $y = x - 2$ (5, 3)
3. $y = -3x$
   $y = -4x + 2$ (2, −6)
4. $y = x - 1$
   $y = -x + 11$ (6, 5)
5. $y = -x$
   $y = 2x$ (0, 0)
6. $y = -x + 3$
   $y = x + 3$ (0, 3)
7. $y = x - 3$
   $y = 2x + 8$ (−11, −14)
8. $y = -x + 6$
   $y = x + 2$ (2, 4)
9. $y = -x + 1$
   $y = x - 4$ $\left(2\frac{1}{2}, -1\frac{1}{2}\right)$
10. $y = -3x + 6$
    $y = x - 2$ (2, 0)
11. $y = 3x - 4$
    $y = -3x - 4$ (0, −4)
12. $y = 2x + 4$
    $y = 3x - 9$ (13, 30)
13. $y = -x + 4$
    $y = x - 10$ (7, −3)
14. $y = -x + 6$
    $y = 2x$ (2, 4)
15. $y = x - 4$
    $y = -2x + 5$ (3, −1)

Lesson 10-6A (Pages 450–451)

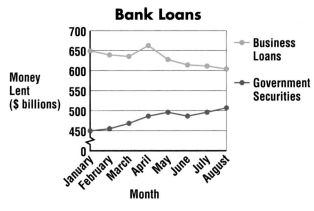
Bank Loans

Use the graph above for Exercises 1–4.

1. Estimate the amount of money banks lent businesses in April. about $660 billion
2. In which month did business loans see its biggest increase? from March to April
3. By about how much did Government Securities increase from March to May? about $20 billion
4. Describe the overall trend in bank lending. There was a decrease in business loans and an increase in government securities.

Lesson 10-6 (Pages 452–455)
Graph each quadratic function. 1–28. See Answer Appendix.

1. $y = x^2 - 1$
2. $y = 1.5x^2 + 3$
3. $f(n) = n^2 - n$
4. $y = 2x^2$
5. $y = x^2 + 3$
6. $y = -3x^2 + 4$
7. $y = -x^2 + 7$
8. $f(n) = 3n^2$
9. $f(n) = 3n^2 + 9n$
10. $y = -x^2$
11. $y = \frac{1}{2}x^2 + 1$
12. $y = 5x^2 - 4$
13. $y = -x^2 + 3x$
14. $f(n) = 2.5n^2$
15. $y = -2x^2$
16. $y = 8x^2 + 3$
17. $y = -x^2 + \frac{1}{2}x$
18. $y = -4x^2 + 4$
19. $f(n) = 4n^2 + 3$
20. $y = -4x^2 + 1$
21. $y = 2x^2 + 1$
22. $y = x^2 - 4x$
23. $y = 3x^2 + 5$
24. $f(n) = 0.5n^2$
25. $f(n) = 2n^2 - 5n$
26. $y = \frac{3}{2}x^2 - 2$
27. $y = 6x^2 + 2$
28. $f(n) = 5n^2 + 6n$

Lesson 10-7 (Pages 456–459) 1–5. See Answer Appendix for graphs.
Find the coordinates of the vertices of each figure after the translation described. Then graph the figure and its translation. 1. $A'(-2, 1)$, $B'(3, 4)$, $C'(6, 1)$ 2. $X'(1, 1)$, $Y'(5, 1)$, $Z'(3, -3)$

1. $\triangle ABC$ with vertices $A(-6, -2)$, $B(-1, 1)$, and $C(2, -2)$, translated by $(4, 3)$
2. $\triangle XYZ$ with vertices $X(-4, 3)$, $Y(0, 3)$, and $Z(-2, -1)$, translated by $(5, -2)$
3. rectangle $HIJK$ with vertices $H(1, 3)$, $I(4, 0)$, $J(2, -2)$, and $K(-1, 1)$, translated by $(-4, -6)$ $H'(-3,-3)$, $I'(0, -6)$, $J'(-2, -8)$, $K'(-5,-5)$
4. rectangle $PQRS$ with vertices $P(-7, 6)$, $Q(-5, 6)$, $R(-5, 2)$, and $S(-7, 2)$, translated by $(9, -1)$ $P'(2, 5)$, $Q'(4, 5)$, $R'(4, 1)$, $S'(2, 1)$
5. pentagon $DGLMR$ with vertices $D(1, 3)$, $G(2, 4)$, $L(4, 4)$, $M(5, 3)$, and $R(3, 1)$, translated by $(-5, -7)$ $D'(-4, -4)$, $G'(-3, -3)$, $L'(-1, -3)$, $M'(0, -4)$, $R'(-2, -6)$

Lesson 10-8 *(Pages 460–463)*
Name the line of symmetry for each pair of figures.

1.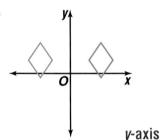
   y-axis

2.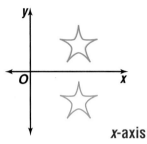
   x-axis

3.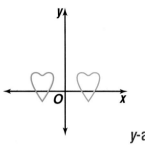
   y-axis

Graph each figure. Then draw its reflections over the x-axis and over the y-axis.

4. △CAT with vertices C(2, 3), A(8, 2), and T(4, −3)
5. trapezoid TRAP with vertices T(−2, 5), R(1, 5), A(4, 2), and P(−5, 2)
6. rectangle ABCD with vertices A(4, −1), B(7, −4), C(4, −7), and D(1, −4)

4–6. See Answer Appendix.

Lesson 10-9 *(Pages 464–467)*
Triangle ABC has vertices A(−2, −1), B(0, 1), and C(1, −1).

1. Graph △ABC. 1, 3. See Answer Appendix.
2. Find the coordinates of the vertices after a 90° counterclockwise rotation. A'(1, −2), B'(−1, 0), C'(1, 1)
3. Graph △A'B'C'.

Rectangle WXYZ has vertices W(1, 1), X(1, 3), Y(6, 3), and Z(6, 1).

4. Graph rectangle WXYZ. 4, 6. See Answer Appendix.
5. Find the coordinates of the vertices after a rotation of 180°. W'(−1, −1), X'(−1, −3), Y'(−6, −3), Z'(−6, −1)
6. Graph rectangle W'X'Y'Z'.

Lesson 11-1 *(Pages 476–479)*
Find the area of each circle to the nearest tenth. 6. 28.3 in² 7. 452.4 in² 8. 201.1 yd²

1.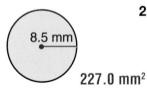
   227.0 mm²

2.
   394.1 m²

3.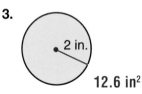
   12.6 in²

4.
   254.5 ft²

5. radius, 4 m 50.3 m²
6. diameter, 6 in.
7. radius, 12 in.
8. diameter, 16 yd
9. diameter, 11 ft 95.0 ft²
10. radius, 5 in. 78.5 in²
11. radius, 19 cm 1,134.1 cm²
12. diameter, 29 mm 660.5 mm²

Find the area of each shaded region to the nearest tenth.

13.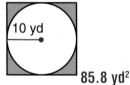
    85.8 yd²

14.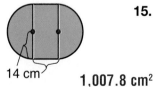
    1,007.8 cm²

15.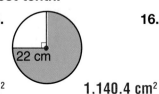
    1,140.4 cm²

16.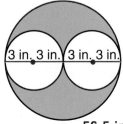
    56.5 in²

Lesson 11-2A (Pages 480–481)
Solve.

1. How many cubes are needed to make this display? **35 cubes**

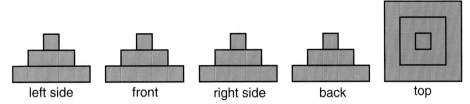

left side   front   right side   back   top

2. How many mirrored tiles would be needed to cover the display in Exercise 1 if each tile is the same size as the face of a cube? (The bottom of the display will not be covered with the tiles.) **61 tiles**

3. How many cubes 3 cm on a side can be packaged in a 12 cm by 9 cm by 6 cm box? **24 cubes**

4. How many cubes are needed to make the display shown in the plan below? **18 cubes**

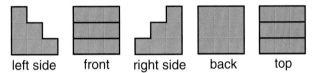

left side   front   right side   back   top

Lesson 11-2 (Pages 482–485) 1–4. See Answer Appendix.
Use isometric dot paper to draw each solid.

1. A rectangular prism that is 4 units high, 5 units long, and 6 units deep.
2. A triangular prism that is 4 units high.
3. A pyramid with a pentagonal base.
4. On grid paper, draw the front view, back view, two side views, and top view of the figure.

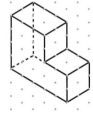

Lesson 11-3 (Pages 486–489)
Find the volume of each solid to the nearest tenth.

1. 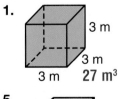 3 m, 3 m, 3 m **27 m³**
2. 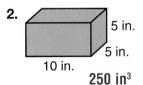 5 in., 5 in., 10 in. **250 in³**
3. 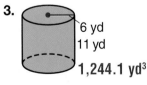 6 yd, 11 yd **1,244.1 yd³**
4. 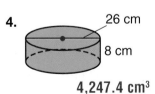 26 cm, 8 cm **4,247.4 cm³**
5. 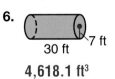 7 mm, 9 mm, 8 mm **504 mm³**
6. 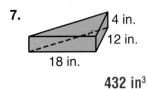 30 ft, 7 ft **4,618.1 ft³**
7. 4 in., 12 in., 18 in. **432 in³**
8. 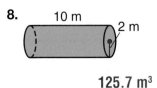 10 m, 2 m **125.7 m³**

Lesson 11-4 (Pages 490–493)
Find the volume of each solid to the nearest tenth.

1. 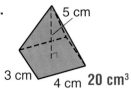 5 cm, 3 cm, 4 cm **20 cm³**
2. 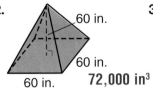 60 in., 60 in., 60 in. **72,000 in³**
3. 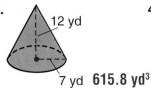 12 yd, 7 yd **615.8 yd³**
4. 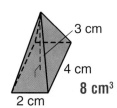 3 cm, 4 cm, 2 cm **8 cm³**

5. 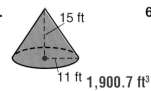 15 ft, 11 ft **1,900.7 ft³**
6. 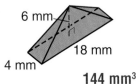 6 mm, 18 mm, 4 mm **144 mm³**
7. 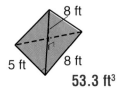 8 ft, 8 ft, 5 ft **53.3 ft³**
8. 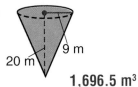 9 m, 20 m **1,696.5 m³**

Lesson 11-5 (Pages 495–498)
Find the surface area of each prism to the nearest tenth.

1. 2 ft, 2 ft, 2 ft **24 ft²**
2. 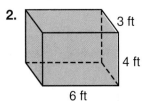 3 ft, 4 ft, 6 ft **108 ft²**
3. 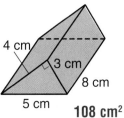 4 cm, 3 cm, 8 cm, 5 cm **108 cm²**
4. 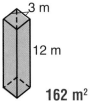 3 m, 12 m **162 m²**

5. 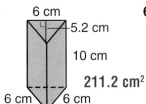 6 cm, 5.2 cm, 10 cm, 6 cm, 6 cm **211.2 cm²**
6. 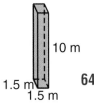 10 m, 1.5 m, 1.5 m **64.5 m²**
7. 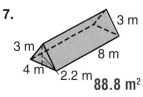 3 m, 3 m, 8 m, 4 m, 2.2 m **88.8 m²**
8. 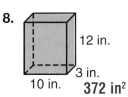 12 in., 3 in., 10 in. **372 in²**

Lesson 11-6 (Pages 499–502)
Find the surface area of each cylinder to the nearest tenth.

1. 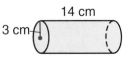 14 cm, 3 cm **320.4 cm²**
2. 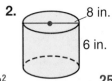 8 in., 6 in. **251.3 in²**
3. 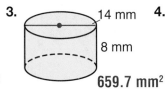 14 mm, 8 mm **659.7 mm²**
4. 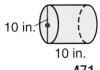 4.2 cm, 12.4 cm **438.1 cm²**

5. 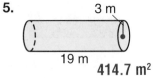 3 m, 19 m **414.7 m²**
6. 8.2 in., 22 in. **1,556.0 in²**
7. 3.6 yd, 14.2 yd **402.6 yd²**
8. 10 in., 10 in. **471.2 in²**

638 Extra Practice

Lesson 11-7 (Pages 504–507)

Analyze each measurement. Give the precision, significant digits if appropriate, greatest possible error, and relative error to two significant digits.

1. 18 min
2. $7\frac{1}{2}$ lb
3. 92.46 m
4. 7 ft
5. 0.067 kg
6. 61.7 cm
7. 8 mm
8. $4\frac{1}{4}$ in.

1–9. See Answer Appendix.

9. Which is a more precise measurement for the length of a pencil, 18 centimeters or 18.0 centimeters? Explain your answer.

10. A wire's length is measured as 0.02 meters.

   a. How many significant digits are there in 0.02? **3**

   b. Suppose the measurement is actually 2.2 centimeters. How many significant digits are there in 2.2? **2**

   c. Compare the preciseness of these two measurements. **See Answer Appendix.**

Lesson 12-1 (Pages 518–520)

Draw a tree diagram to find the number of possible outcomes for each situation. 1–6. See Answer Appendix for tree diagrams.

1. A bakery makes yellow, white, chocolate, or marble cake with a choice of chocolate or vanilla frosting. **8 outcomes**

2. A car comes in white, black, or red with standard or automatic transmission and with a 4-cylinder or 6-cylinder engine. **12 outcomes**

3. A customer can buy roses or carnations in red, yellow, pink, or white. **8 outcomes**

4. A bed comes in queen or king size with a firm or super firm mattress. **4 outcomes**

5. A pizza can be ordered with a regular or deep dish crust and with a choice of one topping, two toppings, or three toppings. **6 outcomes**

6. A woman's shoe comes in red, white, blue, or black with a choice of high, medium, or low heels. **12 outcomes**

Lesson 12-2 (Pages 521–523)

Find each value.

1. 8! **40,320**
2. 10! **3,628,800**
3. 0! **1**
4. 7! **5,040**
5. 6! **720**
6. 5! **120**
7. 2! **2**
8. 11! **39,916,800**
9. 9! **362,880**
10. 4! **24**
11. $P(5, 4)$ **120**
12. $P(3, 3)$ **6**
13. $P(12, 5)$ **95,040**
14. $P(8, 6)$ **20,160**
15. $P(10, 2)$ **90**
16. $P(6, 4)$ **360**
17. $P(7, 6)$ **5,040**
18. $P(9, 9)$ **362,880**

19. How many different ways can a family of four be seated in a row? **24 ways**

20. In how many different ways can you arrange the letters in the word *orange* if you take the letters five at a time? **720 ways**

21. How many ways can you arrange five different colored marbles in a row if the blue one is always in the center? **24 ways**

22. In how many different ways can Kevin listen to each of his ten CDs once? **3,628,800 ways**

Lesson 12-3 (Pages 524–527)

Find each value.

1. $C(8, 4)$ **70**
2. $C(30, 8)$ **5,852,925**
3. $C(10, 9)$ **10**
4. $C(7, 3)$ **35**
5. $C(12, 5)$ **792**
6. $C(17, 16)$ **17**
7. $C(24, 17)$ **346,104**
8. $C(9, 7)$ **36**

9. How many ways can you choose five compact discs from a collection of 17? **6,188 ways**

10. How many combinations of three flavors of ice cream can you choose from 25 different flavors of ice cream? **2,300 combinations**

11. How many ways can you choose three books out of a selection of ten books? **120 ways**

12. How many ways can you choose seven apples out of a bag of two dozen apples? **346,104 ways**

13. How many ways can you choose two movies to rent out of ten possible movies? **45 ways**

Lesson 12-4 (Pages 528–531)

Use Pascal's Triangle to find each value.

1. $C(6, 4)$ **15**
2. $C(6, 2)$ **15**
3. $C(5, 1)$ **5**
4. $C(9, 3)$ **84**
5. $C(3, 2)$ **3**
6. $C(7, 6)$ **7**

Use Pascal's Triangle to answer each question.

7. Four coins are tossed. What is the probability that 2 will show heads and 2 will show tails? $\frac{3}{8}$

8. How many combinations of 7 things taken 3 at a time are possible? **35 combinations**

9. How many different 5-question quizzes can be formed from a test bank of 10 questions? **252 quizzes**

10. How many branches will there be in the last column of a tree diagram showing the outcomes of tossing 4 coins? **8 branches**

Lesson 12-5 (Pages 534–537)

Two socks are drawn from a drawer which contains one red sock, three blue socks, two black socks, and two green socks. Once a sock is selected, it is not replaced. Find the probability of each outcome.

1. a black sock and then a green sock $\frac{1}{14}$
2. a red sock and then a green sock $\frac{1}{28}$
3. a blue sock two times in a row $\frac{3}{28}$
4. a green sock two times in a row $\frac{1}{28}$

There are three quarters, five dimes, and twelve pennies in a bag. Once a coin is drawn from the bag it is not replaced. If two coins are drawn at random, find the probability of each outcome.

5. a quarter and then a penny $\frac{9}{95}$
6. a nickel and then a dime **0**
7. a dime and then a penny $\frac{3}{19}$
8. a dime two times in a row $\frac{1}{19}$

Lesson 12-6A (Pages 538–539)
Solve.

1. Estimate the probability that in a group of 7 people, exactly 5 people will be male. $\frac{1}{6}$

2. In a survey of 130 people, 95 said they exercised regularly and 62 said they watched their diets. If 45 people said they both exercised regularly and watched their diets, how many do neither? **18**

3. Act it out to find the probability that 4 coins tossed in the air all will land showing tails. $\frac{1}{16}$

4. Ryuji has 8 compact discs in his glove compartment. Three are classical, four are rock, and one is country. If he chooses one at random each day, what is the probability that he will choose a rock tape 3 days in a row? $\frac{1}{8}$

Lesson 12-6 (Pages 540–543)
Roll a number cube 50 times. 1–3. See students' work.

1. Record the results.
2. Based on your results, what is the probability of a 5?
3. Based on your results, what is the probability of an odd number?
4. What is the theoretical probability of an odd number? $\frac{1}{2}$

Lesson 12-7 (Pages 546–548)
Use the survey on favorite type of music to answer each question.

1. What is the size of the sample? **250**
2. What is the mode? **light rock**
3. What fraction prefers country music? $\frac{36}{125}$
4. What fraction prefers rap music? $\frac{9}{50}$

Favorite Type of Music	
Country	72
Heavy Metal	41
Rap	45
Light Rock	92

Use the survey on favorite fruit to answer each question.

5. What is the size of the sample? **600**
6. If 7,950 people were to choose one of these three fruits, how many would you expect to choose oranges? **3,975 people**

Favorite Fruit	
Apple	155
Orange	300
Banana	145

Lesson 13-1 (Pages 561–564)

Write a monomial or polynomial for each model.

1. $x^2 - 2x + 1$
2. -5
3. $-2x^2 + 1$
4. $x^2 - 3x$

Model each monomial or polynomial using drawings or algebra tiles. See Answer Appendix.

5. 7
6. $3x$
7. $-2x^2$
8. $-5x + 1$
9. $x + 2$
10. $-2x^2 + 3x$
11. $-x - 4$
12. $x^2 + 2x + 3$
13. $2x^2 - 2$
14. $2x - 5$
15. $-x^2 + 7x$
16. $2x^2 - 3x$

Lesson 13-2 (Pages 565–569)

Name the like terms in each list of terms.   2. $-a, -3a; 2b^2, b^2$

1. $10x, 6y, y^2, 3x$ $10x, 3x$
2. $-a, 2b^2, -3a, b^2$
3. $2m, n, -m, n$ $2m, -m; n, n$
4. $2, 7m, -m^2$ none
5. $3y, 6, 5, y^2$ $6, 5$
6. $4x, 6, 2x, x^2$ $4x, 2x$

Simplify each polynomial using the model.

7. $2x - x^2 + x - 1$ $-x^2 + 3x - 1$

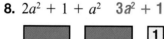

8. $2a^2 + 1 + a^2$ $3a^2 + 1$

Simplify each polynomial. Use drawings or algebra tiles if necessary.

9. $-2y + 3 + x^2 + 5y$
10. $m + m^2 + n + 3m^2$
11. $a^2 + b^2 + 3 + 2b^2$
12. $1 + a + b + 6$ $a + b + 7$
13. $x + x^2 + 5x - 3x^2$
14. $-2y + 3 + y - 2$ $-y + 1$
9. $x^2 + 3y + 3$
10. $4m^2 + m + n$
11. $a^2 + 3b^2 + 3$
13. $-2x^2 + 6x$

Lesson 13-3 (Pages 570–572)

Find each sum. Use drawings or algebra tiles if necessary.

1. $2x^2 - 5x + 7$
   $\underline{+x^2 - x + 11}$
   $3x^2 - 6x + 18$

2. $2m^2 + m + 1$
   $\underline{+(-m^2) + 2m + 3}$
   $m^2 + 3m + 4$

3. $2a - b + 6c$
   $\underline{+ 3a - 7b + 2c}$
   $5a - 8b + 8c$

4. $5a + 3a^2 - 2$
   $\underline{+ 2a + 8a^2 + 4}$
   $7a + 11a^2 + 2$

5. $3c + b + a$
   $\underline{+(-c) + b - a}$
   $2c + 2b$

6. $-z^2 + x^2 + 2y^2$
   $\underline{+ 3z^2 + x^2 + y^2}$
   $2z^2 + 2x^2 + 3y^2$

7. $(5x + 6y) + (2x + 8y)$ $7x + 14y$
8. $(4a + 6b) + (2a + 3b)$ $6a + 9b$
9. $(7r + 11m) + (4m + 2r)$ $15m + 9r$
10. $(-z + z^2) + (-2z + z^2)$ $2z^2 - 3z$
11. $(3x - 7y) + (3y + 4x + 1)$ $7x - 4y + 1$
12. $(5m + 3n - 3) + (8m + 6)$ $13m + 3n + 3$
13. $(a + a^2) + (3a - 2a^2)$ $-a^2 + 4a$
14. $(3s - 5t) + (8t + 2s)$ $5s + 3t$

Lesson 13-4 (Pages 573–576)
Find each difference. Use drawings or algebra tiles if necessary.

1. $5a - 6m$
    $\underline{-\,(2a + 5m)}$
    $3a - 11m$

2. $2a - 7$
    $\underline{-\,(8a - 11)}$
    $-6a + 4$

3. $9r^2 + r + 3$
    $\underline{-\,(11r^2 - r + 12)}$
    $-2r^2 + 2r - 9$

4. $(9x + 3y) - (9y + x)$ $8x - 6y$
5. $(3x^2 + 2x - 1) - (2x + 2)$ $3x^2 - 3$
6. $(a^2 + 6a + 3) - (5a^2 + 5)$ $-4a^2 + 6a - 2$
7. $(5a + 2) - (3a^2 + a + 8)$ $-3a^2 + 4a - 6$
8. $(3x^2 - 7x) - (8x - 6)$ $3x^2 - 15x + 6$
9. $(3m + 3n) - (m + 2n)$ $2m + n$
10. $(3m - 2) - (2m + 1)$ $m - 3$
11. $(x^2 - 2) - (x + 3)$ $x^2 - x - 5$
12. $(5x^2 - 4) - (3x^2 + 8x + 4)$ $2x^2 - 8x - 8$
13. $(7z^2 + 1) - (3z^2 + 2z - 6)$ $4z^2 - 2z + 7$

Lesson 13-5 (Pages 578–581)
Find each product. Express the answer in exponential form.

1. $2^3 \cdot 2^4$ 2^7
2. $5^6 \cdot 5$ 5^7
3. $t^2 \cdot t^2$ t^4
4. $y^5 \cdot y^3$ y^8

Find each product.

5. $a(a + 2)$ $a^2 + 2a$

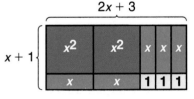

6. $x(2x - 3)$ $2x^2 - 3x$

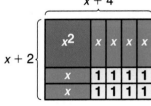

7. $a(a + 4)$ $a^2 + 4a$
8. $m(m - 7)$ $m^2 - 7m$
9. $z^2(z + 3)$ $z^3 + 3z^2$
10. $6x(x + 10)$ $6x^2 + 60x$
11. $3y(5 + y)$ $15y + 3y^2$
12. $2d(d^3 + 1)$ $2d^4 + 2d$
13. $m^3(m^2 - 2)$ $m^5 - 2m^3$
14. $p^5(3p - 1)$ $3p^6 - p^5$
15. $b(9 + 4b)$ $9b + 4b^2$
16. $4t^3(t + 3)$ $4t^4 + 12t^3$
17. $2r^4(5r + 9)$ $10r^5 + 18r^4$
18. $3n^2(6 - 7n^3)$ $18n^2 - 21n^5$

Lesson 13-6 (Pages 583–585)
Find each product.

1. $(x + 1)(2x + 3)$ $2x^2 + 5x + 3$
2. $(x + 2)(x + 4)$ $x^2 + 6x + 8$

Find each product. Use drawings or algebra tiles if necessary.

3. $(r + 3)(r + 4)$ $r^2 + 7r + 12$
4. $(z + 5)(z + 2)$ $z^2 + 7z + 10$
5. $(3x + 7)(x + 1)$ $3x^2 + 10x + 7$
6. $(x + 5)(2x + 3)$ $2x^2 + 13x + 15$
7. $(c + 1)(c + 1)$ $c^2 + 2c + 1$
8. $(a + 3)(a + 7)$ $a^2 + 10a + 21$
9. $(b + 3)(b + 1)$ $b^2 + 4b + 3$
10. $(2y + 1)(y + 3)$ $2y^2 + 7y + 3$
11. $(z + 8)(2z + 1)$ $2z^2 + 17z + 8$
12. $(2m + 4)(m + 5)$ $2m^2 + 14m + 20$
13. $(x + 3)(x + 2)$ $x^2 + 5x + 6$
14. $(c + 2)(c + 8)$ $c^2 + 10c + 16$
15. $(r + 4)(r + 4)$ $r^2 + 8r + 16$
16. $(2x + 4)(x + 4)$ $2x^2 + 12x + 16$

Lesson 13-7A (Pages 586–587)
Solve.

1. Leslie has 6 times as much money as Inazo. If Inazo had $1.20 more, Leslie would have 2 times as much as Inazo. How much money does each person have? **Inazo, $0.60; Leslie, $3.60**

2. Mei-yu bought some pens for $0.89 each and some pencils for $0.19 each. She spent $5.02. How many pens and how many pencils did she buy? **5 pens, 3 pencils**

3. 70 more than a number is twice the number. What is the number? **70**

4. The difference between a number and its double is 28. What is the number? **28**

5. The product of two consecutive numbers is 2,450. What are the two numbers? **49, 50**

Lesson 13-7 (Pages 588–591)
Factor each polynomial.

1. $2x^2 + 4x + 2$ $(2x + 2)(x + 1)$

2. $x^2 + 5x + 4$ $(x + 4)(x + 1)$

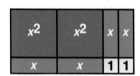

If possible, factor each polynomial. Use drawings or algebra tiles if necessary.

3. $5x + 10$ $5(x + 2)$
4. $x^2 + 8x$ $x(x + 8)$
5. $x^2 + 6x + 1$ not possible
6. $2x^2 + 8x + 6$ $(2x + 2)(x + 3)$
7. $2x^2 + 3x + 5$ not possible
8. $3x^2 + 16x + 5$ $(3x + 1)(x + 5)$
9. $x^2 + x + 4$ not possible
10. $x^2 + 10x + 21$ $(x + 7)(x + 3)$
11. $3x^2 + 22x + 24$ $(3x + 4)(x + 6)$

Mixed Problem Solving

Solve using any strategy.

1. **Money Matters** Out-of-town newspapers cost 60¢, and local papers cost 35¢ each. Bill buys 2 out-of-town papers and 3 local papers. He hands the cashier $3. Should he expect more than 50¢ change? **Yes, he should get 75¢ change**

2. The odometer in Sarah's car registered 68,364.7 miles last week. The odometer registers 71,429.2 miles today. Last week, the number of miles the car traveled was about— **b.**
   a. 900.
   b. 3,000.
   c. 10,000.
   d. 71,000.

3. **School** A section of the school auditorium is set up so that each row has the same number of seats. Cheri is seated in the seventh row from the back and the eighth row from the front of this section. Her seat is the fourth from the right and the seventh from the left. How many seats are in this section?
   140 seats

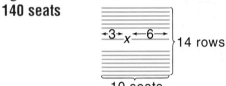

4. David pours 5 gallons of water into a tank of water. Then Carlos adds 8 gallons more to the tank. The tank now contains 32 gallons of water. How many gallons were in the tank before David added water? **19 gallons**

5. Mia arranges her marbles into 2 even columns. Leah puts her marbles into 3 even columns, and Ryan puts his marbles into 5 even columns. Each person has the same number of marbles. What is the least number of marbles that Ryan can have? **30 marbles**

6. **School** When Ruiz walks along the school corridor, he passes through three doorways. The first door is open half the time, the second door is open half the time, and the third door is open two times out of three. What is the probability that all three doors will be open? $\frac{1}{6}$

7. **Design** A memorial waterfall is built of granite cubes using the plans below.

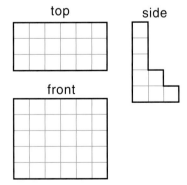

   a. How many cubes of granite are needed? **48 cubes**
   b. If each cube measures 2 feet on an edge, how high is the waterfall? **10 ft**

8. A number plus half the number is 33. Find the number. **22**

Mixed Problem Solving

Solve using any strategy.

1. **Sports** In the Sports Card Collectors' Club, 9 members collect basketball cards, 18 members collect baseball cards, and 15 members collect football cards. Four collect baseball and football cards, 3 collect baseball and basketball cards, 3 collect basketball and football cards, and 1 collects all three types of cards.

   a. Make a Venn diagram of this information.

   b. How many members are in the club?   **33**

   c. How many members collect baseball or football cards?   **29**

   d. How many collect only baseball cards?   **12**

2. Pencils are packed in boxes of 10 and 12. Tori buys 15 boxes of pencils containing a total of 160 pencils. How many of each type does Tori buy?   **10 boxes of 10 pencils; 5 boxes of 12 pencils**

3. **Money** Juan has a mixture of pennies and dimes worth $2.28. He has between 39 and 56 pennies. How many dimes does Juan have?   **18 dimes**

4. Hannah buys 6 boxes of tissues containing 75 tissues each. Mick buys 2 boxes of tissues containing 175 tissues each. Hannah guesses that she has about twice as many tissues as Mick. Is her guess reasonable?   **No, she has 450 tissues and Mick has 350.**

5. **Shopping** The graph shows the average amount of money spent per trip by buyers ages 8 through 17. These shoppers average 12.7 shopping trips a month.

   a. What is the average amount of money a 14-year old would spend shopping in one month?   **$396.24**

   b. Find the average amount an 11-year old would spend in 6 months.   **$1,409.70**

Source: International Mass Retail Association

6. **Money Matters** A model PL-3 shortwave radio costs $12 more than twice the cost of the model PL-1 radio. The PL-3 radio costs $191.98. What is the cost of the PL-1 radio?   **$89.99**

7. Shelby is thinking of a number. Austin is thinking of a number that is 23 more than Shelby's number. Skylar's number is 18 less than Austin's number. Skylar's number is about—   **c.**

   a. 5 less than Shelby's number.

   b. 41 more than Shelby's number.

   c. 5 more than Shelby's number.

   d. 41 less than Shelby's number.

8. Find the sum of whole numbers from 1 to 49.   **1,225**

CHAPTER 1 Test

1. Write 4 · 4 · 6 · 6 · 6 using exponents. $4^2 \cdot 6^3$
2. Evaluate $3^3 \cdot 2^2$. **108**
3. Solve $8x = 88$ if the replacement set is {704, 80, 12, 11}. **11**

Evaluate each expression if $a = 3$, $b = 2$, and $c = 5$.

4. $3(c - a)^2 + b - 12$ **2**
5. $(2c + b) \div a - 3$ **1**

Solve each equation. Check your solution.

6. $k - 10 = 65$ **75**
7. $0.3m = 4.8$ **16**
8. $x + 33 = 43$ **10**
9. $\$2.40 = \$1.85 + b$ **$0.55**
10. $\frac{n}{1.5} = 8$ **12**
11. $10 = 0.8x$ **12.5**
12. $3(n - 5) = 30$ **15**
13. $\frac{a}{7} + 12 = 19$ **49**
14. $100 = 10 + 3x$ **30**

Write each phrase or sentence as an algebraic expression or equation.

15. 7 divided by x $7 \div x$
16. the sum of 5 and y $5 + y$
17. Three times a number is 30. $3n = 30$

Find the perimeter and area of each figure.

18.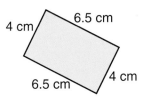
P = 21 cm; A = 26 cm²

19.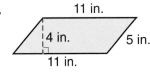
P = 32 in.; A = 44 in²

20.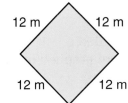
P = 48 m; A = 144 m²

Solve each inequality. Graph the solution on a number line. 21–23. See Answer Appendix.

21. $x + 3 > 15$
22. $11 > 2c - 3$
23. $\frac{z}{3} + 4 \leq 6$

24. Ray has $2,183 in his savings account. He deposits $75. How much will he have in his account after the deposit? **$2,258**
25. Taryn has $8.50 left. She spent $17 and $14.50 at two stores. How much did she start with? **$40**

Chapter 1 Test **647**

CHAPTER 2 Test

1. Find the absolute value of -8.   **8**

2. Graph the set $\{5, -3, -5, 0, 2\}$ on a number line.   **See Answer Appendix.**

Replace each ● with >, <, or = to make a true sentence.

3. 2 ● -19 **>**
4. $|-21|$ ● 21 **=**
5. -816 ● -125 **<**
6. $|-51|$ ● $|-18|$ **>**

Solve each equation.

7. $r = -572 + 58$ **-514**
8. $p = -5(-13)$ **65**
9. $z = 60 + 12$ **72**
10. $m = -105 \div 15$ **-7**
11. $-211 - 127 = d$ **-338**
12. $-789 - (-54) = s$ **-735**
13. $(7)(-10)(4) = h$ **-280**
14. $2q - 12 = -40$ **-14**
15. $g = -109 - (-34)$ **-75**
16. $2x = 14$ **7**
17. $x + 6 = 2$ **-4**
18. $y = 84 + (-13)$ **71**
19. $6t = 38 - (-22)$ **10**
20. $-7 + 22 + (-10) + 16 = s$ **21**
21. $-5(80)(-2) = m$ **800**
22. $k = \dfrac{700}{-100}$ **-7**
23. $y = (-4) - 3$ **-7**
24. $\dfrac{-160}{-5} = x$ **32**

Evaluate each expression if $c = -4$, $m = 5$, and $t = 10$.

25. $-5c + m$ **25**
26. $3ct \div m$ **-24**

Find each sum or difference. If there is no sum or difference, write *impossible*.

27. $\begin{bmatrix} -3 & 0 \\ 5 & -2 \\ -4 & 6 \end{bmatrix} - \begin{bmatrix} 6 & -3 \\ -4 & 6 \end{bmatrix}$ **impossible**

28. $\begin{bmatrix} 3 & -2 \\ 4 & -1 \\ 5 & -3 \end{bmatrix} + \begin{bmatrix} -3 & 8 \\ 0 & 6 \\ -4 & -2 \end{bmatrix}$ $\begin{bmatrix} 0 & 6 \\ 4 & 5 \\ 1 & -5 \end{bmatrix}$

Name the ordered pair for the coordinates of each point graphed on the coordinate plane.

29. A **$(-1, 3)$**
30. B **$(3, 4)$**
31. C **$(-2, -3)$**
32. D **$(2, -2)$**

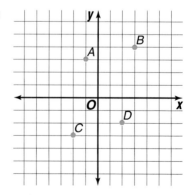

33. **School** Schools were closed in the district due to a winter storm. The temperature dropped 21°F over a three-hour period. If the temperature dropped at an even rate, how many degrees did the temperature fall each hour?   **7 degrees**

CHAPTER 3 Test

Express each ratio or rate in simplest form.

1. 6 misses in 10 tries **3:5**
2. 8 out of 15 **8:15**
3. 9 wins:6 losses **3:2**

Express each rate as a unit rate.

4. 250 miles in 5 hours **50 miles/hour**
5. 180 pages in 6 hours **30 pages/hour**
6. $15 in 4 days **$3.75/day**

Express each ratio or fraction as a percent. Round to the nearest hundredth of a percent.

7. 7 out of 10 **70%**
8. $\frac{3}{4}$ **75%**
9. $\frac{3}{8}$ **37.5%**
10. $\frac{1}{6}$ **16.67%**
11. 2 out of 6 **33.33%**
12. 2:32 **6.25%**

13. *Travel* According to the Travel Association of America, 4 out of 5 vacationers travel to their destination by car, truck, or recreational vehicle. Express the ratio as a percent. **80%**

Express each percent as a fraction in simplest form.

14. 30% $\frac{3}{10}$
15. 8% $\frac{2}{25}$
16. 45% $\frac{9}{20}$

Express each percent as a decimal.

17. 25% **0.25**
18. 8% **0.08**
19. 13.5% **0.135**

Express each decimal as a percent.

20. 0.55 **55%**
21. 0.3 **30%**
22. 0.065 **6.5%**

Solve each proportion.

23. $\frac{3}{7} = \frac{x}{35}$ **15**
24. $\frac{10}{8.4} = \frac{5}{y}$ **4.2**
25. $\frac{3}{2} = \frac{z}{3.4}$ **5.1**
26. $\frac{18}{12} = \frac{24}{y}$ **16**

Compute mentally.

27. 10% of 99 **9.9**
28. 30% of 60 **18**
29. $33\frac{1}{3}$% of 90 **30**

Estimate.

30. 9% of 81 $\frac{1}{10}$ **of 80 or 8**
31. 97% of 16 $\frac{1}{1}$ **of 16 or 16**
32. 46% of 20 $\frac{1}{2}$ **of 20 or 10**

33. *Medicine* There are about 249 million people in the United States. Of these, 37% have type O+ blood. Estimate how many people have O+ blood. $\frac{3}{8}$ **of 240 million people or about 90 million people**

CHAPTER 4 Test

Use the histogram for Exercises 1–3.

1. For which interval is the frequency the greatest? **3–5**
2. How many people were surveyed? **19**
3. How many people spend more than 8 hours per week exercising? **3**

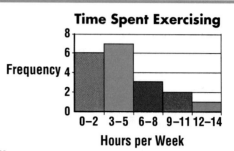

Use the table for Exercises 4–6. 4–6. See Answer Appendix.

On What Day Do Fatal Accidents Occur?							
Day	Sun.	Mon.	Tues.	Wed.	Thurs.	Fri.	Sat.
Accidents	6,590	4,920	4,620	4,920	5,160	6,630	7,560

Source: Carnegie Library of Pittsburgh

4. Make a bar graph of the data.
5. Make a circle graph of the data.
6. What type of display is most appropriate for showing this data in an article proposing more police patrols on the weekends? Explain your choice.

The scores on a 30-point math quiz are 25, 16, 25, 30, 29, 28, 19, 23, 22, 17, 19, 20, 25, 30, and 29.

7. Make a line plot for the data. **See Answer Appendix.**
8. What is the frequency of the score that occurred most often? **3**
9. How many students received a score of 19? **2**
10. How many students received a score of 25 or above? **8**
11. Find the mean, median, and mode of the data. **23.8, 25, 25**

The ages of the customers at Hannah's Bagel Shop are 45, 36, 27, 16, 19, 46, 40, 38, 22, 23, 25, 40, and 17.

12. Find the range. **30**
13. What is the median? **27**
14. What is the upper quartile? **40**
15. Find the lower quartile. **20.5**
16. What is the interquartile range? **19.5**
17. Are there any outliers? If so, name them. **none**
18. *Travel* Would a scatter plot of data describing the gallons of gas used and the miles driven show a *positive, negative,* or *no* relationship? **positive**
19. *Finances* In order to make enough money to meet his budgeted monthly expenses, Hakeem needs to earn an average of $500 a month. Last year, he earned $540, $450, $560, $800, $350, $400, $350, $380, $500, $450, $600, and $200. Did he earn enough last year? If not, how much more would Hakeem have to earn to have an average income of $500? **no; $420**
20. *School* The graph shows the number of students in each grade that had a B average or better. Is the graph misleading? Explain. **See Answer Appendix.**

CHAPTER 5 Test

Find the measure of each angle if a ∥ b and m ∠ 4 = 50°.

1. $m\angle 1$ 130°
2. $m\angle 2$ 50°
3. $m\angle 3$ 130°
4. $m\angle 5$ 130°

5. Can a triangle be scalene and obtuse? Draw a figure to justify your answer.
   Yes; see Answer Appendix for drawing.

Classify each triangle by its angles and by its sides.

6.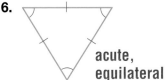
   acute, equilateral

7.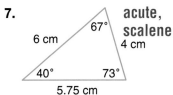
   acute, scalene

8.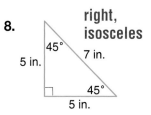
   right, isosceles

9. Draw an isosceles obtuse triangle. **See Answer Appendix.**

10. In $\triangle ABC$, $m\angle A = 35°$ and $m\angle B = 60°$. What is $m\angle C$? 85°

Sketch each figure. Let Q = quadrilateral, P = parallelogram, R = rectangle, S = square, RH = rhombus, and T = trapezoid. Write all of the letters that describe it inside the figure.

11.
    Q, T

12.
    Q, P

13.
    Q, P, R, Rh, S

14. Does a rectangle have line symmetry? If so, draw the lines of symmetry.

15. Does a rhombus have rotational symmetry? **yes**

14. Yes; see Answer Appendix for drawing.

Tell whether each pair of triangles is *congruent, similar,* or *neither*. Justify your answer.

16. 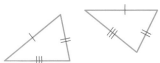   congruent; SSS

17.
    Similar; corresponding angles are congruent.

18. Does the pattern at the right involve a translation or a rotation? **rotation**

19. 30 students; see Answer Appendix for Venn diagram.

19. **Music** A survey of your class shows that 20 students have a portable cassette player and 20 students have a portable CD player. Ten students have both. Make a Venn diagram to help you find out how many students have a portable cassette or CD player.

20. **Business** A corporation is developing a new corporate logo. They can use either the letter X or the letter K. If they want to use the letter with the most lines of symmetry, which will they choose? **X**

CHAPTER 6 Test

Determine whether each number is divisible by 2, 3, 4, 5, 6, 8, 9 or 10.

1. 822   **2, 3, 6**
2. 285   **3, 5**
3. 2,088   **2, 3, 4, 6, 8, 9**

Find the prime factorization of each number.

4. 54 $2 \cdot 3^3$
5. 100 $2^2 \cdot 5^2$
6. 180 $2^2 \cdot 3^2 \cdot 5$

Find the GCF for each set of numbers.

7. 27, 54   **27**
8. 44, 66   **22**
9. 105, 60, 35   **5**

Write each fraction in simplest form.

10. $\frac{12}{16}$ $\frac{3}{4}$
11. $-\frac{36}{54}$ $-\frac{2}{3}$
12. $\frac{36}{81}$ $\frac{4}{9}$

Express each fraction or mixed number as a decimal.

13. $\frac{13}{16}$ **0.8125**
14. $-5\frac{3}{8}$ **−5.375**
15. $\frac{9}{11}$ $0.\overline{81}$

Express each decimal as a fraction or mixed number in simplest form.

16. −1.83 $-1\frac{83}{100}$
17. $0.\overline{67}$ $\frac{67}{99}$
18. $5.\overline{45}$ $5\frac{5}{11}$

The letters of the word "telephone" are written one each on 9 identical slips of paper and shuffled in a bag. A blindfolded student draws one slip of paper. Find each probability.

19. $P(e)$ $\frac{1}{3}$
20. $P(o \text{ or } p)$ $\frac{2}{9}$
21. $P(\text{vowel})$ $\frac{4}{9}$

Find the LCM for each set of numbers.

22. 7, 14   **14**
23. 6, 15, 18   **90**
24. 10, 20, 30   **60**

Replace each ● with <, >, or = to make a true sentence.

25. $\frac{2}{9}$ ● $\frac{5}{27}$ **>**
26. $5\frac{3}{5}$ ● 5.61   **<**
27. −5.89 ● −5.9   **>**

Express each number in standard form.

28. 4.3×10^{-4} **0.00043**
29. 7.93×10^5 **793,000**

Express each number in scientific notation.

30. 65,460,000 6.546×10^7
30. 0.00000057 5.7×10^{-7}

32. **Remodeling** Amanda wants to lay ceramic tiles on her countertop. The countertop is 60 inches long and 48 inches wide. What are the dimensions of the largest square tile that she can use without having to use any partial squares?   **12 in. by 12 in.**

33. **Football** Oscar led the league by completing 50 out of 75 passes he threw. Write the fraction of his throws that were completed as a decimal. $0.\overline{6}$

CHAPTER 7 Test

Solve each equation. Write the solution in simplest form.

1. $-7\frac{3}{7} - 2\frac{5}{7} = m$
2. $\frac{8}{13} + \frac{8}{13} = a$ $1\frac{3}{13}$
3. $r = \frac{9}{14} - \frac{5}{14}$ $\frac{2}{7}$
4. $\frac{3}{8} - \frac{4}{5} = t$ $-\frac{17}{40}$

5. $y = -1\frac{5}{6} + (-8\frac{3}{8})$
6. $-8\frac{5}{9} - 2\frac{1}{6} = b$
7. $-\frac{3}{4} \cdot -\frac{8}{9} = g$ $\frac{2}{3}$
8. $x = -5\frac{1}{3} \cdot 2\frac{2}{3}$ $-14\frac{2}{9}$

9. $h = \left(\frac{4}{9}\right)^2$ $\frac{16}{25}$
10. $j = -4\frac{1}{2}\left(-\frac{2}{3}\right)$ 3
11. $2\frac{4}{5}(10) = d$ 28
12. $f = 12 \times 3\frac{3}{4}$ 45

13. $n = \left(-\frac{4}{5} \cdot \frac{2}{3}\right) \cdot \frac{1}{2}$
14. $c = -\frac{7}{8} \div 1\frac{3}{4}$ $-\frac{1}{2}$
15. $6\frac{1}{6} \div 1\frac{2}{3} = w$ $3\frac{7}{10}$
16. $\frac{3\frac{3}{5}}{10} = k$ $\frac{9}{25}$

17. **Publishing** The width of a page of a newspaper is $13\frac{3}{4}$ inches. The left margin is $\frac{7}{16}$ inch, and the right margin is $\frac{1}{2}$ inch. What is the width of the page inside the margins? $12\frac{13}{16}$ inches

18. **Food** Marcie wants to make enough pudding to serve 16 friends. The recipe that serves 4 requires $1\frac{3}{4}$ cups of milk. How much milk will she need? 7 cups

1. $-10\frac{1}{7}$ 5. $-10\frac{5}{24}$ 6. $-10\frac{13}{18}$ 13. $-\frac{4}{15}$

Identify each sequence as *arithmetic, geometric,* or *neither.* Then find the next three terms.

19. $-10, -6, -2, 2, \ldots$
    A; 6, 10, 14

20. $88, 44, 22, 11, \ldots$
    G; $5\frac{1}{2}, 2\frac{3}{4}, 1\frac{3}{8}$

21. $50, 40, 31, 23, \ldots$
    N; 16, 10, 5

Find the area of each triangle.

	base	height	
22.	11 m	18 m	99 m²
23.	$5\frac{1}{2}$ ft	$6\frac{1}{4}$ ft	$17\frac{3}{16}$ ft²

Find the area of each trapezoid.

	base (*a*)	base (*b*)	height	
24.	10 in.	25 in.	$8\frac{1}{2}$ in.	148.75 in²
25.	3.8 mm	5.3 mm	8.4 mm	38.22 mm²

Find the circumference of each circle described below. Use $\frac{22}{7}$ or 3.14 for π. Round decimal answers to the nearest tenth.

26. The diameter is $6\frac{2}{3}$ yards. $20\frac{20}{21}$ yards

27. The radius is 3.6 meters. 22.6 meters

Solve each equation or inequality. Check your solution.

28. $2\frac{1}{2}w = 4\frac{3}{8}$ $1\frac{3}{4}$
29. $\frac{3}{4} = -\frac{3}{4}x + \frac{9}{16}$ $-\frac{1}{4}$
30. $\frac{a}{1.8} - 7.8 = 11$ 33.84
31. $2.2 < \frac{b}{-10} - 2.4$ $b < -46$
32. $\frac{1}{3}m + 7 \geq 11$ $m \geq 12$
33. $-\frac{3}{4}j - 1 < -5\frac{1}{2}$ $j > 6$

Chapter 7 Test **653**

CHAPTER 8 Test

Write a proportion to solve each problem. Then solve.

1. A car uses 8 gallons to travel 120 miles. How many gallons would be used to travel 80 miles? **Sample answer:** $\frac{8}{120} = \frac{x}{80}$; $5\frac{1}{3}$ gal

2. A muffin recipe calls for 3 cups of flour for 24 muffins. How many cups of flour are needed to make 36 muffins? **Sample answer:** $\frac{3}{24} = \frac{x}{36}$; $4\frac{1}{2}$ cups

Write a percent proportion to solve each problem. Then solve. Round to the nearest tenth.

3. 18 is 25% of what number? $\frac{18}{B} = \frac{25}{100}$; **72**

4. What is 2% of 3,600? $\frac{P}{3,600} = \frac{2}{100}$; **72**

5. 62 is 90% of what number? $\frac{62}{B} = \frac{90}{100}$; **68.9**

Write an equation in the form $RB = P$ for each problem. Then solve.

6. 30 is what percent of 50? $R \cdot 50 = 30$; **60%**

7. Find 45% of 600. $0.45 \cdot 600 = P$; **270**

8. 1 is what percent of 30? $R \cdot 30 = 1$; **3.3%**

Express each percent as a fraction or mixed number in simplest form.

9. 135% $1\frac{7}{20}$

10. $\frac{1}{6}$% $\frac{1}{600}$

Express each percent as a decimal.

11. 0.25% **0.0025**

12. 230% **2.3**

Find each percent of change. Round to the nearest percent.

13. original: 10 **60%**
    new: 16

14. original: 28 **18%**
    new: 23

15. original: 350 **31%**
    new: 240

Find the simple interest to the nearest cent.

16. $300 at 8% for 3 years **$72.00**

17. $1,800 at 6.5% for 6 months **$58.50**

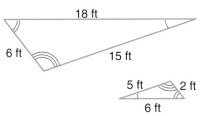

18. Tell whether the pair of polygons are similar. Explain your reasoning. **See Answer Appendix.**

19. A house casts a shadow that is 18 feet long. At the same time, the shadow of a tree in the backyard is 27 feet long. If the house is 24 feet tall, how tall is the tree? **36 ft**

20. A distance on a map is $1\frac{1}{2}$ inches. If the scale on the map is 1 inch = 50 miles, find the actual distance. **75 mi**

Find the coordinates of the image of each point for a dilation with a scale factor of 3.

21. $A(2, 3)$ **(6, 9)**

22. $B(4, 1)$ **(12, 3)**

23. $C(6, 0)$ **(18, 0)**

24. **Budget** Stratman has budgeted $90 a month for entertainment expenses. If this is 15% of his total monthly budget, what is his total monthly budget? **$600**

25. **Travel** It took Tori 8 hours to travel 520 miles. At that rate, how far can she travel in $16\frac{1}{2}$ hours? **$1,072\frac{1}{2}$ mi**

CHAPTER 9 Test

Find each square root.

1. $\sqrt{144}$ **12**
2. $\sqrt{\frac{36}{49}}$ **$\frac{6}{7}$**
3. $-\sqrt{0.16}$ **−0.4**

Estimate to the nearest whole number.

4. $\sqrt{67}$ **8**
5. $\sqrt{510}$ **23**
6. $\sqrt{108.8}$ **10**

Let R = real numbers, Q = rational numbers, Z = integers, W = whole numbers, and I = irrational numbers. Name the set or sets of numbers to which each real number belongs.

7. $\sqrt{14}$ **I, R**
8. $-\sqrt{64}$ **Z, Q, R**
9. 6.1313... **Q, R**
10. $\sqrt{18.743}$ **I, R**

11. State the Pythagorean Theorem. **In a right triangle, the square of the length of the hypotenuse is equal to the sum of the squares of the lengths of the legs. $a^2 + b^2 = c^2$**

Find the missing measure for each right triangle. Round to the nearest tenth.

12. $a = 6$ km; $b = 8$ km **10 km**
13. $b = 20$ in.; $c = 35$ in. **28.7 in.**
14. $a = 1.5$ cm; $c = 2.5$ cm **2 cm**
15. $a = 2.5$ ft; $b = 4.5$ ft **5.1 ft**

Determine whether each triangle with sides of given length is a right triangle.

16. 16 m, 34 m, 30 m **yes**
17. 12 ft, 20 ft, 24 ft **no**

18. A ladder is leaning against a house. The top of the ladder is 16 feet from the ground, and the base of the ladder is 12 feet from the side of the house. How long is the ladder? **20 feet**

19. **Geometry** Find the perimeter of a right triangle with legs of 10 inches and 8 inches. **about 30.8 in.**

20. **Geometry** Find the distance between $R(-2, -2)$ and $S(5, 6)$. Round to the nearest tenth. **10.6 units**

Find the missing lengths. Round decimals to the nearest tenth.

21.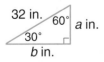
$a = 16; b \approx 27.7$

22.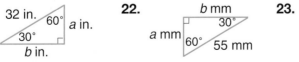
$a = 27.5; b \approx 47.6$

23.
$b = 6; c \approx 8.5$

24.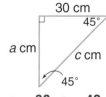
$a = 30; c \approx 42.4$

25. **Construction** A builder wants to build the roof shown at the right. What is the height of the roof? **15 feet**

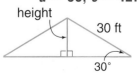

CHAPTER 10 Test

Make a function table for each function. Use −1, 0, and 2 for *n*. Then graph the function. 1–2. See Answer Appendix.

1. $f(n) = -3n$
2. $f(n) = 2n - 4$

Find four solutions of each equation. 3–6. See Answer Appendix.

3. $y = \frac{x}{4} + 2$
4. $y = -4 - 2.5x$
5. $y = 6 - x$
6. $y = 3x - 12$

Graph each function. 7–11. See Answer Appendix for graphs.

7. $y = \frac{x}{3} - 1$
8. $y = x^2 + 6$
9. $y = 2x^2 - 1$
10. $y = -4x + 10$

11. Find the solution of the system $y = 5x + 2$ and $y = 2x + 8$ by graphing. **(2, 12)**

12. Without graphing, determine whether (3, −2) is the solution of the system of equations $y = 2x - 8$ and $y = -3x - 7$. Why or why not?
    No; It is not a solution of the second equation.

Rectangle *ABCD* has vertices *A*(−5, −2), *B*(−2, −2), *C*(−2, −1), and *D*(−5, −1). After a translation, the coordinates of *A*' are (2, 4).

13. Describe the translation using an ordered pair. **(7, 6)**
14. Find the coordinates of *B*', *C*', and *D*'. ***B*'(5, 4), *C*'(5, 5), *D*'(2, 5)**

Triangle *EFG* has vertices *E*(−5, 2), *F*(−2, 3), and *G*(−4, 5).

15. Find the coordinates of the vertices after a reflection over the *y*-axis. ***E*'(5, 2), *F*'(2, 3), *G*'(4, 5)**
16. Find the coordinates of the vertices after a reflection over the *x*-axis.
    ***E*'(−5, −2), *F*'(−2, −3), *G*'(−4, −5)**

Triangle *HIJ* has vertices *H*(4, 4), *I*(1, 2), and *J*(2, 5).

17. Find the coordinates of $\triangle H'I'J'$ after a 180° rotation. ***H*'(−4, −4), *I*'(−1, −2), *J*'(−2, −5)**
18. Find the coordinates of $\triangle H'I'J'$ after a rotation of 90° counterclockwise. ***H*'(−4, 4), *I*'(−2, 1), *J*'(−5, 2)**
19. **Weather** By 6:00 P.M., 3 inches of rain had fallen. For the next three hours, 0.5 inch of rain fell each hour. How many inches of rain fell by 9:00 P.M.? **4.5 in.**
20. **Remodeling** One plumber charges a flat fee of $35 plus $25 per hour. Another plumber charges a flat fee of $20 plus $30 per hour.
    a. When are their charges equal? **3 hours**
    b. If it takes one hour to do a job, who will charge less, the first plumber or the second plumber? **the second plumber**

656 Chapter 10 Test

CHAPTER 11 Test

Find the area of each figure to the nearest tenth.

1. 113.1 m²
2. 50.3 in²
3. 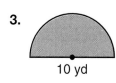 39.3 yd²

4. Draw two views of a rectangular prism whose dimensions are 2 units by 3 units by 4 units. **See Answer Appendix.**

Find the volume of each solid to the nearest tenth.

5. 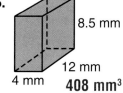 408 mm³
6. 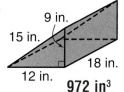 972 in³
7. 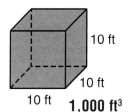 1,000 ft³
8. 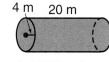 1,005.3 m³

9. 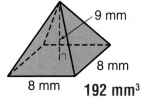 192 mm³
10. 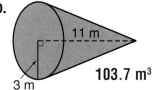 103.7 m³
11. 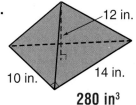 280 in³

Find the surface area of each solid to the nearest tenth.

12. 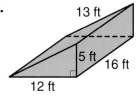 540 ft²
13. 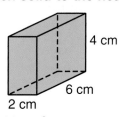 88 cm²
14. 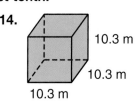 636.5 m²
15. 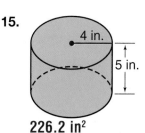 226.2 in²

16. How many significant digits are in 10.4 centimeters? **3**
17. What is the greatest possible error of 10.4 centimeters? **0.05 cm**
18. What is the relative error of 10.4 centimeters? **See Answer Appendix.**
19. A grain silo has a diameter of 20 feet and is 50 feet tall. Another grain silo has a diameter of 30 feet and is 40 feet tall. Which silo has a greater volume? Explain. **See Answer Appendix.**
20. Some cheeses are sealed in wax or a thin plastic film to protect their moisture. Find the surface area of the plastic film on the small cheese wheel. **about 201.1 in²**

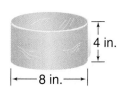

CHAPTER 12 Test

Video cameras come in three formats, VHS, VHS-C, or 8 mm, and with stereo or mono sound. They come with or without remote control.

1. Draw a tree diagram to find the number of possible choices of video cameras.
2. How many of the possible choices are mono video cameras?   **6 choices**

1. See Answer Appendix.

Find each value.

3. $P(6, 3)$ **120**
4. $4!$ **24**
5. $C(10, 5)$ **252**

6. Doug has 8 bowling trophies. How many ways can he arrange 4 of them in a row?   **1,680 ways**
7. How many teams of 5 players can be chosen from 15 players?   **3,003 teams**
8. In how many ways can 6 students stand in line?   **720 ways**

Use Pascal's Triangle to answer each question.

9. How many combinations of 4 baseball cards can be chosen from 6 baseball cards?   **15**
10. What does the 6 in row 4 mean?   **the number of combinations of 4 things taken 2 at a time**

There are 4 blue marbles, 3 red marbles, and 2 white marbles in a bag. Once a marble is selected, it is not replaced. Find the probability of each outcome.

11. a red and then a white marble $\frac{1}{12}$
12. two blue marbles $\frac{1}{6}$
13. Does the probability in Exercise 12 represent dependent or independent events?   **dependent**

Two coins are tossed 20 times. No tails were tossed four times, one tail was tossed eleven times, and two tails were tossed five times.

14. What is the experimental probability of tossing no tails? $\frac{1}{5}$
15. What is the theoretical probability of tossing no tails? $\frac{1}{4}$
16. Why are the experimental probability and theoretical probability of tossing no tails different?   **The theoretical probability is what will probably happen if there are many experiments. With only 20 tries, the experimental probability can be very different.**

Use the survey on favorite flavors of gum to answer each question.

17. What is the size of this sample?   **50**
18. What percent of the sample choose bubble gum?   **40%**
19. The school band plans to sell packages of gum to raise money for new uniforms. How many packages of bubble gum should they order if they plan to sell a total of 800 packages of gum?   **320 packages of bubble gum**

Favorite Flavors of Gum	
spearmint	12
bubble gum	20
fruit	8
peppermint	10

20. Donna reaches into a bag containing 5 McIntosh apples, 3 Golden Delicious apples, and 2 Granny Smith apples. Explain how you could act it out to find the probability that she picks 2 Granny Smith apples.   **See Answer Appendix.**

CHAPTER 13 Test

Model each polynomial using drawings or algebra tiles. 1–2. Answer Appendix.
1. $x^2 + 5x$
2. $-4x^2 - x + 3$

Simplify each expression. Then evaluate if $r = -1$ and $s = 2$.
3. $3r + 2s + 4r - s$ $7r + s; -5$
4. $r - 4s - 5r + 2s$ $-4r - 2s; 0$

5. Identify the polynomial represented by the model at the right. $2x^2 + x + 3$

Simplify each polynomial. Use drawings or algebra tiles if necessary.
6. $2x^2 + 4x + 3x^2 + 5x$ $5x^2 + 9x$
7. $4c^2 + 3c - 3c^2 + c$ $c^2 + 4c$
8. $3x^2 - y - 5x^2 + y$ $-2x^2$
9. $3a^2 + 7a - 4a^2 + 3a - 1$ $-a^2 + 10a - 1$

Find each sum or difference. Use drawings or algebra tiles if necessary.
10. $(9z^2 - 3z) - (5z^2 + 8z)$ $4z^2 - 11z$
11. $(4c^2 + 2c) + (-4c^2 + c)$ $3c$
12. $(-x^2 + 2x - 5) + (x^2 - 6x)$ $-4x - 5$
13. $(5n^2 - 4n + 1) - (4n - 5)$ $5n^2 - 8n + 6$

14. $\begin{array}{r} 5r^2 - 3r + 5 \\ +\ r^2 - 5r - 7 \\ \hline 6r^2 - 8r - 2 \end{array}$

15. $\begin{array}{r} 7y^2 + 4y - 5 \\ -\ (5y^2 + 7y - 10) \\ \hline 2y^2 - 3y + 5 \end{array}$

Find each product. Use algebra tiles or drawings if necessary.
16. $7^3 \cdot 7^5$ 7^8
17. $c^6 \cdot c^8$ c^{14}
18. $8x(x + 3)$ $8x^2 + 24x$
19. $x^3(6x + 5)$ $6x^4 + 5x^3$
20. $(2x + 1)(x + 3)$ $2x^2 + 7x + 3$
21. $(x + 4)(2x + 3)$ $2x^2 + 11x + 12$

Factor each polynomial. Use drawings or algebra tiles if necessary.   24. $(x + 5)(x + 1)$
22. $x^2 + 13x$ $x(x + 13)$
23. $x^2 + 7x + 10$ not factorable
24. $x^2 + 6x + 5$

25. Randi rode her bicycle around a square city block. The length of each side of the block is $(4x + 3)$ feet. What is the total distance she rode? $(16x + 12)$ ft

Getting Acquainted with the Graphing Calculator

When some students first see a graphing calculator, they think, "Oh, no! Do we *have* to use one?", while others may think, "All right! We get to use these neat calculators!" There are as many thoughts and feelings about graphing calculators as there are students, but one thing is for sure: a graphing calculator *can* help you learn mathematics. Keep reading for answers to some frequently asked questions.

What is it?
So what is a graphing calculator? Very simply, it is a calculator that draws graphs. This means that it will do all of the things that a "regular" calculator will do, *plus* it will draw graphs of equations.

What does it do?
A graphing calculator can do more than just calculate and draw graphs. For example, you can program it and work with data to make statistical graphs and computations. If you need to generate random numbers, you can do that on the graphing calculator. If you need to find the absolute value of a number, you can do that too. It's really a very powerful tool, so powerful that it is often called a pocket computer.

What do all those different keys do?
As you may have noticed, graphing calculators have some keys that other calculators do not. The Texas Instruments TI-83 will be used throughout this text.

Graphing Keys

Special Feature Keys

These keys are found on any scientific calculator

These keys allow you to move the cursor up, down, left, and right on the screen.

Basic Keystrokes

- The yellow commands written above the calculator keys are accessed with the [2nd] key, which is also yellow. Similarly, the green characters above the keys are accessed with the [ALPHA] key, which is also green. In this text, commands that are accessed by the [2nd] and [ALPHA] keys are shown in brackets. For example, [2nd] [QUIT] means to press the [2nd] key followed by the key below the yellow [QUIT] command.
- [2nd] [ENTRY] copies the previous calculation so you can edit and use it again.
- [2nd] [QUIT] will return you to the home (or text) screen.
- Negative numbers are entered using the [(-)] key, not the minus sign, [−].
- [2nd] [OFF] turns the calculator off.

Order of Operations

As with any scientific calculator, the graphing calculator observes the order of operations.

Example	Keystrokes	Display
4 + 13	4 [+] 13 [ENTER]	4 + 13 17
5^3	5 [^] 3 [ENTER]	5^3 125
4 (9 + 18)	4 [(] 9 [+] 18 [)] [ENTER]	4(9 + 18) 108
$\sqrt{24}$	[2nd] [√] 24 [ENTER]	√(24 4.8989 79486

Programming on the TI-83

The TI-83 has programming features that allow you to write and execute a series of commands for tasks that may be too complex or cumbersome to perform otherwise. Each program is given a name. Commands begin with a colon (:), which the calculator enters automatically, followed by an expression or an instruction.

When you press [PRGM], you see three menus: EXEC, EDIT, and NEW. EXEC allows you to execute a stored program, EDIT allows you to edit or change a program, and NEW allows you to create a program.

- To begin entering a new program, press [PRGM] [▶] [▶] [ENTER].
- You do not need to type each letter using the [ALPHA] key. Any command that contains lowercase letters should be entered by choosing it from a menu. Check your user's guide to find any commands that are unfamiliar.
- After a program is entered, press [2nd] [QUIT] to exit the program mode and return to the home screen.
- To execute a program, press [PRGM]. Then use the down arrow key to locate the program name and press [ENTER] twice, or press the number or letter next to the program name followed by [ENTER].
- If you wish to edit a program, press [PRGM] [▶] and choose the program from the menu.
- To immediately re-execute a program after it is run, press [ENTER] when Done appears on the screen.
- To stop a program during execution, press [ON] or [2nd] [QUIT].

While a graphing calculator cannot do everything, it can make some things easier and help your understanding of math. To prepare for whatever lies ahead, you should try to learn as much as you can. Who knows? Maybe one day you will be designing the next satellite or building the next skyscraper with the help of a graphing calculator!

Getting Acquainted with the Graphing Calculator

Getting Acquainted with Spreadsheets

What do you think of when people talk about computers? Maybe you think of computer games or using a word processor to write a school paper. But a computer is a powerful tool that can be used for many things.

One of the most common computer applications is a spreadsheet program. Here are answers to some of the questions you may have if you're new to using spreadsheets.

What is it?

You have probably seen tables of numbers in newspapers and magazines. Similar to those tables, a spreadsheet is a table that you can use to organize information. But a spreadsheet is more than just a table. You can also use a spreadsheet to perform calculations or make graphs.

Why use a spreadsheet?

The advantage a spreadsheet has over a simple calculator is that when a number is changed, the entire spreadsheet is recalculated and the new results are displayed. So with a spreadsheet, you can see patterns in data and investigate what happens if one or more of the numbers is changed.

How do I use a spreadsheet?

A spreadsheet is organized into boxes called *cells*. Throughout this text, we will use the Microsoft Excel spreadsheet program. In Excel, the cells are named by a letter, that identifies the column, and a number, that identifies the row. In the spreadsheet below, cell C4 is highlighted.

	A	B	C
1	Width	Length	Area
2	3	4	12
3	2	10	20
4	5	12	60
5	8	14	112

To enter information in an Excel spreadsheet, simply move the cursor to the cell you want to access and click the mouse. Then type in the information and press Enter.

How do I enter formulas?

If you want to use the spreadsheet as a calculator, begin by choosing the cell where you want the result to appear.

- For a simple calculation, type = followed by the formula. For example, in the spreadsheet above, the formula in cell C2 is entered as "=A2*B2." *Notice that * is the symbol for multiplication in a spreadsheet.*

- Sometimes you will want similar formulas in more than one cell. First type the formula in one cell. Then select the cell and click the copy button. Finally select the cells where you want to copy the formula and click the paste button.

- Often it is useful to find the sum or average of a row or column of numbers. Excel allows you to choose from several functions like this instead of entering the formula manually. To enter a function, click the cell where you want the result to appear. Then click on the = button above the cells. A list of formulas will appear to the left. Click the down arrow button and choose your function. Excel will enter a range, which you may alter. For example, to find the average of row 2 of the spreadsheet below, the function chooses to find the average of cells B2, C2, and D2.

	A	B	C	D	E
1	Student	Test 1	Test 2	Test 3	Average
2	Kathy	88	85	91	88
3	Ben	86	89	92	89
4	Carmen	92	86	92	90
5	Anthony	80	88	87	85

The formula for cell E2 is =(B2+C2+D2)/3.

Spreadsheet software is one of the most common tools used in business today. You should try to learn as much as you can to prepare for your future. Who knows? Maybe you'll use what you're learning today as a company president tomorrow!

Selected Answers

CHAPTER 1
Problem Solving and Algebra

Pages 6–7 Lesson 1-1
1. Explore – Identify what information is given and what you need to find. Plan – Estimate the answer and then select a strategy for solving. Solve – Carry out the plan and solve. Examine – Compare the answer to the estimate and determine if it is reasonable. If not, make a new plan. **5.** Sample answer: Oxford Circus – Euston – Old Street **7.** Sample answer: *Explore:* Locate and examine a website and investigate the software used to design homepages. *Plan:* Make a blueprint of how you would like the final page to look. *Solve:* Turn your blueprint into a home page using any available software. *Examine:* Test the code and see if the home page really looks like your blueprint. Make the necessary adjustments. **9.** 4 by 4 by 4 **11.** Sample answer: clockwise from top – 17, 10, 45

Pages 9–10 Lesson 1-2
1. The 3 means to use 2 as a factor 3 times. **5.** $16^2 \cdot 20^2$ **7.** 64 **9.** 288 **11a.** $10 \cdot 10 \cdot 10 \cdot 10 \cdot 10 \cdot 10 \cdot 10 \cdot 10$ **11b.** 484,000,000 **13.** 10^2 **15.** $4^3 \cdot 8^2$ **17.** $12 \cdot 14^2 \cdot 5^3$ **19.** $18^2 \cdot 5^3$ **21.** 1,296 **23.** 1 **25.** 2,000 **27.** 1,000,000 **29.** 22 **31.** 768 **33.** $10^3 \cdot 24^2 \cdot 50$ **35a.** 12^3 **35b.** 12.167 cubic meters **37.** 16 **39.** 1 pen and 2 tablets

Pages 14–15 Lesson 1-3
1. Numerical expressions contain only numbers and algebraic expressions contain numbers and variables. **3.** 33 **5.** 6 **7.** 6 **9.** 44 **11.** 15 **13.** 43,000 **15.** 40 **17.** 8 **19.** 2 **21.** 2 **23.** 37 **25.** 48 **27.** 96 **29.** 13.5 **31.** 9 **33.** 89 **35.** 31 **37.** 8 **39.** 4 **41.** 7 **43a.** y represents the number of years between Ms. Chisholm's and Ms. Braun's elections. **43b.** 24 **45.** 64 **47.** 6 pounds

Page 16 Lesson 1-3B
1. 0; 27; 216 **3.** 48; 192; 1,200 **5.** 4 [STO▶] [X,T,θ,n] [ALPHA] [:] 7 [STO▶] [ALPHA] [Y] [ALPHA] [:] [X,T,θ,n] [ALPHA] [Y] [x^2] [ENTER] 196

Pages 19–20 Lesson 1-4
1. Add when a number is subtracted from the variable and subtract when a number is added to the variable. **3.** Tess is correct. You only need to use an inverse operation to solve if there is an operation performed on the side with the variable. **5.** 7 **7.** 9 **9.** 43 **11.** $247,500 **13.** 4 **15.** 35 **17.** 1 **19.** 27.4 **21.** 83 **23.** 52 **25.** 27 **27.** 14.46 **29.** 86 **31a.** 11,750 pounds **31b.** 4,000 pounds **33.** 4 **35.** 6 **37.** $5^2 \cdot 8^3$

Pages 23–25 Lesson 1-5
1. Multiply; To undo division, you multiply. **3.** $3x = 12; x = 4$ **5.** 5 **7.** 78 **9.** 3.4 **11.** 225 **13.** 6 **15.** 448 **17.** 54 **19.** 0.53 **21.** 200 **23.** 7 **25.** 42 **27.** 183 **29.** 0.55 **31a.** $161 **31b.** $224 **33.** 435 representatives **35.** 1.8 million years **37.** $4^3 \cdot 8^2$

Page 25 Mid-Chapter Self Test
1. $164,700 **3.** 6,912 **5.** 15 **7.** 20.1 **9.** 126

Pages 28–29 Lesson 1-6
1. Sample answer: three more than a number; a number increased by 3 **3.** Chapa is correct. *7 less than y* is translated as $y - 7$ and *7 less y* is translated $7 - y$. **5.** $7 \cdot n$ or $7n$ **7.** $w - 9 = 15$ **9.** $d \div 4 = 7$ **11.** $k + 9$ **13.** $n + 5$ **15.** $16r = 80$ **17.** $6 + b$ **19.** $j + 10$ **21.** $48 - v = 15$ **23.** $2c$ **25.** $n - 8 = 25$ **27.** $n \div 3 = 25$ **29a.** $n + 2,249$ **29b.** $n + 2,249 = 2,425$ **31a.** $s - 9$ **31b.** $2.3a = 103.5$ **33.** 15 **35.** 97

Pages 30–31 Lesson 1-7A
1. They started with the money they had left and added on the amounts they spent in reverse order. **3.** It is easier to work backward than forward when you know a result and need to know the start. **5.** Work the problem forward and make sure the ending number is correct. **7.** Undoing the subtraction first then undoing the multiplication is backward from the order of operations. **9.** $6.72 **11.** $8,100 **13a.** $393 **13b.** $4,882

Pages 35–36 Lesson 1-7
1. Subtract 5 from each side. **3.** Remove 3 counters from each side. Then divide the remaining counters into four groups. The solution is $x = 3$.

664 Selected Answers

Check by substituting 3 counters for each cup and verifying that the result is a true equation. **5.** 12
7. 64 **9.** 4 **11.** 32 clues **13.** 3 **15.** 4
17. 152 **19.** 192 **21.** 0.5 **23.** 0.3 **25.** 0.25
27. 12 **29.** 18.9 **31.** 4 **33.** $\frac{n}{4} - 1 = 7$; 32
35. 355 miles **39.** $c + 3 = 15$ **41.** 92 **43.** 196

Page 37 Lesson 1-7B
1. Multiply the input by 7. Then subtract 2 from the result. **3.** Work backward to solve the two-step equation. **5.** 12

Pages 40–42 Lesson 1-8
1. 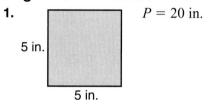 $P = 20$ in.

3. Sample answer: Area is the amount of space a figure takes up, the perimeter is the distance around the figure. **5.** $P = 26$ yd; $A = 40$ yd² **7.** $P = 48$ cm; $A = 128$ cm² **9.** $P = 16$ units; 15 units²
11. $P = 22.6$ m; $A = 29.2$ m² **13.** $P = 20$ m; $A = 23.5$ m² **15.** $P = 16$ units; $A = 12$ units² **17.** $P = 60$ in.; $A = 185$ in² **19.** $144 = 8b$; $b = 18$ yards
23a. 1 in² **23b.** 10,800 boxes **25.** a – $\frac{1}{4}$ unit²; b – $\frac{1}{4}$ unit²; c – $\frac{1}{16}$ unit²; d – $\frac{1}{8}$ unit²; e – $\frac{1}{16}$ unit²; f – $\frac{1}{8}$ unit²; g – $\frac{1}{8}$ unit² **27.** A

Pages 46–47 Lesson 1-9
1. "In excess of" indicates that the speeds are more than 60 mph.

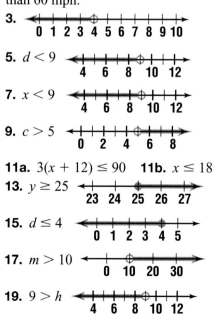

11a. $3(x + 12) \le 90$ **11b.** $x \le 18$

Pages 48–51 Study Guide and Assessment
1. false; exponent **3.** false; inequality **5.** false; second **7.** true **9.** 8 minutes **11.** $5^2 \cdot 6^3$
13. 625 **15.** 322 **17.** 10 **19.** 160 **21.** 50
23. 75 **25.** 2.9 **27.** 4.1 **29.** 294 **31.** 72
33. 12 **35.** $6x$ **37.** $4x = 48$ **39.** 66 **41.** 8
43. $P = 28$ m; $A = 36$ m²
45. $a > 12$

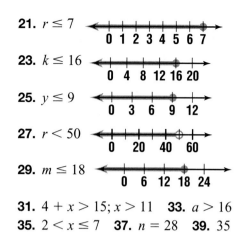

47. $n + 5 > 16$; $n > 11$ **49.** 4 pounds **51.** $6 for each cup; $10 for each bowl

Pages 52–53 Standardized Test Practice
1. C **3.** D **5.** D **7.** B **9.** B **11.** E **13.** 62 pages **15.** $t < 4$ **17.** 20 **19.** $A = 10$ in²; $P = 14$ in.

CHAPTER 2
Algebra: Using Integers

Pages 57–58 Lesson 2-1
1. Sample answer: Draw a number line. Locate the number and draw a dot at that point on the line. Label the point with a letter.

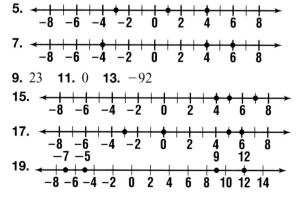

9. 23 **11.** 0 **13.** −92

21. [number line with points at -3, 3, 5, 7, 8, 9]

23. [number line with points at -7, -3, -1, 1, 3, 5, 7]

25. 5 **27.** 12 **29.** 45 **31.** 88 **33.** 4 **35.** 5
37. −3 **39.** 12 **41.** False. For example, $|-4| >$ $|2|$ but $-4 < 2$; and $|-8| > |1|$, but $-8 < 1$. **43.** C
45. 171

Pages 60–61 Lesson 2-2

1. $-3 < 4$; $4 > -3$ **3.** Mariano is correct. **5.** $>$
7. $>$ **9.** $40 > -22$; $-22 < 40$ **11.** = **13.** $>$
15. $<$ **17.** = **19.** $>$ **21.** = **23.** 564, 254, -100, -356, -450 **25.** $1 > -1$; $-1 < 1$
27a. No; $|3| = 3$ and $3 \not< 3$
27b. [number line with points at -5, -3, -1, 1, 3, 5]
29. 6

Pages 64–65 Lesson 2-3

1. Start at the first addend. Move to the left to add a negative number. **3.** -1; no, addition is commutative. **5.** -7 **7.** -5 **9.** 0 **11.** -10
13. -50 **15.** 0 **17.** 11 **19.** -26 **21.** 20
23. 108 **25.** 22 **27.** -130 **29.** -30 **31.** 31
33. 0 **35.** 8 **37a.** $v = -14{,}608 + 5{,}202$
37b. $-9{,}406$ million **39.** no; $|-3| + |5| = 8$, $|-3 + 5| = 2$; $-3 + 5 = 2$ **41.** D **43.** 32

Pages 67–68 Lesson 2-4

1. Sample answer: Add $-13 + 13 = 0$; $-45 + (-55) = -100$; and $15 + 25 = 40$. Then $0 + (-100) + 40 = -60$. **3.** 13 **5.** 2 **7.** 4
9. -7 **11.** 14 **13.** -3 **15.** 12 **17.** 12
19. 214 **21.** 2 **23.** -1 **27a.** $389,100,000 loss **27b.** $-$95,250,000 **29.** B **31.** 64 m²

Pages 71–72 Lesson 2-5

1. Sample answer: The additive inverse of an integer is called its opposite. The opposite of an integer is the same distance from 0, but has the opposite sign. For example, 4 and -4 are additive inverses.

3.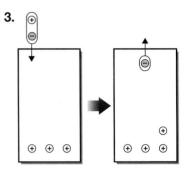

5. -1 **7.** 9 **9.** 0 **11.** 75 **13.** -7 **15.** 0
17. -21 **19.** -25 **21.** -131 **23.** 55 **25.** 78
27. -29 **29.** -412 **31.** 842 **33.** 6 **35.** -18
37. -11 **41a.** Answers will vary. For 1998, it was 6,710 years ago. **41b.** 7,979 years **43.** 47
45. B **47.** 18

Page 72 Mid-Chapter Self Test

1. [number line with points at -9, -5, -3, -1, 1, 3, 5, 7, 9]
3. -2 **5.** 0

Pages 74–75 Lesson 2-6

1. 3 rows, 2 columns **3.** $\begin{bmatrix} -1 & 7 \\ 5 & -2 \end{bmatrix}$

5. $\begin{bmatrix} 0 & 4 & 1 \\ -2 & -3 & -2 \\ 3 & 0 & 3 \end{bmatrix}$ **7.** $\begin{bmatrix} 8 & 4 & 7 & 6 \\ 13 & 2 & 0 & 0 \\ 7 & 5 & 47 & 24 \\ 13 & 12 & 47 & 36 \end{bmatrix}$

9. impossible **11.** $\begin{bmatrix} 11 & -12 \\ 1 & 26 \\ 3 & 28 \end{bmatrix}$ **13.** $\begin{bmatrix} 5 & 5 & 3 \\ 1 & -3 & 10 \end{bmatrix}$

15. $\begin{bmatrix} 16 & 4 & -7 \\ 5 & -1 & 5 \\ 1 & 9 & -12 \end{bmatrix}$ **17.** $\begin{bmatrix} 0 & 8 & -4 \\ 0 & -9 & 10 \\ 1 & 21 & -3 \end{bmatrix}$

19. $\begin{bmatrix} 0 & 10 & 20 \\ 9 & -1 & 48 \\ 3 & -11 & -8 \\ -6 & 34 & 9 \end{bmatrix}$ **21a.** $\begin{bmatrix} 4{,}362 & 5{,}917 \\ 4{,}904 & 4{,}838 \\ 3{,}353 & 3{,}344 \end{bmatrix}$

21b. $\begin{bmatrix} 399 & 554 \\ 157 & 299 \\ 165 & 139 \end{bmatrix}$ **23.** A **25.** more than 700,000

Page 77 Lesson 2-6B

1. $\begin{bmatrix} 5 & 5 & 5 \\ -1 & 11 & 7 \\ 9 & -8 & 2 \end{bmatrix}$ **3.** $\begin{bmatrix} 3 & -5 & -1 \\ 5 & -9 & -5 \\ -3 & 6 & 6 \end{bmatrix}$

5. Yes; $D + E = E + D$.

Page 80 Lesson 2-7

1. Sample answer: $[(-35)(45)](-2) = -1{,}575(-2) = 3{,}150$ or $(-35)[(-2)(45)] = (-35)(-90) = 3{,}150$
3. $(-2)(-4) = 8$

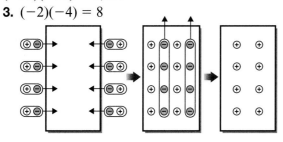

5. 6   7. -54 9. 16   11. -840 13. 48
15. 24   17. 78   19. 70   21. 504   23. 805
25. 441   27. $-2{,}592$ 29. 400   31. 200
33. -60 or 60 heating degree-days   35. y is non-negative. If $y = 0$, then $y^2 = 0$. If y is any other number, y^2 is positive because any two numbers with the same sign have a positive product.
37. -13

Pages 82–83 Lesson 2-8

1. negative   3. Rachel is correct. The divisor is 5 in all of the problems.   5. -40 7. -7 9. -73
11. -3 13. about 2 million tons   15. 2   17. 7
19. -7 21. 28   23. -49 25. -6 27. -12
29. -13 31. -98 33. -8 35. 13   37. 53
39. 8   41a. about $-15{,}533$ 41b. about 6,472,000   45a. Division is not commutative. For example, $9 \div 3 \neq 3 \div 9$.   45b. Division is not associative. For example, $(16 \div 4) \div 2 \neq 16 \div (4 \div 2)$.   47. -37 49. B

Pages 84–85 Lesson 2-9A

1. 9   3. 2   5. $x - 5 = 8$ 7. 6   9. 5   11. -8
13. 2   15. 6

Pages 88–89 Lesson 2-9

1.
$3r + 5 = -10$	
$3r + 5 - 5 = -10 - 5$	Subtract 5 from each side.
$3r = -15$	Simplify.
$\frac{3r}{3} = \frac{-15}{3}$	Divide each side by 3.
$r = -5$	Simplify.
$3r - (-5) = -10$	Rewrite using the additive inverse.
$3r + 5 = -10$	Then follow the same steps as above.

3. Sample answer: Your mistake could be in the solution of the equation or could be in your check. Check both.   5. -3 7. -275 9. -20 11. -9
13. -72 15. -318 17. 58   19. 35   21. -15
23. 36   25. 17   27. -48 29. $-2 - x = 8$; -10
31. $2x + 7 = -21$; -14 33. 40 yards   35. -13
37. 4

Pages 90–91 Lesson 2-9B

1. Crystal and Jackie eliminated the days that they could not play to see which day or days they could.
5. To plan the solution, you need to identify the possibilities for the solution. Then when you solve you can eliminate the ones that are not the solution.
9a. They could take the following buses to Baltimore and home: 8:00 A.M. bus and 3:00 P.M.; 9:00 A.M. and 4:00 P.M.; 10:00 A.M. and 5:00 P.M.; 11:00 A.M. and 6:00 P.M.; 12:00 P.M. and 7:00 P.M.; 1:00 P.M. and 8:00 P.M.   9b. yes   11. A
13. 111.485 mph

Pages 93–95 Lesson 2-10

1. For the x-coordinate, left is negative and right is positive. For the y-coordinate, up is positive and down is negative.   3. Sample answer: x comes before y in the alphabet.   5. $(0, 3)$ 7. $(-3, 2)$
9, 11, 13.

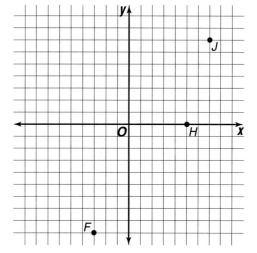

15. $(-5, 5)$ 17. $(2, 1)$ 19. $(0, 5)$ 21. $(-6, -2)$
23. $(0, -7)$ 25. $(2, -1)$

27, 29, 31, 33, 35, 37, 39, 41, 43.

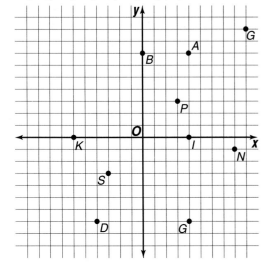

45. It is not a rectangle because the angles are not right.

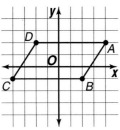

47a. 0, 0; 1, 36; 3, 108; 4, 144; 6, 216
47b.

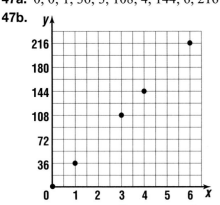

49. $(-6, -3)$ **51.** $\begin{bmatrix} -6 & -2 & 3 \\ 4 & 3 & 5 \end{bmatrix}$ **53.** $<$
55. 207

Pages 96–99 Study Guide & Assessment

1. e **3.** a **5.** g **7.** i **9.** c
11.
13. 11 **15.** $>$ **17.** $=$ **19.** 50 **21.** -8 **23.** 37
25. -18 **27.** 108 **29.** -20 **31.** $\begin{bmatrix} 12 & -4 \\ 5 & 6 \end{bmatrix}$
33. $\begin{bmatrix} -6 & -4 \\ 1 & 10 \\ -2 & -5 \end{bmatrix}$ **35.** 100 **37.** -60 **39.** -30
41. -12 **43.** 15 **45.** -8

47, 49.

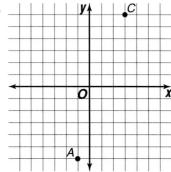

51. -10 points **53.** blue

Pages 100–101 Standardized Test Practice

1. C **3.** A **5.** B **7.** B **9.** D **11.** C **13.** -8
15. -18 **17.** $12x - 8$ **19.** $41\frac{7}{16}$ ft²

CHAPTER 3
Using Proportion and Percent

Pages 105–106 Lesson 3-1

1. Ratios and rates are both comparisons by division. In a rate the units are different; in a ratio the units are the same. **5.** 4 to 5
7. 3 pounds/week **9.** 3:1 **11.** 3:2 **13.** 1:3
15. 1:3 **17.** 3 to 2 **19.** \$0.35/minute
21. 15 students/teacher **23.** \$0.80/pound **25.** $\frac{5}{8}$
27. 3.8 dandelion plants, 17.7 grass plants
29. 6 or -6 **31.** -20 **33.** C

Pages 109–110 Lesson 3-2

1. 25%, $\frac{1}{4}$ **3.**

5. 10% **7.** $\frac{3}{4}$ **9.** 1 **11.** 40% **13.** 90%
15. 84% **17.** 99% **19.** 46% **21.** $\frac{1}{20}$ **23.** $\frac{1}{4}$
25. $\frac{1}{100}$ **27.** $\frac{4}{5}$ **29.** $\frac{77}{100}$ **31.** 70%
33. 23 people **35.** 30% **37.** -8
39. $P = 36$ in., $A = 60$ in²

Pages 112–113 Lesson 3-3

1. reducing fractions and finding cross products
3. Sample answer: assigning a variable; solving equations **5.** yes **7.** 85 **9.** 0.2 **11.** no
13. yes **15.** no **17.** yes **19.** 300 **21.** 15
23. 10.5 **25.** 2.1 **27.** 0.5 cup **31.** $\frac{7}{20}$ **33.** 217

Pages 116–117 Lesson 3-4

1. $\frac{7}{20}$, 0.35, 35%
3. $15\% > \frac{1}{8}$

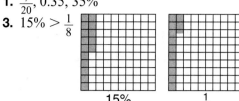

5. 0.09 7. 29% 9. 4.2% 11. 62.5%
13. 15% 15. 0.74 17. 0.01 19. 0.155
21. 0.038 23. 2% 25. 2.6% 27. 66.7%
29. 6.25% 31. 87.5% 33. 62.5% 35. 31.25%
37. 37.5% 39. > 41. = 43. < 45. 40%
47. 40% 49. 12 51. 2 out of 5 53. 81

Page 117 Mid-Chapter Self Test

1. $28/ticket 3. 22.5 5. 6%

Page 119 Lesson 3-4B

1. Except for 2, all are near 1.6. 3. Sample answer: A golden rectangle is a rectangle in which the ratio of the length to the width is 1.6.
5. Sample answer: United Nations building in New York City 7a. 13, 21, 34, 55, 89, 144 7b. After the first few terms in the sequence, the ratio of successive numbers in the Fibonacci sequence is the golden ratio.

Pages 121–122 Lesson 3-5

1. Move the decimal point in the number two places to the left. 3. Elisa; you move the decimal point one place to the left. 5. $13.55 7. 6
9. 450 Calories 11. 0.172 13. 140 15. 6
17. 900 19. 30 21. $3.84 23. < 25. =
27. $1.30 29. about 16,683,700 people
31. 0.395, 0.178 33. −17

Pages 124–125 Lesson 3-6A

1. Sample answer: $0.01 \times 90 = 0.9$ 3. about $3.75 7. $4 9. 4 miles 11. 50 cents
13. 2.7 million people

Pages 128–129 Lesson 3-6

1. $\frac{1}{4}$ of $100 is $25. 3. Darnell; 46% is less than 50%, so the answer will be less than half of 80.
5–27. Sample answers given. 5. about 20%
7. $\frac{7}{8}$ of 64 is 56 9. $\frac{22}{60} \approx \frac{20}{60}$ or 33.3%
11. $\frac{15}{49} \approx \frac{15}{50}$ or 30% 13. $\frac{64}{100}$ or 64%
15. about 12% 17. $\frac{1}{3}$ of 90 or 30

19. $\frac{1}{5}$ of 70 or 14 21. $\frac{2}{3}$ of 9 or 6 23. $\frac{8}{13} \approx \frac{8}{12}$ or 66.7% 25. $\frac{12}{60} = \frac{1}{5}$ or 20% 27. $\frac{9.2}{11} \approx \frac{9}{10}$ or 90% 29. $\frac{1}{3}$ and 150 33. $\frac{60}{206} \approx \frac{60}{200}$ or about 30% 37. 135.8 39. 28

Pages 130–133 Study Guide and Assessment

1. ratio 3. proportion 5. 60% 7. 50% 9. $\frac{1}{4}$
11. 2:3 13. 25 to 9 15. $2.50 per minute
17. 22 miles per gallon 19. 16.5% 21. 56%
23. $\frac{7}{100}$ 25. $\frac{9}{10}$ 27. 4 29. 15 31. 0.90 or 0.9
33. 0.043 35. 70% 37. 65.5% 39. 35%
41. 16.67% 43. 12 45. $21 47–51. Sample answers given. 47. $\frac{8}{100}$ of 100 or 8 49. $\frac{1}{1}$ of 35 or 35 51. $\frac{20}{52} \approx \frac{20}{50}$ or 40% 53. 24 to 65
55. 0.43, 0.16, 0.14, 0.11, 0.06, 0.04, 0.03

Pages 134–135 Standardized Test Practice

1. C 3. A 5. B 7. C 9. A 11. Sample answer: $m + 4 = 3(14)$ 13. 9 grams

CHAPTER 4
Statistics: Analyzing Data

Pages 140–141 Lesson 4-1A

1. Sample answer: Tally marks can be made quickly and each person can add his or her mark without having to erase and add to the total. 5. You would need the lowest and highest numbers in the data set. Knowing those numbers allow you to choose the range of the set. 7. 15
9a. Sample answer:

Allowance	Number of 13- and 14-year-olds
$2.01-3.00	2
$3.01-4.00	1
$4.01-5.00	9
$5.01-6.00	7
$6.01-7.00	3
$7.01-8.00	0
$8.01-9.00	1
$9.01-10.00	2
$10.01-11.00	0
$11.01-12.00	1
$12.01-13.00	0
$13.01-14.00	0
$14.01-15.00	1

9b. $4.01-5.00 11. 144 13. 18-24 years old
15. B

Pages 144–146 Lesson 4-1

1. Sample answer: Because it is more visual, a histogram is more useful than a table when you are trying to show a general trend. Because the individual numbers are shown, a table is more useful when you need to know exact numbers.
5a. Sample answer: incubation times for eggs of different birds
5b. **Incubation Times**

7.

Bicycle Helmet Prices		
Price	Tally	Frequency
$21-30	|||||		6
$31-40	||||| ||	7
$41-50	|||	3

Bicycle Helmet Prices

9. **United Nations Entry**
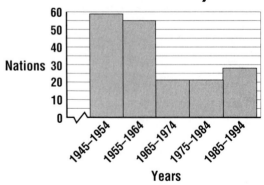

13. Sample answer: $60 ÷ 4 = $15

Page 147 Lesson 4-1B

1. 8 3. The scale for the x-axis is the interval.

Pages 150–151 Lesson 4-2

1. Tokyo Disneyland 3a. 100% 3b. 360°
3c.

Who pays when you date?			
Category	Number	Ratio	Degrees in graph
Boy	1200	0.48	172.8°
Split costs	800	0.32	115.2°
Girl	75	0.03	10.8°
Girl's parents	25	0.01	3.6°
Boy's parents	25	0.01	3.6°
Don't date	375	0.15	54°
Total	2,500	1	360°

3d. **Who pays when you date?**

5. **Medals of Honor**
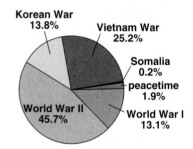

7. **Giving to Charities**
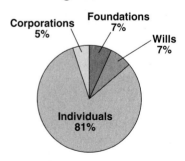

9. No; the percents are not parts of a whole because the categories overlap. 11. A

Pages 154–155 Lesson 4-3

1. Sample answer: Make a number line with appropriate numbers for the data points. Then mark an x for each data point above the appropriate number.

5.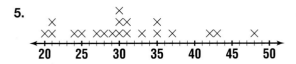

7. twice

9a.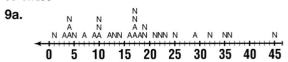

9b. Sample answer: The NFC has won more Super Bowls. The margins tend to be high, with NFC teams usually winning by larger margins than AFC teams.

11.

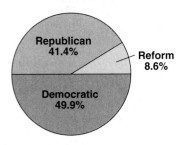

1996 Presidential Votes

13. 324 inches

Page 157 Lesson 4-3B

1. Sample answer: The data are clustered in these intervals. **3.** The north-east quarter of the country, plus California and Florida. Sample answer: There are more birds to be viewed in these areas, there are more people in California than in other states, and there are more retirees with time for leisure activities in Florida. **5.** Maps are more visual than tables. **7.** Sample answer: Both use ranges of numbers to display data. However, a map need not use the same size intervals and a histogram does. Answers may vary. Generally a map is easier to change because there is more flexibility in the data intervals.

Pages 160–161 Lesson 4-4

1. Find the sum of the data and divide by the number of data.
$$\frac{1+3+4+6+6+6+8+9+10+10+14}{11} = 7$$
The data are written in order in the line plot. Since there are eleven pieces of data, the median is the sixth number. The median of the set is 6. In a line plot, you can find the mode by choosing the number that has the most ×s. In this data set, the mode is 6. **3.** Carol is correct. The mean and the median may not be members of the set of data. **5.** 34; 34; 34

7a. $71; $60; no mode **7b.** Sample answer: There is a lot of variation in the prices. The median is a good representative price because half of the prices are above it and half are below. **7c.** Because the value of the new data piece changes the total of the data drastically, the mean is affected most. **9.** 6; 4.5, 3 **11.** 79.3; 79.5; 84 **13.** 9.0; 8.9; 8.3 **15.** 32.1; 33; 33 and 35 **17.** Sample answer: {1, 2, 3, 4, 5, 6, 7} **19.** Sample answer: The mean, median, and mode of {1, 2, 3, 4, 4, 5, 6, 7} are all 4. **21.** Sample answer: This is probably the mean, found by dividing the sum of the attendance figures for each showing of the movie by the number of people in the United States. **23.** Because the value of the new data piece changes the total of the data drastically, the mean is affected most. Adding one very large or very small piece of data will only move the median slightly toward the new number. The mode will not be affected by adding a very large or very small number. **25.** 15

Page 162 Lesson 4-4B

1. 90 **3.** 89

Pages 165–166 Lesson 4-5

1. The measures of variation describe the dispersal of data. The measures of central tendency describe the set as a whole. **3.** 7; 16; 17, 13; 4; no outliers **5.** 38; 52; 57, 48; 9; 22 **7.** 9; 58; 61, 56; 5; no outliers **9.** 22; 37; 40.5, 34; 6.5; 51 **11.** 3.8; 2.9; 3.7, 2.3; 1.4; 6.1 **13.** 52; 62.5; 74.5, 56; 18.5; 104 **15.** 0.7; 0.55; 0.65, 0.25; 0.4; no outliers
17a. Sample answer: {1, 1, 2, 2, 2, 5, 9, 9, 9, 10, 10} and {1, 4, 4, 4, 4, 5, 5, 5, 9, 10, 10}
17b. Sample answer: {1, 2, 5, 7, 9, 10, 12, 14, 15, 17, 22} and {0, 2, 5, 7, 9, 10, 12, 14, 15, 17, 27}
19. 14.9; 17.5; 22

Page 166 Mid-Chapter Self Test

1.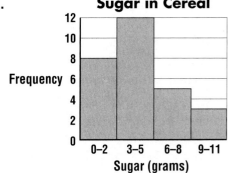

3.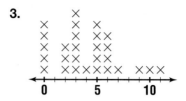

5. 11; 6, 2; 4; none

Page 167 Lesson 4-5B

1. minimum = 29, maximum = 76, upper quartile = 52.5, lower quartile = 36.5, median = 42
3. TI-83: Yes; the box-and-whisker plot for the actresses shows the two outliers using boxes. TI-82: No; the calculator does not evaluate for outliers. The ends of the whiskers are the extreme values.

Pages 169–170 Lesson 4-6

1. Write the data as ordered pairs and graph each ordered pair on a coordinate grid. **5.** negative
7. no relationship **9.** no relationship
11. positive **13.** negative **15.** negative
17. no relationship **19.** positive

21a.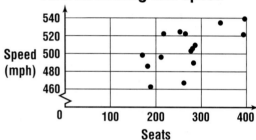

21b. positive **23a.** Sample answer: Both are summer activities, so sales would increase at the same time. **23b.** No. Warm weather could be causing the increase in both skateboard and swim suit sales. **25.** A

Pages 172–173 Lesson 4-7

1. Sample answer: data on the number of immigrants who came to the United States in each decade **5-17.** Sample answers given.
5. histogram or line plot; These are individual data. Which of these is the best choice depends on what the teacher is trying to show. **7.** scatter plot; These are two sets of similar data and you want to show their relationship visually. A scatter plot would allow visual comparison. **9.** circle graph; allows comparison of contribution of each division to the total amount of sales **11.** scatter plot; A trend can be shown and accurate information is needed.
13. line plot; Shows outliers visually and shows concentration of other prices. **15.** histogram or scatter plot; Shows increase visually. **17.** table or bar graph; Data is in categories. **19.** No; a histogram requires that data be divided into intervals. This data is not well suited to intervals. A scatter plot is a better choice.

23.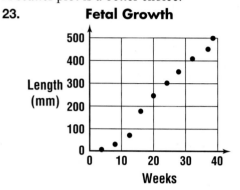

25. $5^2 \cdot 8^3$

Pages 175–177 Lesson 4-8

1. Sample answers: good places – schools, telephone surveys; bad places – computer stores, libraries **3.** No; people at a Gloria Estefan concert will be more likely to choose her than the general public. **5.** No; only people who are concerned enough about the situation will call. **7.** No; people at a movie theater will be more likely to say watching movies than the general public. **9.** Yes; a telephone poll will allow a selection of a number of people with different viewpoints. **11.** No; because Texas has a higher concentration of immigrants than other states and immigrants are more likely to speak more than one language, a sample taken there would not be representative of all Americans. **13.** Yes; if the schools are chosen in different types of neighborhoods, the sample will be representative of teenagers. **15.** No; people who attend a community college will be more likely to have chosen careers that can be pursued with an associate's or bachelor's degree. **17a.** Yes; the space between the tick marks is not consistent with the time between each date. **19a.** The fewer the dentists surveyed, the less reliable the results.
19b. The commercial makes it sound as if the dentists recommended their brand of toothpaste over all others. **21.** 0.048

Pages 178–181 Study Guide and Assessment

1. scatter plot **3.** range **5.** sample **7.** circle graph **9.** The mean is the average of the set and the median is the middle number when the set is arranged in order.   **11.** 7-8 and 9-10 each have 7 data.

13.

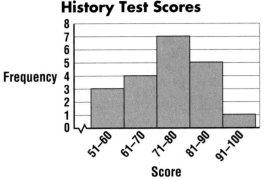

17. 27   **19.** 7.7; 6.8; no mode   **21.** 20; 20; 20 and 21   **23.** 74; 138; 150.5, 123.5; 27; 195
25–27. Sample answers are given.   **25.** positive
27. histogram   **29.** Yes; a good cross section of the population visits a movie theater
33.

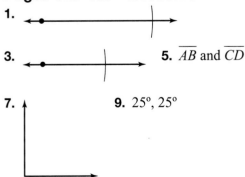

35. The graph on the right is misleading because the intervals are unequal on the vertical scale.

Pages 182–183 Standardized Test Practice
1. D   **3.** C   **5.** D   **7.** B   **9.** D   **11.** A   **13.** 3
15. $2n + 3 = 14$

CHAPTER 5
Geometry: Investigating Patterns

Pages 186–187 Lesson 5-1A
1. ⟷
3. ⟷   **5.** $\overline{AB}$ and $\overline{CD}$
7. ⌐   **9.** 25°, 25°

Pages 191–192 Lesson 5-1
1. Parallel lines are lines that are the same distance apart and never meet.

3.

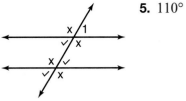

5. 110°

7. 120°   **9.** 100; 80   **11.** 135°   **13.** 138°
15. 68°   **17.** 75°   **19.** 105°   **21.** 75°   **23.** 60
25. 90°, 90°, 90°   **27.** 115°   **29.** C
31.

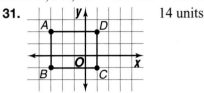

14 units

Page 193 Lesson 5-1B
1.

3. corresponding angles; yes

Pages 194–195 Lesson 5-2A
1. the students who like all three types of music; 8 students
3.

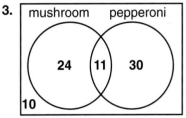

10 customers

5. The numbers in the Venn diagram must have a sum of 75.
7.

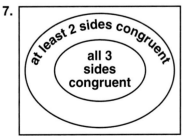

9a. 4 times   **9b.** $21 million   **11.** D

Pages 198–199 Lesson 5-2
1. Both isosceles triangles and equilateral triangles have congruent sides. An isosceles triangle has at least two sides congruent, and an equilateral triangle

has three sides congruent. **3.** Jaali is correct. A triangle has three angles whose sum is 180°. If a triangle had two obtuse angles, then the sum of two of the three angles would already be greater than 180°. **5.** acute, scalene **7.** false **9.** obtuse, isosceles **11.** acute, isosceles **13.** acute, isosceles **15.** 33 **17.** $m\angle A = 65°, m\angle B = 75°, m\angle C = 40°$
19. true

21. 60°, 60°, 60° **25.** 40 **27.** B

Page 200 Lesson 5-3A
1. rectangle, 2, traceable **3.** hexagon, 4, not traceable **5.** octagon, 6, not traceable **7.** Yes; any 4-sided figure with one diagonal will have 2 odd vertices.

Pages 203–204 Lesson 5-3
1. They each have four sides and four angles.
3a. yes **3b.** The angles at each end of either of the parallel sides are congruent to each other.
5. 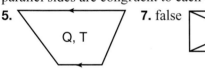 **7.** false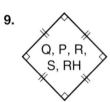

9. **11.**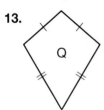

13. 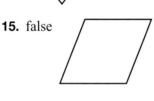 **15.** false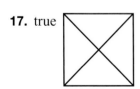

17. true

19. rectangle, square **21a.** 80 **21b.** $m\angle A = 70°, m\angle B = 110°, m\angle C = 70°, m\angle D = 110°$
23. 12 **25.** A

Page 204 Mid-Chapter Self Test
1. 64 **3.** acute, scalene

5. false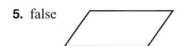

Page 205 Lesson 5-4A
1.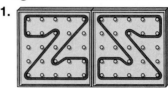

3.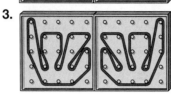

5. Yes, the corresponding sides are congruent. The figures are exactly the same except they are facing opposite directions.

Pages 208–209 Lesson 5-4
1. In line symmetry, half of the figure is the reflection of the other half.
3. 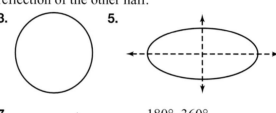 **5.**

7.  180°, 360°

9. 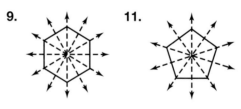 **11.**

13. no lines of symmetry
15. isosceles triangles, equilateral triangles; equilateral triangles
17a.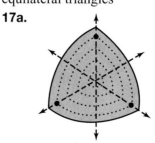

17b. 120°, 240°, 360°

19.

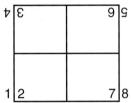

yes; 90°, 180°, 270°, 360°

21. (Numbers on the sides indicate the page numbers on the back of the sheet of paper.)

꜔ε		9ϛ
1 2		7 8

23. A

Pages 211–212   Lesson 5-5

1. Corresponding angles and corresponding sides must be congruent.   **3.** If each of the three sides of one triangle are congruent to each of the sides of another triangle, the triangles are congruent. SSS

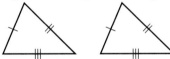

If 2 angles and the included side of one triangle are congruent to 2 angles and the included side of another triangle, the triangles are congruent. ASA

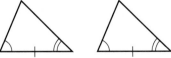

If 2 sides and the included angle of one triangle are congruent to 2 sides and the included angle of another triangle, the triangles are congruent. SAS

5. yes; △EFG ≅ △HIJ; SAS   **7.** yes; the triangles formed by the sections for urban contemporary music and country music; SAS   **9.** no   **11.** yes; △ABD ≅ △CBD; SAS   **13.** 25   **15.** The 4 triangles that form the large triangle in the center appear to be congruent. Three of these triangles are divided into 3 smaller triangles. These smaller triangles appear to be congruent.   **17.** no

Page 213   Lesson 5-5B

1.

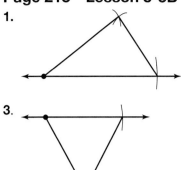

3.

Page 214   Lesson 5-6A

1.   **3.**

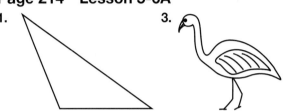

Pages 216–218   Lesson 5-6

1.

3. Helki; since the sum of 3 angles of a triangle is 180°, the unmarked angle in each of the triangles must be 75°. Since corresponding angles are congruent, the triangles are similar.   **5.** Similar; corresponding angles are congruent.   **7.** All of the triangles that form the Sierpinski Triangle are similar. The 4 triangles that form the large triangle are congruent. The 4 triangles that form three of these triangles are congruent, and so on.
9. Similar; corresponding angles are congruent.
11. congruent; ASA   **13.** 29   **15.** Similar; corresponding angles are congruent.   **17.** yes; △ABC ≅ △EFG; SSS

Pages 222–223   Lesson 5-7

1. a tiling of a plane made with copies of the same shape or shapes fit together without gaps or overlaps.   **3a.** 60°, 60°, 60°   **3b.** 6
3c.   6 figures

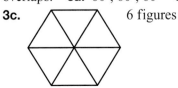

Selected Answers   **675**

3d. The number of figures needed to fill the space equals 360 divided by the measure of the angle.
5. translations **7.**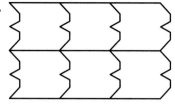

9.   **11.** rotation

15a. translation **15b.** rotation **15c.** rotation
17a. square
17b. translation   **19.** C

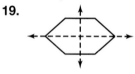

Pages 224–227 Study Guide and Assessment

1. i **3.** g **5.** e **7.** a **9.** If a figure can be turned less than 360° about its center and it looks like the original, then the figure has rotational symmetry. **11.** 127° **13.** right, scalene **15.** 75°
17. 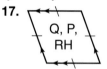 **19.**

21. yes **23.** yes; $\triangle WXY \cong \triangle PQR$; SSS
25. congruent; SAS
27.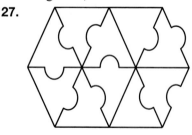

29. 45 students **31.** trapezoids, rectangles

Pages 228–229 Standardized Test Practice

1. A **3.** C **5.** D **7.** B **9.** B **11.** $5a + 5b$
13. 28 **15.** 8

CHAPTER 6
Exploring Number Patterns

Pages 233–234 Lesson 6-1

1. b is a factor of a. **3.** Alonzo; if the sum of the digits is divisible by 9, it is also divisible by 3.
5. 3, 9 **7.** yes **9.** Sample answer: 1,110 **11.** 1 by 40, 2 by 20, 4 by 10, 5 by 8 **13.** 3, 5 **15.** 2, 3, 6, 9 **17.** 3, 5 **19.** 2, 4 **21.** no **23.** yes
25. yes **27.** Sample answer: 1,004 **29.** Sample answer: 2×42, 3×28, 4×21 **31.** Sample answer: $2 \times 8{,}592$, $3 \times 5{,}728$, $4 \times 4{,}296$, $6 \times 2{,}864$, $8 \times 2{,}148$ **33.** Yes; $3 \times 17 = 51$. **35.** A
37. $\dfrac{\$0.32}{1 \text{ ounce}}$

Pages 237–238 Lesson 6-2

1. Both prime and composite numbers are whole numbers. 0 and 1 are neither prime nor composite. Prime numbers have exactly 2 factors, 1 and the number itself. Composite numbers have more than 2 factors.
3.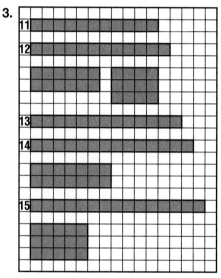

3a. 11, 13 **3b.** 12, 14, 15 **5.** prime **7.** 7^2
9. $3^2 \cdot 5^2$ **11.** prime **13.** composite **15.** neither
17. prime **19.** 5^2 **21.** $3 \cdot 5^2$ **23.** $3^2 \cdot 13$
25. $2^2 \cdot 5^2 \cdot 7^2$ **27.** 3 **29.** 61 **31a.** 128, 256, 512 **31b.** 2^n **33.** $9 = 3^2$, $36 = 2^2 \cdot 3^2$, $100 = 2^2 \cdot 5^2$, $144 = 2^4 \cdot 3^2$ **33a.** All exponents are even.
33b. Yes; in order for a number to be a perfect square, all prime factors must be able to be arranged in pairs. **35.** Similar; the figures are the same shape, but are different sizes. **37.** $4x < 20$; $x < 5$

Page 239 Lesson 6-2B

1. 1, 3, 7, 9 **3.** There is a factor other than 1 that is the same for 10 and each number.

Pages 240–241 Lesson 6-3A

1. Atepa and Curtis have included all the types of pizzas that they can make. **3.** 8 combinations **5.** Sample answer: A sandwich can be made with ham, turkey, American cheese and/or Swiss cheese on white or wheat bread. How many different sandwiches can be made. **7.** 5 packages of 40 and 2 packages of 75 **9.** 10, 5, 4, or 2 people **11a.** bottom right **11b.** bottom left **13.** C

Pages 243–244 Lesson 6-3

1. 1, 2, 4, 5, 10, 20; 20

3. 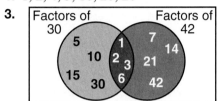 **3a.** 6

3b. 30 = 5·6 = 5·2·3   42 = 6·7 = 2·3·7

5. 27 **7.** 15 **9.** 7 **11.** 21 **13.** 6 **15.** 7 **17.** 12 **19.** 42 **21.** 150 **23.** Sample answer: 7, 14, 21 **25a.** 18 in. **25b.** 7 shelves **27.** $2^3 \cdot 7$ **29.** D

Pages 247–248 Lesson 6-4

1. Sample answer: $\frac{1}{2}$ **3.** rationals **5.** $\frac{1}{9}$ **7.** $-\frac{4}{5}$ **9.** $1\frac{5}{8}$ lb **11.** whole numbers, integers, rationals **13.** whole numbers, integers, rationals **15.** rationals **17.** $\frac{3}{5}$ **19.** $-\frac{1}{3}$ **21.** $\frac{4}{9}$ **23.** $\frac{2}{3}$ **25.** $-\frac{3}{4}$ **27.** $-\frac{9}{20}$ **29.** $\frac{2}{5}$ **31.** $\frac{35}{73}$ **35.** 14 **37.** 3

Page 248 Mid-Chapter Self Test

1. 2, 4, 5, 8, 10 **3.** $2^2 \cdot 17$ **5.** $3^2 \cdot 7$ **7.** 6 **9.** $-\frac{9}{11}$

Pages 251–252 Lesson 6-5

1. 0.012; $\frac{3}{250}$ **3.** Nadia; $\frac{1}{4}$ = 0.25, but $0.\overline{25}$ = 0.252525 **5.** -0.6171717171 **7.** $0.\overline{63}$ **9.** $\frac{33}{50}$ **11.** $-1\frac{5}{33}$ **13.** $0.2\overline{5}$ **15.** $7.0\overline{74}$ **17.** -0.3053053053 **19.** 0.75 **21.** -0.28 **23.** $-5.\overline{6}$ **25.** 12.625 **27.** $\frac{22}{25}$ **29.** $\frac{5}{6}$ **31.** $-1\frac{5}{9}$ **33.** $-5\frac{67}{99}$ **35.** terminating **37.** 5 + 6 ÷ 11 = 5.545454545

39. $\frac{13}{1,000}$ oz **43.** integers, rationals **45.** 12 **47.** C

Pages 254–256 Lesson 6-6

1. Sample answer: An even number picked at random is divisible by 2. **5.** 0 **7.** $\frac{1}{2}$ **9.** 1 **11.** $\frac{5}{6}$; $0.8\overline{3}$ **13.** 0 **15.** 1 **17.** $\frac{1}{8}$ **19.** $\frac{1}{2}$ **21.** $\frac{7}{8}$ **23.** $\frac{2}{11}$ **25.** 1 **27.** $\frac{5}{11}$ **31.** No; the number of ways something can occur cannot be negative. **33a.** $\frac{25}{108}$ **33b.** $\frac{2}{27}$ **33c.** $\frac{2}{27}$ **33d.** $\frac{1}{27}$ **35.** $\frac{4}{7}$ **37.** 80° **39.** 4

Pages 258–259 Lesson 6-7

1. (1) List the multiples of 10 and the multiples of 16. Find the least nonzero number that is a multiple of both numbers. (2) Find the prime factorization of 10 and 16. Multiply all the factors, using the common factors only once. **3a.** 36, 72 **3b.** 36 **5.** 0, t, $2t$, $3t$, $4t$, $5t$ **7.** 90 **9.** 840 **11.** 0, 7, 14, 21, 28, 35 **13.** 0, 14, 28, 42, 56, 70 **15.** 0, 150, 300, 450, 600, 750 **17.** 48 **19.** 100 **21.** 30 **23.** 105 **25.** 1,225 **27.** 340 **29.** yes **31.** when the GCF is 1 **33.** 8:00 A.M. **35.** C **37.** 358

Page 260 Lesson 6-8A

1. The midpoint is the mean of the endpoints. **3.** Add them together and divide by 2.

Pages 263–264 Lesson 6-8

1. 0.3   0.08   0.3 is represented by 30 small squares. 0.08 is represented by 8 small squares.

3. Sample answer: $\frac{1}{2}$, 0.53; Since $\frac{2}{5}$ = 0.4, $\frac{4}{7}$ = $0.\overline{571428}$, and $\frac{1}{2}$ = 0.5, $\frac{1}{2}$ and 0.53 are between the given numbers. **5.** < **7.** = **9.** $\frac{2}{5}$, 0.376, $\frac{3}{8}$, 0.367 **11.** 16 **13.** 90 **15.** < **17.** < **19.** < **21.** < **23.** $-\frac{1}{3}, -\frac{1}{4}, \frac{1}{10}, \frac{1}{9}$ **25.** $-4.75, -4\frac{2}{3}$, $-4.5, -4\frac{2}{5}, -4.1\overline{9}$ **27.** 0.6 **29a.** Q **29b.** S **29c.** P **29d.** R **31.** $\frac{1}{125}$ **33.** No; $0.\overline{4} = \frac{4}{9}$. **35.** Sample answer: 18 **37.** 35

Pages 266–267 Lesson 6-9

1. 36.2 and 0.362 are not greater than or equal to

1 and less than 10   **3.** 488,200   **5.** 234,000,000
7. 5.4×10^{-2} **9.** 14,000 lb   **11.** 8,080,000,000
13. 0.000000075   **15.** 0.0000252   **17.** 202.1
19. 7.67×10^{-3} **21.** 4.004×10^5
23. 3.3×10^{-5} **25.** 7.6×10^3 **27.** 2.5×10^{-8}
29. 3.2×10^{-8} **31.** 7.2×10^6 dollars
33. $7\frac{2}{7}, 7.35, \frac{37}{5}$ **35.** -221

Pages 268–271   Study Guide and Assessment

1. multiple   **3.** simplest form
5. 1   **7.** 3.2×10^4 **9.** Write the digits of the decimal as the numerator. Use the appropriate power of 10 (10, 100, 1000, and so on) as the denominator. Then simplify.   **11.** 5   **13.** 2   **15.** 3   **17.** $3^2 \cdot 7$
19. $2 \cdot 101$ **21.** 3^4 **23.** 14   **25.** 11   **27.** $\frac{5}{7}$
29. $\frac{6}{7}$ **31.** 0.625   **33.** 2.6   **35.** $\frac{13}{20}$ **37.** $\frac{16}{99}$
39. $8\frac{3}{8}$ **41.** $\frac{1}{2}$ **43.** $\frac{2}{3}$ **45.** 30   **47.** 168
49. 350   **51.** <   **53.** <   **55.** 6.48×10^{-5}
57. 5.5×10^{-4} **59.** 4.76×10^{-3} **61.** 0.250
63. $\frac{3}{5}$

Pages 272–273   Standardized Test Practice

1. C   **3.** C   **5.** B   **7.** C   **9.** B   **11.** D   **13.** C
15. I and II

CHAPTER 7
Algebra: Using Rational Numbers

Pages 279–280   Lesson 7-1

1. $\frac{2}{3} - \frac{1}{3} = \frac{1}{3}$
3. $1\frac{1}{2}$

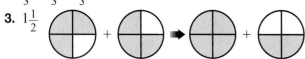

5. $\frac{1}{2}$ **7.** $4\frac{2}{5}$ **9.** $\frac{1}{4}$ **11.** $-1\frac{2}{5}$ **13.** -3
15. -1 **17.** $-3\frac{1}{9}$ **19.** $-\frac{1}{2}$ **21.** 1   **23.** $106\frac{3}{16}$ in. or 8 ft $10\frac{3}{16}$ in.   **25.** $2\frac{3}{4}$ in.   **27.** C

Pages 283–284   Lesson 7-2

1. $\frac{2}{3} + \frac{1}{4} = \frac{11}{12}$
3. $\frac{1}{4}$

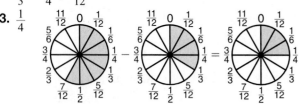

5. 7   **7.** $1\frac{1}{4}$ **9.** $\frac{7}{8}$ **11.** $9\frac{1}{12}$ **13.** $12\frac{1}{8}$
15. $1\frac{1}{24}$ **17.** $-2\frac{1}{10}$ **19.** $6\frac{5}{6}$ **21.** $1\frac{1}{4}$
23. $-9\frac{7}{8}$ **25.** $8\frac{7}{8}$ **27.** $15\frac{1}{4}$ **29.** $-10\frac{23}{40}$
31. $-3\frac{11}{24}$ **33.** $2\frac{5}{24}$ **35.** $66\frac{1}{4}$ in.   **37a.** $a = \frac{4}{5}$, $b = \frac{1}{2}$ **39.** A   **41.** 1,050

Pages 288–289   Lesson 7-3

1. Sample answer: Three out of four rows are shaded to represent $\frac{3}{4}$. Three out of eight columns are shaded to represent $\frac{3}{8}$.   **3.** $\frac{3}{5}$
5. $2\frac{2}{5}$ **7.** $5\frac{19}{25}$
9. $4\frac{1}{4}$ in.; $\frac{5}{8}$ in^2
11. $-\frac{35}{48}$ **13.** $2\frac{1}{4}$ **15.** $-\frac{3}{5}$ **17.** $7\frac{1}{3}$ **19.** $\frac{9}{16}$
21. $-\frac{8}{27}$ **23.** $-\frac{1}{4}$ **25.** $7\frac{1}{2}$ **27.** $\frac{15}{32}$ **29a.** 27 in.
29b. 22 in.   **31.** C   **33.** 12

Pages 291–292   Lesson 7-4

1. $\frac{9}{5}$ **3.** Change $-3\frac{7}{8}$ to an improper fraction and then find the reciprocal.; $-\frac{8}{31}$ **5.** $\frac{8}{7}$ or $1\frac{1}{7}$ **7.** $\frac{1}{4}$
9. $-1\frac{2}{5}$ **11.** $-\frac{3}{2}$ **13.** -5 **15.** $\frac{4}{9}$ **17.** $-\frac{1}{x}$
19. $-\frac{1}{8}$ **21.** $-59\frac{1}{2}$ **23.** $-\frac{5}{11}$ **25.** No; the reciprocal of $-2\frac{1}{2}$ is $-\frac{2}{5}$.   **27.** -1 **29.** $\frac{7}{12}$
31. $36\frac{7}{8}$ in^2 **33.** $19\frac{1}{32}$ **35.** $\frac{1}{2} \times 180 = 90$

Pages 294–295   Lesson 7-5A

1. Santos; He made a rule after seeing several examples.   **3.** $\frac{1}{64}$, 34,200; $\frac{1}{128}$, 39,900   **5.** Sample answer: Looking for a pattern involves determining what you know.   **7a.** 20, 25   **7b.** 21, 15
9a. 25 units   **9b.** 2.25 units   **9c.**
11. 21   **13a.** 10
13b. Change cell B2 to 0.5.

Pages 298–299   Lesson 7-5

1. When the difference between any two consecutive terms is the same, the sequence is arithmetic.   **3.** Joanne is correct. The common difference is -4.   **5.** N; $-11, -14, -16$ **7.** G;

48, −96, 196   **9.** A; 14, $16\frac{1}{2}$, 19   **11.** G; 81, 243, 729   **13.** G; $-\frac{1}{64}, -\frac{1}{256}, -\frac{1}{1,024}$ **15.** A; 36, 40, 44   **17.** G; 162, −486, 1,458   **19.** G; $\frac{7}{9}, \frac{7}{27}, \frac{7}{81}$ **21.** A; $3\frac{1}{6}, 2\frac{5}{6}, 2\frac{1}{2}$ **23.** N; 111, 136, 124   **25.** 4, $7\frac{1}{3}$, $10\frac{2}{3}$, 14   **27.** 74   **29a.** 34, 55, 89; Neither; there is no common difference or common ratio.   **29b.** Sample answer: The numbers are all Fibonacci numbers.   **31.** $33.75   **33.** 24   **35.** 27

Page 299 Mid-Chapter Self Test
1. $\frac{1}{4}$ **3.** $6\frac{2}{5}$ **5.** $-96\frac{1}{4}$ **7.** $\frac{1}{2}(12\frac{4}{5}) = \frac{1}{2} \cdot 12 + \frac{1}{2} \cdot \frac{4}{5} = 6 + \frac{2}{5}$ or $6\frac{2}{5}$ **9.** G; 12, 3, $\frac{3}{4}$

Page 300 Lesson 7-5B
3. Each number is the sum of the previous two numbers.   **5.** Sample answer: All the decimals are about 1.62.

Pages 303–304 Lesson 7-6
1. Sample answer:

3. The height of the parallelogram is half the height of the trapezoid and the length of the base of the parallelogram is the sum of the length of the bases of the trapezoid.   **5.** $b = 2\frac{2}{3}$ ft, $h = 3\frac{3}{4}$ ft, $A = 5$ ft²   **7.** $3\frac{63}{64}$ in²   **9.** 234 ft²   **11.** $b = 3$ ft, $h = 4$ ft, $A = 6$ ft²   **13.** $a = 12$ yd, $b = 18$ yd, $h = 10$ yd, $A = 150$ yd²   **15.** $b = 12$ cm, $h = 5$ cm, $A = 30$ cm²   **17.** $22\frac{1}{2}$ in²   **19.** 33 cm²   **21.** $6\frac{1}{2}$ yd²   **23.** $37\frac{1}{2}$ yd²   **25.** 29.49 m²   **27.** 9 ft²   **29.** Sample answer:

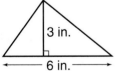

31. $a = 12$ yd, $b = 18$ yd   **33.** The area is quadrupled.   **35.** 8.4×10^{-5} **37.** 27

Page 306 Lesson 7-6B
1. They are the same.   **3.** yes   **5a.** Count the squares.   **5b.** Pick's Theorem works for all of the figures.

Page 308 Lesson 7-7A
1a. diameter   **1b.** circumference   **1c.** Answers will vary. The ratios should be close to $\frac{22}{7}$ or 3.14.   **1d.** Answers will vary.; The ratios should be closer to $\frac{22}{7}$ or 3.14.

Pages 310–311 Lesson 7-7
1. $\frac{22}{7}$ and 3.14   **3.** Sample answer: Divide 22 by 3.14. Then round to the nearest tenth.   **5.** 56.5 in.   **7.** 106.8 cm   **9.** 95.8 m   **11.** 66 mm   **13.** 185.3 cm   **15.** 56.5 ft   **17.** 33 in.   **19.** 60.7 m   **21.** $42\frac{3}{7}$ in.   **23.** 0.325 m   **25.** $9\frac{45}{56}$ or about 9.8 inches   **27.** circumference; height $= 3d$, $C \approx 3.14d$.   **29.** B

Pages 313–314 Lesson 7-8
1. There are 3 sets of $\frac{2}{3}$ to equal 2.   **3.** $\frac{8}{9}$ **5.** $-\frac{5}{12}$ **7.** $\frac{2}{3}$ **9.** $-3\frac{3}{5}$ **11.** $-\frac{3}{20}$ **13.** 6   **15.** 6   **17.** $-2\frac{5}{8}$ **19.** $\frac{9}{25}$ **21.** $29\frac{3}{4}$ feet   **23.** $\frac{m}{n}$ is greater; Dividing by a proper fraction is actually multiplying by an improper fraction.   **25.** B

Pages 316–317 Lesson 7-9
1. Sample answer: $-1\frac{1}{4}a + 4 = 8$ **3.** Sample answer: Esohe's method uses the order of operations in reverse and is probably easier than Priya's method because it requires dividing two fractions.   **5.** 3   **7.** −4.375   **9.** 1.8875   **11.** 60   **13.** $2\frac{3}{10}$ **15.** −3.2   **17.** −14.4   **19.** $-\frac{13}{30}$ **21.** 22.5   **23.** −0.5   **25.** −3.2   **27.** $\frac{1}{21}$ **29.** about 12.9 ft   **31.** $-40°F = -40°C$ **33.** $2^4 \times 3$ **35.** 8

Pages 320–321 Lesson 7-10
1. The process for solving inequalities is the same. If the process includes dividing or multiplying by a negative number, the solution will have the opposite inequality sign.
3. $x < -3$
5. $y \geq -16.5$
7. $h > 4\frac{1}{2}$
9. at least 4 hours
11. $p < 3.17$
13. $y \leq 84$

15. $k < -1\frac{1}{9}$

17. $t < -4\frac{3}{8}$

19. $d > 2\frac{3}{10}$

21. $h \leq \frac{5}{22}$

23. $b \geq 11.4$

25. $g > -9\frac{1}{5}$

27. $n > -8.97$

29. b   31. on days 12–15   33. -15 35. $\frac{1}{6}$

Pages 322–325 Study Guide and Assessment

1. g   3. h   5. j   7. f   9. Sample answer: Rewrite the mixed numbers as fractions. Multiply the numerators and multiply the denominators. Then simplify.   11. $-\frac{1}{2}$ 13. $\frac{1}{15}$ 15. $-11\frac{1}{6}$
17. $-8\frac{1}{3}$ 19. $2\frac{8}{9}$ 21. $-\frac{3}{16}$ 23. $-1\frac{3}{10}$
25. G; 2, 1, $\frac{1}{2}$ 27. A; 32, 20, 8   29. 56 m²
31. 16.3 m   33. 20.4 in.   35. $2\frac{1}{5}$ 37. $\frac{1}{8}$
39. -11.52 41. 3.2
43. $d \geq -21$ 45. $v < 7.6$
47. 40 miles

Pages 326–327 Standardized Test Practice

1. D   3. D   5. A   7. C   9. B   11. D
13. $a = 8$ ft, $b = 14$ ft, $h = 8$ ft, $A = 88$ ft²
15. -484 17. 0.00349   19. yes;

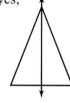

CHAPTER 8
Applying Proportional Reasoning

Pages 331–333 Lesson 8-1

1a. Yes; each ratio represents the time to the cost.
1b. No; one ratio represents the time to the cost and the other represents the cost to the time.   1c. Yes; each ratio represents the cost to the time.
3. Sample answer: $\frac{16}{1} = \frac{56}{p}$; $3\frac{1}{2}$ lb   5. 48 min
7. 43,750 red blood cells   9. Sample answer: $\frac{100}{1} = \frac{170}{m}$; 1.7 m   11. Sample answer: $\frac{12}{96} = \frac{4}{x}$; 32¢   13. Sample answer: $\frac{35}{2.5} = \frac{420}{m}$; 30 min
15. 14,520 ft   17. 335 mi   19a. 21 lb
21. 131,072 bytes   23. Sample answer: $\frac{15}{120} = \frac{25}{p}$, $\frac{120}{15} = \frac{p}{25}$, $\frac{15}{25} = \frac{120}{p}$, $\frac{25}{15} = \frac{p}{120}$ 25. $2\frac{4}{9}$

Page 334 Lesson 8-1B

1. 9 cups cottage cheese, $2\frac{1}{4}$ cups chili sauce, $2\frac{1}{4}$ teaspoons onion powder, $2\frac{1}{4}$ cup skim milk, 27 tablespoons parmesan cheese
3.

| 9 | Tabasco sauce | B2/8 | t. |

Pages 336–338 Lesson 8-2

1. Percent means per one hundred and r is the number out of one hundred.   3. Sample answer: Percentage: Find 20% of 40. $\frac{P}{40} = \frac{20}{100}$; Base: 3 is 40% of what number. $\frac{3}{B} = \frac{40}{100}$; Percent: 5 is what percent of 25? $\frac{5}{25} = \frac{r}{100}$ 5. 5%   7. 38%
9. $\frac{P}{60} = \frac{15}{100}$; 9   11. $\frac{15}{45} = \frac{r}{100}$; 33.3%   13. 14%
15. 37%   17. 20%   19. 52%   21. 82%
23. 50%   25. 18.75%   27. 25%   29. $\frac{P}{66} = \frac{25}{100}$; 16.5   31. $\frac{16}{48} = \frac{r}{100}$; 33.3%   33. $\frac{5}{B} = \frac{26}{100}$; 19.2   35. $\frac{7}{20} = \frac{r}{100}$; 35%   37. $\frac{60}{B} = \frac{15}{100}$; 400
39. $\frac{45}{B} = \frac{35}{100}$; 128.6   41. 10%   43. $48,700
47. 6 cups   49. $\frac{2}{3}$ 51. 40 shrimp/1 pound

Pages 340–341 Lesson 8-3

1. $R \cdot 34 = 28$ 3. $0.47 \cdot 52 = P$; 24.44
5. $R \cdot 90 = 36$; 40%   7. $0.30 \cdot B = 48$; $160
9. 4.5%   11. $R \cdot 66 = 55$; $83\frac{1}{3}$%
13. $0.15 \cdot B = 30$; 200   15. $0.24 \cdot 72$ P; 17.28
17. $\frac{2}{3} \cdot B = 16$; 24   19. $R \cdot 300 = 6$; 2%
21. $R \cdot 80 = 25$; 31.25%   23. $R \cdot 50 = 6$; 12%
25. $R \cdot 80 = 70$; 87.5%   27. $0.08 \cdot B = 54$; $675
29. $25,000   31. about 11 people
33. 1,062,000 tons   35. D

Pages 342–344   Lesson 8-3B

1. Both 100 and 300 are divisible by 20, but they are not divisible by 18 or 22. Therefore, the computations are easier if 20 is used instead of 18 or 22. Sample answer: 25   **3.** vanilla, 15 cups; chocolate 75 cups; strawberry, 45 cups; chocolate chip, 135 cups; peanut butter, 30 cups   **5.** Sample answer: Predict who will win a presidential election. Instead of surveying all voters, you would survey a smaller group that is representative of all voters.
7. about 26 times   **9.** 9 clothespins
11. 50 Calories   **13.** B

Pages 346–347   Lesson 8-4

1. 3 out of every 1,000 farmers in the world live in the U.S.

3a.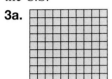

3b. 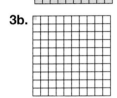   **5.** $1\frac{3}{4}$

7. 1.55   **9.** 1.235   **11a.** 1.17% = 0.0117; 1.06% = 0.0106; 0.92% = 0.0092; 1.08% = 0.0108; 2.21% = 0.0221   **11b.** 1994   **11c.** 1992
13. $1\frac{3}{20}$ **15.** $\frac{3}{5,000}$ **17.** $2\frac{9}{20}$ **19.** $134\frac{47}{100}$
21. $\frac{1}{3}$ **23.** 0.0003   **25.** 0.1025   **27.** 0.00079
29. 0.004   **31.** 1.104   **33.** 104%, 1, 1.04%, 0.4%, 0.04%   **35.** 1.81   **37.** 200% = $\frac{200}{100} = \frac{2}{1} = 2$; therefore 200% means to double.
39. about 51.5 cm   **41.** C

Pages 350–351   Lesson 8-5

1. Find the amount of the change.
3. 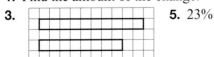   **5.** 23%

7. $200.00   **9.** $16.80   **11.** 40%
13. 67%   **15.** 6%   **17.** $13.05   **19.** $2.69
21. $10.40   **23.** $1,440.00   **25.** $19.60
27. $73.75   **29.** 8%   **31.** $225   **33.** $19.50
35a. about 1.1%   **35b.** decrease   **39.** $\frac{1}{250}$
41. 6 + 7 + 2 = 15, 15 ÷ 3 = 5; yes   **43.** 73

Page 352   Lesson 8-5B

1. sweaters, $22.49; jackets, $27.22; sweatshirts, $18.67; socks, $4.94; T-shirts, $5.99   **3.** $23.59
5.

8	suede jackets	$99.59	(100-B1)/100*B8

Pages 355–356   Lesson 8-6

1. $I = prt$; I represents the amount of interest, p represents the amount of money that is initially deposited in a savings account or the amount of money borrowed, r represents the percent of interest, and t represents the time in years.   **3.** Janay; 8 months equals $\frac{8}{12}$ or $\frac{2}{3}$ year. Sonia forgot to change the time to years.   **5.** $18.40   **7.** $760.52
9. $112.50   **11.** $77.24   **13.** $133.20
15. $840.00   **17.** $294.93   **19.** $421.38
21. $112.50   **23.** 14 years   **25.** $70,200,000,000
27. about 31.6%   **29.** −241

Page 356   Mid-Chapter Self Test

1. 1,176 bushels   **3.** $\frac{63}{84} = \frac{r}{100}$; 75%
5. $0.55 \cdot 86 = R$; 47.3   **7.** $2\frac{17}{20}$ **9.** $140.63

Pages 359–360   Lesson 8-7

1. If 2 polygons have corresponding congruent angles and have corresponding sides that are in proportion, then the polygons are similar.
5. Sample answer: $\frac{10}{5} = \frac{6}{x}$; 3   **7.** $2\frac{1}{2}$ in. by $3\frac{1}{8}$ in.
9. Yes; the corresponding angles are congruent and $\frac{3}{2} = \frac{3}{2} = \frac{3}{2} = \frac{3}{2}$.   **11.** Sample answer: $\frac{5}{3} = \frac{x}{4}$; $6\frac{2}{3}$.
13. Sample answer: $\frac{10}{6} = \frac{5}{x}$; 3   **15.** 8.75 m
17a. yes

17b. Sample answer:  yes

17c. Yes; all angles are right angles, so all angles of one square are congruent to all angles of any other square. Since all 4 sides of a square are the same length, corresponding sides are always in proportion.   **19.** $192   **21.** 9.6

Pages 362–364   Lesson 8-8

1. Indirect measurement means you did not actually measure the distance. Instead you used proportions and other measurements to find the distance.

3. Sample answer: The shadow of a flagpole is 12 feet at the same time the shadow of a nearby sign is 4 feet. If the sign is 8 feet tall, how high is the flagpole? To solve the problem, write the proportion and solve for h.
flagpole's shadow → $\frac{12}{4} = \frac{h}{8}$ ← flagpole's height
sign's shadow → sign's height

5. Sample answer: $\frac{h}{5} = \frac{9}{2\frac{1}{4}}$; 20 ft **7.** Sample answer: $\frac{20}{8} = \frac{x}{5}$; 12.5 km **9.** Sample answer: $\frac{90}{x} = \frac{60}{30}$; 45 mi **11.** Sample answer: $\frac{s}{7} = \frac{26}{10}$; $18\frac{1}{5}$ ft **13.** 630 ft **15.** 2,000 mi **17.** $\frac{451}{1,152}$

Page 365 Lesson 8-8B

1. 30° angle: ratio 1 ≈ 0.5774, ratio 2 ≈ 0.5000, ratio 3 ≈ 0.8660; 60° angle: ratio 1 ≈ 1.7321, ratio 2 ≈ 0.8660, ratio 3 ≈ 0.5000

3. 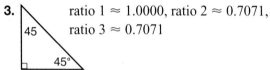 ratio 1 ≈ 1.0000, ratio 2 ≈ 0.7071, ratio 3 ≈ 0.7071

Pages 367–369 Lesson 8-9

1. Some objects are too small or too large to show the actual item in the space available. **3.** Sierra; the scale is 1 inch = 3 feet or 1 inch = 36 inches which is 1:36. **5.** 1 cm = 125 km or 1:12,500,000 **7a.** 16 ft by 18 ft **7b.** 8 ft by 12 ft **7c.** 8 ft by 8 ft **7d.** 8 ft by 10 ft **9.** 10 in. = 305 ft or 1:366 **11.** $2\frac{4}{5}$ in. **13.** 2.6 cm **15.** 25 in. by 35 in. **17.** 60.8475 m

Pages 372–373 Lesson 8-10

1. A dilation is an enlargement or a reduction of an image.

3a.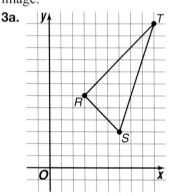

3b. R'(2, 4), S'(4, 2), T'(6, 8)

3c.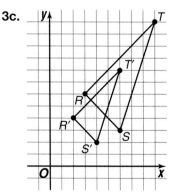

3d. The 2 triangles are the same shape. The dilated figure is smaller than the original. **5.** 1.5
7a. 200%; 50% = $\frac{1}{2}$ so the image would be half the size, but 200% = 2.00 so the image would be twice as large. **7b.** $\frac{3}{4}$ **9.** $B'\left(7\frac{1}{2}, 12\right)$
11. D'(−8, 24), R'(−4, −8), T'(16, 12)

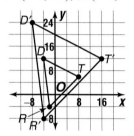

13. $D'(-5, 15), R'\left(-2\frac{1}{2}, -5\right), T'\left(10, 7\frac{1}{2}\right)$

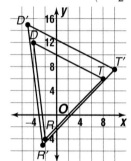

15. 2
17. Sample answer:

19.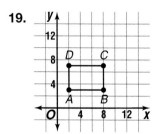

19a. 20 units; 24 units²

19b.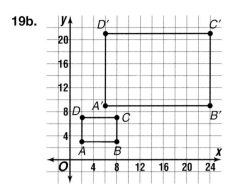

19c. 60 units; 216 units² **19d.** 3:1 **19e.** 9:1
19f. The perimeter of the dilation of a polygon is equal to the perimeter of the original polygon times the scale factor. The area of the dilation of a polygon is equal to the area of the original polygon times the square of the scale factor. **21.** 56 ft **23.** true

Pages 374–377 Study Guide and Assessment

1. d **3.** f **5.** a **7.** g **9.** Sample answer: $\frac{80}{10} = \frac{x}{15}$; $120 **11.** $\frac{P}{18} = \frac{45}{100}$; 8.1 **13.** $\frac{P}{80} = \frac{86}{100}$; 68.8 **15.** $0.66 \cdot 7{,}000 = P$; 4,620 **17.** $0.30 \cdot B = 15$; 50 **19.** $2\frac{3}{20}$ **21.** 0.006 **23.** 25% **25.** 33%
27. 40% **29.** $31.20 **31.** $68.25 **33.** $5\frac{1}{3}$ in.
35. 36 ft **37.** $13\frac{3}{4}$ mi **39.** $22\frac{1}{2}$ mi **41.** (60, 12)
43. (4, 16) **45.** 18 cuts **47.** 107 mi

Pages 378–379 Standardized Test Practice

1. A **3.** A **5.** D **7.** B **9.** A **11.** B **13.** 2
15. 5

CHAPTER 9
Algebra: Exploring Real Numbers

Pages 383–384 Lesson 9-1

1. $-\sqrt{100}$ **3.** When each square root is squared, it equals the perfect square. $5^2 = 25$
5. 9 **7.** -8 **9.** 776 feet **11.** -3 **13.** -6
15. $\frac{2}{3}$ **17.** $-\frac{4}{5}$ **19.** -14 **21.** -1.7
23. $0.5, -0.5$ **25.** $13, -13$; 13 is the principal square root because it is the positive square root.
27. 17 rows of 17 trees in each row **29.** Yes, because the factorization of the product is simply the product of the two perfect square factorizations, and therefore it is also a perfect square.
31. 115.5 cm² **33.** 8.5; 8.5; 6, 12

Page 385 Lesson 9-2A

1. 4 **3.** 12 **5.** You can square numbers by guess and check until you find a number when squared is less than the square root you are finding and then the next greater number when squared is just greater than the square root.

Pages 387–389 Lesson 9-2

1. If three more tiles are shaded, then 12 tiles are shaded with four unshaded. Therefore, $\sqrt{12}$ is closer to $\sqrt{9}$ than $\sqrt{16}$; $\sqrt{12} \approx 3$. **3.** 8 **5.** 12
7. about 2.7 miles **9.** 4 **11.** 4 **13.** 14 **15.** 6
17. 2 **19.** 12 **21.** 9 **23.** about 526 square feet
25a. yes, about 3 mph
25b. Sample answer:
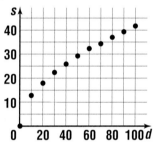
27. 30 **29.** E

Pages 392–394 Lesson 9-3

1. [number line showing $\sqrt{15}$ between 3 and 4] **3.** Their teacher is correct because it depends on what number of which you are taking the square root. Examples, $\sqrt{9} = 3$ and $3 < 9$, but $\sqrt{0.09} = 0.3$ and $0.3 > 0.09$. **5.** Q, R **7.** $\sqrt{20} \approx 4.5$; [number line 2–5 showing $\sqrt{20}$]
9. $-7.7, 7.7$ **11.** W, Z, Q, R **13.** I, R
15. Z, Q, R **17.** 2.4; [number line 0–3 showing $\sqrt{6}$]
19. 7.1; [number line 5–8 showing $\sqrt{50}$] **21.** 10.4; [number line 8–11 showing $\sqrt{108}$]
23. $-8.3, 8.3$ **25.** $-8.7, 8.7$ **27.** $-30.4, 30.4$
29. $\frac{8}{3}$ **31.**

Distance (ft)	Time (s)
128	$\sqrt{8} \approx 2.8$
64	$\sqrt{4} = 2$
32	$\sqrt{2} \approx 1.4$
16	$\sqrt{1} = 1$
8	$\sqrt{\frac{1}{2}} \approx 0.7$

33a. $\sqrt{10} \approx 3.2$ units **33b.** $4\sqrt{10} \approx 12.6$ units **35.** about 700 **37.** E

Page 394 Mid-Chapter Self Test
1. 1 **3.** 1.5 m **5.** 5 **7.** Q, R **9.** I, R

Page 397 Lesson 9-4A
1. 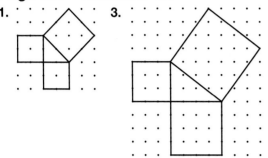 **3.**

5. 32 units **7.** (one small side)² + (another small side)² = (longest side)²

Pages 399–401 Lesson 9-4
1. 100 square units **3.** $6^2 + 12^2 = 36 + 144$ or 180 and $14^2 = 196$. Since $180 \neq 196$, the triangle is not a right triangle. **5.** $9^2 + b^2 = 41^2$; $b = 40$ **7.** $5^2 + 5^2 = c^2$; $c \approx 7.1$ **9.** yes **11.** $12^2 + 9^2 = c^2$; $c = 15$ **13.** $a^2 + 12^2 = 30^2$; $a \approx 27.5$ **15.** $1^2 + 1^2 = h^2$; $h^2 + 1^2 = x^2$; $x \approx 1.7$ **17.** $3^2 + b^2 = 8^2$; $b \approx 7.4$ **19.** $40^2 + b^2 = 41^2$; $b = 9$ **21.** $48^2 + 55^2 = c^2$; $c = 73$ **23.** yes **25.** no **27.** yes **29.** 24 inches **31.** between 18 and 20 inches

33. 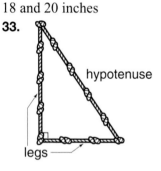 **35.** 44 feet **37.** C

Page 403 Lesson 9-5A
1. Yes, each path is along the faces of the cube, so they are feasible. **3.** There are 49 seats in this section. **7.** 20 people **9.** 9 meters **11.** 11 quarters and 19 nickels **13.** C

Pages 406–407 Lesson 9-5
1. Sample answer: 10-24-26 and 15-36-39 **3.** $x^2 + 21^2 = 30^2$; $x = \sqrt{459} \approx 21.4$ miles **5.** $5^2 + 8^2 = x^2$; $x = \sqrt{89} \approx 9.4$ miles **7.** $x^2 = 9^2 + 18^2$; $x = \sqrt{405} \approx 20.1$ feet; $x^2 = 9^2 + 12^2$; $x = \sqrt{225} = $ 15 feet **9.** $\sqrt{512} \approx 22.6$ feet **11.** about 3.7 feet longer **13.** about 13.2 km **15.** B

Page 409 Lesson 9-5B
1. 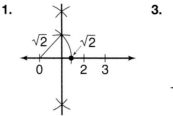 **3.**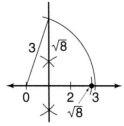

5. Draw a number line. At 1, construct a perpendicular line segment 1 unit in length. Draw a line from O to the top of the perpendicular line segment. Label it c. Open the compass to the length of c. With the tip of the compass at O, draw an arc that intersects the number line to the right of O at A. The distance from O to A is $\sqrt{2}$ units. **7.** Draw a number line. At 1, construct a perpendicular line segment 2 units in length. Draw a line from O to the top of the perpendicular line segment. Label it c. Open the compass to the length of c. With the tip of the compass at O, draw an arc that intersects the number line to the right of O at B. The distance from O to B is $\sqrt{5}$ units. **9.** $(\sqrt{11})^2 + 5^2 = 6^2$; none for 12; $(\sqrt{13})^2 = 3^2 + 2^2$; none for 14; $(\sqrt{15})^2 + 7^2 = 8^2$; $(\sqrt{17})^2 + 8^2 = 9^2$; none for 18; $(\sqrt{19})^2 + 9^2 = 10^2$; $(\sqrt{20})^2 + 4^2 = 6^2$

Pages 411–413 Lesson 9-6
1. 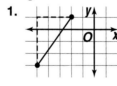 **3.** Both distances are 5 units or about 0.25 miles. Because the Madison Building is at $(4, -3)$ and the Sewall-Belmont House is at $(4, 3)$, the distances are the same. **5.** 7.1 units **7.** about 9.4 miles **9.** 4.1 units **11.** 7.1 units **13.** 4.5 units **15.** 7 units **17.** 4.2 units **19.** 4.2 units

21a. about 8 feet;
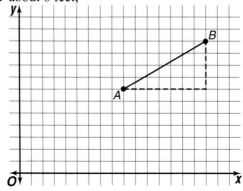

684 Selected Answers

23. B **25.** −384

Pages 416–417 Lesson 9-7

1. $c = 2a$ where a is leg opposite 30° angle
5. $a = 12$ mm; $b \approx 20.8$ mm **7.** $b = 3.2$ m; $c \approx 4.5$ m **9.** $a = 6$ in.; $b \approx 10.4$ in. **11.** $b \approx 34.6$ ft; $c = 40$ ft **13.** 3.75 inches **15a.** $\sqrt{2}, \sqrt{3}, \sqrt{4}$ or $2, \sqrt{5}$ **15b.** yes, the smallest triangle; yes, the fourth triangle **15c.** $\sqrt{6}$ **19.** D

Pages 418–421 Study Guide and Assessment

1. b **3.** c **5.** j **7.** i **9.** Answers will vary. Sample answer: Rational numbers can be expressed in the form $\frac{a}{b}$, where a and b are integers and $b > 0$, but irrational numbers cannot be expressed in that form; rational: 0.75, irrational: $\sqrt{3}$ **11.** 2.5
13. $\frac{2}{3}$ **15.** 1.8 **17.** 7 **19.** 18 **21.** 15
23. Z, Q, R **25.** Q, R **27.** $4^2 + b^2 = 9.5^2$; $b \approx 8.6$ m **29.** 20.6 km
31. 7.8 units; **33.** about 3.6 units

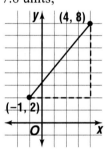 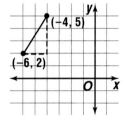

35. $b \approx 10.4$ km; $c = 12$ km **37.** $b = 7$ m; $c \approx 9.9$ m **39.** 9.4 feet **41.** 10 feet

Pages 422–423 Standardized Test Practice

1. B **3.** B **5.** D **7.** C **9.** A **11.** C **13.** −7
15. isosceles triangle **17.** 4.5 units

CHAPTER 10
Algebra: Graphing Functions

Pages 429–431 Lesson 10-1

1. domain, range **3.** Raul is correct. Candace's solution did not follow the order of operations.
5. 4 **7.** 42

9.

n	−5n	f(n)
−4	−5(−4)	20
−2	−5(−2)	10
0	−5(0)	0
$2\frac{1}{4}$	$-5\left(2\frac{1}{4}\right)$	$-11\frac{1}{4}$
5.7	−5(5.7)	−28.5

11. −2 **13.** 0 **15.** 5.84 **17.** −8

19a.

Time (seconds) n	Distance (feet) f(n)
2	2,200
5	5,500
11	12,100
18	19,800

19b. 9.6 seconds **23.** 9.1 feet **25.** B

Page 432 Lesson 10-1B

1.

X	Y = 15X
−3	−45
−2	−30
0	0
1.5	22.5
4.8	72

3.

X	Y = 3 − 2X
−14	31
−3.3	9.6
1.3	0.4
4.5	−6
9.4	−15.8

Pages 434–435 Lesson 10-2

1. Use the n values for x and the $f(n)$ values for y and graph the coordinates (x, y).

3.

n	f(n)	(n, f(n))
−2	3	(−2, 3)
−1	4	(−1, 4)
0	5	(0, 5)
1	6	(1, 6)
2	7	(2, 7)

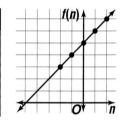

5a.

n	f(n)	(n, f(n))
10	14.8	(10, 14.8)
20	49.6	(20, 49.6)
30	84.4	(30, 84.4)
40	119.2	(40, 119.2)
50	154	(50, 154)

5b.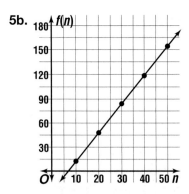

7.

n	f(n)	(n, f(n))
−2	16	(−2, 16)
0	0	(0, 0)
0.5	−4	(0.5, −4)
1	−8	(1, −8)
2	−16	(2, −16)

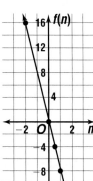

9.

n	f(n)	(n, f(n))
−4	12	(−4, 12)
−1	9	(−1, 9)
0	8	(0, 8)
6	2	(6, 2)
10	−2	(10, −2)

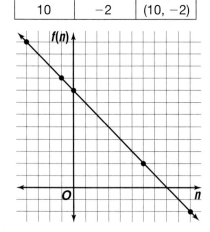

11.

n	f(n)	(n, f(n))
−3	12	(−3, 12)
−2	7	(−2, 7)
−1	4	(−1, 4)
0	3	(0, 3)
1	4	(1, 4)
2	7	(2, 7)
3	12	(3, 12)

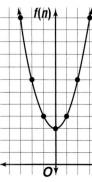

13.

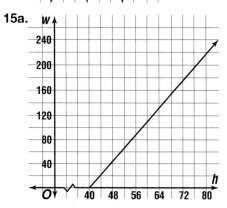

15a.

15b. 132 pounds   **17.** 3,809.48 m

Pages 439–440   Lesson 10-3

1. Sample answers: (−2, −1), (−1, 1), (2, 7), (4, 11)   **5.**

x	y
−4	−2
0	0
5	2.5
16	8

7. Sample answers: (3, 180), (4, 360), (5, 540), (6, 720)   **9.**

x	y
−6	9
−1	1.5
3	−4.5
16	−24

11.

x	y
−5	6
0	7
5	8
10	9

13.

x	y
−3	3
3	5
6	6
8	$6\frac{2}{3}$

15–17. Sample answers given.
15. (−1, 2), (0, 1), (1, 0), (2, −1)   **17.** (0, −57), (2, −27), (4, 3), (6, 33)   **19.** Sample answers: (−459.67, 0), (32, 273.15), (98.6, 310.15)

21.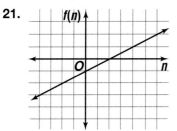

23. 184; 194; 172

15.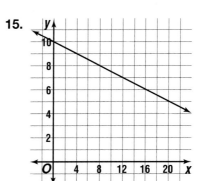

Page 441 Lesson 10-4A

1. Sample answer: yes; each additional washer stretches the rubber band about the same additional distance

3. Sample answer: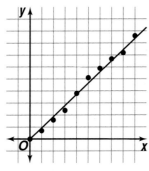

17.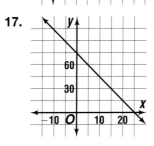

19.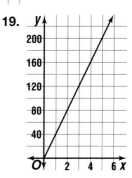

21.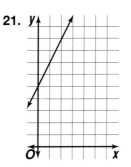

23. Sample answer: {(1, 13), (2, 16), (3, 19), (4, 22)}   **25.** D

Page 445 Lesson 10-4B

1.

Pages 443–444 Lesson 10-4

1. The third point is for a check.   **3.** 5 hours

5.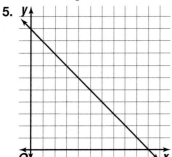

7.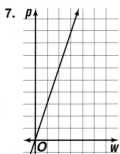

9.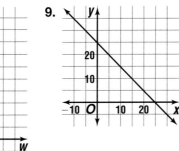

11.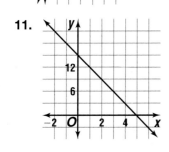

13.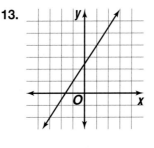

3.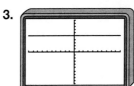

5.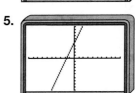

7. perpendicular lines

Pages 448–449 Lesson 10-5

1. two or more equations

3. Sample answer:

5. (2, 2) **7.** No; The lines meet at a point where *x* is negative, so the time when imports and exports has passed. **9.** (1, 2) **11.** no solution **13.** (6, 9) **15.** (4, 3) **17.** (2, 7)

21a.

21b. Yes; about 6 hours after the Fokker leaves.
23. **23a.** They are the same line.

23b. Yes; all of the points on the graph represent solutions. **25.** $2\frac{6}{7}$

Page 449 Mid-Chapter Self Test
1.

n	3n + 5	f(n)
−2	3(−2) + 5	−1
−1	3(−1) + 5	2
0	3(0) + 5	5
3	3(3) + 5	14
5	3(5) + 5	20

3. Sample answers: (0, 1), (1, 3.5), (2, 6), (4, 11)
5. (2, 5)

Pages 450–451 Lesson 10-6A
1. The change in the lumber production from one year to the next is not constant. **5.** *y*-values
7. $58,451 **9.** 15 **11.** Eric - amethyst; Julian - onyx; Louisa - sapphire **13.** A

Pages 453–455 Lesson 10-6
1. 2 **3.** Nicole is correct. The function can be written as $y = \frac{1}{3}x^2$. The highest power in the equation is 2, so the function is quadratic.
5.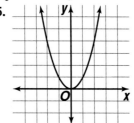

7.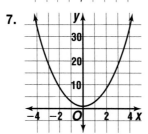

9.

x	y	(x, y)
−3	−9	(−3, −9)
−1	−1	(−1, −1)
0	0	(0, 0)
1.5	−2.25	(1.5, −2.25)
2	−4	(2, −4)

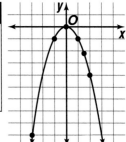

11.

n	f(n)	(n, f(n))
−2	4	(−2, 4)
−1.5	0.5	(−1.5, 0.5)
0	−4	(0, −4)
3	14	(3, 14)
4	28	(4, 28)

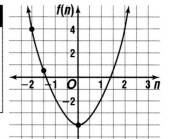

13.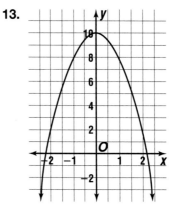

15.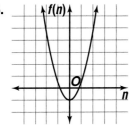

17. 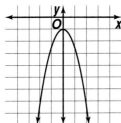 **19.**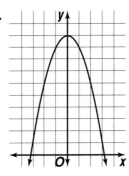

21. $(-2, 8), (0, -4), (1, -1)$ **22.** $(-1.5, -2.75)$, $(3.5, -7.75), (0, 1)$ **23a.**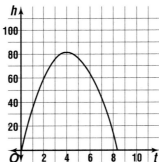

23b. about 80 m **23c.** about 81 m

25.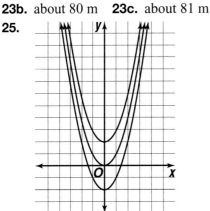

25a. The graphs are all the same shape. The graph of $y = x^2$ has a y-intercept of 0; the graph of $y = x^2 + 2$ has a y-intercept of 2, and the graph of $y = x^2 + (-2)$ has a y-intercept of -2. **25b.** The graphs have the same shape. But the graph of $y = x^2$ has a y-intercept of 0 and the graph of $y = x^2 + c$ has a y-intercept of c. **27.** 3.5%

Pages 458–459   Lesson 10-7

1. slide

3.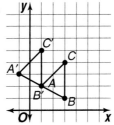

5. $H'(4, -1), J'(7, -4), K'(6, -5), L'(3, -2)$

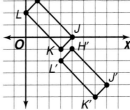

7. yes
9. $P'(-6, 3), Q'(-9, -1), R'(-5, 6)$

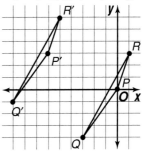

11. $W'(-5, 5), X'(1, 8), Y'(2, 6), Z'(-4, 3)$

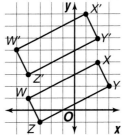

13. $A'(-4, 0), B'(-2, 0), C'(-1, 2), D'(-3, 4), E'(-5, 2)$

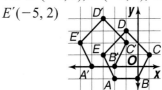

15a. $(2, -5)$ **15b.** $Q'(-8, -3), R'(-2, -5), T'(2, -1)$ **17.** Sample answer: In early years the center moved west, then it moved south and west. **19.** The final position of the figure is the same as the original position of the figure. **21.** D

Pages 462–463   Lesson 10-8

1. Answers may vary. Sample answer: In a mirror, the object seen is the flip of the real object. In a geometric reflection, what is on one side of a line is flipped from what is on the other side of the line.

3.

5a–5b.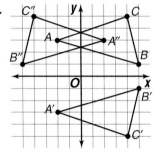

5c. yes; The corresponding sides are all congruent.
7. x-axis   9. y-axis
11.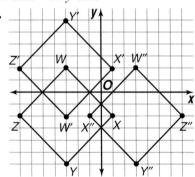

13. 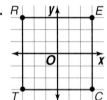   13a.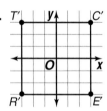

13b.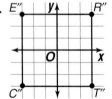

13c. They are all the same figure.   17. Any ordered pair with 0 as its x-coordinate is its own reflection over the y-axis.

19. **Infomercial viewing**

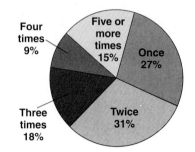

Pages 466–467   Lesson 10-9

1. second quadrant
5.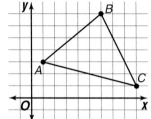

5a.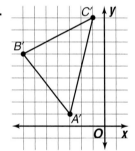

5b.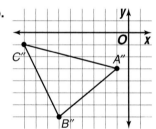

7. yes   9. yes
11. 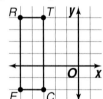   11a.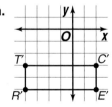

11b. 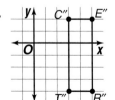   13. (4, −1), (1, −4), and (5, 8)

15a. Mitsubishi and Chevrolet   15b. Mitsubishi:

120°, 240°; Chevrolet: 180° **17.** 2, 4, 10, J, Q, K of hearts; 2, 4, 10, J, Q, K of clubs; 2, 3, 4, 5, 6, 8, 9, 10, J, Q, K, A of diamonds; 2, 4, 10, J, Q, K of spades **19.** $(y, -x)$ **21.** D

Pages 468–471 Study Guide and Assessment

1. domain **3.** system of equations **5.** two **7.** reflection **9.** Substitute the values into the original equation and check to see that a true sentence results.

11.

n	f(n)	(n, f(n))
−3	−8	(−3, −8)
0	1	(0, 1)
1	4	(1, 4)

13. Sample table given.

n	f(n)	(n, f(n))
−2	4	(−2, 4)
0	0	(0, 0)
2	−4	(2, −4)

15.

x	y
−2	−6
4	0
6	2

17. 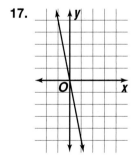 **19.**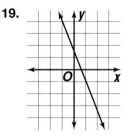

21. (1, 5) **23.** (1, 6)

25. 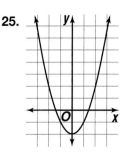 **27.**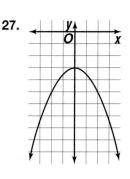

29.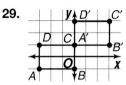

31.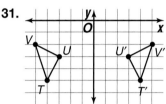

33.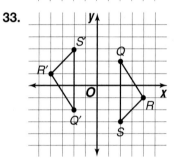

35. A: line symmetry to a horizontal line; E: line symmetry to a vertical line; N: line symmetry to a diagonal line and rotational symmetry for 180°; T: line symmetry to a horizontal line; Z: no symmetry **37a.** Sample answer: about 3.3 million **37b.** from 1950 to 1960

Pages 472–473 Standardized Test Practice

1. B **3.** B **5.** D **7.** A **9.** B **11.** E **13.** 2 **15.** (−1, 0), (4, 5), (8, 1), (3, −4) **17.** 10

CHAPTER 11
Geometry: Using Area and Volume

Pages 478–479 Lesson 11-1

1. Sample answer: Multiply 3 times 10^2. The area is about 300 m². **3.** Alisa; since the diameter is 10 m, the radius is 5 m. Therefore the area of the whole circle equals $\pi \cdot 5^2$ and the area of the semicircle is $\frac{1}{2}\pi \cdot 5^2$.

Selected Answers **691**

5. 530.9 ft² **7.** about 251.93 mm² **9.** 380.1 ft²
11. 1,134.1 m² **13.** 254.5 cm² **15.** 514.2 cm²
17. 8 in. **19.** $\frac{8}{9}$ **21.** 9:1 **23.** C

Pages 480–481   Lesson 11-2A

1. a pre-fabricated unit   **3.** Sample answer: Start with a tower of 3 blocks and 1 block in front of it. Then add a tower of 2 blocks to the right of the original tower, and place 1 block in front of it. Finally, make a tower of 3 blocks to the right of the tower of 2, and place 1 block in front of it.
5. Check to make sure the model satisfies all of the information.   **9.** A
11.

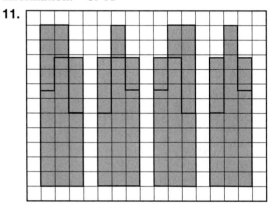

13. C

Pages 484–485   Lesson 11-2

1. Solid lines are used for the edges that you can see and dashed lines are used for the edges you cannot see.   **3.** Both the pentagonal prism and the pentagonal pyramid have a base in the shape of a pentagon. The pentagonal prism has 2 pentagonal bases and the other faces are quadrilaterals. The pentagonal pyramid has 1 pentagonal base and the other faces are triangles.

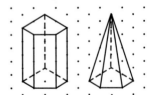

5a. rectangular prism   **5b.** 2 units by 5 units by 3 units   **5c.** 6 faces   **5d.** 12 edges   **5e.** 8 vertices
5f. Sample answer:

7. 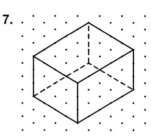 **9.**

11. Lightly draw an oval for the base, draw two edges, and draw another oval for the top. Use dashed lines to show the part you cannot see.

13.

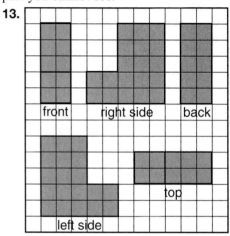

15. Sample answer:

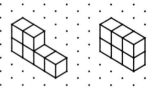

17. 5 faces; 4 faces;

19. A

Pages 488–489   Lesson 11-3

1. The formula $V = Bh$ can be used to find the volume of a rectangular prism, a triangular prism, or a cylinder. Since the area of the base of a rectangular prism equals the length times the width, another formula for the volume of a rectangular prism is $V = \ell wh$. Since the area of the base of a cylinder equals pi times the radius squared, another formula for the volume of a cylinder is $V = \pi r^2 h$.
3c. 18; they are the same.   **5.** 1,125.7 cm³
7. 1,331 in³   **9.** 2,880 cm³   **11.** 5.5 cm³
13. Sample answer: 1,500 ft³   **15a.** 24 units³

15b. 192 units^3 **15c.** 8:1 **15d.** It is multiplied by 8. **17a.** No, the inside dimensions of the refrigerator are $1\frac{5}{12}$ feet by $1\frac{1}{2}$ feet by $3\frac{1}{2}$ feet. Therefore the volume is $1\frac{5}{12} \times 1\frac{1}{2} \times 3\frac{1}{2}$ or $7\frac{7}{16}$ ft^3 which is less than 8 ft^3. **17b.** Double one of the dimensions. **19.** 7,200 ft^3 **21.** B

Pages 492–493 Lesson 11-4

1. Since the area of the base of a cone is πr^2, $V = \frac{1}{3}Bh$ becomes $V = \frac{1}{3}(\pi r^2)h$ or $V = \frac{1}{3}\pi r^2 h$.
3c. They are the same. **3d.** 3 times **3e.** 1:3
5. 121.8 m^3 **7.** 256 in^3 **9.** 56 yd^3 **11.** 85 yd^3
13. Sample answer: 300 m^3 **15.** It is doubled.
17. about 31,672,000 ft^3 **19.** The new height would be $\frac{1}{4}$ of the original height. **21.** about 9%

Page 493 Mid-Chapter Self Test
1. about 38.5 ft^2 **3.** 512 cm^3 **5.** 20.9 m^3

Page 494 Lesson 11-5A
1. Nets of boxes that are rectangular prisms will be made up of rectangles. **3.** Nets of boxes that are rectangular prisms will have 3 sets of congruent shapes.

Pages 496–498 Lesson 11-5

1.
Side	Area
front	ℓh
back	ℓh
top	ℓw
bottom	ℓw
side	wh
side	wh

The sum of the 6 faces is $\ell h + \ell h + \ell w + \ell w + wh + wh$ or $2\ell h + 2\ell w + 2wh$.

3.
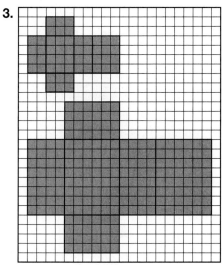

3a. 52 units^2; 208 units^2 **3b.** 1:2 **3c.** 1:4
3d. No; $\frac{1}{2} \neq \frac{1}{4}$. **5.** 630 in^2 **7.** 17.04 m^2
9. 98 m^2 **11.** 1,330 cm^2 **13.** 527.4 cm^2
15. Sample answer: 600 m^2 **17.** Sample answer: Mary is redecorating her room. She wants to cover the 5 sides of an open storage cube with contact paper. If each edge of the cube is 2 ft long, what is the area that needs to be covered with the contact paper? **19.** Block #4; its surface area is 96 in^2 which is less than the surface areas of the other blocks (196 in^2, 112 in^2, and 168 in^2).
21. about 50.3 m^3 **23.** D

Pages 501–502 Lesson 11-6
1.
area of top	πr^2
area of bottom	πr^2
area of curved surface	πdh or $\pi(2r)h$ or $2\pi rh$

The surface area of the cylinder is $\pi r^2 + \pi r^2 + 2\pi rh$ or $2\pi r^2 + 2\pi rh$.

3. Geraldo

Surface Area of Rectangular Prism
top	10 cm^2
bottom	10 cm^2
front	25 cm^2
back	25 cm^2
side	10 cm^2
side	10 cm^2
Total	90 cm^2

Surface Area of Cylinder
top	about 10.2 cm^2
bottom	about 10.2 cm^2
curved surface	about 56.5 cm^2
Total	about 76.9 cm^2

5. 766.9 cm^2 **7.** about 2,136.3 ft^2 **9.** 326.7 cm^2
11. about 169.6 ft^2 **13.** Sample answer: 450 in^2
15. about 227.4 in^2 **17.** Double the radius; consider the expression for the surface area of a cylinder, $2\pi r^2 + 2\pi rh$. If you double the height, you will double the second addend. If you double the radius, you will quadruple the first addend and double the second addend. **19.** D

Pages 505–507 Lesson 11-7
1. No, the length of a horse would be greater than 6 inches. A tape measure would be a better choice for measuring a horse. A 6-inch ruler could be used to measure a paperclip. **3.** Sample answer: An odometer measures length. It measures to the nearest 0.1 mile. It could be used to measure the

distance to school or the distance between 2 cities.
5. The measurement is to the nearest 0.1 km. There are 3 significant digits. The greatest possible error is 0.05 km, and the relative error is $\frac{0.05}{76.4}$ or about 0.00065. **7.** The measurement is to the nearest millimeter. There is 1 significant digit. The greatest possible error is 0.5 mm, and the relative error is $\frac{0.5}{7}$ or about 0.071. **9.** The measurement is to the nearest $\frac{1}{2}$ lb. The greatest possible error is $\frac{1}{4}$ lb, and the relative error is $\frac{1}{38}$ or about 0.026. **11.** The measurement is to the nearest minute. There are 2 significant digits. The greatest possible error is $\frac{1}{2}$ min, and the relative error is $\frac{1}{94}$ or about 0.011. **13.** 34.3 oz; 2 lb is measured to the nearest pound, 34 oz is measured to the nearest ounce, and 34.3 oz is measured to the nearest 0.1oz. Therefore 34.3 is most precise. **15.** 78 in.; the relative error of 17 in. is $\frac{1}{34}$ and the relative error of 78 in. is $\frac{1}{156}$. Since $\frac{1}{34} > \frac{1}{156}$, the relative error of 78 in. is less than the relative error of 17 in. **17.** 10 significant digits **19.** 55 cm² **21.** A

Pages 508–511   Study Guide and Assessment

1. f   **3.** g   **5.** l   **7.** a   **9.** j   **11.** 153.9 m²
13.

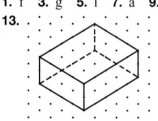

15. 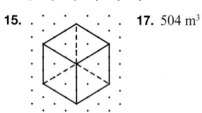   **17.** 504 m³

19. 10 mm³   **21.** 21.4 m²   **23.** 1,633.6 yd²
25. The measurement is to the nearest foot. There are 2 significant digits. The greatest possible error is $\frac{1}{2}$ ft, and the relative error is $\frac{1}{24}$ or about 0.042.
27. The measurement is to the nearest 0.1 m. There are 3 significant digits. The greatest possible error is 0.05 m, and the relative error is $\frac{0.05}{24.2}$ or about 0.0021. **29.** 15 balls **31a.** about 30.5 ft² **31b.** about $6\frac{1}{2}$ trash cans

Pages 512–513   Standardized Test Practice
1. D   **3.** D   **5.** A   **7.** B   **9.** B   **11.** 132 ft²
13. $A'(2, 0), B'(2, -5), C'(-4, -5), D'(-4, 0)$

CHAPTER 12
Investigating Discrete Math and Probability

Pages 516–517   Lesson 12-1A
1.   **3.** 3 ways

Player A	Player B
scissors	scissors
scissors	paper
scissors	stone
paper	scissors
paper	paper
paper	stone
stone	scissors
stone	paper
stone	stone

5. 3 outcomes   **7.** Yes; each player's chance of winning is $\frac{1}{3}$.   **9.** 36 different outcomes
11. 9 ways   **13.** no   **15.** Unfair; player X's chance of winning is $\frac{4}{9}$, and players Y's chance of winning is $\frac{5}{9}$.

Pages 519–520   Lesson 12-1
1. Sample answer: Grandma's Diner has two flavors of ice cream, vanilla and chocolate. They serve the ice cream with one of four toppings; hot fudge, strawberry, butterscotch, and marshmallow. How many different ice cream desserts does Grandma's Diner serve?   **3.** Both a tree diagram and the Fundamental Principle of Counting can be used to find the number of outcomes. A tree diagram actually shows the different outcomes. The Fundamental Principle of Counting does not show the actual outcomes, but it requires less time to solve the problem.   **5.** 40 different T-shirts

7.

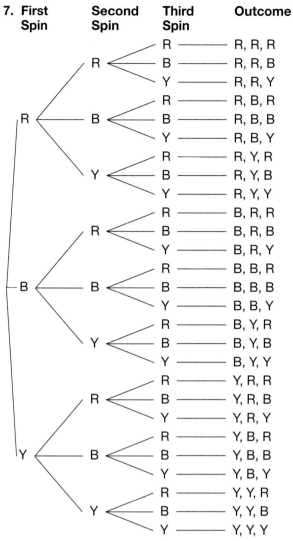

27 outcomes

9.

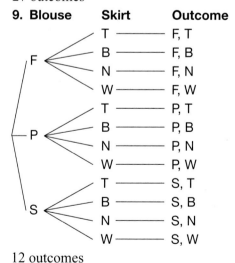

12 outcomes
11. 1,024 outcomes

13a.

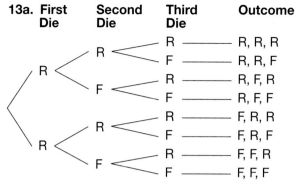

13b. 8 outcomes   **15.** 35,152 call letters
17. 3 significant digits

Pages 522–523   Lesson 12-2

1. Sample answer: How many 2-digit whole numbers can you write using the digits 1, 2, 3, 4, and 5 if no digit can be used twice.   **3a.** 6 ways
3b. They are the same.   **5.** 72   **7.** 2
9. 504 ways   **11.** 336   **13.** 720   **15.** 5,040
17. 3,628,800   **19.** 24,024   **21.** 40,320
23. 720 ways   **25.** 240 ways   **27.** They are the same. $P(n, n) = n \cdot (n - 1) \cdot (n - 2) \cdot \ldots \cdot 2 \cdot 1$ and $P(n, n - 1) = n \cdot (n - 1) \cdot (n - 2) \cdot \ldots \cdot 2$ which are equal.   **29.** 15 yd, 20 yd; 10 yd; 175 yd^2

Pages 526–527   Lesson 12-3

1. $C(20, 6)$ **3.** Montega; the three positions are different, so order is important.   **5.** 56   **7.** 10
9. combination   **11.** 495 ways   **13.** 21   **15.** 210
17. 126   **19.** 286   **21.** permutation
23. combination   **25.** permutation   **27.** 2,598,960 hands   **29.** 253 combinations   **31.** 16 points
33. 840 numbers   **35.** $712\frac{1}{2}$ miles

Pages 530–531   Lesson 12-4

1. Add pairs of number in row 8.
row 8   1   8   28   56   70   56   28   8   1
row 9   1   9   36   84   126   126   84   36   9   1
3a. 1 way; 6 ways; 15 ways; 20 ways; 15 ways; 6 ways; 1 way   **3b.** row 6   **5.** 4   **7.** $\frac{1}{32}$ **9.** 56
11. 10   **13.** 126   **15.** 20 combinations
17. 64 branches   **19.** combination of 4 things taken 2 at a time   **21a.** 4   **21b.** The number of ways is equal to the number in the corresponding position in Pascal's Triangle.   **23.** They are the same. Pascal's Triangle is symmetric. Therefore, $C(9, 0) = C(9, 9)$, $C(9, 1) = C(9, 8)$, $C(9, 2) = C(9, 7)$, and so on.

25.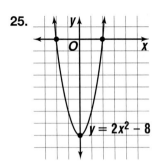

Page 531 Mid-Chapter Self Test

1.

Color	Size	Outcome
black	small	black, small
	medium	black, medium
	large	black, large
	extra large	black, extra large
white	small	white, small
	medium	white, medium
	large	white, large
	extra large	white, extra large
gray	small	gray, small
	medium	gray, medium
	large	gray, large
	extra large	gray, extra large

3. 60 ways **5.** 35 combinations

Pages 532–533 Lesson 12-4B

1. Fibonacci sequence **3.** 30 **5.** Inverted triangles; yes, because when two numbers next to each other have the same factor, the number below them will also have the same factor. **7.** The right and left sides of Pascal's Triangle are ones. The inside numbers are the sum of the two numbers above it. If the rows are numbered from the top starting with 0, each row represents the number of combinations of items corresponding to the number of the row. The sums of the diagonals form the Fibonacci sequence. The product of the 6 numbers surrounding any number is always a perfect square. The multiples of a number form inverted triangles.

Pages 536–537 Lesson 12-5

1. The person is not likely to be 65 years old or older and living in an urban area. Only about 8 out of 100 people would satisfy these conditions. **3.** Tonya; the first spin does not affect the second spin. Therefore the probability is $\frac{2}{5} \cdot \frac{2}{5}$ or $\frac{4}{25}$. **5.** $\frac{1}{12}$ **7.** $\frac{1}{5}$ **9.** $\frac{1}{18}$ **11.** $\frac{3}{20}$ **13.** $\frac{3}{10}$ **15.** $\frac{2}{5}$ **17.** $\frac{2}{9}$ **19.** $\frac{1}{18}$ **21.** $\frac{1}{6}$ **23.** $\frac{1}{27}$ **25.** $\frac{1}{240}$

27. 15 **29.** 120

Pages 538–539 Lesson 12-6A

1. Since there are 6 possible balloons and 6 possible outcomes when a number cube is rolled, a roll of a 1 could represent the balloon Rosita wants. Since there is a 1 in 4 chance of the right wind and 4 possible outcomes for the spinner, a spin of one section could represent the wind Kelsey wanted. By rolling the number cube and spinning the spinner, you can see how many times both favorable outcomes occur out of a total number of tries. **3.** No; the simulation will probably be similar to the computed results. However, simulations will vary. **5.** You must plan a simulation that will correctly represent the situation. **7.** The act it out strategy can find the probability of an event by experimenting. **9.** jacket, $50.00; shirt, $12.50 **11.** 30 students **13b.** No number would guarantee all 3 toys. You could be very unlucky and never get one of the toys.

Pages 541–543 Lesson 12-6

1. If the outcomes cannot be broken down into equally likely events, theoretical probability cannot be calculated. **3.** No; if you draw more marbles, the experimental probability will come closer to the theoretical probability. **5a.** $\frac{1}{5}$ **5b.** $\frac{1}{4}$

7d.

First Coin	Second Coin	Outcome
H	H	H H
H	T	H T
T	H	T H
T	T	T T

7e. $\frac{1}{4}$ **7f.** $\frac{1}{2}$ **9a.** Experimental probability; it is based on what happened in the past. **9b.** 12,000 seeds **13.** D

Page 545 Lesson 12-7A

1. 10, 20, 10 **3.** $\frac{1}{4}, \frac{1}{2}, \frac{1}{4}$

5a.

	F	F
F	FF	FF
f	Ff	Ff

$P(FF) = \frac{1}{2}$, $P(Ff) = \frac{1}{2}$

5b.

	F	F
F	FF	FF
F	FF	FF

$P(FF) = 1$

5c.

	f	f
f	ff	ff
f	ff	ff

$P(ff) = 1$

5d.

	F	f
f	Ff	ff
f	Ff	ff

$P(Ff) = \frac{1}{2}$, $P(ff) = \frac{1}{2}$

Pages 547–548   Lesson 12-7

1. by setting up and solving a proportion
3. Sample answer: If the sample is not random, any predictions will be incorrect. Taking every tenth student from an alphabetical list of students would be an appropriate sampling. Taking a survey of the football players before practice would be an inappropriate sampling.   **5a.** about 760 people
5b. Sample answer: how many T-shirts to stock at future concerts   **7a.** 25 students   **7b.** 48%
7c. $\frac{6}{25}$ **7d.** about 50 votes   **9a.** No; 327 prefer Sunshine Orange Juice and 273 prefer the competitor's juice which is not 2 to 1.   **9b.** Yes; over 50% of the people in one survey did prefer Sunshine Orange Juice.   **9c.** No; another survey may show a different result.   **13.** probability determined by conducting an experiment or simulation

Pages 550–553   Study Guide and Assessment

1. f   **3.** h   **5.** d   **7.** b   **9.** j   **11.** 8 outcomes
13. 12 outcomes   **15.** 120   **17.** 120 numbers
19. 66   **21.** 1,140 groups   **23.** 70 combinations
25. $\frac{1}{4}$ **27c.** $\frac{1}{4}$; $\frac{1}{2}$ **29.** $\frac{9}{50}$
31.

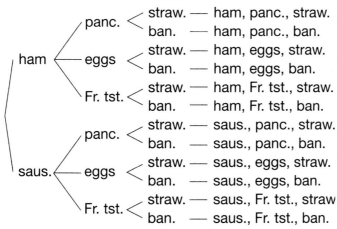

12 outcomes
33. 1,140 combinations

Pages 554–555   Standardized Test Practice

1. A   **3.** C   **5.** B   **7.** E   **9.** C   **11.** C   **13.** 12
15. 1, 2, 3, 4, 6, 8, 9, 12, 18, 24, 36, 72

CHAPTER 13
Algebra: Exploring Polynomials

Page 560   Lesson 13-1A

1. 1, x, x^2 **3.** Sample answer: Each set of tiles will have 3 tiles with areas 1, x, and x^2.
5.

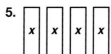

Pages 562–564   Lesson 13-1

1a. polynomial   **1b.** monomial   **1c.** monomial
1d. polynomial   **5.** $-x^2 + 2x - 4$
7.    **9.** -3

11. 70 ft   **13.** $x^2 + 3x$ **15.** $-4x + 2$
17. $2x^2 - x + 4$ **19.**

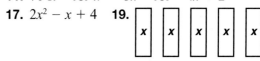

21.

23.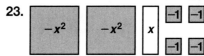

25. 30   **27.** 14   **29.** -4 **31.** 5   **33a.** $2x^2 + 12x$
33b. 32 sq. units   **33c.** rectangular prism if all sides do not have the same measure; cube if $x = 3$
35.

	b
a	ab

   37. 10 units   **39.** 60%

Pages 567–569   Lesson 13-2

1. $2x^2$, $-3x^2$; $-y$, $5y$
3. $-x^2 - x - 3$;

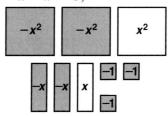

5. none   **7.** $4x + 1$ **9.** $8x - y$; -14

11a. $4q + 8d + 5n$ **11b.** $2.05 **13.** $3x^2, -2x^2$
15. $7a, 10a; 6b, 14b$ **17.** $4y, -2y, -3y; 8, 9$
19. $-2x + 2y$ **21.** $5a + 5$ **23.** $-7y$
25. $3x^2 + 6$ **27.** $17a + 11b; 46$ **29.** $-a + 9b;$
-22 **31.** 52 **33a.** $50r + 100, \$103$
33b. $50r^2 + 150r + 150$ **35a.** $114O + 96G$
35b. $2,232 **37.** 15 **39.** B

Pages 571–572 Lesson 13-3
1. Add like terms.
3. $(x^2 - 2x - 3) + (-2x^2 + 2x + 4) = -x^2 + 1$

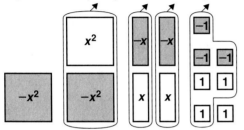

5. $9a + 5b$ **7.** $-x + 6$ **9a.** $(x + 4) +$
$(2x - 3) + (3x + 1) = (6x + 2)$ cm **9b.** 26 cm
11. $6y^2 + 2y + 2$ **13.** $x^2 + 5$ **15.** $16q - r + 1$
17. $12a + 12b$ **19.** $10a^2 + 2a - 3$ **21.** $2x^2 -$
$3x - 10$ **23.** $7g + 3h + 4; 51$ **25.** $15g + h +$
$11; 90$ **27.** 24 **29.** $3x^2 + 21x + 6$
31. $-10a^2 + 15a$ **33.** $3\frac{7}{20}$

Pages 574–576 Lesson 13-4
1. $-2x$
3. $(2x^2 - 3x + 3) - (x^2 + 1) = x^2 - 3x + 2$

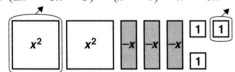

5. $2p^2 - p + 1$ **7.** $2x^2 - 3x + 5$ **9.** $1.03
11. y^2 **13.** $r^2 - 8rt + 2t^2$ **15.** $-5a + 6$
17. $2a^2 - 5a - 9$ **9.** $3x^2 - 3x$ **21.** $2c - 11d -$
$4; -63$ **23.** $9c + 6d - 10; 12$ **25.** $2p^2 + 2pq +$
q^2 **27a.** $-2A + 4B + 3C$ **27b.** 5 **29.** $6y^2 +$
$15y + 3$ **31.** $-\frac{1}{3}$ **32.** B

Page 576 Mid-Chapter Self Test
1. 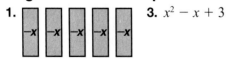 **3.** $x^2 - x + 3$

5. $3y^2 - 5y + 3$ **7.** $4x^2 + 3x + 11$
9. $2a - 7b + 7; -12$

Page 577 Lesson 13-5A
1. $2x + 4$

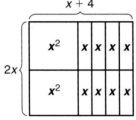

3. $6x + 3$

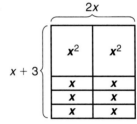

Pages 580–581 Lesson 13-5
1. Multiply $3x$ by $5x$ and multiply $3x$ by 2. Add the results. **3.** $2x(x + 4) = 2x^2 + 8x$

5. n^8 **7.** $b^2 - 2b$ **9a.**

9b. $2x(x + 3); 2x^2 + 6x$ **9c.** 260 ft² **11.** 2^{10}
13. p^{14} **15.** $z^2 - z$ **17.** $b^2 + 3b$ **19.** $2m^3 + 4m^2$
21. $d^8 + 15d^3$ **23.** a^{11} **25.** $15d^2 + 28d$
27. $(3x + 6)$ dollars **29.** $5c + 3d; 21$ **31.** D

Pages 584–585 Lesson 13-6
1. b, c **3.** $4x^2 + 10x + 6$

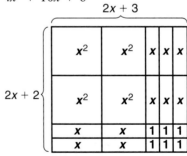

5. $x^2 + 2x - 3$ **7.** $x^2 + 7x + 12$
9. $2x^2 + 7x + 3$ **11.** $b^2 + 10b + 24$
13. $4m^2 + 6m + 2$ **15.** $d^2 + 6d + 9$
17. $3a^2 - 8a - 3$ **19a.** $x^2 + 14x + 40$
19b. $9x^2 + 2x - 48$ **21.** $h^2 + 2mh + m^2$; yes; only h and m will vary from player to player.
23. $t^2 - 4t$ **25.** 306

Pages 586–587 Lesson 13-7A

1. Sample answer: Ramón guessed an equal number of bottles and six-packs; Sheila guessed more than twice as many six-packs than bottles.   **3.** 29, 31
7. $20.30   **9.** $46.20   **11.** 6 postcards, 5 letters
13. D

Pages 589–591 Lesson 13-7

1. Sample answer: You can form a rectangle with the algebra tiles.   **3a.**

3b. $2x + 1$ and $x + 2$; these are the length and the width of the rectangle.   **5.** $x(x + 4)$ **7.** not factorable   **9.** $3(x + 2)$ **11.** $(2x + 1)(x + 2)$
13. $6(3x + 1)$ **15.** $(x + 4)(x + 1)$
17. $(x + 4)(x + 4)$ **19.** $(2x + 1)(x + 3)$
21. $x(3x + 4)$ **23.** $\frac{2x + 1}{x + 2}$ **25a.** $h^2 + 2mh + m^2 = (h + m)(h + m); h + m = 1$
25b. $2 \cdot h^2 + 1 \cdot 2mh + 0 \cdot m^2 = 2h^2 + 2mh$
27. $6k^2 + 5k - 4$ **29.** $\frac{1}{8}$

Pages 592–595 Study Guide and Assessment

1. false; binomial   **3.** true   **5.** false; two
7. Sample answer: A monomial is a number, a variable, or a product of a number and one or more variables.   **9.**

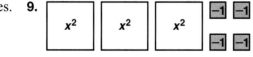

11.   **13.** $5p - 12$

15. $4a - b$ **17.** $12d + 7$ **19.** $3b^2 - b - 4$
21. $2g + 1$ **23.** $s^2 + 5$ **25.** $3p^2 - 4p - 3$
27. 12^{13} **29.** d^{10} **31.** $3z^2 - 12z$ **33.** $2t^7 - 8t^4$
35. $2x^2 - 9x - 5$ **37.** $x^2 - x - 12$
39. $7(2x + 1)$ **41.** $(x + 5)(x + 2)$
43. $(2x + 1)(x + 1)$ **45.** $(4x^2 + 4x + 1)$ yd^2
47. $500 + 1{,}000r + 500r^2$

Pages 596–597 Standardized Test Practice

1. A   **3.** C   **5.** A   **7.** C   **9.** D   **11.** 0.5

Photo Credits

Cover: (bkgd)Terra Forma ©1995 Andromeda Interactive Ltd., (inset)NASA, (hologram)Glencoe photo; **iv v viii** Mark Burnett; **ix** Uniphoto; **x** SuperStock; **xi** Daryl Balfour/Tony Stone Images; **xii** Gary Pearl/Sports Photo Masters; **xiii** E.R. Degginger; **xiv** Ron Kimball; **xv** Jim Mejuto/FPG; **xvi** Paul Dolkos, MODEL RAILROADER Magazine; **xvii** Greg Pease/Tony Stone Images; **xviii** Ralph Wetmore/Tony Stone Images; **xix** Morris Best/Uniphoto; **xx** Ron Berhmann/International Stock; **xxi** Dominic Oldershaw; **2** Dominic Oldershaw; **3** (l)James Nazz/The Image Bank, (r)Elaine Shay; **4** (t)Hulton Getty/Tony Stone Images, (c)Dominic Oldershaw, (b)Uniphoto; **5** L. Smith/H. Armstrong Roberts; **6 7** Timothy Fuller; **8** Paramount/The Kobal Collection; **10** Tom Carroll/International Stock; **11** Focus On Sports; **13** Barros & Barros/The Image Bank; **14** Uniphoto; **15** (t)Reuters/Sue Ogrocki/Archive Photos, (b)Archive Photos; **17** National Archive Photo; **20** Uniphoto; **21** AP/Wide World Photos; **23** (t)David Madison, (b)AP/Wide World Photos; **24** (t)Bob Firth/International Stock, (b)Tom Salyer; **25** Institute of Human Origins; **26** (t)Donna Coveney/MIT, (bl)Uniphoto, (br)Timothy Fuller, (bkgd)Will & Deni McIntyre/Photo Researchers; **27** Jeff Smith/FOTOSMITH; **28** The Dallas Morning News/Evans Caglage; **29** NASA; **30** Timothy Fuller; **32** Don and Susan Sanders; **34** UPI/Corbis-Bettmann; **35** Dominic Oldershaw; **36** Peter Griffith/Masterfile; **37** (l)James L. Amos/Peter Arnold, Inc., (r)Jeff Smith/FOTOSMITH; **38** Newseum/Chris Micelli; **41** AP/Wide World Photos; **42** Washington Stock Photo; **43** courtesy Paramount's Great America; **44** Lewis Carroll, Alice in Wonderland, Grosset & Dunlap, New York, 1996; **46** Bert Sagara/Tony Stone Images; **54** (l)Hugh Rose/Visuals Unlimited, (r)Karen Su/Tony Stone Images; **54-55** Brandon D. Cole/ENP Images; **55** (l)David Madison/Tony Stone Images, (r)Uniphoto; **56** Joe McDonald/Visuals Unlimited; **57** Linda H. Hopson/Visuals Unlimited; **58** Focus On Sports; **59** Mark Junak/Tony Stone Images; **60** Chris Marona/Photo Researchers; **61** Bill Bachmann; **62** Jeff Smith/FOTOSMITH; **64** Everett Collection; **65** SuperStock; **66** Timothy Fuller; **67** SuperStock; **68** Ed Kessler/Zephyr Images; **70 71** Timothy Fuller; **72** Jeff Smith/FOTOSMITH; **73** (bl)Jeff Smith/FOTOSMITH, (others)Timothy Fuller; **75** (t)Brandon D. Cole/ENP Images, (b)SuperStock; **78** AP/Wide World Photos; **81** (t)James Balog/Tony Stone Images, (b)Tim Davis/Tony Stone Images; **82** Joseph Pobereskin/Tony Stone Images; **83** Allsport USA/Rick Stewart; **86** (t)SuperStock, (b)Inga Spence/Tom Stack & Assoc.; **87** Gary Braasch/Tony Stone Images; **88** Corbis-Bettmann; **90** (t)Tim Courlas, (others)Jeff Smith/FOTOSMITH; **92** SuperStock; **94** Glencoe photo; **95** Bill Frakes/Sports Illustrated; **102** (l)Helga Lade/Peter Arnold, Inc., (r)Uniphoto; **102-103** (t)David Sutherland/Tony Stone Images, (b)Tony Craddock/Tony Stone Images; **103** Daryl Balfour/Tony Stone Images; **104** The Pillsbury Company; **106** Manfred Danegger/Peter Arnold, Inc.; **107** (t)Jim Steinberg/Photo Researchers, (b) Rainer Grosskopf/Tony Stone Images; **108** SuperStock; **109** Aaron Haupt; **110** Sandy Roessler/Stock Market; **111** Timothy Fuller; **112** Gerry Ellis/ENP Images; **113** (t)Alan Carey/Photo Researchers, (b)courtesy Nestlé USA; **114** Studiohio; **116** Norbert Wu; **119** (l)Tate Gallery, London/Art Resource, NY, (r)Vic Thomasson/Tony Stone Images; **120** (t)Hulton Getty/Tony Stone Images, (b)People Weekly, ©Time, Inc.; **122** Elaine Shay; **124** (tr)Thomas Veneklasen, (others)Dominic Oldershaw; **126** Thomas Veneklasen; **127** Ed & Chris Kumler; **129** Christopher Bissell/Tony Stone Images; **136-137** Dominic Oldershaw; **137** Aaron Haupt; **138** The Newark Museum/Art Resource, NY; **138-139** (t)Brian Parker/Tom Stack & Assoc., (b)The Newark Museum/Art Resource, NY; **139** Cathlyn Melloan/Tony Stone Images; **140** Jeff Smith/FOTOSMITH; **141** Dominic Oldershaw; **143** Joseph Nettis/Tony Stone Images; **144** John Kieffer/Peter Arnold, Inc.; **145** Joseph Van Os/The Image Bank; **148** Thomas Veneklasen; **149** Dominic Oldershaw; **150** Jeff Smith/FOTOSMITH; **152** (t)Hugh Talman, (b)Dominic Oldershaw, (bkgd)Uniphoto; **153** Jeff Smith/FOTOSMITH; **155** Michael Burr/NFL Photos; **157** Zephyr Images; **158** James Blank/MRTC/Ozark Postcard Co.; **158-159** Paul Niemann/Rushmore Photo & Gifts; **159** Dominic Oldershaw; **160** (t)Hy Peskin/FPG, (b)Dominic Oldershaw; **161** LucasFilm/The Kobal Collection; **163** AP/Wide World Photos; **164** (t)Timothy Fuller, (b)Uniphoto; **165** Everett Collection; **166** Gary Pearl/Sports Photo Masters; **169** Dominic Oldershaw; **171** Alexander Tsiaras/Photo Researchers; **172** Ray Pfortner/Peter Arnold, Inc.; **173** (l)United States Postal Service, (r)Nawrocki Stock Photo; **174** (t)Corbis-Bettmann, (c)Jeff Smith/FOTOSMITH, (b)Timothy Fuller; **175** NBC TV/The Kobal Collection; **176** Brian Parker/Tom Stack & Assoc.; **177** (t)Jeff Smith/FOTOSMITH, (b)Matt Meadows; **184** (l)Timothy Fuller, (r)Uniphoto; **184-185** Thomas Veneklasen; **185** Thomas Veneklasen; **190** Joyce Photographics/Photo Researchers; **192** Barry Slaven/P MODE PHOTOGRAPHY/Visuals Unlimited; **194** Timothy Fuller; **196** Hilary Wilkes/International Stock; **197** Latent Image; **199** (l)Thomas Veneklasen, (r)Jeff Smith/FOTOSMITH; **201** Timothy Fuller; **204** Aaron Haupt; **206** Dominic Oldershaw; **208** Robert Brons/BPS/Tony Stone Images; **209** (t)courtesy NFL, (b)Scott Gilchrist/Archivision; **210** (t)E.R. Degginger, (b)Joe McDonald/Visuals Unlimited; **212** National Council of Teachers in Mathematics; **215** Mark E. Gibson; **216** E.R. Degginger; **217** Matt Bradley/Tom Stack & Assoc.; **218** Willie Holdman/International Stock; **219** (l)Renata Blas, (r)Timothy Fuller, (bkgd)Thomas Veneklasen; **220** *Sky & Water I* by M.C. Escher. ©1997 Cordon Art - Baarn - Holland. All rights reserved.; **222** (t)*Symm. Drawing E25* by M.C. Escher. ©1997 Cordon Art - Baarn - Holland. All rights reserved, (b)United States Postal Service; **223** Thomas Veneklasen; **227** (l)Jeff Smith/FOTOSMITH, (r)Scott Gilchrist/Archivision; **230** (l)Michael Zito/SportsChrome, (r)Ron Sherman/Tony Stone Images; **231** David Madison; **232** Aaron Haupt; **233** Timothy Fuller; **234** Dominic Oldershaw; **239** Jeff Smith/FOTOSMITH; **240** Dominic Oldershaw; **241** Tim Fuller; **242** (t)Disney/The Kobal Collection, (b)John Bramley/Disney/The Kobal Collection; **243** Dominic Oldershaw; **245** J.D. Cunningham/Visuals Unlimited; **246** Archive Photos; **247** AP/Wide World Photos; **248** Uniphoto; **249** David Young Wolff/Tony Stone Images; **250** Earth Imaging/Tony Stone Images; **251** R. Scheiber; **252** Uniphoto; **254 255** Dominic Oldershaw; **256** (l)Ron Kimball, (r)SuperStock; **257** Dominic Oldershaw; **259** Buddy Mays/International Stock; **261** (t)People Weekly ©1996 Donna Terek, (b)Dominic Oldershaw; **262** (l)David Gonzales/NCAA Photos, (r)Dominic Oldershaw; **263** ™ & ©1997 Paramount Parks; **264** (l)Dominic Oldershaw, (r)Rob Nelson/Uniphoto; **265** Disney/Everett Collection; **267** (t)Leonard Lessin/Peter Arnold, Inc., (b)FOXTROT ©1995 Bill Amend. Reprinted with permission of Universal Press Syndicate. All rights reserved; **271** Jeff Smith/FOTOSMITH; **274-275** Tim Davis/Photo Researchers; **275** (t)Aaron Haupt, (b)Dominic Oldershaw; **276** (t)J. Carmichael, Jr./The Image Bank, (c)Dave Jacobs/Tony Stone Images, (b)Greg Vaughn/Tom Stack & Assoc.; **276-277** Richard Ellis/Photo Researchers; **277** (l)R. Lanaud/Jacana/Photo Researchers, (r)Aaron Haupt; **279** Jeff Smith/FOTOSMITH; **281 282** Dominic Oldershaw; **284** SuperStock; **286** Jeff Smith/FOTOSMITH; **287** Gary Holscher/Tony Stone Images; **289** (t)Dominic Oldershaw, (b)CLOSE TO HOME © 1993, 1994 John McPherson/Dist. of Universal Press Syndicate. Reprinted with permission. All rights reserved; **290** (t)NASA/JPL, (b)Derek Berwin/The Image Bank; **292** Kevin Schafer/Tony Stone Images; **293** (t)Barry L. Runk/Grant Heilman Photography, (bl)Angela Seckinger, (br)Dominic Oldershaw, (bkgd)David Phillips/Science Source/Photo Researchers; **294** (tr)Sinclair Stammers/Science Photo Library/Photo Researchers, (others)Jeff Smith/FOTOSMITH; **296** Thomas Veneklasen; **298 299** Jeff Smith/FOTOSMITH; **301** Allsport USA/Nathan Bilow; **306** Timothy Fuller; **309** K. Shimauchi/Photonica; **311** James Randklev/Tony Stone Images; **312** Jeff Smith/FOTOSMITH; **313** Thomas Veneklasen; **314** Ron Kimball; **315** Nils Jorgensen/Rex Features; **316** Rob Tringali Jr./SportsChrome; **317** Jim Mejuto/FPG; **318** (t)Jeff Smith/FOTOSMITH, (b)John Zoiner/International Stock; **319** Jeff Smith/FOTOSMITH; **321** Tim Fuller; **328** (l)Disney/Everett Collection, (c)Christopher Vollker/Shooting Star, (r)CBS/Everett Collection; **328-329** (televisions) Jeff Smith/FOTOSMITH; **329** (l)Dominique Braud/Tom Stack & Assoc., (r)Thomas Veneklasen; **330** (surfer)Jeff Smith/FOTOSMITH, (remote)Dominic Oldershaw, (b)Dominic Oldershaw; **331** Thomas Veneklasen; **332** NASA; **333** Thomas Veneklasen; **335** A & L Sinibaldi/Tony Stone Images; **336** Jeff Smith/FOTOSMITH; **338** (tl)Stockman/International Stock, (tr)Donald L. Miller/International Stock, (b)Dominic Oldershaw; **339** (t)Dominic Oldershaw, (b)Geoff

Butler; **340** Thomas Veneklasen; **341** Tim Davis/Photo Researchers; **342** Jeff Smith/FOTOSMITH; **344** (t)John Elk/Tony Stone Images, (b)Runk/Schoenberger/Grant Heilman Photography; **345** Comstock; **347** O'Brien/International Stock; **349** (racket)Elaine Shay, (ball)Doug Martin; **350** (l)T. Abratis/SportsChrome, (r)Thomas Veneklasen; **351** Boy Scouts of America; **352** Timothy Fuller; **353** Dominic Oldershaw; **354** Aaron Haupt; **355** Uniphoto; **357** The Newark Museum/Art Resource, NY; **359** MAC-1; **360** Paul Dolkos, MODEL RAILROADER Magazine; **362** Dominic Oldershaw; **363** (l)Cosmo Condina/Tony Stone Images, (r)Kunio Owaki/Stock Market; **364** (t)Glenn M. Oliver/Visuals Unlimited, (b)By permission of Johnny Hart and Creators Syndicate, Inc.; **366** (t)ET Archive, London/SuperStock, (b)The Pierpont Morgan Library/Art Resource, NY; **367** (t)Everett Collection, (b)Jeff Smith/FOTOSMITH; **368** SuperStock; **369** Dallas & John Heaton/Westlight; **370** Giraudon/Art Resource, NY; **371** Jeff Smith/FOTOSMITH; **372** K.S. Studio; **373** ©1997, The Art Institute of Chicago, All Rights Reserved; **380** (t)Mark E. Gibson, (c)Michael Agliolo/International Stock, (b)Craig Newbauer/Peter Arnold, Inc.; **381** E.R. Degginger; **382** Patrick Montagne/Photo Researchers; **383** Will & Deni McIntyre/Photo Researchers; **384** (l)Timothy Fuller, (r)Brad Hansel/Great Scott Design; **386** (t)CLOSE TO HOME ©1993, 1994 John McPherson/Dist. of Universal Press Syndicate. Reprinted with permission. All rights reserved, (b)David Tejada/Tony Stone Images; **387** Mark E. Gibson; **388** Greg Pease/Tony Stone Images; **390** (t)E.R. Degginger, (b)David M. Dennis; **392** (t)Warren Faidley/International Stock, (b)Dominic Oldershaw; **393** Jeff Smith/FOTOSMITH; **394** (t)PANTONE®, (b)Adam Jones/Photo Researchers; **395** (t)William B. Folsom/Uniphoto, (c)Curtis Scott/FHP, (b)Dominic Oldershaw, (bkgd)Doug Wilson/Westlight; **399** Eric Hobson/The Ancient Art & Architecture Collection Ltd.; **400** Jeff Smith/FOTOSMITH; **401** (t)Adam Jones/Photo Researchers, (b)Dominic Oldershaw; **402** Dominic Oldershaw; **404** (t)Gloria Kuchinskas/Visuals Unlimited, (b)Dominic Oldershaw; **405** Dominic Oldershaw; **407** MGM/The Kobal Collection; **408** Stock Montage; **411** Jim Pickerell/Westlight; **412** SuperStock; **415** Sylvain Grandadam/Tony Stone Images; **417** Jeff Smith/FOTOSMITH; **424-425** Dominic Oldershaw; **426** Christian Michaels/FPG; **426-427** Timothy Fuller; **428** Disney/Everett Collection; **430** Manuel Denner/International Stock; **431** Timothy Fuller; **433** Hulton Getty/Tony Stone Images; **434** Leverett Bradley/Tony Stone Images; **435** Ed Kashi; **436** (t)Dominic Oldershaw, (bl)Matt Hulbert, (br)Jeff Smith/FOTOSMITH; **437** (t)Phil Schermeister/Tony Stone Images, (b)Dominic Oldershaw; **438** (t)Buddy Mays/International Stock, (c)E.R. Degginger, (b)E.R. Degginger/Photo Researchers; **440** (t)Photo of CA State Science Fair courtesy CA Science Center, (b)HERMAN® is reprinted with permission from Laughingstock Licensing Inc., Ottawa, Canada. All Rights Reserved; **443** (l)David Carriere/Tony Stone Images, (r)Chris Falkenstein; **444** SuperStock; **445** Jeff Smith/FOTOSMITH; **446** Timothy Fuller; **447** (t)Timothy Fuller, (b)Dominic Oldershaw; **449** National Air and Space Museum, Smithsonian Institution, Photo No. A45288-A; **450** Timothy Fuller; **452** (t)Ron Kimball, (b)Tess & David Young/Tom Stack & Assoc.; **454** Ralph Wetmore/Tony Stone Images; **456 459** Timothy Fuller; **460** Jerry Jacka; **462** Susan Harper; **474** (l)The Drachen Foundation, (r)Rich Iwasaki/Tony Stone Images; **474-475** Uniphoto; **475** (l)Mark E. Gibson, (r)Dominic Oldershaw; **476** Nada Pecnik/Visuals Unlimited; **477** Dominic Oldershaw; **478** UPI/Corbis-Bettmann; **480** Dominic Oldershaw; **482** (t)Dominic Oldershaw, (b)George Whiteley/Photo Researchers; **483** Charles R. Belinky/Photo Researchers; **484** SuperStock; **485** Science Photo Library/Photo Researchers; **486** Tribune Media Services, Inc. All Rights Reserved. Reprinted with permission.; **487** Jeff Smith/FOTOSMITH; **488** Craig Aurness/Westlight; **489 490** Dominic Oldershaw; **491** Geoff Butler; **492** Hideo Kurihara/Tony Stone Images; **493** Morris Best/Uniphoto; **495** Collection: Kröller-Müller Museum, Otterlo, The Netherlands ©Tony Smith Estate, courtesy Paula Cooper Gallery, New York; **496** Tom Bean/Tony Stone Images; **497** Lefever/Grushow/Grant Heilman Photography; **499** Uniphoto; **500** Dominic Oldershaw; **501** Tom Raymond/Tony Stone Images; **502** (t)Azzato/International Stock, (b)Thomas Veneklasen; **504** Peter Read Miller/Sports Illustrated; **505** Dominic Oldershaw; **506** (t)Dominic Oldershaw, (b)Steve Elmore/Tony Stone Images; **507** Reprinted with special permission of King Features Syndicate, **514** (t)Allsport USA/Stephen Dunn, (c)Jerry Wachter/Photo Researchers, (b)Phil Degginger/Color-Pic; **515** (t)Phil Degginger/Color-Pic, (b)Uniphoto; **517** Doug Martin; **518** (t)Jeff Smith/FOTOSMITH, (b)Thomas Veneklasen; **521** Thomas Russell; **522** Manuel Denner/International Stock; **524** (l)Steve Gettle/ENP Images, (r)Bill Beatty/Visuals Unlimited; **525** Scott Cunningham; **526** Michel Viard/Peter Arnold, Inc.; **527** Thomas Veneklasen; **528** Hulton Getty/Tony Stone Images; **529 530 531** Jeff Smith/FOTOSMITH; **534** Glencoe photo; **536 537** Timothy Fuller; **538** (l)BLT Productions, (tr)Ron Berhmann/International Stock, (br)Jeff Smith/FOTOSMITH; **539** K.S. Studio; **540** Timothy Fuller; **541** Jeff Smith/FOTOSMITH; **543** Chuck Pefley/Tony Stone Images; **544** (t)E.R. Degginger, (b)James W. Richardson/Visuals Unlimited; **546** Dominique Braud/Tom Stack & Assoc.; **548** E.R. Degginger; **549** (t)MAK-1, (c)Tracy Aiguier, (bl)Timothy Fuller, (br)Jeff Smith/FOTOSMITH, (bkgd)Joseph A. DiChello; **556** Robert Shafer/Tony Stone Images; **557** Ken Frick/International Stock; **558** (l)Allsport USA/David Taylor, (r)Jeff Smith/FOTOSMITH; **558-559** Dominic Oldershaw; **559** John Todd/ABL Photos; **561 562** Jeff Smith/FOTOSMITH; **563** Corbis-Bettmann; **564** John Klein/SportsChrome; **565** Henning Photography; **567 568** Jeff Smith/FOTOSMITH; **569** (t)Dominic Oldershaw, (b)Johnny Johnson; **572** Dominic Oldershaw; **574** Daniel Mainzer; **575** (t)Dominic Oldershaw, (b)Timothy Fuller; **576** (l)Jeff Smith/FOTOSMITH, (r)Timothy Fuller; **579** courtesy Southwest Museum, Los Angeles; **580** Ryan-Beyer/Tony Stone Images; **581** Timothy Fuller; **582** (l)courtesy Phoenix Suns, (r)Timothy Fuller, (clipboard bkgd)Dominic Oldershaw, (floor bkgd)Jeff Smith/FOTOSMITH; **583** Corbis-Bettmann; **585** Mark E. Gibson/Visuals Unlimited; **586** Timothy Fuller; **588** Dominic Oldershaw; **590** Thomas Veneklasen; **591** Dominic Oldershaw; **595** Doug Martin; **599** (t)Uniphoto, (cl)Michel Viard/Peter Arnold, Inc., (cr)Barros & Barros/The Image Bank, (bl)Jeff Smith/FOTOSMITH, (br)Brian Parker/Tom Stack & Assoc.

Teacher's Wraparound Edition: iv v viii Mark Burnett; **x** Superstock; **xi** Daryl Balfour/Tony Stone Images; **xii** Gary Pearl/Sports Photo Masters; **xiii** E.R. Degginger; **ix** Uniphoto; **xiv** Ron Kimball; **xv** Jim Mejuto/FPG; **xvi** J. Saunders/MODEL RAILROADER Magazine; **xvii** Greg Pease/Tony Stone Images; **xviii** Ralph Wetmore/Tony Stone Images; **xix** Morris Best/Uniphoto Picture Agency; **xx** Ron Berhmann/International Stock; **xxi** Dominic Oldershaw; **T1**(t)Jerry Jacka, (c)Manual Denner, (b)Elaine Shay; **T4** Tribune Media Services, Inc. All Rights Reserved. Reprinted with permission.; **T5** Scott Cunningham; **T10** (l)Dominic Oldershaw, (r)BLT Productions; **T11** (t)Scott Cunningham, (b)K.S. Studio; **T12** K.S. Studio; **T13** Aaron Haupt; **T14** K.S. Studio; **T15** Scott Cunningham; **T16** BLT Productions; **T17 T18 T19** K.S. Studio; **T21** Scott Cunningham; **T22 T23** Geoff Butler; **T24** (t)Tim Courlas, (b)George Haling/Photo Researchers; **T30 T31** K.S. Studio; **2A** James Nazz/The Image Bank; **54A** Keren Su/Tony Stone Images; **102A** (t)Uniphoto, (b)Tony Craddock/Tony Stone Images; **138A** ©The Newark Museum/Art Resource, NY; **184A** Uniphoto Picture Agency; **230A** Mike Zito/SportsChrome USA; **276A** Dave Jacobs/Tony Stone Images; **328A** Thomas Veneklasen; **380A** Michael Agliolo/International Stock; **426A** Timothy Fuller; **474A** Mark E. Gibson; **514A** Stephen Dunn/Allsport; **558A** David Taylor/Allsport.

Glossary

A

absolute value (57) The number of units a number is from zero on the number line.

acute (196) An angle with a measure greater than 0° and less than 90°.

addition property of equality (18) If you add the same number to each side of an equation, the two sides remain equal. If $a = b$, then $a + c = b + c$.

additive inverse (70) Two integers that are opposites of each other are called additive inverses. The sum of any number and its additive inverse is zero, $a + (-a) = 0$.

algebraic expression (12) A combination of variables, numbers, and at least one operation.

alternate exterior angles (189) In the figure, transversal t intersects lines ℓ and m. $\angle 1$ and $\angle 7$, $\angle 2$ and $\angle 8$ are alternate exterior angles. If lines ℓ and m are parallel, then these angles are congruent.

alternate interior angles (189) In the figure, transversal t intersects lines ℓ and m. $\angle 3$ and $\angle 5$, $\angle 4$ and $\angle 6$ are alternate interior angles. If lines ℓ and m are parallel, then these angles are congruent.

altitude (40, 301) A segment in a quadrilateral that is perpendicular to both bases, with endpoints on the base lines.

altitude (490) The segment that goes from the vertex of a cone to its base and is perpendicular to the base.

area (39) The number of square units needed to cover a surface enclosed by a geometric figure.

arithmetic sequence (296) A sequence of numbers in which you can find the next term by adding the same number to the previous term.

associative property (8) For any numbers a, b, and c, $(a + b) + c = a + (b + c)$. Also for any numbers a, b, and c, $(ab)c = a(bc)$.

B

bar graph (142) A graphic form using bars to make comparisons of statistics.

bar notation (249) In repeating decimals the line or bar placed over the digits that repeat. Another way to write 2.6363636. . . is $2.\overline{63}$.

base (8) In a power, the number used as a factor. In 10^3, the base is 10. That is, $10^3 = 10 \times 10 \times 10$.

base (40, 301, 302) The base of a parallelogram or a triangle is any side of the figure. The bases of a trapezoid are the parallel sides.

base (335) In a percent proportion, the number to which the percentage is compared. In $\frac{3}{4} = \frac{x}{100}$, the total number of parts, 4, is called the base.

base (483) The bases of a prism are the two parallel congruent faces.

binomial (583) A polynomial with exactly two terms.

box-and-whisker plot (167) A diagram that summarizes data using the median, the upper and lower quartiles, and the extreme values. A box is drawn around the quartile value and whiskers extend from each quartile to the extreme data points.

C

cell (162) The basic unit of a spreadsheet. A cell can contain data, labels, or formulas.

center (309) The given point from which all points on a circle or a sphere are the same distance.

circle (309) The set of all points in a plane that are the same distance from a given point called the center.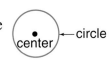

circle graph (148) A type of statistical graph used to compare parts of a whole.

circular cone (490) A shape in space that has a circular base and one vertex.

circular cylinder (487) A cylinder with two bases that are parallel, congruent circular regions.

circumference (309) The distance around a circle.

column (73) Numbers stacked on top of each other in a vertical arrangement.

combination (524) An arrangement or listing of objects in which order is not important.

common difference (296) The difference between any two consecutive terms in an arithmetic sequence.

common ratio (297) The constant factor used to multiply consecutive terms in a geometric sequence.

commutative property (8) For any real numbers a and b, $a + b = b + a$ and $ab = ba$.

compatible numbers (126) Two numbers that are easy to divide mentally. They are often members of fact families and can be used as an estimation.

complementary angles (198) Two angles are complementary if the sum of their measures is 90°.

composite number (235) Any whole number greater than 1 that has more than two factors.

compound event (534) A compound event consists of two or more simple events.

congruent triangles (210) Triangles that have the same size and shape.

converse (399) The converse of the Pythagorean Theorem can be used to test whether a triangle is a right triangle. If the sides of the triangle have lengths a, b, and c, such that $c^2 = a^2 + b^2$, then the triangle is a right triangle.

coordinate (56) A number associated with a point on the number line.

coordinate system (92) A plane in which a horizontal number line and a vertical number line intersect at their zero points.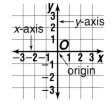

corresponding angles (189) Angles that have the same position on two different parallel lines cut by a transversal. $\angle 1$ and $\angle 5$, $\angle 2$ and $\angle 6$, $\angle 3$ and $\angle 7$, and $\angle 4$ and $\angle 8$ are corresponding angles.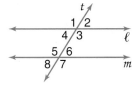

corresponding parts (210) Parts on congruent or similar figures that match.

Counting Principle (518) This principle is used to find the number of outcomes by multiplying. If event M can occur in m ways and is followed by event N that can occur in n ways, then the event M followed by the event N can occur in $m \cdot n$ ways.

cross products (111) The products of the terms on the diagonals when two ratios are compared. If the cross products are equal, then the ratios form a proportion. In the proportion $\frac{2}{3} = \frac{8}{12}$, the cross products are 2×12 and 3×8.

data analysis (154) To study data and draw conclusions from the numbers observed.

density property (260) Between every pair of rational numbers there are infinitely many rational numbers.

dependent events (535) Two or more events in which the outcome of one event does affect the outcome of the other event or events.

diameter (309) The distance across a circle through its center.

dilation (214, 370) The process of reducing or enlarging an image in mathematics.

discount (349) The amount by which the regular price is reduced.

distributive property (34) The sum of two addends multiplied by a number is the sum of the product of each addend and the number. For any numbers a, b, and c, $a(b + c) = ab + ac$ and $(b + c)a = ba + ca$.

divisible (232) A number is divisible by another if the quotient is a whole number and the remainder is zero.

division property of equality (22) If each side of an equation is divided by the same nonzero number, then the two sides remain equal. If $a = b$, then $\frac{a}{c} = \frac{b}{c}$, $c \neq 0$.

domain (428) The set of input values in a function.

E

edge (483) The intersection of faces of a three-dimensional figure.

element (73) Each number in a matrix is called an element.

equation (13) A mathematical sentence that contains the equal sign, =.

equilateral (196) All sides of a figure are congruent.

evaluate (9) To find the value of an expression by replacing the variables with numerals.

event (253) A specific outcome or type of outcome.

experimental probability (540) An estimated probability based on the relative frequency of positive outcomes occurring during an experiment.

exponent (8) In a power, the number of times the base is used as a factor. In 10^3, the exponent is 3.

F

face (483) Any surface that forms a side or a base of a prism.

factor (8) When two or more numbers are multiplied, each number is a factor of the product. In $4 \times 5 = 20$, 4 and 5 are factors.

factorial (522) The expression $n!$ is the product of all counting numbers beginning with n and counting backward to 1.

factoring (588) Finding the factors of a product.

factor tree (236) A diagram showing the prime factorization of a number.

Fibonacci sequence (300) A list of numbers in which the first two numbers are both 1 and each number that follows is the sum of the previous two numbers, 1, 1, 2, 3, 5, 8, . . .

frequency table (142) A table for organizing a set of data that shows the number of times each item or number appears.

function (428) A relation in which each element of the input is paired with exactly one element of the output according to a specified rule.

function table (428) A table organizing the input, rule, and output of a function.

Fundamental Theorem of Arithmetic (236) A property of numbers that states every number has a unique set of prime factors.

geometric sequence (297) When consecutive terms of a sequence are formed by multiplying by a constant factor, the sequence is called a geometric sequence.

graph (56) To draw or plot the points named by those numbers on a number line or coordinate plane.

greatest common factor (GCF) (242) The greatest of the common factors of two or more numbers. The GCF of 18 and 24 is 6.

greatest possible error (504) Half the smallest unit used to make a measurement.

height (40, 301) The length of the altitude of a triangle or a quadrilateral.

histogram (142) A special kind of bar graph that displays the frequency of data that has been organized into equal intervals. The intervals cover all possible values of data, therefore there are no spaces between the bars of the graph.

hypotenuse (398) The side opposite the right angle in a right triangle.

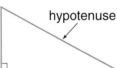

independent events (534) Two or more events in which the outcome of one event does not affect the outcome of the other event(s).

indirect measurement (361) A technique using proportions to find a measurement.

inequality (43) A mathematical sentence that contains $<, >, \neq, \leq,$ or $\geq$.

integer (56) The whole numbers and their opposites.
$$\ldots, -3, -2, -1, 0, 1, 2, 3, \ldots$$

interest (353) The amount charged or paid for the use of money.

interquartile range (163) The range of the middle half of data.

inverse operation (18) Pairs of operations that undo each other. Addition and subtraction are inverse operations. Multiplication and division are inverse operations.

inverse property of multiplication (290) For every nonzero number $\frac{a}{b}$, where a and b do not equal 0, there is exactly one number $\frac{b}{a}$ such that $\frac{a}{b} \times \frac{b}{a} = 1$.

irrational number (391) A number that cannot be expressed as $\frac{a}{b}$, where a and b are integers and $b \neq 0$.

isosceles (196) A triangle that has at least two congruent sides.

least common denominator (LCD) (261) The least common multiple of the denominators of two or more fractions.

least common multiple (LCM) (257) The least of the nonzero common multiples of two or more numbers. The least common multiple of 2 and 3 is 6.

legs (398) The two sides of a right triangle that form the right angle.

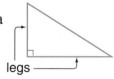

like terms (565) Expressions that contain the same variables, such as $3ab$ and $7ab$.

linear function (442) An equation in which the graphs of the solutions form a line.

line of symmetry (460) A line that divides a figure into two halves that are reflections of each other.

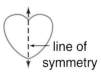

line plot (153) A graph that uses an × above a number on a number line each time that number occurs in a set of data.

line symmetry (206) Figures that match exactly when folded in half have line symmetry.

lower quartile (164) The median of the lower half of data in an interquartile range.

M

markup (349) The difference between the price paid by the merchant and the increased selling price.

matrix (73) A rectangular arrangement of elements in rows and columns.

mean (158) The sum of the numbers in a set of data divided by the number of pieces of data.

measures of central tendency (158) Numbers or pieces of data that can represent the whole set of data.

median (158, 159) The middle number in a set of data when the data are arranged in numerical order. If the data has an even number, the median is the mean of the two middle numbers.

mixed number (278) The sum of a whole number and a fraction. $6\frac{2}{3}$ is a mixed number.

mode (158) The number or item that appears most often in a set of data.

monomial (561) A number, a variable, or a product of a number and one or more variables.

multiple (257) The product of a number and any whole number. 28 is a multiple of 4 and 7.

multiplication property of equality (22) If each side of an equation is multiplied by the same number, then the two sides remain equal. If $a = b$, then $ac = bc$.

multiplicative inverse (290) A number times its multiplicative inverse is equal to 1. The multiplicative inverse of $\frac{2}{3}$ is $\frac{3}{2}$.

N

numerical expression (11) A mathematical expression that has a combination of numbers and at least one operation. $4 + 2$ is a numerical expression.

O

obtuse (196) An obtuse angle is any angle that measures greater than 90° but less than 180°.

open sentence (13) An equation that contains a variable.

opposite (70) Two integers are opposites if they are represented on the number line by points that are the same distance from zero, but on opposite sides of zero. The sum of opposites is zero.

order of operations (11) The rules to follow when more than one operation is used.
1. Simplify the expressions inside grouping symbols; start with the innermost grouping symbols first.
2. Evaluate all powers.
3. Then do all multiplications and divisions from left to right.
4. Then do all additions and subtractions from left to right.

ordered pair (92) A pair of numbers used to locate a point in the coordinate system. The ordered pair is written in this form: (x-coordinate, y-coordinate).

origin (92) The point of intersection of the x-axis and y-axis in a coordinate system.

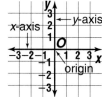

outcome (253, 518) One possible result of a probability event. For example, 4 is an outcome when a number cube is rolled.

outlier (164) Data that is more than 1.5 times the interquartile range from the upper or lower quartiles.

parallel lines (188) Lines in the same plane that do not intersect. The symbol ∥ means parallel.

parallelogram (39, 201) A quadrilateral with two pairs of parallel sides.

Pascal's Triangle (528) A triangular arrangement of numbers in which each number is the sum of the two numbers to the right and to the left of it in the row above.

pentagon (358) A polygon having five sides.

percent (107) A ratio that compares a number to 100.

percentage (335) In a percent proportion, a number (P) that is compared to another number called the base (B).

percent of change (348) The ratio of the amount of increase or decrease to the original amount.

percent of decrease (349) The ratio of an amount of decrease to the previous amount, expressed as a percent.

percent of increase (349) The ratio of an amount of increase to the previous amount, expressed as a percent.

percent proportion (335) Is $\frac{P}{B} = \frac{r}{100}$ where P represents the percentage, B represents the base, and r represents the number per hundred. Also written as $\frac{\text{Percentage}}{\text{Base}}$ = Rate.

perfect square (382) A rational number whose square root is a whole number. 25 is a perfect square because $\sqrt{25} = 5$.

perimeter (38) The distance around a geometric figure.

permutation (521) An arrangement or listing in which order is important.

perpendicular (197) Two lines or line segments that intersect to form right angles.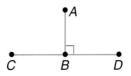

polygon (196) A simple closed figure in a plane formed by three or more line segments.

polynomial (562) The sum or difference of two or more monomials.

population (546) The entire group of items or individuals from which the samples under consideration are taken.

power (8) A number that can be written using an exponent. The power 7^3 is read *seven to the third power*, or *seven cubed*.

precision (504) The precision of a measurement depends on the unit of measure. The smaller the unit the more precise the measurement is.

prime factorization (235) Expressing a composite number as the product of prime numbers. The prime factorization of 63 is $3 \times 3 \times 7$.

prime number (235) A whole number greater than 1 that has exactly two factors, 1 and itself.

principal (353) The amount of an investment or a debt.

principal square root (382) A nonnegative square root. $\sqrt{49}$ indicates the *principal* square root of 49.

prism (483) A three-dimensional figure that has two parallel and congruent bases in the shape of polygons.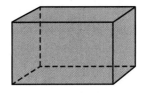

probability (253) The chance that some event will happen. It is the ratio of the number of ways a certain event can occur to the number of possible outcomes.

proportion (111) A statement of equality of two or more ratios, $\frac{a}{b} = \frac{c}{d}, b \neq 0, d \neq 0$.

pyramid (483) A solid figure that has a polygon for a base and triangles for sides.

Pythagorean Theorem (396–398) In a right triangle, the square of the length of the hypotenuse is equal to the sum of the squares of the lengths of the legs. $c^2 = a^2 + b^2$

Pythagorean triple (405) A set of three integers that satisfy the Pythagorean Theorem.

quadrant (92) One of the four regions into which two perpendicular number lines separate the plane.

quadratic function (452) A function in which the greatest power is 2.

quadrilateral (201) A polygon having four sides.

quartiles (163) Values that divide data into four equal parts.

radical sign (382) The symbol used to indicate a nonnegative square root. $\sqrt{}$

radius (309) The distance from the center of a circle to any point on the circle.

random (253) Outcomes occur at random if each outcome is equally likely to occur.

range (428) The set of output values in a function.

range (163) The difference between the greatest number and the least number in a set of data.

rate (105) A ratio of two measurements having different units.

rate (335) In a percent proportion, the ratio of a number to 100.

ratio (104) A comparison of two numbers by division. The ratio of 2 to 3 can be stated as 2 out of 3, 2 to 3, 2:3, or $\frac{2}{3}$.

rational numbers (245) Numbers of the form $\frac{a}{b}$, where a and b are integers and $b \neq 0$.

real numbers (391) The set of rational numbers together with the set of irrational numbers.

reciprocal (290) The multiplicative inverse of a number.

rectangle (38, 201) A quadrilateral with four congruent angles.

reflection (206, 460) A type of transformation where a figure is flipped over a line of symmetry.

relative error (504) The comparison of the greatest possible error with the measurement itself.
$$relative\ error = \frac{greatest\ possible\ error}{measurement}$$

repeating decimal (249) A decimal whose digits repeat in groups of one or more. Examples are 0.181818... and 0.8333.... Using bar notation, these numbers are written as $0.\overline{18}$ and $0.8\overline{3}$

rhombus (201) A parallelogram with four congruent sides.

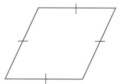

right (196) An angle that measures 90°.

rotation (206, 464) When a figure is turned around a central point.

rotational symmetry (207) A figure has rotational symmetry if it can be turned less than 360° about its center and still looks like the original.

row (73) In a matrix the numbers side-by-side horizontally form a row.

S

sample (174, 546) A randomly-selected group chosen for the purpose of collecting data.

sample space (253) The set of all possible outcomes.

scale (366) The ratio of a given length on a drawing or model to its corresponding length in reality.

scale drawing (366) A drawing that is similar but either larger or smaller than the actual object.

scale factor (370) The ratio of a dilated image to the original image.

scale model (366) A replica of an original object that is too large or too small to be built at actual size.

scalene (196) A triangle with no congruent sides.

scatter plot (168) A graph that shows the general relationship between two sets of data.

scientific notation (265) A way of expressing numbers as the product of a number that is at least 1 but less than 10 and a power of 10. In scientific notation, 5,500 is 5.5×10^3.

selling price (349) The amount a customer pays for an item.

sequence (296) A list of numbers in a certain order, such as, 0 ,1, 2, 3, or 2, 4, 6, 8.

significant digits (504) All of the digits of a measurement that are known to be accurate plus one estimated digit.

similar polygons (357) Two polygons are similar if their corresponding angles are congruent and their corresponding sides are in proportion. They have the same shape but may not have the same size.

similar triangles (215) Triangles that have the same shape but may not have the same size.

simplest form (246) The form of a fraction where the GCF of the numerator and denominator is 1.

simplest form (565) An expression that has no like terms in it. In simplest form $6x + 8x$ is $14x$.

simulation (540) The process of acting out a problem.

solid (482) A three-dimensional figure.

solution (13) A value for the variable that makes an equation true. The solution for $10 + y = 25$ is 15.

spreadsheet (162) A tool used for organizing and analyzing data.

square (38, 201) A parallelogram with all sides congruent and all angles congruent.

square root (382) One of the two equal factors of a number. If $a^2 = b$, then a is the square root of b. A square root of 144 is 12 since $12^2 = 144$.

statistics (142) The branch of mathematics that deals with collecting, organizing, and analyzing data.

substitute (11) To replace a variable in an algebraic expression with a number, creating a numerical expression.

subtraction property of equality (18) If you subtract the same number from each side of an equation, the two sides remain equal. For any numbers a, b, and c, if $a = b$, then $a - c = b - c$.

supplementary angles (190) Two angles are supplementary if the sum of their measures is 180°.

surface area (495) The sum of the areas of all the faces of a three-dimensional figure.

symmetric (460) Figures that can be folded into two identical parts.

system of equations (446) A set of equations with the same variables.

term (565) A number, a variable, or a product of numbers and variables.

term (296) A number in a sequence.

terminating decimal (249) A decimal whose digits end. Every terminating decimal can be written as a fraction with a denominator of 10, 100, 1,000, and so on.

tessellation (220) A repetitive pattern of polygons that fit together with no holes or gaps.

theoretical probability (540) The long-term probability of an outcome based on mathematical principles.

transformation (220) Movements of geometric figures to modify tessellations.

translation (220, 456) A method used to make changes in the polygons of tessellations by sliding a figure horizontally, vertically, or both.

transversal (188) A line that intersects two parallel lines to form eight angles.

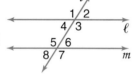

trapezoid (201, 302) A quadrilateral with exactly one pair of parallel sides.

tree diagram (518) A diagram used to show the total number of possible outcomes in a probability experiment.

triangle (196) A polygon having three angles and three sides.

trigonometry (365) The study of triangle measurement.

trinomial (588) A polynomial with three terms.

unit rate (105) A rate with denominator of 1.

upper quartile (163) The median of the upper half of a set of numbers.

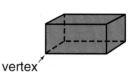

variable (12) A symbol, usually a letter, used to represent a number in mathematical expressions or sentences.

variation (163) The spread in the values in a set of data.

vertex (483) The vertex of a prism is the point where all the faces intersect.

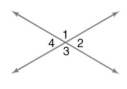

vertical angles (190) Congruent angles formed by the intersection of two lines. In the figure, the vertical angles are ∠1 and ∠3, and ∠2 and ∠4.

volume (486) The number of cubic units needed to fill the space occupied by a solid.

x-axis (92) The horizontal number line which helps to form the coordinate system.

x-coordinate (92) The first number of an ordered pair.

y-axis (92) The vertical line of the two perpendicular number lines in a coordinate plane.

y-coordinate (92) The second number of an ordered pair.

zero pair (63) The result of pairing one positive counter with one negative counter.

Spanish Glossary

absolute value / valor absoluto (57) Número de unidades en la recta numérica que un número dista de cero.

acute / agudo (196) Ángulo que mide más de 0° y menos de 90°.

addition property of equality / propiedad de adición de la igualdad (18) Si sumas el mismo número a ambos lados de una ecuación, los dos lados permanecen iguales. Si $a = b$, entonces $a + c = b + c$.

additive inverse / inverso aditivo (70) Dos enteros que son opuestos mutuos reciben el nombre de inversos aditivos. La suma de cualquier número y su inverso aditivo es cero, $a + (-a) = 0$.

algebraic expression / expresión algebraica (12) Combinación de variables, números y al menos una operación.

alternate exterior angles / ángulos alternos externos (189) En la figura, la transversal t interseca las rectas ℓ y m. El $\angle 1$ y $\angle 7$, así como el $\angle 2$ y el $\angle 8$, reciben el nombre de ángulos alternos externos. Si las rectas ℓ y m son paralelas, estos ángulos son todos congruentes.

alternate interior angles / ángulos alternos internos (189) En la figura, la transversal t interseca las rectas ℓ y m. El $\angle 3$ y el $\angle 5$, así como el $\angle 4$ y el $\angle 6$ reciben el nombre de ángulos alternos internos. Si las rectas ℓ y m son paralelas, estos ángulos son congruentes.

altitude / altitud (40, 301) Segmento en un cuadrilátero que es perpendicular a ambas bases y cuyos extremos yacen en las bases del cuadrilátero.

altitude / altitud (490) Segmento trazado desde el vértice de un cono hasta su base y que es perpendicular a la base.

area / área (39) Número de unidades cuadradas que se requieren para cubrir la superficie encerrada por una figura geométrica.

arithmetic sequence / sucesión aritmética (296) Sucesión de números en que se puede calcular el próximo término sumando siempre el mismo número al término anterior.

associative property / propiedad asociativa (8) Para números a, b y c cualesquiera, $(a + b) + c = a + (b + c)$ y $(ab)c = a(bc)$.

bar graph / gráfica de barras (142) Tipo de gráfica que usa barras para comparar estadísticas.

bar notation / notación de barra (249) En los decimales periódicos, la línea o barra que se escribe encima de los dígitos que se repiten. Otra forma de escribir 2.6363636... es $2.\overline{63}$.

base / base (8) Número que se usa como factor en una potencia. En 10^3, la base es 10, es decir, $10^3 = 10 \times 10 \times 10$.

base / base (40, 301, 302) La base de un paralelogramo o de un triángulo es cualquier lado de la figura. Las bases de un trapecio son sus lados paralelos.

base / base (335) Número con el cual se compara el porcentaje, en una proporción porcentual. En $\frac{3}{4} = \frac{x}{100}$, el número total de partes, 4, recibe el nombre de base.

base / base (483) Las bases de un prisma son sus caras congruentes y paralelas.

binomial / binomio (583) Polinomio que tiene exactamente dos términos.

box-and-whisker plot / diagrama de caja y patillas (167) Diagrama que resume información usando la mediana, los cuartiles superior e inferior y los valores extremos. Se dibuja una caja alrededor de los cuartiles y se trazan patillas que los unan a los valores extremos respectivos.

C

cell / celda (162) Unidad básica de una hoja de cálculos. Las celdas pueden contener datos, rótulos o fórmulas.

center / centro (309) Punto en el plano del cual equidistan todos los puntos de un círculo o de una esfera.

circle / círculo (309) Conjunto de todos los puntos en un plano que equidistan de un punto dado llamado centro.

circle graph / gráfica circular (148) Tipo de gráfica estadística que se usa para comparar las partes de un todo.

circular cone / cono circular (490) Forma espacial que posee una base circular y un vértice.

circular cylinder / cilindro circular (487) Cilindro que posee dos bases circulares paralelas y congruentes.

circumference / circunferencia (309) La distancia alrededor de un círculo.

column / columna (73) Números colocados uno encima de otro en un arreglo vertical.

combination / combinación (524) Arreglo o lista de objetos en que el orden no es importante.

common difference / diferencia común (296) Diferencia entre dos términos consecutivos cualesquiera de una sucesión aritmética.

common ratio / razón común (297) Factor constante que se usa para calcular cualquier término, a partir del segundo, de una sucesión geométrica multiplicando el término anterior por este factor.

commutative property / propiedad conmutativa (8) Para números reales a y b cualesquiera, $a + b = b + a$ y $ab = ba$.

compatible numbers / números compatibles (126) Dos números que son fáciles de dividir mentalmente. Son a menudo miembros de la misma familia de factores y se pueden usar en estimaciones.

complementary angles / ángulos complementarios (198) Dos ángulos son complementarios si la suma de sus medidas es 90°.

composite number / número compuesto (235) Cualquier número entero mayor que 1 que posee más de dos factores.

compound event / evento compuesto (534) Un evento compuesto consiste en dos o más eventos simples.

congruent triangles / triángulos congruentes (210) Triángulos que tienen la misma forma y tamaño.

converse / recíproco (399) El recíproco del Teorema de Pitágoras puede usarse para averiguar si un triángulo es un triángulo rectángulo. Si las longitudes de los lados de un triángulo son a, b y c y si $c^2 = a^2 + b^2$, entonces el triángulo es un triángulo rectángulo.

coordinate / coordenada (56) Número asociado con un punto en una recta numérica.

coordinate system / sistema de coordenadas (92) Plano en el cual se han trazado dos rectas numéricas, una horizontal y una vertical, que se intersecan en sus puntos cero.

corresponding angles / ángulos correspondientes (189) Ángulos que ocupan la misma posición en dos rectas paralelas distintas atravesadas por una transversal. $\angle 1$ y $\angle 5$, $\angle 2$ y $\angle 6$, $\angle 3$ y $\angle 7$, $\angle 4$ y $\angle 8$ son ángulos correspondientes.

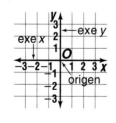

corresponding parts / partes correspondientes (210) Partes de figuras congruentes o semejantes que coinciden.

Counting Principle / Principio Fundamental de Contar (518) Este principio se usa para calcular el número de resultados mediante multiplicación. Si el evento M puede ocurrir de m maneras y es seguido por el evento N, que puede ocurrir de n maneras, entonces el evento M seguido del evento N puede ocurrir de $m \cdot n$ maneras.

cross products / productos cruzados (111) Los productos que resultan de la comparación de los términos de las diagonales de dos razones. Las razones forman una proporción si y sólo si los productos son iguales. En la proporción $\frac{2}{3} = \frac{8}{12}$, los productos cruzados son 2×12 y 3×8.

data analysis / análisis de datos (154) El estudio de datos y la extracción de conclusiones de los números observados.

density property / propiedad de densidad (260) Entre cada par de números racionales hay números racionales infinitos.

dependent events / eventos dependientes (535) Dos o más eventos en que el resultado de uno de ellos afecta el resultado de los otros eventos.

diameter / diámetro (309) Longitud de cualquier segmento de recta cuyos extremos yacen en un círculo y que pasa por su centro.

dilation / dilatación (214, 370) En matemáticas, proceso de reducción o ampliación de una imagen.

discount / descuento (349) La cantidad de reducción del precio normal.

distributive property / propiedad distributiva (34) La adición de dos sumandos multiplicada por un número es igual a la adición del producto de cada sumando por el número. Para números a, b y c cualesquiera, $a(b + c) = ab + ac$ y $(b + c)a = ba + ca$.

divisible / divisible (232) Un número es divisible entre otro si el cociente es un número entero y el residuo es cero.

division property of equality / propiedad de división de la igualdad (22) Si cada lado de una ecuación se divide entre el mismo número no nulo, los dos lados de la ecuación permanecen iguales. Si $a = b$, entonces $\frac{a}{c} = \frac{b}{c}, c \neq 0$.

domain / dominio (428) Conjunto de valores de entrada de una función.

edge / arista (483) Intersección de las caras de una figura tridimensional.

element / elemento (73) Cada número en una matriz recibe el nombre de elemento.

equation / ecuación (13) Enunciado matemático que contiene el signo de igualdad, $=$.

equilateral / equilátero (196) Figura con todos los lados congruentes entre sí.

evaluate / evaluar (9) Calcular el valor de una expresión sustituyendo las variables con números.

event / evento (253) Resultado específico o tipo de resultado de un experimento probabilístico.

experimental probability / probabilidad experimental (540) Probabilidad de un evento que se calcula o estima basándose en la frecuencia relativa de los resultados favorables al evento en cuestión, que ocurren durante un experimento probabilístico.

exponent / exponente (8) Número de veces que la base de una potencia se usa como factor. En 10^3, el exponente es 3.

F

face / cara (483) Cualquier superficie que forma un lado o una base de un prisma.

factor / factor (8) Cuando se multiplican dos o más números enteros, cada número es un factor del producto. En 4 × 5 = 20, 4 y 5 son factores de 20.

factorial / factorial (522) La expresión $n!$ es el producto de los n primeros números de contar, contando al revés.

factoring / factorización (588) Encontrar los factores de un producto.

factor tree / árbol de factores (236) Diagrama que muestra la factorización prima de un número.

Fibonacci sequence / sucesión de Fibonacci (300) Sucesión de números en que los dos primeros términos son iguales a 1 y cada término que sigue es igual a la suma de los dos términos anteriores, 1, 1, 2, 3, 5, 8,... .

frequency table / tabla de frecuencia (142) Tabla que se usa para organizar un conjunto de datos y que muestra cuántas veces aparece cada dato.

function / función (428) Relación en que cada elemento de entrada es apareado con un único elemento de salida, según una regla específica.

function table / tabla de funciones (428) Tabla que organiza las entradas, la regla y las salidas de una función.

Fundamental Theorem of Arithmetic / Teorema Fundamental de la Aritmética (236) Teorema que afirma que todo número entero tiene un conjunto único de factores primos.

G

geometric sequence / sucesión geométrica (297) Sucesión de números en que se puede calcular cualquier término, después del segundo, multiplicando el término anterior por el mismo número.

graph / graficar (56) Dibujar o trazar, en una recta numérica o en un plano de coordenadas, los puntos indicados.

greatest common factor (GCF) / máximo común divisor (MCD) (242) El mayor factor común de dos o más números. El MCD de 18 y 24 es 6.

greatest possible error / error máximo posible (504) Mitad de la unidad más pequeña que se usó para tomar una medida.

H

height / altura (40, 301) La longitud de la altitud de un triángulo o de un cuadrilátero.

histogram / histograma (142) Tipo especial de gráfica de barras que exhibe la frecuencia de los datos una vez que los datos han sido organizados en intervalos iguales. Los intervalos cubren todos los valores posibles de los datos, de modo que no hay espacios entre las barras de la gráfica.

hypotenuse / hipotenusa (398) Lado de un triángulo rectángulo opuesto a su ángulo recto.

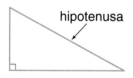

I

independent events / eventos independientes (534) Dos o más eventos en los cuales el resultado de uno de ellos no afecta el resultado de los otros eventos.

indirect measurement / medida indirecta (361) Técnica que usa proporciones para calcular medidas.

inequality / desigualdad (43) Enunciado matemático que contiene $<, >, \neq, \leq$ o $\geq$.

integer / entero (56) Los números enteros no negativos y sus opuestos.
$$\ldots, -3, -2, -1, 0, 1, 2, 3, \ldots$$

interest / interés (353) Cantidad que se cobra o se paga por el uso del dinero.

interquartile range / amplitud intercuartílica (163) El rango de la mitad central de un conjunto de datos.

inverse operation / operaciones inversas (18) Pares de operaciones que se anulan mutuamente. La adición y la sustracción son operaciones inversas. La multiplicación y la división son operaciones inversas.

inverse property of multiplication / propiedad de inverso multiplicativo (290) Para cada número no nulo $\frac{a}{b}$, donde a y b no son cero, existe un único número $\frac{b}{a}$ tal que $\frac{a}{b} \times \frac{b}{a} = 1$.

irrational number / número irracional (391) Un número que no puede escribirse como $\frac{a}{b}$, donde a y b son enteros y $b \neq 0$.

isosceles / isósceles (196) Triángulo que tiene por lo menos dos lados congruentes.

least common denominator (LCD) / mínimo común denominador (mcd) (261) El menor múltiplo común de los denominadores de dos o más fracciones.

least common multiple (LCM) / mínimo común múltiplo (mcm) (257) El menor múltiplo común no nulo de dos o más números. El mínimo común múltiplo de 2 y 3 es 6.

legs / catetos (398) Los lados que forman el ángulo recto de un triángulo rectángulo.

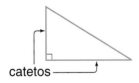

like terms / términos semejantes (565) Expresiones que contienen las mismas variables, como por ejemplo 3*ab* y 7*ab*.

linear function / función lineal (442) Ecuación en que la gráfica de las soluciones es una recta.

line of symmetry / eje de simetría (460) Recta que divide una figura en dos mitades que son reflexiones mutuas.

line plot / esquema lineal (153) Gráfica que usa una recta numérica y un ×, sobre un número en la recta numérica, cada vez que el número aparece en un conjunto de datos.

line symmetry / simetría lineal (206) Exhiben simetría lineal las figuras que coinciden exactamente cuando se doblan.

lower quartile / cuartil inferior (164) Mediana de la mitad inferior de un conjunto de datos, en la amplitud intercuartílica.

markup / margen de utilidad (349) Diferencia entre el precio que paga el comerciante y el precio de venta al consumidor.

matrix / matriz (73) Un arreglo rectangular de elementos en filas y columnas.

mean / media (158) Suma de los números de un conjunto de datos dividida entre el número total de datos.

measures of central tendency / medidas de tendencia central (158) Números que pueden representar el conjunto total de datos.

median / mediana (158, 159) Número central de un conjunto de datos, una vez que los datos han sido ordenados numéricamente. Si hay un número par de datos, la mediana es el promedio de los dos datos centrales.

mixed number / número mixto (278) Suma de un entero y una fracción. $6\frac{2}{3}$ es un número mixto.

mode / modal (158) Número(s) de un conjunto de datos que aparece(n) más frecuentemente.

monomial / monomio (561) Un número, una variable o el producto de un número por una o más variables.

multiple / múltiplo (257) El producto de un número entero por cualquier otro número entero. 28 es múltiplo de 4 y de 7.

multiplication property of equality / propiedad de multiplicación de la igualdad (22) Si cada lado de una ecuación se multiplica por el mismo número, los lados permanecen iguales. Si $a = b$, entonces $ac = bc$.

multiplicative inverse / inverso multiplicativo (290) El producto de un número por su inverso multiplicativo es igual a 1. El inverso multiplicativo de $\frac{2}{3}$ es $\frac{3}{2}$.

numerical expression / expresión numérica (11) Expresión matemática que tiene una combinación de números y por lo menos una operación. 4 + 2 es una expresión numérica.

obtuse / obtuso (196) Ángulo que mide más de 90°, pero menos de 180°.

open sentence / enunciado abierto (13) Ecuación que contiene una variable.

opposite / opuestos (70) Dos enteros son opuestos si, en la recta numérica, están representados por puntos que equidistan de cero, pero en direcciones opuestas. La suma de opuestos es cero.

order of operations / orden de las operaciones (11) Reglas a seguir cuando hay más de una operación involucrada.

1. Primero ejecuta todas las operaciones dentro de los símbolos de agrupamiento, comenzando con los más interiores.
2. Calcula todas las potencias.
3. Luego multiplica y divide, ordenadamente, de izquierda a derecha.
4. Finalmente, suma y resta, ordenadamente, de izquierda a derecha.

ordered pair / par ordenado (92) Par de números que se usa para ubicar un punto en un plano de coordenadas. Se escribe de la siguiente forma: (coordenada x, coordenada y).

origin / origen (92) Punto de intersección axial en un plano de coordenadas.

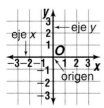

outcome / resultado (253, 518) Uno de los resultados posibles de un experimento probabilístico. Por ejemplo, 4 es un resultado posible cuando se lanza un dado.

outlier / valor atípico (164) Dato o datos que dista(n) de los cuartiles respectivos más de 1.5 veces la amplitud intercuartílica.

parallel lines / rectas paralelas (188) Rectas que yacen en un mismo plano y que no se intersecan. El símbolo ∥ significa paralela a.

parallelogram / paralelogramo (39, 201) Cuadrilátero que posee dos pares de lados paralelos.

Pascal's Triangle / Triángulo de Pascal (528) Arreglo triangular de números en el cual cada número se obtiene sumando los dos números a su izquierda y a su derecha, en la fila superior.

pentagon / pentágono (358) Polígono de cinco lados.

percent / tanto por ciento (107) Razón que compara un número con 100.

percentage / porcentaje (335) Número (P) de una proporción porcentual que se compara con otro número llamado base (B).

percent of change / tanto por ciento de cambio (348) Razón de la cantidad de aumento o disminución comparada con la cantidad original.

percent of decrease / tanto por ciento de disminución (349) Razón de la cantidad de disminución comparada con la cantidad original, escrita como tanto por ciento.

percent of increase / tanto por ciento de aumento (349) Razón de la cantidad de aumento comparada con la cantidad original, escrita como tanto por ciento.

percent proportion / proporción porcentual (335) La proporción $\frac{P}{B} = \frac{r}{100}$ en que P representa el porcentaje, B representa la base y r representa el número por cada 100. También se escribe como $\frac{\text{Porcentaje}}{\text{Base}} = \text{Tasa}$.

perfect square / cuadrado perfecto (382) Número racional cuya raíz cuadrada es un número entero. 25 es un cuadrado perfecto porque $\sqrt{25} = 5$.

perimeter / perímetro (38) La medida del contorno de una figura geométrica cerrada.

permutation / permutación (521) Arreglo o lista en que el orden es importante.

perpendicular / perpendiculares (197) Dos rectas o segmentos de recta que se intersecan formando un ángulo recto.

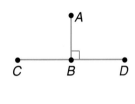

polygon / polígono (196) Figura simple cerrada en un plano, formada por tres o más segmentos de recta.

polynomial / polinomio (562) Suma o diferencia de uno o más monomios.

population / población (546) El grupo total de individuos del cual se toman las muestras que están bajo estudio.

power / potencia (8) Número que puede escribirse usando un exponente. La potencia 7^3 se lee *siete a la tercera potencia* o *siete al cubo*.

precision / precisión (504) La precisión de una medición depende de la unidad de medida. Mientras más pequeña sea la unidad de medida, más precisa es la medición.

prime factorization / factorización prima (235) Manera de expresar un número compuesto como producto de números primos. La factorización prima de 63, por ejemplo, es $3 \times 3 \times 7$.

prime number / número primo (235) Número entero mayor que 1 que sólo tiene dos factores, 1 y sí mismo.

principal / capital (353) Cantidad de dinero invertido o adeudado.

principal square root / raíz cuadrada principal (382) La raíz cuadrada no negativa. $\sqrt{49}$ indica la raíz cuadrada *principal* de 49.

prism / prisma (483) Figura tridimensional que tiene dos bases poligonales paralelas y congruentes.

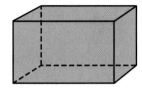

probability / probabilidad (253) La posibilidad de que suceda un evento. Es la razón del número de maneras en que puede ocurrir un evento al número total de resultados posibles.

proportion / proporción (111) Ecuación que afirma la igualdad de dos o más razones, $\frac{a}{b} = \frac{c}{d}$, $b \neq 0$, $d \neq 0$.

pyramid / pirámide (483) Figura sólida que tiene una base poligonal y caras triangulares.

Pythagorean Theorem / Teorema de Pitágoras (396–398) En un triángulo rectángulo, el cuadrado de la longitud de la hipotenusa es igual a la suma de los cuadrados de las longitudes de los catetos. $c^2 = a^2 + b^2$

Pythagorean triple / triplete pitagórico (405) Conjunto de tres enteros que satisfacen el Teorema de Pitágoras.

quadrant / cuadrante (92) Una de las cuatro regiones en que dos rectas perpendiculares dividen un plano.

quadratic function / función cuadrática (452) Función polinómica en que la potencia mayor de la variable es 2.

quadrilateral / cuadrilátero (201) Polígono de cuatro lados.

quartiles / cuartiles (163) Valores que dividen un conjunto de datos en cuatro partes iguales.

R

radical sign / signo radical (382) Símbolo con que se indica la raíz cuadrada no negativa. $\sqrt{}$

radius / radio (309) Distancia desde el centro de un círculo hasta cualquier punto del círculo.

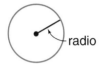

random / al azar (253) Los resultados ocurren aleatoriamente o al azar si cada resultado tiene la misma posibilidad de ocurrir y es imposible predecirlos.

range / rango (428) Conjunto de los valores de salida de una función.

range / rango (163) Diferencia entre los valores máximo y mínimo de un conjunto de datos.

rate / tasa (105) Razón de dos mediciones que tienen distintas unidades de medida.

rate / tasa (335) Razón de un número a 100 en una proporción porcentual.

ratio / razón (104) Comparación de dos números mediante división. La razón 2 a 3 se puede escribir como 2 de cada 3, 2 a 3, 2:3 ó $\frac{2}{3}$.

rational numbers / números racionales (245) Números de la forma $\frac{a}{b}$, donde a y b son enteros y $b \neq 0$.

real numbers / números reales (391) Conjunto de los números racionales junto con el conjunto de los números irracionales.

reciprocal / recíproco (290) El inverso multiplicativo de un número.

rectangle / rectángulo (38, 201) Cuadrilátero cuyos cuatro ángulos son congruentes entre sí.

reflection / reflexión (206, 460) Transformación en que a una figura se le da vuelta de campana por encima de un eje de simetría.

relative error / error relativo (504) Comparación del error máximo posible con la medida misma.

$$\text{error relativo} = \frac{\text{error máximo posible}}{\text{medida}}$$

repeating decimal / decimal periódico (249) Decimal en que los dígitos, en algún momento, comienzan a repetirse en bloques de uno o más números. Por ejemplo, 0.181818... y 0.8333... . Usando notación de barra, estos decimales se escriben $0.\overline{18}$ y $0.8\overline{3}$.

rhombus / rombo (201) Paralelogramo cuyos lados son todos congruentes entre sí.

right / recto (196) Ángulo que mide 90°.

rotation / rotación (206, 464) Cuando se gira una figura en torno a un punto central.

rotational symmetry / simetría rotacional (207) Una figura posee simetría rotacional si se puede girar menos de 360° en torno a su centro, sin que esto cambie su apariencia con respecto de la figura original.

row / fila (73) Los números que están horizontalmente uno al lado del otro en una matriz.

S

sample / muestra (174, 546) Grupo escogido aleatoriamente con el objeto de recoger información.

sample space / espacio muestral (253) Conjunto de todos los resultados posibles de un experimento probabilístico.

scale / escala (366) Razón de una longitud dada en un dibujo o modelo a su longitud real.

scale drawing / dibujo a escala (366) Dibujo que es semejante, pero más grande o más pequeño que el objeto real.

scale factor / factor de escala (370) Razón de la imagen dilatada a la imagen original.

scale model / modelo a escala (366) Réplica de un objeto real el cual es demasiado grande o demasiado pequeño como para construirlo de tamaño natural.

scalene / escaleno (196) Triángulo sin ningún par de lados congruentes.

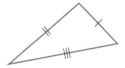

scatter plot / gráfica de dispersión (168) Gráfica que muestra la relación general entre dos conjuntos de datos.

scientific notation / notación científica (265) Manera de expresar números como el producto de un número que es al menos igual a 1, pero menor que 10, por una potencia de diez. En notación científica, $5,500 = 5.5 \times 10^3$.

selling price / precio de venta (349) Dinero que paga un consumidor por un artículo.

sequence / sucesión (296) Lista de números en cierto orden, como, por ejemplo, 0, 1, 2, 3 ó 2, 4, 6, 8.

significant digits / dígitos significativos (504) Todos los dígitos de una medición que se sabe que son exactos, más un dígito aproximado.

similar polygons / polígonos semejantes (357) Dos polígonos son semejantes si sus ángulos correspondientes son iguales y sus lados correspondientes son igualmente proporcionales. Tienen la misma forma, pero pueden no tener el mismo tamaño.

similar triangles / triángulos semejantes (215) Triángulos que tienen la misma forma, pero no necesariamente el mismo tamaño.

simplest form / forma reducida (246) La forma de una fracción en la cual el MCD de su numerador y denominador es 1.

simplest form / forma reducida (565) Una expresión que carece de términos semejantes. La forma reducida de $6x + 8x$ es $14x$.

simulation / simulación (540) Proceso de representar un problema.

solid / sólido (482) Figura tridimensional.

solution / solución (13) Valor de la variable de una ecuación que hace verdadera la ecuación. La solución de $10 + y = 25$ es 15.

spreadsheet / hoja de cálculos (162) Herramienta que se usa para organizar y analizar datos.

square / cuadrado (38, 201) Paralelogramo con todos los lados, y también los ángulos, congruentes entre sí.

square root / raíz cuadrada (382) Uno de dos factores iguales de un número. Si $a^2 = b$, entonces a es una raíz cuadrada de b. La raíz cuadrada no negativa de 144 es 12 porque 12 es un número no negativo y $12^2 = 144$.

statistics / estadística (142) Rama de las matemáticas que trata de la recopilación, organización y análisis de datos.

substitute / sustituir (11) Reemplazar una variable en una expresión algebraica con un número, obteniendo así una expresión numérica.

subtraction property of equality / propiedad de sustracción de la igualdad (18) Si sustraes el mismo número de ambos lados de una ecuación, los lados permanecen iguales. Para números a, b y c cualesquiera, si $a = b$, entonces $a - c = b - c$.

supplementary angles / ángulos suplementarios (190) Dos ángulos son suplementarios si la suma de sus medidas es 180°.

surface area / área de superficie (495) Suma de las áreas de todas las superficies de una figura tridimensional.

symmetric / simétricas (460) Figuras que se pueden doblar en dos partes exactamente correspondientes.

system of equations / sistema de ecuaciones (446) Conjunto de ecuaciones que tienen las mismas variables.

T

term / término (565) Número, variable o producto de números y variables.

term / término (296) Número de una sucesión.

terminating decimal / decimal terminal (249) Decimal cuyos dígitos terminan. Todo decimal terminal puede escribirse como una fracción con un denominador de 10, 100, 1,000, etc.

tessellation / teselado (220) Un patrón repetitivo de polígonos que coinciden perfectamente, sin dejar huecos o espacios.

theoretical probability / probabilidad teórica (540) La probabilidad a largo plazo de un resultado o evento, que se basa en principios matemáticos.

transformation / transformación (220) Movimientos de figuras geométricas con el propósito de modificar teselados.

translation / traslación (220, 456) Método que se usa para hacer cambios en los polígonos que integran un teselado, deslizando una figura horizontal o verticalmente o de ambas maneras.

transversal / transversal (188) Recta que interseca dos rectas paralelas formando así ocho ángulos.

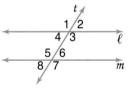

trapezoid / trapecio (201, 302) Cuadrilátero con un único par de lados paralelos.

tree diagram / diagrama de árbol (518) Diagrama que se usa para encontrar y mostrar el número total de resultados posibles de un experimento probabilístico.

triangle / triángulo (196) Polígono de tres lados y con tres ángulos.

trigonometry / trigonometría (365) Estudio de los triángulos.

trinomial / trinomio (588) Polinomio de tres términos.

U

unit rate / tasa unitaria (105) Tasa que tiene un denominador de 1.

upper quartile / cuartil superior (163) La mediana de la mitad superior de un conjunto de números.

V

variable / variable (12) Un símbolo, por lo general, una letra, que se usa para representar números en expresiones o enunciados matemáticos.

variation / variación (163) Dispersión de los valores de un conjunto de datos.

vertex / vértice (483) El vértice de un prisma es el punto en que se intersecan todas las caras del prisma.

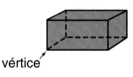

vertical angles / ángulos opuestos por el vértice (190) Ángulos congruentes que se forman de la intersección de dos rectas. En la figura, los ángulos opuestos por el vértice son ∠1 y ∠3, así como ∠2 y ∠4.

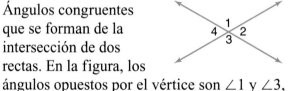

volume / volumen (486) Número de unidades cúbicas que se requieren para llenar el espacio que ocupa un sólido.

X

x-axis / eje x (92) La recta numérica horizontal que ayuda a formar el sistema de coordenadas.

x-coordinate / coordenada x (92) Primer número de un par ordenado.

Y

y-axis / eje y (92) La recta numérica vertical que ayuda a formar el sistema de coordenadas.

y-coordinate / coordenada y (92) Segundo número de un par ordenado.

Z

zero pair / par nulo (63) Resultado de aparear una ficha positiva con una negativa.

Index

Absolute value, 57

Act it out, 538–539

Acute triangles, 196

Addition
equations, 13–15, 17–20
of fractions, 278–284
of integers, 62–68
of matrices, 74
of polynomials, 570–572

Addition property of equality, 18

Additive inverse, 70
using to subtract integers, 70

Algebra
equations, 13–15, 17–25, 32–36,
84–87, 315–317, 339–341, 383,
392, 437–440
evaluating expressions, 8–16, 25,
27–29, 61, 64, 67, 71, 80, 93,
236–237, 280–281, 283, 287,
291, 569, 581
functions, 37, 428–471
solving inequalities, 58, 117
variables, 12
writing equations, 36, 42,
202–203, 256
writing expressions, 68, 106
writing inequalities, 238, 523
See also Applications,
Connections, and Integration
Index on pages xxii–1

Algebraic expression, 12, 27

Algebra tiles
area models from, 560

Alternate exterior angles, 189

Alternative Assessment, 51, 99, 133,
181, 227, 271, 325, 377, 421, 471,
511, 553, 595

Altitude
of circular cones, 490
of parallelograms, 40
of triangles, 301

Angles, 186
alternate exterior, 189
alternate interior, 189
complementary, 198
congruent, 210
corresponding, 189
drawing, 187
measuring, 187
supplementary, 190
vertical, 190

Applications, 2e, 54e, 102e, 138e,
184e, 230e, 276e, 328e, 380e,
426e, 474e, 514e, 558e

Applications, *See* Applications,
Connections, and Integration Index
on pages xxii–1

Appropriate display, 171–173

Area
of circles, 476–479
models for, 40, 476
models with algebra tiles, 560
of parallelograms, 40
Pick's Theorem, 305–306
and probability, 478
of rectangles, 39–40, 588–589
of squares, 40
surface, 495–503
of trapezoids, 302
of triangles, 301–302
using to factor polynomials, 588

Arithmetic sequence, 296–297
common difference, 296

ASA congruence, 210

Assessment Resources, 2c, 2f, 49,
50, 54c, 54f, 97, 98, 102c, 102f,
131, 132, 138c, 138f, 179, 180,
184c, 184f, 225, 226, 230c, 230f,
269, 270, 276c, 276f, 323, 324,
328c, 328f, 375, 376, 380c, 380f,
419, 420, 426c, 426f, 469, 470,
474c, 474f, 509, 510, 514c, 514f,
551, 552, 558c, 558f, 593, 594

Associative property, 8
using to add integers, 66
using to add and multiply rational
numbers, 290

Auditory/Musical Learning Style,
see Multiple Learning Styles

Average, 158

Bar graph, 142–146
histograms, 142–146
interpreting, 4, 129, 144, 157,
171–177, 195, 209, 234, 505,
506, 541, 561, 587
making, 142–143, 394
misleading, 174–176
scale, 142

Bar notation, 249

Base, 8
of circular cones, 490
of circular cylinders, 487
of a parallelogram, 40
percent, 335
of prisms, 483
of trapezoids, 302
of triangles, 301

Binomial, 583
multiplying, 583–585

Box-and-whisker plot, 167
constructing with a graphing
calculator, 167
extreme values, 167
making, 167
outliers, 167
quartiles, 167
lower, 167
upper, 167
scale, 167

Calculator, 9, 12, 114, 118, 127, 136,
236, 250, 252, 260, 266, 274, 339,
365, 394, 424, 447, 453, 477, 503,
522, 532, 533, 556, 578
graphing, 16, 77, 147, 167, 413,
432, 445, 542
pi on, 307, 309
scientific, 578

Calculators, *see Technology*

Capacity, 75

Careers
Computer Game Designer, 436
Former NBA Scout, 582
Graphic Designer, 219
Museum Director, 152
Nutritionist, 549
Pathologist, 293
Police Officer, 395
World Wide Web Inventor, 26

CD-ROM Program, *see Technology*

Center
of a circle, 307, 309

Chapter Projects
Be True to Your School, 185
Consider the Probabilities, 515
Games People Play, 427
Hit or Miss, 559
Home Page Bound, 3
Oldies But Goodies!, 139
Pack Your Bags, 103

Patterns in Nature, 277
Proceed with Caution, 381
Stay Tuned!, 329
Take Me Out to the Ball Game, 231
Up, Up, and Away, 475
Where the Wild Things Are, 55
See also Working on the Chapter Project

Chapter Resources, *see Organizing the Chapter*

Charts, 27, 31, 249, 260, 294, 300, 346, 388, 424, 503

Circle, 307, 309
area of, 476–479
center, 307, 309
circumference, 307, 309
diameter, 307, 309
pi, 307, 309
radius, 307, 309

Circle graphs, 148–151
interpreting, 108, 109, 151, 171–177
making, 126, 127, 129, 148–151, 403, 463
using a calculator to construct, 148

Circumference, 307, 309–310
formula for, 309

Classroom Games, 2e, 54e, 102e, 138e, 184e, 230e, 276e, 328e, 380e, 426e, 474e, 514e, 558e

Classroom Vignettes, 17, 39, 62, 94, 115, 142, 171, 196, 206, 232, 237, 278, 281, 357, 363, 382, 398, 465, 482, 540, 566

Closing Activity
Modeling, 20, 25, 31, 36, 68, 76, 89, 106, 110, 141, 170, 177, 199, 218, 234, 248, 256, 280, 295, 299, 338, 364, 369, 384, 413, 435, 444, 459, 479, 481, 523, 537, 564, 569, 587
Speaking, 10, 42, 47, 58, 65, 72, 80, 113, 129, 146, 161, 166, 192, 195, 204, 223, 244, 252, 264, 289, 292, 311, 317, 341, 351, 356, 373, 389, 417, 431, 455, 463, 489, 498, 507, 531, 548, 585, 591
Writing, 7, 15, 29, 61, 83, 91, 95, 117, 122, 125, 151, 155, 173, 209, 212, 238, 241, 259, 267, 284, 304, 314, 321, 333, 343, 347, 360, 394, 401, 403, 407, 440, 449, 451, 467, 485, 493, 520, 527, 539, 543, 572, 575, 581

Column, 73

Combinations, 524–525

Common difference, 296

Common multiple, 258

Common ratio, 297

Communication, *see Motivating the Lesson*

Commutative property, 8, 290
using to add integers, 66
using to add and multiply rational numbers, 290

Comparing
decimals, 261
fractions, 261
integers, 59–61
percents, 121
rational numbers, 261–264

Compatible numbers, 126

Complementary angles, 198

Composite number, 235

Compound events, 534–537
dependent events, 535–536
independent events, 534–535
probability of, 534–537

Computers, 3, 55, 103, 136, 139, 162, 185, 231, 274, 277, 329, 334, 352, 381, 427, 475, 515, 559
to construct graphs, 127, 274, 515, 556
Internet, 15, 26, 47, 55, 61, 62, 76, 113, 122, 123, 137, 139, 146, 152, 170, 185, 212, 218, 219, 231, 238, 247, 275, 277, 285, 293, 316, 329, 333, 369, 381, 389, 394, 395, 425, 427, 430, 431, 436, 455, 475, 498, 506, 515, 537, 543, 549, 557, 559, 561, 582, 591
software, 3, 149, 185, 231, 277, 425, 427, 475, 559
spreadsheet, 55, 71, 103, 136, 139, 149, 162, 231, 252, 274, 295, 305, 329, 334, 352, 425, 475, 515, 556, 559
See also Applications, Connections, and Integration Index on pages xxii–1

Cone, 482
altitude of, 490
base of, 490
circular, 482
volume of, 490–491

Congruent
angles, 210
figures, 210
sides, 210

Congruent triangles, 210–212
ASA, 210
constructing, 213
corresponding parts of, 210
SAS, 210
SSS, 210

Connections, *See* Applications, Connections, and Integration Index on pages xxii–1

Constructions
of congruent line segments, 186
of congruent triangles, 213

Converse, 399
of Pythagorean Theorem, 399

Cooperative Learning, 136, 274, 424, 425, 556, 557

Cooperative Learning, 37, 84, 85, 118, 119, 156, 186, 187, 193, 200, 205, 213, 214, 239, 260, 300, 305, 306, 307, 308, 365, 385, 396, 397, 408, 409, 441, 494, 503, 516, 517, 532, 533, 544, 545, 560, 577

Coordinate plane, 92
distance on, 410–413

Coordinates, 56
on number line, 56

Coordinate system, 92
graphing points on, 92–93
ordered pairs, 92
origin, 92
quadrants, 92
transformations on, 456–467
x-axis, 92
x-coordinate, 92
y-axis, 92
y-coordinate, 92

Corresponding angles, 189

Corresponding parts, 210

Counting Principle, 518

Critical Thinking, 7, 10, 15, 20, 24, 36, 42, 47, 58, 61, 65, 68, 72, 76, 80, 83, 88, 95, 106, 110, 113, 117, 122, 129, 146, 151, 155, 161, 166, 170, 173, 177, 192, 199, 204, 209, 212, 218, 223, 234, 238, 241, 244, 248, 252, 256, 259, 264, 267, 280, 284, 289, 292, 299, 304, 311, 314, 317, 321, 333, 338, 347, 351, 356, 360, 364, 369, 373, 384, 388, 393, 401, 407, 413, 417, 431, 435, 440, 444, 449, 455, 459, 463, 467, 479, 485, 489, 493, 498, 502, 507, 520, 523, 527, 531, 537, 543, 548, 564, 569, 572, 575, 581, 585, 591

Cross-Curriculum Cue, 4, 63, 108, 148, 188, 249, 286, 330, 390, 437, 504, 518, 561

Cross products, 111, 330

Index **723**

Cultural Kaleidoscope, 9, 69, 121, 163, 196, 245, 361, 398, 464, 505, 518, 565

Customary system, 186

Cylinder, 482
  base of, 487
  circular, 482
  height of, 487
  surface area of, 499–502
  volume of, 487–488

Data
  interquartile range, 164
  outliers, 164
  quartiles, 163
    upper, 163
    lower, 164
  spread of, 163
  variation of, 163

Data analysis, 154
  See also Applications, Connections, and Integration Index on pages xxii–1

Decimals
  bar notation, 249
  comparing, 261
  fractions as, 249, 261
  as fractions, 250–251
  as mixed numbers, 250
  ordering, 261
  percents as, 114, 121
  as percents, 115, 121
  repeating, 249–250
  terminating, 249–250

Decision Making, 19, 28, 60, 82, 122, 128, 144, 160, 198, 216, 233, 251, 298, 316, 355, 367, 392, 429, 453, 478, 501, 526, 536, 589

Deductive reasoning, 294

Denominator
  least common, 261

Density property, 260

Dependent events, 535–536

Diameter
  of a circle, 307, 309

Did You Know?, 5, 19, 59, 73, 107, 160, 206, 235, 290, 301, 313, 335, 349, 382, 411, 438, 477, 500, 540, 570, 583

Dilations, 214, 370–373
  scale factor, 370

Discounts, 349
  rate, 352
  using spreadsheets, 352

Distributive property, 34, 291
  using to find products involving mixed numbers, 291
  using to multiply polynomials by monomials, 579
  using to multiply two binomials, 583

Diversity, 2e, 54e, 102e, 138e, 184e, 230e, 276e, 328e, 380e, 426e, 474e, 514e, 558e

Divisibility patterns, 232–234

Divisible, 232

Division
  divisibility rules, 232–234
  equations, 13–15, 21–25
  estimating,
  of fractions, 312–314
  of integers, 81–84

Division property of equality, 21

Domain, 428

Draw a Diagram, 402–403

Edges, 200, 483

Element, 73

Eliminate possibilities, 90–91

Ellipses, 56

Enhancing the Chapter, 2e, 2f, 54e, 54f, 102e, 102f, 138e, 138f, 184e, 184f, 230e, 230f, 276e, 276f, 328e, 328f, 380e, 380f, 426e, 426f, 474e, 474f, 514e, 514f, 558e, 558f

Equally-likely, 253

Equations, 13
  addition, 13–15, 17–20
  division, 13–15, 21–25
  with fractions, 315–317
  functions written as, 428
  with integer solutions, 84–89
  with irrational solutions, 392
  modeling, 17, 21, 23, 84–85
  multiplication, 13–15, 21–25
  open sentence, 13
  percent, 339–341
  replacement set for, 13
  solutions of, 13
  solving, 13–15, 17–25, 32–36, 84–89, 315–317, 383
  square root, 383
  subtraction, 13–15, 17–20
  two-step, 32–36
  with two variables, 437–440
  using guess-and-check to solve, 45
  writing, 28–29

Equilateral triangles, 196

Error
  greatest possible, 504
  relative, 504

Error Analysis, *see Reteaching the Lesson*

Escher, M. C., 220

Estimation
  area of irregular figures, 126
  with percents, 126–129
  products of fractions, 317
  of square roots, 385–389
  study hints, 149, 358, 399, 443, 465
  using compatible numbers, 126
  using to sketch a circle graph, 126–127

Evaluate, 11–12

Evaluation, *see Assessment Resources*

Events, 253
  compound, 534–537
  dependent, 535–536
  equally–likely, 253
  independent, 534–535

Experimental probability, 540–543
  simulation, 540

Exponents, 8–10
  evaluating expressions containing, 9
  and rational numbers, 287
  writing expression using, 8

Expressions
  algebraic, 12, 27
  evaluating, 9, 11–16
  exponents in, 8–10
  grouping symbols in, 11
  numerical, 11
  substitution in, 11
  using graphing calculators to evaluate, 16
  variables in, 12
  writing, 27–29

Extending the Lesson Activity, 7, 10, 15, 25, 29, 36, 42, 58, 65, 68, 72, 80, 83, 91, 95, 106, 110, 113, 116, 122, 125, 129, 141, 155, 161, 166, 170, 173, 177, 192, 195, 199, 204, 209, 212, 223, 234, 241, 244, 248, 252, 256, 259, 264, 280, 285, 292, 295, 299, 304, 311, 314, 317, 338, 341, 343, 347, 351, 356, 360, 373, 384, 394, 401, 403, 407, 413, 417, 435, 444, 449, 451, 459, 463, 467, 479, 481, 485, 489, 493, 520, 523, 527, 531, 537, 548, 564, 569, 572, 576, 581, 585, 587

F

Faces, 483

Factorial, 522
   on a calculator, 522

Factoring, 588
   trinomials, 588–589

Factors, 8
   greatest common, 242–244
   scale, 370
   of trinomials, 588–589

Factor trees, 236

Fair games, 516

Family Activities, 29, 94, 117, 145, 208, 266, 304, 340, 407, 467, 497, 542, 585

Family Letters and Activities, 2e, 54e, 102,e, 138e, 184e, 230e, 276e, 328e, 380e, 426e, 474e, 514e, 558e
   Math and the Family, 2, 54, 102, 138, 184, 230, 276, 328, 380, 426, 474, 514, 558
   Family Activity, 29, 94, 116, 145, 208, 267, 304, 341, 407, 467, 497, 539, 584

Fibonacci sequence, 300

Figurate numbers, 382

5-Minute Checks, 4, 8, 11, 17, 21, 27, 32, 38, 43, 56, 59, 62, 66, 69, 73, 78, 81, 86, 92, 104, 107, 111, 114, 120, 126, 142, 148, 153, 158, 163, 168, 171, 174, 188, 196, 201, 206, 210, 215, 220, 232, 235, 242, 245, 249, 253, 257, 261, 265, 278, 281, 286, 290, 296, 301, 309, 312, 315, 318, 330, 335, 339, 344, 348, 353, 357, 361, 366, 370, 382, 386, 390, 398, 404, 410, 414, 428, 433, 437, 442, 446, 452, 456, 460, 464, 476, 482, 486, 490, 495, 499, 504, 518, 521, 524, 528, 534, 540, 542, 546, 561, 565, 570, 573, 578, 583, 588

Formulas
   for area of circles, 476
   for area of parallelograms, 40
   for area of rectangles, 40
   for area of squares, 40
   for area of trapezoids, 302
   for area of triangles, 302
   for circumference of circles, 309
   for distance, 21
   for perimeter of parallelograms, 39
   for perimeter of rectangles, 38
   for perimeter of squares, 38
   for simple interest, 353
   in spreadsheets, 71, 162, 295, 305, 334, 352
   for surface area of cylinders, 500
   for volume of cones, 491
   for volume of cylinders, 487
   for volume of prisms, 486
   for volume of pyramids, 491

Four-step plan, 4–7, 34, 79, 87, 127, 197, 243, 262, 282, 300–331, 340, 354, 404, 467–468, 477, 496, 525, 529, 574, 579, 589

Fractions, 245
   adding like, 278–280
   adding unlike, 281–284
   comparing, 261
   decimals as, 250–251
   as decimals, 249, 261
   dividing, 312–314
   least common denominator, 261
   like, 278
   modeling, 278
   multiplicative inverse, 312
   multiplying, 286–289
   ordering, 261
   as percents, 107–108, 115, 121
   percents as, 107–108, 114, 121
   renaming, 281–282
   simplest form, 246
   subtracting like, 278–280
   subtracting unlike, 281–284

Frequency tables, 140–141

Functions, 37, 428–471
   domain, 428
   graphing, 433–435, 442–455
   input, 37, 428
   linear, 442, 445
   machines, 37
   notation, 433
   output, 37, 428
   quadratic, 452
   range, 428
   rule, 428
   tables, 428, 432
   written as equations, 428

Function tables, 428, 432
   using to graph, 433–435

Fundamental Theorem of Arithmetic, 236

G

Games
   fair, 516–517
   unfair, 516–517
   See also Let the Games Begin

Geometric sequence, 296–297
   common ratio, 297

Geometry
   30°-60° right triangle, 365, 414–415
   45°-45° right triangle, 365, 415–416
   acute triangles, 196
   alternate exterior angles, 189
   alternate interior angles, 189
   altitude, 40, 301, 490
   angles, 186–187
   area, 39–40, 301–302, 476–479, 588–589
   ASA congruence, 210
   base, 40, 301–302, 335, 483, 487, 490
   complementary angles, 198
   congruent triangles, 210–212
   construction of congruent line segments, 186
   construction of congruent triangles, 213
   converse of Pythagorean Theorem, 399
   corresponding angles, 189, 357
   corresponding parts, 210, 357
   dilations, 214, 370–373
   distance on the coordinate plane, 410–413
   equilateral triangles, 196
   M. C. Escher, 220
   height, 40, 301, 302
   hypotenuse, 398
   isosceles triangles, 196
   legs, 398
   line of symmetry, 206, 460
   line segments, 186
   line symmetry, 206
   networks, 200
   obtuse triangle, 196
   parallel lines, 188–192
   parallelogram, 39–40, 201
   pentagon, 358
   perimeter, 38–40
   perpendicular, 197
   polygon, 196, 357
   Pythagorean Theorem, 396–407
   Pythagorean triples, 405
   quadrilaterals, 201–204
   rectangle, 38–40, 201
   reflections, 205, 206
   rhombus, 201
   right angles, 197
   right triangles, 196
   rotation, 207, 221
   rotational symmetry, 207
   SAS congruence, 210
   scale, 366, 370
   scale drawing, 366–369
   scale factor, 370
   scale model, 366–369
   scalene triangles, 196
   similar polygons, 357–360
   similar triangles, 215–218
   slide, 220
   square, 38–40, 201
   SSS congruence, 210

supplementary angles, 190
surface area, 495–502
symmetry, 206–209
tessellation, 220–223
transformations, 220–223
translations, 220
transversal, 188
trapezoid, 201
triangles, 196–199
trigonometry, 365
vertical angles, 190
volume, 486–491
See also Applications, Connections, and Integration Index on pages xxii–1

Golden ratio, 118–119

Graphing calculator, 16, 77, 147, 167, 413, 432, 445, 542

Graphs
bar, 142–146
circle, 148–151
coordinate, 92–93, 410–413, 456–467
functions, 433–435, 442–455
histogram, 142–147
integers on the coordinate system, 92–93
integers on number line, 56–60
interpreting, 69, 83, 110, 122, 154, 158, 168–177, 248, 299, 336, 341, 343, 345, 347, 351, 403, 450, 451
irrational numbers, 408–409
line, 570
linear functions, 442–445
line plot, 153–155
making, 142–146, 153–155, 167–170, 295, 515
misleading, 174–177
number line, 44–45, 56–57, 318–319
ordered pairs, 92–93
points, 92–93
points on the coordinate plane, 92
points on number line, 56
quadratic functions, 452–455
reflections, 460–463
relationships, 441
rotations, 464–467
scatter plot, 168–170
systems of equations, 446–449
translations, 456–459
using function tables, 433–435

Greatest common factor, 242–244
using prime factorization to find, 243
using to simplify fractions, 246

Greatest possible error, 504

Guess and check, 586–587

Hands-On Activity, *see Motivating the Lesson*

Hands-on Labs
algebra tiles, 560
area and Pick's theorem, 305–306
basketball factors, 239
constructing congruent triangles, 213
constructing parallel lines, 193
density property, 260
dilations, 214
estimating square roots, 385
fair and unfair games, 516–517
the Fibonacci sequence, 300
function machines, 37
the golden ratio, 118–119
graphing irrational numbers, 408–409
graphing pi, 307–308
graphing relationships, 441
maps and statistics, 156–157
measuring and constructing line segments and angles, 186–187
modeling products, 577
nets, 494
patterns in Pascal's triangle, 532–533
polygons as networks, 200
Punnett squares, 544–545
the Pythagorean Theorem, 396–397
reflections, 205
solving equations, 84
surface area and volume, 503
trigonometry, 365

Hands-On Math, 19, 23, 35, 46, 64, 80, 116, 154, 169, 191, 203, 208, 216, 222, 237, 254, 258, 279, 283, 288, 303, 346, 350, 359, 372, 412, 416, 443, 448, 458, 488, 497, 501, 523, 526, 530, 541, 567, 580, 584, 590

Hands-On Mini-Labs, 17, 21, 32, 40, 44, 63, 69, 78, 107, 127, 153, 168, 188, 190, 202, 207, 215, 220, 235, 257, 278, 286, 301, 344, 348, 358, 371, 410, 414, 415, 442, 447, 456, 460, 464, 486, 490, 495, 499, 521, 524, 529, 535, 540, 566, 570, 573, 583, 588

Height
of a parallelogram, 40
of trapezoids, 302
of triangles, 301

Hexagonal crystal, 482

Histogram, 142–147
intervals, 143

using a graphing calculator to make, 147

Hypotenuse, 398

Improper fractions, 245

Independent events, 534–535

Indirect measurement, 361–364

Inductive reasoning, 294

Inequalities, 43
graphing solutions of, 318–319
on a number line, 44–45
solving, 318–321
symbols for, 43, 59
writing, 44–45

Input, 428

Integers
absolute value of, 57
adding, 62–68
comparing, 59–61
dividing, 81–83
in equations, 86–87
graphing, 56–57
in matrices, 74–75
modeling addition of, 62–63
modeling multiplication of, 78
modeling subtraction of, 69–70
multiplying, 78–80
negative, 56
on number line, 56–60
opposite of, 70
ordering, 60–61
positive, 56
in real number system, 390
subtracting, 69–72

Integration, *See* Applications, Connections, and Integration Index on pages xxii–1

Interactive Mathematics: Activities and Investigations, 2d, 54d, 102d, 138d, 184d, 230d, 276d, 328d, 380d, 426d, 474d, 514d, 558d

Interdisciplinary Investigations
Extra! Extra! Newspapers May Take Over the Planet!, 556–557
Math at the Mall, 424–425
What in the World is "WYSIWYG?", 274–275
What's for Dinner?, 136–137

Interest, 353
simple, 353–356

Internet Connections, 3, 15, 26, 47, 55, 61, 62, 76, 103, 122, 123, 137,

139, 146, 152, 170, 185, 212, 218, 219, 231, 238, 247, 275, 277, 285, 293, 316, 329, 333, 356, 369, 381, 389, 394, 395, 425, 427, 430, 431, 436, 455, 475, 498, 506, 515, 537, 543, 549, 557, 559, 561, 582, 591

Interpersonal Learning Style, see *Multiple Learning Styles*

Interquartile range, 164

Intrapersonal Learning Style, see *Multiple Learning Styles*

Inverse
additive, 70
multiplicative, 290
operation, 18, 21

Inverse property of multiplication, 290

Investigations, *See* Interdisciplinary Investigations

Investigations for the Special Education Student, 12, 87, 126, 175, 190, 242, 296, 335, 415, 464, 477, 534, 573

Irrational numbers, 391
approximating, 391
graphing, 408–409
on number line, 391
in real number system, 391

Isometric crystal, 482

Isosceles triangles, 196

Kinesthetic Learning Style, see *Multiple Learning Styles*

Labs, *See* Hands-On Labs, Problem Solving Labs, Technology Labs, and Thinking Labs

Least common denominator, 261
using to rename fractions, 283

Least common multiple, 257–259
using prime factorization to find, 258

Legs, 398

Leonardo, 300

Lesson Objectives, 2a, 54a, 102a, 138a, 230a, 276a, 328a, 380a, 426a, 474a, 514a, 558a

Lesson Planning Guide, see *Organizing the Chapter*

Let the Games Begin
Absolutely, 61
Architest, 498
Collecting Factors, 238
Estimate and Eliminate, 389
Factor Challenge, 591
Fraction Track, 285
Guess My Rule, 431
Map Sense, 369
Matrix Madness, 76
Parabola Hit or Miss, 455
Per-fraction, 123
Shape Relations, 218
What's Missing?, 333
Win the Lottery, 543
You're the Greatest, 47
You're the Winner...Bar None!, 146

Like fractions,
adding, 278–280
subtracting, 278–280

Like terms, 565

Linear functions, 442, 445
graphing, 442–444
on graphing calculator, 445

Line graphs
interpreting, 570

Line plots, 153–155
data analysis using, 154

Lines
parallel, 188–192
of symmetry, 206

Line segments, 186
constructing congruent, 186
measuring, 186

Line symmetry, 206

Line of symmetry, 206, 460

Logical Learning Style, *see Multiple Learning Styles*

Look for a pattern, 294–295

Lower quartile, 164

Make a List, 240–241

Make a Model, 480–481

Make a Table, 140–141

Making the Connection
Math Connections, 26, 152, 219, 293, 395, 436, 549, 582

Interdisciplinary Connections, 137, 275, 425, 557

Manipulatives, 2c, 2f, 54c, 54f, 102c, 102f, 138c, 138f, 184c, 184f, 230c, 230f, 276c, 276f, 328c, 328f, 380c, 380f, 426c, 426f, 474c, 474f, 514c, 514f, 558c, 558f

Markup, 349
percent of increase, 349

Math Journal, 16, 37, 77, 85, 119, 147, 157, 162, 167, 187, 193, 200, 205, 213, 214, 239, 260, 300, 306, 308, 334, 352, 365, 385, 397, 409, 432, 441, 445, 494, 503, 517, 533, 545, 560, 577

Mathematical Techniques
estimation, 126–129, 149, 358, 399, 443, 465
mental math, 108, 111, 115, 120, 249, 262, 330, 478, 525
number sense, 311, 317
See also Problem Solving

Mathematical Tools
paper/pencil, 9, 120, 477
real objects
algebra tiles, 385, 561–577, 579, 580, 583–585, 588, 590, 591
base-ten blocks, 385
compass, 127, 149, 186, 193, 213, 408, 409, 414, 490
counters, 17, 21, 32, 63, 69, 78, 84, 535, 544, 545
cubes, 486, 488
dot paper, 305, 396, 397, 482–485
equation mat, 17, 21, 32, 84, 85
geoboard and geobands, 205, 305, 396, 397
geomirror, 460
grid paper, 40, 107, 118, 168, 235, 238, 274, 278, 300, 301, 307, 308, 344, 348, 371, 410, 424, 442, 447, 455, 456, 460, 495
hexagonal grid, 532–533
integer mat, 63, 69, 78
number cubes, 255, 280, 516, 517
product mat, 577, 583
protractor, 186, 188, 190, 202, 358, 365, 414, 415, 464
ruler, 168, 186, 202, 214, 286, 358, 365, 410, 414, 415, 441, 442, 490, 494, 499, 556
spinner, 253–255, 455, 520, 536, 537
straightedge, 188, 190, 193, 200, 213, 371, 408, 409, 447, 456, 460

Index **727**

tape measure, 118, 153, 168, 424
technology, *See* Calculators, Computers, Technology Labs, Technology Mini-Labs, and Technology Tips
See also Problem Solving

Math Journal, 6, 7, 9, 40, 57, 88, 93, 105, 109, 112, 172, 211, 244, 263, 291, 310, 337, 362, 399, 439, 462, 466, 484, 506, 519, 547, 562

Math in the Media, 20, 89, 151, 267, 289, 364, 440, 507

Matrix, 73
adding, 74
column, 73
element, 73
on a graphing calculator, 77
rows, 73
subtracting, 74

Mean, 158–166
using spreadsheets to find, 162

Measurement
customary system, 186
greatest possible error, 504
indirect, 361–364
metric system, 186
precision, 504–507
relative error, 504
significant digits, 504–507

Measures of central tendency, 158–161
average, 158–160
mean, 158–160
median, 159–160
mode, 158–160
range, 163

Measures of variation, 163–166
interquartile range, 163
outlier, 164–165
quartiles, 163–164
range, 163

Median, 158–160

Meeting Individual Needs, 2f, 54f, 102f, 138f, 184f, 230f, 276f, 328f, 380f, 426f, 474f, 514f, 558f

Mental math, 120
study hints, 108, 111, 115, 249, 262, 330, 478, 525

Metric system, 186

Mid-Chapter Self Test, 25, 72, 117, 166, 204, 248, 299, 356, 394, 449, 493, 531, 576

MindJogger Videoquizzes, *see Technology*

Misleading
graphs, 174–177
statistics, 174–177

Mixed numbers, 245, 278
as improper fractions, 245

Mode, 158–160

Modeling Mathematics, 2c, 2f, 5, 54c, 54f, 81, 102c, 102f, 138c, 138f, 164, 184c, 184f, 230c, 230f, 276c, 276f, 328c, 328f, 380c, 380f, 397, 426c, 426f, 474c, 474f, 514c, 514f, 558c, 558f, *see Closing Activity*

Monoclinic crystal, 482

Monomials, 561
adding, 570–572
multiplying, 578–581
subtracting, 573–576

Motion geometry, 456

Motivating the Lesson
Communication, 21, 27, 38, 66, 78, 114, 126, 153, 168, 171, 201, 221, 242, 245, 281, 290, 315, 344, 357, 370, 390, 404, 428, 456, 464, 495, 504, 524, 540, 546, 561, 578
Hands-On Activity, 17, 44, 56, 62, 86, 92, 107, 111, 142, 148, 188, 196, 207, 232, 249, 257, 261, 278, 286, 301, 312, 335, 328, 366, 386, 399, 411, 441, 452, 460, 482, 487, 499, 528, 573, 583
Problem Solving, 4, 8, 11, 32, 59, 69, 73, 81, 104, 120, 158, 163, 174, 210, 215, 235, 253, 265, 296, 309, 318, 328, 339, 353, 361, 382, 414, 433, 437, 446, 476, 490, 518, 534, 565, 570, 588

Motivating Students, 26, 152, 219, 293, 395, 436, 549, 582

Multimedia, *see Technology*

Multiples, 257
common, 258
least common, 257

Multiple Learning Styles
Auditory/Musical, 27, 143, 257, 404, 487
Interpersonal, 69, 210, 312, 349, 524, 570
Intrapersonal, 21, 104, 221, 506, 565
Kinesthetic, 11, 73, 201, 366, 410, 521
Logical, 114, 158, 235, 287, 386
Naturalist, 428

Verbal/Linguistic, 153, 163, 265, 446, 583
Visual/Spatial, 66, 107, 290, 361, 457, 500, 528

Multiplication
equations, 13–15, 21–25
of fractions, 286–289
of integers, 78–80
of polynomials, 578–581, 583–585

Multiplication property of equality, 21

Multiplicative inverse, 290

Naturalist Learning Style, *see Multiple Learning Styles*

Nets, 494

Networks, 200
edges, 200
nodes, 200
traceable, 200

Nodes, 200

Nonexamples, 44, 111–113, 171–172, 174, 175, 186, 200, 204, 206, 208, 211, 212, 215, 216, 217, 226, 232, 233, 234, 235, 237, 249, 266, 331, 359, 360, 364, 390–391, 447, 481, 505, 517, 540, 548, 588

Number line
graphing points on, 56–57
inequalities on, 44–45, 318–319
integers on, 56–60
irrational numbers on, 408–409
percents on, 121
pi on, 307–308
real numbers on, 391
square roots on, 386
using to add integers, 62
using to compare integers, 59
using to draw a box-and-whisker plot, 167
using for a line plot, 153–155
using to order integers, 60
whole numbers on, 44

Numbers
compatible, 126
composite, 235
figurate, 382
integers, 390
irrational, 391
prime, 235
rational, 390
real, 390–394
whole, 390

Numerical expression, 11

O

Obtuse triangle, 196
  of 10, 265
  on number line, 386

Open sentence, 13

Operation
  inverse, 18, 21

Opposite, 70

Ordered pairs, 92–93
  graphing, 92–93
  x-coordinate, 92
  y-coordinate, 92

Ordering
  decimals, 261
  fractions, 261
  integers, 60–61
  rational numbers, 261–264

Order of operations, 11–15
  grouping symbols in, 11

Organizing the Chapter, 2b, 54b, 102b, 138b, 184b, 230b, 276b, 328b, 380b, 426b, 474b, 514b, 558b

Origin, 92

Orthorhombic crystal, 482

Outcomes, 253, 518
  counting, 518–520
  Counting Principle, 518
  equally-likely, 253
  random, 253

Outliers, 164

Output, 428

P

Pacing Chart, 2c, 2d, 54c, 54d, 102c, 102d, 138c, 138d, 184c, 184d, 230c, 230d, 276c, 276d, 328c, 328d, 380c, 380d, 426c, 426d, 474c, 474d, 514c, 514d, 558c, 558d

Parallel lines, 188–192
  alternate exterior angles, 189
  alternate interior angles, 189
  constructing, 193
  corresponding angles, 189
  transversal, 188

Parallelogram, 39, 201
  altitude of, 40
  area of, 40
  base of, 40
  height of, 40
  perimeter of, 39
  rectangle, 201
  rhombus, 201
  square, 201

Parents, *See* Family Activities

Pascal, Blaise, 528

Pascal's Triangle, 528–531
  patterns in, 532–533

Patterns, 42, 99, 141
  divisibility, 232–234
  look for a pattern, 284–295
  number, 237, 304, 405, 451

Pentagon, 358
  similar, 358

Percentage, 335

Percent of change, 348–351

Percent of decrease, 349

Percent equation, 339–341
  base, 339
  rate, 339

Percent of increase, 349

Percent proportion, 335–338
  base, 335
  rate, 335

Percents, 107, 114, 335
  base, 335
  on a calculator, 114
  of change, 348–351
  and circle graphs, 108–109, 126–127
  comparing and ordering, 121
  compatible numbers, 126
  decimals as, 115, 121
  as decimals, 114, 121
  of decrease, 349
  discount, 349
  equation, 339–341
  estimation with, 126–129
  finding, 120–123
  fractions as, 107–108, 115, 121
  as fractions, 107–108, 121
  of increase, 349
  large, 344–347
  markup, 349
  modeling, 107, 335, 337, 338, 344
  of a number,
  on a number line, 121
  proportion, 335–338
  rate, 335
  ratios as, 107–108
  selling price, 349
  small, 344–347
  symbol for, 107, 335

Perfect square, 382

Performance Assessment, 51, 99, 132, 181, 227, 271, 325, 377, 421, 471, 511, 553, 595

Perimeter, 38
  of parallelograms, 39
  of rectangles, 38–39
  of squares, 38–39

Permutations, 521–523

Perpendicular, 197
  segments, 197

Pi, 307
  on a calculator, 307, 310
  graphing, 307–308

Pick's Theorem, 305–306

Planning the Chapter, 2c, 2d, 54c, 54d, 102c, 102d, 138c, 138d, 184c, 184d, 230c, 230d, 276c, 276d, 328c, 328d, 380c, 380d, 426c, 426d, 474c, 474d, 514c, 514d, 558c, 558d

Points
  on coordinate plane, 92–93
  on number line, 56–57
  vanishing, 371

Polygon, 196, 357
  as networks, 200
  pentagon, 358
  similar, 357–360

Polynomials, 562
  adding, 570–572
  binomials, 583
  factoring, 588–591
  like terms, 565
  modeling, 561–571, 573–574, 577, 579–580, 583–585, 588, 590
  monomials, 561
  multiplying, 578–581, 583–585
  simplifying, 565–569
  subtracting, 573–576
  term, 565

Population, 174, 546
  general, 174
  sample, 174, 546

Portfolio, 2, 51, 54, 99, 102, 133, 138, 181, 184, 227, 230, 271, 276, 325, 328, 377, 380, 421, 426, 471, 474, 511, 514, 553, 558, 595

Portfolio, 51, 99, 133, 137, 181, 227, 271, 275, 325, 377, 421, 425, 471, 571, 553, 557, 595

Possibilities
  eliminating, 90–91

Power of 10, 265

Powers, 8–10
  base, 8
  exponent, 8
  products of, 578
  of ten, 265

Precision, 504–507
significant digits, 504

Predictions
using sampling, 546–548

Previewing the Chapter, 2a, 54a, 102a, 138a, 184a, 230a, 276a, 328a, 380a, 426a, 474a, 514a, 558a

Prime factorization, 235–238
using to find LCM, 258

Prime numbers, 235

Principal
in simple interest, 353
square root, 382

Principal square root, 382

Prisms, 482–483
bases of, 483
edges of, 483
faces of, 483
surface area of, 495–498
vertices of, 483
volume of, 486–487

Probability, 253–256
and area, 478
certain, 253
compound events, 534–537
dependent events, 535–536
experimental, 540–543
impossible, 253
independent events, 534–535
outcomes, 253
as ratios, 253
simulation, 540
theoretical, 540

Problem Solving, *see Motivating the Lesson*

Problem solving
four-step plan, 4–7, 34, 79, 87, 127, 197, 243, 262, 282, 300–331, 340, 354, 404, 467–468, 477, 496, 525, 529, 574, 579, 589
labs, 30–31, 90–91, 124–125, 140–141, 194–195, 240–241, 294–295, 342–343, 402–403, 450–451, 480–481, 538–539, 586–587
mathematical techniques, 30–31, 90–91, 124–125, 140–141, 194–195, 240–241, 294–295, 342–343, 402–403, 450–451, 480–481, 538–539, 586–587
mathematical tools, 30–31, 90–91, 124–125, 140–141, 194–195, 240–241, 294–295, 342–343, 402–403, 450–451, 480–481, 538–539, 586–587

Problem-Solving Strategies
Act It Out, 538–539
Draw a Diagram, 402–403
Eliminate Possibilities, 90–91
Guess and Check, 586–587
Look for a Pattern, 294–295
Make a List, 240–241
Make a Model, 480–481
Make a Table, 140–141
Reasonable Answers, 124–125
Solve a Simpler Problem, 342–343
Use a Graph, 450–451
Use a Venn Diagram, 194–195
Working Backward, 30–31

Problem Solving Study Hints, 33, 105, 297

Products of powers, 578

Projects, *See* Chapter Projects and Interdisciplinary Investigations

Properties
associative, 8, 290
commutative, 8, 290
distributive, 34, 291
of equality
addition, 18
division, 21
multiplication, 21
subtraction, 18
identity, 290
inverse, of multiplication, 290
of proportions, 111

Proportional Reasoning
area, 479
conversions, 111–113
examples, 111–113
geometry, 357–360
measurement, 361–364
multiplication by a constant, 370
number, 335–338
percent, 115–116, 335–338
perimeter, 360
predictions, 112
price, 348–351
probability, 478
proportion, 111–113, 330–333
ratio, 111–112
recipes, 113, 334, 338
sequences, 296–297
speed density, 332
student-teacher ratio, 105, 106
surface area, 503
volume, 486, 503

Proportions, 111
cross products, 111, 330
percent, 335–338
property of, 111
scale drawings, 366–369
solving, 112–114
using, 330–333

using mental math to solve, 330
using spreadsheets, 334

Punnet squares, 544–545

Pyramid, 482, 483
base of, 491
volume of, 491–492

Pythagorean Theorem, 396–401
converse of, 399
modeling, 396–397, 398, 399
using, 404–407

Pythagorean triples, 405

Quadrant, 92

Quadratic functions, 452
graphing, 452–455

Quadrilaterals, 201
classifying, 201–204
parallelogram, 201
rectangle, 201
rhombus, 201
square, 201
trapezoid, 201

Quartiles, 163
lower, 164
upper, 163

Radical sign, 382

Radius
of a circle, 307, 309

Random, 253

Range, 163, 428

Rate, 104–106
expressed as percents, 107–108
golden, 118–119
percent, 335
rate, 105–106
in simple interest, 353
simplest form of, 104
unit, 105–106

Rational numbers, 245–248
adding, 278–284
comparing, 261–264
decimals, 245, 249–252
density property, 260
dividing, 312–314
exponents and, 287
fractions, 245–246
improper fractions, 245
integers, 245
mixed numbers, 245

multiplying, 286–289
ordering, 261–264
in real number system, 390
subtracting, 278–284

Ratios, 104
as fractions, 104
geometry, 104
measurement, 104
models, 104
number, 104
probability, 253
scale, 370
simplest form, 104

Reading Math Study Hints, 8, 11, 43, 56, 57, 74, 93, 104, 158, 189, 197, 210, 215, 246, 251, 254, 296, 344, 353, 383, 405, 452, 457, 486, 495, 524, 536, 562

Reading Mathematics, 8, 27, 74, 87, 111, 120, 143, 159, 197, 242, 254, 291, 309, 336, 367, 383, 391, 399, 429, 452, 477, 483, 546

Real numbers, 390–394
integers, 390
irrational numbers, 391
on number line, 391
rational numbers, 390
whole numbers, 390

Reasonable answers, 124–125

Reasoning
deductive, 294
inductive, 294

Reciprocal, 290

Rectangle, 38, 201
area of, 39–40
perimeter of, 38–39

Reflections, 205, 206, 460–463
line of symmetry, 460
over x-axis, 461
over y-axis, 461–462
symmetric, 460

Relative error, 504

Repeating decimal, 249–251

Reteaching the Lesson
Activity, 6, 9, 14, 19, 23, 28, 30, 35, 40, 46, 57, 60, 64, 67, 71, 75, 79, 82, 88, 90, 93, 105, 109, 112, 116, 121, 124, 128, 140, 144, 150, 154, 159, 164, 169, 172, 176, 191, 194, 198, 203, 208, 211, 217, 222, 233, 236, 240, 243, 247, 251, 254, 258, 262, 266, 279, 283, 288, 291, 294, 298, 303, 310, 313, 316, 320, 331, 336, 340, 342, 346, 350, 355, 358, 362, 367, 372, 383, 387, 392, 399, 402, 406, 412, 416, 429, 434, 439, 443, 448, 450, 453, 458, 462, 466, 478, 481, 484, 488, 492, 496, 505, 519, 522, 526, 530, 536, 538, 541, 547, 562, 567, 571, 574, 580, 584, 586, 589
Error Analysis, 9, 14, 19, 23, 35, 57, 71, 93, 105, 112, 116, 121, 128, 150, 159, 164, 203, 208, 211, 233, 243, 247, 258, 262, 283, 298, 303, 310, 320, 336, 340, 346, 350, 358, 383, 392, 399, 412, 429, 434, 439, 448, 453, 462, 484, 492, 496, 522, 526, 530, 536, 567, 571, 574

Rhind Papyrus, 399

Rhombus, 201

Right
angles, 197
triangles, 196

Right triangles
30°-60°, 365, 414–415
45°-45°, 365, 415–416
converse of the Pythagorean Theorem, 399
hypotenuse, 398
legs, 398
Pythagorean Theorem, 396–401
special, 414–417
trigonometry, 365

Rotational symmetry, 207

Rotations, 207, 221, 464–467
of 180°, 465
of 90° counterclockwise, 465
point symmetry, 465

Row, 73

S

Sample, 74, 546
population, 546
space, 253
using to predict, 546–548

Sample space, 253

SAS congruence, 210

Scale
for bar graph, 142
for box-and-whisker plot, 167
of drawings, 366
factor, 370
for line plot, 153
of models, 366

Scale drawing, 366–369
maps, 366

Scale factor, 370

Scale model, 366–369

Scalene triangles, 196

Scatter plots, 168–170
negative relationship, 168
positive relationship, 168

School to Career, 2e, 54e, 102e, 138e, 184e, 230e, 276e, 328e, 380e, 426e, 474e, 514e, 558e

School to Career
Advertising, 219
Computer Programming, 436
Health, 293, 549
History, 152
Law Enforcement, 395
Sports Management, 582
Technology, 26

Science and Math Lab Manual, 2e, 54e, 102e, 138e, 184e, 276e, 328e, 426e, 474e, 514e

Scientific calculator, *See* calculator

Scientific notation, 265–267
See also Let the Games Begin

Selling price, 349

Sequences, 296–299
arithmetic, 296–297
common difference, 296
common ratio, 297
Fibonacci, 300
geometric, 296–297
terms of, 296–297

Shih-chieh, Chu, 528

Significant digits, 504–507

Similar polygons, 357–360
corresponding angles, 357
corresponding sides, 357

Similar triangles, 215–218
symbol for, 215

Simple events, 253–256

Simple interest, 353–356
formula, 353
principal, 353
rate, 353
time, 353

Simplest form, 246, 565
of a polynomial, 565

Simulation, 540

Slide, 220

Solids, 482

Solutions, 13
infinite, 437
no, 447

Solve a simpler problem, 342–343

Spatial Reasoning, 38–40, 301–302, 476–479, 486–491, 495, 495–502, 588–589

Speaking, *see Closing Activity*

Spreadsheets, *see Technology*

Spreadsheets, 55, 71, 103, 127, 136, 139, 149, 162, 231, 252, 274, 295, 305, 329, 334, 352, 425, 475, 515, 556, 559

Square, 38, 201
  area of, 40
  perfect, 382
  perimeter of, 38–39
  Punnet, 544–545

Square root, 382
  equations, 383
  estimating, 385–389
  modeling, 382, 385, 386, 387
  perfect squares, 382
  principal, 382
  radical sign, 382

SSS congruence, 210

Standard form, 265

Standardized Test Practice, 52–53, 100–101, 134–135, 182–183, 228–229, 272–273, 326–327, 378–379, 422–423, 472–473, 512–513, 554–555, 596–597

Statistics
  and maps, 156–157
  misleading, 174–177
  population, 174, 546
  sample, 174, 546
  See also Applications, Connections, and Integration Index on pages xxii–1

Study Guide and Assessment, 48–51, 96–99, 130–133, 178–181, 224–227, 268–271, 322–325, 374–377, 418–421, 468–471, 508–511, 550–553, 592–595

Study Hints
  Estimation, 149, 358, 399, 443, 465
  Mental Math, 108, 111, 115, 249, 262, 330, 478, 525
  Problem Solving, 33, 105, 297
  Reading Math, 8, 11, 43, 56, 57, 74, 93, 104, 158, 189, 197, 210, 215, 246, 251, 254, 296, 344, 353, 383, 405, 452, 457, 486, 495, 524, 536, 562
  Technology, 12, 67, 127, 149, 236, 250, 266, 339, 453, 522

Substitute, 11

Subtraction
  equations, 13–15, 17–20
  of fractions, 278–284
  of integers, 69–72
  of matrices, 74
  of polynomials, 573–576

Subtraction property of equality, 18

Supplementary angles, 190

Surface area, 495
  of cylinders, 499–502
  of prisms, 495–498
  and volume, 503

Surveys, 174
  sample population, 174

Symmetric, 460

Symmetry, 206–209, 460
  line, 206
  line of, 206, 460
  point, 465
  reflection, 206
  rotational, 206, 465

System of equations, 446
  graphing, 446–449
  no solutions, 447
  solution of, 446

T

Tables, 7, 20, 29, 42, 44, 55, 57, 60, 61, 67, 68, 72, 73, 75, 76, 89–91, 95, 103, 114, 118, 121, 129, 139, 142–146, 148, 150, 151, 153–159, 166, 170–173, 177, 195, 200, 218, 239, 263, 292, 295, 296, 318, 321, 339, 342, 368, 388, 393, 394, 397, 405, 433, 437, 443, 447, 452, 453, 461, 463, 479, 485, 486, 507, 516, 517, 529, 543, 545, 546, 547, 548, 564, 568, 569, 572, 575, 578, 581
  frequency, 140–146, 164, 166, 256
  function, 428, 432–435

Techniques, *See* Mathematical Techniques

Technology
  Calculators and Spreadsheets, 2f, 16, 54f, 56, 77, 102f, 138f, 162, 184f, 230f, 231, 276f, 328f, 321, 334, 380f, 426f, 432, 474f, 514f, 558f
  CD-ROM Program, 2, 50, 54, 98, 102, 132, 138, 180, 184, 226, 230, 270, 276, 324, 328, 376, 380, 420, 426, 470, 474, 510, 514, 552, 558, 594
  Mindjogger Videoquizzes, 48, 96, 130, 178, 224, 268, 322, 374, 418, 468, 508, 550, 592
  Test and Review Software, 50, 98, 132, 180, 226, 270, 324, 376, 420, 470, 510, 552, 594

Technology Labs
  box-and-whisker plots, 167
  discounts, 352
  evaluating expressions, 16
  finding a mean, 162
  function tables, 432
  histograms, 147
  linear functions, 445
  matrices, 77
  proportions, 334

Technology Mini-Labs, 3, 55, 103, 136, 139, 185, 231, 277, 329, 381, 425, 427, 475, 515, 556, 559

Technology tips, 3, 55, 103, 136, 139, 185, 231, 277, 329, 381, 425, 427, 475, 515, 556, 559

Technology Study Hints, 12, 67, 127, 149, 236, 250, 266, 339, 453, 522

Terminating decimal, 249–250

Terms, 565
  like, 565
  of a polynomial, 565
  of sequence, 296

Tessellation, 220–223
  M. C. Escher, 220
  using rotations, 221
  using transformations, 220
  using translations, 220

Test and Review Software, *see Technology*

Test Practice, 7, 10, 15, 20, 25, 29, 31, 36, 42, 47, 58, 61, 65, 68, 72, 76, 80, 83, 89, 91, 95, 106, 110, 113, 117, 123, 125, 129, 141, 146, 151, 155, 161, 166, 170, 173, 177, 192, 195, 199, 204, 209, 212, 218, 223, 234, 238, 241, 244, 248, 252, 256, 259, 264, 267, 280, 284, 289, 292, 299, 304, 311, 314, 317, 321, 333, 338, 343, 347, 351, 356, 360, 364, 369, 373, 384, 389, 394, 401, 403, 407, 413, 417, 431, 435, 440, 444, 449, 455, 459, 463, 467, 479, 481, 485, 489, 493, 498, 502, 507, 520, 523, 527, 531, 537, 539, 543, 548, 564, 569, 572, 576, 581, 585, 587, 591

Test-Taking Tips, 53, 101, 135, 183, 229, 273, 327, 379, 423, 473, 513, 555, 597

Tetragonal crystal, 482

Theoretical probability, 540

Thinking Algebraically, 11, 59, 92, 104, 149, 210, 265, 297, 312, 315, 340, 354, 387, 433, 505, 519, 562

Thinking Labs
  Act It Out, 538–539

Draw a Diagram, 402–403
Eliminate Possibilities, 90–91
Guess and Check, 586–587
Look for a Pattern, 294–295
Make a List, 240–241
Make a Model, 480–481
Make a Table, 140–141
Reasonable Answers, 124–125
Solve a Simpler Problem, 342–343
Use a Graph, 450–451
Use a Venn Diagram, 194–195
Working Backward, 30–31

Three-dimensional figures, 482–485
bases, 483
cone, 482
cubes, 482
cylinder, 482
drawing, 482–483
edges, 483
faces, 483
hexagonal, 482
isometric, 482
modeling, 480–481
monoclinic, 482
nets, 494
orthorhombic, 482
pyramid, 482, 483
rectangular prisms, 482
solids, 482
surface area of, 495–503
tetragonal, 482
triangular prisms, 482
triclinic, 482
vertices, 483
volume of, 486–493, 503

Tools, *See* Mathematical Tools

Transformations, 220–223
M. C. Escher, 220
rotations, 207, 221
transformations, 220
translations, 220

Translations, 220, 456–459
slide, 220

Transversal, 188

Trapezoid, 201
bases of, 302
formula for area of, 302
height of, 302

Tree diagram, 518
using to count outcomes, 518–519

Triangles, 196
acute, 196
altitude of, 301
area of, 301–302
base of, 301
classifying, 196–199
congruent, 210–212
equilateral, 196
formula for area of, 302

height of, 301
isosceles, 196
obtuse, 196
Pythagorean Theorem, 396–410
right, 196
scalene, 196
similar, 215–218
using Venn diagram to show relationships, 196

Triclinic crystal, 482

Trigonometry, 365

Trinomial, 588
using algebra tiles to factor, 588–589

Two-dimensional figures
congruent, 210–212
nets, 494
parallelogram, 39–40, 201
polygons, 196, 357
quadrilateral, 201–204
rectangle, 38–40, 201
similar, 357–360
square, 38–40, 201
triangle, 196–199

Unfair games, 516

Unit rate, 105–106

Upper quartile, 163

Use a Graph, 450–451

Use a Venn Diagram, 194–195

Vanishing point, 371

Variables, 12

Variation, 163

Venn diagram, 194, 242, 243, 245
of real number system, 390, 391

Verbal phrase, 27

Verbal/Linguistic Learning Style, *see* Multiple Learning Styles

Vertex
of prisms, 483

Vertical angles, 190

Visual/Spatial Learning Style, *see* Multiple Learning Styles

Volume
of circular cones, 490–491
of circular cylinders, 487–488
modeling, 486, 490

of prisms, 486–487
of pyramids, 491
and surface area, 503

Whole numbers
on a number line, 44
in real number system, 390

Work backward, 30–31

Working on the Chapter Project
Be True to Your School, 209, 223, 227
Consider the Probabilities, 543, 548, 553
Games People Play, 444, 449, 467, 471
Hit or Miss, 564, 585, 591, 595
Home Page Bound, 7, 25, 41, 51
Oldies But Goodies!, 146, 161, 173, 181
Pack Your Bags, 106, 113, 129, 133
Patterns in Nature, 248, 252, 264, 271
Proceed with Caution, 388, 393, 413, 421
Stay Tuned!, 338, 351, 377
Take Me Out to the Ball Game, 248, 252, 264, 271
Up, Up, and Away, 485, 502, 507, 511
Where the Wild Things Are, 65, 83, 95, 99

Write a problem, 6, 31, 36, 57, 68, 72, 80, 83, 91, 125, 129, 141, 145, 195, 241, 289, 295, 331, 343, 362, 401, 403, 417, 431, 448, 451, 481, 497, 501, 519, 522, 539, 587

Writing, *see* Closing Activity

x-axis, 92
reflection over, 461

x-coordinate, 92

y-axis, 92
reflection over, 461–462

y-coordinate, 92

Zero pair, 63

Answer Appendix

CHAPTER 1
Problem Solving and Algebra

Pages 6–7, Lesson 1-1

1. Explore – Identify what information is given and what you need to find.
   Plan – Estimate the answer and then select a strategy for solving.
   Solve – Carry out the plan and solve.
   Examine – Compare the answer to the estimate and determine if it is reasonable. If not, make a new plan.

6a.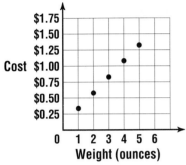

7. Sample answer: *Explore:* Locate and examine a website and investigate the software used to design homepages. *Plan:* Make a blueprint of how you would like the final page to look. *Solve:* Turn your blueprint into a home page using any available software. *Examine:* Test the code and see if the home page really looks like your blueprint. Make the necessary adjustments.

Pages 19–20, Lesson 1-4

4. $x = 3$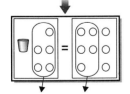

Pages 46–47, Lesson 1-9

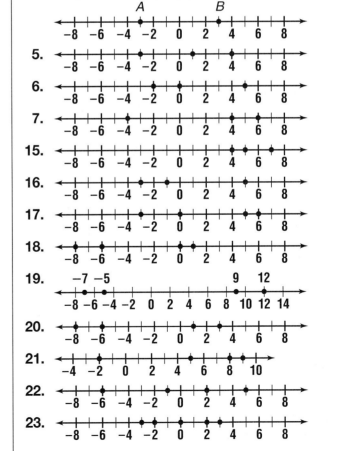

CHAPTER 2
Algebra: Using Integers

Pages 57–58, Lesson 2-1

1. Sample answer: Draw a number line. Locate the number and draw a dot at that point on the line. Label the point with a letter.

2. Sample answer: $|A| = |B|$

AA1

40. They have the same absolute value and opposite signs. Their addition could be modeled by a move left or right to the first addend, followed by a move in the opposite direction back to zero. For the result to be 0, the moves must be of the same length.

42. $s < 9$

Pages 60–61, Lesson 2-2

26. −80 −70 −63 −58 −54

27b.

Pages 71–72, Lesson 2-5

3. $a = 4$

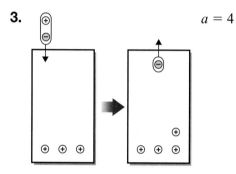

Page 72, Mid-Chapter Self Test

1.

Page 80, Lesson 2-7

1. Sample answer: $[(-35)(45)](-2) = -1{,}575(-2) = 3{,}150$ or $(-35)[(-2)(45)] = (-35)(-90) = 3{,}150$
2. See students' work. The products are the same.
3. $(-2)(-4) = 8$

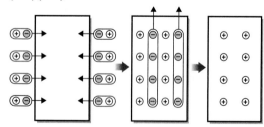

35. y is non-negative. If $y = 0$, then $y^2 = 0$. If y is any other number, y^2 is positive because any two numbers with the same sign have a positive product.

36. Sample answer: $\begin{bmatrix} 3 & -3 & 0 \\ 2 & 5 & 1 \end{bmatrix}$ and $\begin{bmatrix} 3 & -4 \\ 6 & 1 \end{bmatrix}$

Pages 82–83, Lesson 2-8

48.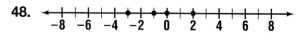

Pages 93–95, Lesson 2-10

1. For the x-coordinate, left is negative and right is positive. For the y-coordinate, up is positive and down is negative.
2. (3, 0): Move right 3, do not go up or down. (0, 3): Do not go left or right, move up 3.

8-13.

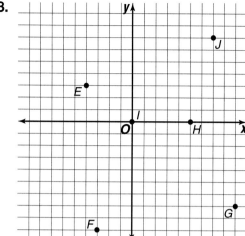

14.

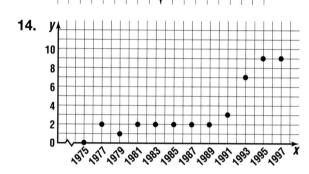

27-44.

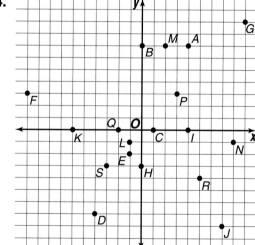

46.

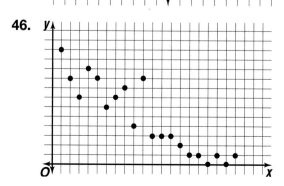

47b.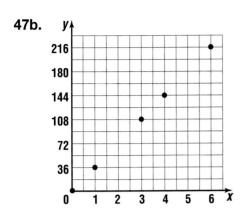

Pages 96–99, Chapter 2
Study Guide & Assessment

11.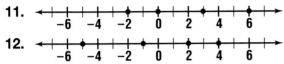

12. (number line from −6 to 6)

46–49.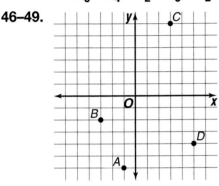

50b. (number line with S, A, T, K, N)

52a. The pattern works for even negatives, but it does not work for odds.

52b. For odd negatives, add 1 instead of subtracting 1. Then divide by 2 and add a 5 to the end. −1 is an exception.

CHAPTER 3
Using Proportion and Percent

Pages 105–106, Lesson 3-1

1. Ratios and rates are both comparisons by division. In a rate the units are different; in a ratio the units are the same.

Page 107, Lesson 3-2, Mini-Lab

1.  **2.**

3.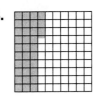

Pages 112–113, Lesson 3-3

1. reducing fractions and finding cross products
2. Sample answer: $\frac{2}{3}, \frac{5}{6}$; cross products are not equal.
3. Sample answer: assigning a variable; solving equations

Pages 116–117, Lesson 3-4

2. Sample answer: For $\frac{13}{20}$, find an equivalent fraction with 100 as the denominator. For $\frac{13}{19}$, solve a proportion.

3. $15\% > \frac{1}{8}$

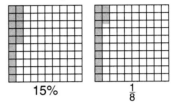

15% $\frac{1}{8}$

Pages 128–129, Lesson 3-6

32. Elements in Earth's Crust

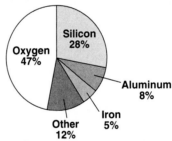

34a. Book Purchases

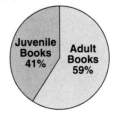

Book Purchases (bar graph: Adult Books 59%, Juvenile Books 41%)

34b. Sample answer: You can see how the two parts make up the whole.

Answer Appendix **AA3**

34c. Sample answer: You can see about how many more adult books were sold.

CHAPTER 4
Statistics: Analyzing Data

Pages 144–146, Lesson 4-1

5b.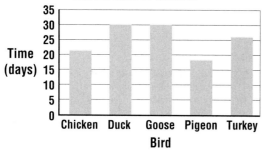

7.

Bicycle Helmet Prices		
Price	Tally	Frequency
$21–30	|||||		6
$31–40	||||| ||	7
$41–50	|||	3

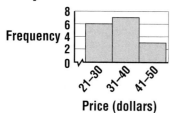

8b.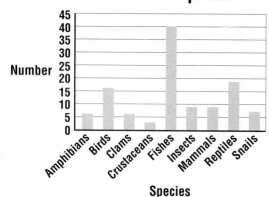

9.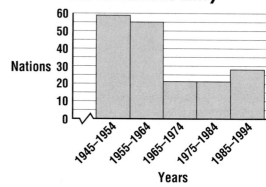

10c.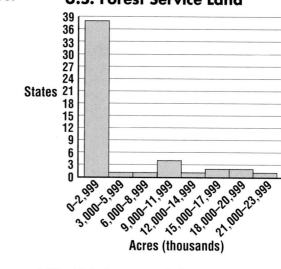

Pages 150–151, Lesson 4-2

4a.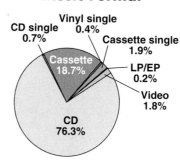

5.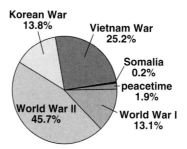

AA4 Answer Appendix

6. Hawaiian Counties

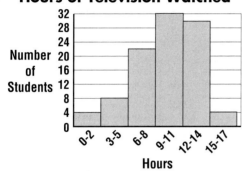

10. Hours of Television Watched

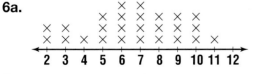

Page 151, Math in the Media

1. Caucasian: 73.7%; African-American: 12.0%; Hispanic: 10.3%; Asian: 3.3%; Native American: 0.7%
2. African-American: 4.424 million; Hispanic: 3.51 million; Caucasian: 52.245 million
3. The number of African-American and Hispanic computer users is 4.424 + 3.51 or 7.934 million. This is much less than the number of Caucasian computer users. The percents added were not found using the same base, so they cannot be added in the way they were in the news report.

Pages 154–155, Lesson 4-3

4a.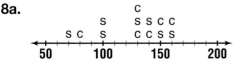

5.

6a.

8a.

8b. Sample answer: The extreme values are 70 and 160 Calories. The chocolate-chip cookies seem to be more clustered than the sandwich cookies. Most of the chocolate-chip cookies are between 130 and 160 Calories per serving.

9a.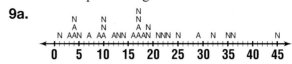

Page 157, Lesson 4-3B

6b. Sample answer: If you are in a state with a large number of tornadoes per year, you would use small ranges to show how many more your state has than others. If you are in a state with a small number of tornadoes per year, you would use larger ranges to make your state appear to have as many tornadoes as other states.

6c. Sample answer: Adjust the ranges so that your state appears to have as few tornadoes as possible. Also see that other states appear to have more tornadoes per year so that they would be better choices for the center.

Pages 160–161, Lesson 4-4

20b. Sample answer: 65; The median and mode are both 65, so it represents the data well.

21. Sample answer: This is probably the mean, found by dividing the sum of the attendance figures for each showing of the movie by the number of people in the United States.

23. Because the value of the new data piece changes the total of the data drastically, the mean is affected most. Adding one very large or very small piece of data will only move the median slightly toward the new number. The mode will not be affected by adding an very large or very small number.

24.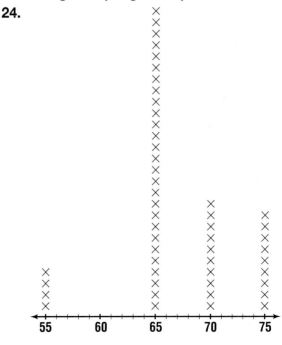

Page 166, Mid-Chapter Self Test

1.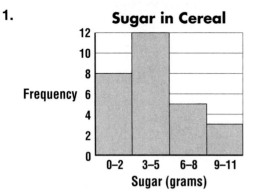

2.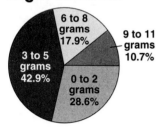

3.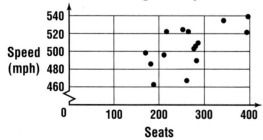

Page 167, Lesson 4-5B

1. minimum = 29, maximum = 76, upper quartile = 52.5, lower quartile = 36.5, median = 42

2. See students' graphs. The box for the actresses is narrower and the values are lower. So the ages of the actresses are younger and more concentrated than those of the actors.

3. TI-83: Yes; the box-and-whisker plot for the actresses shows the two outliers using boxes. TI-82: No; the calculator does not evaluate for outliers. The ends of the whiskers are the extreme values.

Pages 169–170, Lesson 4-6

21a.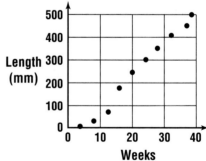

22a. See students' work.

23a. Sample answer: Both are summer activities, so sales would increase at the same time.

23b. No. Warm weather could be causing the increase in both skateboard and swim suit sales.

Pages 172–173, Lesson 4-7

4. The data are in pairs and the scatter plot is visual for viewing quickly on television.

5. These are individual data. Which of these is the best choice depends on what the teacher is trying to show.

6. These are individual data which are being compared to the whole team score.

7. These are two sets of similar data and you want to show their relationship visually. A scatter plot would allow visual comparison.

8. It allows visual comparison of different age intervals.

9. allows comparison of contribution of each division to the total amount of sales

10. It allows quick visual comparison and individual data are not needed.

11. A trend can be shown and accurate information is needed.

12. Allows comparison of two sets of data.

13. Shows outliers visually and shows concentration of other prices.

14. shows proportion of all of computers owned that are Macintosh

15. Shows increase visually.

16. Shows increase visually.

17. Data is in categories.

18. The good and fair prices could be distinguished by symbols and compared easily.

19. No; a histogram requires that data be divided into intervals. This data is not well suited to intervals. A scatter plot is a better choice.

23.

Pages 175–177, Lesson 4-8

6b. Sample answer: The jeans that represent the schools with no restrictions should be a little less than three times as large as the jeans that represent the schools with restrictions. But the jeans are not that much larger.

6c. Yes; The larger jeans that represent the schools with no restrictions give the impression that just a few more schools do not have restrictions than do. This is not consistent with the percents.

7. People at a movie theater will be more likely to say watching movies than the general public.
8. Many apartment complexes do not allow dogs.
9. A telephone poll will allow a selection of a number of people with different viewpoints.
10. People of all different opinions will be in a grocery store.
11. Because Texas has a higher concentration of immigrants than other states and immigrants are more likely to speak more than one language, a sample taken there would not be representative of all Americans.
12. All students will be in a school cafeteria.
13. If the schools are chosen in different types of neighborhoods, the sample will be representative of teenagers.
14. People at a coffee house are more likely to say that coffee is their favorite drink.
15. People who attend a community college will be more likely to have chosen careers that can be pursued with an associate's or bachelor's degree.
16b. The graph on the right is misleading because it makes the preference for Mrs. Field's cookies look larger.
16c. The graph on the right would probably be chosen because it makes the preference for Mrs. Field's cookies look larger.

Pages 178–181, Chapter 4
Study Guide and Assessment

13.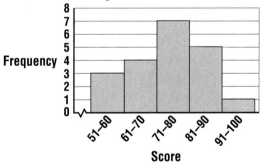

14.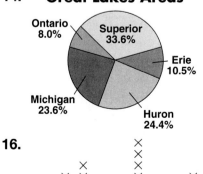

16.

22. 8; 6.5; 8.5, 4.5; 4; no outliers
23. 74; 138; 150.5, 123.5; 27; 195
24. 32; 82; 86.5, 77.5; 9.0; 61, 62
29. Yes; a good cross section of the population visits a movie theater.
30. No; an answer of ice cream is more likely at an ice cream shop.

32.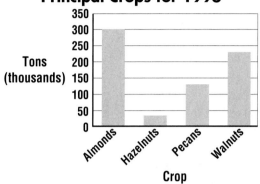

33.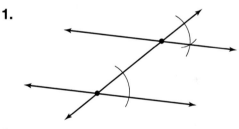

CHAPTER 5
Geometry: Investigating Patterns

Page 193, Lesson 5-1B

1.

2.

Pages 194–195, Lesson 5-2A

1. the students who like all three types of music; 8 students

3.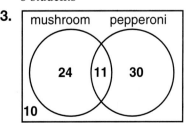

7.

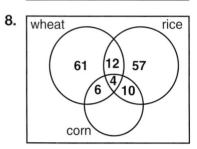

8.

Pages 198–199, Lesson 5-2

24a.  **24b.**

24c. 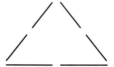 **24d.**

Pages 203–204, Lesson 5-3

15.  **16.**

17. **18.**

22b.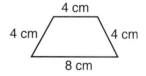

Page 205, Lesson 5-4A

1.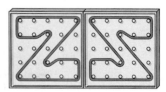

2.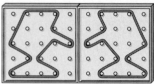

3.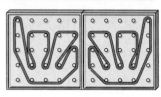

4.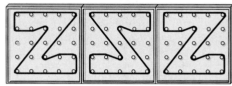

The new image is the same as the original figure.

Pages 208–209, Lesson 5-4

3.

15. isosceles triangles, equilateral triangles; equilateral triangles

21.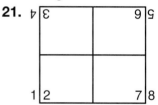

(Numbers on the sides indicate the page numbers on the back of the sheet of paper.)

Pages 211–212, Lesson 5-5, Exercises

1. Corresponding angles and corresponding sides must be congruent.
2. An included angle is the angle between two consecutive sides.

$\angle BAC$ is included between $\overline{AB}$ and $\overline{AC}$.

3. If each of the three sides of one triangle are congruent to each of the sides of another triangle, the triangles are congruent. SSS

If 2 angles and the included side of one triangle are congruent to 2 angles and the included side of another triangle, the triangles are congruent. ASA

If 2 sides and the included angle of one triangle are congruent to 2 sides and the included angle of another triangle, the triangles are congruent. SAS

15. The 4 triangles that form the large triangle in the center appear to be congruent. Three of these triangles are divided into 3 smaller triangles. These smaller triangles appear to be congruent.

16. Sample answer:

Page 213, Lesson 5-5B

1.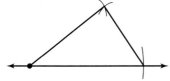

2.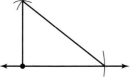

3.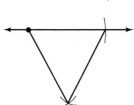

Page 214, Lesson 5-6A

1.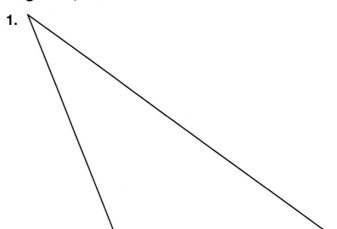

2.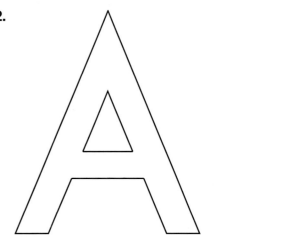

3.

Pages 222–223, Lesson 5-7

4.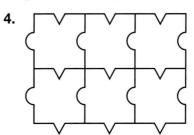

7.
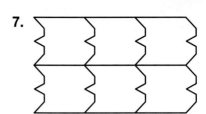

8.
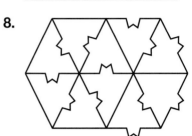

9.
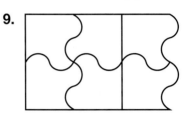

17b. 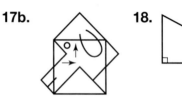 18.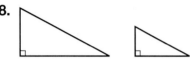

Pages 224–227, Chapter 5
Study Guide and Assessment

26.
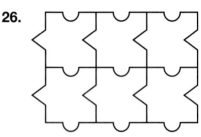

27.
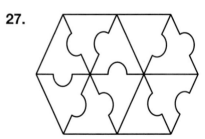

CHAPTER 6
Exploring Number Patterns

Pages 233–234, Lesson 6-1

2. For a number to be divisible by 6, it must be divisible by 2 and 3. However, an odd number is not divisible by 2.

Page 235, Lesson 6-2, Mini-Lab

1.   2.

3.
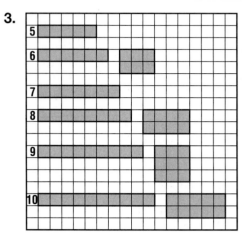

Pages 237–238, Lesson 6-2

2. Sample answers:

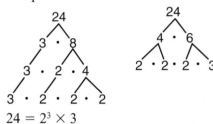

$24 = 2^3 \times 3$

3.
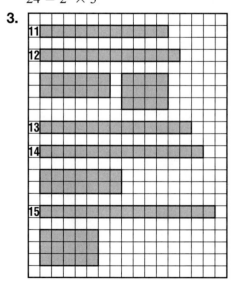

Pages 243–244, Lesson 6-3

3a.
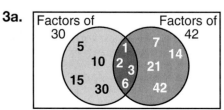

3b.

Pages 251–252, Lesson 6-5

40a. 0.4

40b. 0.363636364

40c. 0.363636363

42a. $\frac{1}{2} = 0.5, \frac{1}{4} = 0.25, \frac{1}{5} = 0.2, \frac{1}{8} = 0.125$

42b. $\frac{1}{3} = 0.\overline{3}, \frac{1}{6} = 0.1\overline{6}, \frac{1}{7} = 0.\overline{142857}, \frac{1}{9} = 0.\overline{1}$

Pages 258–259, Lesson 6-7

1. (1) List the multiples of 10 and the multiples of 16. Find the least nonzero number that is a multiple of both numbers.

(2) Find the prime factorization of 10 and 16. Multiply all the factors, using the common factors only once.

Page 260, Lesson 6-8A

4. Sample answer: 4.26 and 4.27; 4.265; the new number is between the original numbers.

Pages 266–267, Lesson 6-9

1. 36.2 and 0.362 are not greater than or equal to 1 and less than 10.

Page 271, Performance Task

The differences among these decimals are the bars over the repeating digits. For example, 0.83 is a terminating decimal, $0.8\overline{3}$ is a repeating decimal in which the 3 repeats, and $0.\overline{83}$ is a repeating decimal in which 83 repeats. The methods to change the decimals to fractions are shown below.

- $0.83 = \frac{83}{100}$
- Let $N = 0.8\overline{3}$. Then $10N = 8.3\overline{3}$.

$$10N = 8.3\overline{3}$$
$$- \ N = 0.8\overline{3}$$
$$9N = 7.5$$
$$N = \frac{7.5}{9} \text{ or } \frac{5}{6}$$

So, $0.8\overline{3} = \frac{5}{6}$.

- Let $N = 0.\overline{83}$. Then $100N = 83.\overline{83}$.

$$100N = 83.\overline{83}$$
$$- \ N = \ \ 0.\overline{83}$$
$$99N = 83$$
$$N = \frac{83}{99}$$

So, $0.\overline{83} = \frac{83}{99}$.

Since $0.8\overline{3} = 0.833...$ and $0.\overline{83} = 0.838383...,$
$0.83 < 0.8\overline{3} < 0.\overline{83}$.

Using the fractions, $0.83 = \frac{83}{100} = \frac{8,217}{9,900}$, $0.8\overline{3} = \frac{5}{6} = \frac{8,250}{9,900}$, and $0.\overline{83} = \frac{83}{99} = \frac{8,300}{9,900}$. Since $\frac{8,217}{9,900} < \frac{8,250}{9,900} < \frac{8,300}{9,900}$, $0.83 < 0.8\overline{3} < 0.\overline{83}$.

CHAPTER 7
Algebra: Using Rational Numbers

Pages 279–280, Lesson 7-1

3. $1\frac{1}{2}$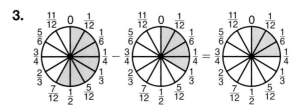

Pages 283–285, Lesson 7-2

3.

Pages 303–304, Lesson 7-6

1. Sample answer:

Pages 320–321, Lesson 7-10

3. 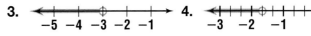   **4.**

5. 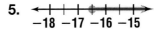   **6.**

7.   **8.**

10. 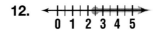   **11.**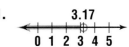

12. 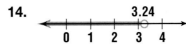   **13.**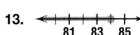

14.

15. 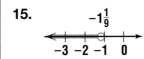   **16.**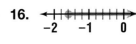

17.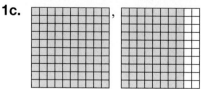
18.
19.
20.
21.
22.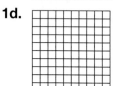
23.
24.
25.
26.
27.

Pages 322–325, Chapter 7
Study Guide and Assessment

42.
43.
44.
45.

CHAPTER 8
Applying Proportional Reasoning

Pages 340–341, Lesson 8-3

10. $0.165 \cdot 60 = P$
11. $R \cdot 66 = 55$
12. $0.50 \cdot B = 75$
13. $0.15 \cdot B = 30$
14. $R \cdot 60 = 18$
15. $0.24 \cdot 72 \; P$
16. $\frac{1}{3} \cdot 420 = P$
17. $\frac{2}{3} \cdot B = 16$
18. $R \cdot 150 = 45$
19. $R \cdot 300 = 6$
20. $0.005 \cdot 3{,}200 = P$
21. $R \cdot 80 = 25$
22. $0.28 \cdot 231 = P$
23. $R \cdot 50 = 6$
24. $0.75 \cdot B = 15$
25. $R \cdot 80 = 70$
26. $0.07 \cdot B = 1.47$
27. $0.08 \cdot B = 54$

Page 344, Lesson 8-4, Mini-Lab

1a.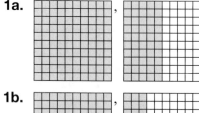

1b.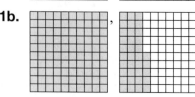

1c.

1d.

1e.

Pages 362–364, Lesson 8-8

1. Indirect measurement means you did not actually measure the distance. Instead you used proportions and other measurements to find the distance.

3. Sample answer: The shadow of a flagpole is 12 feet at the same time the shadow of a nearby sign is 4 feet. If the sign is 8 feet tall, how high is the flagpole?

   To solve the problem, write the proportion and solve for h.

   flagpole's shadow → $\frac{12}{4} = \frac{h}{8}$ ← flagpole's height
   sign's shadow → ← sign's height

Page 365, Lesson 8-8B

1. 30° angle: ratio 1 ≈ 0.5774, ratio 2 ≈ 0.5000, ratio 3 ≈ 0.8660; 60° angle: ratio 1 ≈ 1.7321, ratio 2 ≈ 0.8660, ratio 3 ≈ 0.5000

2. Yes, the corresponding angles are congruent and the sides are in proportion.

3. 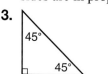 ratio 1 ≈ 1.0000, ratio 2 ≈ 0.7071, ratio 3 ≈ 0.7071

4. Use one of the ratios that relates the given side and one of the missing sides. Then, solve the equation for the missing side. Use another ratio to find the third side.

Pages 372–373, Lesson 8-10

6. $X'(4, 8)$, $Y'(8, 12)$, $Z'(16, 8)$

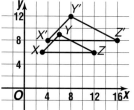

11. $D'(-8, 24), R'(-4, -8), T'(16, 12)$

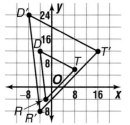

12. $D'(-1, 3), R'\left(-\frac{1}{2}, -1\right), T'\left(2, 1\frac{1}{2}\right)$

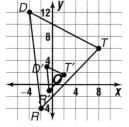

13. $D'(-5, 15), R'\left(-2\frac{1}{2}, -5\right), T'\left(10, 7\frac{1}{2}\right)$

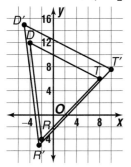

17. Sample answer:

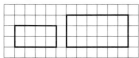

19b.

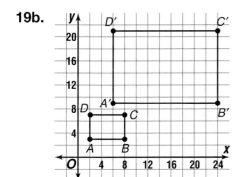

Page 377, Performance Task

The friend is wrong. In this proportion, the first ratio has money over hours and the second ratio has hours over money. A possible correct proportion is $\frac{8}{42} = \frac{24}{h}$. The correct solution is $126. Another possible way to solve this problem is to find the hourly rate by dividing the amount earned by the number of hours worked. Then multiply the hourly rate by 24 hours.

CHAPTER 9
Algebra: Exploring Real Numbers

Pages 383–384, Lesson 9-1

2. $\frac{36}{81}$ is a perfect square because 36 and 81 are both perfect squares and $\left(\frac{6}{9}\right)^2 = \frac{36}{81}$.

3. When each square root is squared, it equals the perfect square.
$5^2 = 25$

29. Yes, because the factorization of the product is simply the product of the two perfect square factorizations, and therefore it is also a perfect square.

30.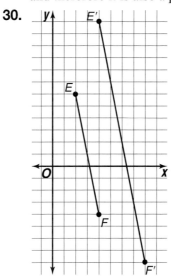

Pages 392–394, Lesson 9-3

17.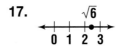
$\sqrt{6}$ between 0 1 2 3

18. $-\sqrt{27}$ between -6 and -4

19. $\sqrt{50}$ between 5 6 7 8

20. $-\sqrt{99}$ between -10 and -8

21. $\sqrt{108}$ between 8 9 10 11

22. $\sqrt{300}$ between 15 16 17 18

36.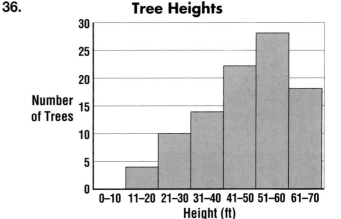
Tree Heights

Page 397, Lesson 9-4A

1.

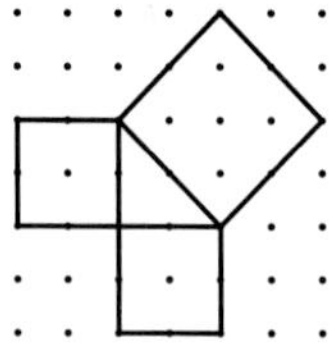

2.

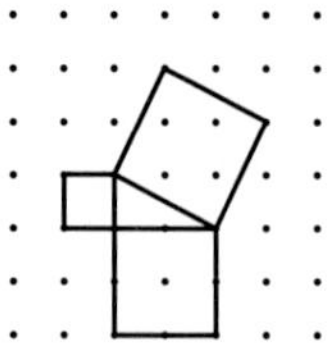

3.

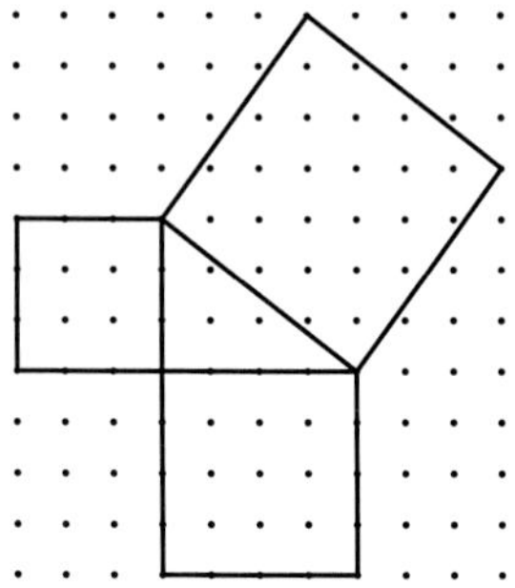

6.

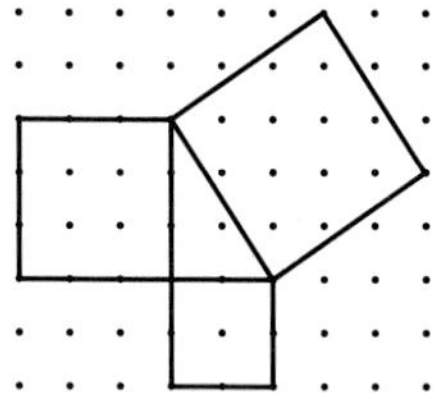

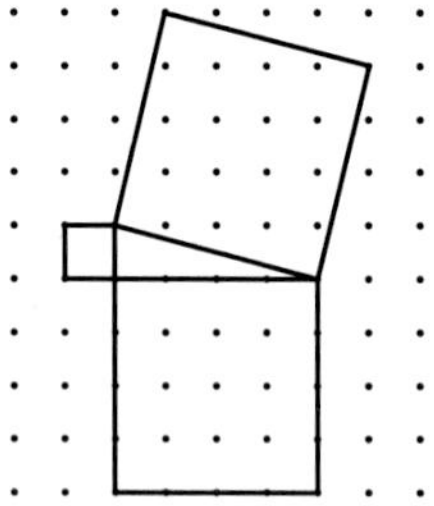

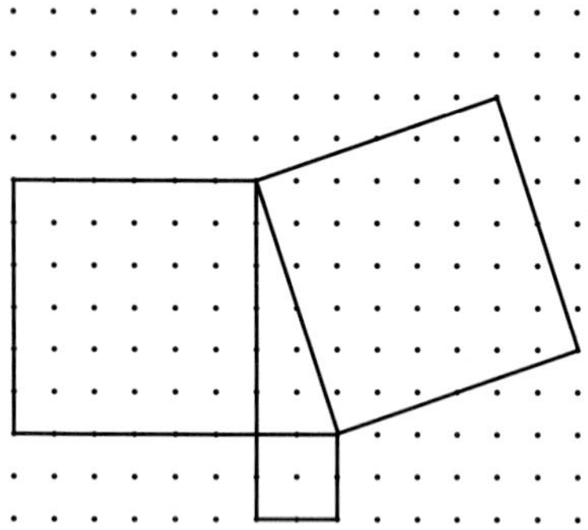

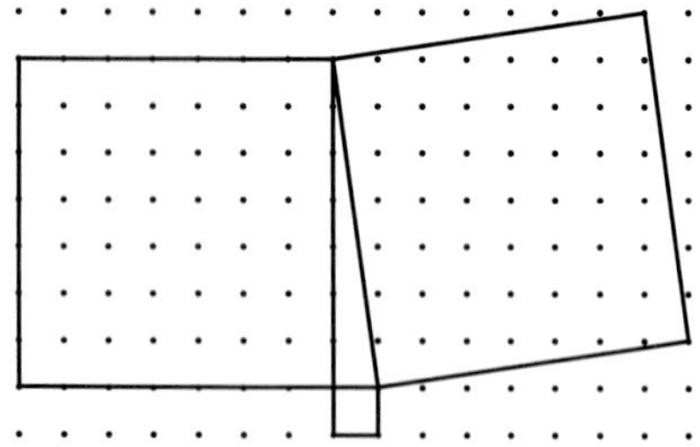

Pages 399–401, Lesson 9-4

2.

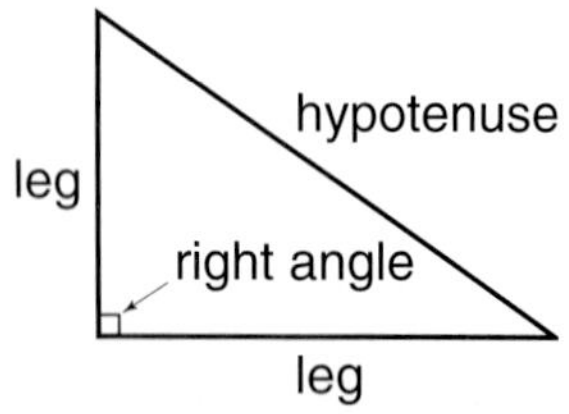

3. $6^2 + 12^2 = 36 + 144$ or 180 and $14^2 = 196$. Since $180 \neq 196$, the triangle is not a right triangle.

Page 403, Lesson 9-5A

2. Answers will vary. Sample answer:

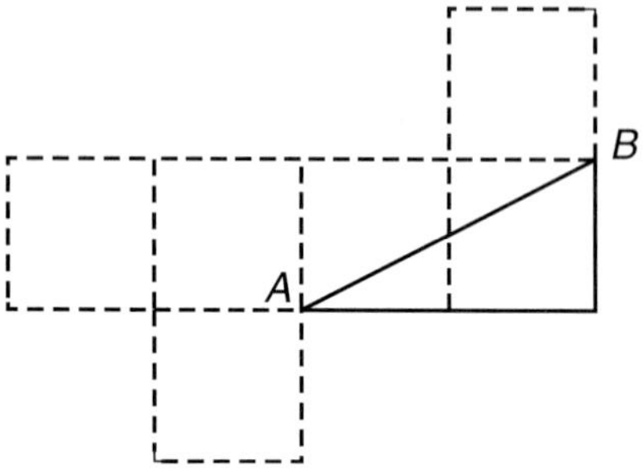

3.

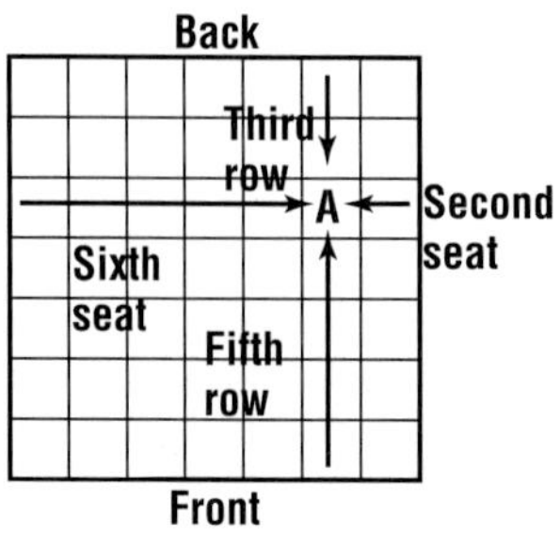

There are 49 seats in the section.

6.

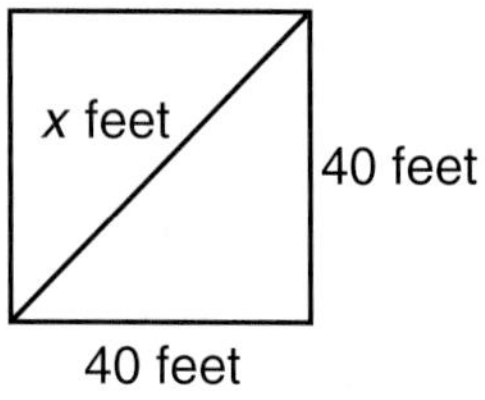

12b. **Grills Sold (millions)**

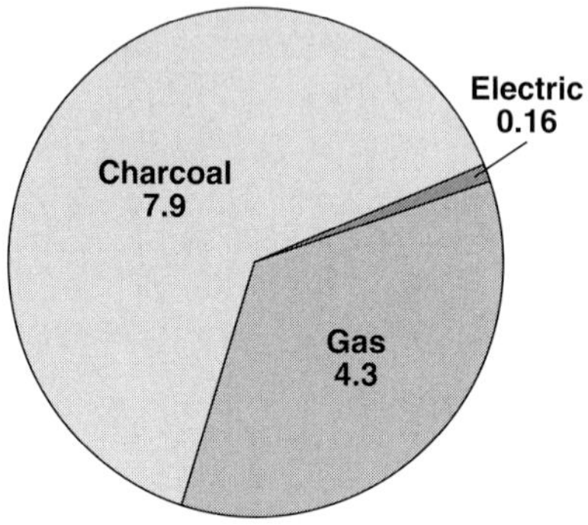

Pages 406–407, Lesson 9-5

14. No, there are many possibilities if given the length of only one side of a right triangle. The two triangles below can represent a right triangle with a side length of 10 cm.

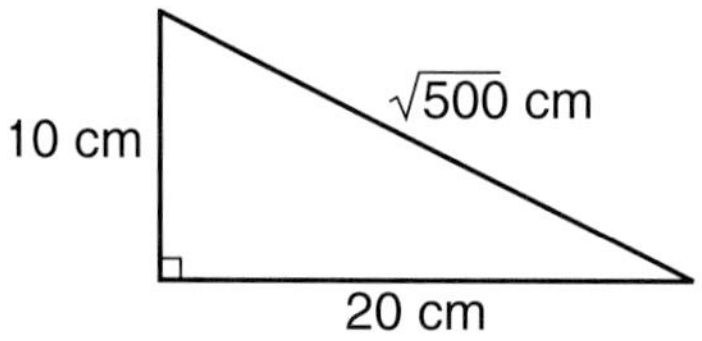

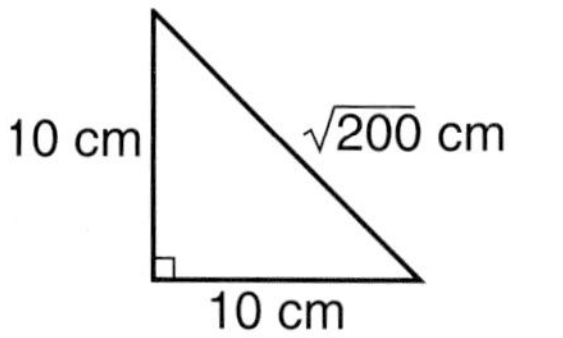

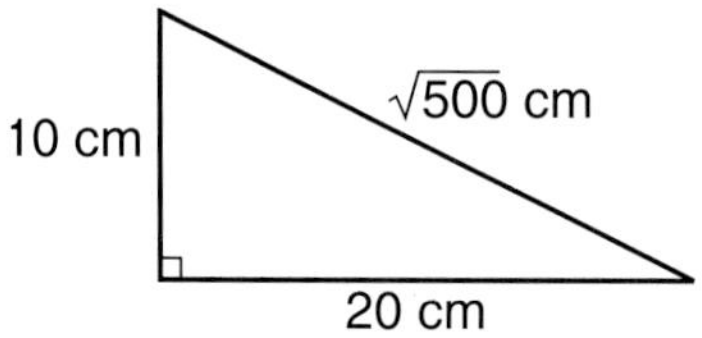

Pages 408–409, Lesson 9-5B

1.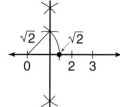
2.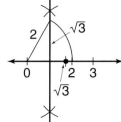
3.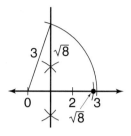
4.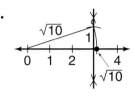

5. Draw a number line. At 1, construct a perpendicular line segment 1 unit in length. Draw a line from 0 to the top of the perpendicular line segment. Label it c. Open the compass to the length of c. With the tip of the compass at 0, draw an arc that intersects the number line to the right of 0 at A. The distance from 0 to A is $\div \sqrt{2}$ units.

6. First method: Draw a number line. At 2, construct a perpendicular line segment 2 units in length. Draw a line from 0 to the top of the perpendicular line segment. Label it b. Open the compass to the length of b. With the tip of the compass at 0, draw an arc that intersects the number line to the right of 0 at B. The distance from 0 to B is $\sqrt{8}$ units.

   Second method: Draw a number line. At 1, construct a perpendicular line segment. Put the tip of the compass at 0. With the compass set at 3 units, construct an arc that intersects the perpendicular line segment. Label the perpendicular leg a. Open the compass to the length of a. With the tip of the compass at 0, draw an arc that intersects the number line to the right of 0 at D. The distance from 0 to D is $\sqrt{8}$ units.

7. Draw a number line. At 1, construct a perpendicular line segment 2 units in length. Draw a line segment from 0 to the top of the perpendicular line segment. Label it c. Open the compass to the length of c. With the tip of the compass at 0, draw an arc that intersects the number line to the left of 0 at B. The distance from 0 to B is $\sqrt{5}$ units.

8. Draw a number line. At 1, construct a perpendicular line segment 2 units in length. Draw a line segment from 0 to the top of the perpendicular line segment. Label it d. Open the compass to the length of d. With the tip of the compass at 0, draw an arc that intersects the number line to the right of 0 at C. The distance from 0 to C is $\sqrt{3}$ units.

9. $(\sqrt{11})^2 + 5^2 = 6^2$; none for 12; $(\sqrt{13})^2 = 3^2 + 2^2$; none for 14; $(\sqrt{15})^2 + 7^2 = 8^2$; $(\sqrt{17})^2 + 8^2 = 9^2$; none for 18; $(\sqrt{19})^2 + 9^2 = 10^2$; $(\sqrt{20})^2 + 4^2 = 6^2$

Pages 411–413, Lesson 9-6

5.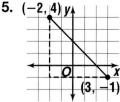
6.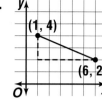
11.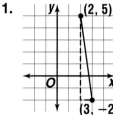
12.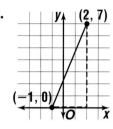
13.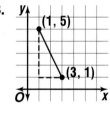
14.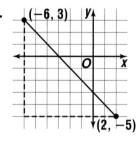
15.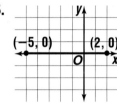
16.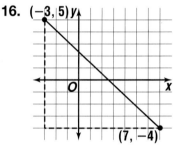
18a.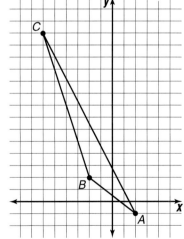

Pages 416–417, Lesson 9-7

14.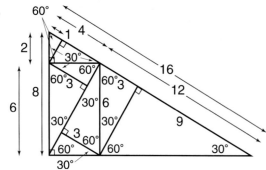

Pages 418–421, Chapter 9
Study Guide and Assessment

30.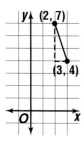
31.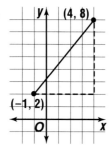
32.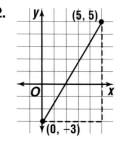
33.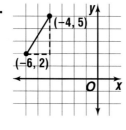

CHAPTER 10
Algebra: Graphing Functions

Pages 429–431, Lesson 10-1

2. Sample answer: A relationship in which one thing depends upon another.

22a.

Pages 434–435, Lesson 10-2

3.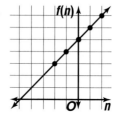

4.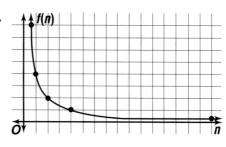

5b.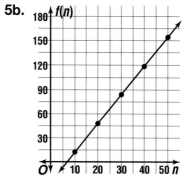
6.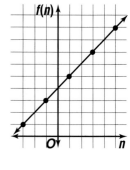

7.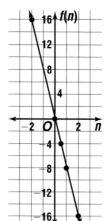
8.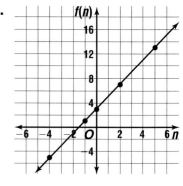

9.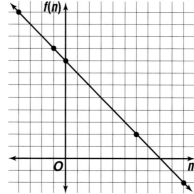

10.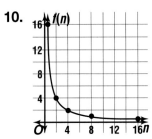

11.

12. See students' tables.
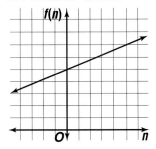

13. See students' tables.
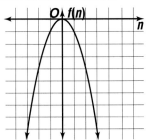

14a. See students' tables.
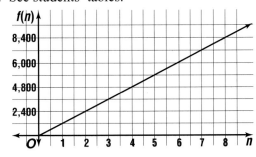

15a. See students' table.
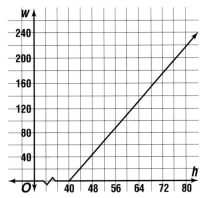

Page 441, Lesson 10-4A

2. Sample answer: The points suggest a line.
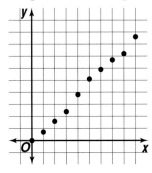

3. Sample answer:
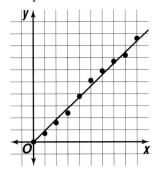

Pages 443–444, Lesson 10-4

4.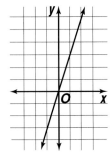

5.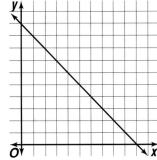

6.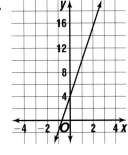

7.

8.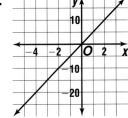

9.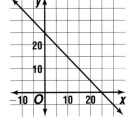

10.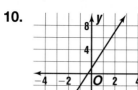
11.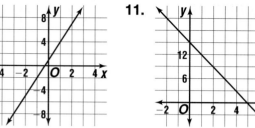
20b.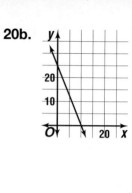
21.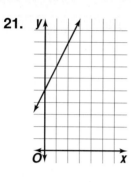

12.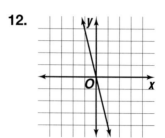
13.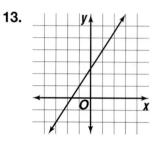
22a.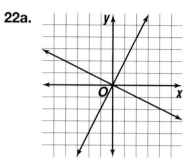

14.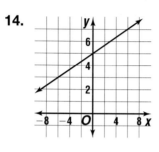

Page 445, Lesson 10-4B

1.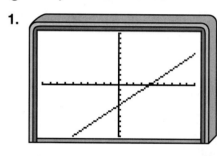

15.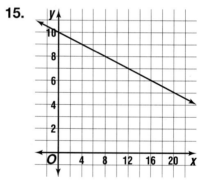

2.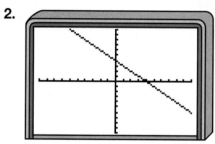

16.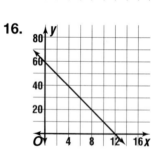
17.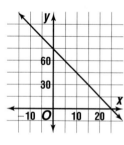

3.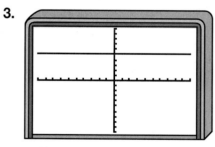

18.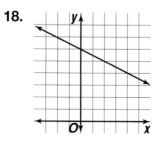
19.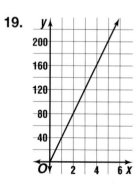

4.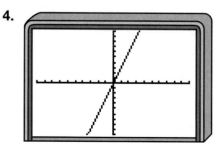

AA18 Answer Appendix

5.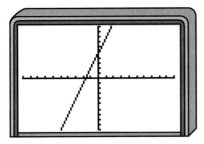

6.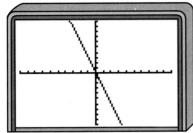

Pages 448–449, Lesson 10-5
3. Sample answer:
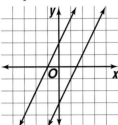
4.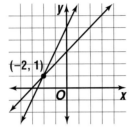

5.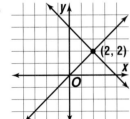
6.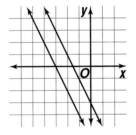

7. No; The lines meet at a point where x is negative, so the time when imports and exports has passed.

8.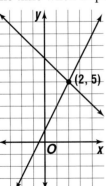
9.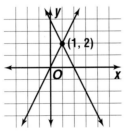

10.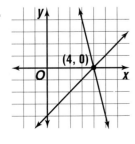
11.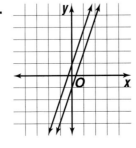

12.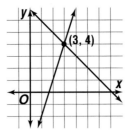
13.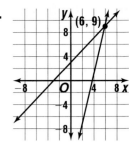

14.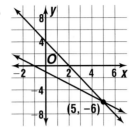
15.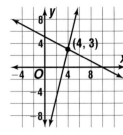

16.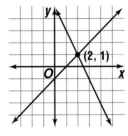
17.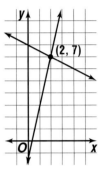

18.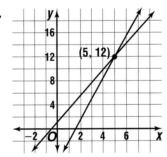

21a.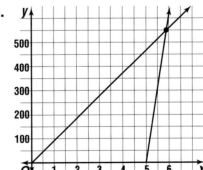

23.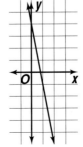

Page 449, Mid-Chapter Self Test

2.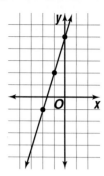

3. Sample answers: (0, 1), (1, 3.5), (2, 6), (4, 11)

4.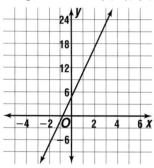

Pages 453–455, Lesson 10-6

3. Nicole is correct. The function can be written as $y = \frac{1}{3}x^2$. The highest power in the equation is 2, so the function is quadratic.

4.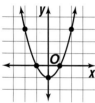

5.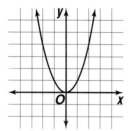

6.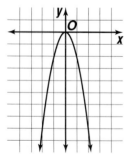

7.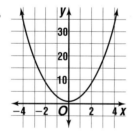

8a.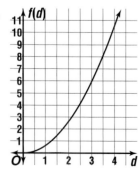

9.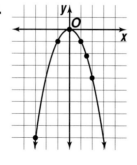

10.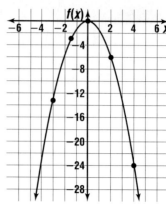

11.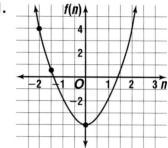

12.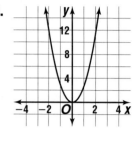

13.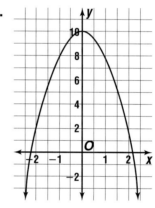

14.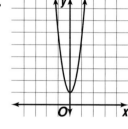

15.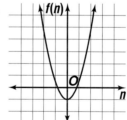

16.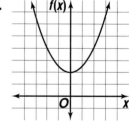

17.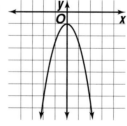

18.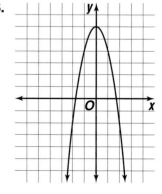

19. 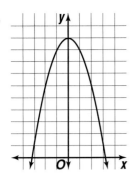   **20.**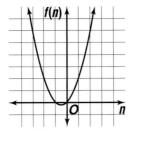

Pages 458–459, Lesson 10-7

8. $E'(1, 1), F'(4, 6), G'(8, 0)$

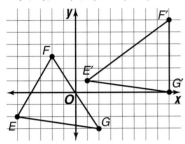

9. $P'(-6, 3), Q'(-9, -1), R'(-5, 6)$

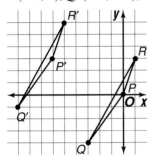

10. $S'(1, 4), Q'(3, 6), A'(1, 8), R'(-1, 6)$

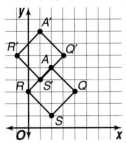

11. $W'(-5, 5), X'(1, 8), Y'(2, 6), Z'(-4, 3)$

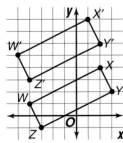

12. $F'(3, 3), G'(1, 0), H'(3, -2), J'(5, 1)$

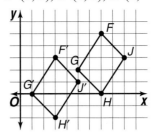

13. $A'(-4, 0), B'(-2, 0), C'(-1, 2), D'(-3, 4), E'(-5, 2)$

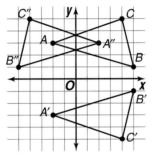

Pages 462–463, Lesson 10-8

5a–5b.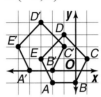

5c. yes; The corresponding sides are all congruent.

10.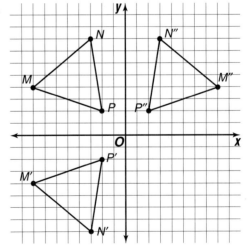

11.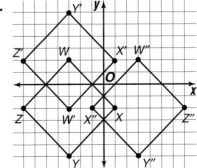

12.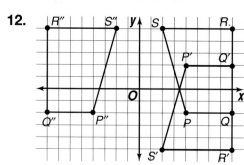

13. 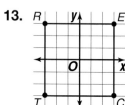 13a.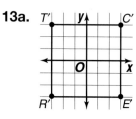

13b.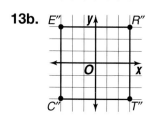

13c. They are all the same figure.

14, 14a.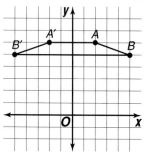

19.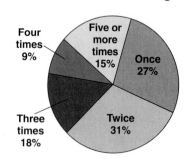

Pages 466–467, Lesson 10-9

5.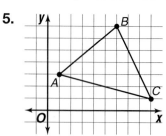

5a. 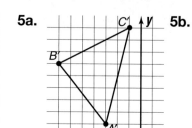 5b.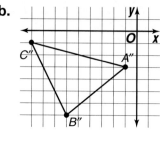

10. 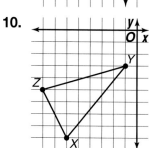 10a.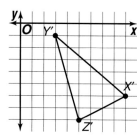

10b. 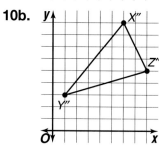 11.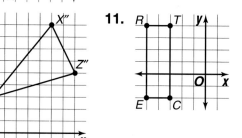

11a. 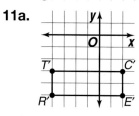 11b.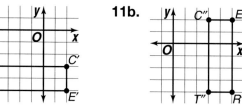

12.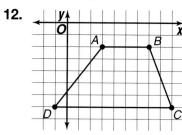

12a.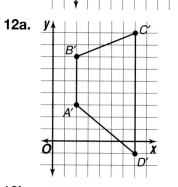

12b.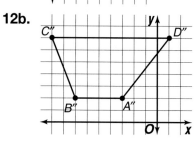

17. 2, 4, 10, J, Q, K of hearts; 2, 4, 10, J, Q, K of clubs; 2, 3, 4, 5, 6, 8, 9, 10, J, Q, K, A of diamonds; 2, 4, 10, J, Q, K of spades

20. 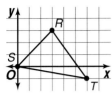 **20a.**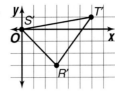

20b.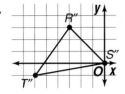

Pages 468–471, Chapter 10
Study Guide and Assessment

11.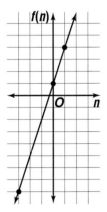

12. Sample table given.

n	f(n)	(n, f(n))
−2	−5	(−2, −5)
0	−4	(0, −4)
4	−2	(4, −2)

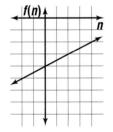

13. Sample table given.

n	f(n)	(n, f(n))
−2	4	(−2, 4)
0	0	(0, 0)
2	−4	(2, −4)

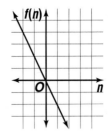

17. 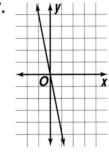   **18.**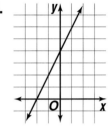

19. 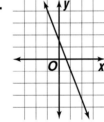   **20.**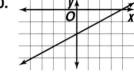

21. 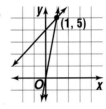   **22.**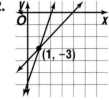

23. 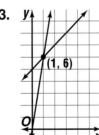   **24.**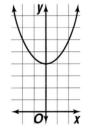

25. 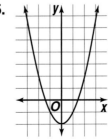   **26.**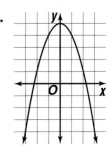

27.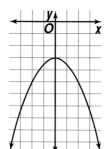

28.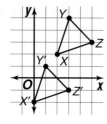

29.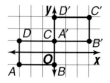

30.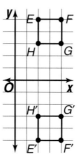

31.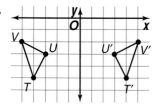

32.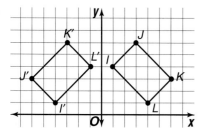

33.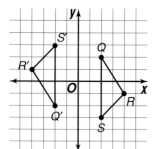

34.

n	f(n)
85	70
87	72
89	74

35. A: line symmetry to a horizontal line; E: line symmetry to a vertical line; N: line symmetry to a diagonal line and rotational symmetry for 180°; T: line symmetry to a horizontal line; Z: no symmetry

36.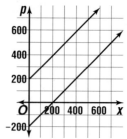

It will never occur since the graphs of their profit equations never intersect.

Page 471, Performance Task

Answers may vary. Possible answer: At one store, there is an initial charge of $5 and then a $2 per movie charge. The other store charges $3 per movie.

Answers may vary. Possible answer: The store that charges $5 and then $2 per movie still charges the initial fee of $5 but now charges $3 per movie. So the graphs never intersect.

CHAPTER 11
Geometry: Using Area and Volume

Pages 480–481, Lesson 11-2A

11.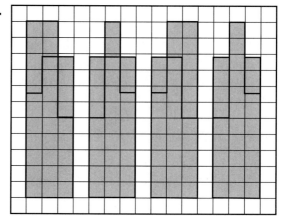

Pages 484–485, Lesson 11-2

3. Both the pentagonal prism and the pentagonal pyramid have a base in the shape of a pentagon. The pentagonal prism has 2 pentagonal bases and the other faces are quadrilaterals. The pentagonal pyramid has 1 pentagonal base and the other faces are triangles.

5f. Sample answer: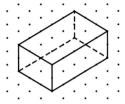

7.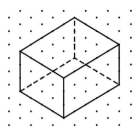

8. 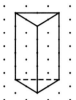   **9.**

10f. Sample answer:

11. Lightly draw an oval for the base, draw two edges, and draw another oval for the top. Use dashed lines to show the part you cannot see.

12. Lightly draw an oval for the base. From a point above the oval draw two edges. Use dashed lines to show the part you cannot see.

13.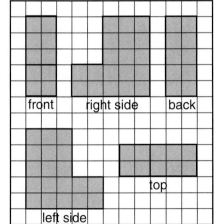

15. Sample answer: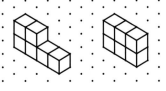

Pages 488–489, Lesson 11-3

1. The formula $V = Bh$ can be used to find the volume of a rectangular prism, a triangular prism, or a cylinder. Since the area of the base of a rectangular prism equals the length times the width, another formula for the volume of a rectangular prism is $V = \ell wh$. Since the area of the base of a cylinder equals pi times the radius squared, another formula for the volume of a cylinder is $V = \pi r^2 h$.

Pages 496–498, Lesson 11-5

3.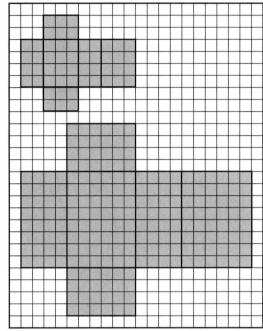

Pages 501–502, Lesson 11-6

17. Double the radius; consider the expression for the surface area of a cylinder, $2\pi r^2 + 2\pi rh$. If you double the height, you will double the second addend. If you double the radius, you will quadruple the first addend and double the second addend.

Pages 505–507, Lesson 11-7

2. Sample answer: The measurements 25 in. and 1 in. both have a greatest possible error of $\frac{1}{2}$ in. The relative error of 25 in. is $\frac{1}{50}$ or 0.02. The relative error of 1 in. is $\frac{1}{2}$ or 0.5.

3. Sample answer: An odometer measures length. It measures to the nearest 0.1 mile. It could be used to measure the distance to school or the distance between 2 cities.

7. The measurement is to the nearest millimeter. There is 1 significant digit. The greatest possible error is 0.5 mm, and the relative error is $\frac{0.5}{7}$ or about 0.071.

8. The measurement is to the nearest 0.001 kg. There are 2 significant digits. The greatest possible error is 0.0005 km, and the relative error is $\frac{0.0005}{0.052}$ or about 0.0096.

9. The measurement is to the nearest $\frac{1}{2}$ lb. The greatest possible error is $\frac{1}{4}$ lb, and the relative error is $\frac{1}{38}$ or about 0.026.

10. The measurement is to the nearest 0.01 m. There are 4 significant digits. The greatest possible error is 0.005 m, and the relative error is $\frac{0.0005}{87.23}$ or about 0.000057.

11. The measurement is to the nearest minute. There are 2 significant digits. The greatest possible error is $\frac{1}{2}$ min, and the relative error is $\frac{1}{94}$ or about 0.011.

12. The measurement is to the nearest 0.1 cm. There are 3 significant digits. The greatest possible error is 0.05 cm, and the relative error is $\frac{0.05}{42.8}$ or about 0.0012.

13. 34.3 oz; 2 lb is measured to the nearest pound, 34 oz is measured to the nearest ounce, and 34.3 oz is measured to the nearest 0.1oz. Therefore, 34.3 is most precise.

14. 305 ft; 100 yd is measured to the nearest yard, and 305 ft is measured to the nearest foot. Therefore, 305 is more precise.

15. 78 in.; the relative error of 17 in. is $\frac{1}{34}$ and the relative error of 78 in. is $\frac{1}{156}$. Since $\frac{1}{34} > \frac{1}{156}$, the relative error of 78 in. is less than the relative error of 17 in.

Pages 508–511, Chapter 11 Study Guide and Assessment

24. The measurement is to the nearest 0.01 mm. There is 1 significant digit. The greatest possible error is 0.005 mm, and the relative error is $\frac{0.005}{0.06}$ or about 0.083.

25. The measurement is to the nearest foot. There are 2 significant digits. The greatest possible error is $\frac{1}{2}$ ft, and the relative error is $\frac{1}{24}$ or about 0.042.

26. The measurement is to the nearest $\frac{1}{3}$ yd or to the nearest foot. The greatest possible error is $\frac{1}{6}$ yd, and the relative error is $\frac{1}{14}$ or about 0.071.

27. The measurement is to the nearest 0.1 m. There are 3 significant digits. The greatest possible error is 0.05 m, and the relative error is $\frac{0.05}{24.2}$ or about 0.0021.

Page 511, Performance Task

Divide 1,017.36 by 81π to find that the height of both pools is about 4 m. So, your friend's pool has a volume of about $\pi(4.5^2)(4)$ or about 254 m³. The volume of your pool is 4 times greater than the volume of your friend's pool.

The surface area of your pool is about 480.7 m² and the surface area of your friend's pool is about 176.7 m². So, you need about 2.7 times more paint than your friend. Since $3(2.7) = 8.1$, you will need about 8 cans of paint.

CHAPTER 12
Investigating Discrete Math and Probability

Pages 516–517, Lesson 12-1A

8.
$1 \times 1 = 1$	$2 \times 1 = 2$	$3 \times 1 = 3$	$4 \times 1 = 4$	$5 \times 1 = 5$	$6 \times 1 = 6$
$1 \times 2 = 2$	$2 \times 2 = 4$	$3 \times 2 = 6$	$4 \times 2 = 8$	$5 \times 2 = 10$	$6 \times 2 = 12$
$1 \times 3 = 3$	$2 \times 3 = 6$	$3 \times 3 = 9$	$4 \times 3 = 12$	$5 \times 3 = 15$	$6 \times 3 = 18$
$1 \times 4 = 4$	$2 \times 4 = 8$	$3 \times 4 = 12$	$4 \times 4 = 16$	$5 \times 4 = 20$	$6 \times 4 = 24$
$1 \times 5 = 5$	$2 \times 5 = 10$	$3 \times 5 = 15$	$4 \times 5 = 20$	$5 \times 5 = 25$	$6 \times 5 = 30$
$1 \times 6 = 6$	$2 \times 6 = 12$	$3 \times 6 = 18$	$4 \times 6 = 24$	$5 \times 6 = 30$	$6 \times 6 = 36$

Pages 519–520, Lesson 12-1

1. Sample answer: Grandma's Diner has two flavors of ice cream, vanilla and chocolate. They serve the ice cream with one of four toppings; hot fudge, strawberry, butterscotch, and marshmallow. How many different ice cream deserts does Grandma's Diner serve?

4. See students' diagrams; outcomes are 1, heads; 1, tails; 2, heads; 2, tails; 3, heads; 3, tails; 4, heads; 4, tails; 5, heads; 5, tails; 6, heads; 6, tails. 12 outcomes

7. See students' diagrams; outcomes are red, red, red; red, red, blue; red, red, yellow; red, blue, red; red, blue, blue; red, blue, yellow; red, yellow, red; red, yellow, blue; red, yellow, yellow; blue, red, red; blue, red, blue; blue, red, yellow; blue, blue, red; blue, blue, blue; blue, yellow, blue; blue, yellow, red; blue, yellow, blue; blue, yellow, yellow; yellow, red, red; yellow, red, blue; yellow, red, yellow; yellow, blue, red; yellow, blue, blue; yellow, blue, yellow; yellow, yellow, red; yellow, yellow, blue; yellow, yellow, yellow. 27 outcomes

8. See students' diagrams; outcomes are H, H, H, H; H, H, H, T; H, H, T, H; H, H, T, T; H, T, H, H; H, T, H, T; H, T, T, H; H, T, T, T; T, H, H, H; T, H, H, T; T, H, T, H; T, H, T, T; T, T, H, H; T, T, H, T; T, T, T, H; T, T, T, T. 16 outcomes

9. See students' diagrams; outcomes are floral, tan; floral, black; floral, navy; floral, white; plaid, tan; plaid, black; plaid, navy; plaid, white; striped, tan; striped, black; striped, navy; striped, white. 12 outcomes

13a. See students' diagrams; outcomes are R, R, R; R, R, F; R, F, R; R, F, F; F, R, R; F, R, F; F, F, R; F, F, F.

Pages 530–531, Lesson 12-4

22. Neither; there are 32 possible outcomes. In 16 of the outcomes, there will be more heads. In the other 16 outcomes, there will be more tails. The probability for each event is $\frac{1}{2}$.

25.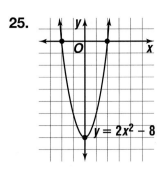

Page 531, Mid-Chapter Self Test

1. See students' diagrams; outcomes are black, small; black, medium; black, large; black, extra large; white, small; white, medium; white, large; white, extra large; gray, small; gray, medium; gray, large; gray, extra large.

Pages 536–537, Lesson 12-5

2. After taking the first event out of the possible outcomes, write the number of remaining possible outcomes in the numerator and the total possible outcomes in the denominator.

4a. See students' diagrams; outcomes are red, red; red, red; red, red; red, yellow; red, yellow; red, red; red, red; red, red; red, yellow; red, yellow; red, red; red, red; red, red; red, yellow; red, yellow; red, red; red, red; red, red; red, yellow; red, yellow; yellow, red; yellow, red; yellow, red; yellow, red; yellow, yellow; yellow, red; yellow, red; yellow, yellow.

Pages 538–539, Lesson 12-6A

1. Since there are 6 possible balloons and 6 possible outcomes when a number cube is rolled, a roll of a 1 could represent the balloon Miko wants. Since there is a 1 in 4 chance of the right wind and 4 possible outcomes for the spinner, a spin of one section could represent the wind Kelsey wanted. By rolling the number cube and spinning the spinner, you can see how many times both favorable outcomes occur out of a total number of tries.

4. Sample answer: Put an X on one of 6 pieces of paper, and place them in a bag. Put an X on one of 4 pieces of paper, and place them in another bag. Pick one piece of paper from each bag. Keep track of how many times you get 2 Xs out of the total number of tries.

Pages 541–543, Lesson 12-6

3. No; if you draw more marbles, the experimental probability will come closer to the theoretical probability.

Pages 547–548, Lesson 12-7

3. Sample answer: If the sample is not random, any predictions will be incorrect. Taking every tenth student from an alphabetical list of students would be an appropriate sampling. Taking a survey of the football players before practice would be an inappropriate sampling.

8d. about 311 rock, 78 jazz, 144 rhythm and blues, 281 country, 43 alternative, 81 oldies, and 61 other

9a. No; 327 prefer Sunshine Orange Juice and 273 prefer the competitor's juice which is not 2 to 1.

9b. Yes; over 50% of the people in one survey did prefer Sunshine Orange Juice.

9c. No; another survey may show a different result.

12. Sample answer: If the questions are not asked in a neutral manner, the people may not give their true opinion. For example, the question "You really don't like Brand X, do you?" might not get the same answer as the question "Do you prefer Brand X or Brand Y?"

13. probability determined by conducting an experiment or simulation

Pages 550–553, Chapter 12 Study Guide and Assessment

31. See students' diagrams; outcomes are ham, pancakes, strawberries; ham, pancakes, bananas; ham, eggs, strawberries; ham, eggs, bananas; ham, French toast, strawberries; ham, French toast, bananas; sausage, pancakes, strawberries; sausage, pancakes, bananas; sausage, eggs, strawberries; sausage, eggs, bananas; sausage, French toast, strawberries; sausage, French toast, bananas.

Page 553, Performance Task

Sample answer: Choose one card from a box containing 3 blue cards, 4 red cards, and 3 green cards. Replace the card and choose another card. Event A is choosing a blue card and event B is choosing a red card. The probability of choosing a blue card and then a red card is $\frac{3}{25}$ or 0.12.

If events A and B are dependent, then the first would not be replaced. The probability of a blue card and then a red card is $\frac{2}{15}$ or $0.1\overline{3}$.

CHAPTER 13
Algebra: Exploring Polynomials

Pages 562–564, Lesson 13-1

18.

19.

20.

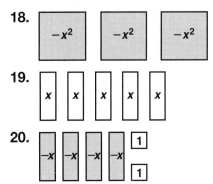

21.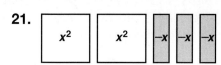

22.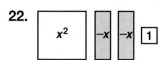

23.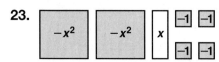

32c.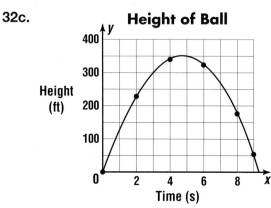

34a.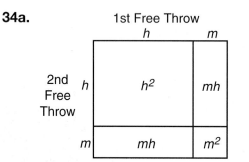

Pages 567–569, Lesson 13-2

3.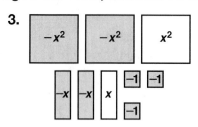

Pages 571–572, Lesson 13-3

3.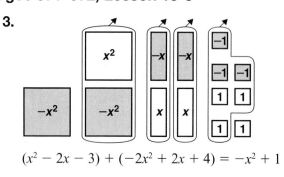

$(x^2 - 2x - 3) + (-2x^2 + 2x + 4) = -x^2 + 1$

Pages 574–576, Lesson 13-4

3.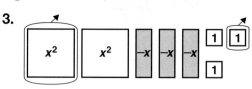

Page 577, Lesson 13-5A

1. 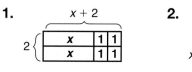   2.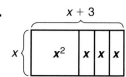

3.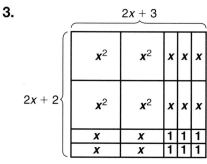

Pages 584–585, Lesson 13-6

3.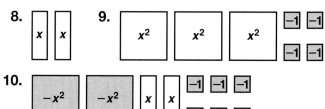

21. $h^2 + 2mh + m^2$; yes; only h and m will vary from player to player.

Pages 592–595, Chapter 13 Study Guide and Assessment

8. 9. 10. 11.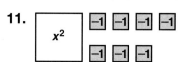

Page 608, Extra Practice, Lesson 1-9

1. 2. 3.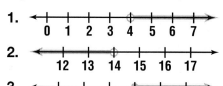

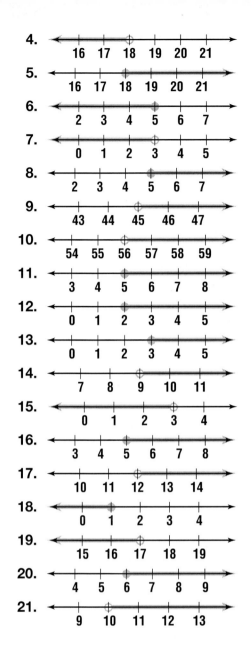

Page 608, Extra Practice, Lesson 2-1

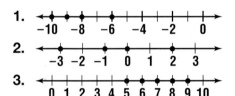

Page 611, Extra Practice, Lesson 2-10

10–19.

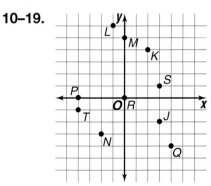

Page 614, Extra Practice, Lesson 4-1A

1a. Sample answer:

Price	Tally	Frequency
$10.01–$20.00	IIII	4
$20.01–$30.00	III	3
$30.01–$40.00	HHI	5
$40.01–$50.00	IIII	4
$50.01–$60.00	II	2

3. Sample answer:

Points	Tally	Frequency
0–4	IIII	4
5–9	HHI I	6
10–14	IIII	4
15–19	III	3
20–24	III	3

Page 615, Extra Practice, Lesson 4-2

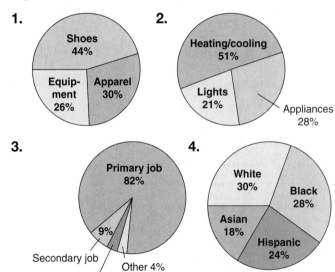

Page 615, Extra Practice, Lesson 4-3

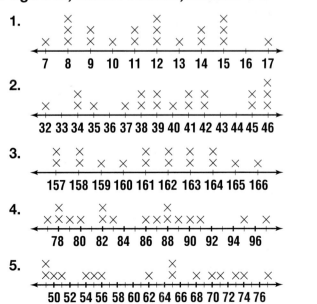

6.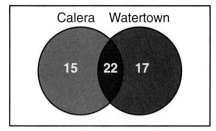

7.

8.

Page 616, Extra Practice, Lesson 4-7

1. Allows for a visual comparison between intervals.
2. Allows comparison of contribution of each person to the total amount of sales.
3. Allows comparison of age group sizes to the total amount of park attendees.
4. A trend can be shown and accurate information is needed.
5. Data is in categories.
6. Allows comparison of number of Pontiac owners to all car owners.

Page 617, Extra Practice, Lesson 5-2A

2a.

Page 619, Extra Practice, Lesson 5-7

1.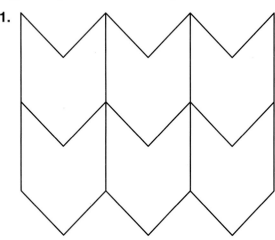

2.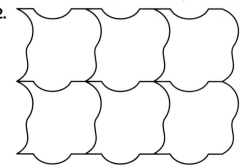

3.

Page 625, Extra Practice, Lesson 7-5

1. 17, 21, 25
2. 162, 486, 1,458
3. 36, 49, 64
4. 27, 9, 3
5. $-18, -23, -28$
6. $5, -5, 5$
7. $10, -3\frac{1}{3}, 1\frac{1}{9}$
8. 26, 29, 32
9. 15, 9, 3
10. $-3, -9, -15$
11. $2, -4, 8$
12. 1, 3, 9
13. 6, 12, 24
14. 20, 27, 35
15. $-3\frac{1}{4}, -3\frac{3}{4}, -4\frac{1}{4}$
16. 21.3, 25.1, 28.9
17. $4\frac{1}{2}, 5\frac{1}{2}, 6\frac{1}{2}$
18. 102, 152, 212
19. $48, -96, 192$
20. 15, 17, 19
21. 6.84, 9.14, 11.44
22. 56, 112, 224
23. 6.6, 10.6, 14.6
24. $-6, 3, -\frac{3}{2}$
25. 108, 324, 972
26. 16, 15, 14
27. $12, 3, \frac{3}{4}$

Page 626, Extra Practice, Lesson 7-10

1.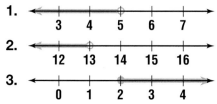
2.
3.
4.
5.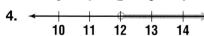
6.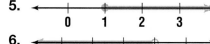
7.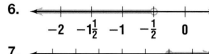

8.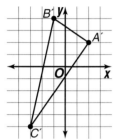
9. (number line 3–7, shaded from 5 onward)
10. (number line with $10\frac{4}{5}, 11, 11\frac{1}{5}, 11\frac{2}{5}, 11\frac{3}{5}$)
11. (number line with $1, 1\frac{1}{21}, 1\frac{2}{21}, 1\frac{3}{21}$)
12. (number line with 11.55, 11.56, 11.57, 11.58)
13. (number line with $4\frac{1}{2}, 5, 5\frac{1}{2}, 6, 6\frac{1}{2}$)
14. (number line with −31, −30, −29, −28, −27)
15. (number line with $1\frac{2}{9}, 1\frac{1}{3}, 1\frac{4}{9}, 1\frac{5}{9}, 1\frac{2}{3}$)

Page 630, Extra Practice, Lesson 8-10

4. A′(2, 2), B′(−1, 4), C′(−3, −5)

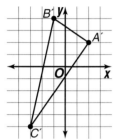

5. A′(1, 1), B′(−5, 2), C′(−1.5, −2.5)

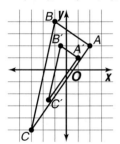

6. A′(4, 4), B′(−2, 8), C′(−6, −10)

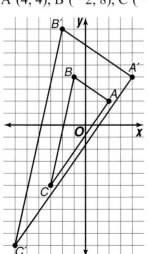

7. A′(6, 6), B′(−3, 12), C′(−9, −15)

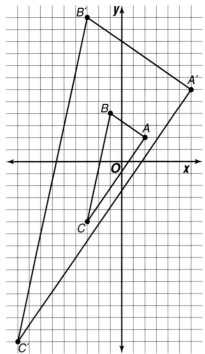

8. A′$\left(\frac{1}{2}, \frac{1}{2}\right)$, B′$\left(-\frac{1}{4}, 1\right)$, C′$\left(-\frac{3}{4}, -\frac{5}{4}\right)$

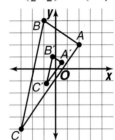

Page 631, Extra Practice, Lesson 9-3

6. $-\sqrt{12}$ (between −4 and −3)

7. $\sqrt{23}$ (between 4 and 5 on 0–7 line)

8. $\sqrt{2}$ (between 1 and 2)

9. $\sqrt{10}$ (between 3 and 4)

10. $\sqrt{30}$ (between −6 and −5)

Page 632, Extra Practice, Lesson 9-6

4.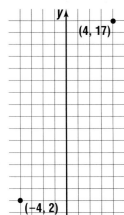
5.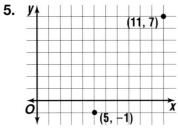
6.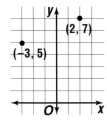
7.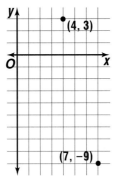
8.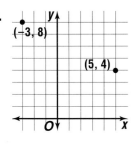
9.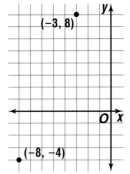
10.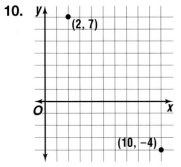
11.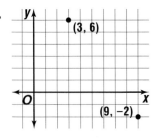

Page 633, Extra Practice, Lesson 10-2

1.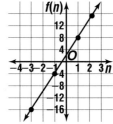
2.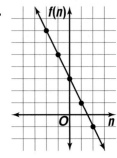
3.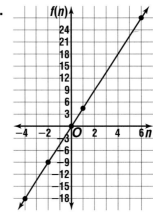
4.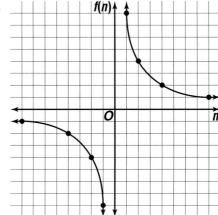
5.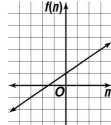
6.
7.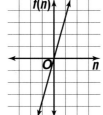
8.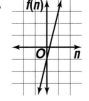

9.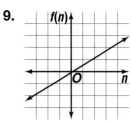

Page 634, Extra Practice, Lesson 10-3

5. $(-2, 11), (-1, 8), (0, 5), (1, 2)$
6. $(-2, -5), (-1, -3), (0, -1), (1, 1)$
7. $(-3, 2), (0, 4), (3, 6), (6, 8)$
8. $(-2, 0.8), (0, 0), (1, -0.4), (2, -0.8)$
9. $(0, -8), (1, 4), (2, 16), (3, 28)$
10. $(0, 2), (4, 5), (8, 8), (12, 11)$
11. $(-2, 1.8), (-1, -0.6), (0, -3), (1, -5.4)$
12. $(-2, 3), (-1, 2), (0, 7), (1, 12)$

Page 634, Extra Practice, Lesson 10-4

1. 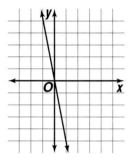 **2.**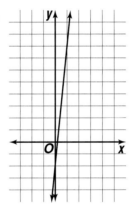

3. 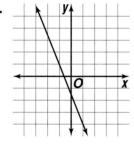 **4.**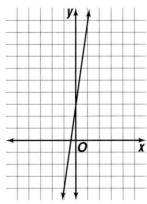

5. 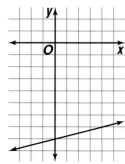 **6.**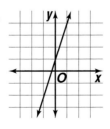

7.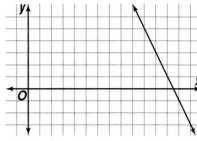

8.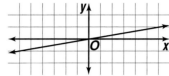

9. 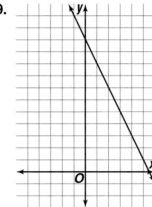 **10.**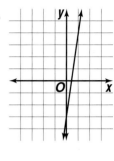

11. 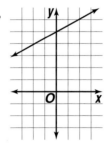 **12.**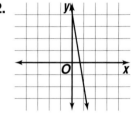

13. 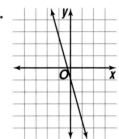 **14.**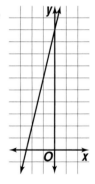

15.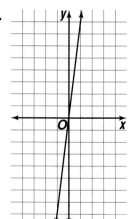
16.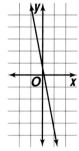

3.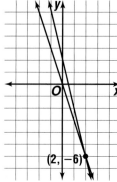
4.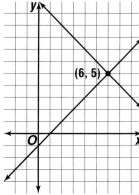

17.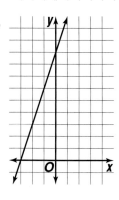
18.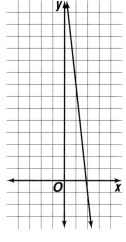

5.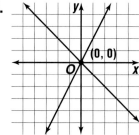
6.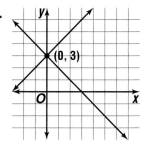

7.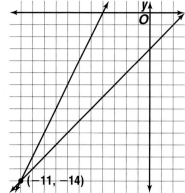

19.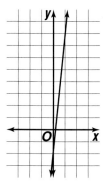
20.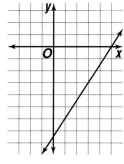

8.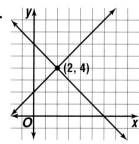
9.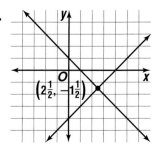

Page 634, Extra Practice, Lesson 10-5

1.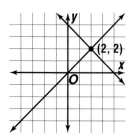
2.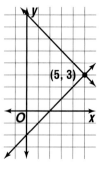

10.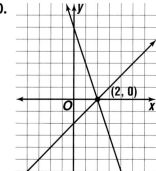
11.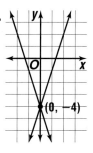

12.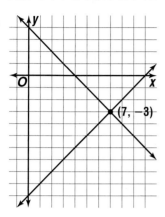

13.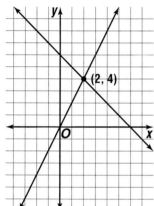

14.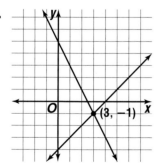

15.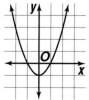

Page 635, Extra Practice, Lesson 10-6

1.
2.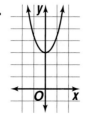
3.
4.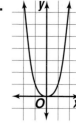
5.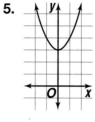
6.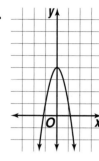
7.
8.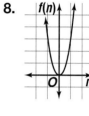
9.
10.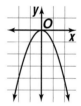
11.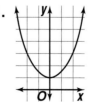
12.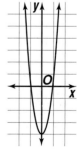
13.
14.

15.
16.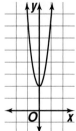
17.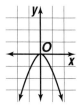
18.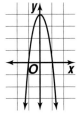
19.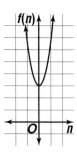
20.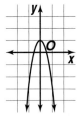
21.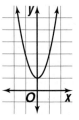
22.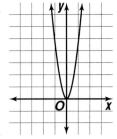
23.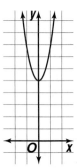
24.
25.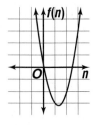
26.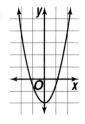
27.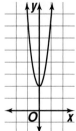
28.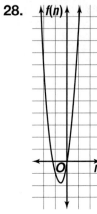

Page 635, Extra Practice, Lesson 10-7

1.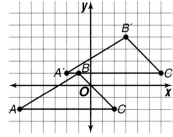
2.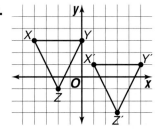
3.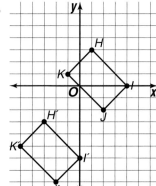
4.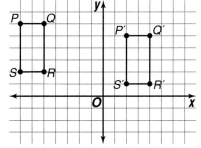

AA36 Answer Appendix

5.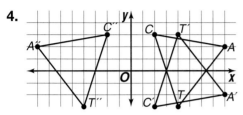

Page 636, Extra Practice, Lesson 10-8

4.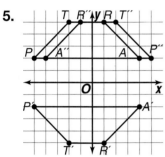

5.

6.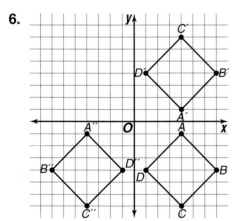

Page 636, Extra Practice, Lesson 10-9

1.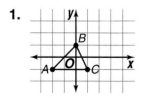
3.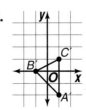
4.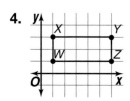
6.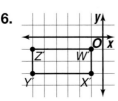

Page 637, Extra Practice, Lesson 11-2

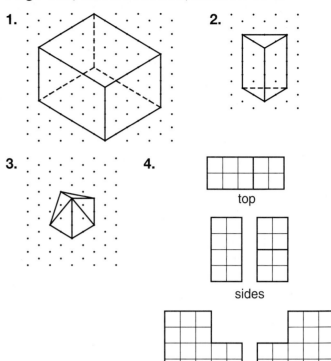

Page 639, Extra Practice, Lesson 11-7

1. The measurement is to the nearest minute. There are 2 significant digits. The greatest possible error is 0.5 minute, and the relative error is $\frac{0.5}{18}$ or about 0.028.

2. The measurement is to the nearest $\frac{1}{2}$ pound. There are 2 significant digits. The greatest possible error is $\frac{1}{4}$ pound, and the relative error is $\frac{0.25}{7.5}$ or about 0.033.

3. The measurement is to the nearest hundredth meter. There are 4 significant digits. The greatest possible error is 0.005 meter, and the relative error is $\frac{0.005}{92.46}$ or about 0.000054.

4. The measurement is to the nearest foot. There is 1 significant digit. The greatest possible error is 0.5 foot, and the relative error is $\frac{0.5}{7}$ or about 0.071.

5. The measurement is to the nearest thousandth kilogram. There are 3 significant digits. The greatest possible error is 0.0005 kg, and the relative error is $\frac{0.0005}{0.067}$ or about 0.0075.

6. The measurement is to the nearest tenth centimeter. There are 3 significant digits. The greatest possible error is 0.05m, and the relative error is $\frac{0.05}{61.7}$ or about 0.00081.

7. The measurement is to the nearest millimeter. There is 1 significant digit. The greatest possible error is 0.5 m, and the relative error is $\frac{0.5}{8}$ or about 0.0625.

8. The measurement is to the nearest $\frac{1}{4}$ inch. The greatest possible error is $\frac{1}{8}$ inch, and the relative error is $\frac{1}{34}$ or about 0.029.

9. 18.0 cm; 18 cm is measured to the nearest centimeter and 18.0 cm is measured to the nearest tenth of a centimeter. Therefore, 18.0 is more precise.

10c. 0.02 is measured to the nearest hundredth of a centimeter and 2.2 is measured to the nearest tenth of a centimeter so 0.02 is more precise.

Page 639, Extra Practice, Lesson 12-1

1. See students' diagrams; outcomes are yellow, chocolate; yellow, vanilla; white, chocolate; white, vanilla; chocolate, chocolate; chocolate, vanilla; marble, chocolate; marble, vanilla.

2. See students' diagrams; outcomes are white, standard, 4-cylinder; white, standard, 6-cylinder; white, automatic, 4-cylinder; white, automatic, 6-cylinder; black, standard, 4-cylinder; black, standard, 6-cylinder; black, automatic, 4-cylinder; black, automatic, 6-cylinder; red, standard, 4-cylinder, red, standard, 6-cylinder; red, automatic, 4-cylinder, red, automatic, 6-cylinder.

3. See students' diagrams; outcomes are roses, red; roses, yellow; roses, pink; roses, white; carnations, red; carnations, yellow; carnations, pink; carnations, white.

4. See students' diagrams; outcomes are queen, firm; queen, super firm; king, firm; king, super firm.

5. See students' diagrams; outcomes are regular, one; regular, two; regular, three; deepdish, one; deepdish, two; deepdish, three.

6. See students' diagrams; outcomes are red, high; red, medium; red, low; white, high; white, medium; white, low; blue, high; blue, medium; blue, low; black, high; black, medium; black, low.

Page 642, Extra Practice, Lesson 13-1

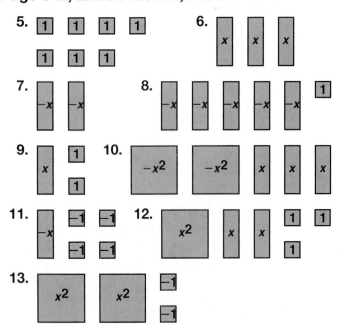

14.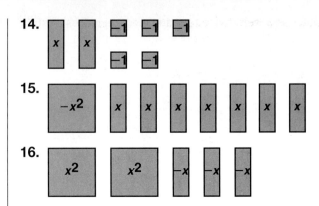

15.

16.

Page 647, Chapter 1 Test

21.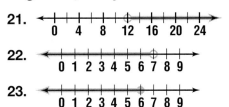

22.

23.

Page 648, Chapter 2 Test

2.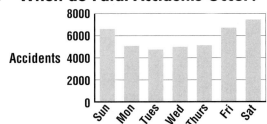

Page 650, Chapter 4 Test

4. **When do Fatal Accidents Occur?**

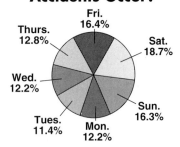

5. **When do Fatal Accidents Occur?**

Fri. 16.4%
Thurs. 12.8%
Sat. 18.7%
Wed. 12.2%
Sun. 16.3%
Tues. 11.4%
Mon. 12.2%

6. Sample answer: A circle graph because it shows it allows the reader to compare the days visually and see that the weekend days have more fatal accidents.

7.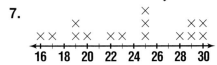

20. The graph is misleading because the scale on the vertical axis is not consistent. It distorts the information. For example, it appears from the bars

that there were twice as many eighth graders as sixth graders, but this is not true.

Page 651, Chapter 5 Test

5. 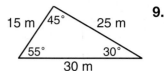 **9.**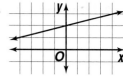

14.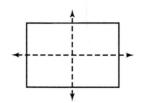

19.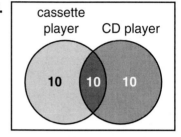

Page 654, Chapter 8 Test

18. Yes; the corresponding angles are congruent and $\frac{18}{6} = \frac{15}{5} = \frac{6}{2}$.

Page 656, Chapter 10 Test

1.
n	f(n)	(n, f(n))
−1	3	(−1, 3)
0	0	(0, 0)
2	−6	(2, −6)

2.
n	f(n)	(n, f(n))
−1	−6	(−1, 6)
0	−4	(0, −4)
2	0	(2, 0)

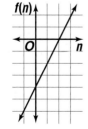

3–6. Sample answers are given.
3. (−4, 1), (0, 2), (4, 3), (8, 4)
4. (−2, 1), (0, −4), (2, −9), (4, −14)
5. (−1, 7), (0, 6), (1, 5), (2, 4)
6. (0, −12), (1, −9), (2, −6), (3, −3)

7. 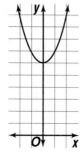 **8.**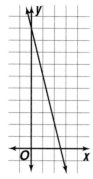

9.  **10.** (graph)

11. (graph with point (2, 12))

Page 657, Chapter 11 Test

4.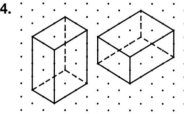

18. $\frac{0.05}{10.5}$ or about 0.0048
19. The one measuring 30 ft by 40 ft; $\pi \cdot 10^2 \cdot 50 \approx 15{,}708.0$ and $\pi \cdot 15^2 \cdot 40 \approx 28{,}274.3$.

Page 658, Chapter 12 Test

1. See students' diagrams; outcomes are VHS, stereo, with remote; VHS, stereo, without remote; VHS, mono, with remote; VHS, mono, without remote; VHS-C, stereo, with remote; VHS-C, stereo, without remote; VHS-C, mono, with remote; VHS-C, mono, without remote; 8 mm, stereo, with remote; 8 mm, stereo, without remote; 8 mm, mono, with remote; 8mm, mono, without remote.

20. Sample answer: Take 10 pieces of paper. Place the pieces of paper in a bag. Pick 2 pieces of paper at random and record the results. Replace the pieces of paper and repeat the simulation numerous times. Use the results to determine the experimental probability that 2 Granny Smith apples are picked.

Page 659, Chapter 13 Test

1.

2.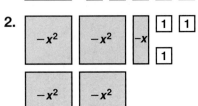

SYMBOLS

Number and Operations

$+$	plus or positive
$-$	minus or negative
$a \cdot b$	
$a \times b$	a times b
ab or $a(b)$	
$\div$	divided by
$\pm$	positive or negative
$=$	is equal to
$\neq$	is not equal to
$<$	is less than
$>$	is greater than
$\leq$	is less than or equal to
$\geq$	is greater than or equal to
$\approx$	is approximately equal to
$\%$	percent
$a:b$	the ratio of a to b, or $\frac{a}{b}$

Geometry and Measurement

$\cong$	is congruent to
$\sim$	is similar to
$\circ$	degree(s)
$\overleftrightarrow{AB}$	line AB
$\overline{AB}$	segment AB
$\overrightarrow{AB}$	ray AB
$\llcorner$	right angle
$\perp$	is perpendicular to
$\parallel$	is parallel to
AB	length of $\overline{AB}$ distance between A and B
$\triangle ABC$	triangle ABC
$\angle ABC$	angle ABC
$\angle B$	angle B
$m\angle ABC$	measure of angle ABC
$\odot C$	circle C
$\overset{\frown}{AB}$	arc AB
π	pi $\left(\text{approximately } 3.14159 \text{ or } \frac{22}{7}\right)$
(a, b)	ordered pair with x-coordinate a and y-coordinate b
$\sin A$	sine of angle A
$\cos A$	cosine of angle A
$\tan A$	tangent of angle A

Algebra and Functions

a'	a prime
a^n	a to the nth power
a^{-n}	$\frac{1}{a^n}$
$\lvert x \rvert$	absolute value of x
$\sqrt{x}$	principal (positive) square root of x
$f(n)$	function, f of n

Probability and Statistics

$P(A)$	the probability of event A
$n!$	n factorial
$P(n, r)$	permutation of n things taken r at a time
$C(n, r)$	combination of n things taken r at a time